Encyclopedia of
Plant Physiology

New Series Volume 11

Editors

A. Pirson, Göttingen
M. H. Zimmermann, Harvard

Hormonal Regulation of Development III

Role of Environmental Factors

Edited by R.P. Pharis and D.M. Reid

Contributors

F.D. Beall D.E. Bilderback M.L. Brenner D. Canary
J.V. Carter D.J. Chalmers W.K. Coleman D.S. Fensom
W.P. Hackett H. Imaseki M.J. Jaffe J.O. Johnson
W.L. Koukkari N.G. Marinos J.P. Mascarenhas T. O'Brien
J.H. Palmer G.F. Pegg R.P. Pharis B.G. Pickard
D.M. Reid F.B. Salisbury H. Smith T.A. Thorpe
D. Vince-Prue R.L. Wample S.B. Warde R.J. Weaver
R.H. Zimmerman

With 121 Figures

Springer-Verlag
Berlin Heidelberg NewYork Tokyo

Editors:
Professor RICHARD P. PHARIS
Professor DAVID M. REID
Department of Biology
The University of Calgary
2500 University Drive N.W.
Calgary, Alberta T2N 1N4/Canada

Planning Volumes 9–11:
Professor N.P. KEFFORD
College of Tropical Agriculture
and Human Resources
University of Hawaii at Manoa
Honolulu, Hawaii 96822/USA

ISBN 3-540-10197-7 Springer-Verlag Berlin Heidelberg New York Tokyo
ISBN 0-387-10197-7 Springer-Verlag New York Heidelberg Berlin Tokyo

Library of Congress Cataloging in Publication Data. Main entry under title: Role of environmental factors. (Hormonal regulation of development; 3) (Encyclopedia of plant physiology; new ser., v. 11) 1. Plant hormones. 2. Botany—Ecology. I. Pharis, R.P., 1937–. II. Reid, D.M., 1940–. III. Beall, F.D. IV. Series. V. Series: Encyclopedia of plant physiology; new ser., v. 11) QK711.2.E5 vol. 11 581.1 s [581.3′1] 84-20208 [QK731] ISBN 0-387-10197-7

Typesetting, printing and bookbinding: Universitätsdruckerei H. Stürtz AG, Würzburg.
2131/3130-543210

Contents

Time-Related Factors and Phenomena

2 Rhythms and Their Relations to Hormones

W.L. KOUKKARI and S.B. WARDE (With 9 Figures)

3 Hormonal Aspects of Phase Change and Precocious Flowering

R.H. ZIMMERMAN, W.P. HACKETT, and R.P. PHARIS (With 5 Figures)

Direction

4 Polarity

W.K. Coleman and T.A. Thorpe (With 1 Figure)

5 Epinasty, Hyponasty, and Related Topics

J.H. Palmer (With 4 Figures)

Light

8 De-Etiolation and Plant Hormones

T. O'Brien, F.D. Beall, and H. Smith (With 7 Figures)

9 Photoperiod and Hormones

D. Vince-Prue

10 Roles of Hormones in Phototropism

B.G. Pickard (With 17 Figures)

Temperature

11 Plant Growth Regulators and Low Temperature Stress

J.V. Carter and M.L. Brenner

Wind and Other Mechanical Factors

12 Wind and Other Mechanical Effects in the Development and Behavior of Plants, with Special Emphasis on the Role of Hormones

M.J. Jaffe (With 14 Figures)

13 Hormonal Control of Wound-Induced Responses

H. Imaseki (With 2 Figures)

14 Water Relations and Plant Hormones

D.M. Reid and R.L. Wample

Organisms

15 Pollen. Symbionts and Symbiont-Induced Structures

J.P. Mascarenhas and D. Canary

16 Pathogenic and Non-pathogenic Microorganisms and Insects

G.F. Pegg (With 2 Figures)

17 Electrical and Magnetic Stimuli

D.S. Fensom (With 9 Figures)

18 Regulators of Plant Reproduction, Growth and Differentiation in the Environment

D.E. Bilderback (With 17 Figures)

Hormones and Plant Ecology

List of Contributors

F.D. Beall
Plant Physiology Research Group
Department of Biology
University of Calgary
Calgary, Alberta T2N 1N4/Canada

D.E. Bilderback
Department of Botany
University of Montana
Missoula, MO 59812/USA

M.L. Brenner
Department of Horticultural Science
and Landscape Architecture
305 Alderman Hall
University of Minnesota
St. Paul, MN 55108/USA

D. Canary
(Deceased)

J.V. Carter
Department of Horticultural Science
and Landscape Architecture
305 Alderman Hall
University of Minnesota
St. Paul, MN 55108/USA

D.J. Chalmers
Irrigation Research Institute
Department of Agriculture
Tatura, Victoria, 3616/Australia

W.K. Coleman
Research Station
Agriculture Canada
P.O. Box 20280
Fredericton, New Brunswick,
E3B 4Z7/Canada

D.S. Fensom
Biology Department
Mount Alison University
Sackville, New Brunswick/Canada

W.P. Hackett
Department of Horticultural Science
and Landscape Architecture
305 Alderman Hall
University of Minnesota
St. Paul, MN 55108/USA

H. Imaseki
Research Institute
for Biochemical Regulation
Faculty of Agriculture
Nagoya University
Chikusa, Nagoya, 464/Japan

M.J. Jaffe
Department of Biology
Wake Forest University
Winston-Salem, NC 27109/USA

J.O. Johnson
Department of Viticulture and Enology
University of California
Davis, CA 95616/USA

W.L. Koukkari
Botany Department
University of Minnesota
220 Biological Sciences Center
1445 Gortner Avenue
St. Paul, MN 55108/USA

N.G. Marinos
School of Biological Sciences
The Flinders University of South
Australia
Bedford Park, 5042/South Australia

J.P. Mascarenhas
Department of Biological Sciences
State University of New York
at Albany
Albany, NY 12222/USA

T. O'Brien
Department of Botany
University of Leicester
University Road
Leicester LEI 7RH/United Kingdom

J.H. Palmer
School of Botany
University of New South Wales
Kensington, 2033/Australia

G.F. Pegg
Department of Horticulture
University of Reading
Reading, RG6 2AU/United Kingdom

R.P. Pharis
Department of Biology
The University of Calgary
2500 University Drive N.W.
Calgary, Alberta T2N 1N4/Canada

B.G. Pickard
Biology Department
Washington University
St. Louis, MO 63130/USA

D.M. Reid
Department of Biology
The University of Calgary
2500 University Drive N.W.
Calgary, Alberta T2N 1N4/Canada

F.B. Salisbury
Plant Science Department UMC 48
Utah State University
Logan, UT 84322/USA

H. Smith
Department of Botany
University of Leicester
University Road
Leicester LEI 7RH/United Kingdom

T.A. Thorpe
Department of Biology
University of Calgary
Calgary, Alberta T2N 1N4/
Canada

D. Vince-Prue
Glasshouse Crops Research Institute
Worthing Road
Littlehampton
West Sussex, BN17 6LP/
United Kingdom

R.L. Wample
Department of Horticulture and
Landscape Architecture
Washington State University
Pullman, WA 99164-6414/USA

S.B. Warde
Biology Department
The Blake School
511 Kenwood Parkway
Minneapolis, MN 55403/USA

R.J. Weaver
Department of Viticulture
and Enology
University of California
Davis, CA 95616/USA

R.H. Zimmerman
Fruit Laboratory
Agricultural Research Service
U.S. Department of Agriculture
Beltsville, MD 20705/USA

List of Abbreviations

A	amplitude
ABA	absicic acid
ACC	1-aminocyclopropane-7-carboxylic acid
ACP	1-(2, carboxyphenyl)-3, phenylpropane-1,3 dione
amitrole	3-amino-s-triazole
AMO-1618	*N*,*N*,*N*, 2-tetramethyl-5-(1-methylethyl)-4-[(1-piperidinylcarbonyl)-oxy]benzenaminium chloride
AMP	adenosine monophosphate
ancymidol	α-cyclopropyl-α-(4-methoxyphenyl)
ATP	adenosine triphosphate
AVG	aminoethoxyvinyl-glycine
B	blue light
BA	6-benzylamino purine (syn: benzyladenine)
BOA	benzothiazole-2-oxyaalocetic acid
CBBP	tributyl (2,4 dichloro-benzyl) phosphonium chloride
CC	constant conditions
CCC (Chlor-mequat; cycocel)	(2-chlorethyl) trimethyl-ammonium chloride
CEPA	2-(chloroethyl) phosphonic acid (ethrel, ethephon)
CK	cytokinins
4-CPA	4-chlorophenoxyacetic acid
cpm	counts per minute
D	dark
DD	continuous dark
2,4-D	2,4-dichlorophenoxyacetic acid
2,6-D	2,6-dichlorophenoxyacetic acid
DIHB	3,5-diiode-5-hydrobenzoic acid
DMSO	dimethyl sulfoxide
DNP	day-neutral plant (for flowering)
dpm	disintegrations per minute
ECM	ectomycorrhiza
EDM	endomycorrhiza
EDTA	ethylene diaminetraacetic acid
EGTA	ethylene-glycol bis-(tetraace-tic acid)
EO	electroosmotic efficiency in water molecules per positive charge or moles H_2O Faraday^{-1}
ET	ethylene
ethephon (ethrel)	(2-chloroethyl) phosphonic acid
FC	fusicoccin
FR	far-red light
GA	gibberellin or gibberellin-like substance
GA_1, GA_2, GA_3 etc.	gibberellin A_1, gibberellin A_2, gibberellin A_3 etc.
GC-FID	gas-liquid chromatography-flame ionization detection, gas liquid chromatography
GMP	guanadine monophosphate
HDT	hormone-directed transport
9-HFCA	9-hydroxyfluorenecarboxylic acid
HIR	high irradiance reaction
IAA	indole-3-acetic acid
IAAsp	IAA-aspartate
IBA	indolebutyric acid
ILA	indole-3-lactic acid

IPA	isopentenyl adenosine
kinetin	6-furfurylamino purine
LD	long day
LDP	long-day plant (for flowering)
LSDP	long-short-day plant (for flowering)
MCPA	[(4-chloro-o-tolyl)oxy]acetic acid
mefluidide	[(N-(2,4-dimethyl-5-(tri-fluore-methyl)amino)phenyl] acetamino
morphactin	2-chloro-9-hydroxyfluorene-9-carboxylic acid butyl ester
MVA	mevalonic acid
NAA	naphthaleneacetic acid
NADH	nicotinamide adenine dinucleotide reduced
NOA	f-naphthoxyacetic acid
NPA	N-1-naphthylphthalanic acid
2ip	isopentenyladenine
PAL	phenylalanine ammonia-lyase
PBA	6-(benzylamino)-9-(2-tetra-hydropyranyl)-9H-purine
PCIB	4-chlorophenoxy-isobutyric acid
Pfr	phytochrome in far-red-absorbing form
picloram	4-amino-3,5,6-trichloro-picolinic acid
Pr	phytochrome in red-absorbing form
R	red light
RNA	ribonucleic acid
RUDPase	ribulosediphosphate carboxylase
SA	salicylic acid
SADA	N-dimethylaminosuccinic acid
SD	short day
SDP	short-day plant (for flowering)
SLDP	short-long-day plant (for flowering)
TIBA	2,3,5-triiodobenzoic acid
TLC	thin layer chromatography
TWF	transcellular water flow
UV	ultra violet light
WL	white light
Ψ	plant water potential
Ψ_1	leaf water potential
Ψ_π	osmotic potential
Ψ_m	matric potential
Ψ_p	pressure potential

Introduction

R.P. PHARIS and D.M. REID

The idea of a separate Encyclopedia volume dealing with the "interrelations of plant hormones with factors in the environment of the plant, and its organs and tissues" originated with N.P. KEFFORD, and we are most appreciative of the help and advice provided by Prof. KEFFORD in the formative stages of this volume. We have thus interpreted "environment" very broadly to include not only factors external to the plant, e.g., gravity, light, temperature, wind, mechanical wounding, water, organisms (including pollen), and magnetic and electric stimuli, but internal factors as well (e.g., nutrients, both inorganic and photoassimilate, direction, and time). In our definition of "hormonal effect", or "hormonal involvement", we have asked our authors to take a broad approach, and to examine not only phenomena that are mediated by the known plant hormones, but to discuss as well a wide variety of processes and events where hormonal involvement is implied through more indirect analyses and observations.

The volume begins with environmental factors internal to the plant; R.J. WEAVER and J.O. JOHNSON thus examine "hormones and nutrients", their interrelationship in movement, accumulation, and diversion. As one studies a plant during its rapid growth phase, and later as maturation and aging proceed, it becomes apparent that time is an environmental cue of great significance, one which may exert a major influence via hormonal messages. Time in the short term, e.g. "rhythmic phenomena", is analyzed in terms of hormonal involvement by W. KOUKKARI and S.B. WARDE. Time in the long term, as plants proceed through the various stages of morphological change, and from a sexually juvenile stage to sexual maturity, is discussed by R.H. ZIMMERMAN, W.P. HACKETT and R.P. PHARIS in a chapter on "juvenility". Although direction within the plant is not often thought of as an environmental factor, analyses by D. CHALMERS (position), W. COLEMAN and T.A. THORPE (polarity) and J.M. PALMER (epinasty and dorsiventrality) tell us of the many developmental events and growth processes that are shaped, via hormonal involvement, by this internal environmental factor.

The environmental factors we traditionally think of as influencing growth of the plant, gravity, light, and temperature, are covered primarily from a hormonal viewpoint. How gravity may mediate plant growth in the broadest sense (root and shoot geotropism, reaction wood formation, plagiotropism and apical dominance) via plant hormones is discussed by B. PICKARD. Light, as a factor that interacts with (and may act through) plant hormones, is covered in three parts: etiolation and intensity effects (T. O'BRIEN, F.D. BEALL and H. SMITH), photoperiod (D. VINCE-PRUE) and phototropism (B. PICKARD). Temperature effects which may involve plant hormones are described by J. CARTER and M. BRENNER.

Less often discussed, but still rather important external environmental factors, wind, mechanical wounding, water, magnetic and electrical stimuli, appear to exert at least some of their influence on plant growth and development via plant hormones. M.J. JAFFE reviews the effects of wind and H. IMASEKI, of wounding. Water, in surfeit, normal, and dearth amounts is dealt with by D.M. REID and R.L. WAMPLE. The seldom-examined area of magnetic and electrical influences, and how they might act via hormones, is covered by D. FENSOM.

A more esoteric group of external environmental factors has also been included. Organisms external to the plant may produce hormones, and/or trigger hormonal production by the plant. J.P. MASCARENAS and D. CANARY examine symbionts and their structures, as well as pollen. D. PEGG covers pathogens, including bacteria, fungi, nematodes and insects. Growth-regulatory substances which are derived from a variety of sources, and are ubiquitous in the external environment of many lower plants, and in the root system of higher plants, are discussed by D. BILDERBACK.

Finally, the above, often seemingly disparate factors are put into a broad "ecophysiological" perspective by F. SALISBURY and N.G. MARINOS, who examine the ecological and evolutionary implications of environmentally mediated hormone-regulated processes.

By necessity there will be overlap with other volumes of the New Series, Encyclopedia of Plant Physiology. For many of the physiological processes that are controlled by the various environmental factors, especially gravity and light, are covered in their own right (e.g., Vol. 7, Physiology of Movements; Vols. 16 A and B, Photomorphogenesis). Hormonal actions are also considered in several chapters of Volume 12 A, Physiological Plant Ecology I, Responses to the Physical Environment. This is by design, as well as by necessity, since each volume should be able to stand alone in introducing and covering its subject material.

As ramified as the reader may think the above approach, there are still environmental factors which are not covered, or covered only briefly. Thus, the reader must look elsewhere with regard to seeds, and the large literature on how they deal with O_2 tension, leaching by water, germination inhibitors in their external environment, etc. Nor could we cover bolting, perennation, aging, and all of the aspects of flowering where plant hormones are key factors in the process in question.

We are especially grateful to Professor M. BOPP for his perusal of the contributed chapters, and for his critical and constructive comments which aided greatly in their revision.

Finally, we express our appreciation for the patience and understanding of our numerous authors as this volume took shape over a very long period. Revision and up-dating has been required more than once, and has been accomplished skillfully and quickly by all concerned.

1 Relation of Hormones to Nutrient Mobilization and the Internal Environment of the Plant: The Supply of Mineral Nutrients and Photosynthate

R.J. WEAVER and J.O. JOHNSON

1 Introduction

Source-to-sink movement of plant assimilates usually involves the transfer of substrate materials from green, photosynthesizing assimilating tissues to growing regions or storage organs. MASON and MASKELL (1928), who measured carbohydrate fractions in fruits, stems, and leaves of cotton at various times during the day, did much to develop the source-to-sink concept for assimilate movement.

Assimilates can come from stored reserves and from recent photosynthate; the former requiring dissolution of the reserves and retranslocation of solute. Directional forces may be involved in mobilization, and mobilizing centers often appear to initiate dissolution and retranslocation of material, even from quite remote sites.

As tissues age they often become weak "sources", exporting assimilates to other sinks, some of which may be newly developing. Plant hormones can hasten senescence and increase the tendency of an organ to give up assimilates.

The partition and distribution of assimilates is of the utmost importance in plant growth and development. Moreover, in crop plants, altering the normal assimilate patterns and distribution with plant hormones offers a tool to increase productivity. Here, the hormone is used to enhance assimilate flow into the plant part that will eventually be harvested.

It is our objective to present as concise a picture as possible of the influence of hormones on translocation patterns of both organic and inorganic assimilates in plants. In doing this we have considered the assimilate sources and sinks, per se, as the major environmental factors, albeit internal, interacting with plant hormones. External environmental factors (i.e., light, temperature, water stress, etc.) have rarely been experimental variables in hormone/translocation experiments, hence are alluded to infrequently. First of all we will discuss some general principles of assimilate movement.

2 Sources, Sinks, and Assimilate Movement in Relation to Morphology

2.1 Definitions of Source and Sink

The source is defined as the plant part that exports, and the sink as the part that imports assimilates. Both definitions are based on direction of transport.

From a morphological viewpoint, certain plant parts, such as leaves, are often sources, whereas other tissues are frequently designated as sinks [fruits, active meristems, roots, and expanding storage organs WARDLAW (1968)]. From a metabolic viewpoint, sources produce assimilates by photosynthesis or from stored materials, whereas sinks utilize assimilates in respiration and growth.

WILSON (1972) suggested that sources and sinks be defined with regard to losses and gains of a particular substance in a certain plant part. Thus, source strength is equal to (source size) × (source activity), or rate of assimilation per plant equals (leaf area) × (rate of assimilation per unit of leaf area). Sink strength equals (sink size) × (sink activity), or absolute growth rate equals (dry weight) × (relative growth rate).

Although WILSON's approach is amenable to experimental measurement, the weight of organ dry matter does not account for respiratory losses; thus, WAREING and PATRICK (1974) noted that only a net sink strength can be obtained. They further suggested that since respiratory losses of imported assimilates may be very high, "sink strength" should refer to its potential capacity to import and accumulate assimilates, whereas "mobilization" should refer to accumulation of dry matter by the sink within the competitive framework of the intact plant.

2.2 Position Centers of Assimilate Production

In most plants lower leaves serve as the main source of assimilates for the roots, upper leaves serve mainly the shoot apex (CRAFTS and CRISP 1971). Leaves in an intermediate position may supply assimilate in either or both directions. Developing leaves can also manufacture assimilates as well as import them, as has been shown for cotton, tomato, soybean, cereals, grasses, and grapes (WARDLAW 1968).

The position of the leaf relative to the shoot apex changes constantly during development (Fig. 1). When a young leaf is less than one-third to one-half expanded, it is a sink for assimilates, with no export. As the leaf enlarges, assimilates are exported to the shoot tip. As more leaves are produced, the movement of assimilates from this first leaf become both apical and basal; finally with further leaf production, movement becomes primarily basipetal (HALE and WEAVER 1962).

When rapidly growing storage organs, flowers, or fruits are produced, assimilate patterns are altered by the intensity and location of the demand (sink), developing fruits being very strong sinks (Fig. 1). Thus, the sink strength of various tissues is constantly changing during growth of the plant.

2.3 Strength of the Sinks

Tissues vary in their demand for assimilates, those that are sinks having one or several mechanisms by which a high demand is created. Carbohydrate depletion may result in a higher concentration gradient, thus enhancing movement

Fig. 1 A–C. Diagram of a rapidly growing grape shoot at three different developmental stages showing main direction of movement of photosynthate. **A** Movement in a very young shoot or shoot tip is apical. **B** In the pre-bloom or bloom stage export is bidirectional from two or three leaves below the shoot tip. Below this region it is basal. **C** After the set of fruit, photosynthate also moves apically into the cluster from leaves blow the cluster. After the rate of shoot growth decreases (several weeks following the set stage), photosynthate moves basally from the tip. (After HALE and WEAVER 1962)

of assimilates to the sinks, and increasing number of completing sinks, or demand sites (WARDLAW 1968). The distance from the site of production may also be a factor (WARDLAW 1968). Increased assimilate translocation to sites of utilization can occur from removal of competing sites, or by applications of plant regulators (WAREING and PATRICK 1974).

In many instances, accumulation of dry matter indicates the competitive ability of a sink to attract assimilates relative to other sink regions (WAREING and PATRICK 1974). However, dry matter increase does not give the potential capacity of a sink to attract assimilates, probably due in part to transport limitation imposed by different types of vascular systems.

2.4 Patterns of Distribution of Assimilates in Relation to Vascular Connections

Distribution patterns of assimilates between sources and sinks may be strongly affected by vascular connections [e.g., PROKOFYEV and SOBOLEV (1957)]. In many plants there is a linearity of transport, and thus a strict limitation on lateral movement of translocating assimilates away from the phloem. It is widely accepted that long-distance movement of the assimilates occurs within the living phloem tissue, especially in the longitudinally oriented sieve tubes, and ^{32}P applied to a sunflower leaf was translocated only to that portion of the inflorescence on the same side of the stem (PROKOFYEV and SOBOLEV 1957). A limitation on lateral movement out of the vascular system has also been shown in cotton (TING 1963), grape (WANG and HSIA 1962), sugarbeet (JOY 1964), rice (ASADA et al. 1960), soybean (THROWER 1962), tomato (McCOLLUM and SKOOG 1960), and numerous other plants (WARDLAW 1968).

In the grape vine, translocation from a given leaf is also often linear; that is, it is restricted to the vertical transport systems more or less in line with the leaf or branch supplying the translocate (SWANSON and EL SHISHINY 1958). However, under certain circumstances this can be modified by the leaves. Lateral transport of 2,4-D and of amitrole in the Tokay shoot only occurred in the apical portion, this uniform lateral distribution of photosynthate was the resulting from anastomoses of the primary phloem at the nodes (LEONARD and WEAVER 1961). A fruit cluster can also divert from its normal vertical path photosynthate moving down the shoot (HALE and WEAVER 1962). In some shoots, the lateral division of ^{14}C-photosynthate was indicated by the presence of radioactivity in axillary buds, the cluster diverting photosynthate laterally from one side of the stem to the other, usually within the vertical distance of one internode. This rapid lateral movement of photosynthate was probably made possible by the development, in secondary growth, of vascular elements which traverse the wide parenchyma rays separating the vascular strands (ESAU 1948).

2.5 The Demand for Assimilates

Reproductive parts are usually stronger sinks than vegetative parts. However, this may vary with stage of development, flowers being a relatively weak sink compared to clusters of developing berries (HALE and WEAVER 1962) which even utilize photosynthate from as far as 4 m (MEYNHARDT and MALAN 1963). A fully expanded leaf usually can no longer import assimilates even when the leaf is in darkness (WARDLAW 1968).

Established shoots generally have priority over roots and buds when there is a deficiency of assimilate. The competition for assimilates among the shoot tip, the leaves at varying stages of development, and expanding stem tissue has already been discussed. To some extent then, the distribution of assimilates can be predicted if one considers leaf position, leaf area, and the growth rates of individual organs.

3 Role of Hormones and Growth Regulators in Assimilate Movement

3.1 Effect of Plant Hormones on Partition of Assimilates

Formerly, partition of assimilates was primarily studied by measuring carbohydrate fractions in stems, leaves, fruits, and other plant parts at various times.

More recently the use of $^{14}CO_2$ or other radioactively labeled assimilate, supplied to plants or plant parts, has become widespread. Autoradiographs made at intervals then reveal assimilate distribution patterns during the course of development, and various counting techniques are also valuable. The effects of hormones on translocation patterns have been well researched with explants, plant parts, partition between roots and shoot, partition within the shoot system, and with intact plants (ROBINSON 1983).

Explants, and such detached parts as leaves and severed hypocotyls, are often used to study hormone-directed transport. However, because of the possible artifacts (e.g., senescense, surging of sieve tubes, and callose formation) of these types of experimental systems, the results from them are difficult to interpret. Auxins, GA's, and cytokinins can effectively promote the power of a sink, whereas inhibitors such as ABA, daminozide, and chormequat often decrease the ability of a sink to attract assimilates, and may cause organs that were acting as sinks to give up assimilates, thus becoming sources.

In earlier work, when certain hormones were applied to plant parts, increased movement of assimilates toward the treated region was noted (ALTMAN and WAREING 1975; MOTHES and ENGELBRECHT 1961; SETH and WAREING 1967). MOTHES and ENGELBRECHT (1961) treated one-fourth of the surface of a detached leaf with kinetin, 4 days later applying a drop of ^{14}C-glycine to another area of the excised leaf. Autoradiographs revealed that in leaves exposed to light or to darkness there was a predominant translocation of radioactivity to the kinetin locus.

Similar results were obtained when a portion of an intact leaf on a plant was treated with BA, and $^{14}CO_2$ was applied to leaves below (QUINLAN and WEAVER 1969) (Fig. 2). PATRICK and WAREING (1973) applied IAA to stumps of dwarf bean (*Phaseolus vulgaris* L. cv. Canadian Wonder) seedlings decapitated above the primary leaves, ^{14}C-sucrose being applied to an epidermal flap on the stem. Within 3 to 6 h ^{14}C-metabolites had moved to the treated stump, localization of ^{14}C being limited to the phloem of treated stumps. Enhanced ^{14}C-metabolite activity also occurred when IAA was applied to one side of the intact internodes. Similar effects of IAA on transport were noted with excised stem segments. This work (PATRICK and WAREING 1973) essentially confirmed the earlier work with decapitated bean plants (MULLINS 1970) or bean hypocotyl segments (ZAERR and MITCHELL 1967). HEW et al. (1967) decapitated the apex of soybean plants (*Glycine max* cv. Comet) and applied IAA or GA_3 to the cut surface. Both the amount and rate of movement of ^{14}C-sucrose from a leaf to the treated area was markedly increased.

MULLER and LEOPOLD (1966a, b) studied kinetin-induced mobilization of ^{32}P in the isolated leaf of corn (*Zea mays* L. hybrid WF 9×38–11). Normal movement was from tip to base. A localized application of kinetin to the leaf caused accumulation of ^{32}P in the area of treatment, as well as a more rapid transport toward the area of kinetin application. When detached blades of oat (*Avena sativa* cv. Victory) were treated with kinetin at the tip, and ^{14}C-2-glycine was applied to the base, no hormone-directed transport was noted; but if ^{14}C-2-glycine was applied to the central part of the blade, there was modest acropetal movement (GUNNING and BARKLEY 1973).

Fig. 2. Autoradiograph showing the distribution of imported ^{14}C within a mature darkened grape leaf following application of 4.4×10^{-3} M BA to right half of the leaf. The left half was treated with water containing wetting agent. The shoot base was ringed and each of the two leaves immediately above the treated one was given 25 μCi $^{14}CO_2$. (After QUINLAN and WEAVER 1969)

GOLDSCHMIDT and HUBERMAN (1974) cut flowers of citrus to shoot peduncles, and then implanted the flowers in an agar sucrose medium supplied with ^{14}C-sucrose, ^{3}H-acetate and ^{14}C-acetate. A characteristic distribution of label among organs for each isotope was obtained, thus allowing them to assess the role of respiration in their experiments (rationale for using both ^{3}H and ^{14}C). Wounding or application of NAA to a single petal shifted the bulk of the ^{14}C-sucrose toward the treated petal, but all petals received some labeled materials. They proposed that each flower organ meristem creates a sink of its own which acts in a typical manner according to its specific endogenous hormonal balance, and that the sink activity of all meristems in concert could result in a sensitive regulatory mechanism which is responsible for the coordination observed in flower development (GOLDSCHMIDT and HUBERMAN 1974).

3.2 Partition Between Roots and Shoots

During most of the growing season, it is important for the plant as an organism that leaf tissue retain or store as much assimilate as possible for subsequent transport to young leaves. Thus, more foliage can be produced which, in turn, can increase the plant's total capacity for growth. The roots, stems, and petioles should be furnished only enough assimilate to support leaves in an efficient arrangement, and produce a root system adequate to supply support, water, and mineral nutrients (WATSON 1971). While from man's viewpoint, it is often desirable to shift assimilates toward a specific portion of the plant, competition between roots and shoots for assimilates is usually strong, and the shoots often predominate.

SHINDY et al. (1973) found that application of BA to grape roots increased the amount of ^{14}C-labeled photosynthate transported to those roots from leaves exposed to $^{14}CO_2$, including amino acids, organic acids, and sugars, indicating that this growth regulator has a very broad mobilizing effect. The BA also altered the relative amounts of label in the amino acid, organic acid, and sugar fractions, possibly due to cytokinin effects on translocation of the compounds from other sources, or synthesis of these substances in the roots themselves, or both.

GINSBURG (1974) studied the effect of GA_3 and chlormequat applied to the entire plant on assimilate distribution in gladiolus in relation to corm growth, labeling a single leaf with $^{14}CO_2$ and following distribution patterns of the assimilates. He found that GA_3 promoted inflorescence growth by directing assimilate movement toward the inflorescence at the expense of the corm. Surprisingly, chlormequat had a similar, but lesser effect. Perhaps the effect of chlormequat was to increase the GA levels in the inflorescence, since chlormequat was found to increase levels of endogenous GA's in pea seedlings (REID and CROZIER 1970).

When the period of rapid inflorescence growth was over and the corm became the main sink, translocation from the ^{14}C-labeled leaf to the corm was promoted by chloremequat, but inhibited by GA_3, relative to control, although total movement to the corm increased for all treatments. It was concluded, therefore, that a decrease in ability of the corm to attract assimilates did not cause decreased assimilate movement into the corm; rather this was caused by increased competition form the inflorescence.

BREEN and MURAOKA (1973) studied effects of IBA on rooting and assimilate distribution in softwood cuttings of Marianna 2624 plum (*Prunus cerasifera* × *Prunus munsoniana*?). Callus on the bases of IBA-treated cuttings became noticeable on day 8 after treatment. A concomitant increase in the accumulation of translocated ^{14}C-photosynthate at the base of the cuttings was observed.

3.3 Assimilate Partition Within the Shoot System

It is often convenient to carry out source-sink studies on the shoot system, especially when there are strong sinks present such as rapidly growing fruits. Because of the difficulty in handling large intact plants, seedlings, or propagules from mature plants are often used. HATCH and POWELL (1971a) noted the effects of several plant hormones on the mobilization of ^{32}P in apple seedlings grown in the greenhouse. Seedlings were defoliated prior to hormone treatment, and the apex removed to avoid competition. Auxins, GA's, and cytokinins could direct the transport of ^{32}P, acropetally, basipetally, and into and out of leaves, depending on which area received hormones, certain combinations being especially effective. This mobilizing effect was less evident in seedlings that had ceased growth and formed terminal buds than in actively growing seedlings. Mobilization of ^{32}P was through the phloem to the sink (cut tip), and required the synthesis of protein. When cycloheximide, an inhibitor of protein synthesis, was added to a decapitated seedling, no mobilization occurred.

In another series of experiments with apple seedlings, HATCH and POWELL (1971b) applied IAA, GA_{4+7} and BA to apple seedlings to determine their influence on translocation of ^{14}C-sorbitol, ^{14}C-glycine, ^{14}C-NAA, ^{3}H-GA_1 and ^{14}C-kinetin. Each hormone mobilized ^{14}C-sorbitol, but only after root competition was eliminated by steam girdling. A mixture of the three growth substances effectively mobilized ^{14}C-glycine acropetally, the translocation being greatly enhanced after root competition was eliminated. ^{14}C-NAA was not mobilized at all. The BA mobilized ^{3}H-GA_1 basipetally. An acropetal movement of ^{14}C-kinetin was induced only with a mixture of GA_{4+7} and IAA. Thus, three growth substances had varying effects in their abilities to mobilize organic substances, including other growth regulators, and these effects were generally muted by the presence of the root system as a competing sink.

WEAVER et al. (1969) sprayed entire shoots of young grape cuttings with BA, 2 days later exposing a lower leaf to $^{14}CO_2$. In the absence of BA, movement of ^{14}C-photosynthate was entirely downward. In shoots that were sprayed with BA, appreciable amounts of ^{14}C-photosynthate moved into the upper parts of the plant, but some moved downward. Thus, as a result of cytokinin treatment, the upper portion of the plant became a sink for assimilates.

When chlormequat (2,000 mg l^{-1}) was used instead of BA, the normal upward movement of ^{14}C-photosynthate from an upper leaf was significantly reduced, only slight movement (bidirectional) of photosynthate occurring (WEAVER et al. 1969). It was not determined, however, whether photosynthesis was reduced or, more probably, translocation inhibited. If the latter, then chlormequat (a growth retardant whose effects can often be counteracted by GA application) greatly decreases the capacity for mobilization of assimilates within the upper portion of the shoot.

3.4 Mobilization of Assimilates into Fruits

Fruits are usually strong sinks and numerous experiments indicate that application of hormones to fruits can increase their power as sinks.

WEAVER et al. (1969) dipped clusters of young Black Corinth grape berries in 4-CPA, GA_3 or BA. After 0.5 h to 5 days the leaves above the clusters were exposed to $^{14}CO_2$, but no growth regulator. Counting and autoradiographic techniques were used at harvest to determine the amount and distribution of activity in the clusters, berries of hormone-treated clusters being heavier than those of controls within 3 days. Total cpm in the fruits were increased by 4-CPA, and the distribution of radioactivity among the sugar, organic acid, and amino acid fractions was usually modified, relative to control, by all hormone treatments. In a time-series experiment (WEAVER et al. 1969) there was an eightfold increase in total cpm, relative to the control, within 6 h of treatment of fruits with GA_3. After 18 h there was a nine-fold and a six-fold increase in counts in tartaric and malic acids, respectively; and in γ-aminobutyric acid, pipecolic acid, and valine, there were increases of 56, 150, and 330%, respectively. Radioactivity in fructose was increased 70% in GA_3-treated clusters, relative to the controls. After 96 h there were only about 1000 cpm g^{-1} fr. wt. in control clusters, compared to about 31,000 cpm in treated clusters. Dipping just portions of clusters into GA_3 also increased the movement of ^{14}C assimilates into the treated portions.

Gibberellin A_3 can also be used to establish competing sinks. Thus, shoots sprayed with GA_3 competed better for ^{14}C-photosynthate against GA_3-treated berry clusters than did control shoots with no GA_3 spray (WEAVER et al. 1969).

Kinetin treatment of citrus fruits alters the normal translocation patterns (KRIEDEMANN 1968) mobilizing labeled assimilates into the fruitlets, irrespective of their position on the shoot.

When BA was applied to peduncles of young fruits of *Vigna unguiculata* L. cv. Adzuki at different stages of ontogeny, no significant effect on the weight increase or the usual gradient of metabolites was observed (ADEDIPE et al. 1976). In contrast, with the cultivar Mala there was a 43% increase in ^{14}C-import (36% increase in weight) over that of the control fruits. Thus, hormonal effects on translocation appear to be under genetic control.

Developing ears of grain in wheat are powerful sinks, although auxin production by *Triticum aestivum* L. cv. Gabo grains does not appear to control assimilate translocation from the flag leaf through the peduncle (WANG and HSIA 1962). Other hormones were not examined. Seeds are rich in a variety of hormones and perhaps assimilate translocation in this instance was controlled by other endogenous regulators.

3.5 Import of Assimilates into Expanded Leaves

Usually fully expanded leaves only export assimilates, and have no import. However, application of hormones, leaf darkening, and/or ringing can cause changes in the patterns of movement.

Leaves of grapes which were nearly fully expanded imported only trace amounts of ^{14}C following photoassimilation of $^{14}CO_2$ by a lower leaf on the same shoot (QUINLAN and WEAVER 1969). However, application of BA resulted in a marked increase in the movement of ^{14}C-photosynthate into these leaves, although older leaves near the shoot base were less responsive to BA treatment. Treatment with BA alone to a leaf had little effect on import of labeled assimilates from leaves immediately adjacent, but when BA-treated leaves were darkened, import of ^{14}C-photosynthate increased. Ringing the base of shoots caused accumulation of high levels of radioactivity in BA-treated dark leaves (Fig. 3), relative to dark controls. However, it should be noted that darkening without the application of BA also increases import of ^{14}C-photosynthate. Accumulation of imported ^{14}C was restricted to the portion of the leaf blade actually receiving BA. A lower percentage of cpm was present in the sugar fraction from BA-treated leaves, with an increased percentage present in the amino

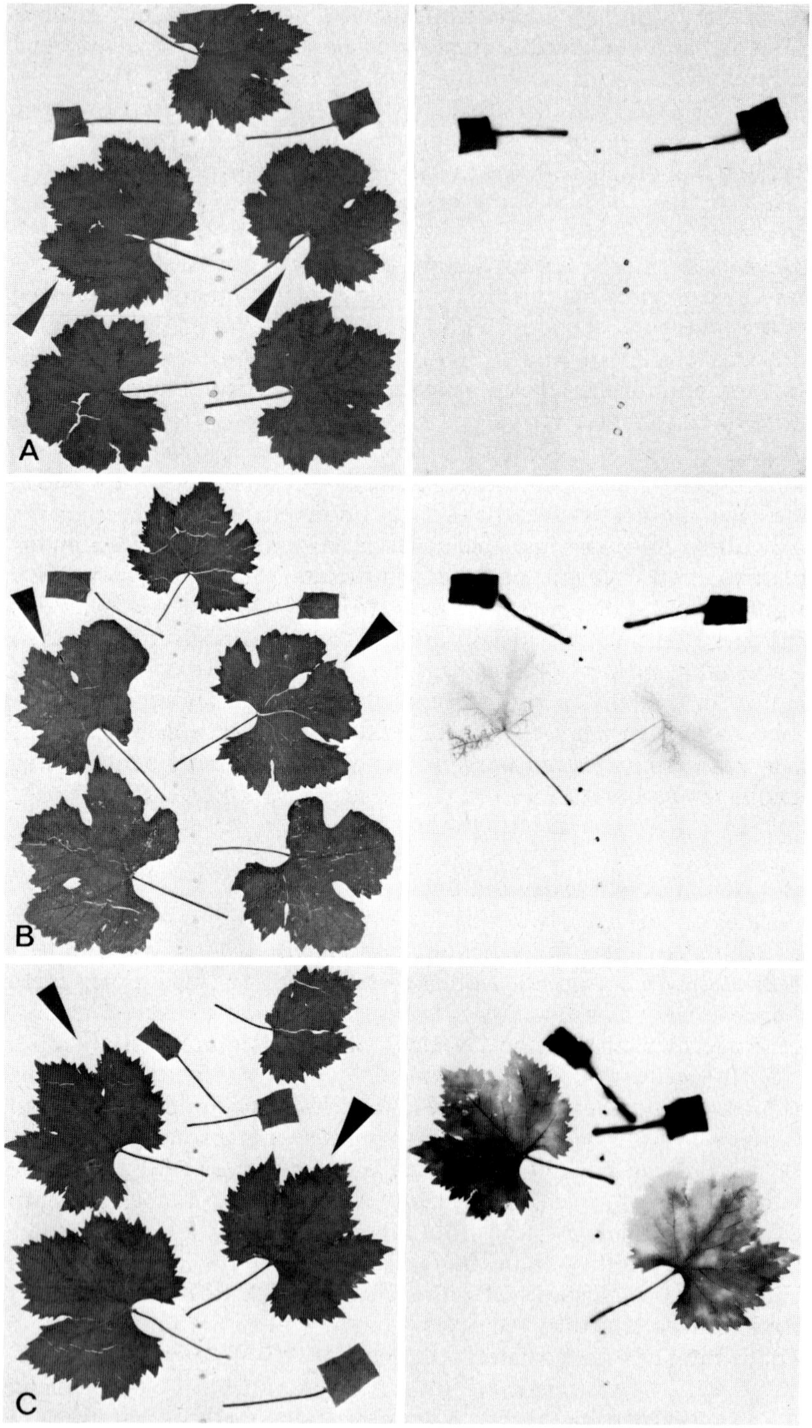
A
B
C

acid fraction. Thus, exogenous cytokinin appears to interact with the light environment to modify both quality and quantity of assimilate translocated within the shoot.

3.6 Export of Assimilates When Leaves or Shoots Are Pretreated with Growth Substances

Leaves. It has been shown that cytokinins and GA's, spotted on a leaf, attract assimilates, and that growth retardants often cause a decrease in the mobilizing power of the treated area (REID and CROZIER 1970; WEAVER et al. 1969) (Fig. 4). With small rooted cuttings of cv. Thompson Seedless grapes, young fully expanded leaves (which were exporting photosynthate upward in the shoot) were dipped momentarily into a solution containing various growth regulators. After 3-h treated leaves were exposed to $^{14}CO_2$, and 6 h subsequent to that shoots were collected, mounted, and autoradiographed. In one experiment, the leaves were dipped in BA at 4000 mg l^{-1} (SHINDY and WEAVER 1970), corresponding shoots being retained as control. In the controls, assimilates were rapidly exported from the leaves, moving acropetally in the stem; but very little assimilate was exported from BA-treated leaves. Similar results were obtained with 1000 mg l^{-1} of GA_3, in lieu of BA. Thus, hormone pre-treatment had virtually stopped export from fully expanded leaves.

When the plant-growth retardant chlormequat (4000 mg l^{-1}) was used, photosynthate moved out acropetally at the same rate as from the untreated leaves (SHINDY and WEAVER 1970). The export of photosynthate from the leaves treated with chlormequat is in agreement with earlier findings (SHINDY and WEAVER 1967; WEAVER et al. 1969) that the retardant tends to counteract the mobilizing power of leaves as a sink.

Shoots. It is often difficult to kill woody plants by foliar application of herbicides, even when the root system is actively growing and is a strong sink. Frequently the aerial portions of plants are killed, but sprouting and regrowth occur from the base of the plant. Experiments conducted with grape indicate that hormonal herbicides sprayed on foliage and stems create sinks, thus inhibiting translocation to the root or storage organs.

Hence, in Ribier grapes high concentrations of 2,4-D and picloram interfered with the basipetal movement of assimilates out of the shoot and into the parent vine (LEONARD et al. 1967). Shoots were bagged and given $^{14}CO_2$ and then the foliage was treated after 1 h with aqueous solutions of 2,4-D or picloram

←

Fig. 3 A–C. Effect of BA, darkening and ringing on movement of ^{14}C-photosynthate into mature grape leaves. Two adjacent lower leaves (*arrowed*) were treated with water containing wetting agent (**A** and **B**), or with 4.4×10^{-3} M BA (**C**). *Arrowed* leaves in (**B**) and (**C**) were darkened and had shoot bases ringed. The two leaves immediately above the arrowed ones were fed with 25 μCi; $^{14}CO_2$. (After QUINLAN and WEAVER 1969)

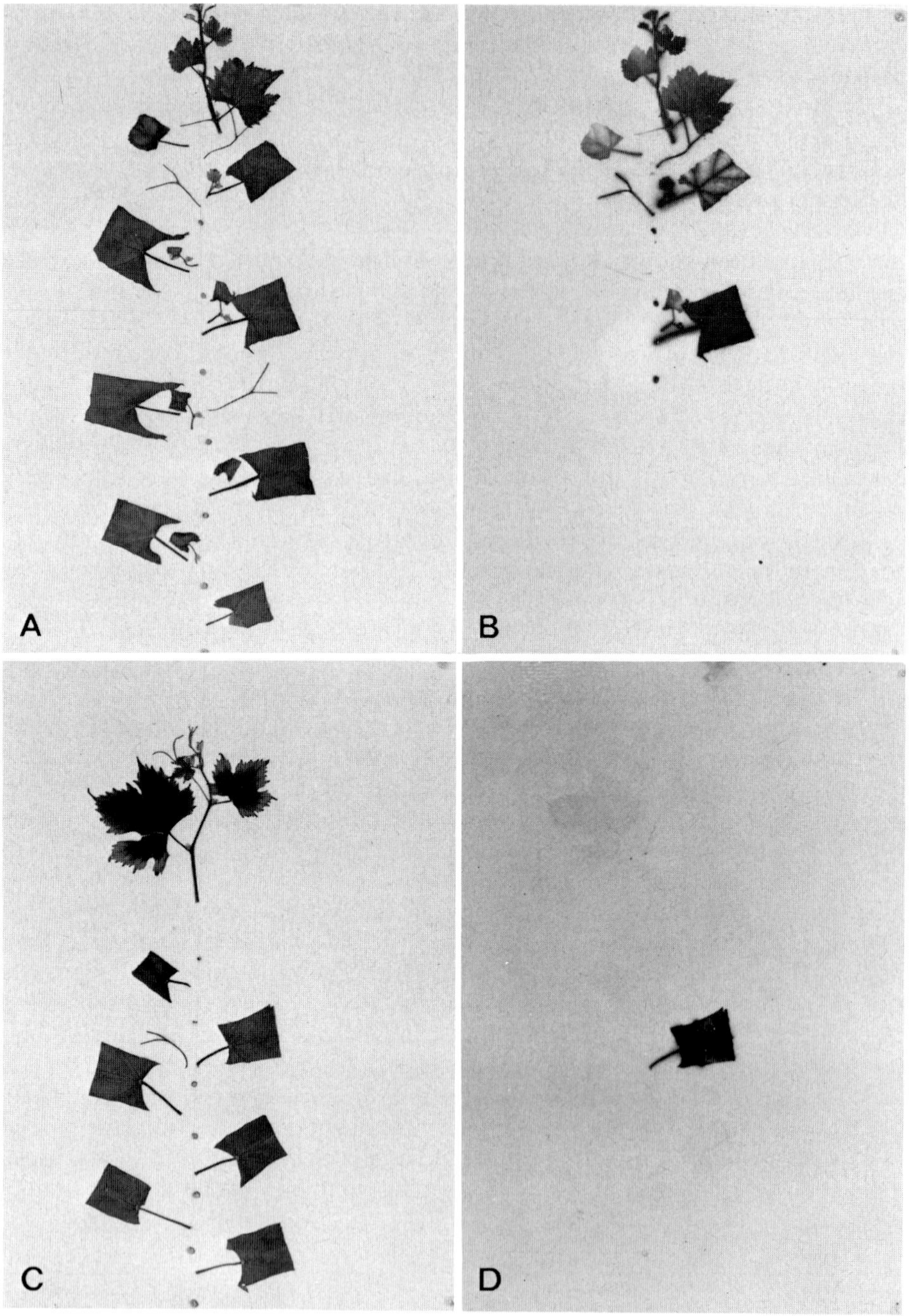

Fig. 4 A–D. Unsprayed grape shoot in which an upper leaf was given $^{14}CO_2$. **A** is a mounted section and **B** an autoradiograph. Note that translocation of ^{14}C is mainly upward. For a shoot sprayed with chlormequat **C** is a mounted specimen and **D** an autoradiograph. Only slight export of ^{14}C from the chlormequat-treated leaf has occurred. (After SHINDY and WEAVER 1970)

(20 ml total) at concentrations of 2,000 to 50,000 mg l^{-1}. Picloram was more effective than 2,4-D in preventing the normal basipetal movement of ^{14}C-photosynthate out of the shoot. However, movement of assimilates within the treated shoots was not prevented by either 2,4-D or picloram, evidenced by the continued transport of ^{14}C-photosynthate into the berry clusters in all treatments. Another striking example of the herbicide not causing symptoms of toxicity in other parts of the plant is the effective killing of the aerial portion of half of a cordon-trained grapevine by 2,4-D without causing injury to the other half (WEAVER 1970). Thus, the alteration of transport patterns by these herbicides appears to be caused by the development of strong sinks.

3.7 Time-Course Studies on Hormone-Induced Movement of ^{14}C Assimilates

Elongated stems of dwarf bean (*Phaseolus vulgaris* cv. Canadian Wonder) were used to study the direct effect of auxin on translocation, the minimum time required to preceive an effect of auxin of ^{14}C-photosynthate accumulation at a decapitated stump being noted (PATRICK and WAREING 1973). Lanolin pastes, with or without IAA, were applied immediately following decapitation, $^{14}CO_2$ being given to one of the primary leaves at various intervals subsequent to decapitation. Three h after each application the treated stumps were radioassayed. Rate of ^{14}C-photosynthate transport increased with time, quantities of ^{14}C-photosynthate being equal at 3 (Fig. 5). But, by h 6, auxin had significantly enhanced transport above that in the controls, a trend which continued to h 9. These results closely correspond to those of (DAVIES and WAREING 1965) for ^{32}P translocation in decapitated stems of pea treated with IAA.

SETH and WAREING (1967), also using decapitated dwarf bean plants, applied IAA, GA_3 or kinetin, all at 1000 mg l^{-1}, to the cut stump. After 12 h ^{32}P

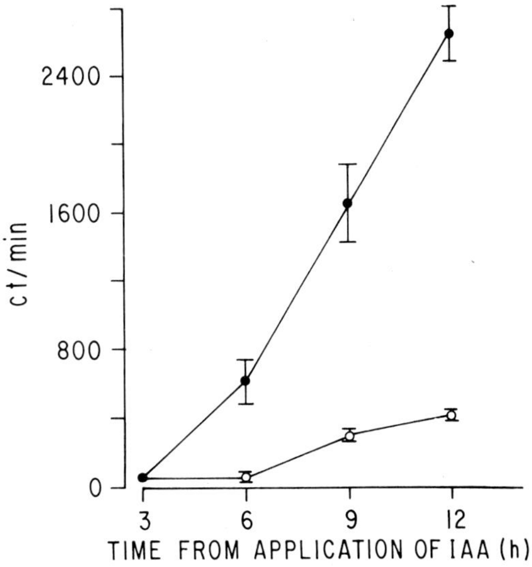

Fig. 5. Time course of ^{14}C-photosynthate transport for IAA-treated (●) and control (o) stems of decapitated plants. Radioactivity (cpm) accumulated in upper 1 to 5 cm of stem stumps at 3-h intervals. (After PATRICK and WAREING 1973)

was applied to the leaf end of the petiole and the radioactivity in the stump counted at intervals of 2, 4, 6, and 12 h. An enhancing effect by IAA was detected within 2 h. No significant effect was noted for GA_3 or kinetin.

In another experiment IAA was applied to decapitated fruit peduncles of dwarf bean, enhancement of the movement of labeled photosynthate from leaves to peduncles being noted (SETH and WAREING 1967).

3.8 Effect of Hormone Concentration on Translocation

For decapitated internodes from young vegetative dwarf bean plants, IAA (from 10 to 1000 mg l^{-1}) increased the movement of ^{32}P to the stump in a linear fashion up to 1000 mg l^{-1} IAA, 10,000 mg l^{-1} depressing ^{32}P transport (SETH and WAREING 1967).

3.9 Interaction of Different Hormones on Hormone-Directed Transport

When kinetin and GA_3 were tested alone for their effectiveness in stimulating transport of ^{32}P using decapitated bean, the results were negative (DAVIES and WAREING 1965; SETH and WAREING 1967). However, using material other than bean, many workers have shown GA_3 and kinetin to be quite effective in stimulating transport of nutrients over short distances, and it is possible that these hormones might also act synergistically in the long-distance transport of metabolites. Thus, SETH and WAREING (1967) attempted to test this possibility on the decapitated bean system. When they applied the three hormones together there was a much greater accumulation of activity than when just IAA plus GA_3, or IAA plus kinetin (K), were applied. The order of effectiveness was:

$$\text{Control} < GA_3 < IAA < IAA + GA_3 < IAA + K < IAA + K + GA_3$$

In another experiment with decapitated bean seedlings internodal stumps were treated with either IAA (1000 mg l^{-1}) or IAA, GA_3 (200 mg l^{-1}), and BA (200 mg l^{-1}) (MULLINS 1970). After 20 h ^{14}C-sucrose was applied to primary leaf petioles. Four h later, internodal segments harvested for radioassay showed a greater accumulation of radioactivity in the internodes of IAA-treated plants than in those of the controls, and the effect of the growth-regulator mixture was greater than that of IAA alone. Although kinetin alone had no effect on long-term transport, when IAA and kinetin were applied together to decapitated fruit peduncles of bean plants (*Phaseolus vulgaris* cv. Canadian Wonder), each at concentrations of 1000 mg l^{-1}, the combination was synergistically effective in promoting greater accumulation of ^{32}P in the area of hormone application (SETH and WAREING 1967). As noted above for kinetin, GA_3 in combination with IAA also gave a synergistic promotion of ^{32}P transport into bean fruit peduncles.

MULLINS (1970) treated decapitated stems of cv. Windsor Longpod bean seedlings with ethylene (0, 500 ppm, and 100%) for 20 h. Then a primary leaf on each plant was exposed to $^{14}CO_2$, and internodes were assayed for radioactivity 5 h later. There were, respectively, five- and tenfold increases in the radioactivity of treated tissues relative to the control. However, when 6 mg of ethrel was applied to freshly decapitated shoots, the movement of ^{14}C-sucrose from a primary leaf was unaffected.

Abscisic acid application to decapitated internodes of bean can inhibit translocation of assimilates from an epidermal flap on a primary leaf petiole (MULLINS 1970). The cytokinin PBA effectively counteracts ABA.

MULLINS (1970) studied the accumulation of ^{14}C-photosynthate in decapitated shoots of bean which had been pretreated with TIBA, an inhibitor of auxin transport. Concentrations of TIBA which inhibited ^{14}C-IAA translocation stimulated protein synthesis, and where IAA was applied after TIBA, the import of ^{14}C-photosynthate was enhanced relative to IAA alone.

Experiments with girdling and its effects on translocation of GA's indicate that GA's are produced in young leaves and roots of sunflower (*Helianthus annuus* L. Russian (KAMIENSKA and REID 1978). Thus, hot wax girdling reduced the endogenous levels of GA's in roots within 3 h of treatment, and after 48 h the levels in some girdled plants were higher than in the nongirdled ones. Girdling also markedly inhibited basal translocation of $^{3}H\text{-}GA_1$ and, to a lesser extent, acropetal movement. After 48 h girdled plants translocated $^{3}H\text{-}GA_1$ from shoots to roots almost as well as in nongirdled controls.

In sugar beet (*Beta vulgaris* L.) there is a good correlation between the level of carbohydrate in the leaf and the rate of photosynthesis (THORNE and EVANS 1964), the accumulation of photosynthate leading to a reduction of photosynthesis (HABESHAW 1973). The rate of photosynthesis can be linked directly to the sinks of the plant, including the roots, through the ability of the sinks to control rates of translocation from the leaves. Hormonal involvement may be direct, by creating or enhancing sinks, or indirect, by stimulating growth.

An example of indirect effects may be the work (HUMPHRIES and THORNE 1964) where CO_2 exchange was studied for fully expanded, detached primary leaves of dwarf bean which had roots on the petioles. The rates of apparent photosynthesis and respiration increased as roots grew, decreased when roots were removed, and increased again as roots regenerated on the petioles. Photosynthesis and respiration decreased when root growth was restricted by kinetin, and were increased when root growth was stimulated by IAA. These and other results (HUMPHRIES and THORNE 1964) indicated that photosynthesis may be positively correlated with the size of the roots.

3.10 Effect of Water Relations on Auxin-Induced Mobilization

PATRICK and WAREING (1976) decapitated one set of dwarf bean seedlings at internode 3, and heat-girdled others at internode 2, 8 cm above the primary node. Lanolin pastes, with or without IAA, were applied immediately below the girdle, or on the stem directly beneath the decapitated stump. Transpiration continued, since the xylem connections with the first trifoliate leaves remained intact and no sign of wilting occurred. Transport of ^{14}C-sucrose and ^{14}C-photosynthate to the stump or girdle regions was similar, IAA treatment enhancing mobilization for both! Thus, auxin-induced mobilization does not appear to be an artifact of any alteration in the water status of the stem that might occur upon decapitation.

3.11 Roots as a Sink

Very few experiments have tested the capacity of roots as a sink for assimilate. One of these (SETH and WAREING 1967) utilized an interesting, but artifact system with decapitated bean plants. Since only a small reduction in the activity of the root sink can markedly increase the amounts of assimilates translocated above the primary node, an experiment was designed to determine whether auxin may increase the flow of assimilates from the primary leaves to the treated stump by inhibiting root activity sufficiently to decrease the demand for photosynthesis. Heat-girdling of decapitated plants 3 to 4 cm below the cotyledonary nodes negated the sink effect imposed by the roots. The enhanced transport in response to IAA treatment was unaffected by the absence of a root sink, although the greater quantities of ^{14}C assimilate moved to the stump in the heat-girdled stems suggest that the presence of a root sink probably does limit the upward flow of photosynthate. It was also apparent that the transport system is not operating at full capacity in the decapitated bean shoot.

3.12 Auxin and Senescence

MULLINS (1970) suggested that in the absence of hormones the decapitated shoot of bean may be a declining system, and that the role of auxin is to maintain metabolite transport merely by preventing senescence. This suggestion was supported by the observation (PATRICK and WAREING 1970) that rapid losses in total protein occurred in control stems. To determine whether protein levels were causally related to the transport ability of stem stumps, bean plants were decapitated, "aged" with lanolin for 48 h, after which fresh lanolin, with or without IAA, was applied to the stump. A harvest of the upper 1.5 cm 12 h later showed no difference in total protein between control and auxin-treated stems. Therefore, auxin-induced mobilization could be examined while total protein levels remained constant. Using this aged system MULLINS (1970) gave either ^{14}C-sucrose or $^{14}CO_2$ to the trifoliate leaves, then measured radioactivity in the upper portions of the treated stumps. Auxin still exerted its normal effect on assimilate mobilization. Thus, the loss in total protein that occurred in the lanolin controls of PATRICK and WAREING (1970) appears to be independent on the mobilization that can be induced by auxin.

3.13 Effect of Hormones on Mineral Nutrient Uptake

There has been little research concerning the effect of hormones on mineral nutrient uptake, compared to that on hormone-directed transport of assimilates in the phloem.

ILAN and his colleagues (ILAN 1971; ILAN and REINHOLD 1963; ILAN et al. 1971) found that IAA inhibited NH_4^+ uptake by segments of sunflower hypocotyls, but increased the absorption of K^+. A small stimulation was found for Li^+ uptake, but there was little or no influence on uptake of Na^+. When kinetin was used instead of IAA, similar effects were obtained on the absorption of monovalent cations by detached cotyledons (incubated with KCl and NaCl) from etiolated seedlings. Thus, both IAA and kinetin affected the selectivity of ion uptake by sunflower plant parts. However, IAA caused no acceleration of K^+ uptake in cotyledons, and kinetin caused no K^+ uptake in hypocotyl sections.

Using leaf discs of sunflower ILAN (1971) showed that the absorption of K^+ was stimulated and that of Na^+ was inhibited by kinetin. Kinetin also promoted K^+ uptake in detached cotyledons from seedlings grown in light, although it had little effect on absorption of Na^+ by these cotyledons (ILAN 1971).

Benzyladenine also caused a higher K^+/Na^+ ratio, this time in cotyledons on intact plants. ILAN (1971) concluded that cytokinins can modify the selectivity of cells of sunflower leaves and cotyledons, increasing the affinity for K^+ relative to Na^+.

BODE (1959), using IAA on tomato plants, found increases of K^+ but no effect of IAA on Ca^{2+} in older leaves.

Investigations were made to determine the effect of GA_3 on ion ratios in plastids, cytoplasm, and vacuoles in short-term experiments in a dwarf maize mutant (*Zea mays*). Following GA_3 treatment the K and Cl content of the cytoplasm increased immediately whereas, the K content of the plastids increased after a 2-h lag phase. However, the Cl content of the chloroplasts increased without a lag phase subsequent to a GA_3 treatment. It was thought that the observed ion content changes were not the "primary action" of the hormone, but were somehow involved in a reaction sequence between the primary site of action of GA_3 at the membrane and the biological reaction membrane (NEUMANN and JANOSSY 1977).

Recent results confirm the movement of specific ions in hormone-directed transport. BERAUD and PENOT (1982) reported that GA_3 and BAP treatments had an attractive effect on ^{32}P phosphate and ^{35}S sulfate but not on ^{36}Cl chloride in detached leaves of *Pelargonium zonale*. However, in 1983 BITTNER and BUSCHMANN found that kinetin treatments reduced the uptake of potassium and calcium ions, with no effect on the uptake of magnesium in seedlings of *Raphanus sativus* L. More recently in 1984 SAKS and ILAN concluded that auxin via the shoot apex regulated the monovalent cation level in cotyledons of sunflower plants.

4 Possible Regulatory Points for Hormone-Directed Transport

As plants evolved from relatively simple organisms to more complex forms, the need for a more elaborate transport system increased. For such a system, some means to regulate the flow of assimilates from regions of manufacture (sources) to areas of utilization in structural requirements and storage (sinks) are needed.

A review of the literature by GIFFORD and EVANS (1981) supports the hypothesis that assimilate distribution is predominately under sink control and may be hormonally mediated, a phenomenon commonly referred to as hormone-directed transport (HDT) (PHILLIPS 1975). Circumstantial evidence for hormonal control over assimilate distribution comes from several sources, including (a) observations (b) correlations between endogenous, exogenous growth regulators and responses (c) studies utilizing excised plants (removal of meristems) where exogenous sources of hormones are substituted for plant meristems, and (d) studies with plant pathogens.

Early observations indicate that hormones often occur in large quantities in rapidly growing tissues such as young seeds, meristems, root tips, etc. Usually these regions are also strong sinks for assimilates, indicating that hormones may be involved in controlling assimilate distribution. If so, it would seem reasonable that the addition of more hormone might strengthen the sink.

Early investigations (see this chapter) have shown a high correlation between exogenous applications of growth regulators and increase in sink strength. A classical example is the work by WEAVER and MCCUNE 1959). Exogenous application of GA_3 to grape berries resulted in a five-fold increase in berry fresh weight. In a current study by WICKS et al. (1985), the application of ABA to grape berries resulted in a significant increase in berry dry weight.

In earlier work NITSCH (1970) found little evidence to support the relationship between dry matter accumulation and endogenous levels of hormones in various tissues. However, recent work by COOMBE and PHILLIPS (1980) has demonstrated a correlation between endogenous levels of ABA and the onset of ripening. According to PATRICK (1976) work on "hormonal compartmentalization within the sink tissues, the relative metabolite-mobilizing potencies of various hormone fractions, and changes in degree of competition for assimilates during plant ontogeny" could provide much information on this subject.

Early studies (see earlier sections of this chapter) and more recent work by PATRICK and WAREING (1982), utilizing decapitated plants with exogenous supplies of growth regulators, have clearly demonstrated that growth regulators may influence assimilate distribution.

Further support for hormonal control of assimilate distribution comes from studies with plant pathogens by PEGG (1981); see also MASCARENHAS and CANARY, Chap. 15, this Vol. When plants are locally infected with fungi or parasitic higher plants, a diversion of assimilates to the site of infection occurs; with parasitic fungi, an increased production of plant hormones occurs at the site of infection (SMITH et al. 1969).

This review section will serve as a brief *introduction* to the evidence for hormonal control of assimilate movement at the level of source-path-sink [also see review by PATRICK (1982)].

4.1 Hormonal Regulation of Assimilate Movement at the Level of Source-Path-Sink

Regulation at the source-end of the phloem pathway may include: (1) source strength; (2) export of compounds; (3) entry of compounds into free space; and (4) phloem loading (LEPP and PEEL 1971b). Regulation at the level of the path (vascular tissues and translocation stream) might be accomplished through stimulation of the phloem transport system some distance from a sink or source (longitudinal transfer within the sieve-tubes) (PATRICK 1976). Regulation at the level of the sink (local effects) might include: (a) stimulating or maintaining local sink activity [e.g., growth (MOORBY 1968), or polymerization of substrate molecules, which creates a continuing demand for assimilates (sink strength)]; (b) indirect stimulation of the rate of photosynthesis, thereby influencing metabolites available for export (source strength); (c) stimulating the accumulation of dry matter by a sink within the competitive framework of other sink regions [mobilizing ability of competing sinks (WAREING and PATRICK 1974)]; and (d) stimulation of the actual processes of phloem unloading. Implicit in HDT is that the sink regulates the level of imported and endogenously produced hormones.

Only the general findings of selected studies published since 1964 are presented which contain evidence supporting the concept of hormonal control of assimilate movement. These findings are based on the following types of studies: (a) time between treatments and response; (b) effects on nonelongating tissues; (c) relationship between the lag period and the distance moved by the hormone; (d) polar auxin transport inhibitors; (e) protein synthesis and nucleic acid metabolism inhibitors; (f) metabolism and accumulation; (g) long-distance transport; (h) distinguishing between hormonal effects on sink strength and phloem transport processes; and (i) rates of photosynthesis, export rates of assimilates, and changes in the mobilizing abilities of competing sinks.

Most of the studies herewithin are based on the use of decapitated plant systems, where exogenous applications of hormones, usually auxin, are substituted for excised meristems (BASALAH and BOOTH 1983; PATRICK 1979a, b; PATRICK and WAREING 1980; SETH and WAREING 1967; WAREING 1978). Assessment of hormonal enhanced transport of sugars, metabolites, and hormone movement are usually based on the detection of exogenously applied labeled (e.g., ^{14}C-sucrose, ^{14}C-IAA) compounds.

4.1.1 Time Between Treatments and Response

Support for the idea that hormone-stimulated transport is independent of sink strength comes from studies where enhanced transport commenced within less than 1.5 h, "short-term experiments" (MULLIGAN and PATRICK 1979; PATRICK

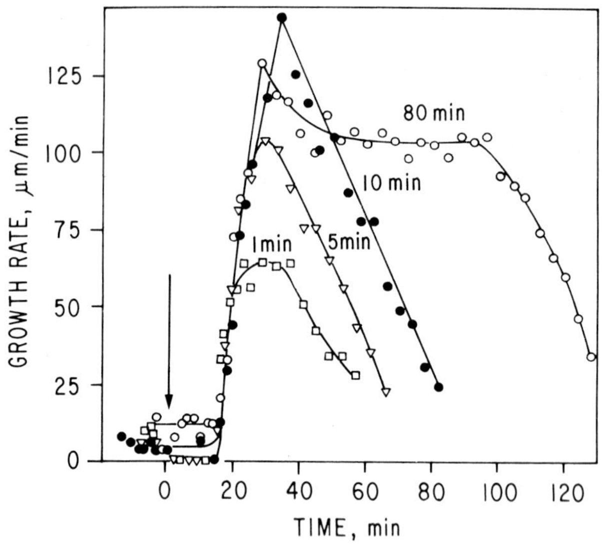

Fig. 6. Kinetics of changes in the growth rate of corn coleoptile sections after the introduction of 10^{-5} M IAA for various time periods, beginning at the *arrow marker*. In each case there was a lag of about 15 min before growth rate increased. Growth rate then decayed after removal of the auxin. (After DELA FUENTE and LEOPOLD 1970)

1979a; PATRICK and WAREING 1980; TURVEY and PATRICK 1979). Thus, how fast the response occurs after a given hormone treatment warrants further investigation.

Time Lag for Growth. Kinetic experiments in which growth versus time was measured are numerous and can be used to ascertain just what a short-term response really is. Time lags of about 15 min for growth of both corn (Fig. 6) (DELAFUENTE and LEOPOLD 1970) and *Avena* (RAY and RUESINK 1962) were noted after IAA applications, the latter using an optical system for continuously monitoring growth. NISSL and ZENK (1969) also used an optical system, and with high concentrations of IAA at high temperatures shortened the lag time to near zero. Since growth enhancement brought about the new synthesis of enzymes (as directed by nucleic acid synthesis) should take longer than a few minutes, hormones may thus act in other, more direct ways to enhance growth, and consideration must be given to the actual length of time needed for growth to occur subsequent to hormone application.

4.1.2 Effects on Nonelongating Tissues

Studies (PATRICK and WAREING 1976), where the magnitude of HDT is insensitive to potential changes in sink strength, and where HDT was demonstrated in (e.g., PATRICK and WAREING 1973, 1980) nonelongating tissues (nongrowing tissues) in conjunction with short-term responses (mentioned above), lend support for hormone-enhanced transport independent of sink strength. However, the possibility exists that hormones may have effects on these "nongrowing tissues", such as the stimulation of synthetic activities which could increase assimilate flow.

Nature of Growth. Advancement in the field of hormone-directed transport will be promoted by research directed toward determining the role of hormones in controlling the initial events of growth. Invaluable in this regard are scientific endeavors with a primary objective of examining the roles of hormones in regulating RNA metabolism in vivo. These studies can be used to provide some insight into just how hormones direct metabolites.

Thus, as early as 1964 a 25 to 30% increase in RNA synthesis was noted (KAY and SHANNON 1964) in the ribosomal fraction of fully elongated soybean hypocotyl cells within 12 h after 2,4-D treatment. Bean (*Phaseolus vulgaris* cv. WINDSOR longpod) internodes pretreated with a mixture of PBA, GA_3, and IAA for 2 to 5 h show an increase in the incorporation of ^{14}C-orotic acid into RNA during the following 2.5 h, well before any increase in elongation could be noted (MULLINS 1970).

Short-term auxin treatments of lentil (*Lens culinaris*) roots resulted in an increase in labeled TB-RNA (RNA which binds strongly to methylated albumen kieselguhr column (PENON 1972). When oat coleoptile segments were incubated in IAA for 1 h, an enhanced incorporation of labeled precursors was noted (CLINE and REHN 1974) into rapidly labeled RNA rich in AMP, a characteristic of polydisperse RNA, or possibly mRNA. The best evidence that auxin enhances the level of specific mRNA is in an experiment (VERMA et al. 1975) where tall Alaska peas were treated with 2,4-D. There was a tenfold increase in buffer-soluble cellulase mRNA per tissue segment within 48 h (VERMA et al. 1975).

Tobacco (*Nicotiana tabacum* cv. Wis 38) was aseptically isolated and grown with and without kinetin or zeatin, on a mineral-sucrose medium. Although there were no short-time (1 d) effects on the incorporation of radioactive precursors into RNA, incubation with zeatin and kinetin enhanced the incorporation of radioactive precursors into r-, t-, and poly-disperse RNA after 4d (ZWAR 1973).

ZURFLUH and GUILFOYLE (1982) reported that auxin treatment increased the levels of translatable messenger RNA for a number of in vitro translation products, the induction occurring within 15 min.

These investigations lend support to the idea that, in time, hormones can elicit responses in nucleic acid metabolism. Future investigations along these lines, coupled with the growing body of evidence indicating that RNA synthesis is a prerequisite for many hormonal responses, should increase our overall understanding of how hormones direct metabolites (JACOBSEN 1977).

In HDT experiments, at least some of the events leading to growth should be reflected in the rates of dark respiration and general metabolism, but unfortunately, measurements of either have been few.

4.1.3 Relationship Between the Lag Period and the Distance Moved by the Hormone

PATRICK (1979a) found a linear relationship (highly correlated) between the lag period, time between enhanced transport at the most basal site of the transport pathway and IAA treatment, and the transport distance. IAA applied to basal portions of decapitated internodes resulted in little effect on acropetal transport. These results taken together suggest that IAA effects are on site(s) beyond the point of hormone application and that IAA-enhanced acropetal transport is due to IAA effects along the entire transport pathway. Furthermore, in the work by PATRICK (1979a) the basipeted velocity of the IAA promoted stimulus, 2.1 cm h^{-1}, was in close agreement with earlier published work (see PATRICK and WOOLEY 1973). In contrast, no relationship between the transport distance and lag period for GA_3 (MULLIGAN and PATRICK 1979) or kinetin-induced (TURVEY and PATRICK 1979) transport response could be demonstrated for decapitated stems of *Phaseolus vulgaris*. These observations lend support to the idea that the hormones, GA, and kinetin act locally at the point of hormone application.

4.1.4 Polar Auxin Transport-Inhibitor Studies

PATRICK (1979a) found that IAA promoted acropetal transport of ^{14}C-assimilate was partially inhibited with ACP 1.55 or eosin, but neither inhibitor affected the control plant's acropetal transport or IAA-promoted basipetal transport. When IAA was applied 2 cm below the site of the inhibitor treatment, IAA-enhanced acropetal transport was restored. It was suggested that the inhibitor reduced basipetal IAA movement, which had been demonstrated in another study (PATRICK and WAREING 1978). The above study (PATRICK 1979a) utilized decapitated plants. However, when PATRICK (1979b) used intact plants and modified the endogenous stem levels of auxin with ACP 1.55, i.e., inhibited basipetal auxin movement, the movement of assimilate was reduced to the shoot apex. These data are interpreted to mean that acropetal transport of photosynthate is related to IAA levels in the stem remote from the source of hormone production (PATRICK 1979b) and hormone application (PATRICK 1979b). Furthermore, when exogenous IAA was introduced via a cut stem, it was found mainly in the stem (HEW et al. 1967). However, MORRIS (1982) reported that TIBA, another auxin transport inhibitor, had no effect on IAA-promoted transport, which confirmed earlier findings by MULLINS (1970). This later study also reported that the accumulation of ^{14}C-photosynthate was enhanced in decapitated internodes pretreated with specific levels of TIBA and subsequently treated with auxin. Other inhibitors of auxin transport, NPA and morphactin, were reported not to inhibit IAA promoted transport (DA CRUZ and AUDUS 1978b). These data (DACRUZ and AUDUS 1978b, MORRIS 1982, MULLINS 1970) suggest hormonal regulation at the local level. However, these data do not allow unequivocal assessment as to which transfer component, i.e., phloem unloading or sink activity, is under hormonal regulation.

Work with inhibitors must be evaluated with care. For example, although TIBA blocks the movement of IAA (CHRISTIE and LEOPOLD 1965), there may be side effects, such as phototoxicity, and inhibition of cell elongation (AUDUS 1959; CHRISTIE and LEOPOLD 1965). DA CRUZ and AUDUS (1978b) has shown that auxin doubled the proportion of ^{14}C incorporated into the soluble fraction compared to water control treatments; however, TIBA significantly reduced incorporation below the water control treatments. This observation and other data (DA CRUZ and AUDUS 1978a) are interpreted to mean that TIBA has inhibitory side effects. Furthermore, concentrations of TIBA that resulted in a reduction of polar auxin transport but showed no effects on IAA-enhanced transport, were associated with an increased incorporation of ^{14}C leucine into proteins. Thus, inhibition, due to TIBA, of the polar transport of endogenous auxin, may retard senescence of internodes (MULLINS 1970).

Hormone-Directed Transport and Senescence. Many experiments that were designed to determine the role of hormones in stimulating assimilate movement must be reevaluated because of possible influences of senescence. MULLINS (1970) suggested that at least in long-term experiments, e.g., 24 h, hormone-directed transport may merely be an artifact of retarding senescence in decapitated shoots. Also, in lanolin control of decapitated bean plants, protein declined to 30 to 50% of intact plant levels, whereas hormone-treated plants maintained high protein levels (PATRICK and WAREING 1970), at least partially through the synthesis of new protein (net incorporation of labeled leucine was five times greater in tissue treated with IAA than in the lanolin control tissue). The protein extracted from the control and from the auxin-treated decapitated bean stems was approximately equal (PATRICK and WAREING 1973), a seeming contradiction to earlier (PATRICK and WAREING 1970) results. However, in the above study (PATRICK and WAREING 1973) 3-day-old decapitated plants retained their ability to respond to auxin, i.e., IAA promoted metabolite transport. However, both experiments compare total protein levels, and these do not indicate possible qualitative changes that may have occurred. Hence, they may be of little significance in evaluating hormone-directed transport. Thus, it is important to determine whether the hormone stimulated overall protein synthesis, or merely prevented its decline. Unfortunately, the literature is somewhat contradictory. Many reports indicate that to maintain protein and RNA levels in leaf organs, RNA and protein synthesis were actually stimulated by hormones (LEOPOLD and KRIEDEMANN 1975). However, the work by TAVARES and KENDE (1970) strongly supports the supposition that the major role of hormones is to retard protein catabolism.

4.1.5 Inhibitor Studies of Protein Synthesis and Nucleic Acid Metabolism (NAM)

Inhibitors of protein synthesis (cycloheximide and puromycin) and NAM (actinomycin), which have no effect on IAA transport in stolons (DA CRUZ and AUDUS 1978a), were reported by DA CRUZ and AUDUS (1978b) to inhibit IAA-promoted accumulation of ^{14}C-assimilates at the distil ends of young growing stolons. However, phloem unloading may be mediated via carrier proteins (see literature cited by PATRICK 1983). Thus, the above evidence supporting the idea that IAA exerts its effect on HDT through regulation at the level of the sink is circumspect.

4.1.6 Metabolism and Accumulation Studies

In a few studies (PATRICK and WAREING 1970, 1976), no changes in sink activity were associated with IAA enhanced transport, thus lending support for the concept that HDT is independent of sink activity. These findings are also supported by investigation using other hormones, e.g., GA_3 and kinetin (cited in review by PATRICK and WAREING 1980) and the finding that IAA depressed sucrose uptake into tissue slices (PATRICK and WAREING 1976). In contrast, MULLINS (1970) reported that subsequent to a 2.5 h pretreatment of decapitated bean internodes with SD 8339, a cytokinin, GA, and IAA there was an increase in protein and RNA synthesis. Therefore, the hormone-enhanced transport of ^{14}C assimilate may have been due to an increase in sink activity (MULLINS 1970). In view of these limited investigations though, the statement that hormone-directed transport exists in the absence of increased local metabolism or growth cannot be considered as a general fact.

4.1.7 Long-Distance Transport

When IAA and labeled sucrose were applied to the same point on intact internodes, no promotive effect on sucrose uptake could be detected (PATRICK and WAREING 1970). It was concluded that IAA effects are only obtained when phloem transport is over a minimum distance. These findings are also supported by later investigations (PATRICK and WAREING 1980).

4.1.8 Distinguishing Between Hormonal Effects on Sink Strength and Phloem Transport Processes

IAA slightly inhibited, kinetin had no effect, and GA_3 stimulated in vitro rates of sucrose uptake by bean stems. IAA and kinetins increased the in vivo apoplast pool size of sugars, and GA_3 yielded mixed effects (PATRICK 1982).

The former and latter data are interpreted to mean that GA_3 exerts its effects at the level of the phloem and of the sink strength; however, IAA and kinetin act principally on the level of phloem transport processes (PATRICK 1982). Furthermore, when GA_3 and kinetin-enhanced transport for decapitated bean plants is restricted to the place of hormone treatment, hormonal regulation is thought to be via phloem unloading, the latter resulting in an increase in the apoplast pool size of sugar. Increases in invertase activity in sink tissues of decapitated plants have been correlated with increases in carbon import and growth subsequent to exogenous applications of hormones (MORRIS 1982), thus lending support for hormonal regulation at the level of phloem unloading.

Further support for hormonal regulation at the level of phloem unloading comes from investigations on abscisic acid (ABA). Abscisic acid has been found in grape berries (*Vitis vinifera* L.) during ripening (COOMBE and HALE 1973, DÜRING 1973, 1974, DÜRING et al. 1978, SCIENZA et al. 1978), and exogenous applications of ABA have been associated with increases in endogenous concentrations of ABA and/or enhanced grape berry ripening (HALE and COOMBE 1974, WICKS et al. 1985) and unpublished data from this laboratory.

In an informative review by DÜRING and ALLEWELDT 1980, the following possible modes of ABA action on *unloading* were advanced: (1) ABA was associated with increases in invertase synthesis which in turn caused the hydrolysis of sucrose, COOMBE 1976; (2) the active proton co-transport of sugars theorized

Fig. 7. Model for sieve-tube unloading and the effect of abscisic acid. (After TANNER 1980)

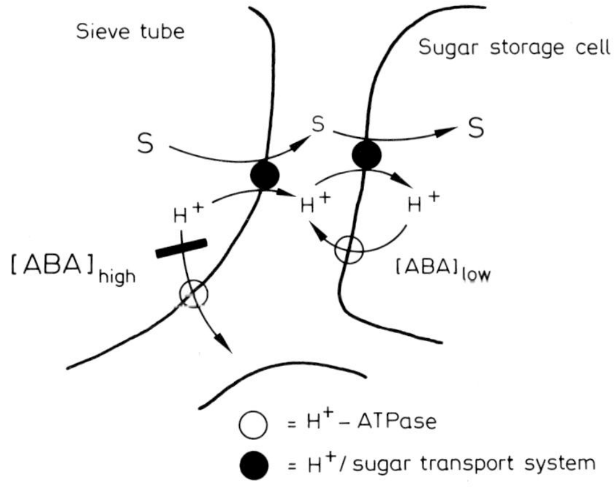

by MALEH and BAKER 1977, and GIAQUINTA 1979, involved ABA which is given credence in light of the fact that ABA affects proton and potassium transport in stomata, VAN STEVENINCK 1976; (3) exogenous applications of ABA to *Vitis vinifera* grapevine leaves increased the retention of assimilates, DÜRING and ALLEWELDT 1980; and (4) that observed increases in ABA in grape berries may be due to increased osmotic potentials which in turn are the result of sugar accumulation (DÜRING and ALLEWELDT 1980).

Some details on the possible role of ABA on phloem unloading have been postulated by TANNER 1980. The process can be envisaged as four consecutive steps as follows: (1) the presence of a high concentration of ABA in the sieve tube would inhibit the H^+-ATPase activity, thereby decreasing the proton motive force (pmf) i.e., the sum of the pH and the membrane potential across the sieve tube plasmalemma; (2) subsequently, sugars and protons (or sugars alone) are transported via the H^+ sugar symport system down the sugar gradient to the free space; (3) the presence of a low concentration of ABA in the sugar storage cell allows a high expression of this H^+-ATPase, i.e., high activity thereby facilitating proton accumulation in the free space and establishing a strong pmf at the plasmalemma of the sugar storage cell; (4) finally the H^+ sugar symport system promotes movement of sugar together with protons from a low sugar concentration in the free space via the plasmalemma into the storage cell with a high sugar concentration (Fig. 7).

4.1.9 Rates of Photosynthesis, Export Rates of Assimilates and Changes in the Mobilizing Abilities Between Competing Sinks

GA_3 (MULLIGAN and PATRICK 1979), or kinetin (TURVEY and PATRICK 1979) treatments to stumps resulted in enhanced acropetal assimilate transport for decapitated, mature tissues and was not associated with changes in rates of net photosynthesis or relative export rates of assimilates from primary leaves. In the above study with GA_3 no reduction in hormone-promoted acropetal ^{14}C-assimilate transport to the area of hormone treatment was detected for de-rooted plants. This data is interpreted to mean that GA_3 does not alter the mobilizing ability of competing sinks. However, UMOESSIEN and FORWARD (1982) reported that GA_3-promoted accumulation of assimilates in young shoots of sunflower was associated with a depletion of assimilates in the roots. While the above studies support the idea of hormonal regulation at the local level, it is difficult to determine which transfer component(s) are being affected by the hormone.

PATRICK and WAREING (1980) reported that GA_3 treatments to decapitated stems, in which there was a source limitation, resulted in little change of the relative rates of ^{14}C export for the source leaf; whereas, in a sink limited situation, GA_3 promoted the export rate. In a review by PATRICK (1982) phloem loading, a carrier-mediated transfer process (GEIGER 1975), was enchanced by the addition of IAA. Auxin promoted transport and enhanced leaf photosynthesis (BIDWELL et al. 1968, BIDWELL and QUONG 1975). Also, according to a recent review by GIFFORD and EVANS (1981), photosynthetic sources do not seem to control the final destination of their exported assimilates. Furthermore, investigations by MORRIS (1977) suggest that auxin produced in shoot meristems is not transported into source leaves in whole plants. However, recently, SAKS and ILAN (1984) reported that labeled IAA applied to the cut stem of sunflower plants was found in the cotyledons. Thus, the need for further studies on possible sink-mediated control via hormonal effects at the source-end of the phloem pathway are needed.

5 Hormonal Regulation of Photosynthate Supply

As our understanding of hormone effects becomes more complete, it is increasingly apparent that photosynthesis is under hormonal control. Several hypotheses have been suggested to explain the role of hormones in regulating photosynthesis (LEOPOLD and KRIEDEMANN 1975). One is concerned with changes in leaf resistance (r_l), i.e., stomatal and cuticular resistances. A second involves the changes in internal leaf resistance (r_r), i.e., biochemical and physical resistances in the leaf to carbon assimilation.

A change in leaf resistance through alteration of the stomatal aperture could regulate photosynthesis, and hormones can affect stomatal aperture. Exogenous applications of kinetin (10^{-6} M) to mature barley leaves rapidly increased stomatal openings (LIVNE and VAADIA 1965). The kinetin-treated leaves had a low CO_2 concentration in the intercellular spaces, and this may have caused stomatal opening. Furthermore, the rate of CO_2 assimilation of barley increased after application of 3×10^{-3} M kinetin (MEYNHARDT and MALAN 1963). Accompanying this was a decrease in leaf resistance ($r_l + r_r$), i.e., an increase in stomatal opening and a reduction in the CO_2 compensation point. In view of this change in compensation point, it is still unclear as to whether photosynthetic regulation by kinetin is mitigated through r_l or r_r (LEOPOLD and KRIEDEMANN 1975). To know this is a key point to better understanding the roles of hormones in regulating photosynthesis. More recently a model by EAMUS and WILSON (1984) describing the interaction of low temperature, IAA, ABA, and CO_2 in the control of stomatal behavior in *Phaseolus vulgaris* has been proposed.

Applications of 3.8×10^{-6} M ABA to wheat leaves rapidly reduced the rate of apparent photosynthesis and simultaneously increased stomatal diffusive resistance (MITTELHEUSER and VAN STEVENINCK 1971). Similarly, the onset of closure of stomates was brought about by ABA in just over 3 min in corn leaves (KRIEDEMANN et al. 1972). LOVEYS and KRIEDEMANN (1974), studying the after-effects of moisture stress, concluded that plant growth regulators (ABA and its metabolites) may influence gas exchange. But, more importantly, these latter investigations lend support to the idea that ABA affects photosynthesis by regulating stomatal opening, not by direct effects at the site of carboxylation. Clearly then, the application of some hormones do elicit a change in r_1.

Evidence indicating that a native hormonal system affecting photosynthesis and/or gas exchange exists is as follows: (1) roots may supply cytokinins to leaves (SKENE and KERRIDGE 1967); (2) the positive effects of cytokinins on stomatal opening, as mentioned above; (3) the changes in stomatal opening that are so well correlated with changes in endogenous levels of ABA (KRIEDEMANN et al. 1972, 1975, LOVEYS and KRIEDEMANN 1974; also see RASCHKE 1982) and (4) recent work by REID and BRADFORD 1984, BRADFORD 1982, with flooding experiments lend further support to hormonal control of the alterations of gas exchange. At least some endogenous hormones, then, appear to be components of a complex regulatory system, regulating gas exchange and/or photosynthesis via r_1.

Other results indicate that hormones may also regulate photosynthesis via r_r. Significant increases in photosynthetic rates of dwarf (d-1) maize plants occurred at high light intensities when plants were sprayed with GA_3 (WAREING et al. 1968). This was confirmed in red clover (*Trifolium pratense* L.), and a positive correlation between high activity of RuBP carboxylase and increased levels of endogenous GA's was also noted (TREHARNE and STODDART 1968). Phaseic acid, a possible product of ABA metabolism (MILBORROW and JONES 1971), had no effect on RuBP carboxylase activity, but did inhibit O_2 evolution, which could indicate reduced photosynthetic electron flow, and thereby decreased CO_2 fixation (KRIEDEMANN et al. 1975).

A third hypothesis refers to hormonal effects on the export rate of assimilates (see Sect. 4.1 p. 21) and on phloem loading at the source (LEPP and PEEL 1971b, MALEH and BAKER 1978).

A fourth hypothesis refers to indirect effects, e.g., changes in sink activity, or changes in leaf architecture that result in more efficient light utilization, thus increasing source activity. Mechanisms here may act individually, or interact to regulate photosynthate supply. MORGAN (1968) postulated that certain events occurring after application of GA_3 to *Festuca arundinacea* were related to increases in total dry weight, which enhanced leaf development, thus leading to increases in relative growth rates. With this knowledge, and the further observation that the immediate effects of stimulated photosynthetic rates by GA_3 can last 6 days (TREHARNE and STODDART 1968), LESTER et al. (1972) recognized the need to measure the effects of GA_3 on net photosynthesis over a longer time interval, i.e., 21 days. During this period no increases in leaf area index were observed, but GA_3 did cause increases in apparent photosynthesis. These were attributed in part to a change in leaf architecture, i.e., elongation of internodes which caused leaf blades to be dispersed more in the horizontal plane, resulting in more efficient light utilization (also see BRADFORD 1983). Apparent photosynthesis in mature leaves was also stimulated, hence GA_3 could have the effect of delaying senescence in old leaves (also see BRADFORD 1983 and literature cited therein). Thus, changes in leaf architecture, or in plant form, are major considerations when drawing conclusions on the effects of hormones in stimulating photosynthesis of the whole plant, or plant community.

A stimulation of photosynthesis may also come about by increasing sink strength via the application of growth regulators ((ALVIM 1960, HALEVY et al. 1964, MONSELISE and HALEVY 1962, TURNER and BIDWELL 1965, WARDLAW 1968) (see Sect. 4). More recently, STARCK (1983), demonstrated that NOA + GA_3 increased sink strength and increased the net photosynthetic rate per leaf of tomato plants.

6 Conclusions

Movement of photosynthetic assimilates to a sink is enhanced by a decrease in available carbohydrates in the sink tissue, thus resulting in a higher concentration gradient toward the sink. The number of competing sinks, and the distance from the site of assimilate production to the sink, also influence mobilization. These "internal environmental factors" are controlled by native plant hormones and appear to be influenced by applied plant-growth regulators.

Fully expanded leaves usually export assimilates, and do not import them. However, hormonal applications, or environmental manipulations (i.e., leaf darkening, and/or ringing) may induce import of labeled materials, or affect the selectivity of ion uptake by plant cells.

Many of the advances made in the area of HDT are based on fully elongated, decapitated plant systems. For such systems, the available evidence supports the supposition that hormones promote assimilate transport. However, it is difficult to assess which components of the source-path-sink are under hormonal regulation and which components contribute most significantly to HDT. This in part is due to the fact that many of the experiments are confounded, for example, by possibly side effects with the use of inhibitors or senescence. It is difficult to extrapolate findings within these to the whole plant system. Thus, the need for further studies utilizing whole plants. More work directed toward elucidating the possible role of hormones in phloem unloading is needed.

At least one cytokinin (kinetin) and ABA have specific effects on stomatal and carboxylation resistances in relation to photosynthetic regulation. Other growth substances associated with increased photosynthesis, e.g., endogenous GA's, whose levels have been shown to be correlated with the activity of RuBP carboxylase, appear to have alternative modes of actions. Phaseic acid, an inhibitor of in vivo photosynthesis, may act at the photochemical level. Also, sink activity may be stimulated, with a consequent rise in source activity. Internally, these hormones interact with both source and sink to regulate the level of source activity (i.e., photosynthesis). Finally, the levels and action of endogenous hormones (as component of a complex regulatory feedback system which regulates photosynthesis) are undoubtedly mediated by external environmental factors.

Acknowledgements. The authors thank Dr. H.B. CURRIER, Dr. D.C. LESTER, Dr. J. DUNNING, Dr. A.S. WICKS, and T.W. JOHNSON for reading the manuscript and making helpful suggestions.

References

Adedipe NO, Fletcher RA, Ormrod DP (1976) Distribution of ^{14}C assimilates in the cowpea (*Vigna unguiculata* L.) in relation to fruit abscission and treatments with benzyladenine. Ann Bot (London) 40:731–737

Altman A, Wareing PF (1975) The effect of IAA on sugar accumulation and basipetal transport of ^{14}C-labelled asimilates in relation to formation in *Phaseolus vulgaris* cuttings. Physiol Plant 33:32–38

Alvim PDT (1960) Net assimilation rate and growth behavior of beans as affected by gibberellic acid urea and sugar sprays. Plant Physiol 35:285–292

Asada K, Konishi S, Kawashima Y, Kasai Z (1960) Translocation of photosynthetic products, assimilated by the top leaf, to the ear of rice and wheat plants. Mem Res Inst Food Sci Kyoto Univ 22:1–11

Audus LJ (1959) Plant growth substances, 2nd edn. Wiley Interscience, New York

Basalah MO, Booth A (1983) The effect of IAA on the distribution of carbon-14 labeled assimilates in *Dahlia variabilis*. Bot Bull Acad Sid 24:11–30

Beraud J, Penot M (1982) Action of phytohormones on ion absorption and transport in leaves of *Pelargonium zonale*. Z Pflanzenphysiol 106:399–400

Bidwell RGS, Quong ECW (1975) Indoleacetic acid effect on the distribution of photosynthetically fixed carbon in the bean plant. Biochem Physiol Pflanzen 168:361–370

Bidwell RGS, Levin WB, Tamas IA (1968) The effects of auxin on photosynthesis and respiration. In: Wight F, Setterfield G (eds) Biochemistry and physiology of plant growth substances. Runge, Ottawa, pp 367–376

Bittner A, Buschmann C (1983) Uptake and translocation of K^+, Ca^{2+} and Mg^{2+} by seedlings of *Raphanus sativus* L. treated with kinetin. Z Pflanzenphysiol 109:181–189

Bode HR (1959) Über den Einfluss des Heteroauxins auf die Kationenzusammensetzung der Blattasche der Tomate. Planta 53:212–218

Bradford KJ (1982) Regulation of shoot responses to root stress by ethylene, abscisic acid, and cytokinin. In: Wareing PF (ed) Plant growth substances. Academic Press, London New York, pp 599–608

Bradford KJ (1983) Involvement of plant-growth substances in the alteration of leaf gas exchange of flooded tomato plants. Plant Physiol 73:480–483

Breen PJ, Muraoka T (1973) Effect of indolebutyric acid on distribution of ^{14}C-photosynthate in softwood cuttings of Marianna 2624 plum. J Am Soc Hortic Sci 98:436–439

Christie AE, Leopold AC (1965) On the manner of triiodobenzoic acid inhibition of auxin transport. Plant Cell Physiol 6:337–345

Cline MG, Rehn MM (1974) Effects of auxin and abscisic acid on polydisperse RNA in *Avena coleoptiles*. Int Conf Plant Growth Subst, Tokyo, pp 582–592

Coombe GB (1976) The development of fleshy fruits. Annu Rev Plant Physiol 27:207–228

Coombe GB, Hale CR (1973) The hormone content of ripening grape berries the effects of growth substance treatments. Plant Physiol 51:629–634
Coombe GB, Phillips PE (1980) Development of the grape berry. III. Compositional changes during veraison measured by sequential hypodermic sampling. UCD Grape and wine centennial Symp, pp 132–136
Crafts AS, Crisp CE (1971) Phloem transport in plants, Ch 3. Freeman, San Francisco
Cruz GS da, Audus LJ (1978a) Studies on the transport of tritiated indole acetic acid in stolons of *Saxifraga sarmentosa*. Ann Bot (London) 42:577–594
Cruz GS da, Audus LJ (1978b) Studies of hormone-directed transport in decapitated stolons of *Saxifraga sarmentosa*. Ann Bot (London) 42:1009–1027
Davies CR, Wareing PF (1965) Auxin-directed transport of radiophosphorus in stems. Planta 65:139–156
Düring H (1973) Abscisinsäure in *Vitis vinifera*-Früchten während der Reife. Naturwissenschaften 60:301–302
Düring H (1974) Abscisinsäure in reifenden Weinbeeren. Vitis 13:112–119
Düring H, Alleweldt G (1980) Effects of plant hormones on phloem transport in grapevines. Ber Dtsch Bot Ges 93:339–347
Düring H, Alleweldt G, Koch R (1978) Studies on hormonal control of ripening in berries of grape vines. Acta Hortic 80:397–405
Eamus DD, Wilson JM (1984) A model for interaction of low temperature, ABA, IAA, and CO_2 in the control of stomatal behavior. J Exp Bot 35:91–98
Esau K (1948) Phloem structure in the grapevine and its seasonal changes. Hilgardia 18:217–296
Fuente RK dela, Leopold AC (1970) Time course of auxin stimulations of growth. Plant Physiol 46:186–189
Geiger DR (1975) Phloem loading. In: Zimmermann MH, Milburn JA (eds) Encyclopedia of plant physiology, Vol I, New Ser. Transport in plants. Springer, Berlin Heidelberg New York, pp 395–431
Giaquinta RT (1979) Phloem loading of sucrose. Plant Physiol 63:744–748
Gifford RM, Evans LT (1981) Photosynthesis, carbon partitioning, and yield. Annu Rev Plant Physiol 32:485–509
Ginzburg C (1974) The effect of gibberellin A_3 and (2-chloroethyl)-trimethyl-ammonium chloride on assimilate distribution in gladiolus in relation to corn growth. J Exp Bot 25:995–1003
Goldschmidt EE, Huberman M (1974) The coordination of organ growth in developing citrus flowers: A possibility for sink-type regulation. J Exp Bot 25:534–541
Gunning, BES, Barkley WK (1973) Kinin-induced directed transport and senescence in detached oat leaves. Nature (London) 199:262–265
Habeshaw D (1973) Translocation and the control of photosynthesis in sugar beet. Planta 110:213–226
Hale CR, Coombe BG (1974) Abscisic acid – an effect on the ripening of grapes. In: Bieleski RL, Ferguson AR, Cresswell MM (eds) Mechanisms of regulation of plant growth. Soc NZ, Wellington, 12:831–836
Halevy AH, Monselise SP, Plaut Z (1964) Effects of gibberellin on translocation and on dry matter and water content in several plant species. Physiol Plant 17:49–62
Hatch AH, Powell LE (1971a) Hormone-directed transport of 32p in *Malus sylvestris* seedlings. J Am Soc Hortic Sci 96:230–234
Hatch AH, Powell LE (1971b) Hormone-directed transport of certain organic compounds in *Malus sylvestris* seedlings. J Am Soc Hortic Sci 96:399–400
Hew CS, Nelson CD, Krotkov G (1967) Hormonal control of translocation of photosynthetically assimilated ^{14}C in young soybean plants. Am J Bot 54:252–256
Humphries ED, Thorne GN (1964) The effect of root formation on photosynthesis of detached leaves. Ann Bot NS 28:391–400
Ilan I (1962) A specific stimulatory action of indolyl-3-acetic acid on potassium uptake by plant cells, with concomitant inhibition of ammonium uptake. Nature (London) 194:203–204
Ilan J (1971) Evidence for hormonal regulation of the selectivity of ion uptake by plant cells. Physiol Plant 25:230–233

Ilan I, Reinhold L (1963) Analysis of the effects of indole-3-acetic acid on monovalent cations. Physiol Plant 16:596–603
Ilan I, Gilad T, Reinhold L (1971) Specific effects of kinetin on the uptake of monovalent cations by sunflower cotyledons. Physiol Plant 24:337–341
Jacobsen JV (1977) Regulation of ribonucleic acid metabolism by plant hormones. Annu Rev Plant Physiol 28:547–564
Joy KW (1964) Translocation in sugar-beet. I. Assimilation of $^{14}CO_2$ and distribution of materials from leaves. J Exp Bot 15:485–494
Kamienska A, Reid DM (1978) The effect of stem girdling on levels of GA-like substances in sunflower plants. Bot Gaz 139:18–26
Key JL, Shannon JC (1964) Enhancement by auxin of ribonucleic acid synthesis in excised soybean hypocotyl tissue. Plant Physiol 39:360–364
Kriedemann PE (1968) An effect of kinetin on the translocation of ^{14}C-labelled photosynthate in citrus. Aust J Biol Sci 21:569–571
Kriedemann PE, Loveys BR, Fuller GL, Leopold AC (1972) Abscisic acid and stomatal regulation. Plant Physiol 49:842–847
Kriedemann PE, Loveys BR, Downton WJS (1975) Internal control of stomatal physiology and photosynthesis. II. Photosynthetic responses to phaseic acid. Aust J Plant Physiol 2:553–567
Leonard OA, Weaver RJ (1961) Absorption and translocation of 2,4-D and amitrole in the shoots of the Tokay grape. Hilgardia 31:327–368
Leonard OA, Weaver RJ, Glenn RK (1967) Effect of 2,4-D and picloram on translocation of ^{14}C-assimilates in *Vitis vinifera* L. Weed Res 7:208–219
Leopold AC, Kriedemann PE (1975) Plant growth and development, 2nd edn. McGraw-Hill, New York
Lepp NW, Peel AJ (1971a) Some effects of IAA and kinetin upon the movement of sugars in the phloem of Willow. Planta 90:230–235
Lepp NW, Peel AJ (1971b) Influence of IAA upon the longitudinal and tangential movement of labelled sugars in the phloem of willow. Planta 97:50–61
Lester DC, Carter OG, Kelleher FM, Laing DR (1972) The effect of gibberellic acid on apparent photosynthesis and dark respiration of stimulated swords of *Pennisetum clandestinum* Hochst. Aust J Agric Res 23:205–213
Livne A, Vaadia Y (1965) Stimulation of transpiration rate in barley leaves by kinetin and gibberellic acid. Physiol Plant 18:658–664
Loveys BR, Kriedemann PE (1974) Internal control of stomatal physiology and photosynthesis. I. Stomatal regulation and associated changes in endogenous levels of abscisic and phaseic acids. Aust J Plant Physiol 1:407–415
Maleh F, Baker DA (1977) Proton co-transport of sugars in phloem loading. Planta 135:297–299
Maleh F, Baker DA (1978) Effect of fusicoccin on proton co-transport of sugars in the phloem loading of *Ricinus communis* L. Plant Sci Lett 11:233–239
Mason TG, Maskell EJ (1928) Studies on the transport of carbohydrates in the cotton plant. I Ann Bot (London) 42:189–253
McCollum JP, Skok J (1960) Radiocarbon studies on the translocation of organic constituents into ripening tomato fruits. Proc Am Soc Hortic Sci 75:611–616
Meidner H (1967) The effect of kinetin on stomatal opening and the rate of intake of carbon dioxide in mature primary leaves of barley. J Exp Bot 18:556–561
Meynhardt JT, Malan AH (1963) Translocation of sugars in double stem grape vines. S Afr J Agric Sci 6:337–338
Milborrow BV (1970) The metabolism of abscisic acid. J Exp Bot 21:17–29
Mittelheuser CJ, Steveninck van RFM (1971) Rapid action of abscisic acid on photosynthesis and stomatal resistance. Planta 97:83–86
Monselise SP, Halevy AH (1962) Effects of gibberellin and Amo-1618 on growth, dry-matter accumulation, chlorophyll content, and peroxidase activity of citrus seedlings. Am J Bot 49:405–412
Moorby J (1968) The effect of growth substances on transport in plants. In: Vardar Y (ed) The transport of plant hormones. Elsevier/North-Holland Biomedical Press, Amsterdam New York, pp 192–206

Morgan DG (1968) Quantitative study of the effects of gibberellic acid on the growth of *Festuca arundinacea*. Aust J Agric Res 19:221–225

Morris DA (1977) Transport of exogenous auxin in two-branched dwarf pea seedlings *Pisum sativum* L. Planta 136:91–96

Morris DA (1982) Hormonal regulation of sink invertase activity: Implications for the control of assimilate partitioning. In: Wareing PF (ed) Plant growth substances. Academic Press, London New York, pp 659–668

Mothes K, Engelbrecht L (1961) Kinetin-induced directed transport of substances in excised leaves in the dark. Phytochemistry 1:58–62

Muller K, Leopold AC (1966a) Correlative aging and transport of P^{32} in corn leaves under the influence of kinetin. Planta 68:167–185

Muller K, Leopold AC (1966b) The mechanism of kinetin-induced transport in corn leaves. Planta 68:186–205

Mulligan DR, Patrick JW (1979) Gibberellic-acid-promoted transport of assimilates in stems of *Phaseolus vulgaris* L. Localized versus remote site(s) of action. Planta 145:233–238

Mullins MG (1970) Hormone-directed transport of assimilates in decapitated internodes of *Phaseolus vulgaris* L. Ann Bot (London) 34:897–909

Neumann D, Janossy AGS (1977) Early responses to gibberellic acid in a dwarf maize mutant (*Zea mays* L.). Planta 137:25–28

Nissl D, Zenk M (1969) Evidence against induction of protein synthesis during auxin-induced initial elongation of *Avena coleoptiles*. Planta 89:323–341

Nitsch JP (1970) Hormonal factors in growth and development. In: Hulme AC (ed) The biochemistry of fruits and their products. Academic Press, London New York, pp 427–472

Patrick JW (1976) Hormone-directed transport of metabolites. In: Wardlaw IF, Passioura JB (eds) Transport and transfer processes in plants. Academic Press, London New York, pp 433–446

Patrick JW (1979a) Auxin-promoted transport of metabolites in stems of *Phaseolus vulgaris* L. Further studies on effects remote from the site of hormone applications. J Exp Bot 30:1–13

Patrick JW (1979b) An assessment of auxin-promoted transport in decapitated stems and whole shoots of *Phaseolus vulgaris* L. Planta 146:107–112

Patrick JW (1982) Hormonal control of assimilate transport. In: Wareing PF (ed) Plant growth substances. Academic Press, London New York, pp 669–678

Patrick JW (1983) Photosynthate unloading from seed coats of *Phaseolus vulgaris* L. General characteristics and facilitated transfer. Z Pflanzenphysol 111:9–18

Patrick JW, Wareing PF (1970) Experiments on the mechanism of hormone-directed transport. In: Carr DJ (ed) Plant growth substances. Springer, Berlin Heidelberg New York, pp 695–700

Patrick JW, Wareing PF (1973) Auxin-promoted transport of metabolites in stems of *Phaseolus vulgaris* L. Some characteristics of the experimental transport systems. J Exp Bot 24:1158–1171

Patrick JW, Wareing PF (1976) Auxin-promoted transport of metabolites in stems of *Phaseolus vulgaris* L. Effects of the site of hormone application. J Exp Bot 27:969–982

Patrick JW, Wareing PF (1978) Auxin-promoted transport of metabolites in stems of *Phaseolus vulgaris* L. Effects remote from the site of hormone application. J Exp Bot 29:359–366

Patrick J, Wareing PF (1982) Hormonal control of assimilate movement and distribution. In: Jeffocat B (ed) Aspects and prospects of plant growth regulators. Monogr 6. Joint DPGRG BPGRG Symp, pp 65–84

Patrick JW, Woolley DJ (1973) Auxin physiology of decapitated stems of *Phaseolus vulgaris* L. treated with indol-3yl-acetic acid. J Exp Bot 24:949–957

Pegg GF (1981) The involvement of growth regulators in the diseased plant. In: Ayres PG (ed) Effect of disease on the physiology of the growing plant. Cambridge University Press, Cambridge, pp 49–83

Penon P (1972) Ribonucleic synthesis and hormone action in lentil roots. Symp Biol Hung 13:325–334

Phillips IDJ (1975) Apical dominance. Annu Rev Plant Physiol 26:341–367

Prokofyev AA, Sobolev AM (1957) On the movement of phosphorus from leaves to seeds. Fiziol Rast 4:12–20

Quinlan JD, Weaver RJ (1969) Influence of benzyladenine, leaf darkening, and ringing on movement of ^{14}C-labeled assimilates into expanded leaves of *Vitis vinifera* L. Plant Physiol 44:1246–1252

Raschke K (1982) Involvement of abscisic acid in the regulation of gas exchange: evidence and inconsistencies. In: Wareing PF (ed) Plant growth substances. Academic Press, London New York, pp 581–590

Ray PM, Ruesink AW (1962) Kinetic experiments on the nature of the growth mechanism in oat coleoptile cells. Dev Biol 4:377–397

Reid DM, Bradford KJ (1984) Effects of flooding on hormone relations. In: Kozlowski TT (ed) Flooding and plant growth. Academic Press, London New York, pp 195–219

Reid DM, Crozier A (1970) CCC-induced increase of gibberellin levels in pea seedlings. Planta 94:95–106

Robinson M (1983) Influence of abscisic acid and ethylene on assimilate distribution in *Gladiolus grandiflorus*. Ann Bot (London) 51:779–785

Saks Y, Ilan I (1984) Hormone mediated regulative action of the sunflower shoot apex on growth and cation level in the cotyledons. Plant Physiol 74:408–412

Scienza A, Miravalle R, Visai C, Fregoni M (1978) Relationship between seed number, gibberellin and abscisic acid levels and ripening in Cabernet Sauvignon grape berries. Vitis 17:361–368

Seth AK, Wareing PF (1967) Hormone-directed transport of metabolites and its possible role in plant senescence. J Exp Bot 18:65–77

Shindy WW, Weaver RJ (1967) Plant regulators alter translocation of photosynthetic products. Nature (London) 214:1024

Shindy WW, Weaver RJ (1970) Export of photosynthate affected when leaves are pretreated with growth substances. Nature (London) 227:301–302

Shindy WW, Kliewer WM, Weaver RJ (1973) Benzyladenine-induced movement of ^{14}C-labeled photosynthate into roots of *Vitis vinifera*. Plant Physiol 51:345–349

Skene KGM, Kerridge GH (1967) Effect of root temperature on cytokinin activity in root exudate of *Vitis vinifera* L. Plant Physiol 42:1131–1139

Smith D, Muscatine L, Lewis D (1969) Carbohydrate movement from autotrophs to heterotrophs in parasitic and mutual symbiosis. Biol Rev 44:17–90

Starck Z (1983) Photosynthesis and endogenous regulation of the source-sink relation in tomato plants. Photosynthetica 17:1–11

Steveninck RFM van (1976) In: Lüttge U, Pittman MG (eds) Transport in plants, vol II, Pt B: tissues and organs. Springer, Berlin Heidelberg New York, pp 307–342

Swanson CA, El Shishiny EDH (1958) Translocation of sugars in the Concord grape. Plant Physiol 33:33–37

Tanner W (1980) On the possible role of ABA on phloem unloading. Ber Dtsch Bot Ges 93:349–351

Tavares J, Kende H (1970) The effect of benzylaminopurine on protein metabolism in senescing corn leaves. Phytochemistry 9:1763–1770

Thorne GN, Evans AE (1964) Influence of tops and roots on net assimilation rate of sugar-beet and spinach beet and grafts between them. Ann Bot NS 28:499–508

Thrower SL (1962) Translocation of labelled assimilates in soybean. II. The patterns of translocation in intact and defoliated plants. Aust J Biol Sci 15:629–649

Ting T (1963) Translocation of assimilates of leaf of main stem in relation to the phyllotaxis in cotton plants. Acta Biol Exp Sin 8:656–663

Treharne KJ, Stoddart JH (1968) Effects of gibberellin on photosynthesis in red clover (*Trifolium pratense* L.). Nature (London) 220:457–458

Turner WB, Bidwell RGS (1965) Rates of photosynthesis in attached and detached bean leaves, and the effect of spraying with indoleacetic acid solution. Plant Physiol 40:446–451

Turvey TM, Patrick JW (1979) Kinetin-promoted transport of assimilates in stems of *Phaseolus vulgaris* L. Localized versus remote site(s) of action. Planta 147:151–155
Umoessien SN, Forward DF (1982) Effect of gibberellic acid on the distribution of products of photosynthesis in sunflower. Ann Bot (London) 50:465–472
Verma DPS, MacLachlan GA, Byrne H, Ewings D (1975) Regulation and in vitro translation of messenger ribonucleic acid for cellulase from auxin-treated pea epicotyls. J Biol Chem 250:1019–1026
Wang FT, Hsia CA (1962) Studies on decapitation and translocation of assimilates in rape (*Brassica napella* Chaix). Acta Bot Sin 10:43–50
Wardlaw IF (1968) The control and pattern of movement of carbohydrates in plants. Bot Rev 34:79–105
Wardlaw IF, Moncur L (1976) Source, sink and hormonal control of translocation in wheat. Planta 128:93–100
Wareing PF (1978) Hormonal regulation of assimilate movement. In: Opportunities for chemical plant regulation. Proc Joint Br Crop Protect Counc Br Plant Growth Reg Group Symp, pp 105–111
Wareing PF, Patrick J (1974) Source-sink relations and the partition of assimilates in the plant. In: Cooper JP (ed) Photosynthesis and productivity in different environments. Cambridge Univ Press, Cambridge, pp 431–499
Wareing PF, Khalifa MM, Treharne KJ (1968) Rate-limiting processes in photosynthesis at saturating light intensities. Nature (London) 220:453–457
Watson DJ (1971) Size structure and activity of the productive system of crops. In: Wareing PF, Cooper JP (eds) Potential crop production. Heinemann, London, pp 76–88
Weaver RJ (1970) Some effects on grapevines of exogenous plant regulators and herbicides. In: Frazier NW (ed) Virus diseases of small fruits and grapevines. Berkeley, Univ Cal, Div Agric Sci, pp 247–254
Weaver RJ, McCune SB (1959) Response of certain varieties of *Vitis vinifera* to gibberellin. Hilgardia 28:297–350
Weaver RJ, Shindy W, Kliewer WM (1969) Growth regulator-induced movement of photosynthetic products into fruits of Black Corinth grapes. Plant Physiol 44:183–188
Wicks AS, Johnson JO, Bonner BA, Weaver RJ (1985) Enhancement of raisin grape ripening *Vitis vinifera* L. cv. Thompson Seedless via exogenous applications of ABA. Am J Enol Vitic (in press)
Wilson WJ (1972) Control of crop processes. In: Rees AR, Cockskull KE, Hand DW, Hord RG (eds) Crop processes in controlled environments. Academic Press, London New York, pp 7–30
Zaerr JB, Mitchell JW (1967) Polar transport related to mobilization of plant constituents. Plant Physiol 42:863–874
Zurfluh LL, Guilfoyle TJ (1982) Auxin-induced changes in the population of translatable messenger RNA in elongating sections of soybean hypocotyl. Plant Physiol 69:332–337
Zwar JA (1973) Effects of cytokinins on the nucleic acids of tobacco pith. J Exp Bot 24:701–710

2 Rhythms and Their Relations to Hormones

W.L. KOUKKARI and S.B. WARDE

1 Introduction

1.1 General

The physical environment in which plants generally grow and develop is not constant. There are fluctuations in the environment, many of which do not occur solely as random or sporadic events. Rather, these changes appear at regular and predictable intervals. For example, each day as the earth rotates on its axis, plants growing out of doors are subjected to alternating spans of light and darkness. Furthermore, depending on the season of the year, the duration of each span changes predictably in relation to the inclination of the planetary axis. This single feature of solar radiation involving the lengths of light and dark spans is but one of many physical events that occur in rhythmic cycles.

In close harmony with the physical cycles of our planet are the natural rhythms of living organisms that inhabit the earth. The pathological injury to growth and development processes that often results from maintaining plants under CC (BÜNNING 1962) may well serve to indicate the interdependence between internal rhythms and external cycles (WENT 1960, WENT 1974).

In spite of being significant and ubiquitous features of life, rhythmic phenomena were generally overlooked until quite recently. Fortunately, the importance of biological rhythmicity is now recognized by plant physiologists (HAMNER 1960, CUMMING and WAGNER 1968, HILLMAN 1976). Even the beginning student can readily conduct experiments (Fig. 1) to show that, when a plant or a portion thereof (e.g., a pair of leaflets) is transferred from a cyclic physical environment to constant conditions (CC), rhythmic processes continue.

1.2 Plant Development

A close relationship exists between rhythmic processes on the one hand and factors which regulate growth and development on the other (see Sect. 7). In a strict sense, growth in plants can be defined as an irreversible increase in volume, involving cell enlargement, but not necessarily accompanied by cell division (THIMANN 1969). Sometimes growth processes are also described as increases in mass, size, and number. Regardless of limitations imposed by definition, both growth (BALL and DYKE 1954, 1957) and cell division (BÜNNING 1973) have been found to be rhythmic in some organisms. However, a word

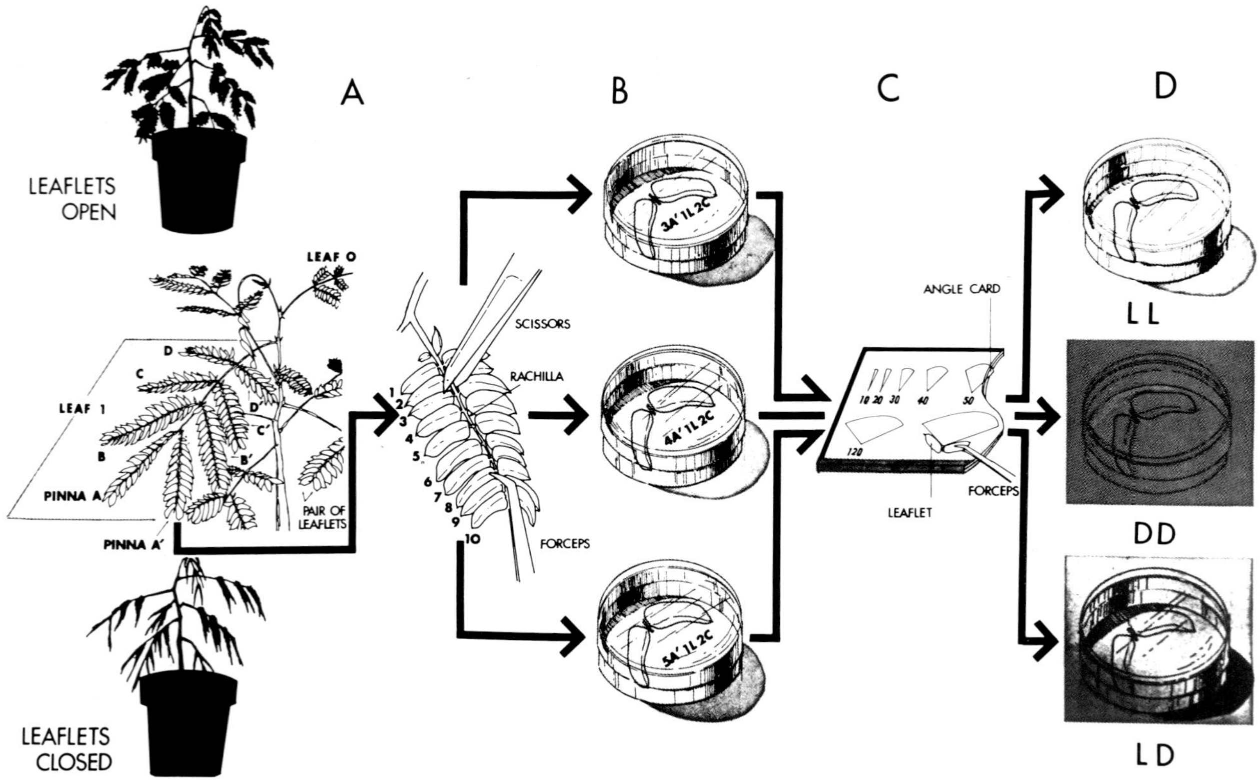

Fig. 1 A–D. Diagrammatic illustration of a procedure for demonstrating the presence of rhythmicity of excised *Albizzia julibrissin* leaflets. **A** First a plant is synchronized in a controlled environmental chamber for ca. 4 d under conditions of alternating light and darkness (e.g., LD 16:8). **B** Pairs of leaflets are excised from a pinna and placed in glass dishes containing a solution, which need only be water. **C** Angles formed by each pair of leaflets as they open and close are measured initially, and at regular intervals thereafter, by comparing the leaflet angles with angle cards inserted between two sheets of clear plastic. **D** Depending upon the experiment, the excised pairs of leaflets may be maintained under conditions of continuous illumination (*LL*) or darkness (*DD*), or alternating light and darkness (*LD*). Similar procedures have been used in many of the hormone and auxin studies described in Sections 5 and 6

of caution is in order regarding the interpretation of cell division or mitosis rhythms in higher plants (HILLMAN 1976) because environmental conditions and experimental procedures have not always been designed to determine the presence or absence of rhythmic endogenicity (see Sect. 2).

The process of differentiation accompanies growth during plant development. Perhaps the classic example of a rhythm-regulating differentiation is the photoperiodic control of flowering (see Sect. 3), in which there is a transition from a vegetative to a reproductive stage. There are numerous other examples clearly showing that morphogenesis and development can, to a large extent, be regulated by the temporal organization of rhythms (cf. SWEENEY 1974c).

Thus far, this introduction has explicitly, though briefly, advanced the concept that rhythms are involved in the developmental processes of plant growth and differentiation. Plant hormones provide one way for mediating these processes of development (cf. PHILLIPS 1971). How plant hormones, their synthetic counterparts, and growth regulators in general relate to rhythms will provide the central theme of this chapter.

2 Bioperiodicities

2.1 Rhythm Characteristics

A rhythm is a change or event which recurs systematically with a specifiable pattern, probability, and frequency (KOUKKARI et al. 1974). Even though an event may possess a rhythmic pattern and be a part of a living system, it is not necessarily a biological rhythm. It is quite possible that a periodic biological event could occur solely in response to an environmental oscillation (e.g., change in light intensity or temperature). Therefore, during rhythm studies, it is important that external factors and synchronizers (also called Zeitgebers, cf. KOUKKARI et al. 1974) are rigorously controlled and manipulated. According to generally accepted standards (ASCHOFF et al. 1965, HALBERG and KATINAS 1973), there are certain criteria that must be fulfilled before a rhythm can be considered to be biological. The rhythm must (1) persist for more than one cycle when isolated from its synchronizer; in other words, the rhythm must be self-sustaining for a time in CC (SWEENEY 1969), (2) be amenable to phase-shifting (see below) when the synchronizer's cycle is shifted (PITTENDRIGH 1954, HALBERG 1969, BÜNNING 1973, KOUKKARI et al. 1974). Restated, biological rhythms are oscillations with properties that do not directly reflect environmental fluctuations or changes (HILLMAN 1976).

Due to variability and dispersion of data on time-course plots, it has sometimes been difficult to quantify rhythm characteristics. Some of these procedural matters have been resolved by computerized statistical techniques which enable more precise evaluation of parameters. According to one such procedure (KOUKKARI et al. 1973), a cosine curve is fitted to the data in order to obtain point and interval estimates of the average value bisecting the cosine curve (mesor),

Table 1. Examples of the free-running circadian periods of four different groups of organisms. Free-running periods may vary according to the type of constant conditions (*) and even among individual organisms of a given species (**) LL: Continuous illumination, DD: Continuous darkness

Group	Organism	Rhythm	Condition	Period (h)	Reference
Vascular plant	*Coleus blumei* × *C. frederici*	leaf movement	DD*	22.67	HALABAN (1968)
	Coleus blumei × *C. frederici*	leaf movement	LL (dim)*	23.06	HALABAN (1968)
	Coleus blumei × *C. frederici*	leaf movement	LL (bright)*	24.79	HALABAN (1968)
Fungus	*Neurospora*	growth	LL (dim red)	22.0	PITTENDRIGH et al. (1959)
Protista	*Gonyaulax polyedra*	bio-luminescence	LL (dim) 19°	23.0	HASTINGS and SWEENEY (1957)
Animal	*Glaucomys volans***	activity	DD	22.97 to 24.35	DE COURSEY (1960)

the location in time of the peak of the cosine curve (acrophase), and the amplitude.

In order to describe, evaluate, and compare rhythms, it is necessary to refer to certain parameters. Three such parameters are amplitude (A), phase (ϕ), and period (τ). Amplitude is one-half the distance from peak to trough of the curve. Changes in A, especially decreases, may be evident in leaf-movement studies when relatively high concentrations of auxin are present (WILLIAMS and RAGHAVAN 1966, MCEVOY and KOUKKARI 1972). ABA may also cause a decrease in A in rhythmic leaf (SKROVE et al. 1982) and petal movements (SCHREMPF 1980).

The term phase is used to denote a particular point or state of an oscillation where a certain event occurs in reference to a fixed time point. The phase is synchronized by various external factors; light and temperature changes are the most important. It has been suggested that plant-growth regulators may also function as synchronizers (VANDEN DRIESSCHE and GLORY 1979). By shifting the Zeitgebers, the phase may be shifted in relation to real time (i.e., the maxima may be delayed or advanced) without altering the other parameters, although several "transients" may occur before the period is re-established.

The period, which is the time necessary to complete a cycle, may be the most important parameter of a rhythm. The period displayed during CC is referred to as the free-running period of a rhythm and usually deviates slightly from the period under entraining conditions (see Table 1). Such deviation supports the concept of endogenicity, particularly in circadian (ca. 24 h) rhythms. The period also provides a systematic criterion by which rhythms may be arranged into three major groups (Table 2): ultradian, circadian, and infradian.

Table 2. Types of rhythms, period ranges, and examples of vascular plants and algae displaying these rhythms

Type	Period	Examples			
		Process	Approximate period	Organism	Reference
Ultradian[a]	<20 h	Auxin transport	25 min	*Zea mays*	HERTEL and FLORY (1968)
Circadian	20–28 h	Coleoptile growth	24 h	*Avena sativa*	BALL and DYKE (1954)
Infradian	>28 h	Gametangia formation	4 d	*Derbesia tenuissima*	PAGE and SWEENEY (1968)
Lunar[b]	ca. 30 d or 15 d	Egg release	16 d	*Dictyota dichotoma*	BÜNNING and MÜLLER (1961)
Circannual[b]	12±2 months	Seed germination	1 yr	*Solanum acaule*	LUDWIG et al. (1982)

[a] According to another scheme (REINBERG 1971), the period range of ultradian rhythms is 0.5 h to 20 h and anything with a period shorter than 0.5 h is referred to as a high frequency rhythm

[b] Lunar and Circannual are not assigned to a fourth type but are included under Infradian

2.2 Ultradian Rhythms

Ultradian rhythms have a frequency greater than one cycle in 20 h (HALBERG 1964, HALBERG and KATINAS 1973) and are referred to as the high (ultra: beyond) frequency rhythms. Many ultradian rhythms have periods that are less than 1 h and can easily be missed unless observations are continuous. Perhaps one reason that they are sometimes not reported is that it has been much easier to ignore high-frequency oscillations or attribute them to noise, sampling error, or poor technique. Although rhythmic phenomena on the order of minutes had been recognized in plants for many years, a sudden surge in studies involving high-frequency oscillations, especially metabolic, commenced in the late 1950's and the 1960's (cf. CHANCE et al. 1973).

There is great diversity in the types of physiological processes that display ultradian oscillations (Table 3). Many ultradian oscillations can be directly associated with growth and/or hormonal regulation of growth (Table 4). Studies designed to monitor stem growth reveal that apical portions of stems may display circumnutations. Circumnutations occur in all three organs of the vascular plant (see JOHNSSON 1979) and have been studied extensively in tendrils (JAFFE and GALSTON 1968). In young *Pisum sativum* L. seedlings, circumnutations (Fig. 2) may be accompanied by a rocking or bobbing movement of the epicotyl and plumule (GALSTON et al. 1964). However, these bobbing movements show a considerable amount of variation among individual seedlings in ampli-

Table 3. Diversity in the types of processes which have ultradian oscillations

Process	Organism	Period	Reference
Protoplasmic streaming	*Physarum polycephalum*	2–2 1/2 min	KAMIYA and NAKAJIMA (1955)
Leaf movements	*Desmodium gyrans*	3 min	BOSE (1927)
Leaf movements	*Gossypium hirsutum*	36 min	MILLER (1975)
Leaf cell viscosity	*Helodea densa*	2–3 h	STÅLFELT (1946)
Transpiration	*Avena sativa*	30 min	JOHNSSON (1973)
Sap flux	*Gossypium areysianum*	20 min	SHERIFF (1974)
Enzyme activity	*Chenopodium rubrum*	12–15 h	DEITZER et al. (1974b)
Enzyme activity	*Pisum sativum*	1–5 h	DUKE and KOUKKARI (1975)
Glycolysis	*Saccharomyces carlsbergensis*	1.8 min	PYE (1969)

Table 4. Examples of ultradian oscillations related to auxins or growth

Process or event	Organism	Plant structure	Period	Reference
Auxin transport	*Avena sativa*	Coleoptile	20–30 min	HERTEL and FLORY (1968)
Auxin transport	*Avena sativa*	Coleoptile	18 min	SHEN-MILLER (1973a)
Auxin transport	*Zea mays*	Coleoptile	27±3 min	SHEN-MILLER (1973a)
Electrical potential differences	*Vicia faba*	Root	5 min	SCOTT (1957)
Electrical potential differences	*Zea mays*	Coleoptile	15 min	STAHLBERG and POLEVOI (1979)
Geotropic nutation	*Carthamus tinctorius*	Hypocotyl	3 1/2–4 h	KARVÉ and SALANKI (1964)
Geotropic nutation	*Carthamus tinctorius*	Hypocotyl	4 h	SHETTY (1969)
Geotropism "Auxins"	*Zea mays*	Coleoptile	18 min	SHEN-MILLER et al. (1970)
Micronutation	*Phaseolus multiflorus*	Stem	12–30 min	HEATHCOTE (1966)
Nutation	*Triticum*	Coleoptile	3 h	ARNAL (1953)
Nutation	*Cuscuta odorata*	Stem	–	BAILLAUD (1953)
Nutation	*Pisum sativum*	Stem	77 min	GALSTON et al. (1964)
Nutation	*Pisum sativum*	Root	–	SPURNÝ (1968)
Phototropic nutation	*Carthamus tinctorius*	Hypocotyl	3 1/2–4 h	KARVÉ and SALANKI (1964)
Leaf movements	*Phaseolus angularis*	Leaf	53 min	ALFORD and TIBBITTS (1971)

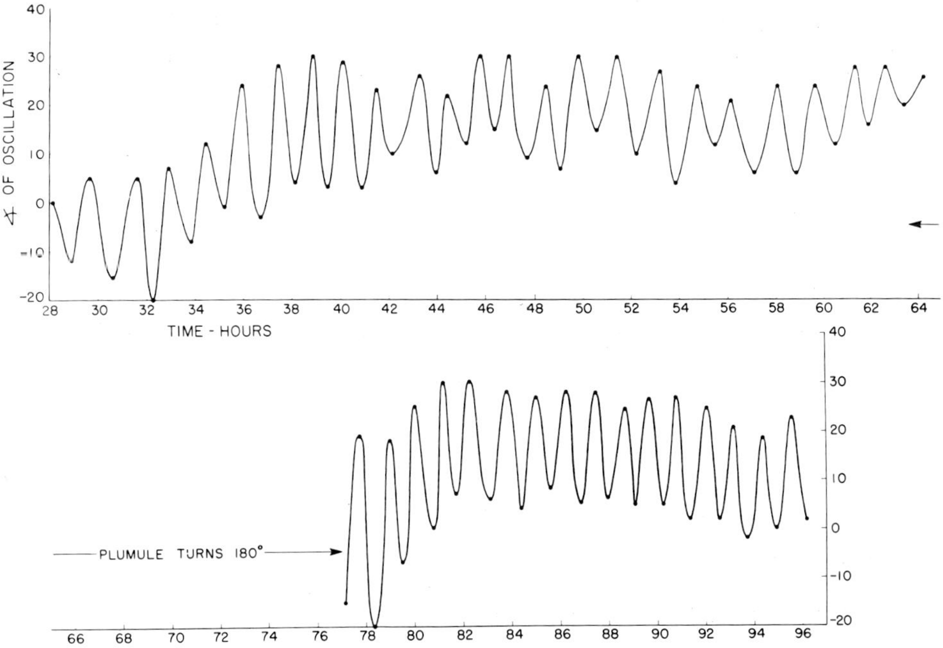

Fig. 2. Ultradian periods in the circumnutations (rotations) of a single etiolated *Pisum sativum* plant maintained under continuous red light following DD. Data obtained by time-lapse photography at 2-min intervals. (GALSTON et al. 1964)

tude and periodicity (Fig. 3). GA may have a pronounced effect on the amplitude of circumnutations (Fig. 4) in *Pisum* (MCCOMB 1962), *Phaseolus vulgaris* L. (BAILLAUD and MONNIER 1960), *Lactuca saligna* L. (TRONCHET and MARCHAL 1960), and *Zinnia elegans* JACQ. (TRONCHET et al. 1960).

Since auxins control cell elongation, at least in part, it is reasonable to expect that auxin levels or transport do not remain constant throughout an organ that is growing in rhythmic spurts. Significant ultradian oscillations involving the transport of auxin have been observed in coleoptiles of *Avena sativa* L. and *Zea mays* L. (HERTEL and FLORY 1968, SHEN-MILLER 1973a). Ultradian periods of approximately 20 min in the basipetal transport of IAA occur in various regions of both *Avena* and *Zea* coleoptiles. Furthermore, oscillations displaying periods of 20–30 min can be initiated by addition of IAA to "hormone-depleted" tissue (Fig. 5). These periods of 20–30 min are very similar in length to those observed in micronutations (HEATHCOTE 1966) and electrical potential differences paralleling auxin transport (NEWMAN 1963), and they are slightly longer than those for geotropism (SHEN-MILLER and MORRIS 1967, SHEN-MILLER et al. 1970, SHEN-MILLER 1973b). Good experimental evidence is lacking for a correlation between ultradian growth patterns and short-period oscillations in hormone levels or transport. Nevertheless, the results obtained from studies conducted in several different laboratories indicate that the hormonal or auxin levels in various sections of a plant display ultradian rhythms (see Sect. 5.1.1).

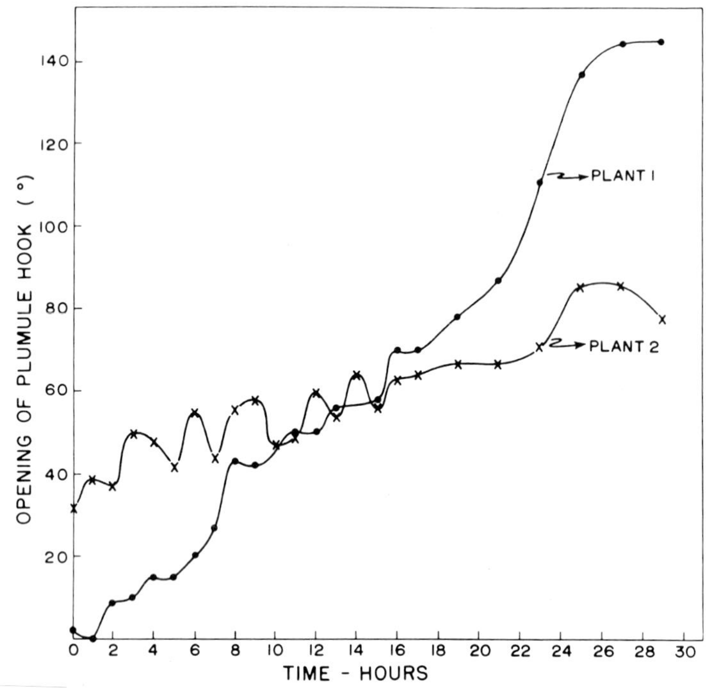

Fig. 3. Variations in *Pisum* plants in the bobbing movement of plumule hook opening during continuous red illumination. (GALSTON et al. 1964)

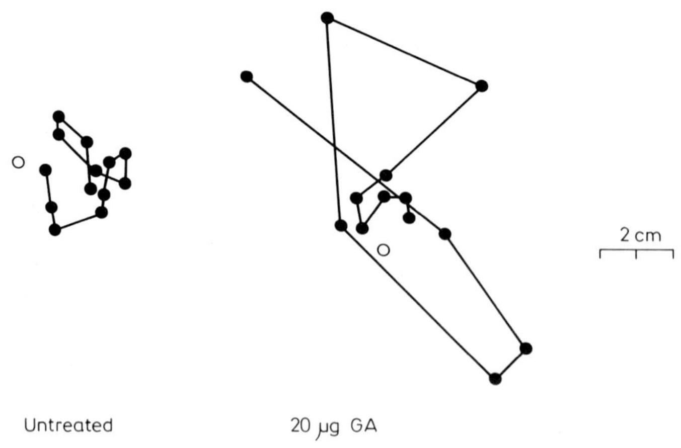

Fig. 4. Circumnutation of the apices of an untreated (*left*) and a GA_3-treated (*right*) *Pisum* seedling as drawn on a glass plate located 6 cm above the apices. (After McCOMB 1962)

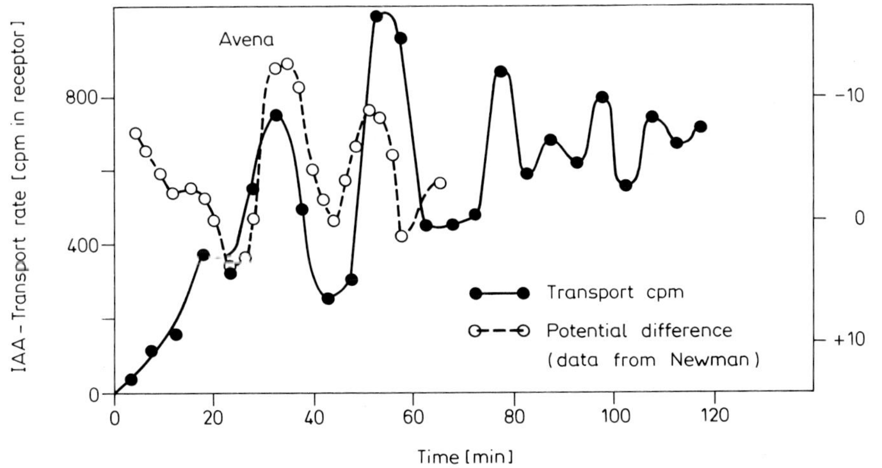

Fig. 5. Transport rate from a single *Avena* coleoptile section. After the sections were cut and drained for 45 min between agar, a donor block containing auxin (^{3}H-IAA) was placed on the apical end at time 0. At indicated time points the sections were transferred to new receptor blocks and the receptors assayed. (Electrical potential differences: NEWMAN 1963; entire figure: HERTEL and FLORY 1968)

2.3 Circadian Rhythms

Circadian rhythms are involved in a variety of physiological processes associated with growth, development, and metabolism (cf. SWEENEY 1963, CUMMING and WAGNER 1968, BÜNNING 1973, QUEIROZ 1974, HILLMAN 1976, KOUKKARI and SOULEN 1981), many of which were described in Volumes II and XVII of the First Series of the *Encyclopedia of Plant Physiology* and again in Volume VII of the New Series. Of the many types of rhythms known, circadian rhythms are probably the most familiar. The term circadian is derived from two Latin words: *circa* (about) and *dies* (day) (HALBERG 1959). These rhythms have a free-running period of ca. 20 to 28 h (HALBERG and KATINAS 1973).

Although the free-running period (Table 1) is determined under CC, the oscillations do not necessarily persist for long durations. In CC the amplitude may damp out (decrease), making the period length difficult to discern. The number of cycles that can be observed during CC varies, and to a certain extent depends upon the environment (LL vs. DD), cultural conditions, type of organism, and process being studied. For example, in the presence of 50 mM sucrose, the number of cycles in excised pulvini of *Samanea saman* (JACQ.) MERR. increases severalfold (SIMON et al. 1976). Likewise, the number of cycles in the flowering response rhythm of *Chenopodium rubrum* L. increases when glucose is added as a supplement (CUMMING 1967).

While it is generally accepted that rhythmic behavior is hereditary (BÜNNING 1973, also cf. KONOPKA and BENZER 1971, KYRIACOU and HALL 1980), circadian rhythms have not always been observed in plants that have been maintained throughout their lifetime in LL or DD, although a single stimulus is all that is necessary to evoke the oscillation (BÜNNING 1960). This feature is well illustrated in *Avena,* where an endogenous growth rhythm can be observed only after the seedlings are transferred from continuous red light to DD (BALL and DYKE 1954). Similarly, *Phaseolus* plants cultured under LL display rhythmic leaf movements after an interruption of the LL by a single dark span of sufficient length, e.g., 18 h (WASSERMANN 1959). *Gossypium hirsutum* L. seedlings grown under LL show little or no variation in response to herbicides (see Sect. 5.3) until a photoperiod triggers sensitivity (RIKIN et al. 1983b).

In many hormone studies, all of the plants have been subjected to cycles of LD and/or temperature changes. When physiological oscillations having a 24-h period are observed under such conditions they may or may not be self-sustained. A rhythm with a 24-h period, observed only under conditions where the environment also oscillates (e.g., LD 12:12), is more appropriately called a daily rhythm (cf. WURTMAN 1967, BÜNNING 1973).

2.4 Infradian Rhythms

Infradian rhythms have periods longer than 28 h (HALBERG and KATINAS 1973). This group includes rhythms with periods that range from slightly more than a day to periods that extend for many years. For example, an interesting 32-year flowering cycle has been reported for *Chusquea abietifolia* GRISEB (cf. SEIFRIZ 1950).

Growth of both temperate-zone and tropical trees may be rhythmic. The shoot growth of *Theobroma cacao* L. maintained on LD 12:12 has periods of 26.6 ± 1.1 days (GREATHOUSE et al. 1971). However, in this study an LL cycle was not included, and plants were fertilized at given intervals, making it more difficult to assess the presence of a true endogenous biological rhythm.

An unusual rhythm, observed under controlled environmental conditions, is that of a coenocytic green alga, *Derbesia tenuissima* CROUAN. The periodicity of gametangial formation is about 4 d (PAGE and KINGSBURY 1968, PAGE and SWEENEY 1968). The significance of this endogenous rhythm in relation to its habitat has not been resolved, since in nature this alga displays a bi- or tri-weekly periodicity.

Circannual rhythms have periodicities more obviously correlated with the environment, since growth and reproduction in most plants reflect seasonal changes and often occur in yearly cycles. It is not easy to carry out experiments to determine the endogenicity of circannual rhythms. Those who have studied rhythmic processes, and have had to maintain constant environmental conditions for only a few weeks, realize the extreme difficulty of providing such conditions for over 2 years. With the stringent requirement for constant environmental conditions, it is little wonder that plants belonging to the family Lemnaceae, such as *Spirodela polyrhiza* L. and *Lemna minor* L., have often been selected for studying circannual rhythms. These tiny aquatic organisms are true vascular plants, but can be readily cultured and maintained in the laboratory (POSNER 1967).

Results from studies with members of the Lemnaceae indicate that circannual rhythms may be present in the formation of winter shoots (HENSSEN 1954), in root growth (PIRSON and GÖLLNER 1953), and in dry weight (BORNKAMM 1966). *Lemna* plants maintained in LL at a constant temperature display a seasonal rhythm in the rate of dry matter produced, with maximum production in the summer and minimum in the winter. Protein content, as well as the protein/carbohydrate ratio, can be correlated with dry weight in that they are both at a maximum in summer. Soluble carbohydrates, on the other hand, reach their highest level in winter. That hormones may have a role in this system is speculative, although it has been demonstrated that cytokinins, in addition to their effects on growth (OVERBEEK and MASON 1968, KOUKKARI and DUKE 1973), influence high starch accumulation in *Lemna* (ERISMANN and FANKHOUSER 1967, TASSERON-DE JONG and VELDSTRA 1971). It has been suggested that these seasonal rhythms in *Lemna* are synchronized by the temperature of the water (BORNKAMM 1966).

The most convincing data in support of the biological endogenicity of circannual rhythms have been provided from studies with seeds, since it is much easier to maintain a relatively constant temperature, even if a compressor or motor momentarily ceases to function, than it is to provide a continuous and uniform source of radiant (light) energy. Humidity can, of course, be a problem, though the structure of seeds makes them less difficult to maintain than succulent tissues. Seeds of *Digitalis lutea* L. and *Gratiola officinalis* L. stored at various temperatures, ranging from −22 °C to 45 °C, show a pronounced circannual rhythm in their capacity to germinate (BÜNNING and MÜSSLE 1951). Seeds of *Solanum acaule* BITT. display a strong rhythmic variation in germination percentages when stored at 4 °C, but only slight variations occur when the seed is stored at 20 °C (LUDWIG et al. 1982). Soaking the *Solanum* seeds in 500 ppm GA_3, combined with low temperature, increases both the speed and percentage of germination. Pretreatment with GA also increases the amplitude, but does not otherwise affect the circannual rhythm.

2.5 Rhythm Interrelationships

There has been a general tendency for plant physiologists to restrict discussions of endogenous rhythms solely to the circadian group. However, all three rhythms may be functionally important (KOUKKARI and SOULEN 1981), regardless of whether the focus is on populations of plants, plant organs, tissues, cells, or even organelles and molecules (Fig. 6). Ultradian and circadian rhythms that directly involve hormones are known to be operative within vascular plants. Some of the ultradian rhythms could conceivably be part of a circadian system, just as some ultradian leaf movements fall within the overall circadian cycle (ALFORD and TIBBITS 1971, MILLER 1975). In addition to structural organization (e.g., cells and tissues), plants also have temporal organization (KOUKKARI and SOULEN 1981). Much of this temporal organization is rhythmic, although during morphogenesis, both single event timers (e.g., a given span of cold required for seed germination) and oscillating timing mechanisms may be present (SWEENEY 1974c).

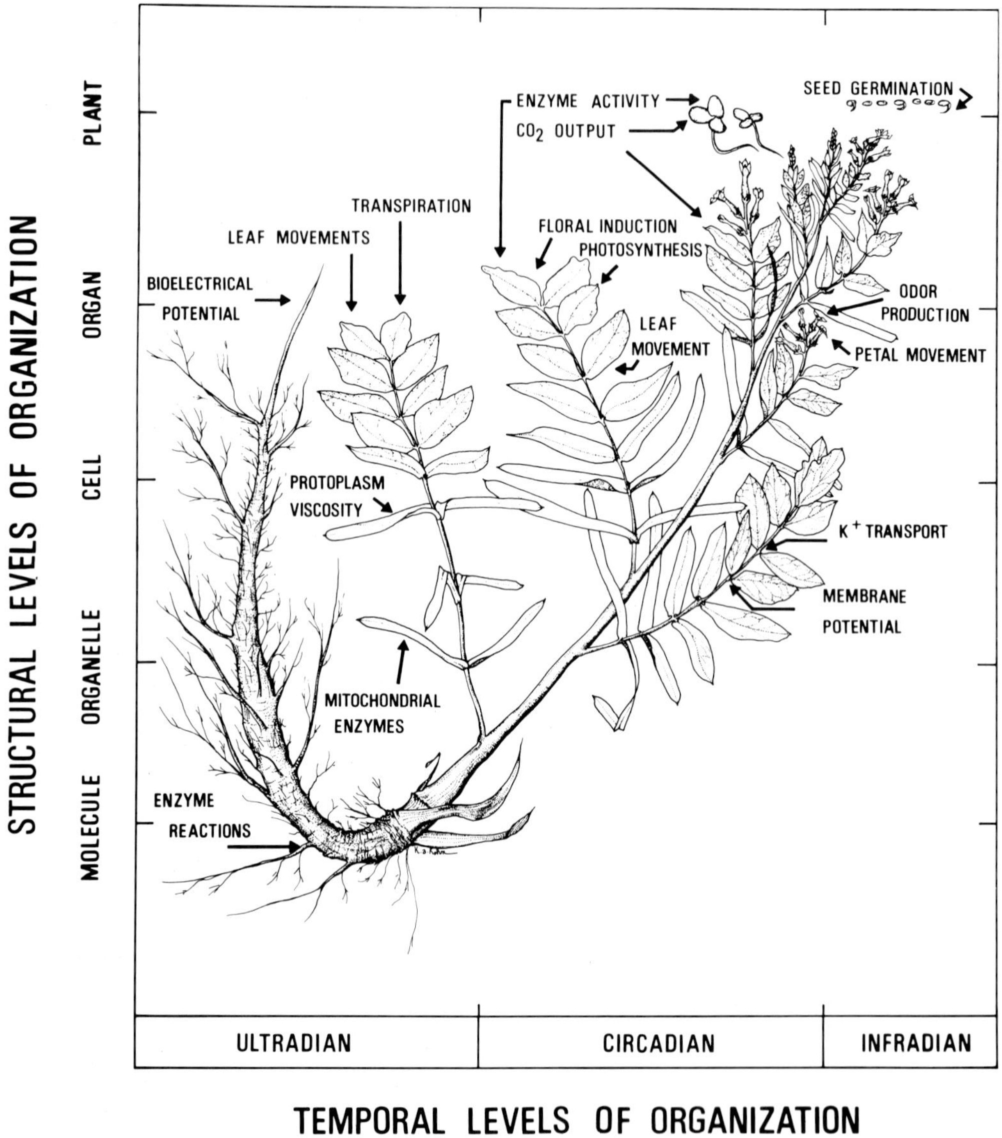

Fig. 6. A diagrammatic illustration of the relationship between structural and temporal levels in the organization of flowering plants. Period lengths for the three groups of rhythms are: ultradian = less than 1 h to 20 h; circadian = 20 h to 28 h; and infradian = over 28 h to years. (KOUKKARI and SOULEN 1981)

3 Photoperiodism

The literature on the relationship between circadian rhythmicity and photoperiodism is confusing (HILLMAN 1976). Yet the rhythmic component of the flowering response of the photoperiodic short-day plant *Chenopodium rubrum* (CUM-

MING et al. 1965, CUMMING 1967, 1969) and the long-day plant *Sinapis alba* L. (KINET et al. 1973) has been superbly demonstrated. A rhythmic component has also been observed in many other species (e.g., HILLMAN 1964, TAKIMOTO and HAMMER 1964, CHORNEY et al. 1970), including Biloxi soybeans *Glycine max* (L.) MERR. (HAMMER 1969), the cultivar that contributed to the introduction of the term photoperiodism (GARNER and ALLARD 1920). Although the interactions between photoperiodism and rhythms are complex, they exist. Florigen, a specific flowering hormone, is viewed by many to be a concept rather than a hormone. Unless a photoperiodic flowering hormone is isolated and characterized, its relationship to rhythmicity can, at best, be speculative.

A hormone that has been known to influence photoperiodic response is GA_3 (cf. LANG 1965, VINCE-PRUE, Chap. 9, this vol.). This hormone also affects the rhythmic sensitivity of flowering in *Chenopodium,* apparently by advancing the phase of the rhythm (KING 1975).

4 Rhythms: Endogenous Hormones

Historically, rhythm studies and hormone research are linked. Many of the investigators listed in the table from a 1937 monograph (WENT and THIMANN 1937) for their work relating to the discovery of the function of auxins also contributed to the study of rhythms (Table 5).

The coleoptile has probably been the most popular plant structure for the study of auxin effects. Growth is by cell elongation, which is under the control of auxin (cf. WENT and THIMANN 1937). Studies with *Avena* coleoptiles have shown that if seedlings are grown under red light and then transferred to DD, an endogenous circadian growth rhythm is induced (BALL and DYKE 1954). The phase of this rhythm can be shifted to any set time by appropriately changing the time at which the seedlings are removed from red light and transferred to DD. Manipulations such as this should eliminate the suggestion of any correlation with fluctuations in "impure" air (see Sect. 5) or other external environmental factors. The phase of the rhythm can also be delayed by hypoxia (BALL and DYKE 1957, WILKINS and WARREN 1963). Coleoptiles of other plants that may indicate rhythmicity in growth-rate include *Triticum vulgare* VILL. cv. Eclipse, *Triticum spelta* L., *Secale cereale* L. cv. Petkus, *Avena fatua* L., and *Hordeum vulgare* L. (BALL et al. 1957). *Avena* coleoptiles also have a CO_2 output rhythm which closely parallels the circadian growth rhythm (BALL et al. 1957).

Of all the rhythms that have been studied in plants, the most commonly observed are those of leaf movements. There are various mechanisms and morphological features associated with leaf movements of plants. In one distinct morphological type, leaf movement results from turgor changes in specialized structures called pulvini (see Sects. 5.1.3 and 6.4). In some plants without pulvini, leaf movement depends on unequal growth rates on opposite sides of the petiole (cf. PFEFFER 1906). The movements of papaya (*Carica papaya* L.) leaves have been attributed to this latter type, and in this species the role of endogenous auxin in rhythmic leaf movements has been convincingly demonstrated (YIN 1941). Using the *Avena* curvature bioassay, auxin levels in petioles

Table 5. From a total of 27 individuals who were listed in a 1937 chart showing the historical interrelations among various lines of research leading toward the discovery of the function of auxins, nine of them are cited in this table because they also contributed to the study of rhythms

Individual[a]	Comments concerning contributions to rhythms[b]
HENRI-LOUIS DUHAMEL DU MONCEAU (1758)	Placed plants in wine cave or covered trunk in closet (CC) and showed that leaf movements persisted and were not controlled by changes in light and temperature
JULIUS SACHS (1863–1887)	Supported endogenicity of periodicity. Showed that oscillations were only modified but not caused by L and D
CHARLES DARWIN (1881)	Described circumnutations and plant movements
WILHELM PFEFFER[c] (1907–1915)	Recorded the persistence of leaf movement rhythms, described interactions between the leaf joint and blade, examined phase-shifts and entrainment, and provided quantitative methods for measuring plant movements
R. DOSTAL[d] (1927, 1941)	Considered the importance of seasonal periodicity and rhythms in plant development
FRITZ KÖGL and A.J. HAAGEN-SMIT (1936)	Studied oscillations in the sensitivity of coleoptiles to auxins in studies ranging from h to weeks
FRITS W. WENT (1937, 1960)	Observed daily oscillations in the sensitivity of coleoptiles to auxin. Demonstrated the importance of environmental cycles and internal timing mechanisms of plants for normal development
FOLKE SKOOG[e] (1938)	Reported on the effects of auxin on the periodicity of rhythmic exudation in *Helianthus*

[a] The year is an approximate date and not necessarily the date of the experiment or publication

[b] General References: WENT and THIMANN (1937), BÜNNING (1960), CUMMING and WAGNER (1968), WARD (1971)

[c] An excellent account of PFEFFER's view of rhythms has been recently summarized (BÜNNING and CHANDRASHEKARAN 1975)

[d] Confirmation of DOSTAL's contributions to the study of rhythms was provided by Dr. MIROSLAV KAMINEK, Czechoslovak Academy of Sciences, Institute of Experimental Botany, Department of Growth and Development, Czechoslovakia (personal communication)

[e] SKOOG et al. (1938)

and leaves of papaya display pronounced circadian variations. The rhythmic movements are related to both the transport of auxin and the histology of the leaf and petiole. For example, auxin may be synthesized uniformly throughout the surface of the lamella, but polar transport favors an accumulation in the basal lobes compared to the tip. Because the basal lobes are connected

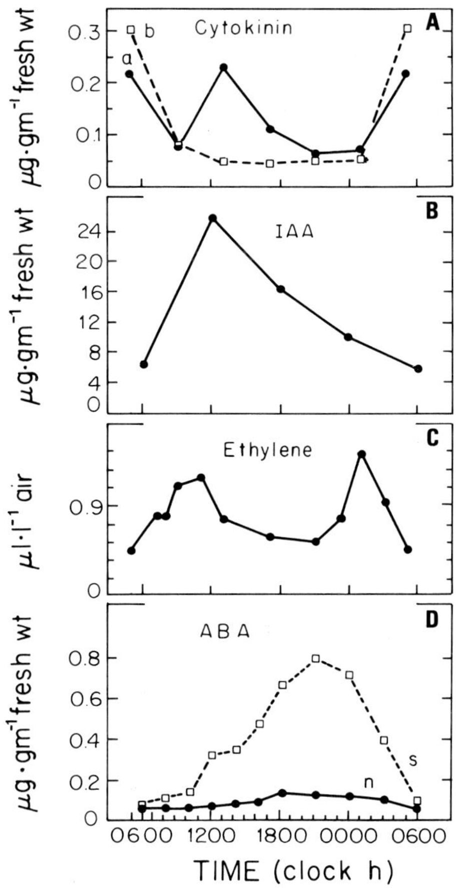

Fig. 7 A–D. Daily oscillations in levels of several hormones. **A** Cytokinin levels in mature leaves during a warm, overcast day (*a*) and during a hot, sunny day (*b*). **B** IAA content in young leaves of coffee (*Coffea arabica* L.). **C** Ethylene concentration in broad bean leaves (*Vicia faba* L.). **D** ABA levels in leaves of stressed (*s*) and non-stressed (*n*) cotton (*Gossypium hirsutum* L.). Poplar (**A**) and coffee (**B**) leaves were collected from field grown plants. Broad bean leaves (**C**) were obtained from plants maintained under LD 16:8 in growth room. Cotton leaves (**D**) were harvested from greenhouse plants. (After **A** HEWETT and WAREING 1973; **B** JANARDHAN et al. 1973; **C** EL-BELTAGY and HALL 1974; **D** MCMICHAEL and HANNY 1977). Data time points on **A**, **C**, and **D** were arranged to match **B**

to the upper side of the petiole, this side receives more auxin and the leaf turns downward. The mechanisms responsible for stopping growth on the upper side may depend on feedback inhibition, auxin destruction, or other regulators. Continued growth of the lower surface causes the petiole to turn up and the leaf to extend to its "day" position (YIN 1941). Differential distribution of auxin in leaves is not unique to papaya and has been known since the early studies on auxins (AVERY 1935).

The levels of growth substances (RAŠKA and HLADÍK 1969) and cytokinins (HEWETT and WAREING 1973) extracted from foliage of trees growing under natural conditions show pronounced daily oscillations (Fig. 7 A). Diurnal fluctuations in IAA levels have been noted in Scots pine (*Pinus sylvestris* L.) seedlings grown under laboratory conditions (SANDBERG et al. 1982). However, these plants were grown under LD 18:6 and the temperature was also varied. It is not known if the hormones in the above studies would display an endogenous rhythm under CC and be amenable to phase-shifting.

When sunflowers (*Helianthus annuus* L.) are grown in a "natural" outside environment, their shoots are turned to the east in the morning, toward the west in the evening, and curve toward the east before sunrise (SHIBAOKA and YAMAKI 1959). To determine if these movements are endogenous, experiments were conducted in which plants grown in pots for 80 d were rotated 180° so that the plant faced west in the morning. The results from a 4-d study show that the entrained movements persist independent of the direction of the sun's movement. In other words, the plants which were rotated 180° moved from west to east. Although many unanswered questions remain, experiments in which hormone levels in *Helianthus* were measured indicate a complex relationship among hormone levels, plant movements, light intensity, and leaf age (SHIBOAKA and YAMAKI 1959).

Circadian oscillations in the concentration of "hormone" (probably IAA) from leaves and petioles of *Kalanchoe blossfeldiana* POELLN. (extracted every 3 h and tested for rooting) have been reported under DD (BECKER 1953) and offer support to the hypothesis that oscillations in hormone concentration display biological rhythmicity. In studies where chromatography has been employed, rather than a bioassay such as rooting, daily rhythms of IAA concentration (Fig. 7B) have been observed in leaves of *Coffea arabica* L. (JANARDHAN et al. 1973), *Hibiscus esculentus* L. and *Piper betle* L. (RAMA DAS et al. 1964).

Single fluctuations in ethylene production consisting of an initial increase followed by a decline have been known for decades to occur during processes such as fruit ripening (BIALE et al. 1954). Recently it has been observed that endogenous ethylene also shows a "daily" oscillation (Fig. 7 C) in leaves of *Caltha polypetala* HOCHST. ex LORENT, *Kalanchoe daigremontiana* HAMET et PERR. (KAPUYA and HALL 1977), *Vicia faba* L. (EL-BELTAGY and HALL 1974) and *Lycopersicon esculentum* MILL. cv. Alicante (EL-BELTAGY et al. 1976). Because the *Lycopersicon* study included experiments conducted under CC, it provides good evidence as to the nature of these daily oscillations. Here, 75-d-old plants, which had been synchronized for 7 d in LD 16:8, were divided into three groups (LD 16:8, LL, DD) and maintained under such regimens for 180 h. The leaves were assayed for ethylene at intervals during three 36-h spans separated by 36 h. A comparison of the results obtained from the three groups indicates the presence of endogenous rhythmicity in the level of ethylene in *Lycopersicon* leaves. More recently, oscillations in ethylene biosynthesis under CC have been demonstrated in cotton (*Gossypium hirsutum*) seedlings (RIKIN et al. 1983a). Rhythmic changes in ethylene levels may well be an endogenous oscillating component within a network of rhythms that constitutes the temporal organization of plants (KOUKKARI and SOULEN 1981), and not merely an oscillation in response to a cyclic environment. The phase of the oscillation could, however, be synchronized by the environment (see Sect. 2.1). Furthermore, the length of the L span under LD may affect the amplitude, since higher levels of ethylene are present under long (e.g., 16 h L) compared to short (e.g., 8 h L) day conditions (KAPUYA and HALL 1977).

No consistent correlation between increased ethylene levels and stomatal resistance has been observed in *Caltha, Kalanchoe,* and *Vicia* (KAPUYA and HALL 1977), yet the phase relationship of ethylene levels between *Kalanchoe* and *Vicia* are inverse, as are the stomatal rhythms.

Ethylene is a gas as well as a hormone, and as such can be grouped with two other gases that also display endogenous rhythmicity: O_2 (SATTER 1979) and CO_2 (WILKINS 1962, HILLMAN 1970, 1971). These gases may have definite interactions in regulatory processes. For example, CO_2 can act as a competitive inhibitor of some of the effects of ethylene (ABELES et al. 1971). This relationship between ethylene and CO_2 may be important in developmental processes and underscores the need for examining both of these gases under a variety of conditions (EL-BELTAGY and HALL 1974, EL-BELTAGY et al. 1976).

When plants are cultured under "normal" or nonstressed conditions, cyclic phenomena, or at least visible indications of their presence, may be minimal (see Sect. 7.1). Endogenous levels of ABA in both field- and greenhouse-grown *Gossypium hirsutum* plants are an example of such a phenomenon. In this species the concentration of ABA under either water-stressed or nonstressed conditions has been shown to increase throughout the day followed by a decrease (MCMICHAEL and HANNY 1977). However, the daily cyclic pattern is much more dramatic in the stressed leaves (Fig. 7 D). A similar situation has been observed in other species, including field-grown pearl millet (*Pennisetum americanum* (L.) LEEKE) (HENSON et al. 1982). Here the peak in leaf ABA level occurs midmorning with a minimum in the afternoon, but the diurnal fluctuations in ABA are more pronounced in the "severely" water-stressed plants.

The concentration of ABA has also been shown to oscillate under CC, suggesting that the rhythm is an endogenous one. In soybean plants (*Glycine max*) grown in controlled-environment chambers, ABA levels are highest at "night" in both plants maintained under LD and those in LL (LECOQ et al. 1983).

It is well known that the level of a particular hormone within a plant is related to the type of organ or tissue. As illustrated in Fig. 7, the levels of a hormone are also dependent upon time. During a normal 24 h cycle, the levels of ABA in leaves may be eight times greater at one time of day than at another. These 24-h oscillations (daily and circadian), along with ultradian rhythms in the levels of hormones, may provide temporal organization (KOUKKARI and SOULEN 1981) during plant development. Rhythmic timing is an integral component, not only of specific hormone-related events, but of metabolic processes and biochemical reactions per se within cells and organelles (HESS and BOITEUX 1971, CHANCE et al. 1973, GOLDBETER and CAPLAN 1976).

5 Rhythms: Exogenous Growth Regulators

Determination of rhythmic fluctuations in endogenous hormone levels is a necessarily destructive procedure. At each time point analyzed, plant material is lost. Any identifiable pattern is based on data from populations rather than from individual plants or organs (cf. SANDBERG et al. 1982). Another approach in determining the functions of hormones and their relation to temporal organization is to observe the results of exogenously applied hormones or growth substances. For the purposes herein, these growth regulators and their relation to rhythms are discussed in three separate sections: (1) auxins (2) ABA, cytokinins, ethylene, and GA's (3) herbicides. Unavoidably, there will be some overlap. For example 2,4-D is, by definition of mode of action or application, both auxin and herbicide.

5.1 Auxins

By the 1930's there was evidence from results of experiments conducted in Europe (KÖGL et al. 1936) and the United States (WENT and THIMANN 1937) that the sensitivity of coleoptiles to the auxin bioassay displayed significant oscillations. However, these oscillations were not observed in certain localities. Because exhaust emission from automobiles and other agents of pollution showed fluctuations relative to traffic and environmental conditions, the effects of air pollution were suspect (HULL et al. 1954). The results from one well-planned study definitely show that air pollution has a significant effect on the sensitivity of the *Avena* test for auxins (HULL et al. 1954). However, it must be noted that the entire procedure for working with coleoptiles during auxin assays may involve the use of red light and other manipulations that could influence rhythms. For example, a growth rhythm can be induced in *Avena* coleoptiles when they are transferred from red light to DD (BALL and DYKE 1954).

The application of auxin to *Avena* coleoptiles has been reported to have approximately the same effect, regardless of whether the hormone is applied when the circadian rhythmic growth rate is high or low. This suggests that the oscillations may not be related to sensitivity of the tissues to auxin (BALL and DYKE 1956). However, a phase-shift delay of this rhythm may be produced by immersing the seedling in either 2,4-D or IAA (BALL and DYKE 1956). Similarly, when *Phaseolus* plants treated with 2,4-D under LD conditions are compared to those transferred to LL or DD prior to treatment, there appear to be indications of phase-shifting. However, the experiments in this study were not of sufficient duration to offer conclusive evidence (cf. CAMPIRANON and KOUKKARI 1976).

Rhythms, or the presence of rhythmic events, can be cited for practically all major physiological processes within plants (KOUKKARI and SOULEN 1981). One of these rhythmic processes is photosynthesis (cf. QUEIROZ 1974), which has been extensively studied in the alga *Acetabularia mediterranea* LAMOUR (SWEENEY and HAXO 1961, VANDEN DRIESSCHE 1970, 1975a, b, c). This single-celled alga, with its numerous temporal components, contains an IAA-like substance (VANDEN DRIESSCHE and DELEGHER-LANGHOR 1975). Exogenously applied IAA induces a pronounced phase-shift advance of the photosynthesis circadian rhythm when applied during the low phase. When applied during the high phase (middle of light span), IAA does not cause a phase-shift (VANDEN DRIESSCHE 1979). Both IAA and morphactins (anti-IAA drugs) effect a phase-shift in the circadian rhythm of chloroplast polysaccharide synthesis (VANDEN DRIESSCHE and GLORY 1979). IAA induces a delay, morphactins induce an advance, and a combination results in an intermediary curve. The effects of IAA on the phases of these rhythms is an excellent example of a rhythmic hormonal response. Within the algal cell, endogenous IAA may well be a major component of temporal organization. It has been hypothesized that during development, IAA functions as an endogenous synchronizer between the organelles of *Acetabularia* (VANDEN DRIESSCHE 1979).

5.1.1 Auxin Transport

Results from studies conducted in several laboratories indicate that the movement of auxins in coleoptiles is rhythmic (NEWMAN 1963, HERTEL and FLORY 1968). When a ^{3}H-IAA donor block is placed on the apical end of an individual section of an *Avena* coleoptile previously depleted of hormone, and the coleoptile (with donor block) is then transferred to new receptor blocks at various intervals, pronounced oscillations in auxin transport are observed (HERTEL and FLORY 1968). The oscillations persist for many cycles and are very similar to previously reported electrical potential measurements (Fig. 5). The periods of these ultradian transport rate rhythms are ca. 20 to 30 min. In another study, ^{14}C-IAA was applied as a 60-s pulse to *Avena* and *Zea* coleoptiles and the radioactivity measured in successive regions (SHEN-MILLER 1973a). Although the velocity of IAA transport was not the same in all regions of a coleoptile, rhythms (ca. 20 min) of basipetal transport were proposed.

5.1.2 Tropisms

Rhythmic oscillations in basipetal transport of auxin have been shown to be present in horizontally, as well as vertically, positioned coleoptiles (SHEN-MILLER 1973b). However, longitudinally split "half coleoptile" sections possess different period lengths. When the velocity of transport along a "geostimulated" coleoptile is measured, regional variations are observed. In lower half-sections, the velocity in the tip region is 41 mm h^{-1}, as compared to 12 mm h^{-1} for the basal region (SHEN-MILLER 1973b).

A lateral movement of auxin in coleoptiles occurs following a brief exposure to unilateral illumination. This effect, classically described in the "theory of Cholodny-Went", results in an uneven distribution of auxin following geotropic or phototropic stimulation. Oscillations in the redistribution of ^{14}C-NAA after phototropic induction have been demonstrated in a series of well-designed experiments (DEHAAN 1969). Auxin in donor blocks is applied to a decapitated *Zea* coleoptile, by splitting the basal end and separating it with a razor blade, with receiver blocks positioned on either side of the blade when necessary. As expected, soon after unilateral photostimulation, more auxin is found on the "dark" side of the coleoptile than on the "illuminated" side. However, after first accumulating on the dark side, the radioactivity is redistributed first to the side that was illuminated and then back again to the dark side, thus displaying pronounced oscillations (Fig. 8). These oscillations are not observed in non-

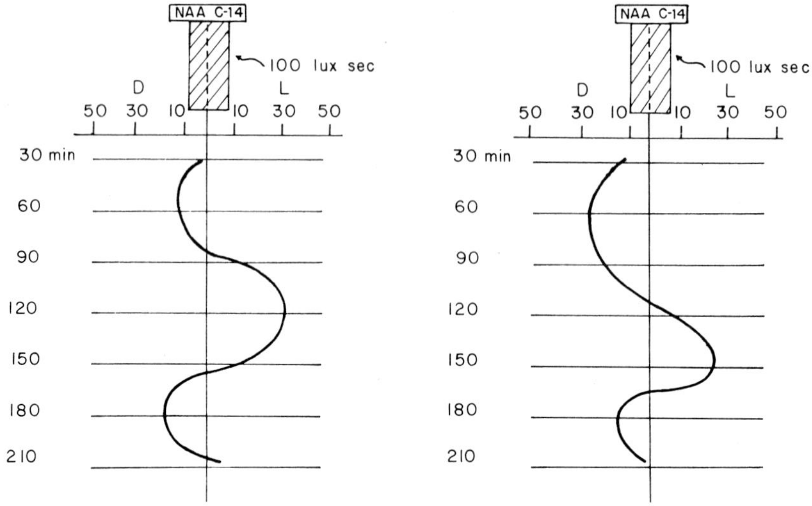

Fig. 8. Time-course of the distribution of ^{14}C-NAA in two *Zea* coleoptiles after unilateral illumination. Sum of radioactivity in both halves (*D* and *L*) calculated as 100%, with the curve illustrating the difference in the percentages. (After DEHAAN 1969)

photostimulated coleoptiles, suggesting that the transverse oscillations in distribution are due to unilateral illumination (DEHAAN 1969).

5.1.3 Leaf Movements

Rhythmic leaf movements, especially for plants possessing pulvini, have been well documented. In addition to phytochrome (FONDEVILLE et al. 1966, HILLMAN and KOUKKARI 1967, JAFFE and GALSTON 1967), auxins, cytokinins, ABA, and GA's also have an effect on nyctinastic leaf movements (FERRI and DE CAMARGO 1950, WILLIAMS and RAGHAVAN 1966, GORDON and KOUKKARI 1972, MCEVOY and KOUKKARI 1972, SATTER et al. 1972, TANADA 1983). A popular method for studying the effects of growth regulators on leaf movements is to place leaflet pairs in dishes containing the growth regulator and to measure the angles formed by leaflets at various times (Fig. 1). In general, auxins, cytokinins, ABA, and GA's either promote opening or inhibit the closure of leaflets in this test. Antiauxins, auxin analogs, and a cytokinin antagonist appear to be ineffective (SATTER et al. 1972, TANADA 1983), suggesting direct hormonal involvement rather than a pH (MCEVOY and KOUKKARI 1972) or osmotic effect. How phytohormones act may be best explained in terms of alterations in membrane permeability (see Sect. 6), since such leaf movements are controlled by turgor changes in the pulvini (or pulvinules).

A different mechanism for leaf movement exists in *Carica papaya* (see Sect. 4). After 2 weeks in LL, nyctinastic movements of papaya leaves damp out, but they resume following the application of IAA to the lobes (YIN 1941). The results strongly support the hypothesis that auxin plays a role in controlling this rhythm.

5.2 Abscisic Acid, Cytokinins, Ethylene, and Gibberellins [1]

ABA is believed to alter ion transport in membranes, particularly the distribution of K^+ and Cl^- (VAN STEVENINCK 1976), and thus might be expected to affect rhythms involving turgor changes (see Sect. 6.4). The effects of exogenously applied ABA on the circadian petal movement of *Kalanchoe blossfeldiana* have been investigated (SCHREMPF 1980). Permanently offered ABA (10^{-8} M to 10^{-4} M) has no effect on period length, but the amplitude damps out at the higher concentrations. At 10^{-6} M, the rhythm can be restored by a light pulse after the ABA is withdrawn. When *Kalanchoe* flowers are given 4-h pulses of 10^{-5} M ABA, phase shifts are observed: treatment at maximum petal closure induces a 5-h delay, while treatment at maximum opening results in a 1-h delay. More recently, the effects of ABA on the circadian leaf movement of *Oxalis regnellii* MIG. have also been examined (SKROVE et al. 1982). Again, no change in the period occurs with permanently offered ABA (10^{-6} M to 10^{-4} M), although in this case there is a slight phase delay. ABA, particularly 10^{-4} M, decreases the amplitude in *Oxalis* as it does in *Kalanchoe*. However, with *Oxalis* 4-h pulses of 10^{-4} M ABA cause both phase advances and delays, the former occurring when the hormone is administered prior to maximum opening and the latter if ABA is applied preceding maximum closure of the leaves.

By themselves ABA and ethylene are not particularly effective in modifying the movements of *Albizzia julibrissin* DURAZZINI leaflets (SATTER et al. 1972), although possible synergisms were not examined. Individually, both IAA and GA_3 influence the leaf movements of *Albizzia* (MCEVOY and KOUKKARI 1972) and *Mimosa pudica* L. (WILLIAMS and RAGHAVAN 1966), and in *Mimosa*, IAA and GA_3 have been observed to act synergistically in decreasing the "dark" closure of leaflets (WILLIAMS and RAGHAVAN 1966).

There is a report that 10^{-6} M GA_3 treatments of 6 h change the rhythmic sensitivity of floral development in *Chenopodium* (KING 1975). Although these results must be confirmed by additional experiments, GA_3 may be capable of inducing a phase advance of the flowering rhythm of *Chenopodium*. GA_3 also causes a marked increase in the amplitude of circumnutation (see Fig. 4) in *Pisum* seedlings (MCCOMB 1962).

The activity of nitrate reductase in catalyzing the reduction of nitrate to nitrite shows circadian rhythmicity in studies conducted with *Glycine* (DUKE et al. 1978), *Capsicum annuum* L. (STEER 1976), *Chenopodium* (COHEN and CUMMING 1974), *Triticum aestivum* L. (UPCROFT and DONE 1972) and *Zea* (DUKE et al. 1978). The oscillatory behavior of enzymatic activity in plants has been fairly well documented (DEITZER et al. 1974a, b, HILLMAN 1976, DUKE et al. 1978). There is experimental evidence that the rhythm parameters of nitrate reductase activity, at least in *Capsicum*, may be influenced by a cytokinin, since kinetin (sprayed on leaves other than those assayed) causes a phase-shift in the activity peaks of nitrate reductase (STEER 1976). Transpiration rates do

1 For another group of substances, reported as turgor regulators, see Addendum on p. 78

not correspond to shifts in nitrate reductase activity. Because cytokinins and nitrate synergistically stimulate nitrate reductase activity (PARKASH 1972, KNYPL 1973), it may be reasoned that the effects of kinetin are due to more than just an increase in the translocation of nitrate (STEER 1976).

5.3 Herbicides

The effectiveness of herbicides in regulating plant growth can sometimes be related to the time of day that the herbicide is applied (Table 6). However, many of the investigations cited in Table 6 were not designed to study biological rhythms and they contain relatively few time points. Because few times were compared (e.g., "early morning" vs. "late evening"), it is difficult to generalize about oscillations and rhythmic behavior. Nevertheless, oscillations have been observed under controlled LD conditions in the response to herbicides of at least four different plants, *Abutilon theophrasti* MEDIC. (DORAN and ANDERSEN 1976, KOUKKARI and JOHNSON 1979), *Gossypium hirsutum* (GOSSELINK and STANDIFER 1967, RIKIN et al. 1983b), *Phaseolus vulgaris* (CAMPIRANON and KOUKKARI 1976), and *Pisum* (WEAVER and NYLUND 1963). Although young *Phaseolus* plants maintained under DD and LL during treatments spanning a single 24-h cycle display a significant oscillation in their sensitivity to 2,4-D (CAMPIRANON and KOUKKARI 1976), the presence of endogenicity has not been fully established (see Sect. 2 for criteria). With *Abutilon,* a significant 24-h rhythm in the response to bentazon was observed under synchronized conditions (KOUKKARI and JOHNSON 1979). This herbicide is more effective in the dark span and less effective in the middle of the light span. Circadian rhythmicity in the response to several herbicides has been satisfactorily demonstrated in the case of *Gossypium* (RIKIN et al. 1983b). Under free-running conditions, cotton seedlings exhibit rhythmical changes in their sensitivity to bentazon, as well as to acifluorfen and fusilade.

Environmental factors could have a strong influence on the effectiveness of herbicide applications in field studies. However, in studies where results from field experiments have been compared to results from experiments conducted in controlled environmental chambers, the most and least effective times of application were quite similar (cf. WEAVER and NYLUND 1963, DORAN and ANDERSEN 1976). With *Phaseolus* and *Pisum* the effects of 2,4-D and MCPA respectively on inhibiting growth (in controlled environments) have been shown to be more pronounced when applied to foliage during the latter part of the light span (WEAVER and NYLUND 1963, CAMPIRANON and KOUKKARI 1976).

Oscillations in the sensitivity of plants to herbicides have been attributed to many factors. Among these are the translocation of photosynthate (WEAVER and NYLUND 1963) and variation in stomatal opening (SCHUSTER 1970), which both display rhythmicity (STÅLFELT 1965, QUEIROZ 1974). In addition, many physiological processes such as photosynthesis, respiration, and transpiration display rhythmicity (HASTINGS et al. 1961, CUMMING and WAGNER 1968, CHIA-LOOI and CUMMING 1972, PALLAS et al. 1974) and are known to be involved in the biochemical effects of herbicides (cf. AUDUS 1976). A rhythm in detoxification of a herbicide within the plant tissue may account for an observed rhythm

Table 6. Plants in which the effectiveness of a herbicide may depend upon the time of day

Plant	Herbicide	Reference
Bean (*Phaseolus vulgaris* L.)	2,4-D [a]	CAMPIRANON and KOUKKARI (1976)
Cocklebur (*Xanthium pensylvanicum* Wallr.)	Bentazon [b]	DORAN and ANDERSEN (1976)
Cotton (*Gossypium hirsutum* L.)	Fluometuron [c] Dicryl [d] EPTC [e]	GOSSELINK and STANDIFER (1967)
	Bentazon [b] Acifluorfen [f] Fluazifopbutyl [g]	RIKIN et al. (1983b)
Crabgrass (*Digitaria* sp.)	Chloroxuron [h]	GOSSETT and RIECK (1970)
Johnsongrass [*Sorghum halepense* (L.) Pers.]	Dalapon [i]	CAULDER and FLETCHALL (1970)
Huisache [*Acacia farnesiana* (L.) Willd.]	Picloram [j] 2,4,5-T [k]	BOVEY et al. (1972)
Macartney rose (*Rosa bracteata* Wendl.)	Picloram [j] 2,4,5-T [k]	BOVEY et al. (1972)
Pea (*Pisum sativum* L.)	MCPA [l]	WEAVER and NYLUND (1963)
Pigweed (*Amaranthus* sp.)	Chloroxuron [h]	GOSSETT and RIECK (1970)
Pricklypear (*Opuntia polycantha* Haw.)	2,4,5-T [k]	SCHUSTER (1970)
Quackgrass [*Agropyron repens* (L.) Beauv.]	Paraquat [m]	PUTNAM and RIES (1968)
Soybean [*Glycine Max* (L.) Merr.]	Chloroxuron [h]	BLACK and WILSON (1969)
Tobacco (*Nicotiana tabacum* L.)	MH [n], _ [o]	SELTMANN and PEEDIN (1972)
Velvetleaf (*Abutilon theophrasti* Medic.)	Bentazon [b]	DORAN and ANDERSEN (1976) ANDERSEN and KOUKKARI (1978) KOUKKARI and JOHNSON (1979)

[a] (2,4-dichlorophenoxy) acetic acid. [b] 3-isopropyl-1 *H*-2,1,3-benzothiadiazin-4(3 *H*)-one 2,2-dioxide. [c] 1,1-dimethyl-3-(α, α, α-trifluoro-*m*-tolyl)urea. [d] 3′,4′-dichloro-2-methyl-acrylanilide. [e] *S*-ethyldipropylthiocarbamate. [f] 5-[2-chloro-4-(trifluoromethyl) phenoxyl]-2-nitrobenzoic acid. [g] Butyl 2-[4-(5-trifluoromethyl)-2-pyridyloxy)phenoxy]propionate. [h] 3-[*p*-(*p*-chlorophenoxy)phenyl]-1,1-dimethylurea. [i] 2,2-dichloropropionic acid. [j] 4-amino-3,5,6-trichloropicolinic acid. [k] (2,4,5-trichlorophenoxy) acetic acid. [l] [(4-chloro-*o*-tolyl)oxy]acetic acid. [m] 1,1′dimethyl-4,4′-bipyridinium. [n] 1,1′dihydro-3,6-pyridazinedione. [o] dimethyldodecylamine acetate.

in sensitivity (RIKIN et al. 1983b). Even for a single herbicide, such as bentazon, both environmental conditions and the stage of plant development influence oscillations in leaf sensitivity. Although daily oscillations in the sensitivity of *Abutilon* leaves to bentazon have been observed under both high and low humidity, they are more evident in the latter case (KOUKKARI and JOHNSON 1979). The presence of moisture that collects on foliage in the form of dew or rain droplets could, depending upon the nature of substance applied, increase or decrease the effectiveness of the herbicide (BEHRENS and ELAKKAD 1976, BEHRENS 1977).

Leaf-movement rhythms have also been suggested as being responsible for changes in the effects of herbicides on plants such as *Abutilon* (ANDERSEN and KOUKKARI 1978) and *Cassia obtusifolia* L. (KRAATZ and ANDERSEN 1978). According to calculations (ANDERSEN and KOUKKARI 1978), the amount of herbicide sprayed from above that comes in contact with a plant that has its leaves extended in a horizontal position is significantly greater than when the leaves are folded vertically. Many plants that are subjected to herbicide sprays show such movements (ANDERSEN and KOUKKARI 1979). Thus, the rhythmic effects of herbicides on such plants could be attributed to rhythmic leaf movements. A point worth noting with respect to the herbicide picloram is that concentrations 10- to 20-fold lower than those necessary to destroy a plant may inhibit leaf movements (MORGAN and BAUER 1970).

When the effects of leaf orientation have been eliminated by applying a solution of bentazon in the form of tiny droplets to precise locations on *Abutilon* leaves, daily oscillations in leaf sensitivity have still been observed (KOUKKARI and JOHNSON 1979). Bentazon inhibits photosynthetic CO_2 fixation (RETZLAFF and FISCHER 1973) and electron transport (BÖGER et al. 1977). Since the peak of the photosynthesis rhythm occurs near the middle of the L span (HASTINGS et al. 1961, PALLAS et al. 1974), the trough in leaf sensitivity to bentazon could be generated by an increase in photosynthetic capacity and a low overall level of bentazon in major intracellular sites (KOUKKARI and JOHNSON 1979).

The nature of the growth regulator is also an important factor. For example, when two substances are compared for controlling suckers in *Nicotiana tabacum* L., one is more effective in the morning, while the other is more effective in the evening (SELTMANN and PEEDIN 1972).

There are some indications that water permeability in subepidermal cells of isolated *Pisum* stem sections may oscillate (LEE and STADELMANN 1972, LEE-STADELMANN and STADELMANN 1979). Such oscillations could account for changes in the susceptibility of plants to chemicals. For example, the herbicide N-1-napthylphthalamic acid may affect the permeability of *Allium cepa* L. epidermal cells to water (FENG 1974). More quantitative results are necessary to substantiate a hypothesis that oscillations in the chemical effects on foliage are related to changes in membrane permeability.

6 Mechanisms

Little is known about the mechanisms responsible for the rhythmic nature of physiological processes that involve, or imply, the participation of plant hormones. There is a limited amount of experimental evidence (compared, for example, to the massive amounts published on auxin action), and consequently models and hypotheses abound. Many of the models and hypotheses that have been proposed for rhythms per se (e.g., PAVLIDIS 1969, NJUS et al. 1974, SWEENEY 1974a, WAGNER et al. 1975, ENGELMANN and SCHREMPF 1980) could account, with certain modifications, for hormonal oscillations as well. In order to provide cohesiveness to this discussion, the subject of mechanisms will focus on phenomena which relate to (1) chemical oscillations (2) transport (3) enzymes, and (4) ions and membranes.

6.1 Chemical Oscillations

Periodicities of chemical reactions that occur in homogeneous solutions have been considered as models for oscillations that take place in biochemical systems (HESS and BOITEUX 1971, HESS 1977). Since hormones are chemicals that may participate in simple reactions (e.g., liberation of H^+, as suggested by the "acid growth" hypothesis), oscillating chemical reactions will be considered here. Some oscillating chemical reactions are fascinating and can be demonstrated easily in the student laboratory (BRIGGS and RAUSCHER 1973, HAWKINS et al. 1975, COOKE 1977). Understanding how they are generated is much more difficult, however. Thermodynamically, chemical oscillations show temporal organization, occur in open systems around a steady state at nonequilibrium conditions, and are capable of displacement (NOYES and FIELD 1974, GOLDBETER and CAPLAN 1976). Sustained stable oscillations require either positive or negative feedback, sometimes combined with cross-catalysis (NICOLIS and PORTNOW 1973).

According to some sources (NICOLIS and PORTNOW 1973, NOYES and FIELD 1974), the oldest reported periodic chemical reaction in a homogeneous solution is the oxidation-reduction reaction in which hydrogen peroxide is decomposed by iodate catalysis (BRAY 1921). Although numerous studies have been devoted to this purely chemical rhythm, in which all initial components can be quantitatively measured, some questions regarding the mechanism remain (NICOLIS and PORTNOW 1973, NOYES and FIELD 1974). Thus, one can appreciate the difficulty in hypothesizing mechanisms for the more complex rhythms in biological systems.

Of the various types of chemical oscillations, those of biological systems observed at the genetic, regulatory, and metabolic levels (HESS and BOITEUX 1971, CHANCE et al. 1973) have provided supportive data for theoretical models. The well-known ultradian oscillations in yeast glycolysis continue for many cycles and are susceptible to phase-shifting (PYE 1969), thus displaying two of the major characteristics of circadian rhythms. It has been proposed that many short-period oscillations could interact to give rise to circadian periods (PAVLIDIS and KAUZMANN 1969, PAVLIDIS 1969, 1971, 1973). Similarly, it is possible that ultradian oscillations involving hormones interact in some manner to produce a circadian cycle (e.g., growth of coleoptiles – see Sect. 5). At the moment, experimental evidence for such hormonal involvement is lacking and other mechanisms appear more probable.

6.2 Transport

The polarity of auxin transport was discovered and quantitatively described over 50 years ago (WENT 1928, VAN DER WEIJ 1932). Classically, auxin has been considered to move in the cytoplasm of the cell, from the tip to the base, where it then enters the next cell by an energy-requiring process (cf. GOLDSMITH 1968). More recently a new hypothesis has been suggested (RUBERY and SHELDRAKE 1973, 1974, RAVEN 1975), descriptively referred to as the "chemiosmotic theory of polar diffusion in cells" (GOLDSMITH 1977). According to the latter hypothesis, transport is thermodynamically "downhill"; metabolic energy is not directly required to transport IAAH (undissociated acid) and IAA, although it is necessary for secretion of H^+ to maintain pH gradients, membrane potentials, and polar permeability (GOLDSMITH 1977). The oscillating electrical potentials that move down the coleoptile (NEWMAN 1963) are in close accord with oscillations in auxin transport in tissue previously depleted of hormone (HERTEL and FLORY 1968). The effects of proton flux can be traced to the membrane level of organization, with changes in electrical potential resulting from possible feedback loops. These changes are reflected by oscillations in the rate of auxin transport. Similarly, the oscillations in the distribution of auxin following phototropic induction (DEHAAN 1969) could arise from asymmetric H^+ secretion by cells on two sides (i.e., illuminated vs. shaded) in response to changing increments in the concentration of auxin (cf. GOLDSMITH 1977), much like the model proposed for the feedback loop in bean roots (JENKINSON and SCOTT 1961).

Oscillations in the response of *Pisum* to the herbicide MCPA may be related to transport. Based on results from studies conducted under controlled environmental conditions, the increased effectiveness of MCPA during the L span may be related to an increase in the rate of its translocation, which in turn is due to an increase in the translocation of photosynthates in L (WEAVER and NYLUND 1963).

Intracellular transport of hormones and translocation of photosynthate certainly play a role in some rhythmic oscillations. However, transport alone is insufficient to explain most rhythmic phenomena.

6.3 Enzymes

An early theoretical model that related enzymes, hormones, and rhythms was based on IAA levels in *Pisum* (GALSTON and DALBERG 1954). This model proposes that oscillations in auxin levels result from the adaptive formation of IAA oxidase. However, IAA oxidase regulation cannot account for the rhythmic growth of *Avena* coleoptiles (BALL and DYKE 1956).

The activity of many enzymes displays rhythmicity (e.g., SULZMAN and EDMUNDS 1972, FROSCH et al. 1973, DEITZER et al. 1974a, b, WAGNER et al. 1974, QUEIROZ 1974, STEER 1976, GORDON and KOUKKARI 1978, DUKE et al. 1978). Moreover, hormones may influence the enzyme status or level in a plant (e.g., DATKO and MACLACHLAN 1968, KASAMO and YAMAKI 1974, VARNER and HO

1976). Nonetheless, a direct interrelationship among enzymes, hormones, and rhythms has not been demonstrated.

6.4 Ions and Membranes

Support for membrane involvement in rhythms has come from a variety of sources (cf. ENGELMANN and SCHREMPF 1980). Among these are studies in which substances such as valinomycin (a K^+ carrier or complexing agent), alcohol, and D_2O have altered rhythms (ENRIGHT 1971a, b, BÜNNING and MOSER 1972, MCDANIEL et al. 1974, SWEENEY 1974b).

A model closely related to chemical oscillations is the physiological model of the feedback-induced oscillations proposed for *Vicia* roots (JENKINSON and SCOTT 1961). Here, a bioelectric current can be generated by growing roots in a weakly conducting salt solution (SCOTT et al. 1955). Results from several electrical potential studies (JENKINSON and SCOTT 1961, JENKINSON 1962a, SCOTT et al. 1955, SCOTT 1957, 1962, SCOTT and MARTIN 1962) reveal an oscillating phenomenon that may be related to auxins. These oscillations, induced by chemical, electrical, or mechanical disturbances of the root's environment, have periods of ca. 5 min. In some experiments, oscillations have been induced by changing the osmotic potential with sucrose or varying the amount of IAA present in the bathing medium in an oscillating manner (i.e., increasing and decreasing the concentration from 0 to 10^{-7} M IAA). These results have provided the basis for a theoretical feedback model (see Fig. 7 in FENSOM, Chap. 17, this Vol.). This proposed feedback loop, with its three time delay constants, has the following sequential characteristics: membrane potential differences that affect the distribution of hormone to active sites (GOLDSMITH 1977); modification of membrane potentials by auxin; and variations in ionic permeability that change the flow of current in and around the root, simultaneously altering the potentials in the external medium. The required energy source for the entire mechanism could be supplied by metabolic reactions (SCOTT 1962). In a model system, application of either auxin (IAA) or an anti-auxin (2,6-D) forces plant potentials to oscillate when the concentrations are low ($<10^{-5}$ M), but inhibits oscillations when the concentrations are high (10^{-5} M). These results, which may be due to competition for active sites (e.g., by enzymes), are more fully explained by JENKINSON (1962b).

Auxin has also been shown to alter the membrane potential in the epidermal cells of corn coleoptiles (STAHLBERG and POLEVOI 1979). IAA induces oscillations with a period of ca. 15 min and an amplitude on the average of 20 mV. It seems probable that these oscillations are caused by an auxin-dependent proton pump on the plasmalemma (STAHLBERG and POLEVOI 1979).

Circadian movements of *Albizzia, Mimosa, Samanea,* and other legume leaves and leaflets are controlled by turgor changes in a specialized region of the leaf called the pulvinus or pulvinule (Figs. 1 and 9). Movement occurs as a result of turgor changes in motor cells located on opposite sides of the pulvinus. It has been proposed that within these cells alterations in membranes and ionic components account for the rhythmic movements (SATTER et al. 1973,

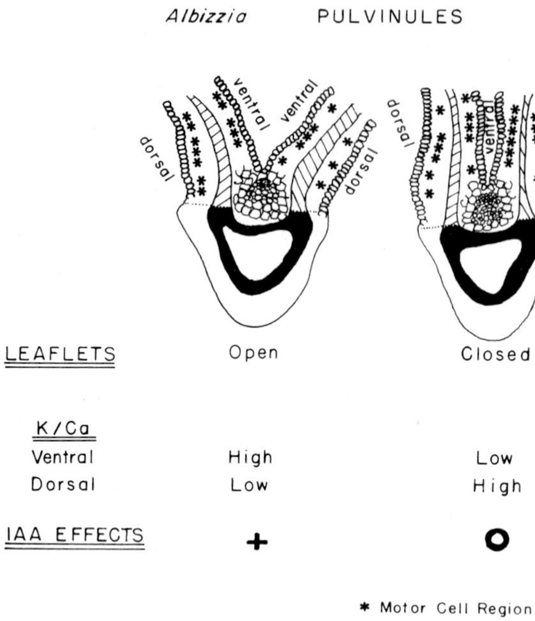

Fig. 9. Diagram of the dorsiventral longitudinal sections of *Albizzia* pulvinules, illustrating their status when leaflets are open and closed, relative K/Ca ratio, and the effects of IAA. (After SATTER et al. 1970a, b, 1972, SATTER and GALSTON 1971b). More recently the terms "extensor" and "flexor" have been substituted for "ventral" and "dorsal". (Cf. SATTER 1979)

1974c, KOUKKARI 1974, RACUSEN and SATTER 1975, SATTER 1979). The subject of leaf movements has also been discussed in Volume 7 of this Series (SATTER 1979, SWEENEY 1979).

Potassium flux, as well as that of calcium and phosphorus, has been measured with an electron microprobe in *Albizzia* and *Samanea* pulvinules or pulvini (SATTER and GALSTON 1971a, b, SATTER et al. 1974a, c, RACUSEN and SATTER 1975). Results from these studies indicate that the rhythmic leaflet movements are accompanied by a rhythmic flux of K^+ in the motor cells on both sides of the pulvinule (Fig. 9). Since the exchange of K^+ ions, which depends on both active and passive transport, does not occur between ventral and dorsal motor cells, and K^+ must be supplied by other cells, the two regions each display rhythmic behavior (SATTER and GALSTON 1973).

Pulvini are not only the active regions for rhythmic movements, but they are also part of the photoreceptive region for phytochrome-controlled nyctinastic movements (KOUKKARI and HILLMAN 1968). Leaflets placed in D close faster if briefly exposed to red illumination than those exposed to far-red light prior to D (FONDEVILLE et al. 1966, HILLMAN and KOUKKARI 1967, JAFFE and GALSTON 1967). Potassium flux is a common feature of both circadian movements and phytochrome-controlled nyctinasty; in the latter case, phytochrome may act as a K^+ gate (SATTER and GALSTON 1971a). Actual membrane potential measurements, such as those obtained from *Samanea* motor cells in DD, show convincing circadian oscillations at the membrane level; phytochrome-controlled changes in transmembrane potentials also have been observed in *Samanea* motor cells (RACUSEN and SATTER 1975).

Leaf movements in *Mimosa* and *Albizzia* may be influenced by exogenously applied growth substances (see Fig. 1 for generalized procedure) such as IAA and GA_3 (WILLIAMS and RAGHAVAN 1966, MCEVOY and KOUKKARI 1972). Other growth substances that may also affect the movement of leaflets include the auxins NAA (MCEVOY and KOUKKARI 1972, SATTER et al. 1972) and 2,4-D (SATTER et al. 1972), and the cytokinin benzyladenine (GORDON and KOUKKARI 1972). Although ethylene effects have also been observed in *Acacia* (MORGAN and BAUER 1970), they were not detected in *Albizzia* (SATTER et al. 1972). ABA, which has been shown to affect turgor in guard cells (TUCKER and MANSFIELD 1971) has been reported to be ineffective in pulvinule cells, as have the antiauxins benzoic acid, 4-chlorophenoxyisobutyric acid, and 2,4-dichlorophenoxyisobutyric acid (SATTER et al. 1972, SATTER and GALSTON 1973).

Exogenously applied growth substances can also affect phytochrome-controlled nyctinasty. It has been shown that far-red light (760 nm) delays nyctinastic closure in *Albizzia* only if red light (660 nm) is given simultaneously or just prior to far-red (cf. TANADA 1983). Several hormones can substitute for this red light requirement. When *Albizzia* pinnules are treated with IAA, ABA, GA, or kinetin and then irradiated with far-red light, leaflet closure is retarded without exposure to red light (TANADA 1983). These hormones or growth substances are effective at concentrations ranging from 0.05 μM to 1.0 μM. Treatment with indole-3-carboxylic acid or the cytokinin antagonist 3-methyl-7-n-pentylaminopyrazol does not delay closure. It is hypothesized that the effect of phytochrome action is to increase the concentration of the endogenous hormone level. The hormone(s) may then interact with an as-yet unidentified far-red absorbing pigment to delay leaflet closure by modification of pulvinule cell membranes (TANADA 1983).

What is the role of endogenous hormones such as IAA in the mechanism of rhythmic leaf movements? In the case of pulvinules, IAA has been envisioned as a "fine tune" control that synchronizes the activities of ventral and dorsal motor cells, while the "rough tuning" is provided by the changes between energetic and non-energetic phases (SATTER and GALSTON 1973). In situations where leaves move in response to growth changes caused by auxins (cf. YIN 1941), the mechanisms may involve the unequal distribution of auxin and the morphological nature of the petiole in relation to different regions or lobes of the lamella (see Sect. 4). The morphological features of the leaf and vascular system that are associated with auxin transport in papaya (YIN 1941) could likewise be present in other plants as well (cf. BRAUNER and ARSLAN 1951). While GA_3 has been shown to affect leaf movements (WILLIAMS and RAGHAVAN 1966, MCEVOY and KOUKKARI 1972), little is known about the mechanism of its action. In *Pisum* seedlings, GA_3 may have an immediate effect on K^+ transport (LÜTTGE et al. 1968).

For a full discussion of membrane models for circadian rhythms, see ENGELMANN and SCHREMPF (1980). Hormonal participation in oscillatory phenomena of plants has been well documented, but more experimental evidence is needed if a direct involvement is to be demonstrated. No single theory has proved satisfactory to explain the basic mechanism of rhythms, and it is possible that more than one mechanism is operative.

7 Role of Rhythms in the Life of the Plant

Endogenous rhythms provide the plant with temporal organization (KOUKKARI and SOULEN 1981) and stability. It is generally accepted that a "universal" mechanism generating circadian rhythms may function at the cellular rather than at the tissue level of organization (cf. SWEENEY 1974a). Nonetheless, interactions or communications between cells, tissues, and even organs may operate synchronously to display or amplify oscillations. Such rhythms have been observed at all levels of organization (KOUKKARI and SOULEN 1981), from metabolic reactions such as the ultradian oscillations in NADH concentration in populations of yeast cells (PYE 1969) to gross movement of organs like the "sleep" movements of the leaves of vascular plants (BÜNNING 1973, ANDERSEN and KOUKKARI 1979).

7.1 Avoidance of Pathological Effects

As a plant grows, develops, matures, and finally reaches senescence during the course of time, it does so under environmental conditions that are cyclic. When plants are experimentally maintained under CC, a single alteration in environmental conditions, such as a change from L to D, or in temperature, may be all that is necessary to evoke a rhythm that was not previously evident (BÜNNING 1956). When an established rhythm becomes dampened under CC (e.g., under LL), a change to another condition of sufficient duration (e.g., 8 h D) is required to reinitiate periodicity (BÜNNING 1962). A more direct relationship between 24-h environmental cycles and endogenous circadian timing can be seen when, by experimental manipulation, an external oscillator is abnormal or absent (WENT 1960, 1962). Under such conditions, pathological symptoms may be observed. For example, when tomato (*Lycopersicon esculentum*) plants are exposed to 4 h D but to various L spans ranging from 4 h to 26 h, the plants maintained under regimens that do not approximate a day (ca. 20 h to 28 h) are injured, whereas LD 16:4, 20:4 and 26:4 regimens are not injurious to the plant (HILLMAN 1956). The injury associated with plants maintained under LL at a constant temperature is eliminated merely by providing a 24-h temperature cycle. Although the pattern of injury and recovery in these tomato experiments is somewhat reminiscent of supra-optimal auxin dosage effects, direct evidence for auxin as the source of injury has not been established. Another interesting example of this phenomenon has been observed in the composite flowers of *Cichorium intybus* L. Under LL, a desynchronization in petal movements occurs and the flowers become deformed (BÜNNING 1962).

7.2 Adaptive Significance

Auxins have been shown to have a significant role in leaf movements – but of what value are such rhythmic movements? There is some evidence that leaf movements may have an adaptive function (DARWIN and DARWIN 1881, SMITH 1974, BÜNNING 1979). One of the more convincing hypotheses to explain the value of such phenomena suggests that a change in position at night protects the leaves from potentially damaging illumination (i.e., moonlight) during critical D spans (BÜNNING 1969, BÜNNING and MOSER 1969). For example, the intensity of moonlight that reaches the upper surface of *Arachis hypogaea* L., *Glycine*, and *Trifolium repens* L. leaves may be reduced by 80–90% when they are in the "night" position. Thus photoperiodic processes are not adversely shifted or altered. Leaf movements may also be important for light perception during twilight intervals (dawn and dusk signals), times when rapid changes occur in light intensity and red/far-red ratios.

The view that a "closed" leaf position at night provides protection from low temperature (DARWIN and DARWIN 1881) has been questioned (SCHWINTZER 1971). However, there are plants, such as a rosette plant in the Andes, where this mechanism may be very important for survival (SMITH 1974). Thus leaf movements could serve various adaptive roles, depending upon the morphology, physiology, and habitat of the species (cf. SATTER 1979).

7.3 Organization in Time

Rhythms appear to provide the organism with some stability whereby certain metabolic events can proceed in time-structured sequences controlled by feedback interactions. If auxin transport occurs according to the chemiosmotic hypothesis (cf. GOLDSMITH 1977) and auxins act via acid growth, then the oscillatory nature of auxin transport under certain conditions (HERTEL and FLORY 1968, DEHAAN 1969) could provide stability in regulating growth by not "driving" any one component (e.g., H^+) beyond or below optimum levels. In more general terms, rhythms may provide for the kinetic optimization of a pathway that, with efficient response times and damping, could allow a system to make adjustments to changing conditions (HESS and BOITEUX 1971).

The periodicities of the rhythmic processes in which hormones, especially auxins, function may be on the order of seconds, hours, days, or months. Not only ultradian, but circadian and infradian rhythms as well, may permit an organism to respond in a predictable manner to the demands of a cyclic and sometimes unpredictable environment.

8 Conclusions

Rhythms are a common component of physiological processes, and they often involve hormones. The relationship among physiological processes, hormones, and rhythms may be indirect, as in the case of coleoptile growth, or direct, as in auxin transport. Hormonal participation is evident in many ultradian (periods <20 h, often min) and circadian (20–28 h) rhythms. Ultradian oscillations that involve hormones include auxin transport, electrical potentials, geotropism, nutations, and leaf movements. Circadian rhythms that may have a hormonal component are represented by events such as growth, development, leaf movements, ion transport, and endogenous levels of growth regulators. There are strong implications of hormonal interactions with rhythmic metabolic processes as well. The responses of plants to growth regulators such as herbicides also display a circadian rhythm in some cases.

Proposed mechanisms or models that contribute to understanding hormone-related rhythmic processes include chemical oscillations, transport, enzyme participation in hormone synthesis and destruction, and the involvement of ions and membranes. Rhythmic phenomena may provide temporal organization, stability, control, and adaptation for plant growth and development in the cyclic environment of our planet.

Note Added in Proof. While this manuscript was in press, a new paper was published on the rhythmic sensitivity of seedlings to several herbicides (Rikin, St. John, Wergin, and Anderson, 1984. Plant Physiol 76:297–300). The oscillations in sensitivity were not detected in seedlings grown under LL from germination. However, in the case of leaf movements, a rhythm has been observed even in seedlings germinated under LL (Hoshizaki and Hamner 1964, Sci 144:1240–1241). Recently we have confirmed the latter observation and have now initiated experiments to examine this matter in relation to the rhythmic effects of growth regulators under CC.

Acknowledgements. The numerous suggestions, comments, and stimulating discussions provided by Thomas K. Soulen, Margo A. Johnson, LeAnne Carlson, Kimberly L. Person, Andrea Gargas, and Robert H. Warde during the preparation of this chapter are gratefully acknowledged. We thank Claire M. Martin for her assistance and patience while typing the various drafts of this manuscript.

References

Abeles FB, Craker LE, Leather GR (1971) Abscission: The phytogerontological effects of ethylene. Plant Physiol 47:7–9

Alford DK, Tibbitts TW (1971) Endogenous short-period rhythms in the movements of unifoliate leaves of *Phaseolus angularis* Wight. Plant Physiol 47:68–70

Andersen RN, Koukkari WL (1978) Response of velvetleaf to bentazon as affected by leaf orientation. Weed Sci 26:393–395

Andersen RN, Koukkari WL (1979) Rhythmic leaf movements of some common weeds. Weed Sci 27:401–414

Arnal C (1953) Recherches sur la nutation des coléoptiles. I. Nutation et croissance. Ann Univ Saraviensis. Scientia 2:92–105

Aschoff J, Klotter K, Wever R (1965) Circadian clocks. In: Aschoff J (ed) Proceedings of the Feldafing summer school. Elsevier/North-Holland Biomedical Press, Amsterdam New York, pp 10–17

Audus LJ (1976) Herbicides. Physiology, biochemistry, ecology, vol I. 2nd edn. Academic Press, London New York

Avery GS Jr (1935) Differential distribution of a phytohormone in the developing leaf of *Nicotiana*, and its relation to polarized growth. Bull Torrey Bot Club 62:313–330

Baillaud L (1953) Action de la température sur la période de nutation des tiges volubiles de cuscute. CR Acad Sci (Paris) 236:1986–1988

Baillaud L, Monnier Y (1960) La circumnutation de la tige d'un *Phaseolus* rendu volubile par l'acide gibbérellique. CR Acad Sci (Paris) 250:4032–4034

Ball NG, Dyke IJ (1954) An endogenous 24-hour rhythm in the growth rate of the *Avena* coleoptile. J Exp Bot 5:421–433

Ball NG, Dyke IJ (1956) The effects of indole-3-acetic acid and 2,4-dichlorophenoxyacetic acid on the growth rate and endogenous rhythm of intact *Avena* coleoptiles. J Exp Bot 7:25–41

Ball NG, Dyke IJ (1957) The effects of decapitation, lack of oxygen, and low temperature on the endogenous 24-h rhythm in the growth-rate of the *Avena* coleoptile. J Exp Bot 8:323–338

Ball NG, Dyke IJ, Wilkins MB (1957) The occurrence of endogenous rhythms in the coleoptiles in various cereal genera. J Exp Bot 8:339–347

Becker T (1953) Wuchsstoff und Säureschwankungen bei *Kalanchöe blossfeldiana* in verschiedenen Licht-Dunkelwechseln. Planta 43:1–24

Behrens R (1977) Influence of dew on the effectiveness of foliar herbicide treatments. Abstract. 1977 Meet Weed Sci Soc Am, p 12

Behrens R, Elakkad M (1976) Influence of simulated rainfall on the effectiveness of foliar herbicide treatments. Abstract. 1976 Meet Weed Sci Soc Am, p 1

Biale JB, Young RE, Olmstead AJ (1954) Fruit respiration and ethylene production. Plant Physiol 29:168–174

Black FS, Wilson HP (1969) Performance of herbicide adjuvant-sprays as affected by the time of day, by the ratio of herbicide to adjuvant, and by the chemical type of the adjuvant. Abstract. 1969 Meet Weed Sci Soc Am, p 1

Böger P, Beese B, Miller R (1977) Long-term effects of herbicides on the photosynthetic apparatus. II. Investigations on bentazone inhibition. Weed Res 17:61–67

Bornkamm R (1966) Ein Jahresrhythmus des Wachstums bei *Lemna minor* L. Planta 69:178–186

Bose JC (1927) Plant autographs and their revelations. Longmans, Green & Co, London New York Toronto

Bovey RW, Haas RH, Meyer RE (1972) Daily and seasonal response of Huisache and Macartney rose to herbicides. Weed Sci 20:557–580

Brauner L, Arslan N (1951) Experiments on the auxin reactions of the pulvinus of *Phaseolus multiflorus*. Rev Fac Sci Univ Istanbul 16 B:257–300

Bray WC (1921) A periodic reaction in homogeneous solution and its relation to catalysis. J Am Chem Soc 43:1262–1267

Briggs TS, Rauscher WC (1973) An oscillating iodine clock. J Chem Educ 50:496

Bünning E (1956) Endogenous rhythms in plants. Annu Rev Plant Physiol 7:71–90

Bünning E (1960) Opening address: biological clocks. Cold Spring Harbor Symp Quant Biol 25:1–9

Bünning E (1962) Mechanism in circadian rhythms: functional and pathological changes resulting from beats and from rhythm abnormities. Ann NY Acad Sci 98:901–915

Bünning E (1969) The adaptive value of circadian leaf movements. In: Menaker M (ed) Biochronometry. Natl Acad Sci, Washington, DC, pp 203–211

Bünning E (1973) The physiological clock, revised 3rd edn. Springer, Berlin Heidelberg New York

Bünning E (1979) Circadian rhythms, light, and photoperiodism: A re-evaluation. Bot Mag (Tokyo) 92:89–103

Bünning E, Chandrashekaran MK (1975) Pfeffer's views on rhythms. Chronobiologia 2:160–167

Bünning E, Moser I (1969) Interference of moonlight with the photoperiodic measurement of time by plants, and their adaptive reaction. Proc Natl Acad Sci USA 62:1018–1022

Bünning E, Moser I (1972) Influence of valinomycin on circadian leaf movements of *Phaseolus*. Proc Natl Acad Sci USA 69:2732–2733

Bünning E, Müller D (1961) How do organisms measure lunar cycles? Z Naturforsch 16b (6):391–395

Bünning E, Müssle L (1951) Der Verlauf der endogenen Jahresrhythmik in Samen unter dem Einfluß verschiedenartiger Außenfaktoren. Z Naturforsch 6b:108–112

Campiranon S, Koukkari WL (1976) Circadian periodic response of *Phaseolus vulgaris* L. to 2,4-dichlorophenoxyacetic acid. Chronobiologia 3:137–148

Caulder J, Fletchall OH (1970) Response of Johnson grass to dalapon applied at different times of day. Abstr Weed Sci Soc Am, p 73

Chance B, Ghosh AK, Pye EK, Hess B (1973) Biological and biochemical oscillators. Academic Press, London New York

Chia-Looi A, Cumming BG (1972) Circadian rhythms of dark respiration, flowering, net photosynthesis, chlorophyll content, and dry weight changes in *Chenopodium rubrum*. Can J Bot 50:2219–2226

Chorney W, Rakosnik E Jr, Dipert MH, Dedolph RR (1970) Rhythmic-flowering response in cocklebur. Bioscience 20:31–32

Cohen AS, Cumming BG (1974) Endogenous rhythmic activity of nitrate reductase in a selection of *Chenopodium rubrum*. Can J Bot 52:2351–2360

Cooke DO (1977) Homogeneous oscillating reactions. Educ Chem 14:53–56

Cumming BG (1967) Circadian rhythmic flowering responses in *Chenopodium rubrum*: Effects of glucose and sucrose. Can J Bot 45:2173–2193

Cumming BG (1969) Circadian rhythms of flower induction and their significance in photoperiodic response. Can J Bot 47:309–324

Cumming BG, Wagner E (1968) Rhythmic processes in plants. Annu Rev Plant Physiol 19:381–416

Cumming BG, Hendricks SB, Borthwick HA (1965) Rhythmic flowering responses and phytochrome changes in a selection of *Chenopodium rubrum*. Can J Bot 43:825–853

Darwin C, Darwin F (1881) The power of movement in plants. Appleton & Co, New York

Datko AH, Maclachlan GA (1968) Indoleacetic acid and the synthesis of glucanases and pectic enzymes. Plant Physiol 43:735–742

DeCoursey PJ (1960) Phase control of activity in a rodent. Cold Spring Harbor Symp Quant Biol 25:49–55

DeHaan I (1969) Oscillations in the redistribution of the growth substance naphthylacetic acid after phototropic induction. Acta Bot Neerl 18:84–94

Deitzer GF, Haertle U, Wagner E (1974a) Frequency patterns of enzyme activities reflecting metabolic control of photoperiodic timing. J Interdiscipl Cycle Res 5:187–198

Deitzer GF, Kempf O, Fischer S, Wagner E (1974b) Endogenous rhythmicity and energy transduction. IV. Rhythmic control of enzymes involved in the tricarboxylic-acid cycle and the oxidative pentose-phosphate pathway in *Chenopodium rubrum* L. Planta 117:29–41

Doran DL, Andersen RN (1976) Effectiveness of bentazon applied at various times of the day. Weed Sci 24:567–570

Driessche Vanden T (1970) Circadian variation in ATP content in the chloroplasts of *Acetabularia mediterranea*. Biochim Biophys Acta 205:526–528

Driessche Vanden T (1975a) Circadian rhythm in the Hill reaction of *Acetabularia*. In: Avron M (ed) Proceedings of the Third International Congress on Photosynthesis, Vol I. Elsevier, Amsterdam, pp 745–751

Driessche Vanden T (1975b) Chloroplast functions are influenced by morphactins. Biochem Physiol Pflanz 168:543–551

Driessche Vanden T (1975c) Circadian rhythms and molecular biology. BioSystems 6:188–201

Driessche Vanden T (1979) Phase-shifting effect of IAA on the photosynthetic circadian rhythm of *Acetabularia*. In: Bonotto S, Kefeli V, Puiseux-Dao S (eds) Developmental biology of *Acetabularia*. Elsevier/North-Holland Biomedical Press Amsterdam New York, pp 195–204

Driessche Vanden T, Delegher-Langohr V (1975) Presence of an auxin-like substance in *Acetabularia*. Protoplasma 83:181

Driessche Vanden T, Glory M (1979) Plant growth regulators as circadian rhythm synchronizers. Chronobiologia 6:167

Duke SH, Koukkari WL (1975) Ultradian oscillations in the activity of two mitochondrial enzymes, glutamate dehydrogenase and malate dehydrogenase, extracted from *Pisum* roots. XII Int Conf Int Soc Chronobiol. Publishing House Washington DC, pp 705–710

Duke SH, Friedrich JW, Schrader LE, Koukkari WL (1978) Cyclic activities of enzymes of nitrate reduction and ammonia assimilation in *Glycine max* and *Zea mays*. Physiol Plant 42:269–276

El-Beltagy AS, Hall MA (1974) Effect of water stress upon endogenous ethylene levels in *Vicia faba*. New Phytol 73:47–60

El-Beltagy AS, Kapuya JA, Madkour MA, Hall MA (1976) A possible endogenous

rhythm in internal ethylene levels in the leaves of *Lycopersicon esculentum* Mill. Plant Sci Lett 6:175–180
Engelmann W, Schrempf M (1980) Membrane models for circadian rhythms. Photochem Photobiol Rev 5:49–86
Enright JT (1971a) Heavy water slows biological timing processes. Z Vergl Physiol 72:1–16
Enright JT (1971b) The internal clock of drunken isopods. Z Vergl Physiol 75:332–346
Erismann KH, Fankhouser M (1967) Change in content of starch, protein and RNA of *Lemna minor* L. under the influence of kinetin (6-furfurylaminopurine). Experientia 23:621–622
Feng KA (1974) Changes of onion epidermal cell permeability due to the treatment with alanap. Physiol Plant 32:311–314
Ferri MG, De Camargo LV (1950) Influence of growth substances on the movement of the pulvini of the primary leaves of bean plants. Acad Bras Cienc An 22:161–170
Fondeville JC, Borthwick HA, Hendricks SB (1966) Leaflet movements of *Mimosa pudica* L. indicative of phytochrome action. Planta 69:357–364
Frosch S, Wagner E, Cumming BG (1973) Endogenous rhythmicity and energy transduction. I. Rhythmicity in adenylate kinase, NAD- and NADP-linked glyceraldehyde-3-phosphate dehydrogenase in *Chenopodium rubrum*. Can J Bot 51:1355–1367
Galston AW, Dalberg LY (1954) The adaptive formation and physiological significance of indoleacetic acid oxidase. Am J Bot 41:373–380
Galston AW, Tuttle AA, Penny PJ (1964) A kinetic study of growth movements and photomorphogenesis in etiolated pea seedlings. Am J Bot 51:853–858
Garner WW, Allard HA (1920) Effect of the relative length of day and night and other factors of the environment on growth and reproduction in plants. J Agric Res 18:553–606
Goldbeter A, Caplan SR (1976) Oscillatory enzymes. Annu Rev Biophys Bioeng 5:449–476
Goldsmith MHM (1968) The movement of plant-growth regulators. Annu Rev Plant Physiol 19:347–360
Goldsmith MHM (1977) The polar transport of auxin. Annu Rev Plant Physiol 28:439–478
Gordon WR, Koukkari WL (1972) The effect of cytokinins and auxins on phytochrome mediated nyctinasty in *Albizzia julibrissin*. Plant Physiol 49 (Suppl) 53
Gordon WR, Koukkari WL (1978) Circadian rhythmicity in the activities of phenylalanine ammonia-lyase from *Lemna perpusilla* and *Spirodela polyrhiza*. Plant Physiol 62:612–615
Gosselink JG, Standifer LC (1967) Diurnal rhythm of sensitivity of cotton seedlings to herbicides. Science 158:120–121
Gossett BJ, Rieck C (1970) Performance of chloroxuron as influenced by spray additives, spray volumes, and early morning versus late afternoon applications. Proc Southern Weed Sci Soc 23:163
Greathouse DC, Laetsch WM, Phinney BO (1971) The shoot-growth rhythm of a tropical tree, *Theobroma cacao*. Am J Bot 58:281–286
Halaban R (1968) The circadian rhythm of leaf movement of *Coleus blumei* × *C. frederici*, a short day plant. I. Under constant light conditions. Plant Physiol 43:1883–1886
Halberg E (1959) Physiologic 24-hour-periodicity; general and procedural considerations with reference to the adrenal cycle. Z Vitamin-Hormon Fermentforsch 10:225–296
Halberg F (1964) Organisms as circadian systems; temporal analysis of their physiologic and pathologic responses, including injury and death. In: Walter Reed Army Inst Res Symp. Medical aspects of stress in the military climate, pp 1–36
Halberg F (1969) Chronobiology. Annu Rev Physiol 31:675–725
Halberg F, Katinas GS (1973) Chronobiologic glossary of the international society for the study of biologic rhythms. Int J Chronobiol 1:31–63
Hamner KC (1960) Photoperiodism and circadian rhythms. Cold Spring Harbor Symp Quant Biol 25:269–277
Hamner KC (1969) *Glycine max* (L.) Merill. In: Evans LT (ed) The induction of flowering. Cornell Univ Press, Ithaca, NY, pp 62–89

Hastings JW, Sweeney BM (1957) On the mechanism of temperature independence in a biological clock. Proc Natl Acad Sci USA 43:804–811
Hastings JW, Astrachan L, Sweeney BM (1961) A persistent daily rhythm in photosynthesis. J Gen Physiol 45:69–76
Hawkins MD, Smith H, Square K (1975) Oscillating chemical reactions. Educ Chem 12:144–146
Heathcote DG (1966) A new type of rhythmic plant movement: micronutation. J Exp Bot 17:690–695
Henson IE, Algarswamy G, Mahalakshmi V, Bidinger FR (1982) Diurnal changes in endogenous abscisic acid in leaves of pearl millet (*Pennisetum americanum* L. Leeke) under field conditions. J Exp Bot 33:416–425
Henssen A (1954) Die Dauerorgane von *Spirodela polyrrhiza* (L.) Schleid. in physiologischer Betrachtung. Flora 141:529–566
Hertel R, Flory R (1968) Auxin movement in corn coleoptiles. Planta 82:123–144
Hess B (1977) Oscillating reactions. Trends Biochem Sci 2:193–195
Hess B, Boiteux A (1971) Oscillatory phenomena in biochemistry. Annu Rev Biochem 40:237–258
Hewett EW, Wareing PF (1973) Cytokinins in *Populus × robusta* Schneid: Light effects on endogenous levels. Planta 114:119–129
Hillman WS (1956) Injury of tomato plants by continuous light and unfavorable photoperiodic cycles. Am J Bot 43:89–96
Hillman WS (1964) Endogenous circadian rhythms and the response of *Lemna perpusilla* to skeleton photoperiods. Am Nat 98:323–328
Hillman WS (1970) Carbon dioxide output as an index of circadian timing in *Lemna* photoperiodism. Plant Physiol 45:273–279
Hillman WS (1971) Entrainment of *Lemna* CO_2 output through phytochrome. Plant Physiol 48:770–774
Hillman WS (1976) Biological rhythms and physiological timing. Annu Rev Plant Physiol 27:159–179
Hillman WS, Koukkari WL (1967) Phytochrome effects in the nyctinastic leaf movements of *Albizzia julibrissin* and some other legumes. Plant Physiol 42:1413–1418
Hull HM, Went FW, Yamada N (1954) Fluctuations in sensitivity of the *Avena* test due to air pollutants. Plant Physiol 29:182–187
Jaffe MJ, Galston AW (1967) Phytochrome control of rapid nyctinastic movements and membrane permeability in *Albizzia julibrissin*. Planta 77:135–141
Jaffe MJ, Galston AW (1968) The physiology of tendrils. Annu Rev Plant Physiol 19:412–434
Janardhan KV, Vasudeva N, Gopel NH (1973) Diurnal variation of endogenous auxin in *arabica* coffee leaves. J Plant Crops 1 (Suppl):93–95
Jenkinson IS (1962a) Bioelectric oscillations of bean roots: Further evidence for a feedback oscillator. II. Intracellular plant root potentials. Aust J Biol Sci 15:101–114
Jenkinson IS (1962b) Bioelectric oscillations of bean roots: Further evidence for a feedback oscillator. III. Excitation and inhibition of oscillations by osmotic pressure, auxins, and antiauxins. Aust J Biol Sci 15:115–125
Jenkinson IS, Scott BIH (1961) Bioelectric oscillations of bean roots: Further evidence for a feedback oscillator. I. Extracellular response to oscillations in osmotic pressure and auxin. Aust J Biol Sci 14:231–247
Johnsson A (1973) Oscillatory transpiration and water uptake of *Avena* plants. I. Preliminary observations. Physiol Plant 28:40–50
Johnsson A (1979) Growth movements not directed primarily by external stimuli. In: Haupt W, Feinleib ME (eds) Physiology of movements. Encyclopedia of plant physiology, new ser vol 7. Springer, Berlin Heidelberg New York, pp 627–646
Kamiya N, Nakajima H (1955) Some aspects of rhythmicity of the protoplasmic streaming in the myxomycete plasmodium. Jpn J Bot 15:49–55
Kapuya JA, Hall MA (1977) Diurnal variations in endogenous ethylene levels in plants. New Phytol 79:233–237
Karvé AD, Salanki AS (1964) A feedback oscillation of a relatively high frequency in hypocotyls of *Carthamus tinctorius* L. Z Pflanzenphysiol 52:98–102

Kasamo K, Yamaki T (1974) Effect of auxin on Mg^{++}-activated and -inhibited ATPases from mung bean hypocotyls. Plant Cell Physiol 15:965–970

Kinet JM, Bernier G, Bodson M, Jacqmard A (1973) Circadian rhythms and the induction of flowering in *Sinapis alba*. Plant Physiol 51:597–600

King RW (1975) Multiple circadian rhythms regulate photoperiodic flowering responses in *Chenopodium rubrum*. Can J Bot 53:2631–2638

Knypl JS (1973) Synergistic induction of nitrate reductase activity by nitrate and benzylaminopurine in detached cucumber cotyledons. Z Pflanzenphysiol 70:1–11

Kögl F, Haagen-Smit AJ, Hulssen CJ van (1936) Über den Einfluss unbekannter äusserer Faktoren bei Versuchen mit *Avena sativa*. Z Physiol Chem 241:17–33

Konopka RJ, Benzer S (1971) Clock mutants of *Drosophila melanogaster*. Proc Natl Acad Sci USA 68:2112–2116

Koukkari WL (1974) Rhythmic movements of *Albizzia julibrissin* pinnules. In: Scheving LE, Halberg F, Pauly JE (eds) Chronobiology. Igaku Shoin, Toyko, pp 676–678

Koukkari WL, Duke SH (1973) Regulating the growth of an aquatic plant: *Lemna perpusilla*. J Minn Acad Sci 39:12–14

Koukkari WL, Hillman WS (1968) Pulvini as the photoreceptors in the phytochrome effect on nyctinasty in *Albizzia julibrissin*. Plant Physiol 43:698–704

Koukkari WL, Johnson MA (1979) Oscillations of leaves of *Abutilon theophrasti* (velvetleaf) and their sensitivity to bentazon in relation to low and high humidity. Physiol Plant 47:158–162

Koukkari WL, Soulen TK (1981) Circadian time structure of vascular flowering plants. In: Kaiser HE (ed) Neoplasms – comparative pathology of growth in animals, plants, and man. Williams & Wilkins, Baltimore, pp 175–184

Koukkari WL, Halberg F, Gordon SA (1973) Quantifying rhythmic movements of *Albizzia julibrissin* pinnules. Plant Physiol 51:1084–1088

Koukkari WL, Duke SH, Halberg F, Lee JK (1974) Circadian rhythmic leaflet movements: Student exercise in chronobiology. Chronobiologia 1:281–302

Kraatz GW, Andersen RN (1978) Response of velvetleaf and sicklepod to herbicide applications at various time of the day. North Centr Weed Conf Proc 33:33

Kyriacou CP, Hall JC (1980) Circadian rhythm mutations in *Drosophila melanogaster* affect short-term fluctuations in the male's courtship song. Proc Natl Acad Sci USA 77:6729–6733

Lang A (1965) Physiology of flower initiation. In: Ruhland W (ed) Encyclopedia of plant physiology Vol XV/1. Springer, Berlin Göttingen Heidelberg, pp 1380–1536

Lecoq C, Koukkari WL, Brenner ML (1983) Rhythmic changes in abscisic acid (ABA) content of soybean leaves. Plant Physiol 72 (Suppl) 52

Lee OY, Stadelmann EJ (1972) Light as a factor in water permeability changes in *Pisum* parenchyma. Plant Physiol 49 (Suppl 62

Lee-Stadelmann OY, Stadelmann EJ (1979) Protoplasmic aspects of drought resistance in *Pisum sativum:* The development of protoplasmic tolerance. In: Goodin JR, Northington DK (eds) Arid land plant resources. Cent Arid Semi-Arid Land Stud, Texas Tech Univ, pp 501–528

Ludwig H, Hinze E, Junges W (1982) Endogene Rhythmen des Keimverhaltens der Samen von Kartoffeln, insbesondere von *Solanum acaule*. Seed Sci Technol 10:77– 86

Lüttge U, Bauer K, Köhler D (1968) Frühwirkungen von Gibberellinsäure auf Membrantransporte in jungen Erbsenpflanzen. Biochim Biophys Acta 150:452–459

McComb AJ (1962) An effect of gibberellic acid on circumnutation. New Phytol 61:128–131

McDaniel M, Sulzman FM, Hastings JW (1974) Heavy water slows the *Gonyaulax* clock: A test of the hypothesis that D_2O affects circadian oscillations by diminishing the apparent temperature. Proc Natl Acad Sci USA 71:4389–4391

McEvoy RC, Koukkari WL (1972) Effects of ethylenediaminetetraacetic acid, auxin, and gibberellic acid on phytochrome-controlled nyctinasty in *Albizzia julibrissin*. Physiol Plant 26:143–147

McMichael BL, Hanny BW (1977) Endogenous levels of abscisic acid in water-stressed cotton leaves. Agron J 69:979–982

Miller CS (1975) Short interval leaf movements of cotton. Plant Physiol 55:562–566

Morgan PW, Bauer JR (1970) Involvement of ethylene in picloram-induced leaf movement response. Plant Physiol 46:655–659
Newman IA (1963) Electric potentials and auxin translocation in *Avena*. Aust J Biol Sci 16:629–646
Nicolis G, Portnow J (1973) Chemical oscillations. Chem Rev 73:365–384
Njus D, Sulzman FM, Hastings JW (1974) Membrane model for the circadian clock. Nature 248:116–120
Noyes RM, Field RJ (1974) Oscillatory chemical reactions. Annu Rev Phys Chem 25:95–119
Overbeek von J, Mason MIR (1968) Dormin and cytokinin: growth regulation of *Lemna*. Acta Bot Neerl 17:441–444
Page JZ, Kingsbury JM (1968) Culture studies on the marine green alga *Halicystis parvula – Derbesia tenuissima*. II. Synchrony and periodicity in gamete formation and release. Am J Bot 55:1–11
Page JZ, Sweeney BM (1968) Culture studies on the marine green alga *Halicystis parvula – Derbesia tenuissima*. III. Control of gamete formation by an endogenous rhythm. J Phycol 4:253–260
Pallas JE Jr, Samish YB, Willmer CM (1974) Endogenous rhythmic activity of photosynthesis, transpiration, dark respiration, and carbon dioxide compensation point of peanut leaves. Plant Physiol 53:907–911
Parkash V (1972) Synergism between cytokinins and nitrate in induction of nitrate reductase activity in fenugreek cotyledons. Planta 102:372–373
Pavlidis T (1969) Populations of interacting oscillators and circadian rhythms. J Theor Biol 22:418–436
Pavlidis T (1971) Populations of biochemical oscillators as circadian clocks. J Theor Biol 33:319–338
Pavlidis T (1973) Biological oscillators: their mathematical analysis. Academic Press, London New York
Pavlidis T, Kauzmann W (1969) Toward a quantitative biochemical model for circadian oscillators. Arch Biochem Biophys 132:338–348
Pfeffer W (1906) The physiology of plants. In: Ewart AJ (ed) A treatise upon the metabolism and sources of energy in plants, vol III. Clarendon, Oxford
Phillips IDJ (1971) The biochemistry and physiology of plant growth hormones. McGraw-Hill, New York
Pirson A, Göllner E (1953) Beobachtungen zur Entwicklungsphysiologie der *Lemna minor* L. Flora 140:485–498
Pittendrigh CS (1954) On temperature independence in the clock system controlling emergence time in *Drosophila*. Proc Natl Acad Sci USA 40:1018–1029
Pittendrigh CS, Bruce VG, Rosensweig NS, Rubin ML (1959) Growth patterns in *Neurospora*. Nature 184:169–170
Posner HB (1967) Aquatic vascular plants. In: Wilt FH, Wessells NK (eds) Methods in developmental biology. Crowell, New York, pp 301–317
Putnam AR, Ries SK (1968) Factors influencing the phytotoxicity and movement of paraquat in quackgrass. Weed Sci 16:80–83
Pye EK (1969) Biochemical mechanisms underlying the metabolic oscillations in yeast. Can J Bot 47:271–285
Queiroz O (1974) Circadian rhythms and metabolic patterns. Annu Rev Plant Physiol 25:115–134
Racusen R, Satter RL (1975) Rhythmic and phytochrome-regulated changes in transmembrane potential in *Samanea* pulvini. Nature 255:408–410
Rama Das VS, Rao JVS, Rao KR (1964) Endogenous auxin and its diurnal rhythm in leaves. Indian J Plant Physiol 7:25–29
Raška Z, Hladík F (1969) Diurnal dynamics of natural growth substances in peach leaves and shoots. Biol Plant 11:60–67
Raven JA (1975) Transport of indoleacetic acid in plant cells in relation to pH and electrical potential gradients, and its significance for polar IAA transport. New Phytol 74:163–172
Reinberg A (1971) La chronobiologie. Recherche 2:242–250
Retzlaff G, Fischer A (1973) Die Beeinflussung der Assimilation verschiedener Pflanzen

durch Bentazon im Vergleich zur Selektivität. Mitt Biol Bundesanst Land-Forstwirtsch 151:179–180
Rikin A, Chalutz E, Anderson JD (1983a) Rhythmical changes in cotton (*Gossypium hirsutum* L.) seedlings: ethylene biosynthesis and cotyledon movement. Plant Physiol 72 (Suppl) 53
Rikin A, St John JB, Wergin WP, Anderson JD (1983b) Rhythmical changes in cotton (*Gossypium hirsutum* L.) seedlings: sensitivity to herbicides. Plant Physiol 72 (Suppl) 175
Rubery PH, Sheldrake AR (1973) Effect of pH and surface charge on cell uptake of auxin. Nature New Biol 244:285–288
Rubery PH, Sheldrake AR (1974) Carrier-mediated auxin transport. Planta 118:101–121
Sandberg G, Oden P, Dunberg A (1982) Population variation and diurnal changes in the content of indole-3-acetic acid of pine seedlings (*Pinus sylvestris* L.) grown in a controlled environment. Physiol Plant 54:375–380
Satter RL (1979) Leaf movements and tendril curling. In: Haupt W, Feinleib ME (eds) Physiology of movements. Encyclopedia of plant physiology, new ser Vol 7. Springer, Berlin Heidelberg New York, pp 442–484
Satter RL, Galston AW (1971a) Phytochrome-controlled nyctinasty in *Albizzia julibrissin*. III. Interactions between endogenous rhythm and phytochrome in control of potassium flux and leaflet movement. Plant Physiol 48:740–746
Satter RL, Galston AW (1971b) Potassium flux: a common feature of *Albizzia* leaflet movement controlled by phytochrome or endogenous rhythm. Science 174:518–519
Satter RL, Galston AW (1973) Leaf movements: Rosetta stone of plant behavior? Bioscience 23:407–416
Satter RL, Marinoff P, Galston AW (1970a) Phytochrome-controlled nyctinasty in *Albizzia julibrissin*. II. Potassium flux as a basis for leaflet movement. Am J Bot 57:916–926
Satter RL, Sabnis D, Galston AW (1970b) Phytochrome-controlled nyctinasty in *Albizzia julibrissin*. I. Anatomy and fine structure of the pulvinule. Am J Bot 57:374–381
Satter RL, Marinoff P, Galston AW (1972) Phytochrome-controlled nyctinasty in *Albizzia julibrissin*. IV. Auxin effects on leaflet movement and K flux. Plant Physiol 50:235–241
Satter RL, Applewhite PB, Kreis DJ Jr, Gaston AW (1973) Rhythmic leaflet movement in *Albizzia julibrissin*. Plant Physiol 52:202–207
Satter RL, Applewhite PB, Galston AW (1974a) Rhythmic potassium flux in *Albizzia*. Effect of aminophylline, cations, and inhibitors of respiration and protein synthesis. Plant Physiol 54:280–285
Satter RL, Geballe GT, Applewhite PB, Galston AW (1974b) Potassium flux and leaf movement in *Samanea saman*. I. Rhythmic Movement. J Gen Physiol 64:413–430
Satter RL, Geballe GT, Galston AW (1974c) Potassium flux and leaf movement in *Samanea saman*. II. Phytochrome-controlled movement. J Gen Physiol 64:431–442
Schrempf M (1980) The action of abscisic acid on the circadian petal movement of *Kalanchoe blossfeldiana*. Z Pflanzenphysiol 100:397–407
Schuster JL (1970) Plains prickly pear control by night applications of phenoxy herbicides. Proc Southern Weed Sci Soc 23:245–249
Schwintzer CR (1971) Energy budgets and temperatures of nyctinastic leaves on freezing nights. Plant Physiol 48:203–207
Scott BIH (1957) Electric oscillations generated by plant roots and a possible feedback mechanism responsible for them. Aust J Biol Sci 10:164–179
Scott BIH (1962) Feedback-induced oscillations of five-minute period in the electrical field of the bean root. Ann NY Acad Sci 98:890–900
Scott BIH, Martin DW (1962) Bioelectric fields of bean roots and their relation to salt accumulation. Aust J Biol Sci 15:83–100
Scott BIH, McAulay AL, Jeyes P (1955) Correlation between the electric current generated by a bean root growing in water and the rate of elongation of the root. Aust J Biol Sci 8:36–46
Seifriz W (1950) Gregarious flowering of *Chusquea*. Nature 165:635–636
Seltmann H, Peedin GF (1972) Application time during the day influences chemical sucker control. Tobacco Sci 16:88
Shen-Miller J (1973a) Rhythmicity in the basipetal transport of indoleacetic acid through coleoptiles. Plant Physiol 51:615–619

Shen-Miller J (1973b) Rhythmic differences in the basipetal movement of indoleacetic acid between separated upper and lower halves of geotropically stimulated corn coleoptiles. Plant Physiol 52:166–170

Shen-Miller J, Morris L (1967) Reciprocity in the geotropic response of gravity-compensated *Avena* coleoptiles. Argonne Nat Lab Biol Med Res Div Annu Rep, ANL-7409, pp 102–104

Shen-Miller J, Noack NG, Baker JE (1970) Kinetics and periodicity of auxin transport, geotropic curvature and growth. Argonne Nat Lab Biol Med Res Div Annu Rep, ANL-7770, pp 108–110

Sheriff DW (1974) A model of plant hydraulics under non-equilibrium conditions. I. Stems J Exp Bot 25:552–561

Shetty GP (1969) Effect of cotyledon injury upon geotropically induced oscillations in hypocotyls of *Carthamus tinctorius* L. Indian J Plant Physiol 11:132–140

Shibaoka H, Yamaki T (1959) Studies on the growth movement of sunflower plant. Sci Pap Coll Gen Educ, Univ Tokyo 9:105–126

Simon E, Satter RL, Galston AW (1976) Circadian rhythmicity in excised *Samanea* pulvini. I. Sucrose–white light interactions. Plant Physiol 58:417–420

Skoog F, Broyer TC, Grossenbacher KA (1938) Effects of auxin on rates, periodicity, and osmotic relations in exudation. Am J Bot 25:749–759

Skrove D, Rinnan T, Johnsson A (1982) Effect of abscisic acid on the circadian leaf movements of *Oxalis regnellii*. Physiol Plant 55:221–225

Smith AP (1974) Bud temperature in relation to nyctinastic leaf movement in an Andean giant rosette plant. Biotropica 6:263–266

Spurný M (1968) Spiral oscillations of the growing radicle in *Pisum sativum* L. Naturwissenschaften 55:46

Stahlberg R, Polevoi VV (1979) Nature of rhythmic oscillations of the membrane potential in corn coleoptile cells. Dokl Acad Nauk USSR 247:1022–1024

Stålfelt MG (1946) The influence of light upon the viscosity of protoplasm. Ark Bot 33:1–17

Stålfelt MG (1965) The relation between the endogenous and induced elements of the stomatal movements. Physiol Plant 18:177–184

Steer BT (1976) Rhythmic nitrate reductase activity in leaves of *Capsicum annuum* L. and the influence of kinetin. Plant Physiol 57:928–932

Steveninck van RFM (1976) Effect of hormones and related substances on ion transport. In: Lüttge U, Pitman MG (eds) Transport in plants II: tissues and organs. Encyclopedia of plant physiology new ser Vol 2/B. Springer, Berlin Heidelberg New York, pp 307–342

Sulzman FM, Edmunds LN (1972) Persisting circadian oscillations in enzyme activity in non-dividing cultures of *Euglena*. Biochem Biophys Res Commun 47:1338–1344

Sweeney BM (1963) Biological clocks in plants. Annu Rev Plant Physiol 14:411–440

Sweeney BM (1969) Rhythmic phenomena in plants. Academic Press, London New York

Sweeney BM (1974a) A physiological model for circadian rhythms derived from the *Acetabularia* rhythm paradoxes. Int J Chronobiol 2:25–33

Sweeney BM (1974b) The potassium content of *Gonyaulax polyedra* and phase changes in the circadian rhythm of stimulated bioluminescence by short exposure to ethanol and valinomycin. Plant Physiol 53:337–342

Sweeney BM (1974c) The temporal regulation of morphogenesis in plants, hourglass and oscillator. In: Basic mechanisms in plant morphogenesis. Brookhaven Symp Biol 25:95–110

Sweeney BM (1979) Endogenous rhythms in the movement of plants. In: Haupt W, Feinleib ME (eds) Physiology of movements. Encyclopedia of plant physiology, new ser vol 7. Springer, Berlin Heidelberg New York, pp 71–93

Sweeney BM, Haxo FT (1961) Persistence of a photosynthetic rhythm in enucleated *Acetabularia*. Science 134:1361–1363

Takimoto A, Hamner KC (1964) Effect of temperature and preconditioning on photoperiodic response of *Pharbitis nil*. Plant Physiol 39:1024–1030

Tanada T (1983) Interaction of phytohormones and far-red irradiation on the nyctinastic closing of *Albizzia julibrissin* pinnules. Physiol Plant 57:42–46

Tasseron-de Jong FG, Veldstra H (1971) Investigations on cytokinins. I. Effects of 6-benzylaminopurine on growth and starch content of *Lemna minor*. Physiol Plant 24:235–238
Thimann KV (1969) The auxins. In: Wilkins MB (ed) The physiology of plant growth and development. McGraw-Hill, New York, pp 1–45
Tronchet A, Marchal J (1960) Action de la gibbérelline sur la croissance et les mouvements de *Lactuca saligna* L. Bull Hist Nat 62:99–100
Tronchet A, Tronchet J, Perney J (1960) Sur les mouvements révolutifs de la tige de *Zinnia elegans* induits par l'acide gibbérellique. CR Acad Sci Paris 250:576–578
Tucker DJ, Mansfield TA (1971) A simple bioassay for detecting "antitranspirant" activity of naturally occurring compounds such as abscisic acid. Planta 98:157–163
Upcroft JA, Done J (1972) Evidence for a complex control system for nitrate reductase in wheat leaves. FEBS Lett 21:142–144
Varner JE, Ho DT (1976) Hormones. In: Bonner J, Varner JE (eds) Plant biochemistry, 3rd edn. Academic Press, London New York, pp 713–770
Wagner E, Frosch S, Deitzer GF (1974) Metabolic control of photoperiodic time measurement. J Interdiscipl Cycle Res 5:240–246
Wagner E, Deitzer GF, Fischer S, Frosh S, Kempf O, Stroebele L (1975) Endogenous oscillations in pathways of energy transduction as related to circadian rhythmicity and photoperiodic control. Biosystems 7:68–76
Ward RR (1971) The living clocks. New English Library, London
Wassermann L (1959) Die Auslösung endogen-tagesperiodischer Vorgänge bei Pflanzen durch einmalige Reize. Planta 53:647–669
Weaver ML, Nylund RE (1963) Factors influencing the tolerance of peas to MCPA. Weeds 11:142–148
Weij van der HG (1932) Der Mechanismus des Wuchsstofftransportes II. Rec Trav Bot Neerl 31:810–857
Went FW (1928) Wuchsstoff und Wachstum. Rec Trav Bot Neerl 25:1–116
Went FW (1960) Photo- and thermoperiodic effects in plant growth. Cold Spring Harbor Symp Quant Biol 25:221–230
Went FW (1962) Ecological implications of the autonomous 24-hour rhythm in plants. Ann NY Acad Sci 98:866–875
Went FW (1974) Reflections and speculations. Annu Rev Plant Physiol 25:1–26
Went FW, Thimann KV (1937) Phytohormones. Macmillan, New York
Wilkins MB (1962) An endogenous rhythm in the rate of carbon dioxide output of *Bryophyllum*. III. The effects of temperature on the phase and period of the rhythm. Proc R Soc London Ser B 156:220–241
Wilkins MB, Warren DM (1963) The influence of low partial pressures of oxygen on the rhythm in the growth rate of the *Avena* coleoptile. Planta 60:261–273
Williams CN, Raghavan V (1966) Effects of light and growth substances on the diurnal movements of the leaflets of *Mimosa pudica*. J Exp Bot 17:742–749
Wurtman RJ (1967) Ambiguities in the use of the term circadian. Science. 156:104
Yin HC (1941) Studies on the nyctinastic movement of the leaves of *Carica papaya*. Am J Bot 28:250–261

Addendum: Turgorins

H. Schildknecht and M. Bopp

In the 19th century Pfeffer stated that specific chemical stimuli must be presumed to play a prominent role in autonomous movement of the Fabaceae. In 1916 Ricca discovered such a substance found in extracts of *Mimosa pudica* L. The substance was subsequently characterized by Fitting (1936), Soltys et al. (1936) and Hesse (1939). The fact that the long-sought Leaf Movement Factor

(MLMF) from *Mimosa pudica* L. is 4-0-[3,5-Dihydroxybenzoic acid]-β-D-glucosid-6′-sulfate, was finally recognized by SCHILDKNECHT et al. 1981, 1983 and named turgorin. It induces leaf movement reaction in *Mimosa pudica* L.

As factors of turgor bending, the Periodic Leaf Movement Factors (PLMF), the turgorins, provide the causal connection between irritation and movement. They are also the chemical basis of nyctinasty. A specific receptor for these substances must exists because of the fact that the turgorins exhibit a distinct structure activity relationship (cf. Table 1).

The turgorins are a widespread group of substances with hormone character. PLMF 1 was found in the following species: *Mimosa pudica, Acacia karroo, A. dealbata, Albizzia julibrissin, Gleditsia triacanthos, Robinia pseudacacia, Samanea saman, Wisteria sinensis, Oxalis stricta, O. deppei, Abutilon grandiflorum, Portulaca oleracea, Vitis gongylodes* (between 1 µg kg^{-1} fw and 3 mg kg^{-1} fw).

Table 1. Structure-activity relationship of naturally occuring turgorins

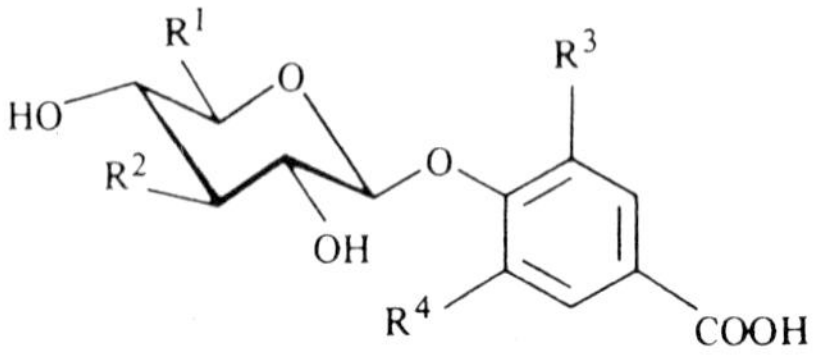

Turgorin	Structure				Activity Minimum conc. [mol/l]
	R^1	R^2	R^3	R^4	
PLMF 1	CH_2OSO_3H	OH	OH	OH	2.3×10^{-7}
PLMF 2	CH_2OSO_3H	OSO_3H	OH	OH	1.9×10^{-7}
PLMF 3	CH_2OSO_3H	OH	OH	H	2.4×10^{-6}
PLMF 4	CH_2OSO_3H	OH	H	H	2.5×10^{-5}
PLMF 5	CH_2OSO_3H	OH	OCH_3	H	2.1×10^{-3}
M-LMF 5	COOH	OH	OH	OH	2.7×10^{-6}

References

Fitting H (1936) Untersuchungen über die chemischen Eigenschaften des Reizstoffes von Mimosa pudica. Jahrb Wiss Bot 83:270–314

Hesse G (1939) Über die Natur der Erregungssubstanz von *Mimosa pudica* L. Biochem Z 303:152

Pfeffer W (1880) Pflanzenphysiology. 1. Aufl., Verlag von Wilhelm Engelmann, Leipzig

Ricca U (1916) Solutione d'un problema di fisiologia: la propagazione di stimulo nella "mimosa". Nuovo G Bot Ital (Nuova Seria) 23:51

Schildknecht H, Bender W (1983) Chemonastisch wirksame Leaf Movement Factors aus *Mimosa pudica* L. Chem-Ztg 107:111–114

Schildknecht H, Schumacher K (1981) Ein hochwirksamer Leaf Movement Factor aus *Acacia karroo*. Chem-Ztg 105:287–290

Soltys A, Unrath K (1936) Über die Erregungssubstanz der Mimosoideen. Biochem Z 284:247

3 Hormonal Aspects of Phase Change and Precocious Flowering

R.H. ZIMMERMAN, W.P. HACKETT, and R.P. PHARIS

1 Introduction

Phase change, the transition from juvenility to maturity, is an intriguing phenomenon in plant development that is observed most widely, but not exclusively, in woody perennial plants. As a plant develops following seed germination, a number of its characteristics change with time, independently but in a more or less coordinated manner. The earlier, juvenile, stage in this development is characterized, among other things, by a general absence of flowering and a ready ability to initiate adventitious roots. Other characteristics, such as leaf shape, phyllotaxy, anthocyanin pigmentation, growth habit, winter leaf retention, and thorniness, may or may not change, depending on plant species.

The transition period for these characteristics tends to be gradual and may occur continuously as the plant matures. Flowering is frequently the key character associated with phase change to maturity, and the transition from the juvenile to the adult phase has usually been defined in terms of the plants attaining the "ability to flower" (e.g., SCHWABE 1976, STOKES and VERKERK 1951, WAREING and FRYDMAN 1976, ZIMMERMAN 1971, 1973).

Once the plant attains this ability to flower, flowering will continue to occur so long as all other requisite environmental conditions are met. If they are not, flowering will be delayed even though the plant is now capable of flowering. Thus, flowering can indicate that the plant has reached the phase of sexual maturity, but may not indicate when the transition occurred. Conversely, lack of flowering may not mean that the plant is still in the sexually immature, juvenile phase (e.g., GREENWOOD 1978). Unfortunately, these definitions and qualifying statements are too simplistic. They may be appropriate for situations where developmental parameters such as size, or number of nodes signal the reaching of sexual maturity with subsequent signals for flowering being environmental in nature (e.g., photoperiod, low temperature). However, they are not completely appropriate for other plants, especially woody perennials for which the attainment of an internal "ability to flower" may be caused by the very environmental conditions that control flowering subsequently. Situations also exist where a fortuitous combination of environmental events may trigger precocious flowering for one year, yet in reality the plant may not have reached sexual maturity, flowering not occurring for many years subsequently [i.e., citrus: (Fig. 1) FURR et al. 1947, HIELD et al. 1966 and some of the conifers SMITH and KONAR 1969, ROSS et al. 1983].

Hence, the definition of the juvenile phase as the period during which the seedling cannot be induced to flower (STOKES and VERKERK 1951) has many

Fig. 1. Terminal flower bud on a 3-month-old Duncan grapefruit seedling. (FURR et al. 1947)

exceptions and is one that cannot be considered as generally valid (LONGMAN 1976). Further evidence of its inappropriateness comes especially from work with several conifer species in the Cupressaceae and Taxodiaceae. Here, seedlings can be readily induced to flower by exogenous application of various GA's when only 2 to 3 months old and just several centimeters high (cf. PHARIS 1974, 1976). These definitions are even less valid in light of recent findings, which indicate that somatic embryos derived from callus from roots of mature *Panax ginseng* plants (CHANG and HSING 1980) and from callus from sexual embryos of date palm (B. TISSERAT, 1983 personal communication) flower in the embryoid stage, whereas sexual embryos do not flower for several years. Whether plants which developed from these somatic embryos retain their ability to flower is not reported.

One may thus conclude that ability to flower is a necessary, but not sufficient, condition to mark the beginning of the adult phase and that the attainment of the ability to flower can be detected only after actual formation of flower primordia. As a working definition, then, one can state that phase change has occurred (1) if a plant flowers, no matter how flowering is induced, *and* (2) if the said plant will then continue to flower in its natural environment without the application of any artificial stimulation that may have originally induced flowering (cf. LONGMAN 1976). Consequently, the conifers alluded to above which were induced to flower by treatment with GA's would be considered mature only if they continued to flower in succeeding years when grown in the same habitat (natural or otherwise) without application of GA's. This definition would not preclude flowering in the juvenile phase induced by hormonal or unusual environmental conditions, since the said flowering would have to continue without benefit of either GA treatment or "unusual" environmental condition for that habitat in which the initial flowering took place.

As a plant grows older and larger, the growth rate slows, the branching structure becomes more complex, the number of terminal growing points increases, and plant vigor declines. WAREING (1959) called these changes aging in contrast to the phase change from juvenile to adult which he called maturation. Seedling plants thus undergo maturation as well as aging. However, one can visualize that some of the internal changes that take place during aging may also affect, perhaps hasten, sexual maturation (PHARIS and MORF 1968).

When mature plants are propagated vegetatively, the resulting plants will generally grow vigorously (a reversal of aging), but usually will not flower for some time (ROSS et al. 1983). This is presumably in part a size effect, but may also be related to the vigor of the root system (COPES 1980). The vigorous growth does not mean that these plants have reverted to the juvenile phase since flowering can still be induced by treatments (cultural/environmental/hormonal) which are ineffective, or nearly so, on seedlings (PHARIS 1976).

Phase change or juvenility has been reviewed or discussed in recent years by a number of authors (ALLSOPP 1967, BRINK 1962, DOORENBOS 1965, KOZLOWSKI 1971, LEOPOLD and KRIEDEMANN 1975, ROBBINS 1964, ROSS et al. 1983, SAX 1962, STOUTEMYER 1967, HACKETT 1980, 1984), and by several participants at the 1975 Juvenility Symposia (BORCHERT 1976, FORTANIER and JONKERS 1976, HACKETT 1976, HEYBROEK and VISSER 1976, SCHWABE 1976, WAREING and FRYDMAN 1976) as has the relation of juvenility to flowering (HIGAZY 1962, JOHNSSON 1949, MATTHEWS 1963, NIENSTAEDT 1961, ROBBINS 1961, ROSS et al. 1983, STOUTEMYER 1962, WAREING 1961, ZIMMERMAN 1972), the influence of juvenility on plant propagation (KESTER 1976, MOTT 1983, STOUTEMYER 1962), and forestry (SWEET 1973, MOTT 1983, SWEET and HARRIS 1976).

A hormonal role is implicit in many of the developmental processes involved in phase change. Plant hormones and growth regulators are known to influence some of the characteristics associated with phase change, but exogenous applications of plant hormones or other growth regulators have not yet been able to bring about a stable change from the juvenile to the adult phase as defined in (1) and (2) above. Reversions from the adult to the juvenile phase can, however, be induced in some plants by GA's, but these effects are temporary in some cases, but very stable in *Hedera helix* (W.P. HACKETT, unpublished data). The effects of plant hormones and growth regulators on phase change and/or precocious flowering and the interactions of these substances with the environment will be discussed below.

2 Factors Affecting Characteristics Associated with Juvenility

2.1 Inability to Flower

Auxin effects on flowering, hence their possible role in sexual phase change, are not clear, since both shortening and lengthening of the juvenile period have been reported. Some years ago, DE ZEEUW and LEOPOLD (1955) observed that NAA, applied during flower-inducing cold treatment, permitted flower initiation in juvenile Brussels sprouts (*Brassica oleracea* var. *gemmifera* L.) seedlings. They suggested that completion of the juvenile phase in nature may result, in part, from the accumulation of a sufficient auxin level at the apical meristem to make it receptive to cold treatment. It is also conceivable that NAA could have stimulated flowering by mobilizing carbohydrates to the shoot tip, since carbohydrate accumulation there at low temperature is correlated with flower formation in other cold-requiring crops, cauliflower (*B. oleracea* var. botrytis L.; SADIK and OZBUN 1968) and broccoli (*B. oleracea* var. botrytis L.; FONTES and OZBUN 1972). In addition, factors such as low light intensity or high temperature that reduce carbohydrate levels delay or prevent flowering, thereby prolonging, in effect, the juvenile phase (HACKETT 1976, SCHWABE 1976). Auxins, usually NAA, alone or in combination with GA's have stimulated flowering (strobilus formation) in young seedlings of several conifers including Japanese larch [*Larix leptolepis* (SIEB. and ZUCC.) GORD.] (HASHIZUME 1967), Sitka spruce [*Picea sitchensis* (BONG.) CARR] (TOMPSETT 1977), Douglas fir [*Pseudotsuga menziesii* (MIRB.) FRANCO] (PHARIS et al. 1976, 1980, ROSS 1976) and lodgepole pine (*Pinus contorta* DOUGL. ex LOUD.; PHARIS et al. 1975, 1976, WHEELER et al. 1980). However, IBA sprayed daily at 50 mg l^{-1} for 2 weeks on a woody angiosperm (*Ribes nigrum* L.) completely inhibited flowering (SCHWABE and AL-DOORI 1973). In the case of the biennials it was, of course, impossible to test criterion (2), but for the woody perennials it was not met, flowering in subsequent years reverting to near pre-treatment levels.

Stimulation of flowering in seedlings of many conifer species by certain of the GA's is the most striking effect of any plant hormone on flower promotion in seedlings (Fig. 2). Since KATO et al. (1959) first reported that GA_3 would promote flowering in cryptomeria [*Cryptomeria japonica* (L.F.) DON], the effect of various GA's on strobilus formation has been noted in seedlings of many species of Cupressaceae and Taxodiaceae (PHARIS and KUO 1977). High levels of GA's are often required; however, response ratios of 1 strobilus: 0.002 µg GA_3 have been obtained on 1- to 2-year-old seedlings of Arizona cypress (*Cupressus arizonica* GREENE; PHARIS 1976). Various methods of application, including sprays, lanolin paste preparations, soil drenches, and injections, have been effective. Strobilus formation in seedlings, or even vegetative propagules of mature scions, of Pinaceae species with GA's was for a number of years more difficult. However, use of the less polar GA_{4+7}, often in combination with other less polar GA's (GA_9 especially; PHARIS et al. 1980, PHARIS and ROSS 1984, ROSS and PHARIS 1976, ROSS et al. 1983, TOMPSETT 1977), has become

Fig. 2. Western red cedar seedling aged 4 months. Initiation of strobili was induced by spraying with gibberellin A_3 at 3 months. The male strobilus is developing at the top of the plant. (PHARIS and OWENS 1966)

a highly effective treatment. The effect of GA_{4+7} can also be synergized with NAA and BA (PHARIS et al. 1980, ROSS and PHARIS 1976, TOMPSETT 1977, WHEELER et al. 1980). Occasionally GA_3 has been successful for Pinaceae family conifers (BLEYMÜLLER 1976, PHARIS and KUO 1977), although results with this "polar" GA are variable and not readily repeatable (CHALUPKA 1978). To date, the most effective treatment for Pinaceae species has involved GA_{4+7} and screening trials have confirmed its efficacy on 16 Pinaceae species in 5 genera (*Larix, Picea, Pinus, Pseudotsuga*, and *Tsuga*; CHALUPKA 1978, HARE 1979, HARE et al. 1979, LUUKKANEN and JOHANSSON 1980, PHARIS and KUO 1977, PHARIS et al. 1980, PHARIS and ROSS 1984, PURITCH et al. 1979, ROSS and GREENWOOD 1979, ROSS et al. 1983, SWEET 1979, WHEELER et al. 1980). Other cultural treatments, such as nondestructive girdling, water stress, high temperature, and nitrate fertilization can enhance the response to these less-polar GA's by seedlings of Pinaceae (CHALUPKA 1981, PHARIS 1976, PHARIS and ROSS 1984, POLLARD and PORTLOCK 1981, ROSS 1976, 1978, ROSS et al. 1983). In sexually mature grafts of *Picea sitchensis,* PHILIPSON (1983) found that heat or drought stress alone would induce flowering, but GA_{4+7} alone would not. However, GA_{4+7} enhanced the stress-induced flowering. Besides the conifers, GA_3 has been reported to stimulate flowering in juvenile plants of other species (ZEEVAART 1983) including *Kalanchoe pinnata* (LAM.) PERS. (WAHDI and MOHAN RAM 1972), *Lunaria biennis* MOENCH, *Salvia occidentalis* SWARTZ, *Silene armeria* L. (HIGAZY 1962), *Rhus typhina* L. (BONNET-MASIMBERT and NITSCH 1969) and *Callistemon viminalis* L. (SINGH 1981), a Myrtaceae species.

In contrast to the above results, GA_3 treatments have reduced or inhibited flowering in apple (*Malus sylvestris* MILL.; GUTTRIDGE 1962, JONKERS 1975, KARNATZ 1963), blackcurrant (SCHWABE and AL-DOORI 1973), peach [*Prunus persica* (L.) BATSCH] (HULL and LEWIS 1959), pear (*Pyrus communis* L.; GRIGGS and IWAKIRI 1961), plum (*Prunus domestica* L.), sweet cherry [*P. avium* (L.) L.], apricot (*P. armeniaca* L.), almond (*P. amygdalus* BATSCH; CRANE et al. 1960), birch (*Betula pendula* ROTH.; ARSHAD 1980) and mango (*Mangifera indica* L.; CHACKO et al. 1976). The question of flowering and GA's in woody angiosperms is reviewed in some detail in PHARIS and KING (1985). Side effects of the GA treatment indicating reversion to juvenile characteristics, such as production of thorns, will be discussed later. Although GA_3 was used in all the trials cited above and GA_{4+7} was used on apple by JONKERS (1975), TROMP (1972), I. THOMPSON, N. LOONEY and R. PHARIS (unpublished data), LOONEY et al. (1978) and GA_{4+7} and GA_9 on birch (ARSHAD 1980), the effects of most GA's have yet to be evaluated. With stem-girdled birch seedlings, ARSHAD (1980) found that the less polar GA_{4+7} was more inhibitory to flowering than GA_3, while GA_9 (also less polar than GA_3) was less inhibitory than GA_3. For birch seedlings, it appears that exogenous applications of both polar and less-polar gibberellins are inhibitory to flower initiation. There is little indication at the present time that GA's will stimulate flowering in juvenile or adult woody angiosperms as they do for both growth phases of the conifers. As noted above for *Rhus* and *Callistemon*, where GA_3 will cause juvenile flowering, in *Clerodendron thomsoniae,* GA_7 gives a slight promotion while GA_3 inhibits flowering in cutting-grown plants (KORANSKI et al. 1979). In *Vitis vinifera*, GA_3 causes the precocious formation of anlagen, the morphogenetic precursor of the tendril and inflorescence, at nodes lower than expected in sexually mature plants (SRINIVASAN and MULLINS 1979, 1980). Whether GA_3 will cause precocious formation of anlagen in juvenile seedlings has not been investigated.

Endogenous GA content has been investigated for different growth phases of several plants. GA-like substances have been determined by bioassay in most cases, although identification of specific GA's has been made in some plants (PHARIS and KUO 1977). In English ivy (*Hedera helix* L.), GA-like substances were higher in 1-cm-long juvenile apices than in adult apices, and were moderately high in roots of juvenile plants (FRYDMAN and WAREING 1973). Since adventitious roots are present on stems of juvenile English ivy, these roots may account for the higher levels of GA-like substances in the juvenile apices. Evidence for synthesis of GA's in, and export from, the roots was presented by FRYDMAN and WAREING (1973). Similarly, SCHWABE and AL-DOORI (1973) detected GA-like substances in roots and lower parts of the stem of blackcurrant. When adventitious roots were induced by air layering near the shoot tip on mature shoots, flower initiation was inhibited. They suggested that GA's exported from these adventitious roots prevented flower initiation. LOONEY et al. (1978) have presented data which indicate that high levels of an endogenous $GA_{3/19}$-like substance during apple flower-bud initiation/differentiation are associated with poor flowering the subsequent spring (ZIESLIN, THOMPSON and PHARIS, unpublished), and also that the metabolism of ^3H-GA_4 of high specific activity is retarded during the initiation/differentiation period in trees which flowered poorly the following spring, but metabolized more rapidly in trees that flowered well. In contrast to these findings, ARSHAD (1980) found that promotion of flowering by stem girdling in birch seedlings was associated with a rapid (3 days after girdling) increase in levels of gibberellin activity in bark above the girdle. He also found that flowering and gibberellin activity in girdled seedlings was enhanced by treatment with the growth retardant chlormequat (CCC or cycocel). Chlormequat alone also promoted flowering slightly and increased levels of gibberellin activity somewhat in bark of nongirdled plants. The chemical identity of the gibberellin activity in birch bark extracts was not investigated. However, treatment of girdled birch seedlings with purified birch bark extracts containing 0.1 or 1.0 μg equivalents of GA_3 (as determined by the lettuce hypocotyl bioassay) inhibited flowering as compared to nontreated but girdled seedlings. Twenty μg of GA_3 or 10 μg GA_{4+7} were required for a comparable inhibition.

In conifers, endogenous GA levels are often high at the time strobilus initiation is occurring and treatments that enhance formation of strobili may also increase the endogenous GA content (KAMIENSKA et al. 1974, PHARIS 1976, 1977, PHARIS and ROSS 1984, ROSS et al. 1983). In Arizona cypress seedlings, increasing daylength brought about a great increase in endogenous GA_3 and other more polar GA's (PHARIS 1976). Increasing daylength is correlated with increased flowering and a decreased need for exogenous GA_3 required to produce a given number of strobili (PHARIS 1976). However, the cultural treatment of nitrogen deficiency which causes precocious flowering in Arizona cypress and cryptomeria (KAMIENSKA et al. 1974; LYR and HOFFMANN 1964) was correlated with a shift from more polar to less polar endogenous GA's in Arizona cypress (KAMIENSKA et al. 1974). Treatments that promote flowering in conifers, such as root pruning, water stress, high temperature, or nitrate fertilization, also increase less-polar GA's or reduce polar GA's (CHALUPKA et al. 1982, PHARIS 1976, 1977, ROSS et al. 1983, PHARIS, ROSS, WEBBER, and OWENS, unpublished data). DUNBERG et al. (1983) have shown that sexually mature propagules of *Picea abies* covered with a clear plastic film (a treatment that promotes flowering) metabolized ^3H-GA_4 more slowly than control plants in the open.

Cytokinins (except as noted above for BA with mature, but nonflowering, propagules of two Pinaceae species) have not, until very recently, been reported to have an effect on modifying the ability of juvenile or mature phase plants to flower. However, in both seedlings as young as 4 weeks, and older propagules of grape (*Vitis vinifera* L.), SRINIVASAN and MULLINS (1978, 1979) have been able to produce (in vivo and in vitro) inflorescences and fruits (with viable seeds) in place of tendrils with the application of a synthetic cytokinin, PBA. Accordingly they (SRINIVASAN and MULLINS 1979) conclude that the tendril is indeed a "weakly differentiated inflorescence" (see discussion under gibberellins on relation of anlagen formation to tendril and inflorescence formation). Hence, the effect of the cytokinin is to promote sexual differentiation (rather than initiate flowering), a situation analagous to the GA_{4+7} effect noted above for conifers in the Pinaceae.

There is little information regarding the ability of abscisic acid (ABA) to promote flowering in juvenile or sexually mature woody plants. EL-ANTABLY et al. (1967) showed that ABA can induce flowering in sexually mature plants of *Ribes nigrum* under noninductive conditions. In birch, ARSHAD (1980) found that ABA enhanced flowering in stem girdled seedlings 6 to 9 months old (see earlier discussion). He also showed that stem girdling dramatically increases ABA levels in buds, leaves and bark of stems above the girdle. Girdled seedlings that were treated with chlormequat had even higher levels of ABA and increased flowering. In mature scions of Sitka spruce, TOMPSETT showed that ABA can reverse the promotion of flowering by gibberellins (TOMPSETT 1977).

Ethylene (as ethephon) sprays (1000 to 2000 mg l^{-1}) stimulated flowering (both number of trees and flowers) two- to sixfold in 3- to 5-year-old seedling apple trees (KENDER 1974), well above the control level of flowering. Similarly, sprays of 1000 to 4000 mg l^{-1} ethepon induced heavy flowering in 3-year-old mango seedlings (CHACKO et al. 1974, 1976) relative to control seedlings. As with GA's applied to young conifers, ethephon treatments to apple and mango probably did not cause phase change, especially since some controls also did flower.

Various plant growth regulators may be useful in deducing possible roles of plant hormones in regulating flowering and phase change. Growth retardants that may inhibit or depress GA synthesis, such as daminozide, chlormequat and CBBP, sometimes stimulate flowering in young fruit trees vegetatively propagated from mature wood. In hopes of stimulating flowering in seedling trees, researchers have applied these growth retardants to apple, crabapple [*Malus hupehensis* (PAMP.) REHD.], grapefruit (*Citrus paradisi* MACF.), pear, and several conifer species (BROOKS 1964, DENNIS 1968, HASHIZUME 1968, HIELD et al. 1966, JONKERS 1973). Although vegetative growth was usually inhibited, flowering was not stimulated except in a few instances, mainly with pear and apple (DENNIS 1968). In sexually mature grafts of *Picea sitchensis*, phosphon D partially counteracted the flower promotive effect of GA_{4+7} (TOMPSETT 1977). There are also reports (CHALUPKA 1979) in which chlormequat (CCC) has been found to enhance flowering of *Picea abies* grafts when used in conjunction with GA_3 which also promoted flowering. TROMP (1972) concluded that the reduction of growth and promotion of flowering by daminozide are independent phenomena. As noted earlier, chlormequat at 10 mg $plant^{-1}$ promotes flowering to a small degree in nongirdled birch seedlings, but to a very substantial degree in stem-girdled seedlings (ARSHAD 1980). Chlormequat causes little or no retardation of extension growth and, as noted earlier, causes an increase in levels of gibberellin activity of the bark. Amo-1618, another growth retardant, inhibited flowering in girdled birch trees at all doses (5–80 mg per plant). Its effect on endogenous gibberellin activity in birch was not investigated (ARSHAD 1980). Similarly, TIBA, an inhibitor of auxin transport, causes earlier flowering in young, vegetatively propagated apple trees (BUKOVAC 1963, STAHLY and PIRINGER 1962) and DAYTON (1966) found that flowering of 9-year-old triploid apple seedlings sprayed with 25 mg l^{-1} TIBA was greatly increased relative to the controls. As noted above for apple and mango which flowered in response to ethephon, the growth retardants and TIBA probably did not bring about phase change, even though they promoted flowering.

Temperature affects precocious flowering in grapefruit seedlings. If the plants are grown at temperatures low enough to induce some dormancy during the winter, flowering from the original terminal bud may occur in as little as 7 months after germination (HIELD et al. 1966), whereas application of a number of growth regulators and other chemicals had no effect on flowering. Perhaps low temperatures influenced the endogenous hormone content/balance of the seedlings, as well as the carbohydrate level, since temperature has been shown to influence the amount and type of cytokinins in grape root exudates (SKENE and KERRIDGE 1967) and of GA-like substances exported from Douglas fir roots (LAVENDER et al. 1973).

2.2 Ability to Initiate Adventitious Roots

Auxin is essential for the initiation of adventitious roots on shoots. The effectiveness of exogenously applied auxin varies with species, time of year, concentration, and growth phase of the plant from which cuttings are taken. Juvenile forms of many plants root easily; nevertheless, auxin still may increase the rate of root formation and the number of roots formed. Auxin usually has much less or no effect on cuttings from mature plants of the same species (HESS 1959; PORLINGIS and THERIOS 1976), an exception being adult Algerian ivy (*Hedera canariensis* WILLD.) (STOUTEMYER and BRITT 1962).

Endogenous auxin content (as measured by bioassay) of juvenile and adult English ivy is virtually the same (HESS 1964), but more auxin and other growth promoters have been found in juvenile than in adult chestnut (*Castanea sativa* MILL.) cuttings (VIEITEZ and VIEITEZ 1976).

Initiation of adventitious roots is generally inhibited by GA's (ROSS et al. 1983). However, very low concentrations (10^{-11} to 10^{-7} M) of GA_3 promoted root initiation in pea cuttings (ERIKSEN 1970, 1971), and in *Pinus radiata* D. DON when supplied coincident with the first observable stage of root initiation (SMITH and THORPE 1975). Also GA_3 may act indirectly to improve rooting by causing reversion to juvenile phase growth (see below) on which adventitious roots are more readily formed.

Cytokinins generally have little effect on root initiation although the balance between the cytokinins and auxins is important since high concentrations of cytokinins can inhibit root initiation (HARTMANN and KESTER 1975) while low concentrations promote the positive effects of auxin on initiation of adventitious roots in pea (ERIKSEN 1974).

No evidence is available that either ABA or ethylene plays a direct role in the difference in rooting ability between the two phases. Growth inhibitors, as measured by the *Avena* coleoptile bioassay, are present in roughly equivalent amounts in juvenile and adult *H. helix* tissue, whether plants are dormant or growing (HESS 1964). In contrast, mature chestnut cuttings contained more growth inhibitors than did cuttings from young (juvenile) seedlings (VIEITEZ and VIEITEZ 1976), although their role is difficult to assess, since the bioassays used did not involve adventitious root formation on the plant studied. PATON et al. (1981) have shown that the G inhibitors of eucalyptus are closely correlated with phase change-related loss of rooting potential in *Eucalyptus grandis* using easy-to-root *Eucalyptus deglupta* as a bioassay for rooting. In *Eucalyptus deglupta*, where rooting potential is not lost with maturation, G inhibitor level is low even in mature tissue.

Rooting cofactors will be discussed later in more detail. Generally, however, there are larger quantities in juvenile than in mature *H. helix* (HESS 1959, 1964) and this correlates well with the greater capacity for forming adventitious roots in the juvenile form; but a similar relationship does not exist for all plants in which rooting ability declines with age of the plant (HEUSER 1976; ZIMMERMAN 1963).

The fact that cuttings from juvenile-phase plants generally initiate adventitious roots more readily than those from mature-phase plants is well established (BRIX and VAN DEN DRIESSCHE 1977, HARTMANN and KESTER 1975). Evidence now available shows that this difference in rooting ability bears no clear or consistent relation to the hormonal or growth-regulator content of the tissue, although purification, separation techniques, and choice of assay often leave much to be desired. Even though auxin applications can sometimes improve rooting of cuttings of mature-phase plants, differences between juvenile and mature phases in capability for forming adventitious roots cannot be eliminated by any exogenously applied growth regulators. Species differences account for some of the difficulty in interpreting the available data. However, the main problem is the lack of comprehensive studies which relate changes in proportion, amount, or metabolism of the known hormones and other relevant growth-regulating substances to the growth phase.

2.3 Other Manifestations of Phase Change

Leaf shape of a number of plants changes in response to GA_3 treatment. In some species, such as *Hedera canariensis* and *H. helix* (ROBBINS 1957, 1960), *Acacia melanoxylon* R. BR. (BORCHERT 1965, TRIPPI 1963e), *Caryota mitis* LOUR. (FISHER 1976) and coconut palm (*Cocos nucifera* L.; SCHWABE 1976), GA_3 treatment causes reversion from the leaf shape of the adult form to that of the juvenile form. Transition from juvenile leaf shape to adult was delayed by GA_3 in *Ipomoea caerulea* KOEN ex ROXB., (NJOKU 1958), hops (*Humulus lupulus* L.; WILLIAMS 1961), and *Centaurea solstitialis* L. (FELDMAN and CUTTER 1970), with other growth regulators having a lesser effect (NJOKU 1958). In contrast, *Eucalyptus melliodora* A. CUNN. ex SCHAUER developed the leaf shape and alternate arrangement characteristic of the adult form sooner following GA_3 treatment than did the controls (SCURFIELD and MOORE 1958), although this might have been a size factor, since GA_3 stimulated shoot elongation of the young plants. Environmental conditions limiting carbohydrate supply (e.g., low light intensity) also delayed changes from juvenile to adult leaf shape in *I. caerulea* (NJOKU 1957).

Thorniness, another manifestation of the juvenile phase, was increased in citrus (COOPER and PEYNADO 1958), pear (GRIGGS and IWAKIRI 1961), almond, apricot, and plum (CRANE et al. 1960) by treatment of mature, fruiting trees with GA_3. Leaf retention, a juvenile characteristic, was extended in *Morus nigra* L. by GA_3 sprays (TRIPPI 1963d).

Thus, for angiosperms, characteristics associated with the juvenile phase generally are induced or prolonged by GA_3 and by environmental conditions limiting growth. The exceptions which occur, however, emphasize our lack of understanding of the processes underlying phase change.

3 Compositional Differences Between Juvenile and Adult Phases

The content, distribution, and form of many substances active in plant metabolism are influenced by environmental factors, and differences in these substances may occur between juvenile- and adult-phase plants. Thus, effects of the environment on these substances may influence the time and rate of phase change. Specific substances which have been examined in juvenile- and mature-phase plants are discussed below.

3.1 Nucleic Acids

Leaves of juvenile and adult forms of several *Pyrus* species contained similar amounts of DNA (ALI and WESTWOOD 1966). DNA contents of mature and juvenile *H. helix* cells have been determined by workers in several laboratories using various methods to examine different tissues. KESSLER and RECHES (1977) and MILLIKAN and GHOSH (1971) found the DNA content of mature phase leaf tissue to be lower than that of juvenile phase leaves on a cellular or dry weight basis. SCHAFFNER and NAGL (1979) analyzed cells of whole buds and leaves and concluded that nuclei of mature cells contained an average of 71% more DNA than those of comparable juvenile tissues. WAREING and FRYDMAN (1976), DOMONEY and TIMMIS (1980), POLITO and ALLIATA (1981), POLITO and CHANG (1984) and HACKETT (cited in STOUTEMYER 1967), report no differences in DNA content per cell between juvenile and mature tissues taken from shoot apices, apical buds, apical meristems and stem callus, respectively. The weight of evidence, then, suggests that the amount of DNA per 2C cell is not different for juvenile and mature tissues in *Hedera*.

Fewer data are available for the RNA content of juvenile and mature tissues. ALI and WESTWOOD (1966) reported less RNA in leaves of mature than juvenile forms of *Pyrus* species. Likewise, MILLIKAN and GHOSH (1971) reported less total, soluble and ribosomal RNA on a dry weight basis in mature than juvenile *Hedera helix* leaf tissue. HACKETT (cited in STOUTEMYER 1967) found higher levels of total RNA per cell in callus derived from juvenile as compared to that derived from mature *Hedera helix* stems. Based on these findings, DOMONEY and TIMMIS (1980) did rRNA hybridization experiments with DNA prepared from juvenile and mature tissue and found no evidence for differences in redundancy in rRNA genes in DNA of the two forms. ROGLER and DAHMUS (1974) used DNA-RNA hybridization techniques to study the RNA populations of juvenile and mature *Hedera* apices and found no qualitative differences between species of RNA in the two forms. Although RNA sequences unique to the juvenile apices could not be detected, RNA from mature apices contained a small proportion of sequences missing from, or present only in very low concentrations, in juvenile apices. Further, some gene sequences transcribed to RNA in mature apices were not active in juvenile ones. Thus, some genes may be active in the mature form, but not in the juvenile form, although the RNA transcribed from these genes would be only a small proportion of the total RNA. They thus concluded that at least part of the molecular basis of phase change, and of GA_3 action in causing reversions from mature to juvenile, could be an alteration in the rate of transcription of certain RNA sequences in mature-form buds.

3.2 Proteins

Both quantity and type of *H. helix* leaf proteins were modified during phase change, whereas senescence of the leaves was reflected mainly in a quantitative decrease in proteins (GHOSH and MILLIKAN 1970). Similar results were noted with leaves of *H. canariensis* and with tissue cultures of juvenile and mature *H. helix* (FUKASAWA 1966). Proteins from slow-growing tissue cultures of juvenile phase plants produced electrophoretic band patterns similar to those from tissue cultures of adult phase plants.

Activity of specific enzymes also varies with changes in growth phase. Catalase activity decreased with maturity in *Robinia pseudoacacia* L. and other enzymes (p-phenol oxidoreductase, amylase, peroxidase) varied with plant age, but could not be related to phase change (TRIPPI 1963c). Cytochrome oxidase, alkaline phosphatase and catalase activities were higher in juvenile *H. helix* than in adult, whereas the reverse was true for ribonuclease and acid phosphatase (GHOSH et al. 1971). Ascorbic acid oxidase and phenolase activities were higher in mature than in juvenile *H. helix* (MILLIKAN and SANIEWSKI 1973), and since auxin content decreased with maturation, the authors suggested that this decrease was inversely related to the higher phenolase activity.

Differences in enzymes or other proteins do not yet provide a basis for separating juvenile and mature phases, variability between species and the limited data base being responsible. While additional data might be useful, other research areas (discussed later) are more likely to yield significant information on phase change.

3.3 Rooting Cofactors

Rooting cofactors were first isolated from *H. helix* and were assayed with mung bean (*Vigna radiata* (L.) WILCZEK) cuttings (HESS 1959). Easy-to-root juvenile-form cuttings contained more rooting cofactors than the difficult-to-root adult-form cuttings (GIROUARD 1969, HESS 1964a, 1967). Since the cofactors were most active in the bioassay in the presence of IAA, HESS (1964a) postulated that they might serve as cofactors of IAA, one of the cofactors being a phenol. Indeed, certain phenols, particularly catechol, were very active in the bioassay (HESS 1964b). These findings fit the theory of BOUILLENNE and BOUILLENNE-WALRAND (1955) that three components are needed for adventitious root initiation, two of which are auxin and ortho-dihydroxy phenolic compounds. Further, GIROUARD (1969) found that the juvenile phase of *H. helix* had a higher phenol content than did the mature phase. Additional evidence on the relation of the cofactors to rooting was developed by grafting the adult form of *H. helix* onto the juvenile (HESS 1960). Following several weeks growth, mature scions reverted to the juvenile form with regard to growth habit and adventitious root formation. After another month, they developed into the mature growth habit once again and flowered; but cuttings made from the scions after they had again developed into the mature form rooted easily and had a rooting cofactor content nearly as high as cuttings from the juvenile form.

Other studies on rooting cofactors showed that the amount present was not necessarily related to plant age or rooting ability (e.g., ZIMMERMAN 1963) but variation in rooting ability was not always linked to phase change in these studies. HACKETT (1970) reported that cofactors extracted from both juvenile and adult *H. helix* stimulated rooting of shoot apices from juvenile *H. helix* but had no rooting effect on adult shoot apices. Since most other evaluations of root-promoting activity in extracts were made on easily rooted cuttings of mung bean, the role, if any, of the rooting cofactors in phase change is not clear. A review of the whole topic of rooting cofactors in relation to juvenility is offered by HEUSER (1976).

3.4 Gibberellins

Gibberellin-like substances occurred in roughly equal quantities in leaves of juvenile and mature *H. helix* (FRYDMAN and WAREING 1973). However, 1-cm shoot tips from juvenile phase plants contained larger quantities of GA-like substances than did mature tips. Adventitious, internodal roots, found only in the juvenile form, contained large quantities of GA-like substances. While these roots could be the major source of the GA-like substances in the shoot tips, definitive proof on this point is lacking (FRYDMAN and WAREING 1973). Further evidence to support the idea that increase in GA-like substances in the shoot tip of juvenile phase plants comes from the adventitious roots is provided by the work of SCHWABE and AL-DOORI (1973) with blackcurrant. In this species, application of GA_3 prevents flowering in mature shoots, and adventitious roots induced to form near the tip of a mature shoot also inhibited flowering, these roots containing significant amounts of GA-like activity (SCHWABE and AL-DOORI 1973).

3.5 Abscisic Acid

HILLMAN et al. (1974) detected ABA in crude extracts of juvenile and mature leaves of *H. helix*. Juvenile-form leaves contained more ABA than adult-form, but data were insufficient to draw conclusions as to any possible relation between the ABA content and the growth phase.

3.6 Cytokinins

HENDRY et al. (1982) reported that buds of juvenile citrus scions accumulated higher levels of polar cytokinins during dormancy than adult scion buds. The authors speculate that the higher level of polar cytokinins in juvenile buds may have resulted in the greater scion growth observed on juvenile plants.

3.7 Sterols

HILLMAN et al. (1975) used thin-layer, column and gas-liquid chromatography and mass spectometry to fractionate, identify and measure the sterols in leaves, flowers and fruits of *H. helix*. Sterol patterns of mature and juvenile leaf tissue were essentially identical with quantities of sterols per 100 gm fresh leaf being: cholesterol, a trace; campesterol, 1 mg; stigmasterol, 9 mg; sitosterol plus spinasterol, 2.5 mg; and stigmast-7-en-3-ol, 0.5–0.7 mg. There were only small quantitative differences between free and esterified sterols. The authors do not comment on any possible involvement of sterols in the transition between juvenile and mature phases in ivy.

4 Juvenile to Adult Phase Change and Its Reversibility

4.1 Influence of Cultural Techniques on Maturation

Manipulation of environmental factors has been used successfully to hasten maturation of both herbaceous and woody plants. The main technique has been to grow plants under long days with high light intensities during all or most of the photoperiod. Control of temperature, water supply, mineral nutrients, and CO_2 has usually been used in various combinations with the longer photoperiods. Controlling these variables has resulted in rapid growth of the plants over long periods of time. The treatments tend to maximize photosynthesis and it is likely that they significantly affect production and movement of plant hormones, although this latter aspect has only been little investigated with specific reference to juvenility. In some instances, these techniques have allowed the plants to proceed directly to flowering. In other cases, specific flower-promotion treatments have had to be applied after the plants have reached a sufficiently large size.

Continuous light combined with minimum temperatures of 18° to 20° C resulted in flowering of peach and camellia seedlings 1 to 3 years earlier than normal (LAMMERTS 1943, 1949). Time to flowering in azalea and azalea-rhododendron hybrid seedlings was halved by growing them under long days (DOORENBOS 1955). Seedlings of birch (*Betula verrucosa* EHRH.; LONGMAN and WAREING 1959; ROBINSON and WAREING 1969), crabapple (ZIMMERMAN 1971), and apple (ALDWINCKLE 1975), have all been brought into flowering in 9 to 18 months by growing them continuously under long days in the greenhouse with minimum temperatures of 15° to 20° C. Apple and crabapple seedlings grown in this way are tall with slender shoots and few or no branches (Fig. 3), lateral flower buds occurring on the main stem, usually above 2 m in height. Seedlings of *Betula* treated as described above flower on lateral branches, however. Similar techniques were used to rapidly produce large seedlings of Japanese larch and blackcurrant (ROBINSON and WAREING 1969). The plants were then subjected to treatments known to promote flowering (e.g., growing the larch horizontally, and placing the blackcurrant in short days) and flowering occurred much sooner than it did in field-grown seedlings.

The idea that seedlings must go through a certain number of cycles of growth and dormancy before undergoing phase change has been held by some workers. POTAPENKO (1939) grew cherry seedlings in large pots so that they went through two cycles of growth and dormancy each year. One-quarter of the plants flowered in the fourth growth cycle, while none of the field-grown controls did so. Although the author presented no quantitative data, photographs of the plants show that those grown for two cycles per year were at least twice as large as the field-grown controls. When SMEETS (1966) attempted to repeat this work in a greenhouse, however, the control seedlings, which were larger than those grown on cycles, flowered first. Results similar to those of SMEETS were obtained with birch (LONGMAN and WAREING 1959) and crabapple (ZIMMERMAN 1971, 1972b) in that plants grown continuously in the greenhouse were larger and flowered sooner than those grown on cycles in the greenhouse or in the field. Thus, it is now clear that seedling size, rather than the number of cycles of growth and dormancy, was the important factor in achieving earlier sexual maturity.

Temperature control has been used mainly to supplement other environmental controls to ensure good growing conditions for plants maintained under long photoperiods (e.g., the maintenance of a minimum temperature to prevent growth cessation). However, use of a carefully controlled temperature regime reduced the juvenile phase in tulips (SISA and HIGUCHI 1967). Thus, a four-part

Fig. 3. *Malus hupehensis* seedling grown in the greenhouse under long days for about 1 year, then given cold treatment to break dormancy. Note flowers in upper portion of seedling along the main stem. (ZIMMERMAN, unpublished)

growing season, each part with a specific temperature, was completed in 200 days. When the seedlings, and then the bulbs derived from them, were grown in a phytotron under this regime, the juvenile phase ended in just over 3 years instead of the normal 6 years.

The relation of earlier flowering to large seedling size has been noted in field-grown seedlings, particularly those of apple and pear (SAURE 1970, VISSER 1964, 1970, WAY 1971, ZIMMERMAN 1977). The same point was made by JONKERS (1971), who had young apple seedlings grown in the field at various latitudes from Finland to Italy. Trees at the more southerly locations, responding to the longer growing seasons and warmer temperatures, grew more rapidly and flowered sooner.

Seemingly in contrast to the earlier flowering with increased size noted above are the striking results obtained by grafting apple seedlings onto certain dwarfing rootstocks, the scions flowering from 1 to 3 or more years earlier than the donor seedlings (CAMPBELL 1961, KARNATZ 1965, KEMMER 1953, KEMMER and THIELE 1954, LOEWEL and SAURE 1963, TYDEMAN 1961, TYDEMAN and ALSTON 1965, VISSER 1973). The younger the seedlings from which the scions were taken, the greater the shortening of the juvenile phase. However, scions taken from older seedlings flowered sooner after grafting, presumably indicating that the older seedlings had progressed more toward phase change. There have, however, been a few reports (FRITZSCHE 1948, SPINKS 1925, ZWINTZSCHER 1964) where dwarfing rootstocks did not result in earlier flowering, with poor growing conditions for the grafted plants being cited as a possible reason (SPINKS 1925).

Dwarfing apple rootstocks most effective in reducing growth of the grafted tree, as measured by stem diameter, are M.9 and M.27 (formerly 3431). Size control can be obtained using M.9 either as a rootstock or as an interstock, with tree size being roughly proportional to the length of the interstock in the latter case. However, M.9 caused precocious flowering in seedlings only when it was used as a rootstock (with or without a vigorous interstock), but not as an interstock on a vigorous rootstock (VISSER 1973). This may indicate that the flower-inducing properties of M.9 reside in the roots, whereas the stem alone can reduce the growth vigor of the grafted tree.

The dwarfing rootstock, M.9, contains fewer GA-like and other growth-promoting substances (assayed by apple leaf discs, dwarf peas, and wheat coleoptiles) (IBRAHIM and DANA 1971, MARTIN and STAHLY 1967) and more growth-inhibiting substances (assayed by wheat coleoptiles and *Lepidium* seed germination) (MARTIN and STAHLY 1967, YADAVA and DAYTON 1972) than vigorous rootstocks such as M.16. Injection of GA_3 into the stem had an increasing effect on apple terminal shoot growth as vigor of the rootstock decreased, and trees on dwarfing rootstocks responded more quickly to injections of ABA by ceasing growth and forming terminal buds (ROBITAILLE and CARLSON 1971). More recently, M.9 as rootstock and M.8, M.9, M.26, and M.7 as interstocks have been found to impede xylem transport of nitrogenous and inorganic sap constituents from the rootstock to scion without affecting water movement (JONES and HOPGOOD 1976, JONES et al. 1975). M.9 interstocks also impede movement of ^{3}H-GA_4 and its metabolites (mainly conjugates) relative to non-dwarfing MM115 interstocks when the former was applied to the rootstock. In the M.9 interstock, which will overgrow, radioactivity (^{3}H-GA_4, free GA metabolites, and GA glucosyl conjugate-like metabolites) is four-fold higher (per g dry wt. tissue) than in MM115 interstocks, which are not dwarfing, and will not overgrow. Further, a higher proportion (12 to 17% higher) of this increased radioactivity in the M.9 interstocks is in the form of free ^{3}H-GA's (precursor GA_4 and its more polar metabolites), relative to the MM115 interstocks, the latter having a higher proportion of ^{3}H-GA glucosyl conjugate-like substances. The tendency for increased free GA's in M.9 exists within the interstock, and around the upper graft union; proportions at the lower graft union are similar for the two interstock types (RICHARDS et al. 1984). The reduction in transport of these sap constituents was roughly proportional to the dwarfing effect of the rootstock or interstocks (JONES and HOPGOOD 1976, JONES et al. 1975) with the site of the reduction in movement being the scion-interstock and/or scion-rootstock union.

Other treatments that can reduce shoot growth to some extent, such as girdling or ringing, scoring, and strangulation, have varying effects on flowering in seedlings. Such treatments stimulated earlier flowering in apple (DENNIS 1968, KEMMER 1953, KEMMER and THIELE 1954, WAY 1971), sweet cherry (FOGLE 1956), citrus (FURR et al. 1947), birch (ARSHAD 1980, LONGMAN 1976), *Thuja plicata* J. DONN ex D. DON (LONGMAN 1976) and red pine (*Pinus resinosa* AIT.) (HOLST 1961) among others. With birch, earlier flowering was obtained with either main stem or lateral branch girdling (ARSHAD 1980). In contrast, no earlier flowering of apple (FRITZSCHE 1948, SPINKS 1925) or eastern white pine (*Pinus strobus* L.; STEPHENS 1964) seedling trees was achieved by girdling branches, although flowers were induced on individually girdled branches of trees that seemed to be in the adult phase (FRITZSCHE 1948, STEPHENS 1964). ROSS and PHARIS (1976) found that girdling, although ineffective by itself in Douglas fir, was a powerful adjunct treatment when given with GA_{4+7}, a similar effect being noted for other conifers (PHARIS 1974, PHARIS and KUO 1977, ROSS et al. 1983).

At least one effect of grafting trees on dwarfing rootstocks and of restricting phloem transport by girdling, scoring, or strangulation is to increase carbohydrate levels of shoots in relation to the rest of the plant (STOLTZ and HESS 1966, TUKEY 1964). A similar situation probably exists in plants grown continuously under long days in the greenhouse (LONGMAN and WAREING 1959, ZIMMERMAN 1971) since these seedlings are grown in containers leading ultimately to restriction of root growth and a very high shoot root ratio. High carbohydrate levels have long been associated with a tendency to flower in fruit trees, especially apples, and treatments known to favor high carbohydrate levels also enhance flowering (DAVIS 1957). The treatments that enhance carbohydrate content of the shoots (including treatment with growth retardants such as daminozide in addition to those treatments discussed above) are even more effective in bringing about flowering in vegetatively propagated plants with scions from mature trees, than when the treatments are applied to seedlings (ROSS 1976, SACHS 1977, ZIMMERMAN 1972).

Alternate interpretations are possible for the results obtained with treatments leading to increased carbohydrate content of the shoots. One interpretation of this evidence is that, for woody angiosperms at least, these treatments prevent or reduce the transport to the shoots of root-produced GA's and/or other substances that promote vegetative growth. Further, the restriction of carbohydrate transport to the roots of plants that have been girdled, or propagated on dwarfing rootstocks, could reduce the production of GA's and other hormones by the roots, thus further reducing amounts which might be moved to the shoot through xylem tissue. Certainly the possibility that an interaction between carbohydrate metabolism and hormone transport affects phase change and flowering merits further study (SACHS 1977).

Nitrogen fertilization and deficiency, P deficiency, high temperature, and water stress have all been reported to stimulate flowering in seedlings and/or young vegetatively propagated trees. The data on these treatments are difficult to interpret since some treatments are not always effective and, when effective, are usually so only on older plants. The topic is reviewed in some detail by JACKSON and SWEET (1972) and briefly for conifers (PHARIS 1974, 1976, ROSS et al. 1983) although the former review is not confined to effects of these treatments on phase change. Water stress, high temperature and nitrate fertilization are being used as successful adjunct treatments to GA_{4+7} application for induction of flowering in seedlings and grafted scions of several conifers (BRIX and PORTLOCK 1982, PHILIPSON 1983, ROSS 1978, ROSS et al. 1981, 1983). PHILIPSON (1983) suggests, based on his experiments with Sitka spruce in which drought and heat stress by themselves or in conjunction with GA_{4+7} promote flowering, that the effects of drought and heat stress may be on reducing root activity. With Douglas fir seedlings, BONNET-MASIMBERT (1982) has shown that irrigation treatments that cause low soil oxygen levels promote flowering both without and with GA_{4+7} treatment. His findings further indicate that only plants with no root growth during the treatment period (whether caused by drought or low soil oxygen tension) produced flowers.

4.2 Reversions in Various Characteristics from Mature to Juvenile Phase

While the only fool-proof method of reversion is sexual reproduction, some cultural and hormonal treatments have been at least marginally successful. Thus, use of nucellar tissue, tissue and organ culture, repeated severe pruning, grafting on juvenile understock, GA_3 treatment, high temperature, and/or low light intensity have all caused reversion from the mature to juvenile phase.

Juvenile plantlets have been regenerated from somatic embryos derived from nucellar tissue of unfertilized ovules of an ancient clone of grapevine (*Vitis vinifera* L. cv. Cabernet-Sauvignon; MULLINS and SRINIVASAN 1976), a situation analagous to that noted for juvenile plantlets obtained by nucellar embryony (*Citrus* and *Mangifera*), (MAHESHWARI 1950), or from *Citrus* nucellar cultures (RANGAN et al. 1968). More recently, somatic embryos of Seyval grape have been produced from callus derived from stem tissue (KRUL and WORLEY 1977) and these embryos developed into plantlets on medium containing no vitamins or growth regulators. In contrast, the establishment of grape plantlets from somatic embryos derived from nucellar tissue required both cytokinin and GA_3. Whether juvenile seedlings obtained from a clone which had undergone both maturation and aging (e.g., Cabernet-Sauvignon is centuries old) constitutes "rejuvenation" or phase reversion is a matter of question if the nucellus is regarded "as an island of juvenility in an otherwise adult organism" (MULLINS and SRINIVASAN 1976). However, it seems more likely that plants of Seyval did undergo rejuvenation since they were derived from mature tissue. Interestingly, when the plants of Seyval (Seyve-Villard 5-276) from somatic embryos fruited in 1979, shoot color and fruit cluster characteristics were uniform among these plants, but neither of these identifying characteristics resembled the typical Seyval plants from which the original explants were obtained. However, both shoot color and fruit cluster characteristics on plants grown from somatic embryos more nearly matched the description of this cultivar (GALET 1956) than do these characteristics on typical source plants (W.R. KRUL, personal communication). Whether the changes with time in these characteristics of Seyval are due to accumulated bud mutations (periclinal chimeras), an infection with a virus or virus-like disease, or to other causes is not known.

With *H. helix*, plantlets regenerated from callus derived from juvenile or mature stems are juvenile in both cases, but the mode of adventive plantlet formation is different for callus from the two phases (embryo for mature callus, bud formation for juvenile callus; BANKS 1979). Returning once again to grape, BARLASS and SKENE (1980) found that plantlets derived via adventitious buds from mature leaf primordia fragments had characteristics identical to seedlings. So perhaps a truly mature meristem cannot be formed de novo. There are only three possible exceptions to this statement reported in the literature. One case (HUHTINEN 1976, WINTON and HUHTINEN 1976) involves adventitious bud formation on callus derived from very early flowering birch seedlings in which the only other clear evidence of the mature condition was the low propensity to form adventitious roots. In this case, plants derived adventitiously retained their ability to flower early, but there were some slight indications of rejuvenation, such as the vegetative character of the terminal bud and a few of the uppermost lateral buds. Therefore, it can perhaps be postulated that some juvenile characteristics were induced even in these plants during callus growth and adventitious bud formation. In another case, SCORZA and JANICK (1980) found that callus derived from the upper parts of *Passiflora suberosa* L. vines produced shoots that flowered for several weeks, but then lost the ability to flower. In a third case, embryoids derived from callus from roots of mature *Panax ginseng* plants flowered in the embryoid stage while still in culture (CHANG and HSING 1980). Whether plants which developed from these embryoids retained their ability to flower is not reported.

The results of recent studies indicate that even though mature meristems are quite stable both in vivo and in vitro, their phase-related characteristics can be modified as a result of length of in vitro culture and number of subcultures involved. Such findings suggest the possible importance of the culture medium and other culture conditions with respect to rejuvenation in vitro. DAVID et al. (1978) have reported the in vitro rejuvenation of *Pinus pinaster* primary meristems. Elongating buds implanted on medium containing the cytokinin, BA, formed new buds from developing needle fascicles (brachyblasts or short shoots). The less elongated the buds implanted, the more juvenile the plantlets developed from the needle fascicle derived buds in terms of foliage characteristics (FRANCLET 1979). This suggests that newly organized meristems are more plastic and subject to rejuvenation. High concentrations of BA and low concentrations of sucrose favored juvenile characteristics (DAVID et al. 1978). Buds taken from cutting propagated plants which had been subsequently hedged and given high levels of fertility gave the highest percentage of successful regeneration (FRANCLET 1979). With *Sequoia sempervirens*, BOULAY (1979) found that when the primary explant was of plagiotropic orientation, the number of orthotropic shoots produced increased with the number of subcultures and orthotropic shoot formation was enhanced by inclusion of activated charcoal in the medium. He also found that there was increasing root formation after a variable number of subcultures and suggests that conditions in the culture produce ontogenetic rejuvenation. FRANCLET (1979) indicates that in vitro-produced plantlets are impossible to distinguish from seedlings in the nursery.

MULLINS et al. (1979) produced "apparent rejuvenation" regarding phyllotaxy and, more consistently, an absence of tendrils through a series of shoot tip subcultures from adult tissue on a sterile, modified White's medium. Some of these rejuvenated plants retained their juvenile state (e.g., lack of tendrils) for up to 3 years (150 nodes), but when transferred to the glasshouse, tendrils appeared within 10 to 15 nodes. Cultured shoot tips treated with GA_3 produced tendrils as did, initially, axillary bud explants treated with BA. However, longer term cultures on medium containing BA did produce shoots lacking tendrils.

FAVRE and GRENAN (1979) pointed out that grapes maintained on a minimal medium (containing only mineral elements and sucrose) will indefinitely retain juvenile characteristics. When the medium is supplemented with vitamins, tendrils, and in some cases, flowers will form. They also state that morphological modifications seem to be a general manifestation of rejuvenation induced by culture in vitro.

Similarly, when strawberries are cultured from shoot tips (0.5 to 2.0 mm in height), the first leaf produced (present as a leaf primordium when the culture is initiated) is trifoliolate and subsequent leaves are monophyllous, a juvenile characteristic normally found only in young seedlings (BOXUS et al. 1977, J.R. MCGREW, personal communication). This occurs whether the medium on which the shoot tips are grown contains cytokinin and auxin or no growth regulators at all. This growth habit can be maintained by frequent subculturing on a medium containing BA and IBA (C. DAMIANO, personal communication), but trifoliolate leaves will develop if more time elapses between subculturing. Shoots with monophyllous leaves form roots on medium containing no cytokinin or auxin or on medium containing a low level of auxin (0.5 to 1.0 mg l^{-1} IBA). Shoots with trifoliolate leaves also form adventitious roots, but the response usually takes a few days longer (ZIMMERMAN 1981). Once the plants are removed from culture and established in the greenhouse, only trifoliolate leaves are formed. The growth habit of strawberry plants produced in this way more closely resembles that of seedlings, i.e., more prostrate leaves and increased runner formation, than of the mature, fruiting plants from which the explants were taken. However, there was no delay in flowering (ZIMMERMAN 1981).

Likewise, thornless blackberries cultured from shoot tips produce monophyllous leaves once a few preformed trifoliolate leaves have grown out (BROOME and ZIMMERMAN 1978) and production of monophyllous leaves in culture has been maintained by repeated subculturing for more than a year (ZIMMERMAN 1981). Shoot cuttings with monophyllous leaves are easily rooted and the resulting plants produce trifoliolate leaves once established in the greenhouse. As with strawberries, there is no delay in flowering, however.

Apple cultivars cultured from shoot tips also show some reversion in leaf shape in tissue culture although it is not as distinct as in the cases described above (SRISKANDARAJAH et al. 1982, ZIMMERMAN 1981). This reversion is manifested by lobed leaves, deep, irregular serrations and thinner leaf blades as found normally in shade leaves.

This apparent reversion to the juvenile phase might also explain the increased rooting of shoots after a certain number of subcultures (SRISKANDARAJAH et al. 1982). TAKENO et al. (1982/83) found that this increase in rooting ability with increasing numbers of subcultures was associated with a reduction in endogenous gibberellin and cytokinin levels. Although the above evidence suggests a rejuvenation of shoots in vitro, apple trees propagated by these in vitro methods do not show delayed flowering in the field (ZIMMERMAN 1981). LYRENE (1981), GUPTA et al. (1981), and ECONOMOU (1982) have reported similar findings for in vitro-cultured blueberry, eucalyptus, and deciduous azalea respectively. In all three cases rejuvenation was measured in terms of increased rooting potential and changes in morphological characteristics. B. MCCOWN (personal communication) has also observed rapid changes to juvenile morphology in cultures of *Thuja occidentalis* and *Cryptomeria japonica*.

Micropropagation in vitro involving adventive bud or embryo formation would be expected to yield rejuvenated plants. Such rejuvenation in plantlets proliferated from pre-existing meristems is less expected and, as indicated earlier, suggests that the small size of the mature explant and the conditions of the culture combine over time (subculture) to induce partial rejuvenation. The fact that in vitro-produced plantlets do not have delayed flowering indicates that the rejuvenation is only a partial one.

Repeated severe pruning of adult *H. helix* plants resulted in production of shoots with juvenile characteristics (ROBBINS 1960). Similar results were observed after adult plants had died back because of poor cultural conditions (ROBBINS 1960) or low winter temperatures (KRANZ 1931). Also, juvenile rooting characteristics in *P. radiata* seedlings have been maintained by hedging (LIBBY et al. 1972). Further, cuttings from cuttings (i.e., second and third generation clones) of *P. radiata* are easier to root than cuttings of the same physiological age taken from trees raised from seedlings (CAMERON 1968).

For *H. helix*, both adult scions grafted on young, juvenile seedlings and adult stocks on which juvenile scions were grafted reverted to the juvenile form (DOORENBOS 1954, HESS 1960). This rejuvenation was stimulated by the presence of leaves on the juvenile form and inhibited by their presence on the adult form. More reversions occurred when the adult form of *H. canariensis* was budded rather than grafted onto juvenile stocks, showing that the amount of adult tissue influenced rejuvenation (STOUTEMYER and BRITT 1961). In this latter study, temperatures above 27° C were more effective in inducing reversion than were lower temperatures. Budding the mature form of *Hevea brasiliensis* (WILLD. ex A. JUSS.) MUELL.-ARG. onto seedlings of *H. brasiliensis* was used to bring about reversion to the easier rooting characteristics of the juvenile form (MUZIK and CRUZADO 1958). Cuttings taken from the resulting shoots were tested for rooting, while buds from the same shoots were budded onto other seedlings. Cuttings from shoots growing from the fourth and fifth buddings rooted while no cuttings from the earlier buddings rooted.

Treatment with GA_3 induces reversion from the adult to the juvenile form in *Hedera* (Fig. 4) (FRYDMAN and WAREING 1974, GOODIN and STOUTEMYER

Fig. 4. Reversion from mature to juvenile form induced by GA_3 treatment of adult *Hedera canariensis variegata* var. *arborescens*. *Left* untreated; *right* treated with GA_3. Note on treated plant one branch with juvenile leaves and aerial roots but orthotropic growth; the apical 45 cm of the second branch has become completely juvenile. Both branches grew from apical buds. *Arrows* show limit of growth before treatment. (ROBBINS 1957)

1961, ROBBINS 1957, 1960, ROGLER and HACKETT 1975a, STOUTEMYER et al. 1961) as well as a number of other species as discussed in Section 2.3 above. Treatment of *Hedera* with GA_3 was effective only at constant temperatures of 15° C or greater (GOODIN and STOUTEMYER 1961). Constant or alternating temperatures of 20° C or greater increased the number of both treated and untreated plants that reverted to the juvenile state (GOODIN and STOUTEMYER 1961, ROGLER and HACKETT 1975a) or speeded the reversion of GA-treated adult *Hedera* (FRYDMAN and WAREING 1974). Temperature had a similar effect on reversion of budded or grafted *H. canariensis* plants that were not treated with GA_3 (STOUTEMYER and BRITT 1961).

Plants growing in low light intensity tend to remain in the juvenile stage (SCHWABE 1976); spontaneous reversion of adult material can also occur (ROGLER and HACKETT 1975a). However, reversion of adult *H. helix* plants treated with GA_3 was not greater at low light intensities (ROGLER and HACKETT 1975a).

In contrast to the results with GA's, IAA, kinetin, ABA, and ethephon did not cause reversion of adult *H. helix* (ROGLER and HACKETT 1975a). However, ABA did counteract the effect of GA_3 on *H. helix*, thus stabilizing plants in the mature form (ROGLER and HACKETT 1975b) and indicating that the balance between GA_3 and ABA, rather than the absolute amount of either, was the important factor in controlling reversion in this species.

The growth retardants chlormequat, daminozide, and ancymidol prevented most spontaneous reversions of adult *H. helix* grown under low light intensity (ROGLER and HACKETT 1975b). While these growth retardants may inhibit GA biosynthesis, they may also have other effects on plant metabolism that could bear on stabilization of the mature phase. For example, chlormequat inhibition of growth of juvenile *H. helix* could be only partially overcome by GA_3 treatment and chlormequat-treated plants had up to tenfold increases in a range of GA-like substances in stem apices with smaller increases in polar GA-like substances from root tissue (FRYDMAN and WAREING 1974).

Anatomical studies of *H. helix* show that apices of juvenile plants have a smaller meristematic area, larger cells in the apex, and cell divisions occurring over a longer part of the shoot for a longer period of time than have adult plants (STEIN and FOSKET 1969). A recent scanning electron microscope study of the shoot apical meristem of mature *H. helix* plants during GA_3-induced rejuvenation (HACKETT and SRINIVASAN 1984) demonstrates that phyllotaxis changes to the juvenile 1/2 configuration before there is any significant reduction in the area of the apical meristem. At a much later time after treatment, there is a reduction in shoot apical meristem size as a result of GA_3-induced rejuvenation of mature plants. These results indicate that GA_3 does not act directly through changes in apex size to influence phyllotaxis and other rejuvenation related morphological characters.

Less cell division occurs in the subapical region in adult plants, resulting in shorter internodes (STEIN and FOSKET 1969). Application of GA_3 causes cell division in the subapical region in some plants (SACHS et al. 1959, SACHS and LANG 1957). This could explain the GA-induced reversions in *Hedera*. However, treatment with growth retardants such as chlormequat does not lead to phase change from juvenile to adult, nor does it inhibit synthesis of GA-like substances in the plant, although it may interfere with their metabolism (FRYDMAN and WAREING 1974). A decline of GA in *Hedera* shoot apices probably precedes maturation; adventitious root production near the shoot tip ceases before maturation under natural conditions, and adventitious roots appear to be a source of GA's in the shoot (FRYDMAN and WAREING 1973, 1974). Further confusing the issue is the fact that *Hedera* which has reverted to the juvenile phase from GA_3 treatment shows increased rooting ability, a rather anomalous situation (see Sect. 2.2, above).

5 Tissue and Organ Culture of Juvenile and Adult Phases

Differences in growth rate for callus cultures established from juvenile and adult phases have been reported for *H. helix* (Fig. 5) and *H. canariensis* (GOODIN 1964, ROBBINS and HERVEY 1970, STOUTEMYER and BRITT 1963, 1965, 1969),

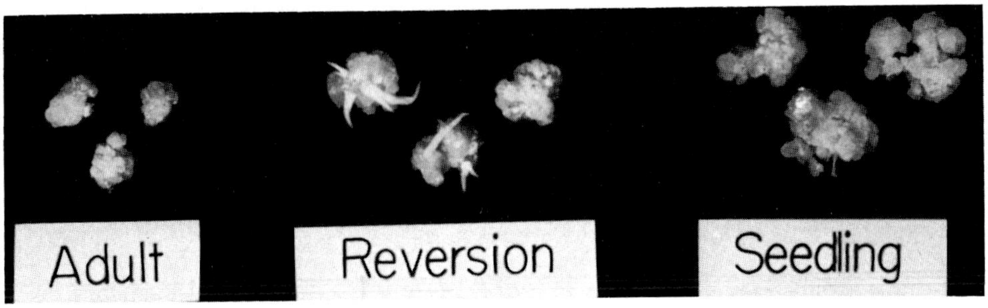

Fig. 5. Tissue cultures of adult (*left*), juvenile reversion from adult (*center*), and seedling (*right*) *Hedera helix*. Note adventitious root formation on callus from juvenile reversion shoots. (Photograph courtesy of V.T. STOUTEMYER)

Aesculus hippocastanum L., *Castanea vulgaris* LAM., and *Robinia pseudoacacia* (TRIPPI 1963a, b). Callus derived from juvenile tissue grew more rapidly in all of these species except for *A. hippocastanum* in which callus from the adult zone proliferated more rapidly. Growth rates of tissue from young *H. helix* seedlings were higher than those from reversions (STOUTEMYER and BRITT 1963), subculturing monthly for more than 3 years not affecting those differences (STOUTEMYER and BRITT 1965, 1969). Cells of callus from juvenile tissue were larger and heavier (fresh weight) than those from adult tissue (STOUTEMYER and BRITT 1965), differences similar to those found in cell size and number for apical meristems of the two growth phases (STEIN and FOSKET 1969). POLITO and ALLIATA (1981) have recently shown that calli derived from shoot apical meristems of juvenile and adult *H. helix* also have characteristically different growth rates, similar to those of callus derived from stem cross-sections.

A minimum temperature of 27° C was necessary for successful subculturing of callus from adult tissue, whereas callus from juvenile tissue would grow at lower temperatures although less vigorously (STOUTEMYER and BRITT 1963, 1965). Callus growth was not affected by light (STOUTEMYER and BRITT 1965).

Callus derived from young seedlings of *H. helix* grew well in the absence of coconut water and NAA, but little proliferation occurred without these components in callus cultures started from reverted juvenile shoots. Portions of some cultures from adult tissue developed an aberrant appearance and a rapid growth rate after several months without transfer. These tissues, when subcultured, required neither coconut water nor NAA. They did require kinetin and casein hydrolysate and attained growth rates comparable to, or greater than, callus from juvenile tissue (STOUTEMYER and BRITT 1969). Whether this represents a reversion in culture is questionable. After adult *H. helix* plants were sprayed with GA_3 to induce reversion, callus cultures were established at weekly intervals from GA_3-treated shoots and from juvenile shoots which had previously reverted from adult plants of the same clone (MILLER and GOODIN 1976). Callus obtained from shoots 3 weeks after GA_3 treatment had growth rates equal to those of callus from juvenile shoots. However, complete reversion of GA_3-treated shoots to the juvenile phase, as judged by standard morphological characters, took at least 10 weeks (MILLER and GOODIN 1976).

6 Effect of Genotype on Length of the Juvenile Period

The length of the juvenile period can vary widely within a species, and is an inherited character. Selection and breeding can thus be used to produce progeny with a shorter juvenile phase. This has been done with apple (SAURE 1970, SCHANDER 1962, SCHMIDT 1970, VISSER 1965, 1976, WAY 1971), pear (VISSER 1976, ZIMMERMAN 1977), *Prunus* species (SCHMIDT 1976), birch (JOHNSSON 1949, STERN 1961), and several *Pinus* species (HEIMBURGER and FOWLER 1969, SCHRÖCK 1949, TEICH and HOLST 1969) among others (cf. ZIMMERMAN 1972 for further discussion on this point).

Although length of the juvenile period is inherited, environmental conditions and cultural techniques interact with genotype to determine just when phase change occurs as noted in Section 4.1 above. While one may speculate that genotypic effects in length of the juvenile phase are mediated, at least partly, through differences in hormonal content and metabolism, this has yet to be examined critically.

7 Summary and Conclusions

Phase change appears to be controlled more by GA's than any other plant hormone, the role of GA's varying greatly with plant species. Although exogenous GA's can induce flowering in conifers, they generally have the opposite effect in woody angiosperms. Gibberellins also induce reversions from adult to juvenile phase in angiosperms with the best-documented examples being the *Hedera* species. Both GA-induced flowering in conifers, and reversion in angiosperms (with the exception of *Hedera*), are temporary phenomena, since without continuing treatment, "normalcy" returns with time. There is evidence that endogenous GA levels may affect or control natural sexual phase change in both conifers and some woody angiosperms, the *effects* of high, or low levels of GA, however, generally being opposite for each group.

Auxins, usually in combination with GA's, stimulate cone bud differentiation in some Pinaceae species, as can, to a lesser degree, exogenous cytokinin; but little evidence of a similar role in angiosperms is available. Except for the possible effects on shoot tips cultured in vitro, there is no indication that auxins or cytokinins play a direct role in other aspects of phase change.

Abscisic acid will tend to stabilize the mature form of *Hedera* once phase change has occurred and appears to be involved in promoting precocious flowering in *Betula* seedlings.

Cytokinins (except as noted above for conifers) and ethylene have no identified role in phase change, although ethylene can stimulate flowering in some angiosperm seedlings that might not normally flower.

Environmental conditions that enhance photosynthesis result in more rapid growth of seedlings, which in turn leads to earlier phase change and flowering in a number of species. Other treatments, such as grafting onto dwarfing rootstocks or girdling, which presumably increase the proportion of photosynthate (carbohydrate) retained in the shoot versus that translocated to the roots, also lead to earlier flowering in many cases. All the treatments that affect carbohydrate level and distribution may also affect production, movement, and probably balance of endogenous plant growth substances. Thus, earlier interpretations of the effects of dwarfing rootstocks or girdling on phase change in terms of carbohydrate accumulation and metabolism need to be re-examined in light of our current knowledge on the probable roles of plant hormones and growth regulators in phase change. The same applies to many other of the stress treatments (nutrient deficiency, water stress, bending, pruning, etc.).

HACKETT (1976) has proposed that phase change may be regulated by hormonal control of the supply of nutritional factors to the shoot apex and partitioning of these factors to different parts of the apex. Considerable evidence from experiments with angiosperms can be marshalled to support this hypothesis (HACKETT 1976, SACHS 1977), and its applicability to conifers is not precluded if one assigns a "mobilizing role" to GA's, auxins and cytokinins in their effects on precocious and enhanced flowering. Recent evidence has been obtained that GA_{4+7} enhances ^{14}C assimilate movement to lateral long shoots (potential female strobili) within the bud of *P. radiata* at the expense of the structural tissue and apical meristematic region (ROSS et al. 1984). Gibberellin A_3 had a similar effect, but only GA_{4+7} caused significant flowering. Hence, in addition to any "mobilizing" effect that GA_{4+7} might have had, there was also a distinct morphogenetic effect, one not possessed by GA_3 (ROSS et al. 1984). Certainly, the interactions of carbohydrates and other nutritional factors with GA's, auxins, and other growth substances need closer examination since the available evidence indicates that these factors can influence some of the biological parameters which are altered in the change from juvenile to mature phase.

Note Added in Proof. The inhibition of flowering in numerous woody angiosperms by GA_3 and $GA_{4/7}$ mixtures (see p. 84) may be due specifically to the presence of the C-1,2 dehydro GA_3 or GA_7, and may be associated mainly with woody angiosperms that are 'spur type' with regard to flowering, and exhibit a pattern of biennial bearing since recent work (Pharis, R.P. and Looney, N.E., UK patent application No. 8502424, 31 Jan., 1985; Looney and Pharis, manuscript submitted to *Planta*) obtained significant flowering increases in a biennial cultivar of apple by applying GA_4 and C-3 epi-GA_4.

References

Aldwinckle HS (1975) Flowering of apple seedlings 16–20 months after germination. HortScience 10:124–126

Ali N, Westwood MN (1966) Nucleic acids as related to juvenility in several *Pyrus* species. Proc Am Soc Hortic Sci 89:123–131

Allsopp A (1967) Heteroblastic development in vascular plants. In: Abercombie M and Brachet J (eds) Advances in morphogenesis. Academic Press, London New York, pp 127–171

Arshad NL (1980) Studies on endogenous growth substances in relation to flowering in seedlings of *Betula pendula* Roth. Ph D Thesis, Univ Wales, Aberystwyth

Banks MS (1979) Plant regeneration from callus from two growth phases of English ivy *Hedera helix* L. Z Pflanzenphysiol 92:349–353

Barlass M, Skene KGM (1980) Studies on the fragmented shoot apex of grapevine. II. Factors affecting growth and differentiation in vitro. J Exp Bot 31:489–495

Bleymüller H (1976) Investigations on the dependence of flowering in Norway spruce [*Picea abies* (L.) Karst] upon age. Acta Hortic 56:169–172

Bonnet-Masimbert M (1982) Influence de l'état d'activité des racines sur la floraison induite par des gibbérellines 4 et 7 chez *Pseudotsuga menziesii* (Mirb.) Franco. Silvae Genet 31:178–183

Bonnet-Masimbert M, Nitsch JP (1969) Induction precoce de la floraison chez une plante ligneuse *Rhus typhina* L. Bull Soc Bot France 116:403–407

Borchert R (1965) Gibberellic acid and rejuvenation of apical meristems in *Acacia melanoxylon*. Naturwissenschaften 52:65–66

Borchert R (1976) The concept of juvenility in woody plants. Acta Hortic 56:21–36

Bouillenne R, Bouillenne-Walrand M (1955) Auxines et bouturage. In: Rep 14th Int Hort Congr Vol I, Veenman en Zonen, Wageningen, pp 231–238

Boulay M (1979) Multiplication et clonage rapide du *Sequoia sempervirens* par la culture in vitro. Micropropagation d'arbres forestiers. Assoc. Foret-Cellulose Domaine de L'Etancon. Nangis, France, pp 49–56

Boxus P, Quoirin M, Laine JM (1977) Large scale propagation of strawberry plants from tissue culture. In: Reinert J, Bajaj YPS (eds) Applied and fundamental aspects of plant cell, tissue, and organ culture. Springer, Berlin Heidelberg New York, pp 130–143

Brink RA (1962) Phase change in higher plants and somatic cell heredity. Q Rev Biol 37:1–22

Brix H, Van den Driessche R (1977) Use of rooted cuttings in reforestation. A review of opportunities, problems and activities. Rep 6, British Columbia For Serv Can For Serv Joint Rep Victoria, pp 1–16

Brix H, Portlock FF (1982) Flowering response of western hemlock seedlings to gibberellin and water-stress treatments. Can J For Res 12:76–82

Brooks HJ (1964) Responses of pear seedlings to N-dimethylaminosuccinamic acid, a growth retardant. Nature (Lond) 203:1303

Broome OC, Zimmerman RH (1978) In vitro propagation of blackberry. HortScience 13:151–153

Bukovac MJ (1963) Chemical promotion of flowering. Am Fruit Grow 83(3):24

Cameron RJ (1968) The propagation of *Pinus radiata* by cuttings. Influences affecting the rooting of cuttings. NZJ For 13:78–89

Campbell AI (1961) Shortening the juvenile phase of apple seedlings. Nature (Lond) 191:517

Chacko EK, Kohli RR, Dore Swamy R, Randhawa GS (1974) Effect of (2-chloroethyl) phosphonic acid on flower induction in juvenile mango (*Mangifera indica* L.) seedlings. Physiol Plant 32:188–190

Chacko EK, Kohli RR, Dore Swamy R, Randhawa GS (1976) Growth regulators and flowering in juvenile mango (*Mangifera indica* L.) seedlings. Acta Hortic 56:173–176

Chalupka W (1978) Effect of growth regulators on the flowering of Scots pine (*Pinus sylvestris* L.) grafts. Silvae Genet 27:62–65

Chalupka W (1979) Effect of growth regulators on flowering of Norway spruce [*Picea abies* (L) Karst.] grafts. Silvae Genet 28:125–127

Chalupka W (1981) Influence of growth regulators and polythene covers on flowering of Scots pine and Norway spruce grafts. Silvae Genet 30:142–146

Chalupka W, Giertych M, Kopciewicz J (1982) Effects of polythene covers, a flower-inducing treatment, on the content of endogenous gibberellin-like substances in grafts of Norway spruce. Physiol Plant 54:79–82

Chang W, Hsing Y (1980) In vitro flowering of embryoids derived from mature root callus of ginseng (*Panax ginseng*) Nature (Lond) 284:341–342

Cooper WC, Peynado A (1958) Effect of gibberellic acid on growth and dormancy in citrus. Proc Am Soc Hortic Sci 72:284–289

Copes DL (1980) Effect of rootstock vigor on leader elongation, branch growth and plagiotropism in 4- and 8-year-old Douglas-fir grafts. Tree Planters Notes 31(1):11–14

Crane JC, Primer PE, Campbell RC (1960) Gibberellin-induced parthenocarpy in *Prunus*. Proc Am Soc Hortic Sci 75:129–137

David H, Isemukali K, Davis A (1978) Obtention de plantes de pin maritime (*Pinus pinaster* Sol) à partir de brachyblastes ou d'apex caulinaires des très jeunes sujets cultivés in vitro. CR Acad Sci Paris Series D 287:245–248

Davis LD (1957) Flowering and alternate bearing. Proc Am Soc Hortic Sci 70:545–556
Dayton DF (1966) Chemical induction of flowering in triploid apple progenies. HortScience 1:92–93
Dennis FG Jr (1968) Growth and flowering responses of apple and pear seedlings to growth retardants and scoring. Proc Am Soc Hortic Sci 93:53–61
De Zeeuw D, Leopold AC (1955) Altering juvenility with auxin. Science 122:925–926
Domoney C, Timmis JH (1980) Ribosomal RNA gene redundancy in juvenile and mature ivy (*Hedera helix*). J Exp Bot 31:1093–1100
Doorenbos J (1954) Rejuvenation of *Hedera helix* in graft combination. Proc K Ned Akad Wet Ser C Biol Med Sci 57:99–102
Doorenbos J (1955) Shortening the breeding cycle of rhododendron. Euphytica 4:141–146
Doorenbos J (1965) Juvenile and adult phases in woody plants. In: Ruhland W (ed) Encyclopedia of plant physiology Vol 15, part 1. Springer, Berlin Heidelberg New York, pp 1222–1235
Dunberg A, Malmberg G, Sassa T, Pharis RP (1983) Metabolism of tritiated gibberellins A_4 and A_9 in Norway spruce, *Picea abies* (L) Karst. Plant Physiol 71:257–262
Economou AS (1982) Chemical and physical factors influencing in vitro propagation of hardy deciduous azaleas (*Rhododendron* spp.) Ph D Thesis Univ of Minnesota
El-Antably HMM, Wareing PF, Hillman J (1967) Some physiological responses to D-L-abscisin. Planta 73:74–90
Eriksen EN (1970) Promotion of root initiation by gibberellin. Roy Vet and Agr Univ (Copenhagen) Yearbook for 1971:50–59
Eriksen EN (1971) Root initiation in pea cuttings: Interactions between tryptophane and gibberellin. Roy Vet and Agr Univ (Copenhagen) Yearbook for 1972, 115–126
Eriksen EN (1974) Root formation in pea cuttings III. The influence of cytokinin at different developmental stages. Physiol Plant 30:163–167
Favre JM, Grenan S (1979) Sur la production de vrilles, de fleurs et de baies, chez la vigne cultivée in vitro. Ann Amélior Plant 29:247–252
Feldman LJ, Cutter EG (1970) Regulation of leaf form in *Centaurea solstitialis* L. I. Leaf development on whole plants in sterile culture. Bot Gaz 131:31–39
Fisher JB (1976) Induction of juvenile leaf form in a palm (*Caryota mitis*) by gibberellin. Bull Torrey Bot Club 103:153–157
Fogle HW (1956) Girdling sweet cherry seedlings for early fruit production. Fruit Var Hortic Dig 10:55–56
Fontes MR, Ozbun JL (1972) Relationship between carbohydrate level and floral initiation in broccoli. J Am Soc Hortic Sci 97:346–348
Fortanier EJ, Jonkers J (1976) Juvenility and maturity of plants as influenced by their ontogenetical and physiological ageing. Acta Hortic 56:37–44
Franclet A (1979) Rajeunissement des arbres adults en vue de leur propagation végétative. In: Micropropagation d'arbres forestiers assoc. Foret-Cellulose. Domaine de L'etancon Nangis, France, pp 3–18
Fritzsche R (1948) Untersuchungen über die Jugendformen des Apfel- und Birnbaumes und ihre Konsequenzen für die Unterlagen- und Sortenzüchtung. Ber Schweiz Bot Ges 58:207–267
Frydman VM, Wareing PF (1973) Phase change in *Hedera helix* L. I. Gibberellin-like substances in the two growth phases. J Exp Bot 24:1131–1138
Frydman VM, Wareing PF (1973) Phase change in *Hedera helix* L. II. The possible role of roots as a source of shoot gibberellin-like substances. J Exp Bot 24:1139–1148
Frydman VM, Wareing PF (1974) Phase change in *Hedera helix* L. III. The effects of gibberellins, abscisic acid and growth retardants on juvenile and adult ivy. J Exp Bot 25:420–429
Fukasawa H (1966) Disc electrophoresis of protein from juvenile and adult specimens of ivy. Nature (Lond) 212:516–517
Furr JR, Cooper WC, Reece PC (1947) An investigation of flower formation in adult and juvenile citrus trees. Am J Bot 34:1–8
Galet P (1956) Cépages et vignobles de France. I. Les vignes américaines. Montpellier: Imp Paul Déhan

Ghosh BN, Glowinkowska A, Millikan DF (1971) Enzymatic activities associated with maturation and senescence in *Hedera helix*. Bull Acad Pol Sci, Ser Sci Biol 19:605–609
Ghosh BN, Millikan DF (1970) Proteins in the leaves of different forms of *Hedera helix* L. Plant Cell Physiol 11:817–818
Girouard RM (1969) Physiological and biochemical studies of adventitious root formation. Extractible rooting cofactors from *Hedera helix*. Can J Bot 47:687–699
Goodin JR (1964) Shoot growth rates as a factor in growth phase transitions in *Hedera*. Proc Am Soc Hortic Sci 84:600–605
Goodin JR, Stoutemeyer VT (1961) Effect of temperature and potassium gibberellate on phases of growth of Algerian ivy. Nature (Lond) 192:677–678
Greenwood MS (1978) Flowering induced on young loblolly pine grafts by out-of-phase dormancy. Science 201:443–444
Griggs WH, Iwakiri BT (1961) Effects of gibberellin and 2,4,5-trichlorophenoxypropionic acid sprays on Bartlett pear trees. Proc Am Soc Hortic Sci 77:73–89
Gupta PK, Mascarenhas AF, Jagannathan V (1981) Tissue culture of forest trees – clonal propagation of mature trees of *Eucalyptus citriodora* by tissue culture. Plant Sci Lett 20:195–201
Guttridge CG (1962) Inhibition of fruit-bud formation in apple with gibberellic acid. Nature (Lond) 196:1008
Hackett WP (1970) The influence of auxin, catechol and methanolic tissue extracts on root initiation in aseptically cultured shoot apices of the juvenile and adult forms of *Hedera helix*. J Am Soc Hortic Sci 95:398–402
Hackett WP (1976) Control of phase change in woody plants. Acta Hortic 56:143–154
Hackett WP (1980) Control of phase change in woody plants. In: Little HA (ed) Control of shoot growth in trees. Proc Inter Union For Res Organ Working parties on xylem physiology and shoot growth physiology. Fredericton, New Brunswick, Can, pp 257–272
Hackett WP (1985) Juvenility, maturation and rejuvenation in woody plants. Hortic Rev 7: (to be published)
Hackett WP, Srinivasan C (1984) *Hedera helix* and *Hedera canariensis*. In: Halevy A (ed) Handbook for flowering. vol III. CRC Boca Raton, FL (to be published)
Hare RC (1979) Promoting flowering in loblolly and slash pine with branch, bud, and fertilizer treatments. Flowering and seed development in trees. Symposium May 1978, US Forest Service, New Orleans, pp 112–121
Hare RC, Synder EB, Schmidtling RC (1979) Longleaf pine flowering in response to nitrogen fertilization; branch girdling, growth substances, and cultivation. US For Serv Gen Tech Rep NC-50, St Paul, Minn, pp 11–16
Hartmann HT, Kester DE (1975) Plant propagation, 3rd edn, Prentice-Hall, Englewood Cliffs, NJ
Hashizume H (1967) Experimental induction of female flowers in young Japanese larch (*Larix leptolepis*, Gordon). J Jpn For Soc 49:405–408
Hashizume H (1968) Chemical regulation of flower-bud formation in conifers. J Jpn For Soc 50:14–16
Heimburger CC, Fowler DP (1969) Precocious flowering in some pines of the lariciones group. Silvae Genet 18:146–150
Hendry NA, Van Staden J, Allan P (1982) Cytokinins in citrus II. Fluctuations during growth in juvenile and adult plants. Scientia Hort 17:247–256
Hess CE (1959) A study of plant growth substances in easy and difficult to root cuttings. Proc Plant Prop Soc 9:39–43
Hess CE (1960) Research in root initiation – a progress report. Proc Plant Prop Soc 10:118–123
Hess CE (1964a) A physiological analysis of root initiation in easy- and difficult-to-root-cuttings. In: Proc 16th Int Hortic Congr vol IV, Editions J Duculot, Gembloux, pp 375–381
Hess CE (1964b) Characterization of the rooting cofactors extracted from *Hedera helix* L. and *Hibiscus rosa-sinensis* L. In: Proc 16th Int Hortic Congr, vol IV. Editions J Duculot, Gembloux, pp 382–388

Hess CE (1967) Root initiation and juvenility – a possible implication of terpene compounds. In: Proc 17th Int Hortic Congr vol III, pp 443–451
Heuser CW (1976) Juvenility and rooting cofactors. Acta Hortic 56:251–261
Heybroek HM, Visser T (1976) Juvenility in fruit growing and forestry. Acta Hortic 56:71–80
Hield HZ, Coggins CW Jr, Lewis LN (1966) Temperature influence on flowering of grapefruit seedlings. Proc Am Soc Hortic Sci 89:175–181
Higazy MKMT (1962) Shortening the juvenile phase for flowering. Meded Landb Hogesch Wageningen 62:1–53
Hillman JR, Young I, Knights BA (1974) Abscisic acid in leaves of *Hedera helix* L. Planta 119:263–266
Hillman JR, Knights BA, McKail R (1975) Sterols of the adult and juvenile forms of ivy, *Hedera helix* L. Lipids 10:542–544
Holst MJ (1961) Experiments with flower promotion in *Picea glauca* (Moench) Voss. and *Pinus resinosa* Ait. In: Recent advances in botany, vol 2, Univ Toronto Press, Toronto, pp 1654–1658
Huhtinen O (1976) Early flowering of birch and its maintenance in plants regenerated through tissue cultures. Acta Hortic 56:243–249
Hull J Jr, Lewis LN (1959) Response of 1-year-old cherry and mature bearing cherry, peach, and apple trees to gibberellins. Proc Am Soc Hortic Sci 74:93–100
Ibrahim IM, Dana MN (1971) Gibberellin-like activity in apple rootstocks. HortScience 6:541–542
Jackson DI, Sweet GB (1972) Flower initiation in temperate woody plants. Hortic Abstr 42:9–24
Johnsson H (1949) Hereditary precocious flowering in *Betula verrucosa* and *B. pubescens*. Hereditas 35:112–114
Jones OP, Hopgood ME (1976) Morphogenetic factors in xylem sap. Effects of dwarfing interstocks on apple xylem sap. Rpt E Malling Res Sta for 1975:66
Jones OP, Hopgood ME, Calder T, Hadlow WCC (1975) Morphogenetic factors in xylem sap. Effects of interstocks and graft unions on apple xylem sap. Rpt E Malling Res Sta for 1974:71
Jonkers H (1971) An international experiment on juvenility in apple. Euphytica 20:57–59
Jonkers H (1973) Effect of SADH on juvenility of apple seedlings. Acta Hortic 34:143–150
Jonkers H (1975) Effect van gibberellinebespuitingen op de jeugdperiode van appelzaailingen. Landbkundig Tijdschr part 87–12, pp 330–333
Kamienska A, Pharis R, Wample RL, Kuo CC, Durley RC (1974) Occurrence and metabolism of gibberellins in conifers. In: Tamura S (ed) Plant growth substances 1973. Hirokawa, Tokyo, pp 305–313
Karnatz A (1963) Tastversuche zur Verkürzung des Primärstadiums bei Apfelsämlingen durch Gibberellin. Erwerbsobstbau 5:8
Karnatz A (1965) Zur Frage des Beginnes der generativen Phase bei Apfelsämlingen. I. Erstblüte. Gartenbauwissenschaft 30:527–540
Kato Y, Fukuhara N, Kobayashi R (1959) Promotion of flower-bud differentiation of conifers by gibberellin. J Jpn For Soc 41:309–311
Kemmer E (1953) Über das primäre und das fertile Stadium bei Apfelgehölzen. Züchter 23:122–127
Kemmer E, Thiele I (1954) Entwicklungsfragen bei Apfelgehölzen. Züchter 24:346–352
Kender WJ (1974) Ethephon-induced flowering in apple seedlings. HortScience 9:444–445
Kessler B, Reches S (1977) Structural and functional changes of chromosomal DNA during ageing and phase change in plants. Chromosomes Today 6:237–246
Kester DE (1976) The relationship of juvenility to plant propagation. Proc Int Plant Prop Soc 26:71–84
Koranski DS, McCown BA, Struckmeyer BE, Beck GE (1979) Gibberellin–growth retardant interactions on the growth and flowering of *Clerodendrum thomsoniae*. Physiol Plant 45:88–91

Kozlowski TT (1971) Growth and development of trees vol I. Academic Press, London New York
Kranz G (1931) Kenntnis der wechselnden Blattform des Efeus und ihrer Ursachen. Flora 25:289–320
Krul WR, Worley JF (1977) Formation of adventitious embryos in callus cultures of Seyval, a French hybrid grape. J Am Soc Hortic Sci 102:360–363
Lammerts WE (1943) Effect of photoperiod and temperature on growth of embryo-cultured peach seedlings. Am J Bot 30:707–711
Lammerts WE (1949) The effect of continuous light, high nutrient level and temperature on flowering of camellia hybrids. Am Camellia Yearbook 1949:53–56
Lavender DP, Sweet GB, Zaerr JB, Hermann RK (1973) Spring shoot growth in Douglas-fir may be initiated by gibberellins exported from the roots. Science 182:838–839
Leopold AC, Kriedemann PE (1975) Plant growth and development 2nd edn. McGraw Hill, New York
Libby WJ, Brown AG, Fielding JM (1972) Effects of hedging radiata pine on production, rooting, and early growth of cuttings. NZJ For Sci 2:263–283
Loewel EL, Saure M (1963) Die Ertragsverfrühung durch schwachwachsende Unterlagen in der Apfelzüchtung. Erwerbsobstbau 5:191–193
Longman KA (1976) Some experimental approaches to the problem of phase change in forest trees. Acta Hortic 56:81–90
Longman KA, Wareing PF (1959) Early induction of flowering in birch seedlings. Nature 184:2037–2038
Looney NE, Kamienska A, Legge RL, Pharis RP (1978) Metabolism of ^{3}H gibberellin A_4 in relation to flower initiation in apple. Acta Hortic 80:105–114
Luukkanen O, Johansson S (1980) Flower induction by exogenous plant hormones in conifers. Silvae Fenn 14:95–105
Lyr UH, Hoffmann G (1964) Über den Einfluss der Mineralsalz-Ernährung auf die Frühfruktifikation von *Cryptomeria japonica* (LFD) Don. Flora 154:189–208
Lyrene PM (1981) Juvenility and production of fast-rooting cuttings from blueberry shoot cultures. J Am Soc Hortic Sci 106:396–398
Maheshwari P (1950) An introduction to the embryology of angiosperms. McGraw-Hill, New York
Martin GC, Stahly EA (1967) Endogenous growth-regulating factors in bark of EM IX and XVI apple trees. Proc Am Soc Hortic Sci 91:31–38
Matthews JD (1963) Factors affecting the production of seed by forest trees. For Abstr 24(1):i–xiii
Miller DR, Goodin JR (1976) Cellular growth rates of juvenile and adult *Hedera helix* L. Plant Sci Lett 7:397–401
Millikan DF, Ghosh BN (1971) Changes in nucleic acids associated with maturation and senescence in *Hedera helix*. Physiol Plant 24:10–13
Millikan DF, Saniewski M (1973) Quantitative differences in ascorbic acid oxidase and phenolase activities and auxin content of different forms of *Hedera helix* L. Plant Cell Physiol 14:213–215
Mott RL (1983) Trees. In: Conger BV (ed) Cloning agricultural plants via in vitro techniques CRC, Boca Raton, FL, pp 217–254
Mullins MG, Srinivasan C (1976) Somatic embryos and plantlets from an ancient clone of the grapevine (cv. Cabernet-Sauvignon) by apomixis in vitro. J Exp Bot 27:1022–1030
Mullins MG, Nair Y, Sampet P (1979) Rejuvenation in vitro: Induction of juvenile characters in an adult clone of *Vitis vinifera* L. Ann Bot 44:623–627
Muzik TJ, Cruzado HJ (1958) Transmission of juvenile rooting ability from seedlings to adults of *Hevea brasiliensis*. Nature (Lond) 181:1288
Nienstaedt H (1961) Induction of early flowering: A critical review of recent literature. In: Recent advances in botany vol 2. Univ Toronto Press, Toronto, pp 1658–1662
Njoku E (1957) The effect of mineral nutrition and temperature on leaf shape in *Ipomoea caerulea*. New Phytol 56:154–171
Njoku E (1958) Effect of gibberellic acid on leaf form. Nature (Lond) 182:1097–1098

Paton DM, Willing RR, Nicholls W, Pryor LD (1981) Rooting of stem cuttings of *Eucalyptus:* A rooting inhibitor in adult tissue. Aust J Bot 18:175–183

Paton DM, Willing RR, Pryor LD (1981) Root-shoot gradients in *Eucalyptus* ontogeny. Ann Bot 47:835–838

Pharis RP (1974) Precocious flowering in conifers: The role of plant hormones. Yale Univ School of Forest and Environ Stud Bull 85:51–80

Pharis RP (1976) Manipulation of flowering in conifers through the use of plant hormones. In: Miksche JP (ed) Modern methods in forest genetics. Springer, Berlin Heidelberg New York, pp 265–282

Pharis RP (1977) Interaction of native or exogenous plant hormones in the flowering of woody plants. In: Schütte HR and Gross D (eds) Regulation of developmental processes in plants. Halle, GDR, pp 344–360

Pharis RP, King RW (1985) Gibberellins and reproductive development in seed plants. Ann Rev Plant Physiol (to be published)

Pharis RP, Kuo CC (1977) Physiology of gibberellins in conifers. Can J For Res 7:299–325

Pharis RP, Morf W (1968) Physiology of gibberellin-induced flowering in conifers. In: Wightman F and Setterfield G (eds) Biochemistry and physiology of plant growth substances. Runge, Ottawa, pp 1341–1356

Pharis RP, Owens JN (1966) Hormonal induction of flowering in conifers. Yale Scientific Mag XLI (2):10–19

Pharis RP, Ross S (1984) Progress in hormonal cone induction in Pinaceae conifers. In: Proceedings Can. Tree Improvement Mtg., Aug 1983. Can For Service Ottawa (to be published)

Pharis RP, Wample RL, Kamienska A (1975) Growth, development, and sexual differentiation in *Pinus,* with emphasis on the role of the plant hormone, gibberellin. In: Baumgartner DM (ed) Proc Symp on Management of Lodgepole Pine Ecosystems. Pullman, Coop Ext Serv, Washington State Univ, pp 106–134

Pharis RP, Ross SD, Wample RL, Owens JN (1976) Promotion of flowering in conifers of Pinaceae by certain of the gibberellins. Acta Hortic 56:155–162

Pharis RP, Ross SD, McMullan EE (1980) Promotion of flowering in the Pinaceae by gibberellins. III. Seedlings of Douglas-fir. Physiol Plant 50:119–126

Philipson JJ (1983) The role of gibberellin $A_{4/7}$, heat and drought in the induction of flowering in Sitka spruce. J Exp Bot 34:291–301

Polito VS, Alliata V (1981) Growth of calluses derived from shoot apical meristems of adult and juvenile ivy (*Hedera helix* L.) Plant Sci Lett 22:387–393

Polito VS, Chang, Y-C (1984) Quantitative nuclear cytology of English ivy (*Hedera helix* L.). Plant Sci Lett 34:369–377

Pollard DFW, Portlock FF (1981) Effects of temperature on strobilus production in gibberellin-treated seedlings of western hemlock. Can For Serv Res Notes 1:21–22

Porlingis IC, Therios I (1976) Rooting responses of juvenile and adult leafy olive cuttings to various factors. J Hortic Sci 51:31–39

Potapenko J (1939) Acceleration of development and fruiting of fruit-tree seedlings. Compt Rend (Dokl) Acad Sci URSS 23:839–842

Puritch GS, McMullan EE, Meagher MD, Simmons CS (1979) Hormonal enhancement of cone production in Douglas-fir grafts on seedlings. Can J For Res 9:193–200

Rangan TS, Murashige T, Bitters WP (1968) In vitro initiation of nucellar embryos in monoembryonic *Citrus*. HortScience 3:226–227

Richards D, Thompson WK, Pharis RP (1984) The influence of apple interstocks on gibberellin uptake and distribution. In: Programme and Abstracts, 24th Mtg Aust Soc Plant Physiol, May, 1984, p 20 (abstract)

Robbins WJ (1957) Gibberellic acid and the reversal of adult *Hedera* to a juvenile state. Am J Bot 44:743–746

Robbins WJ (1960) Further observations on juvenile and adult *Hedera*. Am J Bot 47:485–491

Robbins WJ (1961) Juvenility and the induction of flowering. In: Recent advances in botany, vol 2, Univ Toronto Press, Toronto, pp 1647–1652

Robbins WJ (1964) Topophysis, a problem in somatic inheritance. Proc Am Philos Soc 108:395–403
Robbins WJ, Hervey A (1970) Tissue culture of callus from seedling and adult stages of *Hedera helix*. Am J Bot 57:452–457
Robinson LW, Wareing PF (1969) Experiments on the juvenile–adult phase change in some woody species. New Phytol 68:67–78
Robitaille H, Carlson RF (1971) Response of dwarfed apple trees to stem injections of gibberellic and abscisic acids. HortScience 6:539–540
Rogler CE, Dahmus ME (1974) Gibberellic acid-induced phase change in *Hedera helix* as studied by deoxyribonucleic acid-ribonucleic acid hybridization. Plant Physiol 54:88–94
Rogler CE, Hackett WP (1975a) Phase change in *Hedera helix:* Induction of the mature to juvenile phase change by gibberellin A_3. Physiol Plant 34:141–147
Rogler CE, Hackett WP (1975b) Phase change in *Hedera helix:* Stabilization of the mature form with abscisic acid and growth retardants. Physiol Plant 34:148–152
Ross SD (1976) Differential flowering responses by young Douglas-fir grafts and equisized seedlings to gibberellin and auxin. Acta Hortic 56:163–168
Ross SD (1978) Influences of gibberellins and cultural practices on early flowering of Douglas-fir seedlings and grafts. Proc 3rd World Consultation on Forest Tree Breeding, March 1977. Canberra, CSIRO, Vol 2, pp 997–1007
Ross SD, Greenwood MS (1979) Promotion of flowering in the Pinaceae by gibberellins. II. Grafts of sexually mature and immature *Pinus taeda* L. Physiol Plant 45:207–210
Ross SD, Pharis RP (1976) Promotion of flowering in the Pinaceae by gibberellins. I. Sexually mature non-flowering grafts of Douglas-fir. Physiol Plant 36:182–186
Ross SD, Pharis RP, Bollman M, Sweet GB (1984) Gibberellin $A_{4/7}$ and the promotion of flowering in *Pinus radiata:* Effects on partitioning of photoassimilate within the bud during primordia differentiation. Plant Physiol 76:326–330
Ross SD, Piesch RF, Portlock FT (1981) Promotion of cone and seed production in rooted ramets and seedlings of western hemlock by gibberellins and adjunct cultural treatments. Can J For Res 11:90–98
Ross SD, Pharis RP, Binder WD (1983) Growth regulators and conifers: Their physiology and potential uses in forestry. In: Nickell LG (ed) Plant growth regulating chemicals, vol. II, CRC, Boca Raton, FL, pp 35–78
Sachs RM (1977) Nutrient diversion: An hypothesis to explain the chemical control of flowering. HortScience 12:220–222
Sachs RM, Lang A (1957) Effect of gibberellin upon cell division in *Hyoscyamus*. Science 125:1144–1145
Sachs RM, Bretz CF, Lang A (1959) Shoot histogenesis: The early effects of gibberellin upon stem elongation in two rosette plants. Am J Bot 46:376–384
Sadik S, Ozbun JL (1968) The association of carbohydrate changes in the shoot tip of cauliflower with flowering. Plant Physiol 43:1696–1698
Saure M (1970) Beiträge zur Frühselektion beim Apfel. I. Beziehungen zwischen dem Eintritt der Blühfähigkeit und der Wuchsleistung von Apfelsämlingen. Mitt Obstversuchsringes, Jork 25:156–161
Sax K (1962) Aspects of aging in plants. Annu Rev Plant Physiol 13:489–506
Schaffner RH, Nagl W (1979) Differential DNA application involved in transition from juvenile to adult phase in *Hedera helix* (Araliaceae). In: Nagl W, Hemleben V, Ehrendorfer F (eds) Proc Symp genome and chromatin: organization evolution function. Plant Systematics and Evolution Supplement 2, 105–110
Schander H (1962) Variationsstudie über den Eintritt des Blühbeginns bei Apfelsämlingen. Tagungsber Deut Akad Landw Wiss Berlin 35:111–121
Schmidt H (1970) Beiträge zur Züchtung apomiktischer Apfelunterlagen. II. Dauer der juvenilen Periode in *Malus*-Kreuzungen. Z Pflanzenzücht 63:89–101
Schmidt H (1976) On the inheritance of the length of the juvenile period in interspecific *Prunus* hybrids. Acta Hortic 56:229–234
Schröck O (1949) Die Vererbung der Frühblüte der Kiefer. Züchter 19:247–254
Schwabe WW (1976) Applied aspects of juvenility and some theoretical considerations. Acta Hortic 56:45–56

Schwabe WW, Al-Doori AH (1973) Analysis of a juvenile-like condition affecting flowering in the blackcurrant (*Ribes nigrum*). J Exp Bot 24:969–981

Scorza R, Janick J (1980) In vitro flowering of *Passiflora suberosa* L J Am Soc Hortic Sci 105:892–897

Scurfield G, Moore CWE (1958) Effects of gibberellic acid on species of *Eucalyptus*. Nature (Lond) 181:1276–1277

Singh K (1981) Effect of light and some regulatory substances on morphophysiological and biochemical changes during growth and development of some trees. PhD thesis, Panjab Univ., Chandigarh, India, 160 pp

Sisa M, Higuchi H (1967) Studies on the shortening of the juvenile phase of tulip under the controlled environment. II. On the thermoperiodicity of juvenile tulip. J Jpn Soc Hort Sci 36:315–323

Skene KGM, Kerridge GH (1967) Effect of root temperature on cytokinin activity in root exudate of *Vitis vinifera* L. Plant Physiol 42:1131–1139

Smeets L (1966) Amount of growth and the duration of the juvenile phase in cherry seedlings. In: Proc Balsgard Fruit Breeding Symp., Fjalkestad, Sweden, pp 159–162

Smith DR, Thorpe TA (1975) Root initiation in cuttings of *Pinus radiata* seedlings. II. Growth-regulator interactions. J Exp Bot 26:193–202

Smith WH, Konar RN (1969) Initiation of ovulate strobili in cotyledon-stage seedlings of *Pinus elliottii*. Can J Bot 47:624–626

Spinks GT (1925) The treatment of seedling apple trees to induce early fruiting. J Pomol Hortic Sci 4:141–145

Srinivasan C, Mullins MG (1978) Control of flowering in the grapevine (*Vitis vinifera* L.). Formation of inflorescences in vitro by isolated tendrils. Plant Physiol 61:127–130

Srinivasan C, Mullins MG (1979) Flowering in *Vitis:* Conversion of tendrils into inflorescences and bunches of grapes. Planta 145:187–192

Srinivasan C, Mullins MG (1980) Effects of temperature and growth regulators on formation of anlagen, tendrils, and inflorescences in *Vitis vinifera*. Ann Bot 45:439–446

Srinivasan C, Mullins MG (1981) Induction of precocious flowering in grapevine seedlings by growth regulators. Agronomie 1:1–5

Sriskandarajah D, Mullins MG, Nair Y (1982) Induction of adventitious rooting in vitro in difficult-to-propagate cultivars of apple. Plant Sci Lett 24:1–9

Stahly EA, Piringer AA (1962) Effects of photoperiod, light quality, and two plant growth regulators on growth and flowering of Jonathan apple trees. Proc Am Soc Hortic Sci 81:12–17

Stein OL, Fosket EB (1969) Comparative developmental anatomy of shoots of juvenile and adult *Hedera helix*. Am J Bot 56:546–551

Stephens GR Jr (1964) Stimulation of flowering in eastern white pine. For Sci 10:28–34

Stern K (1961) Über den Erfolg einer über drei Generationen geführten Auslese auf frühes Blühen bei *Betula verrucosa*. Silvae Genet 10:48–51

Stokes P, Verkerk K (1951) Flower formation in Brussels sprouts. Meded LandbHogesch Wageningen 50:141–160

Stoltz LP, Hess CE (1966) The effect of girdling upon root initiation: carbohydrates and amino acids. Proc Am Soc Hortic Sci 89:734–743

Stoutemyer VT (1962) The control of growth phases and its relation to plant propagation. Proc Plant Prop Soc 12:260–264

Stoutemyer VT (1964) Juvenility and flowering potential in woody plants. Am Hortic Mag 43:161–167

Stoutemyer VT (1967) Juvenility in plants, its physiological basis and horticultural significance. In: Proc 17th Int Hortic Congr, Vol III, pp 409–418

Stoutemyer VT, Britt OK (1961) Effect of temperature and grafting on vegetative growth phases of Algerian ivy. Nature (Lond) 189:854–855

Stoutemyer VT, Britt OK (1962) Growth phases and the propagation of *Hedera*. Proc Am Soc Hortic Sci 80:589–592

Stoutemyer VT, Britt OK (1963) Tissue cultures of juvenile and adult specimens of ivy. Nature (Lond) 199:397–398

Stoutemyer VT, Britt OK (1965) The behavior of tissue cultures from English and Algerian ivy in different growth phases. Am J Bot 52:805–810

Stoutemyer VT, Britt OK (1969) Growth and habituation in tissue cultures of English ivy, *Hedera helix*. Am J Bot 56:222–226
Stoutemyer VT, Britt OK, Goodin JR (1961) The influence of chemical treatments, understocks, and environment on growth phase changes and propagation of *Hedera canariensis*. Proc Am Soc Hortic Sci 77:552–557
Sweet GB (1973) The effect of maturation on the growth and form of vegetative propagules of radiata pine. NZJ For Sci 3:191–210
Sweet GB (1979) A physiological study of seed cone production in *Pinus radiata*. NZJ For Sci 9:20–33
Sweet GB, Harris JM (1976) Wood properties of *Pinus radiata:* Seed-grown trees compared with grafts from different-aged ortets. NZJ For Sci 6:114–121
Takeno R, Taylor JS, Sriskandarajah S, Pharis RP, Mullins MG (1982/83) Endogenous gibberellin and cytokinin-like substances in cultured shoot tissues of apple, *Malus pumila* cv. Jonathan, in relation to adventitious root formation. Plant Growth Regulation 1:261–268
Teich AH, Holst MJ (1969) Genetic control of cone clusters and precocious flowering in *Pinus sylvestris*. Can J Bot 47:1081–1084
Tompsett PB (1977) Studies of growth and flowering in *Picea sitchensis* (Bong.) Carr. I. Effects of growth regulator applications to mature scions on seedling rootstocks. Ann Bot 41:1171–1178
Trippi VS (1963a) Studies on ontogeny and senility in plants. II. Seasonal variation in proliferative capacity in vitro of tissues from branches from juvenile and adult zones of *Aesculus hippocastanum* and *Castanea vulgaris*. Phyton 20:146–152
Trippi VS (1963b) Studies on ontogeny and senility in plants. III. Changes in the proliferative capacity in vitro during ontogeny in *Robinia pseudoacacia* and *Castanea vulgaris* and in adult and juvenile clones of *R. pseudoacacia*. Phyton 20:153–159
Trippi VS (1963c) Studies on ontogeny and senility in plants. IV. Activity of some enzymes at different stages of ontogeny and in clones from juvenile and adult zones of *Robinia pseudoacacia*. Phyton 20:160–166
Trippi VS (1963d) Studies on ontogeny and senility in plants. V. Leaf-fall in plants of different age of *Robinia pseudoacacia* and the effect of gibberellic acid on *R. pseudoacacia* and *Morus nigra*. Phyton 20:167–171
Trippi VS (1963e) Studies on ontogeny and senility in plants. VI. Reversion in *Acacia melanoxylon* and morphogenetic changes in *Gaillardia pulchella*. Phyton 20:172–174
Tromp J (1972) Effects of growth regulating substances and tree orientation on growth and flower bud formation in apple. J Hortic Sci 47:525–533
Tukey HB (1964) Dwarfed fruit trees, MacMillan, New York
Tydeman HM (1961) Rootstock influence on the flowering of seedling apples. Nature (Lond) 192:83
Tydeman HM, Alston FH (1965) The influence of dwarfing rootstocks in shortening the juvenile phase of apple seedlings. Ann Rep East Malling Res Sta for 1964 A 48, 97–98
Vieitez E, Vieitez AM (1976) Juvenility factors related to the rootability of chestnut cuttings. Acta Hortic 56:269–274
Visser T (1964) Juvenile phase and growth of apple and pear seedlings. Euphytica 13:119–129
Visser T (1965) On the inheritance of the juvenile period in apple. Euphytica 14:125–134
Visser T (1970) The relation between growth, juvenile period and fruiting of apple seedlings and its use to improve breeding efficiency. Euphytica 19:293–302
Visser T (1973) The effect of rootstocks on growth and flowering of apple seedlings. J Am Soc Hortic Sci 98:26–28
Visser T (1976) A comparison of apple and pear seedlings with reference to the juvenile period. II. Mode of inheritance. Acta Hortic 56:215–218
Wahdi M, Mohan Ram HY (1967) Shortening the juvenile phase for flowering in *Kalanchoe pinnata* Pers. Planta 73:28–36
Wareing PF (1959) Problems of juvenility and flowering in trees. J Linn Soc Lond Bot 56:282–289

Wareing PF (1961) Juvenility and induction of flowering. In: Recent advances in botany, vol 2, Univ Toronto Press, Toronto, pp 1652–1654
Wareing PF, Frydman VM (1976) General aspects of phase change, with special reference to *Hedera helix* L. Acta Hortic 56:57–69
Way RD (1971) Hastening the fruiting of apple seedlings. J Am Soc Hortic Sci 96:384–389
Wheeler NC, Wample RL, Pharis RP (1980) Promotion of flowering in the Pinaceae by gibberellins. IV Seedlings and sexually mature grafts of lodgepole pine. Physiol Plant 50:340–346
Williams IH (1961) Treatment of hops (*Humulus lupulus* L.) with gibberellic acid. Effect on growth and development. Nature (Lond) 191:361–362
Winton L, Huhtinen O (1976) Tissue culture of trees. In: Miksche JP (ed) Modern methods in forest genetics. Springer, Berlin Heidelberg New York, pp 243–264
Yadava UL, Dayton DF (1972) The relation of endogenous abscisic acid to the dwarfing capability of East Malling apple rootstocks. J Am Soc Hortic Sci 97:701–705
Zeevaart JAD (1983) Gibberellins and flowering. In: Crozier A (ed) Biochemistry and physiology of gibberellins, vol 2. Praeger Scientific, pp 333–374
Zimmerman RH (1963) Rooting cofactors in some southern pines. Proc Int Plant Prop Soc 13:71–74
Zimmerman RH (1971) Flowering in crabapple seedlings. Methods of shortening the juvenile phase. J Am Soc Hortic Sci 96:404–411
Zimmerman RH (1972) Juvenility and flowering in woody plants: A review. HortScience 7:447–455
Zimmerman RH (1972) Length of the juvenile period in some apomictic crabapples. HortScience 7:490–491
Zimmerman RH (1973) Juvenility and flowering of fruit trees. Acta Hortic 34:139–142
Zimmerman RH (1977) Relation of pear seedling size to length of the juvenile period. J Am Soc Hortic Sci 102:443–447
Zimmerman RH (1981) Micropropagation of fruit plants. Acta Hortic 120:217–222
Zwintzscher M (1964) Ein Versuch zur Verkürzung der juvenilen Phase junger Obstgehölze. Tagungsber Deut Akad Landw Wiss Berlin 65:21–28

4 Polarity

W.K. COLEMAN and T.A. THORPE

1 Introduction

Polarity can be defined as the bipolar, axiate character of organisms (BLOCH 1965). It is considered the first, visible morphological indication of an internal asymmetrical state within a living system and is a fundamental component of differentiation and spatial organization. Polarity appears to be present in all organisms at some stage of their life cycle and is currently assumed to be an outcome of a biochemical gradient or gradients within a cell or group of cells due to asymmetric environmental stimuli (Fig. 1).

As noted by WOLPERT, polarity is a neglected aspect in developmental studies (WOLPERT 1971, WOLPERT et al. 1971). This situation may have arisen due to a number of causes. For example, the literature appears to vary in its concept of polarity. SINNOTT (1960) referred to polarity as a specific reaction to a specific environment and not a specific cellular characteristic, while KÜHN (1971) recognized a minimum of three types of polarity. Similarly, BLOCK (1965) distinguished three types of morphological polarity: (1) equipolarity, i.e., morphologically similar ends; (2) bipolarity, i.e., morphologically dissimilar ends; and (3) radial polarity.

In this review we will consider polarity as outlined in Fig. 1 and examine some of the manifestations of polarity in single- and multi-celled systems with a view to delineating areas where future research may prove most rewarding in attempting to understand possible causes. Although more is now known about polarity in prokaryotes (see e.g., ADLER 1975, WOLK 1975a, MITCHISON et al. 1976, SHAPIRO 1976) and there are striking similarities between sensory perception in bacteria and in higher plants (ADLER 1975, MACNAB 1979), our review will concentrate on eukaryotic systems.

Previous reviews on the topic of polarity in plants are found in articles by BLOCH (1943, 1965), BÜNNING (1952, 1953), NAKAZAWA (1960, 1962), SINNOTT (1960), BONNER (1963), VON WETTSTEIN (1965), WARDLAW (1968) and QUATRANO (1978). An excellent article by WEISENSEEL on the induction of polarity, particularly in *Pelvetia* and *Fucus* zygotes, is found in a companion volume of this series (WEISENSEEL 1979). STANGE (1984) has recently contributed a lucid discussion of early differentiation events during plant development as an outcome of polarity.

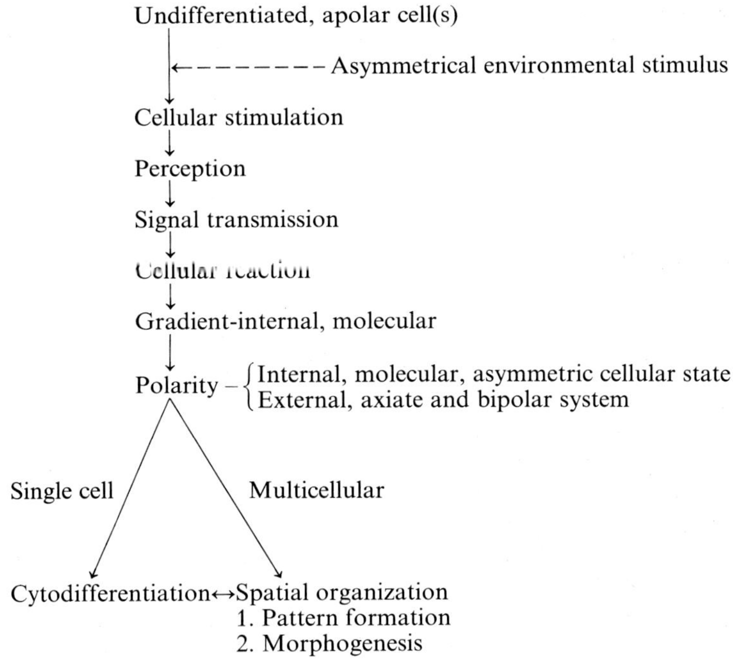

Fig. 1. Hypothetical relationship of polarity to differentiation and spatial organization in organisms

2 Single-Cell Systems

Cellular evolution of the eukaryotic cell has lead to the development of an increased size and internal complexity which has, in turn, allowed an increase in diversity of form. Whether eukaryotes arose (1) through a serial endosymbiosis involving the endocytosis and eventual symbiosis of three to six different prokaryotes, or (2) through the intracellular differentiation of a single prokaryotic species, remains highly speculative in view of the lack of a clear fossil record (e.g., CAVALIER-SMITH 1975). However, STEBBINS' (1974) suggestion that the ability of the cell to become polarized probably evolved in the earliest flagellated eukaryotes seems unlikely in view of the abundant evidence of polarized cellular development among the prokaryotes.

2.1 Fucoid Cells

In *Fucus* and *Pelvetia* zygotes, gradient vectors (see Table 1 of JAFFE 1968; see also WEISENSEEL 1979) such as light, temperature, gravity, pressure, calcium ions, electric current, etc., are capable of polarizing the zygotes which results in a rhizoidal bulge in a highly predictable location within 8–14 h after fertilization. In the absence of environmental gradients, the rhizoidal bulge in *Fucus* zygotes can develop at the point of entry of the spermatozoid. This feature may be generally present in the zygotes of the Fucales (JAFFE 1968, QUATRANO 1974, NAKAZAWA 1975).

The earlier view of WHITAKER (1931–1940, as cited by WARDLAW 1968), that a labile and readily alterable polarity is already established (e.g., during oogenesis) and is subsequently fixed after the reception and integration of environmental signals has been disputed by JAFFE (1968). JAFFE found that approximately half of the *Fucus* zygotes cultured in plane-polarized light initiate rhizoids at two opposite points on the surface of a zygote and in the electrical vector plane of the polarized light. He contends that it is conceptually difficult to explain how a hypothetical invisible axis could rotate in both directions at once and that, if it exists, it could not be a fixed structure rotated by environmental vectors.

A similar polarotropic response has been observed in spores of the fern *Dryopteris felix-mas*. When polarized light is directed toward the spores, the germ tube develops in a direction parallel to the electrical vector of the polarized light (BOPP 1968). This observation suggests that the photoreceptors are organized with respect to each other in such a way that their chromophores lie parallel (SMITH 1975). The nature of the chromophores and the possible method of achieving a precise spatial organization in both the fucoid zygote and fern spore systems are presently unknown, although possible candidates for fixing the photoreceptors include the plasmalemma, microtubules, and/or microfilaments. Further aspects of these phototropic problems can be found in HAUPT's (1965) outline of early work with lower organisms on the photoinduction of polarity and in more recent reviews dealing with problems of photoreception (BRITZ 1979, DENNISON 1979, HÄDER 1979, SHROPSHIRE 1979, WEISENSEEL 1979).

The role of plant hormones in the normal development of fucoid zygotes is unclear. JAFFE (1968) suggests that the tendency of *Fucus* zygotes to germinate toward each other (i.e., the so-called "positive group effect") may be due to a highly localized, short-lived, and diffusible substance ("rhizin") which stimulates rhizoid initiation (see also JAFFE and NEUSCHELER 1969). This substance may be indole-3-acetic acid (IAA) (JAFFE 1968). However, KLEMBE and BENTRUP (1973) disagree with this suggestion on the basis of their finding that IAA had no positive group effect. On the contrary, nucleotides such as AMP and cAMP may partially behave like rhizin although the mechanism is unknown. Interestingly, in *Funaria* spores, the induction of polarity is inhibited by externally applied IAA. This treatment was compared with the effect of colchicine and results in giant and apparently apolar cells (VON WETTSTEIN 1965, the results were confirmed by SCHNEPF 1982).

Electrical potentials have been measured across individuals and populations of *Fucus* and *Pelvetia* zygotes during their development (JAFFE 1968, 1969, JAFFE et al. 1974, NUCCITELLI and JAFFE 1974, 1976). JAFFE hypothesized that the self-generated electric current may be both a cause as well as a consequence of "localization". The developmental phenomena of localization in fucoid zygotes involves the gradual establishment of a polarity which is amplified to the point of determination (or "fixed" irreversibly) with subsequent rhizoidal formation (JAFFE 1968, 1981). JAFFE's hypothesis may be summarized as follows:

1. Hyperpolarization of the cell membrane after fertilization with changes in membrane potential due to KCl accumulation.

2. A localized membrane change occurs in response to a directional or asymmetrical environmental stimulus. The membrane change involves an increased leakiness to certain cations which are normally at a much higher electrochemical potential outside the cell.

3. An entry of these cations (e.g., Ca^{2+}; ROBINSON and JAFFE 1975, 1976, NUCCITELLI and JAFFE 1976) occurs at the prospective rhizoidal site and cation efflux at the opposite end of the cell with the resulting generation of an electric current. In *Pelvetia* zygotes, this current becomes measurable about 1 h after fertilization (NUCCITELLI and JAFFE 1976) and may be due to redistribution of Ca^{2+} pumps and regions of increased membrane leakiness.

4. Cytoplasmic resistance to electric current would generate a cytoplasmic field that is relatively positive under the leaky site of the membrane. This field would, in turn, generate movement of negatively charged cytoplasmic constituents (e.g., vesicles) toward the leaky region, with a resulting increased leakiness of the membrane. This positive feedback loop of self-electrophoresis would tend to further polarize the developing zygote (JAFFE et al. 1974). Actual measurements of single cell current (intracellular in *Fucus* zygotes and externally recorded currents through *Pelvetia* zygotes) indicate that brief periods of no current flow alternate with periods of large flow. These current fluctuations (or "growth pulses") are believed to be due to either a rapid change of a relatively immobile substance (e.g., calcium ions) or to a rapid change in membrane properties (e.g., insertion of a vesicle into a patch of membrane), leading to rapid opening and closing of many ion channels or pumps (NUCCITELLI and JAFFE 1974). JAFFE (1981) has suggested that the calcium ion may be the controlling element of these "developmental" currents.

Actual observations of ultrastructural changes during induction and fixation of polarity due to an applied light gradient in fucoid zygotes were made by QUATRANO (1972) and PENG (1976). Following the determination of polarity (i.e., specification of the future rhizoid site), a localized cytoplasmic accumulation of ribosomes, osmiophilic and fibrillar vesicles, mitochondria and membrane extensions from the nuclear envelope were observed to radiate toward the site of rhizoid formation (QUATRANO 1972). No ultrastructural localizations were observed prior to the fixation of polarity. In a subsequent study, PENG (1976) examined freeze-fracture replicas of the cell membrane of *Pelvetia* zygotes. He observed a redistribution of isolated intramembrane particles (approx. 10 nm wide), as well as apparent insertion of new membrane patches into the cell membrane during polarization.

NOVOTNY and FORMAN (1974) presented evidence to suggest that the localized deposition of the sulfated polyfucose carbohydrate, fucoidin, into the structure of the zygote wall at the future site of the rhizoid was accompanied by the determination of axis polarity. Although these two events were closely associated (i.e., localized deposition and axis commitment), prior sulfation of fucoidin within the Golgi apparatus did not appear to play a role in the establishment of a polar axis (NOVOTNY and FORMAN 1974, CRAYTON et al. 1974, QUATRANO 1978). When a high external osmotic pressure was applied to developing *Fucus* zygotes, no cellular expansion occurred. However, sulfated fucoidin accumulated at the future rhizoid pole similar to the normal situation (VREUGDENHIL et al. 1976). The control of the polar deposition of fucoidin has been discussed by QUATRANO et al. (1979).

These observations support WEISENSEEL's (1979) suggestion that the establishment of cell polarity in fucoid zygotes may involve both the induction of a labile polarity, e.g., the asymmetrical distribution of membrane-bound Ca^{2+} pumps (possibly during oogenesis as originally suggested by WHITAKER) and the reorientation of this labile polarity in response to an external stimulus which has been amplified to the point of determination by a "self-electrophoresis" mechanism.

Although a self-generated electric current may be a crucial and possibly widespread feature of development processes (JAFFE and NUCCITELLI 1977; JAFFE 1981), the influx and efflux of cations as they affect intracellular pH and subsequent metabolic activation, and the induction and stabilization of polarity remain to be elucidated (e.g., see JOHNSON et al. 1976). Nevertheless, the relative ease of fucoid zygote manipulation has led to precise biochemical and biophysical examinations which have, in turn, allowed the construction of a suitable hypothesis or "micro-theory" (BONNER 1963). This hypothesis accounts for the development of polarity at the physiochemical level and postulates that a cytoplasmic electrical potential gradient arises from changes in ion transport through cellular membranes.

2.2 Ochromonas

The flagellated Chrysophycean alga, *Ochromonas danica*, exists as a naked protoplast (SMITH 1955), since there is no indication of a cell wall, sheath, lorica, pellicle or other structure which might function as a cytoskeleton. Nevertheless, *Ochromonas* maintains an asymmetrical, species-specific shape in the vegetative phase of its life cycle. Furthermore, BOUCK and BROWN (1973) describe a close correlation between cell shape and the distribution of two groups of cytoplasmic microtubules. The latter are associated with two distinct intracellular sites or nucleating centers in the anterior region of the cell's cytoplasm (BROWN and BOUCK 1973). They found that cell shape was disrupted by colchicine or high hydrostatic pressure which resulted in spherical cells with disorganized tubules. Release of pressure or washing out of the colchicine resulted in the regeneration of the species-specific shape. Reassembly of the microtubules was correlated with this shape regeneration and the authors suggest that microtubules may play an obligatory role in both developing and maintaining shape in *Ochromonas*. They further hypothesized that nucleating centers temporally and spatially control initiation of tubule assembly – possibly by serving as templates for specifying points of individual microtubule initiation. The mechanism which controls the temporal and spatial organization of microtubule nucleating regions is unknown (HEPLER and PALEVITZ 1974), although a microtrabecular lattice may play a determining role (PORTER and TUCKER 1981). Additional discussions on microtubules and cell shape are found in FILNER and YADAV (1979) and Volume 10, this Series.

2.3 Single Cell Systems of Higher Plants

In comparison with recent advances among the unicellular eukaryotic algae in our understanding of the cellular basis of polarity, progress in solving this fundamental biological problem in higher plants has been minimal.

The vegetative cell of the pollen grain is a useful higher plant system for examining the induction of polarity. In some species this cell is apparently apolar and external factors (e.g., centrifugation) can determine the side from which the pollen tube will emerge (BEAMS and KING 1944b). When pollen grains germinate close together, the tubes tend to emerge so that they are directed away from each other (BEAMS and KING 1944a). This suggests that diffusion of substances from the pollen grain may affect polarity of germination (see MASCARENHAS 1974). Imposed electrical currents also influence pollen tube germination by orienting the outgrowth parallel to the applied field (WEISENSEEL et al. 1975). In SIEVERS and SCHNEPF (1981) one finds very interesting results in comparison of the polarity of germinating pollen grains to other filamentous structures like root hairs and moss protonema. The use of the vibrating electrode has led to the finding that there is a natural transcellular current in pollen grains of *Lilium* (WEISENSEEL 1977). A self-generated current starts to cross the cell 0.5 h after sowing the pollen in a liquid germination medium. The current density rises from zero to ca. 4 μamp cm^{-2} within 3 h and germination begins at the site where current enters the cell. This system needs to be explored further.

Numerous workers have studied the differentiation of xylem tissue in tissue culture (ROBERTS 1969, 1976, TORREY et al. 1971). However, only recently have advances been made in developing a cell suspension culture system possessing such desirable features as (1) initial population of uniform, homogeneous and single cells which would, (2) initiate, in a synchronous manner, a specific cytodifferentiation event [e.g., formation of tracheary elements (ROBERTS 1969)] in response to a specific, externally applied signal. For example, KOHLENBACH and SCHMIDT (1975) and TORREY (1975) have provided indirect evidence that cells in suspension culture differentiate into tracheary elements without intervening mitosis. FUKUDA and KOMAMINE (1980) have produced direct evidence using single cells isolated from the mesophyll of *Zinnia elegans*. TURGEON (1975) also observed that differentiation of wound vessel members can occur without the necessity of DNA synthesis, mitosis or cell division and, hence, no casual association can be inferred between differentiation and cell division. He speculates that cell doubling may be a form of homeostatic mechanism which insures that progenitor cells are reproduced before differentiation occurs and, as a consequence, the integrity of the progenitor cell line is preserved. THORPE (1980), in discussing the importance of cell division for initiation of organized development, points out that tracheary element formation represents the logical culmination of differentiation for all parenchyma cells, and therefore is not an altered pathway of differentiation, but a resumption of the process.

For isolated higher plant cells in culture, cytodifferentiation systems, such as tracheary element formation, are in a very early stage of development. Consequently, specific questions and experimental approaches to investigate the formation of gradients within the cells have not been formulated. Such gradients, caused by asymmetric environmental factors, would lead to the establishment of polarity and subsequent cytodifferentiation.

3 Multiple Cell Systems

Development of the multicellular condition, and particularly the multicellular organism, was probably a frequent event during the evolutionary development of life on this planet (DOBZHANSKY et al. 1977). BONNER (1974) contends that the two basic selection pressures involved the efficiency of energy intake and use of that energy to make more offspring. CAVALIER-SMITH (1975), however, notes that adhesion developed between extracellular layers of complex polysaccharide-protein cell walls, or surface coats, and that this may have been the physical, causal feature which led to multicellularity. Regardless of the phylogenetic origins, however, the multicellular condition has led to increased morphological complexity and the need for intercellular communication to regulate the organism's development. We are concerned in this review with two aspects of multicellular development: (1) pattern formation, or the specification of spatial differences among a population of cells, which leads to overt cell differentiation in either a whole body pattern, or repeated pattern (COOKE 1975), and, (2) morphogenesis, or the generation of form among a cell population, where form is viewed as resulting from a system of polarities set up in the developing organism (SINNOTT 1960).

3.1 Dictyostelium [1]

Cellular slime mold development presents the biologist with a favorable system for studying molecular and genetic aspects of polarity in the multicellular state. As LOOMIS (1975) and SUSSMAN and BRACKENBURY (1976) point out, *Dictyostelium* possesses a number of desirable experimental traits: (1) rapid and developmentally simple life cycle; (2) small size and ease of handling; (3) defined conditions for growth and developmental synchrony which are necessary prerequisites for experimental reproducibility; and (4) isolation and genetic analysis of numerous morphogenic mutants. A comprehensive review of *Dictyostelium* development is available (LOOMIS 1982).

In *Dictyostelium discoideum*, (POFF and WHITAKER 1979) the life cycle alternates between a single-celled amoeba and a multicellular morphogenic phase

1 A more detailed article on this slime mold can be found in Vol. 17, this series (Eds. LINSKENS HF and HESLOP-Harrison J), pp. 309–332. (KONIJN and VAN HAASTERT: Cell interactions in the cellular slime molds)

which is formed by aggregation of numerous amoebae. The aggregate develops either directly into a fruiting body or into a migrating slug and then a fruiting body, depending upon environmental cues perceived by the mass of amoebae (SUSSMAN and BRACKENBURY 1976). The development of an axial polarity in the aggregate, and subsequent migration of the resulting slug, appear to involve a phototactic response in which maintenance (and possibly initiation) of polarity may be due to localized accumulation of cAMP within the aggregate. The presence of a thin protein and polysaccharide sheath (surrounding the slug) which possesses differential permeability is also necessary for maintenance of directional polarity (SUSSMAN and BRACKENBURY 1976).

BONNER (1963) notes that it is impossible to interpret induced slug polarity in terms of the polarity of individual amoebae of the slug, and surgical manipulations of RAPER (RAPER 1940 cited in SINNOTT 1960 and SUSSMAN and BRACKENBURY 1976) would tend to support this idea. Results by STERNFELD and DAVID (1981) have shown that the polarity is predetermined.

Early observations suggested that the developmental fate of individual cells is determined by their positions relative to the polar axis within the slug. The apical cells (i.e., the anterior 20%) differentiate into stalk cells, while the remaining cells become spores. This proportion remains constant regardless of slug size. Furthermore, pre-spore and pre-stalk cells demonstrate reversible development without the necessity of an intervening cell division (BONNER 1974).

Recent evidence suggests that differentiation of the two basic cell types occurs in the late aggregation phase and prior to formation of an axiate slug polarity (HAYASHI and TAKEUCHI 1976). Furthermore, GARROD and FORMAN (1977) have observed an apparently normal pre-stalk pre-spore pattern in spherical aggregates showing no morphological polarity. They suggest that the pre-stalk pre-spore pattern is responsible for generation of an axiate slug polarity, rather than morphological polarity generating the pattern.

On the basis of various observations, it would appear that the following sequence of events is plausible:

1. The developing aggregate possesses a gradient of cAMP, possibly maintained by elaborate feedback regulatory loops (see SUSSMAN and BRACKENBURY 1976, NEWELL 1977, ROOS et al. 1977).

2. A reversible commitment of an individual amoeba to differentiate into one of the two basic cell types may occur in the late phase of aggregation due to its relative position within the aggregate and in response to one or more chemical or biophysical gradients. For example, high cAMP levels are capable of directly inducing apparently uncommitted amoebae to differentiate into stalk cells (BONNER 1974) while NH_3 is proposed to inhibit the formation of stalk cells and promote spore formation by antagonizing cAMP production (GROSS et al. 1981, RUTHERFORD et al. 1982). To date, there appear to be at least four diffusible "morphogens" whose concentrations and distribution may be controlled, at least in part, by the variable thickness of the sheath (GROSS et al. 1981, WILLIAMS 1982). The role of cell sorting, if any, in pattern formation (i.e., the spatial segregation of specific cell types) after commitment to a particular differentiation pathway by individual amoebae is unclear (e.g., BONNER 1974, SUSSMAN and BRACKENBURY 1976, GARROD and FORMAN 1977, GROSS et al. 1981).

3. The aggregate, containing reversibly committed pre-stalk and pre-spore cells, may undergo a morphogenic sequence (involving cell movement) directly into a fruiting body, or it may develop an axiate polarity (i.e., a slug form) in response to external, asymmetrical environmental factors such as light or temperature gradients. Again, we find that the physical and chemical nature of cellular stimulation, perception and signal transmission remain unknown although three general models have been proposed (RUTHERFORD et al. 1982).

3.2 Higher Plant Systems

Plants respond to the external environment through developmental responses, usually by modifying the activity of continuously embryonic regions such as root or shoot apices (SUSSEX 1974). Under extreme environmental conditions where there is a loss of plant tissues or organs, many species regenerate new components. Consequently, the necessity of an efficient intercellular communications system for integrating and regulating the diverse events of a multicellular organism's development is readily apparent.

TOMKINS (1975) speculated that the "necessity" of intercellular communications led to the evolution of relatively metabolically stable hormones or "specific effector molecules" which are secreted from environmentally stimulated cells ("sensor" cells) to more protected cells ("responder" cells) within the multicellular organism. The hormones would apprise responder cells of concentrations of intracellular symbols in sensor cells through a decoding of the metabolic message or hormone into corresponding intracellular symbols (such as the cyclic nucleotides, cAMP or cGMP) within the responder cells. The responder cells might, in turn, initiate or modify specific biosynthetic events in response to this information. However, as TOMKINS (1975) has pointed out, such formulations are essentially speculative and should be regarded, at least for the moment, primarily as "a pedagogical device for organizing a number of facts about cellular control" (p. 760). Certainly the function of cyclic nucleotides as components of a cellular communication network in higher plants remains very hypothetical.

3.2.1 Embryogenesis in Seed Plants

A continuous cycle of positionally transmitted polarity links each generation with the next one in seed plants (STEBBINS 1974). The development of polarity in the zygote appears to be due to the inductive influence of surrounding maternal tissues (i.e., heterogenetic induction; LANG 1974). WARDLAW (1968) observed that polarity of the developing embryo may be due to hormonal and nutritional gradients as well as biophysical components such as osmotic potential of ovule sap and possibly pressure from surrounding tissues. High osmotic potentials due to reducing and nonreducing sugars have been observed in developing ovules and experimental studies using isolated embryo culture (particularly *Capsella*) have documented high osmotic requirements for successful culture of early developmental stages. However, these osmotic requirements may be replaced by specific hormonal mixtures at low concentrations (TORREY 1966). Constantly changing gradients in nutrient levels are also implicated in normal polar development of isolated *Capsella* embryos in culture (MONNIER 1976, 1978). To date, however, successful isolation and culture of *Capsella* embryos smaller than the early globular stage (approximately 50 μ in diameter) has not been achieved.

Recent work by LINTILHAC has presented biologists with a novel approach to the study of mechanical stress as a component (and possible triggering force) for morphogenic events (LINTILHAC and JENSEN 1974, LINTILHAC 1974a, b). LINTILHAC's techniques involved a histological analysis of cotton ovule structure, responses of nucellar tissue in culture with and without the constraints of integuments and photoelastic stress analysis of a two-dimensional plastic model of a sectioned cotton ovule. Comparing cell-wall patterns in the nucellar region of developing ovules with compression and tension stress trajectories in the plastic model gave a high level of coincidence and appeared to be due to pressures generated by the surrounding integuments. LINTILHAC (1974a) observed that integuments ensure that compressive stresses generated during growth of the nucellus do not collapse the embryo sac but pass around it, leaving it as a compression-free space within the growing ovule. The role of these compression and tension stresses in the initiation or maintenance of polarity within the developing zygote is presently unknown. [A better system for studying pressure effects on the plane of cell division may be found in the coenocytic alga, *Caulerpa prolifera*. According to JACOBS (1970), a cross-wall can be induced in this organism by very brief local pressure, followed by a regeneration time of 60 s!].

The ability of somatic tissues in culture to progress through a relatively normal, species-specific, embryogenic sequence (although considerable variation in cellular pattern or segmentation of the somatic embryo is evident) suggests that pressure effects per se may play no direct role in polarized growth and development of the young embryo. Rather, sequential alteration in low concentrations of specific phytohormones appears to be critical for initiation and development of somatic embryos, e.g., auxin (see KOHLENBACH 1978) or gibberellins (NOMA et al. 1982, KOSHIOKA et al. 1983). Phytohormones may play a similar role in zygotic embryo development. Work with *Phaseolus coccineus* had led to the conclusion that the suspensor is the site of phytohormone production, including gibberellins, which are needed for zygotic embryo development (ALPI et al. 1975, 1979, CIONINI et al. 1976). The roles of hormonal or nutritional gradients as they affect polarized development of developing somatic or zygotic embryos remain to be elucidated. Future developments in tissue culture techniques should allow the problem to be explored experimentally. For reviews on somatic embryogenesis see TORREY (1966), WARDLAW (1968), HALPERIN (1969), REINERT (1973), STREET and WITHERS (1974), REINERT and BAJAJ (1977), KOHLENBACH (1978), WETHERELL (1979), and SHARP et al. (1980).

FRY and WANGERMANN (1976) hypothesized that the initiation of polarized auxin transport in the previously apolar globular embryo toward the point of attachment of the suspensor determined the morphological polarity of the embryo. However, as GOLDSMITH (1977) points out, the factors leading to development of the cell asymmetry that produces polar transport remain a mystery.

More specifically, the role played by hormone receptors strategically located in the cell's plasmalemma and, in particular, their ability to be induced or move within the membrane in response to directed hormonal signals and, hence, to initiate an asymmetrical cellular state, remain to be determined (see reviews on hormone receptors by VENIS 1973, KENDE and GARDNER 1976, LIBBENGA 1978, RUBERY 1981).

3.2.2 Growth, Development and Regeneration in Higher Plants

As the embryo grows, it undergoes phenotypic diversity with formation of shoot and root apices as morphologically discrete and relatively autonomously functioning structures. They serve two primary developmental roles: (1) a continuous source of new cells; and (2) as "organizers" or possible control points for differentiation of various organ and cell types in a polar and species-specific pattern (e.g., see reviews on apical organization and possible functions by TORREY 1966, 1972, WARDLAW 1968, GIFFORD and CARSON 1971, CLOWES 1972, 1976, PHILLIPS 1975, LYNDON 1976, HALPERIN 1978).

TORREY (1972) has hypothesized a simple model for polarized development in the plant in which root apices would be primary sites of cytokinin biosynthesis and shoot apices would be primary sites of auxin biosynthesis. Consequently, the earliest polarization in the developing embryo would involve the cellular separation and elaboration of such biosynthetic sites in the cell populations comprising the respective apices. Although differential concentrations of various hormones, their metabolic interconversions and preferential transport pathways (both polar and apolar) among various tissues of root and shoot axes have been demonstrated (e.g., see reviews by SHELDRAKE 1973, TORREY 1976, GOLDSMITH 1977), unequivocal evidence for localized biosynthesis of various hormones in shoot or root apices is absent. Consequently, this hypothesis remains unproven.

According to GREEN (1974), the biological basis for continuous generation of the cylindrical form in tip growth (e.g., root apices) may reside in coordinated gradients in the rate of expansion of the growing surface. This surface stretches and causes the strain alignment of the cellular machinery involved in transverse microfibril synthesis. This is believed to be the ultrastructural basis of elongation in subapical regions of roots. The cause of this "directed strain-generating gradient" is unknown, although GREEN (1974) suggests that in multicellular apices, gradients in phytohormone concentration may be an important component.

Shoot apices are more complicated morphological structures than root apices. They undergo regular and periodic shape changes associated with leaf formation in discrete (i.e., phyllotactic) patterns. Numerous hypotheses have been advanced for this distinctive spatial patterning – often on the basis of various surgical manipulations of shoot apices in vivo and in vitro, localized hormone applications, or through the use of simultated models (e.g., see SNOW and SNOW 1962, WARDLAW 1968, RICHARDS and SCHWABE 1969, SCHWABE 1971, ADLER 1974, WILLIAMS 1974, THORNLEY 1975, LYNDON 1976, MAKSYMOWYCH and ERICKSON 1977a, b, MITCHISON 1977, VEEN and LINDENMAYER 1977). Again, however, we find no concrete evidence for choosing among the various hypothesis – although unspecified, diffusible substances or "morphogens" are often implicated in simple threshold models similar to that proposed for regular spacing of heterocysts in *Anabaena* (FOGG 1949).

The presence of polarity in a tissue system is perhaps best exemplified in organ regeneration. VÖCHTING, as far back as 1878, showed that cuttings of willow stem form roots at their physiological basal ends, and shoot buds at their physiological apical ends, regardless of orientation during rooting. He also observed that bark-ringed cuttings responded as separate cuttings in terms of organ regeneration patterns. Furthermore, irrespective of the size of the cutting, fixed patterns of regeneration were always observed (SINNOTT 1960). Similar fixed patterns of regeneration have also been observed in root cuttings of *Taraxacum* (WARMKE and WARMKE 1950). These and similar findings suggest that once polarity is established in a multicellular system it remains fixed.

The use of tissue and cell cultures promises to be a useful tool in looking at polarity in relation to organized development. The role of concentration gradients in determining the site of xylem tracheid induction was demonstrated in callus of *Syringa vulgaris* (WETMORE and SOROKIN 1955). Xylem elements were formed at specific locations in the callus by placing an excised bud of *Syringa* into a V-shaped notch in the callus. Vascularization was also enhanced by feeding NAA into the notch. The site at which the vascular strands arose could be altered by changing the level of NAA; the higher the concentration, the further away from the notch the xylem arose. Further work (WETMORE and RIER 1963, JEFFS and NORTHCOTE 1967) also implicated sugar levels in determining the ratio of xylem to phloem formed. Similarly, concentration gradients of materials out of the culture medium into the tissue have been implicated in determining loci at which meristemoids and subsequently shoot primordia are induced in tobacco callus (ROSS and THORPE 1973). This evidence was obtained by inverting the callus at different times during culture, which led to the formation of shoot primordia in different positions in the callus.

During regeneration of organs in callus or cell colonies, different species tend to form embryos, or shoot and root primordia independently. Why an entire embryo or half an embryo (i.e., a shoot or root primordium) is formed is unknown. Shoot or root primordium induction requires precise balances of growth regulators in the medium. However, in numerous cases, it is necessary to transfer the tissue to a medium with an altered phytohormonal balance (often a minus phytohormone medium) for full organ development (THORPE 1980). Evidence is now clear that while the initiation of somatic embryos requires high exogenous auxin, their development, in most cases, will take place only in its absence, implying the need for reduced endogenous levels of auxins (KOHLENBACH 1978). During somatic embryogenesis or organogenesis, it is presumably the precise balance of substances, including phytohormones, at particular loci within the cell mass that permits the initiation of organized development. This initiation apparently begins with changes in a single cell within a mass of cells, and the process is associated with the phenomena of (1) cell dedifferentiation, (2) cell interaction and, (3) reaction to specific signals (see THORPE 1980).

Although tissue or organ regeneration remains the best demonstration for the expression of polarity in multicellular systems (both plant and animal), there are problems such as:

1. asynchronous development and imprecise description of organ or tissue siting (i.e., inadequate experimental systems);

2. lack of numerous morphogenic mutants which are amenable to genetic analysis;

3. technical limitations in the evaluation of hormonal or nutritional gradients within the potential "target" cell(s); and,

4. a failure to evaluate theoretical models in terms of actual physiological mechanisms.

In plants, relatively "simple" developmental systems such as xylem induction in *Syringa* callus (WETMORE and SOROKIN 1955), differentiation of wound vessel members in *Coleus* stem explants (ROBERTS 1976) and organogenic responses of epidermal layers in culture (TRAN THANH VAN et al. 1974, TRAN THANH VAN and TRINH 1978; see also HICKS 1980) may provide necessary starting material for future studies on polarity.

4 Models for Polarity

Advances in polarity modeling and subsequent form change or pattern formation in multicellular organisms have been developed in the context of zoological systems. However, SUSSEX (1974) suggests that plants and animals appear to undergo similar basic cellular events during their development. Furthermore, model construction is important if useful questions are to be posed at the molecular and biochemical levels. At the very least, models should provide a stimulus to experimentation by changing the vantage point from which developmental problems are viewed.

Meristematic regions of the plant (or "promeristems"; see SACHS 1972) can be viewed as regulative systems (although of a temporary nature in determinate organs such as the leaf) where the meristems always produce a distinct type form regardless of the number of cellular units or which parts are removed (e.g., see articles on the surgical manipulation of shoot and leaf meristems by SUSSEX 1964, WARDLAW 1968, SACHS 1969, STEEVES and SUSSEX 1972). These studies indicate that intercellular signaling of positional information is a critical aspect of the organism's development (see also MCDANIEL and HSU 1976).

CRICK (1971) observed that there are two basic positioning mechanisms which may be involved in form change or pattern formation as an outcome of the establishment of one or more gradients (see also WEBSTER 1971, WOLPERT et al. 1971, WOLPERT 1981):

1. Random Walk Mechanisms. For example, tissues may set up a gradient of a diffusible "morphogen" (e.g., a hormone) in response to a perturbation from the surrounding environment (i.e., a source-sink diffusion mechanism; SACHS

1978). Within the tissue, each cell would recognize the local concentration of hormone and this absolute level would enable the cell to "discover" its location in the tissue. A homeostatic mechanism for maintaining hormonal concentrations in initially perceived levels may occur through a closed series of feedback loops controlling synthesis, sequestering (e.g., partitioning or conjugation), release, or catabolism of the hormone. Difficulties in this model have been encountered in animal systems when attempts are made to explain the determination of tissue response in terms of absolute value of the gradient (GOODWIN 1971). Nevertheless, a gradient would possess two types of information: (1) it is polarized with respect to direction of slope; and (2) it varies quantitatively down its length. Such gradients could be established in more than one dimension although various considerations (e.g., properties of diffusion and the time available) suggest that gradients may be limited to very localized effects involving small numbers of cells (CRICK 1970, 1971, LAWRENCE 1971). These localized effects would be defined in terms of a "field" which is the set of cells that has its position specified with respect to some set of points (WOLPERT 1971). Hypothetical gradients of diffusible morphogens have been consistently popular in attempts to describe the regulative properties of shoot meristems (e.g., WARDLAW 1968). More recently, the concept of positional information has been applied to various morphogenic events in the plant's life cycle to provide a formal, unifying framework within which one can explain or at least describe such problems as the role of plant growth substances in the interpretation of a cell's positional value (HOLDER 1979). HOLDER (1979) has suggested that plant hormones may be involved in the cell's *interpretation* of its positional value and are not involved in the *specification* of the positional value (i.e., as morphogens). However, his conclusions are based on the observation that high concentrations of hormones are used in in vivo and in vitro experiments and "do not cause gross aberrations in the cellular patterns" (p. 198). Further, if "hormones were establishing positional values, they would need to be capable of producing subtle concentration differences within limited areas of cells to ensure that one cell knew that its value was different from its neighbors, and these would have to be constant from one plant to another in the same species" (p. 198). These comments do not invalidate the concept of plant hormones as morphogens, since we currently lack the necessary quantitative observations on plant hormone distribution within small cells or cell clusters which are undergoing a specific developmental event. The large number of tissue explant and culture systems cited by HOLDER (1979) and WARREN WILSON (1980) would support the latter's interpretation that the hypothetical morphogens possess hormonal characteristics although their true nature is unknown. Consequently, the major problems facing all developmental biologists remain; (1) the nature of diffusible substance(s) involved, and (2) an actual demonstration of the gradient (LAWRENCE 1971, LANG 1974).

2. Signalling Mechanisms. Two types are envisioned. (a) Intrinsic polarity where cells and tissues of the organism possess a pre-established, internal asymmetry or alternatively, a refractory period within each cell which, after reception of a morphogenic signal, prevents that cell from receiving further signals from adjacent cells. Both possibilities would lead to a directed diffusion or transport of the morphogen. A directed or polar transport of various plant hormones (i.e., auxins, GA, cytokinins, and ABA) has been described in various tissues and, in the case of auxins, various mechanisms have been postulated for the necessary cell-to-cell transport on the basis of a pre-existing structural polarity (e.g., see CANDE and RAY 1976, GOLDSMITH 1977). This type of signaling mechanism probably forms the basis of the polar regeneration of specific tissue and organ types. (b) A phase-shift model was developed by GOODWIN and his associates in order to generate an axis along which distance could be measured by cells in a multicellular array (GOODWIN and COHEN 1969, GOODWIN 1971, COHEN 1971, CRICK 1971, WOLPERT et al. 1971). This model emphasized the four-dimensional nature of developmental processes by the introduction of a temporal component. This component is based on at least one self-sustaining oscillatory biochemical process which originates in a "pacemaker" region (i.e., the cell with the highest autonomous frequency) and is propagated as a synchronizing wave(s) through functionally coupled cells. Initiation of a second biochemical event in the pacemaker region, and subsequent propagation at a slower velocity after a defined time interval from the first biochemical event, could generate an axis along which the distance could be measured in terms of the time between the occurrence of the two events. Two potential limitations in this theory are: (1) there is a requirement for an initial heterogeneity in the system in order to site the "pacemaker" region; and, (2) resolution may not be precise enough for divergent differentiation of adjacent cells (e.g., in stomata or trichoblast formation). Nevertheless, the wide-spread occurrence of self-sustaining and oscillating biochemical processes in organisms, and the ability to apply this theory to recently described "pacemaker" cells (i.e., cells producing pulses of cAMP) during the aggregation phase of *Dictyostelium*'s life cycle (ASHWORTH 1971, SUSSMAN and BRACKENBURY 1976, NEWELL 1977) suggest that future experimental approaches based upon hypothetical models may prove very rewarding for the developmental biologist.

5 Conclusions

We continue to increase our knowledge concerning biophysical and biochemical events which surround developmental processes. Nevertheless there is, at present, no convincing evidence that animal or plant hormones are involved in initiation or fixation of polarity in uni- or multi-cellular organisms. In higher plants, use of developmental mutants in simplified systems (e.g., approaches through tissue culture) may facilitate the elucidation of various developmental processes.

The importance of gradients and periodic oscillations in specific biosynthetic pathways as determinants or initiators of polarity remains to be demonstrated. Technical advances are necessary in the development of accurate micromethods for extraction, unequivocal identification and quantitative estimation of specific phytohormones in localized regions of such "fields" as shoot or root apices. Similarly, we require additional evidence for biosynthetic pathways of biologically active growth substances – including intracellular sites of synthesis, compartmentalization, transport, mode of action, inactivation and metabolic interactions with each other.

Cell division may not be an obligate condition for subsequent cellular differentiation in some lower and higher plants. However, for many organisms an asymmetric division is one permissive condition required for establishing a distinct spatial relationship between daughter cells prior to initiation of a specific developmental pathway (STEBBINS 1974, LANG 1974, HEPLER and PALEVITZ 1974, HOLDER 1979). Besides confirming spatial definition upon cells, the primary function of cell division may be to maintain the integrity of the progenitor cell line as noted earlier. Cell division also may be viewed as a necessary prerequisite for the inevitable division of labor required if an organism is to increase in size (BONNER 1963, 1974). Furthermore, resumption of the cell cycle (1) may allow the cell to become exposed to a morphogenic stimulus (possibly a phytohormone or balance of phytohormones) during a sensitive period or phase of the cycle; and/or (2) may be a necessary pre-condition associated with reprogramming of the genetic apparatus during mitosis for the subsequent initiation of a specific developmental pathway (EBERT 1968, HALPERIN 1973).

The physical biochemistry of biological structures will probably provide an ultimate basis for mechanism(s) of polarity – especially during the cellular stimulation phase (see Fig. 1). The importance of template-, enzymatic- and self-assembly in the formation of macromolecular subunits possessing inherent polarity or asymmetry is recognized (HEPLER and PALEVITZ 1974, BOUCK and BROWN 1976). Are phytohormones important in these events? The role of asymmetrical environmental factors in the assembly and alignment of subcellular structures such as microtubules, the microtrabecular lattice, or microfilaments of actin (which may be associated with cellular membranes and involved in cytoplasmic streaming and organelle movement in eukaryotes; see BRITZ 1979, SEITZ 1979, WAGNER 1979) remains to be established and cannot yet be integrated into any discrete model of polarity. Whether positional information in higher plants is perceived by a "cell-edge lattice", such as that associated with microtubule assembly in the elegant morphogenic system of *Azolla* roots (GUNNING and HARDAM 1979; see also KORN 1980), remains to be established. Recent developments in the visualization of cytoskeletal systems using immunofluorescent techniques (see LLOYD et al. 1979, 1980) and high voltage electron microscopy (PORTER and TUCKER 1981) should provide useful additional tools. There remains the possibility that there is no direct role by which expression of genetic information influences polar development (e.g., see SONNEBORN 1970).

Various models have been proposed for developmental processes (see Sect. 4). Attempts also have been made to mathematically describe developmental events in terms of two or three diffusing morphogens which are created and lost according to various hypothetical parameters (e.g., GIERER and MEINHARDT 1972, HERMAN 1972, NICOLIS and AUCHMUTY 1974, BABLOYANTZ and HIERNAUX 1975, OTHMER 1977, OTHMER and PATE 1980, PAPAGEORGIOU 1980, THORNLEY et al. 1980, GIERER 1981, MARKE and KUBÍČEK 1981, MALCHOW and FEISTEL

1982). The developmental biologist must verify that the mathematically described relationships apply to "real-world" situations through determination of specific coefficients or variables encountered within the equations, and must do so without distorting the hypothetical system. As a precautionary note, there is a real danger in hypothesizing, and in theoretical modeling, that researchers will encounter "hypotheson-like" situations – namely, that for every possible hypothesis, the "hypotheson" exhibits precisely those properties predicted by the hypothesis (Barr 1962). Nevertheless, the successful marriage of mathematically described hypotheses with experimental evaluation by interdisciplinary research groups should produce a more favorable climate for achieving an understanding of fundamental processes such as polarity, the apparent first requirement for organized development in plants.

References

Adler J (1974) A model of contact pressure in phyllotaxis. J Theor Biol 45:1–79

Adler J (1975) Chemotaxis in bacteria. Annu Rev Microbiol 29:341–356

Alpi A, Tognoni F, D'Amato F (1975) Growth regulator levels in embryo and suspensor of *Phaseolus coccineus* at two stages of development. Planta 127:153–162

Alpi A, Lorenzi R, Cionni PG, Bennici A, D'Amato F (1979) Identification of gibberellin A_1 in the embryo suspensor of *Phaseolus coccineus*. Planta 147:225–228

Asworth JM (1971) Cell development in the cellular slime mould *Dictyostelium discoideum*. In: Davies DD, Balls M (eds) Control mechanisms of growth and differentiation. SEB Symp XXV. Cambridge Univ Press, Cambridge, pp 27–49

Babloyantz A, Hieraux J (1975) Models for cell differentiation and generation of polarity in diffusion-governed morphogenetic fields. Bull Math Biol 37:637–657

Barr HJ (1962) A new genetical particle: the hypotheson. J Theor Biol 3:514

Beams HW, King RL (1944a) The negative group effect in the pollen grains of *Vinca*. J Cell Comp Physiol 23:39–46

Beams HW, King RL (1944b) Effect of ultracentrifuging on polarity in the pollen grains of *Vinca*. J Cell Comp Physiol 24:109–116

Bloch R (1943) Polarity in plants. Bot Rev IX:261–310

Bloch R (1965) Polarity and gradients in plants: A survey. In: Ruhland W (ed) Encyclopedia of plant physiology vol XV, Pt 1. Springer, Berlin Göttingen Heidelberg, pp 234–274

Bonner JT (1963) Morphogenesis. New York: Atheneum

Bonner JT (1974) On development. Harvard Univ Press, Cambridge, MA

Bopp M (1968) Control of differentiation in fern-allies and bryophytes. Annu Rev Plant Physiol 19:361–380

Bouck GB, Brown DL (1973) Microtubule biogenesis and cell shape in *Ochromonas*. I. The distribution of cytoplasmic and mitotic microtubules. J Cell Biol 56:340–359

Bouck GB, Brown DL (1976) Self-assembly in development. Annu Rev Plant Physiol 27:71–94

Britz SJ (1979) Chloroplast and nuclear migration. In: Haupt W, Feinleib ME (eds) Encyclopedia of plant physiology, New Series vol 7. Springer, Berlin Heidelberg New York, pp 170–206

Brown DL, Bouck GB (1973) Microtubule biogenesis and cell shape in *Ochromonas*. II. The role of nucleating sites in shape development. J Cell Biol 56:360–378

Bünning E (1952) Morphogenesis in plants. Surv Biol Prog 2:105–140

Bünning E (1953) Entwicklungs- und Bewegungsphysiologie der Pflanze. Springer, Berlin Heidelberg New York

Cande WZ, Ray PM (1976) Nature of cell-to-cell transfer of auxin in polar transport. Planta 129:43–52
Cavalier-Smith T (1975) The origin of nuclei and of eukaryotic cells. Nature 256:463–468
Cionini PG, Bennici A, Alpi A, D'Amato F (1976) Suspensor, gibberellin and in vitro development of *Phaseolus coccineus* embryos. Planta 131:115–117
Clowes FAL (1972) The control of cell proliferation within root meristems. In: Miller MW, Kuehnert CC (eds) The dynamics of meristem cell populations. Plenum, New York, pp 133–147
Clowes FAL (1976) The root apex. In: Yeoman MM (ed) Cell division in higher plants. Academic Press, London New York, pp 254–284
Cohen MH (1971) Models for the control of development. In: Davies DD, Balls M (eds) Control mechanisms of growth and differentiation. SEB Symp XXV. Cambridge Univ Press, Cambridge, pp 455–476
Cooke J (1975) Control of somite number during morphogenesis of a vertebrate, *Xenopus laevis*. Nature 254:196–199
Crayton MA, Wilson E, Quatrano RS (1974) Sulfation of fucoidan in *Fucus* embryos. II. Separation from initiation of polar growth. Dev Biol 39:164–167
Crick FHC (1970) Diffusion in embryogenesis. Nature 225:420–422
Crick FHC (1971) The scale of pattern formation. In: Davies DD, Balls M (eds) Control mechanisms of growth and differentiation. SEB Symp XXV. Cambridge Univ Press, Cambridge, pp 429–438
Dennison DS (1979) Phototropism. In: Haupt W, Feinleib ME (eds) Physiology of movements. Encyclopedia of plant physiology, new Ser, vol 7. Springer, Berlin Heidelberg New York, pp 506–566
Dobzhansky TS, Ayala FJ, Stebbins GL, Valentine JW (1977) Evolution. Freeman, San Francisco
Ebert JD (1968) Levels of control: a useful frame of perception. In: Mosona AA, Monroy A (eds) Current topics in developmental biology, vol 3. Academic Press, London New York, pp 15–25
Filner P, Yadav NS (1979) Role of microtubules in intracellular movements. In: Haupt W, Feinleib ME (eds) Encyclopedia of plant physiology new Ser, vol 7. Springer, Berlin Heidelberg New York, pp 95–113
Fogg GE (1979) Growth and heterocyst production in *Anabaena cylindrica* in relation to carbon and nitrogen metabolism. Ann Bot NS 13:241–259
Fry SC, Wangermann E (1976) Polar transport of auxin through embryos. New Phytol 77:313–317
Fukuda H, Komamine N (1980) Direct evidence for cytodifferentiation to tracheary elements without intervening mitosis in a culture of single cells isolated from the mesophyll of *Zinnia elegans*. Plant Physiol 65:61–64
Garrod DR, Forman D (1977) Pattern formation in the absence of polarity in *Dictyostelium discoideum*. Nature 265:144–146
Gierer A (1981) Some physical, mathematical and evolutionary aspects of biological pattern formation. Philos Trans R Soc Lond B 295:429–440
Gierer A, Meinhardt H (1972) A theory of biological pattern. Kybernetik 12:30–39
Gifford EM Jr, Carson G (1971) A shoot apex in seed plants. Bot Rev 37:143–229
Goldsmith MHM (1977) The polar transport of auxin. Ann Rev Plant Physiol 28:439–478
Goodwin BC (1971) A model of early amphibian development. In: Davies DD, Balls M (eds) Control mechanisms of growth and differentiation. SEB Symp XXV. Cambridge Univ Press, Cambridge, pp 417–428
Goodwin BC, Cohen MH (1969) A phase-shift model for the spatial and temporal organization of developing systems. J Theor Biol 25:49–107
Green PB (1974) Morphogenesis of the cell and organ axis – biophysical models. In: Basic mechanisms in plant morphogenesis. Brookhaven Symp Biol 25:166–190
Gross JD, Town CD, Brookman JJ, Jermyn KA, Peacey MJ, Kay RR (1981) Cell patterning in *Dictyostelium*. Philos Trans R Soc Lond B 295:497–508
Gunning BES, Hardman AR (1979) Microtubules and morphogenesis in plants. Endeavour 3:112–117

Häder D-P (1979) Photomovement. In: Haupt W, Feinleib ME (eds) Physiology of movements. Encyclopedia of plant physiology, New Ser, vol 7. Springer, Berlin Heidelberg New York, pp 268–309
Halperin W (1969) Morphogenesis in cell cultures. Annu Rev Plant Physiol 20:395–418
Halperin W (1973) The use of cultured tissue in studying developmental problems. Can J Bot 51:1801–1806
Halperin W (1978) Organogenesis at the shoot apex. Annu Rev Plant Physiol 29:239–262
Haupt W (1965) Perception of environmental stimuli orienting growth and movement in lower plants. Annu Rev Plant Physiol 16:267–290
Hayashi M, Takeuchi I (1976) Quantitative studies on cell differentiation during morphogenesis of the cellular slime mold *Dictyostelium discoideum*. Dev Biol 50:302–309
Hepler PK, Palevitz BA (1974) Microtubules and microfilaments. Annu Rev Plant Physiol 25:309–362
Herman GT (1972) Models for cellular interactions in development without polarity of individual cells. II. Problems of synchronization and regulation. Int J Syst Sci 3:149–175
Hicks GS (1980) Patterns of organ development in plant tissue culture and the problem of organ determination. Bot Rev 46:1–23
Holder N (1979) Positional information and pattern information in plant morphogenesis and a mechanism for the involvement of plant hormones. J Theor Biol 77:195–212
Jacobs WP (1970) Development and regeneration of the algal giant coenocyte *Caulerpa*. Ann NY Acad Sci 175:732–748
Jaffe LF (1968) Localization in the developing *Fucus* egg and the general role of localizing currents. Advan Morphog 7:295–328
Jaffe LF (1969) On the centripetal course of development, the *Fucus* egg, and self-electrophoresis. Dev Biol Suppl 3:83–111
Jaffe LF (1981) The role of ionic currents in establishing developmental pattern. Philos Trans R Soc Lond B 295:553–566
Jaffe LF, Neuscheler W (1969) On the mutual polarization of nearby pairs of fucaceous eggs. Dev Biol 19:549–565
Jaffe LF, Nuccitelli R (1977) Electric controls of development. Annu Rev Biophys Bioeng 6:445–476
Jaffe LF, Robinson KR, Nuccitelli R (1974) Location entry and self-electrophoresis as an intracellular localization mechanism. Ann NY Acad Sci 238:372–383
Jeffs RA, Northcote DN (1967) The influence of indole-3yl-acetic acid and sugar on the pattern of induced differentiation in plant tissue cultures. J Cell Sci 2:77–88
Johnson JD, Epel D, Paul M (1976) Intracellular pH and activation of sea urchin eggs after fertilization. Nature 262:661–664
Kende H, Gardner G (1976) Hormone binding in plants. Annu Rev Plant Physiol 27:267–290
Klembe I, Bentrup FW (1973) Zur Frage nach den Signalstoffen beim Gruppeneffekt keimender Fucuszygoten. Plant Sci Lett 1:315–320
Kohlenbach HW (1978) Comparative somatic embryogenesis. In: Thorpe TA (ed) Frontiers of plant tissue culture 1978. Uni Calgary Print Serv, Calgary, pp 59–66
Kohlenbach HW, Schmidt B (1975) Cytodifferenzierung in Form einer direkten Umwandlung isolierter Mesophyllzellen zu Tracheiden. Z Pflanzenphysiol 75:369–374
Korn RW (1980) The changing shape of plant cells: transformations during cell proliferation. Ann Bot 46:649–666
Koshioka M, Jones A, Koshioka MN, Pharis RP (1983) Metabolism of [^{3}H] gibberellin A_4 in somatic suspension cell cultures of carrot. Phytochemistry 22:1585–1590
Kühn A (1971) Lectures on developmental physiology. Springer, Berlin Heidelberg New York
Lang A (1974) Inductive phenomena in plant development. In: Basic mechanisms in plant morphogenesis. Brookhaven Symp Biol 25:129–144
Lawrence PA (1971) The organization of the insect segment. In: Davies DD, Balls M (eds) Control mechanisms of growth and differentiation. SEB Symp XXV. Cambridge Univ Press, Cambridge, pp 379–390

Libbenga KR (1978) Hormone receptors in plants. In: Thorpe TA (ed) Frontiers of plant tissue culture 1978. Univ Calgary Print Serv, Calgary, pp 325–333
Lintilhac PM (1974a) Differentiation, organogenesis, and the tectonics of cell wall orientation. II. Separation of stresses in a two-dimensional model. Am J Bot 61:135–140
Lintilhac PM (1974b) Differentiation, organogenesis, and the tectonics of cell-wall orientation. III. Theoretical considerations of cell-wall mechanics. Am J Bot 61:230–237
Lintilhac PM, Jensen WA (1974) Differentiation, organogenesis, and the tectonics of cell-wall orientation. I. Preliminary observations on the development of the ovule in cotton. Am J Bot 61:129–134
Lloyd CW, Slabas AR, Powell AJ, McDonald G, Badley RA (1979) Cytoplasmic microtubules of higher plant cells visualised with anti-tubulin antibodies. Nature 279:239–240
Lloyd CW, Slabas AR, Powell AJ, Lowe SB (1980) Microtubules, protoplasts and plant cell shape. Planta 147:500–506
Loomis WF (1975) *Dictyostelium discoideum*. Academic Press, London New York
Loomis WF (ed) (1982) The development of *Dictyostelium discoideum*. Academic Press, London New York
Lyndon RF (1976) The shoot apex. In: Yeoman MM (ed) Cell division in higher plants. Academic Press, London New York, pp 285–314
MacNab RM (1979) Chemotaxis in bacteria. In: Haupt W, Feinleib ME (eds) Physiology of movements. Encyclopedia of plant physiology, new ser vol 7. Springer, Berlin Heidelberg New York, pp 310–334
Maksymowych R, Erickson RO (1977a) Phyllotactic change induced by gibberellic acid in *Xanthium* shoot apices. Am J Bot 64:33–44
Maksymowych R, Erickson RO (1977b) Phyllotaxis in *Xanthium* shoots altered by gibberellic acid. Science 196:1201–1203
Malchow H, Feistel R (1982) Polarity, symmetry, and gradients in a reaction-diffusion system. Stud Biophys 88:125–130
Marke M, Kubíček M (1981) Morphogen pattern formation and development in growth. Bull Math Biol 43:259–270
Mascarenhas JP (1974) The biochemistry of angiosperm pollen. Bot Rev 41:259–314
McDaniel CN, Hsu FC (1976) Position-dependent development of tobacco meristems. Nature 259:564–565
Mitchison GJ (1977) Phyllotaxis and the Fibonacci series. Science 196:270–275
Mitchison GJ, Wilcox M, Smith RJ (1976) Measurement of an inhibitory zone. Science 191:866–868
Monnier M (1976) Culture in vitro de l'embryon immature de *Capsella bursa-pastoris* Moench. Rev Cytol Biol Veg 39:1–120
Monnier M (1978) Culture of zygotic embryos. In: Thorpe TA (ed) Frontiers of plant tissue culture 1978. Univ Calgary Print Ser, Calgary, pp 277–286
Nakazawa S (1960) Nature of protoplasmic polarity. Protoplasma 52:274–294
Nakazawa S (1962) Polarity. In: Lewin R (ed) Physiology and biochemistry of algae. Academic Press, London New York, pp 653–661
Nakazawa S (1975) Physiology of *Fucus*. In: Tokida J, Hirose H (eds) Advance of physiology in Japan. Junk, The Hague, pp 160–170
Newell PC (1977) How cells communicate: the system used by slime moulds. Endeavour, NS 1:63–68
Nicolis G, Auchmuty JFG (1974) Dissipative structures, catastrophes, and pattern formation: a bifurcation analysis. Proc Natl Acad Sci USA 71:2748–2751
Noma M, Huber J, Ernst D, Pharis RP (1982) Quantitation of gibberellins and the metabolism of [^{3}H] gibberellin A, during somatic embryogenesis in carrot and anise cell cultures. Planta 155:369–376
Novotny AM, Forman M (1974) The relationship between changes in cell wall composition and the establishment of polarity in *Fucus* embryos. Dev Biol 40:162–173
Nuccitelli R, Jaffe LF (1974) Spontaneous current pulses through developing fucoid eggs. Proc Natl Acad Sci USA 71:4855–4859
Nuccitelli R, Jaffe LF (1976) The ionic components of the current pulses generated by developing fucoid eggs. Dev Biol 49:518–531

Othmer HG (1977) Current problems in pattern formation. Lect Math Life Sci 9:57–85
Othmer HG, Pate E (1980) Scale invariance in reaction-diffusion models of spatial pattern formation. Proc Natl Acad Sci USA 77:4180–4184
Papageorgiou S (1980) A morphogen gradient model for pattern regulation. 1. Formation of non-repetitive and repetitive structures. Biophys Chem 11:183–190
Peng HB (1976) Polarization of the fucoid egg; galvanotropism and ultrastructure. Ph D Thesis, Purdue Univ West-Lafayette, Ind
Phillips IDJ (1975) Apical dominance. Annu Rev Plant Physiol 26:341–367
Poff KL, Whitaker BD (1979) Movement in slime molds. In: Haupt W, Feinleib ME (eds) Physiology of movements. Encyclopedia of plant physiology, new ser vol 7. Springer, Berlin Heidelberg New York, pp 355–382
Porter KR, Tucker JB (1981) The ground substance of the living cell. Sci Am 244:56–67
Quatrano RS (1972) An ultrastructural study of the determined site of rhizoid formation in *Fucus* zygotes. Exp Cell Res 70:1–12
Quatrano RS (1974) Developmental biology development in marine organisms. In: Mariscal RN (ed) Experimental marine biology. Academic Press, London New York, pp 303–346
Quatrano RS (1978) Development of cell polarity. Annu Rev Plant Physiol 29:487–510
Quatrano RS, Brawley SH, Hogsett WE (1979) The control of the polar deposition of a sulfated polysaccharide in *Fucus* zygotes. In: Subtelny S, Konigsberg IR (eds) Determinants of spatial organization. Academic Press, London New York, pp 77–96
Reinert J (1973) Aspects of organization – organogenesis and embryogenesis. In: Street HE (ed) Plant tissue and cell culture. Univ California Press, Berkeley Los Angeles, pp 338–355
Reinert J, Bajaj YPS (1977) Applied and fundamental aspects of plant cell, tissue, and organ culture. Springer, Berlin Heidelberg New York
Richards FJ, Schwabe WW (1969) Phyllotaxis: a problem of growth and form. In: Steward FC (ed) Plant physiology, a treatise. Academic Press, London New York, pp 79–116
Roberts LW (1969) The initiation of xylem differentiation. Bot Rev 35:201–250
Roberts LW (1976) Cytodifferentiation in plants. Cambridge Univ, Cambridge
Robinson KR, Jaffe LF (1975) Polarizing fucoid eggs drive a calcium current through themselves. Science 187:70–72
Robinson KR, Jaffe LF (1976) Calcium gradients and egg polarity. J Cell Biol 70:37a
Roos W, Scheidegger C, Gerisch G (1977) Adenylate cyclase activity oscillations as signals for cell aggregation in *Dictyostelium discoideum*. Nature 266:259–261
Ross MK, Thorpe TA (1973) Physiological gradients and shoot initiation in tobacco callus cultures. Plant Cell Physiol 14:473–480
Rubery PH (1981) Auxin receptors. Annu Rev Plant Physiol 32:569–596
Rutherford CL, Taylor RD, Merkle RK, Frame LT (1982) Cellular pattern formation: *Dictyostelium discoideum* as a system for a biochemical approach. Trends Biochem Sci 7:108–111
Sachs T (1969) Regeneration experiments on the determination of the form of leaves. Isr J Bot 18:21–30
Sachs T (1972) A possible basis for apical organization in plants. J Theor Biol 37:353–361
Sachs T (1978) Patterned differentiation in plants. Differentiation 11:65–73
Schnepf E (1982) Morphogenesis in moss protonemata. In: Lloyd CW (ed) The cytoskeleton in plant growth and development. Academic Press, London New York, pp 321–344
Schwabe WW (1971) Chemical modification of phyllotaxis and its implications. In: Davies DD, Balls M (eds) Control mechanisms of growth and differentiation. SEB Symp XXV. Cambridge Univ Press, Cambridge, pp 301–322
Seitz K (1979) Cytoplasmic streaming and cyclosis of chloroplasts. In: Haupt W, Feinleib ME (eds) Encyclopedia of plant physiology, new ser, vol 7. Springer, Berlin Heidelberg New York, pp 150–169
Shapiro L (1976) Differentiation in the *Caulobacter* cell cycle. Annu Rev Microbiol 30:377–407

Sharp WR, Sondahl MR, Caldas LS, Maraffa SB (1980) The physiology of in vitro asexual embryogenesis. Hortic Rev 2:268–310

Sheldrake AR (1973) The production of hormones in higher plants. Biol Rev 48:509–559

Shropshire W (1979) Stimulus perception. In: Haupt W, Feinleib ME (eds) Physiology of movements. Encyclopedia of plant physiology, new ser, vol 7. Springer, Berlin Heidelberg New York

Sievers A, Schnepf E, (1981) Morphogenesis and polarity of tubular cells with tip growth. In: Kiermayer E (ed) Cytomorphogenesis in plants. Springer, Wien New York, pp 265–299

Sinnott EW (1960) Plant morphogenesis. McGraw-Hill, New York, pp 10–41

Smith GM (1955) Crytogamic botany, vol 1. McGraw-Hill, New York

Smith H (1975) Phytochrome and photomorphogenesis. McGraw-Hill, London

Snow M, Snow R (1962) A theory of the regulation of phyllotaxis based on *Lupinus albus*. Philos Trans R Soc Lond Ser B 244:483–513

Sonneborn TM (1970) Gene action in development. Proc R Soc Lond B 176:347–366

Stange L (1984) Cellular interactions during early differentiation. In: Linskens HF, Heslop-Harrison J (eds) Encyclopedia of Plant Physiology, new ser Vol 17. Springer, Berlin Heidelberg New York, pp 424–452

Stebbins GL (1974) Evolution of morphogenetic patterns. In: Basic mechanisms in plant morphogenesis. Brookhaven Symp Biol 25:227–243

Steeves TA, Sussex IM (1972) Patterns in plant development. Prentice-Hall, Englewood Cliffs

Sternfeld J, David MN (1981) Cell sorting during pattern formation in *Dictyostelium*. Differentiation 20:10–21

Street HE, Withers LA (1974) The anatomy of embryogenesis in culture. In: Street HE (ed) Tissue culture and plant science. Academic Press, London New York, pp 71–100

Sussex IM (1964) The permanence of meristems: developmental organizers or reactors in exogenous stimuli? In: Meristems and differentiation. Brookhaven Symp Biol 16:1–10

Sussex, IM (1974) Do concepts of animal development apply to plant systems? In: Basic mechanisms in plant morphogenesis. Brookhaven Symp Biol 25:145–151

Sussman M, Brackenbury R (1976) Biochemical and molecular-genetic aspects of cellular slime mold development. Annu Rev Plant Physiol 27:229–265

Thornley JHM (1975) Phyllotaxis. I. A mechanistic model. Ann Bot 39:491–507

Thornley JHM, Warren Wilson J, Colley E (1980) The initiation and maintenance of polarity: mathematical aspects of a proposed control system. Ann Bot 46:713–717

Thorpe TA (1980) Organogenesis in vitro: structural physiological and biochemical aspects. Int Rev Cytol Suppl 11A:71–111

Tomkins GM (1975) The metabolic code. Science 189:760–763

Torrey JG (1966) The initiation of organized development in plants. Adv Morphog 5:39–91

Torrey JG (1972) On the initiation of organization in the root apex. In: Miller MW, Kuehnert CC (eds) The dynamics of meristem cell populations Plenum, New York, pp 1–13

Torrey JG (1975) Tracheary element formation from single isolated cells in culture. Physiol Plant 35:158–165

Torrey JG (1976) Root hormones and plant growth. Annu Rev Plant Physiol 27:435–459

Torrey JG, Fosket DE, Hepler PK(1971) Xylem formation: a paradigm of cytodifferentiation in higher plants. Am Sci 59:338–352

Tran Thanh Van K, Trinh H (1978) Morphogenesis in thin cell layers: concept, methodology and results. In: Thorpe TA (ed) Frontiers of plant tissue culture 1978. Univ Calgary Print Serv, Calgary, pp 37–48

Tran Thanh Van M, Chlyah H, Chlyah A (1974) Regulation of organogenesis in thin layers of epidermal and sub-epidermal cells. In: Street H (ed) Tissue culture and plant science 1974. Academic Press, London New York, pp 101–139

Turgeon R (1975) Differentiation of wound vessel members without DNA synthesis, mitosis or cell division. Nature 257:806–808

Veen AH, Lindenmayer A (1977) Diffusion mechanism for phyllotaxis. Theoretical physio-chemical and computer study. Plant Physiol 60:127–139

Venis MA (1973) Hormone receptor proteins. Curr Adv Plant Sci 2:21–28

Vreugdenhil D, Dijkstra ML, Libbenga KR (1976) Ultrastructure of cell wall of normal and apolar embryos of *Fucus vesiculosus*. Protoplasma 88:305–313

Wagner G (1979) Actomyosin as a basic mechanism of movement in animals and plants. In: Haupt W, Feinleib ME (eds) Physiology of movements. Encyclopedia of plant physiology, new ser vol 7. Springer, Berlin Heidelberg New York, pp 114–126

Wardlaw, CW (1968) Morphogenesis in plants. Methuen, London

Warmke HE, Warmke GL (1950) The role of auxin in the differentiation of root and shoot primordia from root cuttings on *Taraxacum* and *Cichorium*. Am J Bot 37:272–380

Warren Wilson J (1980) A control system for initiating and maintaining polarity. Ann Bot 46:701–711

Webster G (1971) Morphogenesis and pattern formation in hydroids. Biol Rev 46:1–46

Weisenseel MH (1977) Changes in membrane properties and transcellular ion movements in developing plant cells. In: Marre E, Cifferi O (eds) Regulation and cell membrane activities in plant. Elsevier/North Holland Biomedical Press, Amsterdam, pp 267–273

Weisenseel MH (1979) Induction of polarity. In: Haupt W, Feinleib ME (eds) Physiology of movements. Encyclopedia of plant physiology, vol 7. Springer, Berlin Heidelberg New York, pp 485–505

Weisenseel MH, Nuccitelli R, Jaffe LF (1975) Large electrical currents traverse growing pollen tubes. J Cell Biol 66:556–567

Wetherell DF (1979) In vitro embryoid formation in cells derived from somatic plant tissues. In: Hughes KW, Henke R, Constantin M (eds) Propagation of higher plants through tissue culture. Nat Tech Inf Serv, Springfield, pp 102–124

Wetmore RH, Rier JP (1973) Experimental induction of vascular tissues in callus of angiosperms. Am J Bot 50:418–430

Wetmore RH, Sorokin S (1955) On the differentiation of xylem. J Arnold Arbor Harv Univ 36:306–317

Wettstein von D (1965) Die Induktion und experimentelle Beeinflussung der Polarität bei Pflanzen. In: Ruhland W (ed) Encyclopedia of plant physiology, vol XV/1. Springer, Berlin Göttingen Heidelberg

Williams K (1982) Morphogenetic molecules in the multicellular stage of *Dictyostelium discoideum*. Hoppe-Seyler's Z Physiol Chem 363:130

Williams RF (1974) The shoot apex and leaf growth. Cambridge Univ Press, Cambridge, 1974

Wolk CP (1975a) Formation of one-dimensional patterns by stochastic processes and by filamentous blue-green algae. Dev Biol 46:370–382

Wolk CP (1975b) Differentiation and pattern formation in filamentous blue-green algae. Spores 6:85–96

Wolpert L (1971) Positional information and pattern formation. In: Moscona AA, Monroy A (eds) Current topics of developmental biology, vol 6. Academic Press, London New York, pp 183–224

Wolpert L (1981) Positional information and pattern formation. Philos Trans R Soc Lond B 295:441–450

Wolpert L, Hicklin J, Hornbruch A (1971) Positional information and pattern regulation in regeneration of *Hydra*. In: Davies DD, Balls M (eds) Control mechanisms of growth and differentiation. SEB Symp XXV. Cambridge Univ Press, Cambridge, pp 391–415

5 Epinasty, Hyponasty, and Related Topics

J.H. PALMER

1 Introduction

A nastic response is a growth curvature, where the direction of flexure is determined by the morphology of the organ and is not directly related to an external reference, although the response may be induced by an environmental stimulus (BALL 1969; see Sect. 4). A common, readily recognizable form of nastic curvature is epinasty, where in a lateral organ relative growth of the upper or adaxial side[1] increases so that it becomes convex (DE VRIES 1872). In many dicotyledonous species, the growing leaf petiole has a well-developed epinastic capability (CROCKER et al. 1932) (Fig. 1 A), while the entire leaf shows an epinastic response in some monocotyledons, for example narcissus, tulip, and hyacinth (HITCHCOCK et al. (1932). Epinasty is also used to describe curvature in an orthotropic organ such as a stem or pedicel, where the side that initiates the curvature by an acceleration in elongation growth becomes uppermost and the tip of the organ is inclined to form an angle with the vertical. The formation of the terminal hook in the etiolated seedling shoot and in the flower pedicel during its elongation phase are good examples (KALDEWEY 1962, KOHJI et al. 1979, 1981). The complimentary response, hyponasty, where the abaxial side grows most strongly, often occurs in leaf petioles as a sequel to epinasty, when it serves to straighten and restore the petiole's original shape and orientation (Fig. 1 C).

In orthotropic organs, hyponasty removes the effects of a phase of epinastic growth. For example, in many species of dicotyledons exposure of the etiolated seedling to light, particularly R, removes the hooked condition from the hypocotyl (KLEIN et al. 1956) or epicotyl (GOESCHL et al. 1967) by inducing an upward curvature. In flower pedicels, hyponastic curvature below the procumbent floral bud at the time of anthesis results in the opening flower becoming upward-directed (KALDEWEY 1962).

These concepts of epinasty and hyponasty as functions of shoot development originated in the 19th century. The terms have now broadened and are used to describe any rapidly developing organ curvature or change in organ orientation that is not obviously tropic in nature, whether it is part of a developmental process, is environmentally induced, or results from disease or human interference. The operative curvature is sometimes confined to the extreme base of the organ so that the main effect is a change in orientation, rather than curvature of the entire organ. Thus, the thermonastic movements of petals and sepals

1 In this review, "upper" and "lower" are used synonymously with adaxial and abaxial. Generally, but not always, adaxial is equivalent to "dorsal" and abaxial is equivalent to "ventral". (Also see Vol. 7, this Series p 448)

in flowers that open and fold in the morning and evening result from alternating epinastic and hyponastic growth curvatures in the basal region close to the point of attachment to the receptacle. Epinasty has also been used in an incorrect sense to describe the role of apical dominance in inducing inclined growth in lateral shoots, thereby confusing cause with effect (WILSON 1973). A nastic response may become permanent when it is induced during maturity or senescence, whereas in young growing organs it is often a temporary reaction to an environmental stimulus such as nutritional deficiency (ECKERSON 1931), flooding (KAWASE 1974), or lodging (PALMER 1973).

Circadian rhythms of transient epinasty and hyponasty may be induced in growing leaf petioles by the alternation of day and night (SHELL et al. 1974, YIN 1941, YAMANE 1941). In addition to this symptomatic usage, epinasty has been evoked in a theoretical sense to describe a growth tendency, thought to be located in the adaxial half of plagiotropic and diageotropic organs[2] (DE VRIES 1981, KNIEP 1910). In lateral plagiotropic shoots it has been assigned the role of counteracting, either wholly or partially, an assumed innate ability of the shoot to respond to gravity by negative gravitropism so that the resultant growth is at an angle to the vertical. Epinasty was identified with this hypothetical tendency, because of the convex curvatures that were seen to develop in lateral shoots, leaf petioles, and lamina midribs, when potted plants were subjected to gravity compensation on a horizontal clinostat (LARSEN 1962, BROWN et al. 1974, WILSON and ARCHER 1977). Indirect evidence for the existence of this force came from experiments with the intermittent clinostat, using primary shoots of *Tradescantia viridis* and *Zebrina pendula* (RAWITSCHER 1932), where the shoot is normally inclined at an angle to the vertical. Curvatures that appear in symmetrical, orthotropic, or plagiotropic organs following exposure to C_2H_4 or treatment with IAA, NAA or 2, 4-D, have also been assumed to be epinastic, it being postulated that these organs either have a hidden bilateral physiological polarity, or that gravity gradually induces physiological asymmetry, so that the uppermost side of a symmetrical plagiotropic organ becomes physiologically equivalent to the adaxial half of a morphologically asymmetric organ (RAWITSCHER 1932, KALDEWEY 1971). This process of polarization has been termed geo-epinasty. Historical and general aspects of epinasty are reviewed by KALDEWEY (1962) and BALL (1969).

2 Epinasty in Response to Chemical Application

2.1 Ethylene and Ethrel

2.1.1 Leaves and Leaf Petioles

MOLISCH (1911) demonstrated that tobacco smoke and town gas could induce leaf epinasty, thus implicating a gaseous substance; the active principle in town

2 Plagiotropism is used as a general term to describe all forms of growth at an angle to the vertical, including diageotropism where the direction of growth approximates to the horizontal.

gas was later identified as C_2H_4 (HARVEY 1913). CROCKER et al. (1932) exposed 202 dicotyledonous species and cultivars to C_2H_4 and found that 89 developed some degree of leaf epinasty. Sensitivity varied widely between families and genera and between cultivars of the same species, so that it is neither family, generic, nor specific in character. These authors showed that 0.1 $\mu l\ l^{-1}$ C_2H_4 will induce epinasty in some species, while many others are able to respond to 1.0 $\mu l\ l^{-1}$. Higher concentrations of the order of 100.0 $\mu l\ l^{-1}$ are required to induce epinasty in monocotyledons sensitive to C_2H_4 (HITCHCOCK et al. 1932). Of the parts of the dicotyledon leaf, the lamina, petiole, midrib, and lamina veins may all respond to C_2H_4 (HAYES 1981, KAZEMI and KEFFORD 1974), but possibly for ease of observation and measurement, the petiole has been the main subject for investigation. In the tomato, the terminal part of the midrib and the extreme base of the petiole are particularly sensitive to C_2H_4 (KAZEMI and KEFFORD 1974), while in many plants the central region of the petiole responds most actively, the sunflower being a good example (PALMER 1972). Epinasty is not confined to elongating petioles, since mature leaves that have ceased active growth can also respond to C_2H_4, although the curvature usually develops more slowly and is confined to the base of the petiole (Fig. 1A; CROCKER et al. 1932). After exposure to C_2H_4 epinasty gradually disappears from young leaves due to accelerated elongation growth in the lower side of the petiole or midrib. Mature leaves may show only a partial recovery (PALMER 1972). Information about the nature and extent of the recovery process can be obscured by the technique used to record epinasty. Measuring the angle subtended by the petiole or lamina with the stem may fail to record a reduction in the curvature of the petiole or midrib, since this may occur without affecting the angle of insertion of the petiole on the stem. C_2H_4 is now known to be actively synthesized by leaves, stems, flowers, fruits, and other plant organs (ABELES 1973, DENNY and MIDLER 1935, GANE 1934).

Ethrel has been reported to induce leaf epinasty in leaves of bean (EDGERTON and BLANFIELD 1968, HAYES 1981), fig (CRANE et al. 1970), and tomato (KUO and CHEN 1980). The time course of the response to ethrel has not been reported in detail, but in the sunflower ethrel initially induces an epinastic curvature in the petiole that persists for a few days and then gradually disappears, to be replaced by a progressive increase in curvature at the base of the petiole, so that the whole leaf becomes downward-inclined (Fig. 1B). The reason for this change in the nature of the response to ethrel is not known. Possibly it reflects a gradual dimunition in the release of C_2H_4 within the plant. Alternatively, it may be due to persistent epinasty that develops more slowly in the petiole base than in the central region.

2.1.2 Stems

NELJUBOW (1901) showed that in the etiolated pea seedling the epicotyl below the hook region responds to C_2H_4 in the range 0.01–1.0 $\mu l\ l^{-1}$ C_2H_4 by a characteristic inclination to the vertical, due to growth in one side of the epicotyl being more sensitive to ethylene than the other. This response is not seen in the light (GOESCHL and KAYS 1975). Epicotyl elongation is also depressed by

Fig. 1A–C. Examples of epinasty and hyponasty in the leaf petiole of *Helianthus annuus*. **A** *Left plant* control; *right plant* exposed to 10 μl l^{-1} ethylene for 20 h; showing absence of epinasty from very young leaves close to the terminal bud; uniform epinastic curvature of growing leaves; restriction of epinasty to the basal region of aging leaves. **B** Persistent epinasty at the petiole base after straightening has removed curvature from the rest of the petiole, 10 d after application of 10^{-4} M ethrel. **C** *Left plant* control; *right plant* showing petiole hyponasty that replaced the initial epinastic response to 60 μl 10^{-4} M, 2.4-D ester applied to the whole shoot. Photographed 10 d after 2,4-D application

ethylene, so that the plagiotropic orientation must be initiated by growth being more severely inhibited on the side that becomes lower (GOESCHL and KAYS 1975). The angle of inclination increases with ethylene concentration to a limit of 10.0 μl l^{-1}, suggesting that the induced plagiotropic orientation has a gravitational basis. The primary root of the pea seedling also becomes plagiotropic in the presence of C_2H_4, while lateral roots either grow upward at an angle to the horizontal, or at high concentrations become oriented vertically (NELJUBOW 1901).

The terminal region of the hypocotyl or epicotyl of the etiolated dicotyledon seedling is often hook-shaped due to the location of an epinastic curvature in the stem extremity, subadjacent to the apical region. The hook results from newly formed tissue on one side of the stem, the outer, expanding sooner than on the other, so that a sharply curved zone or hook is formed (KANG and BURG 1972). The hooked shape is removed when the expansion phase takes place on the inner side, but the hook entity is retained by younger tissue repeating the process. The hook is thus a standing epinastic wave riding the expansion zone of the plumule. The dynamics of hook maintenance are discussed by SILK and ERICKSON (1978). The terminal region of the etiolated pea epicotyl, including the hook, actively produces C_2H_4 at a rate of 5.0–6.0 µl $kg^{-1}h^{-1}$ (GOESCHL et al. 1967). Light-induced opening of the hook appears to be due to the suppression of this endogenous C_2H_4 production in both bean and pea seedlings (KANG and BURG 1972, KANG and RAY 1969b). Exposure of light-grown pea or lettuce seedlings to C_2H_4 will cause hook formation (BURG and BURG 1968, POOVAIAH and LEOPOLD 1974). In contrast, C_2H_4 has no effect upon illuminated seedlings of mung bean and only slightly affects the hook in dark-grown seedlings (GEE 1977). CO_2 acts as a competitive inhibitor of C_2H_4 action in pea stems, and causes the plumular hook to open in the dark (KANG et al. 1967). Treatment of etiolated seedlings with cobalt chloride, which is known to block C_2H_4 synthesis (LAU and YANG 1976) and responses also induces hook opening (KANG and RAY 1969a). These and other considerations suggest that hook formation and maintenance in the etiolated seedling are commonly a response to C_2H_4 production, while illumination brings about hook opening by suppressing this (KANG et al. 1967). The interaction of C_2H_4 with IAA is considered in Sections 2.2.2 and 5.2.

The normally prostrate habit of the tomato mutant, diageotropica, is replaced by an erect stem when the plant is exposed to C_2H_4 in air (ZOBEL 1973, 1974). ZOBEL proposed that this reflected an impaired ability of the mutant to produce ethylene endogenously, but JACKSON (1979), was unable to detect any differences in C_2H_4 production by mutant petiole or stem tissue, although C_2H_4 production by the roots was about 37% of that of roots for normal plants. He concluded that diageotropica shows an abnormal response to gravity and this condition is removed by exposure to C_2H_4, which presumably raises the C_2H_4 content of the plant to abnormal levels. Increasing the C_2H_4 content of the shoot either by exposure to C_2H_4 or application of ethrel induces the erection, or hyponasty, of shoots in strawberry clover, *Trifolium fragiferum* L. var. Salina (HANSEN and BENDIXEN 1974), *Cupressus arizonica* (BLAKE et al. 1980) and in the peanut (*Arachis hypogaea* L.; ZIV et al. 1976). The tendency of shoots of *Arachis* to become erect in darkness was related to an increase in endogenous C_2H_4 production. Ethrel also increased the level of extractable GA's in the peanut shoot, and it is known that applied GA_3 can induce erection of prostrate shoots of strawberry clover (BENDIXEN and PETERSON 1962). In contrast to these findings, high concentrations of ethrel stimulated the development of plagiotropic rhizomes from the dormant basal buds of lowbush blueberry, *Vaccinium angustifolium*, and converted a proportion of aerial shoots to plagiotropic rhizome-like stems (KENDER et al. 1969).

2.2 Auxins and Herbicides

2.2.1 Leaves

A common response of shoots to auxin application is the development of petiole epinasty (ZIMMERMAN and WILCOXON 1935, HITCHCOCK 1935, FISCHNICH 1935). SOEKARJO (1965) showed that immersion of isolated petiole segments of *Coleus scutellariodes* in IAA solutions induced epinastic curvatures that increased with increasing concentrations, in the range 10^{-9} to 10^{-5} M. That this kind of induced epinasty may be caused indirectly by a gas or vapor produced by the treated shoot was first demonstrated by ZIMMERMAN and WILCOXON (1935). Later the active emanation was identified as C_2H_4 (MORGAN and HALL 1964, ABELES and RUBINSTEIN 1964). It is now known that applied IAA increases endogenous C_2H_4 production in many plant species and in a wide range of tissues (MORGAN 1976), but since IAA-induced epinasty is confined to plants that become epinastic in the presence of C_2H_4 (BORGSTRÖM 1939), it seems probable that applied IAA acts indirectly by stimulating C_2H_4 formation, which is the primary cause of the observed curvature. Evidence to support this was obtained by STEWART and FREEBAIRN (1969), who blocked the ability of tomato petioles to respond to C_2H_4 by exposing the plants to high temperature and showed that applied IAA no longer induced epinasty. PALMER (1979) obtained similar results with sunflower petioles, by pretreating them with silver nitrate solution (Fig. 4B). The increased C_2H_4 production may be of short duration, explaining the rapid recovery from epinasty that is characteristic of young actively growing leaves after IAA treatment (Fig. 4C).

Synthetic auxin-type growth regulators such as NAA and 2,4-D appear to induce epinasty in the same way by the stimulation of endogenous C_2H_4 production (ABELES and RUBINSTEIN 1964, MORGAN and HALL 1962). The epinastic response of leaves of tomato plants to 2,4-D was eliminated by pretreating them with a substituted benzothiadiazole (5-methyl-7-chloro-4-ethoxycarbonylmethoxy-1,2,3-benzothiadiazole) which inhibits responses to C_2H_4 (PARUPS 1973). Further evidence for the existence of an inter-relationship between epinasty and ethylene production is demonstrated by the observation that the chlorophenoxyacetic acid analogs 2,6-D and 2,4,6-T, which are inactive in epinasty tests, do not stimulate C_2H_4 production (RUBINSTEIN and ABELES 1965), and by the failure of IAA to induce petiole epinasty in tomato in the presence of the inhibitor of ethylene synthesis, alpha-aminooxyacetic acid (AMRHEIN and SCHNEEBECK 1980). Other herbicides that have been reported to induce petiole epinasty in receptive plants may do so by stimulating C_2H_4 release, although this has not been demonstrated. These include picloram (MORGAN and BAUR 1970, EISINGER and MORRE 1971), siduron, carbamates, and some triazones. With time, young leaves may recover from herbicide-induced epinasty, if leaf abscission has not been promoted. In the recovery phase, leaves may become more erect than usual. 2,4-D provides a good example (Fig. 1C). 2,4-D-induced C_2H_4 may be a transient phenomenon in cotton (MORGAN and HALL 1962), providing an explanation for the eventual disappearance of leaf epinasty, but not for the development of hyponasty. Since in the case of 2,4-D abnormal growth continues in the recovery phase, it is evident that 2,4-D is still active within the plant. This suggests that plant tissues may have a limited ability to respond to applied chemicals by C_2H_4 production. If in the absence of C_2H_4 production the lower half of the petiole is able to respond to 2,4-D by elongation growth, an explanation for hyponasty in the post-epinastic phase is obtained.

2.2.2 Stems

While curvatures can be induced in green orthotropic stems and lateral branches of some species by the local application of IAA, the response is generally transient. NAA, IBA, and 2,4-D produce more pronounced and more permanent curvatures, but the nature of these has not been investigated in detail (WEAVER 1972), although they are commonly regarded as symptomatic of the application of synthetic auxins or 2,4-D to plant shoots. Their persistence may reflect the relative immunity of these compounds to enzymic degradation, but their ability to induce asymmetric growth in radially symmetrical stem tissue remains unexplained.

The etiolated plumular hook of *Phaseolus vulgaris* L. and *Phaseolus mungo* L. from which the leaves and cotyledons have been removed becomes more epinastic (shut) when treated with IAA in the dark (KLEIN et al. 1957, GEE 1977) and light-induced opening is reduced, as is the response to red light (KLEIN et al. 1967). This effect appears to operate through auxin-induced C_2H_4 production (KANG and RAY 1969b). However, not all plumular hooks respond to applied IAA in this way. The decapitated hook of *Gossypium hirsutum* L. is shut only slightly by IAA, and the rate of light-induced opening is only marginally delayed. In intact seedlings of *G. hirsutum*, IAA promotes opening of the hook in both light and darkness (POWELL and MORGAN 1970).

MÜNCH (1938) found that the application of IAA to one side of the stem stump of a decapitated *Eupatorium cannabinum* shoot resulted in epinastic curvature in the lateral shoot subadjacent to the IAA. SNOW (1945) confirmed this finding for laterals of *Impatiens roylei* and also showed that direct application of IAA to defoliated lateral shoots prevented their erection (hyponasty), a response that normally follows defoliation (LEIKE and GUTTENBERG 1962). The ability of lateral shoots of *Coleus blumei* to develop an epinastic curvature when plants are rotated on a horizontal clinostat depends on the possession of young leaves and apical buds, but if these are removed, then an equivalent epinastic curvature can be induced by IAA applied to the end of the lateral (LYON 1963a). For the lateral shoot of *Solanum andigena* to grow out as a plagiotropic stolon instead of an erect leafy shoot, it must be subject to apical dominance by illuminated leaves and buds (BOOTH 1959). In cuttings, this dominance effect can be replaced by apically applied IAA in lanolin at a concentration of 50 $\mu g\ gm^{-1}$ (KUMAR and WAREING 1972). A condition for the operation of this system is that the inhibitory effect of apical dominance on axillary shoot growth be overcome. This can be achieved by exposure to darkness and high humidity, or by gibberellin application (KUMAR and WAREING 1972). These findings confirm the conclusion of SNOW (1945) that plagiotropic growth is an alternative response to apical dominance, which appears when the lateral is freed from inhibition, but is still subject to some degree of apical influence (CHAMPAGNAT 1961). Conifers respond to removal of the vertical leader shoot, or to girdling between the leader and the laterals, by hyponastic movement of the lateral branches, due to the formation of compression wood on the abaxial or lower side (PHARIS and KUO 1976, WILSON 1973). This is accompanied by increased C_2H_4 evolution from the lateral branch. The amount of extractable C_2H_4 is greatest in the lower half and is correlated with the degree of hyponasty (BLAKE et al. 1980). The application of IAA to the end of the decapitated shoot initially increases C_2H_4 production and the hyponastic response, but with time, C_2H_4 production decreases below that of decapitated control plants and the rate of hyponastic movement declines also. Apical dominance may therefore regulate branch orientation in some conifers by controlling endogenous C_2H_4 production by the branch. It is not known whether the differential formation of compression wood, which is the cause of the change in orientation, is directly affected by C_2H_4 or is controlled in some other way (WILSON and ARCHER 1977, 1981). In contrast to these findings for correlative systems, the direct application of IAA to stolons and rhizomes produces transient curvatures (HANSEN and BENDIXEN 1973) or loss of plagiotropic growth (BENNET-CLARK and BALL 1959).

2.3 Gibberellins

The light-induced unrolling of leaves of grasses and cereals is caused by the relative promotion of expansion growth in the adaxial half of the lamina and is a form of epinasty (BURSTRÖM 1942). The process is promoted by light-induced GA-like substances, but can occur in darkness following gibberellin treatment (Sect. 4.1) (KANG 1979). In many dicotyledons, GA_3 induces hyponasty in leaves by promoting growth of the abaxial half of the petiole and midrib (GRAY 1957, BARBER et al. 1968). Whether this response is brought about by the direct action of GA_3, or indirectly by changing the effectiveness of IAA is not known. GA_3 also reduces the ability of ethylene to induce petiole epinasty by counteracting its inhibitory effect upon cell expansion in the lower half of the petiole (PALMER 1972).

In stems, GA_3 promotes hook opening in the bean (KLEIN 1965), pea (NAKAMURA et al. 1966), and cotton seedling (POWELL and MORGAN 1970). In etiolated pea seedlings GA_3 application does not affect C_2H_4 synthesis (STEWART et al. 1974), but appears to act by antagonizing C_2H_4 action (FUCHS and LIEBERMAN 1968). In a number of conifer species, GA_3 promotes the hyponastic movement of lateral branches, and treatments that reduce endogenous GA production result in epinastic curvatures (PHARIS and KUO 1976). In *Cupressus arizonica*, the hyponastic response of branches to GA_3 is accompanied by a three-to four fold increase in C_2H_4 production. Since exposing plants to C_2H_4 also induces hyponastic branch movement, it has been proposed that GA_3 induces hyponasty in conifers by promoting C_2H_4 synthesis (BLAKE et al. 1980).

2.4 Miscellaneous Chemicals

2.4.1 Brassins

The growth-regulating brassinosteroids, brassinolide, 2α-3α,22,33,-tetrahydroxy-24-methyl-6,7-s-5α-6,7-lactone (GROVE et al. 1979), and its structural isomers such as homobrassinolide induce a pronounced downward inclination in the lamina of the second leaf in rice seedlings. This effect is caused by epinastic curvature in the junction region between the lamina and the leaf sheath, decreasing the angle subtended by the lamina with the sheath (MAEDA 1965). The response which occurs in both intact and isolated leaves is evoked by low concentrations in the range 10–1000 ng plant^{-1} (WADA et al. 1981). Although this lamina response to brassins shows strong synergism with applied NAA and IAA, C_2H_4 does not appear to be involved, since exposure of rice seedlings to C_2H_4 in the range 1–100 µl l^{-1}, and treatment with ethrel, has no effect upon leaf lamina inclination (TAKENO and PHARIS 1982).

2.4.2 Halogenated Phenolics

The halogenated phenolic compounds, 2,6-dichloro, -dibromo- and diiodo-phenol induce severe epinasty when applied to tomato leaf petioles as a 2% mixture in lanolin. With dibromophenol and diiodophenol the induced epinasty persists, whereas normal leaf orientation is gradually restored after dichlorophenol application (WAIN and TAYLOR 1965). It is not known whether this induced epinasty is accompanied by increased C_2H_4 evolution.

2.4.3 Morphactins and Malformin

Morphactins block the gravitropic sensitivity of plant stems (SCHNEIDER 1964). The morphactin Chlorflurenol (methyl-2-chloro-9-hydroxyfluorene-9-carboxylate), when applied to sunflower seedlings, causes the stem to become ageotropic, but in addition induces acute petiole hyponasty (PALMER unpublished).

Petiole epinasty induced by low concentrations of the cyclic peptide Malformin B, is accompanied by increased ethylene production in shoots of *Phaseolus vulgaris*, suggesting that the action of this growth regulator is via regulation of ethylene synthesis and that epinasty is a symptom of this, rather than changed IAA physiology (CURTIS 1968, CURTIS and JOHN 1975).

3 Epinastic and Hyponastic Responses to Parasitic Microorganisms

A number of disease-causing viruses, bacteria, and fungi can induce characteristic changes in shape or orientation in organs of the host plant. Petiole and stem epinasty is induced by cucumber mosaic virus (LEVY and MARCO 1976),

Bacterium solanacearum (GRIEVE 1939), *Bacterium tumefaciens* (LOCKER et al. 1938), *Fusarium oxysporum* (DIMOND and WAGGONER 1953, WELLMAN 1941) and *Pseudomonas solanacearum* (PEGG and CRONSHAW 1976). Hyponastic responses to infection appear to be less common, but the erection of leaves and stems has been reported following infection by the soft rot bacterium *Erwinia phytophthora* (HORSFALL and DIMOND 1959) and by the mycoplasma, Aster Yellows (KUNKEL 1926).

Similarity between disease and ethylene-induced epinasty (HORSFALL and DIMOND 1959) has led investigators to determine C_2H_4 levels in disease infected plants. WIESE and DE VAY (1970) showed that infection of cotton plants by strains of *Verticillium albo-atrum* induced leaf epinasty and increased endogenous C_2H_4 production. ROSS and WILLIAMSON (1951) found a marked increase in C_2H_4 production by shoots of *Physalis floridians*, infected with potato virus, accounting for the development of pronounced leaf epinasty. The C_2H_4 emanation was associated with the appearance of necrotic lesions on the leaves suggesting they may be the site of C_2H_4 synthesis (BALÁZS et al. 1969). In other diseases, petiole epinasty occurs in infected leaves without the appearance of lesions, although C_2H_4 production is increased, for example in tomato plants infected with *Fusarium oxysporum* (DIMOND and WAGGONER 1953). Ethylene release by *Tetragonia expansa* infected with yellow mosaic virus is due to an acceleration of tissue senescence, rather than to induced cell necrosis (GÁBORJÁNYI et al. 1971). Although bacteria, fungi, and other pathogens are often capable of in vitro C_2H_4 production, it is generally held that increased C_2H_4 production by diseased higher plants is wholly of host origin (WILLIAMSON 1950), induced in a similar way to environmental stress, (PEGG 1976, PEGG and CRONSHAW 1976) and Chap. 16, this Vol.

4 Epinasty Induced by Physical Factors

4.1 Light

4.1.1 Wavelength

Leaf epinasty can be induced in *Salvia occidentalis* by exposure to R in a 16-h long-day treatment at an intensity of 370 μW cm^{-2} (MEIJER 1957). Only a brief exposure to R is sufficient to induce downward curvature in the rhizome of *Aegopodium podagraria* (BENNET-CLARK and BALL 1959). In *Dentaria diphylla* R (45 μW cm^{-2}) promotes the opening of the petiolar hook and lamina unrolling in the etiolated leaves that arise from the rhizomatous stem, while FR maintained the dark-formed hook and the rolled lamina (YOPP 1973). FR also induces epinasty in the cotyledons of *Sinapis alba* (BAUMGARTNER and FONDEVILLE 1980). Such R, FR responses suggest the involvement of the phytochrome system in these morphogenetic events. Opening of the hook of dark-grown seedlings of *Phaseolus vulgaris* is also promoted by R and inhibited by FR (WITHROW et al. 1957), and appears to be under the control of the phytochrome

Fig. 2. Change in peroxidase, IAA oxidase and IAA content in the opening hook of *Phaseolus vulgaris*, following exposure to a red light pulse. (After DE GREEF et al. (1977)

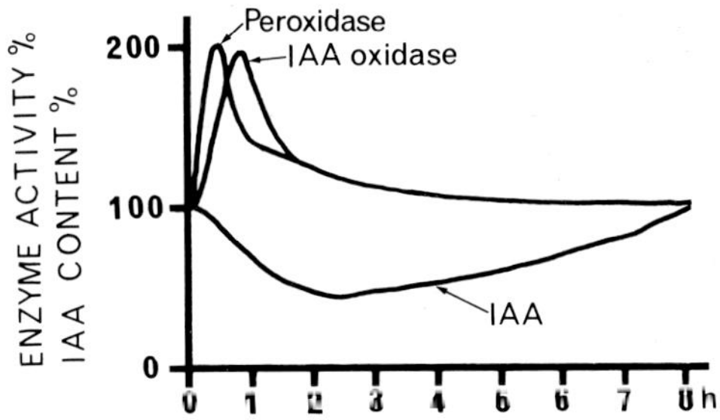

system (KLEIN et al. 1967). Hook opening in etiolated seedlings of *Cucumis sativus* and dodder, *Cuscuta gronovii*, is promoted by both R and FR (PORATH 1977), although for cucumber, FR can become antagonistic to hook opening if it is supplied at the same time as R. In dodder, opening is also promoted by blue light, operating through the "high energy reaction" (KUJAWSKI and TRUSCOTT 1974). In etiolated lettuce seedlings, the hypocotyl remains straight and no hook forms. MOHR and NOBLE (1960) found that R induces a hook to form, so that in lettuce, R promotes hook formation, apparently without changing C_2H_4 production (JANES and LOERCHET 1976). The etiolated bean hook responds to a saturating pulse of R by doubling the relative contents of indol-3-yl-acetate oxidase and peroxidase in 1 h, while there is a concomitant decrease in the endogenous IAA content (DE GREEF et al. 1977). (Fig. 2). The role of C_2H_4 in hook maintenance is considered in Section 2.1.2. Leaf unrolling in grasses and cereals is also promoted by R (VIRGIN 1962) and blue light (DEUTCH 1976; DEUTCH and DEUTCH 1975), and appears to be under the control of the phytochrome system (BEEVERS et al. 1970). The rapid appearance of GA-like substances in etiolated wheat leaves following R irradiation (REID et al. 1968) suggests that GA's may partly regulate leaf unrolling, since this is promoted by applied GA_3 and inhibited by AMO-1618 and CCC (LOVEYS and WAREING 1971).

4.1.2 Intensity

High light intensity is known to induce or enhance plagiotropic tendencies in the prostrate shoots, runners, or rhizomes of a number of plant species (ZIV et al. 1976, BENDIXEN and PETERSON 1962, MAIGE 1901, STAHL 1884) due to the development of a transmissible but not identified stimulus (PALMER 1956, 1958).

High light intensity induces epinastic curvature in the leaf base of *Plantago media*, so that the leaves become prostrate (VEEN and VAN DER MEIJER 1959). In *P. lanceolata*, this response is more prominent in some selection lines than in others (BARBER et al. 1968), and is enhanced by low temperature. Both the high light intensity and low temperature effects can be removed by treatment with GA_3 (BARBER et al. 1968). GA_3 can also block the appearance of the prostrate character in cultivars of *T. fragiferum* (BENDIXEN and PETERSON 1962) and in *Proserpinaca palustris* (WALLENSTEIN and ALBERT 1963).

4.2 Gravity

Disturbing the normal relationship of a plant organ to gravity by inclining the stem or root or rotating it on a clinostat can induce epinastic responses (LEATHER et al. 1972, PALMER 1964, ZIMMERMAN and HITCHCOCK 1938, SALISBURY and WHEELER 1981). Pronounced lateral shoot epinasty occurs when *Coleus* plants are rotated around the stem axis on a horizontal clinostat (LEIKE and GUTTENBERG 1962). These curvatures disappear after about 2 weeks due to the development of an autotropic counter-reaction. Leaves of *Tagetes patula* and *Coleus* leaf petioles also become epinastic during plant rotation on a horizontal clinostat (PFEFFER 1905, TIBBITTS and HERTZBERG 1978), but in the case of *Coleus* the curvature response dies away after a few days and the leaf petioles resume their normal shape. However, tissue growth is stimulated at the base of the petiole, in the region of the abscission zone, leading to an increase in the angle subtended by the petiole with the stem (PALMER, unpublished). These epinastic curvatures have been interpreted in terms of the nullification of the gravitational influence (BROWN et al. 1974, FREYTAG 1930), leading to the movement of auxin toward the normally upper side of the organ (LEIKE and GUTTENBERG 1962). Whether rotation of a plant on the horizontal clinostat reproduces a condition of true weightlessness is debatable (BROWN et al. 1976, TIBBITTS and HERTZBERG 1978). There is evidence that the epinastic curvature seen in leaf petioles of tomato during rotation on the horizontal clinostat is accompanied by increased C_2H_4 (LEATHER et al. 1972). Ethylene production is, however, promoted by simply placing the stems of herbs horizontally (DENNY 1936) or by bending down the branches of fruit trees (LEOPOLD et al. 1972, ROBITAILLE and LEOPOLD 1974). While PALMER (1973) found that the epinastic curvature of leaf petioles of detached inverted leaves coincided with increased ethylene release. If changing the natural orientation of plant organs generally increases C_2H_4 production, the possibility arises that epinastic responses of plants on the horizontal clinostat may reflect the maintenance of the plant in a disoriented state rather than gravity nullification. How gravity may induce polarity in a plant organ inclined to the vertical, so that one side, the upper, can subsequently develop epinastic type reactions (geo-epinasty) is considered by RUFELT (1962), KALDEWEY (1971) and LUNDEGÅRDTH (1966).

4.3 Waterlogging

Leaf epinasty is a common early symptom of root waterlogging. It is often followed by the appearance of other characteristic responses such as reduced stem elongation (WENT 1938, 1943), enhanced leaf senescence (BURROWS and CARR 1969), premature leaf and flower abscission (LLOYD 1914) and adventitious root development (KRAMER 1951). The similarity between these symptoms and those induced by C_2H_4 in plants had been commented upon (TURKOVA 1964), but not until the advent of GC-FID was it possible to measure changes in the endogenous level of C_2H_4 in plant tissues accurately. The involvement of C_2H_4 in responses to waterlogging was then established. KAWASE (1974) found that flooding the roots of potted sunflower plants raised the C_2H_4 content of the shoot. Large increases occurred in the C_2H_4 concentration in the air spaces of bean and tomato leaves during the first 48 h of waterlogging (EL-BELTAGY and HALL 1974, JACKSON and CAMPBELL 1975a). An increase in the C_2H_4 content of the shoot may therefore be a common response to root waterlogging (CLEMENS and PEARSON 1977).

Waterlogged soils actively synthesize C_2H_4, which may attain a concentration equivalent to 8 µl l^{-1} in the soil atmosphere (SMITH and RUSSELL 1969). Such C_2H_4 production depends upon low O_2 concentrations and does not occur in sterilized soil, suggesting that its production is dependent upon microbiological action (SMITH and RESTALL 1971). There is some evidence that the anaerobic environment of waterlogged soil may stimulate C_2H_4 synthesis by the root (KAWASE 1972), although the site of synthesis may be the root rhizosphere, since O_2 is apparently required for C_2H_4 production by flowering plant tissues (MAPSON 1969). Waterlogging restricts the loss of C_2H_4 to the atmosphere by gaseous diffusion and also reduces the emanation of C_2H_4 from root tissues (ZERONI et al. 1977), thus enhancing the retention of C_2H_4 by root tissues. This may be the basis for the "water effect", whereby soil water enhances the effects of soil anoxia (WAMPLE and REID 1975, 1978). The possibility that C_2H_4, originating in the soil or plant roots, may move into the shoot and induce petiole epinasty has been examined following an early demonstration (ZIMMERMANN 1927) that petiole epinasty could be induced in tomato plants by exposing their roots to C_2H_4. JACKSON and CAMPBELL (1975b) found that 10.0 µl l^{-1} ^{14}C-C_2H_4 was taken up by tomato roots and could be detected in stem and leaf tissues after 1 h. They also showed that when the C_2H_4 content of the soil exceeds about 2.0 µl l^{-1}, then sufficient C_2H_4 moves into the shoot system of the tomato plant to exceed the threshold (0.04 µl l^{-1}) required to initiate petiole epinasty. However, the resistance to C_2H_4 movement through the root and stem tissues of *Vicia faba*, and the rate of emanation from stem tissue effectively preclude soil- or root-synthesized C_2H_4 from accumulating in leaf or stem tissues in sufficient concentration to induce petiole epinasty (ZERONI et al. 1977). BRADFORD and DILLEY (1978) reached similar conclusions for dwarf tomato plants. Since all the available evidence indicates that petiole epinasty is a response to an increase in the C_2H_4 content of the shoot, these findings raise the possibility that waterlogging can directly stimulate C_2H_4 production in the shoot. Evidence that this may be the case comes from the finding that, under anaerobic conditions, the roots of a dwarf tomato cultivar showed no change in C_2H_4 evolution, while synthesis of C_2H_4 in the shoot increased from 25 to 70 p mol h^{-1} in the first 20 h (BRADFORD and DILLEY 1978). Normal tomato plants respond similarly, and in this case the possibility that root anoxia is equivalent to isolating the root from the shoot was excluded by the finding that excising tomato shoots did not result in either the accumulation of C_2H_4 or leaf epinasty, although the ability of the leaf petiole to respond to C_2H_4 was retained (JACKSON and CAMPBELL 1975b). If waterlogging directly stimulates shoot C_2H_4 synthesis, the stimulus is unlikely to be small amounts of root C_2H_4 acting autocatalytically, since C_2H_4 synthesis appears to be restricted by root anoxia (JACKSON et al. 1978). Although shoot wilting or the occurrence of water stress has commonly been associated with waterlogging (EL-BELTAGY and HALL 1974), it generally follows, rather than precedes, the occurrence of epinasty (JACKSON and CAMPBELL 1976) and may be entirely absent without restricting the development of shoot C_2H_4 (JACKSON et al. 1978). JACKSON and CAMPBELL (1975b, 1976) also established that the movement into the shoot of transpiration water low in O_2 content does not stimulate leaf epinasty in tomato, and they concluded that the low O_2 content of waterlogged soil directly stimulates the production or release of a root factor, which is not C_2H_4, and that this factor moves into the shoot system, where it induces leaf epinasty. This factor was subsequently identified as ACC. BRADFORD and YANG (1980) established that the flux of ACC in the xylem sap of tomato plants increases appreciably during the first 12 h of flooding, prior to both the rise in shoot C_2H_4 and the development of leaf epinasty. ACC is known to accumulate in plant tissues under anaerobic conditions and to be converted to C_2H_4 when O_2 becomes available. An alternative explanation is that leaf epinasty, under conditions of waterlogging, results from the diminution in a root factor that normally holds in check endogenous C_2H_4 production by the shoot. GA_3 has been shown to reduce the effectiveness of ethylene in petiole epinasty (PALMER 1972), and it is known that the level of GA_3 in the shoot declines rapidly in the tomato plant following flooding of the root system (REID et al. 1969, REID and CROZIER 1971). Waterlogging also reduces the movement of endogenous root cytokinins from the root into the shoot (BURROWS and CARR 1969). It has been demonstrated that treatment of tomato plants with BA suppresses the development of petiole epinasty in subsequent waterlogging (RAILTON and REID 1973, WAMPLE and REID 1978). It is possible that root-synthesized cytokinins may modulate the synthesis of C_2H_4 in the shoot, but supporting evidence is lacking, and application of BA does not reduce the elevated rate of C_2H_4 production by flooded tomato (JACKSON and CAMPBELL 1979) or sunflower plants (WAMPLE and REID 1979). Other difficulties in adopting any hypothesis involving control by a root factor are the findings that severance of the tomato shoot does not induce petiole epinasty, and that epinasty induced in tomato leaves by flooding is not relieved by the promotion of adventitious roots, although these effectively overcome the inhibitory effect of flooding on stem elongation (JACKSON and CAMPBELL 1975b, 1976). Interrupting the cellular connection between the shoot and the roots by stem girdling, while maintaining continuity of the water columns in the nonliving xylem system, results in the rapid development of leaf epinasty (PHILLIPS 1964). The present evidence therefore favors the conclusion that the development of root anoxia during waterlogging generates ACC that is conveyed to the shoot via the transpiration stream, and is there converted to C_2H_4 by aerobic tissues (BRADFORD and YANG 1980).

5 Hormonal Theories

5.1 Lateral Shoots and Leaf Petioles

5.1.1 Auxin Direction

This concept of hormonal control of plagiotropism in lateral shoots developed out of the theory of negative gravitropism. DOLK (1936) showed that the horizontal orientation of coleoptiles established a transverse auxin gradient toward the lower side in the ratio of 7:3. The inclined orientation of plagiotropic organs was attributed to the hypothetical epinastic influence counteracting this downward movement by directing auxin into the upper half, and thereby creating an opposing growth reaction to negative gravitropism. This idea seems to have come from LAAN's finding that when C_2H_4 induced horizontal inclination in the etiolated bean epicotyl, there was a relative increase in the auxin content of the upper half (VAN DER LAAN 1934). Its applicability to plagiotropism came initially from the observation that IAA or IAA analogs could increase epinasty in lateral shoots, either by direct application or indirectly by application to the parent orthotropic shoot (HITCHCOCK 1935, ZIMMERMAN and WILCOXON 1935). Experiments to verify the theory were performed with plants rotating on a horizontal clinostat, since this was thought to enhance the visible epinastic response by nullifying negative gravitropism. It was shown that the epinastic curvature of lateral shoots of *Coleus* required the presence of young leaves in the terminal bud of either the parent or lateral and that this influence could be replaced by 1 mg g^{-1} IAA in lanoline paste (LEIKE and GUTTENBERG 1962). LYON (1963a) confirmed this for *Coleus* and also used the theory to explain the epinastic curvature of leaf petioles on the clinostat (LYON 1963b). In this way, the concept was established that a mobile auxin gradient was an essential element in the epinastic curvature response. The ubiquitous occurrence of epinasty, irrespective of the site of auxin application, was accounted for by the assumption that IAA was rapidly moved into the upper half of the responding organ. This could not, however, explain why the direct application of IAA to the lower half did not stimulate growth there, but induced epinasty instead. No significant difference was found in the auxin content of the upper and lower halves of downward-curving branches of the weeping mulberry, *Morus alba* var. *pendula*, to account for their downward direction of growth (RECHES et al. 1974).

Additional support for the auxin direction theory is claimed from LYON's experiments with plants rotating on a clinostat in which ^{14}C-IAA was applied either to the end of a decapitated lateral shoot (LYON 1963a) or to the surface of the leaf lamina (LYON 1963b). After 40–46 h the distribution of ^{14}C-IAA in the upper and lower half of either the shoot or leaf petiole was determined by chemical extraction. Appreciable radioactivity was always found in each half, but the quantity was greater in the upper half, the excess ranging from 14–145%. A number of deficiencies are apparent in LYON's experimental procedures, which raise doubts about the validity of his findings. While it was reported that lateral shoots and leaves on nonrotated control plants failed to show any epinastic response to applied ^{14}C-IAA, the distribution of ^{14}C in these organs was not determined. Thus the extent to which rotation was responsible for the ^{14}C inbalance cannot be judged. Furthermore, the extraction of ^{14}C was done after the curvature response was completed, so it is not clear whether the reported asymmetry was present while the curvature was developing. It is now known that rotation of plants on a clinostat can stimulate C_2H_4 production, and hence it is difficult to interpret experiments such as LYON's in which IAA application and rotation were combined. Plagiotropic branches of rooted cuttings of Douglas-fir transport ^{14}C-IAA more actively through the lower half of the stem, which is also the site for the compression wood. However, significantly more ^{14}C was located in the upper half of horizontal stems 24 h after removing the apical bud and applying ^{14}C-IAA for 8 h (STARBUCK and ROBERTS 1982).

Early explanations of ethylene-induced leaf epinasty also adopted the auxin direction theory (WENT and THIMANN 1937). Later, LYON (1970) endorsed this, after finding that ^{14}C-IAA applied to the laminae of pepper and tomato leaves could subsequently be extracted in greater amounts from the upper side of the petiole when the parent plant was exposed to C_2H_4 to induce epinasty. However, the region of the lamina to which the ^{14}C-IAA was applied was not specified, although it is known that this largely determines whether its subsequent movement is predominantly through vascular tissue in the upper or in the lower half (YAMANE 1941, YIN 1941). Contrary to LYON's findings, C_2H_4 had no significant directional effect upon the distribution of ^{14}C-IAA in the petiole of *Coleus blumei* × *frederici* while the epinastic response was developing (PALMER 1976). Similarly, epinastic coiling of tendrils of *Passiflora quadrangularis* induced by exposure to CO_2 or to mechanical rubbing is not accompanied by any significant lateral redistribution of ^{14}C-IAA that is moving through the tendril at the time (JUNKER 1977). Despite its continuing popularity, the theory that the hormonal basis for epinasty is the direction of auxin into the upper half of the plagiotropic organ in opposition to the gravitropic influence still lacks any convincing evidence to support it.

5.1.2 Differential Sensitivity

The application of IAA to either the upper or lower surface of the young growing leaf of *Phaseolus vulgaris* induces hyponastic curvature because of the greater sensitivity of the lower half to IAA (HAYES and LIPPINCOTT 1976, LIPPINCOTT and LIPPINCOTT 1971). Such differential sensitivity to IAA is common in leaves, although it more often results in epinasty. Thus the petioles of *Tropaeolum* and *Coleus* show a greater elongation response to IAA when this is applied to the upper side rather than to the lower, suggesting that an inherent differential sensitivity between the tissues of the two halves may be the cause of IAA-induced epinasty (VARDAR 1953). Detailed examination of this effect using tomato leaf petioles showed that the direct application of 2% IAA-lanolin paste to the lower side did not induce any upward curvature or hyponasty, but invariably promoted epinasty (KAZEMI and KEFFORD 1974). In tomato petiole explants, IAA application stimulates endogenous C_2H_4 production as well as inducing epinasty. Since IAA has no additive or synergistic effect upon the magnitude of C_2H_4-induced epinasty, it would appear that epinasty is a direct consequence of the differential sensitivity of tomato petiole tissue to C_2H_4 (KAZEMI and KEFFORD 1974), tissue elongation being promoted in the upper half and inhibited in the lower half, over a large range of C_2H_4 concentrations. While IAA promotes ethylene production in the leaf blade of *Phaseolus vulgaris*, the ensuing epinastic curvature is considered to result from ethylene-induced inhibition of IAA transport from the upper to the lower side of the blade, since ethylene alone does not induce epinastic curvature in the short term (HAYES and LIPPIN-

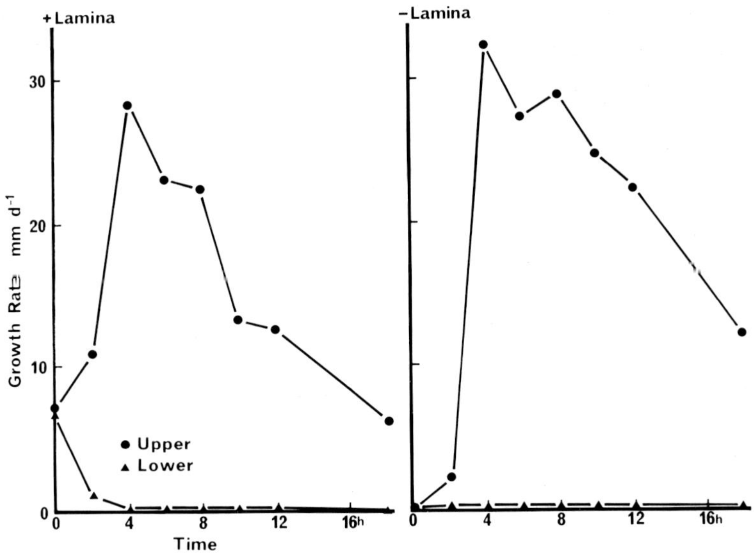

Fig. 3. Differential response to 50 μl l^{-1} ethylene air of the upper and lower half of the intact attached petiole of *Helianthus annuus*. *Left graph* transient promotion of elongation in the upper half and simultaneous inhibition of elongation in the lower. *Right graph* initial zero growth due to prior removal of lamina 1 d before ethylene treatment; showing the retention by each half of the differential response. Rates are calculated from increments of a 50 mm segment marked on the upper and lower surfaces. Values are means for 36 measurements and 5 replicates (PALMER unpublished)

COTT 1981). It has been suggested that the ability of the apical bud to induce epinastic responses in petioles (and also lateral shoots) may be due to endogenous auxin originating in the apical bud and stimulating C_2H_4 production in stem tissues adjacent to the lateral shoot or leaf by diffusion (KAZEMI and KEFFORD 1974). Demonstations that IAA can substitute for the apical influence have relied upon the use of high concentrations of IAA in lanolin paste (SNOW 1945, PALMER and PHILLIPS 1963), and attempts to show that endogenous IAA in the stem raises the rate of C_2H_4 production in the nodal regions have not been successful (BURG and BURG 1968). Hyponastic coiling of tendrils of *Cucumis sativus* is due to the differential production of C_2H_4 by the upper and lower sides under the stimulus of mobile IAA (BANGERTH 1974). Coiling of the tendril of *Passiflora quadrangularis*, induced by contact stimulus, CO_2, or IAA, does not appear to involve C_2H_4 as an intermediary, but is due to the differential sensitivity of the adaxial surface to these agents (JUNKER 1976). C_2H_4-induced petiole epinasty in *Helianthus annuus* is also due to the differential sensitivity of the upper and lower halves. Concentrations of $C_2H_4 > 1.0$ μl l^{-1} air promote elongation in the upper half, but inhibit it in the lower. This differential response is independent of growth factors originating in the lamina, since removing it rapidly halts normal elongation growth, but increases the responsiveness of the upper half to C_2H_4 (PALMER 1972). Time-course studies show that the C_2H_4-induced elongation is transient. Its cessation may be due to

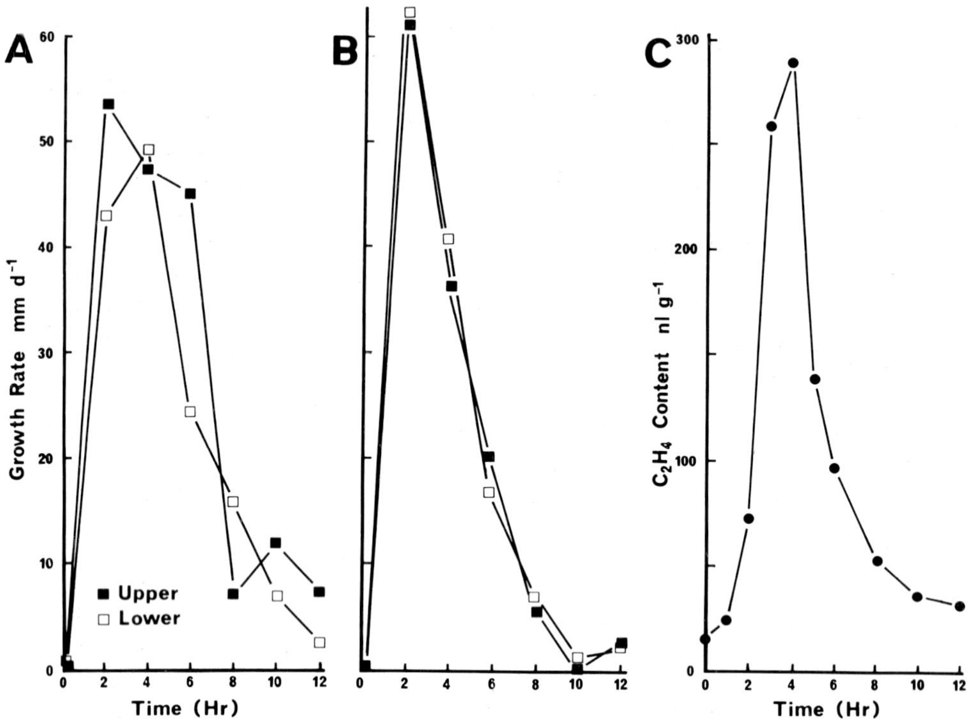

Fig. 4. A, B Elongation response of the upper and lower half of the intact attached petiole of *Helianthus annuus* to 30 µl 10^{-3} M IAA solt., applied to the upper and lower surfaces **A** Lamina removed 24 h before IAA application; *upper half* showing a greater initial response and a second response at 6–8 h. Epinasty appears from 2 h. **B** Lamina removed 24 h before IAA application and petioles pretreated with 10^{-4} M $AgNO_3$ solt; showing identical single response of each half. No epinasty. **C** Ethylene evolution by entire petiole, after application of 60 µl 10^{-3} M IAA solt. Peak in ethylene production coincides with the second response of the *upper half* in **A**. Rates shown in **A** and **B** are calculated from increments of a 50-mm segment marked on the upper and lower surfaces. Values shown in **A** and **B** are means for 28 and 11 measurements respectively and at least two replicates. Values shown in **C** are means for four replicates (PALMER unpublished)

the exhaustion of a co-factor (Fig. 3). This is not likely to be a member of the known growth regulator groups, since the addition of either a cytokinin, IAA, or GA_3, does not prolong the duration of the response (PALMER unpublished). However, if the plant is returned to air and allowed to photosynthesize, then C_2H_4 sensitivity is restored to the petiole after about 12 h. This occurs equally well in petioles from which the lamina has been removed, so that recovery does not depend upon the generation of a lamina factor or the movement of such a factor into the petiole (PALMER 1972). Pretreatment of the plant with GA protects the lower half of the petiole against C_2H_4 inhibition, but does not enhance the relative elongation response of the upper half, explaining why GA_3-treated plants show a diminished epinastic response to C_2H_4 (PALMER 1972).

The sunflower leaf petiole also exhibits asymmetric sensitivity to applied IAA, NAA, and 2,4-D. This can be demonstrated by their application in μg quantities to the upper and lower surfaces of the petiole, epinasty appearing after about 4–6 h. An examination of the elongation response of each half to IAA, for example, shows that the growth rate increases rapidly in both halves during the first 2 h, but then quickly declines in the lower half, while being maintained at a high rate for a further 4 h in the upper half (Fig. 4A). This shoulder coincides with a peak in C_2H_4 production by the petiole tissue (Fig. 4C), suggesting that in the upper half a period of C_2H_4-stimulated elongation may follow the initial response to IAA. Confirmation of this was obtained by observing the growth response to IAA in petioles pretreated with $AgNO_3$ solution to block C_2H_4 responses (PALMER 1979). The second 2–6 h elongation period then disappeared from the upper half, so that both halves showed identical responses to IAA (Fig. 4B). Since there was no longer any differential growth, no epinasty was seen and the petiole remained straight. This finding, and similar ones for cotton (STEWART and FREEBAIRN 1969), suggest that the role of endogenous IAA in petiole epinasty may be one of a mobile agent, directly stimulating growth and indirectly inducing epinasty, via C_2H_4 synthesis, in the same way as has been proposed for growth and hook formation in the etiolated plumule (KANG et al. 1971). This theory of petiole epinasty, however does not explain all kinds of leaf curvature induced by IAA. Lamina hyponasty in bean (LIPPINCOTT and LIPPINCOTT 1971), and midrib epinasty in the tomato leaf (KAZEMI and KEFFORD 1974) appear to be direct responses to IAA resulting from the differential sensitivity of the lamina or midrib tissues to IAA itself. The puzzling observation first reported by CROCKER et al. (1932), that inverted tomato plants do not show leaf epinasty in the presence of C_2H_4, still lacks an explanation based on experimental data, although it is worth recording that C_2H_4-induced petiole epinasty is not generally sensitive to leaf or shoot orientation.

5.2 Plumular Hook Formation

There is no difference in cell number or cell division rate between the two halves of the terminal region of the plumule that would explain in developmental terms the formation of the plumular hook in etiolated seedlings (KANG and BURG 1972), and therefore it is considered to be a product of hormone action, although gravity may play an initiating role in some cases (MACDONALD et al. 1983). A mobile hormone moving preferentially through one side of the plumule and stimulating elongation growth and radial expansion would cause the plumule to bend over to accommodate the inbalance in growth and to develop a hook-like appearance. Since IAA induces reformation of the hook in seedlings exposed to light, and is known to tighten the hook in etiolated seedlings (KANG and RAY 1969a), it has been favored as the hormone that could fulfill this function. However, asymmetry in the distribution of mobile IAA in the hook zone has not been generally demonstrated (KANG 1979), and it now seems more probable that IAA regulates hook formation indirectly by stimulating

C_2H_4 formation (KANG and RAY 1969b). The weight of present evidence for the pea and bean seedling is that the plumular hook results from differential sensitivity of the two sides of the plumule tip to C_2H_4 which is being actively produced by the region where the plumule hooks (KANG et al. 1967, GOESCHL et al. 1967). It is not clear whether the cells that will form the outer side of the hook are insensitive to C_2H_4, or whether their expansion is promoted (KANG 1979), but cell expansion is inhibited in the opposite half of the plumule by C_2H_4, so that growth occurring in the zone is differential. When the inhibited cells of the inner half pass out of the zone of C_2H_4 production, they commence a delayed phase of cell expansion. Consequently, that region of the plumule, the subhook zone, becomes radially symmetrical and straight, or even recurved (KANG et al. 1967). While hook formation and petiole epinasty are similar in that both result from differential responses to C_2H_4, the two phenomena differ in that the hook is a dynamic entity, dependent for its existence upon the formation of new cells by the terminal region, while epinasty occurs in relatively mature tissues that have ceased active cell division (Fig. 1A). The influence that the cotyledons exert in promoting hook formation (GEE and VINCE-PRUE 1976, POWELL and MORGAN 1970) has not been identified in hormonal terms, although there is the strong possibility that it is by IAA production, since it is now established that maintenance of the hook in the flower pedicel of the poppy depends upon the continuous movement of IAA through the pedicel from the site of production in the flower bud (KOHJI et al. 1982). The ability of the primary leaves to induce hook opening in the bean seedling, after their illumination with R, has not yet been integrated into the C_2H_4 theory of hook regulation because of the rapidity of the response (CAUBERGS and DE GREEF 1975). The basal region of the hook, the so-called shank, also promotes hook opening after R (RUBINSTEIN 1971). This control has been attributed to the accelerated production of GA's and the promotion of their acropetal transport (KLEIN 1965). Basally applied kinetin has been shown to reduce the sensitivity of the hook of mung bean to light, suggesting that cytokinins originating in the root may be able to influence hook formation (GEE 1977). Since cytokinins are able to stimulate C_2H_4 production in plumular tissue by synergistic action with IAA (FUCHS and LIEBERMAN 1968), this control may be exerted indirectly via C_2H_4 production (KANG and RAY 1969b). However, the extent to which the rate of C_2H_4 production by the hook region is governed by other hormones has not been precisely determined (KANG et al. 1971).

5.3 Hyponasty

Although C_2H_4-induced hyponasty has been described for plagiotropic shoots of strawberry clover (HANSEN and BENDIXEN 1974), peanut (ZIV et al. 1976), the diageotropica tomato mutant (ZOBEL 1973), and *Cupressus arizonica* (BLAKE et al. 1980), C_2H_4 does not generally induce hyponasty in runners, rhizomes, or plagiotropic aerial shoots, and in leaves and orthotropic green and etiolated shoots it has the reverse effect. It may therefore be that when C_2H_4 induces hyponasty its action is indirect. Possibly it blocks the plagiotropic control mechanism at a basic level so that the shoot reverts to orthotropic growth (JACKSON 1979). Leaf hyponasty that follows after C_2H_4-induced epinasty may be the result of cells in the lower half of the organ accumulating a potential for growth, while their elongation was inhibited during the period of epinasty. When the diminution in C_2H_4 level allows their growth to resume, this potential may then be released, giving relatively greater growth in the lower half and hence upward curvature. Alternatively, the promotion of growth in the upper half during epinasty may exhaust a growth factor, so that elongation becomes less than normal during the post-epinasty phase, leading to relatively greater growth in the lower half, and hence upward curvature (PALMER 1972). Transient leaf blade hyponasty in bean, following low temperature exposure, has been attributed to the stimulation of cell expansion in the lower side of the leaf lamina, following the arrival of IAA that had accumulated in the upper side due to the low temperature inhibition of IAA transport (HAYES and LIPPINCOTT 1981). Although GA_3 induces hyponasty in many plagiotropic organs, and with one exception (BLAKE et al. 1980) is reputed to do this by antagonizing C_2H_4 action, its mode of operation has not been investigated in detail. In the petiole of *Helianthus annuus*, GA_3 is ineffective in promoting growth or hyponasty in the absence of the lamina. It may therefore act by increasing the responsiveness of the petiole tissues to a lamina growth factor which may be IAA (PALMER 1964). Its preferential effect upon cells in the lower half of the petiole could be achieved if their ability to respond to a lamina factor is normally restricted by endogenous C_2H_4 and GA_3 overcomes this.

6 Concluding Remarks

Since the review of epinasty and hyponasty by KALDEWEY in 1962, the role of C_2H_4 as the initiator of hook formation in seedlings, and in almost all forms of petiole epinasty has become recognized, and the differential sensitivity of tissues to C_2H_4, IAA, and GA's is now more widely accepted as the basis for the curvature that characterizes epinasty and hyponasty. It remains to be determined how asymmetric sensitivity to a uniform hormone field could be established. Is it incorporated into the organ by developmental processes during early growth, or is it induced by exposure to polarized environmental influences such as the gravitational field, as suggested by MACDONALD et al. (1983) to account for the induction of the hypocotyl hook? TREWAVAS and JONES (1981) have proposed that differential sensitivity may be created by the presence or absence of specific receptor proteins for growth regulators, the synthesis of these being controlled by a polarized environmental or physiological factor. The time may now be ripe to reappraise the assumption that the epinastic tendency, as distinct from the epinastic response, is an inherent component of all plagiotropic growth. An alternative view is that the type of epinasty manifested by exposure to C_2H_4 is normally latent within orthotropic and plagiotropic stems, but can be evoked by internal processes such as aging or disease, or by environmental factors when they serve to deflect the organ away from an equilibrium position that is separately determined. If epinasty is not normally involved in orientation control, its purpose then needs explanation.

An evident role for the plumular hook is the protection of the terminal bud as the plumule moves through the soil to the surface. Similarly, the hook in the flower pedicel would offer protection to the bud during its passage through the shoot canopy. Leaf epinasty in diseased and waterlogged plants may assist survival by reducing the heat load on the leaf surface during the central part of the day in the way that wilting does for drought-affected plants. Its occurrence in orthotropic plant shoots made prostrate by wind or rain (PALMER 1964) may aid re-erection of the shoot by negative gravitropism and autonomic straightening, by drawing the leaves inward toward the stem, and thus preventing their entanglement with surrounding plants as the stem returns to the vertical. In senescent leaves, epinasty may serve to prevent premature desiccation by reducing transpiration.

This view restricts epinasty to organs in which the tendency can be revealed as a downward curvature; but epinasty has also been implicated in the orientation control of lateral shoots and roots, runners, and rhizomes, which do not normally show epinastic curvature when exposed to C_2H_4 or physiological concentrations of IAA. The evidence for the participation of the epinastic influence in these organs comes mainly from observations on apical dominance and clinostat experiments. There is little evidence that this type of influence, serving to deflect the equilibrium position of the organ away from the vertical, is similar to the epinastic curvatures seen in leaf laminae and petioles, seedling plumules, flower pedicels, and tendrils. KANG (1979) has pointed out that plagiotropic orientation, as distinct from curvature responses, can be accounted for in theory by postulating the occurrence of gravity-induced counter-adaptation that cancels out gravitropism. However, there is no need to relate this concept to auxin movement, since it could equally well be postulated that it creates a gradient of auxin sensitivity across the organ which exactly nullifies the auxin concentration gradient that is supposed to be established in gravitropism. As little is known about the hormonal control of the vertical orientation of orthotropic green shoots (WILKINS 1979), it would be futile at the present time to attempt to translate these theoretical concepts of plagiotropism into physiological terms. Where changes in the orientation of plagiotropic shoots have been produced by hormone application, the response may represent a switch from the operation of one kind of hormone system to another (TREWAVAS 1982), rather than a direct response to the applied hormone.

References

Abeles FB (1973) Ethylene in plant biology. Academic Press, London New York

Abeles FB, Rubinstein B (1964) Regulation of ethylene evolution and leaf abscission by auxin. Plant Physiol 39:963–969

Amrhein N, Schneebeck D (1980) Prevention of auxin-induced epinasty by α-aminooxyacetic acid. Physiol Plant 49:62–64

Balázs E, Gáborjányi R, Toth A, Király Z (1969) Ethylene production in Xanthi tobacco after systemic and local virus infections. Acta Phytopathol 4:355–358

Ball NG (1969) Tropic, nastic and tactic responses. In: Steward FC (ed) Plant physiology: a treatise. vol VA Academic Press, London New York, pp 119–228

Bangerth F (1974) Interaktionen von Auxin und Äthylen bei der thigmotropen Bewegung der Ranken von *Cucumis sativus*. Planta 117:329–338

Barber HN, Halsall DM, Palmer JH (1968) Genetic, environmental and physiological control of leaf orientation in *Plantago lanceolata*. Aust J Biol Sci 21:641–648

Baumgartner N, Fondeville J (1980) Obtention d'épinastie des cotylédons de *Sinapis alba* en éclairement rouge lointain continu. Relations avec le développement de l' hypocotyle et des cotyledons. Physiol Plant 50:153–157

Beevers L, Loveys B, Pearson JA, Wareing PF (1970) Phytochrome and hormonal control of expansion and greening of etiolated wheat leaves. Planta 90:284–294

Bendixen LE, Peterson ML (1962) Physiological nature of gene-controlled growth form in *Trifolium fragiferum* L. II. Auxin-gibberellin relationships to growth form. Plant Physiol 37:245–250

Bennet-Clark TA, Ball NG (1959) The diageotropic behavior of rhizomes. J Exp Bot 10:69–86

Blake TJ, Pharis RP, Reid DM (1980) Ethylene, gibberellins, auxins, and the apical control of branch angle in a conifer, *Cupressus arizonica*. Planta 148:64–68

Booth A (1959) Some factors concerned in the growth of stolons in potato. J Linnol Soc 56:166–169

Borgström G (1939) Theoretical suggestions regarding the ethylene responses of plants and observations on the influence of apple emanations. Kgl Fysiograf Sällskap Lund Förh 9:135–174

Bradford KJ, Dilley DR (1978) Effects of root anaerobiosis on ethylene production, epinasty, and growth of tomato plants. Plant Physiol 61:506–509

Bradford KJ, Yang SF (1980) Xylem transport of 1-aminocyclopropane 1-carboxylic acid, an ethylene precursor, in waterlogged tomato plants. Plant Physiol 65:322–326

Brown AH, Chapman DK, Liu SWW (1974) A comparison of leaf epinasty induced by weightlessness or by clinostat rotation. Bioscience 24:518–520

Brown AH, Dahl AO, Chapmen DK (1976) Limitation on the use of the horizontal clinostat as a gravity compensator. Plant Physiol 58:127–130

Burg SP, Burg EA (1968) Auxin-stimulated ethylene formation: Its relationship to auxin-inhibited growth, root geotropism and other plant processes. In: Wightman F, Settlefield G (eds) Biochemistry and physiology of plant growth substances. Runge, Ottawa, pp 1275–1294

Burrows WJ, Carr DJ (1969) Effects of flooding the root system of sunflower plants on the cytokinin content of the xylem sap. Physiol Plant 22:1105–1112

Burström H (1942) Über Entfaltung und Einrollen eines mesophilen Grasblattes. Bot Not 7:351–361

Caubergs R, De Greef JA (1975) Studies on hook opening in *Phaseolus vulgaris* L. by selective R/FR pretreatments of embryonic axis and primary leaves. Photochem Photobiol 22:139–144

Champagnat P (1961) Dominance apicale, tropisms, epinastie. In: Ruhland W (ed) Encyclopedia of plant physiology vol XIV. Springer, Berlin Göttingen Heidelberg, pp 872–908

Clemens J, Pearson CJ (1977) The effect of waterlogging on the growth and ethylene content of *Eucalyptus robusta* Sm (Swamp Mahogany). Oecologia 29:249–255

Crane JC, Marei N, Nelson MM (1970) Growth and maturation of fig fruits stimulated by 2-chloroethylphosphonic acid. J Am Soc Hortic Sci 95:367–370

Crocker W, Zimmerman PW, Hitchcock AE (1932) Ethylene-induced epinasty of leaves and the relation of gravity to it. Contrib Boyce Thompson Inst 4:177–218

Curtis RW (1968) Mediation of a plant response to malformin by ethylene. Plant Physiol 43:76–80

Curtis RW, John WW (1975) Effect of malformin on phytochrome and ethrel-mediated responses. Plant Cell Physiol 16:719–728

Denny FE (1936) Gravity-position of tomato stems and their production of the emanation causing leaf epinasty. Contrib Boyce Thompson Inst 8:99–104
Denny FE, Midler LP (1935) Production of ethylene by plant tissue as indicated by the epinastic response of leaves. Contrib Boyce Thompson Inst 7:97–102
Deutch B (1976) Barley leaf unfolding under mixtures of red and blue light. Physiol Plant 38:57–60
Deutch B, Deutch BI (1975) Blue light induction of barley leaf unfolding. A phytochrome reaction? Physiol Plant 35:322–327
Dimond AE, Waggoner PE (1953) The cause of epinastic symptoms in *Fusarium* wilt of tomatoes. Phytopathology 43:663–669
Dolk HE (1936) Geotropism and the growth substance. Rec Trav Bot Neerl 33:509–585
Eckerson SH (1931) Influence of phosphorus deficiency on metabolism of the tomato (*Lycopersicum esculentum*). Contrib Boyce Thompson Inst 3:197–217
Edgerton LJ, Blanfield GD (1968) Regulation of growth and fruit maturation with 2-chloroethanephosphonic acid. Nature 219:1064–1065
Eisinger WR, Morre DJ (1971) Growth-regulating properties of picloram, 4-amino-2, 5, 6-trichloropicolinic acid. Can J Bot 49:889–897
El-Beltagy AS, Hall MA (1974) Effect of water stress upon endogenous ethylene levels in *Vicia faba*. New Phytol 73:47–60
Fischnich O (1935) Über den Einfluß von Indolylessigsäure auf die Blattbewegungen und die Adventivwurzelbildung von *Coleus*. Planta 24:552–583
Freytag H (1930) Untersuchungen über die Plagiotropie der Blätter von *Tropaeolum maius*. Planta 12:267–292
Fuchs Y, Lieberman M (1968) Effects of kinetin, IAA and gibberellin on ethylene production, and their interactions in growth of seedlings. Plant Physiol 43:2029–2036
Gáborjányi R, Balázs E, Király Z (1971) Ethylene production, tissue senescence and local virus infections. Acta Phytopathol Hung 6:51–55
Gane R (1934) Production of ethylene by some ripening fruits. Nature 134:1008
Gee HM (1977) The control of hypocotyl hook angle in *Phaseolus mungo* L. The role of growth substances. J Exp Bot 28:367–376
Gee HM, Vince-Prue D (1976) Control of the hypocotyl hook angle in *Phaseolus mungo* L.: The role of parts of the seedling. J Exp Bot 27:314–323
Goeschl JD, Kays SJ (1975) Concentration dependencies of some effects of ethylene on etiolated pea, peanut bean and cotton seedlings. Plant Physiol 55:670–677
Goeschl JD, Pratt HK, Bonner B (1967) An effect of light on the production of ethylene and the growth of the plumular portion of the etiolated pea seedling. Plant Physiol 42:1077–1080
Gray RA (1957) Alteration of leaf size and shape and other changes caused by gibberellins in plants. Am J Bot 44:674–682
Greef de JA, Hoof van R, Caubergs R (1977) Light-induced changes in auxin metabolism during hook opening of etiolated bean seedlings. Biochem Soc Trans 5:1049–1051
Grieve BJ (1939) Epinastic response induced in plants by *Bacterium solanacearum* EFS Ann Bot NS 3:587–601
Grove MD, Spencer GF, Rodwedder WK, Mandava N, Worley JF, Warthen JD, Steffens GL, Flippen-Anderson JL, Cook JC (1979) Brassinolide, a plant-growth-promoting steroid isolated from *Brassica napus* pollen. Nature 281:216–217
Hansen DJ, Bendixen LE (1974) Ethylene-induced tropism of *Trifolium fragiferum* L. stolons. Plant Physiol 53:80–82
Harvey EM (1913) The castor bean plant and laboratory air. Bot Gaz 56:439–442
Hayes AB (1981) The interaction of auxin and ethylene in the maintenance of leaf blade form in *Phaseolus vulgaris* L. var. Pinto. Am J Bot 68:733–740
Hayes AB, Lippincott JA (1976) Growth and gravitational response in the development of leaf blade hyponasty. Am J Bot 63:383–387
Hayes AB, Lippincott JA (1981) The timing of an effect of temperature on auxin-induced hyponastic curvature of the bean primary leaf blade. Am J Bot 68:305–311

Hitchcock AE (1935) Indole-3-n-propionic acid as a growth hormone and the quantitative measurement of plant response. Contrib Boyce Thompson Inst 7:87–95

Hitchcock AE, Crocker W, Zimmerman PW (1932) Effect of illuminating gas on the lily, narcissus, tulip and hyacinth. Contrib Boyce Thompson Inst 4:155–176

Horsfall JG, Dimond AE (1959) The diseased plant. Plant pathology 1. Academic Press, London New York, pp 1–17

Jackson MB (1979) Is the diageotropic tomato ethylene deficient? Physiol Plant 46:347–351

Jackson MB, Campbell DJ (1975a) Ethylene and waterlogging effects in tomato. Ann Appl Biol 81:102–105

Jackson MB, Campbell DJ (1975b) Movement of ethylene from roots to shoots, a factor in the response of tomato plants to waterlogged soil conditions. New Phytol 74:397–406

Jackson MB, Campbell DJ (1976) Waterlogging and petiole epinasty in tomato: the role of ethylene and low oxygen. New Phytol 76:21–29

Jackson MB, Campbell DJ (1979) Effects of benzyladenine and gibberellic acid on the responses of tomato plants to anaerobic root environments and to ethylene. New Phytol 82:331–340

Jackson MB, Gales K, Campbell DJ (1978) Effect of waterlogged soil conditions on the production of ethylene and on water relationships in tomato plants. J Exp Bot 29:183–193

Janes HW, Loerchet L (1976) Effects of red light and ethylene on growth of etiolated lettuce seedlings. Plant Physiol 57:420–423

Junker S (1976) Auxin transport in tendril segments of *Passiflora caerulea*. Physiol Plant 37:258–262

Junker S (1977) Thigmonastic coiling of tendrils of *Passiflora quadrangularis* is not caused by lateral redistribution of auxin. Physiol Plant 41:51–54

Kaldewey H (1962) Plagio- und Diageotropismus der Sprosse und Blätter, einschließlich Epinastie, Hyponastie, Entfaltungsbewegungen. In: Ruhland W (ed) Encyclopedia of plant physiology vol XVII. Springer, Berlin Göttingen Heidelberg, pp 200–321

Kaldewey H (1971) Geoepinasty, an example of gravimorphism. In: Gordon SA, Cohen M (eds) Gravity and the organism. Univ Chicago Press, Chicago, pp 363–339

Kang BG (1979) Epinasty. In: Haupt W, Feinleib ME (eds) Physiology of movements. Encyclopedia of plant physiology, vol VII, Springer, Berlin Heidelberg New York, pp 647–667

Kang BG, Burg SP (1972) Ethylene as a natural agent inducing plumular hook formation in pea seedlings. Planta 104:275–281

Kang BG, Ray PM (1969a) Role of growth regulators in the bean hypocotyl hook-opening response. Planta 87:193–205

Kang BG, Ray PM (1969b) Ethylene and carbon dioxide as mediators in the response of the bean hypocotyl hook to light and auxins. Planta 87:206–216

Kang BG, Yocum CS, Burg SP, Ray PM (1967) Ethylene and carbon dioxide; mediation of hypocotyl hook-opening response. Science 156:958–959

Kang BG, Newcomb W, Burg SP (1971) Mechanism of auxin-induced ethylene production. Plant Physiol 47:504–509

Kawase M (1972) Effect of flooding on ethylene concentration in horticultural plants. J Am Soc Hortic Sci 97:584–588

Kawase M (1974) Role of ethylene in induction of flooding damage in sunflower. Physiol Plant 31:29–38

Kazemi S, Kefford NP (1974) Apical correlative effects in leaf epinasty of tomato. Plant Physiol 54:512–519

Kender WJ, Hall IV, Aalders LE, Forsyth FR (1969) Stimulation of rhizome and shoot growth of the lowbush blueberry by 2-chloroethanephosphonic acid. Can J Plant Sci 49:95–96

Klein RM (1965) Photomorphogenesis of the bean plumular hook. Physiol Plant 18:1026–1033

Klein WH, Withrow RB, Elstad V, Price L (1957) Photocontrol of growth and pigment synthesis in the bean seedling as related to irradiance and wavelength. Am J Bot 44:15–19

Klein WH, Edwards JL, Shropshire W jr (1967) Spectrophotometric measurements of phytochrome in vivo and their correlation with photomorphogenic responses of *Phaseolus*. Plant Physiol 42:264–270

Klein WH, Withrow RB, Elstad VB (1956) Response of the hypocotyl hook of bean seedlings to radiant energy and other factors. Plant Physiol 31:289–294

Kniep H (1910) Über den Einfluß der Schwerkraft auf die Bewegung der Laubblätter und der Frage der Epinastie. Jahrb Wiss Bot 48:1–72

Kohji J, Hagimoto H, Masuda Y (1979) Georeaction and elongation of the flower stalk in a poppy, *Papaver rhoeas* L. Plant Cell Physiol 20:375–396

Kohji J, Nishitani K, Masuda Y (1981) A study on the mechanism of nodding initiation of the flower stalk in a poppy, *Papaver rhoeas* L. Plant Cell Physiol 22:413–422

Kohji J, Hagimoto H, Yamamoto R, Masuda Y (1982) IAA transport and georeaction in the flower stalk of a poppy, *Papaver rhoeas* L. Plant Cell Physiol 23:1329–1336

Kramer PJ (1951) Causes of injury to plants resulting from flooding of the soil. Plant Physiol 26:722–736

Kujawski RF, Truscott FH (1974) Photocontrol of hook opening in *Cuscuta gronovii* Wild. Plant Physiol 53:610–614

Kumar D, Wareing PF (1972) Factors controlling stolon development in the potato plant. New Phytol 71:639–648

Kunkel LO (1926) Studies on aster yellows. Am J Bot 13:646–705

Kuo GG, Chen BW (1980) Physiological responses of tomato cultivars to flooding. J Am Soc Hortic Sci 105:751–755

Laan van der PA (1934) Der Einfluß von Aethylen auf die Wuchsstoffbildung bei *Avena* und *Vicia*. Rec Trav Bot Neerl 31:691–742

Larsen P (1962) Geotropism. An introduction. In: Ruhland W (ed) Encyclopedia of plant physiology, vol VII/2. Springer, Berlin Göttingen Heidelberg, pp 34–73

Lau OL, Yang SF (1976) Inhibition of ethylene production by cobaltous ion. Plant Physiol 58:114–117

Leather GR, Forrence LE, Abeles FB (1972) Increased ethylene production during clinostat experiments may cause leaf epinasty. Plant Physiol 49:183–186

Leike H, Guttenberg H (1962) Die Rolle des Auxins bei der epinastischen Krümmung plagiotroper Seitensprosse von *Coleus blumei* Benth. Planta 58:453–470

Leopold AC, Brown KM, Emerson FH (1972) Ethylene in the wood of stressed trees. Hortscience 7:715

Levy D, Marco S (1976) Involvement of ethylene in epinasty of CMV-infected cucumber cotyledons which exhibit increased resistance to gaseous diffusion. Physiol Plant Pathol 9:121–126

Lippincott BB, Lippincott JA (1971) Auxin-induced hyponasty of the leaf blade of *Phaseolus vulgaris*. Am J Bot 58:817–826

Lloyd FE (1914) Abscission. Ottawa Nat 28:41–43

Locker SB, Riker AJ, Duggar BM (1938) Growth substances and the development of crown gall. J Agric Res 57:21–39

Loveys BR, Wareing PF (1971) The hormonal control of wheat leaf unrolling. Planta 98:117–127

Lundegårdh H (1966) Plant physiology. Oliver & Boyd, Edinburgh

Lyon CJ (1963a) Auxin factor in branch epinasty. Plant Physiol 38:145–152

Lyon CJ (1963b) Auxin transport in leaf epinasty. Plant Physiol 38:567–574

Lyon CJ (1970) Ethylene inhibition of auxin transport by gravity in leaves. Plant Physiol 45:644–646

MacDonald IR, Gordon DC, Hart JW, Maher EP (1983) The positive hook: the role of gravity in the formation and opening of the apical hook. Planta 158:76–81

Maeda E (1965) Rate of lamina inclination in excised rice leaves. Physiol Plant 18:813–827

Maige MA (1901) Recherches biologiques sur les plantes rampantes. Ann Sci Nat Bot Biol Ser 8:11:249–364

Mapson LW (1969) Biogenesis of ethylene. Biol Rev 44:155–187
Meijer B (1957) The influence of light quality on the flowering response of *Salvia occidentalis*. Acta Bot Neerl 6:395–406
Mohr H, Noble A (1960) Die Steuerung der Schließung und Öffnung des Plumula-Hakens bei den Keimlingen von *Lactuca sativa* durch sichtbare Strahlung. Planta 55:327–342
Molisch H (1911) Über den Einfluß des Tabakrauches auf die Pflanze. II. Sitzungsber Kaiserl Akad Wiss Wien 120:813–838
Morgan PW (1976) Effects on ethylene physiology. In: Audus LJ (ed) Herbicides physiology, biochemistry, ecology, 2nd edn. Academic Press, London New York, pp 255–280
Morgan PW, Baur JR (1970) Involvement of ethylene in picloram-induced movement. Plant Physiol 46:655–659
Morgan PW, Hall WC (1962) Effect of 2,4-dichlorophenoxy-acetic acid on the production of ethylene by cotton and grain sorghum. Physiol Plant 15:420–427
Morgan PW, Hall WC (1964) Accelerated release of ethylene by cotton following application of indolyl-3-acetic acid. Nature 201:99
Münch E (1938) Untersuchungen über die Harmonie der Baumgestalt. Jahrb Wiss Bot 86:581–673
Nakamura T, Yamada T, Takahashi J (1966) Effect of gibberellic acid on the growth of the plumular hook section of etiolated pea seedling. Bot Mag Tokyo 70:404–413
Neljubow (1901) Über die horizontale Nutation der Stengel von *Pisum sativum* und einiger anderer Pflanzen. Beih Bot Centralbl 10:128–138
Palmer JH (1956) The nature of the growth response to sunlight shown by certain stoloniferous and prostrate tropical plants. New Phytol 55:346–355
Palmer JH (1958) Studies in the behaviour of the rhizome of *Agropyron repens* (L.) Beauv. I. The seasonal development and growth of the parent plant and rhizome. New Phytol 57:145–159
Palmer JH (1964) Comparative study of the effect of horizontal orientation and the application of indoleacetic acid upon internode extension and petiole orientation in *Helianthus annuus* and the modifying effect of gibberellic acid. Planta 61:282–297
Palmer JH (1972) Roles of ethylene and indole-3-yl-acetic acid in petiole epinasty in *Helianthus annuus* and the modifying influence of gibberellic acid. J Exp Bot 23:733–743
Palmer JH (1973) Ethylene as a cause of transient petiole epinasty in *Helianthus annuus* during clinostat experiments. Physiol Plant 28:188–193
Palmer JH (1976) Failure of ethylene to change the distribution of indoleacetic acid in the petiole of *Coleus blumei* × *frederici* during epinasty. Plant Physiol 58:513–515
Palmer JH (1979) Endogenous ethylene the active principle in IAA and NAA-induced petiole epinasty. 10th Int Conf Plant Growth Substances, Madison, p 54
Palmer JH, Phillips IDJ (1963) The effect of the terminal bud, indoleacetic acid and nitrogen supply on the growth and orientation of the petiole of *Helianthus annuus*. Physiol Plant 16:572–584
Parups EV (1973) Control of ethylene-induced responses in plants by a substituted benzothiadiazole. Physiol Plant 29:365–370
Pegg GF (1976) The involvement of ethylene in plant pathogenesis. In: Heitefuss R, Williams PH (eds) Physiological plant pathology. Encyclopedia of plant physiology. new ser vol 4. Springer, Berlin Heidelberg New York, pp 582–591
Pegg GF, Cronshaw DK (1976) The relationship of in vitro to in vivo ethylene production in *Pseudomonas solanacearum* infection of tomato. Physiol Plant Pathol 9:145–154
Pfeffer W (1905) The physiology of plants, vol III. Translated into English by Ewart AJ. Clarendon Press, Oxford
Pharis RP, Kuo CG (1976) Physiology of gibberellins in conifers. Can J For Res 7:299–325
Phillips IDJ (1964) Root-shoot hormone relations. I. The importance of an aerated root system in the regulation of growth hormone levels in the shoot of *Helianthus annuus*. Ann Bot NS 28:572–584

Poovaiah BW, Leopold AC (1974) Hormone-solute interactions in the lettuce hypocotyl hook. Plant Physiol 55:289–293

Porath D (1977) Hook opening in cucumber seedlings by narrow band red and far-red light. Plant Sci Lett 8:217–222

Powell RD, Morgan PW (1970) Factors involved in the opening of the hypocotyl hook of cotton and beans. Plant Physiol 45:548–552

Railton ID, Reid DM (1973) Effects of benzyladenine on the growth of waterlogged tomato plants. Planta 111:261–266

Rawitscher FL (1923) Epinastie und Geotropismus. Z Bot 15:65–100

Rawitscher FL (1932) Der Geotropismus der Pflanzen. Fischer, Jena Berlin

Reches S, Leshem Y, Wurzburge J (1974) On hormones and weeping: asymmetric hormone distribution and the pendulous growth habit of the weeping mulberry, *Morus alba* var. *pendula*. New Phytol 73:841–846

Reid DM, Crozier A (1971) Effects of waterlogging on the gibberellin content and growth of tomato plants. J Exp Bot 22:39–48

Reid DM, Clements JB, Carr DJ (1968) Red light induction of gibberellin synthesis in leaves. Nature 217:580–582

Reid DM, Crozier A, Harvey BMR (1969) The effects of flooding on the export of gibberellins from the root to the shoot. Planta 89:376–379

Robitaille HA, Leopold AC (1974) Ethylene and the regulation of apple stem growth under stress. Physiol Plant 32:301–304

Ross AF, Williamson CE (1951) Physiologically active emanations from virus-infected plants. Phytopathology 41:431–438

Rubinstein B (1971) The role of various regions of the bean hypocotyl on red light-induced hook opening. Plant Physiol 48:183–186

Rubinstein B, Abeles FB (1965) Relationship between ethylene evolution and leaf abscission. Bot Gaz 126:255–259

Rufelt H (1962) Plagiogeotropism in roots. In: Ruhland W (ed) Encyclopedia of plant physiology vol XVII/2. Springer, Berlin Göttingen Heidelberg, pp 322–343

Salisbury FG, Wheeler RM (1981) Interpreting plant responses to clinostating. I. Mechanical stresses and ethylene. Plant Physiol 67:667–685

Schneider G (1964) Eine neue Gruppe von synthetischen Pflanzenwachstumsregulatoren. Naturwissenschaften 51:416–417

Shell GSG, Lang ARG, Sale PJM (1974) Quantitative measures in leaf orientation and heliotropic response in sunflower, bean, pepper and cucumber. Agric Meteorol 13:25–37

Silk WK, Erickson RO (1978) Kinematics of hypocotyl curvature. Am J Bot 65:310–319

Smith KA, Restall SWF (1971) Occurrence of ethylene in anaerobic soil. J Soil Sci 22:430–443

Smith KA, Russell RS (1969) Occurrence of ethylene and its significance in anaerobic soil. Nature 222:765–771

Snow R (1945) Plagiotropism and correlative inhibition. New Phytol 44:110–117

Soekarjo R (1965) On the formation of adventitious roots in cuttings of *Coleus* in relation to the effect of indoleacetic acid on the epinastic curvature of isolated petioles. Acta Bot Neerl 14:373–399

Stahl E (1884) Einfluß des Lichtes auf den Geotropismus einiger Pflanzenorgane. Ber Dtsch Bot Ges 2:383–397

Starbuck CJ, Roberts AN (1982) Movement and distribution of ^{14}C-indole-3-acetic acid in branches and rooted cuttings of Douglas-fir. Physiol Plant 55:389–394

Stewart ER, Freebairn HT (1969) Ethylene, seed germination and epinasty. Plant Physiol 44:955–958

Stewart RN, Lieberman MV, Kunishi AT (1974) Effects of ethylene and gibberellic acid on cellular growth and development in apical and subapical regions of etiolated pea seedling. Plant Physiol 54:1–5

Takeno K, Pharis RP (1982) Brassinosteroid-induced bending of the leaf lamina of dwarf rice seedlings: an auxin mediated phenomenon. Plant Cell Physiol 23:1275–1281

Tibbitts W, Hertzberg WM (1978) Growth and epinasty of marigold plants maintained from emergence on horizontal clinostats. Plant Physiol 61:199–293
Trewavas AJ (1982) Growth substance sensitivity: The limiting factor in plant development. Physiol Plant 55:60–72
Trewavas AJ, Jones AM (1981) Consequences of hormone-binding studies for plant growth substance research. New Plant Physiol 12:5–8
Turkova NS (1964) Growth reactions in plants under excessive watering. Akad Nauk USSR Leningrad C 42:87–90
Vardar Y (1953) A study of the auxin factor in epinastic and hyponastic movements. Rev Fac Sci Univ Istanbul Ser B 18:317–352
Veen van der R, Meijer G (1959) Light and plant growth. Macmillan, New York
Virgin HI (1962) Light-induced unfolding of the grass leaf. Physiol Plant 15:380–389
Vries de H (1872) Über einige Ursachen der Richtung bilateral symmetrischer Pflanzentheile. Arb Bot Inst Würzburg 1:223–277
Wada K, Marumo S, Ikekawa N, Morisaki M, Mori K (1981) Brassinolide and homobrassinolide promotion of lamina inclination of rice seedlings. Plant Cell Physiol 22:323–325
Wain RL, Taylor HF (1965) Phenols as plant growth regulators. Nature 207:167–169
Wallenstein A, Albert LS (1963) Plant morphology: its control in *Proserpinaca* by photoperiod, temperature and gibberellic acid. Science 140:998–1000
Wample RL, Reid DML (1975) Effect of aeration on the flood-induced formation of adventitious roots and other changes in sunflower (*Helianthus annuus* L.). Planta 127:263–270
Wample RL, Reid DM (1978) Control of adventitious root production and hypocotyl hypertrophy of sunflower (*Helianthus annuus* L.) in response to flooding. Physiol Plant 44:351–358
Wample RL, Reid DM (1979) The role of endogenous auxins and ethylene in the formation of adventitious roots and hypocotyl hypertrophy in flooded sunflower plants (*Helianthus annuus*). Physiol Plant 45:219–226
Weaver RJ (1972) Plant growth substances in agriculture. Freeman, San Francisco
Wellman FL (1941) Epinasty of tomato, one of the earliest symptoms of *Fusarium* wilt. Phytopathology 31:281–283
Went FW (1938) Specific factors other than auxin affecting growth and root formation. Plant Physiol 13:55–80
Went FW (1943) Effect of the root system on tomato stem growth. Plant Physiol 8:51–65
Went FW, Thimann KV (1937) Phytohormones. MacMillan, New York
Wiese MV, Vay De JE (1970) Growth-regulator changes in cotton associated with defoliation caused by *Verticillum albo-atrum*. Plant Physiol 45:304–309
Wilkins MB (1979) Growth-control mechanisms in gravitropism. In: Haupt W, Feinleib ME (eds) Physiology of movements. Encyclopedia of plant physiology, new ser vol 7. Springer, Berlin Heidelberg New York, pp 601–626
Williamson CE (1950) Ethylene, a metabolic product of disease of injured plants. Phytopathology 40:205–208
Wilson BF (1973) White pine shoots: Roles of gravity and epinasty in movements and compression wood location. Am J Bot 60:597–601
Wilson BF, Archer RR (1977) Reaction wood: Induction and mechanical action. Annu Rev Plant Physiol 28:23–43
Wilson BF, Archer RR (1981) Apical control of branch movements in white pine: biological aspects. Plant Physiol 68:1285–1288
Withrow RB, Klein WH, Elstad VB (1957) Action spectra of photo-morphogenic induction and its photoinactivation. Plant Physiol 32:453–462
Yamane G (1941) Untersuchungen über die vertikalen phototropischen Bewegungen der Laubblätter von *Fatsia japonica* vom Standpunkt der Wuchsstofflehre. Jpn J Bot 11:305–326
Yin HC (1941) Studies on the nyctinastic movement of the leaves of *Carica papaya*. Am J Bot 28:250–261

Yopp JH (1973) The role of light and growth regulators in the opening of the *Dentaria* petiolar hook. Plant Physiol 51:714–715

Zeroni M, Jerie PH, Hall MA (1977) Studies on the movement and distribution of ethylene in *Vicia faba* L. Planta 134:119–125

Zimmerman PW, Hitchcock AE (1938) Tropic responses of leafy plants induced by application of growth substances. Contrib Boyce Thompson Inst 9:299–328

Zimmerman PW, Wilcoxon F (1935) Several chemical growth substances which cause initiation of roots and other responses in plants. Contrib Boyce Thomson Inst 7:209–229

Zimmermann W (1927) Die Georeaktionen der Pflanzen. Ergebn Biol 2:116–256

Ziv M, Koller D, Halevy AH (1976) Ethylene and the geotropic response of lateral branches in peanuts (*Arachis hypogaea* L.). Plant Cell Physiol 17:333–339

Zobel RW (1973) Some physiological characteristics of the ethylene-requiring tomato mutant diageotropica. Plant Physiol 52:385–389

Zobel RW (1974) Control of morphogenesis in the ethylene-requiring tomato mutant, diageotropica. Can J Bot 52:735–743

6 Position as a Factor in Growth and Development Effects

D.J. CHALMERS

1 Introduction

When reviewing the subject of this chapter recently, RAVEN and RUBERY (1982) discussed mechanisms of hormone transfer across membranes, the mode of action of receptors and hormone transport. On the latter question, they proposed that while vascular transport of hormones may conduct "temporal" information, transfer of "positional" information must be confined to the polar auxin transport system.

There is no doubt that polar auxin transport is an important positional mechanism at the extremities of the plant. The further apart, however, the extremities become, the less satisfactorily polar auxin transport accounts for what occurs in between. For example, it is a common observation of horticulturists that fruit at the top of a tree grow and mature faster than those lower down. This occurs independently of environmental influences (see Sect. 3.3) and can be shown to be at least partly due to a basipetally mobile factor in the main limb. Because fruit are carried on lateral shoots or spurs some distance from the main limb, polar transport between the main limb and the fruit cannot apply. Further the observed rates of polar transport (10–20 mm h^{-1}) do not realistically account for the transfer of signals from crowns to roots of large trees. Auxin diffuses across membranes relatively freely (GOLDSMITH 1977, RAVEN and RUBERY 1982, RUBERY and SHELDRAKE 1974), is a common constituent of phloem sap (ZIEGLER 1975), and is carried to the roots where it elicits effects (GOODWIN et al. 1978). Quite apart from the fact that it would take 200–400 days to transmit auxin by polar transport from top to bottom of the trunk of the largest trees, it must be doubtful that auxin traveling by that route would survive the journey.

Proponents of the polar transport model (RAVEN and RUBERY 1982, WAREING and ROBERTS 1956), have suggested that phloem transport of IAA may be important in the transfer of temporal rather than spatial signals. This appears to be a dubious distinction, because one presumes that the same receptors will be required and they will be located in space rather than in time. Indeed, temporal signals should be seen as an important source of positional information.

Considering the above points, this discussion will concentrate on mechanisms of propagation and action of the positional signal (excluding polar auxin transport) in large woody plants. I am confident, however, that the meter or so that separates root and shoot apices in herbaceous vascular plants is no less significant a gulf on the scale of molecules, cells, and tissues.

2 The Nature of the Positional Signal

In the terminology of BEER (1975), the propagation and action of the positional signals are probabilistic rather than deterministic processes. O'BRIEN (1982) points out that one cannot predict the behavior of such complex and interactive systems from the analysis of their components. On the other hand, plant growth integrates the signals of nutrient and water supply, apical dominance, correlative growth, tropisms etc. to yield plant form. If the positional signal is the sum of all such influences at a point in the plant, the growth of an organ or tissue at one position in the plant compared to the growth of the same tissue or organ at a different position will give a good quantitative description of the changes in the signal between those positions. The cambium is close to the vascular tissue throughout the nonphotosynthesizing plant. Consequently, growth by the cambial meristem of tree trunks is a clear example of the process wherein the amount the trunk has grown can readily be seen to be greater or less depending on the position of each level in relation to others, and with respect to growth stimuli originating from the crown or the roots.

There are a number of plausible mechanisms by which the positional signal may regulate cellular growth. For example, meristematic activity or cell expansion may be either stimulated or inhibited by a hormone or combination of hormones, or alternatively the positional factor may raise or lower the supply of substrates for division or growth. It could also be possible that a combination of these factors functions simultaneously.

2.1 Compartmentation and the Availability of Growth Substrates

The levels of organic and inorganic growth substrates in buds inhibited by apical dominance have been found to be at least equal to the dominant bud. This led to the conclusion (PHILLIPS 1975) that the presence of an inhibitor or stimulator hormone, not substrate supply, was responsible for correlative inhibition of buds in apical dominance.

The levels of stored carbohydrate in trees fluctuate during the season with changes in the rate of growth (KOZLOWSKI and KELLER 1966). For instance, the level of starch in laterals and roots of peach trees undergoes a marked seasonal change inversely associated with the presence of fruit (Fig. 1). In roots, the concentration of starch drops rapidly to low levels by October (early in the Southern hemisphere spring). In laterals, however, the level of starch declines more slowly, reaching a minimum in summer (mid-December). The reserves of starch in apple bark are also slowly mobilized (PRIESTLEY 1970, 1976), which may suggest that starch reserves contained in the bark in the top of trees in the spring are relatively inaccessible to growing tissue nearby. This conclusion supports the proposition that a block may exist in the utilization of growth substrates by cells in the tree trunk (KOZLOWSI and KELLER 1966).

The fact that there is marked accumulation of starch in bark and laterals of peach trees after fruit is harvested (Fig. 1) shows that, at that time at least,

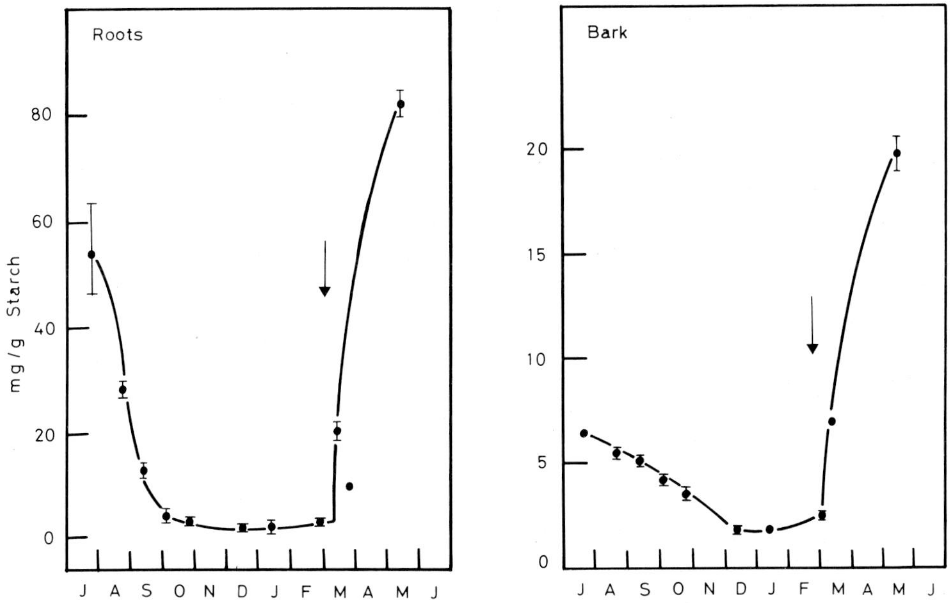

Fig. 1. Changes in the levels of starch in the roots and bark of aerial laterals of peach trees throughout the year. *Arrow* time of fruit harvest

there are more growth substrates available than can be utilized by circumference growth of the limb.

2.2 Some Salient Observations and Experiments on the Control of Cambial Growth

2.2.1 Growth in Relation to Position on the Bole

The amount by which the diameter of the bole of a tree increases at different levels during a season varies. In young trees, and trees growing in the open, the circumference increment tends to be greater toward the base, while the radial increase of the trunk of dominant forest trees is usually greater immediately below the crown and at the base (FARRAR 1961). On the other hand, the width of the xylem rings of suppressed trees decreases with increasing distance from the crown. The growth of the xylem at the base of severely suppressed trees may even stop altogether, resulting in a zone of rings missing in the xylem at the base of the tree (BORMANN 1965).

The positional signal that regulates growth of the cambial meristem at a point along the trunk appears, therefore, to vary with factors such as tree age, vigor, and competition from neighboring trees. In certain circumstances, however, the radial increment of the trunk may increase toward the base. Further, the amount of growth that has occurred at any level of the trunk is

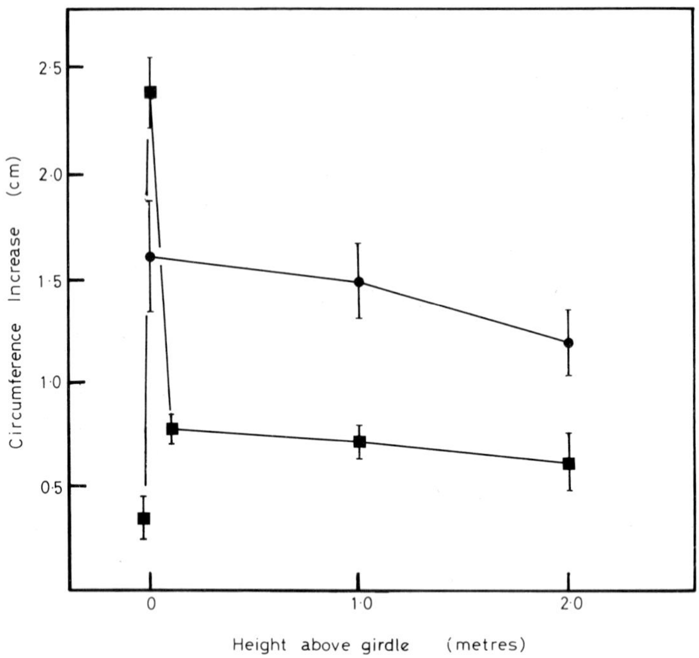

Fig. 2. The seasonal increase in circumference of a peach limb as a function of height above or below a girdle (■), or a reference point on a control limb (●) at the same height above the ground (1.5 m), 1 S.E.

proportional to the increase in cross-sectional area, and when one takes this into account it is probable that most trees that are not suppressed by their neighbors grow more toward the base.

The circumference increment of the limb of a peach tree, below the point of insertion of most of the leaf-bearing laterals, also characteristically increases toward the base (Fig. 2). These data represent a dry weight increase per unit length of limb of 4.56 g cm^{-1} $year^{-1}$ and 3.13 g cm^{-1} $year^{-1}$ at 1.5 m and 2.5 m from the ground respectively. Since assimilate concentration in the phloem appears to decrease with increasing distance from the source (CANNY 1973), it follows that growth is not proportional to assimilate concentration.

Xylem growth, however, may not occur at all at the base of suppressed trees (BORMANN 1965). This is an important observation, for it could indicate that the assimilate in the phloem is depleted before it can reach the lower levels of the tree bole. If this were so, the observation would suggest that assimilate did not reach the roots during those years. When similarly suppressed trees were girdled, however, they had died or were dying within 12 months. BORMANN therefore concluded that the phloem must remain functional in the zone of the missing rings to maintain the supply of assimilates to the roots. If assimilates reached the roots of the suppressed trees, it also indicates that cambial cells could not make use of assimilates passing in the phloem toward the roots. The potential of peach leaves to supply assimilates for growth of the cambium considerably exceeds the requirements for growth by that tissue (Fig. 2). Growth by the cambium itself indicates that the assimilate concentration in the phloem is not the limiting factor in the positional signal, even in conditions which reduce growth of the cambium.

Considering the long distance substrates must travel in the fine conduits of the phloem to reach the roots of large trees, and the ready transfer of substances between the phloem and other tissues (HARDY and POSSINGHAM 1969) it would seem that a mechanism to prevent surrounding tissue from using substances bound for the roots would be essential. In a peach fruit, for example, the seed cannot grow in competition with the fruit, yet to reach the seed the vascular bundle traverses only 1 or 2 cm of the pericarp (CHALMERS and VAN DEN ENDE 1977). The seed only begins to grow when the growth of the fruit declines. Further, the growth rate of the seed decreases immediately when the growth rate of the fruit increases again. When the growth rate of the seed reaches its maximum, the growth rate of the fruit and seed combined has declined to one-tenth of the combined growth rate at the commencement of rapid seed growth. Thus the mechanism facilitating transfer of assimilate through the fruit to the seed is a clear example in which growth by tissues surrounding the vascular system is suppressed.

2.2.2 Growth of the Basal Taper and Buttresses of Tree Trunks

The growth increment of a tree trunk may taper toward the top or the base or even from both ends towards the middle (FARRAR 1961). Furthermore, the taper of the trunk usually increases markedly at the base, and in some tree species and conditions root buttresses may extend well up the trunk from the point at which the roots connect.

It is not logical to propose that the concentration of a growth regulator or substrate exported solely from the crown could account for the variations in trunk growth discussed above. It is simple, however, to envisage how root-derived hormones might form an acropetal concentration gradient. Similarly, a basipetal gradient of hormones from the tree crown is likely to form in the phloem of the trunk below the crown.

There is good evidence that GA's and auxins interact to stimulate cambial growth (see references cited in PHARIS 1976); and there is some evidence that these hormones or their active forms may originate in the roots and the crown respectively (See Sect. 4). These common growth forms of the tree trunk growing in close association with roots are good evidence that a root-derived factor interacts with the products synthesized in the crown to regulate cambial growth. This conclusion might be expected considering the well-established relation between top and root growth.

Wind sway has been shown to promote growth of the lower bole of trees at the expense of upper portions (JAKOBS 1954). Localized ethylene production, stimulated by bending stress, has been implicated in such reaction wood growth (LEOPOLD et al. 1972). More recent work (BLAKE et al. 1980) has shown that localized ethylene production and consequent reaction wood formation of hyponastic (upturning) lateral branches of cypress seedlings, appear to be related to endogenous levels of IAA and GA rather than to bending stress. Therefore, insofar as the basal taper of tree trunks may be regulated by local ethylene levels, it is likely that the rate of ethylene production is regulated by the combined effect of physical factors such as wind sway and the levels of (positional) hormones or substrates.

2.2.3 Inhibition of Cambial Growth

While it seems reasonable to conclude that a positional signal comprised of hormones originating in the roots and the crown stimulates growth of the cambial meristem, a fundamental question remains unanswered: does the positional signal regulate growth and other phenomena directly, or does it act by releasing the tissue from the effect of an inhibitor? For example, ABA is acknowledged to have a characteristically inhibitory role in numerous physiological processes (including growth), is ubiquitous within higher plants, and its endogenous synthesis is readily triggered by such common physiological events as decreased water potential (MILBORROW 1974). Indeed, one might expect that many plant cells could experience conditions during the course of a day (during the growing season) that would stimulate synthesis of ABA. Furthermore, ABA may not be the only physiologically important inhibitor present in plants and, increasingly, research indicates that the balance between stimulator(s) and inhibitor(s) may regulate physiological processes.

While definitive evidence that such a balance regulates growth of the cambium is lacking, there is much circumstantial evidence that this may be so. Obviously, such a mechanism would be ideal to ensure that roots of vigorous young trees received their essential share of assimilates even though the streams of the xylem and phloem contained high levels of growth stimulators. On the other hand, it is hard to believe that the phloem or xylem stream of the suppressed forest trees contained no hormonal growth stimulators at those levels of the bole at which no growth occurred (Sect. 2.2.1).

The growth in circumference of the limbs of heavily cropped and mature peach trees is strongly inhibited during periods of active dry weight increase of the fruit (CHALMERS and WILSON 1978). The time at which growth of the limb is first inhibited coincides with the onset of diurnal limb shrinkage. Thereafter net growth by the cambium is least and often zero when shrinkage is greatest.

Similar experiments on three separate trees gave similar results (CHALMERS and WILSON 1978), except that growth at the top of the limb in the data reported (Fig. 3) was not suppressed during the first rapid stage of fruit growth.

Diurnal limb shrinkage in peach trees appears to be associated with a large increase in water use by cropping compared to noncropping trees (CHALMERS and WILSON 1978). The xylem of woody plants is rigid, however, and shrinkage is confined to the phloem (MOLZ and KLEPPER 1973). The data thus suggest

Fig. 3a–d. Cumulative change in radial diameter of a peach limb, measured by dendrometers at 0800 h (o) and 1600 h (●), the times each day of maximum and minimum limb diameter respectively. Diurnal limb shrinkage was determined by the difference between morning and evening measurements

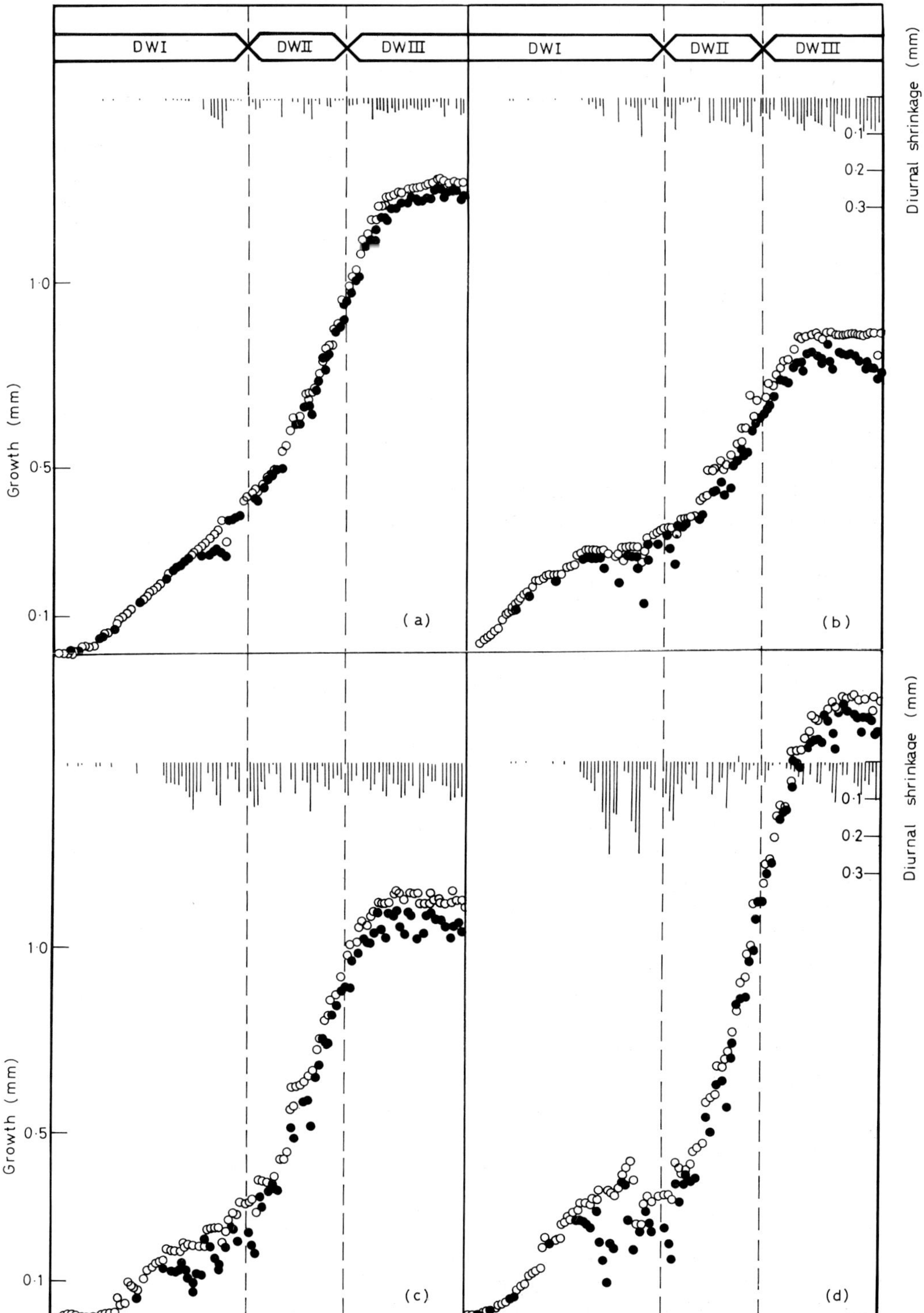

DWI
DWII
DWIII
Diurnal shrinkage (mm)
0·1
0·2
0·3
Growth (mm)
1·0
0·5
0·1
(a)
(b)
(c)
(d)

that a substantial water deficit develops in the rapidly translocating phloem (CHALMERS et al. 1983). Shrinkage (and presumably the water deficit in the phloem) decreased significantly toward the end of the peach limb (CHALMERS and WILSON 1978) and was least in the top zone of the tree where cambial growth is not inhibited during the first stage of rapid fruit growth. This could indicate that growth was inhibited at night by ABA produced during periods of water stress during the day. JENKINS (1974) and LITTLE (1975) have already reported that the inhibitory effect of water stress on the cambial activity of *Abies balsamea* and *Pinus radiata* respectively may be due partly to a build-up of ABA in these tissues, although later work by LITTLE and WAREING (1981) casts some doubt on that conclusion.

HOAD (1973, 1975) has proposed that ABA synthesized in the leaves in response to water stress may be translocated to other tissues. The fact that the limbs of the peach tree grew at the top but not lower down appears to rule out an inhibitor translocated basipetally in the phloem. Thus, if responsible, the inhibitor may be produced and metabolized in situ in the tissues where the shrinkage or water stress occurred. Certainly, ABA levels have been shown to change as rapidly in phloem exudate as they do in leaves of water-stressed lupins (HOAD 1978).

2.2.4 Growth Effects of Ringing or Girdling

Since ringing or girdling completely disrupts the downward flow of the phloem, it will also disrupt the downward transmission of any component of the positional signal which might be carried in the phloem and other tissues outside of the cambium. Ringing experiments, carefully interpreted, could therefore provide definitive information about the positional signal.

The effect of phloem-girdling on meristematic activity and growth above the girdle varies (NOEL 1970). Cambial activity is often interrupted and shoot growth inhibited to some extent, except for the short section of stem immediately above the girdle in which circumference growth is invariably stimulated (Fig. 2). Carbohydrate levels are often increased above the girdle, and this has been thought to be responsible for increased flowering, fruit set, and fruit growth (NOEL 1970), as well as the local stimulation of xylem growth immediately above the girdle (ZIMMERMAN 1960). The circumference growth of the limbs of peach trees above a girdle continues at much the same rate as control limbs, but stops after rapid fruit growth begins, only to recommence at rates similar to nongirdled limbs after harvest (Fig. 4). It has also been shown that increased fruit productivity of older peach trees occurs at the expense of secondary thickening in the limbs and trunk when the sink potential of those tissues declines with increasing age (CHALMERS and VAN DEN ENDE 1975a).

Thus, cambial activity and limb growth may be inhibited above a girdle after fruit begin to grow rapidly and compete for assimilate, or when the supply of stored assimilates is depleted. It is important to note, however, that growth may not be severely inhibited if there is not an active sink, such as a fruit, present to prevent a build-up of assimilates in the tissue (WILSON 1968). When growth of the cambium is inhibited above a girdle, it may well be due to a

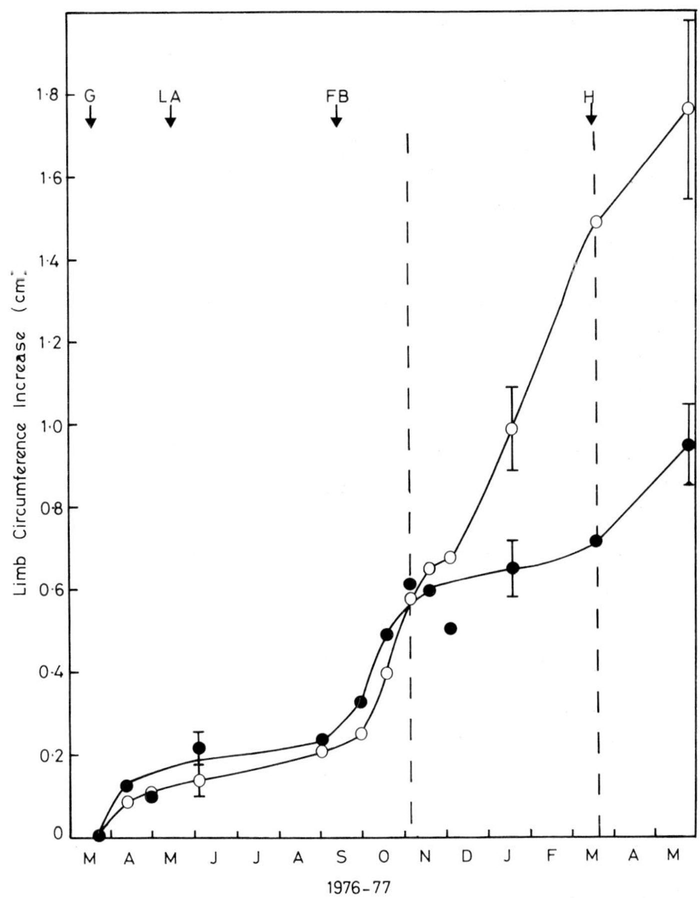

Fig. 4. Cumulative circumference increase of girdled (●) and control (○) limbs of fruiting peach trees (1 S.E.) in relation to time of girdling (*G*), leaf abscission (*LA*), full bloom (*FB*), harvest (*H*) and the period of strong competition from the crop (*between dotted lines*). DANN et al. 1984

reduction in sink strength or competitive ability of the cambium, in relation to other tissues or organs.

Circumference growth of peach tree limbs was severely inhibited below the girdle (Fig. 2), and it appears generally that cambial activity is checked severely and suddenly below a girdle (DANN et al. 1984, KOZLOWSKI 1971, NOEL 1970). Reduced substrate supply does not satisfactorily account for reduced cambial growth beneath a girdle. The mitotic index of the cambium is reduced below, but not above a girdle. On the other hand, thickening of the walls is slightly increased, which suggests no shortage of assimilate (WILSON 1968). Further, in work with peach trees, inhibition was greatest beneath the lower of the two girdles between which all foliage had been removed (DANN et al. 1984).

Ring-girdling of the phloem thus appears to inhibit cambial growth by two quite different mechanisms. Above the girdle the cambial meristem is capable of growth, provided no stronger sinks are present. Below the girdle growth is inhibited, regardless of the availability of substrates. In the former, sink strength of the cambial meristem or growing cells appears to be reduced by the presence of other sinks, while in the latter, meristematic activity may be directly inhibited.

A hormone moving basipetally in the phloem may accumulate above a girdle and, if inhibitory in nature (or at supraoptimal concentrations), could be responsible for reducing the sink strength of the cambium in relation to other competing tissues. Beneath the girdle phloem-borne hormones will be deficient. Reduced growth here would point strongly to a phloem-borne hormone being required for cambial activity. The stimulation of cambial growth that occurs immediately above the girdle contrasts with the general effect above a girdle and appears to be caused differently.

2.2.5 A Possible Mechanism for Girdling Effects and Implications for the Positional Signal

Although not unequivocal, the available evidence strongly supports the hypothesis that IAA is a principal factor in the growth effects of girdling. First, IAA is basipetally mobile (WAREING and ROBERTS 1956). Second, exogenous IAA can initiate or maintain cambial activity provided the cambium is not fully dormant (LITTLE et al. 1981, DENNE and WILSON 1977, LARSON 1960, LITTLE and GONGA 1974). Third, it promotes development in explanted cambial tissue (SAVIDGE 1983) and finally IAA has been positively identified in the cambial region (LITTLE et al. 1978). On the other hand, while many strong correlations between auxin and cambial activity have been found (e.g., AVERY et al. 1937, BONGA and CLARK 1963, LARSON 1964, SHEPHERD and ROWAN 1967) weaker and non-associations have also been reported (JENKINS and SHEPHERD 1974, ZAJACZKOWSKI 1973).

Some recent girdling studies however, may help to explain these discrepancies. Three to 4 weeks after girdling Sitka spruce, LITTLE and WAREING (1981) found auxin had accumulated in the zone immediately above a spring but not an autumn girdle. Within 24 h of girdling peach trees (DANN et al. 1985), auxin accumulated in the same narrow zone above the girdle, but only briefly; it declined in less than 8 days. Cambial growth in that zone, however, was stimulated for an extended period (typically) following a brief lag (Fig. 5).

The results of these two and earlier contradictory experiments can be readily reconciled if one assumes that cambial growth is regulated by auxin turnover rather than auxin concentration (as the data in Fig. 5 suggest). Reduced auxin levels beneath the girdles at all times and in both experiments strongly indicate that auxin supply was intercepted by the girdle. The time-course of the former experiments, however, would not have allowed the authors to detect a transitory rise in auxin concentration above the girdle. Increased auxin level above the spring girdle may simply reflect greater overall auxin transport at that time. The strength and predictability of inhibition of cambial growth that occurs beneath a girdle points to another important conclusion. Auxin is not transported in physiologically significant concentrations in the xylem.

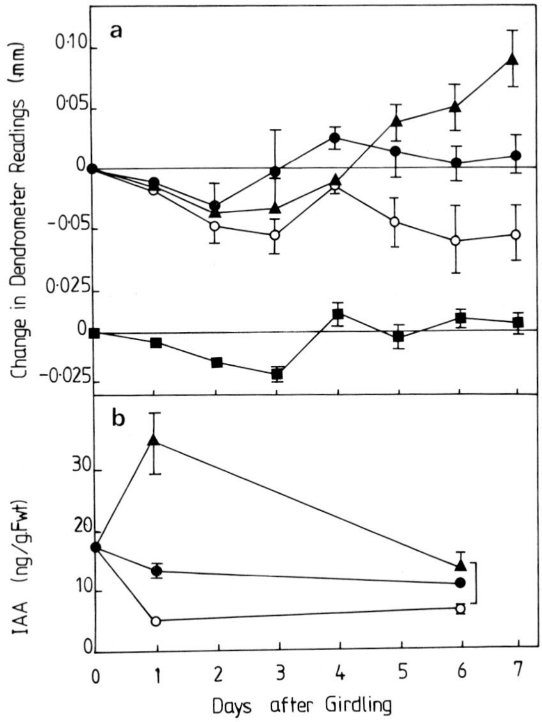

Fig. 5. The effect of phloem girdling of a peach limb in February on **a** cambial growth, measured by dendrometers, and **b** auxin level, estimated by fluorescence of indole pyrone immediately above (▲), 10 cm above (●), and below (○) the girdle (DANN et al. 1985). Growth of a non-girdled limb on the same tree (■)

On the other hand, there is no evidence that auxin accumulates higher up the limb and indeed it may decrease. DANN et al. (1985) report a small (non-significant) decrease while LITTLE and WAREING (1981) observe a decrease in the autumn 20 or so days after girdling. Thus it appears that suppression of the competitive potential of the cambium above a girdle cannot be attributed to supra-optimal levels of auxin. It is possible, however, that the apparent reduction in auxin levels reflects depletion of auxin (and assimilate) in the vascular system associated with cambium, due to competition from the fruit.

3 Regulation of Sinks and Competition by the Positional Signal

The importance of sink activity is regularly acknowledged in studies and reviews of correlative growth phenomena. The rate and direction of phloem transport depends upon the sink activity of the tissues supplied by the vascular system (GEIGER 1975). Hormonal factors alter the sink activity of individual organs and tissues, but the relative sink activity of one cell, tissue, or organ, compared with another, is strongly influenced by the positional signal and needs to be considered here in that light.

3.1 Dependent or Induced Sinks

The sink strength of the cambial meristem that produces phloem and xylem cells appears to be controlled directly (induced?) by the positional signal at that point in the trunk or limb. Since most of the contents of the transpiration and translocation streams reaching a given level of the cambial meristem will pass on to higher or lower (distant) sinks, it is reasonable to propose that the sink capacity of the cells at this given level will be regulated directly by a signal diffusing from the passing vascular streams. Thus the circumference growth of a tree's limb or trunk reflects the minute changes in the positional signal as it reacts with the growing cambium.

3.2 Autonomous or Uninduced Sinks

In contrast, the sink capacity of some plant organs and tissues (e.g., fruit, growth zones associated with developing meristems, some specialized storage tissue, and parasite cankers) is regulated by endogenously produced hormones. Thus, the cells of the peach fruit pericarp develop through quiescent and active stages of growth, clearly controlled by the seed (CHALMERS and VAN DEN ENDE 1975b, 1977), and are independent (more or less) of the influence of the predominating positional signal that controls the activity of surrounding sinks.

3.3 Competition Between Dependent and Autonomous Sinks

The sink strength of competing cambial meristems fluctuates synchronously with the hormone signal that correlates the growth of the roots and the top of the plant. On the other hand, the strength of autonomous sinks will vary according to different priorities determined within the autonomous organ.

The difference in sensitivity to the dominating positional influence will therefore sharply affect the competitive interrelation of dependent and autonomous sinks. Dependent sinks will be favored at times when, or in growth zones where, the positional (correlative) signal is strong, whereas autonomous sinks will benefit from a decline in strength of a positional signal.

Roots, leaves, and apices are usually most active in early spring, and on young plants. Thus young plants produce proportionately more secondary growth of the xylem and phloem, and secondary growth always occurs most rapidly in spring (WAREING 1977). In contrast, seeds and fruit grow fastest at the end of the growing season, while aging (or ailing) trees fruit more prolifically as the strength of their correlative signal subsides.

The dry weight distribution between fruit growth and secondary thickening in a developing peach tree illustrates this shift in assimilate distribution well. Cambial growth is controlled by the main correlative signal linking root and frame growth (CHALMERS and VAN DEN ENDE 1975a). As the vigor of the root system declines with age, the sink strength of the fruit, regulated by its seed (CHALMERS and VAN DEN ENDE 1977), remains relatively constant. This results

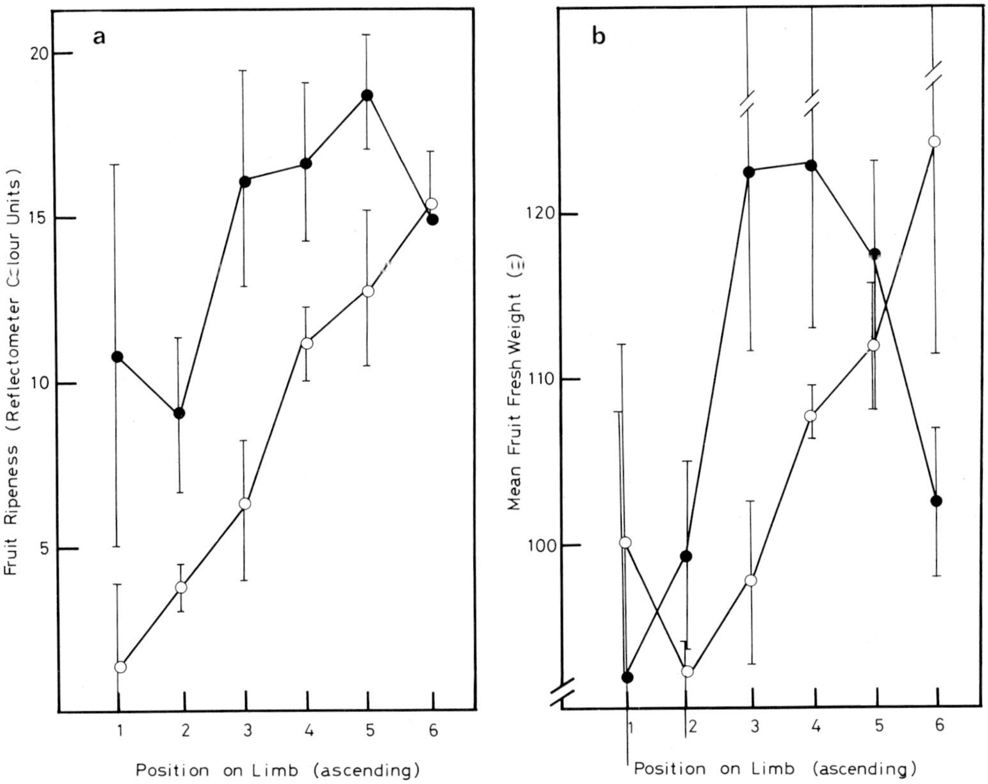

Fig. 6. The average ripeness estimated by color difference **a**, and average weight **b** of fruit harvested at one time from different levels on a control limb (o) and a limb that was girdled between each sample height (●) 1 S.E.

in a major shift in the competitive balance between the two organs, with fruit growth being encouraged.

Similarly, autonomous sinks will be affected less than adjacent dependent sinks by any changes in the hormone signal that are related to the sink position on the plant. Fruit will, however, have less competition and therefore grow best in positions on the tree where positional signals that regulate growth and sink strength of the competing cambial meristem are least strong.

Since the sink strength of the cambium generally increases with increasing distance from the leaves, it follows that the competitive ability of the fruit, in relation to the cambial meristem, will be higher at the top of the tree. A model can thus be derived which predicts the gradations in the rate at which individual fruit grow and mature on the tree (Fig. 6). However, if the phloem is girdled at frequent intervals along the limb to prevent the basipetal translocation of any hormonal substances out of the zones in which they were produced, growth rate and maturation of fruit in the lower part of the tree are markedly enhanced, while higher up fruit growth is at a disadvantage (Fig. 6). These data are further evidence that the competitive ability of the cambium and its consequent influence on the growth of competing sinks nearby, are partly regulated by a basipetally mobile factor in the phloem.

3.4 The Role of Ethylene

Ethylene has been implicated in the regulation of reaction wood formation in the pine (BROWN and LEOPOLD 1973), and it has been suggested that ethylene may have its effect in such situations by altering the levels of other hormones (BLAKE et al. 1980). More commonly, however, the levels of other hormones regulate ethylene production (e.g., ABELES 1973) and in these experiments endogenous ethylene production was stimulated by added IAA and GA_3.

On the other hand, changes in the rates of endogenous ethylene production regulate the rate of dry weight growth (as distinct from fresh weight growth) of peach fruit (CHALMERS and VAN DEN ENDE 1975b, JERIE and CHALMERS 1976). Further, exogenous ethylene applied as ethephon (CHALMERS et al. 1976) or ethylene stimulated the rate of ^{14}C assimilate incorporation by peach fruit

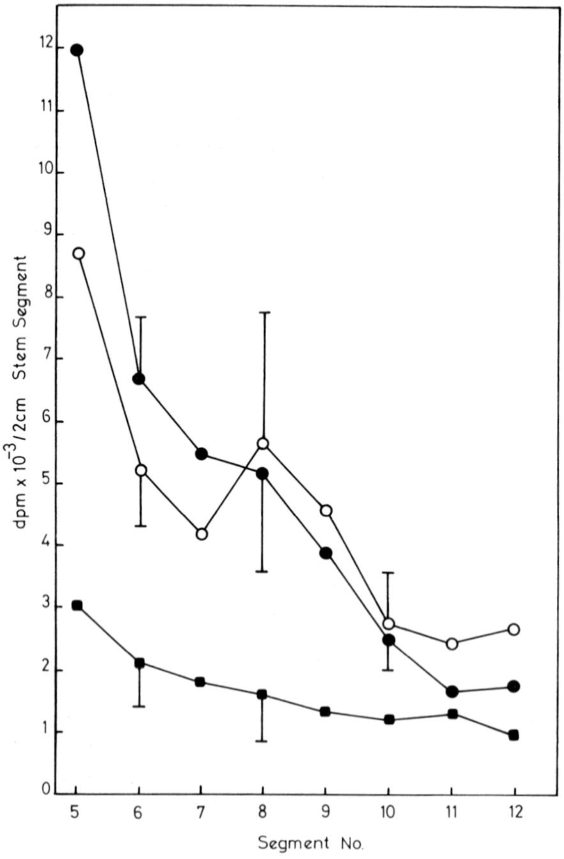

Fig. 7. The amount of label from ^{14}C-sucrose remaining in 2-cm segments of the stem of a peach seedling 24 h after no. 8 segment was painted with 1000 mg l^{-1} (●) or 500 mg l^{-1} (○) ethephon or water (■). ^{14}C-sucrose was applied at the same time as ethephon via a superficial bark incision to segment no. 1 at the base of the plant. All leaves except the apical leaves (which were darkened to create a strong sink) and the basal leaves beneath the point of application of ^{14}C-sucrose were removed

within 8 h of applying the ethephon to the fruit, and $^{14}CO_2$ to nearby leaves. These experiments indicate that ethylene regulates sink strength of peach fruit, and such an explanation could also explain the effects of ethylene on growth by the cambium. In this latter context, ethephon increases the amount of label from ^{14}C sucrose (applied through a bark incision at the base of the stem) that remains in the ethephon-treated and adjacent segments of the woody stem of peach seedlings 24 h after treatment (Fig. 7).

The possibility that growth of the cambial meristem could be simultaneously regulated by factors that control cell division and substrate availability was discussed earlier (Sect. 2). Further it was noted (Sect. 2.2.4) that cambial growth by peach trees was partly regulated by assimilate availability, and that the competitive ability (or sink strength) of the cambium of peach trees could be reduced by girdling to favor the growth of fruit. These observations, together with the proposition that the local ethylene level, mediated by physical factors (LEOPOLD et al. 1972) or endogenous hormone levels (BLAKE et al. 1980), may regulate reaction wood growth, would be consistent with ethylene regulating cambial growth by affecting the sink strength of cambial cells.

4 Propagation of the Positional Signal

The external hormone signal that a cell perceives, and by which it is directed, will exclude those hormones synthesized by the cell itself. Each hormone molecule that escapes from a cell will become part of an overall positional signal, but some tissues obviously contribute more than others to the complement of hormones that comprise the positional signal. If roots regulate (correlate) the growth of the top of a plant, and vice versa (by the limiting effect of the capacity of the root and the top to supply each other with hormones and nutrients), the positional signal at any point in the plant will be composed predominantly of the mixture of hormones diffusing or unloaded from the passing xylem or phloem streams.

4.1 The Descending Component

4.1.1 Sources of Positional Hormones in the Tops of Plants

Although the contribution that auxins produced in apices make to the auxin content of the phloem is uncertain (e.g., see GOODWIN et al. 1978) there are a multitude of apical buds on a tree in the spring that can be seen to exert dominance over the subordinate buds of their respective laterals. Without doubt, this signal propagates in the limbs and trunk, as it is translocated basipetally out of the shoot with assimilates, where it becomes a major part in the positional signal. Indeed in some trees cambial activity ceases at the same time as shoot growth (WORT 1962), indicating that with respect to that system, the hormone contribution of the apex is essential.

In other trees, however, cambial growth continues after shoot growth has ceased (WILSON 1968), probably because of hormones produced by the leaves. The leaves are likely to be the other major contributor to such a descending signal, since the amount of auxin exported by a shoot was found to be substantially greater than that produced by its apex (DIGBY and WAREING 1966). Even during senescence, leaves are the site of continuous physiological activity. They produce and export hormones, as well as assimilates, and thus contribute to a large proportion of the hormones in the phloem. As shoots stop growing and apices become dormant, the relative contribution of leaves to the positional signal should increase.

Leaves appear to synthesize and then export auxin in a discontinuous fashion (DIGBY and WAREING 1966), and they also contain (and therefore probably export) cytokinins and GA's. Leaves are also the major sink for the xylem transpiration steam, and it is likely that they re-export significant amounts of xylem-borne hormones via the phloem. Changes thus caused in the descending signal during the season are bound to be important in the seasonal periodicity of meristematic activity, growth, and other events elsewhere in the plant.

The contribution of other tissues and organs, although uncertain, is likely to be small. Biennial bearing in fruiting crops has been shown to be partly due to inhibitory signals from the seed, suppressing flower bud formation for the following year (CHAN and CAIN 1967). The effect, however, is confined to a brief period of a few weeks after fertilization when the developing fruit requires little assimilate. For most of its development, the seed (and the fruit in particular) is a substantial importer of assimilate and therefore probably not a significant exporter of hormones.

4.1.2 Contribution of the Phloem

Consideration of phloem transport from the point of view of its function of transporting assimilate from a source to a sink obscures the contribution that the phloem makes in propagating the positional signal. The phloem is surrounded by cells which are sinks, all of which presumably succeed to some extent in removing and using some of the phloem contents. The sink activities of these cells appear to create concentration gradients of assimilates in the phloem (CANNY 1973), and one might expect a similar change in the concentration of any hormones traveling with the assimilate, although there is no evidence in this regard. Indeed, the only hormone gradients over substantial distances that have been convincingly demonstrated in plant tissues have been transitional gradients during spring such as occur during the basipetal activation of the cambial meristem (DIGBY and WAREING 1966, SHEPHERD and ROWAN 1967).

4.2 The Ascending Component

4.2.1 Roots as a Source of Hormones

Hormone levels in xylem sap in relation to changes occurring elsewhere in the tree have been thoroughly reviewed (SKENE 1975, GOODWIN et al. 1978).

In short, cytokinins appear to be produced in root meristems, but it is not clear to what extent they may also be produced elsewhere (WAREING 1977). Similarly CROZIER and REID (1970, 1971) have bioassay evidence that root tips may be involved in interconversion, synthesis or release of GA-like substances, certain of which appeared to be exported to the top. However, conclusive evidence regarding the contribution of the root system per se (e.g., as opposed to merely transporting phloem-delivered hormones) to the hormone complement of the xylem is hard to find. It is simple to sample xylem sap in large quantities, however, and as a result we have a good knowledge of what known hormones xylem sap contains when it leaves the root system. Without question root exudates contain cytokinin and GA's, and furthermore the quantity of both is influenced by root environment and activity (SKENE 1975, TORREY 1976). Fewer, but no less convincing studies have detected ABA (TORREY 1976), but in contrast to cytokinins and GA's, evidence that the ABA may have originated in the root system is lacking.

The main difference between the ascending and descending signals with respect to the known hormones appears to be that the roots may not export auxins. Transport of auxin in the roots is predominantly acropetal (BATRA et al. 1975), and although ^{14}C-IAA applied to the shoot apex of pea seedlings was rapidly translocated to the root apex, it was not retranslocated to unlabeled shoots on the same root system (MORRIS 1977). In contrast, when fed to a mature leaf, ^{14}C-IAA moves in small quantities into the unlabeled shoot.

These data, together with the fact that auxin supplied via this phloem is required for cambial activity (see Sect. 2.2.1), suggest that roots do not re-export phloem-borne IAA via the xylem, and may indicate that transfer of IAA from other cells to the xylem in the root and stem does not occur in physiologically significant amounts.

4.2.2 Contribution of the Xylem

Although the phloem certainly carries crucial components of the positional signal to remote target tissues, it is clear that the xylem also carries regulators of cambial meristem growth, and may indeed be mainly responsible for producing their concentration gradients. The classical experiments of STOUT and HOAGLAND (1939) demonstrated that K, Na, and P move rapidly from the xylem to the phloem and associated tissues. Amino acids are similarly rapidly transferred between the xylem and the phloem (SHARKEY and PATE 1975), and there is evidence of such a two-way exchange for ^{14}C-GA_3 and ^{14}C-kinetin (BOWEN and WAREING 1969).

Whether or not xylem-borne hormones penetrate the symplast of the phloem, however, they will diffuse freely throughout the apoplast and hence equilibrate between the xylem and surrounding tissues. Indeed, with the possible exception of auxin, it is not possible to say whether the hormones known to exist in the phloem (see references cited in ZIEGLER 1975) originated in the top of the plant, or came from the roots.

It was pointed out earlier that it is impossible to account for the taper of a tree trunk and its buttress without proposing that stimulatory substances are supplied by the roots. In this context, the factors that control the transfer of solutes between the xylem, and surrounding tissues associated with the cambial meristem are almost certain to be profoundly important in the propagation of the ascending signal.

Hormone molecules that move out of the xylem, and are metabolized in the cambium and other tissues, will reduce the concentration of that hormone in the outer elements of the xylem. Where the xylem first leaves the roots, the concentration of root-derived hormones in the outer elements will be at a maximum; and furthermore the concentration will be constant across the conducting xylem. The hormone concentration in the outer elements of the xylem of a plant with secondary growth will fall most rapidly at the base of the plant until a concentration gradient is established between the outer and inner elements of the xylem, which will then buffer the concentration of hormones in the outer elements against further depletion. The concentration gradient of root hormones at the cambial cells thus formed would be steepest at the base of the plant, and would become more gradual as it ascended the stem or the trunk. It is easy to envisage how such a gradient would closely reflect the growth activity of the cambial meristem that one commonly observes along the stem or trunk.

There is good evidence that hormones move out of the xylem to regulate cambial acitivity and that gradients similar to those proposed above do indeed develop. GREENHAM and BROWN (1957) found during studies of the control of mistletoe on *Eucalyptus* spp. that 2,4-D, applied in numerous shallow holes drilled into the xylem around the base of the host tree, controlled mistletoe more effectively than the same amount of 2,4-D applied in a few deep holes. Conversely, 2,4-D applied into a few deep holes damaged the crown most severely.

Mistletoe almost always grows on the limbs and trunk in the lower part of the crown and, with respect to the solutes in the xylem, is therefore always nearer to the supply moving in the outer elements of the xylem than are the leaves. Since mistletoe is most severely affected when the herbicide is placed into the outer elements of the xylem, while the leaves of the host are most severely affected when 2,4-D is placed into the inner elements, the results indicate: (a) that growth regulators moving in the outer elements of the xylem affect, and are metabolized, by active tissues that they pass; (b) that a higher proportion of growth regulator traveling in the inner elements of the xylem reach the leaves in the crown.

Not surprisingly, such 2,4-D treatments stimulated the subsequent increment of basal area (NICHOLSON 1955). Measurements were made 15 to 30 cm and about 1 m, above the point where the treatments were applied, and the results show (Table 1): (a) cambial growth was stimulated in proportion to 2,4-D concentration, (b) the effect (and therefore the 2,4-D concentration) diminished rapidly as the growth regulator moved up the xylem.

One could expect an endogenous stimulator of cambial growth carried in the xylem from the roots to be distributed and affect cambial growth in a similar way. Indeed WAREING (1977) has suggested that the phase change between distinct juvenile and mature foliage and the suppression of flowering

Table 1. The effect of 2,4-D injection on the increment of basal area during the following growing season (NICHOLSON 1955)

Treatment	Increment of cross-sectional area (%)	
	30 cm above the point of injection	100 cm above the point of injection
Control	0.9	0.8
Low 2,4-D dose	1.2	1.0
Medium 2,4-D dose	3.6	1.5
High 2,4-D dose	4.9	1.1

at the base of stems of blackcurrants (SCHWABE and AL-DOORI 1973) may be evidence of high concentrations of root-produced GA's in that zone.

The second important way in which the xylem will modify the positional signal is by dilution of solutes as the flux of water through the xylem increases with transpiration. The leaf area of deciduous trees increases for 2–3 months in spring. To the extent that transpiration rate reflects the change in leaf area (CHALMERS et al. 1983), one would expect a proportional dilution in the concentration of root-derived hormones. Thus the concentration of hormones in the xylem will decline during the season. This will contribute to the seasonal periodicity of cambial growth and affect the distribution of root derived factors between the leaves and other tissues. LITTLE and WAREING (1981) observed that the changeover from activity to rest in Sitka spruce cambium was due to the development of the inability to respond to basipetally transported IAA. If auxin control of cambial growth requires interaction with a root-derived hormone, the above mechanism could be important.

Seasonal and regional changes in evaporative potential will also affect transpiration by evergreen as well as deciduous trees. FRANCIS (1970) observed that the formation of buttresses on rainforest trees is associated with high relative humidity, and lack of direct sunlight and not with situations subject to wind sway and consequent tension or strain. Low evaporative potential is also common in spring in temperate and cool-temperate regions. It is interesting to speculate whether increased concentrations of root-derived hormones in the xylem resulting from reduced water flux through the xylem contribute to these effects.

5 Conclusion

A tree trunk does not grow in a way that reflects how one might expect the concentration of assimilates in the phloem to change (with distance from the source). In short, the trunk can, and commonly does, grow more at the bottom than the top. Nor, on the other hand, does it appear that growth of the trunk is regulated by hormones exported from the crown or roots separately, since the growth increment may also taper from the top and the bottom toward the middle (FARRAR 1961). Indeed, it is clear that the growth of the tree trunk at all its levels is controlled by the balance of all hormones at that level, including any produced locally in response to either physical or chemical stimuli, as well as those imported from both the crown and the roots.

The concentration of hormones at any point will depend, however, on the activity of the organ that is exporting the hormones, and on the extent to which the hormone has reacted with other tissues along the transport path. Physiologically active tissues will produce and export greater quantities of hormones, while the further the transport stream gets from the source, the lower the hormone concentration is likely to be. The positional signal that cells, tissues, or organs experience at a point within a plant will be determined by the concentration gradients and hormone combinations generated by these mechanisms, and by the relative activity of the two major hormone sources.

One could interpret more growth at the base of a tree trunk, and more growth at the top of a tree trunk to mean that root-derived hormones or top-derived hormones, respectively, have dominated the hormone balance. Thus, conditions that permit vigorous growth and consequently highly tapered trunks are conditions in which the root system becomes the limiting factor for shoot growth. Conversely, trees that have less tapered or more cylindrical trunks have grown relatively more at the top, suggesting that crown-derived hormones limited growth. While for trees in the open, environmental conditions do not usually limit photosynthesis, the root environment combines problems of O_2 and nutrient supply, physical resistance to root growth and soil pathogens or toxins which may limit root activity in relation to top activity, and hence the contribution of the root to the positional signal.

The preceding discussion is not meant to imply, however, that the supply of root-produced hormones to the positional signal would be less in a vigorous than in a nonvigorous tree. On the contrary, the levels of cytokinins and GA's are known to be higher in exudates from vigorous roots (TORREY 1976). This hypothesis simply proposes that availability of root-produced hormones may become the limiting factor in the vigorous situation because the root environment is usually more restricting than the environment above the ground.

While the mechanism by which the positional signal is propagated in the vascular system appears reasonably clear, the mechanism by which it regulates growth and other processes needs much study. Considering their ubiquity it is likely that inhibitors such as ABA play an important part in these regulating processes. There is some evidence they may be produced in situ to function as activity-dampers or suppressors. The positional signal acts by stimulating growth directly or by releasing the tissue (to a regulated extent) from the effect of the inhibitor. The facts that (a) assimilates can be transported over such large distances and (b) that growth is *not* proportional to assimilate concentration even though many compounds exchange readily between the xylem and phloem, suggest that inhibitors may be involved. While further speculation (and research) on the mechanism of action of the positional signal awaits the answer to this question, it is worth observing that the effects of ABA on growth can often be antagonized by IAA and GA (MILBORROW 1974).

The growth of the secondary xylem and phloem in trees has featured prominently in this discussion of positional effects for two important reasons. First, the tree trunk separates the two major sources of hormones that comprise the positional signal, making it the only organ in which the "root-dependent" and "top-dependent" effects on growth may be separated. Further, although processes in all plant cells will be influenced in some way by the positional signal, no other approach illustrates so unequivocally the way in which the positional signal changes with distance from its sources. As CANNY (1973) has pointed out in another context, trees are everywhere, and any hypothesis that cannot explain a tree is useless.

References

Abeles FB (1973) Ethylene in plant biology, Academic Press, London New York

Avery GS, Burkholder PR, Creighton HB (1937) Production and distribution of growth hormone in shoots of *Aesculus* and *Malus*, and its probable role in stimulating cambial activity. Am J Bot 24:51–58

Batra MW, Edwards KL, Scott TK (1975) Auxin transport in roots: Its characteristics and relationship to growth. In: Torrey JG, Clarkson DT (eds) The development and function of roots. Academic Press, London New York, pp 299–325

Beer ST (1975) Platform for change. Wiley & Sons, London

Blake TJ, Pharis RP, Reid DM (1980) Ethylene, gibberellins, auxin and the apical control of branch angle in conifer, *Cupressus arizonica*. Planta 148:64–68

Bonga JM, Clark J (1963) Chromatographic detection of seasonal changes in the location of indole-3-acetic acid in balsam fir. Can J Bot 41:1133–1134

Bormann FH (1965) Changes in the growth pattern of white pine trees undergoing suppression. Ecology 46:269–277

Bowen MR, Wareing PF (1969) The interchange of ^{14}C-kinetin and ^{14}C-gibberellic acid between the bark and xylem of willow. Planta 89:108–125

Brown KM, Leopold AC (1973) Ethylene and the regulation of growth in pine. Can J For Res 3:143–145

Canny MJ (1973) Phloem translocation. Cambridge Univ Press, London New York

Chalmers DJ, Ende van den B (1975a) Productivity of peach trees: Factors affecting dry-weight distribution during tree growth. Ann Bot 39:423–432

Chalmers DJ, Ende van den B (1975b) A reappraisal of the growth and development of peach fruit. Aust J Plant Physiol 2:623–634

Chalmers DJ, Ende van den B (1977) The relation between seed and fruit development in the peach (*Prunus persica* L.). Ann Bot 41:707–714

Chalmers DJ, Wilson IB (1978) Productivity of peach trees: Tree growth and water stress in relation to fruit growth and assimilate demand. Ann Bot 42:285–294

Chalmers DJ, Ende van den B, Jerie PH (1976) The effect of (2-chloroethyl) phosphonic acid on the sink strength of developing peach (*Prunus persica* L.) fruit. Planta 131:203–205

Chalmers DJ, Olsson KA, Jones TR (1983) Water relations of peach trees and orchards. In: Kozlowski TT (ed) Water deficits and plant growth, vo. VII. Academic Press, London New York, pp 197–232

Chan BG, Cain JC (1967) The effect of seed formation on subsequent flowering in apple. Proc Am Soc Hortic Sci 91:63–68

Crozier A, Reid DM (1970) Gibberellin metabolism in the roots of *Phaseolus coccineus* seedlings. In: Carr DJ (ed) Plant growth substances 1970. Springer, Berlin Heidelberg New York, pp 414–419

Crozier A, Reid DM (1971) Do roots synthesize gibberellins? Can J Bot 49:967–975
Dann IR, Wildes RA, Chalmers DJ (1984) Effects of limb girdling on growth and development of competing fruit and vegetative tissues of peach trees. Aust J Plant Physiol 11:49–58
Dann IR, Jerie PH, Chalmers DJ (1985) Short-term changes in cambial growth, endogenous IAA levels and ethylene production in relation to phloem girdling of peach (*Prunus persica* L. Batsch). Aust J Plant Physiol (in press)
Denne MP, Wilson JE (1977) Some quantitative effects of indoleacetic acid on the wood production and tracheid dimensions of *Picea*. Planta 134:223–228
Digby J, Wareing PF (1966) The relationship between endogenous hormone levels in the plant and seasonal aspects of cambial activity. Ann Bot 30:607–622
Farrar J (1961) Longitudinal variation in the thickness of the annual ring. For Chron 37:323–349
Francis WD (1970) Australian rainforest trees. Aust Gov Publ Serv P 12
Geiger DR (1975) Phloem loading. In: Zimmermann MH, Milburn JA (eds) Encyclopedia of plant physiology, Transport in plants, new ser, vol I. Springer, Berlin Heidelberg New York, pp 395–431
Goldsmith MHM (1977) The polar transport of auxin. Annu Rev Plant Physiol 28:439–478
Goodwin PB, Gollnow BI, Letham DS (1978) Phytohormones and growth correlations. In: Letham DS, Goodwin PB, Higgins TJV (eds) Phytohormones and related compounds: A comprehensive treatise, vol II. Elsevier/North-Holland Biomedical Press, Amsterdam New York, pp 215–249
Greenham CG, Brown AG (1957) The control of mistletoe by trunk injection. J Aust Inst Agric Sci 23:308–318
Hardy PJ, Possingham JV (1969) Study on translocation of metabolites within the xylem of grapevine shoots. J Exp Bot 20:325–335
Hoad GV (1973) Effect of moisture stress on ABA levels in *Ricinus communis* L. with particular reference to phloem exudate. Planta 113:367–372
Hoad GV (1975) Effect of osmotic stress on ABA levels on sunflower (*Helianthus annuus*). Planta 124:25–29
Hoad GV (1978) Effect of water stress on ABA levels in white lupin (*Lupinus albus* L.) fruit, leaves and phloem exudate. Planta 142:287–290
Jacobs MR (1954) The effect of wind sway on the form and development of *Pinus radiata* D. Don. Aust J Bot 2:35–51
Jenkins PA (1974) The influence of applied indoleacetic acid and abscisic acid on xylem cell dimensions in *Pinus radiata* D. Don. In: Bieleski RL, Ferguson AR, Cresswell MM (eds) Mechanisms of regulation of plant growth, Bull 12. R Soc N Z, Wellington, pp 737–742
Jenkins PA, Shepherd KR (1974) Seasonal changes in levels of indole-acetic acid and abscisic acid in stem tissues of *Pinus radiata*. N Z J For Sci 4:511–519
Jerie PH, Chalmers DJ (1976) Ethylene as a growth hormone in peach fruit. Aust J Plant Physiol 3:429–434
Kozlowski TT (1971) Growth and development of trees, vol II. Cambial growth, root growth and reproductive growth. Academic Press, London New York
Kozlowski TT, Keller T (1966) Food relations of woody plants. Bot Rev 32:293–382
Larson PR (1960) A physiological consideration of the springwood summerwood transition in red pine. Science 6:110–122
Larson PR (1964) Some indirect effects of environment on wood formation. In: Zimmerman MH (ed) The formation of wood in forest trees. Academic Press, London New York, pp 345–365
Leopold AC, Brown KM, Emerson FH (1972) Ethylene in the wood of stressed trees. HortScience 7:175
Little CHA (1975) Inhibition of cambial activity in *Abies balsamea* by internal water stress: role of abscisic acid. Can J Bot 53:3041–3050
Little CHA, Gonga JM (1974) Rest in the cambium of *Abies balsamea*. Can J Bot 52:1723–1730

Little CHA, Wareing PF (1981) Control of cambial activity and dormancy in *Picea sitchensis* by indol-3-yl-acetic and abscisic acids. Can J Bot 59:1480–1493

Little CHA, Heald JK, Browning G (1978) Identification and measurement of indole acetic and abscisic acids in the cambial region of *Picea sitchensis* (Bong.) Carr. by combined gas chromatography-mass spectrometry. Planta 139:133–138

Milborrow BV (1974) The chemistry and physiology of abscisic acid. Annu Rev Plant Physiol 25:259–307

Molz FJ, Klepper B (1973) On the mechanism of water-stress-induced stem deformation. Agron J 65:304–306

Morris DA (1977) Transport of exogenous auxin in two-branched dwarf seedlings (*Pisum sativum* L) Planta 136:91–96

Nicholson DI (1955) The effect of 2:4-D injections and of mistletoe on the growth of *Eucalyptus polyanthemos*. Leaflet No 69, For Timber Bur Commonw Aust

Noel ARA (1970) The girdled tree. Bot Rev 36:162–195

O'Brien TP (1982) Cell growth and division. In: Smith H, Grierson D (eds) The molecular biology of plant development. Blackwell, Oxford, pp 49–109

Pharis RP (1976) Probable roles of plant hormones in regulating shoot elongation, diameter growth and crown form of coniferous trees. In: Cannell MGR (ed) Tree physiology and yield improvement. Academic Press, London New York, pp 291, 306

Phillips IDJ (1975) Apical dominance. Annu Rev Plant Physiol 26:341–367

Priestley CA (1970) Carbohydrate storage and utilization. In: Luckwill LC, Cutting CV (eds) Physiology of tree crops. Academic Press, London New York, pp 113, 127

Priestley CA (1976) Some effects of ringing branches on the distribution of dry matter in young apple trees (1976). J Exp Bot 27:1313–1324

Raven JA, Rubery PH (1982) Coordination of development: hormone receptors, hormone action and hormone transport. In: Smith H, Grierson D (eds) The molecular biology of plant development. Blackwell, Oxford, pp 28–47

Rubery PH, Sheldrake AR (1974) Carrier-mediated auxin transport. Planta 118:101–121

Savidge RA (1983) The role of plant hormones in higher plant cellular differentiation. II. Experiments with the vascular cambium, and sclereid and tracheid differentiation in the pine, *Pinus contorta*. Histochem J 15:447–466

Schwabe WW, Al-Doori AH (1973) Analysis of a juvenile-like condition affecting flowering in blackcurrant (*Ribes nigrum*). J Exp Bot 24:969–981

Sharkey PJ, Pate JS (1975) Selectivity in xylem to phloem transfer of amino acids in fruiting shoots of white lupin. Planta 127:251–262

Shepherd KR, Rowan KS (1967) Indole acetic acid in cambial tissue of radiata pine. Aust J Biol Sci 20:637–646

Skene KGM (1975) Cytokinin production by roots as a factor in the control of plant growth. In: Torrey JG, Clarkson DT (eds) The development and function of roots. Academic Press, London New York, pp 365–396

Stout PR, Hoagland DR (1939) Upward and lateral movement of salt in certain plants as indicated by radioactive isotopes of potassium, sodium, and phosphorus absorbed by roots. Am J Bot 26:320–324

Torrey JG (1976) Root hormones and plant growth. Annu Rev Plant Physiol 27:435–459

Wareing PF (1970) Growth and its coordination in trees. In: Luckwill LC, Cutting CV (eds) Physiology of tree crops. Academic Press, London New York, pp 1–21

Wareing PF (1977) Growth substances and integration in the whole plant. In: Jennings DH (ed) Soc Exp Biol Symp XXXI. Integration of activity in the higher plant. Cambridge Univ Press, London New York, pp 337–365

Wareing PF, Roberts D (1956) The photoperiodic control of cambial activity in *Robinia pseudacacia*. New Phytol 55:356–366

Wilson BF (1968) Effect of girdling on cambial activity in white pine. Can J Bot 46:141–146

Wodzicki TJ (1971) Mechanism of xylem differentiation in *P. silvestris* L. J Exp Bot 22:670–687

Wort DJ (1962) Physiology of cambial activity. In: Kozlowski TT (ed) Tree growth. Ronald, New York, pp 89–95

Zajaczkowski S (1973) Auxin stimulation of cambial activity in *Pinus silvestris* L. The differential cambial response. Physiol Plant 29:281–287

Ziegler H (1975) Nature of transported substances. In: Zimmermann MH, Milburn JA (eds) Encyclopedia of plant physiology, new ser, vol I. Transport in plants. Springer, Berlin Heidelberg New York, pp 59–100

Zimmermann MH (1960) Longitudinal and tangential movement within the sievetube system of white ash (*Fraxinus americana* L.) Festschr Albert Frey-Wyssling, Beit Z Schwz For 30:289–300

7 Roles of Hormones, Protons and Calcium in Geotropism

B.G. PICKARD

1 Introduction

A variety of hormones and inorganic regulators are involved in geotropism. The evidence for certain of these is extensive and definitive; for others, it is new and incomplete. Because this review is being prepared at a time when there are protests that hormones have little or nothing to do with geotropism (e.g., DIGBY and FIRN 1976, 1979, FIRN and DIGBY 1976, 1977, 1979, 1980, FIRN et al. 1977, FIRN et al. 1978, HALL et al. 1980, FIRN et al. 1981, DIGBY et al. 1982, FIRN 1983, TREWAVAS 1979, 1981, MCINTYRE 1980, 1983, see also WHEELER and SALISBURY 1980, WAREING and PHILLIPS 1981), it will present ample solid evidence to the contrary; but it will also attempt interpretation of still-incomplete data and conjecture about the biochemical bases of gravity reception and the development of hormonal and ionic asymmetries.

In general, the word geotropism will be used narrowly in place of orthogeotropism, or gravity-educed bending toward the vertical; but sometimes it will include bending toward an equilibrium position slightly off vertical. Geotropism toward an equilibrium position markedly away from the vertical will be more strictly designated as plagiogeotropism; diageotropism is a special case of plagiogeotropism in which an organ maintains a horizontal position.

This review will cover both roots and shoots. It will emphasize orthogeotropism of seedling shoots because (1) it is presumably the simple form of geotropism from which diverse plagiogeotropic behaviors of the parts of more mature shoots, and possibly the geotropic behavior of roots, is evolutionarily derived, and (2) it is the best-understood form of geotropism. Where work on hormonal control of plagiogeotropism seems particularly relevant it has been included.

The literature includes reviews by JACKSON and BARLOW 1981, REINHOLD 1978, KANG 1979, KONINGS 1968, PILET 1977, AUDUS 1969, 1971, 1975, 1979, RUFELT 1961, 1962, JOHNSSON 1971, ANKER 1962, KALDEWAY 1962, LARSEN 1962a, b, LYON 1968, GRESSEL and HORWITZ 1982, WILKINS 1966, 1971, 1976a, b, 1977, 1978, 1979, THIMANN 1977, PALMER, this Vol. A book by MERKIS (1973) on the Russian literature is usefully excerpted by BLUZMANAS and DAGIS (1975).

2 Patterns of Response

2.1 Relevance of Kinetic Data

Because the ways in which hormones and inorganic regulators participate in the mediation of geotropism are complicated, information about rates and sites of both georeception and the ensuing response is essential.

2.2 Difficulties of Temporally Separating Sequelae

The convenience of dividing geotropism into stages such as "exposition", "reception", "transmission", "reaction", and "after-effects" (e.g., RAWITSCHER 1932, LARSEN 1962a), coupled with the relative insensitivity of many observations of the onset of the stages, has sometimes tended to lull people into thinking that they can effect a temporal separation of them. For example, in a recent text by WAREING and PHILLIPS (1981), it is stated that reception occurs with a "presentation time" of many minutes and that the response follows slowly.

Yet, even many of the early studies made clear that the minimum stimulation time required to educe detectable curvature can range as low as 2 or 3 min, and in one instance half a minute was sufficient (table in JOHNSSON and PICKARD 1979). In fact, when plots of ultimate curvature resulting from stimuli of increasing duration are extrapolated to the threshold, times of 10–20 s have been noted for both roots and shoots (fig. in JOHNSSON and PICKARD 1979 and VOLKMANN and SIEVERS 1979).

Also, for rapidly elongating roots and shoots, a curvature response can begin almost immediately (Figs. 1 and 3A). Frequently, immediate response is in the "wrong" direction: up for roots, down for shoots. While more often than not a "lag" of 15 to 30 min is observed in lieu of either early normal or transient early "wrong-way" curvature, the rapid responses are of great interest because they indicate that complex secondary events must be underway long before response is usually considered to begin.

Selected frames from two time-lapse movies (Fig. 2A and B) illustrate initial upward (negative) bending as well as the more obvious downward (positive)

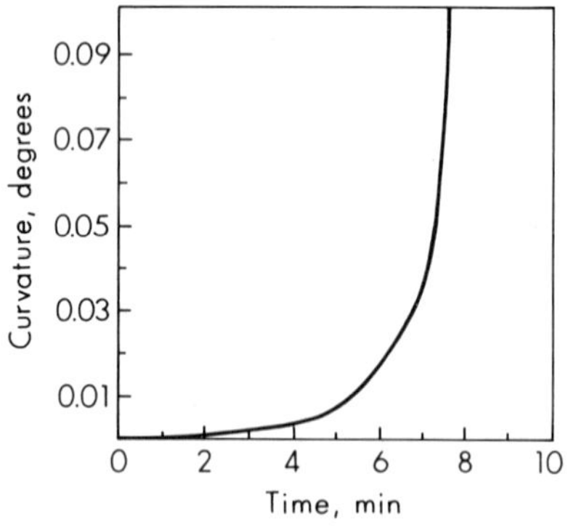

Fig. 1. Early curvature response of a horizontally placed seedling shoot (mesocotyl plus coleoptile) of *Zea*, too slight to be detected by eye but measured with an angular position transducer. (BANDURSKI et al. 1984)

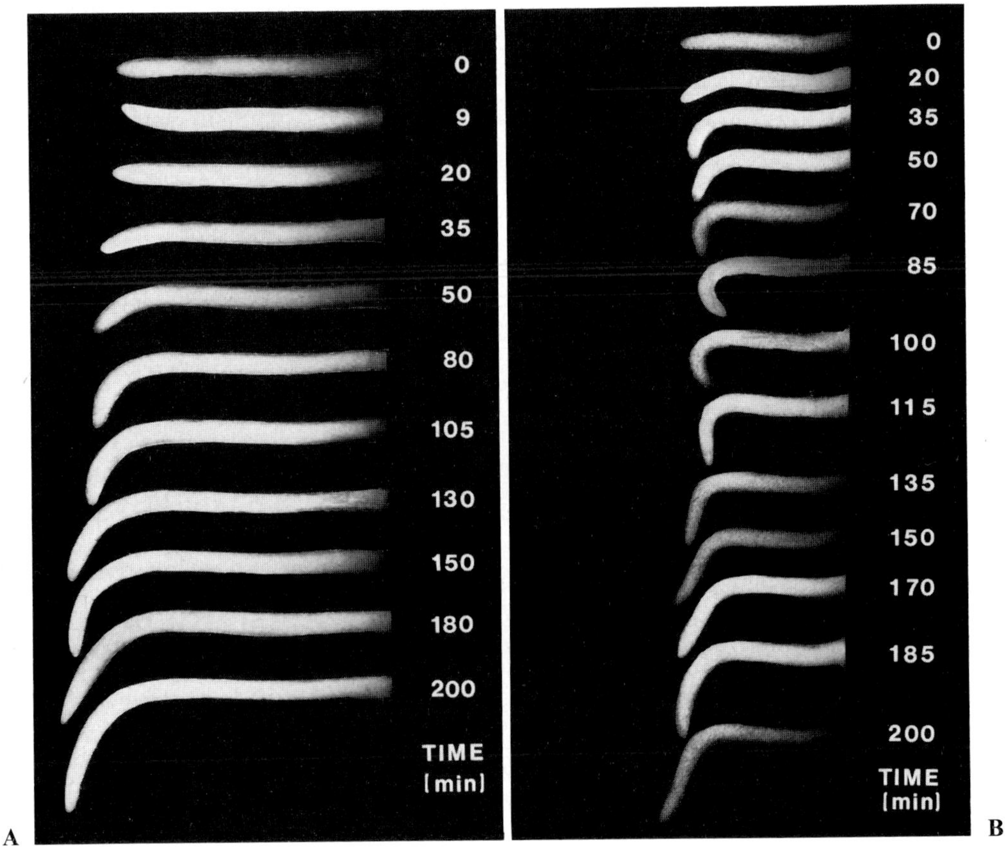

Fig. 2. Selected frames from two unpublished time-lapse movies (**A, B**) by M.L. EVANS and T.J. MULKEY illustrate two different patterns of geotropic response, both in *Zea* roots. Conditions of growth were as described by MULKEY et al. (1984); geotropic response occurred at 22° C with 1.5 W m^{-2} white light (personal communication)

curvature for geotropically stimulated roots of *Zea*. MICHAEL L. EVANS and TIMOTHY J. MULKEY, who provided this illustration, report that they have only seen early negative curvature in roots growing in air saturated with water vapor; 80 or 90% relative humidity seems not to suffice. The two kinds of corn tested had different fractions of early upward benders; for the field corn of Fig. 1 the fraction was about 25%, but for sweet corn the value was somewhat higher. The upward curving roots (Fig. 2A) carry out the main phase of curvature more slowly than do the remainder, and without overshooting the vertical undergo restricted hunting about a plagiogeotropic "position of equilibrium". Roots that bend directly downward (Fig. 2B) attain the vertical rapidly and overshoot it, hunting with large oscillations for a long time.

Although early negative curvature of roots has not generally been remarked in the literature, it has been observed repeatedly that both the time-course

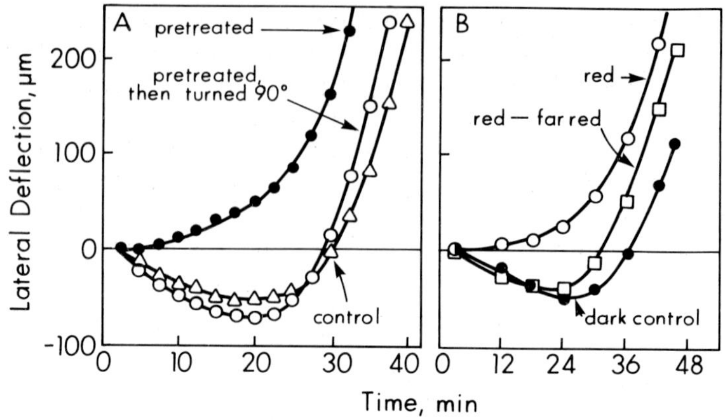

Fig. 3A, B. Pretreatments can determine whether a horizontally placed *Zea* coleoptile bends downward before beginning the major, upward phase of geotropic response, or whether it bends upward monotonically. **A** HILD and HERTEL (1972) physiologically modified coleoptiles by administering 2-min opposing prestimuli in rapid alternation for 1 h. One set was then stimulated by setting it horizontal in the same orientation as for the penultimate prestimulus, while another set was turned 90° and stimulated with the surfaces that had previously been uppermost and lowermost positioned laterally. Curvature response of prestimulated plants differed from controls if the prestimulus vectors were the same and opposite from the stimulus vector, but were indistinguishable if prestimulus vectors were at right angles to the stimulus vecot. **B** HILD (1977) modified plants by pretreating them with 1.1 W m^{-2} R for 3 min; curvature in response to geotropic stimulation administered 2 h later was monotonically upward. The effect of R was largely reversed by immediate 3-min pretreatment with 0.2 W m^{-2} FR; doubtless, complete reversal could have been obtained by administering the light in briefer, more intense doses. The time-courses for both **A** and **B** are relatively slow because the coleoptiles were excised before experimentation

and the equilibrium angle are sensitive to the physical surroundings (e.g. LARSEN 1962b, RUFELT 1957a, b, c, KONINGS 1964, ZINKE 1968).

Early downward (positive) curvature has often been observed with coleoptiles (e.g., BRAUNER and ZIPPERER 1961). It occurs in uprighted (ibid.) or clinostated (PICKARD 1973a) plants, as well as in those maintained horizontal. HILD and HERTEL (1972), who proposed that the early downward curvature may represent an overreaction or "shock" reaction to initiation of a strong stimulus, have shown that it does not take place if a coleoptile is prestimulated by brief opposing accelerations integrating to a zero sum (Fig. 3A). Moreover, it does not occur if coleoptiles are submerged in water (BRAUNER and ZIPPERER 1961). Perhaps most importantly, HILD (1977) has shown that preirradiation with blue or R light can eliminate the downward response, with FR reversing the effect of R (Fig. 3B).

Evidently, early downward curvature is less apt to be observed in hypocotyls and epicotyls, but this might well be because (1) these are often grown in the light and (2) many of the species studied have relatively thick rigid stems, which might provide more mechanical resistance to bending when growth of cells on either side tends to become a little imbalanced.

An attempt to separate reception from subsequent stages was made by BRAUNER and HAGER (1958) with *Helianthus* hypocotyls. They reported that decapitated plants become depleted of auxin and stop growing, but retain the capacity for geoinduction. When they supplied exogenous hormone to stimulated decapitated plants, conspicuous curvatures developed. PICKARD (unpublished) tried to duplicate these experiments but was unable to get repeatedly decapitated hypocotyls to stop growth completely, even after 4–5 days. Correspondingly, though the development of curvature in horizontal decapitated plants was drastically slowed it was not stopped. For coleoptiles, it was already known that a single decapitation does not completely stop growth (e.g., DOLK 1936).

BRAUNER and HAGER also tried to separate primary and secondary processes by stimulating *Helianthus* hypocotyls at 4° C, where they observed no growth. Hypocotyls of plants laid horizontal in the cold did respond when placed upright or on a clinostat in the warmth, however. Again, PICKARD tried to repeat these experiments (unpublished), but under her conditions growth did not stop completely at 4° or even 2° C, and horizontal hypocotyls curved slowly at that temperature. PICKARD carried out an analogous study with *Avena* coleoptiles (1973b), but concluded that (1) the receptor system undergoes rapid progressive loss of sensitivity on deterioration in the cold but once induction occurs it is retained stably, and (2) in spite of the deterioration, the ratio of the threshold for measurable induction to the threshold for measurable negative curvature was about the same at 2° and 25° C. Thus, no reliable way to separate inductive and secondary processes has yet been developed, and this reinforces the view that these processes normally overlap during stimulation.

Evidence that primary and secondary reactions may actually cycle was presented by GÜNTHER-MASSIAS (1928), who with *Lepidium* roots and *Helianthus* hypocotyls obtained larger curvatures from summed brief stimuli than from continuous stimulation of the same net duration. With the roots, a ratio of 10 s stimulation to 10 s rest resulted in a 71% enhancement, and a 10 s: 20 s ratio resulted in 52%. PICKARD (1973a), on the other hand, was unable to observe enhancement with *Avena* coleoptiles. While BÜNNING and GLATZLE (1949) did observe it with coleoptiles, their use of stimulus and rest intervals lasting minutes precludes comparison with the data of GÜNTHER-MASSIAS because it suggests involvement of later stages of response.

Not only are component processes of geotropism difficult to sort out early in the response, but also the main phase of curvature for both roots and shoots can soon be retarded by a counter-reaction. Evidently, primary asymmetries registered in the cellular receptor system are translated rapidly into secondary asymmetries across the organ as a whole, and these may initiate feedback or even disappear. This is best seen when stimulation is terminated artificially: for example, *Avena* coleoptiles responding on the clinostat (Fig. 4A) may achieve maximal curvature as early as 25 min after the end of stimulation (PICKARD 1973a), though net bending sometimes continues for as long as 50 min under similar conditions (GILLESPIE 1959, DOLK 1930/1936, PICKARD 1973a). DOLK showed with decapitated seedlings that all zones of the coleoptile detect as well as respond to gravity (Fig. 5), but curvature is seen earliest in the apical

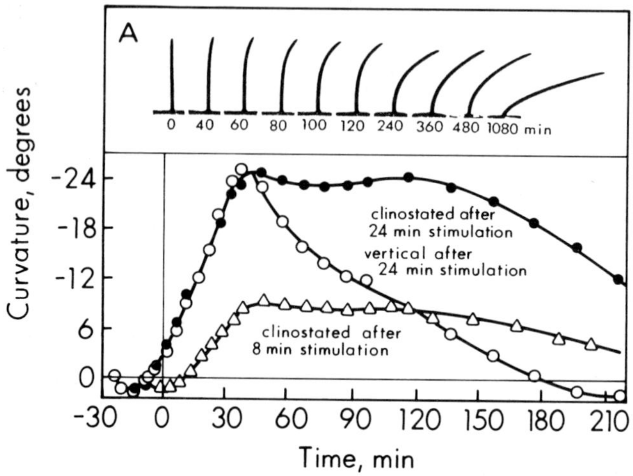

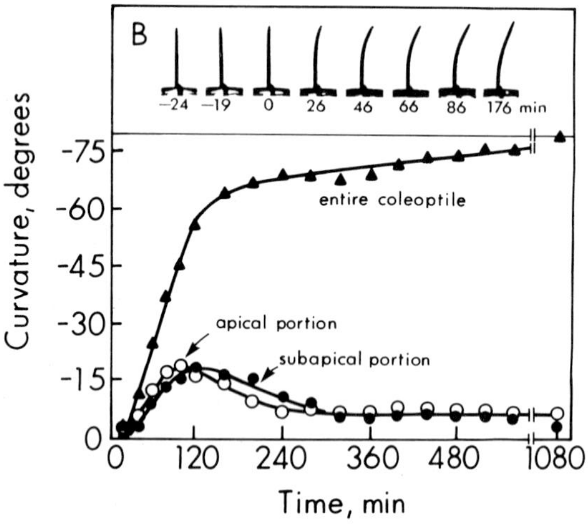

Fig. 4A, B. Time-lapse data for geotropically responding *Avena* coleoptiles (PICKARD 1973a). **A** At time zero, horizontal stimulation of the specified duration was completed and intact plants were placed on the clinostat or upright on the bench. The selected shadowgraphs of a single coleoptile are from the set clinostated after 24 min stimulation. **B** Coleoptiles in a rack were kept horizontal for 18 h; *top trace* shows average net curvatures; *lower traces* show average curvatures of the apical 1–4 mm portions and subapical 6–10 mm portions of the coleoptiles. The selected shadowgraphs are of a single coleoptile

regions (Figs. 5 and 6A; cf. Sect. 2.3). However, during much of the period when net geotropic curvature is still increasing, the tip of an intact seedling actually straightens while the curving zone shifts basally (Fig. 6A). Ultimately, the zone of counter-reaction follows the curving zone, and the entire coleoptile straightens except at its base. Of course, clinostating is omnilateral stimulation, which may well enhance the magnitude of counter-reaction.

In the more natural situation, in which the horizontal plant is neither clinostated nor uprighted, primary roots and shoots also often reduce their net rate of response early, so that the tips either do not attain the vertical or attain it only asymptotically. This is illustrated in Fig. 2B for roots. It is shown in Figs. 4B and 7 for horizontally stimulated, R-pretreated, etiolated coleoptiles of *Avena* and epicotyls of *Pisum;* the reduction typically occurs on attaining a curvature anywhere from 30° to 60°, depending on conditions. It is generally

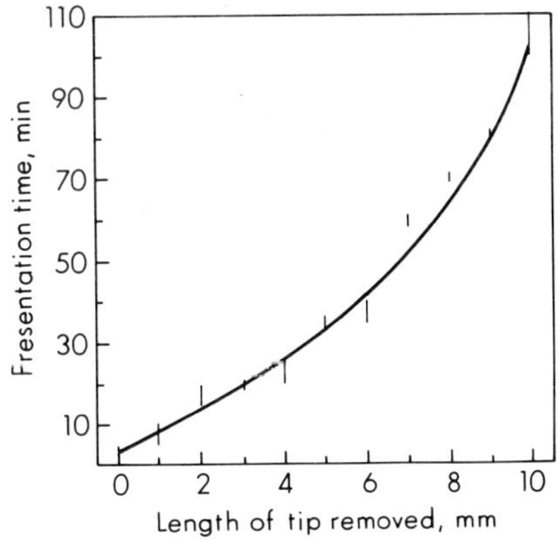

Fig. 5. Distribution of geotropic responsiveness along the shoot. DOLK (1930/1936) removed various lengths of the tip from *Avena* coleoptiles, affixed agar blocks containing a standard quantity of auxin, and determined the presentation time (defined as the stimulus time required to produce just noticeable curvature in half the plants after an optimal interval for curvature development – an index of geotropic sensitivity an order of magnitude less sensitive than the modern minimum effective stimulation time discussed in the text). This test does not separate the extent to which the observed decrease in sensitivity with length of tip decapitated results from loss of ability to receive gravity or to carry out secondary processes such as differential growth, but clearly shows that some degree of capability for geotropism extends at least more than 10 mm down the 30-mm coleoptile

conspicuous even after allowance is made for the continuously decreasing vector of gravitational stimulation, and again typically results from straightening of the tips during basal migration of curvature (e.g., Fig. 4B).

DOLK (1930/36) and others have described this straightening as an "autotropic" response to the initial curvature, and discussed whether it could be due in part to greater depletion of metabolic reserves on the initially faster growing side. While such depletion might contribute, however, work to be described below implicates negative hormonal feedback loops and suggests interferences from a nongeotropic sensory system (Sect. 3).

The counter-reaction varies from species to species and from one set of parameters to another. Often it does not occur at all – e.g., Fig. 2A, and Fig. 11 of RAWITSCHER (1932). This emphasizes that, in seeking hormonal relationships underlying geotropism, one must not expect simplicity. A further argument against simple expectations is that the difference between primary roots and stems and their plagiogeotropic laterals with well-developed counter-reactions must be slight: if a primary root or shoot is damaged, one of the laterals may take its place by adopting a vertical position of equilibrium.

Finally, one of the reasons that the potential rapidity and the complex time-course of geotropism have been emphasized here is that conditions for the response are frequently suboptimal. Of course, when growth is sluggish geotropism is sluggish, and probably its component steps often slow down together; in this case, the several phases of geotropism may be slowed so much as to give a spurious appearance of separation. Frequently, on the other hand, the component processes may be differentially influenced or their coupling may be altered. An already-mentioned example is the differential damage suffered by the induction system at near-freezing temperatures. The history of R irradiation of etiolated plants can certainly influence geotropic responsiveness (e.g., KANG and BURG 1972, HILD 1977, MCARTHUR and BRIGGS 1979; see also Figs. 3 and 7); and since use of dim green or R light during manipulations

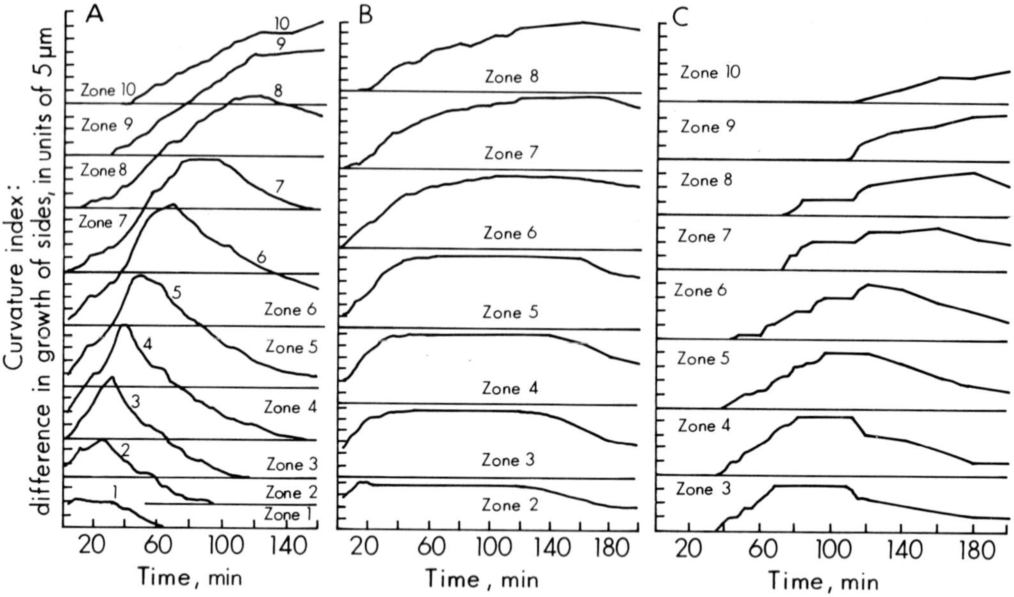

Fig. 6A–C. Time courses for geotropic responses by *Avena* coleoptiles (with root and seed intact), redrawn from DOLK (1930/1936). **A, B** Plants were set horizontal for 30 min, then transferred to a clinostat and photographed every 4 min. Images of coleoptiles were divided into 2-mm zones, starting from the top, and difference in growth rates of the two sides was assessed in 5-µm units. **A** Intact plants, showing curvature and counter-curvature appearing early at the tip, and later at successively more basal zones. **B** Decapitated plants, showing dependence of the counter-reaction on something provided by the tip. **C** Vertical decapitated plants provided asymmetrically with a low amount of auxin for 30 min, starting at time zero, and maintained on the bench. Curvature and counter-curvature are seen to migrate downward, (1) illustrating that auxin redistribution effected geotropically in the tip will migrate basally, and (2) suggesting that sudden asymmetric change of auxin levels is adequate to cause counter-reaction as well as reaction, independent of geotropic stimulation. With a more concentrated supply of auxin, no counter-reaction was evident (data not shown). DOLK recognized that an experiment with auxin-treated, clinostated coleoptiles would have been more definitive, but could not maintain the agar donor-blocks in position during rotation

is apt to vary from experiment to experiment, interpretation of data can be tricky. Similarly, since it is difficult to carry out experiments on hormones without excision of plant parts, the strong inhibition of geotropic curvature by wounding (e.g., WILKINS 1965, PICKARD 1973a) must not be neglected; this is probably due in considerable part to inhibition of elongation (described by CLINE et al. 1974, EVANS and SCHMITT 1975, VESPER and EVANS 1978, EVANS and VESPER 1980), but influences on earlier stages of the geotropic reaction sequence have not been excluded.

In sum, then, the period when induction can be measured without concomitant secondary reactions lasts at most seconds. The development of mediational asymmetries then proceeds hand in hand with induction. Shortly, differential growth may be underway as well. The development of mediational asymmetries

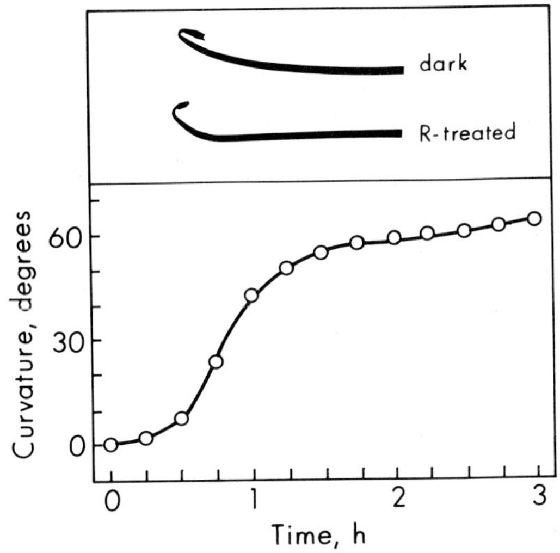

Fig. 7. History of illumination determines the locus of response for etiolated stems of intact *Pisum,* seen here with silhouettes of plants maintained on their sides for 100 min. One plant was kept dark until shadowgraphy, and the other was illuminated briefly with dim R 8 h before stimulation. The time-course of curvature shows the relatively early termination of the rapid bending phase; time-courses are similar for both kinds of plants. (PICKARD and M. HARRISON, unpublished)

due to a given increment of induction is doubtless soon completed; even as they are building up, however, these mediational asymmetries are migrating basipetally. Not only does the differential growth in apical zones diminish as the mediational asymmetries disappear, but also in many cases it reverses as a counter-reaction comes into evidence. The temporal interweaving of these several aspects of geotropism forms so tight a mesh that one must beware of arguments in the literature that because an asymmetry or change in net concentration of some ion, metabolite, or hormone appears early it is the immediate sequel a of the receptive reactions per se; there may well be other asymmetries and changes which are not being assessed with adequate sensitivity. By the same token, late detection of asymmetry or change of a given factor need not necessarily imply that it plays no early role unless adequate sensitivity of assessment can be proven. Furthermore, relative rates and strengths of reception, "transmission", main-phase differential growth, and counter-reaction may be expected to vary greatly between both species and conditions, so that interpretation of molecular data should be grounded in kinetic measurements appropriate to the experimental situation in question.

2.3 Distribution of Receptivity

It has long been believed, on the basis of several types of experiments, that only the tip of a radical or primary root is georeceptive (e.g., LARSEN 1962b, JACKSON and BARLOW 1981). A root begins curving at the tip, and curvature progresses basally (e.g., Fig. 2). Moreover, when its cap is surgically removed, a root loses sensitivity; when the excised tip is affixed in its original position, root sensitivity is partially regained. The surgical technique was refined by JUNIPER et al. (1966); realizing that in corn roots the cap is attached to the body of the root via only a core of cells, they twisted caps gently and snapped them off with minimal damage. Decapped roots continued to elongate but were unable to respond to gravity. After some hours, sensitivity was regenerated at their tips. An excellent review of the functions of the root cap, as well as of evidence that some sensitivity to gravity may exist basal to the cap in certain cases, has been provided by JACKSON and BARLOW (1981).

In contrast, the receptive zone of shoot tissues is typically extensive. It is often strongest at or a short distance from the apical end of a hypocotyl, epicotyl, or coleoptile and grades off more or less gradually. A classical demonstration was discussed in Section 2.2: DOLK (1936), after reviewing a considerable number of cogent earlier papers, determined the average minimum effective stimulation time for *Avena* coleoptiles with different lengths of tip removed and replaced with agar containing growth hormone. The greater the length removed, the greater the stimulation time for a given response or the less the response for a given stimulation time (Fig. 5); while undoubted differences in the capacity for secondary processes precluded strict comparisons, it was clear that (1) receptivity is distributed along the entire length of the tissue, but (2) its effectiveness grades off with distance from the tip. Furthermore, when these experiments are combined with those of Fig. 6 and many others, it is clear that in shoots as in roots, the secondary asymmetries responsible for geotropic curvature and those for counter-reaction shift basipetally.

Sometimes, receptivity is apparently much more nearly uniform along the length of a shoot. For example, although MULKEY et al. (1981) have seen geotropic curvature appear first near the tip of a hypocotyl of *Helianthus* and progress basally (their Fig. 1), KÖHLER (1978) and KÖHLER and DAUM (1979) reported that these hypocotyls begin to bend uniformly along their whole length, and continue to bend uniformly until large curvatures are attained. M.J. VESPER (personal communication) found that *Helianthus* hypocotyls could be converted from uniform to localized curvature by removing the cotyledons. Possibly under normal circumstances the difference is due to different histories of irradiation, as observed with etiolated pea epicotyls. Figure 7, from unpublished experiments, shows typical geotropic curvatures which developed within 100 min after horizontal placement of dark-grown plants and plants irradiated 8 h earlier with R. Curvature is uniformly distributed for the former, but sharply localized for the latter. Time-courses showed that this distinction was valid throughout the development of curvature; and showed also that the sharp bends were displaced farther and farther from the tip as the curvature progressed. Without concomitant measurements of growth distribution and behavior of hormones it is impossible to sort out the precise distributions of gravitational receptivity; certainly, R causes changed distribution of elongation growth in etiolated peas (LEOPOLD 1955, p 39). Whether the distribution of receptivity is unaltered by R and only the distribution of response capability is altered, or whether R educes a variety of changes including both the ability to sense gravity and to carry out secondary reactions, this example illustrates that the extremes of curvature distribution reported in the literature for shoots of different kinds may all be seen in a single species.

Considering that processes of reception and response are better separated in roots than in shoots, it is possible that many ideas about geotropism may meet their most critical test in the root.

2.4 Patterns of Differential Growth

Because it is differential growth which underlies the overlapping phases of the curving response, logic might seem to require next a consideration for representative stimulated organs of the temporal and spatial array of changes in growth of the upper and lower sides. However, available data are both complicated and incomplete, and in themselves make for dull reading. They confuse rather than clarify thought because they are unaccompanied by information on (1) asymmetries of hormones and ionic regulators under identical conditions and (2) depressions and elevations in straight growth rate when concentrations of hormones and ionic regulators are shifted alone and together under identical conditions.

Therefore, this review will proceed to a frontal attack on current knowledge about hormone concentrations, and return briefly to the kinetics of differential growth as a means of obtaining a quick check of ideas and as a critical potential means of keeping book on the significance and completeness of the hormone asymmetries that have been measured.

3 Hormonal Participation

3.1 Generalities About Roots and Shoots

With a knowledge of contemporary hormone physiology, but even without the lore of geotropism, one would make two predictions. Firstly, if mediation of geotropism is hormonal, it must be complex – for no tissue is controlled solely by a single hormone, and different hormones are rate-limiting for growth in different tissues and at different times (cf., e.g., THIMANN 1974, LEOPOLD and NOODÉN, Vol. 10, this Series). Secondly, there must be differences between mediation in root and shoot, though underlying mechanisms would be expected to have many features in common. The requirement for opposing geotropic response itself adds a fundamental constraint to the hormonal control systems, of course, in that the systems in root and shoot must be so poised that the shoot will grow up into the air and the root will burrow down into the soil. Further, although this has not received extensive study, roots and shoots of many plants are coordinated to grow optimally under opposite conditions: shoots expand when moisture is freely available to the roots, ceasing when it is scarce, while roots continue to lengthen as they exhaust soil water in their vicinity (SHARP and DAVIES 1979, PEARSON 1966, EL NADI et al. 1969, HOFFMAN et al. 1971, CALDWELL 1976).

3.2 Shoots

3.2.1 Perspective

The earliest hormone to be implicated in geotropism was auxin: the famous Went-Cholodny theory that auxin limits growth and that tropisms are brought about by its lateral migration made successful predictions about auxin distribution and growth regulation in a number of representative monocot and dicot shoots. In general, the course of early experimentation was dominated by these successes, but at the same time it was seen that auxin does not always limit growth or correlate precisely with geotropic sensitivity. Even for the coleoptile, the primary test system for the Went-Cholodny theory, auxin could not be shown to limit early development (AVERY and LARUE 1938); moreover, when after decapitation of young coleoptiles auxin levels declined only 50%, geotropic response dropped 80% (ibid.). Although cases in which auxin did not limit axial growth of other tissues were not always published, they became part of the lore of the era, putting physiologists on notice that elaboration of the Went-Cholodny theory would ultimately be necessary if its successes were to be generalized.

As it became evident that growth of shoots is often limited by gibberellins, various workers sought to discover whether this group of hormones might act as an alternate mediator of tropisms. It rapidly became clear that, while gibberellins could facilitate curvatures which depended on asymmetric distribution of auxin, they could not by themselves produce tropism; for indeed, unilateral application could not cause curvature (cf. Chap. 10, this Vol.).

The problem was brought to focus by WRIGHT's demonstration (1961) of sequential control of elongation in the *Triticum* coleoptile by gibberellin (GA_3) and auxin (IAA). (Dominant control by cytokinin (kinetin) was also shown for an intermediate period of growth, but its predominance over gibberellin was slight and brief, and it did not invite attention as a potential critical regulator of tropism). While it might initially have seemed reasonable to hypothesize that, for example, redistribution of IAA could mediate shoot geotropism in one species and redistribution of GA in another, it now seemed more credible that there would be only a single, very flexible geotropic mechanism involving both hormones: surely a coleoptile would not switch mechanisms mid-growth! Moreover, any response mechanism with enough complexity to maintain geotropic performance in a plant of a single species as it shifts from dependence on one hormonal control to another is presumably flexible enough to mediate curvature in species for which growth is generally limited by only one of the two hormones. The cogency of this line of reasoning was underlined as more cases of multiple and sequential control (LEOPOLD and NOODÉN Vol. 10, this Series) came to light.

It now appears that (1) lateral redistribution of IAA is a sine qua non of geotropism, (2) asymmetries in GA metabolism and perhaps redistribution are tightly linked with IAA redistribution, and (3) at least as importantly, IAA asymmetry educes a corresponding asymmetry of protons and an opposing redistribution of an inorganic regulator of growth, the calcium ion. Exactly how a given tissue responds to these asymmetries will depend on its physiological condition, but in general, IAA redistribution and the tightly coupled proton asymmetry will tend to slow growth on the upper and enhance it on the lower side, the GA asymmetries will tend to act similarly, and the Ca^{2+} redistribution will tend to inhibit growth on the upper side. It is unlikely that a growing shoot will ever be simultaneously nonresponsive to each of these asymmetries, so that upward geotropic curvature will reliably result from a combination of them.

This section will highlight the experiments which have led to this contemporary generalization of the Went-Cholodny theory, and discuss regulatory asymmetries during the initial lag versus positive phase and during the geotropic counter-reaction.

3.2.2 Auxin

3.2.2.1 Establishing the Occurrence of Lateral Transport

The 1930 experiments of DOLK on geotropic auxin redistribution, analogous to the pioneering experiments of WENT on phototropic redistribution (Chap. 10, this Vol.) are often cited; but it is not always appreciated that the design of these experiments was more conclusive than that of the phototropic ones. As illustrated in Fig. 8A and B, DOLK showed by bioassay that (1) the amounts of auxin diffusing basally from vertical and horizontally stimulated tip sections of both *Avena* and *Zea* coleoptiles are indistinguishable and (2) more auxin emerges from the lower than from the upper sides of horizontal tips of *Zea*. In order to establish that the gravitationally induced asymmetry was due to lateral migration rather than to differential synthesis or release, he further decapitated apical sections and replaced the tips with agar blocks containing a uniformly distributed diffusate of auxin from other coleoptiles (Fig. 8C). Once again, the amount of auxin emerging at the base of horizontal sections was greater on the lower side. Knowing that symmetrically applied auxin can at physiological concentrations stimulate growth in this organ and that asymmetrically applied auxin causes curvature (WENT 1928), DOLK thus built a strong case that the as-yet-unidentified "growth substance" we now know to be IAA (THIMANN 1977, see also GREENWOOD et al. 1972, GREENWOOD et al. 1973, BANDURSKI and SCHULTZ 1974, MCDOUGALL and HILLMAN 1978) is laterally translocated to mediate geotropism.

DOLK's fundamental observations of Fig. 8A and B were repeated by GILLESPIE and BRIGGS (1961), who also collected auxin diffusing from the bases of horizontal upper and lower halves of longitudinally split coleoptile tips and found by bioassay only 53% of the auxin coming from the lower side; this compared with 66% from the lower side of partially split tips (Fig. 8D). This finding extends the argument for lateral migration of auxin by tending to rule out any putative lateral gradation either of conjugation or of basipolar transport capability, which should have led to a significant difference between export of the endogenous store of auxin from the two coleoptile halves. (While splitting the coleoptiles led to some hormone loss during the 3-h diffusion period, nevertheless it is unlikely this would have obscured a substantial difference.)

It could still be wondered if the asymmetry might result from lateral migration of one of two possible kinds of modulator. (1) The putative translocated modulator might linearly control the amount of loss or conjugation within the tissue; this possibility was not tested until the introduction of ^{14}C-IAA (see below). (2) The putative translocated modulator might linearly control

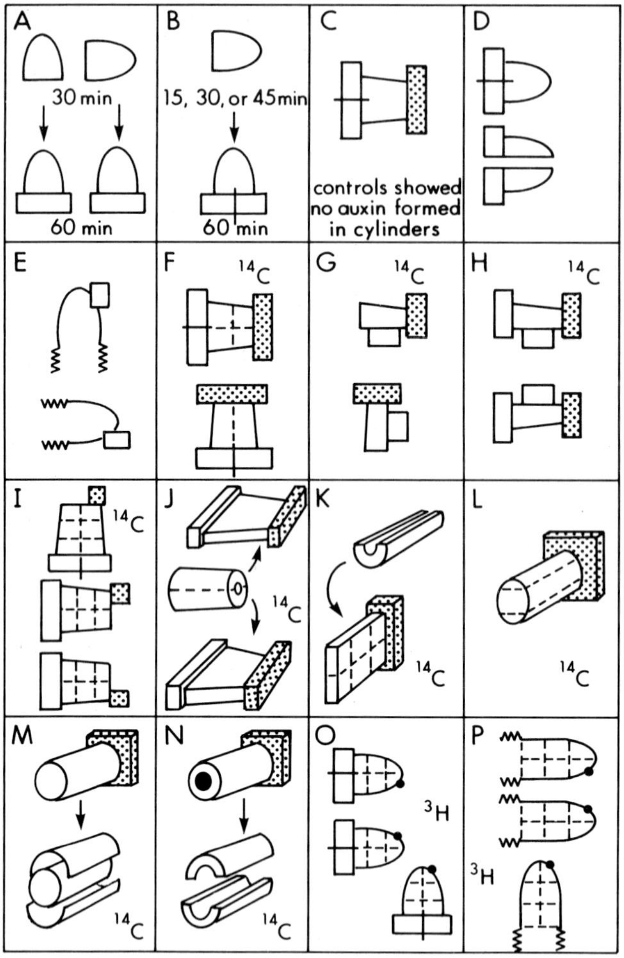

Fig. 8 A–P. Representations of key experiments developing evidence for the lateral transport of IAA. **A, B, C** Dolk 1930/1936. **D** Gillespie and Briggs 1961. **E** Brauner and Appel 1960. **F, G** Gillespie and Thimann 1961, 1963. **H** Hertel and Leopold 1963. **I** Wilkins and Goldsmith 1964. **J, K** Cane and Wilkins 1969. **L** Burg and Burg 1966. **M, N** Iwami and Masuda 1976; stem sections intact during transport for M but cored before transport for N. **O, P** Shaw et al. 1973. All experiments employed coleoptiles except **L–M,** which used dicot stems; **A–E** were with endogenous auxin, **F–P** with exogenous isotopically labeled IAA. The design of early experiments with radioisotopes was strongly influenced by the low specific activity then available for labeled IAA

the capabilities of the polar transport system, thus somehow creating on the lower side a sink toward which auxin from the upper side would diffuse; however, two observations argued against this, even before the era of ^{14}C-IAA studies (see below).

Firstly, suppose that the hypothetically enhanced polar transport on the lower side was indeed strong enough at the tip to drain auxin from the upper side. Then the auxin gradient at the tip would necessarily be the reverse of that in the basal portions of the coleoptile. Yet (excepting during the special case of initial downward response; see Sect. 2.2) the tip bends upward steadily until the relatively late counter-reaction (e.g., DOLK 1936; see Figs. 4 and 6).

Secondly, BRAUNER and APPEL (1960) excised half-tips from coleoptiles to be placed either vertically or horizontally, fitting agar-collecting blocks against the cut surfaces (Fig. 8E). More auxin was bioassayed in blocks from horizontal plants, strongly suggesting an active downward transport mechanism.

In addition to all these experiments with coleoptiles, less extensive but nevertheless convincing demonstrations of auxin asymmetry in horizontal dicot stems were presented by DJIKMAN (1933) for *Lupinus* hypocotyls and by VAN DER LAAN (1934) and BOYSEN-JENSEN (1936) for *Vicia* epicotyls, by BOYSEN-JENSEN (1936) for *Phaseolus* epicotyls, by RUGE (1941) and BRAUNER and BÖCK (1963) for *Helianthus* hypocotyls, and by VAN OVERBEEK et al. (1945) for *Phaseolus* hypocotyls.

All in all, the concept of lateral transport would have seemed well established had not several workers (BÜNNING et al. 1956, GORDON and EIB 1956, REISENER 1957, 1958, REISENER and SIMON 1960, CHING and FANG 1958) tried to repeat the experiments for both geo- and phototropism with a variety of plant parts using ^{14}C-IAA and found no lateral asymmetry in distribution of label. Of course, there are many possible reasons for failure, prominent among which is overloading the tissue with nonradioactive carrier IAA so that the polar transport system is saturated on both the upper and lower or illuminated and shaded sides.

Therefore, GILLESPIE and THIMANN (1961) sought to reproduce the conditions of, e.g., DOLK's experiments as closely as possible but with ^{14}C-IAA substituted for the plants' own auxin. In preliminary experiments with subapical sections from *Avena* coleoptiles, an appropriate difference was indeed detected between upper and lower agar receiver blocks. In more extensive experiments with *Zea* coleoptiles (1963), an auxin differential was found between upper and lower sides of the tissue itself (Fig. 8F). The asymmetry developed was the same over a considerable range of applied and transported IAA. It was additionally shown that when agar collecting blocks were placed lengthwise against halved coleoptile sections transporting ^{14}C-IAA in the upright versus the horizontal position, more radioactivity was collected from the horizontal sections (Fig. 8G). Meanwhile, HERTEL and LEOPOLD (1963) used the closely similar method shown in Fig. 8H with comparable outcome.

Several subsequent studies[1] with isotopically labeled IAA may be roughly summed up as checking whether the observed lateral asymmetry of IAA (1) results from lateral movement of IAA or from lateral gradation of quantities of IAA undergoing basipolar movement, and (2) whether, if lateral movement occurs, it results passively from spillover of IAA from the upper to the lower side or from active transport. Because the possible variations on models invoking differential basipolar transport (cf. Sect. 3.1.3.2 of Chap. 10, this Vol.) are more numerous and intricate than sometimes imagined, the

1 Sophistication of experimentation has increased as isotopically labeled IAA of higher specific activity has become available. At present, IAA labeled with ^{3}H to a level of 30 Ci mmol^{-1} can be obtained commercially. This material should be purified by reverse-phase liquid chromatography immediately before use.

studies have done little to clarify mechanism; but it is nonetheless instructive to examine them, and moreover some important information on the intracellular distribution of gravity receptors has emerged from certain of the experiments.

WILKINS and GOLDSMITH (1964) modified a method of assessing auxin translocation developed by VAN OVERBEEK for phototropism (Chap. 10, this Vol.): as shown in Fig. 8I, ^{14}C-IAA was applied in blocks balanced unilaterally at the apical cut surfaces of vertical coleoptile sections or on the upper or the lower side of horizontal sections. An appropriate time later, the sections were subdivided and amounts of ^{14}C which had moved into the sides lacking donor blocks were compared. More auxin had moved across the midline when the donor block was on the upper side than when it was on the lower side or on one side of the upright controls.

WILKINS (e.g., 1966, 1971, 1979) has supposed that this method is less ambiguous than preceding measurements, claiming that it proves actual movement across the coleoptile, whereas other methods leave open the possibility that basipetal transport is simply enhanced in the lower tissue, resulting in greater uptake at the tip and hence greater accumulation at the base. However, (1) as noted above, experiments with endogenous auxin (Fig. 8D) would seem already to have ruled out this possibility (unless such a laterally graded change in basipetal transport derives from lateral migration of a transport modulator); and, (2) it is hard to imagine why differential uptake due to differential polar transport capability would occur *only* at the tip of a subapical section (e.g., Fig. 8C, F) rather than all along the length, since the position of the cuts is arbitrarily determined and since the uptake mechanism (thought to be diffusive – cf. Sect. 3.1.3.1 of Chap. 10, this Vol. or GOLDSMITH 1977) occurs along the entire length of the coleoptile. After all, the experiments of DOLK (1936) permitted the reasonable conclusion that asymmetry of auxin is enhanced as it travels farther along a section, as is indeed confirmed in the experiments of WILKINS and GOLDSMITH (1964). But if uptake by the lower half from the upper occurs all along the length of the section, this is lateral migration! In fact, (3) the method of WILKINS and GOLDSMITH does not distinguish readily between differential basipolar transport and activated lateral transport. If basipolar transport is indeed more effective on the lower side, any auxin which diffused across from the upper side in WILKINS and GOLDSMITH's experiments would be swept away more rapidly than laterally diffused auxin in vertical controls, thus permitting still more lateral diffusion of auxin. While differential basipolar transport certainly cannot be totally eliminated by the experiments of GILLESPIE and THIMANN in which an agar block rather than tissue acts as a receiver of laterally moving auxin (Fig. 8G), they would seem less susceptible to this interpretation than the experiments of WILKINS and GOLDSMITH themselves.

Further illustrating the tangled thoughts engendered by this subtle problem, in 1963 HERTEL and LEOPOLD tried to settle the issue by showing that movement of ^{14}C-IAA from a horizontal half-section of coleoptile into a laterally applied agar block (cf. Fig. 8H) is inhibited by NPA, an agent known to inhibit polar transport. They inferred from the inhibition that lateral transport resembles polar transport in being actively driven. But inhibition of polar transport alone would prevent accumulation in the block according to a model in which lateral movement is due to increased diffusion from the upper side to a sink on the lower side somehow created by the postulated laterally graded enhancement of polar transport. Insofar as NPA blocks polar transport, it would prevent putative sink formation; and therefore, in the experiment with the half-section, it would eliminate the documented geotropic build-up of IAA in the lower tissue and especially at the lower surface and hence lessen diffusive movement into the agar block.

Ultimately, CANE and WILKINS (1969) concluded that basal transport *is* enhanced on the lower side. They split subapical coleoptile sections and opened them out into sheets, holding them flat between slips of glass, and measured distribution of apically applied ^{14}C-IAA after a suitable transport period (cf. Fig. 8J). If the outer epidermis was oriented upward (or downward) they envisioned that all the cells were transformed into effective "upper" (or "lower") cells in relation to their normal orientation within the organ. Since they found more IAA in the base of the putatively lower section, they concluded that basal transport is enhanced in the lower half of horizontal coleoptiles under normal conditions (though on other grounds they believed this enhancement to be without influence on lateral movement).

There are flaws in this experiment, however, that make it impossible to judge the significance on the changed transport. If auxin migrates to the lower side for whatever reason, the capability of the lowermost tissue for polar transport is indeed critical for net polar transport. The outer epidermis has long been known to be a limiting tissue for auxin-induced growth (WENT and THIMANN 1937, LEOPOLD 1955) and may therefore be supposed capable of more vigorous transport than the parenchyma (cf. IWAMI and MASUDA 1976; see below). About the inner epidermis we have less information, but it might well be an inherently poorer transport tissue than the outer epidermis. Thus, to an important extent the experiment with inverted opened-out coleoptile sections compares contributions to basipetal transport by the outer and inner epidermal layers. Even more critically, however, the experiment compares lowermost tissues with different kinds of manipulative distortion: basipetal transport may well be relatively more efficient on the side compressed during flattening, and less efficient on the side stretched. Compression and stretching will change the spaces available for intercellular diffusion of auxin and hence could alter the rate and polarity of auxin movement. It is also conceivable that the flattening operation, which requires wilting of the tissue if breakage is to be avoided (unpublished experiments with coleoptiles of two dozen corn cultivars by J.W. VAN SAMBEEK, S.E. WILLIAMS and B.G. PICKARD; also personal communication, M.B. WILKINS) could cause some differential damage. The possible magnitude of such effects has not been checked by running appropriate controls.

NAQVI and GORDON (1966) likewise proposed differential change in the basipolar transport system, but believed that lateral movement amounts to accompanying diffusional spillover of auxin from the upper to the lower side. Their experiment is subject to similar yet even stronger criticisms. They compared basal transport of ^{14}C-IAA in longitudinally halved coleoptiles which had been set cut-surface-up or cut-surface-down: while they did find more basipolar transport in the "lower" and less in the "upper" sides, this only suggests that, after all, auxin moved preferentially to the cells at the lower surface would encounter damaged parenchyma in the "upper" sections versus intact outer epidermis in the "lower" ones.

Actually, although the above experiments fail to prove a laterally asymmetric shift in the efficacy of the polar transport system during geotropism, there are at least two theoretical reasons why the occurrence of lateral transport *should* engender asymmetry of polar transport. (1) Stepping up the auxin concentration on the lower side enhances secretion of protons into the apoplast (Sect. 3.3.2) and this in turn enhances diffusion of IAA into the cells; such diffusive entry is thought to be an essential feature of the polar transport system (Chap. 10, this Vol.). (2) Calcium moves in opposition to laterally migrating IAA, and has been demonstrated to stimulate polar transport (Sect. 3.2.7, 3.2.8).

The hypothesis that the primary effect of gravity reception is to simultaneously impede basal auxin transport on the upper side and enhance it on the lower side, so that as a secondary effect auxin spills laterally, is not a simple explanation for lateral auxin transport: as discussed above, to be consistent with the experiments of Fig. 8 it requires the lateral transport of a regulator of polar auxin transport. Such a mobile regulator, if it existed, should have evidenced itself other places in the auxin literature. Thus, the hypothesis seems unnecessarily complicated.

An alternative proposal, also inconsistent with data of Fig. 8, is that basal transport of auxin is changed as a primary consequence of gravity reception by enhancement on the lower side of an organ. This is motivated by an hypothesis that, however gravity might act on a cell, it is only the outer sides of the cell that can detect it: thus, all the cells in the lower half of an organ would make an appropriate response while those in the upper half do not detect gravity at all (cf., e.g., CANE and WILKINS 1969). This idea, intuitively unattractive, would seem discredited in another, very important, experiment by CANE and WILKINS (1969). Opened-out sections of coleoptile (see above) were stimulated by placing them on edge (Fig. 8K). Although this tissue would be unable to discriminate up or down on the basis of morphologically inner versus outer surfaces, net downward lateral movement of applied ^{14}C-IAA was readily elicited. Thus, rotation of a horizontal cell about its polar axis appears not to change its ability to detect gravity; and, since there is no significant "shading" of gravity within the coleoptile, response of an individual cell must be the same whether it finds itself in the upper or lower half of the horizontally placed organ. Thus in geotropism, as in phototropism (Chap. 10, this Vol.), stimulation would appear to result in differential activation of carriers which move auxin across the lateral cell membrane. With respect to the transport process per se, the major difference between the two tropisms would seem to be that following photochemical excitation a long-term change in carrier capability is effected, whereas in geotropism the stimulated state decays shortly after the organ is returned to the vertical (cf. Sect. 2).

Although axial transport seems largely unaffected by shifting an organ from the upright to the horizontal position for periods of 1–3 h, longer stimulation has been reported to bring about a reduction of basal transport (WILKINS and GOLDSMITH 1964). More critically, inversion leads to large, early reductions (HERTEL and LEOPOLD 1963, NAQVI and GORDON 1966, LITTLE and GOLDSMITH 1967, OUITRAKUL and HERTEL 1969). This may well explain the observed reduction in growth rate (e.g., DOLK 1936) accompanying inversion, and more importantly probably explains why a cosinusoidal modifier must be incorporated into the "sine law" in order to explain the dependence of curvature response on stimulation angle between 90° and 180° (e.g., PICKARD 1971, 1973a).

Isotopically labeled auxins have been utilized to confirm the development of asymmetry in dicot stems as well as in coleoptiles: *Helianthus* hypocotyls (eg., GILLESPIE and THIMANN 1963, ABROL and AUDUS 1973a, b, DELA FUENTE and LEOPOLD 1973, MIGLIACCIO and RAYLE 1983, MERTENS and WEILER[2] 1983), *Cucumis* hypocotyls (IWAMI and MASUDA 1976) and *Pisum* epicotyls (BURG and BURG 1966, 1967a, IWAMI and MASUDA 1976).

Furthermore, it has been demonstrated that the gradient of IAA becomes sharpest in the uppermost and lowermost tissues: BURG and BURG showed this by measuring radioactivity in the upper and lower quarters of *Pisum* stem sections (Fig. 8L) and IWAMI and MASUDA extended their work by finding in *Cucumis* a ratio of radioactivity between upper and lower epidermal peels of horizontally held sections equal to 0.23, whereas the ratio was only 0.90 between upper and lower halves of the remaining core of tissue (Fig. 8M). IWAMI and MASUDA also showed by using pre-cored sections that the central tissue is not necessary for the development of asymmetry (Fig. 8N).

Coring the *Zea* mesocotyl after maintaining it horizontal for 15, 30 and 90 min, BANDURSKI et al. (1984) employed a gas chromatographic – mass spectrometric – selected ion monitoring isotope dilution assay to determine an IAA ratio of 0.8 in the cortex and epidermis, but comparable values for the stelar tissue or for the uncored mesocotyl are not available. BANDURSKI et al. suggested that because there is more free IAA per unit fresh weight in the stele than in the cortex and epidermis, the asymmetry also exists in the stele and originates there. It may equally well be speculated that the high concentration of IAA in the central tissue facilitates rapid geotropic loading of IAA into the lower cortical tissue.

2 MERTENS and WEILER reported ratios of 0.7–0.8 in basal agar receiver blocks and of 0.8 in halved tissue, but inexplicably concluded that "it was not possible to detect any significant lateral asymmetry of any of the growth regulators assayed ... in hypocotyls of *Helianthus* ..."

The measurements of BANDURSKI and coworkers also show that in the intact seedling there can be movement of ^{3}H-indole-3-acetyl-*myo*-inositol (as well as ^{3}H-glucose) from the seed to the mesocotyl (personal communication). The inositol ester rapidly equilibrates with IAA. It remains to be worked out the extent to which phloem translocation may be asymmetrically accelerated, phloem unloading may be asymmetric, and symmetrically or asymmetrically unloaded IAA is moved downward by Went-Cholodny transport. However, there is putative, independent evidence with *Vigna* hypocotyls that phloem translocation may be accelerated on the lower side (Sect. 3.2.9.2). The manner of control of IAA levels and asymmetry exhibited by the mesocotyl and the extent to which it can be generalized to other tissues will be an important topic for future research.

3.2.2.2 The Time-Course of Lateral Transport

a) Build Up. Although recent years have seen growing criticism of the Went-Cholodny theory (Sect. 3.2.10), at the same time techniques for measuring lateral transport have been routinely incorporated into experimental designs to probe more deeply into the control of tropisms (e.g. HERTEL et al. 1969, OUITRAKUL and HERTEL 1969, FILNER et al. 1970, SHAW et al. 1973, BANDURSKI et al. 1984). In particular, preliminary measurements of the time-course of asymmetry development promise to open up a much more detailed understanding of geotropism. One such study, that of FILNER et al. (1970), employed the technique schematized in Fig. 8F for 3-mm lengths of *Zea* coleoptiles removed just 2 mm below the apex. ^{3}H-IAA was supplied apically and the sections were equilibrated in the vertical position for 20 min. Then, they were turned horizontal and provided with upper and lower agar receiver blocks, which were replaced every 5 or 10 min. Asymmetry of radioactivity in the receivers was expressed as $(cpm_{lower} - cpm_{upper})/(cpm_{lower} + cpm_{upper})$. Two qualitatively different results were obtained.

In one set of experiments (Fig. 9A), radioactivity was already seen to be moving to the lower side within 5–10 min, when the first measurements were made. By 20 min, the asymmetry index was 3–4 times as great as at 10 min, and by 30 min it had doubled or tripled again. Then, the index declined sharply.

In the second set of experiments, asymmetry also developed rapidly but in the opposite sense (Fig. 9B). Upward transport was evidenced within the first 5 min, but was perhaps already declining by 10 min. Fitting a smooth curve to the data, it would seem that the sign of the asymmetry index reversed between 15 and 20 min and that downward transport of auxin was well established by the end of that period. The fraction of auxin emerging from the lower side continued to increase for the remainder of the experiment, which was terminated at 40 min (just when asymmetry was starting to decline in the first set of experiments).

Simultaneously, work in the same laboratory (OUITRAKUL and HERTEL 1969) employing the technique of Fig. 8G demonstrated early downward movement of an unnatural auxin, naphthaleneacetic acid.

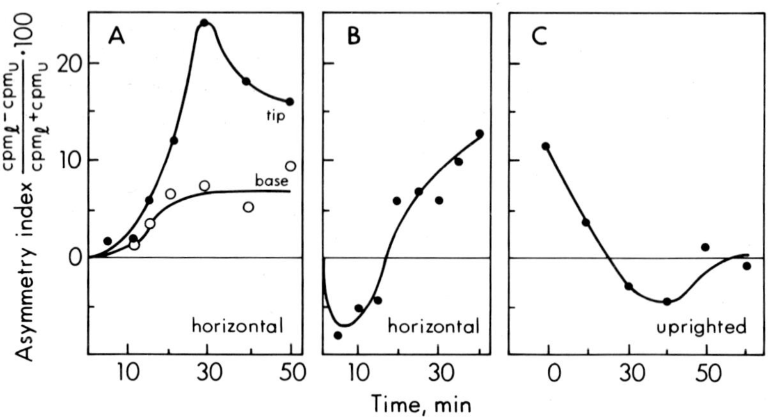

Fig. 9 A–C. Time-courses of lateral transport of IAA in *Zea* coleoptiles (FILNER et al. 1970). **A** *Upper curve* shows early downward movement in apical sections, with maximum at 30 min and decline thereafter; *lower curve* shows less vigorous movement in basal sections which is without an apparent maximum during the period of measurement. **B** Early upward movement, followed by relatively slow development of downward movement. **C** Decline and brief reversal of transport after horizontally stimulated coleoptiles were returned to the vertical

FILNER et al. (1970) focused primarily on their data of Fig. 9B and asserted that transport is transiently upward during the first few minutes. This is consistent with the frequent occurrence of early downward curvature at the coleoptile tip. However, FILNER et al. noted parenthetically that transport might not always be initially upward; for as HILD and HERTEL (1972) and HILD (1977) have painstakingly explored (Fig. 3), the direction of early curvature depends on the previous experience of the coleoptile, and there is generally a slower beginning of upward than downward initial curvature. This is certainly consistent with the relatively slow-starting but nevertheless early monotonic build-up of the positive asymmetry index seen in Fig. 9A.

In order to check this interpretation, HILD and HERTEL (1972) replicated asymmetry tests for coleoptile tissue held horizontal for 10- and 12.5-min response periods after pretreatments (Fig. 3A) known to lead either to early positive curvature or to the slightly slower consistent upward curvature. Using both the assay method of Fig. 8F and that of Fig. 8G, they confirmed that the direction of early transport is indeed oppositely correlated with the direction of early curvature.

The drop in net lateral movement per unit time starting at 30 min in Fig. 9A coincides approximately with the onset of counter-reaction seen by other authors (cf., e.g., Figs. 4 and 6 for the *Avena* coleoptile). While it is not seen in Fig. 9B, perhaps the onset is somewhat variable and the experiment was terminated just a few minutes too soon to display it. The reliability of the data for the decline at about 30 min is not as great as that for the early phase of geotropism, and no comparable decline has been reported by other authors such as WILKINS and GOLDSMITH (1964) or SHAW et al. (1973) (see below). However, the experimental design of WILKINS and GOLDSMITH is not suitable for displaying feedback reactions, for it does not permit IAA levels to build normally on the lower side, and the method of SHAW et al. seems to produce variable results (cf. Chap. 10, this Vol.). Moreover, the kinetics of counter-reaction may well be somewhat variable. The experiments of FILNER et al. (1970) should therefore be repeated, extending the measuring period. Does lateral auxin transport fall to zero at the tip? Reverse direction? How soon does lateral transport decline in more basal tissue? Can conditions be found that will shift the onset of the counter reaction, and do these conditions have a correlated influence on transport? Do closely related cultivars with different counter-reactive tendencies have correlated patterns of lateral transport? Answers to these questions are essential for understanding the relative contributions of auxin and other growth regulators (see below) to development of curvature.

The occurrence of the general downward pattern of early transport evidenced in Fig. 9A was confirmed by SHAW et al. (1973) using a method (Fig. 8O) evolved from that of Fig. 8I: 15-mm coleoptile sections with apices intact were provided with ^{3}H-IAA at a small wound near the tip – on upper or lower sides of horizontal sections or on one side of vertical sections – and diffusate was collected in the tissue and in paired basal receiver blocks. Although the cumulative errors involved in assessing net lateral movement by this method can be large (Sect. 3.1.3.2, Chap. 10, this Vol.) it was clear that net downward movement of auxin had been established within 15 min. Even an indication of slight downward movement within 7.5 min was probably meaningful.

In the same publication, SHAW et al. extended the method of determining lateral transport to seedlings intact except for the tiny apical wound where IAA was applied (Fig. 8P), thus building confidence that experiments with IAA transport in apical and subapical sections do faithfully represent its behavior in intact seedlings.

In the *Zea* mesocotyl, BANDURSKI et al. (1984) assayed endogenous IAA in cortex and epidermis by means of mass spectrometry to show IAA asymmetry within a 15 min horizontal exposure period (Sect. 3.2.2.1). Very recently, BANDURSKI and SCHULZE (personal communication) extended the measurements to show that full asymmetry is developed within 3 min, although no asymmetry can be detected at 1 min. This behavior contrasts with that of the tip of the coleoptile, in which the asymmetry builds gradually over perhaps half an hour (Fig. 9).

b) Decline. Figure 9A and B illustrates decline of lateral transport in near-apical sections maintained horizontal. FILNER et al. (1970) also studied the decay of lateral transport in these sections when they were uprighted after a 20-min horizontal exposure. This decay began as soon as could be measured (Fig. 9C), and the direction of transport apparently reversed at about 25 min and continued upward for some time, peaking within 15 min and subsiding for another 20 min. In interpreting Fig. 9C, it must be kept in mind that transport would have declined shortly, and perhaps reversed, even had the coleoptile sections not been uprighted after only a 20 min horizontal treatment (cf. Fig. 9A, B; and, lacking detailed curvature data for *Zea,* cf. Figs. 4 and 7 and associated references for *Avena*). Had sections been uprighted after only a 10 min treatment, it may be conjectured that downward transport would have continued for another 10–15 min; obtaining a time-course for lateral migration following brief stimulation should be a high research priority.

3.2.2.3 The Time-Course of Auxin-Induced Growth

a) Early Upward Responses. Upward curvature of geotropically stimulated shoots is not visible to the eye until 15–30 min have lapsed. This is consistent with the lags for IAA-induced elongation measured when exogenous IAA is supplied to decapitated shoots or excised shoot segments. However, with the use of high-amplification methods of measuring displacement it has become apparent that upward movement can actually begin within 2–3 min, albeit at a very slow rate (Figs. 1 and 3).

Although the current dogma of auxinology holds firm that elongation lags application of IAA by at least 8–10 min, it is not parsimonious to suppose that earlier upward geotropic response is fundamentally different from the main phase of response. Horizontally placed stems and coleoptiles might be much more natural and vigorous systems for the development and expression of IAA gradients than artificially manipulated tissue provided with IAA. IAA molecules need only travel from the subepidermal to the epidermal cells to establish an effective gradient, and when IAA concentration is increased gradually in an intact plant, the lag for stimulated elongation might perhaps be only a minute or two.

Because proving or disproving the involvement of IAA in early upward curvature may be expected to shed light on a number of aspects of the mechanisms by which elongation is regulated, its investigation should assume high priority.

b) Early Downward Responses. The same workers (Filner et al. 1970, Hild and Hertel 1972) who have attributed early positive (wrong-way) curvature to early upward auxin transport have at the same time considered it due to "overstimulation", analogous to the situation described by Rayle et al. (1970) in which a sudden step-up of IAA concentration causes a transient inhibition of straight growth before the expected enhancement occurs. In order to evaluate the possible role of "overstimulation" by auxin in geotropic curvature, Morath (1972) and Morath and Hertel (1978) as well as Ullrich (1978) have applied IAA asymmetrically and report that a sudden step-up (e.g., 20 fold) on one side of a coleoptile does indeed cause curvature toward the side with the increased concentration.

One argument in favor of "overstimulation" is that transient positive curvatures tend to develop more rapidly than negative ones (Fig. 3), just as transient growth inhibition by sudden step-up of IAA occurs earlier than does promotion by gradual increase of IAA.

Early transport to the upper side is incompatible with such a shock reaction, which in order to effect downward curvature could occur only in response to a sudden build-up of auxin on the lower side. Evidently in order to avoid this contradiction, Hild and Hertel asserted: "One may assume that there is only *one* system – both handling auxin and reacting to auxin and to the transmitted effects of statolithic pressure." While it does not stretch credulity to postulate that both reverse transport and adaptation occur simultaneously, it nevertheless seems impossible that both could play a controlling role in transient early downward curvature in the identical organ. Given that reverse transport has been observed and correlated with "wrong-way" curvature, whereas it has not been shown that geotropic IAA asymmetry can develop fast enough to transiently inhibit growth on the lower side of the plant, it may be speculated that (1) direction of transport is the usual determinant of the direction of early curvature, and (2) conditions causing reversal of transport also promote rapid growth response. The matter urgently requires experimental resolution. In any case, it will be suggested in Section 3.2.8 that the positive curvatures educed by gravity and by laterally applied IAA might be expressions of two rather different processes.

c) Later Responses. DOLK (1936), after painstaking measurement of (1) geotropic curvature and counter-curvature (Fig. 6A) and (2) gravitationally induced lateral asymmetry of auxin, made it plausible that a steady auxin asymmetry could cause not only the curvature, but also the counter-curvature. He applied agar blocks containing a small amount of auxin to decapitated coleoptiles, set the plants upright, and photographed them as he had done for geotropically responding coleoptiles. Figure 6C shows that initially the coleoptiles bent away from the source of auxin. Curvature started at the tip and then migrated basally. As the curvature migrated, counter-curvature began at the tip and it, too, migrated basally. Counter-curvature was not observed, however, when substantially larger amounts of auxin were asymmetrically provided. (Nor was it observed in the *Avena* curvature bioassay for auxin developed by WENT in 1928, but here the coleoptiles were twice decapitated, with an intervening period during which auxin could be drained by the polar transport system and sensitivity could change as shown by VESPER and EVANS 1978).

DOLK suggested that the counter-reaction might be due to asymmetric depletion of materials necessary for growth. In spite of the evidence of Section 3.2.2.2 for the inverse correlation between direction of curvature and direction of auxin transport, and in spite of the obvious experimental flaws in DOLK's experiment that (1) counter-curvature might in part result from counter-induction by gravity as the plant sat on the bench and (2) any effects of wounding are not accounted for, DOLK's explanation that the response is somehow a result of shifting sensitivity of the tissue to a more or less constant supply of auxin has been strongly reinforced by experiments of HATFIELD and LAMOTTE (1984). Noting that growth responses of excised coleoptiles floating or immersed in solution might not be the same as those of intact coleoptiles to which IAA was provided only at the tip, they applied agar blocks to the cut surface of freshly decapitated but otherwise intact *Zea* seedlings and measured coleoptile elongation as a function of time. Figure 10A shows representative results; wheras intact controls grew fairly steadily over an 8-h observation period, plants provided with 0.1 μM IAA blocks (much like plants without IAA added at the cut surface) declined in growth rate for 2–3 h, then increased steadily until they exceeded the intact controls. Application of blocks with 1 μM IAA caused an initial stimulation of growth and then a strong decline starting (within the resolution of the measurements) 1 h after the application. At about 4 h, the growth rate rose sharply and then declined again. Addition of blocks with 32 μM IAA, a very high

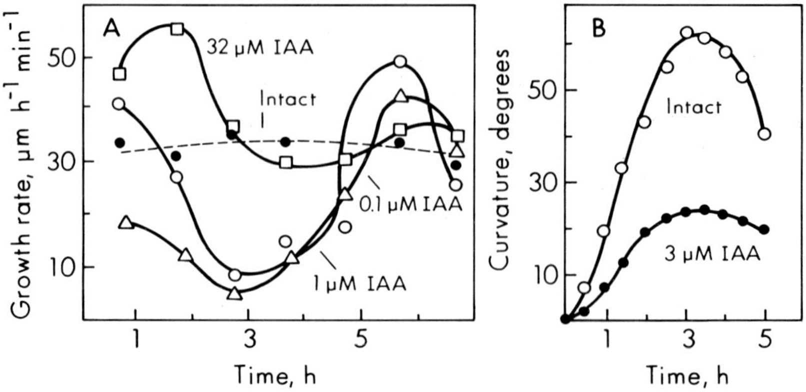

Fig. 10. A When the level of IAA in a *Zea* coleoptile still attached to the rest of the seedling is shifted, elongation rate at first changes in the expected direction but then the tissue undergoes complicated adaptation to the new level. **B** Horizontal decapitated coleoptiles provided with a level of IAA considered similar to the natural level carry out geotropic curvature and counter-curvature with kinetics similar to that of the intact plant. (HATFIELD and LAMOTTE 1984)

concentration, caused strong stimulation followed by a return to control values. These oscillations, though doubtless complicated by a wound response, follow a time-course suggestive of the overall time-course for geotropic curvature under the same experimental conditions (upper curve, Fig. 10B). More specifically, the experiment with 0.1 μM IAA hints at how growth might vary on the upper side of a horizontal coleoptile, while that with 1 μM IAA hints at events on the lower side.

One caveat about the quantitative significance of the experiments of Fig. 10 is appropriate. The agar blocks donating IAA remained on the plants for 8 h, during which the IAA levels were surely diminished (cf., e.g., GILLESPIE and THIMANN 1963). As a check for this, HATFIELD and LAMOTTE replaced donors at 4 h with no change in kinetics of growth, but this check was run only for a single concentration. Clearly, however, the decreasing supply of IAA cannot explain qualitative results such as a rise in growth rate with time.

In an effort to evaluate the possibility of a role of an IAA-induced shift in tissue sensitivity to IAA in geotropic counter-curvature, HATFIELD and LAMOTTE compared the kinetics of geotropic curvature of intact coleoptiles (upper curve, Fig. 10B) with those of decapitated coleoptiles provided with 3.2 μM IAA (lower curve, Fig. 10B). Intact coleoptiles curved 60° in 3 h and net curvature decreased 30% in the next 2 h, whereas decapitated plants curved about 25° in 3 h and lost 20% of their curvature in the next 2 h. Unfortunately, this suggestive experiment proves only that (a physiologically reasonable amount of) IAA is necessary for the counter-curvature (cf. DOLK's experiment of Fig. 6B).

Counter-curvature is not an inevitable consequence of tropistic curvature, as it does not happen when coleoptiles bend toward the light. Prominent among the dissimilarities between geotropism and phototropism which might possibly account for the absence of phototropic counter-curvature is the demonstrated persistence of lateral IAA transport following phototropic induction. Also, probable differences in the pattern of the IAA gradient across the organ, and hence across its several tissues with their differing sensitivities to IAA, should be evaluated in conjunction with data showing that oscillating straight growth (Fig. 10A) and differential growth (DOLK 1936) do not occur with high levels of IAA.

There is no reason to believe that geotropic counter-curvature must be due exclusively either to reversal of transport or to adaptation to shifting auxin levels. Separation of the two contributions will be difficult, for it will depend on detailed kinetic analyses of transport and of shifting tissue sensitivity as a function of distance from the coleoptile apex, and it will be necessary to avoid or separate effects of wounding.

3.2.2.4 Mature Shoots of Grass

The availability of ^{14}C-IAA has made it possible to show that auxin is moved to the lower side of horizontal nodes of flower stalks of *Echinochloa* (OSBORNE 1974) and of node-pulvini of mature shoots of *Avena* (KAUFMAN et al. 1979). Levels of free auxin have long been known to rise during geotropic stimulation of grass nodes (e.g. SCHMITZ 1933, VAN OVERBEEK et al. 1945, BRIDGES and WILKINS 1973, WRIGHT et al. 1978, KAUFMAN et al. 1979), and it is evidently the redistribution of freshly released IAA that permits differential resumption of growth and hence curvature by the mature tissue. Whether the IAA arises by de novo synthesis or by release from conjugates should be settled using definitive new techniques (see BANDURSKI 1980, PENGELLY et al. 1981, ALLEN et al. 1982, EPSTEIN and COHEN 1981, MAGNUS et al. 1980, WEILER 1981, COHEN and SCHULZE 1981). Further, a study of the enzymes controlling the increase in IAA could lead to insight about the links between mediational events and receptor action. It should be tested whether they are activated by a Ca^{2+}-calmodulin complex which could form if Ca^{2+} enters the cytosol either as a direct result of gravitational stimulation or as a secondary result of the redistribution of IAA (cf. Sect. 3.2.7, 3.2.8). Attention should be given to the possible analogy between release of IAA from conjugates in this situation and in the demonstrated (LAU and YANG 1973) suppression of IAA conjugation triggered by cytokinin and possibly mediated by Ca^{2+} entry (see commentary by PICKARD 1984). The grass node-pulvinus might possibly serve as a model for understanding other systems likely to release IAA during geostimulation (consider, e.g., the behavior of floral parts described by RAWITSCHER 1932).

3.2.2.5 Plagiogeotropism

Because plagiogeotropism of a variety of plant parts contributes importantly to the unique form of each species, it is likely that orthogeotropic IAA transport has been evolutionarily modified in a number of ways. As one example, in petioles and lateral branches[3] of plants such as *Coleus* and *Lycopersicon* there is limited evidence that gravitationally induced abaxial transport is opposed by an "endogenous bias" for adaxial transport (review: LYON 1968). The "bias" is expressed conspicuously during omnilateral gravitational stimulation on a clinostat, which results in ET formation, or by exogenous ET (review: ABELES 1973). Such modified transport systems have received relatively little attention; and reviews by KALDEWEY (1971) and KANG (1979), as well as that by LYON, provide access to much of the small literature (see also KOHJI et al. 1982). Recent studies by HAYES and LIPPINCOTT (1976) and HAYES (1981, 1982) on curvature by blades of *Phaseolus* are of particular importance in elucidating the role of IAA transport in ad- and abaxially directed responses and how that transport is modified by ET. Further investigation of mechanisms of IAA distribution, especially in the context of the levels and asymmetries of ET and other regulators involved in plagiogeotropism (Sect. 3.2.3–3.2.8), should be undertaken.

3.2.3 Proton Secretion and Voltage Gradients

3.2.3.1 Perspective

Although it has long been appreciated that gradients of pH develop across a geotropically responding stem (GUNDEL 1933, METZNER 1934) and a role for the gradient in bringing about curvature was suggested (METZNER 1934), it has recently been realized that such gradients occur in the apoplast and are an early and direct expression of IAA asymmetry critical for the mediation of IAA-induced growth. Similarly, the development of trans-organ voltage gradients was measured as early as 1907 by BOSE, yet only recently could an important (but not necessarily the only) component of this response reasonably be ascribed to the proton gradient. Critical experiments to confirm the postulated role of protons are still lacking.

3.2.3.2 Proton Asymmetry

Because H^+ secreted into the cell wall minutes after IAA application plays an important role in mediating the induced growth (e.g., CLELAND 1980, 1982), attention has recently been focused on the participation of H^+ in gravity-induced differential growth. IWAMI and MASUDA (1973) applied acid unilaterally to hypocotyls of *Cucumis* and observed curvature; hence, they suggested that the IAA asymmetry which they later demonstrated to control geotropic response (1974, 1976) acted via correspondingly asymmetric H^+ secretion. Already, GANOT and REINHOLD (1970) had shown that H^+ could be rate-limiting for geotropism in aged hypocotyl sections from *Helianthus*, and at the same time BRIDGES and WILKINS (1973) suggested a critical role for H^+ in geotropism because it, like IAA, could re-establish growth in mature node-pulvini of *Triticum*. (In

3 The distinction between plagiogeotropism and geonasty breaks down for most branches and petioles: the movements and equilibrium positions are partly determined by the position of the organ with respect to gravity and partly by the position with respect to the stem from which it grows (cf., e.g., CROCKER et al. 1932).

mature stem segments of *Avena*, GA_3 has also been reported to enhance H^+ secretion; HEBARD et al. 1976).

The early measurements of the development of H^+ gradients in geotropism were carried out with halved stem tissue (GUNDEL 1933, METZNER 1934), but MULKEY et al. (1981b) were able to visualize the gradients using essentially intact seedlings. After abrading the cuticles of *Zea* coleoptiles and *Helianthus* hypocotyls to permit outward diffusion of H^+, they pressed the organs into plates of agar containing pH indicator dye according to a method worked out by WEISENSEEL et al. (1979) and watched the dye turn color as the pH decreased adjacent to the growing zone. Evidently this method is not highly sensitive, because asymmetry of color change was not remarked until after 90 min for *Zea* and 120 min for *Helianthus;* however, the lower zone of change was much larger than the upper zone by that time. Similar changes were observed for phototropism.

In 1982 WRIGHT and RAYLE showed that applied buffer could inhibit geotropic curvature of *Helianthus* hypocotyls, and in 1983 these studies were extended to make conclusive that H^+ secretion plays an indispensible role in geotropism. Firstly, MIGLIACCIO and RAYLE determined that the buffer did not seriously interfere with lateral redistribution of IAA. Secondly, WRIGHT and RAYLE used (1) PCIB to interfere with IAA action, (2) TIBA to interfere with IAA transport, or (3) vanadate (cf. JACOBS and TAIZ 1980) to inhibit IAA-induced H^+ secretion, and prevented the geotropic H^+ gradient as well as geotropic curvature. The results with vanadate were extended to *Avena* coleoptiles. In accord with expectations, asymmetrically applied IAA could partially overcome PCIB and TIBA but not vanadate inhibitions, whereas asymmetrically applied acid caused curvature in the presence of all three inhibitors. Completing their argument, even after prolonged geotropic exposure the lower side of a hypocotyl did not increase its ability to respond to applied acid; in fact, the upper side responded slightly better than the lower.

3.2.3.3 Electrical Asymmetry

a) Measured Potentials. The development of a potential difference across geotropically responding stems was observed before the role of auxin in geotropism was fully appreciated (e.g., BOSE 1907, BRAUNER 1926, 1927, 1928) and it was early postulated to be intimately related to the reception of gravity. Even after the Went-Cholodny theory became widely accepted, the voltage gradient was suggested as an electrophoretic force driving the migration of IAA (e.g., WENT 1932, SCHRANK 1951). These and other early studies of both gravity- and light-educed potentials (cited, e.g., by WOODCOCK and WILKINS 1969) were not without possibility of artifact and interpretational error, and this section will concentrate on those recent papers that have reliably demonstrated (1) the time-course of development of the voltage gradient, and (2) its dependence on IAA asymmetry.

In order to avoid possible contributions to measured voltage by the various types of salt-bridge required to establish contact between conventional d.c. electrodes and the plant, as well as to avoid any possible influence of the salt-bridge

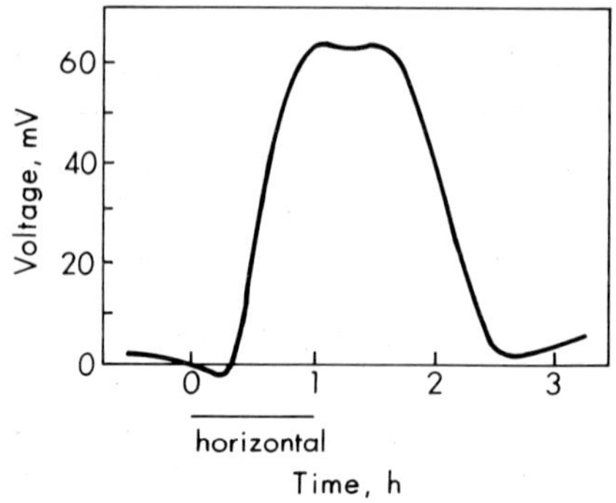

Fig. 11. Development of a transverse electrical potential across a geotropically stimulated *Zea* coleoptile (GRAHM and HERTZ 1962)

on the course of geotropic reaction, GRAHM and HERTZ (1962, 1964, also GRAHM 1964) established a set of plants as one plate of a capacitor and provided a vibrating metal sheet as the other; the alternating voltage generated is roughly proportional to the electric field within the capacitor. Figure 11 illustrates the development of a 65 mV potential across geotropically stimulated *Zea* coleoptiles (GRAHM and HERTZ 1962); a closely similar result was obtained with *Avena* (GRAHM and HERTZ 1964). GRAHM (1964) showed on the one hand that asymmetric IAA distribution gives rise to potentials much like those induced by gravity, and on the other that IAA-depleted coleoptiles do no give rise to such potentials when set horizontal. He also showed that the potentials appeared first at the tip and migrated basally at the same rate as electric potentials shown by NEWMAN (1963) to be associated with basipolar IAA translocation.

GRAHM and HERTZ divided the voltage response into two phases: a lag and a rising phase in which the lower side of the coleoptile became electropositive with respect to the upper. However, they overlooked an apparently consistent small development of negativity on the lower side during the first few minutes of stimulation. Later work with the same equipment (JOHNSSON 1967) did not show such a negative-going phase except at 10° C, where it was not remarked.

More or less confirmatory results were obtained with flowing-solution contacts and conventional electrodes for *Zea* coleoptiles and *Helianthus* hypocotyls by WOODCOCK and WILKINS (1969) and WILKINS and WOODCOCK (1965), although these authors seem to have measured partially shorted voltage differences.

TANADA and VINTEN-JOHANSEN (1980), establishing the excised hypocotyl of a *Glycine* seedling and a closely positioned metal device as (stationary) plates of a capacitor, measured voltage rise across the capacitor starting about 2 min after horizontal placement. The lower surface of the hypocotyl became relatively positive, with the region of maximal bending showing the earliest electrical change. Unaware of the work of, e.g., FILNER et al. (1970) and HILD and HERTEL (1972), they postulated that the early change precedes auxin redistribution.

b) Underlying Ionic Asymmetry. JAFFE and NUCCITELLI (1977) suggested that the geotropic electrical potentials might result from IAA-mediated H^+ secretion. They typically rise with a delay of 10–20 min, corresponding with the similar lag generally preceding detectable increase in H^+ efflux following IAA application (e.g., JACOBS and RAY 1976, CLELAND 1976). They seem generally consistent with the effects of applied IAA and H^+ on straight growth and electrical potentials documented by MIZUNO et al. (1980) and MIZUNO and OKAMOTO (1982) with seedlings of *Vigna*. It would nevertheless be useful to test for the involvement of proton secretion in geotropic electrical potentials by attempting blockage by vanadate.

With regard to the limited data on slight voltage asymmetries developing in only a few minutes, possible interpretations are numerous. Asymmetries such as reported by TANADA and VINTEN-JOHANSEN might result from early development of IAA and H^+ gradients (cf. Sect. 3.2.2.2, 3.2.3.2): the electrical potential across an organ must be a very sensitive indicator of H^+ secretion, for a very small imbalance of ions creates a large voltage asymmetry. Of course, to the considerable extent that charge may be balanced during H^+ secretion, as by entry of K^+ (e.g., TERRY and JONES 1981, CLELAND 1982), the electrical potential is an under-representation of the electrochemical gradient due to H^+.

On the other hand, early subtle voltage asymmetries might more directly index the gravity sensing process. In this case, it should be possible to detect them in shoots drained of auxin before geotropic stimulation.

Early asymmetries have not been specially sought in shoots carrying out early transient downward curvature, and it is not known whether the early transient wrong-way voltage shifts which sometimes appear unremarked in published recordings (Fig. 11 A, JOHNSSON 1967, WOODCOCK and WILKINS 1969) will prove reproducible. Because two hormonal bases for early downward curvature have been suggested, two ways of generating the putative voltage asymmetries must be suggested also. First, if the response is mediated by wrong-way IAA transport, as argued in Section 3.2.2.2, any associated voltage gradients are likely to be due to IAA-stimulated H^+ secretion. Second, if the response is mediated by sudden increase of IAA on the lower side, it is pertinent that MORATH (1972) and MORATH and HERTEL (1978), as well as ULLRICH (1978), have pointed out that when sudden asymmetric step-up of IAA causes unilateral inhibition of growth and consequent bending toward the source of IAA, it simultaneously causes charge polarization. BATES and GOLDSMITH (1983) have shown that sudden step-up of IAA causes a rapid flux of H^+ into the cytosol and concomitant depolarization of the plasmalemma. They suggest that the electrical asymmetries observed during sudden unilateral step-up of IAA result from fluxes of ions associated with membrane depolarization. (Membrane depolarization per se should not be observed with gross extracellular electrodes.)

c) Significance. The contribution to the gross potential apparently made by apoplastic H^+ indexes not only growth-stimulating activity but also a potent electrochemical gradient which could exert some feedback on IAA translocation and could participate in establishing secondary gradients of other regulatory agents (see especially Sect. 3.2.7).

With regard to IAA translocation, although data are limited, it is clear that lowering apoplastic pH depresses dissociation of IAA and thereby encourages more uptake into cells, which tends to boost active export.

It seems likely that the electrical component of the gradient can influence IAA transport directly. MORRIS (1980) has shown that application of a voltage gradient amounting to 180 mV mm^{-1} along the axis of a light-grown *Pisum* stem can completely and reversibly block polar transport, regardless of polarity. This gradient is to be compared with the 65 mV lateral gradient across a *Zea* coleoptile (Fig. 11) of perhaps 2 mm diameter: the geotropic gradient is perhaps a sixth as large. The extent to which geotropic IAA transport limits itself by creating a lateral voltage gradient could be tested if indeed it is possible to prevent the gradient with vanadate (see above); any positive effects of vanadate should be reversible by application of an artificial voltage gradient across the organ. Further, early application of a lateral voltage gradient should inhibit geotropic transport.

Finally, the hyperpolarization of cell membrane potentials occurring during H^+ secretion (references in BATES and GOLDSMITH 1983) deserves further evaluation. The lateral gradient of transmembrane potentials, which influence cellular fluxes of IAA, K^+, Pi, and many other agents, must play a role in differential organ growth.

3.2.4 Gibberellins

3.2.4.1 Young, Orthogeotropic Shoots

The introduction of GA's to the Western world was not accompanied by much attention to the possible roles of this constellation of hormones in geotropism because it was immediately recognized that unilateral application of GA to orthogeotropic shoots commonly studied in the laboratory does not cause curvature. LIBBERT and GERDES (1964) documented this for *Avena* coleoptiles but also explored the considerable extent to which the application of GA_3 can augment IAA-induced curvature (cf. Chap. 10, this Vol). Such experiments have been extended to, e.g., *Cucumis* hypocotyls (IWAMI and MASUDA 1974) and internodes of *Helianthus* (PHILLIPS and HARTUNG 1976) and *Phaseolus* (MEUDT and BENNETT 1978).

Many of the important early experiments were carried out on plagiogeotropic systems, but chronology will be ignored in order to discuss work on the simpler orthogeotropic systems first.

The earliest assessments of the levels of GA's during response by simple orthogeotropic shoots were those of PHILLIPS (1972) and RAILTON and PHILLIPS (1973). Bioassaying material from agar blocks applied for 20 h to the basal surfaces of excised tips of green *Helianthus* epicotyls and etiolated *Zea* coleoptiles, they found that horizontally placed tissue yielded several times as much GA activity as did vertical controls. Further, about an order of magnitude more activity emerged from the lower than from the upper sides. EL-ANTABLY (1975) also reported more extractable GA activity on the lower side of horizontal stem tissue of *Ribes*.

The times allowed for development of asymmetries were relatively large, considering that for horizontal intact seedlings curvature can be induced at 1° per min for *Avena* coleoptiles and 2° per min for *Helianthus* hypocotyls (and presumably epicotyls as well) (e.g., JOHNSSON and PICKARD 1979) and lateral asymmetry of IAA has been detected within 5 min for *Zea* coleoptiles (Fig. 9). Further, it was unclear to what extent GA's diffusing from a cut surface represented conditions in the tissue and to what extent the changes resulted from increased synthesis, release from conjugates, differential formation of conjugates, specific lateral and basal transport, or attraction to metabolic "sinks".

A start at evaluating the possibility of differential transport was made by three sets of workers who applied radioactively labeled GA's to horizontally placed shoots and followed its subsequent distribution. WILKINS and NASH (1974) were unable to detect radioactivity in basal agar collecting blocks even after 24 h when ^{3}H-GA_3 was applied at the apical cut surface of *Zea* coleoptile sections, and no gravity enhancement of downward movement of radioactivity was detected in the horizontal tissue following asymmetric application of donor blocks to the apical cut surfaces in the manner of WILKINS and GOLDSMITH (1964) (Sect. 3.2.2). On the other hand, WEBSTER and WILKINS (1974) reported weak *upward* lateral movement of ^{14}C-GA_3 in *Zea* coleoptiles; they applied the hormone at a small wound in the tip of otherwise intact seedlings using the pipette technique described in Sect. 3.1.3.3 of Chapter 10, this Volume and measured radioactivity in the tissue itself. WEBSTER and WILKINS suggested that careful analysis of the data of WILKINS and NASH reveals subtle evidence of upward transport in their experiments as well. However, the meaning of the putative weak asymmetry is unclear considering the magnitude of the lateral gradient of GA activity reported (albeit for longer stimulation) by RAILTON and PHILLIPS (1973), the absence of data on possible metabolic transformations, and the potential errors of the pipette method (Sect. 3.1.3.3, Chap. 10, this Vol.).

Possible redistribution of ^{3}H-GA_1 in green *Helianthus* epicotyl sections was sought with negative results by PHILLIPS and HARTUNG (1976). (No asymmetry of apically applied ^{14}C-IAA was detected in this study, either; but cf. Sect. 3.2.2). As quite generally found (cf. Sect. 3.2, Chap. 10, this Vol.) symmetric or asymmetric application of GA_1 or GA_3 to sections enhanced elongation but asymmetric application could not cause curvature of upright sections. On the other hand (see above and cf. Sect. 3.2, Chap. 10, this Vol.) exogenous GA did enhance tropistically and IAA-induced curvatures regardless of the point of application. While these results might be considered as evidence against a role for lateral migration of GA_1 (cf. PHILLIPS and HARTUNG 1976), their validity must be questioned because of a concurrent failure to demonstrate lateral translocation of IAA.

All in all, then, this limited evidence concerning lateral movement of GA's in grass coleoptiles and dicot stems appears ambiguous. Moreover, no information on the changes in composition of GA's possibly accompanying the large net increases in these tissues has yet appeared.

3.2.4.2 Mature Shoots of Grass

For the mature shoots of *Avena*, an elaborate behavior of GA's during geotropism was demonstrated by PHARIS et al. (1981). During a 2-day reaction period the stems curved 30°, with both node-pulvini and the basal portions of the internodes participating. At the end of this period, about 97% of the GA activity detected by means of the dwarf rice assay in the horizontally placed node-pulvini was in the lower halves, and about 85% of the GA activity of the associated internodes was in the lower halves. At the same time, net GA activity was depressed in the node-pulvini but was doubled in the internodes relative to vertical controls. The net increase in the internodes was accounted for mainly by an increase in less polar GA's in the lower half, consistent either with increased GA synthesis or with movement of less polar GA's from adjacent node-pulvini or sheaths.

Two days after application of ^{3}H-GA$_4$ the lower sides of both internodes and node-pulvini had accumulated more ^{3}H-GA$_4$ and its acidic metabolites than had the upper sides, while the upper sides had accumulated water-soluble metabolites (presumed to be inactive conjugates).

All these changes are consistent with a role for GA's in the geotropic bending of the mature *Avena* shoot. Indeed, growth of the internode (which has not stopped elongating in plants of the age used by PHARIS et al. (1981) can be stimulated by application of GA$_3$ (KAUFMAN 1965, 1967, see also ADAMS et al. 1973, DAYANANDAN et al. 1976), and GA$_3$ or better GA$_{4/7}$ application can enhance geotropic curvature of the stem. Although GA$_3$ alone will not renew growth in the mature node-pulvinus (see refs above) the effect of GA's on IAA-stimulated growth in vertical mature node-pulvini has not been assessed. Probably, in the mature stem as well as in the coleoptile of *Avena* (Sect. 3.2.4.1), GA's serve to enhance IAA-elicited growth curvatures.

The evidence that upward growth of *Avena* shoots may be enhanced by GA's reinforces the earlier demonstration (see above) of GA$_3$-mediated curving in stems of another grass, *Cynodon* (Bermudagrass). MONTALDI (1969) showed that rhizomes and stolons of intact plants both respond to sprayed GA$_3$ by bending upward. Moreover, he showed that untreated stolon explants hold their apical internodes at angles of 50°–60° with the horizontal by virtue of curvatures in the basal nodes, and application of GA$_3$ increases upward bending. In these nodes, exogenous IAA has little effect on curvature; thus, while (on the basis of accumulated evidence from other plants) IAA asymmetry presumably plays a critical mediational role in establishing asymmetry of GA's, without the latter to act as second mediators upward curvature of *Cynodon* stolons is weak at best.

More detailed description of IAA and GA behaviors in the mature grass shoot would doubtless be rewarding; in addition to its intrinsic interest, this system might serve as a model for the many organs with geotropic and geomorphic responses more intricate than those of the seedling shoot, and might provide some comparative insights into the not uncomplicated seedlings as well.

3.2.4.3 Plagiogeotropic Systems

Concurrent with the work of LIBBERT and GERDES on coleoptiles (Sect. 3.2.4.1) was a strong suggestion that GA's play a critical mediational role in orthogeotropic behavior of a representative conifer, *Cupressus* (PHARIS et al. 1965). Application of growth retardant AMO-1618, an inhibitor of GA synthesis, prevented plagiogeotropic lateral branches of *Cupressus* from assuming the orthogeotropic role of a removed terminal shoot, whereas GA_3 could reverse the inhibition. PHARIS et al. (1972) further found that application of GA_3 could reverse the abnormal positive geotropism of both terminal and lateral shoots growing under very short, 4-h photoperiods; and BLAKE et al. (1980) showed more generally that GA_3, which appears to be the predominant endogenous GA of *Cupressus* (see KAMIENSKA et al. 1974, PHARIS and KUO 1977), shifts branch orientation upward (see more refs., BLAKE et al. 1980). BLAKE et al. (1980) further demonstrated that application of IAA to decapitated terminal shoots can promote or retard upturning of lateral branches, depending on the IAA concentration. Meanwhile, bioassays by MININA et al. (1976) demonstrated more GA activity on the lower than on the upper sides of horizontal young shoots of *Pinus* and of diageotropic branches of *Abies*. While definitive comprehensive assays of endogenous hormones have not yet been accomplished for any georesponsive conifer system, the data are nonetheless still well drawn together by the conjecture of PHARIS et al. in 1965: "it is conceivable that the turning up of the lateral requires auxin in the lateral, and that gibberellin mediates the action of the auxin".

Work on other shoot systems with plagiogeotropic capabilities added to the evidence for tropistic involvement of GA's. HALEVY et al. (1969) demonstrated that several genetic lines of *Arachis* which differed in lateral branch angle each contained the same amount of GA activity gauged by the barley seed assay, but the low-angle runner branches contained higher levels of antagonists of GA function. Applied GA_3 caused the runner branches to grow at a higher angle but had no influence on medium-angle branches, and two growth retardants (CCC and Alar) known generally to reduce the effective level of endogenous GA's caused the medium-angle branches to grow closer to the horizontal, but had little influence on the growth angle of runners. Plants were not responsive to applied auxins or anti-auxins. (Later, ZIV et al. 1976 also showed that the upward growth of runners caused by application of ET is accompanied and perhaps mediated by an increase in endogenous GA activity.)

RECHES et al. (1974) found the GA activity of positively geotropic *Morus* branches to be concentrated in the abaxial tissue, but a material which might be ABA was more abundant adaxially. Physiological control of the putative GA's and ABA has not been evaluated (cf. Sect. 3.2.5.5).

In work described in Section 3.2.6.2, MONTALDI (1969) showed that, whereas application of IAA to diageotropic stolons of *Cynodon* was essentially without influence, GA_3 caused them to bend upward.

Thus, the role of GA's in plagiogeotropic systems may prove to be fundamentally similar to that in orthogeotropic systems, although the plagiogeotropic regulation of the development of GA asymmetry and of the occurrence and asymmetry of promoters and antagonists of GA action may well prove quite elaborate.

3.2.4.4 Overview

In overview, an obligatory role for auxin in the geotropism of shoots is reasonably well established, and in a variety of systems asymmetric IAA must mediate curvature because IAA limits growth. Often, however, GA's are critical regulators of elongation. While GA's have not been as extensively investigated during geotropism, test cases of both IAA-limited and GA-limited organs exhibit GA increases and asymmetries, and one study of the geotropically responding *Avena* shoot has been detailed enough to document massive shifts in the amounts and minor shifts in the metabolism of the GA's in upper and lower tissue. In this particular case, where both a sensitivity of the internode tissue to exogenous GA's and an increase in endogenous GA activity within the lower internode tissue have been documented, it is highly likely that GA's participate in mediation of internode curvature. In the *Helianthus* stem, in which the increases and asymmetries of GA are similarly dramatic, it is also reasonably clear that GA's control elongation and help mediate curvature. Because all evidence suggests that an asymmetry of GA is not normally maintained in the absence of an asymmetry of IAA, GA's presumably transduce and amplify that primary hormonal asymmetry and produce curvature in situations where control of elongation by IAA is negligible or feeble.

While all this heuristic reasoning seems compelling, both more extensive and more intensive investigation of the role of the GA's is urgently needed.

3.2.5 Abscisic Acid

Recent evidence implicates ABA as a stimulator as well as inhibitor of growth in roots, and it has been proposed that it is a critical regulator of their geotropism (Sect. 3.3). ABA can also promote shoot growth (Aspinall et al. 1967, McWha and Jackson 1976, Abou-Mandour and Hartung 1980), and, as was earlier known, can inhibit elongation when concentrations are high. Thus although there is no reason to believe that it plays a general role, it is not unlikely that ABA will be found to participate in shoot geotropism in a variety of circumstances.

In fact, ABA asymmetries have already been reported for a few shoots. For example, a report of asymmetry of putative ABA in the positively geotropic branches of a weeping *Morus* (Reches et al. 1974) suggests that a search for control by ABA might be particularly profitable for plant parts deviating from negative orthogeotropic behavior. In the downward-bending young *Morus* branches, the putative ABA was found in conspicuously greater amount on the upper side; the influence of physiological levels of the substance on elongation of the *Morus* branches was not tested. Auxin and GA bioassays were also carried out. No asymmetry was found in the former, but the straight growth bioassay employed would not be expected to detect the rather small asymmetries normally expected and presumed to mediate the development of secondary asymmetries of other hormones (cf. Sects. 3.2.3, 3.2.4, 3.2.6). GA-like activity was predominant on the abaxial side of the stems, and Reches et al. (1974) speculated that it might be high enough to inhibit rather than stimulate growth in the abaxial tissue. This interpretation of GA activity in weeping branches appears unlikely, however, because the copious literature on GA effects shows few examples of growth inhibition by supraoptimal GA. A promotive effect of asymmetric GA might, however, be overridden by promotion of downward curvature by the opposing asymmetry of putative ABA. Such a method of controlling growth by opposing stimulations might be reasonably efficient for a plant such as *Morus,* which grows rapidly; it is at least as sensible as control by opposing inhibitions.

As a second example, genetic control of lateral branch angle of *Arachis* (peanut) was suggested to be mediated by GA antagonists, one of which cochromatographed with ABA in three solvent systems (HALEVY et al. 1969, cf. Sect. 3.2.3). A running habit, in which lateral branches grow at an angle of 0°–25° with the horizontal, was associated with large amounts of the ABA-like inhibitor plus an inhibitor not detected in erect plants. Even in erect plants there were significant levels of two of the three inhibitors found in running plants; these may help to explain why lateral branches of erect plants assume angles as much as 30° from the vertical. Whether or not the GA antagonists are asymmetric within the branches was not determined.

Further, asymmetry of material cochromatographing in one solvent system with ABA was reported for stems of *Ribes* (black currant) by EL-ANTABLY (1975): twice as much was found on the upper side.

More extensive characterizations of all these ABA-like materials and tests of their growth-controlling activities in the stems from which they are extracted are desirable.

3.2.6 Ethylene

3.2.6.1 Roles in Primary Curvature

It has been proposed in two contexts that ET plays a necessary role in (ortho)-geotropism.

A mutant of *Lycopersicon* requiring ET for vertical growth was described by ZOBEL (1973, 1974). On the basis of measurements suggesting that the mutant synthesized low levels of the gas in response to IAA (although not in response to wounding), ZOBEL postulated that ET is required for some relatively early step in the geotropic process. JACKSON (1979) reported that ET production by mutant and normal plants is similar, and reasoned that the defect in the mutant is associated either with ET receptors or with the geotropic system per se. However, MULKEY (1979) did find differences in ET production, and moreover BRADFORD and YANG (1980) found defective ACC synthase in the mutant, in accord with impaired ET production. The possibilities for the nature and consequences of the genetic lesion in this plant with a vining tendency are broad (cf. Sect. 3.2.6.4) and invite further study.

WHEELER and SALISBURY (1980, 1981) made the second suggestion that ET may be required for the main phase of geotropism; they further specified that the role of ET might be inhibition of cell elongation on the upper sides of stems. They base their proposal on the retardation of geotropic bending of *Xanthium, Ricinus* and *Lycopersicon* stems by treatment with AVG or Co^{2+}, which inhibit ET synthesis, and with Ag^{2+} or CO_2, which block ET action. Several questions must be raised about this provocative study. Firstly, were all the inhibitions significant? There were only three to five plants per experiment, and the total of replicate plants was evidently small. Secondly, the concentrations of inhibitors were high: were the responses specific to ET physiology or indeed to geotropism? For example, older plants were used in the experiments with Ag^{2+} because "younger plants are often burned by the silver solutions" (1981). Thirdly, also raising questions about specificity, the changes effected seem possibly to have differed from inhibitor to inhibitor: for example, in the 1980 paper AVG-treated *Xanthium* plants responded at the same rate as controls for the first 3 h of the experiment but slowed dramatically thereafter; young $AgNO_3$-treated *Xanthium* showed a long lag but thereafter bent at the same rate as controls; older $AgNO_3$-treated *Xanthium* lagged 6–10 h before starting to bend and even then bent much more sluggishly than controls. Fourthly, it was not checked whether exogenous ET could restore geotropic reactivity to AVG- or Co^{2+}-inhibited plants. All in all, the data are at best ambiguous. But even assuming enhancement of geotropism by ET (and there has long been evidence – see above and Sect. 3.2.6.4 – that this can occur in the mutant diageotropica and some other plants), WHEELER and SALISBURY are incautious to suggest (1980) that the gas may play a more critical positive role in orthogeotropism than does auxin.

M. HARRISON (personal communication) has examined the effect of spraying Ag^{2+} on hypocotyls of *Lycopersicon* prior to geotropic stimulation. Application of low levels probably inhibited ET action with reasonable specificity, because the plants remained turgid and healthy. However, geotropism was slowed. Higher levels of Ag^{2+} blocked geotropism completely, but inhibited ET production as well as action and so may possibly have inhibited protein synthesis. Nevertheless, even these high levels acted reversibly, for the plants remained healthy in appearance and partially regained geotropic sensitivity within 24 h. These experiments and similar ones with Co^{2+} tend to support ZOBEL's conclusion from work with the mutant diageotropica that ET is required for some geotropic process controlling IAA transport.

HARRISON has also tested the effects of extra ET on geotropism of *Lycopersicon* hypocotyls, finding a small but significant enhancement for concentrations from 10 nl to 20 µl ET per l air. Plants subjected to 50–100 µl per l responded similarly to controls. The way in which supranormal ET promotes geotropism of *Lycopersicon* hypocotyls has not been determined, but it may be speculated for this particular case that the action of relatively high levels differs from that of very low levels; specifically, it is plausible that added ET acts indirectly by promoting overall elongation. Although the notion that ET is an inhibitor of elongation is based on work with a variety of plants (ABELES 1973), exceptions are known and more will probably be discovered when the effects of very low levels of exogenous ET are better explored. Already, the exceptions include *Phaseolus* and *Xanthium* hypocotyls (GOTÔ and ESASHI 1974), mesocotyls of *Avena* and *Oryza* (SUGE 1971) and stems of some grasses (POOVAIAH and LEOPOLD 1973).

There has been another proposal for how ET might participate in upward curvature in a nonobligatory manner. For geotropically responding flower stalks of *Echinochloa*, OSBORNE (1974) demonstrated elevated ET synthesis in the lower half following increase of IAA in that tissue. Cells in the lower half became elongate and thin-walled, symptomatic of IAA-induced growth, whereas cells in the upper half remained shorter and developed thick walls, suggestive of growth under the influence of ET. OSBORNE suggested that the ET is without effect where it is produced on the lower side because the influence of high levels of IAA there predominates, but that it exerts a large inhibitory effect when it diffuses to the upper side where IAA is depleted. However, later experiments (WRIGHT et al. 1978) indicated that neither added ET nor an inhibitor of ET synthesis modified the geotropic bending response.

3.2.6.2 Counter-Reactive Role

a) Overview. Ethylene and IAA work as balancing members of a feedback system in which IAA stimulates the synthesis of ET, and ET inhibits the synthesis and often the transport of IAA (e.g., BURG and BURG 1966, 1967b, PRATT and GOESCHL 1969, KANG 1979). The parameters of this feedback are taxonomically variable, resulting in considerable diversity of plant behavior (see refs. above). In consequence of the feedback, ET may play a critical role in the geotropic counter-reaction of many plants, causing some petioles and branches to seek an equilibrium position far from the vertical (plagiogeotropism) and slowing the rate at which some primary stems approach the vertical.

b) Effects of Ethylene. In 1934, an elegant but infrequently cited paper of VAN DER LAAN laid the foundation for understanding a wide range of IAA–ET interactions in geotropism. Expanding on work of NELJUBOW (1901, 1911) and others, VAN DER LAAN found by surveying etiolated sprouts of several species that each of these grew thicker and elongated less in the presence of high levels of ET, but that the geotropic responses of their shoots could be divided into three categories. (1) Stems switch from an upward to a horizontal growth habit: this group includes *Vicia* and *Pisum* epicotyls as well as epicotyls of *Ervum, Lathyrus, Tropaeolum* (NELJUBOW 1901), and *Solanum* (SINGER 1903). Because the equilibrium orientation changes slightly as the axis is rotated, the response has a nastic as well as a tropistic component. (2) Stems "... curl like a pig's tail, occasionally growing ... again into the soil"; growth is particularly stunted and abnormal: *Phaseolus* and *Raphanus* hypocotyls provide examples. (3) Geotropism is slow but no abnormalities occur: hypocotyls of *Lupinus* and *Lepidium* as well as coleoptiles of *Avena* are illustrative.

For *Avena* (group 3), VAN DER LAAN showed that ET lowered auxin release at the coleoptile tip to about 65% of control values, but did not change polar auxin transport as a result of either short or long exposures. Because responsivity to applied auxin was essentially unchanged, VAN DER LAAN concluded that the reduction of growth was simply the result of less auxin being distributed to the tissue.

In the case of *Vicia* (group 1), which changes its position of equilibrium in response to ET, auxin production was lowered to about 25% of controls within an hour after application of the gas, and during this interval elongation rate was lowered to 50%. More importantly, in horizontally placed stems ET reversed the normal downward movement. VAN DER LAAN was thus able to account not only for the stubbier growth, but also for its redirection in the gravitational field.

One was left to surmise what happens in plants of the intermediate group (2), for which the direction of growth during ET treatment is not defined with respect to gravity: is the primary geotropic effect of ET simply inhibition of lateral IAA translocation? One single, casually reported experiment (KANG and BURG 1974) with "bean", presumably *Phaseolus,* showed essentially no difference in ^{3}H-IAA asymmetries measured in air and in 100 µl ET per l air. However, in this experiment, ET-treated plants did not curve at all during an 18-h geotropic stimulation, whereas controls in air curved about 45°. The lack of curvature in these ET-treated plants contrasts with the spiraling and random growth reported for ET-treated *Phaseolus* by VAN DER LAAN (1934). Therefore, IAA asymmetries in *Phaseolus* hypocotyls should be tested further using a range of ET concentrations under conditions when spiraling can be observed.

Some of the results for groups 1 and 3 have been confirmed with ^{14}C- and ^{3}H-IAA. BURG and BURG (1966) showed for etiolated *Pisum* epicotyls (group 1) that ET can immediately inhibit lateral IAA migration without appreciably altering basipolar movement, although the latter does subside after a considerable lag. KANG and BURG (1974) found an inverted IAA asymmetry after ET treatment. For coleoptiles (group 3) BURG and BURG (1966) confirmed that lateral as well as basipolar movement is unaffected by short-term ET treatment.

c) Production of Ethylene. The importance of VAN DER LAAN's observations with applied ET was underlined when it became clear that stems produce ET continuously at a low level, and that synthesis can be markedly enhanced by endogenous triggers such as high IAA levels or by numerous environmental perturbations (e.g., ABELES 1973, LIEBERMAN 1979, YANG 1980). Six situations in which ET levels may possibly be high and effective in modifying geotropic orientations will be discussed.

α) Baseline Values. The first is the most general: the "baseline" level may be adequate to retard geotropism. Thus, KANG and BURG (1972) found that hypobaric removal of ET accelerated geotropic curvature of etiolated *Pisum* seedlings by 23%. They also showed that preirradiation with R depressed ET production and, apparently thereby, enhanced geotropism; exogenous ET prevented the enhancement (Fig. 12A). R-pretreated horizontal sections established stronger lateral asymmetry of ^{3}H-IAA than did controls.

β) Stimulation by Auxin. The second condition is that lateral IAA migration, sometimes combined with enhanced IAA synthesis, can change the levels of IAA across the stem enough to increase synthesis of ET in the lower tissue while decreasing it above. This was shown by ABELES and RUBINSTEIN (1964), who placed hypocotyls of *Phaseolus* (VAN DER LAAN's group 2) horizontal for 2 h, bisected them longitudinally, incubated the upper and lower halves in bottles for 18 h, and assessed ET production. Upper and lower halves evolved the gas with a ratio of 45:55 in spite of the presumed obfuscation by symmetric production of wound ET. Similar results were obtained for phototropism. It is a pity that this experiment was performed prior to the development of highly

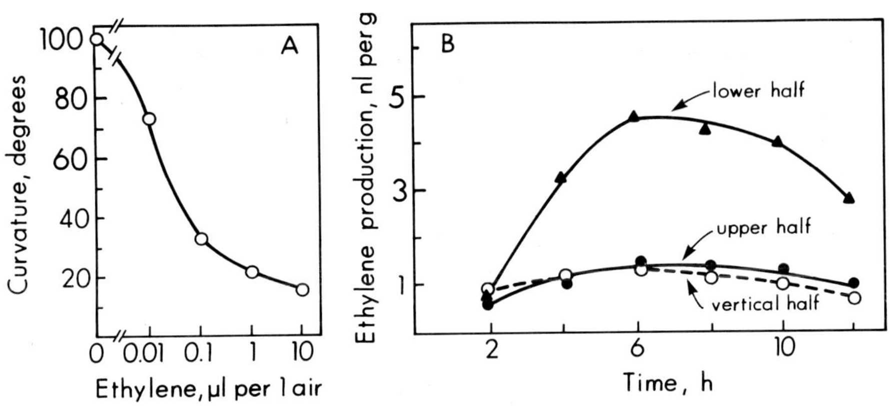

Fig. 12 A, B. Ethylene is thought to retard curving in many plants. **A** Demonstration by KANG and BURG (1972) for sections of etiolated *Pisum* epicotyls. Sections were cut 8 h after a R-pretreatment which reduced ET production, and were geotropically stimulated by floating 18 h in 0.2 µM IAA in a dish provided with the specified level of ET. **B** Peduncles of *Taraxacum* were longitudinally halved by CLIFFORD et al. (1983) and placed in vials in an upright orientation or in positions corresponding to upper and lower halves of an intact horizontal organ. ET was sampled every 2 h, at which time the vial was flushed with fresh air. Production by "upper halves" was closely similar to that by vertical controls, but for "lower halves" rose about four fold. The increase began after 2 h, whereas bending of intact peduncles begins within 25 min

sensitive techniques for measuring ET, for both the 2 h stimulation before halving and the 18 h incubation after cutting are excessively long; moreover, precise measurements of the changing IAA levels would strengthen the experiment.

Recently, CLIFFORD et al. (1983) performed similar experiments with excised *Taraxacum* peduncles. They incubated whole and longitudinally halved sections upright and horizontally, placing the horizontal halves in "upper" and "lower" positions. Removing and measuring accumulated ET every 2 h (Fig. 12 B), they found no differences between treatments during the first interval; but thereafter horizontal peduncle sections produced more ET than vertical ones. The experiments with longitudinal halves indicated that the increase occurred in the lower rather than the upper sides of the sections. Again, the data are most readily accounted for by assuming that increased levels of IAA in the lower tissue stimulate ET production.

Moreover, it is now clear that IAA can enhance ET synthesis with as little a lag as 15–20 min (FRANKLIN and MORGAN 1978) and that low levels of IAA may suffice (e.g., to 0.3 µM external solution; KANG et al. 1971). Under optimal conditions synthesis can probably be enhanced even more rapidly, judging from induction of synthesis by wounding (KENDE and BOLLER 1981).

Thus, it appears that ET synthesis readjusts vigorously enough in response to changing levels of IAA in some plants that (1) growth on the lower side can be inhibited, and (2) lateral transport of IAA can limit itself. It even seems possible that in group 1 plants the level of ET in the lowermost tissue of the apical region of the stem might rise enough during lateral migration to reverse the direction of geotropic IAA transport. For plants of groups 1 and 2 alike, differential growth inhibition and inhibition or reversal of geotropic IAA migration would bring about tip straightening (Sect. 2.2).

IAA-mediated increase of ET production has also been observed within 1.5–2 h in the lower halves of horizontally placed nodes of the grass *Echinochloa,* although in this instance the gas has been shown not to influence the course of geotropic curvature (WRIGHT et al. 1978). This is yet another example of how multiple hormonal changes tend to be linked together even though not every one will have an effect in every tissue.

γ) Stimulation by Friction. Emergence of seedlings through compact, gravelly or heavy soil is a third situation in which ET may slow geotropic bending. GOESCHL et al. (1966) showed that as a growing shoot of *Pisum* rubs through the soil, the friction stimulates ET production with a lag of about 30 min. In addition to retarding geotropism, the ET tends to shift the pattern of stem growth from rapid elongation to increase in girth (see above) and to thicken cell walls (PRATT and GOESCHL 1969, ABELES 1973). In seedlings such as *Pisum,* ET tightens the hook basal to the dragging plumule and inhibits expansion of the still-folded etiolated leaves (see refs. above); in many seedlings, the tightened hook is basal to the inhibited cotyledons. In response to particularly strong friction, *Pisum* seedlings may bend sideways (see refs. above).

Except for the retardation of geotropism in response to light friction, it is obvious that all features of this sensory-response system help the plant to burrow through the soil as rapidly as possible and with the least expenditure of stored resources: only to the extent that a *Pisum* seedling has trouble pushing through its overlay does it recurve the hook and divert energy from elongation to increasing rigidity; and only if this increased rigidity is still inadequate for penetration does it bend sideways to attempt evasion of what could be a local obstacle. Thus, it might seem that the general brake on geotropism by ET is merely a trade-off for the benefits of sturdy growth and, in, e.g., *Pisum* and *Vicia,* of the ability to grow around stones. However, it will be shown in Section 3.2.6.3 how even the counter-reactive inhibition minimizes the time and energy spent in emergence.

δ) Stimulation by Flexure. With established plants, a fourth situation in which ET influences geotropism may come into play: wind can flex the shoot. Flexure produces ET and a host of morphogenetic consequences similar to those of friction (JAFFE and BIRO 1979, JAFFE 1980; see also JAFFE, Chap. 12, this Vol., pp. 444–484). JAFFE and BIRO (1979) found that the ET produced in result of rubbing light-grown *Phaseolus* stems can temporarily halt both geotropism and phototropism. They reversed the inhibitions by hypobaric withdrawal of ET. In alpine or coastal habitats, such inhibition may be important for survival. Persistent winds may lead via ET to stubby, shrubby, plagiogeotropic rather than slender erect growth, permitting plants to huddle near the ground where the windspeed is lowest (cf. KALDEWEY 1962).

Other mechanical disturbances, such as the compressive forces acting on a heavy branch or a toppled sapling, tend to participate in the sturdying morphogenetic responses which may be educed by gravity along with tropism. The ET-mediated production of extra, stronger xylem ("reaction wood") on the lower side of such branches and boles is an excellent example (e.g., LEOPOLD et al. 1972). While these morphogenetic responses fall outside the scope of this chapter, they underscore the importance of ET in geotropism.

ε) Flowering and Fruiting. A fifth circumstance under which ET might control geotropism occurs during flowering and fruiting. Many flowers produce a burst of ET during fading (BURG 1962, PRATT and GOESCHL 1969, ABELES 1973, KENDE and HANSON 1976), and at this time the geotropism of flower stalks alters in some plants (e.g., RAWITSCHER 1932, KALDEWEY 1962). By way of possible example, in some strains of *Taraxacum* it is common to see the previously orthogeotropic peduncle bend downward after pollination and assume a diageotropic position; meanwhile, the apical end of the peduncle bends orthogeotropically so that the head itself is held upright. The closed flower hides in the surrounding vegetation until the seeds have matured for dispersal, and then the peduncle orthogeotropically lifts the head into the air so that its seeds can be caught by the wind. The hormonal basis of these changes has not been elucidated: could *Taraxacum* belong to VAN DER LAAN's category 1, with ET-reversible IAA transport and with increased ET synthesis during seed development? Plagiogeotropism can be regulated in many ways (cf. e.g., Sect. 3.2.6.4), but this possible mode of control in *Taraxacum* deserves test.

η) Stimulation by Gravity Reception. Little effort has been devoted to testing whether ET production may be enhanced by gravitational stimulation without the intervention of IAA. In only one case is it reasonably clear that such enhancement occurs: HARRISON and PICKARD (1984) have shown that *Lycopersicon* seedlings or even isolated hypocotyls briefly double ET production immediately upon horizontal placement, but that the burst lasts no more than 3–4 min. The burst occurs both on the bench top and on the clinostat. HARRISON and PICKARD speculate that it might occur because amyloplasts deform the tonoplast as they settle against and past it. GUY and KENDE (1984a, b) have shown that ACC can accumulate in the vacuole, and that the ET-forming enzyme is associated with the vacuole and may require the integrity of the tonoplast for action. They note that one possible interpretation of their results is that the enzyme is located in the tonoplast; if so, deformation of the tonoplast may transiently increase the ability of the enzyme to attack vacuolar ACC.

In this particular case, ET accelerates rather than retards geotropic bending. If the mechanism of ET production should prove to have some generality, however, it might slow upward curvature of plants such as etiolated *Pisum*.

3.2.6.3 The Rationale for Counter-Reaction

If a fallen dicot stem must correct its growth, the rapidly expanding apical zone reorients without waiting for a slower basal response; but then curvature shifts toward the base due in part to the slow local response there and in part to the continual basipolar movement of now-redistributed IAA. Meanwhile the tip straightens out. By moving the bend as close to the ground as possible, the extending plant maximizes its mechanical stability and requires less length to reach the same height. Also, since the base is generally poorly receptive to gravity, little energy is expended in response to the lateral displacement maintained here where the bend becomes permanent.

ET is only one of the factors that may cause counter-reaction (cf. Sects. 3.2.4.3, 3.2.5, 3.2.8.7) and perhaps often plays a negligible role in it. However, when it does participate, it concomitantly exerts a second mechanically stabilizing influence, thickening the cell walls and the stem as a whole in the zone of curvature where leverage by the out-of-true stem will tend to cause sagging.

Although perhaps less importantly than stabilizing the plant, the counter-reaction also tends to prevent wasteful overshoot. Because reception is proportional to the accelerational vector acting perpendicular to the stem or coleoptile (e.g., PICKARD 1973a), so that the gravitational stimulus is relatively large even for small displacements, overshoot of the vertical would more often occur if differential growth did not reverse at the apex during downward migration of curvature. However, unprevented apical overshoot is common in seedlings, where it is evidently corrected by opposing gravitational stimulation (cf. Fig. 11 of RAWITSCHER 1932).

The evolutionary basis of the ET-mediated counter-reaction is perhaps the close relation between the gravitational and other mechanical sensory systems, as will be explored in Section 3.2.8. Moreover, the evolution of the ET-dependent counter-reaction must have been closely intertwined with that of plagiogeotropism. ET is typically at least partly responsible for the downward component of the plagiogeotropism of many leaf laminae, petioles and branches (see footnote 3 on p 218) (e.g., KANG 1979), and since a lateral branch can in many cases replace the primary stem when it is damaged, the differences between ortho- and plagiogeotropism must be slight. Certainly, any physiological feature which permitted the development of plagiogeotropism must have been at an evolutionary premium, for the plagiogeotropic control of leaf distribution is in most plants critical for optimization of photosynthesis. Additionally, in many plants, vegetative propagation is made possible by the plagiogeotropism of stolons and rhizomes.

3.2.6.4 Plagiogeotropism

Finally, although plagiogeotropism is generally beyond the scope of this chapter, insofar as it is closely related to orthogeotropism it is important to point out that the roles of ET in plagiogeotropic organs may be complicated and diverse. Sometimes, ET leads to downward curvature as one might expect from the above discussion (e.g., *Gossypium,* studied by HALL et al. 1957). Sometimes, however, ET promotes upward bending, as it does in *Lycopersicon* hypocotyls (Sect. 3.2.6.7). BLAKE et al. (1980) found that situations increasing ET production in the conifer *Cupressus* cause upturning of branches, as does exogenous ET, whereas in an atmosphere scavenged of ET branches turn downward. In lateral branches (stolons or runners) of *Trifolium* (HANSEN and BENDIXEN 1974) endogenous and exogenous ET also promote upward curvature; similarly, ET promotes upward plagiogeotropic movement of diageotropic *Arachis* runners

(ZIV et al. 1976). In the lateral branches of each of these three plants, GA's play an important role in negative geotropism (BLAKE et al. 1980, BENDIXEN and PETERSON 1962, ZIV et al. 1976, HALEVY et al. 1969), and in the case of *Arachis* CEPA-released ET increased bioassayed GA activity, whereas applied GA_3 had no effect on ET production. However, in *Cupressus,* applied GA_3 did increase ET production concomitant with its stimulation of upward bending. Each of the studies on these plants has made clear that much more work is required in order to sort out whether ET effects are mediated directly, through promotion of IAA transport, or through controlled synthesis or conjugation of other hormones.

3.2.7 Calcium (and Potassium and Phosphate)

As part of a larger effort to determine whether the action of IAA in a tissue leads to uptake of K^+, BODE (1959) used flame photometry to assess the distribution of K during the geotropism of *Helianthus* hypocotyls. Halving hypocotyls 45, 90 and 120 min after horizontal placement, he found 51% of the K on the lower side, although Na was evenly distributed. Ca, measured only after 120 min, showed a reverse distribution, with 54% on the upper side; Mg was evenly distributed. BODE repeated the asymmetry measurements on the first internodes of *Lycopersicon,* finding 51% of the K but only 50% of the Na on the lower side and 52% of the Ca but only 50% of the Mg on the upper. Because there was a net increase of K in the geotropically responding *Helianthus* tissue and because it occurred even with the plumule and cotyledons removed, BODE concluded that during localized action of auxin K^+ is drawn from both adjacent tissue (in this instance, the upper side) and from the roots.

In the context of the times, BODE's findings caused little stir. The gross asymmetries were slight. K^+ was recognized as a stimulator of growth and Ca^{2+} as an inhibitor, but it had been accepted for some while that the growth differences between upper and lower sides of a shoot were accompanied by metabolic differences, so it seemed reasonable that there should be a variety of lateral gradients in metabolites and electrolytes (such asymmetries have been reviewed by LARSEN 1962b and ANKER 1962).

Nevertheless, ARSLAN-ÇERIM (1966) confirmed BODE's claim of Ca asymmetry by loading hypocotyls with $^{45}Ca^{2+}$ and measuring radioactivity in the two sides. One, 2 and 3 h after horizontal placement she found that about 53, 54 and 57% of the ^{45}Ca was in the upper halves of the hypocotyls when cpm were expressed on a dry weight basis. Vertical controls (for some reason, 5 h rather than 1, 2 or 3 h) contained 50% in each half; hypocotyls set horizontal 3 h in N_2 tended to show reversed distribution but these data were reported to be statistically less reliable.

These results were carefully extended by GOSWAMI and AUDUS (1976), who showed for *Zea* coleoptiles as well as for *Helianthus* hypocotyls that $^{45}Ca^{2+}$ is preferentially distributed to the upper side. Figure 13A illustrates both the extent of the asymmetry and the relatively rapid rate at which it develops in intact *Helianthus* seedlings: asymmetries were nearly fully developed by 1 h. For *Helianthus,* applied $^{42}K^+$ and ^{32}Pi were also found to become asymmetri-

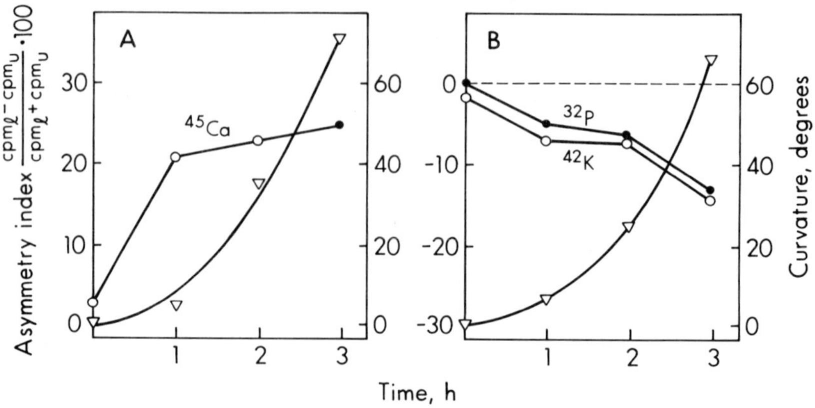

Fig. 13A, B. Development of ionic asymmetries in geotropically bending *Helianthus* hypocotyls was evaluated with radioisotopes by GOSWAMI and AUDUS 1976. **A** Applied ^{45}Ca becomes more abundant on the upper side. It seems likely that better resolution would prove the asymmetry to precede curvature development. **B** Asymmetries of applied ^{42}K and ^{32}P develop in parallel. They oppose that of ^{45}Ca, and develop more slowly

cally distributed, but more slowly, to a lesser extent, and in the opposite direction (Fig. 13B).

Further, GOSWAMI and AUDUS extended BODE's experiments on how the asymmetries might be brought about. Firstly, they established that in *Zea* coleoptiles subjected to phototropically effective illumination, gradients of ^{45}Ca, ^{42}K and ^{32}P develop with much the same time-course as in geotropism. Secondly, they noted that for both geotropism and phototropism, growth curvatures were slight at a time when isotope asymmetries were already relatively large (cf. Fig. 13), which suggests that ion asymmetries may participate in rather than result from differential growth. Thirdly, they applied IAA unilaterally to vertical hypocotyls, and ascertained that early isotope asymmetries comparable to tropistic ones resulted. Fourthly, they found somewhat greater asymmetry in intact seedlings than in sections; since $^{45}Ca^{2+}$ entry was via preequilibration of solution with the roots, this reinforces BODE's proposal that some of the asymmetries could arise from differential vascular supply of the ion. Finally, in both coleoptiles and hypocotyls they blocked geotropic and phototropic IAA redistribution with NPA (cf. ABROL and AUDUS 1973b) and found essentially symmetric isotope distributions. All things considered, then, the asymmetries of Ca, K, and P appear to be linked to IAA asymmetry.

It is easy to imagine that the ^{42}K and ^{32}P asymmetries are the result of growth promotion on the lower side. IAA is thought to stimulate H^+ secretion which can be coupled with K^+ uptake (e.g., CLELAND 1980) and the resulting increase in K^+ might be evidenced as soon as growth is well underway (but cf. ibid.) and if so should become only slightly greater during steady-state growth; certainly, because of dilution during cell expansion, no huge increase should be expected as a final result of growth. Although no explicit coupling between Pi uptake and auxin action is known, nevertheless the requirement

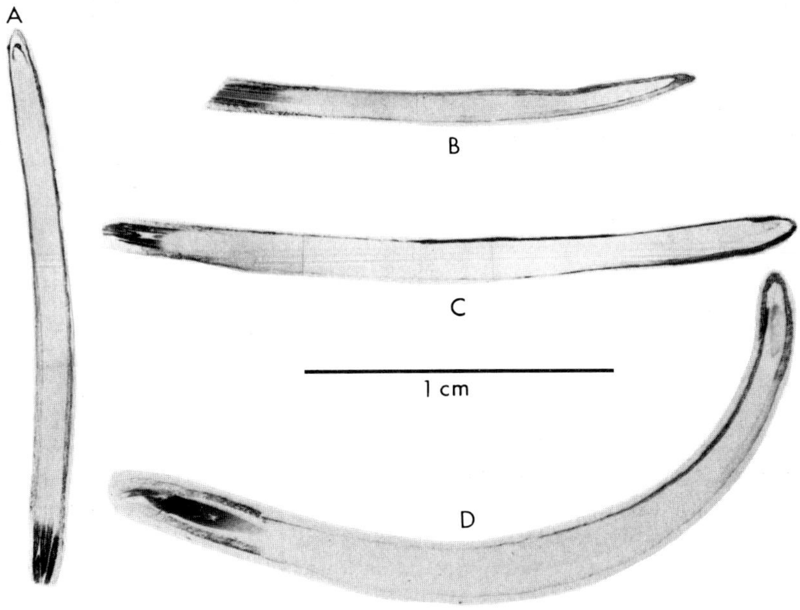

Fig. 14A–D. Light-microscopic localization of Ca in *Avena* coleoptiles by means of precipitation with antimonate; photographs courtesy of R. D. SLOCUM (SLOCUM and ROUX 1983). **A** Vertical control. **B, C, D** Geotropically stimulated for 10, 30, and 180 min, respectively

of growing cells for phosphate is high and uptake will necessarily increase as growth is promoted.

While an asymmetry of K^+ and Pi is probably a participatory nonregulatory concomitant of growth, Ca^{2+} has long been known as a growth regulator. Application of extra Ca^{2+} is inhibitory to auxin-induced growth (e.g. THIMANN and SCHNEIDER 1938a, COOIL and BONNER 1957). Creating a lateral asymmetry of Ca^{2+} across a shoot might certainly be expected to result in curvature. Therefore it is important to know more about the distribution of endogenous Ca^{2+} during geotropism. To this end, SLOCUM and ROUX (1983) treated *Avena* coleoptiles with antimonate in order to precipitate Ca^{2+} and sectioned them to visualize its localization at both the tissue and cellular levels.

The light micrographs of Fig. 14A–D illustrate calcium distribution at the tissue level in coleoptiles held horizontal for 0, 10, 30, and 180 min. Calcium is uniformly distributed at the beginning of the experiment (Fig. 14A), but at 10 min (Fig. 14B) it has already begun to accumulate close to the upper surface, near the tip. At 30 min (Fig. 14C), the accumulation has become heavier and has extended several millimeters in the basal direction. Calcium has been depleted, however, in the lower tissue opposite the upper zone where accumulation is strongest. Further, there is a second zone of calcium accumulation close to the apex and on the lower side. At 3 h (Fig. 14D), the lower apical zone of accumulation remains conspicuous and is associated with the apical, counter-curving region of the coleoptile; the upper zone is associated with the region where negative curvature is maximal and still developing.

Electron micrographs of SLOCUM and ROUX (1983) indicate that the redistributed calcium gathers in the apoplast. It is abundant in all the walls and spaces surrounding the epidermal cells and their parenchymal neighbors.

The large early apoplastic accumulation of calcium in the upper epidermis must quickly inhibit its elongation. Indeed, significant calcium asymmetry must have developed within a few minutes, and the action of Ca^{2+} to block IAA-induced growth may have begun concomitantly; the lag for the measurable expression of Ca^{2+}-inhibition of straight growth may be as brief as 1–2 min (EVANS 1967, CLELAND and RAYLE 1977). SLOCUM and ROUX (1983) emphasize that the calcium asymmetry is seen before upward (negative) growth curvature begins in their coleoptiles; similarly, downward IAA redistribution occurs within minutes of horizontal placement (Fig. 9A), doubtless before upward curvature can occur. It would be worthwhile to evaluate $^{45}Ca^{2+}$ and ^{14}C-IAA or ^{3}H-IAA redistribution simultaneously under conditions educing early downward curvature vs. those educing slower upward curvature (cf. Sect. 3.2.2.2 and 3.2.2.3).

The later accumulation in the apical lowermost tissue, already well developed at 30 min (Fig. 14C), probably plays an important role in the counter-reaction which also begins about that time (Figs. 4, 6). In the experiments of FILNER et al. (1970) illustrated in Fig. 9, net downward IAA transport between 2 and 5 mm from the tip declined sharply at 30 min; is it possible that quite close to the apex, in the region of calcium accumulation, IAA asymmetry had actually reversed? Or was the level of IAA maintained high on the lower side, initiating some feedback loop reversing the asymmetry of growth? Again, in order to determine whether the calcium asymmetries are obligatorily and inversely related to IAA concentrations, whether they are related in some more complicated way, or whether IAA is only one of the natural factors leading to calcium asymmetry, the detailed time-course studies proposed for IAA transport (Sect. 3.2.2.2) should be paralleled by studies with $^{45}Ca^{2+}$, as well as by time-courses of asymmetric growth along the entire coleoptile axis.

GOSWAMI and AUDUS (1976) created opposing gradients of ^{42}K and ^{45}Ca by unilateral application of the diuretic mersalyl to *Zea* coleoptiles, but saw no curvature during a 1-h observation period. They interpreted this as showing that the ionic asymmetries do not of themselves lead to curvature. Yet, only small curvatures resulted within a 1-h period when an auxin was applied unilaterally to the same preparation. Thus, before drawing such a conclusion it would be appropriate to study more reactive plants (cf. Fig. 2 of BRAUNER and HAGER 1958) or to allow a longer opportunity for curvature to develop. More importantly, it would be necessary to show that uniformly applied mersalyl did not itself influence growth. Direct unilateral application of Ca^{2+} might be more instructive in this regard than use of mersalyl.

In both light and electron micrographs of antimonate-stained tissue, SLOCUM and ROUX (1983) observed large amounts of precipitate around the vascular bundles and the stomata, suggesting that some of the calcium that accumulates asymmetrically is delivered in the transpiration stream (in accord with earlier observations; see above). Some must simply be redistributed within the apoplast. SLOCUM and ROUX suggest further that the central vacuoles of the apical 3 mm of the coleoptiles, which contain large amounts of Ca^{2+} before stimulation, may be an important source of the laterally migrating Ca^{2+}. The amount of antimonate precipitated in vacuoles of subapical sections appears greatly diminished during geotropism. This change is reported to be symmetric across the horizontal coleoptile at 3 h, but ROUX and SLOCUM offer no time-course which might provide a clue about how such a change might be initated.

JANISTYN (1973) found that Ca^{2+} is released into the medium when IAA is provided to floating sections of *Zea* coleoptiles; BANGERTH (1979) has reviewed situations in which calcium and IAA appear to move in opposite directions in intact plants; and DE GUZMAN and DELA FUENTE (1981, 1982) and DELA FUENTE (1982) are currently exploring opposing movements in the *Zea* coleoptile. All these observations, coupled with the observation by GOSWAMI and AUDUS (1976, see above) of numerically comparable gross asymmetries of ^{45}Ca in halved coleoptiles following phototropism and unilateral application of IAA as well as geotropism, suggest that geotropic migration of calcium, and possibly the geotropic efflux of calcium from the vacuoles, might be indirect effects of gravitational stimulation which are mediated by IAA.

3.2.8 A Preliminary Model for the Roles of Calcium in Response to Gravity, Friction and Flexure

3.2.8.1 Current Concepts in Calcium Physiology[4]

Remarkably many diverse animal cell functions are controlled by mechnisms which actively maintain the cytosolic level of free Ca^{2+} significantly below normal extracellular levels and which, on receipt of appropriate stimuli, open membrane channels to let extracellular Ca^{2+} enter. Inside the membrane, Ca^{2+} can act in several ways, a prominent one of which is to cooperatively bind to the versatile regulatory protein calmodulin. The complex in turn interacts with cell-specific kinases controlling phosphorylation, and hence activation, of further cell-specific regulatory proteins which initiate the particular function or functions in question (e.g., RASMUSSEN 1983, CHEUNG 1980, GREENGARD 1978).

Calmodulin is present in higher plants, where it has been shown, for example, to regulate a Ca^{2+}-transport ATPase and also NAD kinases (ANDERSON and CORMIER 1978, CORMIER et al. 1980, DIETER and MARMÉ 1980, 1981a, b, RAVEN and RUBERY 1982, MARMÉ and DIETER 1983). Compartmentation of Ca^{2+} between the cytosol and the extracellular space is presumed similar in plant and animal tissues, (e.g., WILLIAMSON and ASHLEY 1982, ROUX and SLOCUM 1982, MARMÉ and DIETER 1983). Also in both plants and animals, mitochondria and perhaps the endoplasmic reticulum compartment Ca^{2+} under the influence of subtle controls; and in plants, chloroplasts compartment Ca^{2+} as well (see refs. above). Perhaps in plant tissues, with wall extension critically controlled by apoplastic Ca^{2+}, the vacuole is an equally important and delicately regulated reservoir of Ca^{2+} (cf. ROUX and SLOCUM 1982).

Section 3.2.8 is an attempt to meld these facts and ideas about calcium with what we know about the geotropism of shoots, and also about responses to mechanical stimuli described in the preceding sections. A model will be propounded in which Ca^{2+} participates in geotropism in a multiplicity of interactive ways, which can be elaborated from seven basic postulates. Although the most extreme case in which all seven postulates are valid will be discussed, it must be emphasized that the postulates are independent and that the validity of any one does not depend on the validity of the others.

3.2.8.2 Reception

It has been argued that, directly or indirectly, it is the pressure exerted laterally either by statoliths or by the entire contents of the cell that mediates the response to gravitational

4 See also MARMÉ D., Calcium transport and function, Vol. 15B, this Series (eds. LÄUCHLI A. and BIELESKI R.L., pp. 599–625, 1983).

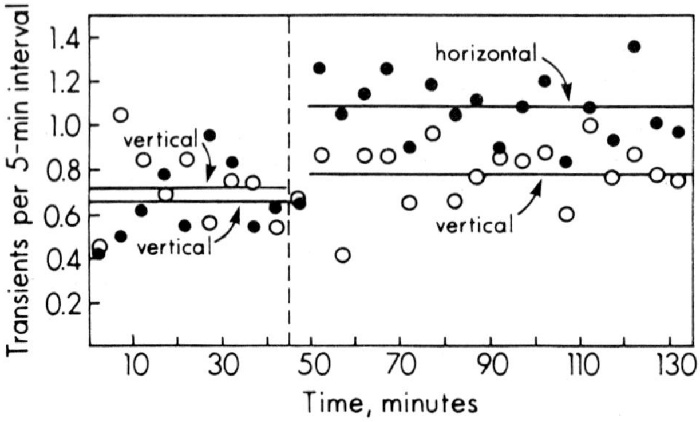

Fig. 15. Extracellularly measured voltage transients elicited by geotropic stimulation in auxin-depleted etiolated *Pisum* stem segments. One of the two sets of vertical segments was swiveled gently to the horizontal position after a 45-min recording period. *t*-tests yielded $p < 10^{-6}$ for comparisons of the stimulated stems with each of the three possible control populations. (PICKARD 1983; experimental technique was essentially as described by PICKARD 1984)

acceleration (e.g., PICKARD and THIMANN 1966, JOHNSSON and PICKARD 1979, SIEVERS and VOLKMANN 1979). Hence, *Postulate 1: The end result of laterally exerted pressure is to deform the plasmalemma on the lower side of a georeceptive cell in such a way that leakage channels for Ca^{2+} open up* (cf. CORMIER et al. 1980). It is agreed that the level of calcium is high in the cell wall, though it is hard to separate contributions from ionically mobile Ca^{2+}, readily exchanged bound calcium, and tightly bound calcium. It is reasonable to assume that total cell wall calcium is in the millimolar range and that free Ca^{2+} is not less than 1% of this amount (cf. CLARKSON and HANSON 1980, ROUX and SLOCUM 1982, MARMÉ and DIETER 1983). Since the estimated cytosolic concentration of free Ca^{2+} is at or below 1 μM and the cytosol is electrically negative with respect to the cell exterior, extracellular Ca^{2+} would respond to this marked electrochemical gradient by entering the cytosol through open Ca^{2+} channels. There are, however, avid binding sites for Ca^{2+} inside the cell, and hence if influx is small the ions should not spread far from their point of entry (cf. HARRIS 1978, HARARAY and BROWN 1984).

During continuing gravitational stimulation, any local rise of cytosolic Ca^{2+} would locally activate plasmalemmal Ca^{2+} expulsion pumps, thereby tending to (1) limit the rise of Ca^{2+} and (2) establish a new, quasi-steady state for the ion in the lower portion of the cell.

Consistent with the postulated necessity for Ca^{2+} to enter the cytosol from the apoplast during gravity reception, SUN and ROUX (1983) found with *Avena* coleoptiles that EGTA can inhibit geotropism without inhibiting elongation growth; possible stimulation by EGTA was not remarked. Geotropic sensitivity was restored by perfusion with solutions containing Ca^{2+}.

In a preliminary test of the idea, PICKARD (1983) found that certain small voltage transients postulated to index the level of cytosolic Ca^{2+} (PICKARD 1984; Sect. 3.2.8.8, 3.2.8.9) appear in horizontally placed etiolated, IAA-depleted *Pisum* stems with a frequency 1.6 times that in vertical controls (Fig. 15). Because of the technical difficulties of assessing cytosolic Ca^{2+} in pea stems directly, the validity of this indirect index may remain equivocal for some time; but also because of these difficulties it is worth some effort to evaluate the rationale by which the signals were postulated (PICKARD 1984, Sect. 3.2.8.8).

3.2.8.3 Activation of an Auxin Carrier

When the level of Ca^{2+} in the cytosol on the lower side of the cell reaches a certain threshold, it cooperatively binds with calmodulin. From this follows *Postulate 2: In georeceptive cells, the calcium-calmodulin complex acts with a special kinase to phosphorylate a plasmalemmal IAA transport protein. In its phosphorylated form the transport protein expels IAA into the apoplast on the lower side of the cell,* thus setting in motion the events leading to lateral migration of IAA (cf., e.g., GOLDSMITH 1977, RUBERY 1980, RAVEN and RUBERY 1982, or Chap. 10, this Vol.).

It seems likely that the geotropic Ca^{2+} leakage channels would be clustered with membrane-associated lateral transport protein and calmodulin; with such a geometry, geotropic Ca^{2+} leakage need not be great to be effective and need not necessarily interfere excessively with other Ca^{2+}-modulated reactions in the same cells.

While there is no evidence in direct support of it, Postulate 2 is consistent with the demonstration by BIRO et al. (1982) that the calmodulin-binding inhibitor chlorpromazine inhibits geotropism of *Avena* coleoptiles at concentrations that do not inhibit elongation (see also BIRO and ROUX 1983). It is also consistent with the findings of DELA FUENTE and LEOPOLD (1973), DELA FUENTE (1981), and DE GUZMAN and DELA FUENTE (1981, 1982) that calcium is required for basipolar translocation of IAA. Although basipolar translocation is much less readily altered by reorientation of cells in the gravitational field than is geotropic translocation, nevertheless both are subject to accelerational induction (e.g., OUITRAKUL and HERTEL 1969).

Postulate 2 could be expressed in terms of methylation rather than phosphorylation. There are many parallels between sensory systems and regulatory systems employing these two mechanisms for activation of proteins (SPRINGER et al. 1979). For simplicity, however, only phosphorylation will be considered here.

3.2.8.4 Electrochemical Migration of Calcium Across the Organ

Normally a large fraction of the apoplastic calcium is bound to the cell wall (see above). However, as demonstrated early by TAGAWA and BONNER (1938), a considerable amount is exchangeable with H^+. Therefore, when IAA redistributed to the lower side of a stem stimulates H^+ secretion, the apoplastic activity of Ca^{2+} will rise. Additionally, the secreted H^+ tends to make the lower tissue electropositive with respect to the upper tissue. Thus, *Postulate 3: Ca^{2+} migrates down its apoplastic electrochemical gradient from the IAA-enriched to the IAA-impoverished side of the stem.*

In the vicinity of the upper epidermis, which becomes depleted of IAA, the apoplast would have a relatively high pH and so the cell walls would bind Ca^{2+} particularly effectively, serving as a strong sink, and thus explaining the heavy accumulation of calcium demonstrated by SLOCUM and ROUX (1983) in that region.

The postulated electrodiffusion of Ca^{2+} does not require a large amount of free ion in the apoplast at any given time. Because Ca^{2+} would tend to leave the electropositive region of the apoplast as soon as it appears, further release of Ca^{2+} from the reservoir of wall-bound Ca^{2+} would be promoted. Thus, it is not unreasonable to imagine that Ca could be depleted from the lower side of a horizontal stem even though the equilibrium between free and wall-bound calcium strongly favors the bound form. This explanation is similar to that proposed by SLOCUM and ROUX (1983), who visualized the calcium migration as due to "an exchange of Ca^{2+} for H^+ along a proton gradient".

The explanation is consistent with the success of GOSWAMI and AUDUS (1976) in creating ion asymmetries with mersalyl (Sect. 3.2.7). It is supported by the report of WRIGHT and RAYLE (1983) that curvature is totally blocked by vanadate (Sect. 3.2.3.2), which inhibits H^+ expulsion pumps: if the IAA gradient alone were directly responsible for Ca^{2+} migration, vanadate would not interfere with it and hence the component of curvature due to Ca^{2+} asymmetry should remain. It could be further tested by evaluating whether a Ca asymmetry fails to develop in horizontal stems or coleoptiles that have been depleted of IAA.

The explanation might be extended to account, at least in part, for the acropetal migration of calcium during basipolar transport of IAA in coleoptiles (DE GUZMAN and DELA FUENTE 1981, 1982).

Finally, while there might be vigorous migration of Ca^{2+} down an electrochemical gradient created by asymmetric H^+ expulsion, some migration might be expected even in a simple electrical field. This may account in part for the numerous influences (e.g., EVERS 1947, LUND and BERRY 1947, LUND et al. 1947, EVERS and LUND 1947, SCHRANK 1947, 1951, ELLIS and TURNER 1978) of imposed voltage gradients on both straight and differential growth. However, H^+ is far more mobile than Ca^{2+} and hence would migrate even more readily. Thus, promotion of growth by electrophoretically displaced H^+ might sometimes outweigh inhibition by electrophoretically displaced Ca^{2+}, and underlie the often-observed (see refs. above) bending of a coleoptile or other plant axis toward the electropositive pole of an external circuit.

3.2.8.5 The Vacuole as a Source of Cytosolic and Apoplastic Calcium

Ca^{2+} accumulates in the central vacuole of some cells. Its concentration there may be 1–10 mM, wherease the cytosolic concentration is thought to be regulated at about 0.1–1 μM (LIN et al. 1977, ROUX and SLOCUM 1982, MARMÉ and DIETER 1983). This reservoir of Ca^{2+} might therefore be a source of cytosolic Ca^{2+} under the control of regulatory agents such as IAA. Hence, *Postulate 4: IAA activates Ca^{2+} channels in the tonoplast and thereby markedly elevates the Ca^{2+} permeability of that membrane.*

It has been suggested that there are receptor sites for IAA on vesicles putatively derived in vitro from the vacuolar membrane (DOHRMANN et al. 1978), but more discriminating studies have cast serious doubt on the source of the vesicles (L. TAIZ, personal communication). It nevertheless seems possible that the efflux is a direct, passive consequence of an as yet undetected binding of IAA to the tonoplast; alternatively, the efflux might be an indirect result of the increase in IAA levels.

If Postulate 4 is valid, cytosolic levels of Ca^{2+} will rise on the lower side of a stem as IAA migrates into it. In consequence, plasmalemmal Ca^{2+} expulsion pumps[5] will be activated on all sides of the cells. As such pumps discharge Ca^{2+} into the apoplast on the lower side of the stem, the ion would participate in the electrochemically driven migration to the upper side, in accord with the argument of Sect. 3.2.8.4.

The early rise in free cytosolic Ca^{2+} should be self-limiting because of the activation of the Ca^{2+} expulsion pumps. However, attainment of a new static equilibrium level of cytosolic free Ca^{2+} will probably be precluded while the plant is executing its geotropic response, due to (1) the asymmetrically shifting levels of intracellular IAA, vacuolar Ca^{2+}, apoplastic Ca^{2+}, and Ca^{2+} pumps as well as of GA's, ET, and apoplastic H^+, and (2) the diminution of the effective gravitational stimulus by curvature.

Ultimately, in those counter-reactive shoots in which the direction of auxin transport reverses, the response should be duplicated with reverse asymmetry in the apical region. Sequential opposing asymmetric responses might account for the symmetric depletion of Ca from the vacuoles in the apical portion of the *Avena* coleoptile observed 3 h after the start of geotropic stimulation by SLOCUM and ROUX (1983).

Postulate 4 concerns geotropism only insofar as IAA is involved in geotropism: it specifies one of the mechanisms of IAA action (PICKARD 1984). The mechanism is flexible because (1) the presumed multifarious responses to cytosolic Ca^{2+} are specific to cell type (refs. in Sect. 3.2.8.1), and (2) the vacuolar concentration of calcium is variable. In fact, some cell types in some plants precipitate almost all of their vacuolar calcium as, e.g., carbonates or oxalates (e.g., ALBERT and KINZEL 1973) which may be enclosed in membranes (SCHÖTZ et al. 1970).

5 Definitive proof of plasmalemmal Ca^{2+} expulsion pumps is lacking for plants. Lincoln TAIZ (personal communication) has evidence consistent with the occurrence of H^+ pumps in the Golgi membranes, and points out that the end results of exchanging Ca^{2+} for H^+ at the plasmalemma and of exchanging Ca^{2+} into Golgi vesicles followed by their exocytosis would be the same.

3.2.8.6 The Early Phase of Curvature

It is plausible that the conditions established by HILD and HERTEL (1972) and HILD (1977) for predisposing coleoptiles to undergo early upward versus downward curvature (Sect. 2.2) operate by influencing the cytosolic Ca^{2+} level. Coleoptiles pretreated with R (under the specific conditions of HILD and HERTEL) bend monotonically upward, whereas controls initially bend downward (Fig. 3). Activation of phytochrome in coleoptile tips has been suggested to cause immediate passive flux of Ca^{2+} into the cytosol from both the apoplast and the mitochondria (ROUX et al. 1981, ROUX and SLOCUM 1982, ROUX 1984). Thus, cytosolic Ca^{2+} might be elevated for a certain period of time following R irradiation of *Avena* coleoptiles (refs. above). When Ca^{2+} levels become high enough to activate a significant amount of calmodulin, however, a reverse cytosolic trend sets in: the activated calmodulin stimulates activity of plasmalemmal Ca^{2+}-expelling pumps (see refs. above and Sect. 3.2.8.1). Cytosolic Ca^{2+} may then drop considerably lower than its level before irradiation.

According to the present model for gravity reception, the nulled accelerational prestimulation used by HILD and HERTEL to prevent early downward curvature (Sect. 2.2) also results in immediate influx of Ca^{2+} followed by activation of Ca^{2+} pumps. The activation should in turn lead to a lowering of the steady-state level in the cytosol. The stimulated influx and corresponding excess activation would be localized to the alternately "lower" sides of each receptive cell, however, so that nulling stimulations in the same and opposite direction as the ultimate stimulation should predispose coleoptiles to respond upward rather than downward, but nulling prestimulation at an angle of 90° to the final stimulation should merely weaken the tendency for downward curving. This in fact is observed (Fig. 3A). The interpretation of nulled accelerational prestimulation is complicated by the opposing preactivations of IAA export proteins in the alternating "lower" sides of the plasmalemma, with the possibility of reducing the pool of unactivated carriers available to respond to the ultimate stimulation. However, the half-life of geotropically activated IAA carriers cannot be long (Sect. 3.2.2.2).

Thus, it may be argued that early upward curvature is correlated with a relatively high initial level of activation of plasmalemmal Ca^{2+} pumps and low cytosolic Ca^{2+} activity, whereas downward bending is correlated with a lower initial level of activation and higher cytosolic activity. But how could the transient "wrong" curvature occur? One possible answer to this question is given by *Postulate 5: When cytosolic levels of* Ca^{2+} *are drastically elevated, the action of lateral plasmalemmal IAA carriers is either inhibited or reversed.* (Basal IAA carriers might be inhibited also.)

Perhaps, when georeceptive Ca^{2+} leakage occurs in cells already being bailed out rapidly by activated plasmalemmal Ca^{2+} export pumps, cytosolic Ca^{2+} on the lower side of each cell attains an elevated but noninhibitory level. In contrast, when georeceptive leakage occurs in the cells with relatively few plasmalemmal Ca^{2+} pumps already activated, Ca^{2+} on the lower side of each cell (at least in the vicinity of IAA transport protein) might transiently rise very high. During such a transient rise, phosphorylation of IAA transport proteins might proceed normally, but their action might be inhibited or reversed.

If, prior to geotropic stimulation, cytosolic Ca^{2+} is elevated well above normal levels, some IAA carriers might continuously be activated on all sides of the cells. Therefore geotropic inhibition might combine with reversal of pumping to result in net upward flux of IAA within each cell, leading to weak upward migration across the tissue. Downward curvature is, after all, only slight (Fig. 3). Meanwhile, the increasing activation of Ca^{2+} export pumps would ensure a gradual lowering of the cytoplasmic Ca^{2+}, and within minutes the malfunction of IAA transport proteins would be relieved and IAA migration would occur in the normal direction.

It has been discussed in Section 3.2.2.3.b how HILD and HERTEL have proposed that there is somehow a common basis for initial wrong-way geotropic curvature of shoots and the initial wrong-way elongation response that can be observed when shoots are subjected to sudden step-up of IAA. It seems at least as likely that there are two categories of wrong-way effects, and recently BATES and GOLDSMITH (1983) have lent plausibility to this notion by suggesting a mechanism which might account for transient growth inhibition and a number of associated events. They have provided evidence that the cytosol may be acidified when a large amount of IAA or other weak organic acid is supplied to a tissue suddenly. The undissociated form of the acid, which is favored in the relatively acidic apoplast, passes readily through the plasmalemma and tends to dissociate in the relatively basic cytosol. Because the ionic form is less soluble in the membrane, the cytosolic concentration of the weak organic acid can build rapidly to a level quite high relative to the apoplast, and the internal pH is postulated to become temporarily imbalanced. Until the imbalance can be corrected, the membrane potential depolarizes. Both internal acidification and such ion fluxes as might be associated with membrane depolarization might indeed plausibly account for transient inhibition of growth, the associated shift in gross voltage distribution, and other effects such as inhibition of cytoplasmic streaming.

It is also possible to imagine, along the lines that HILD and HERTEL have suggested, that Ca^{2+} might play a role in a variety of wrong-way responses. For example, if indeed dramatically elevated IAA concentrations can cause leakage of Ca^{2+} from the vacuole, and if increased cytosolic concentrations of Ca^{2+} activate plasmalemmal Ca^{2+} expulsion pumps, perhaps it can be partly understood how (1) small step-ups in IAA concentration stimulate growth, and may transiently accelerate cytoplasmic streaming (THIMANN and SWEENEY 1937, ULLRICH 1978), whereas (2) sudden large step-ups can transiently inhibit growth and streaming (see refs. above). By analogy with animal systems (DURHAM 1974, SEITZ 1979), a mild increase of cytosolic Ca^{2+} might well stimulate the interaction of plant actin and myosin responsible for streaming[6] (HEPLER and PALEVITZ 1974, ALLEN and ALLEN 1978, KAMIYA 1981), and since the plant is accustomed to small shifts in IAA level there is no reason to expect inhibitory actions on growth. The slight flush of Ca^{2+} would soon disappear through the steady functioning of plasmalemmal Ca^{2+} pumps already in the activated state prior to the step-up. A sudden, unnatural increase of IAA, however, might release a flood of Ca^{2+} from the vacuole. There is little information about the effects of high levels of Ca^{2+} on streaming in higher plants; however, externally applied Ca^{2+} has been observed to accelerate streaming in root hairs of *Hordeum* (VINTILA et al. 1973) and to inhibit it in leaf cells of *Elodea* (FORDE and STEER 1976) and staminal hairs of *Tradescantia* (DORÉE and PICARD 1980). It is known that sudden elevation of Ca^{2+} abruptly stops streaming in Characean algae (e.g., ALLEN and ALLEN 1978, WAGNER 1979, KAMIYA 1981, WILLIAMSON and ASHLEY 1982, KIKUYAMA and TAZAWA 1982). It is also known that the source of Ca^{2+} for streaming stoppage is normally intracellular (W.F. PICKARD 1972, HAYAMA et al. 1979, DORÉE and PICARD 1980). It is uncertain how high cytosolic levels of Ca^{2+} might transiently inhibit growth, but considering the elaborate network of biochemical systems probably regulated by Ca^{2+} several types of mechanism can be imagined. Possibly, membrane "integrity" is disturbed by an unusually high level of Ca^{2+}: sudden external application of Ca^{2+} can cause slight, instantaneous osmotic shrinkage (CLELAND and RAYLE 1977), and certain lanthanide homologs of Ca^{2+} cause rapid turgor loss (PICKARD 1969). Possibly, at high levels of cytosolic Ca^{2+} regulatory proteins are activated and deactivated, either directly or through phosphorylation, in such a pattern as to brake growth. Possibly, some other cytoplasmic system is disturbed. Possibly, with a burst of Ca^{2+} available for expulsion and with an increasing number of plasmalemmal pumps being activated on all sides, efflux of Ca^{2+} into the cell wall would be rapid enough to inhibit growth until, after a delay of a few minutes, enough H^+ is secreted to compete with Ca^{2+} for critical binding sites.

Without further evidence, it seems futile to try to sort out possible effects of H^+ and Ca^{2+} in early wrong-way effects. To a considerable extent, the postulated actions

6 There is no unequivocal evidence on this point: The referenced studies of streaming in Characean algae indicate that Ca^{2+} is exclusively inhibitory but (1) streaming in higher plants is not identical with that in Characeans, and (2) documentation of Ca^{2+} effects in the Characeans is not extensive enough to exclude low-level in vivo stimulation by Ca^{2+}.

of sudden shifts in cytosolic Ca^{2+} and H^+ are compatible and could work in concert. Experimental evaluation of effects of either ion might well proceed faster if possible effects of both are tested together.

Meanwhile, the existence of more than one possible category of explanation of the diverse early wrong-way effects reinforces the suggestion that early transient downward geotropic curvature may have a different origin than early wrong-way responses to applied IAA.

3.2.8.7 A Counter-Reaction

Postulates 4 and 5 imply that there could exist conditions in which parameters favor counter-reaction. If the increase of IAA on the lower side of the stem or coleoptile causes sufficient release of Ca^{2+} from the vacuole into the cytosol, and if plasmalemmal Ca^{2+} pumps do not bail out the cytosol sufficiently rapidly, IAA transport must be activated on all sides of the cell. When this happens, enhanced lateral exchange of IAA may occur. Omnilateral transport would tend to swamp the downward transport more directly stimulated by gravity, and indeed, this may be part of the reason why downward movement of IAA begins to fail at the tip of a *Zea* coleoptile within half an hour (Fig. 9A). Further, if the geotropically elevated concentrations of Ca^{2+} in the lower cytosol actually cause pumping of IAA into the cells (Sect. 3.2.8.6), the direction of net migration would reverse during continued stimulation; this is counter-reaction.

Several parameters might combine to explain why the downward migration of IAA in the basal tissue of the coleoptile (Fig. 9A) does not fail at this time; e.g., the sensitivity to gravity is evidently lower there (DOLK 1936; Sect. 2.3), the responsiveness of the tissue to IAA is axially graded, under at least some conditions the vacuoles sequester less Ca there (SLOCUM and ROUX 1983), and the parameters of reactions involving Ca^{2+} may be a little different.

Ca^{2+} moves laterally during phototropism as well as geotropism (Sect. 3.2.7), but evidently does not participate in a counter-reaction; phototropically induced IAA transport is long-lived (Chap. 10, this Vol.). Moreover, the transient early "wrong" curvatures of geotropism have no known phototropic parallel. Therefore, although the two tropisms may resemble each other in the development of IAA asymmetry, the postulated consequent release of Ca^{2+} from vacuole to cytosol, the expulsion of Ca^{2+} from cytosol to apoplast by plasmalemmal pumps, and the apoplastic migration of Ca^{2+} down the electrochemical gradient formed by IAA-stimulated H^+ expulsion, they probably differ both in the nature of induction of IAA carriers and in its modulation by Ca^{2+}.

3.2.8.8 Membrane Deformation Resulting from Friction and Flexure

Gravitational acceleration is only one form of mechanical stimulation important to plants: the friction between a growing plant and its surroundings results in a variety of responses, as does the bending of a plant by the wind[7] (e.g., GOESCHL et al. 1966, ABELES 1973, JAFFE and BIRO 1979, JAFFE 1980). Membrane deformations would seem a very likely consequence of both friction and flexure; therefore, the above postulates will be extended to build a general model for responses to mechanical stimulation and how they interact.

A close analog of Postulate 1 is *Postulate 6: Friction and flexure deform the plasmalemma in such a way as to permit leakage of Ca^{2+} from the wall space into the cytosol* (cf. PICKARD 1984).

In general, the effects of elevated cytosolic Ca^{2+} depend on cell type. *Postulate 7* specifies: *In tissue responsive to friction and flexure, Ca^{2+} joins cooperatively with calmodu-*

7 Cells also interact mechanically as they grow (e.g., GREEN 1980, LINTILHAC and VESECKY 1981, 1984) and some of the consequences may have biochemical bases similar to those for responses to gravity, friction, and flexure. However, this volume is devoted to responses to external stimuli.

lin and the complex activates a kinase which phosphorylates α- and β-tubulin; this phosphorylated protein polymerizes and causes exocytotic emptying of vesicles that contain a diffusible agent which promotes ethylene synthesis (cf. PICKARD 1984).

When an etiolated *Pisum* epicotyl is rubbed lightly, characteristic voltage transients are educed (PICKARD 1971, 1984). Postulate 7 is inspired by the resemblance of these waveforms to the miniature synaptic potentials triggered by injection of a small amount of Ca^{2+} into the terminal of a nerve cell (MILEDI 1973). It is thought that (1) Ca^{2+} cooperatively binds with calmodulin; (2) a kinase requiring Ca^{2+}-calmodulin phosphorylates α- and β-tubulin, and (3) the phosphorylated tubulins polymerize and bring about exocytotic release of neurotransmitter from small vesicles situated just inside the presynaptic membrane (DELORENZO 1982). The neurotransmitters diffuse across the synaptic cleft and depolarize the synaptic membrane; the depolarization due to exocytosis of a single vesicle is a miniature synaptic potential. If enough vesicles release their contents, as when Ca^{2+} enters the terminal during arrival of an action potential, the depolarization of the postsynaptic membrane will be sufficient to initiate a new action potential (KATZ 1966).

SALTVEIT and DILLEY (1978, 1979) have preliminarily described a diffusible factor from etiolated epicotyls of *Pisum* which can mediate the synthesis of ET. The epicotyls contain calmodulin (e.g., CHARBONNEAU and CORMIER 1979), tubulin, and small vesicles adjacent to the plasmalemma. Perhaps (PICKARD 1984) some of these vesicles contain the factor of SALVEIT and DILLEY, and can release it exocytotically when properly triggered. Perhaps this factor itself depolarizes the plasmalemma, or perhaps this is accomplished by other ions or molecules released from the vesicles; however, in this instance, depolarization itself is likely to be an indicator rather than a mediator of ongoing events. Release of such a diffusible, hormone-like agent could constitute a chemical method of integrating response to stimuli spread over space and time and could amplify the response by causing production of ET throughout a whole population of cells surrounding those actually rubbed or flexed. In turn, of course, the newly synthesized ET undergoes further diffusion, and adds a second stage of hormonal amplification.

3.2.8.9 Crosstalk Between Geotropism and Response to Friction and Flexure

Postulates 1 and 6 provide that leakage of Ca^{2+} into the cytosol is a common feature of the reception of gravity, friction and flexure. Therefore, specificity of accelerational versus frictional and flexural responses must presumably be determined by (1) the distribution of the mediational systems of Postulates 2 and 7, and (2) by the geometry of stimulation.

It was suggested in Section 3.2.8.3 that geotropic Ca^{2+} leakage channels and geotropic IAA transport proteins with their putative kinases might be closely associated. In contrast, frictional and flexural leakage channels might be much more randomly distributed, and the kinase for α- and β-tubulin as well as the tubulins themselves might be located somewhat deeper in the cytosol. It is also possible that the distribution of the mediational systems along the shoot axis and radius might be somewhat different.

The geometry of stimulation may well be the more important of the two factors. Gravitational stimulation is of course unidirectional, whereas the net vectorial component of frictional and flexural stimuli would be expected to vary. Because cell walls are effectively coupled by the middle lamellar glue, it seems possible that even unidirectionally flexing a stem might sometimes distort cells erratically throughout the tissue, and gentle stroking probably distorts cells well removed from the surface of contact. Further, by its very location that surface of contact would seem unable to contribute to asymmetric transport of IAA.

Thus, action by one of the two mediational systems should typically predominate. However, often there must be significant crosstalk between geotropism and thigmomorphogenesis. Sometimes (1) friction or flexure should inhibit response to gravity (directly via cytosolic Ca^{2+} or indirectly via Ca^{2+}-induced ET production), and (2) gravitational stimulation should enhance thigmomorphogenesis. Sometimes, (3) friction and flexure should result in thigmotropism and (4) gravitational stimulation should lead to geomorphogenesis.

As a possible example of (1), JAFFE and BIRO (1979) reported that the upward bending of internodes of *Phaseolus* rubbed just before geotropic exposure lagged an hour behind controls; the inhibition was more than overcome by hypobaric removal of ET.

Perhaps illustrating (3), STARK (1916) caused coleoptiles to bend by stroking them lightly on one side. SCHRANK (1944, 1951) reported not only that gentle unilateral tapping of coleoptiles resulted in thigmotropism toward the stimulus, but also that a potential difference developed across the coleoptile much as in geo- and phototropism; he inferred that the mechanism of bending might be similar. More recently, CLIFFORD et al. (1982) have measured curvature by the *Taraxacum* peduncle in response to a 5–40 min deflection established by attaching a 2-g weight to the tip via a pulley; the bending was away from the pull.

In many vines, of course, either stems or tendrils specialize to curve thigmotropically, although many tendrils curve thigmonastically (e.g, PFEFFER 1900/1906). With clear insight, PFEFFER wrote: "Mechanical agencies probably awaken more or less feeble reactions in all plants, and ... in addition to the special seismonic irritability (of, e.g., vines), other forms of sensitivity to mechanical agencies may be developed. Indeed, all stimulation resulting from movements of water, or from other forms of movement in the plant, may be termed mechanical, while if geotropic irritability is awakened by the changes in the position of the denser particles of the cell it becomes closely related in character to a form of internal contact-irritability."

As regards (4), the early literature provides descriptions of geomorphogenesis (RAWITSCHER 1932), some of which might result from shifting levels of IAA, but some of which might instead be direct consequences of increased cytosolic Ca^{2+}.

An important prediction of the Postulates is that gravity reception should sometimes very slightly stimulate putatively exocytotic voltage transients (Sect. 3.2.8.8) and consequent ET production directly, as well as sometimes via shifting levels of IAA (cf. Sects. 3.2.8.10, 3,2.8.11). The voltage transients are indeed elicited by horizontal placement (Sect. 3.2.8.2); in accord with expectation, the increase over background is relatively small. Available data for ET production, discussed in Section 3.2.6.2.c, are also compatible with this prediction, and closer examination of geotropic ET production should be a high research priority.

3.2.8.10 Auxin Regulation of Ethylene Synthesis via Cytosolic Calcium

If Postulates 6 and 7 are valid, then any agent adequately raising the level of Ca^{2+} in the cytosol of cells responsive to friction and flexure would in turn increase ET synthesis. Postulte 4 specifies that IAA can cause leakage of Ca^{2+} into the cytosol from the vacuole. It follows that a sudden increase of IAA must enhance ET production by initiating the same chain of events as does mechanical stimulation.

If IAA and mechanical stimulation promote ET synthesis by a common mechanism, and if promotion by mechanical stimulation is indeed mediated by vesicle release accompanied by characteristic voltage transients (Sect. 3.2.8.8) then a large step-up of IAA should elicit these transients. As predicted, when a drop of 500 μM IAA was applied to the epidermis of an excised epicotyl of etiolated *Pisum,* within 15 min an electrode recording extracellularly from the site picked up an increase in the frequency of the transients. Within 90 min, the increase was ten-fold for intact shoots or 60-fold for previously decapitated, IAA-depleted plants (PICKARD 1984).

3.2.8.11 Another Counter-Reaction

In Section 3.2.6 it was discussed how ABELES and RUBINSTEIN (1964) showed that, evidently in result of gravity-induced lateral transport, the level of IAA in the lower tissues of a *Phaseolus* section rises high enough to measurably enhance production of ET, which is thought in its turn to inhibit lateral transport in some stems (e.g., *Phaseolus*) and reverse it in others (e.g., etiolated *Pisum*). Such feedback may be responsible for a significant part of the geotropic retardation and counter-reaction in many shoots.

If this feedback occurs in etiolated shoots of *Pisum*, putatively exocytotic voltage transients should become more frequent following horizontal placement of intact shoots, and a small increase has been observed (unpublished data). However, putative contributions from Ca^{2+} leakage during gravity reception and from Ca^{2+} leakage from the vacuole following lateral redistribution of IAA remain unseparated.

It was discussed in Section 3.2.6.3 that a number of benefits of counter-reaction and particularly its extreme manifestation, plagiogeotropism, may have favored its evolution. According to the present model, the evolutionary process would have been readily possible because of the crosstalk between the system educing ET production and that responsible for the geotropic lateral transport of IAA.

3.2.8.12 Alternative Models

The ideas of this section suggest a fairly complicated mesh of regulatory reactions involving IAA, Ca^{2+}, ET and many other agents; yet, if the implicit analogy to regulation in animal systems (cf. RASMUSSEN 1983) has any validity, doubtless the mesh is not complicated enough. The demonstrated flexibility for evolution of second messenger systems in the animal kingdom suggests that, if Ca^{2+} is a second messenger in plants, there will be numerous variations on basic regulatory themes. Thus, the model of this section has not been developed with a strong conviction that it is correct, but rather with a sense that it is the *kind* of model that should be considered, and that it is susceptible to testing which can not only verify or contradict but which can also, in contradicting, provide evidence suggesting alternative models.

While this model was being developed, a model for IAA transport and action was presented by HERTEL (1983). It accounts for the occurrence of early geotropic wrong-way transport of IAA and later counter-reactive transport as consequences of the level of adaptation of the IAA transport system, stipulates that Ca^{2+} moves in opposition to IAA whenever an act of IAA transport occurs, and that Ca^{2+} regulates adaptation. It specifies that cytosolic Ca^{2+} is partially controlled by IAA binding to (and transport across) the tonoplast as well as the plasmalemma.

To do justice to HERTEL's model, a lengthy description would be required; and detailed comparisons with the present model would then be in order. Careful study of the original publication is instead strongly recommended.

Moreover, it is clear that a variety of other models could be elaborated. The present model will have served much of its purpose if it challenges readers to formulate their own hypotheses. Only a few years ago, the construction of sensible molecular models for geotropism seemed an awesome task. Today the ease with which alternative models for molecular events of geotropism can be developed encourages the hope that experimental tests will follow shortly and that models can be replaced or refined. It is likely that a few years will bring a vastly expanded understanding of geotropism.

3.2.9 More Factors and Asymmetries

3.2.9.1 Chemical Substances

The documented geotropic asymmetries of hormones and Ca^{2+}, as well as of growth itself, must be expected to lead to asymmetries of metabolites. A review by LARSEN (1962b) touches on such asymmetries. Because Ca^{2+} is believed to promote the activity of NAD kinase (ROUX and SLOCUM 1982, MARMÉ and DIETER 1983), early changes in the distribution of reducing agents will probably be found. So far, however, the only changes observed in shoots have been after curvature is well underway or completed (FABIAN 1969). Glucose and fructose increase on the lower side of horizontal grass nodes, evidently as a result of increased hydrolysis of sucrose (BRIDGES and WILKINS 1974). Geotropic increases in reducing sugar observed in the lower halves of apical internodes of a variety of plants (see refs. above) probably have a similar origin. The sugars might serve as substrates for respiration and growth, but also CLELAND (1972) has shown that sugars can influence growth in ways more complicated than expected of simple nutrients. Clearly, a complete knowledge of geotropism would require detailed understanding of the metabolism underlying growth and of the feedback loops presumably associated with its geotropic shifts and asymmetries; these feedbacks might be particularly important in the counter-reaction.

Further, it is possible that geotropic gradients of regulatory factors in shoots remain undiscovered. While such gradients might result from the primary IAA asymmetry, they might nonetheless play critical roles in determining parameters of curvature.

3.2.9.2 Phloem Translocation

When differential growth is occurring, supply of raw substrate from the phloem may be expected to occur asymmetrically. However, as discussed in Sect. 3.2.2.1, there is evidence for large changes in the delivery to the mesocotyl of an IAA precursor as well as sugar when corn seedlings are set horizontal, and it seems likely that if intact plants are studied more intensively, a more important role of translocation will become apparent.

Translocation of applied fluorescein is reported to increase unilaterally when dark-grown hypocotyls if *Vigna* are set horizontal (TANADA 1978a). The effect is measurable after excised shoots are induced for as little as 2.5 min, decapitated, and set upside down for 10 min with the apical cut wet by the dye. On the lower side, the dye progresses farther down the stem than in controls. Its fluorescence is visible only in sieve tubes. The effect is blocked by dinitrophenol, and enhanced by provision of extra boron (TANADA 1978b). Although no specifics of method or result are reported, the effect is said to occur in intact seedlings as well as in excised hypocotyls. Controls mounted right side up after induction were not reported. Confirmation of the stimulation of translocation should be attempted using an independent technique of measurement; translocation in short, recently wounded stem segments is suspect. And in spite of its apparent localization in the sieve tubes, dye could have traveled in the xylem undetected due to unfavorable conditons for fluorescence, spreading out into the surrounding tissue and concentrating in the sieve tubes where conditions for fluorescence might happen to be favorable. However, in at least many tissues fluorescein stains the xylem conspicuously, so this counterexplanation is at least as flawed as the initial proposal.

Assuming the putative enhancement to be real, it doubtless favors the increase in growth which is beginning in the lower tissue. It would be worth while to apply NPA in order to prevent the buildup of lateral auxin asymmetry and to see if the asymmetric enhancement of phloem translocation could be sustained without it.

Although TANADA mentions that the enhancement occurs in both detached and intact shoots, it is possible that substrate for growth is less effectively mobilized in the former and that this might account for part of the loss of geotropic performance which typically accompanies excision.

In excised shoots, the differential growth of the vascular parenchyma is less important than that of the epidermis in determining curvature (Sect. 3.2.10); nevertheless, it may be that in intact shoots the postulated supply role of the phloem is so important that it could explain why large sedimentable amyloplasts are so generally found in the vascular parenchyma (as contrasted with the rest of the shoot; for *Vigna* see e.g. HEATHCOTE 1981). Perhaps these amyloplasts indeed function as statoliths, educing increased permeability of the plasmalemma to Ca^{2+} on the outer sides of the lower "statocysts" (cf. Sect. 3.2.9), and perhaps this postulated local increase in cytosolic Ca^{2+} can lead to increased unloading of sugars from the sieve tubes in the lower half of the stem. In view of the evidence that R can increase membrane permeability to Ca^{2+} (Sect. 3.2.9.1), such an idea seems compatible with the also-observed stimulation of translocation by R (TANADA 1978c).

However, other explanations of the effect come readily to mind. As one example, enhanced translocation might be due to early, perhaps localized, development of an IAA asymmetry: although the lateral asymmetry responsible for differential growth is most marked in the outermost tissue of the stem (Sects. 3.2.9.2, 3.2.10), phloem parenchyma is believed to transport IAA vigorously (GOLDSMITH 1977, JACOBS and GILBERT 1983). The apparent involvement of boron in IAA transport (TANG and DELA FUENTE 1982) might thus rationalize its participation in the gravitational stimulation of phloem translocation. As a second example, perhaps an increased call for solutes on the lower side precedes the visible expression of IAA-induced growth, and causes the asymmetric activation of the phloem.

A recent report by JAFFE and LEOPOLD (1984) indicates that deposition (or a tendency toward deposition?) of callose at pit fields and sieve plates in the phloem is enhanced by geotropic stimulation of corn and pea seedlings. An induction period of 5 min is sufficient, but if the plants remain horizontal, further increase can be observed for many hours. The relation of the callose to the enhancement of translocation reported by others is unclear.

3.2.9.3 Transpiration

It has been suggested (MCINTYRE 1980, 1983) that geotropic curvature of shoots may be mediated by transpiration gradients. Since shoot geotropism can occur in water-saturated air and indeed under oil (HUMPHRY 1966, cf. also FRANSSEN et al. 1982) and water (e.g., DEWITT 1957, ANKER 1956, 1958, 1962), asymmetry of transpiration cannot be essential for curvature. Also, it is hard to imagine how asymmetric transpiration could explain the complicated distribution of geotropic curvature in space and time (Sect. 2). Further, because hormonally controlled differential growth seems fully adequate to account for geotropic curvature there is no good reason to invoke asymmetric transpiration. When strong asymmetry of transpiration occurs in nonwoody tissue it must of course result in bending: for example, MCINTYRE (1983) has described a certain bending of *Agropyron* rhizomes that appears to be due to asymmetric arrangement of scale leaves, which evidently protect the underlying epidermis against water loss. Not only is this particular bending unrelated to gravity, however, but also it is nastic rather than tropistic.

3.2.10 Differential Growth, and Recent Cavils About Its Control by Hormones

An ultimate goal of the study of hormones and ions in geotropism is the full accounting for their roles in bringing about differential growth, and hence curvature. While we are far from a quantitative accounting, the asymmetries of regulatory factors and of growth considered in the preceding pages show an encouraging qualitative correspondence.

Nevertheless, lately there has been a spate of papers suggesting or claiming that the apparent dependence of geotropic curvature on the asymmetry of regulatory factors – particularly IAA – is invalid. This point of view, embodied in a fairly large literature listed in Section 1, is well summarized in a 1976 review by DIGBY and FIRN and is repeated in a 1980 review by FIRN and DIGBY and a 1983 review by FIRN. DIGBY and FIRN (1976) pose three major questions.

First: "Is the magnitude of gravity-induced auxin redistribution sufficient to account for the observed growth differential?" A primary reason why DIGBY and FIRN respond negatively is that they tacitly assume that the IAA differentials usually measured for convenience between upper and lower halves of responding stems are based on uniform distributions within the two halves. However, both experimental evidence (Sect. 3.2.9.1) and transport theory (Chap. 10, this Vol., Sect. 3.1.3) suggest that the distribution is graded, with the biggest differences occurring between the upper epidermis and adjacent cortical cells on the one hand and the lower epidermis and its neighboring cells on the other. In one case, the ratio between upper and lower epidermis was about four times that between upper and lower halves of "core" tissues. Because the epidermis is well known to be the tissue generally limiting growth (e.g., PFEFFER 1906, WENT and THIMANN 1937, THIMANN and SCHNEIDER 1938b, BORGSTROM 1939, DIEHL et al. 1939, LEOPOLD 1955, IWAMI and MASUDA 1974, DURAND and RAYLE 1973, CLELAND 1980, PEARCE and PENNY 1983), it is the difference between its uppermost and lowermost regions which is critical for curvature.

DIGBY and FIRN argue that not only are geotropic IAA gradients small, but also that within tissue sections a large increment of IAA is required to produce a small increment in growth. However, short-term assays (CLELAND 1972, VANDERHOEF 1980a, b, BOUCHET et al. 1983) make clear that there are many factors influencing tissue growth in test situations; among them are the effects of excision (EVANS 1973, VESPER and EVANS 1978). Considering that intact plants respond to gravity much more vigorously than do excised stems (Sect. 2.2), perhaps it is not unreasonable to surmise that intact and healthy plants may respond to shifts of IAA more vigorously than do sections.

Moreover, it is critical that IAA acts symmetrically in tests with sections, and without the reinforcement of secondary Ca^{2+} asymmetries. In marked contrast, during growth in response to asymmetric IAA the rapid elicitation of an opposing asymmetry of Ca^{2+} may often dominate in producing curvature. Secondary asymmetries of other factors such as GA's (Sect. 3.2.4) must also be considered.

A second question raised by DIGBY and FIRN is: "Can exogenously applied auxin influence geotropic curvature in a manner predicted by the Cholodny-Went hypothesis?" Considering that the upper and lower epidermal tissues must be the critical places to apply auxin, and that it is not known exactly how much to apply there or even how to apply it there quantitatively in intact plants, this question cannot be answered precisely. However, the success over the years of the classical *Avena* curvature test (WENT 1928) indicates that the qualitative answer is yes. And, unilaterally smearing lanolin paste containing IAA onto intact dicot seedlings is a traditional, almost foolproof, classroom method of producing large curvatures within the lecture hour! DIGBY and FIRN cite two experiments in which gross asymmetries of exogenous IAA were measured, without reference to immobilized and transportable fractions, and side-to-side gradients were found which did not lead to curvature or which were in opposition to that expected for transportable auxin on the basis of observed geotropic bending. Certainly, interpretation of such experiments (cf. Sect. 3.2.2) is too ambiguous to disprove a mediational role for IAA in geotropism.

Third, DIGBY and FIRN ask: "Does the gravity-induced auxin redistribution precede geotropic curvature?" and again answer no. They play down the measurements of HERTEL and collaborators (Sect. 3.2.2.2) showing redistribution of ^{3}H-IAA within only 5–10 min. Many practical limitations preclude demonstrating transport as early as it occurs; limited spatial and temporal resolution of the measurement of isotopically labeled hormone has always frustrated students of tropism. In the absence of real-time information about redistribution, extrapolation is necessary; and if gross asymmetry can be measured within only 5–10 min, gradients between cortex and epidermis must start to develop within some fraction of that time.

Writing before 1982, moreover, DIGBY and FIRN could not have been aware of the evidence of SLOCUM and ROUX (Sect. 3.2.6) for Ca redistribution within 10 min. Growth-inhibiting action of Ca was already known to require only 1–2 min (ibid.).

DIGBY and FIRN emphasize that auxin redistribution must not only precede curvature, but do so by enough time to allow for a growth response. Again, resolution constitutes a major experimental problem. In measurements with excised stem sections, the lag for response to added IAA includes time for it to reach sites of action as well as for the action itself. Ten minutes is a frequently measured lag, yet RAYLE et al. (1970) reported that in half of their experiments (1) with the lipid-soluble methyl ester of IAA, (2) with short (4 mm) sections, or (3) with sudden large step-ups of IAA concentration, lags were only 1–3 min. These experiments are hard to reproduce (e.g., EVANS 1973), but suggest the possibility that the true lag for action may sometimes be brief, particularly for inhibitions due to large increments. However, lags for curvature are generally longer than 10 min (Sect. 2.2), and a 10–15 min lag for IAA action is compatible with the usually reported lags of 15–30 min for upward curvature.

A further argument of FIRN and DIGBY (1980) is based on the erroneous premise that the Went-Cholodny theory requires that gravity reception and auxin redistribution be restricted to the tip of the organ, and that curvature of the lower regions develop only in response to basipolar translocation of the redistributed auxin. Indeed, the fact that curvature often develops first in the tip and only later in the base gave inspiration to the development of the Went-Cholodny theory, but in the very paper in which DOLK published the first solid experimental test of the theory for geotropism he showed that (1) the capability of redistributing auxin is present throughout the coleoptile, although (2) it is strongest at the tip, and (3) curvature in the base is a product of both local redistribution and basipolar migration of auxin already redistributed in apical tissue (Sects. 2.2, 2.3, 3.2.2.1). FIRN and DIGBY suggest that curvature begins uniformly and that its basal migration is an artifact of "autotropism" or counter-reaction which begins at the tip – yet DOLK's extensive analysis of the temporal and spatial distribution of the curvature response took full account of "autotropism", showing that both the initial negative curving and the later counter-reaction begin at the tip of a geotropically responding coleoptile and migrate basally (Sect. 2.2). Qualitatively similar patterns of curvature development have been documented in detail for hypocotyls of *Lepidium* and *Cucumis* (MACDONALD et al. 1983).

Another contention prominent in the discussion of DIGBY and FIRN (1980) is that the Went-Cholodny theory should result in a stimulation of growth on the lower side of a stem exactly balanced by an inhibition on the upper side. While this simplistic interpretation was indeed suggested in the original context of the theory, surely the refinements of 50 years of experimentation on IAA and its control of growth may be taken into account in contemporary discussions of the theory: (1) the lateral gradient of IAA is unlikely to be vertically asymmetric about the midplane of the horizontal stem, (2) the dose-response curve for IAA is not linear, (3) IAA per se is not rate-limiting for the growth of all tissues at all times, (4) IAA redistribution results in positive and negative feedback with other regulators. Thus, turning the argument around, it would be surprising indeed if one could find no diversity in the pattern of differential growth by comparing several species and by varying growing conditions.

One would expect, then, to find cases ranging from primary dependence on inhibition of growth on the upper side to heavy dependence on acceleration of growth on the lower side, with a great many intermediate cases in which inhibition and acceleration are more or less matched. Such cases should be susceptible to further subdivision into those in which curvature begins apically and moves basally and those in which curvature starts more or less evenly along the length of the organ. There should be situations in which counter-reaction is conspicuous and those in which it plays a minimal role. Papers of DIGBY and FIRN (1979) and FIRN and DIGBY (1979), in combination with the data discussed in Section 3.3, indicate that at least many of these possibilities are realized in nature. Of special interest is thorough documentation by MACDONALD et al. (1983) of both accelerated growth on the lower side and retarded growth on the upper side of geotropically curving hypocotyls of *Lepidium* and *Cucumis*, with some compression of certain zones of upper tissue occurring during the response.

In the future it will of course be important to refine assessments of gravity-induced changes in the distribution of hormonal and ionic factors regulating growth as well as of the pattern of tissue sensitivity to these factors. Putting the data together, they must be compared with simultaneous measurements of differential growth and curvature along the length of the stem. Such integrated studies should be made for several species, including different developmental stages and growing conditions for each. These detailed measurements should go a long way toward checking the validity of current concepts of hormonal control. However, the experiments will be tedious and it may be anticipated that limitations in accuracy and resolution will preclude an absolutely complete quantitative accounting for the obviously intricate participation of hormones and ions in geotropism. Indeed, biochemical description of the mechanisms underlying the control of growth may so outpace the bookkeeping method of evaluating the roles of regulatory substances in geotropism that tomorrow's students of tropism will feel no need for a perfectly balanced ledger of regulatory inputs and growth outputs.

3.3 Roots

3.3.1 Perspective

The mechanisms by which elongation is controlled have proved far more elusive for roots than for shoots; some controversy persists over what hormones are primarily responsible. In the literature of geotropism, the classically postulated controlling role for IAA as a critical regulator of differential growth was out of favor for several years, replaced by a set of models in which ABA or other "inhibitors" played key roles (e.g., AUDUS 1975, 1983, WILKINS 1975, 1978). Currently, it is recognized that physiological concentrations of ABA do not always inhibit growth and may stimulate it dramatically, and that high concentrations of ABA do not "swamp" geotropism as would be expected if this substance exerted control over differential growth. New experiments have sharpened the focus on IAA again, and simultaneously on the closely associated calcium ion.

While these experiments appear to have established that movements of IAA and Ca^{2+} are essential, they have not elucidated how they occur and how they are related to each other. Considering the ferment of new ideas, it is not an appropriate time to attempt an exhaustive review of recent work. Moreover, an exhaustive review of earlier literature would be redundant with the thorough and well-reasoned review by JACKSON and BARLOW (1981), which considered among other things the role of the root cap, the distribution and kinetics of curvature, and the evidence concerning possible participation by

IAA and ABA. Therefore, this review will take up where that left off, highlighting work promising to provide a foundation of understanding for root geotropism at a level comparable with that for shoots.

3.3.2 Calcium in the Cap

Recent data indicate that Ca^{2+} migrates downward across the cap of a horizontal root, that it accumulates on the lower side, and that the accumulation ultimately leads to geotropic curvature.

LEE et al. (1983a) recently followed the migration of Ca^{2+} by incorporating isotopically labeled Ca into agar blocks and applying them to the upper or lower sides of horizontal *Zea* roots as well as to the sides of vertical controls. After 45 min, radioactivity was assessed in receiver blocks applied opposite the donors (Fig. 16A). Recovery from receivers was not significantly influenced by root orientation if application was at the zone of elongation or if the cap of the root had been removed. However, the recovery of radioactivity from apical receivers applied to the lower side of intact horizontal roots was enhanced 1.6-fold over that for vertical controls, while recovery from receivers on the upper side was depressed to about half of control values.

The importance of the downward migration for geotropism was demonstrated in two ways (LEE et al. 1983b), as shown in Fig. 16B. First, agar blocks containing EDTA were applied to root tips; this Ca^{2+} chelator eliminated the ability of horizontal roots to respond to gravity. When the EDTA was replaced with $CaCl_2$, responsiveness was restored. Second, blocks containing $CaCl_2$ were applied asymmetrically to root tips [either (1) with caps removed or (2) with EDTA pretreatment]. This asymmetric supply of Ca^{2+} caused vertical roots to curve in the direction of the application. Blocks of the same concentration applied in the zone of elongation caused no curvature.

3.3.3 Auxin

EDWARDS (1983) has shown for *Zea* that asymmetric accumulation of Ca^{2+} at the tip stimulates basipolar movement of IAA in the adjacent meristematic and cortical tissue. She applied agar blocks with ^{3}H-labeled IAA to the apical surface of the meristem of root segments from which the caps had been detached, applied agar blocks containing Ca^{2+} either to the upper or lower surface of the meristematic region, and after 1.5 h assessed the distribution of radioactivity in the basal tissue and in upper and lower receiver blocks applied in the region of elongation. As illustrated in Fig. 17, when Ca^{2+} was supplied via the upper block 54% of the label found in the halved tissue was in the upper portion and 63% of the label found in the agar receivers was in the upper block. But when Ca^{2+} was supplied from below, the distribution was reversed: 56% of the tissue label was on the lower side and 79% of the label in the receivers was on the lower side. Thus, Ca^{2+} applied asymmetrically to the meristematic tissue causes correspondingly asymmetric basipolar movement of IAA. (Incidentally, if the greater asymmetry of ^{3}H-IAA seen when Ca^{2+} was supplied from below rather than above is significant, the experiment suggests that even in the absence of the cap, some downward movement of Ca^{2+} can occur in subja-

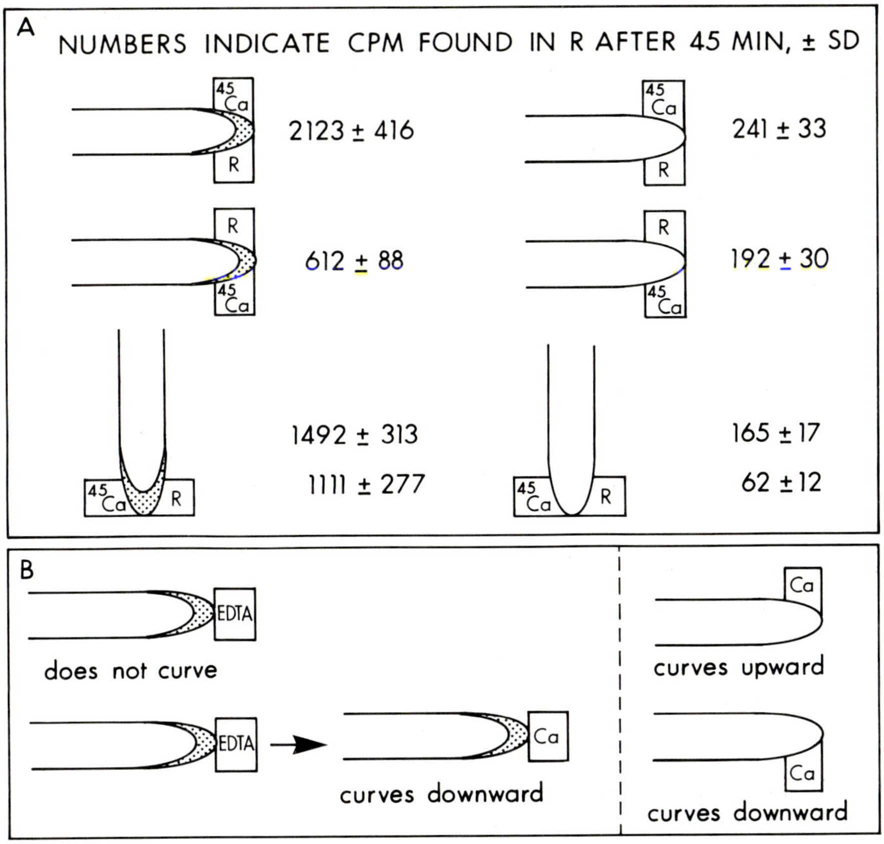

Fig. 16A, B. Demonstration that asymmetry of Ca^{2+} develops in the root cap and is required for geotropic curvature. **A** LEE et al. (1983a) applied agar blocks containing $^{45}Ca^{2+}$ to one side of the tip of intact roots (*left* side of figure) or roots with caps detached (*right* side). At the same time they applied plain agar blocks to the opposite sides; 45 min later, they assessed radioactivity in the receiver blocks with the specified results. For half the vertical controls, movement of ^{45}Ca was assessed from right to left, and for the other half from left to right, as represented respectively by the *upper* and *lower paired values* associated with the *vertical diagrams*. **B** LEE et al. (1983b) applied EDTA (contained in agar blocks) to the tips of intact horizontal roots and found that no curvature occurred even after 3 h, when controls had bent strongly. If after 1.5 h the symmetric EDTA was replaced symmetrically with Ca^{2+}, curvature ensued during the next 1.5 h. Asymmetric application of blocks containing Ca^{2+} to tips of decapped roots caused curvature toward the block, regardless of whether the block was placed on the upper or the lower side

cent meristematic tissue. This suggestion is reinforced by the development of asymmetry in controls to which no Ca^{2+} was applied, and is consistent with data on the distribution of sensitivity to gravity (cf. Sect. 1.3).

Because application of physiologically appropriate concentrations of IAA to one flank of the elongation zone of a vertical root causes bending toward the side of application (e.g., SCHURZMANN and HILD 1980), and because of other evidence discussed by JACKSON and BARLOW (1981), it is reasonable to

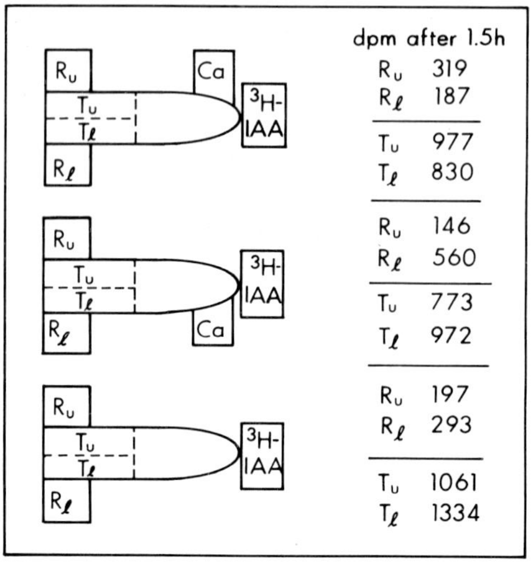

Fig. 17. Demonstration by K.L. EDWARDS that asymmetry of Ca^{2+} at the tip of *Zea* root sections causes corresponding asymmetry of basipetal IAA transport. Donor blocks with 10 nM IAA, labeled with ^3H, were applied to the apex of roots from which caps had been removed; blocks with Ca^{2+} were affixed just basal to the IAA donors on either the upper or the lower side. Sections not provided with Ca^{2+} served as controls; dpm were assessed after 1.5 h in receiver blocks applied to the upper and lower sides at the basal end of the sections; the values are means for the sums of the dpms in sets of 10 blocks. Upper and lower halves of the basal 3 mm of the root sections were also assessed; the dpm are means for sets of 10 samples, expressed per mg dry weight. (Data were provided by EDWARDS; they are published only in abstract. EDWARDS 1983)

assume that the asymmetry of IAA transported basipetally into the cortex of the elongation zone initiates geotropic curvature.

The model of EDWARDS (1983) is new and only tentatively supported. EDWARDS points out (personal communication) that the experiments should be checked, especially with alternate geometries of donor and receiver blocks: they are technically difficult because roots are small, have a moist surface along which films of water move readily, and operate at relatively low levels of IAA. An earlier experimentally based model of KONINGS, evaluated in perspective by JACKSON and BARLOW (1981), retains considerable appeal because of its closer analogy to the mode of geotropic IAA transport demonstrated for shoots. This model specifies that IAA is laterally transported to the lower side of the tip, where it is then preferentially available for loading into the basipetal transport system. The pattern of net geotropic flow of IAA through the apical and the more basal tissues is the same in the two models; they differ only in regard to the region where active control of the flow is exerted.

3.3.4 Connection Between Calcium in the Tip and Development of IAA Asymmetry

It is appropriate to speculate how the asymmetry of Ca^{2+} might come about in the cap and how it educes differential basipetal transport of IAA in the subjacent tissue. As a foundation for building a model, the electrophysiological measurements of BEHRENS et al. (1982) and of BEHRENS, WEISENSEEL, GRADMANN and SIEVERS described by SIEVERS and HENSEL (1982) are pertinent.

BEHRENS et al. (1982) used a vibrating probe to demonstrate a shift in current flow at the root apex within as little as 30 s after a root of *Lepidium* is placed horizontal. Because the orientation of the probe was fixed with respect to gravity rather than to the root, the precise pattern of the gravity-induced change could not be determined (although it could be modeled). However, the demonstration of the shift in ion fluxes encouraged efforts to record membrane potentials from the cells of the root cap (BEHRENS, WEISENSEEL, GRADMANN and SIEVERS, described in SIEVERS and HENSEL 1982). It was found that within seconds after 45° displacement of a root from its vertical position of equilibrium, the plasmalemmas of cells on the lower side of the cap began to depolarize. The depolarization continued for about 2 min, during which time transmembrane potentials dropped from −120 mV to perhaps −80 mV. Then the plasmalemmas began to repolarize, achieving new steady states perhaps a few mV short of the basal transmembrane potential.

The central cells of the cap contain large amyloplasts often considered on the basis of strong correlative evidence (JACKSON and BARLOW 1981) to be statoliths. When a root is set horizontal, these putative statoliths fall within a fraction of a minute to the lateral sides of the cell. There, according to PERBAL (1978), they may come to rest on a single lateral cisternum of endoplasmic reticulum.

It has been argued that the action of statoliths, if amyloplasts truly serve as such, must be to exert pressure on some organelle of the cell (JOHNSSON and PICKARD 1979, SIEVERS and VOLKMANN 1979). If the endoplasmic reticulum of the putative statocytes contains Ca^{2+}, and if the membrane of the lateral cisternum responds to amyloplast pressure by opening Ca^{2+} leakage channels, Ca^{2+} could flood the cytosol on the lower side of the displaced cells. The apical aggregate of endoplasmic reticulum, to which the lateral cisternum would be connected, could serve as a concentration reservoir for the ion and a capacitor for electrical charge. Thus, the lateral cisternum would refill with Ca^{2+} even as leakage to the cytosol continued. As Ca^{2+} flooded the cytosol, it would activate plasmalemmal Ca^{2+} pumps on the lower side of each cell, and (because of the more or less isodiametric cell shape) would depolarize the entire plasmalemma. The pumps would then expel Ca^{2+} into the apoplast, tending to effect lateral transport across the cap and at the same time tending to return the membrane potential toward its original levels. That level would not be achieved, however, as long as the root cap remained out of its equilibrium position, because the continuing statolithic pressure on the endoplasmic reticulum would ensure continuing release of Ca^{2+} to the lower cytosol.

Ca^{2+} expelled to the apoplast would not only tend to re-enter cells and be passed downward (and to some extent basipetally within the curving cell layers of the dome-like cap), but would also diffuse basipetally, activating the polarized IAA transport system in the meristem in the way accomplished by EDWARDS (1983) with exogenous Ca^{2+}. This activation is perhaps analogous to that shown for transport of IAA in shoot tissue by DELA FUENTE and LEOPOLD (1973), DE GUZMAN and DELA FUENTE (1981, 1982) and DELA FUENTE (1981, 1982).

A possible prediction of this model is that the putative pressure-regulated Ca^{2+} channels of the endoplasmic reticulum may also be regulated by IAA, and that the relevant IAA binding site may be blocked by TIBA, NPA and the morphactin 9-HFCA. LEE et al. (1984) showed that these traditional blockers of IAA transport also block Ca^{2+} translocation within the root cap. They speculate (personal communication) that this might occur because the Ca^{2+} movement might be oppositely and perhaps stoichiometrically coupled with IAA transport within the cap. However, such coupling would require that IAA move upward in the geotropically responding cap, and not only has this not yet been observed but also it is hard to imagine how such upward movement could be related to the ultimate stimulation of basipetal translocation of IAA presumed on the basis of the experiments of EDWARDS (1983) to occur in the meristem and elongation zone. While TIBA and NPA have sometimes been considered specific blockers for IAA transport, much of this evidence is based on studies of coleoptiles and shoots and in any case is not based on assessments within root caps.

IAA activation of Ca^{2+} release from endoplasmic reticular stores, with all the postulated consequences, might be an effective regulatory device to assure that a reasonable amount of IAA, which is shipped to the cap via the stele (e.g., BOWEN et al. 1972, SHAW and WILKINS 1974, BATRA et al. 1975), is able to progress basipetally in the cortex to facilitate growth in the zone of elongation.

An alternate suggestion to explain the blocking action of TIBA, NPA and 9-HFCA pretreatments is based on the observation of FELDMAN (1983a) that IAA is necessary in the cap for light-stimulated synthesis of proteins required in some roots for geotropism (Sect. 3.3.8). Possibly, the receptor sites for this action of IAA could strongly bind the three blocking agents.

Considering that the statolith hypothesis for gravity reception by roots is not rigorously established, it is worth noting that it would be possible to construct nonstatolithic models in which membrane pumps for Ca^{2+} are asymmetrically situated in the cell and/or asymmetrically activated in the root. Activation of membrane pumps need not in principle require statoliths; further, the linkage between Ca^{2+} and IAA transport could be envisioned in a variety of ways. Because so many models for the coupling of reception, development of Ca^{2+} asymmetry, and development of IAA asymmetry could be devised, a great deal of testing will doubtless be necessary before any model can be accepted with confidence.

3.3.5 Ethylene

According to early Went-Cholodny theory, auxin accumulates on the lower side of geotropically stimulated roots and shoots alike, but whereas it stimulates shoot growth and causes upward bending, it inhibits root elongation and causes downward bending. The dose-response curves for auxin-induced elongation were postulated to be similar in form but different in parameters: shoots normally operate low in the stimulatory range of auxin, but roots operate high in the stimulatory range, near the inhibitory range (WENT and THIMANN 1937). While the theory was not detailed enough to explain all the observations of subsequent research, and hence fell out of favor for many years, it has now been expanded in such a way that it must be viewed as the best available explanation why stems grow up and roots grow down.

BURG and BURG (1966) proposed that the reason IAA inhibits elongation of stems at high concentrations is that it stimulates the synthesis of ET, which switches the mode of growth from elongation to increase in girth. Extending this proposal to roots, CHADWICK and BURG (1967) provided evidence that in these organs IAA stimulates ET production even at very low levels, thus explaining the early-observed inhibitions by exogenous IAA. They pointed out that accumulation of IAA on the lower side of a root would thus cause local inhibition, and downward growth.

The availability of inhibitors of ET synthesis and action have made it possible to test this theory. Pretreating *Zea* roots with AVG or $CoCl_2$, MULKEY et al. (1982) provided IAA symmetrically and observed enhanced elongation; added asymmetrically, IAA caused curvature away from the site of application (EVANS, personal communication). Geotropism of pretreated roots was strongly negative (ibid.). Thus, it is plausibly via ET that IAA acts as an inhibitor of root growth, and ET is evidently an important agent controlling the rate and equilibrium position of root geotropism.

Such modulation of geotropism by ET is consistent with the plagiogeotropic growth reported by ZOBEL (1974) for roots of the *Lycopersicon* mutant diageotropica.

Moreover, as EVANS has realized (personal communication), the sometimes-observed early transient upward curvature of horizontal placed roots (Fig. 2) can readily be explained if an initially stimulatory level of IAA on the lower side increases gradually through a stimulatory optimum concentration before it builds to an inhibitory level.

Not inconsistent with the notion that roots normally operate a little below the peak of the IAA dose-response curve is the observation by KONINGS and JACKSON (1979) for *Lycopersicon* than 20 nl ET added per l of air stimulates elongation over a 24 h period, whereas 50 nl inhibits it. First, experimental applications of ET are usually above 20 nl ET per l of air, and presumably the increase elicited by geotropic accumulation of IAA on the lower side is adequate to cause production of inhibitory ET levels. Second, 24 h is a very long interval for response, permitting both initial reaction and adaptation. Detailed dose-response curves for any hormone seem always to exhibit subtle complexity.

It appears that ET is an important mediator of environmental alterations of root geotropism. Illumination, for example, can influence geotropism in multiple ways (Sect. 3.3.8), and one is evidently the stimulation of ET synthesis. An inhibitor of ET synthesis and action (DIHB) can sometimes overcome the inhibition of root elongation by light, and in *Lepidium* roots inhibition increases with the amount of ET produced (ROBERT et al. 1975). As a second example, when roots are covered with a film of water, the diffusive escape of ET is retarded; under this circumstance many roots change their equilibrium angle (KONINGS and JACKSON 1979). Indeed, roots are sometimes observed to curl into helices or spirals (personal observation), and this might be a result of abundant ET, just as it is for etiolated *Phaseolus* hypocotyls (Sect. 3.2.6.2.b). In this regard, because ET influences IAA transport in many stems, it seems possible that also in roots ET can influence transport of IAA as well as tissue sensitivity to it.

3.3.6 Protons

In roots as in shoots, at least some of the effects of IAA are mediated by secretion of H^+, and this has been demonstrated specifically for geotropism by MULKEY and EVANS (1981, 1982). Not only does the elongating upper side of geotropically responding *Zea* roots secrete more acid than the inhibited lower side, but also both this secretion and geotropic curvature are blocked by vanadate, thought to be a reasonably specific inhibitor of H^+ secretion. Inhibitors of ET synthesis promote IAA-elicited H^+ secretion concomitant with their promotion of growth (MULKEY et al. 1982).

3.3.7 Secondary Asymmetries: Reaction and Counter-Reaction

The asymmetry of IAA and/or the asymmetry of H^+ secretion it triggers doubtless lead to secondary asymmetries, some of which enhance downward curvature and some of which bring about counter-reaction.

3.3.7.1 Calcium in the Elongation Zone

Employing techniques similar to those used by SLOCUM and ROUX (1983) for coleoptiles, M. DAUWALDER and S.J. ROUX have recently assessed the distribution of Ca in the elongation zone of roots (personal communication). In the dark-grown, geotropically nonreactive *Zea* root, most of the Ca appeared to be localized within the stele. Half an hour after onset of illumination, which triggered geotropic sensitivity and hence downward bending (cf. Sect. 3.3.8), Ca became more abundant in the upper cortex. At this time, curvature was only beginning. Two hours later, when curvature was well developed, accumulation of Ca in the upper cortex was conspicuous.

If Ca^{2+} gradients can be established in response to electrochemical gradients associated with H^+ secretion in elongating tissue of shoots (Sect. 3.2.8.3), they might well develop in the elongation zone of roots by the same mechanism. However, the proton gradient is in the wrong direction to account for DAUWALDER and ROUX's results. A possible basis for the root asymmetry is suggested by a further dissimilarity between roots and coleoptiles: whereas in the upper tissue of coleoptiles Ca^{2+} increases in the cell walls, in the upper tissue of roots it increases within the vacuoles. Perhaps the increased apoplastic H^+ promotes opening of plasmalemmal leakage channels for Ca^{2+}, resulting in movement of Ca^{2+} into the cytosol. Elevated cytosolic Ca^{2+} might lead to activation of Ca^{2+} pumps in the tonoplast, and hence to (1) elevation of vacuolar Ca^{2+} and (2) regulation of cytosolic Ca^{2+}. In this way, cytosolic Ca^{2+} might be maintained at a level suitable for the elongation which must occur in the upper tissue, while cortical apoplastic Ca^{2+} would tend to be depleted. Perhaps Ca^{2+} would then flow down an apoplastic concentration gradient from the stele to the cortex. This loose-knit speculation gives little insight into the

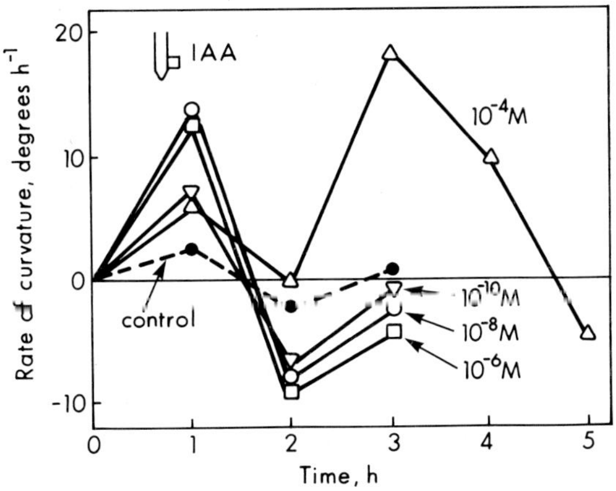

Fig. 18. Unilateral application of an agar block containing IAA to the zone of elongation of a *Zea* root causes curvature and counter-curvature. (SCHURZMANN and HILD 1980)

presumable importance of the Ca^{2+} redistribution for curvature, but does emphasize that study of active and passive mechanisms of controlling Ca^{2+} movement across cellular membranes may be a prerequisite for understanding how the Ca^{2+} redistribution occurs.

3.3.7.2 *Gibberellin and Other Hormones*

It is reasonable to speculate that asymmetries of IAA and H^+ could lead to the asymmetries of regulatory agents such as GA's, just as in shoots. While there is no evidence for a critical role of GA's in root geotropism, a few measurements of asymmetries have been reported; these have been reviewed by JACKSON and BARLOW (1981). A paucity of data prevents further speculation on this matter, but given the elaborate time-courses of curvature and countercurvature sometimes displayed by roots (Sect. 2.2), experimentation would seem worthwhile.

3.3.7.3 *"Adaptation"*

It seems likely that a large part of the geotropic counter-reaction can be accounted for as the result of the shifting sensitivity or adaptation of the tissue to IAA which follows the new IAA distribution. SCHURZMANN and HILD (1980) showed that unilateral application of IAA, even in the absence of geotropic stimulation, can cause fluctuating curvature responses (Fig. 18). Donor block concentrations up to 1 μM cause net bending toward the block during the first hour, net bending away from the block during the second hour, and net bending toward the block again during the third hour. For the very high donor concentration of 100 μM, bending was weakly toward the block for an hour, came to an apparent halt during the next hour, became strongly positive during the next hour, and then became less positive and ultimately strongly negative. Of course, because measurements were made only once per hour, the precise timing of rate maxima and minima is unknown; however, rates of development of geotropic IAA asymmetry and of geotropic counter-reaction under the identical experimental conditions are also unknown.

Adaptation of straight growth of roots to changing levels of IAA has also been studied (GOUGLER and EVANS 1981 and references therein). Controlled comparisons of adaptation to symmetrically and asymmetrically applied IAA would be of interest. Probably the latter is more conspicuous, considering the possibilities for the occurrence of exchanges of substances across the root and possible consequent exaggeration of feedback oscillations.

While it is both common and correct to refer to IAA-elicited shifts in the responsiveness of a tissue to IAA as changes in sensitivity, it is semantically more productive to recognize that the shifts doubtless represent secondary, IAA-elicited changes in levels of one or more specific factors. A good candidate for one such factor is cytosolic Ca^{2+}. Possibly, IAA could increase cytosolic Ca^{2+} by the same mechanism proposed in Section 3.2.8 for shoots. This in turn would have ramifications for control of elongation by IAA. A second candidate is ET, production of which could certainly be viewed as an adaptive phenomenon. Many other candidates can be imagined, but supportive evidence is limited. How changing levels of IAA bring about changing tissue sensitivity to IAA urgently needs exploration.

3.3.8 Light, Inhibitors, and the Effect of Light on Inhibitors

It has been known for a long time that many roots grow ageotropically or plagiogeotropically in complete darkness but become orthogeotropic on receipt of a little light (e.g., KLEMMER and SCHNEIDER 1979, FELDMAN 1984). Ecologically, this makes sense because it permits plants to spread roots widely and establish broad-based anchorage without breaking the soil surface. Conveniently, some varieties of the popular *Zea* root display ageotropic behavior in darkness, whereas some grow downward in both dark and light. This situation has been exploited in searching for processes and substances essential for geotropism.

FELDMAN (1983a) has shown that brief illumination of dark-grown, light-sensitive *Zea* roots causes enhanced protein synthesis in the caps, peaking in 0.5 h and declining to the basal level within 4–5 h. At 0.5 h, at least one particular protein of low molecular weight is enhanced much more dramatically than others (FELDMAN and GILDOW 1984). The enhancement requires not only light but also IAA, as shown by studying its occurrence in detached caps. It appears that the enhancement of protein synthesis and sensitivity to gravity are closely correlated.

Accompanying the enhancement of protein synthesis is the increase of ABA and related "inhibitors" (e.g, JACKSON and BARLOW 1981, FELDMAN 1981a, 1981b, 1982). Particularly important in establishing the increase of ABA has been analysis by mass spectrometry (PILET and RIVIER 1980, 1981). Using *Zea* roots, FELDMAN (1983b) has recently provided evidence that carotenoids, which increase 1.6-fold in result of illumination, are synthetic intermediates for the "inhibitors". One of the carotenoids is violaxanthin. Thus, a possible role of enhanced protein synthesis might be to provide enzymes controlling the synthesis of "inhibitors".

FELDMAN has noted (personal communication) that the production of "inhibitors" might occur in the amyloplasts of the cap (there is suggestive evidence for ABA synthesis in plastids in other systems (WALTON 1980) and that this could explain in part the abundantly observed correlations (e.g., AUDUS 1979) between the condition of the amyloplasts and geotropic sensitivity.

The role or roles of the "inhibitors" is unclear. Their appellation derives from the ability at high concentration to inhibit elongation and to cause stomatal closure. It seems possible, however, that at least one of their perhaps multiple roles in geotropism is stimulatory: they might put the gravity detection or transduction systems into a receptive state for ageotropic roots, or modulate the systems for plagiogeotropic roots. This could explain the important finding of WILKINS and WAIN (1975) that ABA pretreatment of *Zea* roots requiring light for rapid geotropism markedly enhanced geotropism in the dark, in spite of the fact that elongation was not much influenced except by very high concentrations. Thus, ABA does seem likely to be a critical mediator of the effects of light.

Certainly, there is no conclusive evidence that the "inhibitors" exert primary control over the differential elongation of the root. In fact ABA can stimulate as well as inhibit root growth (e.g., ASPINALL et al. 1967, GAITHER et al. 1975, MCWHA and JACKSON 1976, YAMAGUCHI and STREET 1977, ABOU-MANDOUR and HARTUNG 1980, SMITH and HO 1982). MULKEY et al. (1983) have recently assessed the effects of ABA auxanometrically, finding for three cultivars of *Zea* as well as for *Pisum* and *Abelmoschus* that elongation is initially promoted and is inhibited only after 12 h and only by high concentrations. Such inhibition as ABA did exert was further studied in *Zea,* where it was shown to be brought about by delayed induction of ET synthesis (MULKEY et al. 1983). Moreover, unilateral application of ABA to the lower side of horizontal roots decreased geotropic curvature, whereas application to the upper side increased it. This confirms and extends the finding of a number of other authors that unilateral application of ABA fails to cause curvature in a manner consistent with action as a primary mediator of geotropic inhibition of elongation on the lower side of roots (e.g., SUZUKI et al. 1979, SCHURZMANN and HILD 1980, MERTENS and WEILER 1983).

Nevertheless, the level of free ABA can rise in the lower half of the elongation zone of geotropically responding *Zea* roots and diminish in the upper half, according to mass spectrometric determinations of PILET and RIVIER (1981). Also in their experiments, more ABA was found in the lower half of the elongation zone of roots from which the upper half of the root cap had previously been removed. While the authors interpret this as evidence for a primary role of ABA in mediating curvature, and while this interpretation remains a possibility, it is also possible that the asymmetry of ABA is secondary to that of IAA. Arguing for a secondary basis of asymmetry is evidence that ABA is not transported efficiently from root cap to elongation zone (JACKSON and BARLOW

1981), and its distribution from cap to elongation zone is not influenced by Ca^{2+} applied at the root tip (EDWARDS 1983).

Moreover, according to other authors ABA can accumulate on the upper rather than the lower side (e.g., SUZUKI et al. 1979, HARTUNG 1976; see also the review of JACKSON and BARLOW 1981). While these assays were accomplished with less definitive techniques than mass spectrometry, nevertheless they cannot be ignored.

In Section 2 and in Chapter 10, this Volume, it has been extensively discussed how in shoots an asymmetry of IAA might give rise to a wide variety of other asymmetries; and it was speculated that some of these result from the asymmetry of H^+ secretion linked to IAA asymmetry. Perhaps a gradient of ABA could be an example of such a secondary asymmetry for roots. In the upper, acidified portion of the elongation zone ABA would accumulate inside the cells (cf. RUBERY and ASTLE 1982, ASTLE and RUBERY 1983). Such uptake would diminish the local apoplastic concentration, leading to upward diffusion from the apoplast of the lower tissue. Moreover, because a root may exhibit marked counter-reaction as it seeks its position of equilibrium, this proposed mechanism could produce ABA gradients favoring either the upper or lower sides of the root, depending on when the root is sampled. This suggestion could be tested by assessing the time course of ABA distribution in controls and in roots inhibited with vanadate.

Another way in which ABA asymmetry might develop is that, in consequence of asymmetrically shifting IAA levels, the equilibrium between free and conjugated ABA might undergo asymmetric change. Because PILET and RIVIER (1981) assessed only free ABA, currently available data do not address this possibility.

Importantly, then, it does remain plausible that secondarily developed asymmetries of ABA and related substances could enhance positive curvature or elicit counter-reaction, at least in some kinds of roots. Roots of different species or from different environments might well differ in the degree of control of elongation exerted by IAA and ABA, with the secondary asymmetry sometimes being an important effector of geotropic curvature or counter-curvature. However, much more work remains to be done in order to determine whether asymmetry of ABA and related compounds is a general feature of geotropically responding roots.

3.3.9 Summary

Recent experiments on (1) early electrical events elicited by horizontal placement, (2) transduction of Ca^{2+} asymmetry in the cap, (3) the consequent establishment of IAA asymmetry in the elongation zone, and (4) the kinetics of primary and secondary processes by light promise to permit rapid advances in our understanding of root geotropism. For purposes of this review, the data from these experiments have been presented in the context of one of many possible models, but the way in which they actually fit together is far from clear. This or other models should stimulate more precisely targeted experimentation.

4 Concluding Remarks

Historically, studies of tropisms and of growth originated and progressed as a unified effort, but in the intermediate past tended to become more specialized. Recently they have become more interdependent again, and this trend may be expected to continue productively. In particular, investigation of geotropism may well take on greater significance for the understanding of plant growth and development: not only is it convenient to have a precise external trigger for developmental events, but also the study of geotropism promises to yield special insights into how Ca^{2+} might serve as an important regulator and indeed as a second messenger in plants. Moreover, because the basis of geotropic curvature is lateral asymmetry of growth regulators, it appears that many of the feedbacks that occur with temporal overlap and spatial superposition during straight growth may be transiently spread out along and across the plant as the complex reactions and counter-reactions to gravity unfold. Such display may help to pinpoint component regulatory processes for intensive study.

Acknowledgements. The preparation of this review was supported in part by BRSG 507RR07054 awarded by the Biomedical Research Support Grant Program, DRR, NIH; and in part by NAGW-420 from the Space Biology Program, OSS, NASA.

Note Added in Proofs. Reasonably specific inhibition of geotropism of *Avena* coleoptiles by the calcium chelator EGTA has been described by DAYE et al. (1984); the action of the chelator may well be to lower free Ca^{2+} in the apoplast; thus possibly inhibiting both reception of the gravitational stimulus (Sect. 3.2.8.2) and mediation of the curvature response (Sect. 3.2.8.4).

MOORE and SMITH (1984) treated seedlings of *Zea* with fluridone, which prevented synthesis of detectable amounts of ABA, and found that the geotropic response of roots was not impaired. This extends data discussed in Sections 3.3.1 and 3.3.8.

Accumulation of calcium in vesicles derived largely from the endoplasmic reticulum of root caps of *Lepidium* has been assessed by SIEVERS et al. (1984). Additionally, the authors describe in further detail the electrophysiological evidences of Section 3.3.4. On the basis of all their findings, they propose a mechanism for gravity reception similar to the speculations of Section 3.3.4.

An abstract by EDWARDS (1984) provides more information in support of the model described in Section 3.3.3, but EDWARDS announced in presentation of her paper that recent experiments of alternate design were not corroborative. EVANS and LEE (1984) also spoke of recent evidence not published in their abstract; if they apply relatively large amounts of isotopically labeled IAA to the elongation zone they can observe downward movement if and only if the root cap remains in place. It is unclear whether the behavior of exogenous IAA in these experiments resembles the movement of endogenous IAA, or whether it is an abnormal effect secondary to asymmetries established in and by the tip (cf. Sect. 3.3.7). Although the basic tenets of the Went-Cholodny theory at last seem well supported for roots, the proof of the precise mechanism of redistribution of IAA (and now, of Ca^{2+} as well) remains extraordinarily elusive!

References

Abeles FB (1973) Ethylene in plant biology. Academic Press, London New York

Abeles FB, Rubinstein B (1964) Regulation of ethylene evolution and leaf abscission by auxin. Plant Physiol 39:963–969

Abou-Mandour AA, Hartung W (1980) The effect of abscisic acid on growth and develop-

ment of intact seedlings, root and callus cultures and stem and root segments of *Phaseolus coccineus*. Z Pflanzenphysiol 100:25–33
Abrol BK, Audus LJ (1973a) The lateral transport of 2,4-dichlorophenoxyacetic acid in horizontal hypocotyl segments of *Helianthus annuus*. J Exp Bot 24:1209–1223
Abrol BK, Audus LJ (1973b) The effects of N-1-naphthylphthalamic acid and (2-chloroethyl)-phosphonic acid on the gravity-induced lateral transport of 2,4-dichlorophenoxyacetic acid. J Exp Bot 24:1224–1230
Adams PA, Kaufman PB, Ikuma H (1973) Effects of gibberellic acid and sucrose on the growth of oat (*Avena*) stem segments. Plant Physiol 51:1102–1108
Albert R, Kinzel H (1973) Unterscheidung von Physiotypen bei Halophyten des Neusiedlerseegebietes (Österreich). Z Pflanzenphysiol 70:138–157
Allen JRF, Rivier L, Pilet PE (1982) Quantification of indol-3-yl acetic acid in pea and maize seedlings by gas chromatography-mass spectrometry. Phytochemistry 21:525–530
Allen NS, Allen RD (1978) Cytoplasmic streaming in green plants. Annu Rev Biophys Bioeng 7:497–526
Anderson JM,Cormier MJ (1978) Calcium-dependent regulator of NAD kinase in higher plants. Biochem Biophys Res Commun 84:595–602
Anker L (1956) The auxin concentration rule for the geotropism of *Avena* coleoptiles. Acta Bot Neerl 5:335–341
Anker L (1958) The influence of the pH on the growth and geotropism of decapitated *Avena* coleoptiles supplied either with indoleacetic acid or with indoleacetonitrile. Acta Bot Neerl 7:69–76
Anker L (1962) Ortho-geotropism in shoots and coleoptiles. In: Ruhland W (ed) Encyclopedia of plant physiology, vol 17/2. Springer, Berlin Göttingen Heidelberg, pp 103–152
Arslan-Çerim N (1966) The redistribution of radioactivity in geotropically stimulated hypocotyls of *Helianthus annuus* pretreated with radioactive calcium. J Exp Bot 17:236–240
Aspinall D, Paleg LG, Addicott FT (1967) Abscisin II and some hormone-regulated plant responses. Aust J Biol Sci 20:869–882
Astle MC, Rubery PH (1983) Carriers for abscisic acid and indole-3-acetic acid in primary roots: their regional localization and thermodynamic driving forces. Planta 157:53–63
Audus LJ (1969) Geotropism. In: Wilkins MB (ed) Physiology of plant growth and development. McGraw-Hill, London, pp 205–242
Audus LJ (1971) Linkage between detection and the mechanisms establishing differential growth factor concentrations. In: Gordon SA, Cohen MJ (eds) Gravity and the organism. Univ Chicago Press, Chicago, pp 137–151
Audus LJ (1975) Geotropism in roots. In: Torrey J, Clarkson DT (eds) The development and function of roots. Academic Press, London New York, pp 327–363
Audus LJ (1979) Plant geosensors. J Exp Bot 30:1051–1073
Audus LJ (1983) Abscisic acid in root growth and geotropism. In: Addicott FT (ed) Abscisic acid. Praeger, New York, pp 421–477
Avery Jr GS, La Rue GD (1938) Growth and tropic responses of excised *Avena* coleoptiles in culture. Bot Gaz 100:186–199
Bandurski RS (1980) Homeostatic control of concentrations of indole-3-acetic acid. In: Skoog F (ed) Plant growth substances 1979. Springer, Berlin Heidelberg New York, pp 37–49
Bandurski RS, Schultze A (1974) Concentrations of indole-3-acetic acid and its esters in *Avena* and *Zea*. Plant Physiol 54:257–262
Bandurski RS, Schulze A, Dayanandan P, Kaufman PB (1984) Response to gravity by *Zea mays* seedlings. I. Time course of the response. Plant Physiol 74:284–288
Bangerth F (1979) Calcium-related physiological disorders of plants. Annu Rev Phytopathol 17:97–122
Bates GW, Goldsmith MHM (1983) Rapid response of plasma-membrane potential in oat coleoptiles to auxin and other weak acids. Planta 159:231–237

Batra MW, Edwards KL, Scott TK (1975) Auxin transport in roots: its characteristics and relationship to growth. In: Torrey JG and Clarkson DT (eds) The development and function of roots. Academic Press, London New York, pp 299–325

Behrens HM, Weisenseel M, Sievers A (1982) Rapid changes in the pattern of electric current around the root tip of *Lepidium sativum* L. following gravistimulation. Plant Physiol 70:1079–1083

Bendixen LE, Peterson ML (1962) Physiological nature of gene-controlled growth form in *Trifolium fragiferum* L. II. Auxin-gibberellin relationships to growth form. Plant Physiol 37:245–250

Biro RL, Roux SJ (1983) Quantification of calmodulin in isolated organelles from oat by radioimmunoassay and analysis of chlorpromazine binding to calmodulin in vivo. Plant Physiol (Suppl) 72:66

Biro RL, Hale II CC, Wiegand OF, Roux SJ (1982) Effects of chlorpromazine on gravitropism in *Avena* coleoptiles. Ann Bot 50:735–747

Blake TJ, Pharis RP, Reid DM (1980) Ethylene, gibberellins, auxin and the apical control of branch angle in a conifer, *Cupressus arizonica*. Planta 148:64–68

Bluzmanas PI, Dagis IK (1975) Geotropic response of plants. (Review: Merkis AJ). Sov Plant Physiol 22:558–559

Bode HR (1959) Über den Einfluß der Geoinduktion auf die Kationverteilung im Hypokotyl von *Helianthus annuus*. Planta 54:15–33

Borgstrom G (1939) The transverse reactions of plants. CWK Gleerup, Lund

Bose JC (1907) Comparative electro-physiology. Longmans. Green and Co, London

Bouchet M-H, Prat R, Goldberg R (1983) Kinetics of indole acetic acid-induced growth in hypocotyl sections of *Vigna radiata*. Physiol Plant 57:95–100

Bowen MR, Wilkins MB, Cane AR, McCorquodale I (1972) Auxin transport in roots. VIII. The distribution of radioactivity in the tissues of *Zea* root segments. Planta 105:273–292

Boysen-Jensen P (1936) Über die Verteilung des Wuchsstoffes in Keimstengeln und Wurzeln während der phototropischen und geotropischen Krümmung. K Dan Vidensk Selsk Biol Med 13:1–31

Bradford KJ, Yang SF (1980) Stress-induced ethylene production in the ethylene-requiring tomato mutant Diageotropica. Plant Physiol 65:327–330

Brauner L (1926) Über das geoelektrische Phänomen. Kolloidchem Beih Ambronn Festschr 23:143–152

Brauner L (1927) Untersuchungen über das geoelektrische Phänomen. I. Jahrb Wiss Bot 66:381–428

Brauner L (1928) Untersuchungen über das geoelektrische Phänomen. II. Membranstruktur und geoelektrische Effekte. Jahrb Wiss Bot 68:711–770

Brauner L, Appel E (1960) Zum Problem der Wuchsstoff-Querverschiebung bei der geotropischen Induktion. Planta 55:226–234

Brauner L, Böck A (1963) Versuche zur Analyse der geotropischen Perzeption. IV. Untersuchungen über die Auswirkung der Dekapitierung auf den Wuchsstoffgehalt, das Längenwachstum und die geotropische Krümmungsfähigkeit von *Helianthus*-Hypokotylen. Planta 60:109–130

Brauner L, Hager A (1958) Versuche zur Analyse der geotropischen Perzeption. Planta 51:115–147

Brauner L, Zipperer A (1961) Über die Anfangsphasen der geotropischen Krümmungsbewegung von *Avena*-Koleoptilen. Planta 57:503–517

Bridges IG, Wilkins MB (1973) Acid-induced growth and the geotropic response of the wheat node. Planta 114:331–339

Bridges IG, Wilkins MB (1974) The role of reducing sugars in the geotropic response of the wheat node. Planta 117:243–250

Bünning E, Glatzle D (1949) Über die geotropische Erregung. Planta 36:199–202

Bünning E, Reisener HJ, Weygand F, Simon H, Klebe JF (1956) Versuche mit radioaktiver Indolylessigsäure zur Prüfung der sogenannten Ablenkung des Wuchshormonstromes durch Licht. Z Naturforsch 11-B, 363–364

Burg SP (1962) The physiology of ethylene formation. Annu Rev Plant Physiol 13:265–302
Burg SP, Burg EA (1966) The interaction between auxin and ethylene and its role in plant growth. Proc Natl Acad Sci USA 55:262–269
Burg SP, Burg EA (1967a) Lateral auxin transport in stems and roots. Plant Physiol 42:891–893
Burg SP, Burg EA (1967b) Auxin-stimulated ethylene formation: its relationship to auxin-inhibited growth, root geotropism and other plant processes. 6th Int Conf Plant Growth Substances, Ottawa
Caldwell MM (1976) Root extension and water absorption. In: Lange OL, Kappen L, Schulze E-D (eds) Water and plant life. Ecological studies vol 19. Springer, Berlin Heidelberg New York, pp 63–85
Cane AR, Wilkins MB (1969) Independence of lateral and differential longitudinal movement of indoleacetic acid in geotropically stimulated coleoptiles of *Zea mays*. Plant Physiol 44:1481–1487
Chadwick AW, Burg SP (1967) An explanation of the inhibition of root growth caused by indole-3-acetic acid. Plant Physiol 42:415–420
Charbonneau H, Cormier MJ (1979) Purification of plant calmodulin by fluphenazine-sepharose chromatography. Biochem Biophys Res Commun 90:1029–1047
Cheung WY (1980) Calmodulin plays a pivotal role in cellular regulation. Science 207:19–27
Ching TM, Fang SC (1958) The redistribution of radioactivity in geotropically stimulated plants pretreated with radioactive indoleacetic acid. Physiol Plant 11:722–727
Clarkson DT, Hanson JB (1980) The mineral nutrition of higher plants. Annu Rev Plant Physiol 31:239–298
Cleland RE (1972) The dosage-response curve for auxin-induced cell elongation: a reevaluation. Planta (Berl) 104:1–19
Cleland RE (1976) Kinetics of hormone-induced H^+ excretion. Plant Physiol 58:210–213
Cleland RE (1980) Auxin and H^+-excretion: the state of our knowledge. In: Skoog F (ed) Plant growth substances 1979. Springer, Berlin Heidelberg New York, pp 71–78
Cleland RE (1982) The mechanism of auxin-induced proton efflux. In: Wareing PF (ed) Plant growth substances 1982. Academic Press, London New York, pp 23–31
Cleland RE, Rayle DL (1977) Reevaluation of the effect of calcium ions on auxin-induced elongation. Plant Physiol 60:709–712
Clifford PE, Fensom DS, Munt BI, McDowell WD (1982) Lateral stress initiates bending responses in dandelion peduncles: a clue to geotropism? Can J Bot 60:2671–2673
Clifford PE, Reid DM, Pharis RP (1983) Endogenous ethylene does not initiate but may modify geobending – a role for ethylene in autotropism. Plant Cell Environ 6:433–436
Cline MG, Edgerton M, Rehm MM (1974) Accelerated endogenous growth in *Avena* coleoptile segments. Planta 120:213–214
Cohen JD, Schultze A (1981) Double-standard isotope dilution assay. I. Quantitative assay of indole-3-acetic acid. Anal Biochem 112:249–257
Cormier MJ, Anderson JM, Charbonneau H, Jones HP, McCann RO (1980) Plant and fungal calmodulin and the regulation of plant NAD kinase. In: Cheung WY (ed) Calcium and cell function, vol 1. Academic Press, London New York, pp 201–218
Crocker W, Zimmerman PW, Hitchcock AE (1932) Ethylene-induced epinasty of leaves and the relation of gravity to it. Contrib Boyce Thompson Inst 4:177–218
Dayanandan P, Hebard FV, Kaufman PB (1976) Cell elongation in the grass pulvinus in response to geotropic stimulation and auxin application. Planta 131:245–252
Daye S, Biro RL, Roux SJ (1984) Inhibition of gravitropism in oat coleoptiles by the calcium chelator, ethyleneglycol-*bis*-(β-aminoethylether)-N,N′-tetraacetic acid. Physiol Plant 61:449–454
De Guzman CG, Dela Fuente RK (1981) The polarity of calcium movement in young stems and its possible significance to the basipetal secretion of IAA. Plant Physiol (Suppl) 67:2
De Guzman CG, Dela Fuente RK (1982) Further studies on the auxin-promoted acropetal calcium efflux. Plant Physiol (Suppl) 69:57

De Wit JL (1957) Growth substance relations in the *Avena* coleoptile, studied by means of the geotropic response. Acta Bot Neerl 6:1–45

Dela Fuente RK (1981) Is the basipetal secretion of IAA a manifestation of the stimulation-secretion coupling? Plant Physiol (Suppl) 67:2

Dela Fuente RK (1982) The promotion of auxin transport by calcium chelators. Plant Physiol (Suppl) 69:56

Dela Fuente RK, Leopold AC (1968) Geotropism and the lateral movement of auxin. In: Wightman F and Setterfield G (eds) Biochemistry and physiology of plant growth substances. Runge, Ottawa, pp 1129–1143

Dela Fuente RK, Leopold AC (1973) A role for calcium in auxin transport. Plant Physiol 51:845–847

De Lorenzo RJ (1982) Calmodulin in synaptic function and neurosecretion. In: Cheung WY (ed) Calcium and cell function. Academic Press, London New York, pp 272–309

Diehl JM, Gorter CJ, Van Itersen G Jr, Kleinhoonte A (1939) The influence of growth hormone on hypocotyls of *Helianthus* and the structure of their cell walls. Rec Trav Bot Neerl 36:709–798

Dieter P, Marmé D (1980) Calmodulin activation of plant microsomal Ca^{2+} uptake. Proc Natl Acad Sci USA 77:7311–7314

Dieter P, Marmé D (1981a) A calmodulin-dependent, microsomal ATPase from corn (*Zea mays* L). FEBS Lett 125:245–248

Dieter P Marmé D (1981b) Far-red irradiation of intact corn seedlings affects mitochondrial and calmodulin-dependent microsomal Ca^{2+} transport. Biochem Biophys Res Commun 101:749–755

Digby J, Firn RD (1979) An analysis of the changes in growth rate occurring during the initial stages of geocurvature in shoots. Plant Cell Environ 2:145–148

Digby J, Firn RD (1976) A critical assessment of the Cholodny-Went theory of shoot geotropism. Curr Adv Plant Sci 25:953–960

Digby J, Firn, RD, Carrington S (1982) Studies on the differential growth causing tropic curvatures in shoots. In: Wareing PF (ed 1982) Plant growth substances. Academic Press, London New York, pp 519–528

Djikman MJ (1933) A quantitative analysis of the geotropical curvature in dicotyledons. Proc K Ned Akad Wet Amsterdam 36:748–758

Dohrmann U, Hertel R, Kowalik H (1978) Properties of auxin-binding sites in different subcellular fractions from maize coleoptiles. Planta 140:97–106

Dolk HE (1936) Geotropism and the growth substance (translated by Dolk-Hoek F, Thimann KV from Geotropie en Groeistof, 1930). Rec Trav Bot Neerl 33:509–585

Dorée M, Picard A (1980) Release of Ca^{2+} from intracellular pools stops cytoplasmic streaming in *Tradescantia* staminal hairs. Experientia 36:1291–1292

Durand H, Rayle DL (1973) Physiological evidence for auxin-induced hydrogen-ion secretion and the epidermal paradox. Planta 114:185–193

Durham ACH (1974) A unified theory of the control of actin and myosin in nonmuscle movements. Cell 2:123–136

Edwards KL (1983) Calcium elicited asymmetric auxin transport in gravity influenced root segments. In: Halstead T (ed) Program and abstracts of the annual symposium of the NASA space biology program, Arlington. NASA Conf Publ 2299, pp 10–11

Edwards KL (1984) Asymmetric calcium and auxin in root gravitropism. Plant Physiol Suppl 75:170

El-Antably HMM (1975) Redistribution of endogenous indoleacetic acid, abscisic acid and gibberellins in geotropically stimulated *Ribes nigrum* shoots. Physiol Plant 34:167–170

El-Nadi AH, Brouwer R, Locker JT (1969) Some responses of the root and the shoot of *Vicia faba* plants to water stress. Neth J Agric Sci 17:133–142

Ellis HW, Turner ER (1978) The effect of electricity on plant growth. Sci Prog 65:395–408

Epstein E, Cohen JD (1981) Microscale preparation of pentafluorobenzyl esters. Electron-capture gas chromatographic detection of indole-3-acetic acid from plants. J Chromatog 209:413–420

Evans ML (1967) Kinetic studies of the cell elongation phenomenon in *Avena* coleoptile segments. Ph D Thesis, Univ California, Santa Cruz

Evans ML (1973) Rapid stimulation of plant cell elongation by hormonal and non-hormonal factors. Bioscience 23:711–718
Evans ML, Lee J (1984) Relationship of calcium asymmetry to auxin movement in roots of maize. Plant Physiol Suppl 75:177
Evans ML, Schmitt MR (1975) The nature of spontaneous changes in growth rate in isolated coleoptile segments. Plant Physiol 55:757–762
Evans ML, Vesper MJ (1980) An improved method for detecting auxin-induced hydrogen ion efflux from corn coleoptile segments. Plant Physiol 66:561–565
Evers J (1947) Polar bioelectric fields and polar growth under conditions of applied electric fields. History of the problem. In: Lund EJ (ed) Bioelectric fields and growth. Univ Texas Press, Austin, pp 151–161
Evers J, Lund EJ (1947) Reversible decrease in rate of growth by an electric current applied to small region of the root. In: Lund EJ (ed) Bioelectric fields and growth. Univ Texas Press, Austin, pp 197–217
Fabian A (1969) Contributions to the study of geotropism with special reference to the variation of SH-groups in the curvatures. Flora A-160:479–492
Feldman LJ (1981a) Root cap inhibitor formation in isolated root caps of *Zea mays*. J Exp Bot 32:779–788
Feldman LJ (1981b) Light-induced inhibitors from intact and cultured caps of *Zea* roots. Planta 153:471–475
Feldman LJ (1982) Formation and partial characterization of growth inhibitors from cultured and intact root caps. Ann Bot 50:747–756
Feldman LJ (1983a) Light-enhanced protein synthesis in gravitropically stimulated root caps of corn. Plant Physiol 72:833–836
Feldman LJ (1983b) Protein and carotenoid synthesis and turnover in gravistimulated caps. In: Halstead T (ed) Program and abstracts of the annual symposium of the NASA space biology program, Arlington. NASA Conf Publ 2299, p 13
Feldman LJ (1984) Regulation of root development. Annu Rev Plant Physiol 35:223–242
Feldman LJ, Gildow V (1984) Effects of light on protein patterns in gravitropically stimulated root caps of corn. Plant Physiol 74:208–212
Filner B, Hertel R with participation by Steele C, Fan V (1970) Some aspects of geotropism in coleoptiles. Planta 82:123–144
Firn RD (1983) The involvement of gibberellins in geotropism and phototropism. In: Crozer A (ed) The biochemistry and physiology of gibberellins vol 2, pp 375–394
Firn RD, Digby J (1976) Shoot geotropism: new interpretations of old observations. What's New in Plant Physiol 8:1–3
Firn RD, Digby J (1977) The role of the peripheral cell layers in the geotropic curvature of sunflower hypocotyls: a new model of shoot geotropism. Aust J Plant Physiol 4:337–347
Firn RD, Digby J (1979) A study of the autotropic straightening reaction of a shoot previously curved during geotropism. Plant Cell Environ 2:149–154
Firn RD, Digby J (1980) The establishment of tropic curvatures in plants. Annu Rev Plant Physiol 31:131–148
Firn RD, Digby J, Pinsent C (1977) Evidence against the involvement of gibberellins in the differential growth which causes geotropic curvature. Z Pflanzenphysiol 82:179–185
Firn RD, Digby J, Riley H (1978) Shoot geotropic curvature – the location, magnitude and kinetics of the gravity-induced differential growth in horizontal sunflower hypocotyls. Ann Bot 42:465–468
Firn RD, Digby J, Hall A (1981) The role of the shoot apex in geotropism. Plant Cell Environ 4:125–129
Forde J, Steer MW (1976) Cytoplasmic streaming in *Elodea*. Can J Bot 54:2668–2694
Franklin D, Morgan PW (1978) Rapid production of auxin-induced ethylene. Plant Physiol 62:161–162
Franssen JM, Digby J, Firn RD (1982) An investigation of the importance of transpiration gradients in the control of phototropic curvature in coleoptiles of *Avena sativa*. Aust J Plant Physiol 9:583–586

Gaither DH, Lutz DH, Forrence LE (1975) Abscisic acid stimulates elongation of excised pea root tips. Plant Physiol 55:948–949
Ganot D, Reinhold L (1970) The "acid growth effect" and geotropism. Planta 95:62–71
Gillespie BC (1959) An hypothesis for the geotropic curvature of plants. MA Diss, Stanford Univ, Stanford, CA
Gillespie B, Briggs WR (1961) Mediation of geotropic response by lateral transport of auxin. Plant Physiol 36:364–368
Gillespie B, Thimann KV (1961) The lateral transport of IAA-C^{14} in geotropism. Experientia 17:126–129
Gillespie B, Thimann KV (1963) Transport and distribution of auxin during tropistic response. I. The lateral migration of auxin in geotropism. Plant Physiol 38:214–225
Goeschl JD, Rappaport L, Pratt HK (1966) Ethylene as a factor regulating the growth of pea epicotyls subjected to physical stress. Plant Physiol 41:877–885
Goldsmith MHM (1977) The polar transport of auxin. Annu Rev Plant Physiol 28:439–478
Gordon SA, Eib M (1956) Auxin transport in the phototropic response. Plant Physiol 31:xiv
Goswami KKA, Audus LJ (1976) Distribution of calcium, potassium and phosphorus in *Helianthus annuus* hypocotyls and *Zea mays* coleoptiles in relation to tropic stimuli and curvature. Ann Bot 40:49–64
Gotô N, Esashi Y (1974) Stimulation by ethylene of axis and hypocotyl growth in bean and cocklebur seedlings. Physiol Plant 31:204–210
Gougler JA, Evans ML (1981) Adaptation of corn roots to exogenously applied auxin. Physiol Plant 51:394–398
Grahm L (1964) Measurements of geoelectric and auxin-induced potentials in coleoptiles with a refined vibrating electrode technique. Physiol Plant 17:231–261
Grahm L, Hertz CH (1962) Measurements of the geoelectric effect in coleoptiles by a new technique. Physiol Plant 15:96–114
Grahm L, Hertz CH (1964) Measurements of the geoelectric effect in coleoptiles. Physiol Plant 17:186–201
Green PB (1980) Organogenesis – a biophysical point of view. Annu Rev Plant Physiol 31:51–82
Greengard P (1978) Phosphorylated proteins as physiological effectors. Science 199:146–152
Greenwood MS, Shaw S, Hillman JR, Ritchie A, Wilkins MB (1972) Identification of auxin from *Zea* coleoptile tips by mass spectrometry. Planta (Berl) 108:179–183
Greenwood MS, Hillman JR, Shaw S, Wilkins MB (1973) Localization and identification of auxin in roots of *Zea mays*. Planta (Berl) 109:369–374
Gressel J, Horwitz B (1982) Gravitropism and phototropism. In: Smith H, Grierson D (eds) The molecular biology of plant development. Blackwell, Oxford, pp 405–433
Gundel W (1933) Chemische und physikalisch-chemische Vorgänge bei geischer Induktion. Jahrb Wiss Bot 78:623–664
Günther-Massias M (1928) Über die Gültigkeit des Reizmengengesetzes bei der Summation unterschwelliger Reize. Z Bot 21:129–172
Guy M, Kende H (1984a) Ethylene production in *Pisum sativum* and *Vicia faba* protoplasts. Planta 160:276–280
Guy M, Kende H (1984b) Conversion of 1-aminocyclopropane-1-carboxylic acid to ethylene by isolated vacuoles of *Pisum sativum* L. Planta 160:281–287
Hahne I (1961) Untersuchungen zum Problem der Beteiligung des Wuchsstoffes an der geotropischen Induktion bei Koleoptilen. Planta 57:557–582
Halevy AH, Ashri A, Ben-Tal Y (1969) Peanuts: gibberellin antagonists and genetically controlled differences in growth habit. Science 164:1397–1398
Hall AB, Firn RD, Digby J (1980) Auxins and shoot tropisms – a tenuous connection. J Biol Educ 14:195–199
Hall WC, Truchelut GB, Leinweber CL, Herrero FA (1957) Ethylene production by the cotton plant and its effects under experimental and field conditions. Physiol Plant 10:306–317

Hansen DJ, Bendixen LE (1974) Ethylene-induced tropism of *Trifolium fragiferum* L stolons. Plant Physiol 53:80–82
Harary HH, Brown JE (1984) Spatially nonuniform changes in intracellular calcium ion concentrations. Science 224:292–294
Harris P (1978) Triggers, trigger waves, and mitosis: a new model. In: Jeter JR Jr, Cameron IL, Padilla GM (eds) Cell cycle regulation. Academic Press, London New York, pp 75–104
Harrison M, Pickard BG (1984) Burst of ethylene upon horizontal placement of tomato seedlings. Plant Physiol 75:1167–1169
Hartung W (1976) Der basipetale [2-^{14}C] Abscisinsäuretransport in Wurzeln intakter Bohnenkeimlinge und seine Bedeutung für den Wurzelgeotropismus. Planta 128:59–62
Hatfield RD, LaMotte CE (1984) IAA-induced growth responses of decapitated corn seedlings. Indications of two apparent adaptations with a possible role in gravitropism. Plant Physiol 74:302–306
Hayama T, Shimmen T, Tazawa M (1979) Participation of Ca^{2+} in cessation of cytoplasmic streaming induced by membrane excitation in *Characeae* internodal cells. Protoplasma 99:305–321
Hayes AB (1981) The interaction of auxin and ethylene in the maintenance of leaf blade form in *Phaseolus vulgaris* L var Pinto. Am J Bot 68:733–740
Hayes AB (1982) Gravitropic basis of leaf blade nastic curvatures. The Physiologist (Suppl) 25:133
Hayes AB Lippincott JA (1976) Growth and gravitational response in the development of leaf blade hyponasty. Am J Bot 63:383–387
Heathcote DG (1981) The geotropic reaction and statolith movements following geostimulation of mung bean hypocotyls. Plant Cell Environ 4:131–140
Hebard FV, Amatangelo SJ, Dayanandan P, Kaufman PB (1976) Studies on acidification of media by *Avena* stem segments in the presence and absence of gibberellic acid. Plant Physiol 58:670–674
Hepler PK, Palevitz BA (1974) Microtubules and microfilaments. Annu Rev Plant Physiol 25:309–362
Hertel R (1983) The mechanism of auxin transport as a model for auxin action. Z Pflanzenphysiol 112:53–67
Hertel R, Leopold AC (1963) Auxin relations in geotropism of corn coleoptiles. Naturwissenschaften 22:695–696
Hertel R, dela Fuente RK, Leopold AC (1969) Geotropism and the lateral transport of auxin in the corn mutant amylomaize. Planta 88:204–214
Hild V (1977) Wirkung von Vorbestrahlung mit Rot- und Blaulicht auf die geotropische Empfindlichkeit von Maiskoleptilen. Planta 133:309–314
Hild V, Hertel R (1972) Initial phases of gravity-induced lateral auxin transport and geotropic curvature in corn coleoptiles. Planta 108:245–258
Hoffman GJ, Rawlins SL, Garber MJ, Cullen EM (1971) Water relations and growth of cotton as influenced by salinity and relative humidity. Agron J 63:822–826
Humphry VR (1966) The effects of paraffin oil on phototropic and geotropic responses in *Avena* coleoptiles. Ann Bot NS 30:39–45
Iwami S, Masuda Y (1973) Hydrogen-ion induced curvature in cucumber hypocotyls. Plant Cell Physiol 14:757–762
Iwami S, Masuda Y (1974) Geotropic response of cucumber hypocotyls. Plant Cell Physiol 5:121–129
Iwami S, Masuda Y (1976) Distribution of labeled auxin in geotropically stimulated stems of cucumber and pea. Plant Cell Physiol 17:227–237
Jackson MB (1979) Is the diageotropic tomato ethylene deficient? Physiol Plant 46:347–351
Jackson MB, Barlow PW (1981) Root geotropism and the role of growth regulators from the cap: a re-examination. Plant Cell Environ 4:107–123
Jacobs M, Gilbert SF (1983) Basal localization of the presumptive auxin transport carrier in pea stem cells. Science 220:1297–1300

Jacobs M, Ray PM (1976) Rapid auxin-induced decrease in free space pH and its relationship to auxin-induced growth in maize and pea. Plant Physiol 58:203–209

Jacobs M, Taiz L (1980) Vanadate inhibition of auxin-enhanced H^+ secretion and elongation in pea epicotyls and oat coleoptiles. Proc Natl Acad Sci USA 77:7242–7246

Jaffe LF, Nuccitelli R (1977) Electrical controls of development. Annu Rev Biophys Bioeng 6:445–476

Jaffe MJ (1980) Morphogenetic responses of plants to mechanical stimuli or stress. Bioscience 30:241–243

Jaffe MJ, Biro R (1979) Thigmomorphogenesis: the effect of mechanical perturbation on the growth of plants, with special reference to anatomical changes, the role of ethylene, and interaction with other environmental stresses. In: Mussell H, Staples RC (eds) Stress physiology in crop plants. Wiley and Sons, New York, pp 25–69

Jaffe MJ, Leopold AC (1984) Callose deposition during gravitropism of *Zea mays* and *Pisum sativum* and its inhibition by 2-deoxy-D-glucose. Planta 161:20–26

Janistyn B (1973) Indol-3-essigsäure induzierte Calciumionenabgabe bei Maiskoleoptilzylindern. Z Naturforsch 28:777–778

Johnsson A (1967) Relationships between photo-induced and gravity-induced electrical potentials in *Zea mays*. Physiol Plant 20:562–579

Johnsson A (1971) Aspects on gravity-induced movements in plants. Q Rev Biophys 4:277–320

Johnsson A, Pickard BG (1979) The threshold stimulus for geotropism. Physiol Plant 45:315–319

Juniper BE, Groves S, Landau-Schachar B, Audus LJ (1966) Root cap and the perception of gravity. Nature 209:93–94

Kaldewey H (1962) Plagio- und Diageotropismus der Sprosse und Blätter, einschließlich Epinastie, Hyponastie, Entfaltungsbewegungen. In: Ruhland W (ed) Encyclopedia of plant physiology, vol 17/2. Springer, Berlin Göttingen Heidelberg, pp 200–321

Kaldewey H (1971) Geoepinasty, an example of gravimorphism. In: Gordon SA, Cohen MJ (eds) Gravity and the organism. Univ Chicago Press, Chicago London, pp 333–339

Kamienska A, Pharis RP, Wample RL, Kuo CC, Durley RC (1974) Gibberellins in conifers. In: Plant growth substances 1973. Hirokawa, Tokyo, pp 305–313

Kamiya N (1981) Physical and chemical basis of cytoplasmic streaming. Annu Rev Plant Physiol 32:205–236

Kang BG (1979) Epinasty. In: Feinleib ME, Haupt W (eds) Encyclopedia of plant physiology, new ser vol 7. Springer, Berlin Heidelberg New York, pp 645–667

Kang BG, Burg SP (1972) Relation of phytochrome-enhanced geotropic sensitivity to ethylene production. Plant Physiol 50:132–135

Kang BG, Burg SP (1974) Ethylene action on lateral auxin transport in tropic responses, leaf epinasty, and horizontal mutation. In: Plant growth substances 1973. Hirokawa, Tokyo, pp 1090–1094

Kang BG, Newcomb W, Burg SP (1971) Mechanism of auxin-induced ethylene production. Plant Physiol 47:504–509

Katz B (1966) Nerve, muscle and synapse. McGraw-Hill, New York

Kaufman PB (1965) The effects of growth substances on intercalary growth and cellular differentiation in developing internodes of *Avena sativa*. II. The effects of gibberellic acid. Physiol Plant 18:703–724

Kaufman PB (1967) Role of gibberellins in the control of intercalary growth and cellular differentiation in developing *Avena* internodes. Ann NY Acad Sci 144:191–203

Kaufman PB, Bandurski R, Dayanandan P, Koning R, Harrison M (1979) Hormonal regulation of the vegetative geotropic response in *Avena* shoots. Plant Physiol (Suppl) 63:143

Kende H, Boller T (1981) Wound ethylene and 1-aminocyclopropane-1-carboxylate synthase in ripening tomato fruit. Planta 151:476–481

Kende H, Hanson AD (1976) Relationship between ethylene evolution and senescence in morning-glory flower tissue. Plant Physiol 57:523–527

Kikuyama M, Tazawa M (1982) Ca^{2+} ion reversibly inhibits the cytoplasmic streaming of *Nitella*. Protoplasma 113:241–243

Klemmer R, Schneider HA (1979) On a blue light effect and phytochrome in the stimulation of georesponsiveness of maize roots. Z Pflanzenphysiol 95:189–197
Kohji J, Hagimoto H, Yamamoto R, Masuda Y (1982) IAA transport and georeaction in the flower stalk of a poppy, *Papaver rhoeas* L. Plant Cell Physiol 23:1329–1336
Köhler D (1978) „Reizleitung" beim negativen Geotropismus des Sprosses? Z Pflanzenphysiol 86:269–272
Köhler D, Daum M (1979) Autonomy of the parts of a shoot during the geotropic response. Z Pflanzenphysiol 95:91–94
Konings H (1964) On the indoleacetic acid-converting enzyme of pea roots and its relation to geotropism, straight growth and cell wall properties. Acta Bot Neerl 13:566–622
Konings H (1968) Auxin in root geotropism. In: Vardar Y (ed) Transport of plant hormones. North Holland, Amsterdam, pp 237–250
Konings H, Jackson MB (1979) A relationship between rates of ethylene production by roots and the promoting or inhibiting effects of exogenous ethylene and water on root elongation. Z Pflanzenphysiol 92:385–397
Larsen P (1962a) Geotropism. An introduction. In: Ruhland W (ed) Encyclopedia of plant physiology, vol 17/2. Springer, Berlin Göttingen Heidelberg, pp 153–199
Larsen P (1962b) Orthogeotropism in roots. In: Ruhland W (ed) Encyclopedia of plant physiology, vol 17/2. Springer, Berlin Göttingen Heidelberg, pp 34–73
Lau O-L, Yang SF (1973) Mechanisms of a synergistic effect of kinetin on auxin-induced ethylene production: suppression of auxin conjugation. Plant Physiol 51:1011–1014
Lee JS, Mulkey TJ, Evans ML (1983a) Gravity-induced polar transport of calcium across root tips of maize. Plant Physiol 73:874–876
Lee JS, Mulkey TJ, Evans ML (1983b) Reversible loss of gravitropic sensitivity in maize roots after tip application of calcium chelators. Science 220:1375–1376
Lee JS, Mulkey TJ, Evans ML (1984) Inhibition of polar calcium movement and gravitropism in roots treated with auxin-transport inhibitors. Planta 160:536–543
Leopold AC (1955) Auxins and plant growth. Univ California Press, Berkeley, Los Angeles
Leopold AC, Noodén L (1984) Hormonal regulatory systems in plants. In: Scott TK (ed) Hormonal regulation of development. II. From the cell to the whole plant. Encyclopedia of plant physiology, new ser vol 10. Springer, Berlin Heidelberg New York
Leopold AC, Brown KM, Emerson FH (1972) Ethylene in the wood of stressed trees. Hortic Sci 7:175
Libbert E, Gerdes I (1964) Können Gibberelline an tropistischen Krümmungen beteiligt sein? Planta 61:245–258
Lieberman M (1979) Biosynthesis and action of ethylene. Annu Rev Plant Physiol 30:533–591
Lin W, Wagner GJ, Siegelman HW, Hind G (1977) Membrane-bound ATPase of intact vacuoles and tonoplasts isolated from mature plant tissue. Biochim Biophys Acta 465:110–117
Lintilhac PM, Vesecky TB (1981) Mechanical stress and cell-wall orientation in plants. II. The application of controlled directional stress to growing plants; with a discussion on the nature of the wound reaction. Am J Bot 68:1222–1230
Lintilhac PM, Vesecky TB (1984) Stress-induced alignment of division plane in plant tissues grown in vitro. Nature 307:363–364
Little CHA, Goldsmith MHM (1967) Effect of inversion on growth and movement of indole-3-acetic acid in coleoptiles. Plant Physiol 42:1239–1245
Lund EJ, Berry LJ (1947) Polarizable and "non-polarizable" electrochemical systems in the polar cell. Their significance in ion transport and growth. In: Lund EJ (ed) Bioelectric fields and growth. Univ Texas Press, Austin, pp 161–186
Lund EJ, Mahan RI, Hanszen AH (1947) Reversible total inhibition of polar growth in the root by an applied electric current and the nature of this control. In: Lund EJ (ed) Bioelectric fields and growth. Univ Texas Press, Austin, pp 186–197
Lyon CJ (1968) Plagiotropism and auxin transport. In: Vardar Y (ed) Transport of plant hormones. North-Holland, Amsterdam, pp 251–290

MacDonald IR, Hart JW, Gordon DC (1983) Analysis of growth during geotropic curvature in seedling hypocotyls. Plant Cell Environ 6:401–406

Magnus V, Bandurski RS, Schultze A (1980) Synthesis of 4, 5, 6, 7 and 2, 4, 5, 6, 7 deuterium-labeled indole-3-acetic acid for use in mass spectrometric assays. Plant Physiol 66:775–581

Marmé D, Dieter P (1983) Role of Ca^{2+} and calmodulin in plants. In: Cheung WY (ed) Calmodulin and cell function. Academic Press, London New York, pp 263–311

McArthur JA, Briggs WR (1979) Effect of red light on geotropism in pea epicotyls. Plant Physiol 63:218–220

McDougall J, Hillman JR (1978) Analysis of indole-3-acetic acid using GC-MS techniques. In: Hillman JR (ed) Isolation of plant growth substances. Cambridge Univ Press, London, pp 1–25

McIntyre GI (1980) The role of water distribution in plant tropisms. Aust J Plant Physiol 7:401–413

McIntyre GI (1983) Evidence of transpiration-induced curvatures in the rhizome of *Agropyron repens* L Beauv. Ann Bot 51:395–398

McWha JA, Jackson DL (1976) Some growth promotive effects of abscisic acid. J Exp Bot 27:1004–1008

Merkis AI (1973) Geotropic response of plants (in Russian). Vilnius, Mintis

Mertens R, Weiler EW (1983) Kinetic studies on the redistribution of endogenous growth regulators in gravireacting plant organs. Planta 158:339–348

Metzner P (1934) Zur Kenntnis der Stoffwechseländerungen bei geotropisch gereizten Keimpflanzen. Ber Dtsch Bot Ges 52:506–522

Meudt WJ, Bennet HW (1978) Rapid bioassay for auxin. Physiol Plant 44:422–428

Migliaccio F, Rayle D (1983) Relationship between H^+ efflux, auxin redistribution, and shoot gravitropism. Plant Physiol (Suppl) 72:90

Miledi R (1973) Transmitter release induced by injection of calcium ions into nerve terminals. Proc R Soc Lond 183:421–425

Minina EG, Muratov YM, Belyaev VV (1976) Gibberellin-like substances in naturally growing and geoinduced shoots of Siberian fir. Sov Plant Physiol 23:165–167

Mizuno A, Okamoto H (1982) Structure and function of the elongation sink in the stems of higher plants. II. Effects of acids and anoxia on the growth rate and the spatially separate electrogenic ion pumps. Plant Cell Environ 5:131–135

Mizuno A, Katou K, Okamoto H (1980) Structure and function of the elongation sink in the stems of higher plants. I. Effects of anoxia and IAA on the growth rate and the spatially separate electrogenic ion pumps. Plant Cell Physiol 21:395–403

Montaldi ER (1969) Gibberellin – sugar interaction regulating the growth habit of Bermudagrass (*Cynodon dactylon* L Pers). Experientia 25:91–92

Moore R, Smith JD (1984) Growth, graviresponsiveness and abscisic-acid content of *Zea mays* seedlings treated with fluridone. Planta 162:342–344

Morath M (1972) Some early effects of auxin on geotropic exposure of coleoptiles. In: Kaldewey H, Vardar Y (eds) Hormonal regulation in plant growth and development. Verlag Chemie, Weinheim, pp 377–381

Morath M, Hertel R (1978) Lateral electrical potential following asymmetric auxin application to maize coleoptiles. Planta 140:31–35

Morris DA (1980) The influence of small direct electric currents on the transport of auxin in intact plants. Planta 150:431–434

Mulkey T (1979) Physiological and anatomical comparisons between the diageotropica mutant of tomato (*Lycopersicon esculentum* Mill) and its isogenic parent variety VFN-8. MS Thesis, Univ Arkansas, Fayetteville

Mulkey TJ, Evans ML (1981) Geotropism in corn roots: evidence for its mediation by differential acid efflux. Science 212:70–71

Mulkey TJ, Evans ML (1982) Suppression of asymmetric acid efflux and gravitropism in maize roots treated with auxin transport inhibitors or sodium orthovanadate. J Plant Growth Regul 1:259–265

Mulkey TJ, Kuzmanoff KM, Evans ML (1981a) Promotion of growth and shift in the auxin dose/response relationship in maize roots treated with ethylene biosynthesis inhibitors aminoethoxyvinylglycine and cobalt. Plant Sci Lett 25:43–48

Mulkey TJ, Kuzmanoff KM, Evans ML (1981b) Correlations between proton-efflux patterns and growth patterns during geotropism and phototropism in maize and sunflower. Planta 152:239–241

Mulkey TJ, Kuzmanoff KM, Evans ML (1982) Promotion of growth and hydrogen ion efflux by auxin in roots of maize pretreated with ethylene biosynthesis inhibitors. Plant Physiol 70:186–188

Mulkey TJ, Evans ML, Kuzmanoff KM (1983) The kinetics of abscisic acid action on root growth and gravitropism. Planta 157:150–157

Naqvi SM, Gordon SA (1966) Auxin transport in *Zea mays* L coleoptiles. I. Influence of gravity on the transport of indoleacetic acid-2-^{14}C. Plant Physiol 41:1113–1118

Neljubow D (1901) Über die horizontale Nutation der Stengel von *Pisum sativum* und einiger anderer Pflanzen. Beih Bot Zentralbl 10:128–138

Neljubow D (1911) Geotropismus in der Laboratoriumsluft. Ber Dtsch Bot Ges 29:97–112

Newman IA (1963) Electric potentials and auxin translocation in *Avena*. Aust J Biol Sci 16:629–646

Osborne DJ (1974) Auxin, ethylene and the growth of cells. In: Bieleski RL, Ferguson AR, Cresswell MM (eds) Mechanisms of regulation of plant growth. Bulletin 12. R Soc N Z, Wellington, pp 645–654

Ouitrakul R, Hertel R (1969) Effect of gravity and centrifugal acceleration on auxin transport in corn coleoptiles. Planta 88:233–243

Palmer JA (1985) Epinasty, hyponasty and related topics. In: Pharis RP, Reid DM (eds) Hormonal regulation of development. III. Role of environmental factors. Encyclopedia of plant physiology, new ser vol 11. Springer, Berlin Heidelberg New York, pp 139–168

Pearce D, Penny D (1983) Tissue interactions in indoleacetic acid-induced rapid elongation of lupin hypocotyls. Plant Sci Lett 30:347–353

Pearson RW (1966) Soil environment and root development. In: Pierce WH, Kirkham D, Pesek J, Shaw R (eds) Plant environment and efficient water use. Am Soc Agron Soil Sci Soc Am, Madison, pp 95–126

Pengelly WL, Bandurski RS, Schultze A (1981) Validation of a radioimmunoassay for indole-3-acetic acid using gas chromatography-selected ion monitoring-mass spectrometry. Plant Physiol 68:96–98

Perbal G (1978) The mechanism of geoperception in lentil roots. J Exp Bot 29:631–638

Pfeffer WFP (1906) The physiology of plants, vol 2, pp 61–67 and vol 3, p 66 (translated by Ewart EH from Pflanzenphysiologie, 2nd edn, vol 2, 1900) Clarendon, Oxford London

Pharis RP, Kuo CG (1977) Physiology of gibberellins in conifers. Can J For Res 7:299–325

Pharis RP, Kuo CG, Glenn JL (1972) Gibberellins, a primary determinant in the expression of apical dominance, apical control and geotropic movement of conifer shoots. In: Carr DJ (ed) Plant growth substances 1970. Springer, Berlin Heidelberg New York, pp 441–448

Pharis RP, Legge RL, Noma M, Kaufman PB, Ghosheh NS, Lacroix JD, Heller K (1981) Changes in endogenous gibberellins and the metabolism of (^{3}H)-GA$_4$ after geostimulation in shoots of the oat plant (*Avena sativa*). Plant Physiol 67:892–897

Phillips IDJ (1972) Endogenous gibberellin transport and biosynthesis in relation to geotropic induction of excised sunflower shoot-tips. Planta 105:234–244

Phillips IDJ, Hartung W (1976) Longitudinal and lateral transport of [3,4 – ^{3}H]-gibberellin A$_1$ and 3-indolyl (acetic acid-2-^{14}C) in upright and geotropically responding green internode segments from *Helianthus annuus*. New Phytol 76:1–9

Pickard BG (1970) Comparison of calcium and lanthanon ions in the *Avena*-coleoptile growth test. Planta 91:314–320

Pickard BG (1971) Action potentials resulting from mechanical stimulation of pea epicotyls. Planta 97:106–115

Pickard BG (1973a) Geotropic response patterns of the *Avena* coleoptile. I. Dependence on angle and duration of stimulation. Can J Bot 51:1003–1021

Pickard BG (1973b) Geotropic response patterns of the *Avena* coleoptile. II. Induction at low temperature. Can J Bot 51:1023–1027
Pickard BG, with participation by Dutson K, Harrison V, Donegan E (1969) Second positive phototropic response patterns of the oat coleoptile. Planta 88:1–33
Pickard BG (1983) Small gravitationally elicited voltage transients in pea stems. In: Halstead T (ed) Program and abstracts of the annual symposium of the NASA Space biology program, Arlington. NASA Conf Publ 2299, pp 14–15
Pickard BG (1984) Voltage transients elicited by sudden step-up of auxin. Plant Cell Environ 7:171–178
Pickard BG, Thimann KV (1966) Geotropic response of wheat coleoptiles in absence of amyloplast starch. J Gen Physiol 49:1065–1086
Pickard WF (1972) Further observations on cytoplasmic streaming in *Chara braunii*. Can J Bot 50:703–711
Pilet PE (1977) Root georeaction: gravity effect on hormone balance. In: Burke WR, Guyenne TD (eds) Life sciences research in space. ESA Sci Tech Publ Branch, Nordwijk, The Netherlands, pp 223–227
Pilet PE, Rivier L (1980) Light and dark georeaction of maize roots: effect and endogenous level of abscisic acid. Plant Sci Lett 18:201–206
Pilet PE, Rivier L (1981) Abscisic acid distribution in horizontal maize root segments. Planta 153:453–458
Poovaiah BW, Leopold AC (1973) Effects of ethephon on the growth of grasses. Crop Sci 13:755–758
Pratt HK, Goeschl JD (1969) Physiological roles of ethylene in plants. Annu Rev Plant Physiol 20:541–584
Railton ID, Phillips IDJ (1973) Gibberellins and geotropism in *Zea mays* coleoptiles. Planta 109:121–126
Rasmussen H (1983) Pathways of amplitude and sensitivity modulation in the calcium messenger system. Cheung WY (ed) Calcium and cell function, vol 4. Academic Press, London New York, pp 1–61
Raven JA, Rubery PH (1982) Coordination of development: hormone receptors, hormone action, and hormone transport. In: Smith H, Grierson D (eds) The molecular biology of plant development. Blackwell, Oxford, pp 28–48
Rawitscher F (1932) Der Geotropismus der Pflanzen. Fischer, Jena
Rayle DL, Evans ML, Hertel R (1970) Action of auxin on cell elongation. Proc Natl Acad Sci USA 65:184–191
Reches S, Lesham Y, Wurzburger J (1974) On hormones and weeping: asymmetric hormone distribution and the pendulous growth habit of the weeping mulberry, *Morus alba* var pendula. New Phytol 73:841–846
Reinhold L (1978) Phytohormones and the orientation of growth. In: Letham DS, Goodwin PB, Higgins TVJ (eds) Phytohormones and related compounds – a comprehensive treatise, vol 2. Elsevier/North-Holland Biomedical Press, Amsterdam, pp 251–289
Reisener HJ (1952) Versuche zum Geotropismus mit radioaktiver β-Indolylessigsäure. Naturwissenschaften 44:120
Reisener HJ (1958) Untersuchungen über den Phototropismus der Hafer-Koleoptile. Z Bot 46:474–505
Reisener HJ Simon H (1960) Weitere Geotropismus-Versuche mit radioaktiver β-Indolylessigsäure. Z Bot 48:66–70
Robert ML, Taylor HF, Wayne RL (1975) Ethylene production by cress roots and excised cress root segments and its inhibition by 3,5-diiodo-4-hydroxybenzoic acid. Planta 126:273–284
Roux SJ (1984) Ca^{2+} and phytochrome action in plants. Bioscience 34:25–29
Roux SJ, Slocum RD (1982) Role of calcium in mediating cellular functions important for growth and development in higher plants. In: Cheung WY (ed) Calcium and cell function vol 3. Academic Press, London New York, pp 409–444
Roux SJ, McEntire K, Slocum RD, Cedel TE, Hale CC II(1981) Phytochrome induces photoreversible calcium fluxes in purified mitochondrial fraction from oats. Proc Natl Acad Sci USA 78:283–287

Rubery PH (1980) The mechanism of transmembrane auxin transport and its relation to the chemiosmotic hypothesis of the polar transport of auxin. In: Skoog F (ed) Plant growth substances 1979. Springer, Berlin Heidelberg New York
Rubery PH, Astle MC (1982) The mechanism of transmembrane abscisic acid transport and some of its implications. In: Wareing PF (ed) Plant growth substances 1982. Springer, Berlin Heidelberg New York, pp 353–362
Rufelt H (1957a) The course of the geotropic reaction of wheat roots. Physiol Plant 10:231–247
Rufelt H (1957b) Influence of the composition of the nutrient solution on the geotropic reactions of wheat roots. Physiol Plant 10:373–396
Rufelt H (1957c) Influence of temperature on the geotropic reactions of wheat roots. Physiol Plant 10:485–599
Rufelt H (1961) Geotropism in roots and shoots. Annu Rev Plant Physiol 12:409–430
Rufelt H (1962) Plagiogeotropism in roots. In: Ruhland W (ed) Encyclopedia of plant physiology, vol 17/2. Springer, Berlin Göttingen Heidelberg, pp 322–343
Ruge U (1941) Über die geotropische Aufkrümmung decapitierter *Helianthus*-Keimlinge. Planta 132:176–186
Salisbury FB, Wheeler RM (1981) Interpreting plant responses to clinostating. I. Mechanical stresses and ethylene. Plant Physiol 67:677–685
Saltveit ME Jr, Dilley DR (1978) Rapidly induced wound ethylene from excised segments of etiolated *Pisum sativum* L cv Alaska. I. Characterization of the response. Plant Physiol 61:445–450
Saltveit ME Jr, Dilley DR (1979) Rapidly induced wound ethylene from excised segments of etiolated *Pisum sativum* cv Alaska. IV. Requirement of a water-soluble, heat-stable factor. Plant Physiol 64:417–420
Schmitz H (1933) Über Wuchsstoff und Geotropismus bei Gräsern. Planta 19:614–635
Schötz F, Diers L, Bathelt H (1970) Zur Feinstruktur der Raphidenzellen I. Die Entwicklung der Vakuolen und der Raphiden. Z Pflanzenphysiol 63:91–113
Schrank AR (1944) Relation between electrical and curvature responses in the *Avena* coleoptile to mechanical stimuli. Plant Physiol 19:198–211
Schrank AR (1947) Electrical and curvature responses of the *Avena* coleoptile to transversely applied direct current. In: Lund EJ (ed) Bioelectric fields and growth. Univ Texas Press, Austin, pp 217–231
Schrank AR (1951) Electrical polarity and auxins. In: Skoog F (ed) Plant growth substances. Univ Wisconsin Press, Madison, pp 123–140
Schurzmann M, Hild V (1980) Effect of indoleacetic acid, abscisic acid, root tips and coleoptile tips on growth and curvature of maize roots. Planta 150:32–36
Seitz K (1979) Cytoplasmic streaming and cyclosis of chloroplasts. In: Haupt W, Feinleib ME (eds) Physiology of movements. Encyclopedia of plant physiology, new ser vol 7. Springer, Berlin Heidelberg New York, pp 150–169
Sharp RE, Davies WJ (1979) Solute regulation and growth by roots and shoots of water-stressed maize plants. Planta 147:43–49
Shaw S, Wilkins MB (1974) Auxin transport in roots. X. Relative movement of radioactivity from IAA in the stele and cortex of *Zea* root segments. J Exp Bot 25:199–207
Shaw S, Gardner G, Wilkins MB (1973) The lateral transport of IAA in intact coleoptiles of *Avena sativa* L and *Zea mays* L during geotropic stimulation. Planta 115:97–111
Sievers A, Hensel W (1982) The nature of graviperception. In: Wareing PF (ed) Plant growth substances. Academic Press, London New York, pp 497–506
Sievers A, Volkmann D (1979) Gravitropism in single cells. In: Haupt W, Feinleib ME (eds) Physiology of movements. Encyclopedia of plant physiology, new ser vol 7. Springer, Berlin Heidelberg New York, pp 567–572
Sievers AH, Behrens M, Buckhout TJ, Gradmann D (1984) Can a Ca^{2+} pump in the endoplasmic reticulum of the *Lepidium* root be a trigger for rapid changes in membrane potential after gravistimulation? Z Pflanzenphysiol 114:195–200
Singer M (1903) Über den Einfluß der Laboratoriumsluft auf das Wachstum der Kartoffelsprosse. Ber Dtsch Bot Ges 21:175–180
Slocum RD, Roux SJ (1983) Cellular and subcellular localization of calcium in gravistim-

ulated oat coleoptiles and its possible significance in the establishment of tropic curvature. Planta 157:481–492
Smith J, Ho THD (1982) Bimodal response to ABA during barley seed germination. Plant Physiol (Suppl) 69:39
Springer MS, Goy MF, Adler JF (1979) Protein methylation in behavioral control mechanisms and in signal transduction. Nature 280:279–284
Stark P (1916) Experimentelle Untersuchungen über das Wesen und die Verbreitung der Kontaktreizbarkeit. Jahrb Wiss Bot 57:198–320
Suge H (1971) Stimulation of oat and rice mesocotyl growth by ethylene. Plant Cell Physiol 12:831–837
Sun D, Roux SJ (1983) Inhibition of gravitropism in oat coleoptiles by a calcium chelator, and the reversal of this inhibition by calcium. Plant Physiol (Suppl) 72:66
Suzuki T, Kondo N, Fujii T (1979) Distribution of growth regulators in relation to light-induced geotropic responsiveness in *Zea* roots. Planta 145:323–329
Tagawa T, Bonner J (1938) Mechanical properties of the *Avena* coleoptile as related to auxin and to ionic interactions. Am J Bot 25:207–212
Tanada T (1978a) A rapid gravitational effect on the translocation of fluorescein in mung bean hypocotyl sections. Planta 142:221–223
Tanada T (1978b) Boron – key element in the actions of phytochrome and gravity? Planta 143:109–111
Tanada T (1978c) Phytochrome-induced increase of fluorescein translocation in mung bean hypocotyls. Plant Physiol 62:105–106
Tanada T, Vinten-Johansen C (1980) Gravity induces fast electrical field changes in soybean hypocotyls. Plant Cell Environ 3:127–130
Tang PM, dela Fuente RK (1982) Comparison of auxin transport and potassium leakage in calcium and boron deficient sunflower seedlings. Plant Physiol (Suppl) 69:55
Terry ML, Jones RL (1981) Effect of salt on auxin-induced acidification and growth by pea internode sections. Plant Physiol 68:59–64
Thimann KV (1974) Opening lecture. In: Plant growth substances 1973. Hirokawa, Tokyo, pp 3–6
Thimann KV (1977) Hormone action in the whole life of plants. Univ Massachusetts Press, Amherst
Thimann KV, Schneider CL (1938a) The role of salts, hydrogen-ion concentration and agar in the response of the *Avena* coleoptile to auxins. Am J Bot 25:270–280
Thimann KV, Schneider CL (1938b) Differential growth in plant tissues. Am J Bot 25:627–641
Thimann KV, Sweeney BM (1937) The effect of auxins upon protoplasmic streaming. J Gen Physiol 21:123–135
Trewavas AJ (1979) Plant growth substances: what is the molecular basis of their action? What's New in Plant Physiol 10:33–36
Trewavas A (1981) How do plant growth substances work? Plant Cell Environ 4:203–228
Ullrich C-H (1978) Continuous measurement of initial curvature of maize coleoptiles induced by lateral auxin application. Planta 140:201–211
Vanderhoef LN (1980a) Auxin-regulated elongation: a summary hypothesis. In: Skoog F (ed) Plant growth substances 1979. Springer, Berlin Heidelberg New York, pp 90–96
Vanderhoef LN (1980b) Auxin-regulated cell enlargement. In: Leaver CJ (ed) Genome organization and expression in plants. Plenum, New York, pp 159–173
Van der Laan PA (1934) Der Einfluß von Aethylen auf die Wuchstoffbildung bei *Avena* und *Vicia*. Rec Trav Bot Neerl 31:691–742
Van Overbeek J, Davila Olivo G, Santiago de Vasquez EM (1945) A rapid extraction method for free auxin and its application in geotropic reactions of bean seedlings and sugar-cane nodes. Bot Gaz 106:440–451
Vesper MJ, Evans ML (1978) Time-dependent changes in the auxin sensitivity of coleoptile segments. Plant Physiol 61:204–208
Vintila R, Fabian A, Lakar-Keul G (1973) Die Kontrolle der Protoplasmaströmung in den Wurzelhaaren der Gerste (*Hordeum vulgare*) durch ATP, SH-Gruppen und Ca^{2+}-Ionen. Rev Roum Biol Bot 18:287–296

Volkmann D, Sievers A (1979) Graviperception in multicellular organs. In: Haupt W, Feinleib ME (eds) Physiology of movements. Encyclopedia of plant physiology, new ser vol 7. Springer, Berlin Heidelberg New York, pp 573–600
Wagner G (1979) Actomyosin as a basic mechanism of movement in animals and plants. In: Haupt W, Feinleib ME (eds) Physiology of movements. Encyclopedia of plant physiology, new ser vol 7. Springer, Berlin Heidelberg New York, pp 114–126
Walton DC (1980) Biochemistry and physiology of abscisic acid. Annu Rev Plant Physiol 31:453–489
Wareing PF, Phillips IDJ (1981) Growth and differentiation in plants, 3rd edn. Pergamon, Oxford
Webster JH, Wilkins MB (1974) Lateral movement of radioactivity from [^{14}C] gibberellic acid (GA_3) in roots and coleoptiles of *Zea mays* L seedlings during geotropic stimulation. Planta 121:303–308
Weiler EW (1981) Radioimmunoassay for pmol-quantities of indole-3-acetic acid in plant extracts. Planta 153:319–325
Weisenseel MH, Dorn A, Jaffe LF (1979) Natural H^+ currents traverse growing roots and root hairs of barley (*Hordeum vulgare* L). Plant Physiol 64:512–518
Went FW (1928) Wuchstoff and Wachstum. Rec Trav Bot Neerl 25:1–116
Went FW (1932) Eine botanische Polaritätstheorie. Jahrb Wiss Bot 76:528–557
Went FW, Thimann KV (1937) Phytohormones. Macmillan, New York
Wheeler RM, Salisbury FB (1980) Gravitropism in stems may require ethylene. Science 209:1126–1128
Wheeler RM, Salisbury FB (1981) Gravitropism in higher plant shoots. I. A role for ethylene. Plant Physiol 67:686–690
Wilkins H, Wain RL (1975) Abscisic acid and the response of the roots of *Zea mays* L seedlings to gravity. Planta 126:19–23
Wilkins MB (1965) Red light and the geotropic response of the *Avena* coleoptile. Plant Physiol 40:24–34
Wilkins MB (1966) Geotropism. Annu Rev Plant Physiol 17:376–408
Wilkins MB (1971) Hormone movement in geotropism. In: Gordon SA, Cohen MJ (eds) Gravity and the organism. Univ Chicago Press, Chicago pp 107–112
Wilkins MB (1975) The role of the root cap in root geotropism. Curr. Adv. in Plant Sci 13:317–328
Wilkins MB (1976a) Identification of auxin in growing plant organs and its role in geotropism. In: Sutherland N (ed) Perspectives in experimental biology, vol 2. Pergamon, Oxford New York, pp 77–78
Wilkins MB (1976b) Gravity-sensing guidance systems in plants. Sci Prog Oxford 63:187–217
Wilkins MB (1977) Gravity and light-sensing guidance systems in primary roots and shoots. In: Jennings DH (ed) Integration of activity in the higher plant. Proc Symp Soc Exp Biol 31:275–335
Wilkins MB (1978) Gravity-sensing guidance systems in roots and shoots. In: Controlling factors in plant development. Bot Mag Tokyo (Special Issue) 1:255–277
Wilkins MB (1979) Growth-control mechanisms in gravitropism. In: Haupt W, Feinleib ME (eds) Physiology of movements. Encyclopedia of plant physiology, new ser vol 7, Springer, Berlin Heidelberg New York, pp 601–626
Wilkins MB, Goldsmith MHG (1964) The effects of red, far-red, and blue light on the geotropic response of coleoptiles of *Zea mays*. J Exp Bot 15:600–615
Wilkins MB, Nash LJ (1974) Movement of radioactivity from [^{3}H]GA_3 in geotropically stimulated coleoptiles of *Zea mays*. Planta 115:245–251
Wilkins MB, Woodcock AER (1965) Origin of the geoelectric effect in plants. Nature 208:990–992
Williamson RE, Ashley CC (1982) Free Ca^{2+} and cytoplasmic streaming in the alga *Chara*. Nature 296:647–651
Woodcock AER, Wilkins MB (1969) The geoelectric effect in plant shoots. I. The characteristics of the effect. J Exp Bot 20:156–169
Wright LZ, Rayle DL (1982) Inhibition of shoot geotropism by neutral buffers. Plant Physiol 69:278–279

Wright LZ, Rayle DL (1983) Evidence for a relationship between H^+ excretion and auxin in shoot gravitropism. Plant Physiol 72:99–104
Wright M, Mousdale DMA, Osborne DJ (1978) Evidence for a gravity-regulated level of endogenous auxin controlling cell elongation and ethylene production during geotropic bending in grass nodes. Biochem Physiol Pflanz 172:581–596
Wright STC (1961) A sequential growth response to gibberellic acid, kinetin and indolyl-3-acetic acid in the wheat coleoptile (*Triticum vulgare* L). Nature 190:699–700
Yamaguchi T, Street HE (1977) Stimulation of the growth of excised culture roots of soya bean by abscisic acid. Ann Bot 41:1129–1133
Yang SF (1980) Regulation of ethylene biosynthesis. Hortscience 15:238–243
Zinke H (1968) Versuche zur Analyse der positiven und der negativen geotropischen Reaktionen von Keimwurzeln. Planta 82:50–72
Ziv M, Koller D, Halevy AH (1976) Ethylene and the geotropic response of lateral branches in peanuts (*Arachis hypogaea* L). Plant Cell Physiol 17:333–339
Zobel RW (1973) Some physiological characteristics of the ethylene-requiring tomato mutant diageotropica. Plant Physiol 52:385–389
Zobel RW (1974) Control of morphogenesis in the ethylene-requiring tomato mutant, diageotropica. Can J Bot 52:735–742

8 De-Etiolation and Plant Hormones [1]

T. O'BRIEN, F.D. BEALL, and H. SMITH

1 Introduction

Light, perceived and transduced by a number of photoreceptors, is known to regulate or modify many aspects of plant growth and development. Beginning with seed germination and ending with fruiting and senescence, light can regulate processes which can also be modified by plant hormones. This has been taken as circumstantial evidence that light may be exerting its influence via plant hormones, and considerable research effort has been directed toward elucidating the role of plant hormones in light-mediated processes. The two areas of light and plant hormone research that have received the most attention are de-etiolation and photoperiodism (see VINCE-PRUE, Chap. 9 this Vol.). While the emphasis, by necessity, of this chapter is on plant hormones in de-etiolation, we include relevant work on other aspects of photomorphogenesis and plant hormones.

The phenomenon of de-etiolation is extremely complex, with the plant undergoing a wholesale change in pattern of development, a change which is usually complete within 48 h under continuous high-irradiance WL. A number of photoreceptors are undoubtedly involved – phytochrome, the blue-absorbing photoreceptor (cryptochrome), protochlorophyll(ide) and, in the final stages, chlorophyll itself. De-etiolation would seem to require a reorganization of hormonal balances and actions throughout the plant, but it has proven extremely difficult to determine precisely what these changes are. The literature contains many reports in which the hormone work has been first class, but the photocontrol aspects most uncritical. Similarly, much good photomorphogenesis experimentation is invalidated through a naive approach to the problems of hormone isolation, identification, and quantification. Perhaps the majority of work on photoreceptors and hormones must be viewed with caution since it is uncritical in both areas.

This review, consequently, is extremely selective. We attempt here to collect together the most important work bearing on the relationship between light and hormones. Inevitably, the emphasis is placed on phytochrome, since this is the photoreceptor at present best understood. The available information is diffuse and sporadic, and we have not attempted to draw out a "model" of hormone action in de-etiolation; such an attempt would certainly be futile. On the other hand, there are one or two promising areas of research, and we hope this review gives them the appropriate prominence. Readers are also

1 In this chapter the term "hormone" is used as a convenient operational expression, equivalent to "plant-growth regulator", but without implying any assumptions as to function.

referred to the review by deGreef and Fredericq in Vol. 16A of this Encyclopedia (1983) which overlaps with this chapter, but is complementary in scope.

2 Auxins

Many of the morphogenic events which occur during de-etiolation can also be affected by auxin, and this has often been considered as circumstantial evidence for the involvement of auxin in vivo as a natural regulator of photomorphogenic phenomena. Liverman and Bonner (1953) showed that IAA was required for the far-red (FR) reversal of the inhibitory effect of R on the growth of excised coleoptiles. Similarly, Vince (1968) found that exogenously supplied 2,4-D suppressed light-induced anthocyanin synthesis in excised dark-grown sorghum internodes. Furuya and Torrey (1964) showed that phytochrome-controlled lateral root initiation in peas required the presence of exogenous IAA, and, perhaps at a more fundamental level, Tanada (1968) observed that the phytochrome-dependent attachment of barley root tips to glass required a low exogenous supply of IAA.

Because of the well-documented inhibition of coleoptile extension which occurs during de-etiolation, and the important role assigned to auxin in the promotion of coleoptile growth, considerable attention has been paid to the possibility that light brings about de-etiolation by reducing the availability or effectiveness of auxin in the irradiated tissue. Specifically, light effects on auxin transport and metabolism (and more recently on auxin binding) have been investigated. It must be pointed out, however, that in no instance has it been shown that the ultimate developmental response observed is mediated by a direct effect of light on auxin; where attempts have been made to relate apparent light–auxin interactions to subsequent photomorphogenic events, the results have been found not to fit a simple scheme for the control of development by light via auxin. In the specific case of the cereal coleoptile, it should also be remembered that light both inhibits and promotes cell extension, depending on the physiological age of the coleoptile, and whether it is illuminated as an intact seedling or as a coleoptile segment.

2.1 Light and Auxin Transport

Experiments by Briggs (1963) gave the first indications that light might affect auxin transport. Two h of R reduced auxin diffusion out of corn coleoptile tips by about 50%, and this reduction in diffusion rate gradually disappeared during 3 h of subsequent darkness. FR-reversibility of this effect was not tested. Similar effects of long-term light treatments on auxin diffusion were also observed (Muir 1970) with *Avena* coleoptile tips, and in this instance the effect did seem to be FR-reversible (Table 1). Sherwin and Furuya (1973) investigated light effects on auxin uptake by intact, etiolated, rice coleoptiles using short pulses of R (3 min) and FR (10 min). Absorption of ^{14}C-IAA by the coleoptiles was apparently bi-phasic, and each phase showed R/FR-revers-

Table 1. a Effect of R and FR irradiation on the content of diffusible auxin in the apex of the *Avena* coleoptile. **b** Reversibility of the effect of R irradiation on the content of diffusible auxin in the apex of the *Avena* coleoptile. (After MUIR 1970)

	Light treatment	Auxin diffusion ng IAA-equivalents/apex/3 h
a	24 h dark	0.22
	12 h dark, 12 h R	0.08
	12 h dark, 12 h FR	0.19
b	24 h dark	0.25
	4 h FR, 20 h R	0.08
	4 h R, 20 h FR	0.23

Table 2. R/FR reversibility of the effect of light upon the amount of ^{14}C-IAA in rice seedlings at (**a**) 3 h and (**b**) 5 h after irradiation. 10 M ^{14}C-IAA solidified with 1.5% agar was applied as a cylinder to the tips of intact coleoptiles immediately following irradiation. (After SHERWIN and FURUYA 1973)

a Uptake of ^{14}C-IAA (cpm $\times 10^{-3}$), 3 h after irradiation

Dark	FR	R	R/FR	R/FR/R
66.5	75.6	43.3	73.3	50.1

b Uptake of ^{14}C-IAA (cpm $\times 10^{-3}$), 5 h after irradiation

Dark	FR	R	R/FR	R/FR/R
77.2	74.5	122.7	80.6	107.1

ibility, suggesting phytochrome control (Table 2). It was concluded that light effects on auxin transport might be an important factor in the regulation of photomorphogenesis, and that the photoreceptor involved seemed to be phytochrome. RAJAGOPAL and BULARD (1975), using 3–14 min irradiations, also observed apparent phytochrome control of auxin diffusion from *Avena* coleoptiles, but in this case the effect on auxin secretion did not seem to be related to subsequent coleoptile growth. The most effective wavelengths for reducing coleoptile growth were the blue (B) and ultraviolet (UV) regions, which also reduced auxin diffusion from the coleoptile tip. Neither the B nor the UV effects on auxin diffusion or coleoptile growth were FR-reversible (Table 3).

A major difficulty with these auxin diffusion experiments, of course, lies in determining why the amount of auxin diffusing from the isolated coleoptile tips varies with phototreatment. Rather than being an expression of effects on auxin transport, it may be that specific phototreatments modify binding of auxin within the irradiated tissue (WALTON and RAY 1981), or alter rates of synthesis or degradation therein, as diffusion experiments are based on bioassays which only measure biologically active auxins. In the work quoted above (RAJAGOPAL and BULARD 1975), for example, it is quite possible that although phytochrome may modify auxin transport, it is B and UV light, acting perhaps

Table 3. Effect of light quality on growth and diffusible auxin content of *Avena* coleoptiles 12 to 13 h after equilateral irradiation of intact seedlings at the specified wavelengths. (After RAJAGOPAL and BULARD 1975)

Light treatment	pg IAA equivalents/apex/3 h		Coleoptile height (mm ± SE)	
	Expt. 1	Expt. 2	Expt. 1	Expt. 2
dark control	200	181	34.0 ± 0.21	28.9 ± 0.42
FR (730 nm)	235	215	33.5 ± 0.40	27.4 ± 0.34
R (660 nm)	88	76	33.0 ± 0.53	27.0 ± 0.30
R/FR	215	206	32.6 ± 0.34	28.0 ± 0.34
blue (455 nm)	160	143	27.9 ± 0.42	23.0 ± 0.31
UV (380 nm)	124	103	22.0 ± 0.32	19.4 ± 0.34

through promotion of auxin degradation, which produce an effect on auxin levels that subsequently induces a morphogenic response. Thus, it is important that such studies of light effects on auxin diffusion are accompanied by investigations of auxin metabolism rates and binding capacities in the tissue involved.

2.2 Light and Auxin Metabolism

Red-light-treated tissue has been shown many times to contain reduced amounts of extractable auxin, relative to dark-grown material, and this has led to the idea that upon de-etiolation, elongation rates decrease due either to increased degradation/conjugation or lowered rates of synthesis of auxin. HILLMAN (1949) had already shown that application of high levels of exogenous IAA could prevent R-induced growth inhibition, and MUIR (1974) also noted that treatment of *Avena* coleoptiles with tryptophan (a precursor of IAA) eliminated the inhibitory effect of R on coleoptile growth. Reduction induced by R in extractable (BLAAUW-JANSEN 1959) and diffusible (MUIR 1974) auxin was found in *Avena* and a detailed study of the response using *Phaseolus vulgaris* seedlings has also been made (FLETCHER and ZALIK 1964, 1965, EVANS 1975, TILLBERG 1974a, b). FLETCHER and ZALIK (1964) measured auxin levels after one photoperiod, and the lengths of various plant parts after seven photoperiods, using plants grown under a range of light qualities (Table 4). A direct relationship was thus demonstrated between auxin content after one cycle and plant height after seven cycles. Under R, plants were shortest, and had lowest endogenous auxin levels. This effect could be partially removed by adding FR. However, TILLBERG (1974b) showed that in bean seedlings the level of IAA increased after de-etiolation relative to plants maintained in the dark.

Subsequent studies (FLETCHER and ZALIK 1965), using exogenously supplied ^{14}C-IAA, showed that the IAA was metabolized into three distinct fractions, the relative proportions of these being related to light quality (Table 5). Fraction II was tentatively identified as indolealdehyde, a postulated degradation product of IAA which appears during in vitro photo-oxidation of IAA (GAL-

Table 4. Effect of light of several spectral bands on elongation and auxin content of bean plants, as measured by the *Avena* first internode test. (After FLETCHER and ZALIK 1964)

	Spectral range	Length (cm)			Auxin g IAA equiv. plant
		Hypocotyl	Epicotyl	Petiole	
—	dark	25.6	14.5	1.0	13.31
393–463	B	21.6	13.5	7.2	11.92
580–655	R	16.9	8.0	3.1	9.04
610–705	"deep" R	18.2	10.1	4.7	—
643–738	R+FR	20.0	11.5	7.0	10.96

Table 5. Effect of light of several spectral bands on the recovery of radioactivity from bean seedlings injected with ^{14}C-IAA. (After FLETCHER and ZALIK 1965)

Light treatment	Percent label recovered			Total
	Fraction I	Fraction II	Fraction III	
dark	81	11	2	94
B	75	12	4	91
R	41	31	7	79
R+FR	66	13	5	84

STON and HILLMAN 1961). These results therefore provided further evidence for a photomorphogenic response involving light effects on auxin metabolism.

In the dwarf pea variety Progress No. 9, R has been shown to inhibit the growth of etiolated internodes, and also to cause a shift to higher IAA concentrations in the IAA dose/response curve for excised sections (RUSSEL and GALSTON 19619). This was interpreted as being consistent with R-enhanced IAA degradation. Similarly, MUIR (1970), using normal and dwarf pea varieties, found that the diffusible auxin content of the subapical internode in light-treated plants was greatly reduced for dwarf varieties, relative to normal varieties or dark-grown dwarves. At the same time the level of IAAsp, an inactive conjugate of IAA, and one of its proposed degradation products, was three times higher in light-treated dwarf varieties relative to light-treated normal plants, and was not detectable in dark-grown plants. These results also argue strongly for R-enhanced auxin catabolism, though in rice coleoptiles (SHERWIN and FURUYA 1973) and bean seedlings (TILLBERG 1974b), IAAsp levels were found not to vary with phototreatment.

Further evidence for light-induced auxin degradation comes from work on light effects on hypocotyl extension in seedlings of *Lupinus alba,* where growth inhibition caused by continuous WL was largely overcome by application of exogenous IAA, whereas IAA itself caused growth inhibition in etiolated seedlings (ACTON and MURRAY 1974). It was suggested that light-treated plants contained depressed levels of auxin, and so responded to added IAA by extension, whereas addition of exogenous auxin to etiolated material produced supra-optimal levels, leading to growth inhibition.

More recent work by BANDURSKI et al. (1977) has emphasized, for light effects on endogenous auxin levels, a concept which is already familiar for the regulation of GA levels, i.e., control of the balance between free and conjugated forms of the hormone, only one of which forms is physiologically active in any particular growth response. Thus, BANDURSKI et al. (1977) have found that a 27-s light exposure, sufficient to produce a 30% reduction in growth rate of etiolated seedlings of *Zea mays,* caused a decrease of 40% in the concentration of free IAA in the seedling, and an increase in the content of esterified IAA 90 min after the light flash. Such a system has obvious parallels in the changes in ratio of "conjugated" and "free" GA which have been observed following light treatment of etioplast suspensions (COOKE and KENDRICK 1976), and may represent a general phenomenon whereby certain environmental stimuli control development by alteration of the ratio of free to conjugated hormones. A further point from this work is that it would not now seem to be possible to draw simple conclusions concerning absolute endogenous auxin levels from agar diffusion experiments (especially if only the free auxin is transportable), since conjugated auxin may represent a significant reservoir of hormone in the tissue which can again be made readily available rapidly, without the need for further synthesis, in response to the appropriate external stimulus.

The most important objection to the hypothesis that light-induced growth inhibition is a result of enhanced auxin degradation lies in the failure to establish a satisfactory mechanism by which this might be achieved, perhaps because most schemes have ignored the possibility that conjugates are the major catabolites of endogenous auxin. GALSTON (1970) proposed a scheme for photocontrol of development via light effects on phenolic cofactors of peroxidases involved in IAA-degradation, the "IAA-oxidase" enzyme system. Red light produces high levels of kaempferol glycosides in the stems of etiolated pea plants, and quercetin glycosides in the leaves (MUMFORD et al. 1964, BOTTOMLEY et al. 1966). Stem elongation in these plants is inhibited by R, and leaf expansion is promoted (GALSTON et al. 1964). It has also been shown that kaempferol glycosides promotes peroxidase-induced IAA degradation, whereas quercetin glycosides inhibit the enzyme system (FURUYA 1962). However, the critical observation correlating light treatment with enhanced endogenous IAA-oxidase activity in stems has not been made. In fact, where such results are reported, R has been shown either to depress, or to have no effect on, IAA-oxidase activity, which seems difficult to reconcile with the apparent effect of R in reducing endogenous IAA levels. Thus, GALSTON and HILLMAN (1957) demonstrated R inhibition of IAA-oxidase activity in pea stems, an effect which was also found to be FR-reversible. More recently, RAJAGOPAL and BULARD (1975) have shown that in etiolated *Avena* coleoptiles, IAA-oxidase activity is unaffected by R, whereas UV, which reduces auxin secretion from the coleoptile and also inhibits coleoptile growth, actually suppresses IAA-oxidase activity by about 30%.

In the few studies involving light-grown plants, the addition of FR to background WL (TUCKER 1976) or monochromatic FR (MUIR 1974) resulted in increased amounts of bioassay-detectable IAA-like activity. However, these workers did not analyze the source (synthesis or release from conjugate) of the increased IAA. Such systems, utilizing light regimes to manipulate the growth and development of light-grown plants, hold considerable promise for elucidating the role of all plant hormones in light-regulated growth and development.

The role of auxin metabolism in photomorphogenesis thus remains obscure at the present time; a more complete understanding of the in vivo significance of peroxidase-induced IAA degradation, enhanced conjugate formation, the mechanism and function of auxin transport, and the role of auxin-binding proteins in regulating available hormone levels is necessary before sufficiently detailed schemes can be formulated.

3 Gibberellins

The most fully understood relationship between photomorphogenesis and plant growth regulators is the effect of light on GA's. This topic has been extensively reviewed (SMITH 1980, REID 1983), and as a consequence this review we will only emphasize a few aspects. Evidence implicating GA's in the de-etiolation process comes principally from two experimental systems: (a) light effects on cereal leaf growth; and (b) photo-controlled stem extension.

3.1 Cereal Leaf Growth

One of the earliest observations linking phytochrome with GA's was made by SCOTT and LIVERMAN (1957), who observed that R-stimulated expansion of bean leaf discs could be simulated by treatment with GA_3. REID et al. (1968) observed a similar effect in the unrolling of barley leaves, and found that 30 min R produced a large transient increase in the GA content of irradiated leaf segments, peaking after 15 min of subsequent darkness. This increase could be reduced considerably by treatment of the segments with CCC or AMO-1618 (inhibitors of GA synthesis), arguing for at least a partial involvement of R-induced GA synthesis in the leaf unrolling process. REID and CLEMENTS (1968) showed further that the R-induced GA increase could be blocked by actinomycin-D and chloramphenicol, implicating RNA- and protein-synthesis in the response. Another effect of R in barley seemed to be to enhance rates of GA catabolism, since in homogenates of etiolated barley leaves the conversion of $^3H\text{-}GA_9$ into other acidic ^3H-substances was increased by R (REID et al. 1972).

A similar system was observed to operate in wheat leaves, where unrolling is again stimulated by R. Exogenously supplied GA_3 will substitute for R to induce unrolling (BEEVERS et al. 1970) (but cf. Sect. 5: Cytokinins). In this system, 5 min R was found to induce a rapid and transient increase in endogenous

GA, which reached a maximum after 10 min subsequent darkness (BEEVERS et al. 1968). This GA increase did not occur if FR was given at the end of the R irradiation (BEEVERS et al. 1970).

LOVEYS and WAREING (1971a) made a comparative study of the two experimental systems, wheat and barley. In wheat, where a short (5 min) R pulse was used to induce the transient GA increase, the response was unaffected by AMO-1618 or CCC. In barley, however, where 30 min R pulses had been used, the subsequent GA increase could be significantly reduced with AMO-1618 or CCC. Further, in the wheat leaf system, the effect of 5 min R on increased acidic "free" GA was mirrored by a corresponding decrease in the level of highly water-soluble conjugates, and protein-bound GA. It was suggested that two distinct effects were operative:

1. a rapid R-induced release of GA from a bound to a free form; and
2. a subsequent effect of R stimulating GA synthesis, and turnover.

To bring in a note of caution, LOVEYS and WAREING (1971b) have shown that when chromatograms of extracts of etiolated leaves are assayed simultaneously by both the lettuce hypocotyl test and by the effects upon the unrolling of etiolated wheat leaf sections, only one narrow zone of the chromatogram is effective in stimulating leaf unrolling, whereas a much wider zone shows GA-like activity in the lettuce hypocotyl bioassay. This and other work (REID et al. 1972) suggest that metabolism of existing pools of active GA precursors may be an important source of quickly produced biologically active GA's, an area which had received little attention.

The leaf unrolling system was further defined by SMITH and co-workers (EVANS and SMITH 1976a, b, HILTON and SMITH 1980), who demonstrated a R-induced increase in GA levels in a preparation of partially purified etioplasts from etiolated barley leaves. Irradiation with FR following R prevented this increase (Fig. 1). Further work (EVANS and SMITH 1976b) showed that phytochrome was associated with etioplasts and indicated that it seemed to be located at the etioplast envelope. This has obvious implications in terms of a potential role for the photoreceptor in the control of membrane permeability, though the 5-min time lag between the end of R treatment and the onset of a detectable increase in GA levels suggests that phytochrome itself is not directly involved in GA-transport across the etioplast membrane.

Recently, HILTON (1984) has attempted to obtain evidence on the hypothesis that phytochrome controls GA-efflux from barley etioplasts by regulating the permeability of etioplast membranes to gibberellins. Crude and partially purified etioplast preparations were layered onto three-step sucrose gradients, given R treatment, and centrifuged so that the intact plastids would band in the lower sucrose layer. If R caused GA-efflux, then an increase in GA-like activity would be expected in the upper layer. The limited results presented show that GA-like activity is increased (relative to dark controls) in both the upper (soluble) layer and the lower (particulate) layer. The author interprets this in terms of there being no effect on GA-efflux, and that R acts through increasing GA-extractability. The increases in GA-like activity in the upper layer are attributed to plastid breakage. This interpretation is plausible, but in the face of a conspicuous lack of other evidence on the integrity of the plastids, it is difficult to accept;

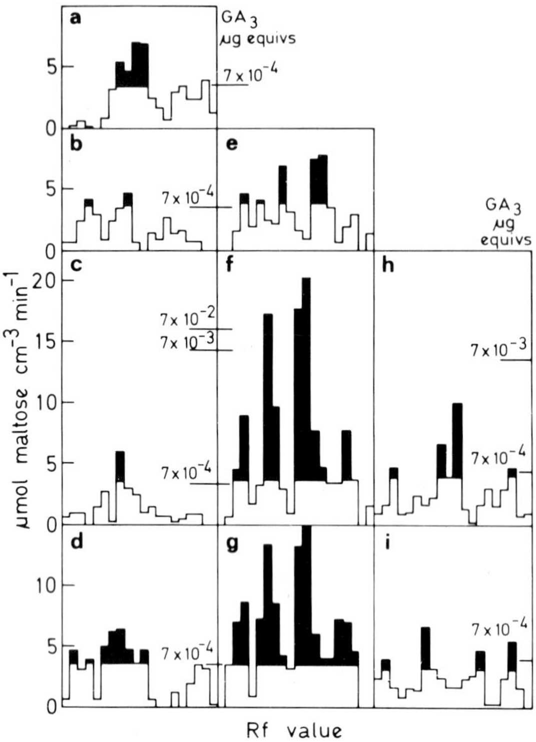

Fig. 1a–i. Thin layer chromatography of acidic, ethyl acetate soluble fractions of extracts from partially purified suspensions on intact etioplasts from barley leaves, following α-amylase assay. **a** Zero time in dark; **b** 5 min dark; **c** 10 min dark; **d** 15 min dark; **e** 5 min R; **f** 5 min R and 5 min dark; **g** 5 min R and 10 min dark; **h** 5 min R and 5 min FR; **i** 5 min R and 10 min FR. *Shaded portions* significant promotion of P<5%. (Evans and Smith 1976a)

it is equally likely, as proposed by Evans and Smith (1976a), that R causes gibberellin interconversions within the etioplasts, leading to both increased GA-efflux, and increased internal GA-levels. A crucial test of etioplast integrity would be to determine the distribution on the gradient of a plastid soluble protein, for example, RuBPCase, which is present in high concentration in barley etioplasts; that such tests were not carried out renders satisfactory interpretation of the reported results impossible.

Increases in GA levels induced by R were also observed (Cooke and Saunders 1975a) in etioplast-enriched preparations from etiolated wheat leaves (Fig. 2). However, the very low levels of GA activity present in these preparations emphasize the importance of using a more sensitive assay in studies on the etioplast, to obtain accurate, detailed information about the changes in hormone levels which occur [see Reeve and Crozier, Vol. 9, this Series (1980)].

The question obviously arises as to whether the surges in GA observed on de-etiolation are causally important, or whether they might be merely indirect effects of photomorphogenesis. Some evidence which would seem to argue against a direct functional involvement of GA in leaf unrolling is derived from a comparison of the time-courses of: (a) R-induced GA efflux from etioplasts; and (b) escape from photoreversibility of the leaf-unrolling response.

Cooke and Saunders' data (1975b) seems to indicate that the time-courses for GA increases in wheat leaves and for escape from photoreversibility of the wheat leaf-unrolling phenomenon, were all of a similar order (ca. 5 to

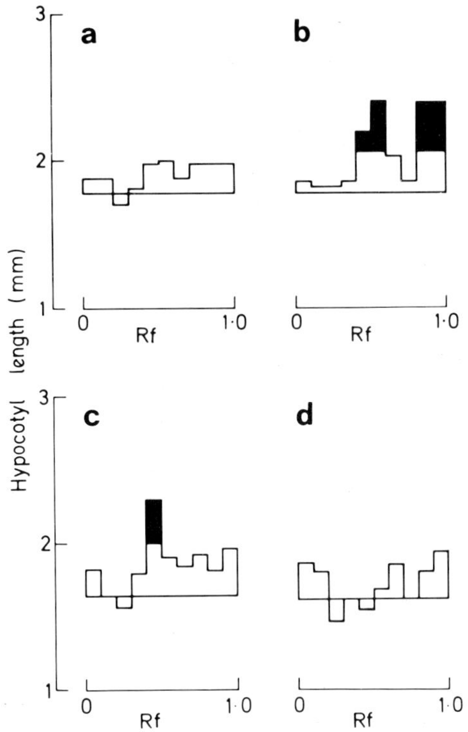

Fig. 2a–d. Lettuce hypocotyl assays of thin layer chromatograms of the acidic, ethyl acetate soluble fractions of etioplast preparations obtained from wheat leaves. **a** Dark control; **b** 5 min R and 15 min dark; **c** 5 min R and 15 min FR; **d** 15 min FR. *Shaded portions* significant promotion at P<1%. (COOKE and SAUNDERS 1975a)

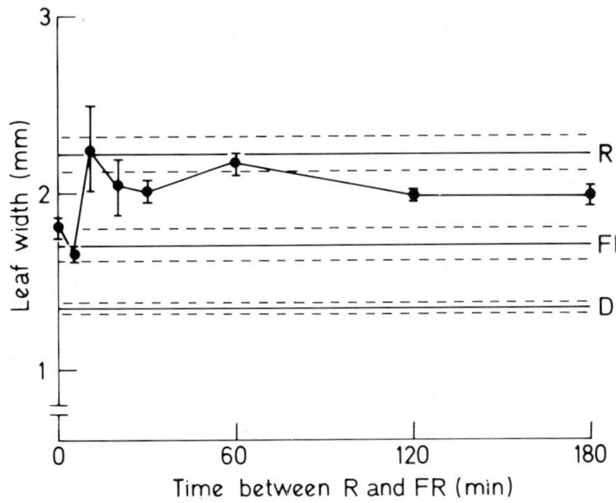

Fig. 3. Effect of an increasing period of darkness between R and FR on wheat leaf unrolling. Leaf width was measured after 24 h in darkness following irradiation. (From data of COOKE and SAUNDERS 1975b)

10 min), allowing the possibility of a cause-and-effect relationship between GA efflux and R-induced unrolling. The magnitude of the unrolling response in this system, however, was very small (Fig. 3).

By comparison, a time-course of escape from photoreversibility of leaf unrolling in barley (SMITH 1975, SUNDQUIST and BRIGGS 1982) showed that the unrolling response could be significantly decreased by FR given up to 1.5 to

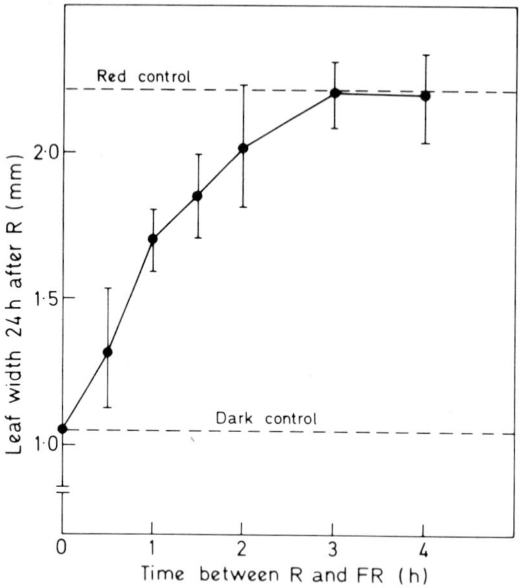

Fig. 4. Effect of an increasing period of darkness between R and FR on barley leaf unrolling. Leaf width was measured 24 h after R irradiation. (SMITH 1980)

2 h after a saturating R irradiation, i.e., long after the initial increase in GA levels had presumably occurred (Fig. 4). This observation throws considerable doubt on the hypothesis that GA changes are causally involved in the developmental response to light, and is an obvious problem requiring further experimentation.

However, it may be that the initial R irradiation of the etiolated leaves induces a series of co-ordinated reactions, all of which are necessary for the expression of leaf unrolling, and some of which remain photoreversible after other associated responses are complete.

A recent claim has been made that phytochrome controls the amount of gibberellin-like substances in the chloroplasts of light-grown barley leaves (HILTON 1983). Seedlings were grown for 7 days in four specially constructed light cabinets designed to provide equal amounts of photosynthetically active radiation (PAR = 400–700 nm), but with differing amounts of added FR (700–800 nm) so as to vary the R:FR ratio and thus phytochrome photoequilibrium (i.e., $P_{fr}/(P_r+P_{fr})$. [See SMITH (1982) and SMITH and MORGAN (1983) for full discussion of this general approach to the study of the role of phytochrome in light-grown plants]. Under these conditions, barley leaf expansion is inversely proportional to $P_{fr}/(P_r+P_{fr})$, although such data are not presented by HILTON (1983). Crude chloroplast preparations from equal weights of laminae were extracted for GA-like substances and bio-assayed by the barley amylase assay. Each treatment ($P_{fr}/(P_r+P_{fr})$ ranging from 0.24 to 0.69) yielded similar amounts of GA-like activity. To "correct" for incomplete recovery of chloroplasts, the GA data were normalized on the basis of extracted chlorophyll levels, inducing the author to claim that the amounts of gibberellin per chloroplast were directly related to estimated phytochrome photoequilibrium. There are two major criticisms of this work. Firstly, extractable chlorophyll level is not

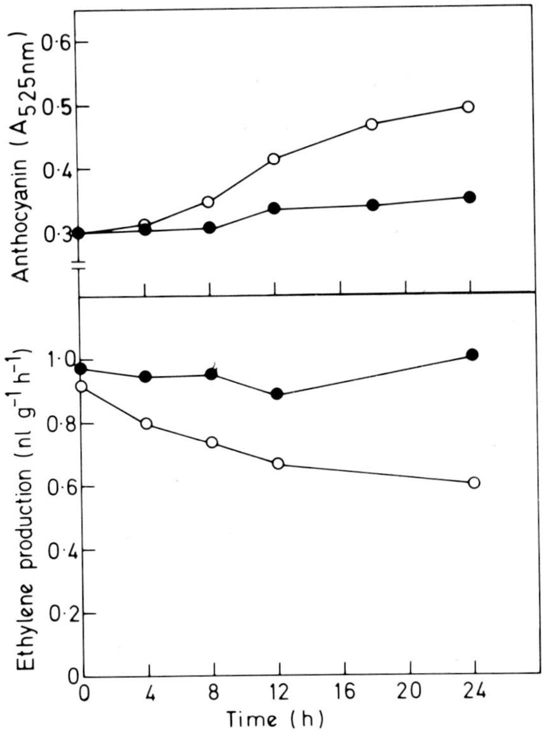

Fig. 5. Kinetics of anthocyanin synthesis and ethylene production in etiolated red cabbage seedlings, following 5 min R. Ethylene production was measured using excised 1 cm apical sections, cut at various times following illumination. Values are calculated from the total ethylene produced during a 12-h incubation period

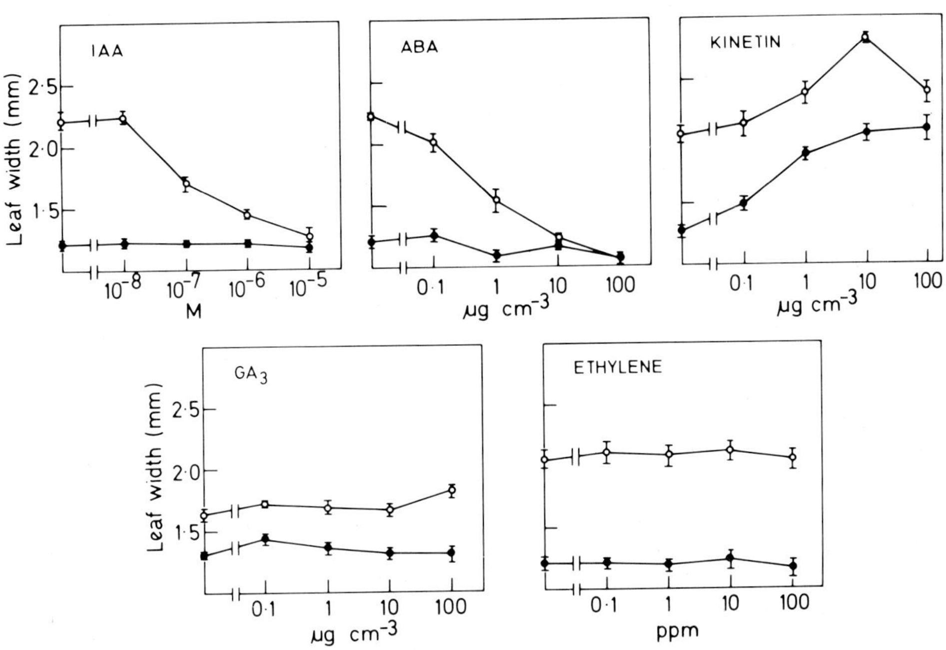

Fig. 6. Effects of various growth hormones on wheat leaf unrolling. ○ 30 min R; ● dark

an acceptable quantitative marker for chloroplast numbers: 7-days' growth in differing R:FR ratios will certainly affect the chlorophyll amount per chloroplast. There is no real substitute for counting chloroplasts directly; after all, counting chloroplasts is even easier than measuring chlorophyll contents. The second criticism is of the conclusions. The author proposes that the changes in chloroplast gibberellin levels may be responsible for the growth differences, but leaf extension growth is inversely proportional to estimated $P_{fr}/(P_r+P_{fr})$, whilst the claimed GA levels are directly proportional to $P_{fr}/(P_r+P_{fr})$. The author fails to resolve the logical contradiction, and it would seem prudent to await independent confirmation of these findings.

The role of phytochrome in regulating the gibberellin levels of cereal leaf etioplasts is an intriguing topic which at first sight appears to hold considerable promise for understanding at least one aspect of the photocontrol of development. Furthermore, the response is the only independently confirmed cell-free phytochrome-mediated phenomenon yet available. All of the work that has been done to date, however, may be criticized for inadequate methodology, particularly with respect to the isolation, purification, and characterization of the plastid preparations used. Although it seems clear now that whatever is happening is a function of the etioplasts, rather than of some other organelles, no group has yet applied the necessary stringent biochemical criteria of purity, intactness and viability customarily required of those seeking to characterize a particular in vitro response. Furthermore, the reported levels of both phytochrome and gibberellins in etioplasts are extremely close to the detection limits of the methods used, and it will be necessary to use more sensitive techniques (i.e., immunochemistry) to make further progress. On these grounds, therefore, it would be wise at this stage to regard the hypothesis that phytochrome controls cereal leaf enlargement through modulation of etioplast gibberellin levels as extremely tentative.

3.2 Photocontrol of Stem Extension

Another phenomenon which has been used to study the possible involvement of plant hormones in the photocontrol of development is the stem extension response, which appears to be controlled, at least in part, by phytochrome. Most workers have used normal and dwarf varieties of plants, usually peas. Peas may be an unfortunate choice of experimental material for the study of light and GA's, as recent results with single gene mutants (POTTS et al. 1982, REID 1983) suggest that while some dwarfs are GA-deficient, others may be GA-response mutants, which makes comparisons between tall and dwarf peas less clear.

The normal and dwarf varieties grow at similar rates in darkness, but on transfer to light, the dwarf characteristic becomes evident, R being most effective for inducing this transition. KENDE and LANG (1964) examined GA levels in R-treated and dark-grown dwarf peas (var. Progress No. 9), and isolated two major fractions of GA activity. No difference was found in the amounts of either fraction from dark-grown or R-treated material. When the GA-zones were reapplied to peas in R or darkness, it was found that Fraction II was highly active in inducing stem extension in both R-treated and dark-grown material. Fraction I (which may be GA_{20}) was active in increasing stem extension in dark-grown peas, but was inactive in R. These results were interpreted as indicating that R reduces stem extension rate by reducing the sensitivity of the plant to specific, rate-controlling endogenous GA's. It should be noted, however, that the results could also be interpreted as evidence for a differential effect of R on the metabolism of certain GA's.

The idea of an effect of light on "sensitivity" to endogenous GA received some support from the work of MUSGRAVE et al. (1969) which showed that the accumulation of ^{3}H-GA_1 or ^{3}H-GA_5 in the apical region of dwarf pea (Progress No. 9) stems was reduced in R-treated material, compared with etiolated tissue. Furthermore, a lower diffusion of ^{3}H-GA_5 from apical sections occurred in the dark than in R. It was inferred from these results that light modifies the binding of GA in elongating tissue. Because a 3-h lag was found to exist before the effect of light on GA accumulation became measurable, it was further suggested that light inhibits the synthesis of GA receptor sites.

KÖHLER and LANG (1963) could find no change in GA levels in Alaska peas (a "normal" variety) following R-irradiation, but were able to isolate an inhibitor of stem extension. This too showed no differences in levels between dark-grown or R-treated normal peas. In the dwarf variety Progress No. 9, however, R-irradiation produced much higher levels of inhibitor than were found in either the R-treated or dark-grown normal variety, or in the dark-grown dwarf. The inhibitory substance was also able to reduce the response of the dwarf variety to exogenous GA_3.

Thus, there seemed to be two possible effects of R with respect to GA-mediated stem extension:

1. R lowers the sensitivity of the plant to GA; or
2. R induces the appearance of an inhibitor of GA-induced stem extension (see Sect. 6).

No effect of light on GA levels per se could be found. However, in a study of de-etiolation in Schnabel, another normal pea variety, KÖHLER (1965) found that light-treated peas contained ten times *more* GA than etiolated seedlings. Furthermore, this R-induced increase was later shown to be FR-reversible (KÖHLER 1966). Soon after this observation was made, JONES (1967) confirmed the original findings of KENDE and LANG (1964): no quantitative or qualitative differences could be found between GA diffusates or extracts from normal (Alaska) or dwarf (Progress No. 9) peas, whether grown in the dark or in the light.

In an extensive study (NESKOVIC and KONJEVIC 1974) of the effects of light on GA levels in dark-grown Alaska pea internodes, a rapid (5 min) R pulse induced an increase in GA, detectable after 20 min, followed by a decrease to an almost undetectable level in the following 2 h. This effect was not FR-reversible, though FR reduced the R-induced inhibition of internode elongation.

CROZIER and AUDUS (1968), investigating light effects on GA levels in *Phaseolus multiflorus* (now *coccineus*) seedlings, isolated two GA-like zones – *Pha-*

Table 6. Levels of endogenous GA-like substances extracted from light- and dark-grown seedlings of *Phaseolus multiflorus*. (CROZIER and AUDUS 1968)

	g GA_3-equivalents/1000 seedlings	
	Phaseolus I	*Phaseolus* II
Light-grown	1.2	0.73
Dark-grown	0.27	3.6

seolus I and II. Light-grown plants were found to contain more *Phaseolus* I and less *Phaseolus* II than dark-grown seedling (Table 6). It appears that *Phaseolus* II may represent a dihydroxylated, and thus more active, GA, and that light may inhibit its production. BOWN et al. (1975) extended this work with light- and dark-grown seedlings of *Phaseolus coccineus*. Here light-grown seedlings contained less endogenous GA than dark-grown plants and were responsive to exogenously supplied GA while dark-grown plants were not. Further, light-grown seedlings metabolized exogenously supplied ^3H-GA_4, ^3H-GA_5 and ^3H-GA_{20} more rapidly, although no evidence was found for qualitative differences in GA metabolites between light- and dark-grown seedlings. The implication is that the process of de-etiolation involves an increased metabolic conversion of GA's to inactive products. In this respect, the responses to *light* with regard to *Phaseolus* GA's appear to be consistent with the observations made on cereal leaves.

4 Ethylene

Two major factors controlling the rate of ethylene synthesis in plants have been identified to date: auxin and light. In the case of light, it is found that the most effective wavelengths are those to which the phytochrome system is particularly responsive. CRAKER et al. (1973) produced an action spectrum for the induction of ethylene production in etiolated and excised sorghum which is similar to many reported HIR responses (SMITH 1975).

In etiolated seedlings ethylene is produced primarily in the apex. In etiolated pea seedlings ethylene production in the apex decreased after exposure to 60 s R, the decrease being detectable 2.5 h after treatment; production of ethylene reached a minimum rate 9 h after irradiation (GOESCHL et al. 1967). Most importantly, the time course of inhibition of ethylene production matched closely that of R-induced plumular expansion, though it preceded the latter phenomenon by about 1 h. Both R-induced plumular expansion and inhibition of ethylene production showed partial FR reversibility (Table 7). SAMIMY (1978) showed a similar effect in etiolated soybean seedlings. In this study reduced ethylene production was correlated with hypocotyl elongation and was fully reversible by FR.

Table 7. Reversal by FR of R-induced changes in ethylene production and plumular expansion in etiolated pea seedlings. (GOESCHL et al. 1967)

Light treatment	Ethylene production $10^6 \times mm^3$ plant^{-1} h^{-1}	Plumular growth mg h^{-1} mg^{-1} tissue
dark	52±7	0.011±0.002
R	10±2	0.074±0.009
FR	22±3	0.033±0.006
R/FR	19±3	0.037±0.006

There is a similar effect in bean hooks, where 30 min R induces hook opening and also reduces ethylene evolution, both phenomena being FR-reversible (KANG and RAY 1969a). Further, R-induced hook opening is completely inhibited by applying 0.2 ppm ethylene (KANG and RAY 1969a). The plumular hook will open in darkness if the seedlings are placed in a hypobaric chamber where ethylene is removed as it appears; the hook never formed in darkness if the seedling was continuously grown under these conditions (BURG 1973). It was concluded that ethylene serves as a natural growth regulator mediating the hypocotyl hook-opening response to light in bean seedlings (KANG and RAY 1969a).

Later work (KANG and BURG 1972a) demonstrated the existence of another apparent phytochrome/ethylene interaction controlling the de-etiolation process in peas. HENSHALL and GOODWIN (1964) had previously shown that photoinduced carotenoid synthesis in the shoot apex of etiolated peas was phytochrome-mediated; 5 min R induced carotenoid synthesis, and this effect was FR-reversible. This R treatment also reduced the rate of ethylene production at the apex (BURG and BURG 1968). KANG and BURG (1972a) then found that synthesis of carotenoids in the shoot apex of R-treated pea seedlings could be completely inhibited by 1 ppm ethylene. It was further demonstrated that seedlings maintained under hypobaric conditions in darkness accumulated carotenoids more rapidly than similar plants growing at atmospheric pressure.

Similarly, in etiolated *Brassica oleracea* seedlings, R first reduced endogenous ethylene production, then subsequently induced the formation of anthocyanin (KANG and BURG 1973) (Fig. 5). Application of 100 ppm ethylene inhibited anthocyanin formation in illuminated material, and hypobarism induced anthocyanin formation in darkness. The results again argue for the involvement of ethylene in a phytochrome-mediated response.

The effect of light on morphogenesis via changes in ethylene production seems, however, not to be a general phenomenon. In etiolated lettuce (*Lactuca sativa* cv. Grand Rapids) seedlings, either 24 h R or an increase in ethylene concentration induced plumular hook-closure even under hypobaric conditions, and when seedlings were exposed to ethylene and R simultaneously, their effects were found to be additive, suggesting independent actions on hook closure (JANES et al. 1976).

By a similar analysis, BÜHLER et al. (1978) demonstrated that light and ethylene acted independently in the control of anthocyanin accumulation in *Sinapis alba*. Nevertheless, where light and ethylene effects on a de-etiolation response have been shown to be associated with the appropriate light effects on ethylene production itself (KANG and BURG 1973, KANG and RAY 1969a), then the preliminary conclusions that light modifies development via the hormonal action of ethylene is reasonable.

The effect of light on ethylene production in nonetiolated tissue is more complex. A number of workers (GEPSTEIN and THIMANN 1980, DELAAT et al. 1981, KAO and YANG 1982) have shown that white light, following a dark period, decreases ethylene production, while other workers (GRODZINSKI et al. 1982, 1983, BASSI and SPENCER 1982) have shown light stimulates ethylene production. However, as KAO and YANG (1982) have shown, light stimulates ethylene production only when a CO_2 source is included in the incubation, which was not done in the work of GEPSTEIN and THIMANN (1980) and DE LAAT et al. (1981). Further, KAO and YANG (1982) have suggested that CO_2 modulates the activity of the enzyme which converts ACC to ethylene, whereby treatments which decrease CO_2 (light, photosynthetic inhibitors) decrease ethylene production. Light quantity has also been shown to affect ethylene production (GRODZINSKI et al. 1982). Increasing PAR resulted in increased photosynthesis, measured as O_2 evolution, and increased ethylene release in the presence of bicarbonate in the incubation medium. Without bicarbonate, increasing PAR decreased ethylene release. One report (EREZ 1977) has demonstrated that, in light-grown plants, ethylene production is affected by the R/FR ratio of the incident light, low ratios promoting ethylene production. In *Marchantia* thalli, when FR is given in conjunction with WL there is a CO_2-dependent response, in the absence of CO_2, FR promotes ethylene production and conversely in the presence of CO_2 (VEROUSTRAETE et al. 1982). These aspects of light modulation of ethylene production deserve further experimentation.

KANG and BURG (1972b) investigated another system where phytochrome seems to modify ethylene production, as well, perhaps, as auxin transport, a situation which may prove to have important consequences for other apparent phytochrome–auxin interactions. It was found that the geotropic sensitivity of subapical sections from etiolated pea stems was enhanced by exposure to 5 min R, and that the development of geotropic curvature in these sections was closely correlated with decreasing ethylene production. It was also shown that geotropic curvature of such stem sections was very sensitive to applied ethylene; 10 ppb inhibited bending by 20 to 30%, and 100 ppb by about 70%. Geotropic sensitivity was greatly enhanced by hypobaric treatment. Since ethylene is known strongly to inhibit lateral transport of auxin (BURG and BURG 1967, LYON 1970), the effect of R on distribution of ^{3}H-IAA across the stem sections was studied. The increased geotropic sensitivity following R treatment was accompanied by increased lateral migration of ^{3}H-IAA. The suggestion was made that R enhances geotropic sensitivity primarily by reducing ethylene production in the geosensitive tissue, allowing the introduction of lateral auxin transport.

The apparently close relationship between light perception and subsequently modified ethylene production might be questioned, however, when it is recalled that the other principal factor known to alter ethylene production, along with light, is auxin. Application of exogenous auxin is capable of inducing ethylene synthesis after a lag of 30 to 60 min, and inhibitor studies suggest that this process requires de novo synthesis of the appropriate enzymes (SAKAI and IMASEKI 1971, KANG et al. 1971), which has been shown to be ACC synthetase (YOSHII and IMASEKI 1981). It might therefore be argued that apparent effects of light on ethylene production could, in fact, be achieved via light effects on auxin.

In the bean hook-opening system, it has been suggested that light might act principally by reducing the level of endogenous auxin in the tissue, leading to a decrease in ethylene production. Evidence cited in support of this hypothesis includes:

1. the opposing effects of auxin and R on hook opening;
2. the promotion of hook opening in removing the cotyledons and plumule (KLEIN et al. 1956), which are the major source of auxin for hook tissue, and
3. the induction of opening by PCIB, an auxin antagonist, in darkness.

However, a R treatment sufficient to induce substantial hook opening has been shown not to change significantly the amount of diffusible auxin present in the hook tissue, whereas a similar R treatment reduced ethylene evolution by about 25% (KANG and RAY 1969b). It was concluded that although auxin was evidently capable of modifying ethylene evolution, the effect of light on hook opening was probably mediated by a direct effect on ethylene synthesis.

In addition to light exerting an effect of ethylene production via auxins, other plant hormones can also modify ethylene production. ABA has been shown to inhibit auxin-induced ethylene production (KONDO et al. 1975, YOSHII and IMASEKI 1981) and the latter suggested that ABA exerts its effect at a different regulatory point than IAA. Cytokinins have been reported to stimulate ethylene production in etiolated mung beans (YOSHII and IMASEKI 1981) and inhibit ethylene production and delay the senescence of carnation petals (MOR et al. 1983).

The foregoing discussion highlights the complex interactions of light, plant hormones, and the resultant plant responses. Treatments which modify ethylene production, including light, can also modify internal CO_2 levels and the transport and/or metabolism of plant hormones, all of which may result in changes in ethylene production. Obviously, it is an area of research that must be approached with caution.

5 Cytokinins

Leaf unrolling, a de-etiolation phenomenon which, as described in Section 3.1, seems to involve large changes in GA levels, is also known to be sensitive to cytokinins. Kinetin has been shown (POWELL and GRIFFITH 1960) to promote the expansion of bean leaf discs in a manner similar to R, and application of kinetin induced unrolling of etiolated wheat leaf segments (BEEVERS et al. 1970), a phenomenon also inducible by GA or R. KANG (1971), re-investigating the effect of applied growth regulators in the wheat leaf-unrolling phenomenon, found no significant effect of GA_3, although kinetin induced unrolling in darkness (Fig. 6).

When etiolated wheat leaves are placed in WL, synthesis of chlorophyll is induced, but only after a lag phase of 3 to 4 h. This lag phase can be reduced by pretreatment of the tissue with a short burst of R (VIRGIN 1961). BEEVERS further showed that pretreatment of etiolated leaves with 10 mg l^{-1} kinetin could substitute for R in reducing the lag phase in chlorophyll formation; GA_3 at the same concentration was ineffective. LONGO et al. (1981) suggest that cytokinins and light act together in the formation of functional plastids in etiolated watermelon cotyledons; however, no consideration was given to light quality and quantity.

Endogenous cytokinin levels in attached and detached mature poplar leaves, exhibited a large transient increase in activity following short periods of exposure to R; 5 min R induced an approximately tenfold increase in cytokinin levels, peaking after 30 min subsequent darkness (HEWETT and WAREING 1973). This effect is closely analogous to that observed with endogenous GA levels in etiolated wheat leaves, although it is not strictly a de-etiolation phenomenon since the poplars were light-grown.

Similarly, red light has been shown to increase cytokinin-like activity in some light-sensitive seeds during germination (TAYLOR and WAREING 1979, VAN STADEN and WAREING 1972). In lettuce seed, however, VAN STADEN (1973) showed that treatments promoting germination (RL and GA) decreased cytokinin activity. KÖHLER et al. (1980) demonstrated that blue light was more effective than red in promoting an increase in cytokinin-like activity in *Amaranthus* seedlings.

6 Abscisic Acid and Other Inhibitors

Since the observation by KÖHLER and LANG (1963) that R-irradiated dwarf peas contained high levels of an unknown inhibitor of stem extension, and the subsequent identification of ABA as a potent naturally occurring inhibitor of stem growth, many investigations have been made to determine the relationship between light-induced inhibition of stem extension, and levels of endogenous ABA.

KAYS and KENDE (1971), using optical rotatary dispersion, found no significant difference in ABA contents of stems of 48 h R-treated or completely dark-grown (Progress No. 9) seedlings. This result was subsequently confirmed by GLC for both tall and dwarf pea cultivars grown in darkness or in R (BARNES 1972); nor was there any difference between the ABA contents of stems of the tall and dwarf plants (BARNES 1972, HIGGINS and BONNER 1974).

Table 8. Length of shoots and ABA-content of 11-day-old pea seedlings, irradiated from days 7 to 11 with the light program indicated. (After DÖRFFLING 1973)

Light treatment	Final shoot length (mm) at day 11	ABA content	
		ng plant^{-1}	ng g^{-1} wt.
R/FR	103.9±3.4	3.4±0.7	5.8±1.1
FR/R	60.2±0.8	3.4±0.8	6.0±1.2

DÖRFFLING (1973), also using GLC, again found no significant difference in the ABA content of pea shoots for plants given either:

1. 5 min R followed by 5 min FR; or
2. 5 min FR followed by 5 min R

every hour, from days 7 to 11 following germination. Final plant height and ABA content, measured on day 11, are shown in Table 8. LOVEYS (1979) found that, in green plants grown under different R:FR ratios, ABA levels did increase in those treatments resulting in less stem elongation. However, the increase in ABA did not occur until 2 to 3 d after the elongation response.

BURDEN et al. (1973) measured the levels of both ABA and xanthoxin, another potent growth inhibitor, in etiolated and R-irradiated dwarf pea seedlings (cv. Meteor). Once again, ABA concentrations were found to be very low, and the differences between treatments were not significant. R-irradiated material, however, contained five times as much xanthoxin, on a dry weight basis, as dark-grown material. Further, it was found that two major xanthophylls present in pea seedlings, lutein and violaxanthin, increase markedly after R-treatment. It was suggested that the dwarfing effect of R on etiolated seedlings might involve enhancement of xanthoxin levels, perhaps by increasing the availability of the precursor violaxanthin. The enhancement of xanthoxin levels by R was confirmed by ANSTIS et al. (1975), who also demonstrated that the increase in xanthoxin associated with R treatment is paralleled by a decrease in growth rate of the irradiated seedlings (Fig. 7).

HIGGINS and BONNER (1974) also isolated an inhibitor from peas similar to xanthoxin which increased in response to light and was also found to be present in greater amounts in dwarf peas relative to the tall cultivar. More recently, HASAGAWA et al. (1983) have isolated and characterized an inhibitor from peas, which they have named pisumin, and based on its partitioning and chromatographic properties have suggested that it is different from previously isolated inhibitors. They demonstrated that pisumin increased markedly in light and decreased when transfered to darkness.

A number of new inhibitors have also been isolated from radish seedlings (HASEGAWA et al. 1982, HASEGAWA and HASE 1978, 981, HASEGAWA and MIYAMOTO 1980). These new inhibitors are very active, inhibiting radish and lettuce

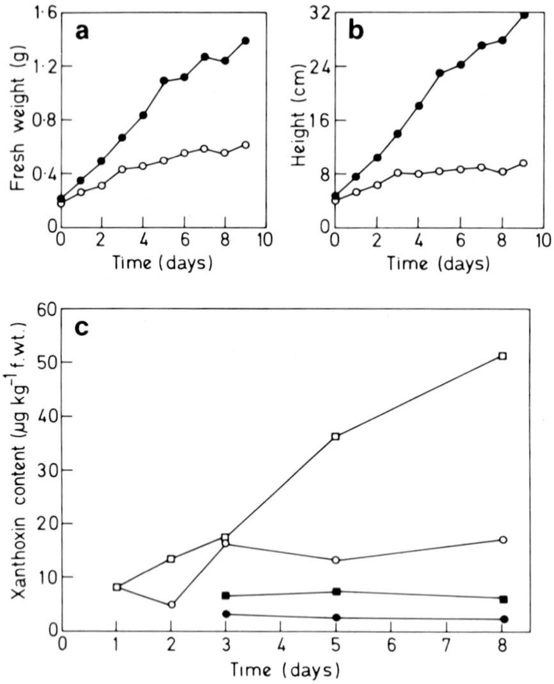

Fig. 7. **a** Fresh weight increase for R- or dark-treated dwarf pea seedlings (cv. Meteor). ● Darkness; ○ 1 min R per 29 min dark. **b** Increase in height for R- or dark-treated dwarf pea seedlings (cv. Meteor). ● Darkness; ○ 1 min R per 29 min dark. **c** Xanthoxin content of Meteor pea seedlings grown in complete darkness or R. □ trans, trans-xanthoxin; R; ■ trans, trans-xanthoxin, darkness; ○ cis, trans-xanthoxin; R, ● cis, trans-xanthoxin; darkness. Time-scale refers to days after R-irradiation commenced. (ANSTIS et al. 1975)

hypocotyl elongation at less than 10^{-6} M (raphanusanin) or 0.03 mg l^{-1} (raphanusol A and B). Furthermore, all of the inhibitors were shown to increase in response to red light reaching levels at least twice that in the dark.

7 Concluding Remarks

It is significant that no attempts have yet been made to study the changes and fluxes in more than one hormone in any given experimental material. Since there is evidence, of varying reliability, implicating at least auxins, GA's, ethylene, cytokinins and various inhibitors in a range of de-etiolation phenomena, it is most important to undertake the task of studying the effects of light on all of these hormones, simultaneously, in one tissue. At all times, however, it will be crucially important to test whether observed effects of light on hormone levels, rates of synthesis, sites of transport, etc., precede (or are simultaneous with) and are *quantitatively* related to the ultimate growth response. Only rigorous demonstration of such quantitative relationships will provide the necessary reliable leads towards a final resolution of the possible role of hormones as intermediaries of photoreceptor function in de-etiolation.

References

Acton GJ, Murray PB (1974) The roles of auxin and gibberellin in reversing radiation inhibition of hypocotyl lengthening. Planta 117:219–226

Anstis PJP, Friend J, Gardner DCJ (1975) The role of xanthoxin in the inhibition of pea seedling growth by red light. Phytochemistry 14:31–35

Bandurski RS, Schulze A, Cohen JD (1977) Photo-regulation of the ratio of ester to free indole-3-acetic acid. Biochem Biophys Res Commun 79:1219–1223

Barnes MF (1972) Abscisic acid in pea shoots. Planta 104:182–184

Bassi PK, Spencer MS (1982) Effect of carbon dioxide and light on ethylene production in intact sunflower plants. Plant Physiol 69:1222–1225

Beevers L, Loveys B, Pearson JA, Wareing PF (1970) Phytochrome and hormonal control of expansion and greening of etiolated wheat leaves. Planta 90:286–294

Blaauw-Jansen G (1959) The influence of red and far-red light on growth and phototropism of the *Avena* seedling. Thesis, Univ Utrecht, Netherlands

Bottomley W, Smith H, Galston AW (1966) Flavonoid complexes in *Pisum sativum*. III. The effect of light on the synthesis of kaempferol and quercetin complexes. Phytochemistry 5:117–123

Bown AW, Reeve DR, Crozier A (1975) The effect of light on the gibberellin metabolism and growth of *Phaseolus coccineus* seedlings. Planta 126:83–91

Briggs WR (1963) Red light, auxin relationships and the phototropic responses of corn and oat coleoptiles. Am J Bot 50:196–207

Buhler B, Drumm H, Mohr H (1978) Investigations on the role of ethylene in phytochrome-mediated photomorphogenesis. I. Anthocyanin synthesis. Planta 142:109–117

Burden RS, Firn RD, Hiron RWP, Taylor HF, Wright STC (1971) Induction of plant-growth inhibitor xanthoxin in seedlings by red light. Nature 234:94–96

Burg SP (1973) Ethylene in plant growth. Proc Natl Acad Sci USA 70:591–597

Burg SP, Burg EA (1967) Auxin-stimulated ethylene formation; its relationship to auxin-inhibited growth, root geotropism and other plant processes. In: Wightman F, Setterfield G (eds) Biochemistry and physiology of plant growth substances. Runge, Ottawa, pp 1275–1294

Burg SP, Burg EA (1968) Ethylene formation in pea seedlings: its relation to the inhibition of bud growth caused by indole-3-acetic acid. Plant Physiol 43:1069–1074

Cooke RJ, Kendrick RE (1976) Phytochrome-controlled gibberellin metabolism in etioplast envelopes. Planta 131:303–307

Cooke RJ, Saunders PF (1975a) Phytochrome-mediated changes in extractable gibberellin activity in a cell-free system from etiolated wheat leaves. Planta 123:299–302

Cooke RJ, Saunders PF (1975b) Photocontrol of gibberellin levels as related to the unrolling of etiolated wheat leaves. Planta 126:151–160

Craker LE., Abeles FB, Shropshire W (1973) Light-induced ethylene production in sorghum. Plant Physiol 51:1082–1083

Crozier A, Audus LJ (1968) Distribution of gibberellin-like substances in light and dark grown seedlings of *Phaseolus multiflorus*. Planta 83:207–217

De Greef JA, Frederick H (1983) Photomorphogenesis and hormones. In: Shropshire W, Mohr H (eds) Photomorphogenesis. Encyclopedia of plant physiology, New Ser vol 16. Springer, Berlin Heidelberg New York, pp 401–427

DeLaat AMM, Brandenburg DCC, Van Loon LC (1981) The modulation of the conversion of 1-aminocyclopropane-1-carboxylic acid to ethylene by light. Planta 153:193–200

Dörffling K (1973) Regulation of internode growth in pea seedlings by light and abscisic acid. Z Pflanzenphysiol 70:131–137

Erez A (1977) The effect of different portion of the sunlight spectrum on ethylene evolution in peach (*Prunus persica*) apices. Physiol Plant 39:285–289

Evans A (1975) Light and plant development: Proc 22nd Univ Nott. Easter School in Agric Sci 1975. Butterworths, London, pp 129–141

Evans A, Smith H (1976a) Localisation of phytochrome in etioplasts and its regulation in vitro of gibberellin levels. Proc Natl Acad Sci USA 73:138–142
Evans A, Smith H (1976b) Spectrophotometric evidence for the presence of phytochrome in the envelope membranes of barley etioplasts. Nature 259:323–325
Fletcher RA, Zalik S (1964) Effect of light quality on growth and free IAA content in *Phaseolus vulgaris*. Plant Physiol 39:328–331
Fletcher RA, Zalik S (1965) Effects of light of several spectral bands on the metabolism of radioactive IAA in bean seedlings. Plant Physiol 40:549–552
Furuya M (1962) Thesis, Yale University
Furuya M, Torrey JG (1964) The reversible inhibition by red and far-red light of auxin-induced lateral root initiation on isolated pea roots. Plant Physiol 39:987–991
Galston AW (1970) The relation of flavonoids and peroxidase activity to the control of growth by phytochrome and homones. Am J Bot (Abstr) 57:764
Galston AW, Hillman WS (1957) Inductive control of indoleacetic acid oxidase activity by red and near infrared light. Plant Physiol 32:129–135
Galston AW, Hillman WS (1961) Encyclopedia of plant physiology, vol 14. Nucleic acids in proteins and plants. Parthier B, Boulter D (eds) Springer, Berlin Heidelberg New York, pp 647–670
Galston AW, Tuttle AA, Penny PJ (1964) A kinetic study of growth movements and photomorphogenesis in etiolated pea seedlings. Am J Bot 51:853–858
Gepstein S, Thimann KV (1980) The effect of light on the production of ethylene from 1-aminocyclopropane-1-carboxylic acid by leaves. Planta 149:196–199
Goeschl JD, Pratt HK, Bonner BA (1967) An effect of light on the production of ethylene and the growth of the plumular portion of etiolated pea seedlings. Plant Physiol 42:1077–1080
Grodzinski B, Boesel I, Horton RF (1982) Ethylene release from leaves of *Xanthium strumarium* L. and *Zea mays* L. J Exp Bot 33:344–354
Grodzinski B, Boesel I, Horton RF (1983) Light stimulation of ethylene release from leave of *Gomphrena globosa* L. Plant Physiol 71:588–593
Hasegawa K, Hase T (1978) Light growth inhibition and growth inhibitors in Sakurajima radish seedlings. Plant Cell Physiol 19:1077–1083
Hasegawa K, Hase T (1981) Raphanusol B: A growth inhibitor of light-grown radish seedlings. Plant Cell Physiol 22:303–306
Hasegawa K, Miyamoto K (1980) Raphanusol A: a new growth inhibitor of light-grown radish seedlings. Plant Cell Physiol 21:363–366
Hasegawa K, Shihara S, Iwagawa T, Hase T (1982) Isolation and identification of a new growth inhibitor, raphanusanin from radish seedlings and its role in light inhibition of hypocotyl growth. Plant Physiol 70:626–628
Hasegawa K, Koreeda M, Hase T (1983) A new growth inhibitor, pisumin, involved in light inhibition of epicotyl growth of dwarf peas. Plant Physiol 72:391–393
Henshall JD, Goodwin TW (1964) The effect of red and far-red light on carotenoid and chlorophyll formation in pea seedlings. Photochem Photobiol 3:243–247
Hewett EW, Wareing PF (1973) Cytokinins in *Populus × robusta* (Schneid): light effects on engogenous levels. Planta 114:119–129
Higgins TJV, Bonner BA (1974) Natural inhibitors in normal and dwarf peas. J Exp Bot 25:705–714
Hillman WS (1949) Photoperiodism and related phenomena in plants and animals. Am Assoc Adv Sci Washington DC, pp 181–196
Hilton JR (1983) The influence of phytochrome photoequilibria on gibberellin-like substances and chlorophyll content of chloroplasts of *Hordeum vulgare* L. New Phytol 95:545–548
Hilton JR (1984) Further analysis of the in vitro phytochrome mediation of the amount of gibberellin-like substances in etioplast-enriched suspensions isolated from *Hordeum vulgare* L. leaves. New Phytol 96:161–166
Hilton JR and Smith H (1980) The presence of phytochrome in purified barley etioplasts and its in vitro regulation of biologically-active gibberellin levels in etioplasts. Planta 148:312–318

Janes HW, Loercher L, Frenkel C (1976) Effects of red light and ethylene on growth of etiolated lettuce seedlings. Plant Physiol 57:420–423

Jones JF, Kende H (1979) Auxin-induced ethylene biosynthesis in subapical stem sections of etiolated seedlings of *Pisum sativum* L. Planta 146:649–656

Jones RL (1967) Agar diffusion techniques for estimating gibberellin production by plant organs; the discrepancy between extractable and diffusible gibberellins in pea. In: Wightman F, Setterfield G (eds) Biochemistry and physiology of plant growth substances. Runge, Ottawa, pp 73–84

Kang BG (1971) Phytochrome-controlled leaf unrolling and protein synthesis. Plant Physiol 47:352–356

Kang BG, Burg SP (1972a) Involvement of ethylene in phytochrome-mediated carotenoid synthesis. Plant Physiol 49:631–633

Kang BG, Burg SP (1972b) Relation of phytochrome-enhanced geotropic sensitivity to ethylene production. Plant Physiol 50:132–135

Kang BG, Burg SP (1973) Role of ethylene in phytochrome-induced anthocyanin synthesis. Planta 110:227–235

Kang BG, Ray PM (1969a) Ethylene and carbon dioxide as mediators in the response of the bean hypocotyl hook to light and auxins. Planta 87:206–216

Kang BG, Ray PM (1969b) Role of growth regulators in the bean hypocotyl hook-opening response. Planta 87:193–205

Kang BG, Newcomb W, Burg SP (1971) Mechanism of auxin-induced ethylene production. Plant Physiol 47:504–509

Kao CH, Yang SF (1982) Light inhibition of the conversion of 1-aminocyclopropane-1-carboxylic acid to ethylene in leaves is mediated through carbon dioxide. Planta 155:261–266

Kays SE, Kende H (1971) The level of (+)-abscisic acid in dwarf pea shoots. Naturwissenschaften 58:524–525

Kende H, Lang A (1964) Gibberellins and light inhibition of stem growth in peas. Plant Physiol 39:435–440

Klein WH, Withrow RB, Elstad VB (1956) Response of the hypocotyl hook of bean seedlings to radiant energy and other factors. Plant Physiol 31:289–294

Köhler D (1965) Über den Gibberellingehalt von Zwerg- und Normalerbsen in Rotlicht und die Wirkung von Chlorocholinchlorid auf das Wachstum der Erbsen. Planta 65:218–224

Köhler D (1966) Die Abhängigkeit der Gibberellinproduktion von Normalerbsen vom Phytochromsystem. Planta 69:27–33

Köhler D, Lang A (1963) Evidence for substances in higher plants interfering with the response of dwarf peas to gibberellin. Plant Physiol 38:555–560

Köhler KH, Dörfler M, Göring H (1980) The influence of light on the cytokinin content of *Amaranthus* seedlings. Biol Plant 22:128–134

Kondo K, Watanabe A, Imaseki H (1975) Relationships in actions of indoleacetic acid, benzyladenine and abscisic acid in ethylene production. Plant Cell Physiol 16:1001–1007

Liverman J, Bonner J (1953) The interaction of auxin and light in the growth responses of plants. Proc Natl Acad Sci USA 39:905–916

Longo CP, Longo GP, Lampugnani M, Rossi G, Servettaz O (1981) Light and fusicoccin as tools for discrimination among responses of cotyledons to cytokinins. In: Guérin J, Péaud-Lenoel C (eds) Metabolism and molecular activities of cytokinins.Springer, Berlin, pp 261–266

Loveys BR (1979) The influence of light quality on levels of abcisic acid in tomato plants and evidence for a novel abscisic acid metabolite. Physiol Plant 46:79–84

Loveys BR, Wareing PF (1971a) The red-light-controlled production of gibberellin in etiolated wheat leaves. Planta 98:109–116

Loveys BR, Wareing PF (1971b) The hormonal control of wheat leaf unrolling. Planta 98:117–127

Lyon CJ (1970) Ethylene inhibition of auxin transport by gravity in leaves. Plant Physiol 45:644–646

Mor Y, Spiegelstein H, Halevy AH (1983) Inhibition of ethylene biosynthesis in carnation petals by cytokinin. Plant Physiol 71:541–546
Muir RM (1970) The control of growth by the synthesis of IAA and its conjugation. In: Carr DJ (ed) Plant growth substances 1970. Springer, Berlin Heidelberg New York, pp 96–101
Muir RM (1974) Effect of red light on coleoptile growth. Plant Physiol 54:286–288
Muir RM, Zhu L (1983) Effect of light in the control of growth by auxin and its inhibitor(s) in the sunflower. Physiol Plant 57:407–410
Mumford FE, Smith DH, Heytler PG (1964) The effect of red light on the flavonoid composition of etiolated pea plumules. Biochem J 91:517–522
Musgrave A, Kays SE, Kende H (1969) In vivo binding of radioactive gibberellins in dwarf pea shoots. Planta 89:165–177
Neskovic M, Konjevic R (1974) The non-reversible effects of red and far-red light on the content of gibberellin-like substances in pea internodes. J Exp Bot 25:733–739
Potts WC, Reid JB, Murfet IC (1982) Internode length in *Pisum*. I. The effect of the Le/le gene difference on endogenous gibberellin-like substances. Physiol Plant 55:323–328
Powell RD, Griffith MM (1960) Some anatomical effects of kinetin and red light on discs of bean leaves. Plant Physiol 35:273–275
Rajagopal R, Bulard C (1975) Auxin balance in the *Avena* coleoptile: effect of light-quality. Physiol Vég 13:1–11
Reed WA (1971) Investigations on the rapid phytochrome-induced inhibition of *Tropaeolum* stem elongation. Plant Physiol 47:Abstr 10
Reid DM (1983) Gibberellins and phytochrome. In: Crozier A (ed) The biochemistry and physiology of gibberellins. vol 2. Praeger, New York, p 452
Reid DM, Clements JB (1968) RNA and protein synthesis: pre-requisites of red light-induced gibberellin synthesis. Nature 219:607–609
Reid DM, Clements JB, Carr DJ (1968) Red light induction of gibberellin synthesis in leaves. Nature 217:580–582
Reid DM, Tuing MS, Durley RC, Railton ID (1972) Red light-induced conversion of ^3H-GA_9 into other gibberellin-like substances in homogenates of etiolated barley leaves. Planta 108:67–75
Reid JB (1983) Internode length in *Pisum*. Do the internode length genes affect growth in dark-grown plants? Plant Physiol 72:759–763
Russell DW, Galston AW (1969) Blockage of gibberellic acid of phytochrome effects on growth, auxin responses, and flavonoid synthesis in etiolated pea internodes. Plant Physiol 44:1211–1216
Sakai S, Imaseki H (1971) Auxin-induced ethylene production by mungbean hypocotyl segments. Plant and Cell Physiol 12:349–359
Samimy C (1978) Effect of light on ethylene production and hypocotyl growth of soybean seedlings. Plant Physiol 61:772–774
Scott RA, Liverman JL (1957) Control of etiolated bean leaf-disc expansion by gibberellins and adenine. Science 126:122–124
Sherwin JE, Furuya M (1973) A red-far red reversible effect on uptake of exogenous indole-acetic acid in etiolated rice coleoptiles. Plant Physiol 51:295–298
Smith H (1975) Phytochrome and photomorphogenesis. McGraw-Hill, London, pp 33–158
Smith H (1980) Gibberellins in photomorphogenesis. In: Brit Plant Growth Regulator Group Monograph #5, Gibberellins
Smith H (1982) Light quality, photoperception and plant strategy. Annu Rev Plant Physiol 33:481–518
Smith H, Morgan DC (1983) The function of phytochrome in nature. In: Shropshire W, Mohr H (eds) Encyclopedia of plant physiology (NS), vol 16. Springer, Berlin, pp 491–517
Sundquist C, Briggs WR (1982) The effect of delta-ALA on red-light-induced unrolling of dark-grown barley leaf sections. Physiol Plant 54:131–136

Tanada T (1968) A rapid photoreversible response of barley root tips in the presence of IAA. Proc Natl Acad Sci USA 59:376–380
Taylor JS, Wareing PF (1979) The effect of light on the endogenous levels of cytokinins and gibberellins in seeds of Sitka spruce (*Picea sitchensis* Carriere). Plant Cell Environ 2:173–179
Tillberg E (1974a) Levels of indole 3yl-acetic acid and acid inhibitors in green and etiolated bean seedlings (*Phaseolus vulgaris*). Physiol Plant 31:106–111
Tillberg E (1974b) Occurrence of endogenous indole 3yl-aspartic acid in light and dark-grown bean seedlings (*Phaseolus vulgaris*). Physiol Plant 31:271–274
Tucker DJ (1976) Effects of far-red light on the hormonal control of side shoot growth in the tomato. Ann Bot 40:1033–1042
van Staden J (1973) Changes in endogenous cytokinins of lettuce seed during germination. Physiol Plant 23:222–227
van Staden J, Wareing PF (1972) The effect of light on endogenous cytokinin levels in seeds of *Rumex obtusifolius*. Planta 104:126–133
Veroustraete F, Fredericq H, Van Wiemeersch L, De Greef J (1982) Specific photoregulation by phytochrome of epinasty and light-induced ethylene production in *Marchantia polymorpha*. Photochem Photobiol 35:261–264
Vince D (1968) Growth and anthocyanin synthesis in excised *Sorghum* internodes. I. Effects of growth-regulating substances. Planta 82:261–279
Virgin HI (1961) Action spectrum for the elimination of the lag phase in chlorophyll formation in previously dark-grown leaves of wheat. Physiol Plant 14:439–452
Walton JD, Ray PM (1981) Evidence for receptor function of auxin-binding sites in maize. Red light inhibition of mesocotyl elongation and auxin binding. Plant Physiol 68:1334–1338
Yoshii H, Imaseki H (1981) Biosynthesis of auxin-induced ethylene. Effects of indole-3-acetic acid, benzyladenine and abscisic acid on the endogenous levels of 1-aminocyclopropane-1-carboxylic acid (ACC) and ACC synthetase. Plant Cell Physiol 22:369–379

9 Photoperiod and Hormones

D. VINCE-PRUE

1 General Concepts of Photoperiodism

1.1 Range of Responses

Photoperiod has been shown to affect the tissue levels of hormones in all of the major categories. More difficult to assess is the precise role of these variations in controlling the many responses of plants to daylength. The great range of photoperiodic behavior (see VINCE-PRUE 1975) precludes discussion of the hormonal regulation of all daylength-controlled phenomena and this review, therefore, concentrates on selected responses in three major areas:

a) vegetative growth (stem elongation),
b) dormancy phenomena (the formation of storage organs and resting buds),
c) reproductive behavior (the formation of floral primordia).

1.2 The Role of Leaves

The first experimental evidence for the idea that hormones are formed in particular daylength cycles was the discovery that photoperiodic perception can take place in the leaves, whereas the target sites are usually situated elsewhere and often at some distance. KNOTT (1934) was the first to demonstrate that a stimulus from leaves exposed to LD caused initiation of floral primordia at the terminal growing point (of *Spinacia oleracea* L. cvs. Old Dominion and Virginia Savoy). HAMNER and BONNER (1938) subsequently showed that SD treatment of a single leaf of *Xanthium strumarium* L. would cause the development of flower buds even when all the other leaves remained in LD. Vegetative responses such as bulbing, tuberization, dormancy, and cold acclimation have also been shown to depend on the daylength experienced by the leaves (HAMNER and LONG 1939, ESASHI 1960, HEATH and HOLDSWORTH 1948, HURST et al. 1967, WAXMAN 1957, NITSCH 1959). These, and other experiments involving exposure of selected plant parts or treatment of detached leaves prior to grafting [e.g., in *Perilla crispa* (THUNB.) TANAKA (ZEEVAART 1958)], have established that, in general, the daylength signal is perceived by the leaves and it is immaterial whether or not the site of response is subjected to the controlling photoperiod. Consequently, control by daylength must involve the generation in the leaves of one or more transmissible stimuli which are transported to the target sites where particular responses are evoked. The stimulus generated by daylength thus exhibits the classical properties of a hormone.

Results from a variety of experiments have shown that the floral stimulus is exported from the leaves through living tissue and probably mainly through the phloem. These include girdling, localized heating or cooling, and narcotic treatment of the petiole (LANG 1965). Grafting experiments indicate that a functional phloem connection is usually necessary for (or at least facilitates) transport of the floral stimulus, which does not move without some tissue connection (WITHROW and WITHROW 1943, ZEEVAART 1958, WELLENSIEK 1970, ZEEVAART 1979). Movement only in living tissue was also reported for the tuber-forming stimulus in *Helianthus tuberosus* L. (HAMNER and LONG 1939) and potato [*Solanum tuberosum* L. (GREGORY 1956)].

Studies on the rate of transport have yielded conflicting results. Rates of ca. 1–3 mm h^{-1} were estimated for SDP *Pharbitis nil* CHOISY and *Perilla* (IMAMURA and TAKIMOTO 1955, CHAILAKHYAN 1940) and LDP *Lolium temulentum* L. (EVANS and WARDLAW 1966). Later experiments with *Pharbitis,* however, indicated a much faster rate of transport of 24–37 cm h^{-1} (KING et al. 1968) or 50 cm h^{-1} (TAKEBA and TAKIMOTO 1966), comparable with the rate of movement of assimilates. Apparent differences between transport of the stimulus in *Pharbitis* and *Lolium* led to the suggestion that the stimuli exported from their leaves are not identical (EVANS 1969). However, the results of grafting experiments indicate that the floral hormone of at least some LDP and SDP may be similar, if not identical, (see p. 334) and there is no good evidence for the suggestion that the graft-transmissible stimulus may differ from the immediate transmissible product of photoperiodic induction in the leaf (ZEEVAART 1971c).

In addition to the generation in appropriate photoperiods of a stimulus which promotes flowering, leaves in inappropriate (noninductive) daylengths may produce a transmissible inhibitor of flowering, although there is evidence that these leaves can also act as "sinks" for substances required in flower evocation and/or development (see LANG 1965, KING and ZEEVAART 1973). A few plants will flower in noninductive cycles if the leaves are removed, indicating that inhibitory substances originating in the leaves act at the apex; examples include the LDP *Hyoscyamus niger* L. (LANG and MELCHERS 1943, LANG 1965) and the SDP strawberry, *Fragaria* × *ananassa* DUCH. (THOMPSON and GUTTRIDGE 1960). Transmissible inhibitors of flowering are also indicated for *Lolium* where leaves in SD reduced the effect of a leaf in LD, even when they were inserted below it, and did not affect movement of assimilates from the LD leaf (EVANS 1960, EVANS and WARDLAW 1966). Removal of the inhibitory SD leaves at various times suggested that transport of the inhibitor preceded transport of the floral stimulus. In *Chenopodium rubrum* L., however, an inhibitory effect occurred after 30 h dark even though the floral stimulus arrived at the apex after 22.5 h; consequently an inhibitory effect arising in the leaf after 24 to 30 h dark appeared to be translocated to the apex (KING 1975). Inhibitory and promotive substances are also indicated for other daylength-controlled processes such as the induction of cold-hardiness (IRVING and LANPHEAR 1967, 1968).

The production of transmissible promotors and inhibitors of flowering has also been demonstrated in grafting experiments (LANG et al. 1977). Flowering of the day-neutral tobacco, *Nicotiana tabacum* L. cv. Trapezond was accelerated (compared with Trapezond-Trapezond grafts) when grafted to the SDP tobacco cv. Maryland Mammoth (grafts in SD) or to the LDP *N. sylvestris* L. (grafts in LD); these results indicate a transmissible flower promoter from both the SDP Maryland Mammoth and the LDP *N. sylvestris* in appropriate daylengths. However, when the grafts were maintained under daylengths inappropriate for the donor, there was little delay in flowering of Trapezond receptors grafted to Maryland Mammoth, whereas when grafted to *N. sylvestris,* the Trapezond receptors remained vegetative and their growth was dwarfed.

Thus a potent inhibitor of flowering appears to be produced in *N. sylvestris,* which seems not to be present in Maryland Mammoth. Donor plants of *Hyoscyamus* under SD were also extremely inhibitory to flowering of Trapezond receptors, as might be expected from the results of the defoliation experiments with *Hyoscyamus* cited above. To gain some insight into the interaction of the floral stimulus and the inhibitor, double grafts were made with varying numbers of *N. sylvestris* and Maryland Mammoth leaves: the results indicated a rather simple "stoichiometric" interaction, most probably in the shoot meristems (LANG 1980).

Grafting experiments have shown that some SDP also produce a floral inhibitor. When leaves of the SDP *Coleus fredericii* were taken from plants in LD, they delayed flowering of both *C. fredericii* and the day-neutral *C. blumei* (JACOBS 1980). Similarly, the SDP Maryland Mammoth delayed flowering when grafted to the LDP *N. sylvestris,* although not when grafted to the day-neutral variety Trapezond (LANG 1980).

A considerable body of physiological evidence thus indicates that control by daylength may be effected by a balance between the production of a transmissible promotive substance(s) for the particular phenomenon and an inhibitory substance(s), one or the other predominating. This conclusion is supported by the spectral studies discussed below.

1.3 Photoperception and the Photoperiodic Mechanism

1.3.1 Long-Night Processes

In photoperiodism, plants respond to the duration and timing of light and dark periods in the daily cycle. For several species it has been shown that the measurement of time is primarily concerned with the duration of the dark period; long days coupled with long nights give a SD-type of response, while short days with short nights give a LD response (HAMNER and BONNER 1938, HAMNER 1940). Thus, the onset of darkness appears to initiate a process which, after a sufficiently long period, results in a long night (i.e., SD) response. If the dark period is too short, no response occurs.

The discovery that the effect of a long night may be annulled when it is interrupted near the middle with a brief (<30 min) exposure to light (see VINCE-PRUE 1975) emphasized further the importance of the reactions that occur in darkness. When a short exposure to light (as a "night-break") is given at the appropriate time, usually several hours into the dark period, the dark response is prevented. The action spectrum for this night-break effect is similar to that for phytochrome-mediated responses (PARKER et al. 1946, 1950, SAJI et al. 1983). The R/FR reversibility phenomenon has been demonstrated for many night-break responses, confirming that phytochrome is the controlling pigment. These include the inhibition of flowering in SDP and the promotion of flowering in LDP (DOWNS 1956); the control of dormancy and the prevention of cold acclimation (VINCE-PRUE 1975, 1984, WILLIAMS et al. 1972, MCKENZIE et al. 1974): and the inhibition of tuber formation (ESASHI 1966, BATUTIS and EWING 1982).

The response to R in the night-break effect indicates the presence of the red-light absorbing form of phytochrome, P_r. Until fairly recently it was generally accepted that this P_r was formed from P_{fr} (established during daylight) by reversion during the intervening hours of darkness or that P_{fr} was lost by decay, the appearence of P_r requiring new synthesis of phytochrome. A problem with this interpretation is that, in light-grown plants, most of the P_{fr} is relatively stable in darkness ($T_{\frac{1}{2}}$=7–9 h at 25° C) and the synthesis of new phytochrome occurs much more slowly than in dark-grown plants (HEIM et al. 1981). Moreover, the stable P_{fr} has been shown to be required for flowering in several SDP (SAJI et al. 1982, KING et al. 1982 and see VINCE-PRUE 1983a). Consequently, in relation to flowering, the precise changes that occur in phytochrome during an inductive dark period are not yet resolved. Several hypotheses have been put forward to explain the observed responses; these include (1) separate pools of phytochrome with different functions and different rates of P_{fr} loss in darkness and (2) a reaction of phytochrome in the night-break response which requires turnover between the P_{fr} and P_r forms. For a detailed consideration of the actions of phytochrome in photoperiodic responses and their relationship with time measurement see VINCE-PRUE (1982, 1983a, b) and LUMSDEN and VINCE-PRUE (1984). However, irrespective of the mechanism through which phytochrome controls these photoperiodic responses, it is clear that after a particular night length is achieved, reactions occur (unless P_{fr} is re-established in the leaf) which lead to the observed long night response.

1.3.2 Long-Day Processes

In many LD responses, light appears not only to counteract the effect of darkness but also to have a positive effect which requires long daily exposures (VINCE-PRUE 1979, 1983a). In the LDP *Lolium,* for example, the inhibitory effect of a long night can partly be counteracted by giving a relatively short night-break with R at the appropriate time (VINCE 1965) or by anaerobic conditions (EVANS 1962), but a much greater flowering response is obtained by exposing plants to a single long photoperiod. Flowering in many other LDP is also accelerated considerably more by long daily photoperiods than by a night-break of 0.25 to 0.5 h, which is generally considered sufficient to saturate the photoconversion of P_r to P_{fr} (see VINCE 1970, VINCE-PRUE 1975, 1983a).

The spectral characteristics of such LD responses are well documented (EVANS 1971, VINCE 1969, VINCE-PRUE 1975, 1976, 1979). In contrast to the

Table 1. Examples of responses of plants to daylength. (After VINCE-PRUE 1976; Table 22.2, p. 362)

	Red before SD	Red after SD	SD
Light-dominant responses[a]:			
Lolium temulentum Ba 3081 (LDP) % flowering	100	0	0
Fragaria × ananassa (SDP) % flowering	20	100	100
Picea abies % dormant plants	41	100	100
Long-night responses[b]:			
Fuchsia hybrida cv. Lord Byron (LDP) days to anthesis	30.4	31.7	Vegetative
Kalanchoe blossfeldiana (SDP) % flowering	0	0	100
Weigela florida % dormant plants	0	0	100

[a] In light-dominant responses, the effect of an 8–9-h day-extension with R light given immediately after a SD in sunlight resembles that of SD, whereas when day-extension with R precedes the SD, a different response is obtained. The prevention of a long-night response may contribute to the effect of long photoperiods in some cases, but this cannot be determined from these results.

[b] In long-night responses both types of sequence produce similar results, which differ from those in SD.

typical phytochrome photoconversion action spectrum (λ_{max} ~640–660 nm) obtained with short night-breaks, the long night-breaks of several hours which are necessary for maximum flowering in many LDP are more effective when mixtures of R and FR light are given (see VINCE-PRUE 1975, 1979, 1981, 1983a) and the λ_{max} of the action spectrum occurs at about 700–710 nm (SCHNEIDER et al. 1967, BLONDON and JACQUES 1970, IMHOFF et al. 1979). Although the R and FR regions of the spectrum are effective, R/FR reversibility has not been demonstrated. The evidence for phytochrome as the photoreceptor pigment is thus not conclusive.

A characteristic feature of LD responses is that the optimum ratio of R/R+FR light varies during the course of the light period, being higher in the first than in the second half of a 16-h daily photoperiod. For example, the sequence 8 h sunlight – 8 h R – 8 h D was essentially ineffective in promoting flowering of *Lolium* Ba 3081, whereas the sequence 8 h R – 8 h sunlight – 8 h D was strongly promotive (VINCE 1965). Clearly, flowering is here not determined solely by events which occur in darkness since the duration of darkness was the same in both sequences and was less than the critical night length for this strain of *Lolium*. Such sequences (Table 1) can be used to distinguish between those responses which are controlled primarily by a long-night process and those which also depend on a LD process: the latter have been termed "light–dominant" responses (by HILLMAN in VINCE-PRUE 1979). The difference is clearly seen in the flowering responses of the LDP, *Fuchsia hybrida* VOSS. cv. Lord Byron and *Lolium* Ba 3081. In *Fuchsia,* flowering is prevented by a long-night process and plants flowered when the dark period was shortened to 8 h, irrespective of the spectral sequence during the LD. In *Lolium,* on the other hand, a light-dominant process promoted flowering, which occurred in 16-h photoperiods only when 8 h R preceded the 8 h day in sunlight; shortening the dark period to 8 h did not result in flowering if the spectral sequence during the LD was inappropriate. In *Lolium,* the changing sensitivity to R and FR light has been shown to depend on a change in the response to P_{fr}, which is inhibitory to flowering at certain times in the daily cycle (HOLLAND and VINCE 1971): however, in *Hordeum vulgare* Wintex, a change in sensitivity to a "high–irradiance reaction" of phytochrome has been proposed (DEITZER et al. 1979). Both plants show a rhythmic change in sensitivity to FR with a 24-h periodicity.

Examples of both light-dominant and long-night processes are found in LDP and SDP (Table 1). Flowering in many SDP appears to depend primarily on a long-night, promotive process but in strawberry (VINCE-PRUE and GUTTRIDGE 1973), flowering appears to be controlled by a light-dominant inhibitory process. Similarly, in most LDP there is evidence for the generation of a flower-promoting stimulus in LD but *Fuchsia* is an example of an LDP where the control of flowering is primarily via an inhibitory effect of SD/long-night cycles (VINCE-PRUE 1976). Dormancy phenomena in woody plants also exhibit both kinds of response (see Table 1 and p. 323). This type of spectral analysis may provide some insight into whether promotive or inhibitory stimuli are likely to be involved in daylength-controlled processes, although in *Pisum sativum,* where a light-dominant response appears to predominate under natural photoperiods (see *Pisum sativum* in VINCE-PRUE 1979), long days lead to the suppression of production of an inhibitor of flowering (PROEBSTING and HEFTMAN 1980).

The exact mechanism of action of phytochrome in photoperiodism is not understood but it is known that a consequence of this action is the generation in the leaf of transmissible stimuli which ultimately give rise to the overt physiological responses. The possible nature of these daylength-controlled stimuli and their function in selected photoperiodic responses is considered in the remainder of this chapter.

2 Vegetative Growth: Stem Elongation

Photoperiodism is commonly associated with reproductive changes, but many aspects of vegetative growth are also profoundly affected by daylength. While some of these vegetative responses are closely linked with the induction of flowering, they are not necessarily dependent upon it. Even where there is a close association under natural conditions, vegetative and floral responses can often be separated experimentally [e.g., stem elongation and flowering in *Silene armeria* L. by AMO-1618 treatment (CLELAND and ZEEVAART 1970)]. There are also many vegetative responses to daylength that are clearly independent of flowering, such as the induction of dormancy and the formation of storage organs. Although the induction of flowering is by far the most widely studied aspect of photoperiodism, it is appropriate in a hormonal context to consider first the vegetative phenomena, since the hormones which control some of these processes are known, whereas the identity of the floral stimulus engendered by inductive photoperiods remains an enigma.

Plants growing in LD are often considerably taller than their counterparts in SD and/or have much longer petioles. The differences in stem length can arise from the cessation of extension growth under SD [e.g., as with the onset of SD-induced dormancy in some woody species (see Sect. 3.2)]. Or elongation may be reduced under SD (relative to LD) in plants (woody and otherwise) which do not show SD-induced dormancy. Within this latter group it is possible to distinguish two categories of response:

- *rosette plants* where stem elongation normally accompanies flowering: these are mostly LDP;
- *caulescent plants* of various photoperiod response groups for flowering, where internode length is increased by LD.

2.1 Rosette Plants

In rosette LDP, flowering and stem elongation usually occur more or less together following transfer to LD (LANG 1965). Because of this close relationship, it has been suggested that the same hormonal stimulus, possibly a GA, controls both processes. In many rosette LDP, treatment with GA_3 in SD leads to both stem elongation and flowering, as, for example, in *Silene armeria, Samolus parviflorus* RAF. and *Hyoscyamus niger* (LANG 1957). Furthermore, treatment with the plant-growth retardants chlormequat and AMO-1618, which are known to inhibit endogenous GA biosynthesis (although both may also affect other metabolic processes) suppressed both flowering and stem elongation in *Samolus* (BALDEV and LANG 1965). However, despite their close association, the two processes of stem elongation and floral initiation can be separated in many plants by appropriate treatments. Stem elongation without flowering can be induced in a number of species by GA_3 application (for example in *Hieracium aurantiacum* L. and *H. pratense* TAUSCH (PETERSON and YEUNG 1972) and, in *Hyoscyamus,* by 11-h photoperiods at 20° C (LANG and MELCHERS 1943). Flowering without elongation was induced (in LD) by AMO-1618 in *Silene* (CLELAND and ZEEVAART 1970), by chlormequat in radish, *Raphanus sativus* L. cv. Miyashige-sofuto (SUGE and RAPPAPORT 1968), and (in SD) by high temperature in *Rudbeckia bicolor* NUTT. (MURNEEK 1940). Further evidence that the two phenomena depend on separate processes comes from the genetic analysis of *Silene* strains which has demonstrated that GA-induced stem elongation and flower formation are determined by two separate genes (WELLENSIEK 1973). Thus, although they are closely related, the photoperiodic control of stem elongation and flower initiation in at least some rosette LDP can be separated and probably result from different metabolic events.

2.1.1 Gibberellins

The content of endogenous GA's is generally higher in LD than in SD irrespective of photoperiodic classification (Table 2). Even though workers have used different extraction and assay techniques, and have not always restricted their extractions to leaves or apical buds, they have generally found that the transfer of LDP from SD to LD causes an increase in both GA content and rate of metabolism (ZEEVAART 1976). For example, in *Silene* after 2 to 6 LD, the increase in GA (especially diffusible GA) occurring in the region of the shoot apex was considered to be an important factor in the stimulation of elongation, as there was little increase in stem growth when GA synthesis was blocked with AMO-1618 (CLELAND and ZEEVAART 1970). The increase occurred largely in the stem tip (which included young leaves); there was little diffusible GA from mature leaves and their extractable GA was less in LD than in SD as measured by bioassay.

More significant than photoperiodic effects on total GA's may be those on GA metabolism. In *Silene*, ^{3}H-GA_5 was more rapidly metabolized in LD than in SD (VAN DEN ENDE and ZEEVAART 1971) and in the closely related plant, *Agrostemma githago*, substantial differences in GA metabolism between LD and SD have been reported (JONES and ZEEVAART 1980a). In *Agrostemma*, LD induced a large but transitory increase in a zone which co-chromatographed with GA_5 and GA_{20}, and this increase coincided with the onset of rapid elongation. Subsequently (JONES and ZEEVAART 1980b), this zone was found to contain GA_{17}, GA_{19}, GA_{20} and GA_{44}, while GA_1 and GA_{53} were identified in other bioassay-active zones. Transient peaks of GA_{17}, GA_{19}, GA_{20} and GA_{44} after 8 LD coincided with the onset of rapid elongation, while the GA_1 maximum occurred 4 days later. AMO-1618 studies indicated that for all (except GA_{53}), the rate of metabolism was higher after 12–16 LD than in SD. More direct evidence for these changes in metabolism was obtained by applying ^{3}H-labeled GA's (JONES and ZEEVAART 1982). In *Agrostemma*, the complexity of photoperiod-dependent changes in at least 7 GA's illustrates the difficulties of relating these to physiological responses. The authors concluded that the induction of elongation in *Agrostemma* by LD requires more than qualitative or simple quantitative changes in GA's. However, the changes in GA_{20} may be particularly important in view of results with spinach (see below) and the observations that GA_{20} elicited elongation in *Agrostemma* in SD and overcame AMO-1618 inhibition of elongation in LD (JONES and ZEEVAART 1980a).

In spinach, photoperiodic effects on GA metabolism have been intensively studied using a variety of techniques. Although the total GA content was not affected (ZEEVAART 1971a), LD conditions led to a decline in GA_{19} and an increase in GA_{20} and GA_{29} (METZGER and ZEEVAART 1980a, c). The increased levels of GA_{20} were accompanied by increases both in its rate of synthesis (from GA_{19}) and its conversion to the inactive GA_{29} (METZGER and ZEEVAART 1982): consequently, a large, rapidly turning over pool of GA_{20} was present in LD but not in SD. The proposed interconversion pathway from GA_{53} to GA_{29} was examined further by applying deuterium-labeled GA_{53} to spinach leaves and following its metabolism in LD and SD (GIANFAGNA et al. 1983).

After 2–6 SD, GA_{53} was metabolized to GA_{44} and GA_{19} while, after 2 LD, GA_{20} was the only metabolite found. This study supported the proposed $GA_{53} \rightarrow GA_{44}$, $GA_{19} \rightarrow GA_{20}$ pathway in spinach and afforded further evidence for the dependence of GA_{20} production on LD. In spinach, GA_{20} application has also been shown to mimic LD in causing petiole elongation (METZGER and ZEEVAART 1980c); moreover, the rise in GA_{20} in LD preceded the onset of stem elongation (METZGER and ZEEVAART 1980c). Consequently, there is good circumstantial evidence that the availability of GA_{20} is a limiting factor for stem extension in spinach in SD. Earlier experiments (METZGER and ZEEVAART 1980b) indicated that GA_{20} occurs in shoots but not in roots of spinach, and it was also found in phloem exudates but not in xylem. GA_{20} may, therefore, be synthesized in leaves and transported to the apex via the phloem. This would be consistent with the location of daylength perception in the leaf and supports the concept that $GA_{19} \rightarrow GA_{20}$ is under photoperiod control in the leaves; the mechanism for such control is, however, unknown, as is the mechanism for the increased metabolism of $GA_{20} \rightarrow GA_{29}$ in LD shoots.

As well as affecting the rate of metabolism and the content of GA's, LD may also increase the sensitivity to a given amount of GA. For example, both *Silene armeria* and spinach responded more to a given dose of GA_3 in LD than in SD (CLELAND and ZEEVAART 1970, ZEEVAART 1971a). Other examples of plants which have a greater sensitivity to applied GA_3 in LD include the winter annual, *Thlapsi arvensis,* where GA_3 was able to substitute for cold in inducing elongation (METZGER 1983); and strawberry, where stem elongation occurs in response to GA_3 application even though strawberry is normally a rosette plant in both SD and LD (TAFAZOLI and VINCE-PRUE 1978).

Although a SDP, the promotion of petiole elongation and the regulation of flowering in strawberry appear to depend on LD processes (VINCE-PRUE and GUTTRIDGE 1973, VINCE-PRUE et al. 1976). Exogenous GA_3 increased petiole length but only partly replaced the LD stimulus (GUTTRIDGE 1970) and it was concluded from the analysis of the interaction between daylength and applied GA_3 that factors other than GA's are important in the endogenous system. Nevertheless, growth-retardant application suppresses petiole elongation (GUTTRIDGE 1969), and the possibility is not excluded that the LD stimulus is a GA other than GA_3. A GA tentatively identified as GA_{19} (from bioassay and the R_f TLC) increased in the young leaves and crown after 9 SD plus night-breaks compared with the level in SD alone (UEMATSU and KATSURA 1983).

In conclusion, GA's appear to be closely associated with photoperiodic effects on stem and petiole elongation in rosette herbaceous plants. LD may increase the total content of GA, but more often influences the rate of metabolism between GA species, leading to transient increases in some of them; the pool of GA_{20} may be particularly important for these physiological responses in some plants. LD also increases the sensitivity of elongation responses to applied GA.

2.1.2 Inhibitors

It has been suggested that flower formation and stem elongation may be prevented under SD by high levels of endogenous inhibitors. An increased inhibitor content could also explain the observed decrease in response to a given dose of GA when applied in SD. However, despite an increase in the growth-inhibitory activity of extracts from *Spinacia* grown in SD (WAREING and EL-ANTABLY 1970), chemical analysis showed that one of the inhibitors (ABA) was actually two to three times higher in LD. The xanthoxin level remained unchanged (ZEEVAART 1971b, 1974, Table 2). A similar situation was found in chrysanthemum [*Chrysanthemum morifolium* (RAMAT) HEMSL. (SENGUPTA et al. 1974)]. The increased inhibitory activity of SD extracts may arise from a decrease in GA's, some of which may co-chromatograph with ABA and other inhibitors, thereby reducing the inhibitor's effect in certain bioassays (LENTON et al. 1972).

2.2 Caulescent Plants

Many caulescent plants develop longer internodes in LD than in SD, irrespective of their photoperiodic classification for flowering. The response appears to be quantitative, internode length being a function of the daily duration of light (DOWNS and BORTHWICK 1956a, VINCE-PRUE 1977). Spectral composition is also important and, with long daily light periods, elongation increases as the R/R+FR ratio decreases to that which sets the photoequilibrium of phytochrome at $P_{fr}/P \sim 0.3$ (MORGAN and SMITH 1976, VINCE-PRUE 1977). A further reduction in the R/R+FR ratio (for example by giving a day extension for several hours with FR alone) results in shorter internodes (VINCE-PRUE 1977, SATTER and WETHERELL 1968). Stem elongation is, therefore, a function both of daylength and of the R/R+FR content of the light.

There appear to have been few, if any, studies of hormonal effects or contents in such plants, nor is it known whether the effect of daylength is perceived in the leaves to result, as with other photoperiodic phenomena, in the production of a transmissible stimulus. Although the semi-quantitative response to photoperiod duration, together with the requirement for a low R/R+FR ratio, resembles the light-dominant response for the control of flowering, there is no evidence for a change in sensitivity to the R/R+FR ratio with time of day, and so it is probable that the mechanism differs from that which controls flowering.

3 Dormancy Phenomena

3.1 Storage Organs

Vegetative storage organs arise by lateral swellings of stems (tubers, corms), roots (root tubers) and leaves (bulbs). Their formation is usually accompanied by senescence and death of the rest of the plant and, once formed, they enter into a state of dormancy and function as perennating organs. In many species the formation of storage organs is under the control of daylength and, with the major exception of bulbs in *Allium*, it is usually favored by exposure to SD (see VINCE-PRUE 1975). The photoreceptor for the perception of daylength is phytochrome, and the mechanism is probably very similar to that for the photoperiodic control of flowering (ESASHI 1966, NITSCH 1965, BETRIA and MONTALDI 1974, VINCE-PRUE 1975, BATUTIS and EWING 1982). Leaves perceive the daylength signal (HAMNER and LONG 1939, ESASHI 1960, HEATH and HOLDSWORTH 1948, HAMMES and BEYERS 1973) and, therefore, are presumed to generate a stimulus which evokes tuberization and bulbing in the responsive region. As with flowering, this stimulus has been shown to be graft-transmissible in potato, where grafting induced stem pieces to noninduced ones led to tuber formation on the latter (GREGORY 1956). However, there is also evidence for the production (generally in LD) of an inhibitory stimulus (ESASHI 1961) and several authors have concluded that the formation of storage organs is consequent on a particular balance of inductive and inhibitory substances. In potato, the complexity of the responses of stolons and tubers to increasing induction by SD also suggests that the stimulus formed under daylength control may be a complex one (EWING and WAREING 1978). Although assimilates are mobilized into developing storage organs (for example, a high concentration of soluble carbohydrate has been found to be associated with tuberization in potato (BORAH and MILTHORPE 1962), excess assimilate alone does not induce their formation (KATO 1965) and, indeed, exposure to the appropriate daylength has induced bulb formation in onion plants which were so heavily shaded that many died (HEATH and HOLLIES 1965). In grafting experiments between sunflower (*Helianthus annuus* L.) and Jerusalem artichoke (*H. tuberosus*), tubers developed on the latter when leaves of the sunflower donor were maintained in SD, even though sunflower itself cannot form tubers; control grafts with leaves in LD did not form tubers (NITSCH 1965). Thus the formation of a tuber-evoking stimulus in leaves exposed to appropriate daylengths is not restricted to plants that are capable of forming tubers. On the other hand, the difference between clones of potato that can tuberize under LD and those that cannot resides in the ability of the leaf to form the stimulus under LD, as demonstrated by grafts between the two types (EWING and WAREING 1978). The grafting experiments between sunflower and Jerusalem artichoke indicate that no unique substances are involved and that specificity lies in the target cells. It seems likely that control by daylength is achieved through changes in known growth regulators, since in many cases these have been found effective (see below) in promoting or inhibiting the development of storage organs, but unknown factors have also been postulated (WAREING 1982).

3.1.1 Cytokinins

Tuber formation in vitro has been promoted by cytokinin (CK) in potato (PALMER and SMITH 1969, SMITH and PALMER 1970, MAUK and LANGILLE 1978), *Dioscorea batatas* DECN., *Ullucus tuberosus* Caldas (ASAHIRA and NITSCH 1968) and *Helianthus tuberosus* (COURDUROUX 1967), and CK has been isolated from tuber tissue of both *H. tuberosus* and potato (NITSCH and NITSCH 1960, TIZIO 1960). Cytokinin application was also found effective in inducing the formation of root tubers in *Ipomoea batatas* LAM. (SPENCE and HUMPHRIES 1972), bulbils in *Dioscorea bulbifera* (UDEBO 1971) and aerial bulblets in onion (THOMAS 1972).

The general effectiveness of CK's for plants with different photoperiodic requirements and different types of storage organs suggests that they may be of widespread importance in the initiation of storage organ growth. *Dahlia pinnata* CAV. cv. Choot Hashani may be an exception, since treatment of leaf cuttings with CK did not affect tuberization (BIRAN et al. 1974). The response of intact potato plants to applied CK's is also rather variable; although a foliar spray with kinetin had earlier been reported to increase tuber yield (TAKASHIMA et al. 1962, the exogenous application of BA was without effect (BADIZADEGAN et al. 1972).

It is by no means established, however, that the daylength-controlled stimulus is a CK. There is some evidence that the amount of endogenous CK exported from leaves can be affected by daylength (PHILLIPS and CLELAND 1972) and, in potato, a cis-zeatin riboside (ZR), the amounts of which varied with daylength, has been identified (MAUK and LANGILLE 1978). Following transfer to SD, there was a large increase in ZR in the belowground parts within 4 days: this increase preceded visible tuber formation, which was noted after 8 days. ZR also promoted tuberization in isolated stolons. However, although similar (but smaller) changes occurred in the aboveground tissues, the time course studies were not sufficiently detailed to determine whether or not the ZR originated in the induced leaves. Moreover, the cultivar used by MAUK and LANGILLE did not have an obligate SD requirement. When tested with a clone of *Solanum andigena,* which forms tubers only in SD, cuttings from LD plants failed to form tubers in response to either kinetin or BA application (WAREING 1982). Consequently, while there is evidence suggesting a role for CK's in tuber formation in potato, this is not clearly associated with photoperiodic effects.

In onion, transfer to LD resulted in a very large increase in the amount of CK's that could be extracted from the leaf blade: a maximum was reached on the 29th day but, by the 42nd day, there were no detectable CK's, as in the SD controls (LERCARI and MICHELI 1981). However, the relationship of these photoperiodically induced changes in the extractable CK levels in the leaf with the bulbing response to LD is obscure, since bulbing ceased when plants were returned to SD, even after the endogenous CK's had already fallen to the SD level.

3.1.2 Inhibitors

The evidence concerning the function of inhibitory substances in tuber and bulb induction is also inconclusive, although some experiments have indicated a role for ABA in tuber initiation. In potato, ABA was identified in tubers (MILBORROW 1967) and, when applied to leaves of *Solanum andigena,* ABA increased the number of plants which formed tubers in LD (EL-ANTABLY et al. 1967). Subsequently, other workers reported that ABA promotes tuberization in whole plants, in 1-node stem cuttings (MENZEL 1980) and when applied to stolon tips (KRAUSS and MARSCHNER 1976). The application of ABA also promoted the formation of tuberous roots in *Dahlia pinnata* (BIRAN et al. 1974) and *Helianthus tuberosus* (CHARNAY and COUDUROUX 1972). In contrast, SMITH and RAPPAPORT (1969) were unable to induce tuber formation in potato by spraying plants with ABA or by application to stolons in vitro (although it did inhibit stolon growth, which is a pre-requisite for tuber formation). Furthermore, ABA was strongly antagonistic to kinetin-induced tuberization in vitro (PALMER and SMITH 1969). The application of ABA also completely failed to induce the formation of aerial tubers in *Begonia evansiana* ANDR. under LD (HASHIMOTO and TAMURA 1969) and, as with potato, suppressed cytokinin-induced tuber swelling (ESASHI and LEOPOLD 1968).

Recent studies have provided further support for the proposal that ABA may be required for tuber initiation in the SD-requiring *Solanum andigena* although, as with CK's, there is little evidence that the SD effect itself operates by changing the transport or amount of the inhibitor. When 1-node cuttings with one leaf are taken from SD-induced plants, the axillary bud develops directly into a tuber (WAREING and JENNINGS 1980). When the leaf is removed, the axillary bud develops into an erect shoot; the application of ABA can replace the effect of the SD leaf. ABA cannot, however, cause tuber formation in cuttings taken from plants in LD, either with or without leaves. There is evidence that ABA is produced in leaves and transported out via the phloem: however, although ABA can be extracted from potato leaves, there are no differences in the amounts present in LD and SD (WAREING and JENNINGS 1980), nor were any differences detected in the amount of ABA transported out. WAREING (1982) has concluded that ABA is an essential part of the tuber stimulus produced by the leaf but that this "leaf effect" is equally supplied by LD or SD leaves of potato and can even be supplied by a tomato leaf which is effective in inducing tuber development when grafted onto a SD-induced stem cutting of *S. andigena*. Treatments that reduce endogenous ABA levels such as a continuous nitrogen supply (which also increased GA) inhibited tuber formation in SD (KRAUSS and MARSCHNER 1982), supporting further the proposal that ABA is part of the tuber-forming stimulus. It is probable that ABA promotes tuber formation by inhibiting stolon extension growth, since other treatments that inhibit extension growth also promote tuberization in induced cuttings (WAREING and JENNINGS 1980). It is worth noting that, in contrast to potato, SD has been reported to increase the level of endogenous ABA-like inhibitors in the leaves of *Dahlia*, where the application of exogenous ABA also promotes tuber formation (BIRAN et al. 1974). However, such an effect of daylength on the level of ABA is not supported by results from other plants (Table 2) and so may be an artifact.

3.1.3 Gibberellins

Endogenous GA's may be inhibitors of SD-promoted tuber formation. Stimulation of tuber development by the application of growth retardants to intact plants has been recorded for *Dahlia* (daminozide, BIRAN et al. 1972), *Begonia* (chlormequat, NAGAO and OKAGAMI 1966), artichoke (AMO 1618, COURDUROUX 1966) and potato (chlormequat, HAMMES 1971, HUMPHRIES and DYSON 1967). In contrast, the application of GA_3 prevented or reduced the growth of tubers in potato (in natural LD, BOOTH and LOVELL 1972), *Dahlia* (in LD or SD, BIRAN et al. 1974) and *Begonia* (in SD, NAGAO and OKAGAMI 1966,

OKAGAMI et al. 1977). Removal of the tip and young leaves of potato resulted in the development of tubers in noninductive photoperiods, while GA_3 application nullified this response (HAMMES and BEYERS 1973). Endogenous GA levels were found to be higher in LD (RAILTON and WAREING 1973) and the onset of tuberization was associated with a decrease in GA (OKAZAWA 1960, 1967). High endogenous GA levels in plants continuously supplied with nitrogen were also associated with the inhibition of tuberization in SD (KRAUSS and MARSCHNER 1982). Thus at least part of the effect of daylength on tuber formation may be mediated through changes in GA's, perhaps originating in the young leaves. However, although these young leaves may directly perceive the daylength signal, the effect of their removal, in potato, could not be imitated by exposing them to SD (HAMMES and BEYERS 1973). Tuberization in potato was enhanced by exposing mature leaves to SD, but an even greater effect was obtained by exposing the whole plant. It is clear that tuberization is not exclusively controlled by the young leaves, whether via GA and/or other substances.

The involvement of GA's with LD-promoted processes, such as bulbing, has been little studied. Application of GA_3 enhanced bulbing of *Hordeum bulbosum* L. under suboptimal photoinduction, but was ineffective without some exposure to LD: a non-GA component was inferred (OFIR 1976). However, in onion, the amount of GA-like substances in extracts from inner leaves was less after 26 LD than in SD (see BREWSTER 1977).

3.1.4 Auxins

Auxins may be involved in the LD-induced bulbing of onion. Transfer to LD resulted in a transient increase in extractable auxin from the leaf base, with a peak after about 5 LD (CLARK and HEATH 1962). In seedling sections without roots, immersion in IAA at 10^{-4} M led to bulbing in the dark; without IAA, bulbing occurred only in LD. Auxin extracted from the leaf bases of plants in LD substituted for IAA in this test, and it was concluded that the photoperiodic induction of bulbing was mediated via an increased level of endogenous auxin in LD. However, foliar sprays of 2,4-D at 20 mg l^{-1} retarded bulb formation in inductive daylengths (see BREWSTER 1977).

3.1.5 Ethylene

The application of ethylene changes the orientation of cell expansion to produce lateral enlargement rather than extension growth (BURG and BURG 1966). Ethylene is, therefore, of potential interest in relation to tuber initiation, which involves cessation of stolon elongation and the radial expansion of cells, and in bulbing, which involves lateral swelling of leaf cells.

In onion, application of the ethylene-releasing substance, CEPA, induced bulbing in noninductive conditions (LEVY et al. 1973), being effective when applied as a leaf spray, a soil drench, or by direct injection into the root growing zone. Seedlings with roots and sheaths constantly immersed in a CEPA solution of 1.5 mg l^{-1} began to bulb after only 7 days. As with LD treatment, CEPA inhibited the growth and emergence of leaves, as well as inducing bulb development. As high concentrations of exogenous auxin are known to stimulate ethylene evolution in treated tissues, it is possible that auxin-induced bulb formation in onion is related to ethylene formation. However, the relationship to bulbing of daylength effects on endogeneous auxin and ethylene levels does not appear to have been investigated.

Studies with potato tubers have produced inconsistent results but do not, in general, indicate a unique role for ethylene in tuber formation. Treatment of potato plants was reported to promote the development on stolons of subapical swellings devoid of starch (CATCHPOLE and HILLMAN 1969), but no effect was found in a subsequent experiment (LANGILLE 1972). In vitro treatment of stolons with CEPA did not induce tuberization although there was some generalized swelling of the stolon (PALMER and BARKER 1973). However, at 5 $\mu l\, l^{-1}$, ethylene strongly depressed kinetin-induced tuberization (MINGO-CASTEL et al. 1974), and the promotion by CO_2 (an ethylene antagonist) of tuber formation on isolated potato stolons (MINGO-CASTEL et al. 1974) suggests that endogenous ethylene may have an inhibitory role in tuberization.

In *Dahlia,* ethylene evolution from leaf cuttings was increased from 0.5 to 1.5 $nl\, g^{-1}\, h^{-1}$ following transfer from LD to SD; the increase occurred only in samples taken after 2 to 3 weeks in SD, the rate subsequently declining to that in LD (BIRAN et al. 1972). The enhanced rate of ethylene production preceded the onset of tuberization and was thought to be associated with the induction or initiation of tuber growth. Treatment with CEPA enhanced tuberization in LD but tuber production was substantially greater in SD. Tuberization is also stimulated in *Dahlia* by ABA (see p. 320).

3.1.6 Steroids

Transfer to SD resulted in changes in the sterol composition of leaves of the wild potato during the period of tuber induction, and the application of inhibitors of steroid biosynthesis prevented tuber formation (BAE and MERCER 1970). However, the relationship of steroid metabolism in leaves to the daylength regulation of tuber induction has not been studied.

3.1.7 Conclusions

The available evidence suggests that there may be no unique tuber- or bulb-evoking substances, but rather that the regulation of storage organ development involves interactions between several hormonal factors – some of which are influenced by photoperiod. The inhibition of tuber initiation by LD appears to result from increases or changes in GA's, and there is some evidence that both CK's and ABA are required for tuber initiation. In potato, ABA appears to originate in the leaves but the amount is not influenced by photoperiod: in contrast, CK's do increase in SD but have not unequivocally been shown to originate in the leaves. It is clear that an additional and as yet unidentified "tuber-promoting" stimulus is produced in leaves exposed to SD. Attempts to extract such a tuber-inducing factor from shoots of SD-induced plants of *Solanum andigena* have been unsuccessful (WAREING 1982).

The LD promotion of bulbing may be associated with auxin and/or ethylene metabolism in onion but, in other plants (e.g., in *Hordeum*), GA partially replaces the LD effect. Cytokinin metabolism may be implicated in LD bulbing, as with SD-dependent processes.

3.2 Resting Buds

In many woody species, particularly angiosperms of the temperature zones, the formation of resting buds is under photoperiodic control. As with the production of storage organs, this type of dormancy is usually favored by SD (see VINCE-PRUE 1975), although at least one example of LD-induced dormancy has been recorded, namely that in *Lunularia cruciata* (L.) DUM. a desert liverwort (SCHWABE and NACHMONY-BASCOMBE 1963). Again like storage organs, the photoperiodic mechanism that controls the formation of dormant buds seems to be basically the same as that which controls flowering. Phytochrome is the photoreceptor (VINCE-PRUE 1975, 1984, WILLIAMS et al. 1972, MCKENZIE et al. 1974), and the leaves are the photoreceptive sites (NITSCH 1959, WAXMAN 1957). In *Betula pubescens* EHRH., however, only the daylength to which the apical bud is exposed seems to be important and there may be no transmissible effect (WAREING 1954). A number of additional responses are associated with the formation of resting buds; these include leaf fall in deciduous species; decreased cambial activity (WAREING 1951); and increased cold hardiness (HURST et al. 1967). The cold-hardiness effect has been shown to be under the control of phytochrome. Although the leaves do not themselves become cold-hardy, they perceive the daylength signal (HURST et al. 1967) and produce a stimulus which appears to be transported through the phloem (FUCHIGAMI et al. 1971). Sometimes the resumption of growth by dormant buds is accelerated by LD (WAREING 1953, 1954, DOWNS and BORTHWICK 1956a, b), although usually the breaking of bud dormancy requires a period of low temperature. When the low temperature requirement is fully saturated, daylength often has little or no effect (see VINCE-PRUE 1975).

As with flowering, there may be two types of photoperiodic regulation of bud dormancy (VINCE-PRUE 1976, 1984). The first is a long-night response which induces dormancy and is prevented by a night-break given at the appropriate time or by any day-extension treatment which shortens the dark period to below the critical length. The second requires a long daily exposure to light and is not very sensitive to a night-break; this light-dominant type of response prevents dormancy. As discussed previously (p. 312), an 8 h day extension with R is much more effective when given before the main sunlight period than when given immediately following it. The two types of response are illustrated in Table 1.

In *Weigela florida* BUNGE the control of dormancy depends on a dormancy–promoting process which occurs during a long dark period, suggesting formation of a dormancy–inducing substance. In contrast, for first-year seedlings of *Picea abies* (L.) KARST, events which occur during the long photoperiod are important and dormancy is not controlled by night length alone; consequently the production of inhibitors of dormancy may be inferred. An indication of the relative importance of dormancy-inhibiting and dormancy-inducing substances in any particular case might be obtained from this approach, but few results are available. However, the responses of a large number of species to LD and to night-breaks were examined by NITSCH and SOMOGYI (1958) and their results also indicated two classes of behavior. A night-break of 30–60 min was just as effective as a LD in preventing the induction of dormancy in several cases [e.g., in *Rhus typhina* L., *Tsuga canadensis* (L.) CARR.], indicating that control was primarily by night length. In other species, however, night-breaks were almost without effect and long daily light periods were needed for continued growth (e.g., in *Juniperus communis* L., *Cercis canadensis* L.), indicating that there was a specific promotion of growth by LD in addition to any effect which resulted from preventing dark reactions.

It is clear from the foregoing discussion that photoperiodic regulation of dormancy may result from the production by leaves of dormancy-inducing and/or dormancy-preventing substances with the possibility that one or the other may predominate in particular species. Similarly, hardiness promoters and/or inhibitors are thought to be produced in leaves, but it is not known whether they are the same substances as those which control dormancy.

3.2.1 Growth Inhibitors

The association of growth inhibitors with bud dormancy has been recognized for many years and an increase (as measured by bioassay) in the growth-inhibiting activity of extracts from tissues of plants entering dormancy has been recorded for a range of species (Table 2). Evidence that these inhibitors originate in the leaves comes from studies with sycamore (*Acer pseudoplatanus* L.; WAREING 1959). After transfer to SD, an increase in inhibitory activity was detected in leaf extracts after only two SD cycles, and this was followed by an increase of inhibitors in apex extracts after five cycles. This increase in "inhibitor" content of the apical bud preceded the retardation of growth. Inhibitors have been detected in transport systems and there was a pronounced increase in ether-soluble inhibitors in both phloem and xylem of willow (*Salix viminalis* L.) during entry to dormancy (BOWEN and HOAD 1968).

These results raise two questions which still remain largely unanswered. Firstly, does the change in the growth-inhibitory activity of the extract result from a change in the content of inhibitor(s) and/or of growth promoter(s) (which may co-chromatograph)? Secondly, is the growth inhibitor involved in the endogenous regulation of bud dormancy?

In sycamore, the inhibitory effect was found to be largely due to the presence of ABA (originally termed "dormin", ROBINSON and WAREING 1964). Similarly, the inhibitor in the sap of willow was identified as ABA, which was the main inhibitory component present (BOWEN and HOAD 1968, ALVIM et al. 1976). Abscisic acid has also been identified in other dormant tissues such as buds of apricot (RAMSEY and MARTIN 1970). Present evidence indicates, however, that ABA content in leaves and buds (as determined by GLC) is actually decreased by exposure to SD in sycamore, *Acer rubrum* L., *Betula pubescens* (LENTON et al. 1972), *B. lutea* MICHX. (LOVEYS et al. 1974), *B. papyrifera* MARSH (POWELL 1976) and *Salix viminalis* (ALVIM et al. 1978). These results are consistent with those obtained for herbaceous plants such as chrysanthemum (SENGUPTA et al. 1974), *Phaseolus vulgaris* L. (BENTLEY et al. 1975), strawberry (PLANCHER and NAUMANN 1978), and *Spinacia* (ZEEVAART 1974). In other cases, daylength had no effect (Table 2). A problem with determinations of ABA content relates to the large increases resulting from water stress – a factor which has not always been appreciated and which may have caused spurious results in some cases. For example, more ABA may be synthesized in response to water stress in LD compared with SD, perhaps because of the thinner leaves (PLANCHER and NAUMANN 1978). However, recent experiments in which the water stress factor has been eliminated have confirmed that SD does not increase ABA content (PLANCHER and NAUMANN 1978, PHILLIPS et al. 1980). The increased inhibitory activity of extracts from leaves and buds exposed to SD must, therefore, have resulted from changes in inhibitors other than ABA such as naringenin or prunin (EREZ and LAVEE 1966) or from changes in growth-promoting substances which co-chromatograph with ABA in some solvent systems.

Even though ABA content is not increased by SD, it may still be a component of the endogenous system regulating dormancy if it is always exported from leaves to buds (PHILLIPS and HOFFMANN 1979), thus causing dormancy unless antagonized by compounds whose concentration varies with daylength. This possibility can be explored by considering effects of ABA applications. The induction of dormancy in LD by (±) ABA has been reported for sycamore, *B. pubescens*, and blackcurrant (*Ribes nigrum* L.) (EL-ANTABLY et al. 1967). Typical bud scales were formed in all cases and the buds were fully dormant. Except at very high concentrations (up to 1000 mg l^{-1} for *Acer palmatum* THUNB. and *Cornus florida* L. CATHEY 1968b), a daily foliar spray was not effective and it was necessary both to immerse the youngest expanded leaf in ABA and to spray the apical region of the shoot. Similarly, leaf-applied ABA did not induce bud dormancy in *Betula* and *Alnus glutinosus* GAERTN. under LD (HOCKING and HILLMAN 1975), but only a small proportion of the leaf-applied ABA reached the apical region unaltered. ABA application in LD also increased low temperature hardiness in *Acer negundo* L. (IRVING and LANPHEAR 1968) and caused the formation of dormant structures (turions) in *Spirodela polyrrhiza* L. (PERRY and BYRNE 1969) and *Myriophyllum verticillatum* L. (in marginally inductive conditions only, WEBER and NOODEN 1976). However, (±) ABA applications were reported to be without effect in causing dormancy in *Weigela florida* (from Table 1, a species in which dormancy is presumed to be induced by a long night) and *Catalpa bignonioides* WALT., nor did ABA induce cold-hardiness in *Cornus stolonifera* MICHX. (CATHEY 1968b, FUCHIGAMI et al. 1971) or the formation of normal winter buds in *Acer rubrum* (PERRY and HELLMERS 1973). Furthermore, even where it is effective in causing the formation of resting buds, ABA does not always entirely mimic SD (CATHEY 1968b, EL-ANTABLY et al. 1967, IRVING 1969).

Table 2. Effects of daylength on the content of endogenous growth regulators.

Plant	Photoperiod response	Content increased by: LD	Content increased by: SD	Method[a]	Tissue(s)	Comment	Source
Auxins							
Allium cepa	LD→bulbs	×		B	Leaf bases	Co-chromatographed with IAA	CLARK and HEATH (1962)
Begonia × cheimantha	SDP	×		B	Leaves		HEIDE (1967, 1968)
Cucumis sativus Beit Alpha	SD→♀		×	B	Shoot tips		RUDICH et al. (1972b)
Festuca arundinacea		×		B	Shoot	31/17° C. No difference at 22/11° C	YEH et al. (1976)
Glycine max Biloxi	SDP	×		C	Shoot tips	Free IAA	VLITOS and MEUDT (1954)
Rhus typhina	SD→dormancy	×		B	Shoot tips		NITSCH (1957)
Zea mays	SDP	×		B	Shoot tips	SD and auxin→♀	HESLOP-HARRISON (1964)
Cytokinins							
Allium cepa	LD→bulbs	×		B	Leaf blades	Maximum after 29 LD; then fell to SD value	LERCARI and MICHELI (1981)
Begonia × cheimantha	SDP		×	B	Leaves		HEIDE and SKOOG (1967)
Dactylis glomerata	SD→reduced growth		×	B	Shoot		MENHENETT and WAREING (1977)
Myriophyllum verticillatum	SD→turions	×		B			WEBER and NOODEN (1976)
Perilla frutescens	SDP		×	B	Xylem sap	Increase after about 20 SD; flower buds visible after 18 SD	BEEVER and WOOLHOUSE (1973)

Phaseolus vulgaris P47	SD prevent flower bud abscission		×	B	Xylem sap		NEWTON et al. (1980) MORGAN and MORGAN (1984)
Solanum tuberosum	SD→tubers		×	B/C	Stolons and shoots	Cis-zeatin riboside	MAUK and LANGILLE (1978)
Xanthium strumarium							
(1)	SDP		×	B	Phloem-aphid honeydew	Plants flowering (SD) or vegetative (LD)	PHILLIPS and CLELAND (1972)
(2)		×		B	Buds: leaves: root exudate	Change after 1 × SD	HENSON and WAREING (1974, 1977b)
Ethylene							
Cucumis sativus							
Beit Alpha	SD→♀		×	C	Shoot tips	Ethylene enhances ♀	RUDICH et al. (1972a)
Matsunomidori	SD→♀	×		C	Leaves	Ethylene enhances ♀	TAKAHASHI et al. (1983)
Higan-fushinari	LD→♀	×		C	Leaves	Ethylene enhances ♀	TAKAHASHI et al. (1983)
Dahlia hybrida Choot Hashani	SD→tubers		×	C	Leaf cuttings	Ethylene increase preceded tuberization and subsequently decreased	BIRAN et al. (1972)
Gibberellin							
Acer negundo	SD→cold hardiness	×		B	Leaves	GA_3 decreased hardiness in SD	IRVING (1969)
Acer pseudoplatanus	SD→dormancy	×		B	Shoot tips		LENTON et al. (1972)
Agrostemma githago	LDP	×		B/C	Shoots	Transient increases in GA_{17}, GA_{19}, GA_{20}, GA_{44}, GA_1	JONES and ZEEVAART (1980a, b)
Avena sativa	LDP	×		B	Leaves		CHAILAKHYAN (1968)
Begonia × cheimantha	SDP	×		B	Leaves		HEIDE (1967)
Betula pubescens	SD→dormancy	×		B	Shoot tips		DIGBY and WAREING (1966)
Bryophyllum daigremontianum	LSDP	×		B/C	Shoot tips: leaves	GA_{20}. LD→SD increased GA more than continuous LD	ZEEVAART (1969b, c, 1973), GASKIN et al. 1973

[a] B = bioassay; C = chemical or physical method

Table 2 (continued)

Plant	Photoperiod response	Content increased by LD	Content increased by SD	Method[a]	Tissue(s)	Comment	Source
Gibberellin (continued)							
Cupressus arizonica	LD ♂, SD ♀	×		B/C	Green shoot	Plants juvenile (vegetative); >LD=>GA_3. Exogenous GA causes flow ering, less GA required under LD to induce ♂ flowering	Pharis (1976), Table III
Fagopyron esculentum	DNP	×		B	Leaves		Chailakhyan (1968)
Fragaria × ananassa	SDP. LD→petiole elongation	×		B/C	Young leaves and crown	Possibly GA_{19} (TLC)	Uematsu and Katsura (1983)
Hyoscyamus niger	LDP	×		B			Lang (1960)
Nicotiana sylvestris	LDP	×		B	Leaves	Different GAs present in LD	Grigorieva et al. (1971), Lozhnikova et al. (1982)
Nicotiana tabacum Maryland Mammoth	SDP	×		B	Leaves	No evidence for new GAs in LD	Grigorieva et al. (1971), Lozhnikova et al. (1982)
Perilla crispa	SDP	×		B	Leaves		Chailakhyan (1968)
Pisum sativum G	LDP			B/C	Shoot tips	GA_E (=GA_{19}) synthesised only in SD	Proebsting et al. (1978), Davies et al. (1982)
Prunus persica	SD→dormancy	×		B	Leaves		Kojerzenko and Majko (1967)
Rudbeckia bicolor	LDP	×		B	Leaves		Chailakhyan (1968)
Salix viminalis	SD→dormancy	×		B	Phloem sap		Hoad and Bowen (1968)
Silene armeria	LDP	×		B	Shoot tips	Increase mainly in diffusible GAs	Cleland and Zeevaart (1970)
Spinacia oleracea Savoy Hybrid 612	LDP			B/C	Shoot	Change in GAs present, no change in total GA	Zeevaart (1971a), Metzger and Zeevaart (1980a, c, 1982)

Solanum andigena	SDP	×		B	Leaves	Decrease after 2 × SD cycles. Two new GA zones in LD	RAILTON and WAREING (1973)
Trifolium pratense	LDP		×	B	Leaves and shoots	Diffusible GA. Extractable GA transiently higher in LD	STODDART and LANG (1968)
Vicia faba	DNP	×		B	Leaves		CHAILAKHYAN (1968)
Inhibitors							
Acer pseudoplatanus							
(1)	SD→dormancy		×	B	Shoot tips; leaves	Change after 2 SD in leaves	PHILLIPS and WAREING (1959)
(2)		×		C	Shoot tips	ABA	LENTON et al. (1972)
(3)				C	Young/old leaves	ABA; free and bound. LD = SD. No water stress	PHILLIPS et al. (1980)
Acer rubrum	SD→dormancy	×		C	Shoot tips	ABA	LENTON et al. (1972)
Betula lutea	SD→dormancy	×		B/C	Leaves	ABA	LOVEYS et al. (1974)
Betula papyrifera	SD→dormancy	×		C	Shoot tips	ABA	POWELL (1976)
Betula pubescens							
(1)	SD→dormancy		×	B	Shoot tips; leaves	Inhibitor β	EAGLES and WAREING (1964), DIGBY and WAREING (1966)
(2)		×		C	Shoot tips	ABA	LENTON et al. (1972)
Chrysanthemum morifolium Indianapolis White	SDP	×		C	Leaves; shoot tips	ABA. (B – more inhibitor in SD)	SENGUPTA et al. (1974)
Dahlia hybrida Choot Hashani	SD→tubers		×	B	Leaves	ABA-like inhibitors	BIRAN et al. (1974)
Fragaria × ananassa	SDP	×		C	Leaves; shoot tips	ABA. Night-break	PLANCHER and NAUMANN (1978)
Lolium temulentum	LDP			C	Shoot apices: leaves	ABA, xanthoxin; LD = SD	KING et al. (1977)

[a] B = bioassay; C = chemical or physical method

Table 2 (continued)

Plant	Photoperiod response	Content increased by LD	Content increased by SD	Method[a]	Tissue(s)	Comment	Source
Inhibitors (continued)							
Malus hupehensis	SD→dormancy			C	Shoot tips	ABA; LD=SD	Powell (1976)
Nicotiana sylvestris	LDP		×	B	Leaves	ABA-like	Lozhnikova et al. (1982)
Nicotiana tabacum Maryland Mammoth	SDP		×	B	Leaves	ABA-like	Lozhnikova et al. (1982)
Perilla crispa	SDP			C	Leaves	ABA; LD=SD	Purse (1984)
Phaseolus vulgaris P47	buds abscise in LD	×		B/C	Leaves: stems: buds	ABA: abscising buds have higher ABA	Bentley et al. (1975) Morgan and Zehni (1980), Morgan and Morgan (1984)
Salix viminalis							
(1)	SD→dormancy		×	B	Phloem sap: xylem sap	Co-chromatographs with ABA: a second inhibitor in xylem sap decreased in SD	Bowen and Hoad (1968)
(2)		×		C	Shoot tips	ABA; plants in field	Alvim et al. (1978)
(3)				B/C	Shoot tips: xylem sap	LD=SD. ABA, inhibitor β; plants in controlled environments	Alvim et al. (1979)
Solanum andigena	SD→tubers			C	Leaves	ABA; LD=SD	Wareing and Jennings (1980)
Spinacia oleracea Savoy Hybrid 612	LDP	×		C	Shoot	ABA. No change in xanthoxin	Zeevaart (1971b, 1974)
Vitis vinifera	DNP			C	Leaves	ABA; LD=SD	Alleweldt and During (1972)
Xanthium strumarium	SDP			C	Leaves	ABA; LD=SD	Zeevaart and Kivilaan (1971)

[a] B=bioassay; C=chemical or physical method

The exact status of ABA as an endogenous regulator of dormancy remains unresolved. It now seems unlikely that photoperiodic control of entry to dormancy involves simply changes in ABA content (LOVEYS et al. 1974, ALVIM et al. 1978, 1979, PHILLIPS et al. 1980). However, most extracts have been from tissue macerates and there appear to have been no studies of subcellular compartmentation, nor have any detailed studies of ABA metabolism been made in relation to dormancy. For example, light enhanced the glucosylation of ABA and also induced the appearance of both free and bound phaseic acid. Under LD, amounts of free and bound phaseic acid were not significantly different, but SD leaves contained less of the bound form (PHILLIPS et al. 1980). It has also been observed that, during winter, the proportion of esterified ABA in birch buds progressively increased as dormancy was broken (HARRISON and SAUNDERS 1975), although there was no association between total ABA levels and emergence from dormancy in buds of sycamore (PHILLIPS and HOFFMANN 1979) and apple (BORKOWSKA and POWELL 1982–1983). The possibility that, as with GA, photoperiod might alter the sensitivity to ABA, has also largely been overlooked. In apple, where entry to dormancy is not dependent on daylength (WAREING and SAUNDERS 1971), such changes can clearly occur, since growing buds responded much less to exogenous applications of ABA than did nongrowing buds (BORKOWSKA and POWELL 1982–1983). After 18 weeks of exposure to cold, bud sensitivity to ABA was also drastically reduced.

3.2.2 Gibberellins

The involvement of GA in dormancy phenomena is fairly well established. Following transfer to SD, a fall in GA content (as assayed by dwarf pea or lettuce hypocotyl tests) has been recorded in shoot tips of *B. pubescens* (DIGBY and WAREING 1966) and sycamore (LENTON et al. 1972), and in leaves of peach (KOJERZENKO and MAJKO 1967). GA-like substances also greatly decreased in apical buds of *Salix pentandra* before the cessation of apical growth and shoot-tip abscission under natural conditions, although changes in photoperiod were small at this time (JUNTTILA 1982). Moreover, the application of GA may prevent the induction of dormancy in SD for some woody angiosperms. In *Salix pentandra,* GA_3 prevented apical growth cessation in SD, especially when applied to the apex (JUNTTILA 1981). In *Rhus typhina,* a single application of 10 μg GA_3 in lanolin delayed dormancy in SD and resulted in growth equivalent to LD controls during a 2-week test period; however, GA_3 also increased growth of plants growing in LD, although to a lesser extent (NITSCH 1957). Finally, interactions have been observed between exogenously applied GA and ABA in the regulation of bud dormancy, with the inhibitory effect of ABA being partly, but not entirely, overcome by GA_3 (EL-ANTABLY et al. 1967, WAREING 1969).

Very similar results have been obtained in cold-hardiness studies. An inhibitor of cold-hardiness appears to be produced in LD leaves (IRVING and LANPHEAR 1967): the application of GA_3 decreases hardiness in SD (IRVING and

LANPHEAR 1968) and the content of GA-like substances is higher in leaves under LD (IRVING 1969). Moreover, growth retardants such as chlormequat and daminozide can increase cold-hardiness in several plants (ALDEN and HERMANN 1971, REID et al. 1974), suggesting the possible involvement of endogenous GA's. [On the other hand, a fall in gross GA content was not strongly correlated with increase in cold-hardiness of winter wheat cv. Kharkov (REID et al. 1974)]. However, hardiness promoters appear also to be produced by leaves. Although these have not been positively identified, ABA can induce cold-hardiness in some plants (IRVING 1969), but not as effectively as exposure of leaves to SD. The present evidence, however, is that the ABA content of tissues is either the same, or lower in SD compared with LD (see Sect. 3.2.1 above).

Although evidence is limited and detailed metabolic studies remain to be done, a tenable hypothesis is that the induction by daylength of bud dormancy is partly achieved through changes in the content and/or composition of GA's in the buds which, in LD, counteract the dormancy-inducing effect of inhibitory substances (e.g., ABA, naringenin, prunin) from the leaves. The LD-induced apical senescence in pea, which may be a related phenomenon, has been shown to be associated with changes in GA metabolism. A biologically active GA (GA_E), which prevents senescence, is synthesized only in SD: in LD, there is a low rate of conversion to this GA and senescence occurs (PROEBSTING et al. 1978). It was suggested that, in this system, GA_{20} was metabolized to the active GA_E (proposed as GA_1), while in LD the inactive GA_{29} was formed from GA_{20}. However, more recent work on G_2 pea (DAVIES et al. 1982) could not characterize GA_1 in vegetative tissue, but did characterize GA_{19} from the SiO_2 partition column fraction which will contain both GA_1 and GA_{19}. DAVIES et al. (1982) thus conclude that GA_E is most probably GA_{19}. A build-up of GA_{19} in G_2 pea would be consistent with results in spinach (METZGER and ZEEVAART 1980a, c, 1982, GIANFAGNA et al. 1983) where GA_{19} was high under SD, and its metabolism to GA_{20} was inhibited, relative to LD. It is also not clear that the GA's which increased in the shoot tips of G_2 pea originated in the leaves. It may be that a stimulus from the leaves is acting to increase the synthesis, activity, or metabolism of GA's in the apical region.

From a consideration of the photoperiodic mechanism, one might also expect an increase in a dormancy promoter in at least some species. Since physical measurements of ABA show no change or a decrease under SD conditions, ABA cannot be the dormancy-inducing compound which is expected to change in amount with daylength. Once again, as with tuber formation, one is driven to the conclusion that, while the LD substances are probably GA's, identity of the SD (long–night) substance remains an enigma.

Irrespective of whether it is induced by photoperiod or develops autonomously, once dormancy has been fully established, the resumption of bud growth will normally not occur until the buds have experienced a period of low temperature. In many species, and particularly in conifers, neither GA nor LD will substitute for the low temperature requirement during deep dormancy, although both may be effective before dormancy is fully established or when it is partly broken.

3.2.3 Cytokinins

Cytokinins are not directly implicated in the induction of bud dormancy in woody plants by daylength but there is some evidence that changes in CK levels are associated with the breaking of dormancy. For example, CK activity was absent from dormant buds, but increased after dormancy was broken, reaching a maximum shortly before bud break (DOMANSKI and KOZLOWSKI 1968, HEWETT and WAREING 1973) and a similar increase in sap CK occurred prior to bud break in willow (ALVIM et al. 1976). Moreover, application of synthetic CK's will stimulate growth of dormant buds in some woody species (CHVOJKA et al. 1962, WEAVER 1963, ANGRISH and NANDA 1982) and of dormant turions in *Spirodela* (CZOPEK 1964) and *Myriophyllum* (WEBER and NOODEN 1976). Changes in endogenous CK's may, however, be involved in the SD induction of overwintering turions in some vascular hydrophytes. The application of CK prevented turion formation in *Myriophyllum* (WEBER and NOODEN 1976) and counteracted the effect of ABA to induce turion formation in *Spirodela* (STEWARD 1969). Since CK activity decreased during development and CK content is higher in LD (Table 2), it was suggested that the induction of turions in *Myriophyllum* by SD, involves a decrease in endogenous CK's, as well as an increase in acidic inhibitors (WEBER and NOODEN 1976).

4 Reproductive Behavior

4.1 Initiation of Floral Primordia

By far the most widely studied aspect of daylength control is that exerted over the switch from vegetative to reproductive growth. There is good physiological evidence for the existence of a floral stimulus which moves from leaves to the receptive buds and there causes floral differentiation. In photoperiodically sensitive plants, the generation of this stimulus is under daylength control, but the results from grafting experiments suggest that the stimulus may be common to plants of different photoperiodic classes, including day-neutral plants.The term "florigen" for the hypothetical flower-forming hormone was proposed by CHAILAKHYAN (1936).

4.1.1 Florigen

Although proof of the existence of a specific floral hormone awaits its extraction and identification, there is a great deal of circumstantial evidence that, when they are in appropriate daylengths, leaves produce a transmissible substance capable of evoking flowering at receptive shoot apices (see, for example, LANG 1965, VINCE-PRUE 1975). This evidence includes: the small amount of leaf tissue that needs to be exposed to appropriate daylengths for floral induction; repeated transfer of the stimulus by grafting; failure to flower if the induced leaf is removed after a single inductive cycle; and the possibility of summation of inductive cycles even when noninductive cycles are intercalated between them, as, for example, in the SDP *Impatiens balsamina* L., where as many as 16 LD can be intercalated between SD without altering the number of SD required (KRISHNAMOORTHY and NANDA 1967). Moreover, the results of several cross-grafting experiments indicate that there may be a floral hormone which is capable of evoking flowering in plants of all photoperiodic classes in a wide range of species. At least 35 successful graft transfers of flowering, including those between different photoperiodic response types, have been listed (LANG 1965, ZEEVAART 1976). In the Crassulaceae, for example, all photoperiodic response types are available and it has been possible to transmit the floral stimulus in various graft combinations between the SDP *Kalanchoe*, the LDP *Sedum*, the LSDP *Bryophyllum*, and the SLDP *Echeveria* (ZEEVAART 1978, 1982), indicating that the final outcome of photoperiodic induction in all four response types is the production of transmissible stimuli that are similar, if not identical.

However, there are several cases where the floral stimulus does not appear to be interchangeable between closely related species (VAN DE POL 1972, GRIESEL 1963, ZEEVAART 1958, JACQUES 1973). This does not necessarily indicate different stimuli, since differences in the rate of production and persistence and perhaps different sensitivities at the receptor site could account for negative results. Similarly, although there are different mechanisms to preserve the flowering state, this appears not to be due to different stimuli. In some plants (*Xanthium*; *Silene*; *Bryophyllum*; green *Perilla ocymoides*), the floral stimulus appears to have self-perpetuating properties and shoots that are induced to flower by grafting can themselves elicit flowering when grafted to a noninduced receptor (indirect induction, ZEEVAART 1976). In the red *Perilla crispa*, a leaf induced by SD continues to produce floral stimulus indefinitely, but a flowering receptor shoot cannot itself function as a donor (ZEEVAART 1958). However, the stimulus is readily transmitted between *P. crispa* and *P. ocymoides* and, moreover, *P. crispa* donor leaves bring about indirect induction in *P. ocymoides*, indicating that the differences in behavior are the result of different internal conditions rather than of different stimuli (ZEEVAART 1984).

If such a ubiquitous *florigen* does exist, the question then arises as to why attempts to identify it have met with so little success. One problem undoubtedly lies in the lack of a suitable assay. Both excised stem tips and intact plants have been used as test material but there are difficulties, such as penetration and the possible presence of inhibitory compounds, associated with both approaches. As ZEEVAART (1979) has pointed out, it is crucial that the extract is active in the species from which it was obtained, while a similar extract from vegetative plants is not active. The floating pondweed *Lemna* spp. is often used, and offers a conveniently simple system; however, results may be difficult to interpret because of the many chemicals that can elicit a flowering response and/or modify photoperiodic behavior in this plant.

Assays of crude extracts made with organic solvents have generally yielded only negative results or, if positive, they have been hard to duplicate (CLELAND 1975, ZEEVAART 1976, 1979, CLELAND and BEN-TAL 1983). Also the crude extracts were not amenable to further purification for identification of the active fraction (LINCOLN and CUNNINGHAM 1964). The most promising extracts were made from flowering plants of the SDP *Xanthium;* these caused some flowering when applied to noninduced plants of *Xanthium* (LINCOLN et al. 1961, HODSON and HAMNER 1970) or to the SDP *Lemna paucicostata* (HODSON and HAMNER 1970). Extracts from vegetative *Xanthium* plants were inactive, but preparations active on *Xanthium* were obtained from flowering plants of the DNP sunflower (LINCOLN et al. 1962) and from the fungus *Calonectria rigidiuscula* (BERK & BR.) SN. & H. (LINCOLN et al. 1966). The active substance was acidic, but further purification resulted in loss of activity (LINCOLN and CUNNINGHAM 1964) and its chemical identity remains unknown. CARR (1967), and later HODSON and HAMNER (1970), found that the addition of GA_3 to *Xanthium* extracts improved the flowering response of *Xanthium* but, when applied to *Lemna,* the extracts alone were more effective (HODSON and HAMNER 1970). However, the extracts did not appear to contain all of the necessary materials for normal inflorescence development in *Xanthium,* since male inflorescences were abnormal and female inflorescences absent. CHAILAKHYAN (1982) has also reported successful results with extracts from lyophilized Maryland Mammoth tobacco (SDP): when tested on the SDP, *Chenopodium rubrum,* extracts from flowering plants elicited a marked flowering response, while those from vegetative plants were without effect. Some of the partially purified fractions were also tested for their effect on flowering in stem explants from the DNP, Trapezond tobacco. Two of the fractions more than doubled the number of flowering buds, while the third gave some increase. Samples from leaves of vegetative plants depressed flowering in the receptors. These results confirm that the extracts have florigenic activity in tobacco, although it must be noted that the Trapezond explants were already in a flowering condition (50% of the buds in the controls were flower buds) and that there are no reports that the extracts can induce flowering in Maryland Mammoth itself.

These approaches indicate that it is possible to prepare extracts that can cause flowering in test plants under noninductive conditions but, without better reproducibility and further identification of the active components, the significance of the results remains in doubt.

The long-distance transport of the floral stimulus appears to take place in the phloem (see LANG 1965, VINCE-PRUE 1975) and, therefore, analysis of the phloem contents has been attempted using various approaches such as extracts of phloem sap, diffusates from leaves into agar, and honeydew from phloem-feeding aphids (CLELAND 1975). Both the phloem exudates and the honeydew approaches have yielded some positive results but they have not yet been very conclusive.

Phloem exudates from induced leaves of the SDP, *Perilla crispa,* elicited some flowering response when tested on excised *Perilla* shoot tips under weak continuous light (ZEEVAART 1979), but analysis of the neutral, acidic and basic fractions by GLC revealed no significant differences in fractions obtained from induced and noninduced plants (ZEEVAART 1984). PURSE (1984) has also tested phloem exudates from red-leaved *Perilla* on shoot tip explants of the same species. Phytotoxicity of the phloem exudates was a problem limiting the concentration that could be used but flowering was promoted by exudates from SD leaves in a few cases, although not always. The active constituent(s) was not identified but ABA and sucrose were excluded as being inadequate to account for the flowering response. CLELAND and BEN-TAL (1983) have reported negative results from phloem exudates of *Perilla*. Using a different approach, CHAILAKHYAN (1982) has investigated the effect of substances diffusing from shoot explants taken from stems of vegetative or reproductive plants of the DNP, *Nicotiana tabacum* Trapezond. Donor explants were cultivated on agar for 3–7 days and receptor explants were then placed on the same media. There was no stimulation of flowering in the vegetative explants, although the number of flower buds was increased when flowering receptors were cultivated after flowering donors. Taken over all, therefore, the results from the phloem exudate and diffusion approaches have not been very encouraging.

Honeydew from aphids feeding on vegetative or flowering plants of the SDP *Xanthium* has been tested on the LDP *Lemna gibba* L. G_3 and, following thin layer chromatography of an acidic ethyl acetate fraction, one zone of flower-inducing and two zones of flower-inhibiting activity were detected. The promoting substance was identified as salicylic acid (SA), and it was also shown that authentic SA would induce a good flowering response in *L. gibba* and in at least two other species of *Lemna* (CLELAND 1974a, CLELAND and AJAMI 1974). SA had no effect on flowering in *Xanthium* when given alone or in combination with GA and/or kinetin, nor was there any difference in the SA content of honeydew from aphids feeding on flowering or vegetative plants of *Xanthium* (CLELAND 1974a, 1975); consequently a role for SA in the control of flowering in *Xanthium* is unlikely. However, flowering can be induced by SA in the SDP *Impatiens balsamina* L. in strictly noninductive conditions (NANDA et al. 1976). In this plant, it has been suggested that SA and other phenols may affect flowering through effects on IAA oxidase and so on IAA levels (KUMAR and NANDA 1981).

Salicylic acid is effective in many different members of the family Lemnaceae (see CLELAND et al. 1982) and in the closely related aquatic plant, *Pistia stratiotes* (PIETERSE 1982). It is also able to induce flowering in both LD and SD species of *Lemna* (CLELAND and TANAKA 1979), which is one of the properties expected of a true florigen. The effect is quite specific, being limited to SA, benzoic acid and a small number of closely related compounds (CLELAND 1974b, WATANABE and TAKIMOTO 1979). There are no published reports of the occurrence of SA in the Lemnaceae. However, benzoic acid has been identified in both *L. paucicostata* and *L. gibba,* although there was no evidence that the amount present was correlated in any way with flowering in either species (FUJIOKA et al. 1983). The effect of SA on flowering in *Lemna* is strongly daylength-dependent. Optimal levels of SA caused a shift in the critical daylength in both LD and SD species (CLELAND and TANAKA 1979). In the LDP *L. gibba,* the critical daylength was reduced from about 10 to 8 h, while in the SDP *L. paucicostata,* it was increased from 14 to 15 h. This suggests that SA is interacting directly with the photoperiodic induction mechanism. This conclusion is further substantiated by the fact that, although SA remains in the plant for at least 5 days, its effect is lost in less than a day (CLELAND 1982). It has, therefore, been suggested that the action of SA is related to its uptake and that it may be causing its effect at the level of the cell membrane (CLELAND and BEN-TAL 1983) or by chelation (WATANABE et al. 1981).

The general conclusion, therefore, is that SA affects photoperiodic induction in the frond rather than being a transmissible floral stimulus, which acts at the apex.

4.1.2 Flower-Inhibiting Substances

While much of the inhibitory effect of leaves in noninductive cycles can be explained in terms of interference with transport of the floral stimulus to the apex (CHAILAKHYAN and BUTENKO 1957, KING and ZEEVAART 1973) or with its production in the leaf (SCHWABE and WIMBLE 1976), there is also a good deal of evidence pointing to the generation of a transmissible substance which is inhibitory to flowering (see p. 309).

The idea of a specific floral inhibitor(s) was proposed almost as early as that of a floral promoter (LANG 1980, JACOBS 1980, SCHWABE 1984). Extracts of such specific flower-inhibiting substances, or "anti-florigens" have, however, been rewarded with no more success than "florigens". Crude extracts from LD leaves of *Kalanchoe* were inhibitory to flowering when injected into SD leaves, whereas when extracts from SD leaves were injected, flowering was not significantly different from the controls (SCHWABE 1972). The inhibitory component has not yet been characterized, but it has been suggested that 2-flavan-3-ol fractions of relatively low molecular weight may be involved (SCHWABE 1984). A report that gallic acid was the endogenous inhibitor of *Kalanchoe* (PRYCE 1972) could not be confirmed (SCHWABE and WIMBLE 1976). An inhibitory LD effect in *Hibiscus cannabinus* cv. South Selected has also been transmitted by injection into the cotyledons of the SDP *Pharbitis nil* (REN et al. 1982b). The inhibitory effect of the *Hibiscus* extract could not be simulated by known hormones but, in *Kalanchoe,* there was some inhibition of flowering when ABA or xanthoxin was injected into the leaves in SD (SCHWABE 1972). In *Kalanchoe,* the evidence suggests that the inhibitor acts in the leaf to prevent production of the floral stimulus (SCHWABE and WIMBLE 1976). Such an inhibitor is likely to be different from the graft-transmissible "anti-florigen" that interferes with the action of "florigen" at the shoot apical meristems. There are no reports of the isolation and chemical identification of "anti-florigens" (LANG 1980).

4.1.3 Gibberellins

Gibberellins are the only group of chemicals which evoke flowering when applied to a wide range of plant species. However, the role of endogenous GA in the control of the flowering process by daylength is far from being understood. Applied GA's cause flowering most commonly in LDP which grow as rosettes in SD, but they can also substitute for an inductive photoperiod requirement in plants of other photoperiodic groups. They may act at the apex or in the leaves, and they may inhibit, promote, or be ineffective in flowering. Some examples of the effects of GA on flowering in plants of various response groups are given in Table 3. Action of GA at the apex to promote flowering is indicated

Table 3. Examples of effects of gibberellin application on floral initiation in plants of different photoperiodic response groups. (After VINCE-PRUE 1975, Table 6.1, pp. 218–220)

Long day plants	
Gibberellin promotes initiation of flowers in short days:	
Lolium temulentum	EVANS (1964)
Samolus parviflorus	LANG (1957)
Gibberellin inhibits initiation of flowers in long days:	
Lemna gibba G3	CLELAND and BRIGGS (1969)
Fuchsia hybrida cv. Lord Byron[a]	SACHS et al. (1967)
Gibberellin promotes stem elongation in short days, but does not induce flowering:	
Blitum virgatum	JACQUES (1968)
Hieracium aurantiacum	PETERSON and YEUNG (1972)
No effect of gibberellin on elongation or flowering:	
Anagallis arvensis[a]	CHOUARD (1957a)
Hieracium floribundum	PETERSON and YEUNG (1972)
Short day plants	
Gibberellin promotes initiation of flowers in long days:	
Impatiens balsamina	NANDA et al. (1967)
Zinnia elegans	SAWHNEY and SAWHNEY (1976)
Gibberellin inhibits or delays initiation of flowers in short days:	
Fragaria × ananassa (strawberry)	THOMPSON and GUTTRIDGE (1959)
Lemna paucicostata	GUPTA and MAHESHWARI (1970)
No effect of gibberellin on initiation:	
Pharbitis nil (some promotion in SD)	OGAWA and ZEEVAART (1967)
Xanthium strumarium (some promotion in SD)	LANG (1965)
Plants with dual photoperiodic requirements	
Gibberellin promotes initiation in short days:	
Bryophyllum daigremontianum (LSDP)	ZEEVAART and LANG (1963)
Coreopsis grandiflora (SLDP)	KETELLAPPER and BARBARO (1966)
Gibberellin promotes initiation in long days:	
Scabiosa succisa (SLDP)	CHOUARD (1957b)
Gibberellin inhibits initiation:	
Cestrum nocturnum (LSDP)	SACHS (1969)

[a] Caulescent plant

for the LDP *Rudbeckia* (CHAILAKHYAN 1958) and to inhibit flowering for the LDP *Fuchsia* (SACHS et al. 1967), while action in the leaves was clearly indicated for the LSDP *Bryophyllum* where GA promoted flowering in SD (ZEEVAART 1969a).

The content of endogenous GA's has usually been found to be higher in LD than in SD, and this seems to be independent of the photoperiodic class for flowering (Table 2). CHAILAKHYAN and LOZHNIKOVA (1966) reported that a SD followed by a night-break was the equivalent of a LD for the production of GA's in LDP (*Nicotiana sylvestris; Rudbeckia*), SDP (*Xanthium;* tobacco cv. Maryland Mammoth) and DNP (tobacco cv. Trapezond). The content of both diffusible and extractable GA's may be increased in LD, but not always in parallel (CLELAND and ZEEVAART 1970, STODDART and LANG 1968). It has also been shown that LD leaves may export more GA in the phloem (HOAD and BOWEN 1968).

The metabolism of GA's is also influenced by photoperiod. In some higher plants, a series of GA's have been identified in the pathway $Ga_{53} \rightarrow Ga_{44} \rightarrow GA_{19} \rightarrow GA_{20} \rightarrow GA_1 \rightarrow GA_8$, with GA_1, being presumed as the active form. Minor modifications of this sequence have been found; in spinach, for example, the active form appears to be GA_{20}, which is converted to the inactive GA_{29}. As already discussed in Section 2.1.1, in the LDP spinach, photoperiod is a controlling factor in the conversion of GA_{19} to GA_{20}, which increases in LD; metabolism of GA_{20} to GA_{29} is also faster in LD. Somewhat similar changes occur in another LD rosette plant, *Agrostemma githago*. Differences in GA metabolism between LD and SD are also indicated by results of GRIGORIEVA et al. (1971), who reported the presence of three GA-like substances in the leaves of the LDP *N. sylvestris* grown in LD; extracts containing them were effective in evoking flowering of the LDP *Rudbeckia* in noninductive cycles. In SD the leaves produced GA's which were inactive for flower induction in *Rudbeckia*. In *Trifolium*, a substance co-chromatographing with GA_4, GA_5 or GA_7 appeared during floral induction in LD (STODDART 1966). But in the G_2 strain of *Pisum sativum*, ^3H-GA_9 was metabolized by leaves to more polar GA's in SD but not in LD (PROEBSTING and HEFTMANN 1980). Thus, in many LDP, the GA content is increased in LD, even if only transiently, and there is often an increase in GA metabolism together with the formation of new GA's.

The LD-induced changes in GA's may coincide with or precede the morphological events of floral initiation in LDP, and a role in flowering has been postulated. However, in rosette LDP, interpretation is complicated by the relationship between flower formation and stem elongation, which usually occur together in LD. Both phenomena can be induced by GA application, but it now seems likely that they depend on separate processes (ZEEVAART 1976). The role of endogenous GA's in LD-induced stem elongation has already been discussed in Section 2.1.1. A role in flowering is less certain. The application of the growth–retardant, AMO-1618, completely suppressed stem growth in *Silene*, yet flower formation under inductive photoperiods proceeded normally, with endogenous GA below the limit of detection in the d_5 corn bioassay (CLELAND and ZEEVAART 1970); consequently, in *Silene* at least, flower formation can proceed in the absence of detectable levels of endogenous GA. On the other hand, growth–retardant application depressed both elongation and flowering in *Samolus* and GA_3 reversed these effects; hence the higher GA level in LD appears to be necessary for flowering in this plant (BALDEV and LANG 1965). A comparison of the pattern of differentiation at the apex of *Rudbeckia* following GA_3 application in SD with that induced by transfer to LD indicated that the immediate responses were different: GA_3 increased the mitotic activity of the medullary zone of the apex whereas, in LD, mitotic activity was stimulated in the central zone (CHAILAKHYAN et al. 1969). These and other results led CHAILAKHYAN (1975a, b) to conclude that, in the shoot apex, the sites of action of GA_3 and the floral hormone are different, although both may be required for floral evocation.

Where GA is required for flowering in LDP, its action may be in the leaves to cause production of the floral stimulus (ZEEVAART 1984). WARM (1980) working with the annual LDP, *Hyoscyamus niger*, has concluded that applied GA has two separate effects: to cause production of the floral stimulus in the leaves and so cause flowering, and to act at the shoot apex to cause stem elongation. In neither of these effects is GA considered to be the equivalent of the floral stimulus itself. Action in the leaf has also been demonstrated in the LSDP *Bryophyllum*. When leaves of de-budded plants were treated with GA_3 in SD, these plants could act as donors and induce flowering in receptor plants kept in LD. Since GA_3 application does not cause flowering in *Bryophyllum* in LD it appears that, in the presence of GA, leaves in SD produce a floral stimulus which is translocated to receptor plants or to the apex (ZEEVAART 1969a). The application of chlormequat suppressed flowering in response to the change from LD to SD in *Bryophyllum*, indicating that endogenous GA's from LD leaves are necessary for SD to function in floral induction (ZEEVAART and LANG 1963). It is not yet clear, however, which endogenous GA's are active. Endogenous GA's were higher in LD than in continuous SD, and increased further when plants were transferred from LD to SD (ZEEVAART 1969b, c). The increase was mainly in GA_{20} (GASKIN et al. 1973, ZEEVAART 1973) but neither GA_{20}, nor any of its metabolites (mainly GA_{29} and 3-epi-GA_1) had high biological activity when applied to leaves (GASKIN et al. 1973, DURLEY et al. 1975).

Gibberellins may also be necessary for flowering in some SDP. Flowering in the SDP, *Pharbitis nil* CHOISY was inhibited by the application of chlormequat during an inductive night and the inhibition was prevented by GA_3. These effects occurred at the shoot apex and ZEEVAART (1970) concluded that the action of GA_3 was to stimulate cell division in the apex when the floral hormone was present. GA application had no effect in strictly noninductive conditions: however, it has been reported to enhance flowering in *Pharbitis* in suboptimal photoperiods (REN et al. 1982a), and also to increase the response to low temperature (OGAWA 1981b). Recently, a new endogenous growth factor synthesized in response to long dark periods has been obtained from a dwarf strain of *Pharbitis*. This strain produces only 2–4 flowers per plant following a single inductive dark period in contrast to the normal cultivar, Violet, which produces about 7 (SIMMONS and COULTER 1979, OGAWA 1981a). Extending the dark period brings about a large increase in the number of flower buds in the dwarf strain, and this additional duration of darkness can be replaced by GA_3 (SIMMONS and COULTER 1979). As with applied GA_3, the endogenous compound stimulated subapical elongation and enhanced floral formation, but did not induce flowering in noninductive conditions (COULTER 1983). The substance is presumed to be a hormone in that the photoperiod signal responsible for its synthesis is perceived by the leaves, but the responses occur at the shoot apex.

Gibberellin application does not normally cause flowering in SDP, with a few notable exceptions. The most studied case is *Impatiens balsamina*, which is an obligate SDP requiring a minimum of 2 SD cycles for initiation of flower buds. Gibberellins A_3, A_{13}, A_{4+7} and even (−)-kaurene are able to induce flower buds under strictly noninductive conditions (NANDA and KRISHNA-

MOORTHY 1967, NANDA et al. 1969). Subthreshold photoinduction (one SD) can be added to subthreshold GA (a single application of GA_3 or GA_{13}, which alone did not cause flowering); the resultant flowering occurred earlier than that obtained by continuous treatment of plants with GA in LD (NANDA and JINDAL 1975). For the additive effect to occur it was necessary to make the GA application after the single SD, although several LD could be intercalated without losing the effect of the subsequent GA treatment. Thus GA application was effective with SD only after the photo-induced stimulus had been synthesized. Perhaps, as in *Pharbitis* (ZEEVAART 1970), GA may be needed for action of the floral hormone at the apex and with prior application this effect is lost. There is some support for this suggestion from the fact that the second SD cycle can be substituted for by LD of 16 h or less (SAWHNEY et al. 1972). However, it is worth emphasizing that in *Impatiens,* as in *Zinnia elegans* (SAWHNEY and SAWHNEY 1976), GA treatment can completely substitute for SD induction and may (as is suggested for some LDP and *Bryophyllum*) act through production of the floral stimulus. An interaction with the photoperiodic induction process in the leaf may be indicated for three qualitative SD species of *Amaranthus* (*A. caudatus* f. albiflorus, *A. caudatus* f. caudatus and *A. tricolor* var. tristis). In all three, GA_{13} application promoted flowering and reduced the critical night length by 1–2 h (as in *Lemna paucicostata* with SA): GA_3 and $GA_{4/7}$ increased elongation but did not affect flowering, while GA_{13} had no effect on elongation (KOHLI and SAWHNEY 1979). However, GA was applied to the apex and the authors assumed that its action occurred there.

In some SDP the increased amounts of GA in LD may be involved in the *inhibition* of flowering. Application of GA_3 will inhibit flowering in at least 14 different SDP when applied in inductive photoperiods (see VINCE-PRUE 1975). In strawberry, a large amount of GA_3 completely prevented floral initiation in SD although, once initiation was completed, GA_3 accelerated development of the flower (GUTTRIDGE and THOMPSON 1964, PORLINGIS and BOYNTON 1961, THOMPSON and GUTTRIDGE 1959, TAFAZOLI and VINCE-PRUE 1978). Furthermore, a transmissible inhibitor of flowering has been shown to be produced in LD (GUTTRIDGE 1959a, b). However, the role of endogenous GA's and their relationship with the LD inhibitor remain in doubt, since neither chlormequat nor daminozide is more than marginally promotive of flowering (GUTTRIDGE 1969). Moreover, although a wide spectrum of GA's (e.g., A_1, A_2, A_3, A_4, A_5, A_7, A_9) inhibit flowering, they also cause elongation of the main stem (GUTTRIDGE and THOMPSON 1964); in LD alone, the main stem remains as a rosette. However, apart from the aberrant elongation of the stem, the effects of GA's are very similar to those of LD for petiole growth, stolon formation inhibition of floral initiation and promotion of flower growth (GUTTRIDGE 1970, TAFAZOLI and VINCE-PRUE 1978).

There are also a few species of LDP in which flowering is *inhibited* by GA_3 application. In *Fuchsia* this inhibitory effect occurred only when GA_3 was applied to the apical bud before, or at least concurrently with the arrival of the floral stimulus from the leaves. Later applications were without effect. The repeated application of daminozide promoted initiation in SD, and also increased the effect of subinductive LD treatment. The simultaneous application of GA_3 and daminozide under SD prevented the expected enhancement of flowering (SACHS et al. 1967). Thus, in this LDP, there is evidence for an inhibitory role of GA's in flowering, as there is for other woody angiosperms whose photoperiod requirements are not known (see ZIMMERMAN, Chap. 3, this Vol.). Furthermore, endogenous GA's may actually decrease during LD in *Fuchsia* (SACHS et al. 1967). The spectral responses of *Fuchsia* also suggest that an inhibitory effect of SD (i.e., long-night) cycles is a major component of the control system for flowering (Table 1). It remains to be proved that the endogenous SD inhibitor in *Fuchsia* is a GA, but there is good evidence that an endogenous GA is inhibitory for flowering in the LDP *Pisum sativum*. The strain G_2, a late–flowering line, is an obligate LDP. In SD, ^{3}H-GA_9 was metabolized to more polar GA's but this was blocked by exposure to a single LD, which also induced flowering. Moreover, when reproductive plants were reverting to the vegetative state, they regained the ability to synthesise polar GA's from ^{3}H-GA_9 (PROEBSTING and HEFTMAN 1980). It was concluded that flowering was inhibited in this strain of pea by a polar (on SiO_2 partition columns) gibberellin (GA_E) produced in SD. This polar GA [which is now presumed to be GA_{19} (DAVIES et al. 1982)] is also an anti–senescence factor (see p. 332).

A role for gibberellins in the control of flowering in conifers is now well established. A variety of biologically active GA's, including GA_3, has been found to enhance flowering in members of the Cupressaceae and Taxodiaceae while the "less polar" $Ga_{4/7}$ mixture (having only one hydroxyl group) is the most effective treatment in the Pinaceae (ROSS et al. 1983). A variety of stress treatments enhances the response to $GA_{4/7}$ application (BRIX and PORTLOCK 1982, PHILIPSON 1983, ROSS et al. 1983). Evidence that endogenous GA's participate in the control of flowering comes from studies of GA metabolism using cultural treatments known to induce flowering. For example, in *Picea abies* (L.) KARST., the less polar ^{3}H-GA_4 was metabolized more slowly in plants covered with clear plastic film (DUNBERG et al. 1983) suggesting that the stimulating effect on flowering of this cultural tratment may come about through an increased level of endogenous less polar GA's. Other cultural treatments that enhance flowering have also been shown to increase the levels of endogenous GA's of a less polar nature (ROSS et al. 1983) which, in the Pinaceae, are most effective in causing flowering.

Conifers have not been reported to be sensitive to photoperiod for the induction of flowering, although there are some effects of daylength on sex determination. In several Cupressaceae species, for example, LD promote maleness and SD promote femaleness (PHARIS et al. 1970, OWENS and PHARIS 1971) and there is limited evidence that daylength also affects sex in the Pinaceae (LUUKKANEN 1981, ROSS et al. 1983). The application of GA_3 has been shown to affect sex determination in Taxodiaceae and Cupressaceae species, with a high concentration increasing the number and proportion of female cone buds in LD. However, endogenous GA concentration has not been clearly implicated in sex determination, and it has been suggested that the high concentration of exogenous GA may increase femaleness by increasing the concentration of exogenous auxin (ROSS et al. 1983).

From the foregoing discussion it is clear that, while GA's influence flowering in a wide range of plants, their endogenous role is by no means well understood. In general there appear to be increased amounts of GA in LD, and this may be associated with the regulation of flowering by daylength. There is some evidence for inhibitory action in certain SDP (e.g., strawberry). On the other hand, the increased endogenous GA levels induced by LD may be necessary for synthesis of floral hormone in leaves of the LSDP *Bryophyllum*. In rosette LDP, the increased GA in LD is necessary for stem elongation, which normally accompanies flowering, although in *Silene*, at least, it has been established that GA's may have no direct role in floral initiation. Growth-retardant studies with other rosette LDP do, however, suggest a role for endogenous GA's, perhaps in the leaves. The greater sensitivity of the flowering response in several plants also indicates that GA may be having an effect on flowering distinct from that on elongation. For example, in both the SDP *Pharbitis* and the LDP *Hyoscyamus*, low doses of GA promoted flowering but did not increase elongation (OGAWA 1977, WARM 1980). Similarly, flowering in *Amaranthus* was promoted only by GA_{13}, which did not enhance elongation. The possibility that particular GA's have a specific role in flowering remains.

4.1.4 Abscisic Acid and Xanthoxin

The observation that (±) ABA application was able to mimic the effect of SD on flowering in some photoperiodically sensitive plants suggested that daylength effects might, in part, be attributed to changes in endogenous ABA content. Flowering was stimulated in the SDP blackcurrant, *Chenopodium, Pharbitis,* strawberry and *Lemna* (EL-ANTABLY et al. 1967, HIGHAM and SMITH 1969). ABA also induced flowering when applied to excised shoot tips of *Perilla* in LD, although the flowers were abnormal, resembling those produced with threshold induction (PURSE 1984). In the LDP, *Spinacia, Lolium temulentum, Petunia hybrida,* and carnation *Dianthus caryophyllus,* ABA delayed or inhibited flowering (EL-ANTABLY et al. 1967, EVANS 1966, CATHEY 1968a, b). However, in later experiments under strictly noninductive conditions, ABA was not able to induce flowering in SDP, although it enhanced the flowering response of *Pharbitis* and *Chenopodium* when plants were partly induced (HARADA et al. 1971, KREKULE and HORAVKA 1972, KREKULE and KOHLI 1981). Furthermore, the application of ABA was inhibitory to flowering in the SDP *Kalanchoe* (SCHWABE 1972). It seems unlikely, therefore, that endogenous ABA is generally involved in the process of floral induction in SDP.

Although (±)-ABA was inhibitory to flowering in the LD species mentioned, many others were not affected (CATHEY 1968b) and ABA seems to have no general function as an inhibitor of flowering in LDP. Indeed, the endogenous levels of ABA were higher in LD than in SD in the LDP *Spinacia* (ZEEVAART 1974) and were unaffected by daylength in perennial ryegrass, *Lolium perenne* L. (HANKS 1974) and *Lolium temulentum* (KING et al. 1977). Since the contents of ABA and xanthoxin in leaves and shoot apices of *Lolium temulentum* were not changed in any consistent way by daylength, it was concluded that the photoperiod response, which in this plant has an inhibitory component from SD leaves (EVANS 1960), is not due to either of these two inhibitors [although (±)-ABA injection into *Lolium* is inhibitory to flowering, and water stress both inhibited flowering and increased the content of endogenous ABA in the apex (KING and EVANS 1977)]. In *Spinacia,* ABA levels were higher in LD while xanthoxin remained unchanged, and there appeared to be no function for either substance in the photoperiodic control of flowering in this plant (ZEEVAART 1974) despite the inhibitory effect of applied (±)-ABA (EL-ANTABLY et al. 1967). ZEEVAART (1974) suggested that the higher ABA in LD might be a consequence of the thinner leaves produced under LD, these being more susceptible to water stress. In some South American cultivars of *Phaseolus vulgaris* L., however, the higher level of ABA in LD may be important in the regulation of flower development, since buds abscise in LD and this effect of daylength is mimicked by the application of ABA. Further evidence for the importance of endogenous ABA is the fact that a daylength-insensitive cultivar had the lowest content of ABA in LD while the most sensitive cultivar had the highest (MORGAN and ZEHNI 1980, MORGAN and MORGAN 1984).

The postulated role of ABA as a promoter of flowering in SDP is also not consistent with physical measurements of endogenous ABA levels, which were higher in LD than in SD in the SDP chrysanthemum (SENGUPTA et al. 1974) and strawberry (PLANCHER and NAUMANN 1978) and did not differ in extracts from induced and non-induced leaves of *Perilla* (PURSE 1984). There seems little evidence, therefore, to suggest that changes in endogenous ABA or xanthoxin concentration function in the photoperiodic regulation of floral initiation. It, of course, remains possible that endogenous ABA (or any other inhibitor) might play a role in flowering if it were produced at a constant rate irrespective of daylength, its effect being overcome in LD by the presence of increased amounts of other compounds. LOZHNIKOVA et al. (1982) have, for example, suggested that flowering may be affected by the relative amounts of ABA and GA: these workers reported that the LD induction of flowering in *Nicotiana sylvestris* was accompanied by an increase in GA and a decrease in ABA while, in the SDP Maryland Mammoth tobacco, SD induction led to a reduced level of GA and an increase in ABA in the leaves.

4.1.5 Cytokinins

Cytokinins have been shown to induce flowering in only a few plants. Application of kinetin reduced the number of SD required for induction in the SDP, *Perilla* (LONA and BOCCHI 1957), and shortened the critical night length in *Pharbitis* (OGAWA and ZEEVAART 1967). Moreover, the application of BA resulted in flowering in dark-grown plants given only a few minutes of R before transfer to an inductive night; control plants without BA were essentially vegetative (OGAWA and KING 1979). Kinetin also reduced the inhibitory effect of FR given at the beginning of an inductive night in *Pharbitis* and, to a lesser extent, overcame the inhibitory effect of a R night-break. It has been shown, however, that the application of CK may increase assimilate transport to the stem apex and so the effect on flowering is probably indirect, resulting from an increased movement of floral stimulus and/or assimilate under marginally inductive conditions (OGAWA and KING 1979).

The application of a CK has also been shown to evoke flowering in other plants but the site of action was not established, nor were possible effects on transport patterns considered. Short-day plants induced to flower in noninductive daylengths include *Wolffia microscopica* (VENKATARAMAN et al. 1970) and *Lemna paucicostata* HEGELM. (GUPTA and MAHESHWARI 1970), but the wide range of chemicals that can cause flowering in the Lemnaceae has already been mentioned. Exogenous application of BA to the SDP, chrysanthemum, cv. Pink Champagne, caused some flowering in LD when SD controls remained vegetative; there was a strong synergism between GA and CK in their effect on flowering in this plant, the combination of GA_5 with BA being highly effective (PHARIS 1972). Cytokinin application to apical buds promoted the development in LD of *Bougainvillaea glabra* CHOISY cv. San Diego Red inflorescences, which are initiated but normally fail to develop further under these conditions (HACKETT and SACHS 1968, TSE et al. 1974). Cytokinin treatment also stimulated the development of flower buds of *Phaseolus* which would otherwise have abscised in LD (BENTLEY et al. 1975). In both cases, CK may have caused an increased movement of metabolites to the flowers. *Arabidopsis thaliana* (L.) HEYNH. appears to be the only LDP in which flowering has been induced in noninductive photoperiods by CK application (MICHNIEWICZ and KAMIENSKA 1965).

Studies of endogenous cytokinins do not yet afford evidence for a role in the photoperiodic regulation of flowering. The CK content (probably zeatin) in leaves of the SDP *Begonia* × *cheimantha* EVERETT was increased by SD (HEIDE and SKOOG 1967). Similarly, transfer to LD resulted in a fall in n-butanol soluble CK's (mainly zeatin and zeatin riboside) in shoots of *Dactylis glomerata* L., which was detectable after two cycles and continued over 5 days. Following the reciprocal transfer from LD to SD, there was some evidence for an increase in CK's (MENHENETT and WAREING 1977). A greater CK content was also found in SD in xylem sap of *Perilla* (BEEVER and WOOLHOUSE 1973) and phloem sap of *Xanthium* (PHILLIPS and CLELAND 1972). CK also increased in SD in the xylem sap of *Phaseolus vulgaris* P-47 (NEWTON et al. 1980). In this variety, flower development requires SD and is promoted by CK application, suggesting that endogenous CK's (together with ABA, which causes flower abscission and increases in LD) may be involved in the control of flower development (MORGAN and ZEHNI 1980, MORGAN and MORGAN 1984).

In contrast to these results, transfer of the SDP *Xanthium* to SD led to a substantial decrease in the amount of extractable CK (VAN STADEN and WAREING 1972) and this decrease could be effected by a single SD (HENSON and WAREING 1974). The induction of flowering in *Xanthium* and the reduction in CK content were similar in several respects; both were effected by a single SD, an increase in length of the dark period gave a greater response, the effect of the long dark period could be at least partly negated by a night-break with R, and the leaf was the site of photoperiodic perception (HENSON and WAREING 1977b). As with flowering, the decrease in CK was an irreversible "induction" phenomenon (VAN STADEN and WAREING 1972). Furthermore, conditions which resulted in suboptimal rates of floral development (e.g., night-break treatments, near-threshold dark periods, and exposure of only a few leaves to SD) also gave intermediate levels of CK (HENSON and WAREING 1977b). Thus, for *Xanthium,* there is a strong correlation between floral induction and the lowered levels of CK brought about by SD conditions. However, a causal relationship appears unlikely since bark-ringing (which maintains high levels of CK in SD) had little effect on flowering.

Although CK levels in *Xanthium* decrease in SD, a direct effect of daylength on metabolism in leaves is unlikely. The capacity for CK synthesis in leaves is inadequate to sustain normal levels in the absence of a root system, neither is there evidence for any effect of daylength on (a) the rate or pattern of CK metabolism in leaves, nor (b) levels of endogenous CK's in detached leaves (HENSON and WAREING 1977a). It is not clear to what extent export to other plant parts is involved in these lowered CK levels. Increase in content of honeydew from aphids feeding on phloem sap was greater in flowering than in vegetative plants of *Xanthium (*PHILLIPS and CLELAND 1972), indicating the possibility of more rapid export. However, an influence of SD to increase CK export from leaf laminae would not explain the lower CK levels in the buds themselves, unless utilization is substantially stimulated by the SD treatment (HENSON and WAREING 1977b). Thus, HENSON and WAREING (1977b) concluded that the influence of daylength on CK content of leaves and buds of *Xanthium* is possibly mediated by a shoot to root signal from the leaves in SD, thus reducing the amount of CK exported from the roots. Such a reduction in root exudate CK's from SD plants was observed earlier (HENSON and WAREING 1974) and was also seen in *Chenopodium rubrum* (KREKULE 1979). The signal from SD leaves, like the floral stimulus, appears to move in the phloem since bark-ringing prevented the SD effect. Once again the evidence points to a stimulus of unknown composition, originating in leaves in appropriate daylengths and causing changes in activity at selected target sites. A similar conclusion might be drawn regarding the control of CK production in *Phaseolus vulgaris* P-47 (where SD increased the amount of CK present in xylem sap), since the transmissible stimulus that prevents bud abscission must originate in the leaves which have been shown to be the site of perception for this SD effect (MORGAN and MORGAN 1984). Although only circumstantial, the evidence indicates that a stimulus from the leaves in SD increases CK export from the roots and that this increased CK supply to the buds may prevent their abscission.

The lowered CK content in induced plants of *Xanthium* may be uncommon and is not consistent with the reported effects of CK's to enhance flowering in some other SDP. However, CK's may be responsible for the root-mediated suppression of flowering that has been observed in several plants (KREKULE 1979). For example, excision of the roots enhanced flowering in the quantitative SDP *Chenopodium polyspermum* in noninductive conditions and the addition of zeatin to the buds mimicked the presence of the roots and reduced flowering: zeatin also counteracted the effects of inductive SD (SOTTA 1978). Similar results have been obtained in the LDP *Scrophularia arguta* SOL. (KREKULE 1979). Since, in *Chenopodium,* sucrose acted like zeatin, it is possible that, as in *Pharbitis,* CK is affecting flowering through an effect on assimilate transport. However, it is difficult to understand how an enhanced movement of assimilate to the bud could inhibit flowering. It has been suggested that, in these plants, flowering is controlled by a balance between a floral stimulus coming from the leaves and inhibitory substances (CK's) from the roots (MIGINIAC 1978).

In some plants, one of the earliest events observed at the shoot apex following photoperiod induction is a transitory increase in the mitotic index and any treatment that abolishes this increase also abolishes flowering. When CK's are added to the stem apex of the LDP *Sinapis alba,* they cause an increase in mitotic activity similar to that seen after a single LD (BERNIER et al. 1977). However, the LD induces flowering, while the CK treatment does not. It has been suggested, therefore, that the floral stimulus in *Sinapis* consists of at least two components, one of which is a CK that stimulates mitotic activity. Work is clearly needed on CK levels in relation to flowering, with particular reference to their origin (leaves or roots) and the influence of daylength on their synthesis, metabolism and movement.

4.1.6 Auxin and Ethylene

In early studies, auxin application was shown to be inhibitory to flowering in a number of SDP; the inhibition appeared to be localized in the leaf and was thought to be due to interference with synthesis of the flowering hormone, to an increase in its breakdown, or perhaps to an effect on its transport (see LANG 1965, VINCE-PRUE 1975). In more recent studies with another SDP, *Chenopodium rubrum,* the effect occurred at the apex via the suppression of axillary bud differentiation (KREKULE 1979). In *Chenopodium* and in the LDP *Brassica campestris* L., auxin application both inhibited flowering and promoted apical dominance (KREKULE and SEIDLOVA 1977, SEIDLOVA 1980). One of the earliest changes that occurs with the onset of flowering in many plants is the activation of the axillary meristems, and so a decrease in the level of auxins in the apical meristem may be associated with the transition to flowering. However, there is considerable disagreement concerning the need for a reduced level of auxin. Although floral events have been linked with a drop in auxin level (KREKULE 1979, KUMAR and NANDA 1981, GASPAR 1980), auxin application may promote flowering in some LDP especially in subthreshold photoperiods (LIVERMAN and LANG 1956, EVANS 1964). However, the endogeneous auxin content of leaves shows no consistent trend with photoperiod (Table 2), and its role if any, in the photoperiodic regulation of flowering remains obscure.

Within recent years, some of the effects of applied auxin have been attributed to an increase in the release of ethylene. ABELES (1967) observed that ethylene prevented flowering in *Xanthium* (16 h dark) and attributed the inhibitory effect of applied auxin in this plant to the release of ethylene. Ethylene also inhibited SD-induced flowering in *Perilla, Pharbitis,* and chrysanthemum (ZHADANOVA 1950, SUGE 1972, ROGERS and TSIA 1966) but induced flowering in another SDP, *Plumbago indica* L. (NITSCH and NITSCH 1969). In the facultative SDP, chrysanthemum cv. Polaris, CEPA delayed flowering in LD as well as in SD, the number of additional leaves formed being the same in both daylengths (COCKSHULL and HORRIDGE 1978); in this plant, therefore, the delaying effect of ethylene is probably independent of daylength. Ethylene can also cause flowering in the DNP pineapple, other members of the Bromeliaceae (BURG and BURG 1966), and in apple (WILLIAMS 1972). In the SDP *Pharbitis,* the inhibitory effect of ethylene occurred in the cotyledons during the second half of the night (SUGE 1972), whereas auxin exerted its inhibitory effect only in the first half (OGAWA and ZEEVAART 1967), suggesting that the two hormones may be acting differently. It was proposed that ethylene may affect the inductive process since it did not cause production of an inhibitor nor did it modify transport of the floral stimulus.

The few studies on the effect of daylength on the rate of endogenous ethylene evolution are noted in Table 3.

4.1.7 Steroids

The possibility that steroids might be involved in the control of flowering was tested several years ago by applying a range of compounds which block cholesterol synthesis. Two of these compounds suppressed flowering in both *Xanthium* and *Pharbitis* (BONNER et al. 1963). There were also reports that floral initiation was evoked in noninductive photoperiods by the application of steroids [*Chrysanthemum* (SDP), BISWAS et al. 1967; *Callistephus chinensis* (LDP), CHOUARD 1936]. However, more recent studies on the metabolism of the steroid precursor, mevalonic acid, in *Xanthium* have revealed no differences between plants in LD or SD for leaf, petiole or bud extracts, although the mevalonic acid was rapidly metabolized and its metabolites transported from leaves to buds (BLEDSOE and ROSS 1978).

4.1.8 Conclusions

Florigen continues to remain a physiological concept rather than a chemical entity. Its elusive nature, together with the observed effects of naturally occurring growth regulators, has led several workers to question whether it is a single substance (CLELAND 1975, 1978, EVANS 1969, 1971). Certainly, other morphogenetic processes, such as the initiation of roots and buds on callus, require the interaction of two or more known hormones. Similarly, floral evocation may depend on a particular ratio and/or temporal sequence of endogenous growth regulators which are not themselves unique for flowering. If florigen is such a complex of substances, then much more research is needed on the influence of daylength on concentration, transport, and metabolism of the various growth regulators associated with the flowering situation.

Despite the failure to characterize florigen, there is much circumstantial evidence for the existence of such a hormone which is uniquely needed for floral initiation and which is effective in a range of different plants (LANG 1965, EVANS 1971, VINCE-PRUE 1975, KING 1976, ZEEVAART 1976). There are a number of reasons which could account for the failure, so far, to isolate such a substance; these include lack of suitable assay methods, use of improper extraction solvents, difficulties of re-introduction, lability, subthreshold concentrations in extracts, the presence of inhibitors from noninduced leaves of test plants, or rapid breakdown in such leaves, and failure of phloem loading in leaves in noninductive daylengths. Inhibitors, or the balance between inhibitors and promoters of flowering, appear to be particularly important in some plants [e.g., in strawberry (GUTTRIDGE 1959a, b), *Hyoscyamus* (LANG and MELCHERS 1943) and *Lolium* (EVANS 1960)]. A further possibility is that action of the floral hormone requires the presence of a complementary stimulus such as GA.

CHAILAKHYAN's original single substance concept was replaced by a hypothesis which proposed that flowering is regulated by a stimulus consisting of two components, namely GA's and a hypothetical substance or group of substances, called anthesins (CHAILAKHYAN 1975a, b). It was supposed that LDP produce anthesins in all daylengths but GA's only in LD, while SDP produce sufficient GA's in any daylength, but anthesins only in SD. The scheme could also account for plants with dual photoperiodic requirements if it is assumed that LD are needed for GA production and SD for anthesin production, although it does not explain why the obligatory sequence should be different in LSDP and SLDP. The hypothesis that GA's limit the flowering of LDP in SD, while anthesins limit the flowering of SDP in LD, requires that donors of one photoperiodic type should evoke flowering in receptors of the other when both are in noninductive conditions. Few such grafts have been made, and negative results have been obtained in at least four cases (LANG 1965, ZEEVAART 1958). Both LANG and ZEEVAART observed, however, some flowering response in Maryland Mammoth tobacco receptors (SDP maintained in LD) when grafted to vegetative donors of *N. sylvestris* (LDP maintained in SD). Additional results have been obtained in LSDP, *Bryophyllum,* where one part of the graft was in LD and the other in SD; flower buds formed on the receptor shoots which were situated between the LD and SD parts and were thought to be evoked by the influx to the receptor of GA's from LD- and anthesins from SD-treated leaves (CHAILAKHYAN 1975a, b). An alternative explanation was offered by ZEEVAART (1976), who proposed that GA's from LD leaves moved to SD leaves, thereby resulting in the production of "florigen" which subsequently moved into the intervening receptor shoots. Furthermore, it has been unambiguously shown that LD and SD induction must be perceived by one and the same leaf in the LSDP, *Cestrum nocturnum* L. (SACHS 1969) and probably also in *Bryophyllum* (ZEEVAART 1969d) whereas the anthesin/GA hypothesis proposed by CHAILAKHYAN assumes that both substances are acting at the apex. Thus the anthesin/GA hypothesis of flowering does not seem to be generally applicable to different photoperiodic response types. Even so, the possibility that one or more additional hormones may be needed for the synthesis, transport or action of "florigen" is not excluded. An appropriate supply of hormones appears to be necessary for initiation, differentiation, and growth of various floral parts and, in many species, these further stages of floral development are also under photoperiodic control (see VINCE-PRUE 1975, CLELAND and BEN-TAL 1983).

Several hypotheses for the control of flowering by hormones have been put forward during the past 50 years. These include control by a single specific flower-promoting substance or "florigen"; a relatively simple interaction between two substances, "florigen" and "anti-florigen"; a complex multi-component stimulus with different factors being limiting in different conditions; and a nutrient-diversion theory that assumes only a secondary role for hormones. If there is a specific "florigen", its identity and even the class of substance involved remains a mystery. The idea of a multi-component stimulus is attractive and there are several lines of evidence that point in this direction. Perhaps one of the most convincing arguments for a relatively nonspecific stimulus is the enormous range of responses that are under daylength control, indicating that specificity may be a function of the target site. Nevertheless, for several of the responses discussed in this chapter, there is evidence for an unidentified stimulus originating in SD leaves. It is clear that efforts should continue toward the possible discovery of new hormonal substances, but there should also be a detailed investigation with the techniques now available of the influence of daylength on the known hormones in relation to selected processes under photoperiod control.

References

Abeles FB (1967) Inhibition of flowering in *Xanthium pennsylvanicum* by ethylene. Plant Physiol 42:608–609

Alden J, Hermann RK (1971) Aspects of the cold hardiness mechanism in plants. Bot Rev 37:37–142

Alleweldt G, Düring H (1972) Einfluss der Photoperiode auf Wachstum und Abscisinsäuregehalt der Rebe. Vitis 111:280–288

Alvim RW, Hewett EW, Saunders PF (1976) Seasonal variation in the hormone content of willow. I. Changes in abscisic acid content and cytokinin activity in the xylem sap. Plant Physiol 57:474–476

Alvim RW, Thomas S, Saunders PF (1978) Seasonal variation in the hormone content of willow. II. Effect of photoperiod on growth and abscisic acid content of trees under field conditions. Plant Physiol 62:779–780

Alvim RW, Saunders PF, Barros RS (1979) Abscisic acid and the photoperiodic induction of dormancy in *Salix viminalis* L. Plant Physiol 63:774–777

Angrish R, Nanda KK (1982) Dormancy and flowering process in reproductive buds of *Salix babylonica* cultured in vitro. Z Pflanzenphysiol 106:263–269

Asahira T, Nitsch JP (1968) Tuberization in vitro: *Ullucus tuberosus* et *Dioscorea*. Bull Soc Bot (France) 115:345–352

Badizadegan M, Tafazoli E, Kheradnam M (1972) Effect of N^6-benzyladenine on vegetative growth and tuber production in potato. Am Potato J 49:109–116

Bae M, Mercer EI (1970) The effect of long- and short-day photoperiods on the sterol levels in the leaves of *Solanum andigena*. Phytochemistry 9:63–68

Baldev B, Lang A (1965) Control of flower formation by growth retardants and gibberellin in *Samolus parviflorus* a long-day plant. Am J Bot 52:408–417

Batutis EJ, Ewing EE (1982) Far-red reversal of red light effect during long-night induction of potato (*Solanum tuberosum* L.) tuberization. Plant Physiol 69:672–674

Beever JE, Woolhouse HW (1973) Increased cytokinin from root systems of *Perilla frutescens* and flower and fruit development. Nature 246:31–32

Bentley B, Morgan CR, Morgan DG, Saad FA (1975) Plant growth substances and effects of photoperiod on flower bud development in *Phaseolus vulgaris*. Nature 256:121–122

Bernier G, Kinet J, Jacqmard A, Havelange A, Bodson M (1977) Cytokinin as a possible component of the floral stimulus in *Sinapis alba*. Plant Physiol 60:282–285

Betria AI, Montaldi ER (1974) Light effects on bulb differentiation and leaf growth in *Cyperus rotundus* L. Phyton 32:1–8

Biran I, Gur I, Halevy AH (1972) The relationship between exogenous growth inhibitors and endogenous levels of ethylene, and tuberization of dahlias. Physiol Plant 27:226–230

Biran I, Lesham B, Gur I, Halevy AH (1974) Further studies on the relationship between growth regulators and tuberization of dahlias. Physiol Plant 31:23–28

Biswas PK, Paul KB, Henderson JHM (1967) Effects of steriods on chrysanthemum in relation to growth and flowering. Nature 213:917–918

Bledsoe CS, Ross CW (1978) Metabolism of mevalonic acid in vegetative and induced plants of *Xanthium strumarium*. Plant Physiol 62:683–686

Blondon F, Jacques R (1970) Action de la lumière sur l'initiation florale du *Lolium temulentum* L: spèctre d'action et rôle du phytochrome. CR Acad Sci Ser D 270:947–950

Bonner J, Heftman JE, Zeevaart JAD (1963) Suppression of floral induction by inhibition of steriod biosynthesis. Plant Physiol 38:81–88

Booth A, Lovell PH (1972) The effect of pre-treatment with gibberellic acid on the distribution of phytosynthate in intact and disbudded plants of *Solanum tuberosum* L. New Phytol 71:795–804

Borah MH, Milthorpe F (1962) Growth of the potato as influenced by temperature. Indian J Plant Physiol 5:53–72

Borkowska B, Powell LE (1982/1983) Abscisic acid relationships in dormancy of apple buds. Sci Hortic 18:111–117

Bowen MR, Hoad GV (1968) Inhibitor content of phloem and xylem sap obtained from willow (*Salix viminalis* L.) entering dormancy. Planta 81:64–70

Brewster JL (1977) The physiology of the onion. Hortic Abst 47:17–23

Brix H, Portlock FT (1982) Flowering response of western hemlock seedlings to gibberellin and water-stress treatments. Can J For Res 12(1):76–82

Burg SP, Burg EA (1966) Interaction between auxin and ethylene and its role in plant growth. Proc Natl Acad Sci USA 55:262–269

Carr DJ (1967) The relationship between florigen and the flower hormones. Ann NY Acad Sci 144:305–312

Catchpole AH, Hillman J (1969) Effect of ethylene on tuber initiation in *Solanum tuberosum* L. Nature 223:1387

Cathey HM (1968a) Response of *Dianthus caryophyllus* L. (carnation) to synthetic abscisic acid. Proc Am Soc Hortic Sci 93:560–568

Cathey HM (1968b) Response of some ornamental plants to synthetic abscisic acid. Proc Am Soc Hortic Sci 93:693–698

Chailakhyan M KH (1936) New facts in support of the hormonal theory of plant development. Dokl Akad Nauk USSR 12:443–447

Chailakhyan M KH (1940) Translocation of the flowering hormones across various plant organs. III. Across the root. Dokl Akad Nauk USSR 27:373–376

Chailakhyan M KH (1958) Hormonale Faktoren des Pflanzenblühens. Biol Zentralbl 77:641–662

Chailakhyan M KH (1968) Flowering hormones of plants. In: Wightman F, Setterfield G (eds) Biochemistry and physiology of growth substances. Runge Press, Ottawa, pp 1317–1340

Chailakhyan M KH (1975a) Substances of plant flowering. Biol Plant 17:1–11

Chailakhyan M KH (1975b) Forty years of research on the hormonal basis of plant development – some personal reflections. Bot Rev 41:1–29

Chailakhyan M KH (1982) Hormonal substances in flowering. In: Wareing PF (ed) Plant growth substances 1982. Academic Press, London New York, pp 645–655

Chailakhyan M KH, Butenko RG (1957) Movements of assimilates of leaves to shoots under different photoperiodic conditions of leaves. Fiziol Rast 4:450–462

Chailakhyan M KH, Lozhnikova UN (1966) Reaction of the interruption of darkness on plant gibberellins. Fiziol Rast 13:833–841

Chailakhyan M KH, Kakhidse NT, Milyeva EL, Gukasyan IA, Yanina LI (1969) Effects of daylength and gibberellins on growth rate, flowering, and complex differentiation in bi-colored coneflower. Fiziol Rast 16:392–399

Charnay D, Courduroux JC (1972) Acide abscissique et tubérisation in vitro de bourgeons de topinambour (*Helianthus tuberosus* L. var D19). CR Acad Sci (Paris) 275:2351–2354
Chouard P (1936) Nouvelles recherches experimentals sur l'action de la folliculine sur floraison des Reines-Marguerites et de plusiers autres plantes. CR Seances Soc Biol 122:823–825
Chouard P (1957a) Diversité des méchanismes des dormances, de la vernalisation et du photopériodisme revelée notamment par l'action de l'acide gibbérellique. Mem Soc Bot Fr 1956–1957:51–64
Chouard P (1957b) La journée courte ou l'acide gibbérellique comme succedanées du froid pour la vernalisation d'une plante vivace en rosette, le *Scabiosa succisa* L. CR Hebd Seances Acad Sci 245:2520–2522
Chvojka L, Travnick M, Zakourilova M (1962) The influence of stimulating doses of 6-benzylaminopurine on awaking apple buds and on their consumption of oxygen. Biol Plant 4:203–206
Clark JE, Heath OVS (1962) Studies in the physiology of the onion plant. V. An investigation into the growth substance content of bulbing onions. J Exp Bot 13:227–249
Cleland CF (1974a) Isolation of flower-inducing and flower-inhibiting factors from aphid honeydew. Plant Physiol 54:899–903
Cleland CF (1974b) The influence of salicylic acid on flowering and growth in the long-day plant. *Lemna gibba* G_3. In: Bieleski RL, Ferguson AR, Cresswell MM (eds) Mechanisms of regulation of plant growth. R Soc NZ, Wellington, pp 553–557
Cleland CF (1975) The florigen enigma. New Plant Physiol 7/6:1–4
Cleland CF (1978) The flowering enigma. Bio Sci 28:265–269
Cleland CF (1982) The chemical control of flowering – a status report. In: Wareing PF (ed) Plant growth substances. Academic Press, London New York, pp 635–644
Cleland CF, Ajami A (1974) Identification of the flower-inducing factor isolated from aphid honeydew as being salicylic acid. Plant Physiol 54:904–906
Cleland CF, Ben-Tal Y (1983) Hormonal regulation of flowering and sex expression. In: Meudt WJ (ed) Strategies of plant reproduction. BARC Symp 6. Allanheld, Osmun, Totowa, pp 157–180
Cleland CF, Briggs WR (1969) Gibberellin and CCC effects on flowering and growth in the long-day plant, *Lemna gibba* G_3. Plant Physiol 44:503–507
Cleland CF, Tanaka O (1979) Effect of daylength on the ability of salicylic acid to induce flowering in the long-day plant *Lemna gibba* G_3 and the short-day plant *Lemna paucicostata* 6746. Plant Physiol 64:421–424
Cleland CF, Zeevaart JAD (1970) Gibberellins in relation to flowering and stem elongation in the long-day plant *Silene armeria*. Plant Physiol 46:392–400
Cleland CF, Tanaka O, Fledman LJ (1982) Influence of plant growth substances and salicylic acid on flowering and growth in the Lemnaceae (duckweeds). Aquat Bot 13(1) (Spec Iss) 3–20
Cockshull KE, Horridge JS (1978) 2-Chloroethanephosphonic acid and flower initiation by *Chrysanthemum morifolium* Ramat. in short days and in long days. J Hortic Sci 53:89–90
Coulter MW (1983) Isolation of a dark-induced hormone from *Pharbitis nil*. Plant Physiol 72 (Suppl):85, No 481
Courduroux JC (1966) Mécanisme physiologique de la tubérisation du topinambour. Bull Soc Fr Physiol Veg 12:213–232
Courduroux JC (1967) Etude du mécanisme physiologique de la tubérisation chez le topinambour (*Helianthus tuberosus* L.). Ann Sci Nat Bot 8:215–355
Czopek M (1964) Action of kinetin and gibberellic acid and red light on the germination of turions of *Spirodela polyrrhiza*. Bull Acad Pol Sci Cl II 12:177–182
Davies PJ, Emshwiller E, Gianfagna TJ, Proebsting WM, Noma M, Pharis RP (1982) The endogenous gibberellins of vegetative and reproductive tissue of G_2 peas. Planta 154:266–272
Deitzer GF, Hayes R, Jabben M (1979) Kinetics and time dependence of the effect of far-red light on the photoperiodic induction of flowering in Wintex barley. Plant Physiol 64:1015–1021

Digby J, Wareing PF (1966) The relationship between endogenous hormone levels in the plant and seasonal aspects of cambial activity. Ann Bot 30:607–622

Domanski R, Kozlowski TT (1968) Variations of kinetin-like activity in buds of *Betula* and *Populus* during release from dormancy. Can J Bot 46:397–403

Downs RJ (1956) Photoreversibility of flower initiation. Plant Physiol 31:279–284

Downs RJ, Borthwick HA (1956a) Effect of photoperiod upon the vegetative growth of *Weigela florida* var variegata. Proc Am Soc Hortic Sci 68:518–521

Downs RJ, Borthwick HA (1956b) Effects of photoperiod on growth of trees. Bot Gaz 117:310–326

Dunberg A, Malmberg G, Sassa T, Pharis RP (1983) Metabolism of tritiated gibberellins A_1 and A_9 in Norway Spruce, *Picea abies* (L.) Karst. Effects of a cultural treatment known to enhance flowering. Plant Physiol 71:257–262

Durley RC, Pharis RP, Zeevart JAD (1975) Metabolism of (^{3}H) gibberellin A_{20} under long- and short-day conditions. Planta 126:139–149

Eagles CF, Wareing PF (1964) The role of growth substances in the regulation of bud dormancy. Physiol Plant 17:697–708

El-Antably HMM, Wareing PF, Hillman J (1967) Some physiological responses to DL abscisin (dormin.). Planta 73:74–90

Ende van den H, Zeevaart JAD (1971) Influence of daylength on gibberellin metabolism and stem growth in *Silene armeria*. Planta 98:164–176

Erez A, Lavee S (1969) Prunin identification, biological activity and quantitative change in comparison to naringenin in dormant peach buds. Plant Physiol 44: 342–346

Esashi Y (1960) Studies on the formation and sprouting of aerial tubers in *Begonia evansiana* Adr. IV. Cutting method and tuberizing stages. Sci Rep Tohoku Univ 26:239–249

Esashi Y (1961) Studies on the formation and sprouting of aerial tubers in *Begonia evansiana* Andr. V. Antagonistic action of long-days to short-day response. Plant Cell Physiol 2:117–127

Esashi Y (1966) The relation between red and blue or far-red lights in the night interruption of the photoperiodic tuberization of *Begonia evansiana*. Plant Cell Physiol 7:405–414

Esashi Y, Leopold AC (1968) Regulation of tuber development in *Begonia evansiana* by cytokinin. In: Wightman F, Setterfield R (eds) Biochemistry and physiology of plant growth substances. Runge, Ottawa, pp 923–941

Evans LT (1960) Inflorescence initiation in *Lolium temulentum* L. II. Evidence for inhibitory and promotive processes involving transmissible products. Aust J Biol Sci 13:429–440

Evans LT (1962) Inflorescence initiation in *Lolium temulentum* L. III. The effect of anaerobic conditions during photoperiodic induction. Aust J Biol Sci 15:281–290

Evans LT (1964) Inflorescence initiation in *Lolium temulentum* L. V. The role of auxins and gibberellins. Aust J Biol Sci 17:24–35

Evans LT (1966) Inhibitory effect on flower induction in a long-day plant. Science 151:107–108

Evans LT (1969) (ed) The nature of flower induction. In: The induction of flowering. MacMillan, Melbourne, pp 457–580

Evans LT (1971) Flower induction and the florigen concept. Ann Rev Plant Physiol 22:365–394

Evans LT, Wardlaw IF (1966) Independent translocation of ^{14}C-labelled assimilates and of the floral stimulus in *Lolium temulentum*. Planta 68:310–316

Ewing EE, Wareing PF (1978) Shoot, stolon and tuber formation on potato (*Solanum tuberosum* L.) cuttings in response to photoperiod. Plant Physiol 61:348–353

Fuchigami LH, Evert DR, Weiser CJ (1971) A translocatable cold hardiness promoter. Plant Physiol 47:164–167

Fujioka S, Yamaguchi I, Murofushi N, Takahashi N, Kaihara S, Takimoto A (1983) Flowering and endogenous levels of benzoic acid in *Lemna* species. Plant Cell Physiol 24:235–239

Gaskin P, MacMillan J, Zeevaart JAD (1973) Identification of gibberellin A_{20}, abscisic acid, and phaseic acid from flowering *Bryophyllum daigremontanum* by combined gas chromatography-mass spectrometry. Planta 111:347–352

Gaspar T (1980) Rooting and flowering, two antagonistic phenomena from a hormonal point of view, Joint DPGRG & BPGRG Symp. Aspects and prospects of plant growth regulators. Monogr 6:39–49

Gianfagna T, Zeevaart JAD, Lusk WJ (1983) Effect of photoperiod on the metabolism of deuterium-labelled gibberellin A_{53} in spinach. Plant Physiol 72:86–89

Gregory LE (1956) Some factors for tuberization in the potato plant. Am J Bot 43:281–288

Griesel WO (1963) Photoperiodic responses of two *Cestrum* species and non interchangeability of their flowering hormones. Plant Physiol 38:479–482

Grigorieva NYa, Kucherov VF, Lozhnikova VN, Chailakhyan MKH (1971) Endogenous gibberellins and gibberellin-like plants: a possible correlation with photoperiodic response. Phytochemistry 10:509–517

Gupta S, Maheshwari SC (1970) Growth and flowering of *Lemna paucicostata* II. Role of growth regulators. Plant Cell Physiol 11:97–106

Guttridge CG (1959a) Evidence for a flower inhibitor and vegetative growth promoter in the strawberry. Ann Bot 23:351–360

Guttridge CG (1959b) Further evidence for a growth-promoting and flower-inhibiting hormone in strawberry. Ann Bot 23:612–621

Guttridge CG (1969) *Fragaria*. In: Evans LT (ed) The induction of flowering. MacMillan, Melbourne, pp 247–267

Guttridge CG (1970) Interaction of photoperiod, chilling and exogenous gibberellic acid on growth of strawberry petioles. Ann Bot 34:349–364

Guttridge CG, Thompson PA (1964) The effect on growth and flowering of *Fragaria* and *Duchesnea*. J Exp Bot 15:631–646

Hackett WP, Sachs RM (1968) Experimental separation of inflorescence development from initiation in *Bougainvillea*. Proc Am Soc Hortic Sci 192:615–621

Hammes PS (1971) Tuber initiation in the potato (*Solanum tuberosum* L.) Agroplantae 3:73–74

Hammes PS, Beyers EA (1973) Localization of the photoperiodic perception in potatoes. Potato Res 16:68–72

Hamner KC (1940) Interrelation of light and darkness in photoperiodic induction. Bot Gaz 101:658–687

Hamner KC, Bonner J (1938) Photoperiodism in relation to hormones as factors in floral initiation and development. Bot Gaz 100:388–431

Hamner KC, Long EM (1939) Localization of the photoperiodic perception in *Helianthus tuberosus*. Bot Gaz 101:81–90

Hanks GR (1974) Abscisic acid levels and daylength in perennial ryegrass. Z Pflanzenphysiol 71:63–66

Harada H, Bose RK, Cheruel J (1971) Effects of four growth regulating chemicals on flowering of *Pharbitis nil*. Z Pflanzenphysiol 64:267–269

Harrison MA, Saunders PF (1975) The abscisic acid content of dormant birch buds. Planta 123:291–298

Hashimoto T, Tamura S (1969) Effects of abscisic acid on the sprouting of aerial tubers of *Begonia evansiana* and *Dioscorea batatas*. Bot Mag (Tokyo) 82:69–75

Heath OVS, Holdsworth M (1948) Morphogenic factors as exemplified by the onion plant. Symp Soc Exp Biol 2:326–350

Heath OVS, Hollies MA (1965) Studies in the physiology of the onion plant. VI. A sensitive morphological test for bulbing and its use in detecting bulb development in sterile culture. J Exp Bot 16:124–144

Heide OM (1967) The auxin level of *Begonia* leaves in relation to their regeneration ability. Physiol Plant 20:886–902

Heide OM (1968) Auxin level and regeneration of *Begonia* leaves. Planta 81:153–159

Heide OM, Skoog F (1967) Cytokinin activity in *Begonia* and *Bryophyllum*. Physiol Plant 20:771–780

Heim B, Jabben M, Schäfer E (1981) Phytochrome destruction in dark- and light-grown *Amaranthus caudatus* seedlings. Photochem Photobiol 34:89–93

Henson IE, Wareing PF (1974) Cytokinins in *Xanthium strumarium*: a rapid response to short-day treatment. Physiol Plant 32:185–187

Henson IE, Wareing PF (1977a) Cytokinins in *Xanthium strumarium* L.: the metabolism of cytokinins in detached leaves and buds in relation to photoperiod. New Phytol 78:27–33

Henson IE, Wareing PF (1977b) Cytokinins in *Xanthium strumarium* L.: some aspects of the photoperiodic control of endogenous levels. New Phytol 78:35–45

Heslop-Harrison J (1964) The control of flower differentiation and sex expression. In: Régulateurs naturels de la croissance végétale. 5th Int Conf Plant Growth Substances. CNRS Paris, pp 597–609

Hewett EW, Wareing PG (1973) Cytokinins in *Populus × robusta*; changes during chilling and bud burst. Physiol Plant 28:393–399

Higham BH, Smith H (1969) The induction of flowering by abscisic acid in *Lemna perpusilla* 6746. Life Sci 8:1061–1065

Hoad GV, Bowen MR (1968) Evidence for gibberellin-like substances in phloem exudate from higher plants. Planta 82:22–32

Hocking TJ, Hillman JR (1975) Studies in the role of abscisic acid in the initiation of bud dormancy in *Alnus glutinosa* and *Betula pubescens*. Planta 125:235–242

Hodson HK, Hamner KC (1970) Floral inducing extracts from *Xanthium*. Science 167:384–385

Holland RWK, Vince D (1971) Floral initiation in *Lolium temulentum* L.: the role of phytochrome in the responses to red and far-red light. Planta 98:232–243

Humphries EC (1963) Effects of gibberellic acid and CCC on growth of potato. Rothamsted Exp Stn Ann Rep 89

Humphries EC, Dyson PW (1967) Effect of a growth inhibitor, N-dimethylaminosuccinamic acid (B9) on potato plants in the field. Eur Potato J 10:116–126

Hurst C, Hall TC, Weiser CJ (1967) Reception of the light stimulus for cold acclimation in *Cornus stolonifera* Michx. Hortic Sci 2:164–166

Imamura S, Takimoto A (1955) Transmission rate of photoperiodic stimulus in *Pharbitis nil*. Bot Mag 68:260–266

Imhoff CH, Lecharny A, Jacques R, Brulfert J (1979) Two phytochrome dependent processes in *Anagallis arvensis* L.: flowering and stem elongation. Plant Cell Environ 2:67–72

Irving RM (1969) Characterization and role of an endogenous inhibitor in the induction of cold hardiness in *Acer negundo*. Plant Physiol 44:801–805

Irving RM, Lanphear FO (1967) The long-day leaf as a source of cold-hardiness inhibitors. Plant Physiol 42:1384–1388

Irving RM, Lanphear FO (1968) Regulation of cold hardiness in *Acer negundo*. Plant Physiol 43:9–13

Jacobs WP (1980) Inhibition of flowering in short-day plants. In: Skoog F (ed) Plant growth substances. Springer, Berlin Heidelberg New York, pp 301–309

Jacques M (1968) Diversité et caractéristiques des processus de l'induction florale chez deux Chénopodiacées. C R Hebd Seances Acad Sci Ser D 267:1592–1595

Jacques M (1973) Transfert par voie de greffage du stimulus photopériodique. C R Acad Sci 276D:1705–1708

Jones MG, Zeevaart JAD (1980a) Gibberellins and the photoperiodic control of stem elongation in the long-day plant *Agrostemma githago* L. Planta 149:269–273

Jones MG, Zeevaart JAD (1980b) The effect of photoperiod on the levels of seven endogenous gibberellins in the long-day plant *Agrostemma githago* L. Planta 149:274–279

Jones MG, Zeevaart JAD (1982) Effect of photoperiod on metabolism of [^{3}H] gibberellins A_1, 3-*epi*-A_1, and A_{20} in *Agrostemma githago* L. Plant Physiol 69:660–662

Junttila O (1981) Effects of different gibberellins on elongation growth under short-day conditions in seedlings of *Salix pentandra*. Physiol Plant 53:315–318

Junttila O (1982) Gibberellin-like activity in shoots of *Salix pentandra* as related to the elongation growth. Can J Bot 60:1231–1234

Kato T (1965) Physiological studies on the bulbing and dormancy of onion plants. V. The relations between the metabolism of carbohydrates, nitrogen compounds and auxin and the bulbing phenomenon. J Jpn Soc Hortic Sci 34:187–195

Ketellapper HJ, Barbaro A (1966) The role of photoperiod, vernalization and gibberellic acid in floral induction in *Coreopsis grandiflora* Nutt. Phyton 23:33–41

King RW (1975) Multiple circadian rhythms regulate photoperiodic flowering responses. Can J Bot 53:2631–2638

King RW (1976) Implications for plant growth of the transport of regulatory compounds in phloem and xylem. In: Passioura J (ed) Transport and transfer processes in plants. Academic Press, London New York, pp 415–431

King RW, Evans LT (1977) Inhibition of flowering in *Lolium temulentum* L. by water stress: a role for abscisic acid. Aust J Plant Physiol 4:225–233

King RW, Zeevaart JAD (1973) Floral stimulus movement in *Perilla* and flower inhibition caused by non-induced leaves. Plant Physiol 51:727–738

King RW, Evans LT, Wardlaw IF (1968) Translocation of the floral stimulus in *Pharbitis nil* in relation to that of assimilates. Z Pflanzenphysiol 59:377–388

King RW, Evans LT, Firn RD (1977) Abscisic acid and xanthoxin contents in the long-day plant *Lolium temulentum* L. in relation to daylength. Aust J Plant Physiol 4:217–223

King RW, Schäfer E, Thomas B, Vince-Prue D (1982) Photoperiodism and rhythmic response to light. Plant Cell Environ 5:395–404

Knott JE (1934) Effect of a localized photoperiod on spinach. Proc Am Soc Hortic Sci 31:152–154

Kohli RK, Sawhney S (1979) Promotory effect of GA_{13} on flowering of *Amaranthus* – a short day plant. Biol Plant 21:206–213

Kojerzenko IE, Majko TK (1967) On the role of natural gibberellins in the photoperiod reaction of peach and vine. Dokl Akad Nauk USSR 177:720–723

Krauss A, Marschner H (1976) Einfluss von Stickstoffernährung und Wuchsstoffapplikation auf die Knollen-Induktion bei den Kartoffelpflanzen. Z Pflanzenern Bodenkd 2:143–155

Krauss A, Marschner H (1982) Influence of nitrogen nutrition, daylength and temperature on contents of gibberellic and abscisic acid and on tuberization in potato plants. Potato Res 25:13–21

Krekule J (1979) Stimulation and inhibition of flowering. In: Champagnat P, Jacques R (eds) Physiologie de la floraison. CNRS, Paris, pp 19–57

Krekule J, Horavka B (1972) The response of short-day plant *Chenopodium rubrum* L. to abscisic acid and gibberellic acid treatment applied at two levels of photoperiodic induction. Biol Plant 14:254–259

Krekule J, Kohli RK (1981) The condition of the apical meristem of seedlings responsive to a promotive effect of abscisic acid on flowering in the short-day plant *Chenopodium rubrum*. Z. Pflanzenphysiol 103:45–52

Krekule J, Seidlova F (1977) *Brassica campestris* as a model for studying the effects of exogenous growth substances on flowering in long-day plants. Biol Plant 19:462–468

Krishnamoorthy HN, Nanda KK (1967) Effect of intercalated long days and light interruption of dark period on flowering, extension growth and senescence of *Impatiens balsamina*. Physiol Plant 20:760–770

Kumar S, Nanda KK (1981) Effect of gibberellic acid and salicylic acid on the activity and electrophoretic pattern of IAA-oxidase during floral induction in *Impatiens balsamina*. Biol Plant 23:328–334

Lang A (1957) The effect of gibberellin upon flower formation. Proc Natl Acad Sci USA 43:709–717

Lang A (1960) Gibberellin-like substances in photoinduced and vegetative *Hyoscyamus* plants. Planta 54:498–504

Lang A (1965) Physiology of flower initiation. In: Ruhland W (ed) Encyclopedia of plant physiology, vol XV, Pt 1. Springer, Berlin Göttingen Heidelberg, pp 1380–1536
Lang A (1980) Inhibition of flowering in long-day plants. In: Skoog F (ed) Plant growth substances. Springer, Berlin Heidelberg New York, pp 310–322
Lang A, Melchers G (1943) Die photoperiodische Reaktion von *Hyoscyamus niger*. Planta 33:653–702
Lang A, Chailakhyan MKH, Frolova IA (1977) Promotion and inhibition of flower formation in a day-neutral plant in grafts with a short-day plant and a long-day plant. Proc Natl Acad Sci USA 74:2412–2416
Langille AR (1972) Effects of (2-chloroethyl) phosphonic acid on rhizome and tuber formation in the potato *Solanum tuberosum* L. J Am Soc Hortic Sci 97:305–308
Lenton JR, Perry VM, Sauners PF (1972) Endogenous abscisic acid in relation to photoperiodically induced bud dormancy. Planta 106:13–22
Lercari B, Micheli P (1981) Photoperiodic regulation of cytokinin levels in leaf blades of *Allium cepa* L. Plant Cell Physiol (Tokyo) 22(3):501–505
Levy D, Kedar N, Karacinque R (1973) Effect of ethephon on bulbing of onion under non-inductive photoperiod. HortSci 8:228–229
Lincoln RG, Cunningham A (1964) Evidence for a florigenic acid. Nature 202:559–561
Lincoln RG, Mayfield DL, Cunningham A (1961) Preparation of a floral initiating extract from *Xanthium*. Science 133:756
Lincoln RG, Mayfield DL, Hutchins RO, Cunningham A, Hamner KC, Carpenter BH (1962) Floral induction of *Xanthium* in response to application of an extract from a day-neutral plant. Nature 195:918
Lincoln RG, Mayfield DL, Cunningham A (1966) Florigenic acid from a fungal culture. Plant Physiol 41:1079–1080
Liverman JL, Lang A (1956) Induction of flowering in long-day plants by applied indoleacetic acid. Plant Physiol 31:147–150
Lona F, Bocchi A (1957) Effetti morfogenetici ed organo-genetici provocati dalla cinetine (kinetin) su piante erbacee in condizioni esterne controllate. N G Bot Ital 64: 236–246
Loveys BR, Leopold AC, Kriedemann PE (1974) Abscisic acid metabolism and stomatal physiology in *Betula lutea* following alteration in photoperiod. Ann Bot 38:85–92
Lozhnikova VN, Krekule J, Seidlova F, Bavrina TV, Chailakhyan MKh (1982) The balance of gibberellins and abscisins in tobaccos during the process of photoperiodic induction. Sov Plant Physiol 29:185–190
Lumsden PJ, Vince-Prue D (1984) The perception of dusk signals in photoperiodic time-measurement. Physiol Plant 60:427–432
Luukkanen O (1981) Effects of gibberellins GA_4 and GA_7 on flowering in Scots pine grafts. Silva Fenn 15:359–365
Mauk CS, Langille AR (1978) Physiology of tuberization in *Solanum tuberosum* L. Cis-zeatin riboside in the potato plant: its identification and changes in endogenous levels as influenced by temperature and photoperiod. Plant Physiol 62:438–442
McKenzie J, Weiser CJ, Burke MJ (1974) Effects of red and far-red light on the initiation of cold hardiness acclimation in *Cornus stolonifera* Michx. Plant Physiol 53:783–789
Menhenett R, Wareing PF (1977) Effects of photoperiod and temperature on the growth and cytokinin content of two populations of *Dactylis glomerata* (Cocksfoot). New Phytol 78:17–25
Menzel CM (1980) Tuberization in potato at high temperatures: responses to gibberellins and growth inhibitors. Ann Bot 46:259–265
Metzger JD (1983) Gibberellins and cold-induced stem elongation in *Thlapsi arvense* L. Plant Physiol 72 Suppl No 467:82
Metzger JD, Zeevaart JAD (1980a) Identification of six endogenous gibberellins in spinach shoots. Plant Physiol 65:623–626
Metzger JD, Zeevaart JAD (1980b) Comparison of the levels of six endogenous gibberellins in roots and shoots of spinach in relation to photoperiod. Plant Physiol 66:679–683

Metzger JD, Zeevaart JAD (1980c) Effect of photoperiod on the levels of endogenous gibberellins in spinach as measured by combined gas chromatography-selected ion current monitoring. Plant Physiol 66:844–846

Metzger JD, Zeevaart JAD (1982) Photoperiodic control of gibberellin metabolism in spinach. Plant Physiol 69:287–291

Michniewicz M, Kamienska A (1965) Flower formation induced by kinetin and vitamin F treatment in long-day plant (*Arabidopsis thaliana*) grown in short-day. Naturwissenschaften 52:623

Miginiac E (1978) Some aspects of regulation of flowering: role of correlative factors in photoperiodic plants. Bot Mag (Tokyo) Spec Iss 1:159–173

Milborrow BV (1967) The identification of (+)-abscisin II ((+)-dormin) in plants and the measurement of its concentrations. Planta 76:93–113

Mingo-Castel AM, Negm FB, Smith OE (1974) Effect of carbon dioxide and ethylene on tuberization of isolated potato stolons cultured in vitro. Plant Physiol 53:798–801

Morgan DC, Smith H (1976) Linear relationship between phytochrome photoequilibrium and growth in plants under simulated natural radiation. Nature 262:210–212

Morgan DG, Morgan CB (1984) Photoperiod and the abscission of flower buds in *Phaseolus vulgaris*. In: Vince-Prue D, Cockshull KE, Thomas B (eds) Light and the flowering process. Academic Press, London New York, pp 227–240

Morgan DG, Zehni MS (1980) The effect of light breaks in the dark period on flower-bud development in varieties of *Phaseolus vulgaris* L. Ann Bot 46:37–42

Murneek AE (1940) Length of day and temperature effects in *Rudbeckia*. Bot Gaz 102:269–279

Nagao M, Okagami N (1966) Effect of (2-chloroethyl) trimethylammonium chloride on the formation and dormancy of aerial tubers of *Begonia evansiana*. Bot Mag (Tokyo) 79:687–692

Nanda KK, Jindal RK (1975) Summation of the effect of sub-threshold gibberellin treatment and sub-threshold photo-induction of flowering of *Impatiens balsamina*. Indian J Plant Physiol 18:109–114

Nanda KK, Krishnamoorthy HN (1967) Photoperiodic studies in growth and development of *Impatiens balsamina*. II. Floral bud initiation, flower opening and extension growth. Planta 72:338–343

Nanda KK, Anaradha TA, Lal K (1967) Floral induction by gibberellic acid in *Impatiens balsamina* L., a qualitative short-day plant. Planta 76:367–370

Nanda KK, Toky KL, Lata K (1969) Effects of gibberellins A_3, A_{4+7} and A_{13} and of (−)-kaurene on flowering and extension growth of *Impatiens balsamina* under different photoperiods. Planta 86:134–141

Nanda KK, Kumar S, Sood V (1976) Effect of gibberellic acid and some phenols on the flowering of *Impatiens balsamina*, a qualitative short-day plant. Physiol Plant 38:53–56

Newton C, Morgan CB, Morgan DG (1980) Evaluation of a bioassay for cytokinins using soybean hypocotyl sections. J Exp Bot 31:721–729

Nitsch C, Nitsch JP (1969) Floral induction in a short-day plant, *Plumbago indica* L. by 2-chloroethane phosphonic acid. Plant Physiol 44:1747–1748

Nitsch JP (1957) Growth responses of woody plants to photoperiodic stimuli. Proc Am Soc Hortic Sci 70:512–525

Nitsch JP (1959) Réactions photopériodiques chez les plantes ligneuses. Bull Soc Bot France 106:259–287

Nitsch JP (1965) Éxistence d'un stimulus photopériodique non spécifique capable de provoquer la tubérisation chez *Helianthus tuberosus* L. Bull Soc Bot France 112:333–340

Nitsch J, Nitsch C (1960) Le problème de l'action des auxines sur la division cellulaire: présence d'un cofacteur de division dans le tubercle de topinambour. Ann Physiol Vég 2:261–268

Nitsch JP, Somogyi L (1958) Le photopériodisme des plantes ligneuses. Ann Sci Nat Hortic France 16:466–490

Ofir M (1976) Interaction of gibberellin with photo-induction in the initiation of the dormant phase in vernalized *Hordeum bulbosum*. Aust J Plant Physiol 3:827–832

Ogawa Y (1977) Difference in effectiveness of various gibberellins on shoot elongation and flower formation in *Pharbitis nil* Chois. J Jpn Soc Hortic Sci 46:117–122
Ogawa Y (1981a) Stimulation of the flowering of *Pharbitis nil* Chois. by gibberellin Plant Cell Physiol 22:675–681
Ogawa Y (1981b) Effect of chilling at germinating stage on the flowering response of *Pharbitis nil* Choisy. J (Jpn) Soc Hortic Sci 49:102–106
Ogawa Y, King RW (1979) Establishment of photoperiodic sensitivity by benzyladenine and a brief red irradiation in dark grown seedlings of *Pharbitis nil* Chois. Plant Cell Physiol 20:115–122
Ogawa Y, Zeevaart JAD (1967) The relations of growth regulators to flowering. In: Imamura S (ed) Physiology of flowering in *Pharbitis nil*. Jpn Soc Plant Physiol, pp 107–119
Okagami N, Esashi Y, Nagao M (1977) Gibberellin-induced inhibition and promotion of sprouting in aerial tubers of *Begonia evansiana* and in relation to photoperiodic treatment and tuber storage. Planta 136:1–6
Okazawa Y (1960) Studies on the relationship between tuber formation of potato and its natural gibberellin content. J Crop Sci Soc Jpn 29:121–124
Okazawa Y (1967) Physiological studies on the tuberization of potato plants. J Fac Agric Hokkaido Univ Sapporo 55:267–336
Owens JN, Pharis RP (1971) Initiation and development of western red cedar cones in response to gibberellin induction under natural conditions. Can J Bot 49:1165–1175
Palmer CE, Barker WG (1973) Influence of ethylene and kinetin on tuberization and enzyme activity in *Solanum tuberosum* L. Ann Bot 37:85–93
Palmer CE, Smith OE (1969) Effect of abscisic acid on elongation and kinetin-induced tuberization of isolated stolons of *Solanum tuberosum* L. Plant Cell Physiol 10:657–664
Parker MW, Hendricks SB, Borthwick HA, Scully NJ (1946) Action spectrum for the photoperiodic control of floral initiation of short-day plants. Bot Gaz 108:1–26
Parker MW, Hendricks SB, Borthwick HA (1950) Action spectrum for the photoperiodic control of floral initiation of the long-day plant *Hyoscyamus niger*. Bot Gaz 111:242–252
Perry TO, Byrne OR (1969) Turion induction in *Spirodela polyrrhiza* by abscisic aid. Plant Physiol 44:754–785
Perry TO, Hellmers H (1973) Effects of ABA on growth and dormancy of two races of red maple (*Acer rubrum*). Bot Gaz 134:283–288
Peterson RL, Yeung EC (1972) Effect of two gibberellins on species of the rosette plant *Hieracium*. Bot Gaz 133:190–198
Pharis RP (1972) Flowering of *Chrysanthemum* under non-inductive long-days by gibberellins and by N^6-benzyladenine. Planta 105:205–212
Pharis RP (1976) Probable roles of plant hormones in regulating shoot elongation, diameter growth and crown form of coniferous trees. In: Cannel MGR, Last FT (eds) Tree physiology and yield improvement. Academic Press, London New York, pp 291–306
Pharis RP, Ruddat MDE, Glenn JL, Morf W (1970) A quantitative requirement for long day in the induction of staminate strobili by gibberellin in the conifer *Cupressus arizonica*. Can J Bot 48:653–658
Philipson JJ (1983) The role of gibberellin $A_{4/7}$, heat and drought in the induction of flowering in Sitka spruce. J Exp Bot 34:291–302
Phillips DA, Cleland CF (1972) Cytokinin activity from the phloem sap of *Xanthium strumarium* L. Planta 102:173–178
Phillips IDJ, Hoffmann A (1979) Abscisic acid (ABA), ABA esters and phaseic acid in vegetative terminal buds of *Acer pseudoplatanus* during emergence from dormancy. Planta 146:591–596
Phillips IDJ, Wareing PF (1959) Studies in dormancy of sycamore. II. The effect of daylength on the natural growth inhibitor content of the shoot. J Exp Bot 10: 504–514
Phillips IDJ, Miners J, Roddick JG (1980) Effect of light and photoperiodic conditions on abscisic acid in leaves and roots of *Acer pseudoplatanus* L. Planta 149:118–122

Pieterse AH (1982) A review of chemically induced flowering in *Lemna gibba* G_3 and *Pistia stratiotes*. Aquat Bot 13 Spec Iss: 21–28
Plancher B, Naumann WD (1978) Influence of water supply and day length on abscisic acid content of *Fragaria ananassa*. Gartenbauwissenschaft 43: 126–136
Pol van de PA (1972) Floral induction, floral hormones and flowering. Meded Landbouwhogesch Wageningen 72: 1–89
Porlingis IC, Boynton D (1961) Growth responses of the strawberry plant, *Fragaria chiloensis* var ananassa to gibberellic acid and to environmental conditions. Proc Am Soc Hortic Sci 78: 261–269
Powell LE (1976) Effect of photoperiod on endogenous abscisic acid in *Malus* and *Betula*. Hortic Sci 11: 498–499
Proebsting WM, Heftmann E (1980) The relationship of (^{3}H) GA_9 metabolism to photoperiod-induced flowering in *Pisum sativum* L. Z Pflanzenphysiol 98: 305–309
Proebsting WM, Davies PJ, Marx GA (1978) Photoperiod-induced changes in gibberellin metabolism in relation to apical growth and senescence in genetic lines of peas (*Pisum sativum* L.). Planta 141: 231–238
Pryce RJ (1972) Gallic acid as a natural inhibitor of flowering in *Kalanchoe blossfeldiana*. Phytochemistry 11: 1911–1918
Purse JG (1984) Phloem exudate of *Perilla crispa* and its effects on flowering of *P. crispa* shoot explants. J Exp Bot 35: 227–238
Railton IDM, Wareing PF (1973) Effects of daylength on endogenous gibberellins in leaves of *Solanum andigena* L. Changes in levels of free acidic gibberellin-like substances. Physiol Plant 28: 88–94
Ramsay J, Martin GC (1970) Isolation and identification of a growth inhibitor in spur buds of apricot. J Am Soc Hortic Sci 95: 574–577
Reid DM, Pharis RP, Roberts DWA (1974) Effects of four temperature regimens on the gibberellin content of winter wheat cv Kharkov. Physiol Plant 30: 53–57
Ren XC, Lian HK, Cao GY (1982a) The possible role of GA_3 in promotion of flowering in SDP *Pharbitis nil*. Acta Phytophysiol Sin 8: 134–140
Ren XC, Zhang J-Y, Luo W-H, Jin S-P (1982b) Flower inhibitory effects of long days preceding the inductive short days on SDP *Hibiscus cannabinus* cv. South Selected. Acta Phytophysiol Sinica 8: 214–221
Robinson PM, Wareing PF (1964) Chemical nature and biological properties of the inhibitor varying with photoperiod in sycamore, *Acer pseudoplatanus*. Physiol Plant 17: 314–323
Rogers MN, Tsia BOS (1966) Effects of ethylene in the atmosphere on photoperiodic responses of chrysanthemums. Proc 17th Int Congr Hortic 1, Abstr No 471
Ross SD, Pharis RP, Binder WD (1983) Growth regulators and conifers: their physiology and potential uses in forestry. In: Nickell LG (ed) Plant growth regulating chemicals. CRC Press, Boca Raton
Rudich J, Halevy AH, Kedar N (1972a) Ethylene evolution from cucumber plants as related to sex expression. Plant Physiol 49: 998–999
Rudich J, Halevy AH, Kedar N (1972b) The level of phytohormones in monoecious and gynoecious cucumbers as affected by photoperiod and ethephon. Plant Physiol 50: 585–590
Sachs RM (1969) *Cestrum nocturnum* L. In: Evans LT (ed) The induction of flowering. Mac Millan, Sydney, pp 424–434
Sachs RM, Kofranek AM, Shyr SY (1967) Gibberellin-induced inhibition of floral initiation in *Fuchsia*. Am J Bot 54: 921–929
Saji H, Furuya M, Takimoto A (1982) Spectral dependence of the night-break effect on photoperiodic floral induction in *Lemna paucicostata* 441. Plant Cell Physiol 23: 623–629
Saji H, Vince-Prue D, Furuya M (1983) Studies on the photoreceptors for the promotion and inhibition of flowering in dark-grown seedlings of *Pharbitis nil* Choisy. Plant Cell Physiol 24: 1183–1189
Satter RL, Wetherell DF (1968) Photomorphogenesis in *Sinningia speciosa* cv. Queen Victoria. II. Stem elongation: interaction of a phytochrome controlled process and a red requiring energy dependent reaction. Plant Physiol 43: 961–967

Sawhney S, Sawhney N (1976) Floral induction by gibberellic acid in *Zinnia elegans* under non-inductive long days. Planta 131:207–208

Sawhney S, Sawhney N, Nanda KK (1972) Partial substitution by long days of short days required for floral induction in *Impatiens balsamina*. Plant Cell Physiol 13:1113–1115

Schneider MJ, Borthwick HA, Hendricks SB (1967) Effects of radiation on flowering of *Hyoscyamus niger*. Am J Bot 54:1241–1249

Schwabe WW (1972) Flower inhibition in *Kalanchoe blossfeldiana*. Bioassay of an endogenous inhibitor and inhibition by (±) abscisic acid and xanthoxin. Planta 103:18–23

Schwabe WW (1984) Photoperiodic induction – flower-inhibiting substances. In: Vince-Prue D, Cockshull KE, Thomas B (eds) Light and the flowering process. Academic Press, London New York, pp 143–153

Schwabe WW, Nachmony-Bascombe S (1963) Growth and dormancy in *Lunularia cruciata* (L) DUM. II. The response to daylength and temperature. J Exp Bot 14: 353–378

Schwabe WW, Wimble RH (1976) Control of flower initiation in long- and short-day plants – a common model approach. In: Sunderland N (ed) Perspectives in experimental biology, vol II, Botany. Pergamon, Oxford New York, pp 41–57

Seidlova F (1980) Sequential steps of transition to flowering in *Chenopodium rubrum* L. Physiol Veg 18:477–487

Sengupta SK, Rogers MN, Lorah EJ (1974) Effects of photoperiod and ethephon treatment on abscisic acid levels in *Chrysanthemum morifolium* Ramat. J Am Soc Hortic Sci 99:416–420

Simmons HA, Coulter MW (1979) Responses of normal and dwarf *Pharbitis nil* to an extended dark period and gibberellic acid. Plant Cell Physiol 20:1671–1673

Smith OE, Palmer CE (1970) Cytokinin induced tuber formation on stolons of *Solanum tuberosum*. Physiol Plant 23:599–606

Smith OE, Rappaport L (1969) Gibberellins, inhibitors and tuber formation in the potato, *Solanum tuberosum* L. Am Potato J 46:185–191

Sotta B (1978) Interaction du photopériodisme et des effets de la zéatine, du saccharose et du l'eau dans la floraison du *Chenopodium polyspermum*. Physiol Plant 43:337–342

Spence J, Humphries EC (1972) Effect of moisture supply, root remperature and growth regulators on photosynthesis of isolated rooted leaves of sweet potato (*Ipomoea batatas*). Ann Bot 36:115–121

Staden van J, Wareing PF (1972) The effect of photoperiod on levels of endogenous cytokinins in *Xanthium strumarium*. Physiol Plant 27:331–337

Steward GR (1969) Abscisic acid and morphogenesis in *Lemna polyrrhiza* L. Nature 221:61–62

Stoddart JL (1966) Studies on the relationship between gibberellin metabolism and daylength in normal and non-flowering red clover (*Trifolium pratense* L.). J Exp Bot 17:96–107

Stoddart JL, Lang A (1968) The effect of daylength on gibberellin synthesis in leaves of red clover (*Trifolium pratense* L.). In: Wightman F, Setterfield G (eds) Biochemistry and physiology of growth substances. Runge, Ottawa, pp 1371–1383

Suge H, (1972) Inhibition of photoperiodic floral induction in *Pharbitis nil* by ethylene. Plant Cell Physiol 13:1031–1038

Suge H, Rappaport L (1968) Role of gibberellins in stem elongation and flowering in radish. Plant Physiol 43:1208–1214

Tafazoli E, Vince-Prue D (1978) A comparison of the effects of long days and exogenous growth regulators on growth and flowering in strawberry, *Fragaria* × *ananassa* Duch. J Hortic Sci 53:255–259

Takahashi H, Saito T, Suge H (1983) Separation of the effects of photoperiod and hormones on sex expression in cucumber. Plant Cell Physiol 24:147–154

Takashima S, Nakamura K, Naito M, Hirose T (1962) Studies on the growth of potato by the foliar spray of growth regulating substances (Japanese). Sci Rep Kyoto Prefect Univ Agric 14:37–43

Takeba G, Takimoto A (1966) Translocation of the floral stimulus in *Pharbitis nil*. Bot Mag (Tokyo) 79:811–814

Thomas TH (1972) Stimulation of onion bulblet production by N^6-benzyladenine. Hortic Res 12:77–79

Thompson PA, Guttridge CG (1959) Effect of gibberellic acid on the initiation of flowers and runners in the strawberry. Nature 184:72–73

Thompson PA, Guttridge CG (1960) The role of leaves as inhibitors of flower induction in strawberry. An Bot 24:482–490

Tizio R (1960) Présence de kinines dans périderme de tubercles de pomme de terre. C R Hebd Seances Acad Sci Paris 262:868–869

Tse ATY, Ramina A, Hackett WP, Sachs RM (1974) Enhanced inflorescence development in *Bougainvillea* San Diego Red by removal of young leaves and cytokinin treatment. Plant Physiol 54:404–407

Udebo AE (1971) Effect of external supply of growth substances on axillary proliferation and development in *Dioscorea bulbifera*. Ann Bot 35:159–163

Uematsu Y, Katsura N (1983) Changes in endogenous gibberellin level in strawberry plants induced by light breaks. J Jpn Soc Hortic Sci 51:405–411

Venkataraman R, Seth PN, Maheshwari SC (1970) Studies on the growth and flowering of a short-day plant, *Wolffia microscopica*. I. Growth aspects and induction of flowering by cytokinins. Pflanzenphysiol 62:316–327

Vince D (1965) The promoting effect of far-red light on flowering in the long-day plant *Lolium temulentum*. Physiol Plant 18:474–482

Vince D (1969) The regulation of flowering in long-day plants. Acta Hortic 14:91–95

Vince D (1970) Lighting techniques for the control of flowering. Proc 18th Int Congr Hortic 5:169–180

Vince D (1972) Phytochrome and flowering. In: Mitrakos K, Shropshire W (eds) Phytochrome. Academic Press, London New York, pp 257–291

Vince-Prue D (1975) Photoperiodism in plants. McGraw-Hill, London New York

Vince-Prue D (1976) Phytochrome and photoperiodism. In: Smith H (ed) Light and plant development. Butterworths, London, pp 347–369

Vince-Prue D (1977) Photocontrol of stem elongation in light-grown plants of *Fuchsia hybrida*. Planta 133:149–156

Vince-Prue D (1979) Effect of photoperiod and phytochrome in flowering: time measurement. In: Champagnat P, Jacques R (eds) Physiologie de la floraison. CNRS, Paris, pp 91–127

Vince-Prue D (1981) Daylight and photoperiodism. In: Smith H (ed) Plants and the daylight spectrum. Academic Press, London New York, pp 223–242

Vince Prue D (1982) Phytochrome and photoperiodic physiology. In: Brady J (ed) Biological time-keeping. Cambridge Univ Press, Cambridge, pp 101–117

Vince-Prue D (1983a) Photomorphogenesis and flowering. In: Shropshire W, Mohr H (eds) Encyclopedia of plant physiology new ser vol XVIb. Springer, Berlin Heidelberg New York, pp 457–490

Vince-Prue D (1983b) The perception of light – dark transitions. Philos Trans R Soc London Ser B 303:523–536

Vince-Prue D (1984) Contrasting types of photoperiodic response in the control of dormancy. Plant Cell Environ 7:(in press)

Vince-Prue D, Guttridge CG (1973) Floral initiation in strawberry: spectral evidence for the regulation of flowering by long-day inhibition. Planta 110:165–172

Vince-Prue D, Guttridge CG, Buck MW (1976) Photocontrol of petiole elongation in light-grown strawberry plants. Planta 131:109–114

Vlitos AJ, Meudt W (1954) The role of auxin in plant flowering. III. Free indole acids in short-day plants grown under photo-inductive and non-photo-inductive daylengths. Contrib Boyce Thompson Inst 17:413–417

Wareing PF (1951) Growth studies in woody species. III. Further photoperiodic effects in *Pinus sylvestris*. Physiol Plant 4:41–56

Wareing PF (1953) Growth studies in woody species. V. Photoperiodism in dormant buds of *Fagus sylvatica* L. Physiol Plant 6:692–706

Wareing PF (1954) Growth studies in woody species. VI. The locus of photoperiodic perception in relation to dormancy. Physiol Plant 7:261–277

Wareing PF (1959) Photoperiodism in seeds and buds. In: Withrow RB (ed) Photoperiodism and related phenomena in plants and animals. AAAS, Washington, pp 73–87

Wareing PF (1969) Germination and dormancy. In: Wilkins MB (ed) Physiology of plant growth and development. McGraw-Hill, London, New York, pp 605–644

Wareing PF (1982) Hormonal control of stolon and tuber development, expecially in the potato plant. In: Wareing PF (ed) Plant growth substances. Academic Press, London New York, pp 181–195

Wareing PF, El-Antably HMM (1970) The possible role of endogenous growth inhibitors in the control of flowering. In: Bernier G (ed) Cellular and molecular aspects of floral induction. Longman Group, London, pp 285–303

Wareing PF, Jennings AMV (1980) The hormonal control of tuberization in potato. In: Skoog F (ed) Plant growth substances. Springer, Berlin Heidelberg New York, pp 147–153

Wareing PF, Saunders PF (1971) Hormones and dormancy. Annu Rev Plant Physiol 22:261–288

Warm E (1980) Effect of phytohormones and salicylic acid on flowering and stem elongation in the long-day plant *Hyoscyamus niger*. Z Pflanzenphysiol 99:325–330

Watanabe K, Takimoto A (1979) Flower-inducing effects of benzoic acid and some related compounds in *Lemna paucicostata* 151. Plant Cell Physiol 20:847–850

Watanabe K, Fujita T, Takimoto A (1981) Relationship between structure and flower-inducing activity of benzoic acid derivatives in *Lemna paucicostata* 151. Plant Cell Physiol 22:1469–1479

Waxman S (1957) The development of woody plants as affected by photoperiodic treatments. Ph D Thes, Cornell Univ, Ithaca, NY

Weaver RJ (1963) Use of kinin in breaking rest in buds of *Vitis vinifera*. Nature 198:207–208

Weber JA, Nooden LD (1976) Environmental and hormonal control of turion formation in *Myriophyllum verticillatum*. Plant Cell Physiol 17:721–731

Wellensiek SJ (1970) The floral hormones in *Silene armeria* and *Xanthium strumarium*. Z Pflanzenphysiol 63:25–30

Wellensiek SJ (1973) Gibberellic acid, flower formation and stem elongation in *Silene armeria*. Neth J Agric Sci 21:245–255

Williams BJ Jr, Pellett NE, Klein RM (1972) Phytochrome control of growth cessation and initiation of cold acclimation in selected woody plants. Plant Physiol 50:262–265

Williams MW (1972) Induction of spur and flower bud formation in young apple trees with chemical growth retardants. J Am Soc Hortic Sci 97:210–212

Withrow AP, Withrow RB (1943) Translocation of the floral stimulus in *Xanthium*. Bot Gaz 104:409–416

Yeh RY, Matches AG, Larson RL (1976) Endogenous growth regulators and summer tillering of tall fescue. Crop Sci 16:409–413

Zeevaart JAD (1958) Flower formation as studied by grafting. Meded Landbouwhogesch Wageningen 58:1–88

Zeevaart JAD (1969a) The leaf as a site of gibberellin action in flower formation in *Bryophyllum daigremontianum*. Planta 84:339–347

Zeevaart JAD (1969b) Gibberellin-like substances in *Bryophyllum daigremontianum* and the distribution and persistence of applied gibberellin A_3. Planta 86:124–133

Zeevaart JAD (1969c) Changes in the gibberellin content of *Bryophyllum daigremontianum* in connection with floral induction. Neth J Agric Sci 17:215–220

Zeevaart JAD (1969d) *Bryophyllum*. In: Evans LT (ed) The induction of flowering. MacMillan, Melbourne, pp 435–456

Zeevaart JAD (1970) Gibberellins and flower formation. In: Bernier G (ed) Cellular and molecular aspects of floral induction. Longman, London, pp 335–344

Zeevaart JAD (1971a) Effects of photoperiod on growth rate and endogenous gibberellins in the long-day rosette plant spinach. Plant Physiol 47:821–827

Zeevaart JAD (1971b) ($\pm$)-Abscisic acid content of spinach in relation to photoperiod and water stress. Plant Physiol 48:86–90

Zeevaart JAD (1971c) Lack of evidence for distinguishing florigen and flower hormone in *Perilla*. Planta 98:190–194

Zeevaart JAD (1973) Gibberellin A_{20} content of *Bryophyllum daigremontianum* under different photoperiodic conditions as determined by gas-liquid chromatography. Planta 114:258–288

Zeevaart JAD (1974) Levels of (+)-abscisic acid and xanthoxin in spinach under different environmental conditions. Plant Physiol 53:644–648

Zeevaart JAD (1976) Physiology of flower formation. Ann Rev Plant Physiol 27:321–348

Zeevaart JAD (1978) Transmission of the floral stimulus between four different photoperiodic response types in the *Crassulaceae*. Plant Physiol 61 Suppl:14, No 72

Zeevaart JAD (1979) Perception, nature and complexity of transmitted signals. In: Champagnat P, Jacques R (eds) Physiologie de la floraison. CNRS, Paris, pp 59–90

Zeevaart JAD (1982) Transmission of the floral stimulus from a long-short-day plant, *Bryophyllum daigremontianum*, to the short-long-day plant *Echeveria harmsii*. Ann Bot 46:549–552

Zeevaart JAD (1984) Photoperiodic induction, the floral stimulus and flower-promoting substances. In: Vince-Prue D, Cockshull KE, Thomas B (eds) Light and the flowering process. Academic Press, London New York, pp 137–142

Zeevaart JAD, Kivilaan T (1977) Abscisic acid in the short-day plant, *Xanthium strumarium*. In: Plant research. Annu Rep MSU/AEC Plant Res Lab Michigan State Univ, East Lansing

Zeevaart JAD, Lang A (1963) Suppression of floral induction in *Bryophyllum daigremontianum* by a growth retardant. Planta 59:509–517

Zhadanova LP (1950) Significance of the gaseous regime for the passage of the light stage in plants. Dokl Akad Nauk USSR 70:715–718

10 Roles of Hormones in Phototropism

B.G. PICKARD

1 Introduction

Because sunlight is the foremost environmental requirement of higher plants, processes which help seedling shoots emerge from the soil into the air, and which direct further development so as to optimize the intensity of light received by the leaves, are critical for plant survival. It is thus natural that plants, in addition to using sunlight as a source of energy, use it as a sensory cue to guide growth. It is the shorter wavelengths of the photosynthetic action spectrum to which the guidance systems respond, whereas the longer wavelengths often modulate the sensitivity or effectiveness of guidance.

In accord with the abundant evidence that growth is regulated by hormones, the asymmetric growth by which orientation is generally achieved is thought to be due to asymmetric hormone distribution. Hypotheses that light might cause differential growth by more direct means have been proposed, and because the plant kingdom is rich with variety and the number of species subjected to physiological scrutiny is small, such hypotheses should not be forgotten. However, for those higher plant systems which have been examined with care, it has become clear that the orientation of growth with respect to light, or phototropism, is indeed brought about by a gradient in hormone concentration.

The only hormone that has been proven to mediate phototropism in shoots of vascular plants is auxin. The means by which its asymmetric distribution is brought about will be the main topic of this chapter, and the prospect that the action of redistributed IAA is reinforced by a secondary distribution of apoplastic Ca^{2+} will also be evaluated. Tropisms and related movements of algae, fungi and sporelings of the lower vascular plants will be excluded because little can be said about the possibility that they are under hormonal control. Furthermore, knowledge about hormonal involvement in the occasionally observed phototropism of certain roots is too scanty to justify discussion. Regrettably few publications have been devoted to the phototropism of leaves and floral parts since BRAUNER's review (1959) in the first series of this Encyclopedia. However, reports on leaves by WAINWRIGHT (1977) and SCHWARTZ and KOLLER (1978, 1980), VOGELMANN and BJÖRN (1983), WERK and EHLERINGER (1984), and on tendrils by JAFFE (1970) are worth special notice, as is a brief review by SMITH (1984). In Volume 7 of the present series, DENNISON (1979) has provided a comprehensive review of phototropism in which studies on the action spectrum and the chemical identity of the pigment are discussed at some length. Recently there have appeared numerous reviews and analyses on the nature of the pigment (BRIGGS 1976, SONG 1980, DE FABO 1980, STRASSER and BUTLER

1980, SCHMIDT 1980, HARTMAN and SCHMID 1980, LÖSER and SCHÄFER 1980, SCHMID 1980, NINNEMAN 1980, SENGER and BRIGGS 1981; see also JESAITIS et al. 1977, SCHMIDT et al. 1977, MANABE and POFF 1978, GRESSEL 1979, HERTEL et al. 1980, LEONG and BRIGGS 1981, LEONG et al. 1981, VIESTRA and POFF 1981, BRITZ et al. 1979, SCHMIDT 1983, SENGER and BRIGGS 1981, VIESTRA and POFF 1981a). The coverage of these reviews, as well as the continuing usefulness of earlier ones (GALSTON 1959, 1974, THIMANN 1964, 1967, 1977, THIMANN and CURRY 1961, CURRY 1969, GOLDSMITH 1968, BRIGGS 1963a, 1964, BALL 1969, HAUPT 1970, 1979, WILKINS 1977, WAREING and PHILLIPS 1978, REINHOLD 1978), make it appropriate to limit the present discussion to the dependence of the curvature response on conditions of illumination and to the mediational processes.

2 Phototropic Phenomena

2.1 Overview: Dark-Grown Seedlings

Happily, it appears that the basic processes of phototropism may be essentially similar in most or all dark-grown seedlings. Exploratory dose-response plots have been prepared for at least 18 species (Table 1) and all show comparable patterns. Representative plots for radish seedlings (EVERETT 1974, STEYER 1967) show that very small amounts of light can produce vigorous phototropism (Fig. 1). All authors agree that for small unilateral exposures response depends primarily on the total dose rather than on how it is delivered. Thus, at a fixed wavelength irradiation with 10 W m^{-2} for 0.1 s yields the same curvature as 1 mW m^{-2} for 17 min. Even though such a phototropic exposure may be as brief as a fraction of a second, the bending it initiates typically continues steadily for about an hour at 20° to 25° C. By this time, counter-induction of the curving plants by gravity begins to take effect and net curvature may diminish, although this can be avoided experimentally by permitting response to develop on a clinostat. In the absence of net geotropic stimulation, bending continues for several hours (WENT and THIMANN 1937, SHEN-MILLER and GORDON 1967).

In nature, this large and prolonged reaction to the sensitive summing of all available light plays an important role while the seedling is still deeply buried in the ground. As the shoot's geotropic responses guide it upward, dim sunlight may transiently seep through cracks in the soil as the sun wheels across the sky. These tiny bits of light will tend to provide cues about the easiest or most direct path to full sunshine: the seed may have become buried under a small stone, or the soil directly above it may have become deeper or more compacted than that to one side, so that straight up may not be the most effective way to grow. The large response to the faint light overwhelms the geotropic response long enough to carry the tip of the seedling closer to the crack or thin region in the soil, where further phototropic cues may well be available, or where in their absence geotropism will take over again to lift the tip out of the ground.

Table 1. Index of dose-response data for seedlings

Seedling category	Species	Reference
Dark-grown grasses	*Avena sativa*	BRIGGS (1960)[a], ZIMMERMAN and BRIGGS (1963a), STEYER (1967), PICKARD et al. (1969), BLAAUW and BLAAUW-JANSEN (1970a, b)
	Hordeum distichon	STEYER (1967)
	Secale cereale	STEYER (1967)
	Triticum aestivum	STEYER (1967)
	Zea mays	BRIGGS (1960) CHON and BRIGGS (1966)
Dark-grown dicots	*Agrostemma githago*	STEYER (1967)
	Brassica napus	STEYER (1967)
	B. oleracea	STEYER (1967)
	Convolvulus tricolor	STEYER (1967)
	Cucumis sativus	STEYER (1967)
	Helianthus annuus	DIEMER (1961), STEYER (1967), FRANSSEN and BRUINSMA (1981)
	Lens culinaris	STEYER (1967)
	Lepidium sativum	STEYER (1967)
	Linum usitatissimum	STEYER (1967)
	Raphanus sativus	STEYER (1967), EVERETT (1974)
	Sinapis alba	STEYER (1967)
	Vicia villosa	STEYER (1967)
Light-grown dicots	*Cucumis sativus*	SHUTTLEWORTH and BLACK (1977)
	Helianthus annuus	SHUTTLEWORTH and BLACK (1977), FRANSSEN and BRUINSMA (1981)
	Raphanus sativus	EVERETT (1974)
	Phaseolus aureus	BRENNAN et al. (1976)

[a] Due to abundance of recent studies, earlier less complete ones are omitted.

As the dose of light increases, the sensitive response fails (Fig. 1B). Still further increase of dose – often provided experimentally as an increase of light duration at a fixed, relatively high intensity – elicits a less sensitive response which is dependent on both the duration and the intensity of irradiation, but not as a simple function of total dose. Like the sensitive response, this one achieves its fullest development in about 1 h under ordinary conditions, but continues for perhaps 3 h on a clinostat (Fig. 2).

In nature, this less sensitive response assumes guidance of the plant as the seedling nears the surface, and maintains its role long after the shoot has pushed into the air. Whereas the sensitive response, if not inactivated, would waste energy in over-reaction to movement of the sun across the sky or of clouds across the sun, or even to the shifting shadows of a leafy canopy, the less

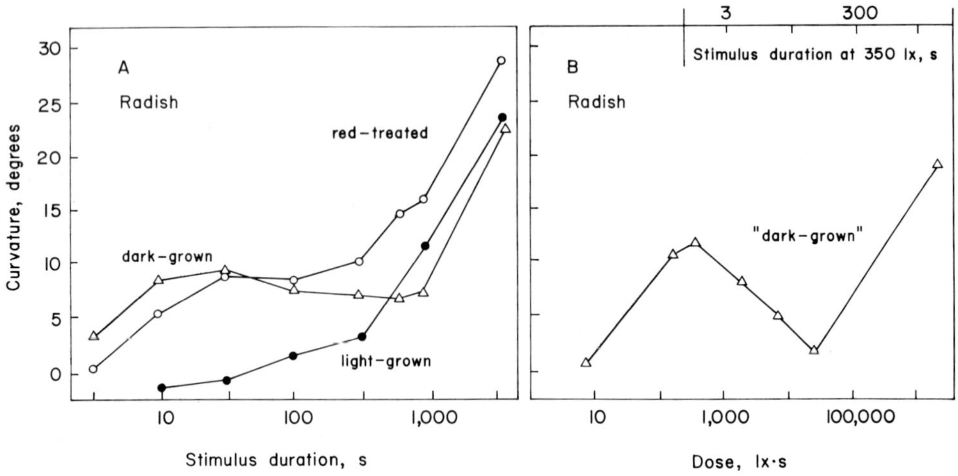

Fig. 1 A, B. Dose-response curves for radish (*Raphanus sativus*). **A** Data from EVERETT (1974). Irradiance 6 mW m^{-2} at 460 nm. Curvature was allowed to develop for 90 min following stimulation, except for the last sets of plants for each curve, which were measured after only 30 min. Seedling age, 4 d. For R-treated plants, 6 mW m^{-2} was given for 1 d following 3 d growth in dark. For "white"-light-grown plants, 1-d-old dark-germinated seedlings were transferred to continuous 6.5 klx fluorescent light. **B** Data from STEYER (1967). Illuminance for first two points of curve, 8 lx; for remainder, 350 lx; unfiltered incandescent source. Curvature development was permitted for 75 min following stimulation except for the last set of plants, which was measured after only 45 min. Seedling age, 2 d. Plants were handled under R work-light.

It is suggested that the initial peak, shallow dip, and final rise in the curve for dark-grown plants in **A** and the initial peak, deep trough, and final rise in **B** are comparable, although conditions in the two experiments are quite different. Moreover, the response curve for light-grown plants appears similar to the final rise in the curve for dark-grown plants

sensitive response permits the plant efficiently to seek a photosynthetically optimal light intensity. Further, when the plant ultimately attains full sunshine, over-reaction to gradients of bright light might needlessly jeopardize the mechanical stability which is gained from more or less upright growth. Thus, at light intensities adequate for rapid photosynthesis, the plant is better off with a phototropic response weak enough to permit geotropic interaction. Under these conditions (Fig. 3, and Sect. 2.3) the strength of phototropism is dramatically diminished.

Although it is only blue light (B) toward which the plant will grow, red light (R) modifies the response pattern dramatically. Information on the complex effects of R is restricted to a small number of species, but it may be generalized that R shifts the peak of the dose-dependent response toward higher doses, whereas it greatly increases the sensitivity of the intensity- and duration-dependent response.

The time elapsed between sequential R and B must be known in order to predict the modifications; for oat coleoptiles at least 90 min must pass after R is provided if maximal effects are to be obtained, and it does not matter whether R is continued or discontinued during that period. For oat coleoptiles,

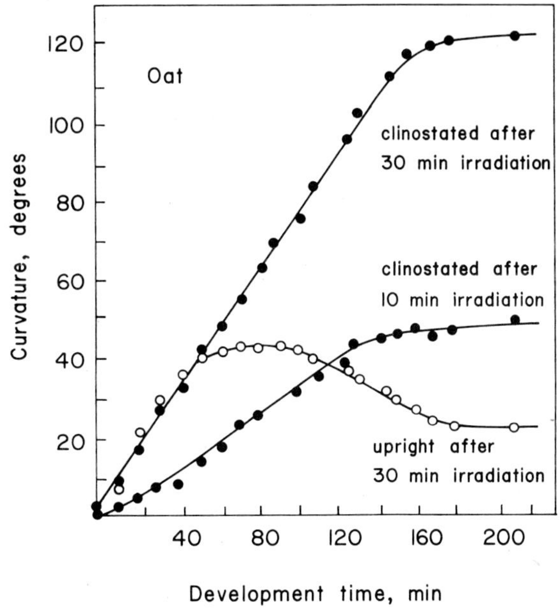

Fig. 2. Comparison of curvature developed on and off the clinostat by R-pretreated oat (*Avena sativa*) coleoptiles. (Data from PICKARD et al. 1969). Irradiance 0.32 W m^{-2} B.

at least, R can dramatically increase the maximum of the dose-dependent response as well as shifting its sensitivity (e.g., BLAAUW and BLAAUW-JANSEN 1964, 1970a).

In nature, R which reaches seedlings not quite emerged from the ground prepares them to meet relatively prolonged illumination; i.e., it serves as a cue that the seedling is near the soil surface. Small amounts of B can also modulate response to subsequent stimuli but, as will be discussed later, their effects do not endure nearly as long as those of R, and the underlying mechanisms appear to be dissimilar (e.g., BLAAUW and BLAAUW-JANSEN 1964, 1970a–b).

2.2 Overview: Green Shoots

Studies on green plants have been relatively few. Casual observation in garden, field, and forest has raised the question whether in some plants the site of photoreception might be in the leaves rather than in the stem.

The sunflower is an obvious candidate for such a plant. LAM and LEOPOLD (1966) removed the apices from young seedlings, covered one cotyledon of each pair with metal foil, and set the plants in white light. Each bent in the direction of the illuminated cotyledon, and this was attributed to decreased growth on the side underneath it. (The extent to which hormonal contributions from an intact apex might override asymmetries imposed by the cotyledons was not studied.) However, DIEMER (1961) was able to cut away both the apex and cotyledons from sunflower hypocotyls, supply auxin at the cut surface, and demonstrate a response to unilateral illumination which was almost equal to

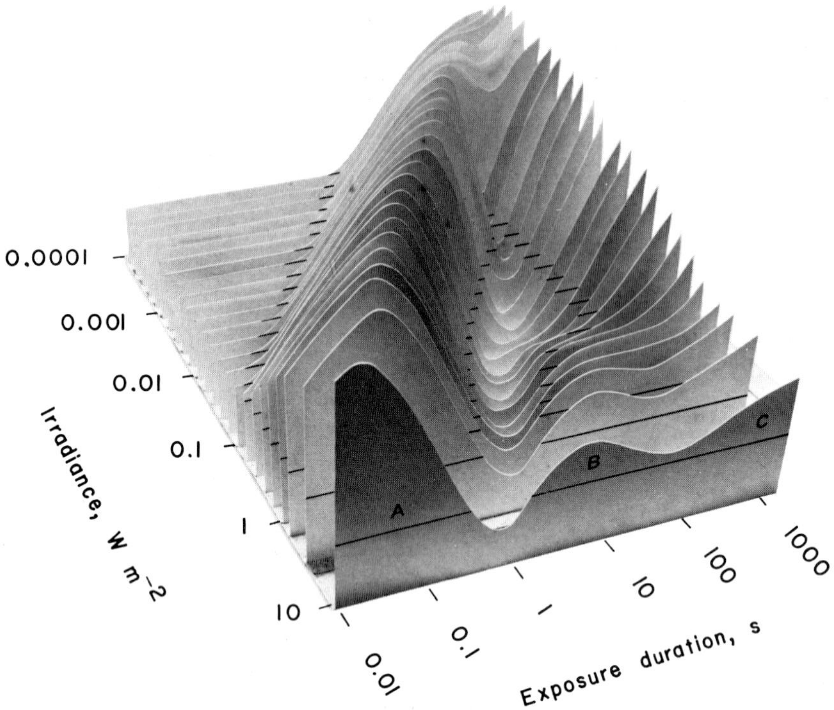

Fig. 3. Curvature of R-pretreated excised oat coleoptiles as a function of log irradiance and log exposure time. (BLAAUW and BLAAUW-JANSEN 1970a). *Letters of first panel* of graph indicate response regions *A*, *B* and *C*; between *A* and *B* is designated the zone of negative curvature and between *B* and *C* is the indifferent zone. Height of A peak is about 25°

that of intact controls. Some phototropic curvature of the trimmed stem was possible even in the absence of applied hormone (see also BRUINSMA et al. 1975). Moreover, sunflower seedlings studied by FRANSSEN and BRUINSMA (1981) responded equally with or without both tip and cotyledons masked, and responded almost as much with tip and cotyledons removed. Thus, the phototropic role of sunflower cotyledons appears to be secondary under normal conditions.

A primary role for the hypocotyl and secondary one for the apex and cotyledons also emerged from a study of mung bean (BRENNAN et al. 1976): (1) masking a single cotyledon depressed phototropic curvature little if at all, (2) masking the apex cut the response in half and (3) masking the stem alone caused complete failure (Table 2). Even in the dark the apex and cotyledons appear to contribute necessary materials, for excision of either prevented response (Table 2); but lack of evidence on hormonal supplementation or on the effect of wounding in mung bean prevents conclusions.

The maximal role which cotyledons can play in cucumbers and sunflowers has been clarified using "de-etiolated" seedlings, in which cotyledons differentially illuminated by white light can cause bending of the hypocotyl toward the illuminated side. SHUTTLEWORTH and BLACK (1977) showed that the response

Table 2. Effect upon phototropic curvature of masking or removing parts of light-grown mung bean seedling. Plants were exposed to 2 h of unilateral light. Leaves were parallel to light rays in treatment 2, but all others were perpendicular. (Data from BRENNAN et al. 1976)

Treatment	Average curvature, degrees
1. Unmasked, leaves perpendicular	25.2 ± 5
2. Unmasked, leaves parallel	20.2 ± 6
3. Primary leaves masked	18.7 ± 3
4. Tip masked	11.2 ± 3
5. Stem masked	0
6. Tip or leaves removed	0

is due to R rather than the more typically phototropically active B wavelengths. In fact, when they exposed only one cotyledon of a seedling to B, the plant actually underwent slight curvature away from that cotyledon. These results are consistent with an earlier study (BLACK and SHUTTLEWORTH 1974) in which growth inhibition by white light was shown to be transmitted from cucumber cotyledons to the hypocotyl.

If asymmetric influences of cucumber cotyledons were eliminated either by wrapping both in foil or exposing both fully, seedlings unilaterally illuminated with B bent toward it; response vs. irradiance is shown in Fig. 4A and B. The response of sunflowers is somewhat similar (Fig. 4C, 10C) but in B alone shows pronounced oscillation during phototropic exposure (Fig. 10C). R suppresses the oscillation (Fig. 4C), but suppression need not be pronounced (Fig. 10A, and see also FRANSSEN and BRUINSMA 1981). Quite possibly, the oscillation results from a geotropic counter-reaction.

SHUTTLEWORTH and BLACK (1977) are reluctant to consider that the asymmetric growth accomplished by completely darkening one cotyledon and brightly illuminating its partner with red or white lights is "true phototropism", for the difference between the intensities of light falling on the two cotyledons in nature is generally small. Moreover, as discussed by BRENNAN et al. (1976) this difference is apt to be an unreliable guide to the direction of incident light because it depends on the axillary angle of the cotyledons, and therefore may in principle lead to bending either toward or away from the sun. In fact, it is likely that this complex dependence enhances the tendency of seedlings to nutate. The practical distinction made by SHUTTLEWORTH and BLACK (1977) is thus an important one, even though insofar as the bending response partly depends on the direction from which light strikes the plant it technically meets the criterion of a tropism.

Discrimination of the vector of incident light occurs in leaves of *Lavatera cretica*, apparently by means of dichroic photoreceptors oriented in the membranes of cells in the vascular bundles (SCHWARTZ and KOLLER 1978). These leaves are sun-trackers, in which laminar orientation is regulated by pulvinar turgor. Effects of localized applications of the auxin transport inhibitor TIBA

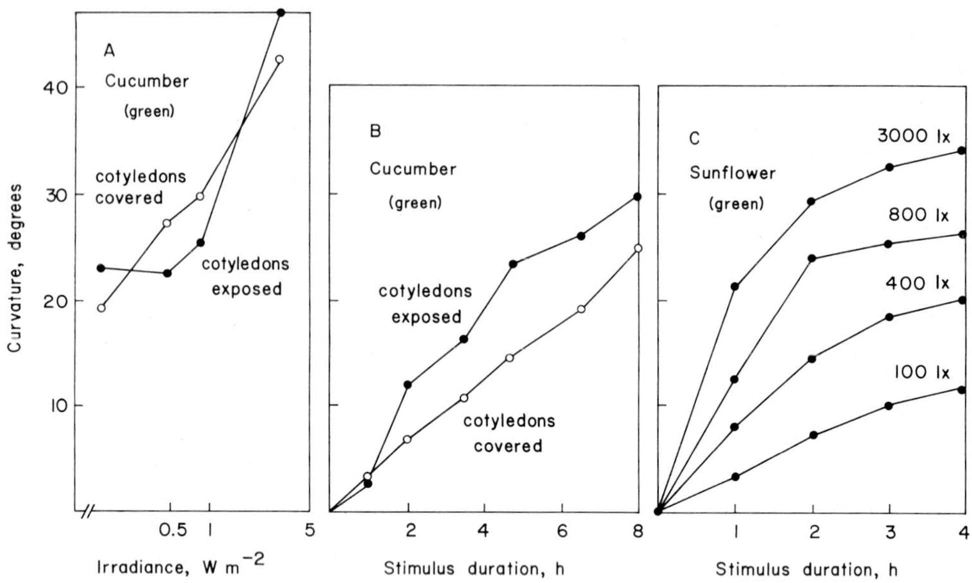

Fig. 4. Increasing response of two green dicot seedlings with increasing light intensity (**A, C**) and stimulus duration (**B, C**). **A, B.** Data for "de-etiolated" cucumber (*Cucumis sativus*) exposed to B, from SHUTTLEWORTH and BLACK (1977). Note response is not substantially affected by irradiation of cotyledons. **A** Exposure duration 8 h. **B** Irradiance 1 W m^{-2}. **C** Data for light-grown sunflower (*Helianthus annuus*) from DIEMER (1961)

support the hypothesis that asymmetries in turgor result from asymmetric transport of auxin into the pulvinus from the veins.

If such a vector-generated asymmetry of auxin transport were to be maintained all the way down the petiole into an elongating stem, phototropism could result there instead of or in addition to phototropism by the leaves. However, in *Lavatera* seedlings such an effect does not predominate under ordinary conditions (SCHWARTZ and KOLLER 1980). As in more commonly studied plants the hypocotyl independently senses the light gradient. A negative or positive contribution by the cotyledons to hypocotyl bending can, however, be demonstrated by masking the hypocotyl and constraining asymmetrically illuminated cotyledonary laminae from movement, so that their excitation persists an unusually long time.

In general, contributions to stem phototropism by true leaves remain unassessed. With sunflower, if all true leaves but one were trimmed from a small vegetative plant in the dark the stem did not bend; whereas in the light it bent nastically away from the leaf (SHIBAOKA and YAMAKI 1959). Thus, the phototropic contributions of sunflower true leaves *might* be important, and possibly primary – but this experiment by no means excludes a primary role for the stem, especially under natural circumstances. Because normal phototropic movements in older plants became entrained to oscillate diurnally from east to west (SHIBAOKA and YAMAKI 1959), the multi-leaved sunflower is an interesting but by no means simple system for model studies of phototropism.

In sum, on the basis of present evidence with seedlings it appears that the general role of the apex and leaves in phototropism of the stem is to contribute factors involved in growth. Although unequal illumination of cotyledons or leaves can in some plants under some circumstances cause tropistic bending, the phototropic receptor system which normally controls axial orientation is in the hypocotyl or epicotyl itself. Doubtless, as more kinds of plants are studied, numerous evolutionary modifications and elaborations will be described, but a common mechanism underlies the phototropism of stems of diverse light-grown seedlings.

Furthermore, one is tempted by the limited evidence to speculate that phototropism in green shoots may be basically the same as the low-sensitivity response of dark-grown plants. When phototropic curvature is assessed as a function of duration of illumination at fixed intensity for both light-grown and dark-grown seedlings of radish (Fig. 1A), the two plots appear similar for long stimuli (EVERETT 1974). For green cucumber seedlings, Fig. 4A and B shows that curvature increases fairly steadily with time of exposure for as long as 8 h at 1 W m^{-2} B, and increases monotonically with intensity (SHUTTLEWORTH and BLACK 1977); sunflower (Fig. 4C) similarly shows increase of curvature with both time and intensity of exposure to white light (DIEMER 1961). Both plants show high- and low-sensitivity responses when etiolated (STEYER 1967), and it may be assumed that the absence of data on a high-sensitivity response in the green seedlings means that they do not exist. Clearly, more elaborate and directly comparable data are needed for evaluation of the speculation, although even when they are available, detailed comparisons of green and etiolated plants may be complicated by differences in hormonal contents, in ways in which they respond to hormones, and in differing levels of screening pigment.

2.3 A Relatively Thorough Dose-Response Study

2.3.1 General Aspects

The etiolated oat coleoptile is perhaps the most thoroughly studied phototropic system, both at the level of hormone participation and of gross response. Because coleoptile phototropism appears very similar to that of dark-grown seedling stems (Table 1), it may serve as a generally useful model system.

The complexity of the dose-response relations is astonishing. Some knowledge of this complexity is a prerequisite to proper study of hormonal mediation. Moreover, an appreciation of whether the complexity arises from two or more closely related interacting photosystems or from a single system modulated by its own history of illumination will aid in appraising whether there might be more than one basic mechanism of hormonal mediation. The accumulation of data in the literature seems to justify development of a unified outlook on dose-response relations, and the rest of this section will be devoted to that task. Readers not interested should move directly to Section 3.

Until fairly recently, students of phototropism were content to study the dose-response domain by taking limited transects along its time or intensity axes, or even by measuring response as a function of total dosage without regard to individual variations in time and intensity. Fortunately, BLAAUW and BLAAUW-JANSEN (1970a) have produced a series of plots of curvature response vs. stimulus time and an intersecting series of plots of curvature vs. stimulus intensity. Their coleoptiles were treated with a small amount of R 2 h before phototropic exposure. Combining all plots, they produced a three-dimensional representation of dose-response relations over a fairly large domain (Fig. 3).

Preliminary to a discussion of the four or five distinctive regions of this topographical display, an important caveat must be set forth: the coleoptiles seem relatively nonresponsive. For example, in a separate study (PICKARD et al. 1969) R-pretreated coleoptiles exposed unilaterally for 40 min to about 0.4 W m^{-2} B curved 68°, whereas similarly treated coleoptiles curved 15° or 20° for BLAAUW and BLAAUW-JANSEN (1964, 1970a, b). The low sensitivity may be in part attributable to the use of excised coleoptiles, for excision prior to gravitational stimulation reduces geotropic responses by about 40% (WILKINS 1965). More critically, however, the small curvatures may be due to diminution of developing phototropism by geotropism, as can be surmised from the time-lapse records in Fig. 2. Coleoptiles of plants clinostated after a 30 min exposure to about 0.4 W m^{-2} B curved steadily for almost 160 min, whereas curvature rates of identically stimulated plants placed on the bench top began to decline at 40 min and reversed sign after 80 min (PICKARD et al. 1969). Significantly, BLAAUW and BLAAUW-JANSEN did not clinostat their coleoptiles, and shadowgraphed them after 100 min. Curvature of nonclinostated coleoptiles at 100 min is only one-third that of clinostated coleoptiles at 180 min (Fig. 2). Unfortunately, the discrepancy increases with the effectiveness of the stimulation, since geotropic counter-induction on the bench varies with the sine of the angle of curvature response (cf. PICKARD 1973), which increases continuously as curvature increases.

Also, topographic contours in portions of the domain where curvature is slight are somewhat uncertain, as a small displacement of one or two points in a region where the terrain is low and changing slope can greatly influence the interpretational construction. The seriously concerned reader might want to try his own hand at fitting a curve to the data points of Fig. 7 in the 1970a paper.

BLAAUW and BLAAUW-JANSEN must not be judged harshly because of these quantitative uncertainties. The systematic representation of dose-response relations was a tedious task too formidable to attract others. Since one can process at least ten times more coleoptiles a day if they are excised and left to curve on the bench top instead of retained intact and permitted to curve on a clinostat, their method was practicable even though somewhat imprecise.

2.3.2 Regions of the Dose-Response Domain

Angling out from the low end of the logarithmic time axis and the high end of the logarithmic intensity axis in Fig. 3 is a response ridge which peaks at about 0.12 J m^{-2}. Curvatures of this region are designated as A-type responses (BLAAUW and BLAAUW-JANSEN 1970a); this term replaces the "first positive curvature" of earlier authors. Over a fairly long time interval, A-type responses depend on the product of exposure time and intensity. If R had not been administered to the coleoptiles 1.5 h before phototropic exposure, the A ridge would not have risen so high and would have been found at a lower total dose (BLAAUW and BLAAUW-JANSEN 1964, 1970a and refs. therein).

For the longest tested exposures, another major ridge rises; curvature in this region is termed C-type (BLAAUW and BLAAUW-JANSEN 1970a) and corresponds to the "second positive curvature" of most recent workers. In the C response zone, curvature depends on both duration of irradiation and on intensity, but not simply on the duration-intensity product. If R had not been administered to the coleoptiles before phototropic exposure, the C response would have arisen later (e.g., BLAAUW and BLAAUW-JANSEN 1970a) but the response domain for coleoptiles untreated with R has not been fully mapped.

The A ridge in this log-log plot of the response domain drops off more steeply than it rises, and for the higher intensities ends in a conspicuous trough. This is the zone of negative curvature. Pretreatment with R enhances the depth and breadth of the negative curvature trough.

Between the zone of negative curvature and the C ridge lies a region in which either zero or very small positive curvature occurs; this is termed the indifferent zone. Looming out of the indifferent zone at the highest intensities is the B ridge. It is seen only if the coleoptiles are continuously pre-irradiated with R for a full 2 h before phototropic stimulation (BLAAUW and BLAAUW-JANSEN 1970a); in this respect it differs from the A ridge which is maximally influenced by 2 min of 4 nW m^{-2} R given 90 min before phototropic exposure. The B ridge evidently corresponds to the "second positive curvature" of DU BUY and NUERNBERGK (1934); its existence had been doubted by many recent authors, who unwittingly created semantic confusion by using the term "second positive" to describe the zone which DU BUY and NUERNBERGK named "third positive". In view of the discordant interpretations of the historically accepted terms, the vocabulary of BLAAUW and BLAAUW-JANSEN will be used herein.

2.3.3 A Unified View

It has long been thought that, in spite of the variety of dose-response behaviors, the photoreceptor system employs a single pigment. This view was strengthened by the observation (EVERETT andTHIMANN 1968) that the action spectrum for C-type curvature closely resembles that for the previously studied A curvature.

However, some workers have assumed that the pigment is divided into two or more noninterchangeable populations, differing perhaps in the molecules or molecular environment with which they are associated, or perhaps in grosser aspects of their distribution within the tissue. For example, this assumption was fundamental to the mathematical model of ZIMMERMAN and BRIGGS (1963b), and it was implicit in the reasoning of PICKARD et al. (1969).

It currently appears that models involving two noninterchangeable pigment populations may not be required to explain dose-response behavior. BLAAUW and BLAAUW-JANSEN (1970b) have developed an ingenious scheme in which the two major peaks of sensitivity result from a dark-adapted and a light-adapted form of the same pigment system. Implicit in their scheme, as in almost all models of phototropism, is the assumption that the dose-response plots directly and simply reflect transformations of the photoreceptor system – something difficult to prove but devoutly to be hoped, since if secondary complications contribute significantly to the shape of the plots, current information is inadequate to support even the loosest theory.

For the A-type curvature, BLAAUW and BLAAUW-JANSEN accepted the standard (cf. CURRY 1969, BRIGGS 1963a) explanation: the shape of the rising side of the dose-response curve coupled with obedience to the reciprocity law suggests a finite quantity of pigment which is not regenerated rapidly. The response undoubtedly depends on the light gradient within the coleoptile (MEYER 1969b, PICKARD et al. 1969); but it is difficult to prove whether it results from the summed actions of individual cells or from integrated comparison of the intensities within each cell in the gradient (PICKARD et al. 1969). The former idea seems the simpler[1]. In any case, with increasing dosage the pigment population

1 If integration across the coleoptile does occur, it seems likely that symplastic continuity would be required; yet both phototropism and geotropism can occur in coleoptiles in which most cells have been plasmolyzed (DRAKE and CARR 1978).

There appears to be a simple, definitive experiment which might test the possibility of integration across the coleoptile. In order to understand it, one must be familiar with the material of Section 3.1. HOCKING and WILKINS (WILKINS 1977) slit sections of coleoptile, opened them out, and flattened them between glass slips. They applied ^{14}C-IAA to the apical surface of the resulting sheet of tissue. One half was irradiated with B, the other shaded. There was no stimulation of lateral movement of ^{14}C-IAA into the dark half. This result is to be expected if integration is lacking and each cell responds directly to its own light gradient. If indeed the actions of cells are summed in series, a large auxin gradient would develop if the sheets were illuminated edge-on with a type C dose. The gradient should be even larger than that found in cylindrical sections, because of the additional cells in series within the light gradient. Any auxin gradient would be most apparent if the outer edges of the sheet were compared: it would thus be more appropriate to longitudinally quarter the sheet prior to assessing IAA than to halve it.

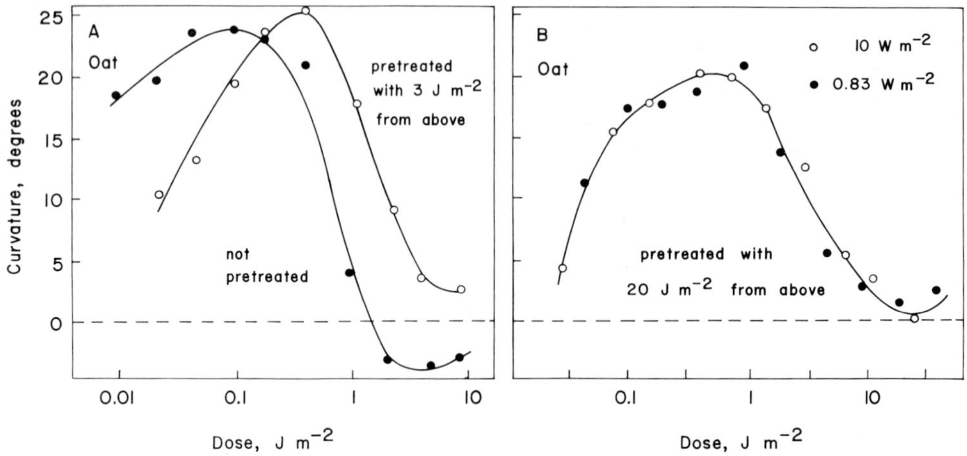

Fig. 5. Dose-response curves for oat coleoptiles following pre-irradiation with (**A**) 3 J m^{-2} or (**B**) 20 J m^{-2} B from above and a brief dark period; total treatment time 20 min. (Data from BLAAUW and BLAAUW-JANSEN 1970b). **A** Curve is shifted to higher doses by pre-irradiation. **B** Following pre-irradiation, response depends only on dose in a range for which response normally depends on the irradiance at which a dose is administered

becomes completely activated, first on the illuminated side and then on the side in shadow. Such saturation could explain the rise of the curvature response to a maximum and partially account for its fall to zero. Overlap with the mechanism of negative curvature (see below) could also influence the shape of the falling slope of the A ridge.

It is for the C-type curvature that BLAAUW and BLAAUW-JANSEN offered valuable new insight. They hypothesized that after initial activation *the photoreceptor system is regenerated to a light-adapted state*. In other words, it recovers to a state of lower sensitivity. Thus once the recovery process has begun, pigment can be activated a second time, but in order to produce a given response a considerably higher dose is required.

BLAAUW and BLAAUW-JANSEN drew from data of Fig. 3 two conclusions about parameters of the C response which (though they will be challenged later) will serve as a basis for the presentation of their analysis. Firstly, recovery begins 4 min after the start of illumination, this time being the baseline of the C ridge. Secondly, the rise of the C ridge is more or less linear with the logarithm of exposure duration. A series of five experiments supports the light adaptation hypothesis. In the first, a type C irradiation with B at about 10^{-2} W m^{-2} (cf. Fig. 3) was similarly (but not identically) effective if delivered unilaterally for 10 min, or if given from above for 2 to 9 min and from the side for the remainder of a 10-min period. The vertical irradiation is of course without tropistic effect. In the second, all but the first 1 or 2 min of the overhead irradiation could be replaced by a dark period without diminishing the curvature. In the third, when a 1-min overhead exposure was separated from a 1-min lateral exposure by dark intervals varying from 0 to 40 min, the effectiveness of the waiting period between exposures was fully realized at perhaps 5 min, and persisted to more than 20 min.

In the more critical fourth experiment, for which the first three served as preparation, coleoptiles were irradiated from above for 1 min at an intensity of 30 nW m^{-2} and were next maintained for 20 min in the dark. Then, as shown in Fig. 5A, unilateral irradiations at the same intensity were administered for increasing lengths of time in order to vary the dose, and the response plot thus created was compared with a standard A-type dose-response plot. In agreement with the hypothesis, the coleoptiles which received the vertical pre-illumination produced the same general shape of response curve

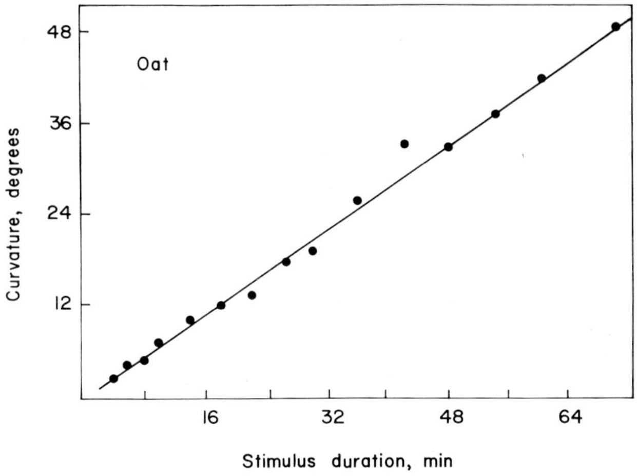

Fig. 6. Linear dependence of type C response on stimulus duration. (Data from PICKARD et al. 1969). Irradiance was 40 mJ m^{-2} "white" light. Correlation coefficient for fitted line is 0.996

as the controls, but it was shifted to higher doses. (The sharper rise seems not to be a repeatable feature of the experiment – cf. Fig. 5A with 5 B. The loss of negative curvature will be discussed later.)

If, after recovery to a desensitized state, the response mechanism is analogous to type A tropism, the dependence on intensity and duration of exposure should simplify to a dependence on dose. The fifth experiment confirms this simple dependence: when exposure time was varied for light administered at two different intensities, the "reciprocity law" or "dose law" was in effect (Fig. 5B).

In sum, after an initial type A activation of the pigment system there is a short lag following which pigment gradually returns to a light-receptive state. This state might be a second, inefficient, light-receptive form of the pigment, or might involve the original form of pigment in a different environment. In any case, at some time between the original activation and the full recovery, the state of the pigment system is changed in such a way that overall sensitivity is lowered. Ultimately, the original level of sensitivity is evidently restored, but this was explored only in a preliminary manner.

The reasoning and data of BLAAUW and BLAAUW-JANSEN are compelling. Moreover, their hypotheses can be generalized to a more powerful form. As already noted, the desensitized state persists for perhaps 20–30 min without apparent change. If pigment recovery is evidenced about 4 min after the start of irradiation, perhaps during protracted C-type stimulation the pigment could recycle many times. Thus, their hypothesis might be reformulated as follows: *after initial activation the photoreceptor system will be regenerated to a relatively stable light-adapted state, and in that state it can be reactivated and regenerated repetitively.*

Accordingly, continued pigment cycling would result in an approach toward steady-state conditions, and a more or less linear time-dependence should be expected for C-type phototropism during uninterrupted irradiation. Although BLAAUW and BLAAUW-JANSEN apparently observed a logarithmic rather than linear time-dependence (Fig. 3), they only studied a span of a few minutes; and PICKARD et al. (1969) did find linear behavior over a broad range (Fig. 6). The difference might arise in part from geotropic counter-induction in Fig. 3 and its prevention by clinostating in Fig. 6; probably ridge C in Fig. 3 falls off at spuriously low values.

A precise value for the presumed latency period for recovery of photoreceptor is important for evaluation of the cycling hypothesis: the briefer the lag, the more likely it is that cycling will reach a steady state during experimental exposure periods (but the more restricted will be the region for which a special schedule of irradiation can bring out reciprocity). BLAAUW and BLAAUW-JANSEN remarked that the C ridge begins

Table 3. ^{3}H-IAA distribution data of GARDNER et al. (1974) for coleoptiles. Transport periods either 1 or 2 h

	0.75 s Irradiation			60 min Irradiation		
	Curvature	Lateral movement	Reduction of basipetal transport	Curvature	Lateral movement	Reduction of basipetal transport
In situ						
Oat						
Red	+	0	...	+	0	...
Dark	+	+	+[a]	+	0	+[b]
Corn						
Red	+	0	...	+	+	0
Dark	+	+	0	+	+	0 (?)
Excised						
Oat						
Red	...	...	...	...	...	...
Dark	0[c]	+	...	0	0	...
Corn						
Red	0	0 in tissue	...	0	+ in tissue	...
		+ in receiver	...	0	0 in receiver	...
Dark	0	+ in tissue	...	...	...	...
		+ in receiver				

[a] Inhibited if IAA was applied to the illuminated side, not if applied to shaded side.
[b] Inhibited if IAA was applied to either side.
[c] Could failure of excised coleoptiles to curve be due in part to reduced availability of water?

to rise after a lag of 4 min, but judging from Fig. 3 it may at very high intensity rise at 2 min or less (and cf. Table 3 in BLAAUW and BLAAUW-JANSEN 1970b). However, the lags for initiation of C curvatures and for continued pigment recycling may well be different, and unfortunately there are no data on the latter.

A lag independent of intensity implies that desensitization of the pigment system as a whole must be triggered in all-or-none fashion by the onset of illumination rather than quantitatively related to the molecule-by-molecule activation of the pigment population. In fact, further data (BLAAUW and BLAAUW-JANSEN 1970b) argue in the opposite direction: the extent of desensitization as well as of activation is a function of total dose. It thus seems likely that activation of an individual unit of the pigment system initiates events which lead to the unit's own recovery and light adaptation. In this perspective (cavalierly ignoring the occurrence of negative curvature), the C response would shift toward shorter times as a function of increasing intensity of exposure. The shift might amount to 1 or 2 min within the intensity range of Fig. 3, but because C curvature occurs late along the logarithmic time axis such a shift would not be effectively displayed. Moreover, at relatively low intensities the system might begin to undergo C response before the entire population had participated in A response.

Since the dose-response map is not refined enough to permit evaluation of interpretations such as these, it would be worthwhile to obtain more detailed data. Use of responsive coleoptiles clinostated during curvature development and use of linear as well as logarithmic time axes would permit better assessment of the baseline as a function of intensity

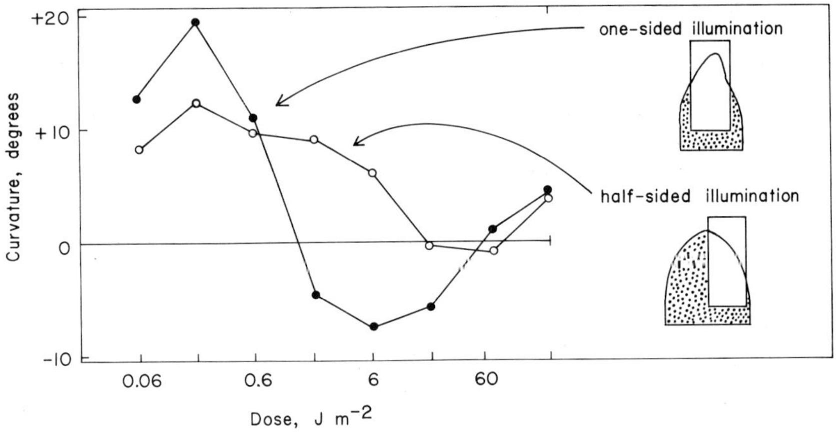

Fig. 7. Failure of negative curvature to appear following irradiation of only half of one side of an oat coleoptile. (Data from MEYER 1969b). Standard unilateral (one-sided) irradiation resulted in curvature perpendicular to the plane of the paper, whereas half-sided irradiation produced curvature in the plane of the paper

as well as elucidation of the true pattern of slope of the C response. The time and intensity range of Fig. 3 should be extended. Extensive dose-response maps for unilateral exposures which follow a range of initial symmetric illuminations would be equally helpful.

However, it is doubtful that ideas about the C response can be developed effectively without consideration of the negative and B-type responses which, because they are not apt to be of any significance in nature, must be the incidental result of the properties of the A–C system.

How, then, might negative curvature be explained? In the absence of any data on the steps immediately following photoreception, one cannot eliminate the possibility that the negative response is due not to the photochemistry per se, but rather to a "shock reaction" attendant on very sudden saturation of the receptor system. However, MEYER (1969b) found that illumination of only one half of the coleoptile with doses that would normally yield negative curvatures produces only positive curvatures (Fig. 7). Therefore, it seems more likely that negative curvatures result from two-hit inactivation of some of the pigment in the front flank while most pigment in the rear remains activated. Such bleaching is, of course, more apt to occur at high than at low intensities, because of the greater probability that a second photon will strike before the energy of a photoactivated pigment molecule is transferred by secondary reactions. This idea was initially entertained by BRIGGS (1960), although later discounted (ZIMMERMAN and BRIGGS 1963a, b) on the basis of certain mathematically modeled dose-response curves.

MEYER (1969a) proposed that negative curvature is due to a different chain of reactions than are responsible for positive type A curvature, because following pre-irradiation from above the coleoptile requires a longer dark period to recover the capacity for negative than for positive response. However, this behavior is in excellent accord with the generalization of the theory of BLAAUW and BLAAUW-JANSEN. In regaining susceptibility for induction of positive curvature, recovery of pigment (usually inseparable from loss of sensitivity) is all that is required. In regaining susceptibility for induction of negative curvature, the coleoptile would first have to pass through a period of low sensitivity when it would be capable of type C response. Then it would have to return toward its original sensitivity, regaining the capability for both negative and type A responses simultaneously. It will be recalled that the data of BLAAUW and BLAAUW-JANSEN (1970b) indicate return of sensitivity in the dark following desensitization (and cf. BRIGGS 1960).

The existence of B curvature might be easy to understand. The rate of pigment recovery from the activated and inactivated states might be different. Possibly, recovery of photoreceptive capability and loss of sensitivity need not be tightly linked. Since in continuously and strongly irradiated plants the A response is activated and inactivated very early, perhaps the pigment molecules begin to recover before or during the onset of system desensitization. Recovery might not be precisely synchronous, and light absorption by the recovered fraction of the pigment population might tend toward saturation even as ongoing recovery made more pigment available for absorption. Hence, the response plot would not peak sharply, but would spread out as a low ridge. The less synchronous the recovery the stronger would be the tendency toward early saturation, and the lower and flatter would lie the ridge. Unfortunately, the resolution of Fig. 3 is not adequate to evaluate such possibilities, which in any case must receive ultimate test by molecular studies.

Finally, the decline of the slope of the C ridge at very high intensity deserves comment. It is readily imagined to be due either to a tendency toward saturation on both sides of the coleoptile, or to bleaching, or to both factors combined (cf. PICKARD et al. 1969). Alternatively, it might result from increasing desensitization of the pigment system (cf. BLAAUW and BLAAUW-JANSEN 1970b).

Detailed comparative studies of several species (both with and without red pretreatment; cf. BLAAUW and BLAAUW-JANSEN 1970a) might well clarify some of these problems. Presently available response curves (Table 1) suggest that oat may be atypical in that it is the only species with which negative curvature has been observed; is phototropism in other species simpler, or does it merely seem so due to a different balance of parameters or due to inadequate mapping of their dose-response domains? Moreover, a comparative study of the C-type response (STEYER 1967) showed quantitative differences between species. Intensive work with three or four carefully chosen species might permit considerable refinement of ideas about the falling phase of the A response, negative curvature, and pigment system recovery, desensitization, and resensitization.[2]

2.3.4 Some Unexplained Problems

The basic observation underlying BLAAUW and BLAAUW-JANSEN's explanation of C-type curvature is that the photoreceptor system of oat coleoptiles displays a low sensitivity during its recovery from an initial exposure to light. These authors characterize plants during the recovery period as being in a special state. However BRIGGS (1960), using coleoptiles of corn prepared under R, documented steady recovery of responsiveness (presumed to indicate recovery of pigment) without an apparent stable intermediate state of low sensitivity (Fig. 8). Without such a stable period, reciprocity such as shown in Fig. 5B would obtain only roughly and over a small range. If the situation is peculiar to corn, critical predictions about differences between the dose-response domains for

---

2 Some years ago STEINER (1969a) provided provocative evidence that phototropic and polarotropic responses in moss and fern spores and sporelings and in algae and fungi might share multiphasic dose-response behavior with the coleoptile. The particular organ he studied, the protonema of *Dryopteris*, seems unusually complicated because it bends toward both red and blue light (1969b). Yet, his findings suggested that there may be relatively simple organs which bend toward blue light only and which are much better suited than the coleoptile for the large scale replication of experiments required for further testing of models for the reception process. In a structure such as the *Dryopteris* protonema there appears to be a much more direct coupling between differential activation of the photosystem and differential growth, and the protonema lacks the formidable problems presented by the coleoptile with its complex spatial overlay of receptor, mediational, and response systems (cf. Sect. 2.3.4). The photochemical implications and putative generality of the multiphasic dose-response curve are currently under investigation (e.g., SHROPSHIRE 1980, LÖSER and SCHÄFER 1980, SCHROTT 1980, refs. therein).

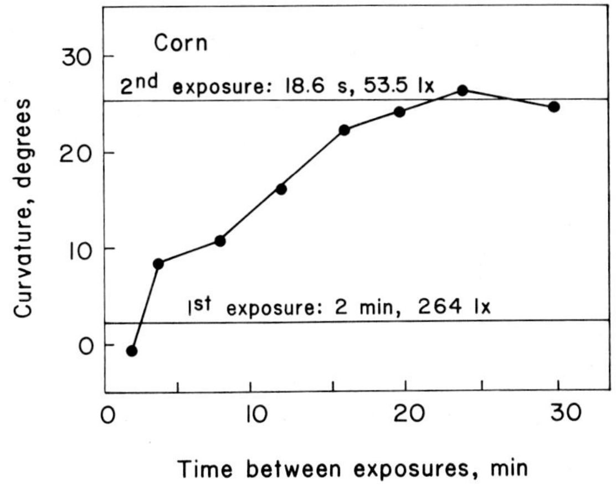

Fig. 8. Effect of preliminary inactivating exposure (2 min at 264 lx, an "indifferent zone" dosage) on response to subsequent type A exposure (18.6 s at 53.5 lx). Control responses to the two exposures are indicated by *straight lines*. Recovery of sensitivity appears complete within 25 min. (Data from BRIGGS 1960)

coleoptiles of the two kinds of Gramineae could be made and tested. On the other hand, the initial exposures of BRIGGS and of BLAAUW and BLAAUW-JANSEN were different: did this cause significant differences in subsequent desensitization? It is important to resolve this apparent discrepancy.

A second problem involves the distribution of the A- and C-type response systems. It will be recalled that phototropism theories can be broken into two categories: those assuming a single population of photoreceptor units (which may, however, exist in more than one state) and those involving more than one population, i.e., a population responsible for A tropism and restricted to the extreme tip and a population responsible for C tropism and distributed throughout the upper portion of the coleoptile. The model of BLAAUW and BLAAUW-JANSEN of course falls into the former category, and they have reviewed evidence (1970b; see also CURRY 1969) in support of a single population. There are two ways to approach this problem directly: cut off the tips before illumination, or mask them.

Both DU BUY (1933) and POHL (1960) found that decapitated oat coleoptiles yielded a type A response only after so-called physiological tip regeneration, although the coleoptiles responded to type C stimulation immediately after decapitation if provided with auxin.

In contrast, NAQVI et al. (1972) claimed to observe type A response in both decapitated oat and corn coleoptiles supplied with auxin-laden agar blocks, and BRAUNER (1922) and REINDERS (1934) found type A curvature in oats either if illumination was delayed to permit physiological tip regeneration or if tips were reserved in the dark during illumination of the stump and then immediately replaced. Presumably the sole effect of the replaced tip was to supply auxin at a physiologically appropriate level, although REINDERS expressed the view that "stumps with auxine-agar have not the normal sensitivity."

Masking the apical 500 μm of oat coleoptiles, ARISZ (1914) and REINDERS (1934) obtained curvature from type A doses. Type A curvature was also obtained by MEYER (1969b) after irradiating more than 350 μm below the tip, and the peak of the response curve did not appear to be displaced toward higher doses. However, light entering the coleoptile undergoes an amazing amount of internal scattering and channeling (MANDOLI and BRIGGS 1982a, b, 1983), and a gradient of somewhat reduced intensity could well have reached the apical cells. The slight displacement of MEYER's dose-response curve that would occur if this happened might be undetectable on the logarithmic plot of her widely separated data points.

Considering the critical significance of these data for the BLAAUW and BLAAUW-JANSEN hypothesis, and considering the evident delicacy of the determinations, a more comprehensive study of decapitated coleoptiles provided with auxin in agar and irradiated

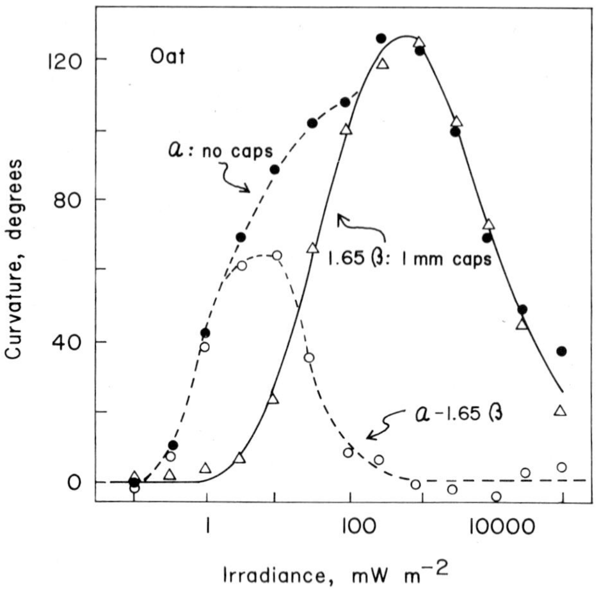

Fig. 9. Dependence of response on intensity of B (45 min exposure) for capped coleoptiles and controls. Values for capped plants have been multiplied by a normalizing factor of 1.65. (Data from PICKARD et al. 1969)

briefly over a range of intensities would seem essential. Even so, the existing data are mildly supportive.

On the other hand, one cannot dismiss two kinds of data supporting the idea of two spatially differentiated pigment populations. Firstly, there are different distributions of capacity for lateral movement of ^{14}C-IAA during type A and type C responses of corn coleoptiles treated with dim R (PICKARD and THIMANN 1964). For an A response, lateral movement was essentially confined to the uppermost cells, whereas for a C response lateral migration occurred throughout the tip (PICKARD and THIMANN 1964), and indeed the lateral asymmetry of ^{14}C-IAA developed in sections with 2 mm removed was about one-fourth that in intact tips. It should be noted that the intact tip effects a strong asymmetry of endogenous auxin before basipetal movement carries the hormone to the elongating subapical regions, which thus can only enhance an already marked gradient: in consequence, A- and C-type curvatures are visually almost indistinguishable in intact R-treated corn coleoptiles. Secondly, PICKARD et al. (1969) found that oat coleoptiles with masked apices yielded a plot of curvature vs. light intensity which was simpler in shape than that obtained with unmasked plants and which had a higher threshold (Fig. 9). Assuming that the response of the masked coleoptiles was of the C type and that the apical portion of the C system was prevented from participation, they multiplied the masked response curve by a factor to match its peak height with that of the unmasked curve and found that the curves overlapped for the high range of intensities. On subtraction of the masked curve from the unmasked curve there remained in the lower intensity range a small peak resembling the A type response (Fig. 9). While other methods of manipulating the data can be envisioned, there seem to be none which readily permit the conclusion that the base and tip responses are qualitatively the same.

In spite of the formal conflict between the two-population model of e.g. PICKARD et al. and the one-population model of BLAAUW and BLAAUW-JANSEN, it is possible that both could be revised and fused to better accommodate the data. Suppose that (1) pigment at the very apex (or at the regenerated physiological tip) behaves according to BLAAUW and BLAAUW-JANSEN, and (2) the basal pigment is always in a state associated with C type response. This condition would be fully compatible with the data of PICKARD et al. and might only *appear* irreconcilable with the data of BLAAUW and BLAAUW-JANSEN, in that action at the tip would dominate curvature, and an influence of the base would be hard to discern.

Since the effects of R pretreatment are still imperfectly described, perhaps some of the apparent conflicts of data may be understood in terms of differing exposures of plants to R. Doubtless, some R effects are expressed at the level of the photoreceptor system, whereas others are expressed at the level of the capability for secondary reactions. However, it may be that one critical effect of R is to shift a certain fraction of the photoreceptor system from the A to the C state, thus raising the threshold for induction of A-type curvature and lowering it for C type (The influence of R on the magnitude of the A peak might be secondary; cf., e.g., BRIGGS 1963a). Further, any effect of R pretreatment on the capacity for regeneration of the physiological tip (and putative associated development of type A responsiveness) might lead to variable results with decapitated plants.

Certainly, the above considerations argue for further research on the differences between response by tips and basal regions. As a basis, a set of experiments analogous to that of Fig. 9 but with irradiation fixed at about 0.1, 1, and 10 mJ m^{-2} and with time varied over a wide range would be useful. One test of the above interpretations that would emerge from such data is whether type C irradiations of the tip and base differ in producing reciprocity of response in the latter case but not in the former.

In evaluating the intricacies of tip and base responses, the existence of internal light-guiding paths demonstrated for coleoptiles and stems by MANDOLI and BRIGGS (1982a, 1983) should be kept in mind. While it is not obvious that such paths will figure prominently in models of phototropic behavior, in that they do not cause swamping of the light gradients due to absorbance, they must be taken into account for detailed predictions; and, as the authors note (1983), "Further data on the geometrical patterns of light generated within plant tissues may reveal ... surprises."

In summary, increasingly detailed information about the dose-response patterns of coleoptiles has led to increasingly sophisticated interpretations, but it is not clear how the puzzle-pieces will fit together into a single picture. One is tempted to speculate that the several regions of the dose-response domain will ultimately be shown to result from the operation of a single photosystem, and therefore probably a single mediational system. However, convenient as it would be to understand the photochemical basis of the dose-response domain before embarking on a study of the role of hormones, historically the investigations of the two topics have been interwoven. Therefore, it is appropriate to shift our focus to the headway made in understanding hormonal participation in phototropism.

3 Hormonal Basis of Phototropism

3.1 Auxin

3.1.1 Dark-Grown Seedlings

The early discovery of the role of auxin in phototropism was inextricably linked to the discovery of auxin itself. Almost every elementary text of plant biology describes the decisive experiment in which WENT (1928) lightly pressed tips of oat coleoptiles onto the blade of a razor, fitted blocks of agar on either side against the cut base of the tissue, illuminated the assembly perpendicular to the razor, and after a suitable response period bioassayed the blocks for auxin by setting them asymmetrically on decapitated coleoptiles and measuring resultant curvature. The ratio of auxin in blocks from the lighted vs. shaded sides was about 25:75 for a type A dose. The total amount of auxin recovered, while not in every instance quite as high as that obtained from dark controls, was far too great to permit an interpretation that the asymmetry resulted from photodestruction of auxin. WENT proposed that within the tip auxin moved from the illuminated side to the shaded side unter the influence of a type A dose, and moved in the reverse direction following a dose eliciting negative curvature. Associated experiments indicated the continuous operation of a polar transport system which carried the asymmetrically distributed auxin into the lower regions of the coleoptile, resulting in differential growth and therefore curvature.

The only flaw in WENT'S experiment was the slight apparent reduction of auxin yield when illuminated tip assemblies were compared with controls. That this difference was either a sampling error or a result of the slightly more extensive manipulation to which the illuminated assembly was subjected was amply shown for type A tropism by BRIGGS et al. (1957), who used the larger coleoptiles of corn. Further, these workers prevented the appearance of the auxin gradient by splitting the tips completely and separating the halves of each with a slip of glass. This provided unequivocal evidence that either auxin or a substance intimately involved in auxin synthesis or distribution actually does move from one side to the other during normal phototropism.

WILDEN (1939) early demonstrated asymmetry of auxin in oats following what seems to have been type C irradiation, and newer results of BRIGGS (1963c) with corn seemed to show definitively that both A and C curvatures result from lateral migration of auxin in that organism. Already in 1938 ASANA had confirmed that the negative curvature of oats is mediated by migration of auxin toward the light. The general validity of the lateral migration theory was shown in 1932–1933 by VAN OVERBEEK with light-grown seedlings of radish. Moreover, (as will be discussed in Sect. 3.1.2) VAN OVERBEEK extended the evidence supporting lateral migration by applying auxin either to the lighted or shaded side of the decapitated radish hypocotyl and showing that far more crossed from the lighted to the shaded side than vice versa. Thus, for a number of years WENT'S idea seemed soundly established and broadly applicable to seedling plants. Experiments on phototropism were all the more convincing because a parallel series of experiments on gravitationally stimulated seedling shoots provided equally impressive evidence for a key role for lateral migration in geotropism (REINHOLD 1978).

However, despite this large body of supportive data, the lateral migration theory met two challenges. First, ^{14}C-IAA applied to coleoptiles failed to undergo significant phototropic redistribution (BÜNNING et al. 1956, GORDON and EIB 1956, REISENER 1958, see also SHEN-MILLER and GORDON 1966). Barring flaws in experimental technique, these experiments could only indicate that the early bioassays of auxin had been grossly misinterpreted.

Impressed by the simple elegance of the original bioassay methods, PICKARD and THIMANN (1964) re-investigated the fate of applied ^{14}C-IAA, taking care to copy as closely as possible the conditions under which work with endogenous auxin had yielded positive results. Measuring ^{14}C-IAA both in agar blocks placed at the base of unilaterally illuminated coleoptiles and in the tissue itself, they showed A-type gradients closely resembling those obtained by bioassay. Because of its apparent localization in the extreme apex of the coleoptile, all work with A-type response was necessarily carried out with intact tips, but lateral asymmetry of label was demonstrated with both tips and subapical sections for C-type response. With sections, it was possible to show that all the radioactivity provided at the beginning of the experiment was recovered at the end for both illuminated assemblies and controls, ruling out creation of the gradient by photodestruction. Subsequently, DELA FUENTE and LEOPOLD (1968) confirmed that phototropism enhances lateral auxin movement by applying ^{14}C-IAA to one side of corn coleoptiles and showing that unilateral light enhanced the amount that moved to the opposite side. Thus, the first challenge to WENT's theory of lateral migration was unsuccessful.

Because determinations of the lateral distribution of auxin are still being made on other plants, sometimes with divergent results, it is worth considering seven experimental problems which can result in a failure to detect asymmetry under conditions in which it would normally occur. (1) The region of the tissue sensitive to the stimulus must not be destroyed prior to the test stimulation or redistribution perid. (2) Entry of exogenous IAA must be restricted to the apical region of the sensitive tissue. (3) The timing of application of exogenous IAA must be compatible with the timing of induced redistribution. (4) Exogenous IAA must not increase the total auxin level in the tissue above the endogenous level of transportable auxin in controls by enough to saturate the transport system. (5) Ideally, only auxin in the transport system should be measured; auxin "bound" in the tissue (either chemically reacted or sequestered in a pool inaccessible to the transport system) before redistribution could be effected will mask asymmetry of basipetally moving auxin in tissue segments. (6) It cannot be assumed that the asymmetry is linear across the tissue; rather, the gradations are probably minimal in the central tissue and dramatic at the lighter and particularly at the darker sides. Thus, for stems with a great deal of pith, the perhaps uniformly distributed auxin in that tissue may swamp asymmetry of auxin which would be dramatic if sections of only epidermal and subepidermal tissue were compared. Each of these problems has plagued one or more attempts to detect auxin asymmetry during phototropism.

The second challenge to the theory came from workers who postulated or measured loss of auxin during phototropic experiments (see REINERT 1953, 1959, GALSTON 1959). After additional experiments in which full recovery of either endogenous (BRIGGS et al. 1957, BRIGGS 1963c, PICKARD and THIMANN 1964) or exogenous (PICKARD and THIMANN 1964) auxin was documented, as described above, this challenge dwindled. It was easy to understand how nonphysiological loss could confuse results under certain conditions, because IAA is indeed photolabile (e.g., MENSCHICK et al. 1977). Moreover, BRIGGS (1963b) expanded on suggestive early experiments to show that light received by the phytochrome system, incapable even at very high doses of causing phototropic curvature, could halve the amount of auxin diffusing out of apical sections of corn coleoptiles. The decrease might well be due to conjugation of the IAA (cf. BANDURSKI et al. 1977). Since phytochrome absorbs weakly in the B as well as strongly in the R region of the spectrum (SMITH 1975), and since BRIGGS' coleoptiles were sensitive to extremely small irradiations, it appears that nonphototropic diminution of free auxin may obfuscate hormone distribution measurements unless the phytochrome system is activated equally in experimental and control plants.

These two nonphototropic phenomena plausibly account for all observations of auxin disappearance during phototropism, and reinforce the view that tropism is due to migration rather than destruction or chemical transformation. Thus,

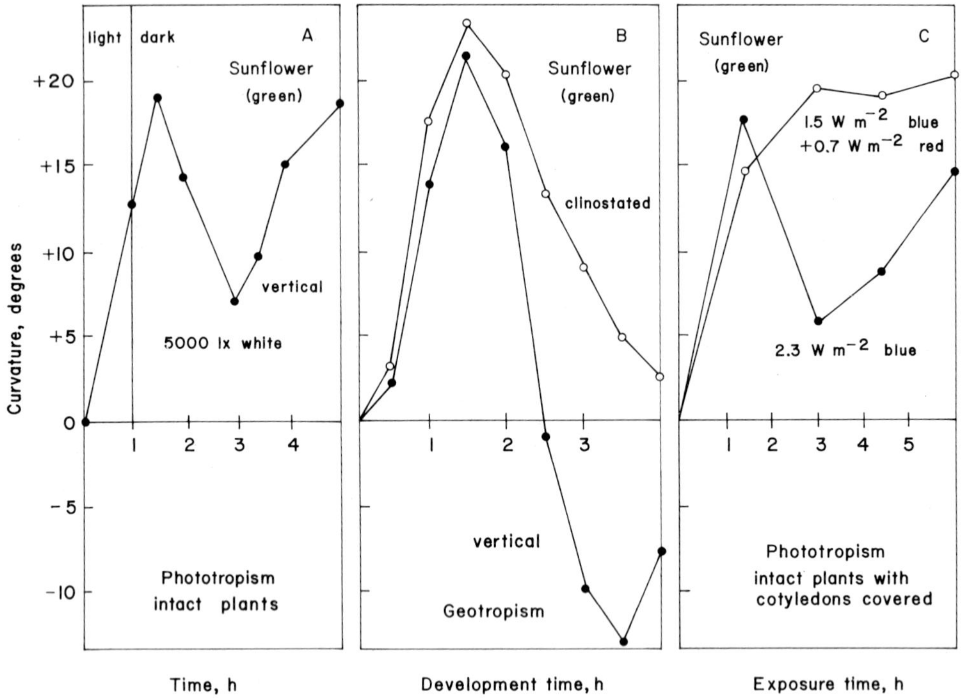

Fig. 10 A–C. Oscillation of the tropistic responses of sunflower seedlings. **A** Phototropism in response to 1 h of 5000 lx "white" light. (Data from DIEMER 1961). **B** Geotropism on bench-top vs. clinostat following 10 min horizontal exposure. (Data from BRAUNER and HAGER 1958). **C** Phototropism by de-etiolated seedlings with cotyledons covered during continuous exposure to B (~440 nm) vs. mixed B and R (~660 nm) light. R eliminates the oscillation. (Data from SHUTTLEWORTH and BLACK 1977)

lowering of the free auxin level may accompany and influence phototropism, but does not cause it.

All in all, one can sympathize even if not totally agree with WENT'S (1974) evaluation of the early work and criticisms of it: "This evidence was completely clear cut, yet ... it was attacked until work with radioactive indoleacetic acid fully confirmed our results. The fascination of radioisotope work was finally able to overcome prejudices against equally exacting pre-isotope work."

3.1.2 Light-Grown Dicots

3.1.2.1 Auxin in the Stem

It is perhaps unfortunate that the sunflower has been selected by modern authors as a model system for the study of hormonal relations in the phototropism of green seedlings. Circumnutations in the sunflower seedling can be pronounced (e.g., JOHNSSON 1971), and following cessation of the initial stimulus both phototropism and geotropism can be complicated by conspicuous counter-reactions

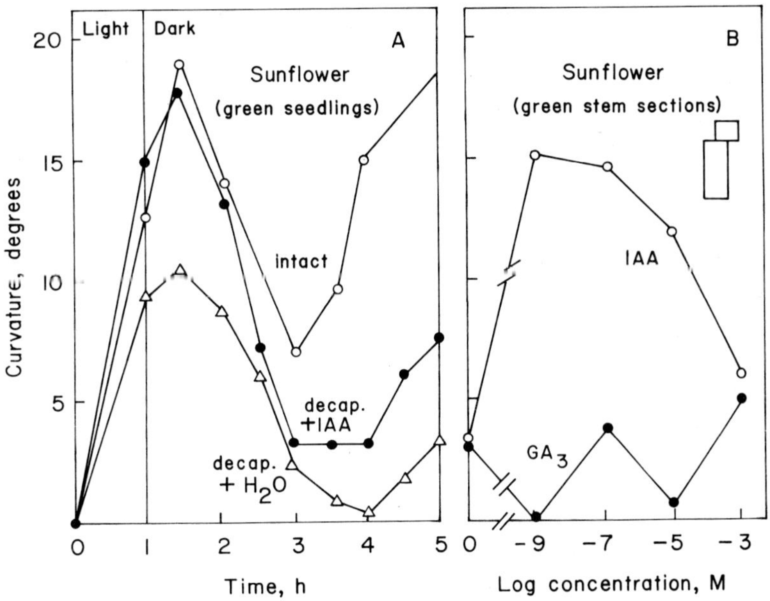

Fig. 11. A Following decapitation sunflower hypocotyls lose phototropic responsiveness, but it can be restored by IAA. Data obtained with decapitated seedlings exposed for 1 h to 5000 lx "white" light. (DIEMER 1961). **B** Curvature of sunflower epicotyl sections following unilateral application of agar blocks containing IAA, contrasted with lack of curvature following application of blocks with GA_3. (Data from PHILLIPS and HARTUNG 1976)

which cause oscillation of the curvature response, as illustrated by data of DIEMER (1961) and of BRAUNER and HAGER (1958) in Fig. 10A and B. SHUTTLEWORTH and BLACK (1977) showed that in B oscillation can occur even while the light remains on, though it is suppressed if R is provided concurrently. The dose-response relations for this effect of R have not been investigated.

In spite of these and other difficulties, some provocative information about possible involvement of auxin has been gained. For example, DIEMER (1961) showed that following decapitation hypocotyls become less sensitive to unilateral light, and that exogenous auxin can restore full sensitivity (Fig. 11A). She also found that a lanolin paste of auxin applied to one side causes differential growth. PHILLIPS and HARTUNG (1976) applied agar blocks containing auxin to one side of a decapitated hypocotyl, and they too observed differential growth (Fig. 11B). BRUINSMA et al. (1975) measured auxin in the illuminated and shaded halves of hypocotyls after a 4-h unilateral illumination with 1.1 W m^{-2} blue light, but neither results of bioassay nor of spectrofluorometric assay provided evidence of asymmetry (see also BRUINSMA 1977). FRANSSEN and BRUINSMA (1981) present confirmatory spectrofluorometric data and mention without presentation of data results of a confirmatory bioassay. Now, it is possible that lateral migration of auxin does not occur in sunflower hypocotyls. However, most workers have assessed such tiny differences in auxin as are to be expected between the two halves of a stem by means of the sensitive Went curvature bioassay, which features a more or less linear range of response; BRUINSMA'S laboratory used a much less sensitive logarithmic straight growth assay. Also, specificity of their spectrofluorometric assay may be equivocal, and complications in quantitation cannot be ruled out. Furthermore, while no asymmetric auxin distribution has yet been measured for phototropically stimulated sunflower hypocotyls, RUGE (1941) reported a geotropic asymmetry of endogenous auxin, and subsequently geotropic asymmetries of applied ^{14}C-IAA and ^{14}C-2,4-D have been reported by several authors (Chap. 7, this Vol.). It is not parsimonious to suppose that in sunflower hypocotyls phototropism and geotropism are mediated by entirely different mechanisms, but of course supposition is no substitute for experiment. Clearly, more work must be done on this plant. In view of the apparently atypical complexity of its response patterns, however, comprehensive measurements on a more representative kind of seedling should perhaps be given priority, so that mediational mechanism in sunflower can be studied on a comparative basis as has been done for general response relations (recall Sect. 2.2).

The morphology of the hypocotyl must be taken into account in the design of any experiments on its phototropic mechanisms, but such considerations may be particularly critical for thick hypocotyls such as that of the sunflower. Firstly, it is widely believed that, in stems of many plants, cells in the epidermal and subepidermal layers tend to limit growth of the organ as a whole. It is thought that this is because they show a different responsiveness to auxin than those in the core (e.g., WENT and THIMANN 1937, IWAMI and MASUDA 1974, cf. FIRN and DIGBY 1977), and curvature in the traditional split stem bioassay for auxin (WENT and THIMANN 1937) is presumably due to this differential responsiveness. In the specific case of sunflower, MENTZE et al. (1977) have shown that if the outer two or three layers of cells are removed, sections of hypocotyl will not elongate when auxin is applied. Of course, a wound response is not excluded; however, the main explanation is probably that the core of the sunflower hypocotyl is not particularly responsive to auxin.

Secondly, the role of the central parenchyma in auxin translocation is thought to be slight (IWAMI and MASUDA 1974, 1976, SACHS 1975, GOLDSMITH 1977, WANGERMANN and WITHERS 1978), and even if the pith participates in lateral redistribution, the auxin gradient there must surely be negligible. BURG and BURG (1967) addressed an analogous problem in their study of geotropic auxin asymmetry in etiolated peas: they demonstrated that the relative difference between auxin in upper and lower quarters was much more striking than that between upper and lower halves. Furthermore, in a later study of phototropism KANG and BURG (1974) showed a marked difference between illuminated and shaded quarters of etiolated pea stems.

It is also worth noting that both phototropic and geotropic sensitivity may be greatest at the apical end. When working with sections, it is tempting to cut them below the tapered bulges where the cotyledons emerge, but sensitivity may be lost in so doing. Could this account for the inability of VAN OVERBEEK (1936) and PHILLIPS and HARTUNG (1976) to detect auxin asymmetry?

Finally, sunflower might present a special physiological problem in addition to these morphological ones: phenylacetic acid has been reported to constitute a sizable fraction of the natural auxin in sunflower (SCHNEIDER and WIGHTMAN 1974), although its efficacy in this system is unknown.

A more suitable model system might be radish. Preliminary dose-response studies have already been carried out (Fig. 1), and more importantly, the extensive bioassays of VAN OVERBEEK (1932, 1933) provided strong evidence that lateral movement of auxin does occur during phototropism of light-grown radish

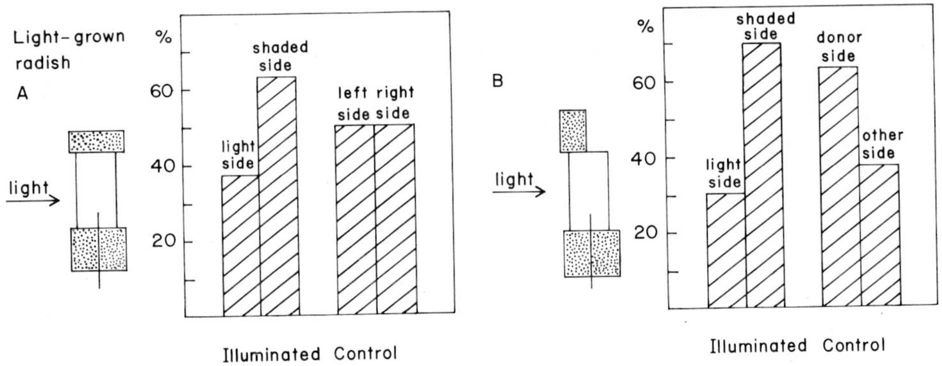

Fig. 12A, B. VAN OVERBEEK's demonstration (1932, 1933) that auxin moves toward the darker side of an asymmetrically illuminated section of light-grown radish hypocotyl. *Sketches* indicate how auxin was supplied via apical blocks of agar and collected in paired agar receiver blocks; it was assayed by means of the *Avena* curvature test. *Histograms* show the percent of collected auxin found in each of the paired receivers

seedlings. At the same time, his measurements of elongation as a function of omnilateral illumination substantiate a significant role for light-induced change in responsiveness to auxin (a form of "light-growth reaction"), though whether this change is capable of independently leading to tropistic response is open to doubt. Because VAN OBERBEEK's experiments have been neglected, it is pertinent to consider them at some length.

Seedlings grown in a glasshouse were removed to a dark room for "adaptation" the night before use. They were highly reactive: during a 2-h exposure to bright unilateral light, they curved about 70°.

Bioassaying agar blocks placed at the bases of excised cotyledons, VAN OVERBEEK first showed that light-grown cotyledons provide auxin to the hypocotyl in amounts about five times higher than do dark-grown cotyledons. Long-term return of plants to the dark resulted in a reduction of cotyledonary auxin production (particularly in older seedlings), but darkening cotyledons for 3 h did not measurably influence the amount of auxin collected during that period.

Secondly, he excised sections from the hypocotyls and provided auxin via agar blocks at the apical ends while collecting auxin in basal blocks. Bioassay of the basal blocks after 1-, 2-, or 3-h periods showed that equal amounts were transported in light and dark.

Thirdly, collecting auxin from the two halves of phototropically stimulated hypocotyl sections (Fig. 12A), he found a greater flow of auxin down the darker side.

Fourthly, he asked whether the difference in auxin flow might arise from a decrease in the capability for transport on the lighted side matched by an increase on the dark side, without resultant lateral movement. Mounting an auxin-donating block on the side of the section to be illuminated (Fig. 12B), he compared auxin collected under the donor block with that collected on the opposite side, and in turn compared these values with amounts collected from similar assemblies which had been kept in the dark. Unequivocally, there is a net movement of auxin toward the darker side of a unilaterally illuminated hypocotyl (Fig. 12B).

GIMMLER has proposed an attractive mechanism for phototropism during asymmetric illumination of green stems (personal communication) which might supplement lateral translocation of IAA. Within the chloroplast, light-induced proton transport from the stroma into the thylakoids results in the stroma becoming alkaline with respect to the cytoplasm. Protonated IAA should cross the chloroplast membrane freely, but must dissociate in the aqueous phases on either side to an extent dependent on pH. Within the alkaline stroma more IAA will dissociate, and its concentration will build in the chloroplast because of the predominance there of the anionic form. Thus, the pool of IAA able to participate in cytoplasmic activities, or in exchange across the plasmalemma, is diminished.

Under symmetric conditions, analogous sequestering of ABA, which has a pK_a and solubility properties similar to those of IAA, has already been demonstrated in leaves of spinach (HEILMAN et al. 1980; see also KAISER and HARTUNG 1981). GIMMLER notes that the differential sequestering of ABA in unilaterally illuminated stems would lower the cytoplasmic level of this "inhibitor" on the illuminated side, thus opposing the effect of the differential sequestering of IAA in tissues where ABA inhibits elongation and combining with it where ABA stimulates (cf. ASPINALL et al. 1967, MCWHA and JACKSON 1976, ABOU-MANDOUR and HARTUNG 1980).

Because the net magnitude of these potentially modifying effects will depend on very many factors, an experimental assessment of its quantitative contribution to phototropism under a range of conditions is required.

3.1.2.2 Auxin from Leaves

VAN OVERBEEK's demonstration that radish cotyledons provide auxin to the stem (Sect. 3.1.2.1) has been duplicated for a wide variety of cotyledons and leaves; young leaves are a rich source of the hormone. While VAN OVERBEEK demonstrated that auxin production by radish cotyledons is equivalent for some hours in white light and darkness, and that in radish seedlings it is the stem in which lateral asymmetry of auxin arises, the evidence of Section 2.2 raises the question whether there might be circumstances under which release of auxin from cotyledons or leaves into the stem is influenced rapidly by light, so that if they were differentially illuminated some phototropic bending might result. Indeed, the finding (see above) that ABA is partitioned from the cytoplasm into the chloroplasts during photosynthesis of spinach leaves indicates one mechanism by which availability of IAA, which probably partitions similarly, could be modified. The extent to which release of IAA by a petiole or by individual veins within it is localized to the tissue directly below will, of course, vary with species and with stage of plant and leaf development, since parenchymal IAA transport is thought to be strongest in the cells surrounding the vascular tissue (SACHS 1975, GOLDSMITH 1977, WANGERMANN and WITHERS 1978) and since there are characteristic species differences and developmental progressions

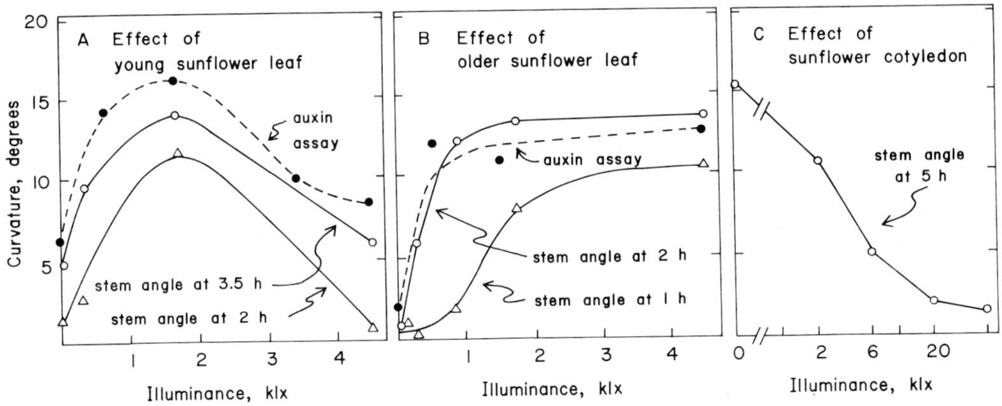

Fig. 13A–C. Potential of leaves and cotyledons to cause differential growth of sunflower stems. **A, B** and **C** show curvature of epicotyl or hypocotyl resulting from one young leaf, one older leaf, or one cotyledon following removal of its opposite prior to test illumination from above. **A** and **B** also show bioassayed auxin which diffused from comparable excised leaves into agar blocks; the auxin assay results are expressed in degrees curvature of *Avena* test plants and for convenience the same scale is used for both this curvature and that of sunflower stems. (**A** and **B** SHIBAOKA and YAMAKI 1959; **C** LAM and LEOPOLD 1966)

in the connections between vascular bundles of the leaf and stem (e.g., ESAU 1965).

Two studies on sunflower speak directly to this issue. The first, by SHIBAOKA and YAMAKI (1959), indicates a strong dependence of auxin production by the first leaf above the cotyledon on intensity of white light and on age (Fig. 13A and B). If all leaves but one were removed before plants were subjected to light from above, the resultant curvature of the epicotyl away from the leaf depended on intensity in much the same way as did auxin production measured by the *Avena* curvature bioassay. With young leaves (Fig. 13A) the responses to illumination peak dramatically with intensity, whereas in older ones (Fig. 13B) they reach a plateau. Imagining it unlikely that auxin production would be both enhanced by weak white light and inhibited by strong light, SHIBAOKA (1961) sought and found a diffusible growth inhibitor from brightly illuminated leaves; but its identity and role have not been established (cf. MUIR and ZHU 1983 and refs. therein).

Further, the demonstration by SCHWARTZ and KOLLER (1978; see Sect. 2.2) that at least some leaves can respond to the vector of incident light as well as to its intensity serves notice that sunflower leaves should be tested for their ability first to discriminate light direction per se and second to relay this information to the epicotyl. It is likely that several mechanisms can contribute to phototropism of shoots, and the relative importance of each in a given situation must be sorted out before the hormonal basis of the net response can be satisfactorily understood.

The second study, by LAM and LEOPOLD (1966), used cotyledons rather than regular leaves of 10- to 14-day-old sunflowers. A hypocotyl with the epicotyl and one of the cotyledons removed bent away from the remaining cotyledon less rather than more as light intensity increased (Fig. 13C). Release of endogenous auxin from decapitated seedlings cut just below the cotyledonary node was twice as great in the dark as in white light (about 20 klx?). Chromatography of diffusate obtained in darkness indicated the presence of both auxin and a material which inhibited auxin transport. Though the amount of endogenous auxin released by the cotyledons was low in the light, applied ^{14}C-IAA was more effectively moved into the hypocotyl from lighted than from darkened cotyledons. Unfortunately, the method of application did not allow an estimate of dilution by endogenous IAA, the concentration of which was apparently lower in illuminated cotyledons. When ^{14}C-IAA was applied either to the shaded or to the illuminated cotyledon, the ratio within the stem of radioactivity under the donor cotyledon to that under the opposite cotyledon was sensibly the same. Viewed in the perspective of subsequent experiments by SHUTTLEWORTH and BLACK (Sect. 2.2) these results cannot be considered to demonstrate that cotyledons normally bring about tropism. Rather, they show that the level of auxin available for transport out of sunflower cotyledons is depressed by light, and hint that the ability to transport auxin out of the cotyledons might be enhanced by light.

Thus, in spite of numerous gaps in knowledge, the sum of past work indicates that in many plants auxin supplied by leaves and cotyledons is phototropically redistributed in stems, evidently with other as yet unidentified agents from the leaves and cotyledons modifying responsiveness. Light-induced changes in auxin output occur to varying extent in different species, probably in general affecting the expression of phototropic induction more conspicuously than they impose systematic asymmetry. Indeed, it is likely that somewhat arbitrary light-induced asymmetries in auxin flow from the cotyledons and leaves often superimpose a wobble or nutation on the growth directed by the geo- and photoreceptors of the stem.

There is a conspicuous need, then, to better establish the phenomena of photoreception, to better define the relative contributions of the several processes in leaf and stem which may contribute to curvature of the stem toward the light, to better map the changes in pattern and intensity of auxin flow during response, and to identify any inhibitors which might be involved and clarify their role. Current knowledge gives a very incomplete accounting of processes mediating and modifying phototropism. Nevertheless the evident importance of lateral transport of auxin within the coleoptile and stem has stimulated a great deal of research, the understanding of which is basic for phototropic theory and to which we now turn.

3.1.3 Mechanism of Auxin Transport

3.1.3.1 Basipetal Transport

Lateral movement of auxin in growing coleoptiles and stems is tropistically superimposed on the normal basipetal flow. To understand this redistribution, it is, of course, necessary to understand the mechanism of basipetal flow. Fortunately, a theory is emerging which seems capable of accounting for many, if not all, of the observations in polar movement which have accumulated over the years, and which has already proven to have considerable power of prediction (RUBERY and SHELDRAKE 1973, 1974, RAVEN 1975, CANDE and RAY 1976, GOLDSMITH 1977, DAVIES and RUBERY 1978, RUBERY 1977, 1978, 1979, 1980, EDWARDS and GOLDSMITH 1980, SUSSMANN and GOLDSMITH 1981, GOLDSMITH et al. 1981, HERTEL et al. 1983). An abbreviated presentation of the theory will be given below.

The theory specifies two kinds of path for the passage of IAA across the plasmalemma; net movement along both paths occurs down electrochemical gradients.

The first path involves diffusion, both simple and carrier-mediated, across the membrane. Outside the cell, the pH tends to be low, and a considerable fraction of the IAA is undissociated. Because this uncharged form is relatively soluble in the plasmalemma, it can diffuse into the cell freely; however, the occurrence of a specific carrier in the plasmalemma accelerates the transfer. Once in the more or less neutral cytoplasm the IAA tends to dissociate, thus maintaining the concentration gradient for the uncharged species. In result, the cell is a trap for IAA. (The partitioning of IAA between cytoplasm and central vacuole is similar to that between cytoplasm and the exterior, so that the vacuolar concentration of IAA is low; but because the vacuole is large it maintains a considerable quantity of IAA.)

The second path involves carriers which bind anionic IAA. These might function in several ways (cf. references cited above for discussion). Whatever the specific properties of the carriers, they tend to effect a net efflux of anion out of the cell, down both the chemical and electrical gradients.

Earlier modeling by LEOPOLD and HALL (1966) showed that striking polar transport can be achieved in a tissue if there is even a slight, systematic, directional bias in secretion by each cell. In their model, if the rate of basal export from each cell is as little as 1.01 times that of apical export, auxin moves basipetally through the tissue. Thus, if anion carriers were to release IAA^- into the apoplast preferentially at the basal end of each cell, auxin would increase in concentration there, and entrance of the undissociated form into the cell below would be promoted. Over a path many cells in length, the flow would appear both polarized and more rapid than diffusion through a homogeneous medium.

Direct evidence for the presumed carrier of anionic auxin and information on its distribution in stem tissue has recently been obtained by JACOBS and GILBERT (1983). They treated sections of pea stem with a fluorescent monoclonal antibody raised against a protein binding naphthylphthalamic acid, a reasonably specific inhibitor of auxin transport. Fluorescence could be observed at the basal ends of cells in the vascular parenchyma, a tissue known to transport auxin well. Further use of this or of similar antibodies promises not only to permit detailed mapping of the distribution of the auxin transport protein but also to facilitate its isolation and biochemical characterization.

3.1.3.2 Two Proposals for the Origin of Lateral Asymmetry

Although many mechanisms have been proposed for the origin of phototropic auxin asymmetry, only two are now seriously entertained. According to one model, lateral polarization of movement would be closely analogous to basipetal polarization, but under environmental rather than morphogenetic control. It will be called the "lateral carrier model", or MI. The model in its generalized form has evolved over many years with contributions from many authors. According to the other model, asymmetric blue light would cause a correspondingly asymmetric inhibition of basipetal transport. Auxin would tend to accumulate in or above the zone of inhibition, slowly in the darker side of the stem or coleoptile and more rapidly in the brighter side. As auxin became more concentrated in the brighter half, it would diffuse into the shaded half, where it would be preferentially exported to more basal tissue. Thus, *below* the inhibited zone, cells on the darker side would receive more auxin than cells on the lighter side. This model will be called the "axial carrier model", or MII. It was originated by GEORGE CURRY (personal communication, December 6, 1961; publication, 1969).

Both models are compatible with current views on the mechanism of basipetal transport; but discrimination between them is difficult because, whereas MI yields qualitatively decisive predictions, the behavior of MI is so strongly dependent on selection of parameters that its predictions could at one extreme overlap those of MI.

The lateral carrier model (MI) will be considered first. In unilateral light, the gradient across a single cell (cf. Sect. 2.3.3, and footnote 1, p. 375) is quite small, but since the photoreceptive system is probably associated with the plasmalemma (e.g., BRIGGS 1976, HERTEL et al. 1980) the maximum gradient of light within each cell could be transduced to a gradient of photochemical excitation. Moreover, since cell walls doubtless act both as optical diffusers and as neutral density filters, it is possible that significant drops in intensity and hence in excitation occur between cells. This excitation gradient could be transduced into a gradient in action of membrane-situated carriers in many ways; two categories of possibilities in accord with the model for basipetal transport described in Section 3.1.3.1 will be discussed.

i) There could be integration within the plasmalemma of each cell. Such integration could occur in one of three ways: (α) Carriers might always be present in the lateral walls, the light gradient across a cell causing inhibition of carrier activity predominantly on the brighter side. (β) Light might enhance the total number or the total activity of carriers, with the gradient causing displacement of carriers toward the darker side. (γ) Light might simply cause displacement of carriers toward the darker side. Simple pictorial representations of these three versions of Model MI/i are shown in the upper left corner of Fig. 14.

ii) The amount of activity of carrier per unit area of lateral membrane could be directly related to local photochemical excitation. Like category (i), category (ii) can be formally subdivided into inhibitory and excitatory versions: (α) Carriers might be present at all times, with local light intensity determining the extent of their inhibition. (β) Carrier activity might be enhanced by light, with light intensity determining the extent of their activation. In either case, according to the basic transport model, the small gradients to be expected across the cells would have a large effect in polarizing the tissue.

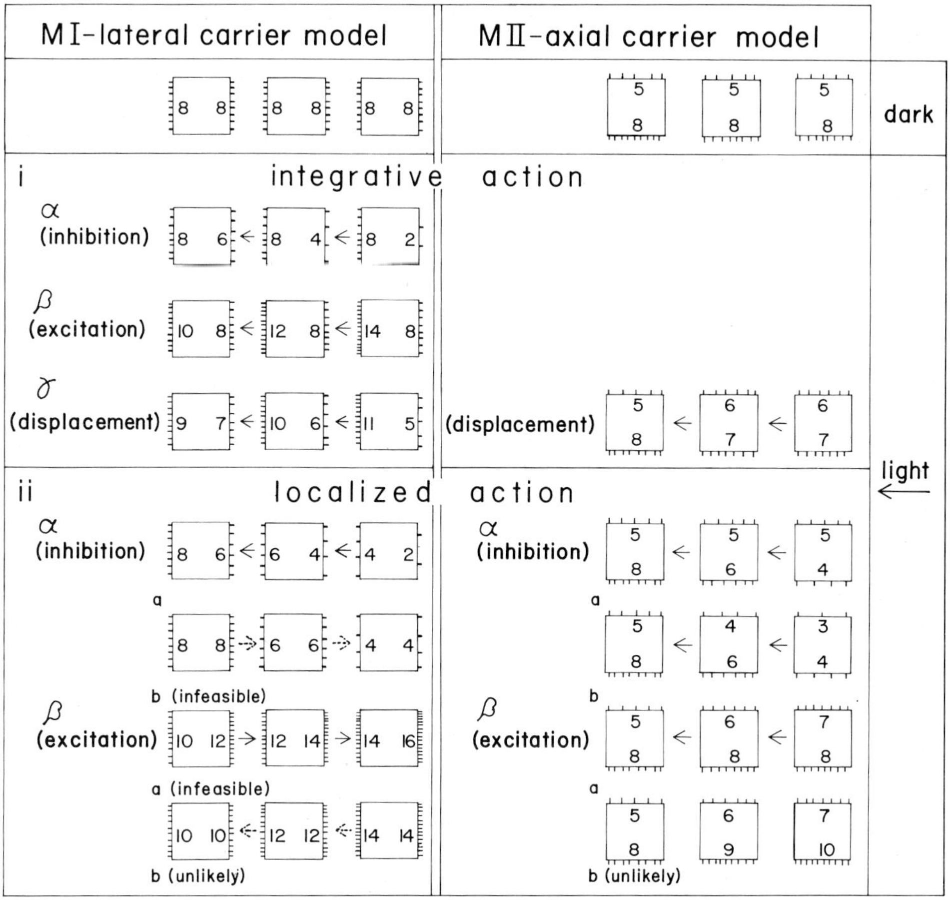

Fig. 14. Models for lateral movement of auxin in phototropism, based on current transport theory. All models assume that the gradient of light exerts its control at the cellular level, rather than by integration across the tissue. Each *row of three squares* represents three cells in a tissue; spikes on the sides of the squares represent IAA^- carriers (or carrier capability) and *numbers inside the squares* redundantly indicate the number of carriers (or the extent of carrier capability). The carrier system is considered responsible for movement of auxin (as IAA^-) out of cells. (Whether or not this facilitated diffusion is energized is immaterial to the models.) It is assumed that movement into cells is largely by diffusion of IAA, and that permeability to IAA is uniformly high. *Arrows between squares* indicate direction of net lateral transport. The numbers of depicted carriers are intended only as an aid in visualizing transport trends and are not quantitative predictions of carrier density. In particular, the actual light gradient across a row of three cells must be very small, but dramatic shifts in carrier abundance or capability are indicated for emphasis. For excitatory models, there need not be active lateral carriers in the unstimulated, dark state, but for illustration a number greater than zero has been assigned. Additional variations on the major modeling themes as well as combinations of mechanisms could also be constructed

Figure 14 illustrates that these two schemes MI/ii/a and b are diametrically opposed. However, the direction of net transport in either scheme would depend on whether fall-off of light occurred mainly across the protoplasm or the cell walls. If there is considerably greater intracellular extinction, net transport would be from the brightly lit toward the dimly lit cells in scheme a (see Fig. 14, case MI/ii/α-a), but toward the brightly lit cells in scheme β (see case MI/ii/β-a); therefore scheme β-a seems unrealistic and can be discarded. If most of the absorption occurs in the cell walls, the direction of net transport would be reversed in schemes α and β (see Fig. 14, MI/ii/α-b and β-b). However, net transport would probably be established only with considerable inefficiency so these two schemes also seem unlikely, and of course α-b operates in the wrong direction. Thus scheme MI/ii/α-a, in which there is nondirected, omnilateral export in the dark and differential inhibition in unilaterally illuminated plants, is the only version of Model MI/ii worth further consideration.

Schemes MI/i/α and MI/ii/α depend on a net decrease in the lateral export of auxin from cells; in both, therefore, phototropic stimulation would cause a transient increase in the amount of auxin inside the cells and hence available for basipetal transport. Schemes MI/i/β and MI/ii/β, on the other hand, both depend on a diversion of auxin into the lateral transport system; that is, an increase in lateral exchange. Phototropic stimulation would thus tend to cause a transient decrease in auxin available for basipetal transport. In every case the transient shift in the amount of auxin undergoing basipetal transport would be small, and as the system soon approached a steady state basipetal transport would return to normal. This assumes, of course, that the basipetal transport system is well below its level of saturation, as is probably the usual case in intact plants (cf. GILLESPIE and THIMANN 1963).

As the basipetal system of carriers approaches saturation with auxin, predicting the transient effects of phototropically active light becomes impracticable. On the one hand, transients should diminish. On the other hand, the tendency of an induced lateral transport system to transfer auxin to the darker side of the stem or coleoptile, perhaps increasing the concentration there beyond the saturation level for basipetal transport, might play a dominant role: if so, there would be a small, long-term reduction in net basipetal transport.

MI also predicts that transients in the rate of basipetal export would occur at the onset of omnilateral illumination. No net asymmetry of lateral auxin transport would be induced, but according to the localized action schemes carriers would be affected even in perfectly uniform light, whereas the integrative action mechanism would operate due to the small opposing gradients of light from the outside to the inside of the plant.

Model MII, in which the effectiveness of basipetal transport is inhibited preferentially on the brighter side of the stem or coleoptile, need not invoke gradients within single cells; rather it is enough that inhibition be graded between cells. Like MI, MII can be subdivided into several categories (right side of Fig. 14); but since these yield overlapping predictions, this model will be discussed only in its general form.

According to MI, unilateral light would lead to a substantial pile-up of auxin on the brighter side of the tip of the stem or coleoptile. Hence (unless the inhibition all took place *extremely* close to the apex) the tip should curve away from the light. Meanwhile, the supply of auxin initially moving to the tissue below would be diminishing more on the brighter than darker side, and this differential loss would tend to cause subapical curving toward the light. But because photoreception occurs throughout the tip in C-type phototropism and under some circumstances in A-type as well, the consequent distributed increase in resistance would result in an initial auxin pile-up on the brighter side, tending to cause curving away from the light. The overall pattern of early curvature might thus be S-shaped or even completely negative. With time, the gradient in the tip would lead to a steady spill-over of auxin to the darker side, perhaps resulting in strong curvature toward the light. The important predictions of MII about S-shaped and negative curvatures are not borne out by experiment: such bending has not been reported except in or near the restricted negative zone of the dose-response map of oat coleoptiles (see photographs in ZIMMERMAN and BRIGGS 1963a).

Model MII predicts a transient dip in net export of auxin into the base of the stem or coleoptile. The extent of the recovery cannot be predicted without more information on the parameters of transport than is available. However, complete recovery seems unlikely, even though the carriers of the inhibited axial transport system would operate with increasing frequency as auxin accumulated. Nevertheless it is not impossible that parameters in the plant are so nicely mated that the return would be complete for practical purposes.

3.1.3.3 Experiments on Net Basipetal Transport

Although Models MI and MII overlap in their predictions about basipetal transport, a substantial number of recent publications (restricted for the most part to coleoptiles) have aimed at discriminating between them on the basis of the presence or absence of light-induced changes in longitudinal auxin movement. The data are ambiguous but deserve examination.

Large long-term shifts in the amount of auxin undergoing basipetal transport in coleoptiles can be ruled out by the aggregate of data discussed in Sections 3.1.1 and 3.1.2; experiments of BRIGGS (1960) and of PICKARD and THIMANN (1964) in which amounts of auxin indistinguishable from each other were collected from illuminated and control coleoptiles over a 3-h period are particularly cogent. However, recognizing that a transient change might result in only a trivial difference in yield during such a long period, THORNTON and THIMANN (1967) followed the time-course of basipetal transport of ^{14}C-IAA from subapical sections by removing and replacing collector blocks every 5 min. Omnilateral type C stimulation with B lasted 15 min. There was indeed a small dip in output of auxin from the bases of the coleoptile sections, but it persisted less than 30 min (Fig. 15A). Raising the IAA concentration to nonphysiological levels resulted in a decrease of much longer duration (Fig. 15A). It was concluded that these results did not permit a choice between MI and MII. In an experiment of this design, as in phototropism under natural circumstances, activation of a lateral transport system might not only temporarily divert auxin from the basipetally flowing stream (as the laterally polarized channels become filled), but also might bring about an inwardly directed movement of auxin from the epidermis and its underlying cells. In general, cells in the peripheral layer or layers of stems tend to be more responsive to auxin than those within, perhaps

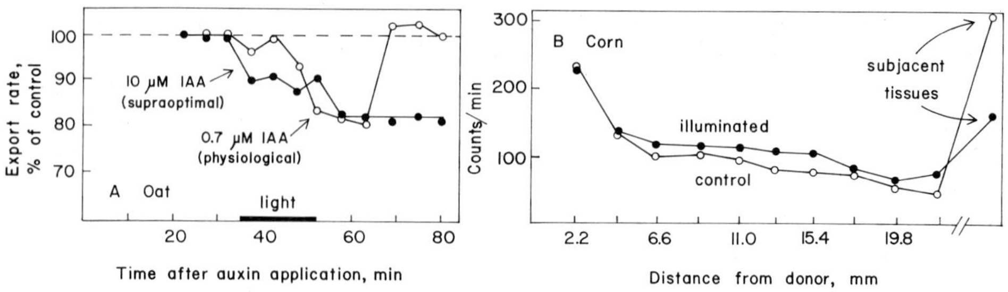

Fig. 15A, B. Two experiments on the influence of blue light on net basipetal transport. **A** (Data from THORNTON and THIMANN 1967). ^{14}C-IAA was supplied via agar donor blocks to the apical end of inverted 4-mm sections of oat coleoptiles. At the basal ends collector blocks were renewed every 5 min to assess ^{14}C-IAA movement. About 300 W m^{-2} of B was provided for 15 min from a zenith angle of 45°, controls were exposed only to a R work-light. One trance shows a transient decrease in transport when a physiological level of IAA is provided (from a donor block with 0.7 μM), while the other shows a more prolonged inhibition of transport when IAA is supraoptimal (donor at 10 μM). **B** (Data from NAQVI and GORDON 1967). Six corn seedlings were bilaterally illuminated with 1 klx "white" light; the apical mm of the coleoptile was removed 30 min later and replaced by an agar block containing 1.1 μM ^{14}C-IAA. After 90 min coleoptiles were sliced and corresponding segments pooled as were all tissues below 22 mm. Slightly more radioactivity seemed to appear in the upper 22 mm of illuminated than control seedlings, and less in tissues below 22 mm

tend to transport auxin more effectively, and tend to limit the growth of the stem as a whole (cf. Sect. 3.1.2.1; PICKARD Chap. 7, this Vol.). Should lateral transport be given a net inward direction by blue light, some auxin would be transferred to a portion of the organ in which basipetal transport might be less efficient.

NAQVI and GORDON (1967) also studied photoinhibition of basipetal transport. Intact corn seedlings were bilaterally illuminated with a type A dose of white light, their tips were removed immediately, and agar blocks containing ^{14}C-IAA were applied to the cut surfaces. After a transport period of 1.5 or 2 h, 10 sections of 2.2 mm length were excised from the apical end of each plant and these plus the "remainder of the seedling" were assayed for radioactivity: slightly more was retained in the remaining upper 22 mm of the coleoptile in the light than in the dark, with less moving into the subjacent tissue (Fig. 15B). The effect, not seen before 1.5 h, intensified between 1.5 and 2 h. The interpretation was that "the primary hormonal phenomenon in phototropism is a light-induced impairment in the basipetal transport of auxin."

There are at least five problems with these experiments. (1) The retention was delayed until 1.5 h, long after the curvature would normally begin to develop. (2) If the effective zone of inhibition is well below the tip, early curvature should show a distinctive S-shape (Sect. 3.1.3.2); unfortunately, there were no controls for the measurement of curvature. (3) BRIGGS et al. (1957) showed that, under similar experimental conditions, lateral movement of auxin in response to type A doses of white light occurred only in the very tip of the corn coleoptile. NAQVI et al. (1972) have counter-claimed that freshly decapitated corn coleoptiles are indeed responsive to type A doses of light (cf. Sect. 2.3.4) but presented only meager data. (4) ^{14}C radioactivity was not proved uniquely associated with IAA. (5) Except for the "remainder of the seedling", differences between experimental and control tissues are small; and for the "remainder" there is lack of information about the evidently large and variable amounts of coleoptile, mesocotyl, and root (and perhaps also seed?) included and about their quenching of the counting scintillant. Thus, these experiments yield no decisive results for phototropism.

SHEN-MILLER et al. (1969) studied basipetal transport following illumination of up to 15 s duration with white and monochromatic light. A dose-response curve for the effect of white light on basipetal transport during a subsequent 2-h period appears to show: (i) an impairment at low total dose, purported to correspond to type A response; (ii) a small enhancement at intermediate dose, purported to correspond to the negative response; and (ii) a large inhibition at high illuminance, purported to correspond to type C response. However, three problems beset the interpretation of the data. (1) The maximum duration of exposure was 15 s. The dose-response plot of Fig. 3, which accords well with earlier dose-response data, indicates that significant type-C response is quite unlikely for so brief an illumination. The apparent shifts in basipetal transport, then, do not relate to the usually observed pattern of phototropism. (2) There were only four coleoptiles per treatment, and "the individual datum points for the illuminated tissues do not differ significantly from the dark controls"; nor was it made clear at what level of significance the collective data differ from the controls. (3) ^{14}C-IAA was applied via 1.5% agar in plastic tubes slipped over the tip of the coleoptile. Thus, most of the auxin entered below the level at which lateral transport demonstrably occurred in experiments on type A tropism by PICKARD and THIMANN (1964).

Further experiments (SHEN-MILLER et al. 1969) compared phototropic curvature in unilateral light with basipetal transport under balanced illumination on the basis of both spectral response and dose-dependence at 455 nm B. Curvature and inhibition of transport did not correspond in a detailed way as dose was varied in the type A range; for example curvature was about 70% maximal before inhibition of transport became evident, and was maximal when transport was inhibited only 25%. Nevertheless, the inhibition of transport during a 1-h period and the appearance of curvature during a 1.5-h period following illumination peaked at roughly the same dose. Similarly, the spectral response plots both peaked in the blue, even though curvature showed a broad, flat peak (dissimilar to that of other workers – cf. DENNISON 1979) and inhibition of transport showed two major peaks rather than one. Again, however, the relevance of the transport data of SHEN-MILLER et al. for phototropism must be questioned. In these experiments, ^{14}C-IAA dissolved in a solution of detergent (0.1% Tween-20) was applied to the tips with a camel-hair brush. Although the amount of carrier IAA per coleoptile averaged 2.6 ng, the volume and concentration are not clearly specified. If the concentration used in this method was the same as that in the tubing reservoir method (see above), the volume was about 500 nl. This must be compared with a similar application of 100–200 nl in earlier experiments (SHEN-MILLER and GORDON 1966) in which

^{14}C-IAA was symmetrically distributed in apical coleoptile sections after type A irradiation and emerged only slightly asymmetrically into basally positioned agar receiver blocks. PICKARD (unpublished observations, 1966), using a 100 nl drop of inert dye in lieu of the ^{14}C-IAA solution, observed with a microscope that the drop remained localized on the crest-like tip of the coleoptile for a moment or two, bulging at its lower rim due to the pull of gravity. It then disappeared rapidly, as if by evaporation. That it did not immediately evaporate became evident when, repeatedly, a ring of crystallized dye was discovered about 1.5 mm below the tip. Evidently, the drop flowed downward from the apex and under the influence of both gravity and surface tension formed a symmetric ring of liquid which ultimately dried, leaving the conspicuous dye band. This experiment, coupled with the weak asymmetries found by SHEN-MILLER and GORDON (1966), suggests that most of the IAA is entering the coleoptile below the zone where type A asymmetry has been shown to develop (cf. Sect. 3.1).

In sum, even when all the experiments are taken together, it is still impossible to be certain what factors contribute to shifts in basipetal auxin transport during or after illumination or just how the small shifts observed when IAA is applied at physiological levels relate to phototropism.

How differential inhibition of basipetal transport might be brought about has been the subject of another set of papers. An ingenious proposal was set forth by HAGER and SCHMIDT (1968a, b): a tiny fraction of the plant's auxin was postulated to photo-oxidize to 3-methylene-oxindole, a product which impairs polar auxin transport. However, in a search for decarboxylation of 1-^{14}C-IAA applied to oat coleoptile tips, MENSCHICK et al. (1977) found neither a shift in decarboxylation rate upon irradiation with light nor a significant difference between irradiated assemblies and controls. Thus, earlier reports of indistinguishable auxin recoveries in light and dark (Sects. 3.1.1, 3.1.2) were reinforced.

NAQVI (1972a) and NAQVI and ENGVILD (1974) also proposed that irradiation leads to the production of an inhibitor of basipolar transport. They speculated that the inhibitor is ABA, and that it arises from the photolysis of the carotenoid violaxanthin. NAQVI and ENGVILD further supplied ABA and ^{14}C-IAA to decapitated coleoptiles, reporting that the spatial distribution of ABA inhibition of IAA transport resembles that due to light (NAQVI and GORDON 1967; see above). However, measurements of violaxanthin, related compounds, and total carotenoid (VALADON and MUMMERY 1971) showed no correlation with light treatment.

In order to dispel any doubt that auxin will move laterally along a concentration gradient within a shoot, some workers have established such gradients in the absence of phototropic stimulation and measured subsequent distribution of IAA. DELA FUENTE and LEOPOLD (1968) simulated putative asymmetric phototropic resistance to basipetal auxin movement by laterally incising sunflower hypocotyls, finding both IAA accumulation apical to the cuts and net lateral IAA movement. NAQVI (1972) paired donor blocks containing either 1.1 or 1.7 μM IAA on the right and left halves of the split apical surfaces of corn coleoptile sections; in each case, IAA in one or both members of the pair was radioactively labeled (Fig. 16). After a 90-min transport period, distribution of ^{14}C-IAA which had passed into and through the tissue was assessed: a

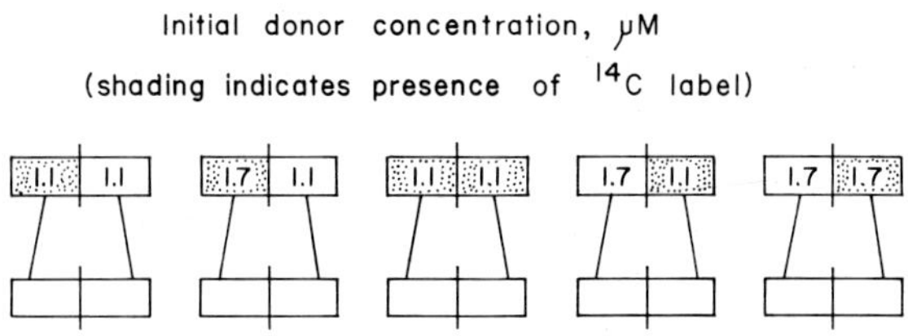

Final distribution of label, $\frac{cpm_{right}}{cpm_{left} + cpm_{right}} \times 100$

Tissue	11 %	18 %	50 %	94 %	88 %
Receivers	15 %	21 %	50 %	91 %	85 %

Fig. 16. A demonstration by NAQVI (1972b) that auxin can move laterally down a concentration gradient in unirradiated 12-mm coleoptile sections. Gradients were established by placing a pair of agar blocks, each provided with a different concentration of labeled or unlabeled auxin, at the apical cut surface; note the two controls with paired blocks of matched concentrations. After 90 min, radioactivity was assessed in corresponding halves of the tissue sections and corresponding basal agar collecting blocks; ratios are presented for each assembly

greater percentage moved across the midline if it came from a donor of higher concentration. Opposite results in either experiment would have invalidated the hypothesis that polar transport is differentially impaired in phototropism. However, the suggestions that these results support MII are not justified: they are as compatible with the concept of laterally polarized carriers as with differentially inhibited basipolar carriers.

Finally, some quite surprising results of GARDNER et al. (1974) on the lateral movement of ^{3}H-IAA in coleoptiles of both oat and corn must be considered. They made three comparisons: (1) dark-grown vs. R-pretreated coleoptiles; (2) excised vs. in situ coleoptiles; and (3) coleoptiles exposed to type A vs. type C irradiations. According to their evaluation, "... different species may use different mechanisms under different conditions". Their findings are summarized in Table 3; there are nearly as many kinds of results as there are combinations of variables. It should be specially noted that for type A responses lateral movement of label was found in all coleoptiles handled only in the dark or in green light, regardless of whether or not the coleoptiles were intact or excised. In contrast, oat (but not corn) coleoptiles pretreated for 2 h with R failed to show lateral movement of label, in spite of the fact that they curved. For C-type responses, oats did not exhibit lateral movement of label under any of the conditions tested, whereas in intact corn seedlings lateral movement was evident in both R-treated coleoptiles and those maintained in the dark. Strong inhibition of basipetal transport was observed for type C tropism of oats in the dark, but was not seen under these circumstances for corn.

The possible origin of these differences must be examined. Firstly, neither MI nor MII yield simple predictions for the amount of basipolar movement during lateral deflection. By either model, variation of conditions could certainly influence the strength of lateral movement, as well as the degree of inhibition of basipolar transport. This is highlighted by the demonstration by THORNTON and THIMANN (see above) that at physiological levels of IAA equilateral type C irradiation produces a small transient decrease in basipetal efflux, whereas at higher concentration the depression of net transport is more prolonged.

Secondly, the experimental method of GARDNER et al. had unusually high potential for variability. ^{3}H-IAA was introduced just below the apex by pricking the surface of the coleoptile with a solution-filled pipette of approximately 0.1 mm diameter. A coleoptile was next irradiated either on its pricked or unpricked side. After a 1- or 2-h transport period, each coleoptile was divided into 5-mm segments and these were halved for assessment of radioactivity (sometimes receiver blocks were also measured). The average percentage of radioactivity moved across the midline in coleoptiles irradiated on the unpricked side was subtracted from that in oppositely irradiated coleoptiles, with the difference defined as "net lateral movement".

Because the tip of the coleoptile is small and irregularly shaped, pricking with a pipette will damage a significant, variable number of cells. Even cells some distance from those actually broken open may be expected to show signs of injury (e.g., MOLLENHAUER et al. 1960, MACNICOL 1976, VAN SAMBEEK and PICKARD 1976a, b). The problem might be particularly severe with coleoptiles of oat, which are smaller than those of corn; notably, four out of five failures to observe lateral movement were with oats. Furthermore, pricks which damage cells too far below the apex or permit downward drainage of solution might decrease net apparent light-induced movement of label across the midline if light and auxin are supplied from the same side, or increase it if light and auxin are supplied from opposite sides. Both effects would minimize "net lateral movement". Thus GARDNER et al. are unjustified in espousing multiple mediational mechanisms. The evidence favoring lateral movement of auxin in coleoptiles (Sect. 3.1.1) is powerful and a convincing case for exceptional behavior remains to be made.

Although the results described in this section do not come close to pinpointing a mechanism for phototropically induced lateral movement of auxin, one tends to narrow the possibilities: the early, small, transient decrease in basipetal transport observed by THORNTON and THIMANN (1967) is compatible with only half of the feasible categories of models MI and MII, as indicated in Table 4.

The transient decrease in basipetal transport can be predicted with all versions of MII, with inhibition deepening markedly as dose increases within the A and C ranges. However, because some decrease in transport capability would be expected on the darker as well as on the brighter side, recovery from the initial transient decrease should be incomplete. Insofar as transport appeared to recover completely when IAA was supplied at a physiological level in the experiment of THORNTON and THIMANN, the data do not accord well with MII. Further, while it is easy to propose molecular mechanisms for MII, it is hard to see how they could operate over the full range of light intensity or of light dose to which plants are observed to respond. Also, the S-shaped curvatures predicted by MII (at least for the C-type response) are not observed. Thus, the totality of evidence does not offer strong support for MII at this time.

Of the several versions of MI, MI/i/β remains likely. In this scheme, light polarizes the cell membrane via a transcellular gradient of excited pigment, and the polarization leads in turn to activation of carriers on the darker side. How such a polarization might occur is obscure; the concept at present is physiological rather than molecular. A favorable feature of MI/i/β is that it could operate over large ranges of light intensity and dose, in spite of the

Table 4. Predictions of feasible categories of models described in Section 3.1.3.2

Category of model	Predicted shift in basipolar transport (at physiological auxin levels)	Compatible with transient decrease observed by THORNTON and THIMANN (1967)?
MI (lateral carrier)		
i (integrative action)		
α	Transient increase	No
β	Transient decrease	Yes
γ	Little change	No
ii (localized action)		
α–a	Transient increase	No
MII (basipetal carrier)		
i (integrative action)	Transient decrease with return to a level somewhat below starting value	Possibly
ii (localized action)	Transient decrease with return to a level somewhat below starting value	Possibly

almost trivial gradients that can be imposed across the individual cells or even, in many plants, across the tissue as a whole.

Although the data of NAQVI and GORDON (1967) for the type A response of corn appear unrelated to phototropism (see above), if applicable their finding of a late decline in net basipetal transport would eliminate both MI and MII.

Thus, while the predictions of the several versions of MI and MII listed in Table 4 have some overlap, these versions do not as a group predict all possible results and the overlap of individual versions is limited enough that further work might permit convincing discrimination between them.

In particular, it appears that the work of THORNTON and THIMANN (1967) could be extended with profit. Use of sections from larger, sturdier coleoptiles of corn instead of oat should reduce errors, and selection of appropriate type C doses on the basis of an extensive dose-response map should optimize the response. Relatively brief irradiation (which should cause a transient depression of auxin export followed by a considerable return) could be compared with continuous irradiation (which should cause a depression with relatively little return). Also, unilateral and omnilateral exposures could be compared.

However, sophisticated measurements of net basipetal transport may do little more than provide further heuristic weighting of the relative likelihood of MI or MII. Also, although MI and MII appear to have general validity, they have been developed consistent with the current model of basipolar transport. Ultimately, a different experimental approach would seem to be required, such as molecular description of the transport mechanism.

An encouraging beginning for such a study has been made by JACOBS and GILBERT (1983), who with fluorescent monoclonal antibodies have visualized basipolar auxin carriers in the vascular parenchyma of pea stems (Sect. 3.1.3.3). In principle, similar methods could be applied to phototropic auxin carriers; and once the carriers can be tagged their isolation and hence characterization might be possible.

3.2 Gibberellins

Although control of elongation by auxin has been demonstrated for a wide variety of species, it has long been known that GA's share the control and sometimes dominate it. Even in coleoptiles, control by GA is more conspicuous than control by IAA at an early stage of development (e.g., BOOTHBY and WRIGHT 1962). In shoots for which addition of IAA leads to little or no increase in growth (e.g., SILK and JONES 1975, STUART 1976), it may be questioned whether lateral auxin transport could play a decisive role in either geo- or phototropism.

However, until recently there were very few efforts to seek asymmetry of GA's in phototropically stimulated shoots because it had not proved possible to produce curvature by applying GA_3 to one side of a stem (cf. MEUDT and BENNETT 1978). Two experiments of LIBBERT and GERDES (1964) with oat coleoptiles were particularly compelling. Firstly, not only did they show that laterally applied GA_3 could not produce curvature in unirradiated plants, but also they showed it did enhance curvature caused by asymmetric IAA (Fig. 17A). Secondly, they showed that GA_3 alone was without phototropic effect: coleoptiles were decapitated twice in order to reduce the level of endogenous auxin, then

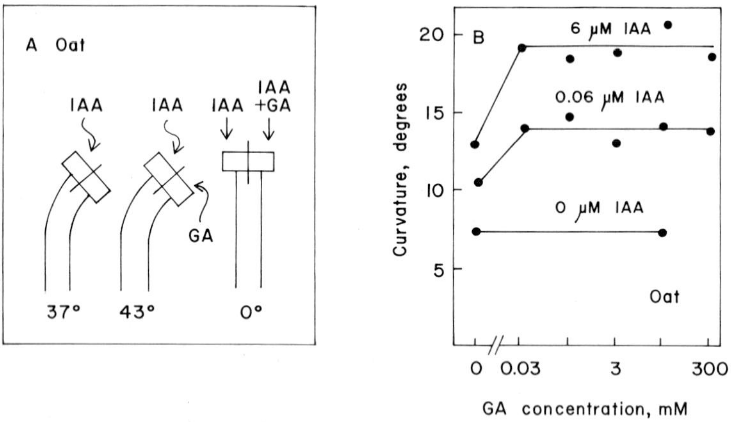

Fig. 17. A demonstration by LIBBERT and GERDES (1964) that GA_3 enhances curvature caused by phototropic (**B**) or artificially created (**A**) IAA asymmetry in oat coleoptiles, yet is incapable of causing curvature when supplied asymmetrically itself (**A**). In **A** an agar block containing 3 μM GA enhanced curvature away from an opposite block containing 6 μM IAA, but one-sided application of GA caused no curvature even in the presence of uniformly applied IAA. In **B** effectiveness of a 160-lx, 2-h illumination was enhanced by added GA only in the presence of added IAA

subjected to type C irradiation. Symmetric addition of GA_3 did not increase curvature, whereas 6 μm IAA doubled it. Addition of 30 μM GA along with the IAA, however, tripled curvature (Fig. 17B).

Nevertheless, PHILLIPS found a dramatic difference in the levels of GA-like activity diffusing from the lighted and shaded sides of phototropically stimulated sunflower shoot tips (1972b) and from the upper and lower sides of geotropically stimulated tips (1972a). The respective ratios were 1:8 and 1:10. Moreover, the total yield of activity was three times higher for geotropically stimulated shoot tips than for controls. Similar results were found for tips of geotropically stimulated corn coleoptiles, in which auxin asymmetry could also be confirmed (RAILTON and PHILLIPS 1973).

Concurrent experiments by SCOTT and MOST (1972) seemed to bear on PHILLIPS' (1972a, b, RAILTON and PHILLIPS 1973) findings: the longitudinal movement of $^3H\text{-}GA_1$ in sugar cane stems is weak and nonpolar unless IAA is supplied simultaneously. In the second internode, added IAA resulted in a three- to fourfold increase in basipetal movement of $^3H\text{-}GA_1$ while acropetal movement remained the same; in the basal portion of the third internode there were increases of 15-fold and three- to fourfold, respectively. Thus, during tropism, an increase of auxin in the lower or shadowed side of the stem or coleoptile might stimulate basipetal GA movement in that region.

Even between two young internodes of the same plant, SCOTT and MOST found a large difference in the stimulation of basipetal $^3H\text{-}GA_1$ movement by IAA: thus it might be expected that the effect would vary greatly between organs and between species, and perhaps even in the same tissue under different experimental conditions.

What effect such a putative tropistic increase in basipetal GA movement might have on the concentration of GA or on elongation in the lower or shadowed side cannot be known without more direct measurement. Also, it is not apparent whether lateral auxin gradients can enhance lateral movement of GA. There is too little information to be sure whether the observed increase in basipetal movement reflects in vivo translocation processes or merely a facilitation of the release of GA into agar collection blocks. Finally, it is possible that the asymmetric tropistic increases in GA activity resulted in part or in whole from new synthesis or metabolic transformation.

Examination of $^{14}C\text{-}GA_3$ distribution has been made for phototropically responding etiolated peas (KANG and BURG 1974). While application of GA_3 to excised stems conspicuously enhanced curvature in response to a 4-h exposure to blue light, no asymmetry of ^{14}C was detected in the tissue. On the other hand, lateral movement of ^3H-IAA was clearly evident. These important experiments reinforce and extend the earlier ones on oats (LIBBERT and GERDES see above). In both, added GA appears to enhance auxin-induced asymmetry of elongation. Similar enhancement of straight growth is known in a variety of systems (KATSUMI and KAZAMA 1978).

To date, the few experiments on the participation of GA's in phototropism have been designed to elucidate only one or two facets of their movement and metabolism in a particular situation. There is a strong possibility that, at least in some tissues, phototropic auxin redistribution leads to polarized GA movement or differential GA metabolism, which might themselves contribute to curvature. The role of GA's may be critical when auxin redistribution occurs in tissue for which it is not growth-limiting. Thus, a much more systematic study is required: close comparisons should be made between stimulated and unstimulated plants of several species, and should include assay of endogenous activity, identification of auxin and GA's by definitive means, and studies of the fates of radioactively labeled appropriate GA's and IAA, applied together as well as separately.

3.3 Calcium and Other Agents and Effects

Auxin and GA's are not the only substances which may become asymmetrically distributed during tropistic response, and many things can limit the rate of differential growth. Better knowledge of such agents and their functions should lead to a more detailed understanding of the time-courses and magnitudes of curvature, as well as explain many anomalous tropistic situations.

In particular, phototropically established gradients of Ca^{2+} may well be as important as those of auxin in mediating curvature. GOSWAMI and AUDUS (1976) extended earlier work on lateral gradients of ions during tropism, showing that the type C phototropic response of corn coleoptiles is accompanied by a shift of K and P toward the shaded side and a reverse shift of Ca. Similar differential distribution occurred during the geotropic response of sunflower hypocotyls. Comparisons of time-courses for the development of ion asymmetries and differential growth during natural stimulation, during stimulation under inhibitory conditions, and following asymmetric application of auxin indicated that the ion asymmetries were a consequence of auxin asymmetry.

Evidently there have been no subsequent studies on the development and significance of these phototropic ion asymmetries, but for the geotropism of coleoptiles (Chap. 7, this Vol.) there is considerable evidence that Ca asymmetry develops early and plays a key role in controlling differential growth. Closer evaluation of the phototropic asymmetry is essential.

Moreover, there is considerable evidence and extensive speculation that the intracellular distribution of Ca^{2+} plays a role analogous to that of a hormone in many sensory systems, particularly those mediated by the R-receptor phytochrome (ROUX 1984) and also those mediated by B-receptors (SCHMIDT and ECKERT 1976). Thus, another important priority for future research is investigation of whether intracellular partitioning of Ca^{2+} plays a role in phototropic induction or in the adaptive transition from the A to the C type response. But currently, there are few if any data on possible changes in the intracellular distribution of Ca^{2+} during the phototropism of higher plants; this and the close association of any such postulated effects with reception exclude the subject from the defined scope (Sect. 3) of the present review.

Because IAA stimulates secretion of protons into the cell wall (CLELAND and RAYLE 1978, TERRY and JONES 1981), an asymmetry of apoplastic acid was sought and found in phototropically curving *Zea* coleoptiles and *Helianthus* hypocotyls by MULKEY et al. (1981). As they discuss, this asymmetry must be an important mediational link in the development of the IAA-induced curvature.

Additionally, the pH gradient doubtless influences the capability for curvature in several ways. For example, because low external pH promotes the uptake of IAA across the plasmalemma into the cytoplasm, it must be expected that IAA will tend to enhance its own uptake and transport; this has indeed been observed (POOLE and THIMANN 1964, GOLDSMITH 1977). Similarly, other agents which affect intra-or extracellular pH may influence transport and hence tropism. A possible example of rate limitation by acid is the finding (GANOT and REINHOLD 1970) that symmetrically applied acid restored the ability of aged sections of sunflower hypocotyl to carry out geotropism better than did auxin. Normally, it may be presumed, the extracellular space is acidic enough to permit transport of auxin. During aging, the levels of auxin and extracellular acid might fall, and added acid could perhaps prime the transport of the remaining auxin, which in turn might permit development of secondary asymmetries (but cf. counter-arguments of REINHOLD 1978, GANOT and REINHOLD 1970, REINHOLD and GANOT 1972).

Differential acidification of the cell walls may be expected to bring about redistribution of a variety of compounds. Among them is ABA, although no phototropic asymmetry of this compound has yet been reported. Increased wall acidity on the lighted side would increase the amount of ABA partitioning into the cytoplasm there, creating an apoplastic concentration gradient across the plant and consequent net ABA movement out of the shaded side (cf. KAISER and HARTUNG 1981). To the extent that ABA inhibits elongation (Sect. 3.1.2.1), such an effect would tend to enhance phototropic curvature. Additionally, ABA partitions into photosynthesizing chloroplasts because of a pH gradient, and therefore the cytoplasmic level of ABA must decrease preferentially on the illuminated side of a green stem during C-type stimulation (the "Gimmler effect", Sect. 3.1.2.1). Such a decrease might tend to counteract positive phototropism, but it would also lead to net diffusion of ABA from the shadowed to the lighted side. Bulk extraction of ABA (and ABA metabolites) from halved stems could thus indicate asymmetry different from that prevailing in the cytoplasm.

Taking into account the widespread occurrence of Mitchell-type transport mechanisms (e.g., POOLE 1978, KINRAIDE and ETHERTON 1980), actively as well as passively maintained distributions of numerous dissociable hormones and metabolites are probably influenced by pH gradients between the cytoplasm, its organelles, and its extracellular surround. Moreover, a lateral gradient in metabolism and growth would itself effect lateral gradients in many molecules. Interpretation of secondary asymmetries, even though accurately determined, may prove difficult. It will doubtless be a long time before we can account in detail for the time-courses of tropistic curvatures.

Xanthoxin (a compound related to ABA) occurs asymmetrically in sunflower stems responding to type C stimulation: FRANSSEN and BRUINSMA (1981) reported about twice as much in the lighted as in the shaded halves. Applying the compound to the plant, they reported inhibition of growth. Moreover, they grew seedlings under different conditions and reported that xanthoxin content was correlated with phototropic sensitivity. Since they could find no lateral gradient of auxin (Sect. 3.1.2.1), they concluded that xanthoxin, rather than

IAA and GAA, regulates sunflower phototropism, but they did not determine whether its asymmetry was primary or secondary.

As might be expected, endogenous and exogenous ethylene can influence phototropic curvature. Working with peas, KANG and BURG (1974) both applied ethylene and removed the endogenous gas by hypobaric treatment. In this case removal had no effect. Added ethylene dramatically reduced phototropic curvature, but because it did not alter the redistribution of ^{3}H-IAA they concluded that it played no role in the phototropic mechanism per se, merely diminishing the ability of the cells to elongate. With common beans, JAFFE and BIRO (1979) also observed inhibition of phototropic curvature by elevated ethylene levels. BRENNAN et al. (1976) found that phototropic curvature of mung bean was decreased by hypobaric treatment; while they suggested that further studies might implicate ethylene in the phototropic mechanism they adduced no evidence for a direct effect. HUMPHREY (1980) found that exogenous ethylene slowed the phototropic response of oat coleoptiles but accelerated that of rice coleoptiles, in parallel with its effects on elongation. His conclusion that ethylene acts indirectly accords with the view of KANG and BURG.

All in all, while a tremendous number of agents and processes may contribute to differential growth and may, under some circumstances, limit it, postulation of very complicated primary mechanisms or of a large variety of primary mechanisms seems unnecessary at this time.

Just as growing knowledge about the complex participation of a wealth of factors in phototropism has not yet required abandoning the simple idea of primary control by asymmetric auxin, the resuscitation of the old hypothesis that phototropism is primarily caused by effects of B unrelated to lateral migration of auxin (GORDON and DOBRA 1972, ELLIOT and SHEN-MILLER 1976, TOWERS and ABEYSEKERA 1984) is not warranted at this time. GALSTON (1959) and THIMANN (1967, 1977) are among those who have earlier discussed cogent arguments why phototropism is unlikely to be brought about by "direct" effects of B on growth.[3] On the other hand, it is probable that subtle secondary influences of B on the magnitude of phototropic curvature will be further documented. B influences a large variety of cellular processes, such as plastid redistribution and cytoplasmic streaming (e.g., GALSTON 1959, BLATT and BRIGGS 1980a, b). It would be surprising indeed if B could not both induce lateral migration of auxin and influence the expression of growth-dependent curvature which normally follows.

Finally, FIRN and DIGBY (1980) and TREWAVAS (1981) propose that all evidence for the participation of hormones in phototropism is spurious. Because

3 COSGROVE (1981) has defended the inverse hypothesis that suppression of growth by B is unrelated to B-elicited changes in auxin transport associated with phototropism. His conclusion is appealing because B does influence many cellular processes, but his arguments are premature, in that they are based on comparisons of his careful dose-response studies of growth suppression in cucumber with scattered data for other plants from the phototropic literature. They ignore the possibility that B might cause a rapid shift of auxin from the epidermis, which tends to limit organ extension, to the inner tissues (Sect. 3.1.3.3). Considering the complexities of the kinetics and dose-response relations of both phototropic curvature and auxin translocation (Sects. 2.3, 3.1), realistic evaluation must await a more detailed and systematic comparison of effects of B on growth, phototropism, and auxin distribution within a single species.

their arguments deal more heavily with geotropism than phototropism (but see FIRN et al. 1983), discussion of them is relegated to the chapter on that topic.

3.4 Conclusion

The assemblage of the available facts into a unified picture should not blind us to the possibility that the depiction is suppressing some inadequately appreciated set of data, or lull us to neglect conflicting data that may appear in the future. The diversity of mechanisms which appear to coordinate phototropic and polarotropic responses in phanerogamic leaves, cryptogamic sporelings, and various organs of algae and fungi (e.g, DENNISON 1979) and the geotropism of various higher plant organs (REINHOLD 1978, PICKARD Chap. 7, this Vol.) remind us that a rich array of tropistic devices is not only formally conceivable but more importantly has been evolutionarily realized. To assert that all aerial axes of higher plants have identical primary phototropic mechanisms would be foolish, because relatively few species have been studied in much detail and because available data for certain species do not conclusively support the current theory. Moreover, it seems clear that secondary mediational processes will prove variable.

Notwithstanding, the concept that lateral movement of auxin is the primary mediational event in the phototropism of young shoots is well documented for some plants and has successfully systematized a great amount of experiment, observation, and theory to date: it promises to remain useful into the indefinite future.

Notes Added in Proof. A recent detailed assessment of growth distribution during type A phototropic curvature of *Zea* coleoptiles (IINO and BRIGGS 1984) deserves attention because it shows that phototropic curvature can be based on symmetric redistribution of growth from the lighted to the shaded side of the organ, beginning at the tip and migrating basipetally. Such balanced redistribution of growth is not a necessary consequence of the Went-Cholodny redistribution of auxin, which appears to initiate a variety of nonlinear processes contributing to the regulation of growth. Certainly, however, it would not be surprising if equal increases and decreases of growth occurred under some phototropic conditions in seedling shoots of some species. Because a number of authors have mistakenly doubted the Went-Cholodny theory on the grounds that phototropic response of many shoots is due to unequally opposed changes in growth rate on the lighted and shaded sides (e.g. FIRN and DIGBY, Sect. 3.3, and refs. cited by IINO and BRIGGS; also see Chap. 7, this Vol.), the work of IINO and BRIGGS may be expected to play a prominent role in abating controversy about the theory.

MCINTYRE and BROWNE (1984) have shown that asymmetry of photosynthesis by *Helianthus* cotyledons can lead under some circumstances to asymmetry of transpiration and hence of water potential in the hypocotyl. Doubtless, such asymmetry can lead to "phototropic" bending in the laboratory; but, as BRENNAN et al. have pointed out (Sect. 2.2), in the field the difference between illumination of paired cotyledons is not a good indicator of the direction of incident light.

Further, FISHER and WRIGHT, and LISTER and FISHER have suggested that suntracking by leaves of *Lavatera* can result from uneven distribution of photosynthesis across the blade with consequent gradients in turgor across the pulvinus. It is premature to compare their preliminary data with the experiments of Sect. 2.2, but their reports should stimulate interest in the possibly varied mechanisms by which leaves optimize their positions.

References

Abou-Mandour AA, Hartung W (1980) The effect of abscisic acid on growth and development of intact seedlings, root and callus cultures, and stem and root segments of *Phaseolus coccineus*. Z Pflanzenphysiol 100:25–33

Arisz WH (1914) Adjustment to light in oats. Proc K Ned Akad Wet Amsterdam, vol 16, part 2, pp 615–628

Asana RD (1938) On the relation between the distribution of auxin in the tip of the *Avena* coleoptile and the first negative phototropic curvature. Ann Bot NS 2:955–957

Aspinall D, Paleg LG, Addicott FT (1967) Abscisin II and some hormone-regulated plant responses. Aust J Biol Sci 20:869–882

Ball NG (1969) Tropic, nastic and tactic responses. In: Steward FC (ed) Plant physiology, a treatise, vol VA. Academic Press, London New York, pp 119–228

Bandurski RS, Schultz A, Cohen JD (1977) Photo-regulation of the ratio of ester to free indole-3-acetic acid. Biochem Biophys Res Commun 79:1219–1223

Blaauw OH, Blaauw-Jansen G (1964) The influence of red light on the phototropism of *Avena* coleoptiles. Acta Bot Neerl 13:541–552

Blaauw OH, Blaauw-Jansen G (1970a) The phototropic responses of *Avena* coleoptiles. Acta Bot Neerl 19:755–764

Blaauw OH, Blaauw-Jansen G (1970b) Third positive (C-type) phototropism in the *Avena* coleoptile. Acta Bot Neerl 19:764–776

Black M, Shuttleworth JE (1974) The role of the cotyledons in the photocontrol of hypocotyl extension in *Cucumis sativus* L. Planta 117:57–66

Blatt MR, Briggs WR (1980a) Blue-light-induced cortical fiber reticulation concomitant with chloroplast aggregation in the alga *Vaucheria sessilis*. Planta 147:355–362

Blatt MR, Briggs WR (1980b) Actin and cortical fiber reticulation in the siphonaceous alga *Vaucheria sessilis*. Planta 147:363–375

Boothby D, Wright STC (1962) Effects of kinetin and other plant growth regulators on starch degradation. Nature 196:389–390

Brauner L (1922) Lichtkrümmung und Lichtwachstumreaktion. Z Bot 14:497–547

Brauner L (1959) Phototropismus und Photonastie der Laubblätter. In: Ruhland W (ed) Encyclopedia of plant physiology, vol 17/1. Springer, Berlin Göttingen Heidelberg, pp 492–529

Brauner L, Hager A (1958) Versuche zur Analyse der geotropischen Perzeption. I. Planta 51:115–147

Brennan T, Gunckel JE, Frenckel C (1976) Stem sensitivity and ethylene involvement in phototropism of mung bean. Plant Physiol 57:286–289

Briggs WR (1960) Light dosage and phototropic response of corn and oat coleoptiles. Plant Physiol 35:951–962

Briggs WR (1963a) The phototropic response of higher plants. Annu Rev Plant Physiol 14:311–352

Briggs WR (1963b) Red light, auxin relationships, and the phototropic responses of corn and oat coleoptiles. Am J Bot 50:196–207

Briggs WR (1963c) Mediation of phototropic responses of corn coleoptiles by lateral transport of auxin. Plant Physiol 38:237–247

Briggs WR (1964) Phototropism in higher plants. In: Giese AC (ed) Photophysiology. Academic Press, London New York, pp 223–271

Briggs WR (1976) The nature of the blue light photoreceptor in higher plants and fungi. In: Smith H (ed) Light and development. Butterworths, London, pp 7–18

Briggs WR, Tocher RD, Wilson JF (1957) Phototropic auxin redistribution in corn coleoptiles. Science 126:210–212

Britz SJ, Schrott E, Widell S, Briggs WR (1979) Red light-induced reduction of a particle-associated b-type cytochrome from corn in the presence of methylene blue. Photochem Photobiol 29:359–365

Bruinsma J (1977) Hormonal regulation of phototropism in dicotyledonous seedlings. In: Pilet PE (ed) Plant growth regulation. Springer, Berlin Heidelberg New York, pp 218–225

Bruinsma J, Karssen CM, Benschop M, van Dort JB (1975) Hormonal regulation of phototropism in the light-grown sunflower seedling, *Helianthus annuus* L (1975): Immobility of endogenous indoleacetic acid and inhibition of hypocotyl growth by illuminated cotyledons. J Exp Bot 26:411–418

Bünning E, Reisener HJ, Weygand F, Simon H, Klebe JF (1956) Versuche mit radioaktiver Indolylessigsäure zur Prüfung der sogenannten Ablenkung des Wuchshormonstromes durch Licht. Z Naturforsch 11-B:363–364

Burg SP, Burg EA (1967) Lateral auxin transport in stems and roots. Plant Physiol 42:891–893

Cande WZ, Ray PM (1976) Nature of cell-to-cell transfer of auxin in polar transport. Planta 129:43–52

Cleland RE, Rayle DL (1978) Auxin, H^+-excretion, and cell elongation. Bot Mag Tokyo (Special Issue) 1:125–139

Cosgrove DJ (1981) Rapid suppression of growth by blue light. Plant Physiol 67:584–590

Curry GM (1969) Phototropism. In: Wilkins MB (ed) The physiology of plant growth and development. McGraw-Hill, London, pp 245–273

Davies PJ, Rubery PH (1978) Components of auxin transport in stem segments of *Pisum sativum* L. Planta 142:211–219

De Fabo E (1980) On the nature of the blue light photoreceptor: still an open question. In: Senger H (ed) The blue light syndrome. Springer, Berlin Heidelberg New York, pp 187–197

Dela Fuente RK, Leopold AC (1968) Lateral movement of auxin in phototropism. Plant Physiol 43:1031–1036

Dennison DS (1979) Phototropism. In: Haupt W, Feinleib ME (eds) Physiology of movements. Encyclopedia of plant physiology, new ser, vol 7. Springer, Berlin Heidelberg New York, pp 506–566

Diemer R (1961) Untersuchungen des phototropischen Induktionsvorganges an *Helianthus*-Keimlingen. Planta 57:111–137

Drake G, Carr DJ (1978) Plasmodesmata, tropisms, and auxin transport. J Exp Bot 29:1309–1318

Du Buy HG (1933) Über Wachstum und Phototropismus von *Avena sativa*. Rec Trav Bot Neerl 30:798–925

Du Buy HG, Nuernbergk E (1934) Phototropismus und Wachstum der Pflanzen. II. Ergeb Biol 10:207–322

Edwards KL, Goldsmith MHM (1980) pH-dependent accumulation of indoleacetic acid by corn coleoptile sections. Planta 147:457–466

Elliott WM, Shen-Miller J (1976) Similarity in dose responses, action spectra and red light responses between phototropism and photoinhibition of growth. Photochem Photobiol 23:195–199

Esau K (1965) Vascular differentiation in plants. Holt, Rinehart and Winston, New York

Everett ML (1974) Dose-response curves for radish seedling phototropism. Plant Physiol 54:222–225

Everett M, Thimann KV (1968) Second positive phototropism in the *Avena* coleoptile. Plant Physiol 43:1786–1792

Firn R, Digby J (1977) The role of the peripheral cell layers in the geotropic curvature of sunflower hypocotyls: a new model of shoot geotropism. Aust J Plant Physiol 4:337–347

Firn RD, Digby J (1980) The establishment of tropic curvatures in plants. Annu Rev Plant Physiol 31:131–148

Firn R, Digby J, Macleod K, Parsons A (1983) Phototropism: patterns of growth and gradients of light. What's New in Plant Physiol 14:29–32

Fisher FJF, Wright M (1984) The dependence of suntracking in *Lavatera cretica* L. upon carbon dioxide availability. (Abstr.) Annual Meeting, Canadian Soc Plant Physiologists, p 16

Franssen JM, Bruinsma J (1981) Relationships between xanthoxin, phototropism, and elongation growth in the sunflower seedling *Helianthus annuus* L. Planta 151:365–370

Galston AW (1959) Phototropism of stems, roots and coleoptiles. In: Ruhland W (ed) Encyclopedia of plant physiology, vol 17/1. Springer, Berlin Göttingen Heidelberg, pp 429–529
Galston AW (1974) Plant photobiology in the last half-century. Plant Physiol 54:427–436
Ganot D, Reinhold L (1970) The "acid growth effect" and geotropism. Planta 95:62–71
Gardner G, Shaw S, Wilkins MB (1974) IAA transport during the phototropic responses of intact *Zea* and *Avena* coleoptiles. Planta 121:237–251
Gillespie B, Thimann KV (1963) Transport and distribution of auxin during tropistic response. I. The lateral migration of auxin in geotropism. Plant Physiol 38:214–225
Goldsmith MHM (1968) The transport of auxin. Annu Rev Plant Physiol 19:347–360
Goldsmith MHM (1977) The polar transport of auxin. Annu Rev Plant Physiol 28:439–478
Goldsmith MHM, Goldsmith TH, Martin MH (1981) Mathematical analysis of the chemiosmotic polar diffusion of auxin through plant tissues. Proc Natl Acad Sci USA 78:976–980
Gordon SA, Dobra WA (1972) Elongation responses of the oat shoot to blue light, as measured by capacitance auxanometry. Plant Physiol 50:738–742
Gordon SA, Eib M (1956) Auxin transport in the phototropic response. Plant Physiol (Suppl) 31:xiv
Goswami KKA, Audus LJ (1976) Distribution of calcium, potassium, and phosphorus in *Helianthus annuus* hypocotyls and *Zea mays* coleoptiles in relation to tropic stimuli and curvatures. Ann Bot 40:49–64
Gressel J (1979) Blue light photoreception. Photochem Photobiol 30:749–754
Hager A, Schmidt R (1968a) Auxintransport und Phototropismus. I. Die lichtbedingte Bildung eines Hemmstoffes für den Transport von Wuchsstoffen in Koleoptilen. Planta 83:347–371
Hager A, Schmidt R (1968b) Auxintransport und Phototropismus. II. Der Hemmechanismus des aus IES gebildeten Photooxidationsproduktes 3-Methylenoxindol beim Transport von Wuchstoffen. Planta 83:372–386
Hartmann E, Schmid K (1980) Effects of UV and blue light on the biopotential changes in etiolated hypocotyl hooks of dwarf beans. In: Senger H (ed) The blue light syndrome. Springer, Berlin Heidelberg New York, pp 221–237
Haupt W (1970) Phototropismus. Fortschr Bot 32:168–175
Haupt W (1979) Bewegungen. Fortschr Bot 41:153–160
Heilman B, Hartung W, Gimmler H (1980) The distribution of abscisic acid between chloroplasts and cytoplasm of leaf cells and the permeability of the chloroplast envelope for abscisic acid. Z Pflanzenphysiol 97:67–78
Hertel R, Jesaitis AJ, Dohrmann U, Briggs WR (1980) In vitro binding of riboflavin to subcellular particles from maize coleoptiles and *Cucurbita* hypocotyls. Planta 147:312–319
Hertel R, Lomax TL, Briggs WR (1983) Auxin transport in membrane vesicles from *Cucurbita pepo* L. Planta 157:193–201
Humphrey VR (1980) Effect of ethylene on the phototropic response of *Avena* and *Oryza* coleoptiles. Naturwissenschaften 67:198
Iino M, Briggs WR (1984) Growth distribution during first positive phototropic curvature of maize coleoptiles. Plant Cell Environ 7:97–104
Iwami S, Masuda Y (1974) Geotropic response of cucumber hypocotyls. Plant Cell Physiol 5:121–129
Iwami S, Masuda Y (1976) Distribution of labeled auxin in geotropically stimulated stems of cucumber and pea. Plant Cell Physiol 17:227–237
Jacobs M, Gilbert SF (1983) Basal localization of presumptive auxin transport carrier in pea stem cells. Science 220:1297–1300
Jaffe MJ (1970) On heliotropism of tendrils of *Pisum sativum:* a response to infrared irradiation. Planta 92:146–151
Jaffe MJ, Biro R (1979) Thigmomorphogenesis: the effect of mechanical perturbation on the growth of plants, with special reference to anatomical changes, the role of ethylene, and interaction with other environmental stresses. In: Mussell H, Staples RC (eds) Stress physiology in crop plants. Wiley and Sons, New York, pp 25–69

Jesaitis AJ, Heners PR, Hertel R, Briggs WR (1977) Characterization of a membrane fraction containing B type cytochrome. Plant Physiol 59:941–947

Johnsson A (1971) Aspects on gravity-induced movements in plants. Q Rev Biophys 4:277–320

Kaiser WM, Hartung W (1981) Uptake and release of abscisic acid by isolated photoautotrophic mesophyll cells, depending on pH gradients. Plant Physiol 68:202–206

Kang BG, Burg SP (1974) Red light enhancement of the phototropic response of etiolated pea stems. Plant Physiol 53:445–448

Katsumi M, Kazama H (1978) Gibberellin control of cell elongation in cucumber hypocotyl sections. Bot Mag Tokyo (Special Issue) 1:141–158

Kinraide TB, Etherton B (1980) Electrical evidence for different mechanisms of uptake for basic, neutral, and acidic amino acids in oat coleoptiles. Plant Physiol 65:1085–1089

Lam S-L, Leopold AC (1966) Role of leaves in phototropism. Plant Physiol 41:847–851

Leong T-Y, Briggs WR (1981) Partial purification and characterization of a blue light-sensitive cytochrome-flavin complex from corn membranes. Plant Physiol 67:1042–1046

Leong TY, Vierstra R, Briggs W (1981) Blue light–sensitive cytochrome–flavin complex form corn coleoptiles. Further characterization. Photochem Photobiol 34:697–703

Leopold AC, Hall OF (1966) A mathematical model of polar auxin transport in plants. Plant Physiol 41:1476–1480

Libbert E, Gerdes I (1964) Können Gibberelline an tropistischen Krümmungen beteiligt sein? Planta 61:245–258

Lister GA, Fisher FJF (1984) The action-spectrum of suntracking in *Lavatera cretica* L. (Abstr) Annual Meeting, Canadian Soc Plant Physiologists, 1984, p 16

Löser G, Schäfer E (1980) Phototropism in *Phycomyces*: a photochromic sensor pigment? In: Senger H (ed) The blue light syndrome. Springer, Berlin Heidelberg New York, pp 244–250

Macnicol PK (1976) Rapid metabolic changes in the wounding response of leaf discs following excision. Plant Physiol 57:80–84

Manabe K, Poff KL (1978) Purification and characterization of the photoreducible b-type cytochrome from *Dictyostelium discoideum*. Plant Physiol 61:961–966

Mandoli DF, Briggs WR (1982a) Optical properties of etiolated plant tissues. Proc Natl Acad Sci USA 79:2902–2906

Mandoli DF, Briggs WR (1982b) The photoperceptive sites and the function of tissue light–piping in photomorphogenesis of etiolated oat seedlings. Plant Cell Environ 5:137–145

Mandoli DF, Briggs WR (1983) Physiology and optics of plant tissues. What's New in Plant Physiol 14:13–16

McIntyre GI, Browne KP (1984) The role of water in the light-induced growth inhibition and phototropic curvature of the sunflower hypocotyl. (Abstr) Annual Meeting, Canadian Soc Plant Physiologists, 1984, p 16

McWha JA, Jackson DL (1976) Some growth promotive effects of abscisic acid. J Exp Bot 27:1004–1008

Menschick R, Hild V, Hager A (1977) Decarboxylierung von Indolylessigsäure im Zusammenhang mit dem Phototropismus in *Avena*-Koleoptilen. Planta 133:233–228

Mentze J, Raymond B, Cohen JD, Rayle DL (1977) Auxin-induced H^+ secretion in *Helianthus* and its implications. Plant Physiol 60:509–512

Meudt WJ, Bennet HW (1978) Rapid bioassay for auxin. Physiol Plant 44:422–428

Meyer AM (1969a) Versuche zur Trennung von 1. positiver und negativer Krümmung der *Avena*koleoptile. Z Pflanzenphysiol 60:135–146

Meyer AM (1969b) Versuche zur 1. positiven und zur negativen phototropischen Krümmung der *Avena*koleoptile. I. Lichtperception und Absorptionsgradient. Z Pflanzenphysiol 60:418–433

Mollenhauer HH, Whaley WG, Leech JH (1960) Cell ultrastructure responses to mechanical injury. J Ultrastruct Res 4:473–481

Muir RM, Zhu LJ (1983) Effect of light in the control of growth by auxin and its inhibitor(s) in the sunflower. Physiol Plant 57:407–410

Mulkey TJ, Kuzmanoff KM, Evans ML (1981) Correlations between proton-efflux patterns and growth patterns during geotropism and phototropism in maize and sunflowers. Planta 152:239–241
Naqvi SM (1972a) Possible role of abscisic acid in phototropism. Z Pflanzenphysiol 67:454–456
Naqvi SM (1972b) Lateral transport of indoleacetic acid-2-^{14}C in *Zea mays* L coleoptile sections. Physiol Veg 10:495–501
Naqvi SM, Engvild KC (1974) Action of abscisic acid on auxin transport and its relation to phototropism. Physiol Plant 30:283–287
Naqvi SM, Gordon SA (1967) Auxin transport in *Zea mays* coleoptiles. II. Influence of light on the transport of indoleacetic acid-2-^{14}C. Plant Physiol 42:138–143
Naqvi SM, Ansari R, Shere SM (1972) Phototropic response of decapitated coleoptiles. Pak J Bot 4:51–53
Ninneman H (1980) Blue light photoreceptors. Bioscience 30:166–170
Phillips IDJ (1972a) Endogenous gibberellin transport and biosynthesis in relation to geotropic induction of excised sunflower shoot tips. Planta 105:234–244
Phillips IDJ (1972b) Diffusible gibberellins and phototropism in *Helianthus annuus*. Planta 106:363–367
Phillips IDJ, Hartung W (1976) Longitudinal and lateral transport of [3,4-^{3}H]gibberellin A_1 and 3-indolyl(acetic acid-2-^{14}C) in upright and geotropically responding green internode segments from *Helianthus annuus*. New Phytol 76:1–9
Pickard BG (1973) Geotropic response patterns of the *Avena* coleoptile. I. Dependence on angle and duration of stimulation. Can J Bot 51:1003–1021
Pickard BG, Thimann KV (1964) Transport and distribution of auxin during tropistic response. II. The lateral migration of auxin in phototropism of coleoptiles. Plant Physiol 39:341–350
Pickard BG with participation by Dutson K, Harrison V, Donegan E (1969) Second positive phototropic response patterns of the oat coleoptile. Planta 88:1–33
Pohl R (1960) Beiträge zum Phototropismus der *Avena*-Koleoptile. Phyton (Horn) 15:145–157
Poole RJ (1978) Energy coupling for membrane transport. Annu Rev Plant Physiol 29:437–460
Poole RT, Thimann KV (1964) Uptake of indole-3-acetic acid and indole-3-acetonitrile by *Avena* coleoptile sections. Plant Physiol 39:98–103
Railton ID, Phillips IDJ (1973) Gibberellins and geotropism in *Zea Mays* coleoptiles. Planta 109:121–126
Raven JA (1975) Transport of indoleacetic acid in plant cells in relation to pH and electrical potential gradients, and its significance for polar IAA transport. New Phytol 74:163–172
Reinders DE (1934) The sensibility for light of the base of normal and decapitated coleoptiles of *Avena*. Proc K Ned Akad Wet Amsterdam 37:308–314
Reinert J (1953) Über die Wirkung von Riboflavin und Carotin beim Phototropismus von *Avena*-Koleoptilen und bei anderen pflanzlichen Lichtreizreaktionen. Z Bot 41:103–122
Reinert J (1959) Phototropism and phototaxis. Annu Rev Plant Physiol 10:441–458
Reinhold L (1978) Phytohormones and the orientation of growth. In: Letham DS, Goodwin PB, Higgins TJV (eds) Phytohormones and related compounds – a comprehensive treatise, vol II. Elsevier/North-Holland Biomedical, Amsterdam, pp 251–289
Reinhold L, Ganot D (1972) Asymmetric "acid growth" response following gravistimulus. In: Carr DJ (ed) Plant growth substances 1970. Springer, Berlin Heidelberg New York, pp 726–730
Reisener JH (1958) Untersuchungen über den Phototropismus der Hafer-Koleoptile. Z Bot 46:474–505
Rubery PH (1977) The specificity of carrier-mediated auxin transport by suspension-cultured crown gall cells. Planta 135:275–283
Rubery PH (1978) Hydrogen ion dependence of carrier-mediated auxin uptake by suspension-cultured crown-gall cells. Planta 142:203–206

Rubery PH (1979) The effects of 2,4-dinitrophenol and chemical modifying reagents on auxin transport by suspension-cultured crown-gall cells. Planta 144:173–178

Rubery PH (1980) The mechanism of transmembrane auxin transport and its relation to the chemiosmotic hypothesis of the polar transport of auxin. In: Skoog F (ed) Plant growth substances 1979. Springer, Berlin Heidelberg New York, pp 50–60

Rubery PH, Sheldrake AR (1973) Effect of pH and surface charge on cell uptake of auxin. Nature New Biol 244:285–288

Rubery PH, Sheldrake AR (1974) Carrier-mediated auxin transport. Planta 118:101–121

Ruge U (1941) Über die geotropische Aufkrümmung decapitierter *Helianthus*-Keimlinge. Planta 32:176–186

Sachs T (1975) The induction of transport channels by auxin. Planta 127:201–206

Schmid GH (1980) Conformational changes caused by blue light. In: Senger H (ed) The blue light syndrome. Springer, Berlin Heidelberg New York, pp 198–204

Schmidt W (1980) Artificial flavin/membrane systems; a possible model for physiological blue light action. In: Senger H (ed) The blue light syndrome. Springer, Berlin Heidelberg New York, pp 212–220

Schmidt W (1983) The physiology of blue-light systems. In: The biology of photoreception, Symposium XXXVI, Soc for Exptl Biol. Cambridge University Press, Cambridge London New York, pp 305–330

Schmidt W, Hart J, Filner P, Poff KL (1977) Specific inhibition of phototropism in corn seedlings. Plant Physiol 60:736–738

Schneider EA, Wightman F (1974) Metabolism of auxin in higher plants. Annu Rev Plant Physiol 25:487–513

Schrott EL (1980) Dose-response and related aspects of carotogenesis in *Neurospora crassa*. In: Senger H (ed) The blue light syndrome. Springer, Berlin Heidelberg New York, pp 309–318

Schwartz A, Koller D (1978) Phototropic responses to vectorial light in leaves of *Lavatera cretica* L. Plant Physiol 61:924–928

Schwartz A, Koller D (1980) Role of the cotyledons in the phototropic response of *Lavatera cretica* seedlings. Plant Physiol 66:82–87

Scott TK, Most BH (1972) The movement of growth hormones in sugar cane. In: Kaldewey H, Vardar Y (eds) Hormonal regulation in plant growth and development. Verlag Chemie, Weinheim, pp 57–67

Senger H, Briggs WR (1981) The blue light receptor(s): primary reactions and subsequent metabolic changes. In: Smith CK (ed) Photochemical and photobiological reviews, vol 6. Plenum Press, New York, pp 1–38

Shen-Miller J, Gordon SA (1966) Hormonal relations in the phototropic response. III. The movement of C^{14}-labeled and endogenous indoleacetic acid in phototropically stimulated *Zea* coleoptiles. Plant Physiol 41:59–65

Shen-Miller J, Gordon SA (1967) Gravitational compensation and the phototropic response of oat coleoptiles. Plant Physiol 42:352–360

Shen-Miller J, Cooper P, Gordon SA (1969) Phototropism and photo-inhibition of basipolar transport of auxin in oat coleoptiles. Plant Physiol 44:491–496

Shibaoka H (1961) Studies on the mechanism of growth-inhibiting effect of light. Plant Cell Physiol 2:175–197

Shibaoka H, Yamaki T (1959) Studies on the growth movement of sunflower plant. Sci Papers Coll Gen Ed, Univ Tokyo, vol 9, pp 105–126

Shropshire W JR (1980) Carotenoids as primary photoreceptors in blue-light responses. In: Senger H (ed) The blue light syndrome. Springer, Berlin Heidelberg New York, pp 172–186

Shuttleworth JE, Black M (1977) The role of cotyledons in phototropism of de-etiolated seedlings. Planta 135:51–55

Silk WK, Jones RL (1975) Gibberellin response in lettuce hypocotyl sections. Plant Physiol 56:267–272

Smith H (1975) Phytochrome and photomorphogenesis. McGraw Hill, London

Smith H (1984) Plants that track the sun. Nature 308:774

Song PS (1980) Spectroscopic and photochemical characterization of flavoproteins and carotenoproteins as blue light photoreceptors. In: Senger H (ed) The blue light syndrome. Springer, Berlin Heidelberg New York, pp 157–171
Steiner AM (1969a) Dose response behavior for polarotropism of the chloronema of the fern *Dryopteris filix-mas* (L) Schott. Photochem Photobiol 9:493–506
Steiner AM (1969b) Action spectrum for polarotropism in the chloronema of the fern *Dryopteris filix-mas* (L) Schott. Photochem Photobiol 9:507–513
Steyer B (1967) Die Dosis-Wirkungsrelationen bei geotroper und phototroper Reizung: Vergleich von Mono- mit Dicotyledonen. Planta 77:277–286
Strasser RJ, Butler WL (1980) Interactions of flavins with cytochrome C and oxygen in excited artificial systems. In: Senger H (ed) The blue light syndrome. Springer, Berlin Heidelberg New York, pp 205–211
Stuart DA (1976) Gibberellic acid and elongation in lettuce hypocotyl sections. Dissertation, Univ California, Berkeley
Sussman MR, Goldsmith MHM (1981) Auxin uptake and action of N-1-naphthylphthalamic acid in corn coleoptiles. Planta 151:15–25
Terry ME, Jones RL (1981) Effect of salt on auxin-induced acidification and growth by pea internode sections. Plant Physiol 68:59–64
Thimann KV (1964) Phototropism. Photochem Photobiol 3:463–469
Thimann KV (1967) Phototropism. In: Florkin M, Stotz EH (eds) Comprehensive biochemistry, vol 27. Elsevier, Amsterdam, pp 1–29
Thimann KV (1977) Hormone action in the whole life of plants. Univ Massachusetts Press, Amherst
Thimann KV, Curry GM (1961) Phototropism and phototaxis. In: Florkin M, Mason HS (eds) Comparative biochemistry, vol I. Academic Press, London New York, pp 243–309
Thornton RM, Thimann KV (1967) Transient effects of light on auxin transport in the *Avena* coleoptile. Plant Physiol 42:247–257
Towers GHN, Abeysekera B (1984) Cell wall hydroxycinnamate esters as UV–A receptors in phototropic responses of higher plants – a new hypothesis. Phytochem 23:951–952
Trewavas A (1981) How do plant growth substances work? Plant Cell Environ 4:203–228
Valadon LRG, Mummery RS (1971) Relationship between the inhibitory effects of 5,6-epoxyxanthophylls and phototropism. Physiol Plant 24:363–368
van Overbeek J (1932) An analysis of phototropism in dicotyledons. Proc K Ned Akad Wet Amsterdam 35:1325–1335
van Overbeek J (1933) Wuchsstoff, Lichtwachstumreaktion und Phototropismus bei *Raphanus*. Rec Trav Bot Neerl 30:537–626
van Overbeek J (1936) Light growth response and auxin curvatures of *Avena*. Proc Natl Acad Sci USA 22:421–415
Van Sambeek JW, Pickard BG (1976a) Mediation of rapid electrical, metabolic, transpirational, and photosynthetic changes by factors released from wounds. I. Variation potentials and putative action potentials in intact plants. Can J Bot 54:2642–2650
Van Sambeek JW, Pickard BG (1976b) Mediation of rapid electrical, metabolic, and photosynthetic changes by factors released from wounds. III. Measurements of CO_2 and H_2O flux. Can J Bot 54:2662–2671
Vierstra RD, Poff KL (1981) Mechanism of specific inhibition of phototropism by phenylacetic acid in corn seedling. Plant Physiol 67:1011–1015
Vierstra RD, Poff KL (1981a) Role of carotenoids in the phototropic response of corn seedlings. Plant Physiol 68:798–801
Vierstra RD, Poff KL, Walker EB, Song P-S (1981) Effect of xenon on the excited states of phototropic receptor flavin in corn seedlings. Plant Physiol 67:996–998
Vogelmann TC, Björn LO (1983) Response to directional light by leaves of a sun–tracking lupine (*Lupinus succulentus*). Physiol Plant 59:533–538
Wainwright CM (1977) Sun-tracking and related leaf movements in a desert lupine (*Lupinus arizonicus*). Am J Bot 64:1032–1042
Wangermann E, Withers LA (1978) Auxin transport characteristics and cellular ultrastructure of different types of parenchyma. N Phytol 81:1–17

Wareing PF, Phillips IDJ (1978) The control of growth and differentiation in plants, 2nd edn. Pergamon, Oxford
Went F (1928) Wuchstoff und Wachstum. Rec Trav Bot Neerl 25:1–116
Went FW (1974) Reflections and speculations. Annu Rev Plant Physiol 25:1–26
Went FW, Thimann KV (1937) Phytohormones. MacMillan, New York
Werk KS, Ehleringer J (1984) Non–random leaf orientation in *Lactuca serriola* L. Plant Cell Environ 7:81–87
Wilden M (1939) Analyse der positiven und negativen phototropischen Biegungen. Planta 30:286–288
Wilkins MB (1965) Red light and the geotropic response of the *Avena* coleoptile. Plant Physiol 40:24–34
Wilkins MB (1977) Gravity and light-sensing guidance systems in primary roots and shoots. In: Symposium XXXI Soc Exp Biol. Integration of activity in the higher plant. Cambridge Univ Press, Cambridge, pp 275–335
Zimmerman BK, Briggs WR (1963a) Phototropic dosage-response curves for oat coleoptiles. Plant Physiol 38:248–253
Zimmerman BK, Briggs WR (1963b) A kinetic model for phototropic response of oat coleoptiles. Plant Physiol 38:253–261

11 Plant Growth Regulators and Low Temperature Stress

J.V. CARTER and M.L. BRENNER

1 Introduction

Plants respond to temperature stresses in a bewildering variety of ways. The extent to which hormones are involved in these responses is still under active investigation. The practical benefits to be wrought from a more complete understanding of this interrelationship are potentially great since even partial alleviation of crop damage due to chilling and freezing would result in greater production and, ultimately, more people being fed.

Studies suggesting a connection between plant hormone levels and low temperature stress extend back to the late 1940's when in the U.S.S.R. experiments were carried out showing that treating overwintering plants with auxins affected their frost resistance (VASIL'YEV 1961, pp 150–151). However, it has been within the past 15 years that significant advances have occurred in this area, owing to the development of sensitive analytical instrumentation and techniques.

This review is especially concerned with the possible role(s) that endogenous plant hormones may play in the processes by which many plants acclimate to low temperatures. Studies concerned with exogenous application of natural or synthetic plant growth regulators to protect against low temperature stress will also be dealt with, especially when their results bear on the endogenous mechanism(s) of action.

2 Freezing Stress – Background Information

Plant cold hardiness is a complex phenomenon about which much remains to be learned in spite of the large amount of research already done (for reviews, see WEISER 1970, LEVITT 1980, STUSHNOFF 1972, 1974)[1]. During late summer and early fall a series of signals – decreasing photoperiod, changes in light quality, and cool days and cooler nights – trigger many chemical and physiological changes in plants capable of becoming cold-hardy. During this initial stage of cold hardiness, called the cold acclimation period, starch declines in concentration, for example, with an accompanying increase in the levels of simple sugars such as glucose and sucrose. Free amino acid levels rise, due principally to substantial increases in proline and glutamic acid. Ascorbic acid has been

1 General aspects of cold and chilling resistance can be found in this Encyclopedia Vol. 12A, Chap. 12: P.L. STEPONKUS, Responses to Extreme Temperatures, Cellular and Sub-Cellular Bases, pp. 372–402 and Chap. 13: W. LARCHER and H. BAUER, Ecological Significance of Resistance to Low Temperatures, pp. 403–437 (1981)

found to increase in concentration in some plants during acclimation. In some cases unsaturated hydrocarbon chains in lipids are increased at the expense of a decrease in the level of more highly saturated chains. Among those compounds whose concentrations are sensitive to the acclimation process must be listed several of the plant-growth hormones.

After acclimating by exposure to short days and subzero temperatures, hardy plants can withstand quite low temperatures. This period, after acclimation is complete, can be considered the middle stage of the cold-hardiness phenomenon. During late winter and early spring increasing daylength, higher temperature, and changing light quality combine to initiate the final stage of cold hardiness, the deacclimation period. It is difficult to differentiate between physiological processes which initiate growth and those which bring about deacclimation. Although deacclimation commences well before visible signs of new growth, it is certainly possible that the physiological changes inherent in the early stages of deacclimation are also essential for new growth.

Research interrelating plant-growth substances with cold hardiness has mainly focused on acclimation. There has been a small amount done on deacclimation and very little on the middle stage of cold hardiness. However, because of the kinship between deacclimation and new growth, as well as between the first two stages of cold hardiness and dormancy, the effects of plant-growth regulators on the timing of new growth and dormancy may well provide legitimate information about the interrelations between plant-growth regulators and cold hardiness.

3 Dormancy and Cold Acclimation

The onset of winter dormancy is very closely related to the onset of cold hardiness. Conditions which cause acclimation to begin, such as low temperatures, short days, and water stress, also lead to growth cessation, or dormancy in temperate-zone plants. It is thus very difficult, if not impossible, to separate the onset of dormancy from acclimation. Consequently, studies of growth-regulator effects on dormancy are pertinent to our subject.

Since the term dormancy is used in several ways, it is necessary to indicate the definition which will be used in this article. Simply put, a dormant tissue is tissue that fails to grow due to either physiological or environmental conditions. It is well known that a chilling requirement must be met for certain plants which have entered dormancy to resume active growth. After fulfillment of the chilling requirement, the plant may still not grow, due to unfavorable environmental conditions. Various terms have been advanced to make clear whether the plant is dormant because of physiological or environmental factors. Some scientists call the first state "dormancy" and the second state "rest". Others use "true dormancy" and "imposed dormancy". Still another nomenclature uses "organic dormancy" and "forced dormancy". "Deep dormancy" has been used to describe the first state. The term rest has caused confusion, since it is used by many writers as a synonym for dormancy, thus encompassing both states, but by others to denote only the dormant state imposed by environmental factors, after the chilling requirement has been met.

It is, of course, an oversimplification to subdivide dormancy into only two states. Rather, it should be recognized that there is a continuous gradation, all the way from the first slight slowing of growth due to a change in temperature or daylength or water stress, to complete cessation of growth and then the inverse of the sequence (LEVITT 1980).

Throughout this review the term "dormancy" will include all stages in the life cycle of a plant in which growth has ceased. "Deep dormancy" will apply to the physiological state which begins in the fall, requiring a certain amount of chilling for completion. "Imposed dormancy" will refer to the state in which the plant is not growing due to unfavorable environmental conditions.

FUCHIGAMI et al. (1982) and KOBAYASHI et al. (1983) have described the various rhythmic cycles of growth tissue that woody perennial species undergo in a "degree growth stage (°GS) model." The °GS model contains five point events: (1) Spring Bud Burst is when vegetative growth first occurs from a dormant bud. (2) Maturity Induction Point occurs when plants become sensitive to environmental factors such as daylength. (3) Vegetative Maturity occurs when bud growth is no longer stimulated upon leaf removal. (4) Maximum Rest (referred to as deep dormancy in this review) is the point when buds need the maximum time of chilling or the highest concentration of gibberellic acid in order to resume growth. (5) End of Rest (end of deep dormancy) occurs when the chilling requirement is fulfilled. This model, which can be described by a sine curve, shows the sequence of changes that occur from growth through to deep dormancy.

Acclimation to cold occurs in two stages, the first initiated by short days (FUCHIGAMI et al. 1971, IRVING and LANPHEAR 1967b); the second develops when low temperature conditions succeed the completion of the first stage (IRVING and LANPHEAR 1967b, 1967c, WEISER 1970). FUCHIGAMI et al. (1982) suggest in their °GS model that the first stage of acclimation occurs after the maturity induction point, but before vegetative maturity. The second stage of acclimation occurs after vegetative maturity. The rate of cold acclimation during this second stage is increased as the acclimating temperature is lowered from 20° to 5° (FUCHIGAMI et al. 1982).

4 Interrelations of Hormones with Freezing Stress

4.1 Translocatable Factors in Cold Acclimation

In the late 1960's and early 1970's, several important experiments were performed which demonstrated that plants capable of acclimating to below-freezing temperatures actively produce translocatable materials which can either help or hinder the acclimation process.

IRVING and LANPHEAR (1967c) identified leaves exposed to long days (the long-day leaf) as an apparent source of cold hardiness inhibitors. They placed box elder (*Acer negundo* L.) seedlings outdoors in September; by October 4th plants under natural conditions had hardened to a killing point of −18° C. However, plants exposed to the same outdoor temperatures and maintained in a long day (LD) photoperiod with supplemental lighting for 6 h at the end of the day did not harden until November 2nd when the foliage was killed, at which time the killing point dropped from −7° to −23.5° C in 2 weeks. They obtained a similar, although not so dramatic, effect with doublefile virburnum (*Viburnum plicatum tomentosum* THUNB.).

By wrapping one limb on each of five *Viburnum* plants with aluminum foil from 16:00 to 08:00 h each day for 6 weeks, while illuminating the rest of each of the plants in a LD regimen, IRVING and LANPHEAR (1967c) were able to demonstrate that in this plant the short day (SD) leaf is a source of translocatable hardiness-promoting factors. Both sides of these SD/LD plants hardened to an average killing point of −17° C, while plants exposed to only LD had a killing point of −12° C, and those plants exposed to only SD had a killing point of −22° C. They also noted that defoliation of a LD branch in a SD/LD plant removed the cold-hardiness inhibition produced in that plant.

The University of Minnesota group independently arrived at similar conclusions. Working with Haralson apple (*Pyrus malus* L.), HOWELL and WEISER (1970) found that leaves exposed to SD produced a translocatable factor(s) which stimulated cold acclimation of the bark. In 1967 and 1968 they performed double-branched plant experiments and found, as did IRVING and LANGPHEAR (1967c), that the factor(s) produced on the SD side of the split plant induced cold acclimation in the LD portion. Both branches of the divided plants acclimated to the same hardiness level, which was intermediate between that produced by SD and LD treatments.

In a related study on red osier dogwood (*Cornus sericea* M.) FUCHIGAMI et al. (1971) found that removing leaves from plants on a SD, low temperature (LT) regimen inhibited their acclimation; the plants needed at least 4 weeks of SD prior to defoliation if they were to acclimate. Further, they found that if leaves were removed from some branches, but not from others, the branches with and without leaves both acclimated as well as nondefoliated plants. Results from this study support the concept of the SD leaf as a source of a hardiness-promoting factor(s), in agreement with IRVING and LANPHEAR (1967c), but conflict with other studies implicating LD leaves as a source of a hardiness-inhibiting factor(s). This disagreement could be due to the different plants used in these studies or sampling of different points of the plant's growth stage.

The results obtained in girdling experiments performed by these two groups also appear to be in disagreement. In their double-branched plant experiment using *Viburnum,* IRVING and LANPHEAR (1967c) found that girdling the LD branch of a LD/SD plant failed to prevent hardening of the LD branch, indicating that either the translocatable hardiness promotor moved through the xylem or that the girdle produced hardiness. However, the SD side of this LD girdled/SD combination became significantly more hardy than did similar plants without the girdle. In their study using red-osier dogwood, FUCHIGAMI et al. (1971) reported that girdling the phloem between foliated and defoliated branches did prevent hardening of the defoliated branches, which would imply that the hardiness-promoting factor produced in SD dogwood leaves is transported through the phloem and not through the xylem. These authors point out that the disagreement between their results and those of IRVING and LANPHEAR (1967c) is more apparent than real, since in the latter study both the LD girdles and the SD branches received the same prolonged exposure to LT (4 weeks at 4.4° C + 1 week at −1.1° C and 1 day each at −2.2°, −3.3°, −4.4°, and −5.5° C), which as several studies have shown can fully acclimate woody plants exposed to LD.

TUMANOV and coworkers (1976) have found that a hardiness-promoting factor(s) is produced in leaves of winter wheat (*Triticum aestivum*) and is translocated to the tillering nodes. Plants hardened fully when leaves were exposed to as little as 15 min light (4500 lx, type unspecified) per day for 30 days while being kept at 2° C, while those kept in total darkness, but grown at the same temperature, were significantly less hardy. They found that tillering nodes which were kept in the dark, while the rest of the plant to which they were attached was lighted normally, survived to −18° C. In a companion experiment plants were treated identically, except that the tillering nodes were excised from the rest of the plant; these tillering nodes were all killed by exposure to −10° C. They concluded that a substance (or substances) was produced in the lighted leaves which translocated to the tillering nodes, located underground, allowing them to develop cold hardiness. They also showed that 6-benzylaminopurine (BAP), when added to the nutrient solution at 4 mg l^{-1} of plants kept in the dark, enabled these plants to absorb as much sucrose from the nutrient solution as occurred in plants exposed to light. The hardiness level of the BAP-treated plants was −20° compared to −16° C for the dark controls; the lighted controls survived −23° C. They also reported that winter wheat treated with gibberellin showed a sharply decreased capacity to harden.

TIMMIS and WORRALL (1974), working with Douglas-fir (*Pseudotsuga menziesii*) seedlings, performed a double-branched plant experiment in which photoperiod was held constant while the two parts of the "double-branched" plants received different temperature treatments. Warm branches were held at 21° ± 1° C, chilled branches were kept at 1.5° ± 0.5° C, and 8 h of mixed incandescent and fluorescent light (4280 lx) was supplied to all parts of all the seedlings. After 8 weeks, chilled branches were 9° C more hardy than warm branches on the same plant, but significantly less hardy (by about 5° C) than branches of wholly chilled plants, showing that some translocatable factor was involved which inhibited hardening. This might also be explained as a hardiness promoter being diluted by the warm branch, which is not making as much promoter and has amassed more dry weight. Also, warm branches acquired the same degree of hardiness whether on a plant with a chilled branch or on a wholly warm plant, showing that the effect of chilling was not translocated. These results do not conflict with those from the articles discussed above; rather, they emphasize the complexity of the acclimation process. IRVING and LANPHEAR (1967c) reported that LD, LT treated box elder plants produce cold hardiness inhibitor(s); their work did not demonstrate that this inhibitor(s) was translocatable. TIMMIS and WORRALL (1974) have shown that a hardiness inhibitor(s) can be produced in Douglas-fir under SD, HT conditions which can translocate and interfere with the hardening of a SD, LT-treated branch.

4.2 Evidence that Translocatable Factors Are Hormones

What is the hardiness-promoting factor? In his review, WEISER (1970) discussed this topic, suggesting that it is likely to be a hormone or complex of hormones, and mentioned that the extensive metabolic changes involved in cold acclimation imply this hormone (or hormones) ..." exerts its regulatory function on nucleic acid transcription". He further pointed out that elucidation of the hardiness promoting factor "deserved concentrated research attention". Although progress has been made, final identification of the factor(s) has yet to be accomplished.

In an attempt to learn more about the nature of the hardiness-promoting and inhibiting factor(s) for which they had earlier found evidence, IRVING and LANPHEAR (1968) focused on gibberellins (GA) and abscisic acid (ABA). They determined endogenous GA-like and ABA-like levels by methanol extraction followed by paper chromatography and a bioassay based on percent germination of lettuce seed. After establishing that exogenous GA_3 application inhibited cold-hardiness induction in box elder, while ABA and the synthetic growth retardants AMO and daminozide promoted cold hardiness, they carried out a bioassay experiment to see whether short days promote hardiness through the action of an inhibitor such as ABA. They tested extracts from LD and SD leaves in their performance on the lettuce seed germination test, comparing the results with a similarly separated solution containing 25 mg l^{-1} each of GA_3 and ABA. They found the SD extract to be richer in inhibitors while the LD extract had more germination-promoting activity. The inhibiting zones had R_F values very similar to those of ABA under the same chromatographic conditions and the germination-promoting areas of the chromatograms were very similar to the peak produced by GA_3. IRVING and LANPHEAR (1968) interpreted these results as being consistent with high gibberellin activity produced by exposing box elder leaves to LD being responsible for hardiness inhibition, whereas SD-treated leaves promote hardiness through build-up of an inhibitor which counteracts gibberellin activity. Although they find in their experiments that exogenously applied ABA acts as a hardiness promoter and behaves similarly to the natural hardiness promoter(s), they are correct in not suggesting that one of the natural hardiness promoters is ABA until more rigorous chemical and physiological experiments are completed. They do suggest GA as the inhibitor or one of the inhibitors, since AMO, one of the synthetic growth retardants which they found promoted hardiness, is known to interfere with GA biosynthesis in other plant systems.

Further work by FUCHIGAMI et al. (1971) was directed toward identification of translocatable hardiness-promoting and inhibiting factors which had already been shown to exist in red-osier dogwood (FUCHIGAMI et al. 1971) and in apple (HOWELL and WEISER 1970). They exposed test plants to 10^{-4} M ABA and to GA_3 (concentration not given) both separately and in combination, for 4 weeks, using treatment methods similar to those of IRVING with box elder (1969a, b). They found no measurable increase in hardiness in ABA-treated plants, but surprisingly reported a growth *increase* in these plants compared to nontreated controls. This finding, contrary to the commonly found growth inhibition produced by the application of ABA, makes it seem likely that the treatment method was not successful in introducing ABA into the test plants. They did find that GA_3 reduced hardiness. Hardiness-promoting and/or inhibiting factors were not isolated from the test plants. Thus their positive result, that GA_3 reduced hardiness, must be classified as suggestive rather than conclusive evidence concerning the nature or existence of hardiness-inhibiting factors, a conclusion they evidently shared.

PROEBSTING (1983) has further examined the nature of a translocatable factor in relation to the onset of bud dormancy induced by short days. He conducted a series of experiments on double-branched red-osier dogwood plants. Growth of the terminal bud on one of the branches, which had been defoliated and maintained in LD, is suppressed by exposing the foliated branch to SD. The reciprocal treatment of maintaining the defoliated branch in SD conditions and the foliated in LD promoted growth. Application of GA_3 to plants with SD-exposed leaves reversed the growth suppression, while ancymidol treatment caused formation of an apparent dormant terminal bud on plants exposed to LD. Application of ABA to LD leaves had no impact on growth of the terminal bud. These experiments support the hypothesis that gibberellins function as a translocatable promoter of shoot growth in LD. Assuming the °GS model of FUCHIGAMI et al. (1982) is correct in associating the onset of dormancy with cold acclimation, then gibberellins may function as acclimation inhibitors.

IRVING (1969a) tried to characterize the hardiness-promoting factor(s) more definitively. He extracted SD leaves of box elder and compared the effect of the extract on the germination of lettuce seeds with that of ABA, after paper chromatography in four different solvent systems. In each solvent system the inhibitor present in the extract behaved very similarly to ABA. He then treated LD plants with both ABA and the extract taken from SD leaves, and found an increase in cold hardiness with both. Abscisic acid applied to SD plants did not increase their hardiness, but he did not report the length of the SD treatment prior to ABA application; it is possible that the photoperiod treatment had already increased the cold-hardiness level past the point at which ABA might have had an effect. This work provides suggestive evidence that at least in box elder ABA acts as a hardiness promoter. Also in this paper IRVING determined the levels of growth promoters and inhibitors in extracts from box elder leaves exposed for 4 weeks to the following three treatments: SD, LD, and LD + 5° nights (LD + 5° N). He used both the lettuce (*Lactuca sativa*) seed and normal corn (*Zea mays*) growth bioassays. There was more GA-like activity in the LD extract than in either the LD + 5° N or the SD extracts; both the latter treatments induced cold hardiness. The extracts from the SD and LD + 5° N-treated leaves contained more ABA-like activity than did the LD extract. He concluded that hardiness development seems more closely correlated with a buildup of ABA-like activity than with a reduction in GA-like activity. Examination of his data indicates to us, however, that hardiness development in box elder correlates with both an increase in ABA-like activity and a decrease in GA-like activity.

BRAY (1982) applied modern analytical methods to determine whether endogenous ABA in red-osier dogwood plants relates to cold acclimation. She used high performance liquid chromatography (HPLC) followed by gas chromatography on a capillary column coupled to an electron capture detector. The level of ABA in leaves of plants in SD conditions decreases significantly compared to plants maintained in LD. The decrease in ABA in the leaves is the consequence of higher catabolism and export. From BRAY's work, the earlier reports of IRVING and LANPHEAR (1968), and IRVING (1969a) could be reinterpreted. Their observations of greater inhibitory material in leaves of SD-treated plants may be from the occurrence of ABA metabolites rather than ABA.

Though BRAY (1982) observed that SD conditions result in increased ABA transported in both the xylem and phloem, she found that elevated levels of ABA moving within the plant during SD conditions will not result in dogwood attaining cold hardiness. This conclusion was based on an experiment in which roots of one set of dogwood plants were maintained at 5°, while another set of plants had their roots treated at 20° C; the shoots of both sets of plants were maintained at 21° in SD conditions. Root pressure exudate from the roots treated at 5° C had threefold more ABA than those from the higher temperature treatment.

4.3 Effects of Hormones on Cold Acclimation

RICHMOND's group in Israel made a major contribution to this area. In work done on two cultivars of alfalfa (*Medicago sativa*) (RIKIN et al. 1976, WALDMAN et al. 1975) they have focused on GA and ABA activities and effects as related to cold acclimation. RIKIN et al. (1976) examined the relationships between the ABA/GA ratio and both morphology and cold hardiness using cv. Ranger, which is acclimatable, and cv. Hairy Peruvian, nonacclimatable. Under LD and high temperatures (HT), nonacclimating conditions, both cultivars produced elongated shoots and were not cold-hardy. Under SD and LT cv. Ranger produced a rosette and showed improved cold hardiness while cv. Hairy Peruvian still produced an elongated shoot and no cold hardiness. When cv. Hairy Peruvian was grown in a nutrient solution containing ABA it did develop rosette growth and improved cold hardiness. When GA was fed to cv. Ranger seedlings grown under SD and LT, the environmental effects were nullified; instead of rosette growth and hardiness development they exhibited elongated shoots and no cold resistance. It was concluded that the morphological change from elongated shoot to rosette is related to development of cold hardiness and that cv. Hairy Peruvian does not cold-acclimate because it lacks a triggering mechanism for ABA increase and/or GA decrease under inducing conditions.

In a related article, WALDMAN et al. (1975) described further studies on the same two alfalfa cultivars. They extracted and determined the activities of ABA and GA in seedlings exposed to various treatments. When cv. Ranger and cv. Hairy Peruvian were exposed to SD, LT, conditions which induce cold hardiness, ABA activity was not increased in either cultivar; the effect was on GA activity. Shoots of cv. Ranger seedlings grown under SD and LT showed almost no GA_3-like activity; under LD and HT they had relatively high levels. GA_3-like activity in cv. Hairy Peruvian was not affected by changing day length or temperature, but remained higher than cv. Ranger under inducing conditions. Significantly, making the nutrient solutions 4 μM in ABA produced a marked decline in GA_3-like activity in both cultivars. In the case of cv. Ranger, acclimation caused the ABA/GA ratio to increase solely by reducing GA activity, whereas in box elder (IRVING 1969a) the increase in this ratio with acclimation was apparently caused primarily by an increase in ABA activity with an accompanying apparent decrease in GA activity. WALDMAN et al. (1975) hypothesized that ABA application induces cold hardiness by stopping GA synthesis. It is also possible that it blocks GA translocation or displaces an equilibrium between some form of conjugated GA and free GA.

DAIE and CAMPBELL (1981) observed that when tomato (*Lycopersicon esculentum*) plants were transferred from a 25°/15° C day/night temperature regime to 10°/5°, 15°/10°, 35°/25°, or 45°/35° C day/night conditions, ABA content in the plant tissue significantly increased within 12 h. Their ABA analytical procedures relied on gas chomatography fitted with a packed column and an electron capture detector. The most stressful temperature treatments (10°/5° and 45°/35° C) caused the largest increase in ABA content. The increase was for both free ABA and conjugated ABA during the 72 h in which the plants were maintained in these altered temperature regimes. There were no differences in leaf water potential as a result of the various temperature treatments. DAIE and CAMPBELL concluded that the change to stressful temperature conditions resulted in an increase of ABA synthesis similar to that which occurs from other types of stressful conditions.

CHEN et al. (1983) found a significant increase of endogenous ABA in *Solanum commersonii* when cultured shoots were placed in cold-acclimating conditions compared to similar shoots maintained in warm conditions. Interestingly, the increase in ABA content was observed only on the 4th day after placing the cultured shoots in the acclimating environment. The increase in ABA was preceded by an increase in osmotic concentration of the leaf tissue 2 days earlier. Soluble proteins increased in the leaf tissue by the 6th day, which is when cold hardiness increased. ABA added to the culture medium promoted cold acclimation of shoots maintained in warm (25°) conditions. Addition of cycloheximide blocked the cold acclimation of shoots treated with ABA or placed in acclimating conditions. Similar tests were done on *S. tuberosum,* a potato species that does not cold-acclimate. No changes in osmotic concentration, ABA, or soluble protein were observed during the culture period for either *S. commersonii* maintained in the warm conditions or in the leaves of *S. tuberosum* maintained at either temperature. CHEN et al. (1983) suggested that cold conditions cause an increase in osmotic potential in the *S. commersonii* leaves which, in turn, leads to an increase in ABA. The increased ABA may promote those processes which are required for an increase in cold hardiness.

Recently, CHEN and GUSTA (1983) reported that addition of 7.5×10^{-5} M ABA to the medium of cultured suspension cells significantly increased the cold hardiness of wheat, rye, and bromegrass. The ABA-treated cells maintained at 20° C exhibited freezing resistance to below −30° C, while control treatments could only tolerate −7° to −8° C. Suspension cells acclimated at 2° for 7 days attained freezing resistance to −17° to −20° C. However, the combination of 7 days of cold acclimation and ABA was not significantly different from the ABA treatment. CHEN and GUSTA (1983) hypothesize that ABA may be substituting for low temperature induction of cold acclimation.

Taken together, these papers indicate that one of the hardiness-promoting factors, if not ABA, is very similar to ABA in both chemical (chromatographic) and biological (various bioassays) behavior. Also emerging, however, is the importance of the ratio of hardiness-promoting to hardiness-inhibiting factors. It is this ratio which evidently increases in a hardy plant under environmental conditions which induce cold acclimation. The work by WALDMAN et al. (1975) indicates the complexity of the control mechanism, since it demonstrates that extractable ABA activity can remain constant under environmental conditions and yet cold hardiness can still develop. What caused the decrease in GA activity which they found? Was it another hardiness-promoting factor? Or is it possible that any increase in ABA content that occurred was too transient to be observed? This would be similar to the relatively brief increase in ABA content observed by CHEN et al. (1983).

Perhaps some plants have adapted mechanism(s) for shutting off GA synthesis which do not require ABA for an on-off switch, while in others such a switch is necessary. If a plant such as alfalfa cv. Ranger was grown under LD and HT in a nutrient solution with elevated concentrations of both ABA and GA, one could perhaps better determine the importance of their concentration ratio to the development of cold hardiness. Certainly the work of CHEN and GUSTA (1983) demonstrates that addition of ABA to the culture medium promotes the cold acclimation of other species.

The conclusion that ABA is a hardiness-promoting factor must remain tentative, since its record is not unblemished. As was mentioned earlier, FUCHIGAMI et al. (1971) found no increase in cold hardiness of red-osier dogwood when ABA was applied; rather they found a growth increase in these plants above that of the controls. In a study of the effect of growth regulators on cold hardiness of citrus seedlings YOUNG (1971) found that ABA applied as a spray at 1 and 10 ppm did not increase hardiness. At ABA concentration levels of 100 and 500 ppm hardiness was decreased. However, this treatment also caused partial defoliation and increased leaf ethylene production, which makes identification of the cause of the decrease difficult.

In a study of the interrelations of various kinds of stress BOUSSIBA et al. (1975) subjected tobacco plants (*Nicotiana sylvestris*) to four different stresses: leaf dehydration, mineral deprivation, salination, and boron toxicity. All these stresses resulted in increased ABA levels in the leaf, as determined by extraction and gas chromatographic analysis. Leaves of the stressed plants all showed an increased resistance to subzero temperatures. The ABA increase was always associated with declining leaf transpiration. When the salinated or mineral-deprived plants were transferred back to pre-stressed conditions their ABA levels declined, as did their cold hardiness. Direct application of ABA to whole plants as well as to excised leaves caused the same increase in cold hardiness, except during the summer months. They state, however, that in many other experiments on barley (*Hordeum vulgare*), wheat, and alfalfa, ABA application increased cold hardiness throughout the year.

If, as previous studies suggest, ABA is a translocatable hardiness-promoting factor in at least some plants, it is also evident that there are interrelationships among photoperiod, temperature, and the concentrations of at least two hormone classes – ABA and gibberellins – the balancing of which results in initiation of the complex series of physiological changes that together constitute the acclimation process. Thus, identification of a single substance as "the hardiness-promoting factor" may be an impossibility.

Application of growth regulators exogenously to affect hardiness can cause radically different results depending on the time of application, which in field studies implies a dependence on both daylength and temperature (FUCHIGAMI 1977). Some growth regulators have totally opposing effects, depending upon their concentrations. For example, DROR et al. (1976) found that kinetin applied to detached tobacco leaves or to wheat and barley plants in the range of 10^{-8} M caused little or no reduction of their frost resistance, while in the range of 10^{-4} to 10^{-7} M it caused a significant increase in resistance. Kinetin at 10^{-3} M caused a decrease in resistance. Similar results were obtained with zeatin and benzyladenine.

Gibberellins have been found to decrease cold hardiness in several plants when applied at certain times of the year. PROEBSTING and MILLS (1974) reported that GA_3 applied to Bing cherry trees (*Prunus avium* L.) prior to late June had little effect on fruit bud or wood hardiness. Application on August 22 resulted in much more damage than that produced in trees sprayed on August 1, in agreement with the concept of vegetative maturity as described by FUCHIGAMI (1977).

Kharkov 22 MC winter wheat has been found to be less cold-hardy when grown in a nutrient solution containing GA_3 or GA_7 (ROBERTS 1971). Further work by the same group (REID et al. 1974) showed that, although winter wheat plants grown at 2° had much less GA activity than those grown at 20° C, plants grown at 6° had almost the same level of GA as the 20° C plants, and yet showed the dwarfing of winter wheat induced by low temperature as well as a moderate level of cold hardiness. They conclude that the negative correlation of GA levels with cold hardiness in winter wheat is weak. However, their results show that plants grown at 2° had almost no GA activity and were decidedly more hardy than those grown at 6° C. Thus their work is in accord with the conclusion that achievement of full hardiness is accompanied by reduction of GA activity to very low levels. WÜNSCHE (1966) also noted that GA_3-treated Starke winter wheat plants were slightly less cold-hardy than controls. The site of greatest damage in the GA_3-treated plants was the base of the youngest leaf. KORNS (1959) reported that soaking Kharkov winter wheat seeds for 24 h in 100 ppm GA_3 solution before seeding decreased the vigor and extent of survival of plants grown from the treated seeds. Other plants whose cold hardiness has been shown to decrease due to exogenous application of GA include alfalfa (PAQUIN et al. 1976), citrus (YOUNG 1971), and *Chrysanthemum morifolium* (BRUCKEL and WIGGANS 1969).

To emphasize that the relationship between GA and cold hardiness is complex, we now mention work which has shown a positive correlation between exogenous GA and cold hardiness. EDGERTON (1967) applied potassium gibberellate (KGA_3) as a spray at 50 ppm concentration to peach trees (*Prunus persica*) in July. He found that in both the November and February freezing tests the KGA_3-treated flower buds had increased cold hardiness over the control plants although the hormone lowered the number of buds per length of stem significantly. The GA_3-promoted increase in hardiness of the treated flower buds was correlated with a delay in flower primordia differentiation. He also noted that small buds with only partially developed primordia were the ones which best survived the November and February freezing tests. MODLIBOWSKA (1965) sprayed 1-year-old pear trees (*Pyrus communis*) with 500 ppm GA_3 during April and May using four different application schedules and reported a smaller percentage of the next year's flowers killed in three of the four treatments as compared to the nontreated set of controls. Application of KGA_3 to mature Elberta peach trees by PROEBSTING and MILLS (1964) at concentrations between 80 to 1000 ppm in late August produced a small increase in bud cold hardiness, determined by controlled freezing tests. Even though the increase was small (0.5° to 2° C) the response translated into significantly improved yields. November application reduced hardiness. As will be discussed later, a careful consideration of the timing of these applications may allow reconciliation of these results with those previously discussed.

TUMANOV et al. (1970) observed that the concentrations of GA-like substances in buds and bark of black currant (*Ribes nigrum*) and birch (*Betula papyrifera*) remain low from October to January and that there is a sharp increase in February and March before decline to normal levels in April. They found that spraying 100 mg l^{-1} of GA (presumably GA_3) on plants in October produced a higher extractable GA level during these periods, but the concentration dynamics were unchanged. At the time of spraying the plants were dormant and the GA spray did not lessen their frost resistance. If, however, GA was applied in the summer – sprayed 7 times with 200 mg l^{-1} GA in July and August – there was a sharp reduction in frost resistance after a hardening regimen. In blackcurrant the control withstood −40° while the GA-treated plants survived only −5° C; with birch the control withstood −60° and the GA-treated plants only −5° C. TUMANOV considers the main significance of GA is to help wintering woody plants pass from deep to imposed dormancy in the second half of winter, without noticeably reducing the frost resistance of plants.

There is some evidence suggesting that indoleacetic acid (IAA) levels are affected not only by low temperature but also by GA. BOLDUC et al. (1970) followed the activity of the enzyme IAA oxidase in winter wheat cv. Knox. The activity of this enzyme, inversely correlated to the IAA levels in the wheat seedlings, was increased tenfold during a 40-day treatment at 2° C. Addition of GA_3 also stimulated IAA oxidase development both during room temperature germination and during the 40-day cold treatment. Further, they reported that AMO inhibits IAA oxidase development in the cold. This is in accord with the early report by TUMANOV and TRUNOVA (1958) that auxin decreases on hardening of coleoptiles and that exogenous application of IAA reduces the ability of juvenile tissues to cold acclimate. As noted by HAARD (1973), this suggests that auxin metabolism is an integral part of the cold acclimation process.

VASIL'YEV (1961, pp. 150–151) cites some early work done on the relation between auxins and cold hardiness. He summarizes (p. 150) a 1947 study in which treatment of the plant koksaghyz (no latin name given) with heteroauxin in the fall caused a sharp decrease in frost resistance during the winter. VASIL'YEV (pp. 150–151) also mentions a study done in 1948 in which lemon (*Citrus limon*) branches sprayed with a "higher-than-usual concentration" of α-naphthylacetate stopped growing and gained some hardiness. We have not found any published work which deals with the endogenous level of auxins as a function of the degree of cold hardiness of a plant which can acclimate to freezing temperatures.

Published work implies that cytokinins do not play an important role in the initial part of cold acclimation. DUMBROFF and BROWN (1976), working with 2-year-old sugar maple (*Acer saccharum*) seedlings, exposed their test plants to normal overwintering conditions from early November through April. The cytokinin-like material in lateral roots and stems remained below control group levels until early March, as measured by the cucumber (*Cucumis sativus*) cotyledon greening bioassay. It then rose steadily and peaked in mid-April, behavior which has been noted in other trees as well (HEWETT and WAREING 1974). Again the lesson to be learned is that cytokinins apparently play a minor role in the onset of hardiness, but a cautionary word is in order, since in neither of these studies was the cytokinin activity examined during the initial cold acclimation period. Thus it remains to be conclusively demonstrated that cytokinins are not involved in cold acclimation. The work by SKENE (1972), showing that grape vine (*Vitis vinifera* L.) canes collected in winter had only slight cytokinin activity when compared to growing canes, supportive of the apparent noninvolvement of cytokinins in cold acclimation, is also inconclusive in this regard.

Ethylene is not thought to play a direct role in cold acclimation (FUCHIGAMI et al. 1971), but there is suggestive evidence that it has an indirect effect on the depth of hardiness achieved. PROEBSTING and MILLS (1973, 1976) treated sweet cherry with ethephon, a compound which is hydrolyzed to ethylene after being absorbed by plant tissue, and found that late summer sprays at 100 and 500 ppm increased fruit bud winter hardiness, delayed bloom and improved survival from spring frosts. The average increase in bud hardiness over and above naturally attainable levels was estimated at only 1° to 2° C, a small but nevertheless significant effect. The response of peach was not as positive.

Recent reports on the use of ethephon to promote cold hardiness (HOLUBOWICZ et al. 1982, RAESE 1983) have documented that fall applications of ethephon to fruit trees has been consistently detrimental. On the other hand, CGA-15281 has been found to be the plant-growth regulator that most consistently (69% of the tests) increases cold hardiness of 1-year-old apple trees (RAESE 1983). This is surprising, since CGA-15281 is also believed to act as an ethylene-releasing agent. The means by which ethylene exerts this effect is not known, and to our knowledge no hypotheses have been made.

4.4 Exogenous Application of Synthetic Growth Regulators that Affect Cold Hardiness

A casual reading of the literature leaves the impression that the chemical industry has provided a bewildering array of compounds for plant scientists intent on fine-tuning various growth aspects of economically important plants. However true this may be in general, in the specific area of low temperature stress most of the compounds that have imparted cold hardiness may be classified as growth retardants. It is significant that all of these compounds have been implicated as gibberellin antagonists in one plant system or another or as ethylene-releasing agents. Of course it should not be concluded that they all function in this manner in all plant systems.

It should be emphasized that the effect of these compounds on plant cold hardiness must be intimately associated with their growth retarding abilities. In each of the articles reviewed here in which aspects of growth were monitored in addition to hardiness levels, increases in cold hardiness were accompanied by growth retardation.

The synthetic growth retardant which has seen the widest usage for frost protection is CCC. Its effect on cold hardiness has been examined in winter wheat, rape (*Brassica napus* L.), citrus, box elder, alfalfa, potato, cabbage (*Brassica oleracea* L.), tomato, pear, and barley.

Applying 4×10^{-3} M CCC to soil, WÜNSCHE (1966) found an increase of less than 1° in the frost resistance of Starke variety winter wheat. KHISAMUTDINOVA et al. (1975) reported that treating Albidum-114 winter wheat seeds with CCC before sowing results in deeper-than-usual hardening of the tillering nodes and increases frost resistance. Concentrating on energy metabolism and water status as affected by CCC, they found that CCC-treated plants have higher rates of photosynthesis and respiration during acclimation than do test plants.

ROBERTS (1971), in his study of the effects of CCC on Kharkov 22 MC winter wheat, uncovered an interesting concentration effect. Plants grown for 3 weeks at 21° C with 3×10^{-4} or 3×10^{-5} M CCC in the nutrient solution show an increased ability over controls to survive exposure to −5° C for 58 h, while 3×10^{-3} M CCC has a toxic effect that is manifested in plant death 1 to 3 weeks after exposure. This toxicity is somehow related to the cold treatment, since plants grown with 10^{-2} M CCC for 4 weeks, but not subjected to −5° for 48 h, did not die. As the author points out, this result underscores the necessity of considering the ultimate survival of test plants in any such study. Short-term visual examination would have given completely misleading results. He also found that the effectiveness of CCC treatment in enhancing cold hardiness of winter wheat is inversely correlated with plant age.

Frost hardiness of winter rape has been extensively studied by KACPERSKA-PALACZ and coworkers. In one facet of this study (KACPERSKA-PALACZ and EGIERSZDORFF 1972), the effect of CCC on hydration and frost and desiccation hardiness of winter rape was examined. It was found that including CCC in the nutrient solution of rape plants grown in water culture increases frost resistance of the leaves and makes them more resistant to desiccation injury. They also reported that plants grown in hydroculture either receiving CCC or exposed to 5° C so as to cause them to cold harden actually have more water in leaf tissue than the control plants, grown at 20° C. This result shows clearly that CCC does not provide frost protection by causing dehydration. However, these authors also found that the CCC treatment causes an increase in the intercellular volume, enhances the rate of water loss of rape leaves exposed to a desiccant, and allows the plant to survive a larger water deficit. So although CCC does not cause dehydration, it renders the plant more easily dehydratable and more capable of surviving dehydration. CHROMINSKI et al. (1969) have also shown that CCC increases hardiness of winter rape. They applied CCC as a foliar spray, 6 g l^{-1}, at levels of 1 to 16 kg ha^{-1} and found that 2 kg ha^{-1} was the minimum level at which significant protection was provided; only 32% as many plants were killed as in the control plots receiving no CCC.

YOUNG (1971) looked at the effects on cold hardiness of citrus seedlings of several growth regulators, both natural and synthetic, applied as sprays. He found no effect with CCC at levels of 500, 1000, 2000, and 3000 ppm; no effect with daminozide at 500, 1000, 2000, and 5000 ppm; no effect with ethephon at 5, 100, and 500 ppm. The daminozide used was the Alar formulation which contains 85% active daminozide, and it is not clear whether this factor was taken into account in determining the above concentrations.

We have so far covered an example of a plant which shows no change in hardiness level when treated with CCC – citrus – and a plant which has its cold hardiness increased by applying CCC with no accompanying low-temperature or short-day treatment – winter rape. The study of IRVING (1969b) on box elder further demonstrates the variety of responses by plants to synthetic growth regulators. He found that CCC, AMO, and daminozide are all ineffective at increasing the frost resistance of this plant as long as it is exposed to long days. However, applying any of the three growth regulators each week for 5 weeks, followed by 3 weeks of hardening at 4° C, causes a significant increase in frost resistance compared to controls receiving the cold treatment but not the growth retardants. He also reported that none of the three is effective at preventing or slowing down dehardening caused by exposure of hardened plants to 21° C for 5 days.

Manipulation of cold hardiness of alfalfa by synthetic growth retardants has been observed by several workers. A representative article, containing references to other studies, is that of PAQUIN et al. (1976), who applied CCC, daminozide and phosphon to alfalfa in the field and found that cold hardiness can be enhanced by all three, with widely varying results depending principally upon time of application and temperature. Such dependence is similar to that found for box elder by IRVING (1969b). PAQUIN et al. (1976) found CCC inferior to daminozide and phosphon for enhancing cold hardiness in alfalfa.

CHEN and LI (1976) experimented with CCC and daminozide on various potato (*Solanum*) species, attempting to increase their cold hardiness. They determined that CCC sprayed at 2000 ppm increased the hardiness of *S. acaule* and *S. commersonii* each by 1° C but had no effect on *S. tuberosum*. Daminozide had no effect on the hardiness of any of the three species.

KACPERSKA-PALACZ et al. (1969) sprayed cabbage with CCC (2000 ppm) and daminozide (4000 ppm) and found that the effectiveness of the treatment depended on photoperiod and temperature to which the plant had previously been exposed. They grew their plants in growth chambers and measured several variables associated with hardening: morphological changes, tissue water content, leaf succulence, osmotic potential of cell sap, carbohydrate content, nitrogen-containing compounds, and frost resistance. Both CCC and daminozide were found to have no effect on the level of investigated components of nonhardened plants – those exposed to long days and moderate temperatures. Plants in the hardening conditions (SD, LT), when treated with either CCC or daminozide, were reported to have enhanced leaf succulence (ratio of fresh weight of leaf blades to their surface area), increased cell sap osmotic potential, and increased content of water-soluble proteins and nonprotein nitrogen-containing compounds. The treated plants in the hardening conditions were more resistant to frost damage than were the controls, and CCC was found to be more effective than daminozide. The authors offer several alternative hypotheses to explain the enhanced effect of these growth retardants when applied to plants exposed to SD/LT as compared to the same plants under LD/HT: (1) LD/HT conditions cause a large increase in dry weight which could "dilute" the growth retardants. Also, under these conditions the retardants may move rapidly from leaves to roots, where they are not active. (2) The activity of the growth retardants may be greater at lower temperatures. (3) The retardants may express their activity by enhancing rates of reactions initiated in plants by SD/LT conditions.

MARTH (1965) also experimented with CCC and daminozide on cabbage and found increased frost resistance. The chemicals were sprayed on young field plants in the fall and exposed to outdoor winter conditions with the lowest temperature being −19° C. Growth was significantly suppressed by the retardant treatments. All the treated plants survived and 90% of them resumed growth when brought into a greenhouse the next spring, whereas 40–60% of the control plants were killed by winter frost, and those that survived had damaged terminal buds.

Other plants which have shown an increase in frost resistance accompanied by growth retardation when treated with CCC are: tomato (MICHNIEWICZ and KENTZER 1965), pear (MODLIBOWSKA 1965), and winter barley (BELZILE et al. 1973).

CALDER et al. (1973) examined the effects of daminozide on the physiology of alfalfa and white clover (*Trifolium repens* L. cv. Ladino). As well as a suppression of growth, they noted an improvement in the frost resistance of both plants, this improvement being temporary in the case of alfalfa. In clover, daminozide can induce a frost-hardy condition independent of temperature hardening; this result was repeated in successive years. Daminozide has also been found effective in reducing raspberry (*Rubus strigosus*) cane dieback due to freezing (GRANGER and HOGUE 1968). Application in June and July was much more effective than in August, in apparent contradiction to the fairly general result that application during autumn when days are shorter and cooler gives more frost protection.

IRVING and LANPHEAR (1968), in their study of the cold hardiness of box elder, applied daminozide and AMO and found that both chemicals produce measureable increases of cold hardiness under LD conditions besides decreasing growth. They assumed that this effect resulted from counteracting endogenous GA.

Phosphon and daminozide were applied to *Chrysanthemum morifolium* cultivars Ann Ladygo and Joanette by BRUCKEL and WIGGANS (1969), after the plants were partially hardened by exposure to SD/LT. Joanette suffered slightly less injury from exposure to −9.7° C when treated with either daminozide or phosphon than did control plants, while neither chemical afforded Ann Ladygo any enhanced resistance to frost damage, a result which is not surprising if the SD/LT treatment had already suppressed growth.

In concluding this section on exogenous application of growth regulators to enhance cold hardiness, the work by FUCHIGAMI (1977, FUCHIGAMI et al. 1982) provides some clarification. Overwintering plants harvested before they have reached "vegetative maturity" suffer various injuries during the subsequent winter, whereas if they are vegetatively mature the plants can be defoliated, harvested, stored outdoors or transplanted without ill effects. The time period over which vegetative maturity develops is rather sharp, at least in the case of red-osier dogwood which in a 1-week period drops from 90% stem dieback to no stem dieback following defoliation. The date varies according to environmental conditions, and an empirical model for predictive purposes has been described (KOBAYASHI et al. 1983). This stage of development may well provide the key to the sometimes conflicting results obtained when growth regulators are applied exogenously to affect cold hardiness of woody perennials. Especially relevant here is the GA work reviewed in the previous section. Application of growth promoters before vegetative maturity should either not affect hardiness or should promote it by producing more vigorous, healthy plants. Note that in those studies in which GA_3 was found to promote hardiness, it was probably applied prior to vegetative maturity. After this stage has been reached, application of growth promoter would be counter-productive to cold acclimation, tending to bring the plants back into active growth. By the same reasoning, growth retardants should either have little effect or reduce the ability of the plant to achieve full hardiness when applied before vegetative maturity, but should enhance hardiness development after maturity has been achieved.

5 The Relationship Between Chilling and Growth Regulators

The injury suffered by many tropical and subtropical plants, and some from temperate zones, which results when they are exposed to temperatures between 0° and about 12° C, is now most commonly referred to as chilling injury (LYONS 1973). Such economically important crops as citrus, avocado (*Persea americana*), banana (*Musa sapientum*), pineapple (*Ananas comosus*), tomato, and apple (actively growing) are susceptible to this injury. Considerable work has been done on both causes and effects of chilling injury, as can be seen in the review by LYONS. However, changes in natural plant hormone levels during exposure of chilling-sensitive plants to the aforementioned temperature range has received only a fraction of the attention which has been focused upon the relationship between plant hormones and subzero temperature stress. Further, although synthetic growth regulators have been applied to some chilling-sensitive plants, most application has been made to protect against freezing, not chilling temperatures.

In one study on the effect of chilling on plant hormone content, transfer of cocklebur (*Xanthium strumarium* L.) from a warm greenhouse to a growth chamber with 10° days and 5° C nights produced an increase in ABA content of the leaves (RASCHKE et al. 1976). This increase was found to be particularly large in young, incompletely expanded leaves. Levels of ABA were determined by a combination of TLC and analytical GC; bioassays were not performed. It was further shown that this chilling-induced increase in ABA level was not primarily due to any decrease in water content caused by the temperature treatment. However, water content is not a particularly good measure of water status; had they followed osmotic or water potential, which may have changed, their conclusions might have been different.

RIKIN and RICHMOND (1976) studied the effect of exogenously applied ABA on 3-day-old cucumber seedlings exposed to a chilling regime of 1.5° for 24 h in the dark, RH 80–85%. They reported that a 10^{-4} M ABA spray reduced leakage of cell contents to 50% of that by control plants. Such leakage is a typical chilling injury. Water loss was decreased, and the ABA-treated plants were able to regain full turgor after chilling-induced wilting much more rapidly than control plants. Morphological differences between ABA-treated and non-ABA-treated chilled plants were found to exist 8 weeks after resumption of pre-experimental conditions. The control plants had less new growth than the ABA-treated plants, suggesting that chilling may have damaged the apical buds of the control plants. Coupling these results with their finding that high relative humidity during chilling greatly reduces injury led the authors to hypothesize that ABA may afford protection against chilling injury by improving plant water belance.

MARKHART (1984) pretreated soybean (*Glycine max*) plants by placing their roots in 50 μM ABA or water for 1 h before exposing the roots of the plants to chilling conditions of 10° C. The ABA-treated plants, unlike the control plants, did not wilt. Since the ABA-treated plants had lower leaf stomatal resistance and slightly higher leaf water potential, MARKHART concluded that the ABA increased root hydraulic conductivity. He suggests ABA may act by protecting membranes from damage normally imparted by chilling conditions.

In a related study RIKIN et al. (1976) tested the possibility that root salination and water deprivation, both of which stresses are known to elevate ABA levels, induced increased chilling resistance in cucumber seedlings. Pretreating roots with 0.10 and 0.15 M NaCl solutions was found to increase chilling resistance, as well as to increase endogenous levels of ABA. The same pattern was observed when the cucumber seedlings were deprived of water. The authors conclude that the increase in chilling resistance in cucumber seedlings resulting from salt/or water stress is an example of cross-adaptation resulting at least in part from modification of hormonal balance and involving, in particular, an increase in endogenous ABA.

The activity of IAA oxidase from green bananas was shown by HAARD (1973) to have a complex temperature dependence. Between 15° and 30° C the enzyme-catalyzed oxidation of IAA has a low activation energy, while between 5° and 15° C the process has a high activation energy. This indicates that IAA might accumulate when banana is exposed to chilling temperatures. The author concludes that this build-up may be related to chilling unjury. However, IAA activity is presumably governed by many other processes such as synthesis, conjugation, and receptor sites, so to conclude that auxin activity within the tissue may be determined by IAA oxidase activity is tenuous.

Increased ethylene production by plants exposed to chilling temperatures is evidently a fairly general phenomenon. COOPER et al. (1969) found that exposure of grapefruit (*Citrus pardisi*), orange (*C. sinensis*), and tangerine (*C. reticulata*) trees to temperatures between 0° to 10° C caused enhanced ethylene formation in the airspace between peel and fruit. This increase in ethylene concentration was always associated with loss of chlorophyll from leaves and fruit, allowing carotenoid pigment to be seen. These workers also carried out this study on three avocado varieties, chilling-sensitive Simpson and Booth 7 and chilling-tolerant Lula. The two chilling-sensitive varieties accumulated more than twice as much ethylene in the fruit when chilled at 5° C for 3 days after harvest than when held at 20° C for the same period, while Lula had from 20 to 100 times as much ethylene when kept at 20° C than when kept at 5° C for 3 days.

Holding harvested Bosc pears (*Pyrus communis*) at 5° or 10° C for 7 days before increasing the temperature to 20° C causes an almost immediate increase in ethylene production, according to SFAKIOTAKIS and DILLEY (1974), accompanied by ripening. Holding the fruit at a constant 20° results in very low ethylene concentrations. Somewhat surprisingly, 0° was a less effective chilling temperature than 5° or 10° C. The metabolic process which is evidently induced by chilling temperatures is greater at 5° than at 10° and slower yet at 0° C.

The relationship between ethylene produced by chilling and by water stress was examined by WRIGHT (1974) in bean (*Phaseolus vulgaris* L. cv. Canadian Wonder). Seedlings of this plant chilled at 5° C and 45% relative humidity lose fresh weight, produce ethylene at an elevated rate when warmed to 25° C, and leak electrolytes at an increased rate when placed in water. Since ethylene formation is not increased by exposure to 5° C when the plants are fully hydrated, WRIGHT concludes that it is the partial dehydration accompanying chilling which is the primary cause of enhanced ethylene formation. Strengthening this conclusion is his finding that rates of ethylene formation during the 1st h of exposure to 20% relative humidity increases for leaves at 25° as well as at 5° C.

One exception to the lack of examples of applied growth regulator increasing tolerance to chilling conditions is a recent report by TSENG and LI (1984). They found that root application of mefluidide imparted protection for cucumber and corn to chilling conditions of 0° to 3° C. Interestingly, TSENG and LI found that application of mefluidide above the optimum concentrations for the two plants provided no protective effect. The mechanism by which mefluidide causes the protection from chilling injury is not known.

6 The Relationship Between Deacclimation and Growth Regulators

Cold hardiness is multi-faceted, one of the least-understood facets being the rapid deacclimation and subsequent reacclimation which can take place in the field. Warm spells, which initiate this cycle, are more common in late winter.

Such temperature fluctuations can be quite rapid and the resulting damage to commercially important tree crops has often been substantial.

Suggestive evidence of a connection between hormone levels and sensitivity to deacclimation conditions in early spring has existed for several years. PROEBSTING and MILLS (1964) showed that spraying GA_3 solutions on mature Elberta peach trees in August and September of the previous year delayed blossoming in the spring by up to 7 days. Similar results have been obtained by several other workers. Although hardiness levels were not followed prior to bud burst, it is reasonable to suspect that the delay of growth was accompanied by a delay in the onset of deacclimation.

IRVING and LANPHEAR (1967a), in a study primarily aimed at the relationship between dormancy and deacclimation (or dehardening), found that exogenous ABA significantly inhibited deacclimation. They also examined the effect of GA_3. Working with box elder and doublefile *Viburnum*, they developed a regimen under which the plants could become fully cold-hardy and yet remain dormant since their chilling requirement was not satisfied. This was achieved by by a 6-week treatment of 4.4° in darkness for 6 h day^{-1}. Their description does not make clear what the conditions were during the other 18 h of each day; the temperature was raised, presumably to 21° C, but the photoperiod during this treatment is not specifically stated. Plants which were kept at 4.4° C for 24 h day^{-1} during the 6-week treatment reached the same hardiness level but were nondormant, having had their chilling requirement satisfied. Two groups of plants from each genera were acclimated, being divided between the two hardening treatments, and then held at 21° C continuously for 7 days. This treatment at 21° C significantly deacclimated the nondormant plants of both species, while the dormant doublefile *Viburnum* plants were deacclimated to a lesser degree and the dormant box elder plants were not deacclimated at all.

After GA_3 application to both groups of box elder plants and deacclimation at 21° for 7 days the plants had the same killing points as had been found in the same experiment without GA_3. However, it was found that GA_3 treatment did break dormancy of the dormant plants. Thus breaking dormancy by GA_3 treatment did not speed deacclimation, whereas breaking dormancy by exposure to low temperature did.

The effect of ABA on deacclimation was examined by placing freshly cut, nondormant box elder stems in either water or a 25 ppm ABA solution for 9 days at 21° C. The killing point of the stems in plain water increased by 13° but there was only a 6.7° C increase in killing-point temperature for the ABA-treated plants. The authors assumed that ABA retarded deacclimation by either stimulating or reintroducing dormancy. Their ABA supply was unfortunately insufficient to allow testing of this assumption.

In further work on box elder IRVING (1969b) again examined the effect of GA_3 on deacclimation and in addition experimented with AMO and CCC. Test plants were given 5 weeks of long days (sic) and 4.4° C nights followed by 3 weeks of darkness at 4.4° C, thus partially acclimating them and rendering them nondormant. The plants were then deacclimated by holding them at 21° C for 5 days. The photoperiod during this 5-day period was not specified. One group was irrigated with water and three other groups were irrigated with

1000 ppm solutions of AMO, CCC, or GA_3. The average killing temperature for plants receiving only water was −15° compared to −22° C, which was the killing temperature of a separate group receiving the same acclimating treatment but no deacclimating treatment. The killing temperatures of the AMO- and CCC-treated plants did not differ significantly from that of the group irrigated with plain water. However, in the GA_3-treated group the killing temperature was elevated to −11.4° C. Similar results were found when these growth regulators, and also daminozide, were applied by dipping plant stems in 1000 ppm solutions rather than as a soil drench. Thus in this study GA_3 increased the deacclimation rate of hardy, nondormant plants as opposed to their earlier finding (IRVING and LANPHEAR 1967a). The conclusion was that single applications of growth inhibitors immediately prior to onset of deacclimation conditions are an ineffective means of increasing survival. IRVING suggested that multiple applications of these compounds at earlier dates may be the only means of retarding development enough to provide adequate protection.

These results are similar to those reported by TUMANOV et al. (1970) for blackcurrant and birch. As previously mentioned, they sprayed dormant plants from both species six times with a 100 mg l^{-1} GA solution, keeping them under short days and warm temperatures during treatment. Specific photoperiod and temperatures were not mentioned in the article. This treatment did not cause deacclimation. They also examined the effect of GA on plants in transition from deep to imposed dormancy, using slips cut from mature willow trees [*Salix alba*(?)] in February. One batch of slips was held in water for 1 day, another batch in a 100 mg l^{-1} GA solution; this treatment did not cause deacclimation, but when slips were exposed to GA for a period of 1 month at 2° C during this part of the winter (GA concentration and photoperiod not specified) they were killed at −40°, while control slips survived at −60° C.

The first part of their results is similar to that reported by IRVING and LANPHEAR (1967a): one-time GA application does not cause deacclimation of dormant woody plants. The result from the willow study is not at odds with what was found by IRVING and LANPHEAR (1967a) and by IRVING (1969b), but one wonders whether a noticeable deacclimation of the willow slips might not have occurred prior to 1 month after GA application commenced, particularly since nondormant *Acer* plants had their deacclimation enhanced by only 5 days of GA_3 application (IRVING 1969b).

When these results are coupled with the previously mentioned finding by TUMANOV et al. (1970) that natural gibberellin levels rise to a maximum in late February and March, when plants are generally more susceptible to deacclimation, it is apparent that there is at least strong suggestive evidence that gibberellin levels play an important role in the deacclimation process.

The study by TIMMIS and WORRALL (1974) on Douglas-fir showed that a translocatable deacclimation factor is produced in this plant during flushing. As mentioned in an earlier section they used a "double-branched" experimental design in which each branch of a seedling could be exposed to a different temperature. All branches in a particular experiment had the same light exposure. In one of their experiments plants were exposed to natural daylength in a greenhouse. From December to the following June chilled branches were cooled during a 15-h period coinciding approximately with night at 3°±2° C, while warm branches were maintained at night temperatures of 16.5°±2° C. Day temperatures ranged from 18° to 30° C and were the same for all branches. Swelling of buds and subsequent flushing occurred only on those branches which had been chilled and were on a plant the other branch of which had been in the warm environment. This result led to the hypothesis that flushing is not only dependent on dormancy being broken by adequate chilling, but also on a translocatable factor(s) produced during warm nights. They found that although all-warm plants had not deacclimated 6 weeks after flushing began, warm branches on plants having the other branch chilled (and flushing) had deacclimated significantly. This led to the hypothesis that a second translocatable factor is produced during the expansion of new shoots on chilled branches which can cause deacclimation of older foliage on the same plant. Further, because the old foliage on the chilled branch beneath the new growth did not deharden, it appears that low night temperature inactivates this factor. An experiment was carried out which indicated that the flush-promoting factor produced in warm branches is not gibberellin. No attempt was made to identify the deacclimation-producing factor, but sound arguments were presented that these translocatable factors are hormones. The authors state that, while the initial deacclimation of Douglas-fir is triggered only by temperature, their results indicate that final deacclimation is dependent on factors produced in and translocated from expanding shoots.

It appears that the substantial effort which has been made to understand the nature and extent of hormonal involvement in bud burst may also be applicable to deacclimation. For example, HEWETT and WAREING (1974) have shown that cytokinin levels in poplar (*Populus robusta* SCHNEID) vegetative buds, nondetectable in December and January, reach their maximum in February and March, closely coinciding with bud break. Similar concentration dynamics have been observed for cytokinins in sugar maple by DUMBROFF and BROWN (1976), the peak occurring just before bud burst. At present, work has not been done to assess the role cytokinins play in deacclimation. Based on what has been found out about their importance in bud burst, and the fact that their concentration levels peak during the time when deacclimation occurs, such a study certainly seems worthwhile.

In summary, the deacclimation process appears at least in part to be triggered by plant hormones. Gibberellins have been implicated as the most important hormone group involved in deacclimation. More work in this area could provide significant gains in both basic understanding of the process and the more practical aspects of avoiding spring frost damage.

7 Summary

Resistance to low temperature stress in the plant kingdom is of paramount importance. Evolutionary pressure has produced a broad spectrum of physiological responses to low temperature exposure. Research to date indicates that plant hormones are intimately involved in many, if not all of these responses.

Cold acclimation, the initial phase of plant response to low temperature, is closely related to the onset of winter dormancy. However, since hardy plants can maintain their hardiness after they have passed from deep to imposed dormancy, it is clear that cold hardiness cannot be simply equated to deep dormancy. It appears safe to say that, for plants capable of acclimating, whenever growth stops, acclimation begins, and perhaps more importantly, acclimation does not begin unless growth slows or stops. It is also clear that once a plant begins to grow, its cold hardiness rapidly diminishes. This coupling of growth and deacclimation is somewhat more tenuous, however, since deacclimation can begin before visible signs of growth develop.

Studics have shown that there are translocatable factors which both promote and inhibit cold acclimation in a variety of plants. ABA has been at least tentatively identified as a hardiness promoter while the inhibitor(s) has been found to behave very much like GA_3.

It appears that it is the ratio of promoters to inhibitors which is the most important factor in the hormonal component of acclimation rather than the level of any individual hormone, and different plants regulate this ratio in different ways in response to stimuli which initiate acclimation. At present, other plant hormones such as cytokinins and auxins have not been found to play prominent roles, whereas ABA and gibberellins apparently do. In some cases application of GA_3 to fruit trees during active growth has led to an enhancement of freezing resistance during the following cold season. These conflicting results are apparently resolved by the concept of vegetative maturity described by FUCHIGAMI, pointing out that the timing of application of growth regulators is crucial to the resulting effect.

Synthetic growth retardants have been applied to a number of plant species by many workers seeking to protect against low temperature damage. The most widely tested chemical for promotion of hardiness has been CCC; also commonly applied were AMO, phosphon, daminozide and ethephon. Results vary widely depending upon such factors as timing, concentration, and application method. However, growth retardation resulting from use of these compounds has almost always been accompanied by enhanced cold hardiness. It remains to be resolved whether the effect on cold hardiness of growth retardants is direct or indirect. The mechanism of action of growth retardants with respect to cold hardiness may well be different for woody perennials than for herbaceous plants since the former go through a vegetative maturity stage after which they can more successfully overwinter, whereas herbaceous plants do not.

Comparatively little work has focused on changes in plant hormone levels of chilling-sensitive plants during exposure to low temperature, but that which has been done indicates that ABA levels are tied to chilling stress. RICHMOND's group has been especially active in this area, showing not only that exogenous ABA can reduce chilling injury but also that manipulations which increase endogenous ABA can increase chilling resistance.

There is suggestive evidence that gibberellins are actively involved in the deacclimation process. Exogenous application of synthetic growth retardants during late winter and early spring to retard deacclimation has met with little success; IRVING and LANPHEAR (1967a) did find that exogenous ABA was able to retard deacclimation in box elder, however. Although the work that has been done in trying to understand the involvement of hormones in deacclimation is of high quality, it appears a neglected area. The considerable body of research on hormonal involvement in bud burst may, however, be applicable to deacclimation.

References

Belzile L, Paquin R, Therrien HP (1973) Effects of CCC on the development of cold resistance by winter barley cult. Dover. Can J Plant Sci 53:31–36

Bolduc RJ, Cherry JH, Blair BO (1970) Increase in indoleacetic acid oxidase activity of winter wheat by cold treatment and gibberellic acid. Plant Physiol 45:461–464

Boussiba S, Rikin A, Richmond AE (1975) The role of ABA in cross-adaptation of tobacco plants. Plant Physiol 56:337–339

Bray EA (1982) The involvement of abscisic acid in the cold acclimation of *Cornus sericea* L. Ph D Thesis, Univ Minnesota

Bruckel DW, Wiggans SC (1969) Effects of certain growth regulators on low temperature injury of cultivars of *Chrysanthemum morifolium*. HortScience 4:220–221

Calder FW, Canham WD, Fensom DS (1973) Some effects of Alar-85 on the physiology of alfalfa and Ladino clover. Can J Plant Sci 53:269–278

Chen HH, Li PH, Brenner ML (1983) Involvement of abscisic acid in potato cold acclimation. Plant Physiol 71:362–365

Chen P, Li PH (1976) Effect of photoperiod, temperature, and certain growth regulators on frost hardiness of *Solanum* species. Bot Gaz 137:105–109

Chen THH, Gusta LV (1983) Abscisic acid-induced freezing resistance in cultured plant cells. Plant Physiol 73:71–75

Chrominski A, Belt H, Michniewicz M (1969) Effect of CCC on frost resistance, yield and seed quality indexes of winter rape. Rocz Nauk Roln 95:191–197

Cooper WC, Rasmussen GK, Waldon ES (1969) Ethylene evolution stimulated by chilling in *Citrus* and *Persea* sp. Plant Physiol 44:1194–1196

Daie J, Campbell WF (1981) Response of tomato plants to stressful temperatures. Plant Physiol 67:26–29

Dror Z, Rikin A, Richmond AE (1976) Hormonal regulation of plant resistance to subzero temperatures. II. Influence of treatment with cytokinins. Isr J Bot 25:96

Dumbroff EB, Brown DCW (1976) Cytokinin and inhibitor activity in roots and stems of sugar maple seedlings through the dormant season. Can J Bot 54:191–197

Edgerton LJ (1967) Some effects of gibberellin and growth retardants on bud development and cold hardiness of peach. Proc Amer Soc Hortic Sci 88:197–203

Fuchigami LH (1977) Determining maturity and predicting harvest dates of deciduous plants. Proc Woody Ornamentals Winter Storage Symp 17–21

Fuchigami LH, Weiser CJ, Evert DR (1971) Induction of cold acclimation in *Cornus stolonifera* Michx. Plant Physiol 47:98–103

Fuchigami LH, Weiser CJ, Kobayashi K, Timmis R, Gusta LV (1982) A degree growth stage (°GS) model and cold acclimation in temperate woody plants. In: Li PH, Sakai A (eds) Plant hardiness and freezing stress. Academic Press, New York, pp 93–116

Granger RL, Hogue EJ (1968) Effect of Alar on winter hardiness of raspberries. Can J Plant Sci 48:100 – 101

Haard NF (1973) Chilling injury of green banana fruit: kinetic anomalies of IAA oxidase at chilling temperatures. J Food Sci 38:907–908

Hewett EW, Wareing PF (1974) Cytokinin changes during chilling and bud burst in woody plants. In: Bieleski RL, Ferguson AR, Cresswell MM (eds) Mechanisms of regulation of plant growth. Bulletin 12, R Soc NZ Bull 12:693–701

Holubowicz T, Pieniazek J, Khamis MA (1982) Modification of frost resistance of fruit plants by applied growth regulators. In: Li PH, Sakai A (eds) Plant hardiness and freezing stress. Academic Press, New York, pp 541–559
Howell GS, Weiser CJ (1970) The environmental control of cold acclimation in apple. Plant Physiol 45:390–394
Irving RM (1969a) Characterization and the role of an endogenous inhibitor in the induction of cold hardiness in *Acer negundo*. Plant Physiol 44:801–805
Irving RM (1969b) Influence of growth retardants on development and loss of hardiness of *Acer negundo*. J Am Soc Hortic Sci 94:419–422
Irving RM, Lanphear FO (1967a) Dehardening and the dormant condition in *Acer* and *Viburnum*. Proc Am Soc Hortic Sci 91:699–705
Irving RM, Lanphear FO (1967b) Environmental control of cold hardiness in woody plants. Plant Physiol 42:1191–1196
Irving RM, Lanphear FO (1967c) The long day leaf as a source of cold hardiness inhibitors. Plant Physiol 42:1384–1388
Irving RM, Lanphear FO (1968) Regulation of cold hardiness of *Acer negundo*. Plant Physiol 43:9–13
Kacperska-Palacz A, Egierszdorff S (1972) Effects of cold hardening and CCC treatment on hydration and frost and desiccation hardiness of plant tissues. Bot Gaz 133:355–360
Kacperska-Palacz A, Blaziak M, Wcislinska B (1969) The effect of growth retardants CCC and B-9 on certain factors related to cold acclimation of plants. Bot Gaz 130:213–221
Khisamutdinova VI, Vasil'yev IM, Kuz'mina GG, Vershinin AA (1975) Effect of CCC on energy metabolism and the state of water in winter wheat during the process of hardening. Sov Plant Physiol 22:917
Kobayashi K, Fuchigami LH, Weiser CJ (1983) Modeling cold hardiness of red-osier dogwood. J Am Soc Hortic Sci 108:376–381
Levitt J (1980) Responses of plants to environmental stresses, 2nd ed, vol 1. Academic Press, New York
Lyons JM (1973) Chilling injury in plants. Ann Rev Plant Physiol 24:445–466
Markhart AH III (1984) Amelioration of chilling-induced water stress by abscisic acid-induced changes in root hydraulic conductance. Plant Physiol 74:81–83
Marth PC (1965) Increased frost resistance by application of plant growth-retardant chemicals. J Agric Food Chem 13:331–333
Michniewicz M, Kentzer T (1965) The increase in frost resistance of tomato plants through application of CCC. Experientia 21:230–231
Modlibowska I (1965) Effects of CCC and GA on growth, fruit bud formation and frost resistance in one-year-old pear trees. Nature 208:503–504
Paquin R, Belzile L, Willemot C, St.-Pierre J-C (1976) Effects of some growth retardants and gibberellic acid on the frost resistance of alfalfa (*Medicago sativa*). Can J Plant Sci 56:79–86
Proebsting EL Jr, Mills HH (1964) Gibberellin-induced hardiness responses in Elberta peach flower buds. Proc Am Soc Hortic Sci 85:134–140
Proebsting EL Jr, Mills HH (1973) Bloom delay and frost survival in ethephon-treated sweet cherry. HortScience 8:46–47
Proebsting EL Jr, Mills HH (1974) Time of gibberellin application determines hardiness response of Bing cherry buds and wood. J Am Soc Hortic Sci 99:464–466
Proebsting EL Jr, Mills HH (1976) Ethephon increases cold hardiness of sweet cherry. J Am Soc Hortic Sci 101:31–33
Proebsting WM (1983) Transmissible factors regulating shoot growth of *Cornus sericea* L. Z Pflanzenphysiol 112:191–198
Raese JT (1983) Conductivity tests to screen fall-applied growth regulators to induce cold hardiness in young Delicious apple trees. J Am Soc Hortic Sci 108:172–176
Raschke K, Pierce M, Popiela CC (1976) ABA content and stomatal sensitivity to CO_2 in leaves of *Xanthium strumarium* L. after pretreatments in warm and cold growth chambers. Plant Physiol 57:115–121

Reid DM, Pharis RP, Roberts DWA (1974) Effects of four temperature regimens on the gibberellin content of winter wheat *cv.* Kharkov. Physiol Plant 30:53–57
Rikin A, Richmond AE (1976) Amelioration of chilling injuries in cucumber seedlings by abscisic acid. Physiol Plant 38:95–97
Rikin A, Blumenfeld A, Richmond AE (1976a) Chilling resistance as affected by stressing environments and ABA. Bot Gaz 137:307–312
Rikin A, Dror Z, Richmond AE (1976b) Hormonal regulation of plant resistance of sub-zero temperatures. I. Influence of environmental conditions on the response to abscisic acid (Abst.). Isr J Bot 25:96
Roberts DWA (1971) The effect of CCC and gibberellins A_3 and A_7 on the cold hardiness of Kharkov 22MC winter wheat. Can J Bot 49:705–711
Sfakiotakis EM, Dilley DR (1974) Induction of ethylene production in Bosc pears by postharvest cold stress. HortScience 9:336–338
Skene KGM (1972) Cytokinins in the xylem sap of grape vine canes: changes in activity during cold storage. Planta 104:89–92
Stushnoff C (1972) Breeding for cold hardiness (symposium). Breeding and selection methods for cold hardiness in deciduous fruit crops. HortScience 7:10–22
Stushnoff C (1974) Cold hardiness of woody plants. 19th Int Hortic Congr, Warsaw
Timmis R, Worrall J (1974) Translocation of dehardening and bud-break promoters in climatically 'split' Douglas-fir. Can J For Res 4:229–237
Tseng MJ, Li PH (1984) Mefluidide protection of severely chilled crop plants. Plant Physiol 75:249–250
Tumanov II, Trunova TI (1958) The effect of growth processes on the capacity for hardening. Sov Plant Physiol 5:108–117
Tumanov II, Kuzina GV, Karnikova LD (1970) Effect of gibberellins on the period of dormancy and frost resistance of plants. Sov Plant Physiol 17:737–745
Tumanov II, Trunova TI, Smirnova NA, Zvereva GN (1976) Role of light in development of frost resistance of plants. Sov Plant Physiol 23:109–114
Vasil'yev IM (1961) Wintering of plants (Levitt J, ed of English translation) Roger and Roger, Washington DC, pp 150–151
Waldman M, Rikin A, Dorvat A, Richmond AE (1975) Hormonal regulation of morphogenesis and cold-resistance. II. Effect of cold acclimation and of exogenous abscisic acid on gibberellic acid and abscisic acid activities in alfalfa seedlings (*Medicago sativa* L.). J Exp Bot 26:853–859
Weiser CJ (1970) Cold resistance and injury in woody plants. Science 169:1269–1278
Wright M (1974) The effect of chilling on ethylene production, membrane permeability and water loss of leaves of *Phaseolus vulgaris*. Planta 120:63–69
Wünsche U (1966) Influence of CCC and gibberellin A_3 on frost hardiness of winter wheat. Naturwissenschaften 53:386–387
Young R (1971) Effect of growth regulators on citrus seedlings cold hardiness. J Am Soc Hortic Sci 96:708–710

12 Wind and Other Mechanical Effects in the Development and Behavior of Plants, with Special Emphasis on the Role of Hormones

M.J. Jaffe

"I hear the wind among the trees
playing celestial symphonies;
I see the branches downward bent,
like keys of some great instrument."
– Longfellow, A Day of Sunshine

1 Introduction

There are three types of visible responses that are available to plants that have been mechanically stimulated: thigmonasty, thigmotropism, and thigmomorphogenesis (Fig. 1). The prefix "thigmo" indicates "touch". A nastic response is one in which the direction of the movement is determined by anatomical constraints, and not by the direction of the stimulus. An additional characteristic of almost all nastic responses is that they are freely reversible, and may therefore be said to have the property of elasticity. The duration of the recovery period is characteristically longer than that of the motor response period. A tropic response is typically a bending movement, the direction of which is always determined by the vector of the stimulus. Tropisms often involve assymetric growth on one side of an organ, which results in its leaning toward (positive tropism) or away from (negative tropism) the stimulus vector. Tropic responses are sometimes, but not always, irreversible, and may be said to often have the property of plasticity. A morphogenetic response is one which affects the form of the plant, involves growth and/or differentiation, and is irreversible.

It should be pointed out that these definitions say nothing about the cellular mechanisms. For example, both geotropism and epinasty are caused by the asymmetric growth of cells on one side of an organ, and yet are classified differently because the direction of the first movement is determined by the direction of the stimulus, whereas the direction of the second is determined by the anatomy of the plant. In contrast, nyctinasty of various leaves is caused by contractions of motor cells, as is thigmotropism of the stamens of *Portulaca grandiflora,* Hook (Jaffe et al. 1977). Thus, the designations of tropic and nastic responses are meaningful only on the level of the whole plant, and seem to have little relevance at the cellular level. This will be discussed more thoroughly in the last section, but for the present, it will be useful to divide the responses into the three categories described above.

In analyzing the kind of systems to be considered here, it is worthwhile thinking of the mechanism in terms of a *sensory* component coupled to a *responding* component by one or more *transduction* steps. In general, this will be the framework upon which each of the systems to be discussed will be structured. It is hoped that this type of "systems analysis" will aid in an understanding of the various plant organs in their responses to mechanical stimulation.

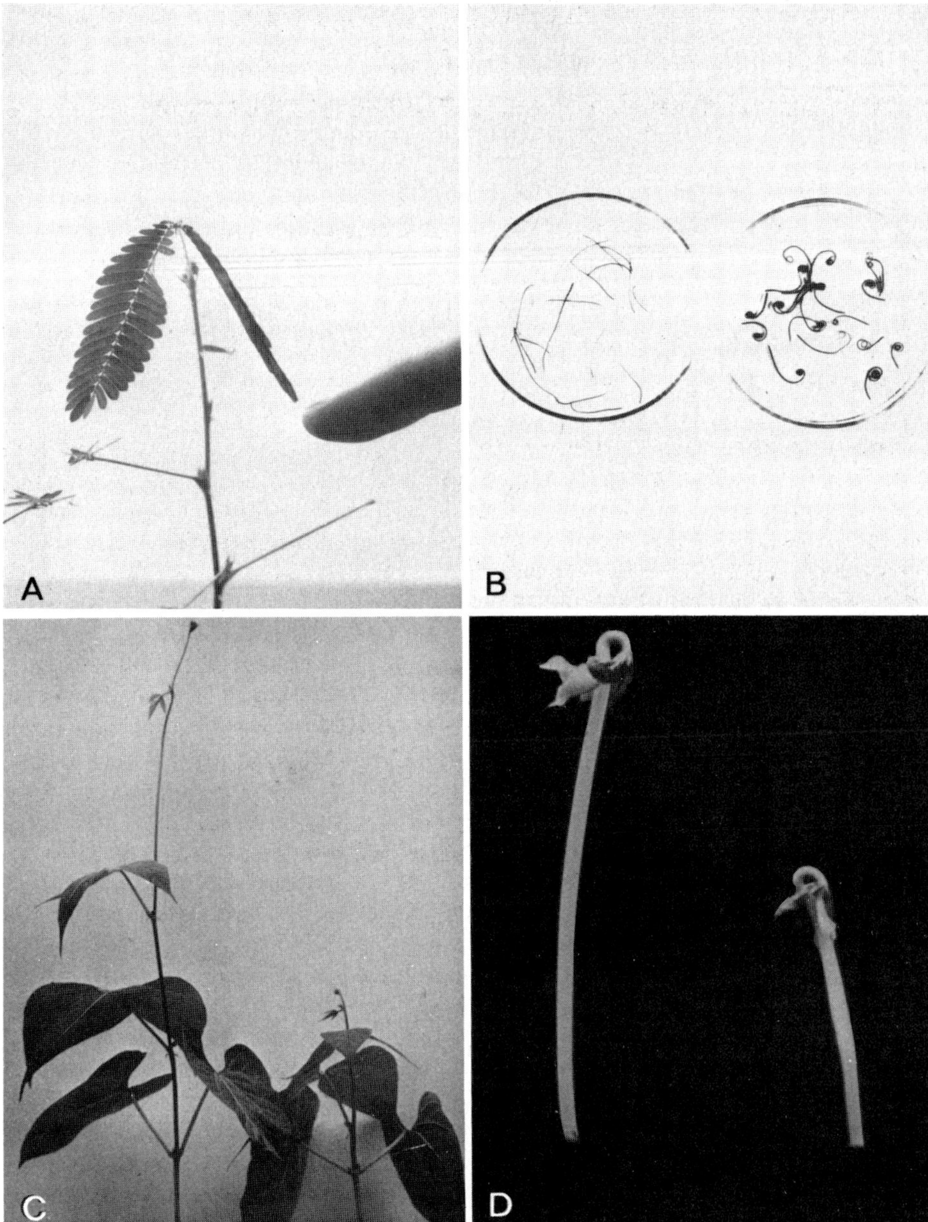

Fig. 1 A–D. Different kinds of plant responses to touch. Thigmonasty is shown in the sensitive mimosa (**A**), and excised pea tendrils (**B**); thigmomorphogenesis in beans can be seen in light-grown (**C**) and etiolated (**D**) plants. In each case, the unstimulated control is on the *left*, and the response to touching is on the *right*

The components of the environment that can produce mechanical perturbations in plants are many and varied [e.g., wind, pressure (roots in soil or shoots growing up through soil), animals brushing through the foliage, etc.]. The interrelationships of the various phytohormones, as well as specialized endogenous "factors" with the environmental stimuli in effecting changes in plants will be discussed. It should also be noted that other changes in metabolism due to mechanical perturbations are also briefly discussed, in order to place the information concerning hormones in the broadest possible context.[1]

2 Thigmonasty

2.1 The Thigmonastic Leaves of the Sensitive Mimosa and the Venus' Fly Trap

2.1.1 Sensitive Mimosa

The thigmonastic movements of the leaf of *Mimosa pudica* L. constitute one of the most interesting phenomena in the field of plant physiology. The well-known external features of these movements consist of: the folding together upward of the pinnule pairs; the lateral approach of the subpetioles to each other; and the sudden falling of the main petiole (Fig. 1 and 2). The thigmonastic response takes from 0.5 to 3 s, and recovery of the resting position takes about 16 min (BOSE 1913). When a leaf or pinnule is only slightly mechanically displaced, thigmonasty occurs. There seem to be no specially adapted sensory cells on the leaves, although large hairs, called "stimulators" by HABERLANDT (1914) may serve to assist by catching on the surface of a moving object, (e.g., an animal's fur). The sensory system for mechanical stimulation seems rather to reside in the pulvinus itself, in which, as in the fulcrum of a lever, there can be amplification of small displacements at the end of the leaf. The epidermal cells of the pulvinus are capable of a great deal of deformation (Fig. 1) and may be the sensory cells. If, however, the end of the leaf is traumatically wounded by being held in a flame or by being crushed, a nastic response will also occur. If the intensity of the stimulus is great enough, leaves above the stimulated leaf will also move, with a specific interval occurring between each movement. Other plants exhibiting leaf thigmonasty are listed in Table 5, together with their response and recovery time. It is generally believed that the mechanism is similar in all cases (HABERLANDT 1914), but satisfactory evidence for this is still lacking. Whatever the sensory system may be, there seems to be evidence for both kinds of biological integrative mechanisms in the sensitive mimosa: action potentials and hormones.

In a very interesting experiment, PFEFFER (1901), later elaborated by RICCA (1916) and by HIRSCH (1953), demonstrated that when the tip of a leaf is trauma-

1 For comparison the reader is referred to the relevant Chap. 15 of Vol. 12A of this Encyclopedia: P.S. NOBEL, Wind as an Ecological Factor, pp. 475–500 (1981)

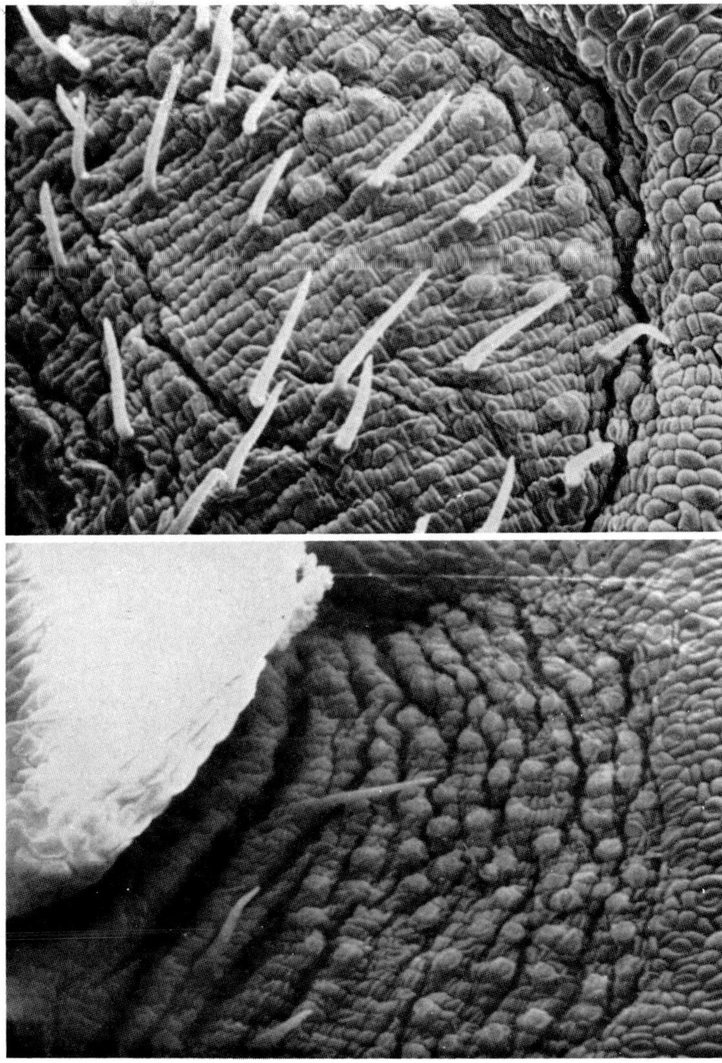

Fig. 2. Upper (adaxial) surface of the tertiary pulvinus of *Mimosa pudica* before (*top*) and after (*botton*) mechanical stimulation and movement. The crescent groove at the left of each picture is where the pulvinus (*top*) joints the lamina (*bottom*). Note the wrinkling of the epidermal cells of the closed pulvinus (× 150). (JAFFE unpublished)

tized, a stimulatory principle moves up the stem. The supertending internode was cut across and the two cut halves held apart by a tightly fitting glass tube containing water. After trauma and closing of the subtending leaf, a greenish turbidity could be seen to issue from the lower cut end and diffuse upward through the water. The rate of migration of this turbidity was computed, and the time calculated for it to reach the next supertending pulvinus. At precisely the predicted time, the leaf of that pulvinus fell (Fig. 3). The active principle

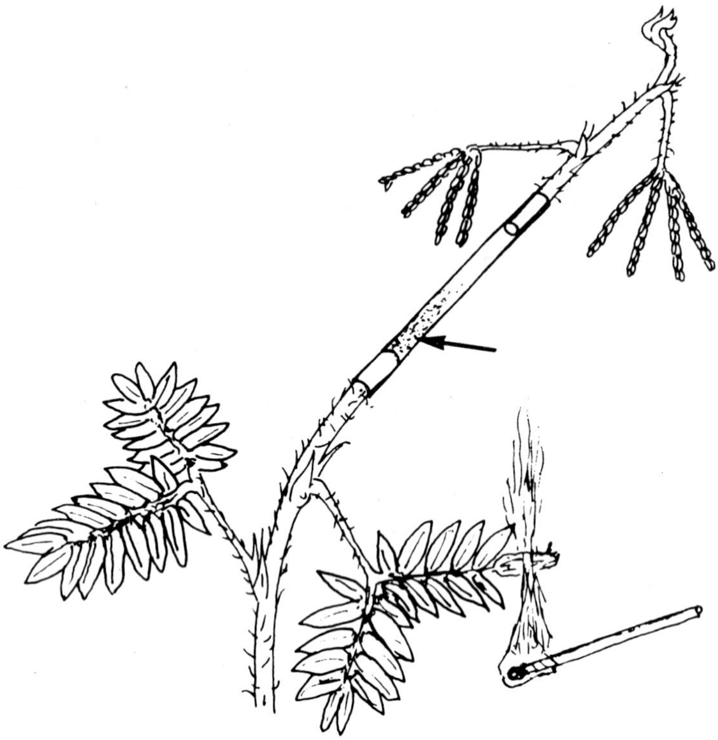

Fig. 3. *Mimosa* preparation for the observation of touch-induced turbidity (*arrow*) through the glass tube. (After HIRSCH 1953)

has not been identified, but HIRSCH (1953) has shown that extracts of *Mimosa* leaves will induce leaf movement when added to the soil. UMRATH (1972) has postulated than an excitatory substance triggers action potentials of the *Mimosa* when leaves are traumatized, and indeed, he has shown that such a substance can be isolated from *Mimosa* extracts, and used to stimulate other motor systems, such as guard cells of other species (UMRATH 1934). It is not known if the putative hormone described by RICCA (1916) and HIRSCH (1953) is the same as UMRATH's so-called "excitatory substance" (UMRATH 1972). The identity of these active principles is also unknown, however HESSE et al. (1957) have analyzed the eluate from cut *Mimosa* petioles. They concluded that the putative *Mimosa* hormone is highly sensitive to oxygen and that it behaves like a "reductone". It was further suggested that meso-inositol may be a precursor of the active principle. Another substance, mimosine, has been extracted and isolated from *Mimosa,* and identified as the amino acid 3-hydroxy-4-keto-pyridylalanine. Although mimosine has been shown to oppose the effect of auxin in bacteria (SUDA 1960), there is no direct evidence that it may be involved in thigmonasty of *Mimosa.* The only exogenously applied plant hormone that has been shown to induce epinasty in *Mimosa* is IAA. As early as 1936 IAA was applied in a lanolin paste to one side of the pulvinus, and pronounced

Fig. 4. The effect of IAA on leaf movement of mimosa. *Upper* treated with tap water; *lower* treated with 10^{-5} M IAA. (After WATANABE 1971)

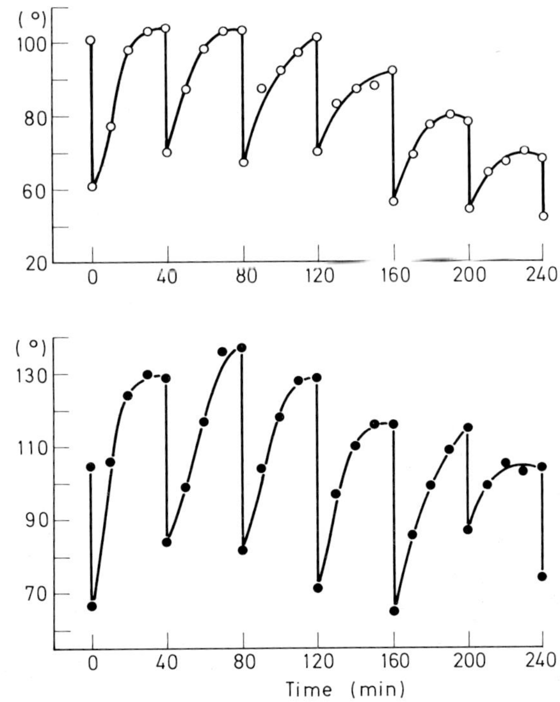

movement of the petioles and leaftlets resulted, (BURKHOLDER and PRATT 1936). More recently, WATANABE (1971) showed that IAA increases not only the amount of thigmonastic movement of *Mimosa* leaves, but also the amount and rate of recovery (Fig. 4). On the basis of cytochemical electron microscopy, he concluded that IAA increases the activity of a ouabain-sensitive ATP-ase on the surface of the motor cell tonoplast. The study of the roles of plant hormones in *Mimosa* thigmonasty is remarkable for its scarcity, and should provide a rich field for future investigations.[2]

The role of action potentials in thigmonasty of *Mimosa* leaves has been reviewed elsewhere (SIBAOKA 1966) and will not be discussed here in any detail. Several different types of action potentials which are propagated basipetally have been reported which travel from the point of stimulus down the rachilla and the rachis and then upward along the stem. They undoubtedly reflect the propagation of sensory information within the plant to the motor cells. These "medullary" pulvinal cells appear to be modified parenchyma cells (TORIYAMA 1953) and contain a very large central vacuole and a smaller contractile vacuole. TORIYAMA and JAFFE (1972) have suggested that Ca^{2+} act as an intracellular hormone, or "second messenger" during contraction of the motor cells, shown schematically in Fig. 5. Upon reception of the electrochemical stimulus (action potential), Ca^{2+} are released from the contractile vacuole to the central vacuole (TORIYAMA and JAFFE 1972) the contractile vacuole shrinking to less than half

2 See also addendum to Chap. 2, this Vol., p. 77

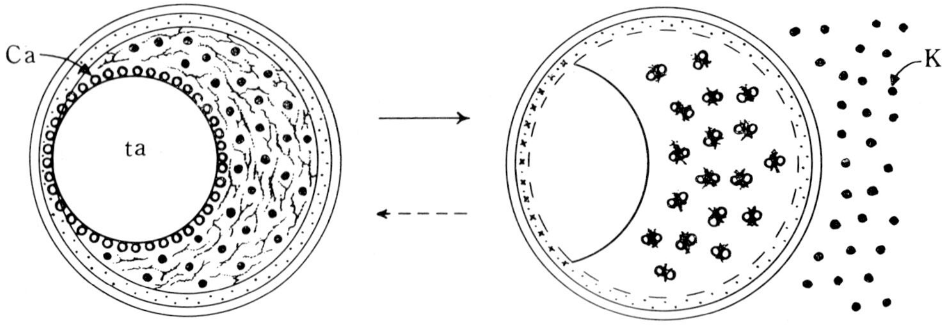

Fig. 5. Scheme of the function of the motor cell of abaxial half of mimosa primary pulvinus. →: indicates bending movement; --→: indicates recovery phase of bending movement. (After TORIYAMA and JAFFE 1972)

of its resting size. It is suggested that Ca^{2+} bind to microfibrils in the central vacuole (which contract or "clump up"), the microfibrils being connected to, and continuous with the tonoplast (TORIYAMA and SATO 1968), which WATANABE (1971) has shown to contain an ATP-ase. BISWAS and BOSE (1972) have demonstrated a contractile ATP-ase, which is activated by Ca^{2+} or Mg^{2+} that can be extracted from *Mimosa* pulvini. They have further shown that the contractile moiety is separable from the Ca^{2+}-activated moiety. Thus, the contractile moiety has characteristics that are very similar to the actomyosin found in animal muscle, and since ATP has been shown to be used up in the pulvinus during contraction (LYUBIMOVA et al. 1964), the proposed mechanism is reminiscent of that of the muscle cell, with the contractile vacuole taking the place of the sarcoplasmic reticulum, and the putative contractile microfibrils playing the role of the muscle cells' actomyosin contractile fibers.

As the microfibrils in the tonoplast contract, K^+ move from the central vacuole to the cytoplasm and thence out of the motor cells (TORIYAMA 1955, ALLEN 1969). In a manner that is perhaps similar to the K^+ flux-related turgor change in stomatal guard cells, the efflux of K^+ is related to the efflux of water from the *Mimosa* motor cells. Since the cells on the other side of the pulvinus retain their turgor, the leaf is rapidly forced to bend.

The slower recovery of the leaf seems to be due to a reverse of the above events in the motor cells. The contractile vacuole scavenges the Ca^{2+}, returning to its resting configuration, while the microfibrils in the central vacuole relax. Potassium ions move into the motor cells as does water, and full turgor is restored, lifting the leaf.

Although the sensitive *Mimosa* is one of the most intensively studied of the plant systems that respond to mechanical stimulation, our knowledge of the mechanisms of both its sensory and motor functions is still incomplete. In particular, a systematic study of the roles that phytohormones may play, and the identification of the stimulatory substance are high on the list of priorities of possible areas of future research.

2.1.2 Venus' Fly Trap

The thigmonastically responsive leaf blade is divided by the midrib into two lobes, each having three sensory hairs on the adaxial surface and surrounded by relatively long spines at right angles to the rim. When the hairs are touched, the lobes become slightly concave as they snap together, while the perimeter spines intercalate with one another to effectively bar the escape of trapped insects. The sensory hairs, the fine structure of which have been elucidated by WILLIAMS and MOZINGO (1971), display a true receptor potential of about 5 to 20 mV when they are stimulated (JACOBSON 1965). Thus, they are true sensory organs and their indented layer cells, which are deformed by mechanical stimulation, are the sensory cells (BENOLKEN and JACOBSON 1970). If the hairs are touched twice within 20 s, an action potential on the order of 50 mV follows the receptor potential. It spreads in concentric waves across the leaf blade at the rate of from 6 to 17 cm s^{-1}, and closure of the trap occurs 0.5 to 1 s after the spread of the second action potential (SIBAOKA 1966). The production of the receptor potential may be thought of as the sensory function of this system and the action potential as part of its transduction mechanism. Adenosine triphosphate may be the native source of energy for the motor response, since its endogenous level drops by 32% during closure, and exogenous ATP (but not AMP) increases the rate of closure by 38% (JAFFE 1973a).

Although the Venus' fly trap system has many features in common with that of the sensitive *Mimosa,* no studies have been made of the possible integrative role of plant hormones or intracellular "second messengers". In one preliminary study, however, LEA (1976) showed that an extract of Venus' fly traps will cause muscle contraction (according to the frog abdominis rectus muscle assay system) and that the active principle may be lysophosphatic acid. At present, our understanding of this function is limited to the electrochemical events described above.

2.2 Thigmonastic Flower Parts

Various flower parts of some 10 or 12 species show motor responses to touch. The stamens of *Cynara scolymus* L. and *Centurea jacea* L. are very elastic and may be stretched to twice their usual length and return to their original size when released. When the filaments are touched they immediately contract so that the anther touches the stigma (SACHS 1882). The flower of *Opuntia* spp. contains a ring of short stamens surrounding a central style. When a pollinator enters the flower, it touches the filaments and the stamens immediately curve inward, trapping the insect (TOUME 1899). It is assumed that the insect brushes against the anthers, causing pollen to be deposited on its body. Similarly thigmonastically sensitive stamens can be found in species of *Mammillaria, Echinocactus,* and *Echinocereus* (MAC DOUGAL 1908). The relatively thick filaments of the stamens of *Berberis vulgaris* L. are sensitive on their adaxial side only, and when they are touched there, they bend inward (SACHS 1882). Portions of the style or stigma are sensitive in species of *Martynia louisiana* MILL. and *Mimulus luteus* L. (HANSGIRG 1890). When one of the two lobes of the style of *M. luteus* is touched, it gradually moves from the horizontal to the vertical position in about 6 s. In the flower, the other lobe is usually pressed up against an interior petal wall, and is already vertical. Thus, the movement of a pollinator within the flower may have the effect of causing the two lobes to press their inner surfaces vertically together.

In general, it is assumed that the motor responses of plant sexual organs are adaptations to the behavior of pollinating insects. There have been essentially no studies in these systems on the integrative functions of either action potentials or phytohormones.

2.3 Aquatic Thigmonastic Plants

2.3.1 Aldrovanda

The classical and most comprehensive work to date of this system is that of ASHIDA (1934), and the brief mention here is derived from his work. *Aldrovanda vesiculosa* L. is a small aquatic plant containing series of leaf traps arranged in whorls. When it is touched, the two lobes of the leaf snap shut with a similar rate and movement to the Venus' fly trap. The rate of closing is about 5 cm s^{-1} and the recovery rate is about 5 cm per 45 min. Since the leaf itself only has to move about 3 mm, the trap will close in about 60 ms. No studies have been done to test the possible role of either action potentials or phytohormones as integrative mechanisms.

2.3.2 Nematode Trapping Fungi

Although the most common type of fungal nematode trap is an adhesive hyphal network found in *Arthrobotrys oligospora* FRES. and others, some of the predacious Moniliales depend on constricting rings. These structures (Fig. 6) capture their prey by the garroting action of the ring cells which, by swelling, grasp the animal in a strangle-hold from which it cannot escape. The ring is composed of three curved cells, thigmonastically sensitive on their inner edges, and attached to the mycelium by a short stalk. When a nematode enters the ring, its body rubs against the inner sides of the ring cells and, after a lag period of a few seconds, the cells swell to three times their previous volume in less than 0.1 s, deeply indenting the body of the animal. Trophic hyphae then invade its body, and consume the contents (DUDDINGTON 1968).

Three types of possible chemical messengers are implied. First, the nematode *Pancyrellus redivivus* induced ring formation in four species of the predacious genus *Dactylella*. The putative substance may be thought of as a pheromone which the authors have designated endogenous nermin and which would tend to encourage trap formation in soil water containing high populations of nema-

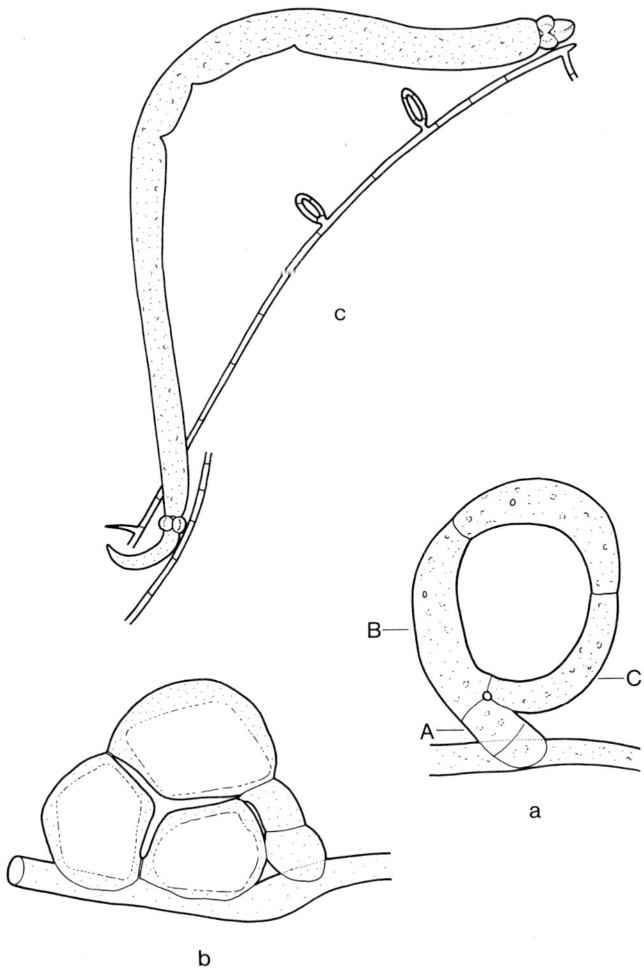

Fig. 6a–c. The constricting trap of *Dactylella bembicodes* Drechsler. **a** (× 525), a mature trap, showing the upper stalk cell (*A*), and the first (*B*), second and thrid (*C*) loop cells. **b** (× 525) a loop, the three cells of which have expanded, throwing the trap without catching a nematode. **c** (× 110) a large nematode trapped by both head and tail ends. The animal's body was completely filled with fungal hyphae. (After COUCH 1937)

todes. Second, BALAN and GERBER (1972) have shown that the predacious fungus *Arthrobotrys dactyloides* Drechsler is capable of exuding a nematode-attracting substance. This pheromone is capable of luring the animal to the trap where it can be killed. The third way in which an integrative factor might play a role is after the mechanical stimulus has been received. It is possible that a "second messenger" or an intracellular hormone might be involved, or that action potentials might play a role, but to my knowledge, no studies have been done to test for either of these possibilities.

2.4 Plant Tendrils

The older literature dealing with the physiology of plant tendrils has been reviewed up to 1967 (JAFFE and GALSTON 1968b), and except where specifically pertinent, the older literature will not be dealt with here. Tendrils are hair-like organs, often modified leaves which usually respond thigmonastically when their ventral or underside is rubbed. When mechanically stimulated, they coil around the support, but if the source of the stimulus is soon removed, they uncoil (Fig. 1). In many tendrils, the sensory function seems to be performed by specialized tactile protuberances or pits on the epidermal layer, or in some cases simply a particular surface (like "knitting") which is sensitive to touch (TRONCHET 1964, JUNKER and REINHOLD 1975), which seem to be characterized by especially thin cell walls. These are dramatically apparent in the tendrils of *Eccromorcapus scaber* (Fig. 7), where TRONCHET (1964) has claimed that the deformation of the tactile papillae is the cause of the subsequent coiling. BOSE (1927) and ZELTNER (1931–1932) both demonstrated that an applied electrical current could substitute for the mechanical stimulus in the induction of contact coiling, suggesting that action potentials might be involved as part of the transduction step. Such a supposition has been confirmed in tendrils of *Cucumis melo* L. by UMRATH (1934), who found that within milliseconds after mechanical stimulation, an action potential of 60 to 100 mV propagates down the tendril at the rate of 0.04 cm s^{-1}. REINHOLD et al. (1972), using techniques similar to UMRATH's, were able to detect rhythmic bursts of bioelectric potentials in the tendrils of *Parthenocissus tricuspidata* (SIEB and ZUCC) Planch, but did not report a confirmation of UMRATH's findings. However, since *Parthenocissus* tendrils do not exhibit contact coiling, it is unlikely that these observations can be considered pertinent to that phenomenon.

A somewhat clearer picture can be seen concerning the role of phytohormones in mediating contact coiling. JAFFE and GALSTON (1966a) showed that physiological concentrations of exogenous 2,4-D or IAA enhanced contact induced coiling in pea tendrils. Such an enhancement might, of course, be purely fortuitous and have nothing to do with the mechanism of coiling. However, REINHOLD (1967) was able to demonstrate that dipping the tips of *Marah fabaceus* L. tendrils into auxin solutions could substitute for touch in inducing coiling (Fig. 8). In an extension of this work (REINHOLD et al. 1972) tendrils of *Parthenocissus* were induced to produce adhesive discs, a response normally caused by mechanical stimulation (DARWIN 1876). They also showed that treatment with either mechanical stimulation or auxin induces lignification of tendril tissues by 5 days after treatment, although it is unlikely that lignification at 5 d after mechanical stimulation need necessarily have anything to do with the coiling response itself. If auxin is involved in coiling, the question is how. BORESCH (1938) suggested that there is an asymmetrical distribution of auxin in the tendril and that this results in curvature in much the same way as in geotropism or phototropism. In order for this to be part of the coiling mechanism, there would have to be more auxin on the dorsal (convex) side than on the ventral (concave) side so as to make the dorsal side grow faster. However, the mechanism of coiling is not that simple. FITTING (1903) marked tendrils

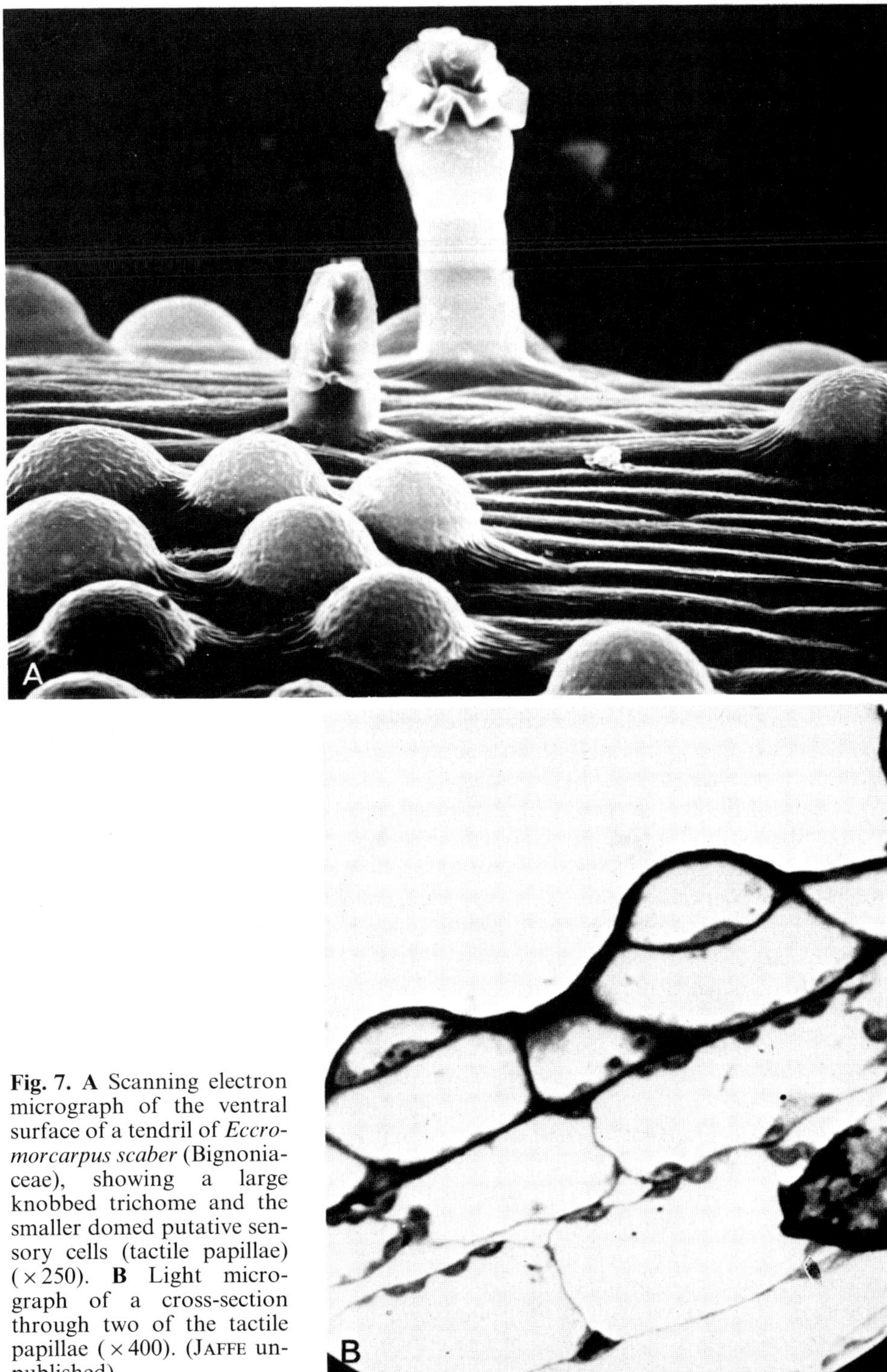

Fig. 7. A Scanning electron micrograph of the ventral surface of a tendril of *Eccromorcarpus scaber* (Bignoniaceae), showing a large knobbed trichome and the smaller domed putative sensory cells (tactile papillae) (×250). **B** Light micrograph of a cross-section through two of the tactile papillae (×400). (JAFFE unpublished)

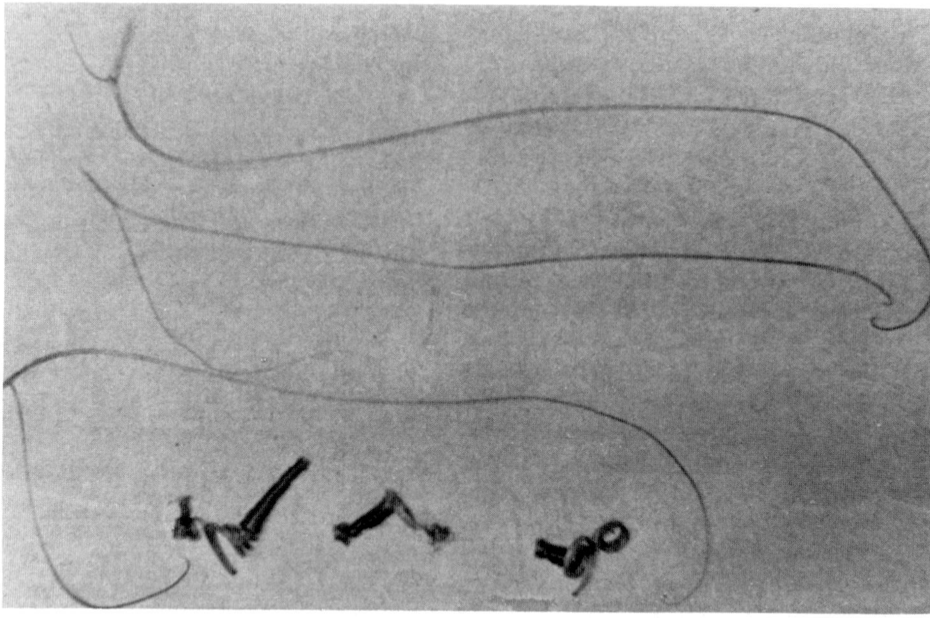

Fig. 8. Tendrils of *Marah fabaceus* after floating for 15 h on water (*top*) or on IAA at 100 mg l^{-1} (*bottom*). (After REINHOLD 1967)

of many species 0.5 to 2.5 cm behind the tips and then rubbed them. He found that in almost all species tested, the dorsal side expanded, whereas the ventral side contracted within a few minutes after stimulation. This was confirmed with pea tendrils by JAFFE and GALSTON (1966a), who showed that in the first 30 min after rubbing, the dorsal side elongates 0.18 mm but the ventral side contracts 0.12 mm. Thus, it is a combination of ventral contraction and dorsal elongation that produces the first rapid curvature in mechanically stimulated tendrils. In order to see if auxin accentuated the contraction or elongation, pea tendrils were symmetrically treated with 5×10^{-7} M IAA (JAFFE and GALSTON 1966a). They found that dorsal elongation increased to 0.23 mm in the first 0.5 h, but that the ventral side also showed elongation to the extent of +0.27 mm over the −0.12 mm (see above), or a net elongation of +0.15 mm. Since FITTING (1903) observed ventral contraction within a few minutes after stimulation, and since exogenous auxin abolishes normal ventral contraction, it seems unlikely that auxin distributed asymetrically on the dorsal and ventral sides is responsible for the early stages of contact coiling. It is possible, however, that auxin is involved in two other ways. One clue to the first of these is to be found in REINHOLD's paper dealing with tendrils of *Marah fabaceus* L. She shows very clearly that dipping the tip of the tendril into an auxin solution will rapidly induce coiling (Fig. 8), a finding confirmed by JAFFE (1975) for tendrils of eight species. *Cucurbita moschata* mechanically stimulated tendrils coiled twice as fast during the first hour than auxin-induced coiling, but after $1^1/_2$ h, the auxin-treated tendrils continued to coil, whereas the rubbed plants

Table 1. Specific ^{14}C-IAA binding ability of gourd (*Cucurbita moschata* L.) tendril extracts. (After JAFFE 1975)

Treatment	Specific binding (%)[a]
Basal quarter of resting tendrils	0.06[b] ±0.08[c]
Apical quarter of resting tendrils	0.35 ±0.09
Apical quarter of tendril allowed to coil on the vine	0.06 ±0.06
Apical 1 cm of resting tendrils	0.26 ±0.09
Subapical 2 cm of resting tendrils	0.02 ±0.04

[a] 0.1 mM IAA was used as the nonradioactive antagonist

[b] $\%\ \text{specific binding} = \dfrac{\text{radioactivity in total supernatant}}{\text{radioactivity in total supernatant fluid} + \text{pellet (dpm)}} \times 100$

Note: The apparent minus value of the first datum is not significantly different from 0.00

[c] S.E. of $\bar{x}$

did not (JAFFE 1975). Thus, auxin may be involved in the sustained asymmetrical growth occurring around a support, but not in the early rapid contraction phase, as first suggested by REINHOLD (1968). In fact, auxin may play a role in the very early events of contact coiling, since only a few seconds of dipping the tip into auxin suffice to induce coiling, suggesting that very few auxin molecules are needed to start the response (JAFFE 1975). Only the active auxins IAA, NAA, or 2,4-D cause this effect, 2,6-D and TIBA inhibiting IAA-induced coiling (JAFFE 1975). JAFFE (1975) also showed that $^{14}C_2$-IAA was translocated basipetally, an observation confirmed by JUNKER (1976), both noting no change in the amount of translocated $^{14}C_2$-IAA after either mechanically induced- or auxin-induced coiling. JUNKER (1977) who applied $^{14}C_2$-IAA to the tips of tendrils of *Passiflora quadrangularis* L., and rubbed the tendrils, also failed to detect any difference from the initial dorsal: ventral ratio of ^{14}C (43:57) as a possible consequence of stimulus or of coiling. However, an interesting observation has been made concerning the tendril tip which is sensitive to auxin. HERTEL et al. (1972) have shown that auxin can be specifically bound to a putative receptor which can be recovered in a membrane-rich fraction from various plants. He also showed that this binding exhibits the same kinetic properties as does auxin action in these plants. JAFFE (1975) was subsequently able to demonstrate that specific auxin-binding activity could be extracted from the tips but not the remainder of uncoiled tendrils, and that extracts of tips of coiled tendrils no longer exhibited this specific binding property (Table 1). It was thus concluded that upon mechanical stimulation, auxin may act as a trigger by being released at the tip, thence becoming bound to specific receptor molecules (JAFFE 1975).

Auxin is then capable of being translocated both ventrally and dorsally, but at no greater rate on either side (JUNKER 1977), the basipital rate being 1.45 cm h^{-1} (JUNKER 1976). Since no more label is translocated in coiling than in resting tendrils, the effect of translocatable auxin may be due to differential dorsiventral sensitivity to auxin, rather than to differential dorsiventral distribution.

REINHOLD (1967) noted that the symmetric application of ethylene induced a slight coiling. Subsequently, JAFFE (1970) applied the ethylene precursor 2-chloroethylphosphoric acid (ethrel) asymetrically by painting it onto the ventral or dorsal surfaces of pea tendrils in the presence of DMSO. The ethylene treatment mimicked mechanical stimulation. Ventral application of ethrel induced coiling, (as does ventral rubbing) but dorsal application of ethrel had no effect, (as also dorsal rubbing), whereas ethrel applied dorsally inhibited coiling due to ventrally applied ethrel, (as coiling due to ventral rubbing was inhibited by dorsal rubbing). Further, tendrils coiling due to mechanical stimulation produced three times as much ethylene during the first hour than did resting tendrils (JAFFE 1970). Thus, the production of ethylene may be responsible for contact coiling, a possibility given further support by BANGERTH (1974). He measured ethylene production in tendrils of *Cucumis sativus* L., and found that auxin-stimulated ethylene synthesis in the ventral half of the tendrils was several times greater than in the dorsal half. As noted above, there is equal translocation of ^{14}C auxin on both sides of resting and coiling tendrils, no distribution of IAA (JAFFE 1976, 1977), and no increase in total translocated ^{14}C during coiling (JAFFE 1975).

Thus, it seems that mechanical stimulation may sensitize the ventral side so that translocating auxin induces ethylene production; and that the localized ethylene production is, in turn, responsible for the curvature. Afterward, if the stimulating support remains, the sustained growth curvature may be due to continued dorsi-ventral asymmetry of auxin transport, perhaps triggered by the activity of auxin receptors at the tip.

The effects of other phytohormones on thigmonasty in tendrils have only been studied in very little depth. Gibberellic acid does not induce curvature in tendrils of *Cucumis sativum* and *Marah fabaceus*, (REINHOLD 1967, JAFFE and GALSTON 1968a), respectively. However, thigmonasty of tendrils of *Pisum sativum* was enhanced by GA_3 in the dark and retarded by it in the light; a small retarding effect of kinetin in the light also being noted (JAFFE and GALSTON 1966a).

There have, of course, been many studies of the role of phytohormones on growth and circumnutation in tendrils (BOYER 1972, 1973), but since these reports may not deal directly with the phenomenon of thigmonasty, they will not be considered here.

To complete the picture of the physiology of contact coiling in tendrils, the following observations are noted. After reception of the mechanical stimulus, probably by modified epidermal cells (TRONCHET 1964, JUNKER and REINHOLD 1975), the ATP content of the tendrils decreases rapidly (JAFFE and GALSTON 1966a) as does the activity of a contractile ATPase (JAFFE and GALSTON 1967). Both factors begin to recover as coiling ceases and their correlation with the coiling process is probably causally significant since exogenous ATP induces coiling in light-starved tendrils (JAFFE and GALSTON 1966b). These properties of the tendril appear to be related to the contraction phase of contact coiling because, during the first 30 min of coiling, ^{14}C-sodium acetate, ^{14}C-sucrose, and electrolytes efflux more rapidly from coiling, than from resting tendrils (JAFFE and GALSTON 1968a). Fifteen minutes after mechanical stimulation, H^+

ions efflux more rapidly and K^+ ions less rapidly from coiling tendrils than from those at rest (JAFFE and GALSTON 1968a). That this is related to ventral contraction is shown by the following observations by JAFFE and GALSTON (1968a). When ^3H_2O is taken up by resting tendrils, it leaves the ventral side during the early stages of coiling. It does not go to the dorsal side, but passes basipetally down the axis of the tendril and presumably into the subtending petiole and thence to the stem. Thus, the conclusion may be drawn that ventral contraction is due to turgor loss on that side caused by water efflux from the ventral motor cells. This in turn causes curvature due to greater turgor in cells of the dorsal side. It is probable that the ^3H_2O loss from the motor cells follows the observed efflux of solutes.

2.5 Sundew

Although tentacles of the sundew plant (*Drosera* spp.) are normally thought of as chemotropic organs (LLOYD 1942), they also respond to touch. Thus, stimulation by a live insect or a clean inert object will induce thigmonasty toward the center of the leaf, the curvature being completed within 2 to 4 min (WILLIAMS and PICKARD 1972a). Although there have been no critical studies involving phytohormones, bioelectric potentials have been observed, and put forward as a putative integrative mechanism. In an elegant series of papers, WILLIAMS and PICKARD (1972a, b, 1974) showed that after mechanical stimulation a receptor potential of about −15 mV can be detected at the mucilage head. This is followed by a train of action potentials having an amplitude of about 20 mV. The action potentials occurred at a frequency dependent on the magnitude of the receptor potential, and, similar to those of peripheral nerves, were propagated basipetally, possibly via plasmadesmata.

3 Thigmotropism

We will see that thigmomorphogenesis is fairly widespread among vascular plants and thigmonasty (as noted above) is quite widespread. Thigmotropism, on the other hand, has not been observed too often. Other than the tendrils of a few species which have been found to be thigmotropic (JAFFE 1976b), plant systems responding to direction of the stimulus have only been reported in the few cases noted below.

3.1 Thigmotropic Roots and Shoots

DARWIN (1876) was the first to report the effects of mechanical stimulation on root growth. He demonstrated that when a tiny piece of mica was affixed to one side of the tip of a pea root, its touch caused the root to curve away from the stimulus. This negative tropism was so pronounced, that eventually the root could be forced to grow in a complete circle.

Only one case of thigmotropism in shoots has been reported (SCHRANK 1944, 1945). He stimulated the side of the tip of *Avena sativa* coleoptiles with a rapidly vibrating reed and observed growth toward the direction of the stimulus. This positive thigmotropism was strong enough to overcome gravitropic curvature, and was accompanied by bioelectric potentials (SCHRANK 1944).

3.2 Thigmotropic Stamens

It has been known since the time of SACHS (1882) that the stamens of *Portulaca grandiflora* are positively thigmotropic (Fig. 9). The first critical study of this system was made by IWANAMI (1962a, b, c, d). He partially characterized the movement with regard to rates of response and recovery, nature of the optimum stimulus, and several other parameters. Later, JAFFE et al. (1977) thoroughly characterized the response and tested the effects of various environmental factors. The stamens exhibit positive thigmotropism when the filaments are gently touched or pushed. The bending response can be detected within 0.2 s and is complete after 2 or 3 s. It is not dependent on the presence of the style,

Fig. 9A–C. The thigmotropic stamens of *Portulaca grandiflora* Hook. The pictures show: **A** the appearance of the flower in the open position; **B** with the petals tied back and ready for stimulation with a glass rod; **C** after stimulation and responding thigmotropically (taken with stroboscopic light, 6 flashes s^{-1}). (After JAFFE et al. 1977)

Table 2. Effects of hormones and cyclic nucleotides on stamen movement in *Portulaca grandiflora*. (After JAFFE et al. 1977)

Hormone	Concentration	Total movement angle (degrees)
Buffer control	–	23.0 ± 2.0[a]
GA_3	1 μM	19.0 ± 1.3
IAA	0.1 μM	36.9 ± 1.9
	1 mM	14.3 ± 3.6
Ethrel	50 ml l^{-1}	15.3 ± 2.7
Cyclic AMP	50 nM	50.9 ± 1.9
	50 μM	0.0 ± 0.0
Cyclic GMP	50 nM	5.5 ± 1.3
	50 μM	30.7 ± 2.6

[a] S.E. of $\bar{x}$

and can be stopped by touching the other side of the stamen. If stimulated stamens have moved on one side of the flower, stamens on the other side of the flower stimulated subsequently will not respond. After moving, stamens take about 5 min to return to their original position, and 6 to 8 min to recover full ability to respond again.

The ability of the stamens to respond follows a pattern during the daylight hours, with an activity peak in the afternoon. Light does not modulate the response, but temperature does, with the response having a Q_{10} of 1.98. Respiratory inhibitors block thigmotropism, and exogenous ATP enhances it (JAFFE et al. 1977), suggesting a role for ATP as a native energy source for the response.

When exogenous phytohormones were applied to the filaments (Table 2) GA_3 had no effect, whereas 0.1 μM IAA significantly increased movement and 1 mM IAA significantly inhibited it. The latter effect may have been due to stimulation of ethylene production, since exogenous ethrel had the same inhibitory effect. Although it is not known if the cyclic nucleotides are active as secondary messengers in plants, interesting results were obtained when the filaments were treated with them prior to mechanical stimulation, 50 μM c-AMP completely inhibiting the thigmotropic response, 50 nM enhancing it (Table 2). Cyclic GMP displayed the reverse pattern, since 50 nM inhibited, and 50 μM promoted movement (JAFFE et al. 1977). Although it is premature to draw conclusions from these data, the fact that the concentration effects of c-AMP and c-GMP are reversed is not incompatible with the suggestion that, if they are parts of the natural mechanism, they may function in a complementary-antagonistic fashion, as has been demonstrated in animal systems (GREENGARD 1975).

3.3 Thigmotropism in Fungi

Very interesting directional growth responses have been observed in rust and powdery mildew fungi. When the spores of these fungi germinate on the surface

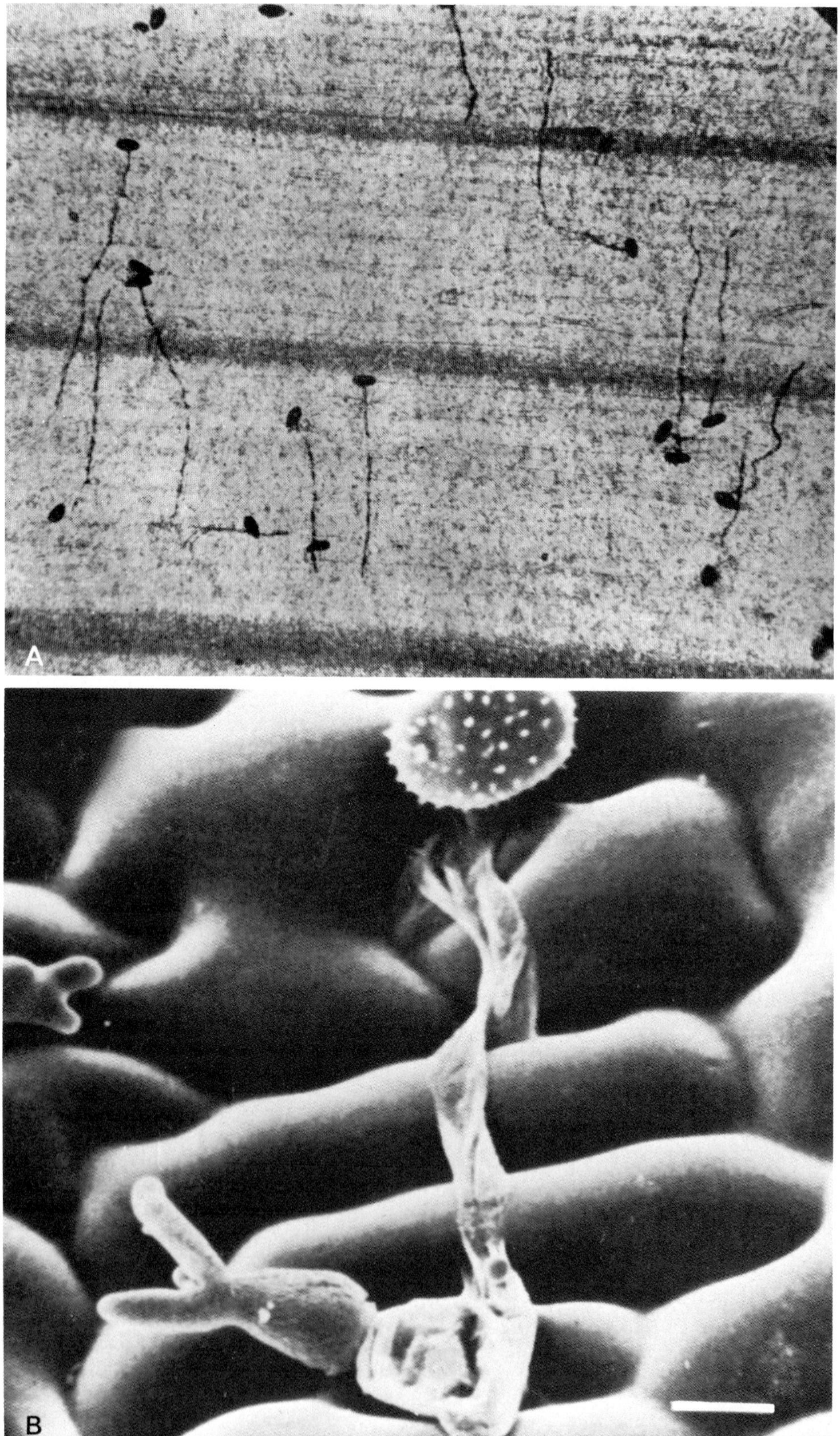
A
B

of a leaf, the selection of the direction of the growth of the hyphal germ tube is determined by mechanical contact with the leaf surface (DICKINSON 1949a). Both negative thigmotropism (growth away from the surface) and diathigmotropism (growth along the surface) have been observed (DICKINSON 1971). Similar responses by the fungus were obtained with surfaces of collodian replica films (e.g., serial replica) (DICKINSON 1949b), indicating that the response is due to contact rather than to a chemical found on the surface of the leaf.

A particularly interesting thigmotropic response was first reported by JOHNSON in (1934). Studying the growth of germ tubes of urediospores of *Puccinia graminis* tritici Eriks, he noted that most of them grew transversely across the wheat leaves, in a direction at right angles to the veins. The penetration of the surface hypha is via a stomatal aperture, and the stomata are arranged in longitudinal rows on the leaf. However, the stomata of two adjacent rows are rarely opposite one another, so a germ tube growing directly across a leaf is bound to encounter a stoma eventually, if it continues to grow in a straight line (Fig. 10). The study of this phenomenon has been extended by WYNN (1976), using the bean rust fungus, *Uromyces phaseoli* Pers., and to show that no chemical stimuli were involved, he used serial replicas of the surfaces of bean leaves. The germ tubes grew at right angles to the direction of the long axis of the cell replicas of bean and wheat leaves (Fig. 10). No physiological or biochemical mechanisms of this response are currently known.

4 Thigmomorphogenesis

4.1 Thigmomorphogenesis in Fungi

As noted above, pathogenesis in leaf-infecting powdery mildews and rusts involve appresorium formation at the end of the germ tube when it reaches a stomate. For *Sclerotinia sclerotiorum* Lib., this differentiation was due to mechanical contact (PURDY 1958), although a carbon source was also necessary. WYNN (1976) found that contact with a stomatal aperture of a serial replica of a bean leaf was sufficient to induce appresorium formation, in *U. phaseoli*, but not on a replica of wheat leaves, a nonsusceptible host. In an extension of this kind of study, STAPLES et al. (1976) have examined thigmomorphogenetic induction of appresoria in the soybean anthracnose fungus, *Colletotrichum truncatum* (Schweinitz) Andrus et Moore. They found that when the second round of nuclear division is blocked by the inhibitors of DNA synthesis, fluorodeoxyuridine or hydroxyurea, appresoria do not form. In addition, when DNA-induced RNA synthesis was blocked by actinomycin D, appresorium formation

◄

Fig. 10. A Germinating urediospores of *Puccinia graminis* tritici Eriks., showing germ tubes elongating transversely on a wheat leaf. (After JOHNSON 1934). **B** Scanning electron micrograph showing growth of the germ tube of *Uromyces phaseoli* Pers. for 12 h on a replica of Pinto bean leaf surface. Line = 10 μm. (After WYNN 1976)

was reversibly inhibited. They thus concluded that mechanical stimulation induces the formation of appresoria by altering the nucleic acid programming of the germ tube nucleus.

4.2 Thigmomorphogenesis in Vascular Plants

Only recently has a direct effort been made to elucidate thigmomorphogenesis; so much of the literature on the topic is anecdotal. That is, the observations dealing with the responses of plant growth and development to mechanical stimulation were made accidentally while an investigator was studying something else and are parenthetically buried in the report of the primary study. Nevertheless, in searching the literature, one is able to find a number of papers that report responses in a wide variety of vascular plants. Table 3 shows that 39 species from 18 families all have been reported to respond to some form of mechanical stimulation.

The primary thigmomorphogenetic response is one of vertical growth retardation coupled with increased radial growth (Figs. 1, 11) (JAFFE 1973a, b). In most species tested, shaking, bending, rubbing, or blowing on the plant induces a retardation of elongation. The ubiquity of this response is significant. It should be stressed that thigmomorphogenesis is as common a response as other, much more widely studied plant-growth responses, such as geotropism or photomorphogenesis.

4.2.1 The Ecological Significance of Thigmomorphogenesis

From an ecological point of view, the mechanical stimulus itself may come from a variety of sources. The rubbing of plants caused by the daily passage of animals or machines (BURDEN and RANDERSON 1972), vibration of the soil caused by machinery, or even the frequent rubbing by animals to relieve an itch, as well as certain horticultural practices such as the tying of trunks or branches, might all act as adequate stimuli to produce a developmental change (NEEL and HARRIS 1971, TURGEON and WEBB 1971). Trails through vegetation may form in response to such stimuli. In addition, the two environmental conditions that are probably the most important causes of thigmomorphogenesis are bending by the wind and the mechanical rubbing by soil particles as the root grows down or laterally, or as the shoot grows up. Thus, plants exposed to bending by natural winds (NEEL and HARRIS 1971, JACOBS 1954) or to wind treatment in a growth chamber (NEEL and HARRIS 1971, JAFFE 1976a) grew less than did controls. Wind sway is a powerful causal factor in thigmomorphogenesis (WESTING 1965, 1968). LARSON (1965) found that exposure of *Latrix laricina* (DUROI) K. KOCH to natural winds caused a pronounced downward drift of incremental growth toward the base of the tree, usually at the expense of the upper stem parts. At the same time, elongation of free-swaying trees was reduced. Trees held erect with guy wires did not show either of these responses. In a similar study, JACOBS (1954) showed that free-swaying trees exhibited greater radial growth in the lower trunk and in the upper roots than those stayed with guy wires. Such trees, therefore, had a broader base than stayed trees. He found that after 10 years the increased radial growth due to sway could be detected to a height on the trunk of 7.6 to 9.1 m.

The effect of sway at a height of 1.2 m was greater in a heavily thinned test area than in a well-stocked test area. This "wind-break" effect has been field tested by PELTON (1967), who showed that wheat yields within the sheltered area were 24–43% above control yields. It may be suggested, therefore, that motion due to wind be regarded as a factor in the dwarfing characteristics of such habitats as the alpine tundra and in the reduced and procumbent forms of plants on coastal dunes. Such a hypothesis has been sporadically tested in a number of laboratories with various kinds of wind tunnels, and WADSWORTH (1959, 1960) found that when *Brassica campestris* L. seedlings were less than 1 cm tall, their relative growth rate increased with continuous wind speed; when they were between 1 and 4 cm tall, 1.1 km h^{-1} was the optimum wind speed; and plants from 4 to 7 cm were progressively retarded as wind speed increased. Similarly, WHITEHEAD (1962) found that dry weight, internode length, and leaf area of seedlings of *Helianthus giganteus* L. decreased as continuous wind speed increased from 7.6 to 52.8 km h^{-1}. He concluded that the plants became more xeromorphic as wind speed increased, to the benefit of their water economy. Comparable results were obtained by VENNING (1949), who showed that wind-swayed celery plants (*Apium graveolens* L.) were shorter than the controls, with thicker, less flexible petioles. MARTIN and CLEMENTS (1935) showed that 6–8 weeks exposure to continuous winds of 8, 16, and 24 km h^{-1} decreased the leaf area and elongation of sunflower plants. Plants grown under the higher wind velocities exhibited the gnarled and twisted appearance characteristic of plants growing in windy habitats. Both continuous and gusty winds will induce thigmomorphogenesis. Working with young bean plants, JAFFE (1976a) showed that ten 10-s gusts of 17 km h^{-1} d^{-1} wind would retard elongation by as much as 34% over a period of 10 days.

Since the mechanical effect of wind is to flex the aerial plant parts, several laboratories have tried shaking or rubbing plants, either by hand, or by motor. Thus, shaking cotton plants by hand or with a stick reduced plant height, and number of flowers and bolls (Fig. 11) (COAD and MCGEHEE 1917, BAILEY and TEMPLETON 1926, FRIZZELL et al. 1960), but merely walking through the rows had no effect. Flexing has effected thigmomorphogenesis in a number of different species (Table 3).

BÜNNING (1940) studied the effects of mechanical stimulation on etiolated *Phaseolus multiflora* L. plants, noting that rubbing retarded elongation of the epicotyl by 50% as did light. He concluded that as the plant pushed up through the soil it was mechanically stimulated. Similarly, when plants were grown through glass beads, etiolated pea shoots of *Pisum sativum* (GOESCHL et al. 1966), and barley or maize roots (BARBER and GUNN 1974) exhibited decreased elongation. STEUCEK and GORDON (1975) demonstrated that there may be profound varietal differences in the ability to respond to mechanical stimulation. The elongation of Blue Boy variety of wheat, which resists lodging, was retarded. Wheat, however, is probably not the best species in which to study thigmomorphogenesis, since the retardation response is at best 11 to 14% (JAFFE 1973b,

Table 3. Reported occurrence of thigmomorphogenesis in plant species (JAFFE unpublished data)

Family	Species	Light habit	Plant part stimulated	Type of stimulus	Response	References
Anacardiaceae	*Pistacia chinensis*	Lt.	Wh. Pl.	Shaking and flexing by hand	Decr. elong. and incr. rad. gr.	MITCHELL (1975)
Compositae	*Bidens pilosa*	Lt.	Cot.	Pricking	Inhib. of axillary bud	DESBIEZ (1973)
	Chrysanthemum sinense	Lt.	Wh. Pl.	Platform shaking or br. rubbing	Decr. elong.	HAMNER et al. (1974)
	Tagetes patula	Lt.	Wh. Pl.	Br. rub.	Decr. elong.	JAFFE (1973b)
				Leaves showed epinasty	Increased leaf area, dry weight	SUDA (1960)
	Xanthium strumarium	Lt.	Leaves	Br. rub.	Lvs. yellowed, died	SACHS (1882)
	Helianthus annuus	Lt.	Wh. Pl.	Wind tun.	Decr. elong., rad. gr., and lf area	WHITEHEAD and LUTI (1962), JUNKER (1976)
Cruciferae	*Brassica napus*	Lt.	Wh. Pl.	Wind tun.	>4 cm decr. elong; >1 cm incr. growth	VICTOR and VANDERHOEF (1975)
	Sinapis alba	Dk.	Wh. Pl.	Shaking	Decr. elong.	BÜNNING et al. (1948)
Cucurbitaceae	*Bryonia dioica*	Lt.	Wh. Pl.	Rub. stem	Incr. rad. gr.; decr. elong.	BOYER (1967)
				Br. rub.	Decr. elong.	JAFFE (1973b)
	Cucumus sativus	Lt.	Wh. Pl.	Br. rub.	Decr. elong.	JAFFE (1973b)
	Cucurbita melopepo	Lt.	Wh. Pl.	Rub. stem	Incr. rad. gr.; decr. elong.	TRONCHET (1964)
	Cucurbita pepo	Lt.	Wh. Pl.	Br. rub.	Decr. elong.	JAFFE (1973b)
Euphorbiaceae	*Ricinus communis*	Lt.	Wh. Pl.	Br. rub.	Decr. elong.	JAFFE (1973b)
Gramineae	*Avena sativa* var. Victory	Dk.	Coleop.	Rub.	Incr. elong.	EVANS and RAY (1969)
	Hordeum vulgare	Lt.	Wh. Pl.	Restraint of plant	Incr. root efflux of amino acids and carbohy	BARBER and GUNN (1974)
				Br. rub.	Decr. elong.	JAFFE (1973b)
	Secale cereale	Lt.	Wh. Pl.	Br. rub.	Decr. elong.	JAFFE (1973b)
	Triticum aestivum	Lt.	Wh. Pl.	Bent by wind	Decr. yield	NEEL and HARRIS (1971)
				Br. rub.	Decr. elong.	JAFFE (1973b)

	Zea mays	Lt.	Wh. Pl.	Wind tun.	Incr. rad. gr., decr. elong., incr. leaf size	WATANABE (1971)
				Restraint of plant	Incr. root efflux of amino acids and carbohy	BARBER and GUNN (1974)
				Br. rub.	Decr. elong.	JAFFE (1973b)
Hamamelidaceae	*Liquidambar styraciflua*	Lt.	Wh. Pl.	Bent, shk. by hand	Incr. rad. gr., decr. elong., induc. of terminal buds	MITCHELL et al. (1977)
Leguminosae	*Glycine max.* var. Wayne	Lt.	Hypocotyl sections	Restraint of plant	Incr. rad. enlgmt.	VENNING (1949)
			Wh. Pl.	Rub. or vib.	Incr. rad. gr., decr. elong., inhib. of photosyn. CO_2 fixation	MILBURN (1971)
	Mimosa pudica	Lt.	Wh. Pl.	Rub. of lvs.	Decr. elong. of stem, smler throns, decr. # inflores and seeds.	BÜNNING et al. (1948)
				Br. rub.	Decr. elong.	JAFFE (1973b)
	Pisum sativum	Dk.	Wh. Pl.	Restraint of plant	Decr. elong.	GOESCHL et al. (1966)
		Lt.	Wh. Pl.	Br. rub.	No effect	JAFFE (1973b)
	Phaseolus multiflorus	Dk.	Wh. Pl.	Restraint of plant	Decr. elong.	BÜNNING (1940)
	Phaseolus vulgaris	Lt.	Wh. Pl.	Br. rub.	Decr. elong., incr. rad. gr., retarded flow	JAFFE (1973b, 1976a)
			Excised Inter	Br. rub.	Decr. elong.	JAFFE (1976a)
	Vicia faba	Dk.	Wh. Pl.	Shake	Decr. elong.	BÜNNING et al. (1948)
Liliaceae	*Lilium longiflorum*	Lt.	Wh. Pl.	Br. rub.; restraint of plant	Decr. elong.	HIRAKI and OTA (1975)
Malvaceae	*Gossypium hirsutum*	Lt.	Wh. Pl.	Shk. by hand, slt. rub.	Decr. elong and yield	COAD and MCGEHEE (1917), FRIZZELL et al. (1960)

Abbreviations: Lt. = light grown; Dk. = dark grown, etiolated; Wh. Pl. = whole plant; Cot = cotyledon; Lf. = leaf; Br. = brief; Rub. = rubbing; Shk. = shaking; Decr. elong. = decreased elongation; Incr. rad. gr. = increased radial growth; Resp. = response; Flow. = flowering; Induc. = induction; Vib. = vibrating; Tun. = tunnel

Table 3 (continued)

Family	Species	Light habit	Plant part stimulated	Type of stimulus	Response	References
Myrtaceae	*Eucalyptus sideroxylon*	Lt.	Wh. Pl.	Bendg., shk. by hand	Decr. elong., incr. rad. gr.	MITCHELL et al. (1977)
Pinaceae	*Larix laricina*	Lt.	Wh. Pl.	Bent by wind	Incr. rad. gr., decr. elong.	LARSON (1965)
	Pinus radiata	Lt.	Wh. Pl.	Bent by wind	Incr. rad. gr.	JACOBS (1954)
Pinaceae	*Pinus strobus*	Lt.	Branch	Tying 180° bend	Incr. rad. gr.	BROWN and LEOPOLD (1973)
Plantaginaceae	*Plantago coronopus*	Lt.	Wh. Pl.	Trampling	Incr. standing crop becus rosette	BURDEN and RANDERSON (1972)
Plumbaginaceae	*Armeria maritima*	Lt.	Wh. Pl.	Trampling	Suppressed tussock formation and flowering	BURDEN and RANDERSON (1972)
Rosaceae	*Prunus persica*	Lt.	Wh. Pl.	Tying branch 90° bend	Nongrowing branches so no change in gr.; incr. ethylene	LEOPOLD et al. (1972)
	Pyrus malus	Lt.	Wh. Pl.	Tying branch 90° bend	Nongrowing branches so no change in gr.; incr. ethylene	LEOPOLD et al. (1972)
Solanaceae	*Lycopersicum esculentum*	Lt.	Wh. Pl.	Br. rub.; vib. plt.	Decr. elong., incr. rad. gr., epinastic foliage	JAFFE (1973b), MILBURN (1971)
	Solanum melongena	Lt.	Wh. Pl.	Br. rub.	Decr. elong.	JAFFE (1973b)
Umbelliferae	*Apium graveolens*	Lt.	Wh. Pl.	Wind tun., vib. plt.	Decr. elong.; incr. rad. gr.	VAN SAMBECK et al. (1976), WADSWORTH (1960)
Ulmaceae	*Zelkova serrata*	Lt.	Wh. Pl.	Bendg., shk. by hand	Decr. elong.; incr. rad. gr.	MITCHELL et al. (1977)

Fig. 11. Comparison of shaken (*right*) and unshaken (*left*) cotton plants. Adapted from COAD and MCGEHEE (1917)

STEUCEK and GORDON 1975). The effect of the mechanical stimulation of soil on root growth and development has been studied by BARLEY (1962). He found that mechanical stress induced slower elongation, but more lateral root production.

The effect of rubbing on young plants of *Bryonia dioica* L. has been tested by BOYER (1967), who found that when the internodes were rubbed thrice daily with the fingers, the growth of the plants was reduced at least to 58% of the unstimulated controls (BOYER 1967). She further found that growth retardation resulted when either the upper or lower internodes were rubbed. The growth of the petioles and shoots of *Cucurbita melopepo* L. was found to decrease when they were rubbed or shaken, with the young tissue being the most responsive (TURGEON and WEBB 1971), and etiolated oat hypocotyls were stimulated to elongate faster when they were rubbed (EVANS and RAY 1969). MITCHELL et al. (1975) showed that plants placed on a vibrating platform also had their growth retarded, and HAMNER et al. (1974) have used this technique to reduce the "legginess" of greenhouse-grown chrysanthemum plants, thereby increasing their saleability. Similarly, HIRAKI and OTA (1975) stroked easter lily (*Lilium longiflorum* THUNB.) plant leaves lightly with a brush each day. They found that stem elongation was significantly inhibited as part of an epinastic response.

In addition to the usual effect of growth retardation, several studies have shown developmental effects. Thus, NEEL and HARRIS (1971) reported that shaken trees of *Liquidambar styraciflura* L. produced only terminal buds. Apical dominance is also affected in plants of *Bidens pilosus* L. and DESBIEZ (1973) found that stimulating one of the cotyledons of a decapitated seedling quickly induced the inhibition of the axillary bud on that side of the plant. Micromolar amounts of lithium reversed this effect, whereas potassium or sodium mimicked it (DESBIEZ and THELLIER 1975). Mechanically manipulating young cocklebur *Xanthium pennsylvanicum* WALLR. leaves inhibited growth by 30%, and senescence was hastened (SALISBURY 1963). The fact that growth stops immediately following mechanical stimulation in a vining plant such as *Bryonia dioica* (JAFFE 1973b) or *P. vulgaris* (JAFFE 1976a) may be of adaptive advantage. A similar response was observed in circumnutating pea tendrils (JAFFE 1972), and may be interpreted as a mechanism to keep a plant from circumnutating away from a potential support that it has contacted.

Thigmomorphogenetic responses in woody species may be related to the formation of reaction wood (WESTING 1965, 1968, SCURFIELD 1973). When a gymnosperm tree trunk is bent or tipped from the vertical, more wood forms on the bottom side, and if a board is cut from such wood it will expand. Such expansion in the intact tree would tend to bring the tree to an erect position, if that is possible. When an angiosperm tree is bent or tipped, the reaction wood forms on top and a band cut from it tends to shrink, which, if it happened in nature, would bring the trunk back toward the vertical. This is known to be a response of the developing wood cells to the direction of the gravitational field rather than a response to the stresses and strains established in the trunk (WILSON and ARCHER 1977). Thus is might be quite independent of thigmomorphogenesis, which is often observed following mechanical stimulation and in the absence of any reorientation of the plant. In some locations, however, sustained wind may reorient a plant in relation to gravity as well as provide mechanical stimulation (WILSON and ARCHER 1977). Hence, the formation of reaction wood and thigmomorphogenesis may interact in a given plant in nature.

4.2.2 The Time Course of Thigmomorphogenesis

The detailed kinetics of the response to mechanical stimulation has been studied by JAFFE (1973b, 1976a, b) in *B. dioica, Cucumis sativus, M. pudica,* and *P. vulgaris.* After rubbing, there is an immediate increase in the growth rate that lasts 2 or 3 min. Following this, growth stops completely for 15 to 45 min. Then the growth rate resumes at about one half the pre-stimulus rate. If the plant is not stimulated again, it will regain its original growth rate within 2 to 3 d, depending on the species (JAFFE 1973b, 1975). Thus, the thigmomorphogenetic response is a rapid one, the first measurable reaction occurring instantaneously. Since no injury is involved, the plant is free to resume the normal rate of growth after several days.

4.2.2.1 Histological Studies

Several laboratories have examined the histological changes that occur during thigmomorphogenesis. BÜNNING (1940) compared the effects of light and rubbing on the growth of etiolated *P. multiflora* and *Vicia faba*. He was able to show that epidermal and cortical cells did not elongate as much as controls, but had greater diameters. His published drawing (BÜNNING et al. 1948) clearly shows cuticular thickening and thickening of the cortical cell walls due to mechanical stimulation. GOESCHL et al. (1966) also reported that mechanically stimulating etiolated peas caused about 70% decrease in length and 50% increase in diameter of cortical parenchyma cells, the changes closely paralleling the decreased epicotyl growth. NEEL and HARRIS (1971), working with *Liquidamber*, showed that vessel members were about 30% shorter in mechanically stimulated plants than in controls.

If environmental mechanical stress tends to produce a plant more capable of withstanding that stress, one might expect that the strengthening elements in the stem would be enhanced. Such has been shown to be the case in a number of studies. Mechanical stimulation induces a thickening of collenchyma cell walls in celery and *Datura stramonium* L. (VENNING 1949, WALKER 1957, 1960) and the number and thickness of schlerenchyma rows and fiber bundles in corn (WHITEHEAD and LUTI 1962). A histological study of light-grown *P. vulgaris* entirely confirms and extends the previous workers' results (JAFFE and BIRO 1979). Thus, two conclusions may be drawn on the cellular basis of thigmomorphogenesis: Firstly, both retardation of elongation and thickening of the internodes can be partially accounted for by correlative dimensional changes in cell enlargement. Changes in cell division are partially responsible for radial enlargement. Secondly, wall thickening (cutin on the epidermal cells and possibly cellulose on the cortical cells) follows mechanical stimulation.

4.2.3 Integrative Mechanisms in Thigmomorphogenesis

4.2.3.1 Action Potentials

Whatever the adaptive values of thigmomorphogenesis, the underlying mechanism is only beginning to be elucidated. Both DESBIEZ (1973) and PICKARD (1971) have shown that nonsynchronous action potentials of low amplitude follow mechanical stimulation of stem tissue. DESBIEZ (1973) postulated a primary action by a sodium-potassium ATPase in her system (the decrease in growth on the stimulated side of the cotyledon is inhibited by the postulated ATPase inhibitor, oubain). In a correlative study using detailed kinetics of thigmomorphogenesis in *P. vulgaris*, JAFFE (1976b) showed that during the brief initial period of stimulated growth, the electrical resistance of the tissue decreased within seconds after stimulation. This effect can also be detected at the cut base of excised plants as a decrease of the electrical resistance of the solution bathing them. The electrical resistance of the stem tissue recovers during the first 3 min after stimulation.

4.2.3.2 The Role of Ethylene

PICKARD (1971) suggests that touch-stimulated action potentials may be involved in ethylene production and that ethylene may inhibit internode elongation. Such a suggestion is supported by the observations that rubbing tomato plants induces

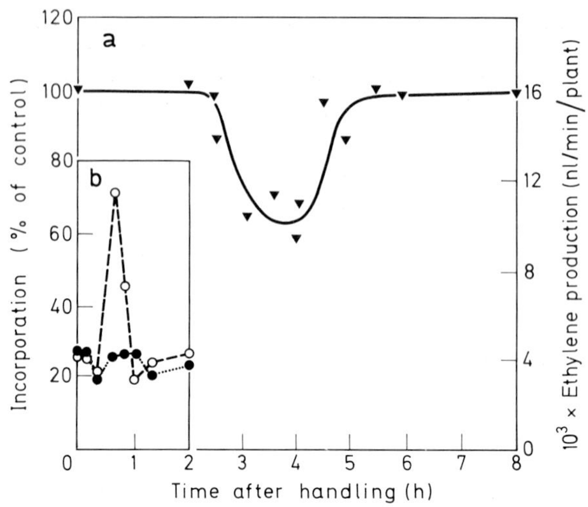

Fig. 12. Effect of handling etiolated pea plants on [1-^{14}C] glycerol incorporation into phospholipids, and on ethylene production. (After IRVINE and OSBORNE 1973)

epinasty in the leaves near the site of stimulation (JAFFE 1973b), and that rubbing bean internodes causes them to elongate less and to increase markedly in diameter (JAFFE 1976a). Both of these latter responses can also be caused by ethylene (ABELES 1973).

GOESCHL et al. (1966) first implicated ethylene in a response to mechanical stimulation, finding that when etiolated peas were mechanically stimulated by growth through glass beads, or by the pressure of a foam neoprene stopper, internode elongation decreased, stem diameter increased, and significant increases in ethylene production by the tissue occured. Ethylene production has also been found to increase when lily plants are mechanically stimulated (HIRAKI and OTA 1975), the greatest increase being observed after 3 days. IRVINE and OSBORNE (1973) studied this effect in etiolated peas and have shown that a burst of ethylene evolution occurs between 25 and 60 min after mechanical stimulation. Since this happens just at the time that growth resumes at an attenuated rate, ethylene production seems to be related to the long-term slowing of growth, but not to the immediate responses, which seem to be membrane-related (JAFFE 1976b; Fig. 12).

Exogenous ethylene simulates mechanical stimulation in bean plants (Table 4) (JAFFE unpublished), and ethylene is also implicated in mechanical stress responses in woody plants. BROWN and LEOPOLD (1973) and LEOPOLD et al. (1972) stressed branches of a conifer and several woody angiosperms by tying them in arcs. In all cases the tissues responded first by producing increased amounts of ethylene, and then by increasing their radial growth. Since hypobaric conditions significantly nullify mechanically stimulated growth retardation (JAFFE and BIRO 1979), it appears that ethylene mediation is part of the chain of causal events linking mechanical stimulation and growth retardation. Also, both exogenous ethylene and mechanical stimulation decreased the rate of incorporation of 1-^{14}C-glycerol into the phospholipids of etiolated pea stems (IRVINE and OSBORNE 1973) (Fig. 12). Since phospholipids are an important component of

Table 4. Effect of mechanical stimulation and plant hormones on the growth of young bean plants, and on the thickness of the first internode. Hormones were applied to the bud once daily in a 50 µl drop. The first true internode of the stimulated plants was rubbed ten times, once each day. The duration of the experiment was 5 d. (After JAFFE and BIRO 1979)

Additive	Change in total length (mm)		Change in diameter of first internode (mm)	
	Control	Rubbed	Control	Rubbed
Control (H_2O)	72±5[b]	39±2[b]	0.30±0.08[b]	0.92±0.09[b]
1 µM GA_3	99±12	77±6	0.28±0.06	0.74±0.16
1 µM ABA	42±4	32±2	0.21±0.04	0.28±0.06
10 µM IAA	46±3	30±1	0.19±0.06	—
44 mM ethrel[a]	37±1	13±1	0.57±0.05	0.93±0.05

[a] Enough to produce about 5 µl ethylene g^{-1} 24 h^{-1}. A series of concentrations of each hormone was tried, but only the most successful one is shown here
[b] S.E. of $\bar{x}$

the endomembrane system, this could be a mechanism by which decreases in cell growth result from mechanical stimulation.

4.2.3.3 The Role of Auxin

Auxin may also be involved in thigmomorphogenesis. BOYER (1967) has shown that auxin is not detectable in the internodes of mechanically stimulated *Bryonia* plants, although it is in the controls (Fig. 13). Further, VICTOR and VANDERHOEF (1975) found that mechanical stimulation reverses auxin-promoted elongation in soybeans *Glycine max* (L.) Merr; but auxin *increases* in tissues that are forming reaction wood (SCURFIELD 1973). Exogenous auxin retards elongation but does not cause radial expansion in intact *P. vulgaris* plants (Table 4).

4.2.3.4 The Role of Translocation and Transpiration

In addition to any possible involvement of hormones in the response to mechanical stimulation, translocation in general may be affected. This has been demonstrated with *Ricinus communis* L. plants by MILBURN (1970, 1971), who found that a 5-day pretreatment of thrice-daily mechanical stimulation resulted in a significant increase of phloem exudation from subsequently tapped plants. If, however, control plants were tapped immediately after a massage treatment, or if rubbing was done during exudation, exudation stopped immediately. MILBURN (1970) observed no effect of mechanical stimulation on the growth of *Ricinus,* but under other conditions, such a response can be demonstrated (JAFFE 1973b).

Transpiration is also affected by mechanical stimulation. MARTIN and CLEMENTS (1935) report that either 2 h or 6 to 8 weeks of continuous 24 km h^{-1} wind decreased the rate of transpiration, but that the requirement for water increased by 50%. The stomates tended to close when stimulated by wind,

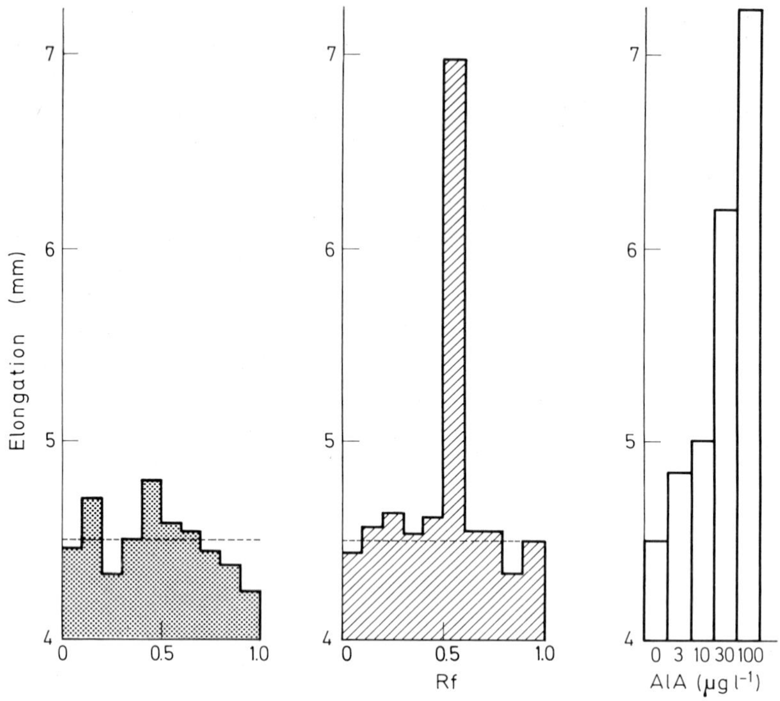

Fig. 13. Histograms of the biological activity (on *Avena* mesocotyl elongation) from chromatograms of extracts of control nonstimulated stems (*center*), and of stimulated stems (*left*). The figure on the *right* indicates the growth effects of different concentrations of IAA on *Avena* mesocotyls. (After BOYER 1973)

and the decrease of transpiration was greater at night than during the day. Similar results were found with sunflower by WHITEHEAD (1962), with tomato by MITCHELL et al. (1977), and with beans by JAFFE (unpublished). Stimulation of the leaves produced a greater effect on transpiration than stimulation of stems (MITCHELL et al. 1977).

4.2.4 Photosynthesis, Respiration, and Metabolism

There have been relatively few studies of the energetics of thigmomorphogenesis. HUNT et al. (1967) used a wind tunnel to study the effects of wind on photosynthesis and evapo-transpiration in *Medicago sativa* and *Dactylis glomerata* (orchard grass). They found that photosynthesis, but not transpiration, increased with wind speed. Thus, plants were apparently able to use water more efficiently under the highest wind regime, although temperature or other environmental parameters could be confounding this effect; but shaking the plants inhibits photosynthetic CO_2 fixation according to MITCHELL et al. (1977). Obviously, more work needs to be done in this area to understand the effects of wind on photosynthesis.

The situation regarding respiration is more clear. GODWIN (1935) and AUDUS (1935) independently showed that bending or rubbing cherry laurel leaves or twigs increased the rate of respiration by as much as threefold. This increase was not due to incidental variations in the light, temperature, or humidity. Such an effect was even found with soft potato tuber tissue (PARKER 1935). Clearly, handling is an important factor when performing respiration experiments! Later, TODD et al. (1972) showed that wind speeds of 13 to 26 km h^{-1} also increase, by 20 to 40%, the CO_2 released by plants of wheat, barley, oats, maize, sorghum, beans, soybeans, peas and *Magnolia grandiflora* L. The respiration rate returned to normal when the wind ceased. These effects were noted even when leaf flutter was largely prevented by fastening the leaf, suggesting to the authors that the response is subcellular and not due to gross movement of the leaves. The slight motions possible with fastened leaves suggest that the phenomenon is a typically sensitive thigmomorphogenetic response. BARBER and GUNN (1974) have shown that the exudation of both amino acids and carbohydrates was greater from stimulated barley and maize roots than from controls, a conspicuous exception being asparagine; efflux of it from control roots was four times that from stimulated roots.

4.2.5 Interaction of Mechanical Stimulation with Other Environmental Cues

The interaction of phytochrome and touch-induced growth retardation in etiolated bean seedlings was tested by JAFFE (1976a). Both red light and mechanical stimulation retarded hypocotyl elongation, but far-red could not reverse thigmomorphogenesis, nor could rubbing reverse the effect of far-red light. An interaction between the two systems has, however, been reported (GREGORY-SOUTHWORTH and KLEIN 1974). Etiolated bean seedlings were given 2-s red irradiations followed by mechanical stimulation in the dark. Weight increase of the leaves 24 h later varied by up to 40% in a complex pattern which was reproducibly dependent on the force of the stimulation and the exact interval between light and shaking.

Several clear-cut interactions have been observed between gravity and the mechanical stimulation of bean seedlings. When 11-day-old bean seedlings were placed horizontally, their upward geotropic bending is retarded by 50% in rubbed plants. This effect is lost after 5 h. Similarly, the stem–petiole angle is larger and the petiole–leaf angle is smaller in rubbed clinostated plants than in control clinostated plants. Thus, there is probably interaction between gravity and wind sway, which should be examined further.

5 Conclusions

Although responses to mechanical stimulation found among plants are of many different types, still it is possible to generalize about the mechanisms that are involved (Fig. 14, Table 5). It seems probable that a physical perturbation of the plant is required for the sensory function. In all of the cases involving vascular plants, a flexing or bending of the sensory organ is involved, as shown

Table 5. Various mechanosensitive systems, and various factors and properties associated

Phenomenon	Plant	Organ	Response[a]	Recovery[a]
I Thigmonasty	*Mimosa pudica*	Pulvinus	3	960
	Biophytum sensitivum	Pulvinus	1	180
	Neptunia oleracea	Pulvinus	180	3600
	Dionaea muscipula	Leaf	1	>24 h
	Drosera spp.	Tentacle	180	?
	Aldrovanda	Leaf	0.04	3600
	Pisum sativum	Tendril	1920	14,400
	Cucumis sativum	Tendril	5400	18,000
	Marah fabaceus	Tendril	?	?
	Various nematode-trapping fungi	Ring trap	0.1	?
II Thigmotropism	*Avena sativa*	Coleoptile	<300	
	Pisum sativum	Root		
	Various rust and powdery mildew fungi	Germ tube	<24 h	?
	Portulaca grandiflora	Stamen	5	360
III Thigmomorphogenesis	*Bryonia dioica*	Stem	900	3 d
	Phaseolus vulgaris	Stem	360	3 d
	Lycopersicon esculentum	Stem	900	3 d

[a] In s, unless otherwise stated. [b] ? = not known. [c] + = present in the system.

by IWANAMI (1962c) for *Portulaca* stamens. Pressure on the tip of the anther, while holding the base steady, results in the filament bending like a bow, and the resulting thigmotropic curvature is in the direction of the convex side of the bow. The significance of IWANAMI's experiment lies in the fact that the organ was not rubbed, but bent. The same kind of flexing stimulus can be demonstrated in as different a system as thigmomorphogenesis in beans (JAFFE 1976a). When the flexing stimulus is given, one side of the organ stretches and the other side is compressed, and in the tropic systems, it is the stretched side that subsequently contracts to produce a curvature.

The first event that can be recorded after stimulation is a change in bioelectrical properties, whether changes in electrical resistance (JAFFE 1976b), or in resting or action potentials (SIBAOKA 1966, UMRATH 1934, SCHRANK 1945, PICKARD 1971, WILLIAMS and PICKARD 1972b). Where they have been looked for, receptor potentials have often been found to precede action potentials (SIBAOKA 1966, WILLIAMS and PICKARD 1972a, JACOBSON 1965). Such induced electrical activity may be thought of as part of the transduction method, coupling the stimulus to the utilization of energy in the motor response.

with their functions (JAFFE unpublished)

Receptor potential	Action potential	ATP Utilized or otherwise involved	Involved in mechanism: Auxin	Involved in mechanism: Ethylene	References
?	+	+	+	?	BOSE (1913), SIBAOKA (1966), WATANABE (1971), LYUBIMOVA et al. (1964)
?[b]	+	?	?	?	BOSE (1913), SIBAOKA (1973)
?	?	?	?	?	BOSE (1913)
+	+	+	?	?	JAFFE (1973a), JACOBSON (1965), SIBOAKA (1966)
+	+	?	?	?	WILLIAMS and PICKARD (1972a)
?	?	?	?	?	ASHIDA (1934)
?	?	+	+	+	JAFFE (1970), JAFFE and GALSTON (1966a, b, 1967)
?	+	?	+	+	BANGERTH (1974), UMRATH (1934)
?	?	?	+	+	REINHOLD (1967)
?	?	?	?	?	DUDDINGTON (1968)
?	+	?	?	?	SCHRANK (1944, 1945)
?	?	?	?	?	DARWIN (1881)
?	?	?	?	?	JOHNSON (1934), PURDY (1958), WYNN (1976)
?	?	+	+	+	IWANAMI (1962b), JAFFE et al. (1977)
?	?	?	+	?	BOYER (1972, 1973), JAFFE (1973b)
?	+	?	+	+	JAFFE (1976a, b), PICKARD (1971)
?	+	?	+	+	MILBURN (1971), PICKARD (1971)

While the propagation of action potentials is probably part of the integrative mechanism, phytohormones have also been implicated. UMRATH (1972) postulated an "excitatory substance" to be involved in a number of rapidly responding systems, which may or may not be the same as the putative principle described by RICCA (1916) and HIRSCH (1953). In a most interesting paper, VAN SAMBEEK et al. (1976) have shown that an unknown substance or substances is released into the transpiration stream of traumatized plants which seems to move together with action potentials. This principle, which they have named Ricca's factor can induce action potentials and is stable in air. Thus, it cannot be the same substance reported by HESSE et al. (1957).

Insofar as they have been studied, certain phytohormones may also be involved as integrative factors. Thus, auxin has been shown to induce coiling in tendrils and potentiate the response of *Portulaca* stamens and *Mimosa* leaves. Ethylene has been found to induce coiling in tendrils and initiate thigmomorphogenesis in beans, and much more ethylene is produced by both systems when they are mechanically stimulated. It does not seem probable that the mimicking effect of ethylene and the production of ethylene in response to mechanical

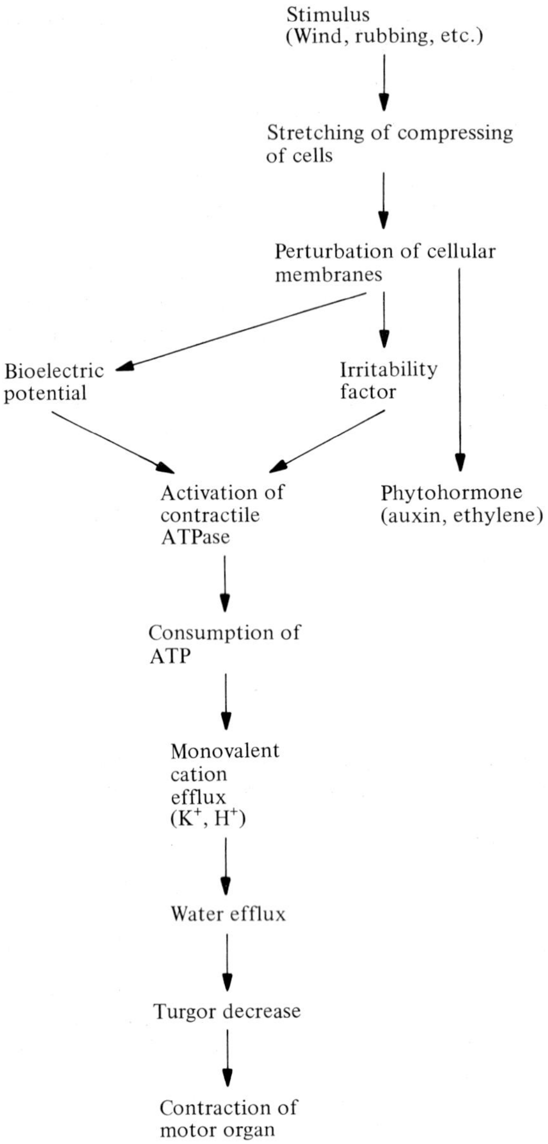

Fig. 14. Generalized outline of the events found to occur in most systems responding to mechanical stimulation

perturbation is merely fortuitous. Rather, it seems likely that ethylene is an internal mediator, or "trigger" of these systems.

In all systems that have been studied, exogenous ATP can potentiate the response, and endogenous ATP is used up during the response. These observations, which indicate that ATP is the native source of energy for the response, are supported by the presence of contractile "actomyosin-like" ATP-ase in several of the systems (pea tendrils and sensitive mimosa).

The rapid turgor drop in certain parts of the responding organ, seems to be accompanied by ion effluxes from the appropriate cells (e.g., H^+, K^+ in pea tendrils and *Mimosa* motor cells). The water efflux responsible for the turgor drop may be in response to a solute gradient set up by the ion efflux.

In all cases, the initial response to mechanical stimulation, be it nastic, tropic, or morphogenetic, is very rapid, minutes, seconds, or less. The recovery phase, on the other hand, is typically much longer, and may involve a reversal of the response.

Thus, the accumulated experience with these "thigmo" systems indicates that they have many processes in common and are probably developments arising in their variety from a fundamental mechnism. That the thigmoresponse system probably involves changes at the membrane level, is supported by current work in several laboratories. Thigmomorphogenesis in vascular plants is a good model for all of the systems, since its early events parallel those of the tropic and nastic responses.

Acknowledgment. This work was supported in part by grant No. PCM-77-24798 from the National Science Foundation (USA) and by grant No. NSG-7352 from the National Aeronautical and Space Administration (USA).

References

Abeles FB (1973) Ethylene in plant biology. Academic Press, London New York

Allen RD (1969) Mechanisms of seismonastic reaction in *Mimosa pudica*. Plant Physiol 44:1101–1107

Ashida J (1934) Studies on the leaf movement of *Aldrovanda vesiculosa* L. Mem Coll Sci Kyoto Imp Univ Ser B 9:141–244

Audus LJ (1935) Mechanical stimulation and respiration rate in the cherry laurel. New Phytol 34:386–402

Bailey MA, Templeton JA (1926) A note on the abnormal behavior of cotton plants when subjected to handling. Emp Cotton Grow Rev 3:477

Balan J, Gerber NN (1972) Attraction and killing of the nematode *Panagrellus redivivus* by the predacious fungus *Arthrobytrys dactyloides*. Nematologica 18:163–167

Bangerth F (1974) Interaktionen von Auxin und Äthylen bei der thigmotropen Bewegung der Ranken von *Cucumis sativus*. Planta 117:329–338

Barber DA, Gunn KB (1974) The effect of mechanical forces on the exudation of organic substances by the roots of cereal plants grown under sterile conditions. New Phytol 73:39–45

Barker J (1935) A note on the effect of handling on the respiration of potatoes. New Phytol 34:407–408

Barley KP (1962) The effects of mechanical stress on the growth of roots. J Exp Bot 13:95–110

Benolken RM, Jacobson SL (1970) Response properties of a sensory hair excised from Venus' flytrap. J Gen Physiol 56:64–82

Biswas S, Bose DM (1972) An ATPase in sensitive plant *Mimosa pudica* I. Purification and characterization. Arch Biochem Biophys 148:199–207

Boersch K (1938) Personal communication to G. Borgstrom, 1939, and quoted by the latter in: The transverse reactions of plants. Williams & Norgate, London

Borgstrom G (1939) The transverse reactions of plants. Williams & Norgate, London

Bose JC (1913) Researches on irritability of plants. Longmans, Green & Co, New York

Bose JC (1927) Plant autographs and their revelations. MacMillan, New York

Boyer N (1967) Modifications de la croissance de la tige de Bryone (*Bryonia dioica*) à la suite d'irritations tactiles. CR Acad Sci Paris 264:2114–2117

Boyer N (1972) Movement périodique de circumnutation et croissance des milles. Ann Sci Nat Bot Biol Veg 13:417–475

Boyer N (1973) Corrélations de croissance dans les plantes à milles. Ann Sci Nat Bot Biol Veg 14:301–366

Brown KM, Leopold AC (1973) Ethylene and the regulation of growth in pine. Can J For Res 3:143–145

Bünning E (1940) Über die Verhinderung des Etiolements. Ber Dtsch Bot Ges 59:2–9

Bünning E, Haag L, Timmermann G (1948) Weitere Untersuchungen über die formative Wirkung des Lichtes und mechanischer Reize auf Pflanzen. Planta 36:178–187

Burden RF, Randerson PF (1972) Quantitative studies of the effects of human trampling on vegetation as an aid of the management of semi-natural areas. J Appl Ecol 9:439–458

Burkholder PR, Pratt R (1936) Leaf movements of *Mimosa pudica* in relation to light. Am J Bot 23:46–52

Coad DR, McGehee TF (1917) Collection of weevils and infested squares as a means of control of the cotton boll weevil in the Mississippi delta. USDA Bull 546:1–51

Colla S (1937) Die kontraktile Zelle der Pflanzen. Protoplasma 30:1–163

Couch JN (1937) The formation and operation of the traps in the nematode-catching fungus *Dactylella bembicodes* Drechsler. J Mitchell Soc 53:301–309

Darwin C (1876) The movements and habits of climbing plants, 2nd edn rev, Appleton, New York

Darwin C (1881) The power of movement in plants. Appleton, New York

Desbiez MO (1973) Précisions sur le mécanisme des corrélations entre bourgeons cotyledonaires induits par des stimulation méchaniques du cotylédon chez *Bidens pilosus* L. Z Pflanzenphysiologie 69:174–180

Desbiez MO, Thellier M (1975) Lithium inhibition of the mechanically induced presence between cotyledonary buds. Plant Sci Lett 4:315–321

Dickinson S (1949a) Studies in the physiology of obligate parasitism. I. The stimuli determining the direction of growth of the germ tubes of rust and mildew spores. Ann Bot NS 13:89–104

Dickinson S (1949b) Studies in the physiology of obligate parasitism. II. The behavior of the germ-tubes of certain rusts in contact with various membranes. Ann Bot NS 13:219–236

Dickinson S (1971) Studies in the physiology of obligate parasitism. VII. An analysis of fungal responses to thigmotropic stimuli. Phytopathol Z 70:62–70

Duddington CL (1968) Fungal parasites of invertebrates. 2. Predacious fungi. In: Ainsworth GC, Sussman AS (eds) The fungi, an advanced treatise. Academic Press, New York, Vol III, pp 239–251

Evans ML, Ray PM (1969) Timing of the auxin response in coleoptiles and its implications regarding auxin action. J Gen Physiol 53:1–20

Feder WA, Everard COR, Wootton LMO (1963) Sensitivity of several species of the nematophagous fungus *Dactylella* to a morphogenic substance derived from free-living nematodes. Nematologica 9:49–54

Fitting H (1903) Untersuchungen über den Haptotropismus der Ranken. Jahrb Wiss Bot 38:545–634

Frizzell JL, Brown LC, Waddle BA (1960) Some effects of handling on the growth and development of cotton. Agron J 52:69–70

Galun E (1959) The cucumber tendril – a new test organ for gibberellic acid. Experentia 15:184–185

Godwin H (1935) The effect of handling on the respiration of cherry laurel leaves. New Phytol 34:403–406

Goeschl JD, Rappaport L, Pratt HK (1966) Ethylene as a factor regulating the growth of pea epicotyls subjected to physical stress. Plant Physiol 41:877–884

Greengard P (1975) Cyclic nucleotides, protein phosphorylation, and neuronal function. In: Drummond GI, Greengard P, Robinson GA (eds) Advances in cyclic nucleotide research vol 5. Raven, New York, pp 585–601

Gregory-Southworth A, Klein AO (1974) Light-induced sensitivity of bean seedlings to mechanical stimulation. Plant Physiol 53 Suppl:46

Haberlandt G (1914) Physiological plant anatomy. (Translated from the 4th German Ed by M. Drummond) MacMillan, London

Hamner PA, Mitchell CA, Weiler TC (1974) Height control in greenhouse chrysanthemum by mechanical stress. HortScience 9:474–475

Hertel R, St. Thomson K, Russo VEA(1972) In-vitro auxin binding to particulate cell fractions from corn coleoptiles. Planta 107:325–340

Hesse G, Banerjee B, Schildknecht H (1957) Die Reizbewegungen der Mimosoideen and ihre Hormone. Experientia 13:13–19

Hiraki Y, Ota Y (1975) The relationship between growth inhibition and ethylene production by mechanical stimulation in *Lilium longiflorum*. Plant Cell Physiol 16:185–189

Hirsch P (1953) Eine empfindliche Pflanze. Kosmos 6:266–268

Hunt LA, Impens II, Leman ER (1967) Preliminary wind studies on photosynthesis and evapotranspiration of forage stands. Crop Sci 7:575–578

Irvine RF, Osborne DJ (1973) The effect of ethylene on [$1-{}^{14}C$] glycerol incorporation into phospholipids of etiolated pea stems. Biochem J 136:1133–1135

Iwanami Y (1962a) The movements of the stamens of *Portulaca grandiflora*. I. Bot Mag (Tokyo) 75:133–139

Iwanami Y (1962b) The movements of the stamens of *Portulaca grandiflora*. II. Bot Mag (Tokyo) 75:289–295

Iwanami Y (1962c) The movements of the stamens of *Portulaca grandiflora*. III. Bot Mag (Tokyo) 75:331–335

Iwanami Y (1962d) The movements of the stamens of *Portulaca grandiflora*. IV. Bot Mag (Tokyo) 75:371–376

Jacobs MR (1954) The effect of wind-sway on the form and development of *Pinus radiata* D. Don. Aust J Bot 2:35–51

Jacobson SL (1965) Receptor response in Venus' fly-trap. J Gen Physiol 49:117–129

Jaffe MJ (1970) Physiological studies on pea tendrils. VII. Evaluation of a technique for the asymmetrical application of ethylene. Plant Physiol 45:631–633

Jaffe MJ (1972) Physiological studies on pea tendrils. VIII. The relationship of circumnutation to contact coiling. With a description of a laboratory intervalometer using integrated digital circuits. Physiol Plant 26:73–80

Jaffe MJ (1973a) The role of ATP in mechanically stimulated rapid closure of the Venus' fly trap. Plant Physiol 51:17–18

Jaffe MJ (1973b) Thigmomorphogenesis: The response of plant growth and development to mechanical stimulation. With special reference to *Bryonia dioica*. Planta 114:143–157

Jaffe MJ (1975) The role of auxin in the early events of the contact coiling of tendrils. Plant Sci Lett 5:217–225

Jaffe MJ (1976a) Thigmomorphogenesis: A detailed characterization of the response of beans (*Phaseolus vulgaris* L.) to mechanical stimulation. Z Pflanzenphysiol 77:437–453

Jaffe MJ (1976b) Thigmomorphogenesis: Electrical resistance and mechanical correlates of the early events of growth retardation due to mechanical stimulation in beans. Z Pflanzenphysiol 78:24–32

Jaffe MJ, Biro R (1979) Thigmomorphogenesis: The effect of mechanical perturbation on the growth of plants, with special reference to anatomical changes, the role of ethylene, and interaction with other environmental stresses. In: Mussel H, Staples RC (eds) Stress physiology in crop plants. Wiley & Sons, New York, pp 25–59

Jaffe MJ, Galston AW (1966a) Physiological studies on pea tendrils. I. Growth and coiling following mechanical stimulation. Plant Physiol 41:1014–1025

Jaffe MJ, Galston AW (1966b) Physiological studies on pea tendrils. II. The role of light and ATP in contact coiling. Plant Physiol 41:1152–1158

Jaffe MJ, Galston AW (1967) Physiological studies on pea tendrils. III. ATPase activity and contractility associated with coiling. Plant Physiol 42:845–847
Jaffe MJ, Galston AW (1968a) Physiological studies on pea tendrils. V. Membrane changes and water movement associated with contact coiling. Plant Physiol 43:537–542
Jaffe MJ, Galston AW (1968b) The physiology of tendrils. Annu Rev Plant Physiol 19:417–434
Jaffe MJ, Gibson C, Biro R (1977) Physiological studies of mechanically stimulated motor responses of flower parts. I. Characterization of the thigmotropic stamens of *Portulaca grandiflora* Hook. Bot Gaz 138:438–447
Johnson T (1934) A tropic response in germ tubes of urediospores of *Puccinia graminis* tritici. Phytopathology 24:80–82
Junker S (1976) Auxin transport in tendril segments of *Passiflora caerulea*. Physiol Plant 37:258–262
Junker S (1977) Thigmonastic coiling of tendrils of *Passiflora quadrangularis* is not caused by lateral redistribution of auxin. Physiol Plant 41:51–54
Junker S, Reinhold L (1975) A scanning electron microscopic survey of the surface of sensitive tendrils. J Microsc Biol Cell 23:175–180
Larson PR (1965) Stem formation of young *Larix* as influenced by wind and pruning. For Sci 11:412–424
Lea HW (1976) A muscle contracting substance from a plant's closing fly-trap. Planta 129:39–41
Leopold AC, Brown KM, Emerson FH (1972) Ethylene in the wood of stressed trees. HortScience 7:175
Lloyd FE (1942) The carnivorous plants. Ronald, New York
Lyubimova MN, Demyanovskaya NS, Fedorovich IB, Itomlenskite IV (1964) Participation of ATP in the motor function of the *Mimosa pudica* leaf. (English transl) Biokhimiya 29:663–667
MacDougal DT (1908) Practical text-book of plant physiology. Longmans Green & Co New York
Martin EV, Clements FE (1935) Effect of artificial wind on growth and transpiration of *Helianthus annuus*. Plant Physiol 10:613–636
Milburn JA (1970) Phloem exudation from castor bean: Induction by massage. Planta 95:272–276
Milburn JA (1971) An analysis in the response of phloem exudation on application of massage to *Ricinus*. Planta 100:143–154
Mitchell CA, Severson CJ, Wott JA, Hamner PA (1975) Seismomorphogenic regulation of plant growth. J Am Soc Hortic Sci 100:161–165
Mitchell CA, Dostal HC, Seipel TM (1977) Dry weight reduction in mechanically dwarfed tomato plants. J Am Soc Hortic Sci 102:605–608
Neel PL, Harris RW (1971) Motion-induced inhibition of elongation and induction of dormancy in *Liquidambar*. Science 173:58–59
Pelton WL (1967) The effect of a windbreak on wind travel, evaporation and wheat yield. Can J Plant Sci 47:209–214
Pfeffer W (1901) Pflanzenphysiologie, 2nd edn. Engelmann, Leipzig
Pickard BG (1971) Action potentials resulting from mechanical stimulation of pea epicotyls. Planta 97:106–115
Purdy LH (1958) Some factors affecting penetration and infection by *Sclerotinia sclerotiorum*. Phytopathology 48:605–609
Reinhold L (1967) Induction of coiling in tendrils by auxin and carbon dioxide. Science 158:791–793
Reinhold L, Sachs T, Vislovska L (1972) The role of auxin in thigmotropism. In: Carr DJ (ed) Plant growth substances 1970. Springer, Berlin Heidelberg New York, pp 731–737
Ricca U (1916) Solution d'un problème de physiologie. La propagation de stimulus dans la sensitive. Arch Ital Biol 65:219–232

Sachs J (1882) Text-book of botany, morphological and physiological, 2nd edn. (S.H. Vines) Clarendon Press, Oxford
Salisbury F (1963) The flowering process. MacMillan, New York
Schrank AR (1944) Relation between electrical and curvature responses in the *Avena* coleoptile to mechanical stimuli. Plant Physiol 19:198–211
Schrank AR (1945) Effect of mechanical stimulation on the electrical and curvature responses in the *Avena* coleoptile. Plant Physiol 20:344–358
Scurfield G (1973) Reaction wood: Its structure and function. Science 179:647–655
Sibaoka T (1966) Action potentials in plant organs. Symp Soc Exp Biol 20:49–74
Sibaoka T (1973) Transmission of action potentials in *Biophytum*. Bot Mag Tokyo 86:51–61
Staples RC, Laccetti L, Yaniv Z (1976) Appressorium formation and nuclear division in *Colletotrichum truncatum*. Arch Microbiol 109:75–84
Steucek GL, Gordon LK (1975) Response of wheat (*Triticum aestivum*) seedlings to mechanical stress. Bot Gaz 136:17–19
Suda S (1960) On the physiological properties of mimosine. Bot Mag Tokyo 73:142–147
Todd GW, Chadwick DL, Tsai S-D (1972) Effect of wind on plant respiration. Physiol Plant 27:342–346
Toriyama H (1953) Observational and experimental studies of sensitive plants, I. The structure of parenchymatous cells of pulvinus. Cytologia 18:283–292
Toriyama H (1955) Observational and experimental studies of sensitive plants. VI. The migration of potassium in the primary pulvinus. Cytologia 20:367–377
Toriyama H, Jaffe MJ (1972) Migration of calcium and its role in the regulation of seismonasty in the motor cell of *Mimosa pudica* L. Plant Physiol 49:72–81
Toriyama H, Sato S (1968) Electron microscope observation of the motor cell of *Mimosa pudica* L. II. On the contents of the central vacuole of the motor cell. Proc Jpn Acad 44:949–953
Toumey JW (1899) Sensitive stamens in the genus *Opuntia*. Asa Gray Bull 7:35
Tronchet A (1964) Quelques aspects de la sensibilité de vrilles. P V Mem Acad Sci Belles-Lett Arts Besancon 175:23–39
Turgeon R, Webb JA (1971) Growth inhibition by mechanical stress. Science 174:961–962
Umrath K (1934) Über die elektrischen Erscheinungen bei thigmischer Reizung der Ranken von *Cucumis melo*. Planta 23:47–50
Umrath K (1966) Durch Erregungssubstanz ausgelöste Spaltöffnungsschließbewegung. Z Pflanzenphysiol 55:217–223
Umrath K (1972) Erregungsleitung bei jungen Pflanzen von *Mimosa pudica* und ihre vermutete Beziehung zur Erregungssubstanz. Ber Dtsch Bot Ges 85:451–457
Van Sambeek JW, Pickard BG, Ulbright CE (1976) Mediation of rapid electrical, metabolic, transpirational, and photosynthetic changes by factors released from wounds. II. Mediation of the variation potential by Ricca's factor. Can J Bot 54:2651–2661
Venning FD (1949) Stimulation by wind motion of collenchyma formation in celery petioles. Bot Gaz 110:511–514
Victor TS, Vanderhoef LN (1975) Mechanical inhibition of hypocotyl elongation induces radial enlargement. Plant Physiol 56:845–846
Wadsworth RM (1959) An optimum wind speed for plant growth. Ann Bot NS 23:195–199
Wadsworth RM (1960) Effect of artificial wind on growth rates of plants in water culture. Ann Bot NS 24:200–211
Walker WS (1957) The effect of mechanical stimulation on the collenchyma of *Apium graveolens* L. Proc Iowa Acad Sci 64:177–186
Walker WS (1960) The effects of mechanical stimulation and etiolation on the collenchyma of *Datura stramonium*. Am J Bot 47:718–724
Watanabe S (1971) Ouabain and IAA effects on *Mimosa* pulvinus. Artes Liberales 8: 75–80
Westing AH (1965) Formation and function of compression wood in gymnosperms. Bot Rev 31:381–480

Westing AH (1968) Formation and function of compression wood in gymnosperms II. Bot Rev 34:51–78
Whitehead FH (1962) Experimental studies of the effect of wind on plant growth and anatomy. II. *Helianthus annuus*. New Phytol 67:59–62
Whitehead FH, Luti R (1962) Experimental studies of the effect of wind on plant growth and anatomy. I. *Zea mays*. New Phytol 61:56–58
Williams ME, Mozingo HN (1971) The fine structure of the trigger hair in Venus' fly trap. Am J Bot 58:532–539
Williams SE, Pickard BG (1972a) Receptor potentials and action potentials in *Drosera* tentacles. Planta 103:193–221
Williams SE, Pickard BG (1972b) Properties of action potentials in *Drosera* tentacles. Planta 103:222–240
Williams SE, Pickard BG (1974) Connections and barriers between cells of *Drosera* tentacles in relation to their electrophysiology. Planta 116:1–16
Wilson BF, Archer RA (1977) Reaction wood: induction and mechanical action. Annu Rev Plant Physiol 28:23–43
Wynn WK (1976) Appressorium formation over stomates by the bean rust fungus: Response to a surface contact stimulus. Phytopathology 66:136–146
Zeltner H (1931–1932) Über Elektronastik und andere Reizbewegungen der Ranken. Z Bot 25:97–172

13 Hormonal Control of Wound-Induced Responses

H. IMASEKI

1 Introduction

Wounding is a frequent, but irregular event imposed on plant life by the natural environment. Since higher plants are unable to move to better environments to escape from animal browsing or severe winds, mechanisms that overcome the wound effect are essential to survival of the individual, and the species as well. Wounding is defined herein as a mechanical process which destroys cells in a specific area of tissue. It thus breaks cell to cell continuity in a multicellular plant so that cells, or at least one side of cells which were previously in contact with other cells, are now exposed. In most cases, this will include loss of part of a tissue or organ. The reactions which occur in response to wounding are so diverse that we presently cannot integrate all of these reactions into a cogent series of physiological processes. However, the fundamental physiological outcome of the wound response is regeneration of part or all of the functions which were previously shared by the damaged or lost cells, tissues or organs (LIPETZ 1970). This does not necessarily mean regeneration of the complete lost structure, however (LIPETZ 1970). Thus, formation of protective materials in cells near the cut surface, initiation of cell proliferation, regeneration of vascular elements, or rooting at the base of shoot cuttings are common physiological responses to wounding. In a broad sense, one may also include in the wound response the lateral bud growth which occurs when the apical portion of the central axis is removed or damaged, a phenomenon well-known as the breaking of apical dominance.

Wound-response phenomena are supported by biochemical changes in otherwise quiescent unwounded cells near the wound, the cells becoming quite active metabolically (KAHL 1974, 1978, VAN STEVENINCK 1979). Apparently, the wound surface, or adjacent wound-affected cells generate a signal(s) which is transmitted through at least several cell layers inducing these quiescent cells to become active (BORCHERT and MCCHESNEY 1973, URITANI 1961, 1971, 1976). The chemical entity of the signal generated by wounding has been a matter of great interest, but remains unknown with only a few exceptions (RYAN 1974, TRESHOW 1955a, b). However, a currently acceptable concept is that one or more of the known plant hormones plays at least a part in the signaling process. Emphasis will thus be placed on the effects of exogenously applied hormones on wound reactions at both physiological and biochemical levels, and on the changes of plant hormone levels associated with wounding.

2 Formation of Protective Materials

2.1 Suberization or Cutin Formation at the Wound Surface

The intact plant is covered with cell layers which are differentiated to protect the inner cells from desiccation and microbial invasion. Wounding exposes these inner cells and their normal response is to produce and deposit protective materials in order to seal off the wound. Most of the natural protective material deposited in the epidermal cells of aerial parts of the plant is the lipoidal polymeric substance, cutin (EDELMAN and HALL 1964), and for cells of underground organs is suberin (KOLATTUKUDY et al. 1975). When tomato fruits, bean pods or jade plant (*Crassula*) leaves are wounded by removing the original cuticle-containing layers, and the organ is allowed to heal on the plants, new protective layers are produced. Comparative analysis of the lipid components of hydrogenolysis products of the newly formed protective materials revealed, rather surprisingly, that when wounded, aerial parts of plants whose natural protective material was cutin produced and deposited suberin, but not cutin (DEAN and KOLATTUKUDY 1976). The potato tuber periderm contains suberin as its natural protective material, and wounding of it induces parenchymatous cells to produce the same material, suberin (DEAN and KOLATTUKUDY 1976). Thus, DEAN and KOLATTUKUDY (1976) concluded that suberization might be the fundamental process involved in wound-healing of all plant tissues.

Suberization of the wound surface of potato tubers occurs after about a 4-day lag period, but it is severely inhibited if the wound surface is washed for 3 h immediately after wounding with large amounts of water (SOLIDAY et al. 1978). The suppression of suberization was also observed when the washing was done at any time within the 4-day lag period. Washing after day 4 had no effect on the suberization process. An interpretation of this observation by SOLIDAY et al. (1978) is that the factor, easily removed by a 3-h washing and responsible for induction of suberization, is generated at or near the wound; but the continued presence of the factor at or near the wound is needed until induction of the enzymes involved in suberization occurs (SOLIDAY et al. 1978). They also examined a possibility that the factor might be one, or a combination of plant hormones. Short-term treatment of potato tuber slices with ABA enhanced suberization at the wound surface of the slices, while treatment with cytokinin severely inhibited suberin formation. The inhibition of suberin formation caused by washing could be partially reversed when ABA was applied to potato slices washed for 3 h within 1 day after wounding. Inhibition of suberization caused by washing slices between 2 and 4 days after wounding was not reversed by ABA treatment. It was also shown that the washing released endogenous ABA, but the release was observed only when the slices were washed within 2 days after cutting. Therefore, ABA may participate in a positive manner in the early phases of suberin formation. The factors which can be washed off the wound surface between day 2 to 4 after wounding are, however, probably not ABA. It is thus presumed that the suberin-inducing factor is formed as a result of ABA action at early stages (SOLIDAY et al. 1978). This was more directly proven by adding ABA to potato tuber tissue cultures. Suberization was stimulated without significant inhibition of growth (Soliday et al. 1978). A point which should be raised here, however, is that although potato tuber tissue contains significant amounts of ABA (72 ng g^{-1} fr. w.), endogenous ABA apparently does not stimulate suberization unless the tissue is wounded. Thus, a mechanism which brings endogenous ABA into action after wounding must be sought in order to understand the hormonal relations of wound-induced suberin formation.

2.2 Lignification

Lignification in response to wounding is in most cases associated with xylem element development near the wound (this will be discussed in a later section), but there are several examples of lignification on the wound surface which are independent of the induced xylem differentiation. Parenchyma cells of sweet potato root do not contain a significant amount of lignin when examined by the phloroglucinol-HCl test, a conventional test for lignins (Imaseki et al. 1968). The wound surface of the root tissue, however, gradually deposits substances which give an intense red color with the phloroglucinol-HCl test (Imaseki et al. 1968). Induced lignification on the wound surface has also been reported in radish roots (Kahl 1973).

The regulatory mechanism of lignification and the role of plant hormones in it are not clear. In natural lignin formation the degree of lignification in the reaction wood which develops in leaning stems of both angiosperms and gymnosperms is correlated with the endogenous levels of auxin (Wardrop 1971, Wilson and Archer 1977). Lignification on the wound surface in swede root slices is stimulated by ethylene treatment of the slices (Rhodes and Wooltorton 1973). Cytological examination of control and treated slices revealed that lignification was confined to cells near the wound surface. However, when a whole swede root was incubated in air containing ethylene for 6 days, parenchyma cells of secondary xylem throughout the root were uniformly lignified (Rhodes and Wooltorton 1973). Thus, ethylene enhances lignin formation in swede root cells regardless of whether or not the wound is imposed. However, it is clear that wounding, together with ethylene, has some specific effect on lignin formation or deposit since, when wounded, the slices deposited lignins only in cells in the vicinity of the wound. Stimulation of lignin formation in swede root slices by ethylene was shown to result from enhanced synthesis of lignin precursors as a result of increased PAL activity, a key enzyme in the biosynthesis of cinnamic acid derivatives (Rhodes and Wooltorton 1971). Increased supply of lignin precursors together with hydrogen peroxide enhances lignin formation in *Phleum pratense* and several other species (Stafford 1965, Mäder and Füssl 1982, Mäder and Amberg-Fisher 1982). Wounding generally promotes ethylene production (Abeles 1972, 1973), and ethylene is known to increase PAL in several plants (Hyodo and Yang 1971, Imaseki et al. 1968, Riov et al. 1969). Thus, although wound-induced ethylene usually increases lignification in the cells near the wound surface, this is not always true. For example, wound-induced lignification of sweet potato root cells was greatly inhibited by ethylene (IMASEKI and URITANI unpublished data), although synthesis of cinnamic acid derivatives was stimulated by it (IMASEKI et al. 1968).

3 Wound-Induced Cell Division and Its Hormonal Control

Technically, almost all tissue cultures of plant cells originate from excised tissue fragments from various sources, and the success of establishment is dependent upon hormonal as well as nutritional balances in the culture medium. Since the hormonal aspects of callus induction (YEOMAN 1970) and cell culture (STEWARD et al. 1970) are not within the scope of this chapter, only a few examples of cell proliferation which may be related to wounding will be dealt with here.

Wounded tissues of Jerusalem artichoke tubers respond to auxin in two different ways, depending on their "physiological state" (ADAMSON 1962, MASUDA 1965). If thin slices of the tuber are incubated immediately after wounding on a suitable medium containing in auxin, cells on the wound periphery start to divide synchronously after a lag period of about 8 h (YEOMAN et al. 1965, YEOMAN and EVANS 1967). These newly meristematized cells eventually develop into callus. If, however, the slices are first incubated in water without auxin, or with GA_3, cells in the pretreated slices rapidly expand without cell division in response to auxin added later (MASUDA 1966). Thus, during the pretreatment of the tuber slices in water, a metabolic alteration occurs so that cells originally capable of dividing in response to auxin lose their dividing ability, but attain an expansion potential. This metabolic alteration is triggered by wounding, and involves RNA and protein synthesis (MASUDA 1966, SKOOG and SCHMITZ 1972, SPERLING and LATIES 1963). The biochemical sequence that occurs during water pretreatment is not known, but it was shown that the process is not dependent on auxin (MASUDA 1966, SETTERFIELD 1963, SPERLING and LATIES 1963), but is enhanced by GA_3 (MASUDA 1965) or cyclic AMP (KAMISANA et al. 1973). MASUDA (1966) also showed that the increase in RNA and protein content continued only when the slices pretreated with water were further incubated with auxin. It should be mentioned here that the increased synthesis of RNA and protein after wounding is generally observed in almost all plant tissues regardless of whether they are responsive to auxin or to other plant hormones (BYRNE and SETTERFIELD 1977, CHERRY 1968, ELLIS and MACDONALD 1967, IMASEKI et al. 1968, LEAVER and KEY 1967, WATANABE and IMASEKI 1973).

YEOMAN et al. (1968), using cylindrical tissue plugs of Jerusalem artichoke tuber, analyzed four factors that would cause the initiation of cell division in peripheral cells in the presence of auxin: (1) greater availability of O_2, (2) more rapid release of CO_2, (3) greater availability of nutrients and (4) more rapid release of a volatile inhibitor. However, none of these possibilities fits their results, obtained under various conditions. They did, however, observe that autolysis products of cells near the wound surface interacted with auxin to stimulate cell division. Cutting apparently exerts an unusual stress on cells adjacent to the destroyed cells, and it is probable that this stress disrupts cellular organization of these adjacent cells, thus leading to autolysis. This hypothesis was supported by the fact that the proportion of dividing cells in tissue segments with different surface–volume ratios increased as the surface area relative to volume increased. Similar, but more limited cell divisions on the wound periphery have also been observed in potato tuber slices incubated without auxin and nutrients (LANGE et al. 1970). BORCHERT and MCCHESNEY (1973), using 2-mm thick potato tuber slices, analyzed spatial distribution of the wound response. As discussed earlier, suberization takes place in the walls of cells at the wound surface. Cell divisions accompanied by massive degradation of starch were limited to two or three subperipheral cell layers located immediately below the outermost, suberizing cell layers (BORCHERT and MCCHESSNEY 1973). Enhanced synthesis of RNA and protein occurred in much broader region below the wound surface (BORCHERT 1978, BORCHERT and MCCHESNEY 1973).

WATANABE and IMASEKI (1973) observed an induced DNA synthesis in nuclei of 1-mm-thick potato tuber slices which had been incubated in water without auxin. In contrast to other tissues, such as Jerusalem artichoke tuber (VAN STEVENINCK 1961, 1972), tobacco pith (PATAU et al. 1957) and carrot root slices (STEWARD et al. 1970), it is clear that in potato tuber, wounding per se is enough to induce DNA synthesis. In fact, 1-mm-thick potato slices doubled their DNA content within 3 days of cutting (WATANABE and IMASEKI 1973). Distribution of cells in which DNA synthesis occurred were fairly random, but even after 1 week's incubation period, cells division, if any, occurred only in peripheral cells and most of the cells in which DNA synthesis occurred did not divide. BORCHERT and MCCHESNEY (1973) explained this by speculating that normal wound healing does not take place in very thin slices. If this interpretation is correct, then for cell division to occur it may require the metabolic support of cells adjacent to those that would divide (e.g., located two to three cells layers from the wound surface). The possibility that endogenous auxin is involved in the initiation of DNA synthesis in potato tuber slices was tested using the anti-auxin, p-chlorophenoxyisobutyric acid, but it did not significantly suppress the induced DNA synthesis (unpublished data).

The wound-induced DNA synthesis in potato tuber slices is accompanied by wound-induced protein synthesis at two periods after wounding (WATANABE and IMASEKI 1976). The DNA synthesis that occurs at 36 h after wounding was completely blocked when the slices were supplied with cycloheximide immediately after wounding (Fig. 1). Sensitivity of the tissue slices to cycloheximide decreased in two steps when the time of application of the inhibitor was delayed. When the tissue slices were allowed to incubate without cycloheximide for the first 6 h, and protein synthesis after h 6 was inhibited by cycloheximide, the rate of DNA synthesis was about 25% relative to the untreated tissue slices at h 36. Tissue sensitivity toward cycloheximide did not change from h 6 to about h 30, i.e., until 6 h immediately before DNA synthesis begins, but it quickly dropped after h 30. These results indicate that protein synthesized within 6 h after wounding (Phase I protein) is essential to initiate DNA synthesis and protein synthesized in the 6 h immediately before DNA synthesis (Phase II protein) is needed for the full development of DNA synthesis capacity. That Phase I

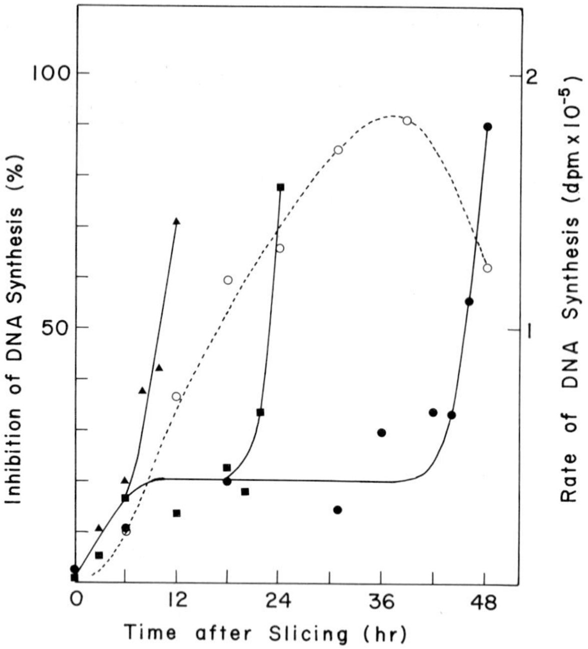

Fig. 1. Effect of delayed application of cycloheximide on the induction of DNA synthesis. Cycloheximide was applied to discs at the indicated time and the rate of DNA synthesis (e.g., ^{3}H-thymidine incorporation during a 2-h incubation) was determined 12 h (▲), 24 h (■) and 48 h (●) after slicing. DNA synthesis is shown as percent of the respective control rate. Time-course changes in the rate of DNA synthesis for control discs (○) was also included. (WATANABE and IMASEKI 1976)

protein synthesis was initiated on wounding was shown when the slices were treated with cycloheximide for 6 h followed by 6 h of washing and incubation in cycloheximide-free water: initiation of DNA synthesis was delayed some 12 h, to h 48.

The wound-induced DNA synthesis in potato tuber slices was not affected by exogenous auxin and GA_3, but was inhibited by ABA and partially inhibited by ethylene (SATO et al. 1976). Ethylene was effective *only* when supplied during the several hours shortly after the end of Phase I protein synthesis (SATO et al. 1976). Reduction of endogenous ethylene levels increased DNA synthetic rate, thus the ethylene may indeed have a regulatory function in wound-induced DNA synthesis.

That wounding alone induced massive cell proliferation was also reported in *Stizolobium hassjoo* (Fabaceae) epicotyls by KOMAMINE and his colleagues. When etiolated seedlings of *S. hassjoo* were decapitated far below the elongating zone and the basal portions were further incubated in the dark, massive cell division took place at the cut end of epicotyl, resulting in a callus-like outgrowth (KOMAMINE et al. 1963). It was also found that the base of the 2-cm-long stem segments excised from the matured part of epicotyls similarly developed the outgrowth. Application of IAA in lanolin paste to the cut surface did not significantly affect the outgrowth; but if the stems are cut at the elongation zone where natural auxin content should be higher, the outgrowth did not develop. However, this does not mean that auxin is not involved in initiation of the outgrowth. MOROHASHI et al. (1969) reported that the cut surface of the matured portion of *S. hassjoo* epicotyls accumulated a significant amount of auxin before the outgrowth became visible. The accumulation of auxin at the wounded area was explained by a decreased IAA-oxidizing activity, rather than an increased synthetic rate of auxin after wounding. The increased auxin level at the wounded area of matured epicotyls may be one of the causal factors that induce cell proliferation in *S. hassjoo,* since if the cut surface of the young portion of the stem (which does not form the outgrowth by cutting alone) is supplied with IAA, an outgrowth similar to that formed in the matured stem cut end occurred.

4 Organized Wound Response

4.1 Vascular Element Differentiation

It has been extensively documented that wounds which sever vascular continuity induce differentiation of xylem and phloem elements from nearby parenchyma cells (JACOBS 1970, ROBERTS 1969). If the wound is made to a whole plant, near vascular elements eventually form functional vascular strands replacing the severed vascular function. Most studies on redifferentiation of vascular elements or strands around the wound aimed primarily at understanding the mechanism of primary vascular cytodifferentiation of procambial cells in apical meristems, but little attention has been paid to the effect of wounding on the process. A question pertinent to this problem is: "does wounding have a specific effect on vascular differentiation?"

When a wound was made in the stem of *Coleus,* tomato, or pea by a horizontal half-way cut or by a wedge-shaped cut, the most remarkable characteristic of the vascular regeneration around the wound is that the regeneration occurs in a basipetal direction (ALONI 1976, ALONI and JACOBS 1977, SACHS 1968b, THOMPSON and JACOBS 1966). Xylem as well as phloem regeneration starts at parenchyma cells above the wound and proceeds downward to the upper edge of the wound, followed by a further progress of regeneration circling around the wound. The process may follow the possible basipetal transport flow-path of a stimulating substance. Vascular strands regenerate as if they are attracted by the existing strands below the wound. The vascular regeneration is significantly reduced if leaves above the wound are removed, suggesting that leaves above the wound provide promotive substances which flow basipetally. Using *Coleus* internodes, JACOBS (1952) showed that auxin could replace the function played by leaves above the wound. When IAA in lanolin paste was given to the cut end of a decapitated internode which was wounded in the middle, regeneration of xylem and phloem strands around the wound occurred as in an intact but similarly wounded plant (THOMPSON 1966). A series of experiments carried out by JACOBS and his colleagues clearly established that the wound-induced vascular regeneration is strictly regulated by auxin concentration coming from the distal part of the wound, and the morphogenetic pattern is determined by the polarity of auxin transport (JACOBS 1952, 1956, JACOBS and MACCREADY 1967, LAMOTTE and JACOBS 1963). Simple, but novel experiments by SACHS (1968a, b, 1969) and ALONI (1976) indicate that the effect of wounding to induce additional vessel strands is to sever the basipetal flow of auxin at the wound, so that the increased auxin level in the part above the wound initiates the cytodifferentiation of parenchyma cells (see also ROBBERTSE and MCCULLY 1979, RANA and GAHAN 1983). That auxin induces vessel element differentiation in undifferentiated cells has also been shown with callus tissue to which auxin was supplied locally (COMER 1978, JEFFS and NORTHCOTE 1967), and in woody plants (WAREING 1958). Recent histological study with pith parenchyma explants of *Coleus blumei* revealed that cell division induced by auxin occurred before parenchyma cells differentiated into vessel members (COMER 1978).

It seems unlikely that there is any specific effect of wounding on the vascular regeneration. However, for light-grown pea plants (BENAYOUN et al. 1975), a wound made just below the second photosynthetic leaf, that is in the fourth internode from below, promoted xylem differentiation in the third internode (below), as well as in the wounded internode. Thus, the effect of wounding appeared in a part far distant from the wound. This cannot be simply explained by a severed polar transport of auxin which leads to increased auxin concentration around the wound. Wounding appears to release some stimuli which act synergistically with auxin to promote further vascular differentiation (see also BEHNKE and SCHULZ 1980, GERSAN and SACHS 1980).

Vessel element differentiation is variously modified by other plant hormones such as cytokinins (DALESSANDRO 1973, DALESSANDRO and ROBERTS 1971, FOSKET and ROBERTS 1964, FOSKET and TORREY 1969, MIZUNO et al. 1971), and gibberellins (ALONI 1979, GLASZIOU 1969, ROBERTS and FOSKET 1966), but detailed accounts of these effects will not be dealt with here.

4.2 Root Formation on Stem Cuttings

The initiation of adventitious roots on stems is apparently a response to wounding. Numerous physiological as well as morphological studies to characterize the mechanism underlying in this process have presented several theories (COOPER 1938, SOEKARJO 1965, VAN OVERBEEK and GREGORY 1945, WENT 1938), but a general concept is that the effect of cutting the stem may be similar to that of wounding on vascular regeneration. Thus, an accumulation of root-forming substances occurs at the cut base. These had formed in leaves and buds and moved to the base in a polar and basipetal manner. Evidence that the root-forming substance is auxin (probably IAA) was presented by THIMANN and KOEPFLI (1935), and similar evidence is available from a number of experiments with a variety of tissues (cf. WARMKE and WARMKE 1950). A historical as well as a physiological account of this problem was described by THIMANN (1972, 1977).

There are, however, a few indications showing that the interaction of wounding with natural auxin has a specific effect on adventitious root formation of stem cuttings. For example, the degree of wounding at the base of stem is related to the degree of rooting (THIMANN 1972) and when the base of stem cuttings is longitudinally cut 2 to 3 mm up from the base, it increases the number of adventitious roots (THIMANN 1977). Results probably related to this effect of wounding on root formation are found with pea epicotyl sections where natural phenolic substances such as caffeic and chlorogenic acid interacted with auxin and stimulated adventitious root formation (TOMASZEWSKI 1963), a similar interaction being observed in aseptically cultured shoot apices of *Hedera helix* (HACKETT 1970). Phenolic substances are widely distributed in higher plants, and it is well established that wounding or injury greatly stimulates their biosynthesis and accumulation near the wounded or injured area (TRESHOW 1955b). It is thus probable that wound-induced phenolic substances interact with auxin to stimulate root initiation.

A different aspect of the wound effect on adventitious root formation was noted by SHIBAOKA and his colleagues. Shoot cuttings with a pair of primary leaves from light-grown seedlings of *Phaseolus mungo* produced four adventitious roots at the cut base, when incubated in water, and protrusion of roots occurred after a 36-h lag period. When the base of their stems was treated with inhibitors of nucleic acid synthesis (e.g., 2-thiouracil or 5-fluorouracil) for the first 24-h period followed by incubation in water, the number of roots increased three- to fivefold (SHIBAOKA et al. 1967). The additional roots formed in the upper part of stem rather than at the base. Treatment with actinomycin D, 5-bromodeoxyuridine and mitomycin C also increased root formation (ANZAI et al. 1971). The stimulating effect of 2-thiouracil was completely negated by simultaneous addition of uracil or uridine, that of 5-fluorouracil by orotic acid and that of 5-bromodeoxyuridine by thymidine. Thus, if nucleic acid synthesis near the wound is blocked, root formation in the upper part is stimulated. They interpret these results as indicating that in intact plants the root system suppresses the ability of pericycle cells to differentiate into roots. If, however, the root system is removed, pericycle cells on the stem are released from the suppression and roots form at the base. This, in turn, induces again the suppressing mechanism in the upper part of the stem. Since cell division and growth of roots require nucleic acid synthesis, a supply of nucleic acid inhibitors blocks root formation. Thus, the mechanism which suppresses root formation on the upper part of the stem does not develop, but at least partially determined cells accumulate. Then, when the cuttings are released from inhibition by transferring the treated cuttings to water, root formation starts in these accumulated, predetermined cells and more roots form in the upper part of stem. In fact, during the period when the shoot cuttings are treated with the inhibitors, root formation does not occur, and the actual lag periods from the end of inhibitor treatment to the emergence of roots were about 30 to 35 h, the same as control, regardless of the duration of treatment with the inhibitors. This is basically an extension of the "basal dominance" theory proposed by SOEKARJO (1965), combined with auxin function. It is apparent that death or disorganization of cells at the base of shoot cuttings, imposed by wounding and further enhanced by metabolic inhibitors, has a specific effect, since killing cells near the cut base of *Coleus* internode by acid or alkali treatment greatly stimulated the number of roots formed (SOEKARJO 1965).

The stimulative effect of IAA on root initiation was studied in detail with *Azukia angularis* (Fabaceae) epicotyl cuttings without leaves and buds which, in contrast to mung bean shoot cuttings, produce adventitious roots only in a plane about 1 mm above the cut surface (MITSUHASHI et al. 1969). If IAA is supplied to the base of epicotyl cuttings for 48 h, followed by incubation in water, the number of roots formed at the base is increased threefold. *Azukia* epicotyls contain 12 vascular bundles, 6 large ones alternating with 6 smaller bundles. Adventitious roots develop from pericycle cells between the two adjacent vascular bundles (MITSUHASHI-KATO et al. 1978a, b). The effect of IAA on stimulating root formation in *Azukia* is divided into the two phases, as observed in pea stem cuttings some 50 years ago by WENT (1936). In *Azukia* cuttings, the first phase occurs up to h 24, the second phase from 24 to 48 h after cutting. While IAA given only during the first phase had no effect, IAA given only during the second phase doubled the number of roots, relative to control cuttings. However, if IAA was given during both the first and second phases, the number of roots increased with increasing concentrations of IAA applied during the first phase. Clearly then, the first phase is sensitive to IAA. The IAA effect in the first phase is not specific to IAA, but can be replaced by the anti-auxin, p-chlorophenoxyisobutyric acid and 2,4,6-trichlorophenoxyacetic acid only when the second phase is treated with IAA, in the absence of which treatment these anti-auxins inhibit root formation. The effect of IAA in the second phase could not be replaced by the anti-auxins.

Extensive anatomical examination of the initial process leading to root primordium formation (MITSUHASHI-KATO et al. 1978a, b) revealed that a root primordium was initiated by the transverse cell division of pericycle cells, first observable at 10 h after cutting. All the daughter cells produced by the transerve 1st division divided again, 37% of them transversely (transverse 2nd division) 63% longitudinally (longitidinal 1st division). About half of the daughter cells from the transverse 2nd division underwent longitudinal division (longitudinal 2nd division); these were observed from 18 to 22 h after cutting. Cells in this newly formed meristematic cell cluster thus continued division without an increase in gross volume of the cluster up to about 30 h, then the cluster started to enlarge, protruding a root. The development of a root primordium was separated into five stages; (A) only cells formed by the transverse 1st division; (B) cells finished the transverse 2nd division, but not the longitudinal 1st division; (C_1) cells formed by the longitudinal 1st or 2nd division; (C_2) cell clusters formed by further divisions without a volume increase; (D) cell clusters which have increased their volume by expansion of cells, with roots starting to protrude outside (MITSUHASHI-KATO et al. 1978a, b). Analysis of frequency changes of root primordia after cutting (Fig. 2) clearly indicates that at all possible sites on the basal part of the stem, cutting alone triggered off the first cell division (after a lag period of 8 h), and all the primordia had entered into Stage B by h 20. However, the number of primordia proceeding to Stages C_1, C_2, and D was limited, and only a very few entered Stage D by h 36; these eventually became visible roots. When IAA was applied to the 1st phase (0 to 24 h) it shortened the lag period for the transverse 1st division and hastened the formation of Stage A primordia. A remarkable effect of IAA treatment was to increase

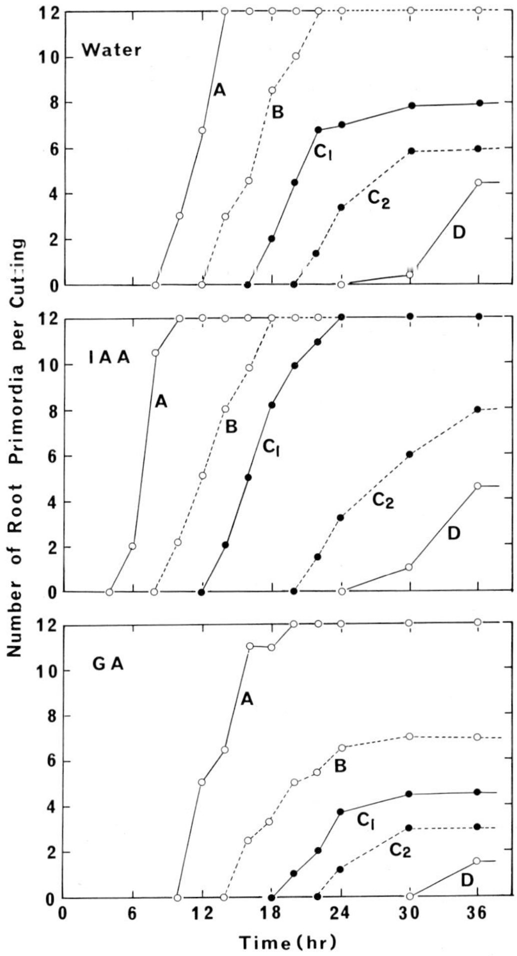

Fig. 2. Developmental processes of root primordia in the basal parts of *Azukia* stem cuttings treated with water, IAA, and GA_3. Number of root primordia which had progressed past each of stages A, B, C_1, C_2 and D were plotted against time after cutting. (After MITSUHASHI-KATO et al. 1978)

the number of primordia that entered Stage C_1 from Stage B, and by h 24 all the primordia at the 12 possible sites proceeded to Stage C_1. Also, in the presence of IAA during the 2nd phase (24–48 h), Stage B primordia developed into roots at h 36. In its absence roots were produced only from Stage D primordia. Both transverse and longitudinal divisions were inhibited by GA_3 and it significantly suppressed the Stage A to Stage B progression, thus decreasing the number of primordia developing into roots (MITSUHASHI-KATO et al. 1978a, b).

Wounding, then, directly induces the process occurring before the transverse 1st division takes place, i.e., within 8 h of cutting. If actinomycin D, p-fluorophenylalanine or 2,4-DNP is supplied in the 1st phase followed by IAA treatment in the 2nd phase, rooting is stimulated (MITSUHASHI-KATO et al. 1978a). For *Azukia* cuttings, where the "basal dominance" theory is not applicable, this may indicate that a mechanism that suppresses root formation in the stem develops during the 1st phase. Since wounding induces formation of only a small number of roots, it is likely that it partially inhibits development of the "suppressing mechanism". It is also apparent (Fig. 2) that the process of root formation on stem cutting is regulated at two steps, one at initiation of the transverse first division, the other during the progression of stages; the positive effects of plant hormones are greatest on the latter steps, and these determine the number of roots finally formed (Mitsuhashi-Kato et al. 1978a, b).

5 Biochemical Activation of Wound-Affected Cells

Metabolic activation induced by wounding in higher plant tissues, particularly in storage tissues, has been extensively studied from various points of view (Glasziou 1969, Kahl 1974, van Steveninck 1961). Since simple incubation of tissue slices under aerobic conditions for a day or so dramatically increases various metabolic activities, the phenomenon has provided a useful system to study salt uptake (Rains 1969, van Steveninck 1961, 1972), respiration (Adams and Rowan 1970, apRees 1966, Click and Hackett 1963, Laties 1962), protein synthesis (Bacon et al. 1965, Edelman and Hall 1965, Kahl 1971), and nucleic acid synthesis (Duda and Cherry 1971, Kahl 1971, Leaver and Key 1967, Sato et al. 1976, 1980, Watanabe and Imaseki 1973, 1976).

5.1 Factors Affecting Wound-Induced Metabolic Changes

Since the metabolic changes induced in response to wounding are rapid and often transient, it has often been postulated that the wound "stimuli" must be mediated by some chemical factors which are produced or released by wounding, and that these play a crucial role in changing the regulatory state of cell metabolism (Gayler and Glasziou 1964, Glasziou 1969, Imaseki et al. 1968). However, except for a few cases, experimental evidence is lacking for this hypothesis. "Wound hormone", first proposed by Haberlandt (1921), was a term assigned to substances that are produced or released by wounding and cause adjacent uninjured cells, to begin dividing. Such a substance, traumatic acid, was isolated from crushed bean pods (English et al. 1939) and was claimed to induce cell division of mesocarp cells of split young bean pods. However, its physiological activity as a cell-division inducer has since been questioned (Skoog and Schmitz 1972). Even so, there was an indication that traumatic acid did increase enzyme activity of Jerusalem artichoke tuber slices (Palmer 1968), although this was not related to increased cell division.

Formation of a substance structurally related to traumatic acid (1-decene-1,10-dicarboxylic acid) occurs upon maceration of green leaves (Hatanaka et al. 1977), linolenic acid liberated from phospholipids or glycerides being oxidatively split at the 12–13 C double bond, yielding cis-3-hexenal and 12-oxo-cis-10-dodecenoic acid, the latter being further isomerized to 12-oxo-trans-10-dodecenoic acid (traumatic half-aldehyde). It is quite possible that traumatic half-aldehyde is oxidized to the dicarboxylic acid (traumatic acid). ZIMMERMAN and COUDRON (1979) have recently shown that traumatic half-aldehyde, as well as traumatic acid, induced intumescence in young bean pods, and that bean pods contained the half-aldehyde but not traumatic acid as the endogenous constituent. Wounding greatly enhanced formation of traumatic half-aldehyde, which was converted by auto-oxidation to traumatic acid in the presence of oxygen. Traumatic half-aldehyde also stimulated elongation growth of cucumber hypocotyls, thus ZIMMERMAN and COUDRON (1979) concluded that the half-aldehyde is a natural "wound hormone".

A clear demonstration of the presence of "wound hormone" in another sense has been presented by RYAN and his colleagues (GREEN and RYAN 1972, 1973). When a tomato or potato leaf is wounded by bruising or an insect bite, proteinase inhibitors are accumulated in leaves throughout the plant within just a few hours. These inhibitors are active only on animal and microbial proteinases, but not on plant proteinases (BALLS and RYAN 1963, WALKER-SIMMONS and RYAN 1977a). Their accumulation apparently functions as a protection mechanism against insects which feed on the plant, since an insect bite on part of a single leaf induces accumulation of the inhibitors in large quantities in other unwounded leaves (GREEN and RYAN 1972, RYAN 1978). It was also shown that the wounded leaf released water-soluble "proteinase inhibitor-inducing factor" (PIIF), which was rapidly transported to other unwounded parts where it caused the inhibitors to accumulate (GREEN and RYAN 1973, RYAN 1974). Since it behaves as a hormone in the traditional sense, RYAN (1974) applied the term "wound hormone" to the inducing factor. A similar wound response has been observed in several other species, including *Medicago sativa* and *Nicotiana tabacum*. The PIIF seems nonspecific to the species which produces the factor and the factor from the wounded tomato leaves was as effective on potato or tobacco plants as their own inducing factor (WALKER-SIMMONS and RYAN 1977b).

Accumulation of proteinase inhibitors as affected by PIIF occurred in vacuoles (SHUMWAY et al. 1970, 1976, WALKER-SIMMONS and RYAN 1977) accompanied by a net loss of other leaf proteins (GUSTAFSON and RYAN 1976). The PIIF from tomato leaves seems to be a polysaccharide, as PIIF activity is always associated with polysaccharide fractions during purification (RYAN 1978). RYAN et al. (1981) noticed a similarity in sugar compositions between tomato leaf PIIF and polysaccharide fragments released from sycamore cell walls by a fungal endo-α-1,4-polygalacturonase, and these polysaccharide fragments were tested for their possible PIIF activity. They found that a large pectic polysaccharide fragment called rhamnogalacturonan I, which was solubilized from the purified walls of suspension-cultured sycamore cells, had a potent PIIF activity. Although there are a number of differences between tomato leaf PIIF and rhamnogalacturonan I in molecular size and quantitative composition of component sugars, similarities of PIIF to cell-wall-associated methylated pectic substances in their physical properties led RYAN et al. (1981) to propose that PIIF may be released from cell walls after wounding.

As it is assumed that there are numerous kinds of polysaccharidases in plant cells, which are normally separated from their substrate polysaccharides by the cellular membrane system, wounding would cause rapid partial hydrolysis of these polysaccharides through the loss of the membrane permeability or the partial disintegration of the membrane system. THEOLOGIS and LATIES (1981) found that rapid breakdown of membrane lipid occurred in potato tuber after slicing. An interesting feature of this finding is that the membrane lipid breakdown elicited by slicing proceeded to a depth of at least 10 mm from the cut surface. The partial breakdown of membrane lipid may change membrane permeability and may also release some glycoproteins localized in the membranes. Polysaccharide fragments or glycoproteins possibly released by wounding are expected to be of various kinds, but their component structure is contained in every cell of a plant. They would have affinity for the parent polysaccharides or the glycoproteins present in other nonwounded cells, and their interaction with the cell surface components of unwounded quiescent cells could activate the cells. RYAN et al. (1981) reported that the PIIF activity was associated with a particular kind of polysaccharide fragment. Therefore, it is possible that biochemical wound responses of various kinds which occur in apparently intact cells located away from the directly damaged cells may be triggered by different sorts of polysaccharide fragments or glycoproteins which may be released by autolysis and transported through the vascular system or apoplastic channels. This possibility was also discussed by YEOMAN et al. (1968). The presence of hormonal factors of this type will open a new avenue to studies of wound responses at biochemical as well as at molecular levels.

A possible controlling function of plant hormones in wound-induced metabolic changes has been suggested based on the results of hormone application to wounded tissues, or thin slices (IMASEKI et al. 1968, MITSUHASHI-KATO et al. 1978b). As described in preceding sections, plant hormones definitely affect a number of physiological phenomena occurring after wounding; it is thus logical that plant hormones might regulate metabolic reactions underlying the wound responses. EDELMAN and HALL (1964) observed that invertase activity in Jerusalem artichoke tuber slices during aerobic incubation was significantly stimulated by exogenous GA_3, but inhibited by IAA. Prolonged treatment of beet root slices with IAA, 2,4-D or kinetin suppressed wound-enhanced respiratory rate, phosphate uptake and invertase activity, but all of these were significantly stimulated by GA_3 (PALMER 1968, 1970a). An opposite effect for auxin was, however, observed in wound-induced acid phosphatase (PALMER 1970b) and invertase (RUTHERFORD 1971) activities in Jerusalem artichoke tuber slices. The stimulation of acid phosphatase activity by IAA appeared rapidly (4 h), 0.3 mM IAA resulting in about a threefold increase.

Activities of different enzymes in the same tissue slices behave differently by a growth substance. Thus, wound-enhanced invertase activity in sugarcane stem segments excised from immature tissue was not affected, whereas peroxidase activity was severely suppressed by GA_3 (GLASZIOU et al. 1968). Auxin (NAA) stimulated the wound-induced increase in invertase, but it suppressed the peroxidase increase (GAYLER and GLASZIOU 1964). There is one case where a growth hormone promotes or inhibits wound-enhanced enzyme activity, depending on precisely when the hormone is applied to the tissue after wounding. CHERRY (1968) showed that GA_3 applied to Jerusalem artichoke tuber slices within 16 h of slicing stimulated invertase activity (by h 24), but suppressed the increase occurring during 24 to 48 h after wounding.

Activation of RNA synthesis is a universal response of wounded tissues (DUDA and CHERRY 1971, KAHL 1971, LEAVER and KEY 1967, SATO et al. 1980). However, hormonal factors related to the wound-induced activation of RNA synthesis are not clear. Although simple wounding activates RNA synthesis and polysome formation in potato tuber tissue (KAHL 1971, SATO et al. 1976, 1980), application of GA_3 at 0.1 μM to the tuber slices further enhanced the activation of RNA synthesis in a time-dependent way, i.e., the degree of the enhancement increased up to 24 h after wounding, then declined (WIELGAT and KAHL 1979a). The GA_3 effect was observed not only in ribosomal RNA but also in transfer RNA and messenger-like 18S RNA. Activation of polysome formation by wounding was also stimulated by GA_3 application to the slices, though the size distribution of polysomes was not much affected by GA_3. GA_3 also activated chromatin-bound DNA-dependent RNA polymerase activity and increased the template availability of chromatin in wounded potato tuber (WIELGAT and KAHL 1979b). IAA and 2,4-D did not show such an effect. Although GA_3 enhanced the activation of both ribosomal RNA synthesis and polysome formation (thus indicating enhancement of messenger RNA synthesis), the chromatin-bound RNA polymerase stimulated by GA_3 was RNA polymerase I, which transcribes ribosomal RNA, but not RNA polymerase II which transcribes messenger RNA (WIELGAT and KAHL 1979b). In potato tuber tissue, wounding induced alteration of relative composition of coding properties of messenger RNA (ISHIZUKA et al. 1981). Gel electrophoretic analysis of in vitro translation products of messenger RNA extracted from wounded potato tuber indicated that several messenger RNA's which were not detected in nonwounded tissue started to be synthesized after wounding. Gibberellin, may thus regulate wound response by affecting the switching mechanism of genome activation.

5.2 Wound-Induced Change in Hormone Levels

In Jerusalem artichoke tuber, where GA_3 will stimulate invertase activity, levels of endogenous GA-like substances increased as a result of tissue excision prior to the normal wound-enhanced increase in invertase activity (BRADSHAW and EDELMAN 1969) and increases in the amount of invertase were correlated with increases in the level of endogenous GA-like substances. Potato tubers also increased in GA-like substances after wounding (RAPPAPORT and SACHS 1967) and this increase was considered to be related to metabolic activation associated with the breaking of bud dormancy that was triggered by bud excision (CLEGG and RAPPAPORT 1970).

While auxin has been shown to influence wound responses, the possible involvement of endogenous auxin in regulating biochemical changes after wounding is not clear, primarily because it has been little studied. Observations by TANAKA and URITANI (1977), however, provide interesting information on this aspect. When roots of sweet potato, cv. Kokei 14, were transversely cut in 2-cm thick slices and incubated, accumulation of polyphenolic substances occurred in the wounded area of the proximal side of the root piece, but only little accumulation was found on the wound surface of the distal side. Activities of enzymes related to polyphenol synthesis, PAL and trans-cinnamic acid-4-hydroxylase also increased more on the proximal side than on the distal side. Although IAA application to the cut surface of the proximal side did not further enhance the increase in polyphenol accumulation, IAA supplied to the cut surface of the distal side significantly increased polyphenol accumulation in the distal surface. Acropetal transport of ^{14}C-IAA in tissue plugs of sweet potato root occurred at a very slow rate and it was also found that IAA content (as measured by a spectrofluorometric method) in the proximal side increased from about 5 ng g^{-1} fresh weight to about 11 ng g^{-1} fresh weight during the 20 h incubation (TANAKA and URITANI, personal communication, TANAKA et al. 1979). These results indicate that in this particular variety of sweet potato root, IAA produced at the proximal surface in response to wounding migrates acropetally at a low rate, perhaps thereby stimulating cells near the proximal surface to increase activities of enzymes involved in polyphenol synthesis. A similar polyphenol accumulation also occurred on the distal side, but in the narrow layers of the cut surface. They speculated that production of IAA might also occur in the distal surface, but its acropetal polar movement would limit the cells affected by IAA to the narrow wound surface.

The gaseous growth regulator, ethylene, has a pronounced effect on the metabolic changes induced by the wounding of plant tissues. When excised blocks of sweet potato were incubated in air containing a small amount of ethylene, the wound-induced increase in peroxidase and polyphenol oxidase activities was greatly stimulated (GAHAGAN et al. 1968, IMASEKI 1970, IMASEKI et al. 1968a, STAHMANN et al. 1966). It was further shown that the increase in PAL, chlorogenic acid and O_2 uptake were stimulated by ethylene at the same time (IMASEKI et al. 1968d). The tissue slices are very sensitive to ethylene, responding to a concentration as low as 0.01 ppm. The effect of ethylene is reversible, thus stimulation of the peroxidase increase was observed only when the tissue slices were incubated in ethylene:air mixtures; removal of ethylene from the ambient air resulted in an immediate reduction in the rate of peroxidase increase (IMASEKI et al. 1968a). Ethylene production in intact roots was so low that an accurate measurement of ethylene produced could not be made, but within 1 h of cutting the tissue slices produced enough ethylene to have been responsible for a stimulation of the metabolic rates (IMASEKI et al. 1968a, b). Stimulation of PAL activity by ethylene has also been found in flavedo slices of grapefruit (RIOV et al. 1969), pea epicotyls (HYODO and YANG 1971) and swede root slices (RHODES and WOOLTORTON 1971), while ethylene partially suppressed an increase in PAL activity in potato slices (HYODO and YOUNG 1974).

Increased rate of ethylene production after wounding has been recognized in a large number of plant tissues (ABELES 1972, 1973, LIEBERMAN 1979). Wounding, as well as other types of tissue injury, e.g., by toxic chemicals (ABELES and ABELES 1972, IMASEKI et al. 1968c) or invading pathogens (IMASEKI et al. 1968b, STAHMANN et al. 1966, WILLIAMSON 1950) generally caused a significant rise of the rate of evolution shortly after injury. When the immature fruits of *Citrus unshiu* were cut into pieces, the increased ethylene evolved was related to the number of pieces per unit weight of tissue, i.e., the degree of wounding (HYODO 1977a, 1978), similar results being obtained with albedo tissue of the mature fruits, which had produced only trace amounts of ethylene before excision (HYODO 1977b, HYODO and NISHINO 1981).

HANSON and KENDE (1976) proposed a hypothesis that may be applicable to the mechanism of wound-induced ethylene production, namely that increased ethylene production in senescing tissue of morning glory flower results from an increased supply of methionine to a cellular compartment containing the ethylene-synthesizing system. The increased supply of methionine is due either to a breakdown of compartmentation that segregates the methionine-synthesizing system from its substrate homocysteine or to increased synthesis of homocysteine. In fact, permeability of cell membrane increases in senescent flower tissue (HANSON and KENDE 1975) and cell membrane potentials do change after wounding (KOPOWITZ et al. 1975). As discussed earlier, wounding most probably caused disintegration of cellular organization in cells adjacent to the destroyed cells. Thus, it is possible that the ethylene-synthesizing system in wound-affected cells is mixed with its substrate owing to the loss of compartmentation. This hypothesis, of course, needs further experimental verification.

It has been shown that the rate of ethylene production is under the control of cell membranes (IMASEKI and WATANABE 1978, MATTOO et al. 1977, ODAWARA et al. 1977) and an ethylene synthesizing-enzyme system in apple fruits was reported to be localized in the cell wall–cell membrane complex (MATTOO and LIEBERMAN 1977). However, more recent experiments have revealed that apple protoplasts devoid of cellulose also produced ethylene, indicating that the cell wall was not involved in the ethylene-synthesizing enzyme system (ANDERSON et al. 1979, also IMASEKI, unpublished data).

A noticeable characteristic of wound-induced ethylene production is the transient burst after wounding. When juvenile rib segments of morning glory flowers were wounded, ethylene production rose within 1 h, reached a maximum ten times the control rate, and then declined (HANSON and KENDE 1976, KENDE and HANSON 1976). Recent observation by SALTVEIT and DILLEY (1978) clearly indicated that wound ethylene production followed a remarkably consistent pattern regardless of plant species. They showed that in segments excised from the elongating zone of etiolated pea epicotyls there was a definite lag period of 26 min. Production rate of ethylene then increased linearly up to 56 min, decreased to a minimum at 90 min, and increased again to form a second peak at 131 min. The pattern of wound-induced ethylene production was only slightly different between dicots and monocots, or between the young and mature parts of a stem. This particular pattern of wound-induced ethylene production was thought to result from negative feedback control by endogenous ethylene (SALTVEIT and DILLEY 1978).

Knowledge of ethylene biosynthesis was greatly advanced by re-discovery of 1-aminocyclopropane-1-carboxylic acid (ACC) as an immediate precursor of ethylene in a sequence of reactions leading up to ethylene from methionine (ADAMS and YANG 1979, LÜRSSEN et al. 1979). Methionine is converted to ACC only in ethylene-producing tissues, but not in those tissues which do not produce ethylene (ADAMS and YANG 1979, BOLLER et al. 1979, JONES and KENDE 1979, YOSHII et al. 1980, YU and YANG 1979). However, a number of plant tissues do produce a large amount of ethylene if ACC is exogenously supplied, regardless of their inherent ability to produce ethylene (CAMERON et al. 1979, YOSHII et al. 1980, YU et al. 1979). Apparently, the enzyme system which converts ACC to ethylene is constitutive, and the cellular content of ACC determines the rate of ethylene production. In other words, the reaction which forms ACC is the rate-limiting step in the ethylene biosynthetic reactions from methionine (IMASEKI et al. 1982, YANG 1983, YANG et al. 1982). ACC is formed from SAM by the enzyme, ACC synthase, for which an in vitro assay system has been established (BOLLER et al. 1979, YU et al. 1979). In many cases, the increase or decrease in ACC synthase activity in tissues parallels the endogenous content of ACC and the rate of ethylene production in the tissue (IMASEKI et al. 1982, JONES and KENDE 1979, KENDE and BOLLER 1981, YANG 1983, YANG et al. 1982, YOSHII and IMASEKI 1981, YU and YANG 1979). One of the remarkable characteristics of ACC synthase is its instability in cells. Its apparent half-life in cells has been estimated to be 25 to 30 min in auxin-treated tissues (IMASEKI et al. 1982, YOSHII and IMASEKI 1982) and tomato fruit (KENDE and BOLLER 1981).

Wound-induced ethylene production also was shown to be coupled with the increase in ACC content and ACC synthase activity of the wounded tissues (KENDE and BOLLER 1981, KONZE and KWIATKOWSKI 1981, YU and YANG 1979, 1980). The increased rate of ethylene production in excised Valencia orange albedo tissue and excised columella of unripe tomato fruit was shown to follow the increase in ACC content in the tissues. A similar relationship was also found in mung bean hypocotyls treated with a toxic level of Cu^{2+} (10 mM) (YU and YANG 1979). In these cases, development of the ACC synthase activity in the wounded tissues is inhibited by the presence of cycloheximide. Thin slices of fruit of *Cucurbita pepo* do not produce appreciable amount of ethylene, but when they were incubated under aerobic conditions, they started to produce large amounts of ethylene after a 6-h lag period. This increase in the ethylene production rate was also supported by a parallel increase in ACC synthase activity, the latter being inhibited by cycloheximide (HYODO et al. 1983). These studies clearly indicate that wound-induced ethylene production is supported by wound-induced de novo synthesis and accumulation of ACC synthase in the affected area. Once synthesis of ACC synthase stops or slows, ethylene production rapidly decreases because the cellular ACC synthase activity is quite labile and inactivated rapidly, whereas conversion of ACC to ethylene continues until ACC is no longer available. Conversion of ACC to ethylene was recently found to be a minor portion of the metabolic fate of ACC, and a major portion of ACC formed was conjugated with malonate to form 1-(malonylamino)-cyclopropane-1-carboxylic acid (AMRHEIN et al. 1981, 1982, HOFFMAN et al. 1982, YANG et al. 1982). The conjugate formation rapidly decreases endogenous free ACC, thus leading to exhaustion of ACC available for ethylene synthesis.

In vegetative tissues, synthesis of ACC synthase has been shown to be regulated by other plant hormones. Auxin and cytokinin stimulate, but ABA inhibits synthesis of ACC synthase (JONES and KENDE 1979, YOSHII and IMASEKI 1981, IMASEKI et al. 1982). Roles of these hormones in wound-induced ethylene production, however, are not clear. Water stress is known to increase ABA content in the stressed tissues (WRIGHT 1978), but also causes a transient surge of ethylene production (APELBAUM and YANG 1971, WRIGHT 1979). WRIGHT (1979) and MCKEON et al. (1982) reported that when wheat leaves were treated with ABA followed by water stress, ethylene production that occurred after water stress was lower in the ABA-treated leaves than in the stressed leaves without ABA treatment. WRIGHT (1979) suggested that the decline in ethylene production that occurred after an initial surge was controlled by ABA that accumulated after water stress. However, the timing of the increases in ethylene production and ABA accumulation was almost the same.

Synthesis of ACC synthase is also controlled by a negative feedback repression by the ethylene produced. Development of wound-induced ethylene-producing activity in excised albedo tissues of grapefruit was nearly completely inhibited when the albedo tissues were incubated with ethylene (20 $\mu l\, l^{-1}$) (RIOV and YANG 1982). Removal of exogeneous ethylene initiated an increase in ethylene production in a manner similar to that in the control tissues. However, the ACC synthase activity in the ethylene-treated albedo tissues was reduced to about 31% of the control tissue. RIOV and YANG (1982) noticed that the inhibitory degree of the applied ethylene on the ACC synthase activity was too small to account for the greater inhibitory effect on the ACC content, and interpreted this to mean that ethylene also inhibited the activity of ACC synthase. Repression of synthesis of ACC synthase by ethylene was also reported in auxin-treated stem segments of mung bean (IMASEKI et al. 1982, YOSHII and IMASEKI 1982). When hypocotyl segments were incubated with IAA, ACC synthase activity in the tissue increased markedly and the increase paralleled increases in both ACC content and ethylene production rate. However, when the tissue was incubated with IAA in the presence of AVG, the increase in the enzyme activity was enhanced several-fold, while ethylene production was nearly completely abolished by AVG. This enhancement in enzyme activity by AVG, however, was diminished when the tissue was incubated in 0.3 to 10 $\mu l\, l^{-1}$ ethylene. The ACC synthase reaction was not inhibited by ethylene being added to the reaction mixture. Therefore, the apparent enhancement of synthesis of ACC synthase by AVG is thought to be due to release of the tissue from the repressive effects of ethylene, which are, however, only partial. As the effective concentration of ethylene was estimated to be less than 0.1 $\mu l\, l^{-1}$ (YOSHII and IMASEKI 1982), and wound-induced ethylene production of a relatively low degree is observed in plant tissues on which partial cuts are made (SALTVEIT and DILLEY 1978), it is possible that a negative feedback mechanism may be operating.

Although environmental stresses other than wounding are not within the scope of this chapter, it should be mentioned that a variety of stresses, including drought, flooding, and physical contact greatly affect ethylene synthesis. Many of the morphological and physiological responses to these stresses have been shown to be controlled by altered ethylene production by the tissues (ABELES 1973, LIEBERMAN 1979). Epinasty occurred in leaf petioles of tomato plants after a flooding of the root system resulted in increased ethylene production in leaves and petioles (BRADFORD and YANG 1980). A working hypothesis would thus be that flooding the root system causes anaerobiosis in roots, which inhibits conversion of ACC to ethylene and, moreover, stimulates formation of ACC. ACC, then, is transported through the xylem stream up to the terrestrial plant parts, where ACC is then converted to ethylene since oxygen is available, and the ethylene-forming enzyme is present constitutively.

As discussed in this section, physiological and morphological responses to wounding are associated in many cases with changes in hormonal content or balance. However, even if a change of a hormonal content is proven to be a cause of a specific wound response, the mechanism of the wound response is not simply explained by the change in hormone content/balance. The question still remains: what initiates the change in the hormonal balance? For example, wound-induced ethylene production is primarily supported by wound-induced synthesis and accumulation of ACC synthase. This mechanism may thus involve selection of a specific gene that is activated on wounding.

6 Conclusion

Stress biology, which deals with a wide variety of responses of organisms to environmental stimuli, includes wounding as an environmental factor. It is a subject to which plant physiologists and biochemists are giving more and more attention. As we have discussed in this chapter, a wound given to a higher plant gives rise to significant physiological (e.g., functional) as well as biochemical alteration, not only in the wounded area, but throughout the plant. However, we have scarcely any knowledge of the chemical and physical factors that must be involved in the signal transfer between the wound stimulus and biochemical changes. An understanding of the mechanisms of wound responses at physiological, biochemical, and molecular levels should lead us to a better understanding of survival of the whole plant in nature. Furthermore, many wound responses appear to include activation of genes which are otherwise never expressed in the life of the intact plants. Studies on the specific effect of wounding or, in more general terms, stress, may be one approach to resolving current problems in plant physiology, such as mechanisms of hormonal control of specific gene expression, the role of membranes in metabolic control, and cell–cell interactions. Such studies should, of course, include parallel, normal (unwounded) plant systems.

Acknowledgments. The auther expresses his sincere thanks to Dr. YOSHIYUKI TANAKA (Nagoya University) and Dr. HIROH SHIBAOKA (Osaka University) for discussion and providing unpublished data.

References

Abeles AL, Abeles FB (1972) Biochemical pathway of stress-induced ethylene. Plant Physiol 50:496–498

Abeles FB (1972) Biosynthesis and mechanism of action of ethylene. Ann Rev Plant Physiol 23:259–292

Abeles FB (1973) Ethylene in plant biology. Academic Press, New York London, pp 87–108

Adams DO, Yang SF (1979) Ethylene biosynthesis: identification of 1-aminocyclopropane 1 carboxylic acid as an intermediate in the conversion of methionine to ethylene. Proc Nat Acad Sci USA 76:170–174

Adams PB, Rowan KS (1970) Glycolytic control of respiration during aging of carrot root tissue. Plant Physiol 45:490–494

Adamson D (1962) Expansion and division in auxin-treated plant cells. Can J Botany 40:719–744

Aloni R (1976) Polarity of induction an pattern of primary phloem fiber differentiation in *Coleus*. Am J Botany 63:877–889

Aloni R (1979) Role of auxin and gibberellin in differentiation of primary phloem fibers. Plant Physiol 63:609–614

Aloni R, Jacobs WP (1977) Polarity of tracheary regeneration in young internodes of *Coleus* (Labiatae). Am J Botany 64:395–403

Amrhein N, Schneebeck D, Skorupka H, Tophof S, Stockigt J (1981) Identification of a major metabolite of the ethylene precursor, 1-aminocyclopropane-1-carboxylic acid in higher plants. Naturwiss 68:619–620

Amrhein N, Breuing F, Eberle J, Skorupka H, Tophof S (1982) The metabolism of 1-aminocyclopropane-1-carboxylic acid. In: Wareing PF (ed) Plant growth substances 1982. Academic Press, pp 249–258

Anderson JD, Lieberman M, Stewart RN (1979) Ethylene production by apple protoplasts. Plant Physiol 63:931–935

Anzai T, Shibaoka H, Shimokoriyama M (1971) Increases in the number of adventitious roots caused by 2-thiouracil and 5-bromodeoxyuridine in *Phaseolus mungo* cuttings. Plant Cell Physiol (Tokyo) 12:695–700

Apelbaum A, Yang SF (1981) Biosynthesis of stress ethylene induced by water deficit. Plant Physiol 68:594–596

apRees T (1966) Evidence for the widespread occurrence of induced respiration in slices of plant tissue. Australian J Biol Sci 19:981–990

Bacon JSD, MacDonald IR, Knight AH (1965) The development of invertase activity in slices of the root of *Beta vulgaris* L. washed under aseptic conditions. Biochem J 94:175–182

Balls AK, Ryan CA (1963) Concerning a chymotryptic inhibitor from potatos and its binding capacity for the enzyme. J Biol Chem 238:2976–2982

Behnke H-D, Schulz A (1980) Fine structure, pattern of division, and course of wound phloem in *Coleus blumei*. Planta 150:357–365

Benayoun J, Aloni R, Sachs T (1975) Regeneration around wounds and the control of vascular differentiation. Ann Botany (London) 39:447–454

Boller T, Herner RC, Kende H (1979) Enzymatic formation of an ethylene precursor, 1-aminocyclopropane-1-carboxylic acid. Planta 145:293–303

Borchert R (1978) Time course and spatial distribution of phenylalanine ammonia-lyase and peroxidase activity in wounded potato tuber tissue. Plant Physiol. 62:789–793

Borchert R, McChesney JD (1973) Time course and localization of DNA synthesis during wound healing of potato tuber tissue. Develop Biol 35:293–301

Bradford KJ, Yang SF (1980) Xylem transport of 1-aminocyclopropane-1-carboxylic acid, an ethylene precursor in waterlogged tomato plants. Plant Physiol 65:322–326

Breadshaw MJ, Edelman J (1969) Enzyme formation in higher plant tissue: The production of a gibberellin preceding invertase synthesis in aged tissue. J Exp Botany 20:87–93

Burg SP, Clagett CO (1967) Conversion of methionine to ethylene in vegetative tissue and fruits. Biochem Biophys Res Commun 27:125–130

Byrne H, Setterfield G (1977) Activation of ribosomal and messenger RNA synthesis in excised Jerusalem artichoke tuber slices. Planta 136:203–210

Cameron AC, Fenton CAK, Yu Y, Adams DO, Yang SF (1979) Increased production of ethylene by plant tissues treated with 1-aminocyclopropane-1-carboxylic acid. Hort Sci 14:178–180

Cherry JH (1968) Regulation of invertase in washed sugar beet tissue. In: Wightman F, Setterfield G (eds) Biochemistry and physiology of plant growth substances. The Runge Press, Ottawa, pp 417–431

Clegg MD, Rappaport L (1970) Regulation of bud rest in tuber of potato, *Solanum tuberosum* L. VI Biochemical changes induced in excised potato buds by gibberellic acid. Plant Physiol 45:8–13

Click RE, Hackett DP (1963) The role of protein and nucleic acid synthesis in the development of respiration in potato tuber slices. Proc Nat Acad Sci US 50:243–250

Clutter M (1960) Hormonal induction of vascular tissue in tobacco pith in vitro. Science 132:548–549

Comer AE (1978) Pattern of cell division and wound vessel member differentiation in *Coleus* pith explants. Plant Physiol 62:354–359

Cooper WC (1938) Hormones and root formation. Botan Gaz 99:599–614

Dalessandro G (1973) Hormonal control of xylogenesis in pith parenchyma explants of *Lactuca*. Ann Botany (London) 37:375–382

Dalenssandro G, Roberts LW (1971) Induction of xylogenesis in pith parenchyma explants of *Lactuca*. Am J Botany 58:378–385

Davies E, Schuster A (1981) Intercellular communication in plants: Evidence for a rapidly generated, bidirectional transmitted wound signal. Proc Natl Acad Sci 78:2422–2426

Dean BB, Kolattukudy PE (1976) Synthesis of suberin during wound-healing in jade leaves, tomato fruit, and bean pods. Plant Physiol 58:411–416

Duda CT, Cherry JH (1971) Chromatin- and nuclei-directed ribonucleic acid synthesis in sugar beet root. Plant Physiol 47:262–268

Edelman J, Hall MA (1964) Effect of growth hormones on the development of invertase associated with cell walls. Nature 201:296–297

Edelman J, Hall MA (1965) Enzyme formation in higher plant tissues. Development of invertase and ascorbate oxidase activities in mature storage tissue of *Helianthus tuberosus* L. Biochem J 95:403–410

Ellis RJ, MacDonald IR (1967) Activation of protein synthesis by microsomes from aging beet disks. Plant Physiol 42:1297–1302

English J Jr, Bonner J, Haagen-Smit AJ (1939) The wound hormones of plants. II. The isolation of a crystalline active substance. Proc Natl Acad Sci USA 25:323–329

Fosket DE, Roberts LW (1964) Induction of wound-vessel differentiation in isolated *Coleus* stem segments in vitro. Am J Bot 51:19–25

Fosket DE, Torrey JG (1969) Hormonal control of cell proliferation and xylem differentiation in cultured tissues of *Glycine max* var. Biloxi. Plant Physiol 44:871–880

Gahagan HE, Holm RE, Abeles FB (1968) Effect of ethylene on peroxidase activity. Physiol Plant 21:1270–1279

Gayler KR, Glasziou KT (1964) Plant enzyme synthesis: Hormonal regulation of invertase and peroxidase synthesis in sugar cane. Planta 84:185–194

Gersan, Lips, Sachs T (1980) Effects of wounding on transport in phloem. J Exp Bot 31:783–789

Glasziou KT (1969) Control of enzyme formation and inactivation in plants. Annu Rev Plant Physiol 20:63–88

Glasziou KT, Gayler KR, Waldron JC (1968) Effects of auxin and gibberellic acid on the regulation of enzyme synthesis in sugarcane stem tissue. In: Wightman F, Setterfield G (eds) Biochemistry and physiology of plant growth substances. Runge, Ottawa, pp 433–442

Green TR, Ryan CA (1972) Wound-induced proteinase inhibitor in plant leaves: A possible defense mechanism against insects. Science 175:776–777

Green TR, Ryan CA (1973) Wound-induced proteinase inhibitor in tomato leaves. Some effects of light and temperature on the wound response. Plant Physiol 51:19–21
Gustafson G, Ryan CA (1976) Specificity of protein turnover in tomato leaves. Accumulation of proteinase inhibitors, induced with the wound hormone, PIIF. J Biol Chem 251:7004–7010
Haberlandt G (1921) Wundhormone als Erreger von Zellteilungen. Beitr Allg Bot 2:1–53
Hackett WP (1970) The influence of auxin, catechol, and methanolic tissue extracts on root initiation in aseptically cultured shoot apices of juvenile and adult forms of *Hedera helix*. J Am Hortic Sci 95:398–402
Hanson AD, Kende H (1975) Ethylene-enhanced ion and sucrose efflux in morning glory flower tissue. Plant Physiol 55:663–669
Hanson AE, Kende H (1976) Methionine metabolism and ethylene biosynthesis in senescent flower tissue of morning glory. Plant Physiol 57:528–537
Hanson AE, Kende H (1976) Biosynthesis of wound ethylene in morning glory flower tissue. Plant Physiol 57:538–541
Hatanaka A, Kajiwara T, Sekiya J, Kido Y (1977) Formation of 12-oxo-trans-10-dodecenoic acid in chloroplasts from *Thea sinensis* leaves. Phytochemistry 16:1827–1829
Hoffman NE, Yang SF, McKeon T (1982) Identification of 1-(malonyl-amino)cyclopropane-1-carboxylic acid, an ethylene precursor in higher plants. Biochem Biophys Res Commun 104:765–770
Hyodo H (1977a) Ethylene production and respiration of Satsuma mandarin (*Citrus unshiu* Marc.) fruit harvested at different stages of development. J Jpn Soc Hortic Sci 45:427–432
Hyodo H (1977b) Ethylene production by albedo tissue of Satsuma mandarin (*Citrus unshiu* Marc.) fruit. Plant Physiol 59:111–113
Hyodo H (1978) Ethylene production by wounded tissue of citrus fruit. Plant Cell Physiol 19:545–551
Hyodo H, Nishino T (1981) Wound-induced ethylene formation in albedo tissue of citrus fruit. Planta Physiol 67:421–423
Hyodo H, Yang SF (1971) Ethylene-enhanced synthesis of phenylalanine ammonia-lyase in pea seedlings. Plant Physiol 47:765–770
Hyodo H, Yang SF (1974) The effect of ethylene on the development of phenylalanine ammonia-lyase in potato tuber disks. Z Naturforsch 71:76–79
Hyodo H, Tanaka K, Watanabe K (1983) Wound-induced ethylene production and 1-aminocyclopropane-1-carboxylic acid synthase in winter squash fruit. Plant Cell Physiol 24:963–969
Imaseki H (1970) Induction of peroxidase activity by ethylene in sweet potato. Plant Physiol 46:172–174
Imaseki H, Watanabe A (1978) Inhibition of ethylene production by osmotic shock. Further evidence for membrane control of ethylene production. Plant Cell Physiol (Tokyo) 19:345–348
Imaseki H, Asahi T, Uritani I (1968a) Investigations on the possible inducers of metabolic changes in injured plant tissues. In: Hirai T, Hidaka Z, Uritani I (eds) Biochemical regulation in diseased plants and injury. Phytopathol Soc Jpn, Tokyo, pp 189–201
Imaseki H, Teranishi T, Uritani I (1968b) Production of ethylene by sweet potato roots infected by the black rot fungus. Plant Cell Physiol 9:769–781
Imaseki H, Uritani I, Stahmann MA (1968c) Production of ethylene by injured sweet potato root tissue. Plant Cell Physiol 9:757–768
Imaseki H, Uchiyama M, Uritani I (1968d) Effect of ethylene on the inductive increase in metabolic activities in sliced sweet potato roots. Agric Biol Chem 32:387–389
Imaseki H, Yoshii H, Todaka I (1982) Regulation of auxin-induced ethylene biosynthesis in plants. In: Wareing PF (ed) Plant growth substances 1982. Academic Press, London, pp 259–268
Ishizuka M, Sato T, Watanabe A, Imaseki H (1981) Alteration of coding properties of polysome-associated messenger RNA in potato tuber slices during aging. Plant Physiol 68:154–157

Jackson MB, Campbell DJ (1976) Production of ethylene by excised segments of plant tissue prior to the effect of wounding. Planta 129:273–274
Jacobs WP (1952) The role of auxin in differentiation of xylem around a wound. Am J Botany 39:301–309
Jacobs WP (1956) Internal factors controlling cell differentiation in the flowering plants. Am Naturalist 90:163–169
Jacobs WP (1970) Regulation and differentiation of sieve tube elements. Intern Rev Cytol 28:239–273
Jacobs WP, MacCready CC (1967) Polar transport of growth regulators in pith and vascular tissues of *Coleus* stems. Am J Botany 54:1035–1040
Jeffs RA, Northcote DH (1967) The influence of indol-3-ylacetic acid and sugar on the pattern of induced differentiation in plant tissue culture. J Cell Sci 2:77–78
Jones JF, Kende H (1979) Auxin-induced ethylene biosynthesis in subapical stem sections of etiolated seedlings of *Pisum sativum* L. Planta 146:649–656
Kahl G (1971) Synthesis of rRNA, tRNA and other RNA-species concomitant with polyribosome formation in aging potato tuber slices. Z Naturforsch 26b:1058–1064
Kahl G (1971) Activation of protein synthesis in aging potato tuber slices. Z Naturforsch 26b:1064–1067
Kahl G (1973) Genetic and metabolic regulation in differentiating plant storage tissue cells. Botan Rev 39:274–299
Kahl G (1974) Metabolism in plant storage tissue slices. Botan Rev 40:263–314
Kahl G (1978) Biochemistry of wounded plant tissues. De Gruyter, New York, p 680
Kamisaka S, Sakurai N, Masuda Y (1973) Auxin-induced growth of tuber tissue of Jerusalem artichoke VIII. Role of cyclic AMP in the action of auxin, cytokinin and gibberellic acid. Plant Cell Physiol (Tokyo) 14:183–193
Kende H, Boller T (1981) Wound ethylene and 1-aminocyclopropane-1-carboxylate synthase in ripening tomato fruit. Planta 151:476–481
Kende H, Hanson AD (1976) Relationship between ethylene evolution and senescence in morning glory flower tissue. Plant Physiol 57:523–527
Kolattukudy PE, Kronman K, Poulose AJ (1975) Determination of structure and composition of suberin from the roots of carrot, parsnip, rutabaga, turnip, red beet, and sweet potato by combined gas-liquid chromatography and mass spectrometry. Plant Physiol 55:567–573
Komamine A, Sato M, Shimokoriyama M (1963) Physiological studies on the outgrowth of the epicotyl in *Stizolobium hassjoo* I. Properties of the outgrowth. Botan Mag (Tokyo) 76:130–137
Konze JR, Kwiatkowski MK (1981) Rapidly induced ethylene formation after wounding is controlled by the regulation of 1-aminocyclopropane-1-carboxylic acid synthesis. Planta 151:327–330
Koopowitz H, Dhys R, Fosket DE (1975) Cell membrane potentials of higher plants: Changes induced by wounding. J Exp Botany 26:131–137
Lamotte CE, Jacobs WP (1963) A role of auxin in phloem regeneration in *Coleus* internodes. Develop Biol 8:80–98
Lange H, Rosenstock G, Kahl G (1970) Induktionsbedingungen der Suberinsynthese und Zellproliferation bei Parenchymfragmenten der Kartoffelknolle. Planta 90:109–118
Laties GG (1962) Controlling influence of thickness on development and type of respiratory activity in potato slices. Plant Physiol 37:679–690
Leaver CJ, Key JL (1967) Polyribosome formation and RNA synthesis during aging of carrot-root tissue. Proc Nat Acad Sci US 57:1338–1344
Lieberman M (1979) Biosynthesis and action of ethylene. Ann Rev Plant Physiol 30:533–591
Lieberman M, Kunishi AT, Mapson LW, Wardale DA (1965) Ethylene production from methionine. Biochem J 97:449–459
Lieberman M, Kunishi A, Mapson LW, Wardale DA (1966) Stimulation of ethylene production in apple tissue slices by methionine. Plant Physiol 41:376–382
Lipetz J (1970) Wound-healing in higher plants. Intern Rev Cytol 27:1–28
Lürssen K, Naumann K, Schröder R (1979) 1-Aminocyclopropane-1-carboxylic acid –

a new intermediate of ethylene biosynthesis in higher plants. Z Pflanzenphysiol 92:285–294
Mäder M, Amberg-Fischer V (1982) Role of peroxidase in lignification of tobacco cells. I. Oxidation of nicotinamide adenine dinucleotide and formation of hydrogen peroxide by cell wall peroxidases. Plant Physiol 70:1128–1131
Mäder M, Füssl R (1982) Role of peroxidase in lignification of tobacco cells. II. Regulation by phenolic compounds. Plant Physiol 70:1132–1134
Masuda Y (1965) Auxin-induced growth of tuber tissue of Jerusalem artichoke. I. Cell physiological studies on the expansion growth. Botan Mag (Tokyo) 78:417–423
Masuda Y (1966) Auxin-induced growth of tuber tissue Jerusalem artichoke. II. The relation to protein and nucleic acid metabolism. Plant Cell Physiol (Tokyo) 7:75–91
Mattoo AK, Lieberman M (1977) Localization of the ethylene-synthesizing system in apple tissue. Plant Physiol 60:794–799
Mattoo AK, Baker JE, Chalutz E, Lieberman M (1977) Effect of temperature on the ethylene-synthesizing systems in apple, tomato and *Penicillium digitatum*. Plant Cell Physiol 18:715–719
McKeon TA, Hoffman NE, Yang SF (1982) The effect of plant-hormone pretreatments on ethylene production and synthesis of 1-aminocyclopropane-1-carboxylic acid in water-stressed wheat leaves. Planta 155:437–447
Mitsuhashi M, Shibaoka H, Shimokoriyama M (1969) Morphological and physiological characterization of IAA-less-sensitive and IAA-sensitive phases in rooting of *Azukia* cuttings. Plant Cell Physiol 10:867–874
Mitsuhashi-Kato M, Shibaoka H, Shimokoriyama M (1978a) Anatomical and physiological aspects of developmental processes of adventitious root formation in *Azukia* cuttings. Plant Cell Physiol 19:393–400
Mitsuhashi-Kato M, Shibaoka H, Shimokoriyama M (1978b) The nature of the dual effect of auxin on root formation in *Azukia* cutting. Plant Cell Physiol 19:1535–1542
Mizuno K, Komamine A, Shimokoriyama M (1971) Vessel element formation in cultured carrot-root phloem slices. Plant Cell Physiol 12:823–830
Morohashi Y, Komamine A, Shimokoriyama M (1969) Physiological studies on the outgrowth of the epicotyl in *Stizolobium hassjoo* VI. Changes in the IAA content and the activity of IAA destruction in the decapitated epicotyls of etiolated *Stizolobium* and *Vicia* seedlings. Bot Mag (Tokyo) 82:110–120
Odawara S, Watanabe A, Imaseki H (1977) Involvement of cellular membrane in regulation of ethylene production. Plant Cell Physiol 18:567–575
Palmer JM (1968) The effect of some plant growth substances on the induction of enzymic activities in thin slices of plant tubers. In: Wightman F, Setterfield G (eds) Biochemistry and physiology of plant growth substances. Runge, Ottawa, pp 401–415
Palmer JM (1970a) The influence of growth-regulating substances on the development of enhanced metabolic rates in thin slices of beetroot storage tissue. Plant Physiol 41:1173–1178
Palmer JM (1970b) The induction of phosphatase activity in thin slices of Jerusalem artichoke tissue by treatment with indoleacetic acid. Planta 93:53–59
Patau K, Das ND, Skoog F (1957) Induction of DNA synthesis by kinetin and indoleacetic acid in excised tobacco pith tissue. Physiol Plant 10:949–966
Rains DW (1969) Sodium and potassium absorption by bean stem tissues. Plant physiol 44:547–554
Rana MA, Gahan PB (1983) A quantitative cytochemical study of determination for xylem element formation in response to wounding in roots of *Pisum sativum*. Planta 157:307–316
Rappaport L, Sachs M (1967) Wound-induced gibberellins. Nature 214:1149–1150
Rhodes MJC, Wooltorton LSC (1971) The effect of ethylene on the respiration and on the activity of phenylalanine ammonia-lyase in swede and parsnip root tissue. Phytochemistry 10:1989–1997
Rhodes MJC, Wooltorton LSC (1973) Stimulation of phenolic acid and lignin biosynthesis in swede root tissue by ethylene. Phytochemistry 12:107–118
Riov J, Yang SF (1982) Autoinhibition of ethylene production in citrus peel discs. Plant Physiol 69:687–690

Riov J, Monselise SP, Kahan RS (1969) Ethylene-controlled induction of phenylalanine ammonia-lyase in *Citrus* fruit peel. Plant Physiol 44:631–635

Roberts LW (1969) The initiation of xylem differentiation. Botan Rev 35:201–250

Roberts LW, Fosket DE (1966) Interaction of gibberellic acid and indoleacetic acid in the differentiation of wound vessel members. New Phytologist 65:5–8

Robbertse PJ, McCully M (1979) Regeneration of vascular tissue in wounded pea roots. Planta 145:167–773

Rutherford PP (1971) Inhibition by actinomycin D of water uptake and invertase and hydrolase activities induced in Jerusalem artichoke tuber tissue discs by treatment with 2,4-dichlorophenoxyacetic acid. Phytochem 10:1469–1473

Ryan CA (1974) Assay and biochemical properties of the proteinase inhibitor-inducing factor, a wound hormone. Plant Physiol 54:328–332

Ryan CA (1978) Proteinase inhibitors in plant leaves: A biochemical model for pest-induced natural plant protection. Trends Biochem Sci July, 148–150

Ryan CA, Bishop P, Pearce G, Darvill AG, McNeil M, Albersheim P (1981) A sycamore cell wall polysaccharide and a chemically related tomato leaf polysaccharide possess similar proteinase inhibitor-inducing activities. Plant Physiol 68:616–618

Sachs T (1968a) The role of the root in the induction of xylem differentiation in peas. Ann Botany (London) 32:389–399

Sachs T (1968b) On the determination of the pattern of vascular tissue in peas. Ann Botany (London) 32:781–790

Sachs T (1969) Polarity and the induction of organized vascular tissues. Ann. Botany (London) 33, 263–275

Sachs T, Cohen D (1982) Circular vessels and the control of vascular differentiation in plants. Differentiation 21:22–26

Sakai S, Imaseki H (1972) Ethylene biosynthesis: Methionine as an in vivo precursor of ethylene in auxin-treated mung bean hypocotyl segments. Planta 105:165–173

Saltveit ME Jr, Dilley DR (1978) Rapidly induced wound ethylene from excised segments of etiolated *Pisum sativum* L. cv. Alaska I. Characterization of the response. Plant Physiol 61:447–450

Saltveit ME Jr, Dilley DR (1978) Rapidly induced wound ethylene from excised segments of etiolated *Pisum sativum* L. cv. Alaska II. Oxygen and temperature dependency. Plant Physiol 61:675–679

Sato T, Watanabe A, Imaseki H (1976) Effect of ethylene on DNA synthesis in potato tuber slices. Plant Cell Physiol 17:1255–1262

Sato T, Ishizuka M, Watanabe A, Imaseki H (1980) Synthesis and properties of polysomal RNA of potato tuber slices in the early stage of aging. Plant Cell Physiol 21:137–147

Setterfield G (1963) Growth regulation in excised slices of Jerusalem artichoke tuber tissue. Cell differentiation. Symp Soc Exp Biol 17:98–127, Univ Press, Cambridge

Shibaoka H, Anzai T, Mitsuhashi M, Shimokoriyama M (1967) Interaction between heliangine and pyrimidines in adventitious root formation of *Phaseolus* cutting. Plant Cell Physiol 8:647–656

Shumway LK, Rancour JM, Ryan CA (1970) Vacuolar protein bodies in tomato leaf cells and their relationship to storage of chymotrypsin inhibitor I protein. Planta 93:1–14

Shumway KK, Yang V, Ryan CA (1976) Evidence for the presence of proteinase inhibitor I in vacuolar protein bodies of plant cells. Planta 129:151–165

Skoog F, Schmitz RY (1972) Cytokinins. In: Steward FC (ed) Plant physiology vol VIB. Academic Press, New York London, pp 181–213

Soekarjo R (1965) On the formation of adventitious roots in cuttings of *Coleus* in relation to the effect of indoleacetic acid on the epinastic curvature of isolated petioles. Acta Bot Neerl 14:373–399

Soliday DL, Dean BB, Kolattukudy PE (1978) Suberization: Inhibition by washing and stimulation by abscisic acid in potato disks and tissue culture. Plant Physiol 61:170–174

Sperling E, Laties GG (1963) The dependence of auxin-induced growth on auxin-independent metabolic changes in slices of storage tissue. Plant Physiol 38:546–550

Stafford HA (1965) Factors controlling the synthesis of natural and induced lignins in *Phleum* and *Elodea*. Plant Physiol 40:844–851

Stahmann MA, Clare BG, Woodbury W (1976) Increased disease resistance and enzyme activity induced by ethylene and ethylene production by black rot-infected sweet potato tissue. Plant Physiol 41:1505–1512

Steward FC, Ammirato PV, Mapes MO (1970) Growth and development of totipotent cells. Some problems, procedures and perspectives. Ann Bot 34:761–787

Tanaka Y, Uritani I (1977) Polarity of production of polyphenols and development of various enzyme activities in cut-injured sweet potato root tissue. Plant Physiol 60:563–566

Tanaka Y, Uritani I (1979) Polar transport and content of I-3-AA in wounded sweat potato root tissues. Plant Cell Physiol 20:1087–1096

Theologis A, Laties GG (1981) Wound-induced membrane lipid breakdown in potato tuber. Plant Physiol 68:530–538

Thimann KV (1972) The natural plant hormones. In: Steward FC (ed) Plant physiology. Academic Press, New York London, pp 95–100

Thimann KV (1977) Hormone action in the whole life of plants. Univ Massachusetts Press, Amherst, pp 188–203

Thimann KV, Koepfli JB (1935) Identity of the growth-promoting and root-forming substances of plants. Nature 135:101

Thomas B, Hall MA (1975) The effect of growth regulators on wound-stimulated callose formation in *Salix viminalis*. Plant Sci Lett 4:9–15

Thompson NP (1966) Vascular regeneration and long-distance transport of indole-3-acetic acid in *Coleus* stems. Plant Physiol 41:1106–1112

Thompson NP, Jacobs WP (1966) Polarity of IAA effect on sieve-tube and xylem regeneration in *Coleus* and tomato stems. Plant Physiol 41:673–682

Tomaszewski M (1963) The mechabism of synergistic effects between auxin and some natural phenolic substances. In: Régulateurs naturels de la croissance végétale. CNRS, Paris, pp 335–351

Treshow M (1955a) Physiology and anatomical development of tomato fruit tumor. Am J Botany 42:198–202

Treshow M (1955b) The etiology, development, and control of tomato fruit tumor. Phytopathol 45:132–137

Uritani I (1961) The role of plant phenolics in disease resistance and immunity. In: Johnson G, Geissman TA (eds) Proceedings of symposium on biochemistry of plant phenolic substances. Colorado State University

Uritani I (1971) Protein changes in diseased plants. Ann Rev Phytopathol 9:211–234

Uritani I (1976) Protein metabolism. In: Heitefuss R, Williams PH (eds) Encyclopedia of plant physiology, New Series, vol 4. Springer, Berlin Heidelberg New York, pp 509–525

Van Overbeek J, Gregory LE (1945) A physiological separation of two factors necessary for the formation of roots on cuttings. Am J Botany 32:336–341

Van Steveninck RFM (1961) The "lag-phase" in salt uptake of storage tissue. Nature 190:1072–1075

Van Steveninck RFM (1972) Inhibition of the development of a cation accumulatory system and of Tris-induced uptake in storage tissues by N^6-benzyladenine and kinetin. Plant Physiol 27:43–47

Van Steveninck RFM (1975) The "washing" or "aging" phenomenon in plant tissues. Ann Rev Plant Physiol 26:237–258

Walker-Simmons M, Ryan CA (1979a) Immunological identification of proteinase inhibitors I and II in isolated tomato leaf vacuoles. Plant Physiol 60:61–63

Walker-Simmons M, Ryan CA (1979b) Wound-induced accumulation of trypsin inhibitor activities in plant leaves. Survey of several plant genera. Plant Physiol 59:437–439

Wardrop AB (1971) Occurrence and formation in plants. In: Sarkanen KV, Ludwig CH (eds) Lignins, occurrence, formation structure and reactions. Wiley-Interscience, New York, pp 19–41

Wareing PF (1958) Interaction between indole acetic acid and gibberellic acid in cambial activity. Nature 181:1744–1745
Warmke HE, Warmke GL (1950) Role of auxin in differentiation of root and shoot of *Taraxacum* and *Cichorium*. Am. J Bot 37:272–280
Watanabe A, Imaseki H (1973) Induction of deoxyribonucleic acid synthesis in potato tuber tissue by cutting. Plant Physiol 51:772–776
Watanabe A, Imaseki H (1976) Induction of deoxyribonucleic acid synthesis in potato tuber slices. Role of protein synthesis. Plant Physiol 57:568–571
Went FW (1936) The dual effect of auxin on root formation. Am J Bot 26:24–29
Went FW (1938) Specific factors other than auxin affecting growth and root formation. Plant Physiol 13:55–80
Wielgat B, Kahl G (1979a) Enhancement of polyribosome formation and RNA synthesis of gibberellic acid in wounded potato tuber tissue. Plant Physiol 64:863–866
Wielgat B, Kahl G (1979b) Gibberellic acid activates chromatin-bound DNA-dependent RNA polymerase in wounded potato tuber tissue. Plant Physiol 64:867–871
Williamson CE (1950) Ethylene, a metabolic product of diseased or injured plants. Phytopathology 40:205–208
Wright STC (1978) Phytohormones and stress phenomena. In: Letham DS, Goodwin PB, Higgins TVJ (eds) Phytohormones and related compounds – a comprehensive treatise, vol II. Elsevier/North-Holland Biomedical Press, Amsterdam New York, pp 495–536
Wright STC (1979) The effect of plant growth-regulator treatments on the levels of ethylene emanation from excised turgid and wilted wheat leaves. Planta 144:177–188
Yang SF (1983) Mechanism and regulation of ethylene biosynthesis. In: Akazawa T, Asahi T, Imaseki H (eds) The new frontiers in plant biochemistry. Jpn Sci Soc Press and Nijhoff/Junk, pp 133–151
Yang SF, Hoffman NE, NcKeon T, Riov J, Kao CH, Yung KH (1982) Mechanism and regulation of ethylene biosynthesis. In: Wareing PF (ed) Plant growth substances 1982. Academic Press, London New York, pp 239–248
Yeoman MM (1970) Early development in callus cultures. Int Rev Cytol 29:383–409
Yeoman MM, Evans PK (1967) Growth and differentiation of plant tissue cultures II. Synchronous cell divisions in developing callus cultures. Ann Bot 31:323–332
Yeoman MM, Dyer AF, Roberson AI (1965) Growth and differentiation of plant tissue cultures I. Changes accompanying the growth of explants from *Helianthus tuberosus* tubers. Ann Bot 29:265–276
Yeoman MM, Naik GG, Robertson AI (1968) Growth and differentiation of plant tissue cultures III. The initiation and pattern of cell division in developing callus cultures. Ann Bot 32:301–313
Yoshii H, Imaseki H (1981) Biosynthesis of auxin-induced ethylene. Effects of indole-3-acetic acid, benzyladenine and abscisic acid on endogenous levels of 1-aminocyclopropane-1-carboxylic acid (ACC) and ACC synthase. Plant Cell Physiol 22:369–379
Yoshii H, Imaseki H (1982) Regulation of auxin-induced ethylene biosynthesis. Repression of inductive formation of 1-amino-cyclopropane-1-carboxylate synthase by ethylene. Plant Cell Physiol 23:639–649
Yoshii H, Watanabe A, Imaseki H (1980) Biosynthesis of auxin-induced ethylene in mung bean hypocotyls. Plant Cell Physiol 21:279–291
Yu Y, Yang SF (1979) Auxin-induced ethylene production and its inhibition by aminoethoxyvinylglycine and cobalt ion. Plant Physiol 64:1074–1077
Yu Y, Yang SF (1980) Biosynthesis of wound ethylene. Plant Physiol 66:281–285
Yu Y, Adams DO, Yang SF (1979) Regulation of auxin-induced ethylene production in mung bean hypocotyls. Role of 1-aminocyclopropane-1-carboxylic acid. Plant Physiol 63:589–590
Yu Y, Adams DO, Yang SF (1979) 1-Aminocyclopropane-1-carboxylate synthase, a key enzyme in ethylene biosynthesis. Arch Biochem Biophys 198:280–286
Zimmerman DC, Coudron CA (1979) Identification of traumatin, a wound hormone, as 12-oxo-trans-10-dodecenoic acid. Plant Physiol 63:536–541

14 Water Relations and Plant Hormones

D.M. REID and R.L. WAMPLE

1 Introduction

Water is required for all aspects of plant growth and development, the control of rates of water loss and water uptake is of vital importance to the organism. Plants exert considerable control over their water status through morphological and behavioral adaptations. Typical morphological adaptations are a thicker epicuticular wax layer and deeper root system. Short-term control of water status by changes in stomatal aperature, and long-term strategies of dormancy or ephemeral growth habit also increase plant survival.

Since the effects of water stress on development are so general, one might expect changes in hormonal metabolism and water stress-induced modifications of growth and development to be highly correlated. Indeed, there is much data to show that water stress affects hormone levels and that exogenous application of these compounds will alter metabolism and the course of development. However, this is not unequivocal evidence that endogenous hormones are a link between changes in water potential and growth, and such correlations could be fortuitous. One can devise schemes explaining many of the effects of water stress without involving the action of hormones.

Do hormones act as metabolic messengers of changes in water potential? What is the evidence that hormones are essential in the response of plants to water stress? Hormones might have two roles; to act as fine turners of the response to changes in water potential, and as messengers of alterations in water status. The aim of this chapter is to examine the evidence that hormones might be so involved.

Plants are rarely exposed to the precise amount of water they require for optimal growth. Most commonly they have insufficient water (drought), however they are often exposed to excess water (waterlogging or flooding). This chapter will explore the role of hormones in the responses of plants to drought, and their role in response to flooding.[1]

It is difficult to give a precise definition of "drought" and when one should consider a plant to be stressed. A broad definition is when there is any reduction in water availability below that required for maximum growth. With this definition one could consider most plants to be droughted for much of their lives.

1 In addition to several titles from Volume II A of this Encyclopedia (cited in this chapter), the reader may also consult Volume 12 B (1982), Physiological Plant Ecology II, Water Relations and Carbon Assimilation (eds. LANGE OL, NOBEL PS, OSMOND CB, ZIEGLER H), in particular: Chapter 9, BRADFORD KJ, HSIAO TC, Physiological Responses to Moderate Water Stress, pp. 264–324 and Chapter 14, CRAWFORD RMM, Physiological Response to Flooding, pp. 453–477

This is demonstrated by comparing the higher productivity of plants grown in hydroponic and aeroponic conditions to those in the normal agricultural environment (HUBICK et al. 1982). For a more detailed examination of whether or not a plant is stressed the reader is directed to LEVITT's (1980) concepts of "stress" and "strain". In this chapter we will consider a plant as responding to stress (i.e., strained) when its growth or development is significantly different from the behavior it exhibited immediately before the imposition of the stress. We recognize that such a definition is simplistic and covers LEVITT's elastic (reversible) and plastic (irreversible) strains. Although not attempted here, it might be useful to examine the role of elastic and plastic strains separately.

2 Insufficient Water

2.1 Causes of Drought

There is an enormous range of conditions and treatments that affect tissue water potential (Ψ) and cause a plant to develop symptoms typical of a water deficiency. A lack of soil moisture, low relative humidity and saline soils are obvious causes. Fumigation with SO_2 (MAJERNICK and MANSFIELD 1970), hydrogen fluoride (POOVAIAH and WIEBE 1973), water loss from wounding, reduced xylem conductivity caused by pathogens (MACHARDY and BECKMAN 1973), sudden chilling of the roots (BENGTSON et al. 1977), increased osmotica around the roots (MIZRAHI et al. 1970) and elevated temperatures all can have an effect.

Salt-stressed plants may show symptoms of drought, but are actually subjected to the twin stresses of very negative soil water potentials and salt toxicity, and the situation is complicated by the differing abilities of various species to deal with these stresses. Some halophytes might not show toxicity to a specific ion since they may exclude, extrude the salt ions, compartmentalize, or chemically inactivate that ion. Alternatively, the plant might be capable of osmoregulation by salts and organic solute accumulation within living cells and thus avoid drought stress, but could still be adversely affected by the salt ions (FLOWERS et al. 1977).

Even a healthy, well-watered plant will undergo rapid changes in Ψ. A passing cloud obscuring the sun can have an effect (STANSELL et al. 1973), while diurnal fluctuations in leaf water potential (Ψ_l) are common (JORDAN and RITCHIE 1971). Growth itself affects overall plant water potential. Longer roots and taller shoots increase the resistance of the water transport pathway. As a cell enlarges its osmotic (Ψ_π), matric (Ψ_m), and pressure (Ψ_p) potentials are all in a state of flux. In essence, rapid growth creates a situation similar to drought stress. Understandably, modeling such a dynamic system and its control mechanisms is complex.

2.2 Some Problems in Interpreting the Literature

There are many problems in interpreting the literature concerning the mechanisms of drought-induced stress. A frequent problem is the use of a wide range of species with different responses and sensitivities to water stress. Furthermore, one should not assume that different types of stress all act through the same mechanism or initially on the same organs. For instance, although it is possible to cause water-deficiency stress in leaves by submerging the roots in a polyethylene glycol (PEG) solution (SINGH et al. 1973a, b), this might also, even in aerated solutions, cause a flooding response (WAMPLE and REID 1975). It may also be incorrect to assume that different methods of inducing water stress, such as root chilling, PEG, soil moisture depletion and hot, dry air flows will all result in the same influence on Ψ_p, endogenous ABA levels, and control of stomatal aperture. Exposure to air pollution can modify stomatal transpiration (MAJERNIK and MANSFIELD 1970, UNSWORTH et al. 1972, BISCOE et al. 1973) by affecting stomatal opening, but also might increase dessication by altering the rate of cuticular wax deposition (BYSTROM et al. 1968).

Another problem is that experimental material subjected to sudden, intense stress undergoes catastrophic reductions of Ψ. This is most unnatural, and as HSIAO (1973) points out, water stress usually develops gradually. On the rare occasions in nature when drought causes severe wilting, the plants will often die, or lose most or all their foliage. In research material experiencing abnormally rapid reductions in Ψ, metabolic changes could happen so quickly that it would be difficult to observe their precise sequence, thereby obscuring the key to early metabolic effects of drought stress.

With the onset of a gradual increase of stress, a leaf might have time to import some vital component needed in its response to stress, such as cytokinins from the roots. Rapidly imposed stress might not allow enough time for this response to occur. Thus, studies on rapidly wilted material are perhaps not appropriate to investigate the basis of the more common condition of gradual drought.

Stress can also be applied for different "durations" and at various "intensities", and the latter can be gradually stepped up or down during the course of the experiment. Such variations in technique make it difficult to compare the results from different studies.

The use of detached leaves, while convenient, leads to problems of discovering how water stress affects the whole plant. DÖRFFLING et al. (1980) have shown, while studying the role of ABA in water-stressed plants, that results with attached leaves are different from those with detached leaves. As mentioned above, detached organs are removed from the sources of compounds that might be needed in their response to stress (HATRICK and BOWLING 1973).

Leaf age or stage of development is also important, since older leaves may show a different degree of response than younger leaves (RASCHKE and ZEEVAART 1976, SIVAKUMARAN and HALL 1978a).

Out of necessity most work is carried out using pot-grown plants in artificial environments. Such plants differ from wild or agricultural plants due in part to restricted space for root growth, constant growth environment and relative lack of prestress. JORDAN and RITCHIE (1971) found that field-grown cotton plants were able to extract water from apparently droughted soil and had stomates which were less sensitive to decreased water potential than greenhouse-grown plants.

A final problem is prestress (treatment prior to the experiment) which is often not well controlled or even reported. There is increasing literature (see later) showing that prestressed (hardened) plants react very differently to stress than do nonhardened plants.

2.3 Control Points for Water Loss and Gain

The major site of water loss is through the stomata. In some species an appreciable amount of water loss occurs through the cuticle and this can be a major problem in rust-infected plants (DUNIWAY and DURBIN 1971). Presumably increased water loss would occur in any plant with a thin or damaged cuticle (HALL and JONES 1961), such as one exposed to wind and abrasion by soil particles (MARTIN and JUNIPER 1970). Abrupt salinization of the soil could cause a transient water loss through the roots.

The prime site of water gain is the root, although occasionally water absorption can occur through the leaves following heavy dew (MEIDNER 1954, 1968, KAPPEN et al. 1979, 1980). The major routes of water flow into and through the roots have been discussed by various workers (DAINTY 1969, NEWMAN 1976, PASSIOURA 1981).

The main driving force of water flow through plants is "passive" hydrostatic absorption caused by transpiration. "Active" osmotic absorption of water [the force producing root pressure and xylem exudate in a detopped plant (KRAMER 1974)] is of much less importance, especially in tall trees. In active uptake, the roots act somewhat as osmometers and the inhibitory effects of low soil temperatures, anoxia, and metabolic poisons, could all affect water absorption by slowing uptake and accumulation of ions. The effect of low temperature might also reduce hydraulic conductivity of the endodermal membrane (MILBURN 1979), and poisons could modify membrane structure and thus affect the pathways of water movement (STUART 1973).

The relative importance of active versus passive water uptake and the problem of how to separate the role of the flux of solutes into roots from the flow of water is complex and far from understood (DAINTY 1969, GREACEN et al. 1972, DALTON et al. 1975, FISCUS 1975, KAUFMANN 1976).

Although the transpirational loss of water by a plant is usually the most important driving force for water absorption by roots, one must also consider the processes such as root pressure and other factors that may alter the Ψ of the xylem and other root cells, especially when transpiration is slow. Thus there are a multitude of endogenous factors that influence water movement into roots and throughout the plant.

One difficulty in studying Ψ of actively growing systems is the dynamic nature of the cell-wall elasticity (E) and cell volume (V). It is also difficult to separate the effects of hormones on Ψ and growth from their influence upon ion movement. The different results obtained with experiments studying the effects of hormones on excised roots as compared to intact roots might be explained by the operation of transpiration in the intact system. Transpiration will affect long-distance transport of ions. The high levels of carbohydrate and different quantities of hormones in the intact system could also influence rates of water flow.

Table 1. Some endogenous factors that influence rates of exit (A) and entry (B) of water

A	B
Number and size of leaves Density and size of stomata Reponsiveness of stomata Orientation of leaves and leaf-rolling Methods of reducing heat load Quantity and distribution of wax in cuticles Dormancy and ability to control growth rate Rates of leaf tip die back, abscission and senescence Ψ of leaves Ability of osmoregulate (Ψ_π) Succulence, cell structure and volume Rates of cell expansion Cell wall elasticity Number and diameter of xylem elements and frictional resistance of xylem	Ease by which new lateral and root hairs can be formed Spatial distribution of roots Rate of root elongation Degree of suberization Integrity of endodermis Ψ of roots Ability to osmoregulate (Ψ_π) and control intra- and intercellular distribution of osmatica Rate of movement of assimilate from shoot to root Rates of ion transport (Js) Hydraulic permeability and conductivity (Lp) Degree of soil–root contact Root/shoot size ratio Number and diameter of xylem elements

Hormones have been shown to affect many of the above points and also those factors listed in Table 1. Thus, it is at any one of these events or points that one could expect to find an effect of hormones if they are to have any direct role to play in maintaining the water status of the plant. The obvious examples would be the influence of exogenous ABA on the stomatal closure and hydraulic conductivity of roots. These two effects are easily observed in the laboratory, which may explain the lack of information on the less dramatic, but perhaps important effects of hormones on some of these other control points.

Rates of water efflux can be controlled by changing the rate of leaf formation, growth, and in extreme circumstances by drought-induced abscission or dormancy. Leaf orientation, turning away from the rays of the sun, could also be of use to lessen heat load and evapotranspiration (RAWSON 1979). The effective leaf area can be reduced in grasses by leaf-rolling. Reduction in growth rate, thus lowering the requirement for water, is also used as a mechanism of drought avoidance.

A particularly effective technique would be to increase water use efficiency and hold back the initiation of important events (such as formation of reproductive structures) until water is available in the appropriate quantity.

With regard to water uptake, the main control point would be the efficiency of water absorption by roots. Direction and rate of growth, numbers, positioning, lateral and density of root hairs are all important. The major resistances are the root endodermis and to a lesser extent the hypodermis. Growth of roots away from saline zones would conserve water. Osmotic adjustment (producing more negative Ψ_π) may also be another long-term strategy. All will have a significant effect on the rates of water movement (see Sect. 2.5).

2.4 Sequence of Responses to Water Stress

2.4.1 The Initial Response

What are the earliest responses to drought? This has been discussed at some length by HSIAO (1973) and ZIMMERMANN (1978) and will be considered only briefly here. The first major response is probably a reduction in turgor pressure. This would cause a slowing of cell expansion and might coincidentally activate a turgor pressure-sensing mechanism. ZIMMERMANN (1978) proposed that a sensing mechanism is located in membranes and that alterations in their thickness affect their electromechanical properties. This would then transform a mechanical stress into a chemoelectrical one, perhaps eventually regulating ion transport between various organelles and even different cells. As a consequence of these changes, metabolism, growth, and development could be affected. Indeed, as has been observed by everyone who has studied the responses of plants to drought, even small reductions in water potential have profound effects on growth and development, some occurring within minutes, but others taking days.

The turgor pressure hypothesis was criticized by ITAI and BENZONI (1976) on the basis that "it separates the known response to water stress from the additional information which can be deduced from the known response to other stress, such as heat, salinity, mineral deprivation, or boron toxicity, where the turgor remains unchanged ... All these stresses are characterized by hormonal changes and other symptoms similar to water stress. It is likely, therefore, that an alternative hypothesis based on the hormonal balance as the primary effect and other responses as secondary ones is more appropriate." We agree that most stresses seem to alter hormone levels and thus may affect growth and development. However, the advantage of the turgor pressure hypothesis is that it puts forward a simple, convincing model for the initial, rapid sensing and transduction of altered water potential which then could influence hormone levels. All stresses might eventually influence hormone metabolism but each stress may have its own initial sensor. An analogy would be that different forms of stress result in increased ethylene production, but the sensor could differ in each case.

2.4.2 Later Responses

A change in water availability, especially a severe or sudden one, initially affects turgor, increasing or slowing cell expansion. This triggers a number of secondary alterations in metabolism, presumably including transcription, translation, and hormone content. The third level of stress-induced changes will be modifications in substrate production, rate and direction of transport, osmotic potential, all eventually leading to alterations in growth and development (fourth level). The plant physiologist examining the effects of water stress below level 3 finds interpretation of the results complicated by the number and interaction of events that have occurred by this time. A good example of this is the difficulty of deciding how drought affects photosynthesis. Is photosynthesis showed by stomatal closure resulting in a reduced availability of CO_2, or by increases in ABA and so on?

In the next section a summary is given of some of the data showing that water stress affects the endogenous amount of hormones. As we see it, hormones might act in two ways: (a) The level of endogenous hormone present at the time of stress might modify the effects of the stress, perhaps by controlling processes such as the movement of water or ions into the roots. (b) Some time later there might be secondary changes in the production of that or other hormones which modify the effects of subsequent stress and initiate new directions in growth and development.

2.5 Drought-Induced Changes in Levels of Hormone: Effects on Stomatal Functioning

Since the detailed mechanics of guard cell movement have been dealt with elsewhere (RASCHKE 1975a, 1979), we will only consider the role and effects of plant hormones on stomatal function.

2.5.1 Gibberellins and Stomates

There is little evidence that changes in the endogenous levels of gibberellins (GA) are directly involved in the control of stomatal function. AHARONI et al. (1977) found a dessication-induced decline in GA activity 6 h after the onset of dessication of lettuce leaves, and in pea seedlings wilting alters the rate of GA interconversion (TAYLOR and RAILTON 1977). The latter finding suggests that wilt-induced increases in abscisic acid (ABA) influence GA metabolism. However, the reported effects of exogenous ABA on GA levels and metabolism seem to depend upon the species being examined and need further study (NADEAU et al. 1972, DURLEY et al. 1976).

Although exogenous GA_3 enhanced transpiration of barley (LIVNE and VAADIA 1965) this was a minor effect, while GA was ineffective in detached oat leaves (LUKE and FREEMAN 1968). HORTON (1979) using isolated epidermal strips peeled from leaves of *Vicia faba* found that GA had little influence on stomatal opening, nor did it interfere with the ABA inhibition of opening (see also HORTON 1971, TUCKER and MANSFIELD 1971). In contrast, AHARONI et al. (1977) showed that GA pretreatment of detached lettuce leaves increased water loss from dessicated tissues but had little effect on fully turgid leaves.

2.5.2 Auxins and Stomates

DARBYSHIRE (1971a, b) and MILLS and TODD (1973) have observed stress-induced changes in IAA-oxidase activity and there is a report of effects on auxin-like activity (TAL and IMBER 1970). There is as yet no conducive evidence of a direct influence of endogenous auxin on stomatal function. Although the stomates of some dicotyledonous species respond to synthetic auxins (FERRI and LEX 1948, BRADBURY and ENNIS 1952, ZELITCH 1965), not all do (TUCKER and MANSFIELD 1971), and monocotyledonous species appear to be unresponsive (PLAYER 1950, MANSFIELD 1967). More recently PEMADASA (1979 and 1982) and SNAITH and MANSFIELD (1982) have found that an exogenous supply of various auxins can affect stomatal movement in *Commelina communis* L. In view of these studies, estimates of endogenous auxins in drought-stressed plants are now essential.

2.5.3 Ethylene and Stomates

Some droughted plants produce more ethylene than nonstressed controls (MCMICHAEL et al. 1972, ADATO and GAZIT 1974, BEN-YEHOSHUA and ALONI 1974, EL BELTAGY and HALL 1974, GUINN 1976, WRIGHT 1977), and osmotic stress has a similar stimulatory effect (HALL et al. 1977, SIVAKUMARAN and HALL 1978b).

Little is known about the causes of increased ethylene production. In some cases the rate of production of the gas is in two stages (EL BELTAGY and HALL 1974). The early stage is perhaps primarily due to the lowered water potential. The latter stage might be due to the stress-induced changes in overall growth and development, since variations in ethylene evolution seem to follow marked alterations in growth (HUXTER et al. 1979). It is possible that stomatal closure slows the diffusion of ethylene to the surrounding air (ABELES 1973, pp 39–41); and it is the ethylene in this air that some workers monitor. Results may therefore depend upon whether the ethylene being measured is that given off by the plant, or that which is vacuum-extracted from the tissue.

Ethylene supplied at a wide range of concentrations had little effect on stomatal movement in *Zea mays, Pisum sativum* (PALLAGHY and RASCHKE 1972) and *Vicia faba* (EL BALTAGY and HALL 1974). A possible ABA and ethylene interaction (Sect. 2.7) suggests that the gas could affect stomatal functioning by modification of ABA production. However, WRIGHT (1977) has shown that in water-stressed excised wheat leaves, the rise in ABA occurred more rapidly than ethylene and concluded that ethylene levels did not influence ABA production.

Reports that the ethylene-releasing agent, 2-chloroethyl-phosphonic acid, will open (VITAGLIANO 1975) or close (BROWNING 1974) stomates must be interpreted carefully since it is rare that synthetic growth regulators are specific in their action. More recently PALLAS and KAYS (1982) found ethylene to inhibit photosynthesis in *Arachis hypogaea* L. Although the gas did not affect ribulose 1,5-bisphosphate carboxylase activity, leaf conductance to water vapor was lowered, it was argued that ethylene might be affecting stomatal opening. Little or no effect was observed on six other species (PALLAS and KAYS 1982).

There are a few reports that exogenous ethylene might reduce the numbers of stomates per leaf by inhibiting their development (FUNKE et al. 1938, REID and WATSON in press).

2.5.4 Cytokinins and Stomates

Changes in endogenous cytokinins caused by drought and osmotic stress are reported to influence stomatal behavior.

ITAI and VAADIA (1965) first found that reduced water supply lowered cytokinin content of xylem fluid and BROWNING (1973) reported that irrigation of droughted plants led to a re-establishment of normal cytokinin content. Osmoticum around roots had similar effects (ITAI and BENZIONI 1976), and other workers found that dessicated lettuce leaves showed lower levels of cytokinin than did turgid tissue (AHARONI et al. 1977). Cytokinin content has been observed to drop within 15 min of the onset of drying in tobacco leaves (ITAI and VAADIA 1971), and the rates of recovery to nonstressed levels were slow (ITAI et al. 1968). Osmotic stress has now been shown to reduce endogenous cytokinin content in various species (BOUSSIBA and RICHMOND 1976, SIVAKUMARAN and HALL 1978b, WALTER and DUMBROFF 1981).

Detached leaves of oats, barley, bean, and tobacco transpire more rapidly when treated with cytokinin (LIVINE and VAADIA 1965, MEIDNER 1967, LUKE and FREEMAN 1967, MIZRAHI et al. 1970, PALLAS and BOX 1970, KIRKHAM et al. 1974). On the other hand, in some species cytokinin will not cause stomatal opening in epidermal strips (HORTON 1971, TUCKER and MANSFIELD 1971, OGUNKANMI et al. 1973). If cytokinins only affect guard cell movement in intact leaves, one could argue, as has RASCHKE (1975a) that the cytokinins are not acting directly on the guard cells but are slowing processes such as senescence. However, there are reports of a stimulation of stomatal opening by several cytokinins in leaf epidermal strips (DAS et al. 1976, JEWER and INCOLL 1980) and others showing a promotive effect of kinetin on transpiration in excised leaves and epidermal strips of a grass (*Anthephora pubescens*; INCOLL and WHITELAM 1977). The latter authors suggest that the previously reported ineffectiveness of cytokinins might be due to a concentration requirement or structural charateristic of the cytokinin used. It is also possible that a presumptive site of cytokinin action could be damaged during isolation of the epidermis. KIRKHAM et al. (1974) found that addition of cytokinins to salt-stressed *Phaseolus vulgaris* decreased stomatal resistance, lowered leaf turgor pressure and increased the symptoms associated with osmotic stress. Thus, the normal response of droughted or salt-stressed plant (less cytokinin production) might be a useful adaptation to these stresses. The question of how the cytokinins exert their influence on guard-cell movement needs much more study.

2.5.5 ABA and Stomatal Closure

The debate of exactly how the drought-induced increases in ABA fit into the events between the initial onset of water stress and subsequent stomatal closure is far from over. However, there is a strong evidence showing a relationship between these two events.

a) Wilting and moderately severe water stress cause increased in ABA levels in wheat leaves (WRIGHT and HIRON 1969, WRIGHT 1969). A very wide range of species act in this manner (HOAD 1973, LOVEYS and KRIEDEMAN 1973, LARQUE-SAAVEDRA and WAIN 1974, BEARDSELL and COHEN 1975, BOROHOV et al. 1976, RASMUSSEN 1976, BLAKE and FERRELL 1977). Even plants found in aquatic environments, and thus presumably not well adapted to drought, show a wilt-induced build of ABA (MILBORROW and ROBINSON 1973).

b) Less severe water stress can also produce a rise in ABA. A decrease in Ψ_l of as little as 2 bar can stimulate ABA production (MOST 1971, WRIGHT 1977). With a greater drop in Ψ_l, a more substantial increase occurs, suggesting a two-phase response (WRIGHT 1977). The latter, steeper increase might be similar to the threshold which was observed by ZABADAL (1974). There has been discussion of whether or not there is a threshold Ψ below which ABA levels do not rise, and above which there is a sudden increase. Although SIVAKUMARAN and HALL (1978b) found no evidence for it in *Euphorbia lathyris* or *Vicia faba*, it is possible that it might only be seen in certain species or to a different extent in various species. QUARRIE and JONES (1979) have shown that there are considerable differences in the amount of endogenous ABA in 26 wheat genotypes grown under similar conditions. The observation of a sharp threshold might depend upon the manner in which the stress is imposed (sudden versus slow). PIERCE and RASCHKE (1980) observed the changes of ABA content together with Ψ, Ψ_π and turgor of the tissue of three species. They found that water stress always increased ABA content, which rose sharply as turgor approached zero, but the water potential needed to achieve this varied from leaf to leaf, depending upon the history of previous stress of the individual plant. They concluded that "the leaf water potential per se does not control ABA content" and also showed that there was not a clear relationship between ABA level and osmotic potential. For these and other persuasive reasons, they argued that the build-up of ABA was controlled by turgor. This agrees with the concept of a turgor pressure-sensing mechanism being the initial trigger of all the responses to a period of drought (HSAIO 1973, ZIMMERMAN 1978).

c) Consistent with the idea that ABA is involved with guard-cell functioning is the recent finding by WEILER et al. (1982) that ABA levels can increase more rapidly in epidermal cells than in mesophyll cells. Also in agreement with this is the observation that ABA catabolism can occur in the leaf epidermis (SINGH et al. 1979). Metabolism has also been observed in the cytoplasm of mesophyll cells (HARTUNG et al. 1980), and the distribution of ABA between the chloroplasts and cytoplasm might be controlled by differences in chloroplast pH (HEILMANN et al. 1980). Thus although ABA degradation occurs in the cytoplasm (HARTUNG et al. 1980) it appears that ABA synthesis takes place in chloroplasts since lyzed preparations of that organelle will incorporate mevalonic acid into ABA (MILBORROW 1974). Also chloroplasts which contain ABA (RAILTON et al. 1974) seem to contain an appreciable amount of the total leaf ABA (LOVEYS 1977). LOVEYS (1977) proposed that the hormone might (especially during water stress) leak out of the mesophyll plastids, move through the mesophyll cytoplasm to the epidermis and finally to the site of action in the guard cells. In agreement with this, WELLBURN and HAMPP (1976) and HARTUNG et al. (1980) demonstrated that chloroplasts are permeable to ABA. Furthermore, ABA can be taken up by the epidermal cells of *Commelina* and accumulate in the guard cells (ITAI et al. 1978, WEYERS and HILLMAN 1979).

Minor problems with this hypothesis include the low rate of incorporation of MVA into ABA by lyzed chloroplasts (MILBORROW 1974) and there appear to have been few attempts to follow up on the intracellular site(s) of ABA synthesis and movement.

d) Water stress has a marked influence over the regulation of the rates of the ABA biosynthesis. This has been discussed by WALTON (1980), SEMBDNER et al. (1980) and MILBORROW (1981), and will not be dealt with in length here. It does seem as if the overall drought-stimulated increase in the hormone in leaves is the result of de novo synthesis rather than the release from a bound form or conversion of an immediate precursor (MILBORROW and NODDLE 1970, MILBORROW 1978). However, the levels of bound ABA (sugar conjugates) might be of importance in the recovery from stress (HIRON and WRIGHT 1973). Based on (1) his own data (MILBORROW 1979, 1981), (2) that of LOVEYS' (1977) indicating that ABA moves out of the chloroplasts during drought, and (3) the ideas of HARRISON and WALTON (1975), who studied the rapid metabolism of ABA to phaseic and dihydrophaseic acids in wilted leaves, MILBORROW (1981) proposes that the high concentration of ABA in chloroplasts inhibits further synthesis in that organelle. Water stress somehow makes the plastids leaky to ABA, which eventually moves to the guard cells. The reduction in chloroplastic ABA level allows synthesis to restart. MANSFIELD et al. (1978) have proposed that increased plastid leakiness to ABA is the results of drought-stress-induced synthesis of all-trans-farnesol (OGUNKANMI et al. 1973, WELLBURN et al. 1974). Applications of this compound cause stomatal closure (FENTON et al. 1977) and alter the structure of the outer chloroplast envelope (WELLBURN et al. 1974, FENTON et al. 1976). This led MANSFIELD et al. (1978) to propose that all-trans-farnesol controls plastid permeability and increases ABA synthesis by lowering chloroplastic ABA levels. This would result in higher whole plant ABA levels typical of drought-stressed plants. The question still remains of how water stress, and lowered turgor, affect farnesol synthesis.

e) An exogenous supply of ABA reduced transpiration in cuttings of *Abies balsamea* and *Picea glauca* (LITTLE and EIDT 1968), and it will initiate stomatal closure in *Vicia faba* epidermal strips (HORTON 1971) and in excised cereal leaves (MITTELHEUSER and VAN STEVENINCK 1969). Other terpenoids, all-trans-farnesol (WELLBURN et al. 1974) and vomifoliol (STUART and COKE 1975), will also affect stomatal functioning (see reviews by RASCHKE 1975a, WALTON 1980, MANSFIELD and DAVIES 1981).

All the earlier experiments were done with high levels of ABA, and one could, of course, conclude that the exogenous ABA might be exerting a pharmacological effect. However, KRIEDEMANN et al. (1972) carried out experiments indicating that only a doubling of ABA content was needed to initiate stomatal closure in detached leaves of three species. More recently it has been calculated that in *Commelina* closure is complete when the guard cells have taken up no more than 5.9 mmol ABA per stomatal complex (WEYERS and HILLMAN 1979). On some occasions (RASCHKE and ZEEVAART 1976) there is a good negative correlation between endogenous ABA and stomatal conductance.

f) As predicted by the hypothesis that an increase in endogenous ABA is a factor in stomatal closure, the rate of response of guard cells to applied ABA can be rapid. CUMMINS et al. (1971) reported that barley stomata started to close within 5 min and were fully closed by 30 min and in *Zea mays* there was an initial lag of only 3 min (KRIEDEMANN et al. 1972).

g) Some of the most convincing evidence of the involvement of ABA comes from the studies on the wilty mutant of *Lycopersicon esculentum* var. flacca. The mutant contains little ABA and shows morphological and stomatal abnormalities, most of which (especially stomatal behavior) can be overcome by exogenous ABA (IMBER and TAL 1970). These plants also show imbalances in the levels and metabolism of other hormones, but it would seem as if the lack of ABA is the primary defect (TAL et al. 1979).

h) The observation that ABA prevents K^+ influx into guard cells, which normally precedes stomatal opening, also supports the idea of a direct effect of ABA (MANSFIELD and JONES 1971, HORTON and MORAN 1972). ABA influences the movement of malate (DITTRICH and RASCHKE 1977, VAN KIRK and RASCHKE 1978), K^+ and protons in a range of tissues (MARRÈ 1977, KASAMO and SHIMOMURA 1981), it will inhibit auxin-induced H^+ excretion (RAYLE 1973, RAYLE and JOHNSON 1973) and affects K^+ and Cl^- efflux from guard cells (MACROBBIE 1981, WEYERS and HILLMAN 1980). ABA may inhibit H^+ expulsion (RASCHKE 1975b), and thus alter K^+ transport by some effect on membrane-bound ATPases, which may be associated with stomatal control (FUJINO 1967).

i) Indirect evidence that the small quantities of endogenous ABA might be involved in stomatal functioning in epidermal strips of *Vicia faba* comes from the work of TAYLOR et al. (1981) who have shown that an exogenous supply of ABA will counteract the stomatal opening caused by SO_2. Their work suggests that low levels of SO_2 might act by blocking the normal action of endogenous ABA which is tending to promote closure.

j) Petiole girdling inhibits ABA translocation out of soybean leaves and causes stomatal closure, thereby suggesting that increased ABA might trigger closure (SETTER et al. 1980, 1981). Their work also indicated that the build-up of photosynthate was not responsible for the closing of the stomates (SETTER et al. 1980).

Despite the evidence in paragraphs (a) to (j) there are inconsistencies with the idea that the elevated levels of ABA in droughted plants are responsible for initiating stomatal closure.

a) A variety of *Lupinus luteus* which responded to water deficit in the usual manner with higher ABA levels, was unusual in that stomates of both epidermal peels and intact leaves were unresponsive to exogenous ABA (LANCASTER et al. 1977). WEYERS and HILLMAN (1979) argued that the failure to obtain complete closure might have been due to the use of excess KCl in the solutions upon which the epidermal strips were floated. This, however, does not explain the observation that the ABA was not effective in the intact leaves, which were able to close their stomates in response to wilting (LANCASTER et al. 1977).

b) HSIAO (1973) and WALTON (1980) have both pointed out that the evidence for ABA as the main link between water stress and stomatal closing is only circumstantial. There is still some doubt that the drought-induced rise in ABA occurs rapidly enough to cause stomatal closure, especially in quickly wilted plants. A number of authors have reported that some guard cells react before there is a measurable increase in ABA (BEARDSELL and COHEN 1975, BENGTSON et al. 1977, WALTON et al. 1977, DÖRFFLING et al. 1980). Furthermore, isolated epidermal strips of *Vicia faba* (LOVEYS 1977) or *Commelina communis* (DÖRFFLING et al. 1980) are unable to synthesize ABA in response to water stress and yet respond to water stress by closing their stomates. In isolated leaves of *Commelina,* stomata closed within 5 min after the imposition of stress. This suggests that even if ABA is synthesized and/or released from the mesophyll chloroplasts, it is unlikely that there would be sufficient time for it to travel to its presumptive site of action in the guard cells. To counter this argument, it has been suggested that some ABA might first be produced in the epidermal cells (WEILER et al. 1982).

Frequently, research on ABA and stomatal movements has been carried out with detached, rapidly wilted leaves. Perhaps under these circumstances there is a significant change in leaf physiology from those of an intact plant, and thus it is unwise to attempt to relate the changes in ABA levels with stomatal resistance. In circumstances such as this, the control mechanism for stomatal closure might not involve ABA, but ABA levels rise independently in response to the lowered turgor. Once the hormone reaches an appropriate concentration, it could reinforce the initial mechanism that closed the stomates and prolong this condition. Plants which undergo a slow water stress will generally have sufficient time for ABA to be synthesized, translocated, and to act in the guard cells. This might even include ABA synthesis in roots and translocation to the leaves in the transpiration stream (DAVISON and YOUNG 1973). Furthermore, a detached leaf would not have this source of ABA or cytokinins and it is known that detached leaves do not always respond to water stress and ABA in exactly the same manner as attached leaves (KRIEDEMANN et al. 1972, DÖRFFLING et al. 1980).

Comparison of data from different workers is further complicated by the changing sensitivity of guard cells to water stress and ABA with, developmental age (JORDAN et al. 1975, RASCHKE and ZEEVAART 1976), physiological state (TURNER 1975, RASCHKE 1979) and even between ab- or adaxial leaf surface (JORDAN et al. 1975).

Much of the work attempting to correlate endogenous changes of ABA with the stomatal opening and closing, has simply looked at the overall alterations of ABA concentration in the whole leaf or the entire shoot. What we really need to know is: what are the changes occurring in the guard cells themselves? The more recent work studying the ABA levels in the epidermis itself (LOVEYS 1977, SINGH et al. 1979, DÖRFFLING et al. 1980, WEILER et al. 1982), and the movement of ABA in the epidermis to the guard cells (ITAI et al. 1978, WEYERS and HILLMAN 1979) is a start in this direction. In all such experiments with isolated bits of tissue, it would be necessary to remember that being "isolated" they may respond to stress differently than when in situ.

Data is also needed on the mechanism of action of ABA on the movement of ions through guard-cell and root membranes; for instance, how does it affect K^+/H^+ exchange or malate concentration? A useful approach to this type of question might be to examine the effects of a variety of environmental stimuli that are known to affect stomatal movements; such as light, humidity, drought, flooding, CO_2 concentration, and so on. Would they all interact with ABA in the same way? Do they all elicit the same response mechanisms? Would they and ABA all have the same influence on H^+/K^+ exchange or malate concentration? For example, RASCHKE (1975b) has reported on an intriguing relationship between ABA and CO_2 on the stomatal response in which ABA or water stress made stomates more sensitive. Utilization of such responses could be a useful experimental tool in the understanding of the mechanisms of both ABA and CO_2 action.

The difficulty in relating variations in the level of ABA with stomatal resistance is exacerbated by three factors; (a) rather high levels of ABA are found in some leaves (RASCHKE and ZEEVAART 1976) but only very small increases appear to be needed to close stomates (KRIEDEMANN et al. 1972, WEYERS and HILLMAN 1979); (b) the redistribution of stored ABA, rather than a synthesis, might be the initial controlling factor (BEARDSELL and COHEN 1975). Thus, it might be difficult to obtain a good correlation between stomatal opening and the total ABA content of the leaf. This brings us to the interesting question of what all this ABA is doing if only a small portion is required for stomatal functioning? As described in Sections 2.6 and 2.7, this hormone could be involved in other aspects of plant physiology which allow it to further influence water status.

2.5.6 ABA and Stomatal Opening

The high ABA level produced after a period of water stress eventually drops to almost pre-drought concentration (HIRON and WRIGHT 1973). The rate of recovery of ABA to the prestress level is variable and has been reported to be 2 days in *Zea mays* (BEARDSELL and COHEN 1975) and peas (DÖRFFLING et al. 1974), 1 day in *Vitis vinifera* (KRIEDEMANN and LOVEYS 1974) and sorghum (BEARDSELL and COHEN 1975) and as short as 3 h in wheat (BENGTSON et al. 1977) and *Xanthium* (ZEEVAART 1980). Recovery time also depends upon the severity and length of time for which the stress was imposed (BENGTSON et al. 1977). There might also be a difference between the rate of recovery in detached versus attached leaves. Perhaps this is due to the inability of ABA to be exported from the isolated leaf (SETTER et al. 1981).

In intact plants the ABA concentration rarely falls to the pre-stress level and is often twice the original level (HIRON and WRIGHT 1973, KRIEDEMANN and LOVEYS 1974). Successive periods of stress cause the basal amount of ABA in turgid tissue to rise steadily (LIN et al. 1978, ACKERSON 1980). If this residual amount of ABA, or the slow decline in ABA following a period of stress has any effect on stomatal function, the history of water stress of a plant would be important in its subsequent survival.

What is the evidence that these post-stress levels of ABA influence stomatal opening? Stomatal functioning can be impaired for some time (often days) after the relief of the stress and resumption of normal turgor (STÅLFELT 1955). This "after-effect" is reflected in reduced transpiration and photosynthesis (LOVEYS and KRIEDEMANN 1973). In some cases there is a fair correlation between fall of ABA levels and resumption of normal transpiration rates (BOUSSIBA and RICHMOND 1976, DÖRFFLING et al. 1977). However, this is not always so, and ABA levels may return to near-normal values more quickly than does transpiration (LOVEYS and KRIEDEMANN 1973, BEARDSELL and COHEN 1975, BENGTSON et al. 1977, DÖRFFLING et al. 1977, 1980). Perhaps in these cases although the total amount of ABA in the leaf is low, there could be appreciable quantities in the guard cells. Alternatively, KRIEDEMANN et al. (1974) suggested that since only a doubling of ABA content is needed to close stomates (KRIEDEMANN et al. 1972), that sufficient ABA is present to prevent complete opening for some days.

Another possibility is that perhaps those few molecules of ABA that are active at presumptive sites on the guard cell membranes have some long-term effect on those sites, even after they have been metabolized or transported away. This implies that either ABA pre-treatment or water stress should alter the sensitivity of the guard cells to further ABA application. Although this has been observed, the alteration in sensitivity is not always in the same direction (KRIEDEMANN et al. 1972, DAVIES 1978, ACKERSON 1980). ACKERSON (1980) found that stomates became more sensitive to exogenous ABA after they had been subjected to repeat cycles of water stress.

Repeated episodes of water stress will "harden" or "precondition" the stomatal response, such that the Ψ_1 needed to initiate closure is more negative and therefore presumably less sensitive to drought stress (MCCREE 1974, BROWN et al. 1976, JONES and TURNER 1978). Part of this response is due to osmoregulation in leaves (BROWN et al. 1976, JONES and TURNER 1978). The question of how much of the pre-stressing effects on stomatal sensitivity are due to stress-induced changes in ABA or osmoregulation is still unanswered.

2.5.7 Summary of Section 2.5

Of the five classes of hormones, it appears most likely that ABA and related compounds play the major role in stomatal functioning, with cytokinins having a less direct effect. However, in more indirect ways, all the hormones can influence stomatal behavior by their influence on nutrient flow, ion uptake, photosynthetic rate, and growth. For instance, control of either photosynthetic rate or direction of photosynthate transport could affect osmotic potential, energy for osmoregulation, and ion transport.

Hormones could serve as permissive factors with their presence being necessary for proper functioning of processes such as membrane permeability, involved in guard-cell operation. As an example, consider the absolute requirement for IAA in the Tanada-effect, in which red light brings about rapid alterations in the permeability of root membranes (TANADA 1968). The ability of auxins to affect stomatal movement (PEMADASA 1982) and to modify the response of stomates to ABA (SNAITH and MANSFIELD 1982) is interesting in this regard, and looks well worth following up.

2.6 Water Uptake and Movement Through Roots

Many studies on the effects of drought on roots and transport in roots have used isolated roots. Although this greatly simplifies experimental procedure, it complicates the extrapolation of results to the whole plant. For instance, transpiration can affect ion transport in roots (SUTCLIFFE 1962, LÜTTGE 1969, PITMAN 1975) and there are some interesting models of root–shoot cooperation involving hormone control over water relations (CRAM and PITMAN 1972, PITMAN 1972, LÜTTGE and HIGINBOTHAM 1979). Thus, when one reads the literature concerning effects of drought on water and ion relations in roots, these considerations must be born in mind.

Water flow into roots could be influenced by any of the factors mentioned in Table 1 B.

Compared to the shoots, there is relatively little work on the effects of drought on roots. Nevertheless, drought is known to influence many of the factors in Table 1 B. Long-term drought stress normally causes reduced growth of all plant organs. However, shorter periods of water stress can result in root elongation, apparently by a repartitioning of solutes from leaves to roots allowing for a maintenance of turgor and cell expansion (SHARPE et al. 1979). There are also reports of solute accumulation (osmotic adjustment) in roots of droughted plants (GREACEN and OH 1972, OSONUBI and DAVIES 1978). Some plants subjected to salinity also exhibit osmotic adjustment and increased root growth (see Sects. 2.7.1 and 2.7.7). These adaptive responses to stress may be a very useful feature (SHARP et al. 1979).

The importance of new root growth has been demonstrated by ACEVEDO (1975) and is discussed by CALDWELL (1976) and HSIAO et al. (1976). Another feature which requires further investigation is root shrinkage and swelling. KAUFMANN (1976) suggests that the importance of root contact with the soil is underestimated with regard to water absorption. Expanding roots are in contact with moist soil, but roots which have been subjected to drying reduce their diameter by as much as 60% and also become more suberized (KAUFMANN 1968, HUCK et al. 1970). Larger root systems increase water-collecting ability, and this is a trade-off between shoot and root growth; root growth being more important during water shortage. Reduced leaf expansion would also slow water loss by transpiration.

Other morphological features useful in water absorption are the production of new lateral roots and a control over their spatial distribution. SHARPE and DAVIS (1979) suggest that there could be an increase in the numbers of secondary roots and PASSIOURA (1976), points out that cereals with deep seminal and shallow nodal roots have the ability to obtain water from the top soil in wet years and lower horizons in dry periods. Deep root systems can be very effective in extracting moisture during drought (JORDAN and RITCHIE 1971).

Could any of these processes be controlled by drought-induced alterations in hormone level in the roots?

The rate of xylem exudation from de-topped root systems varies greatly and is affected by environmental conditions (REID et al. 1969). It has long been known that there is a diurnal fluctuation of this exudation (GROSSENBACHER 1939, VAADIA 1960) and that roots exhibit a variable resistance to water flow (STOKER and WEATHERLEY 1971). These observations, together with the reported inhibitory effects of metabolic inhibitors on root pressure and flow, support the idea that the physiological state of the root (and perhaps its hormone content) is important in controlling xylem flow.

2.6.1 Drought and Hormone Levels in Roots

A prerequisite for a hormonal control over water flow into roots would be either a drought-induced alteration in sensitivity to endogenous hormones, or changes in their levels. As to the former, nothing is known. Even with regard to the latter, there is less data on roots than shoots. Nevertheless, the content of cytokinin-like compounds in xylem sap is reduced by withholding irrigation or by the addition of osmotica (ITAI and VAADIA 1965, ITAI et al. 1968). Irrigation after water stress brings about a rise in cytokinin activity in the xylem fluid of *Coffea arabica* (BROWNING 1973). Reduced hormone levels in the xylem sap might not reflect a reduction of overall concentration in the root tissues. During stress conditions roots could conserve hormone levels by reducing export. Even if the quantity of hormones is reduced, the question is, how this is achieved? Roots seem to be a major site of cytokinin production (KULAEVA 1962, KENDE 1965, BURROWS and CARR 1969). Does a more negative Ψ of the root thus directly inhibit cytokinin synthesis, or is the lack of substrates from drought-stressed leaves the cause of lowered hormone production?

Water stress will cause an increase in the quantity of ABA in roots (MILBORROW and ROBINSON 1973, WALTON et al. 1976) and in xylem fluid (HOAD 1975). Using tall and dwarf *Helianthus* HOAD (1975) investigated the idea that ABA in the xylem sap might not be synthesized in the root, but in the leaves. He found a longer lag in tall as compared to dwarfs, before he could detect elevated levels of ABA in the xylem sap. He concluded that this was caused by the longer transport distance for ABA from the presumptive site of synthesis in the leaves of taller plants and therefore that the roots might not be an important locus of ABA production. Reports of ABA in phloem exudate (HOAD 1967, BELLANDI and DÖRFFLING 1974a, ZEEVAART 1977) support this idea.

On the other hand, the work of WALTON et al. (1976) suggests that mild water stress brought about an increase in ABA synthesis in roots of *Phaseolus vulgaris*. Their experiments in which the phloem was blocked, and those with excised plants, indicated that the rise in ABA levels did not require transport of the hormone from the shoots, and might be produced in root tissues. The site(s) of ABA biosynthesis, directions and rates of transport, and the influence of water stress, still require much study.

There is little information on the effects of drought or osmotic stress on GA, auxin and C_2H_4 metabolism and production by roots.

2.6.2 Effects of Hormones on Nutrient Redistribution and Ψ of Roots

Not much is known of the hormonal controls of drought-induced redistribution of solute from shoots to roots and there is debate over the precise effect of drought on phloem loading, source–sink relationships and long-distance translocation (REINHOLD 1975, WARDLAW 1980). Phloem does remain functional during water stress (WARDLAW 1968) and two studies (SHARPE and DAVIES 1979, SHARPE et al. 1979) suggest that in water-stressed *Zea mays*, solutes move from older leaves to actively growing zones such as roots. This result fits well with the frequent observation that droughted plants lose their older leaves and, prior to abscission, mobilize nutrients to the rest of the plant. These processes could be affected by the relative levels of hormones in the various organs. Although the export of cytokinins from roots is reduced by water stress (ITAI et al. 1968), it is possible that hormone levels in roots are maintained. This would cause a relative increase in cytokinin quantity in roots as compared to older leaves having a twofold effect; increasing rate of senescence in these leaves and increasing the relative sink strength of roots. SHINDY et al. (1973) showed that benzyladenine treatment of roots resulted in an enhanced transport of ^{14}C photosynthate to roots. This flow of carbohydrate could affect root Ψ and Ψ_π and stimulate root growth.

Auxins, C_2H_4, GA, and ABA all affect long-distance transport (PATRICK 1976). Again there is little information on effects on roots or whether the hormones are acting directly on the transport process or on overall cell growth.

2.6.3 Effects of Hormones on Ion Transport

It is now generally recognized that auxins, cytokinins, GA, and ABA all affect active ion transport (VAN STEVENINCK 1976). Most work with roots has been with ABA and cytokinins. There is little evidence showing drought-induced changes in GA or auxins in roots. Nonetheless, the quantity of these hormones present in roots prior to drought could influence the response to stress in view of the effects of IAA and GA on secretion/uptake processes in root tissues. Application of GA_3, increased K^+ and Na^+ leakage from slices of beetroot (VAN STEVENINCK 1976), and permeability of a liposome model membrane (WOOD and PALEG 1972, WOOD et al. 1974). Although the responses of root tissue to auxins are different, and not as distinct as those with coleoptile and shoot tissue, IAA can stimulate pigment leakage and inhibit Na^+ uptake by beetroot (PICKLES and SUTCLIFFE 1955). Endogenous levels of hormones are rarely taken into account by researchers in this area.

It is clear that ABA and cytokinins have a strong influence on ion transport and water permeability of roots. Ion transport across the root into the xylem and permeability of root tissues to water have been estimated by measuring the rates of xylem exudation by roots of decapitated plant and the ionic content of that fluid. Besides isolated root systems, the results of the effects of root-applied hormone on xylem flow as measured by techniques such as a magnetohydrodynamic sap flow meter (SHERIFF 1972) would be useful. Unfortunately, whole plant studies are complicated by the effects of ABA and cytokinin (CK) on shoots (i.e., on transpiration). However, CRAM and PITMAN (1972) found that ABA increased ^{36}Cl accumulation in roots of intact barley seedlings. KARMOKER and VAN STEVENINCK (1979) found a similar stimulatory effect with ^{36}Cl and ^{22}Na, but showed an inhibition of long-distance transport to shoots of *Phaseolus vulgaris*. They observed that ABA inhibited K^+ transport into the xylem in intact plants but in excised roots it was stimulated. KARMOKER and VAN STEVENINCK (1979) concluded that ABA exerts a strong control over ion flux into the xylem and proposed that in intact plants, ABA causes stomatal closure, resulting in a positive pressure of xylem fluid opposing root pressure.

They then invoked ZIMMERMANN's (1978) hypothesis of a pressure-sensing mechanism that turns off further K^+ transport.

In decapitated root systems, ABA can affect exudation rate in less than 6 min after application (GLINKA 1980) and ion release into the xylem within 30 min (PITMAN and WELLFARE 1978). While ABA has been shown to inhibit these processes (CRAM and PITMAN 1972, PITMAN et al. 1974a, b, ERLANDSSON et al. 1978) it can promote these processes in others (GLINKA and REINHOLD 1971, TAL and IMBER 1971, VAN STEVENINCK 1972, GLINKA 1973, 1977, COLLINS and KERRIGAN 1974, KARMOKER and VAN STEVENINCK 1978). In addition, a stimulation of ion transport followed by an inhibition has been observed in excised roots of *Hordeum vulgare* (PITMAN et al. 1974a, PITMAN and WALLFARE 1978) and in intact *Phaseolus vulgaris* seedlings (KARMOKER and VAN STEVENINCK 1979). Thus the response to ABA is complex, and depends upon such factors as temperature and nutrients in the bathing medium (PITMAN et al. 1974a).

Cytokinins also affect ion and water movement in roots and tend to oppose the effect of ABA. In isolated *Zea mays* roots, ABA increases water permeability, while kinetin reduces it (COLLINS and KERRIGAN 1974). They also found ABA treatment to promote K^+ flux into the xylem, which was inhibited by the cytokinin. Ca^{2+} flux was reduced by ABA, while kinetin had the opposite action (COLLINS and KERRIGAN 1974). There are other reports showing various effects of these hormones on water flow and ion transport in roots and other tissues (ILAN 1971, ILAN et al. 1971, BENZIONI et al. 1974, HONG and SUCOFF 1976). The work of HONG and SUCOFF (1976) is of particular interest, as it indicates that the cytokinin production of the root tip might influence water and ion flow into roots.

The mode and locus of action of cytokinins and ABA on ion and water movement in roots is obscure. Also the effect of altered ion concentrations on water flow, or the influence of more rapid water movement into roots on transport of ions is far from understood. After cutting off the external supply of ions to isolated *Helianthus annuus* roots, GLINKA (1980) found that ABA-treated plants had higher levels of ions in the xylem exudate, and concluded that ABA had enhanced the flow of ions from vacuoles to the symplast. He also found a greatly increased water flow to the xylem and that the relationship between time in ABA and exudation rate was biphasic. The initial increase occurred in 0–20 min and the second after 80 min. GLINKA (1980) suggested that the first increase in xylem flow was due to the release of ions from vacuoles into the cytoplasm and then into the xylem.

PITMAN and WELLFARE (1978) used intact and isolated roots from *Hordeum vulgare* to measure rates of ion transport and exudation, and thus calculated values for hydraulic conductivity. In their system ABA inhibits ion transport across roots and into the xylem and causes reduced guttation from leaves in intact plants and xylem exudation from excised roots. Whether this reduced transport was due to lower salt transport or to root permeability was further studied. They found that although hydraulic conductivity was initially increased by ABA, it decreased after 90 min to control levels. At this time, when there was no effect of ABA on hydraulic conductivity, the rates of exudation and salt transport were slowed. Thus the effect of ABA on exudation rate appeared to be by inhibition of salt transport rather than hydraulic conductivity. PITMAN and WELLFARE (1978) also proposed that ABA may have two effects; a rapid stimulation of hydraulic conductivity and a slow inhibition of salt transport (also see GLINKA 1980). They emphasized the importance of the levels of ABA and variable temperature.

Is there any direct connection between the effects of ABA and cytokinins? More data is needed to resolve this question. GLINKA (1980) found that the effect of kinetin treatment on K^+ release in sunflower xylem fluid was variable, while the effect of ABA was reproducible. He proposed that the two hormones acted independently. ITAI and BENZIONI's (1976) data suggested that there might be a connection between the cytokinins and ABA. They showed that ABA-treated *Phaseolus* plants had lower levels of cytokinin-like substances in xylem exudate than controls while kinetin treatment enhanced the catabolism of ^3H-ABA in tobacco leaves.

It is clear that exogenous ABA, cytokinins and water stress can affect water and ion movement into roots and up the xylem. Thus endogenous ABA and cytokinin levels in roots may be important when studying the affects of water stress on roots and whole plants.

Many reports of conflicting effects of hormones might exist because various workers have used different species and a wide range of experimental conditions, and have not paid close attention to the condition of plant growth prior to the experiment. Additional information is needed on the rates of penetration of exogenous cytokinins and ABA and a knowledge of the intra- and intercellular sites of hormone biosynthesis and metabolism in roots.

2.7 Other Mechanisms Controlling Water Status of the Plant

There are other mechanisms by which a plant might control entry or loss of water (Table 1), or by other means increase its (or its progenies) chances of survival.

2.7.1 Reduced Growth Rate

A partial or total cessation of growth gives a water-stressed plant a simple mechanism for lowering water demand. Small reductions in Ψ, as seen in droughted plants, slow cell expansion (BOYER 1968), cell-wall synthesis (CLELAND 1967) and cell division (KIRKHAM et al. 1972). Cell division may be more sensitive than enlargement (KIRKHAM et al. 1972). Drought reduces overall growth, with leaves usually being the most sensitive, roots intermediate, and stem growth least affected (GATES 1955, WANGERMANN 1961, JARVIS and JARVIS 1963). Much of the recent work has been with the effects on leaves, with little study of stem cells. The apical meristem appears to be most drought-resistant of all plant tissues (BARLOW et al. 1977). Perhaps this is because the apex is not well connected to the vascular system, does not transpire, is a strong sink for nutrients and assimilates, and maintains its Ψ_π (BARLOW et al. 1980).

Inhibition of growth could be achieved in many ways, but lowered turgor seems to be the most important and rapid influence (BOYER 1970, ACEVEDO et al. 1971). A lack of photosynthate caused by partial closure of the stomates (review by HSIAO 1973, and Sect. 2.7.6), will secondarily slow cell expansion and growth. Extensive diversion of nitrogen to proline (KEMBLE and MACPHERSON 1954, STEWART et al. 1966) and glycine betaine (HANSON and NELSON 1978) would reinforce this effect. However, there is little subtle long-term control here. The role of hormones, as fine tuners of the responses of plants to water stress, might be found in the following. They could modify CO_2 fixation by controlling stomatal functioning; they could also directly interfere with the enzymes of CO_2 fixation (TREHARNE and STODDART 1968, POSKUTA et al. 1972, WELLBURN et al. 1973, SANKHLA and HUBER 1974 and 1979), as ABA and GA have been shown to influence this process. Exogenous ABA promotes proline accumulation (ASPINALL et al. 1973), and thus reduces the supply of nitrogen available for growth. Perhaps stress-induced increases in endogenous ABA have the same effect.

There are reports suggesting that ABA (reviews in MILBORROW 1974, WALTON 1980) and ethylene (ABELES 1972, LIEBERMAN 1979) inhibit growth and development. For example, ABA in the cambial region of woody species (LITTLE et al. 1978) could affect cambial growth and development (LITTLE 1975) and bud break (LITTLE and EIDT 1968). Exogenous ABA affects morphological development of wheat leaves (QUARRIE and JONES 1977) and inhibits overall growth of *Lemna minor* (OVERBEEK et al. 1968), elongation of *Avena* coleoptiles (REHM and CLINE 1973) and *Pisum* stem cells (WARNER and LEOPOLD 1971) and light-stimulated leaf expansion of *Phaseolus vulgaris* (VAN VOLKENBURGH and DAVIS 1983). Other work (HUBICK, YEUNG and REID unpublished) also shows that ABA and drought will slow both cell division in root and shoot apicies and developing leaves of *Helianthus annuus*.

Thus, considering their potential mode of action, drought-induced increases of ABA and ethylene and reductions in cytokinins are consistent with the idea that changes in the levels of these hormones are one cause of reduced growth in stressed plants. For instance, drought inhibits growth of many tissues and species (HSIAO 1973), and in water-stressed plants ABA accumulates (WRIGHT and HIRON 1969). BELLANDI and DÖRFFLING (1974a, b) found the apical buds were sinks for exogenous ABA. Furthermore the shoot apex in *Xanthium* is a very active ABA producer (RASCHKE and ZEEVAART 1976). ABA appears to inhibit cell enlargement (MILBORROW 1974) and bud growth (TUCKER and MANSFIELD 1973, BELLANDI and DÖRFFLING 1974b), and will, for example, inhibit growth of *Lemna* (TILLBERG et al. 1980). Thus one could simply conclude that the stress-induced increases in ABA are the main cause of a slowing of growth during and after drought.

It is unlikely that ABA is the only factor that inhibits growth, and in some cases ABA is not an effective inhibitor. DAMPTEY et al. (1978) found that the application of ABA to corn seedlings did not slow stem elongation as much as did water stress. In some situations ABA can even promote growth (MCWHA and JACKSON 1976), and as shown by HARTUNG and FUFER (1981), the degree of inhibitory or stimulatory effect depends upon tissue age.

It has been argued that there is still insufficient evidence that drought-induced changes in the ABA level are directly responsible for the reduction in growth rate (WALTON 1980), or as pointed out by KING (1976) that the evidence so far "fails to establish final causality in the control of growth". Perhaps the initial mechanism by which drought slows growth is reduced turgor; then, secondarily, other factors, such as lack of assimilates, become more important. Hormones such as ABA and cytokinins might modify these secondary processes. Perhaps too much emphasis has been placed on work with exogenous application of growth-regulatory compounds or on studies in which workers have examined "levels" of hormones rather than studying their metabolism or the alterations in sensitivity of tissue to hormones (TREWAVAS 1981). In addition, looking at the levels of only one hormone, for example ABA, in the absence of estimations of antagonistic compounds, such as GA, possibly located in the same cellular compartment, might be of little use. To attempt to explain a water-stressed reduction in growth or the induction/maintenance of dormancy on the basis of an alteration in the pool sizes of one hormone is too simplistic.

Nevertheless, in view of the wide range of effects of hormones on the processes listed in Table 1, it seems fair to argue that changes in endogenous levels of hormones must have some effect on the eventual responses of a plant to drought. For instance, drought tends to increase root/shoot ratio (MAXIMOV 1929, READ and BARTLETT 1972, MEYER and BOYER 1981), and conversely, well-watered plants have relatively small root systems (HUBICK et al. 1982). Plants with a small area of transpiring shoot tissue and a large root system have obvious drought-stress avoidance capabilities. It would be interesting to speculate on the role of hormones as factors that could influence the relative growth rate of shoot versus root, perhaps by controlling the direction and rate of translocation. The works of RICHARDS (1976) and RICHARDS and ROWE (1977) which discuss the importance of the root, and root-produced hormones in the control of root–shoot interaction and water uptake is worth following up.

Overall translocation is moderately insensitive to water shortage (WARDLAW 1980). However, it would seem that there are subtle shifts in the control of the direction of source–sink flow of assimilates during drought which allow those parts of the plant most vital to its survival to continue growth at the expense of less important organs. ASPINALL (1965) has reported a diversion of nutrients toward reproductive structures. Chaneling of assimilates from surface roots sensitive to drought to rapidly growing deep roots that are capable of reaching new water supplies could also be of utmost importance, as would alterations in the ability of roots to absorb water. Hormones could aid indirectly in these decisions by promoting senescence and abscission of old parts of the plant. They could also have a more direct influence on the mobilization of assimilates (PATRICK 1976, WEAVER and JOHNSON Chapt. 1, this vol.). Such nutrients can be used to maintain metabolism in the more drought-resistant tissues, thus allowing for their continued growth. The increase in Ψ_m would also aid in the abilities of organs, especially roots, to grow (MEYER and BOYER 1981) and to continue to absorb, retain, and transport water.

Little is known of the precise mode of action of hormones on source–sink relationships in plants (WEAVER and JOHNSON, this vol.). In view of the effects of drought on promoting senescence, and thus the export of assimilates from leaves, it is surprising how little work has been done with the two hormones that are considered to be the stress hormones, ABA and ethylene.

Another interesting growth response to drought and salt stress is a decrease in apical dominance. Plants in highly saline areas become prostrate (UNGAR 1967). *Zea* plants, on release from water stress, increase the number of axillary inflorescences (DAMPTEY et al. 1978). ABA treatment of *Chenopodium rubrum* increases the rate of lateral branch production (KREKULE and HORAVKA 1972), and a number of workers have discussed the possibility that alterations in the levels of cytokinins and ABA influence apical dominance (MULLINS and OSBORNE 1970, TUCKER and MANSFIELD 1973, BELLANDI and DÖRFFLING 1974b, TUCKER 1978). Since the level of these hormones is affected by water stress, there might be a connection, especially in view of the fact that during stress ABA moves to the main shoot apex (KING and EVANS 1977, ZEEVAART 1977).

Even if ABA is involved in prostrateness, it seems likely that other hormones are also involved. Auxins could influence correlative inhibition of lateral buds by controlling ABA level (ELIASSON 1975, TUCKER 1978). On the other hand, work with flacca tomato mutant (TAL et al. 1979) examines the relationship between various hormones and suggests that the level of ABA may control auxin metabolism and thus potentially the metabolism of ethylene and cytokinins. In the same paper, the authors put forward a scheme of postulated hormone interactions that nicely underlines the point that our understanding of the role of hormone interactions in morphogenesis is still at a primitive stage.

2.7.2 Root Growth and Development

Root mass and its distribution in the soil is vital in the maintenance of the ability to take up water. KUMMEROW (1980) argues that roots possess fewer adaptive structures which can be used to resist or tolerate drought than do shoots, and suggests that roots survive water stress by their plastic behavior. For example, *Adenostoma fasciculatum* can grow successfully either in shallow soils with a large, spreading, shallow root system or in deep soils with a deep root growth form (HELMERS et al. 1955, KUMMEROW et al. 1977).

The effects of water stress on root growth in trees depends upon the season, spring growth being less affected, while in the summer the roots are sensitive to drought (ZAHNER 1978). Root elongation is usually inhibited by low Ψ_{soil} (FEILDING 1955). However, although Ψ_{tissue} and Ψ_{soil} are obviously important, the increased mechanical impedance of dry soil (ZIMMERMAN and KARDOS 1961) also restricts root elongation (ZAHNER 1968, HSIAO 1973).

It is advantageous for a root system to respond to drying top soil by producing roots which penetrate the densely packed dry soil into moister lower horizons. An alternative strategy is to develop a shallow but very extensive root system enabling the plant to take advantage of frequent but light showers that might not reach deep into the soil (MEYER and BOYER 1981). Water stress can also affect the direction of root growth (BILAN 1960) and the number of laterals (BROWN and LACATE 1961). Recently TESKEY and HINCKLEY (1981), working with *Quercus alba,* have found that as the Ψ_{soil} steadily becomes more negative, the elongation of existing roots slows. However, they conclude that the main determining factor of root mass is the total number of growing roots rather that the rate of root elongation, and that this is an important factor in drought resistance. When growth is inhibited by dry, tightly packed soil (ZIMMERMAN and KARDOS 1961), there can also be a promotion of root initiation leading to an increase in the number of roots (BARLEY 1962).

It is possible to imagine a role for hormones as intermediaries between the sensing of the stress and the eventual alteration in growth and development. Water and osmotic stress reduce cytokinin export from roots with a concomitant rise in ethylene and ABA (Sect. 2.5). Ethylene might well be involved in a thigmomorphogenetic sensing of the soil particles by roots (BARLEY 1962, GOESCHL et al. 1966, JAFFE, Chap. 12, this Vol.) and also affect root initiation. Consistent with this is the finding that ethylene inhibits root elongation (CHADWICK and BURG 1970, SMITH and ROBERTSON 1971) and under certain circumstances promotes root initiation (FABIJAN et al. 1981b). Also, low cytokinin (HEIDE 1965, RAILTON and REID 1973, FABIJAN et al. 1981b) and elevated ABA (CHIN et al. 1969) and ethylene concentrations (FABIJAN et al. 1981b) can promote the formation of new roots. In addition the appropriate level of ABA stimulates root elongation (GAITHER et al. 1975), and brief periods of water stress increase root formation on pea cuttings (RAJAGOPAL and ANDERSEN 1980). However, the situation is far from clear-cut, as there are reports of ABA inhibition of root growth and root initiation (HEIDE 1968, PILET 1968). The precise endogenous concentration of the hormone required, and the sensitivity of root tissue are little-researched factors.

For the deep, rapidly elongating roots, hormones could, by controlling translocation, direct the products of photosynthesis to that organ. Drought-induced alterations in hormone levels might also be responsible for controlling direction of root growth by modifying tropistic behavior. Furthermore, changes in the ability of the roots to absorb water, by alterations in root-hair density and hydraulic conductivity, could result in increased root growth. As discussed in Sects. 2.6.3 and 2.7.1, some of these processes could be affected by hormones.

Much less is known of root physiology than of shoot and it seems to be a case of "out of sight, out of mind". We do not have sufficient information on the effects of drought on hormone metabolism in root tissue.

2.7.3 Leaf Morphology and Behavior

Even a brief water shortage slows leaf expansion (BEGG and TURNER 1976, RAWSON et al. 1980), partly due to turgor loss. However, even after watering, some of the inhibition persists (BOYER 1970), suggesting that other factors such as hormone levels might be involved. Water stress also inhibits the production of leaf primordia in shoot apices of barley (NICHOLLS and MAY 1963, HUSAIN and ASPINALL 1970), sunflower (MARC and PALMER 1976, YEGAPPAN et al. 1980) and tobacco (CLOUGH and MILTHORPE 1975). Cell division is also reduced (YEGAPPAN et al. 1982).

Arid-zone species have evolved xeromorphic structures to control transpiration and heat load, aiding their survival during drought. Details of these are given by PARKER (1968) and include: the number of stomatal pores, cuticle and epidermal cell-wall thickness, stomatal size, leaf size, orientation, number, leaf-rolling, and, in extreme situations, abscission. Any or all of the hormones could influence these processes.

In wheat, ABA and water stress decrease mean leaf cell size, reduce numbers of stomata and leaf area and increase numbers of trichomes (QUARRIE and JONES 1977). The authors propose that ABA "may have an important role in determining the pattern of leaf development".

A stress-induced rise in ethylene (Sect. 2.5.3) might reinforce the inhibitory effects of ABA on leaf expansion, since ethylene will slow leaf growth (FUNKE et al. 1938, BURG and BURG 1968, GOESCHL and PRATT 1968, ABELES and HEGGESTAD 1973).

QUARRIE and JONES (1977) pointed out that reduced turgor and limited assimilate might also be important. Also the reduction in cytokinin export from the roots to the shoot (ITAI and VAADIA 1971) could have some importance here. The action of cytokinins often opposes that of ABA (MIZRAHI et al. 1970, 1971), suggesting that ABA and cytokinins act antagonistically in these situations.

It might be worthwhile to study the possible role of hormones in drought-induced leaf rolling in xerophytic plants. This response can reduce transpiration rate (PARKER 1968) and occurs when bulliform cells lose turgor. Could these cells react to water stress and ABA as do guard cells?

Another response of leaves to severe drought is abscission (OPPENHEIMER 1960, KOZLOWSKI 1976) and increased rates of senescence (YEGAPPAN et al. 1982). The metabolic cost of this must be great. Older leaves usually abscise before younger ones. ABA accelerates (CRAKER and ABELES 1969), and the ethylene enhances abscission (JACKSON and OSBORNE 1970, JORDAN et al. 1972, BEN-YEHOSHUA and ALONE 1974, LIPE and MORGAN 1973, ADATO and GAZIT 1974). Thus it seems possible that with increases in stress, ethylene promotes leaf senescence. However, other hormones may be involved, since the altered levels of other growth regulators in water-stressed plants may influence ethylene evolution. Depending upon the tissue, exogenous ABA can either stimulate (leaves) or inhibit (fruit) ethylene release (COOPER et al. 1968). Also depending upon the species, ABA has been shown to inhibit or increase ethylene production (CRAKER and ABELES 1969, GERTMAN and FUCHS 1972, MAYAK and HALEVY 1972). In two elegant papers, WRIGHT (1979, 1980) concludes that the balance of auxins, cytokinins, GA, and ABA in the plant at the time of the water stress could be a major factor in controlling ethylene levels. This and other evidence suggests that leaf senescence is associated with alterations in a range of hormones (CHIN and BEEVERS 1970). Whether or not these changes initiate senescence or merely follow them is not known. WRIGHT (1980) also proposes that the eventual decline in ethylene evolution, observed some hours after the onset of water stress, is related to prior accumulations of ABA. The effect of ABA was not straightforward, as ABA first stimulated and then inhibited ethylene release in turgid leaves. In senescing rose petals, ABA and ethylene may control each other's synthesis (MAYAK and HALEVY 1972) in a feed-back system somewhat similar to auxin–ethylene interactions (BURG and BURG 1966).

These complications aside, stress-induced ethylene evolution is sometimes positively correlated with, and occurs prior to, abscission in droughted plants, and may be a factor in promoting abscission. PRISCO and O'LEARY (1972, 1973) argue that in the leaves of salt-stressed *Phaseolus vulgaris,* the enhancement of senescence is brought on by hormonal imbalance (reduction of cytokinins and increase in ABA). These workers also show that the commonly observed increase in leaf succulence of salt-stressed plants is probably not caused by a lowered level of cytokinins (PRISCO and O'LEARY 1973).

SAMET and SINCLAIR (1980) report that endogenous ABA in soybean leaves rises rather late during leaf aging. It seems as if the increase in ABA was not the cause of leaf senescence, but could be more instrumental in abscission. With regard to leaf abscission, there is little evidence that ABA is important. For instance, DAVENPORT et al. (1977) found little correlation between ABA transport and stress-induced abscission in cotton. They did, however, demonstrate that the reduced capacity of stressed petioles to transport IAA might be involved (DAVENPORT et al. 1980).

An interesting study linking ABA to leaf morphology is that of ANDERSON (1978), in which he found that treatment with exogenous ABA converts the submerged leaf type to the floating form in *Potamogeton nodosus*. The latter develop a different shape, stomata on the adaxial surface, and a thick waxy cuticle, none of which are seen in the submerged type and are presumably advantageous in the drier, above-water environment. He also showed that GA and cytokinins tended to reverse the effect of ABA. The influence of ABA, GA, IAA, and cytokinins on wax synthesis was studied by CURRIE et al. (1978) who showed no effect of GA or ABA, while IAA stimulated and kinetin inhibited the process.

2.7.4 Flowering and Reproductive Development

These processes are affected by reductions in Ψ_{soil}, can be modified by the application of hormones, and are associated with changes in hormone content. Drought reduces spikelet initiation in wheat (BARLOW et al. 1977), and MORGAN (1980) has shown that drought and exogenous ABA both reduce seed set in wheat. Alterations in the supply of ABA and GA (THORPE and MEIER 1973), cytokinins and auxins (STEWARD 1968), and ethylene (HUXTER et al. 1979) can modify a range of developmental processes such as shoot formation and root initiation (EBELL 1967, GREENWOOD 1981). An exogenous supply of GA (PHARIS et al. 1980) promoted flowering in water-stressed Pinaceae, and stressed seedlings showed changes in the metabolism of endogenous GA (PHARIS 1977). PHARIS (1981) also found that a combination of drought and elevated temperature promoted flowering of *Tsuga heterophylla*. Alterations in ABA can affect flowering and there have been reports of ABA treatment modifying the rate and time of floral initiation, and influorescence growth (EL-ANTABLY and WAREING 1966, EVANS 1969, ASPINALL and HUSAIN 1970, KREKULE and ULLMANN 1971, KINET et al. 1975, KING and EVANS 1977, QUARRIE and JONES 1977, DAMPTEY et al. 1978, MORGAN 1980). KREKULE and HOROVKA'S (1972) data suggests that ABA is more effective on the post-inductive stages of flowering than on inductive stages.

In many cases the application of ABA exerts an effect similar to water stress (MORGAN 1980); however, it is unwise to try to explain anything as complex as flower initiation by postulating an alteration in one hormone. A specific hormone might be the key controller of one of the many steps in this series of events. In a water-stressed plant, two stages at which changes in the balance of hormones could exert an influence of flowering and reproductive development are a direct effect on the initiation of the flower primordia, and in the diversion of assimilates to a developing reproductive structure (ASPINALL 1965, CHALMERS, Chap. 6, this Vol., WEAVER and JOHNSON, Chapt. 1, this Vol.). Some of the arguments for and against the idea that flowering might be controlled by either a single hormone or by a balance of hormones, have been discussed at length by VINCE-PRUE (1975) and ZEEVAART (1976).

An interesting development is that various tissues grown in vitro respond to the Ψ_π of the medium in that small changes in Ψ_π can stimulate organogenesis (DHINDSA et al. 1975, GRANATEK and COCKERLINE 1978, BROWN and THORPE 1980). Whether or not this is mediated through alterations in hormone metabolism is not known.

2.7.5 Summary of Sections 2.7.1 to 2.7.4

In the period immediately following an episode of stress one might expect to see an orderly sequence of changes in the metabolism of all the hormones; the correct sequence leading to a series of specific hormonal balances, each one modifying a certain part of the overall developmental pattern. The plasticity (TREWAVAS 1979) and multivariate response of plants (STEWARD and KRIKORIAN 1971) to environmental stimuli, with the possibility of alterations in hormonal metabolism occurring together with variations in sensitivity to those hormones (TREWAVAS 1981), make it difficult to sort out exactly what is happening.

2.7.6 Water Stress and CO_2 Assimilation

Although the rate of CO_2 fixation does not directly affect the water status of the plant, it plays such a vital role in growth that it is mentioned here. Photosynthesis influences water relations by producing the substrates that allow the plant to grow and osmoregulate, and it has been known for a long time that the assimilation of CO_2 is reduced by water stress (BOYER 1976). The quantity of CO_2 fixed is controlled by processes such as stomatal functioning, leaf area development and modes of CO_2 fixation, respiration, photorespiration, and translocation.

The major effect of drought-induced stomatal closure seems to be in reducing CO_2 supply. However, water deficiencies appear to have other inhibitory effects on photosynthesis (BOYER 1971, FRY 1972), such as chloroplast ultrastructural changes (GILES et al. 1974), and some decreases in RuBP carboxylase activity (O'TOOLE et al. 1977). LAWLOR and FOCK (1975) have found that as sunflowers are subjected to increasing stress, photorespiration does not decline as much as does photosynthesis, suggesting that photorespiration allows the plant to consume photosynthetically produced reductant and to protect the photosynthetic apparatus from damage (KRIEDEMANN and DOWNTON 1981).

Both water and osmotic stress seem to promote a shift from a C_3 to CAM-type of metabolism in facultative CAM species (WINTER and VON WILLERT 1972, WINTER and LÜTTGE 1976), while in halophytes increasing salt stress appears (on the basis of change in δ^{13} values) to cause decreasing dependence on RuBP carboxylase and higher PEP carboxylase activity (GUY et al. 1980). On the other hand GUY et al. (1980) also suggest that this change in δ^{13} values could merely reflect an increase in stomatal diffusional resistance. Water stress also slows the flux of recent photosynthetically fixed carbon towards amino acids and sugar synthesis, but increases the utilization of stored reserves for this purpose (LAWLOR and FOCK 1975).

Many of these points in metabolism could be potential sites of hormone action. How might drought achieve these effects? The degree of stomatal opening is an important factor, in that it restricts availability of CO_2. Following stomatal closure, there will be rapid adjustments in internal CO_2 and O_2 concentrations which will alter the ratio of RuBP oxygenase to RuBP carboxylase activities (FORRESTER et al. 1966, LAWLOR 1978). The closing of the stomates will also alter transpiration and leaf temperature. The problem of determining the major control points is difficult because the prime effector, the stomatal complex, is itself controlled by a series of feedback loops (RASCHKE 1975a) involving the levels of CO_2, Ψ, and carbohydrate level. Is there any evidence that endogenous hormones can control any of these processes?

We have reviewed evidence showing that drought and osmotic stress stimulate ABA and ethylene, and reduce cytokinin levels. Thus the primary influence of hormones on CO_2 fixation may be as controllers of stomatal opening and closing (Sects. 2.5.1 and 2.5.2) by ABA and biosynthetically related compounds, and to a lesser extent cytokinins. Hormones could be but one of many factors modulating early stomatal closure and also the subsequent sensitivity of the stomates to further periods of drought. Although CUMMINS et al. (1971) concluded that the main effect of ABA on photosynthesis was stomatal closure, hormones can affect aspects of CO_2 assimilation other than through a control of the stomates (SANKHLA and HUBER 1974). *Lemna* has nonfunctional stomates, and thus ABA cannot close them. Exogenous ABA inhibited CO_2 fixation in this genus (BAUER et al. 1976), and has been shown to inhibit this process within 4 h of ABA application (TILLBERG et al. 1981). The latter authors argue that the rapidity of action suggests that the effects of the ABA are "closely associated with the primary target for ABA action and that ABA was directly inhibiting some photosynthetic process." The target could be an ABA-related reduction of RuBP carboxylase activity since ABA lowers the activity of this enzyme in various species (POSKUTA et al. 1972, WELLBURN et al. 1973, SANKHLA and HUBER 1979, RYĆ and LEWAK 1980). The Hill reaction may also be inhibited (BAUER et al. 1976). On the other hand, MCLAREN and SMITH (1977) observed a small increase in photosynthetic activity in *Lemna* treated with ABA. As well, MAWSON et al. (1981) using isolated mesophyll cells, found no effect of

ABA on photosynthesis, and this system is similar to *Lemna* in that the effect of stomates is eliminated. It is difficult to explain how ABA can have this range of effects. Perhaps it is due to the mode of application of ABA or the methods of tissue treatment. However, such comparisons between the action of a hormone on systems as different as the fronds of a hydrophyte and enzymatically isolated cells from the leaves of mesophytes could be misleading. Nevertheless, in those plants where ABA is inhibitory to the photosynthetic process, it seems likely that drought can lower plant productivity and CO_2 fixation by raising ABA level, thus causing stomatal closure and/or reducing the activity of the CO_2-fixing enzymes.

There have been suggestions that cytokinins are involved in regulating photosynthesis. Application of this growth regulator tends to oppose the action of ABA on stomatal closure (Sect. 2.5) and in some situations enhances transpiration (LIVNE and VAADIA 1965). Plants with water-stressed roots produce less cytokinin (ITAI and VAADIA 1971), and thus one would expect that with lowered cytokinin levels photosynthesis might be reduced as a result of stomatal closure.

In *Phaseolus vulgaris* cytokinin treatment will stimulate the rate of photosynthesis (ADEDIPE et al. 1971) and the activity of RuBP carboxylase (TREHARNE et al. 1970). Cytokinins seem to be root-synthesized (KULAEVA 1962, KENDE 1965, BURROWS and CARR 1969). Root removal will, amongst other things, slow growth (CROZIER and REID 1971) lower cytokinin production (RICHARDS and ROWE 1977), and also reduce RuBP carboxylase production (FEIERABEND 1969). In addition FEIERABEND showed that exogenous cytokinin would overcome the effects of root removal, which agrees with the findings of HUMPHRIES and THORNE (1964) and THORNE and EVANS (1964), who showed that the size of the root systems may control photosynthetic rate and shoot growth (RICHARDS and ROWE 1977). When all the evidence is considered, CARMI and KOLLER's (1978) conclusion that root-produced cytokinins are one of the factors that control the rate of photosynthesis, seems reasonable.

Thus the effects of stress on roots might in part influence photosynthetic rate by alterations in cytokinin production and export to shoots. Data (SIVAKUMARAN and HALL 1979, AHARONI et al. 1977) showing that, after water stress is relieved, cytokinin and GA levels tend to return to normal, support the idea that some of the effects of stress could be related to the action of these two hormones.

There has only been limited work investigating the role of GA in stress and photosynthesis. Net assimilation rate is increased after application of GA (COULOMBE and PAQUIN 1959, ALVIM 1960b). As described in Section 2.5.1, the effects of GA on stomatal functioning are unclear, however GA_3 partially overcomes the negative effects of salt stress on photosynthesis and carbohydrate metabolism (HUBER and SANKHLA 1973, HUBER et al. 1974, STRACK et al. 1975). GA_3 will also promote RuBP carboxylase activity (TREHARNE and STODDART 1968) but CCC had both promotive and inhibitory effects, depending upon its concentration and other conditions. Although CCC can effectively inhibit GA biosynthesis, these effects of CCC are not altogether unexpected, in view of the action of this compound in other systems in which it can sometimes promote GA levels or affect the permeability of tissues (REID and CROZIER

1970, FABIJAN et al. 1981a). Although the work of TREHARNE et al. (1970) confirms the stimulatory influence of GA on RuBP carboxylase, the hormone did not seem to affect the synthesis of the enzyme.

It would be interesting to observe the effects of GA on various photosynthetic processes in GA-deficient mutants. As yet there is no unequivocal evidence of a rapid and direct effect of endogenous GA on photosynthesis.

There are no convincing data showing a close relationship between ethylene and auxins and any process concerned with CO_2 fixation.

In summary, research investigating the role for hormones in CO_2 assimilation is still in its early stages, and most of this work has been concerned with experiments using exogenous hormones. Much more work now needs to be done in which correlations are made between endogenous levels and metabolism of hormones (using mutants with lesions in hormone biosynthesis) and the numerous metabolic processes involved in carbon balance in the stressed plant. Applying large amounts of any growth regulator to a stressed plant might merely demonstrate the pharmacological effects of that substance.

2.7.7 Osmoregulation

Plants that are drought- or salt-stressed often accumulate high levels of certain ions and organic compounds, thus lowering their Ψ_π (osmoregulation). Ions are accumulated from exogenous sources while the organic compounds are synthesized in vivo. Depending upon the species, drought can cause a remarkably large build-up of glycine betaine (HANSON and NELSON 1978) and/or proline (KEMBLE and MACPHERSON 1954, SINGH et al. 1973a and b); halophytes show a similar response (STEWART and LEE 1974, STOREY and WYN JONES 1975).

Other compounds may also have some involvement in osmoregulation. CUTLER and RAINS (1978) and CUTLER et al. (1977) have found evidence for a build-up of organic acids and sugars in water-stressed cotton plants.

The rates of carbohydrate metabolism, utilization, and transport of stressed plants are all important in the control of overall tissue Ψ_π (TURNER 1975, BROWN et al. 1976, HSIAO et al. 1976, JONES and TURNER 1978, ACKERSON 1981, ACKERSON and HEBERT 1981, MEYER and BOYER 1981). Structural changes in the leaf might also be a determinant of leaf–water relations (CUTLER and RAINS 1978).

In halophytes it seems clear that the uptake of ions and the generation of organic compounds has an important role in allowing the plant to decrease its Ψ below that of the soil. Osmoregulation may also be useful in maintaining root and shoot elongation (Sect. 2.7.2). This ability to take up water in saline soils is essential for survival but occurs at considerable metabolic cost. In *Triglochin maritima*, over 10% of the shoot dry weight can be free proline (STEWART and LEE 1974). This represents a large amount of nitrogen not available for protein synthesis, and may have negative effects on growth. On the other hand, high levels of proline and glycine betaine do not inhibit the activities of certain enzymes as much as does NaCl. This suggests that it is advantageous for a halophyte to adjust its Ψ_π with these types of compatible solutes rather than a toxic one (STEWART and LEE 1974, OSMOND 1978, WYN JONES et al. 1977).

Discussion of whether or not the ability to accumulate proline and betaines is advantageous to drought-stressed as opposed to salt-stressed plants other than to remove excess toxic nitrogenous substances and reducing power has occurred often (STEWART and HANSON 1980, ASPINALL and PALEG 1981, WYN JONES and STOREY 1981). The verdict is far from in. Even though we do not know the precise utility and mode of action of all the compounds that accumulate in stressful conditions, it seems likely that they are useful to the plant. What role do hormones have in their accumulation or subsequent metabolism?

ABA promotes proline accumulation in a few species (ASPINALL et al. 1973, HUBER 1974, EDER and HUBER 1977, RAJAGOPAL and ANDERSEN 1978), but not in all (WAMPLE and BEWLEY 1975, ASPINALL and PALEG 1981). The build-up of other osmotica is also stimulated by treatment with ABA; osmotic stress and ABA cause an increase in glycine betaine in *Pennisetum typhoides* and ABA raises the concentration of amino acids in *Lemna* (ANDRES and SMITH 1976). The role of stress-induced changes of endogenous ABA is uncertain, since all the above work has been carried out with exogenous ABA. It would be interesting to follow the detailed biochemical kinetics of ABA, proline, and glycine betaine metabolism simultaneously during a period of stress and recovery and determine the locus of hormone action (i.e., synthesis, turnover, or some more general effect on amino acid metabolism).

There are few reports suggesting the involvement of other hormones in proline metabolism. In salt-stressed *P. typhoides,* GA tended to overcome the effects of NaCl stress on various enzymes of amino acid metabolism (HUBER and SANKHLA 1973) and in wheat, GA treatment will counteract the effects of salinity on α-amylase activity (SARIN and NARAYANA 1968). GA_3 treatment reduces proline levels in barley while CCC increases them (SINGH et al. 1973a). CCC also tends to protect against drought injury (HALEVY and KESSLER 1963). However, drought (SIMON 1974) and CCC can also increase membrane leakiness (FABIJAN et al. 1981a), and effect synthesis of membrane components (DOUGLAS and PALEG 1974). Nevertheless, in view of the effects of (1) GA on starch hydrolysis in barley seeds (PALEG 1960, YOMO 1960), (2) the observation that osmotic stress has a negative feed-back effect on this process (JONES and ARMSTRONG 1971), and (3) that GA enhances production of invertase synthetase (EDELMAN and HALL 1964, KAUFMANN 1968, GAYLER and GLASZIOU 1969), it is tempting to hint at a possible role of GA in osmoregulation by sugar accumulation.

Ethylene treatment (using ethrel) causes proline accumulation (EDER et al. 1977) and kinetin has the opposite effect (EDER and HUBER 1977). Although such reports are intriguing, much more work needs to be done before a convincing case can be made to relate stressed-induced hormone changes and proline or betaine metabolism.

3 Excess Water

In addition to drought stress, another form of stress less frequently studied, but of considerable importance, is excess water or flooding. Flooding or waterlogging can occur during partial or total submersion of plants in running-aerated water or stagnant-deoxygenated water. The studies that will be reviewed encompass most of the possible types of flooding and range from natural situations to laboratory-controlled experiments.

Economically, flooding may have a considerable impact on both the local and national level. An example is the extensive flooding of the Missouri and Mississippi River delta regions of the United States which destroyed a major portion of the U.S. soybean crop (KENNEDY and KRINARD 1974). Because water is the predominant limiting factor in the cultivation of land, irrigation has become a common practive in many countries. However, irrigation has not been without its detrimental effects. In Western Canada it is estimated that well over 60,000 acres of land are permanently waterlogged and highly saline due to leakage from irrigation canals (Alberta Inst Agric 1971). This leakage is known to reduce crop yields (HAGAN et al. 1967). It has also been shown that partial soil saturation or high water table levels will reduce productivity of crops such as soybean, millet, corn, and cabbage by as much as 50–75% (WILLIAMSON and KRIZ 1970). In some cases as little as 24 h of this condition is enough to cause such crop losses (KRAMER 1969, WILLIAMSON and KRIZ 1970). Clearly, whether flooding results from either natural or man-caused conditions it may result in significant economic losses.

3.1 Morphological Effect of Flooding

Morphologically, the extent and type of response of plants to flooding is species-specific, such as the formation of aerenchyma in stems and roots of several species (BROWN 1947, EVANS and ELBERT 1960, BARBER et al. 1962, COULT 1964, GREENWOOD 1967). The formation of knees or knee-roots by *Taxodium* has been associated with a survival mechanism during flooding, but has also been seen on trees growing in well-drained soils (DEN UYL 1961). Pneumatophores or negatively geotrophic roots produced by *Anthocleista nobilis* and *Xylopia stadtii* and other species were originally thought to function only in gas exchange but have recently been shown to contribute photosynthetically (GILL 1970), and might therefore supply the roots with oxygen, produced by photosynthesis, for respiration.

The more often observed morphological effects of flooding are: reduced shoot elongation, chlorosis, leaf epinasty, accelerated senescence, and abscission of leaves, hypertrophy of the hypocotyl and stem, adventitious root formation, lenticel formation, decreased dry weight accumulation, reduced root elongation, suppression of root hair formation, and death of part or all of the original root system. These effects of flooding have been reported for woody plants (HUNT 1951, DEN UYL 1961, MCALPINE 1961, MINORE 1968, HOOK et al. 1971, MCMINN and MCNABB 1971, HOOK and BROWN 1973, CRAWFORD 1974, BUMBIERIS 1976, CLEMENS and PEARSON 1977), while the flooding tolerance of woody species has been reviewed by GILL (1974). There are also numerous reports of these effects occurring in flooded herbaceous plants (SARTORIS and BELCHER 1949, JACKSON 1955, 1956a, b, FULTON and ERICKSON 1964, PHILLIPS 1964a, b, BURROWS and CARR 1969, CRAWFORD and TYLER 1969, REID et al. 1969, YEE et al. 1969, BOLTON and ERICKSON 1970, HEINRICHS 1970, WILLIAMSON 1970,

FORSYTHE and PINGHINAT 1971, PLACE et al. 1971, REID and CROZIER 1971, KAWASE 1972a, 1974, SELMAN and SANDANAM 1972, WRIGHT and HIRON 1972, CARCIA-NOVA and CRAWFORD 1973, RAILTON and REID 1973, REID and RAILTON 1974a, JACKSON and CAMPBELL 1975a, b, WAMPLE and REID 1975, 1978, 1979, DUTHION and MINGEAU 1976).

3.2 Causes of Flood-Induced Morphological Changes

There are a number of reasons why flooding will indirectly have many effects on plant growth. There are physical effects of water on soil texture and leaching of nutrients. More importantly the reduced level of oxygen in the flood water will inhibit processes dependent upon that gas, resulting in changes in the soil ecosystem such as reductive rather than an oxidative soil chemistry, and replacement of aerobic microflora and fauna by anaerobes. There will also be very direct effects of anoxia and leaching on the metabolism of the submerged root tissue itself. Various aspects of these influences have been reviewed by a number of authors (BERGMAN 1959, GRABLE 1966, KRAMER 1969, GILL 1970, REID 1977, HOOK and CRAWFORD 1978, LEVITT 1980).

The most obvious way in which long-term flooding affects plant growth appears to be through a reduction in oxygen availability for respiration and electron transport (KRAMER 1969); the primary consequences of this being a build-up in the products of anaerobic respiration, lack of ATP and a reduction of root growth (see review of HOOK and CRAWFORD 1978). However, most of these metabolic symptoms of anoxia only appear rather slowly and do not satisfactorily explain the rapid appearance of phenomena such as adventitious root primordia and stem hypertrophy (WAMPLE and REID 1975, 1978, FABIJAN et al. 1981b), or wilting (KRAMER 1969).

There might, however, be other more subtle and rapid mechanisms by which growth is affected by flooding. These could start to act shortly after the inundation with water, but before oxygen content falls to a level low enough to inhibit overall electron transport or other major oxygen-requiring processes. Such mechanisms might act as messengers of oncoming anoxia, thus allowing the plant to initiate procedures that might in some cases prevent flood-induced damage. For instance, production of adventitious roots, aerenchyma and stem hypertrophy might allow for an increasing movement of gases such as oxygen and ethylene (JACKSON 1955, KAWASE and WHITMOYER 1980, BLAKE and REID 1981). Leaf and petiole epinasty could reduce heat load and help alleviate subsequent water stress (BRADFORD and YANG 1981). Some of the early warning signals which produce these effects could be: (1) increases in ethylene (KAWASE 1971, JACKSON and CAMPBELL 1975a, WAMPLE and REID 1979, see Sect. 3.7 for a discussion on the role of oxygen in this process). (2) leaching and the movement of certain ions from roots (DRAKEFORD and REID 1984). (3) an unidentified process sensitive to only slight reductions in oxygen level around the roots.

Quite apart from the consideration above regarding the relative importance of the various environmental factors that initially produce responses in the plant, the appearance of a flooded plant gives obvious clues as to some of the links between the perception of the flood water and the eventual effects on growth and development. The idea that some of these links are hormonal in nature comes from the known effects of hormones on one hand, and the appearance of a flooded plant on the other. Put simplistically, a flooded plant looks as if it has undergone increases in levels of ethylene, abscisic acid and auxin and decreases in GA and cytokinins. In fact, many years ago WENT (1928, 1938) suggested that some of the inhibitory effects of flooding on growth might be due to a reduction in the export of hormonal-like substances to the shoots. This concept was subsequently supported by the work of JACKSON (1956a, b), who showed that the export of metabolic toxins by flooded roots was not the cause of the perturbation in growth. The first evidence that flooding might cause change in certain hormones was reported by PHILLIPS (1964a, b).

With the exception of ethylene, the hormonal responses of woody plant species to flooding has not, to our knowledge, been examined in much detail. On the other hand, the hormonal responses of herbaceous plants have been studied more and will be reviewed here on the basis of five classes of plant hormones (cytokinins, ABA, GA, auxins, ethylene).

3.3 Cytokinins

BURROWS and CARR (1969) reported a reduction in the level of cytokinin-like activity in the bleeding sap of flooded sunflower plants which recovered if flooding was terminated within 72 h. As determined by a comparison of detached leaves from control and flooded plants, these authors suggested that is was not just a reduction in cytokinins and nutrients, supplied by the root to the shoots, which caused increased chlorosis. Hence, they proposed either the production of one or more compounds which stimulated chlorophyll degradation or that the cytokinins were being preferentially translocated to younger leaves, resulting in accelerated chlorosis of older leaves.

Most subsequent reports concerning cytokinins and flooding have dealt with the application of cytokinin-like substances to flooded plants in an attempt to maintain these plants in an approximately nonflooded state of development (SELMAN and SANDANAM 1972, RAILTON and REID 1973). This work demonstrated that foliar application of benzyladenine (BA) to tomato plants reduced chlorosis, petiole epinasty, and adventitious root formation and restored shoot growth to near-normal. The minor differences between these two studies may reflect different cultural techniques (hydroponics versus soil) or cultivars used. REID and RAILTON (1974b) extended this work to show that BA applications could restore the endogenous level of gibberellins in flooded tomato plants to control levels.

However, BA applied to flooded sunflower plants did not prevent hypocotyl hypertrophy, nor did it restore shoot growth to control levels (WAMPLE and REID 1979). Based on the rather small reductions in cytokinin-like activity in extracts of bleeding sap of flooded sunflower plants during the first 48 h (BURROWS and CARR 1969) and the rather rapid and irreversible effects of flooding during the first 48 h (WAMPLE and REID 1975, 1978), it seems clear that other factors (hormonal and/or metabolic) are involved. JACKSON and CAMPBELL (1975b) support this hypothesis by two different experiments in which they demonstrate (1) a lack of similarity between flooded and nonflooded plants which have had all or a portion of their roots removed, thus removing the source of cytokinins, and (2) the high level of sensitivity of plants with healthy, nonflooded roots to low levels (0.04 ppm) of ethylene and subsequent development of characteristics similar to flooding. Recently, BRADFORD (1983b) has shown that leaf conductance and photosynthetic rates, which are reduced in flooded tomato plants (BRADFORD 1983a) could be maintained at control levels by BA treatment. There was no attempt to determine if this treatment influenced the endogenous level of other hormones. These findings do not rule out the long-term influence of reduced levels of cytokinins in flooded plants on development, but suggest that with improvements in the techniques for extraction, purification, and quantification of cytokinins that further work will provide us with a better understanding of the role they play in flooded plants.

3.4 Abscisic Acid

It is generally accepted that plants subjected to stress of various forms will show changes, frequently increases, in the levels of endogenous ABA. Such increases are most commonly associated with droughted plants and are discussed elsewhere in this chapter.

It seems reasonable to relate changes in ABA levels in flooded plants to leaf chlorosis and senescence. There are several reports concerning changes in ABA in flooded plants. WRIGHT and HIRON (1972) found a fivefold increase in extractable ABA equivalents in leaves of flooded dwarf bean plants (*Phaseolus vulgaris*) after 5 days. Measurements of ABA levels before or after this time period were not made. SIVAKUMARAN and HALL (1978a, b) reported nearly a ninefold increase in ABA levels extracted from leaves of young (50 to 104 d) *Euphorbia lathyris* after 4 days of flooding while older plants (150 d) showed a slower, less dramatic increase (twofold increase after 16 days). They also showed a decrease in ABA in the younger plants after the apparent peak at 4 days while the older (150 d) plants showed no change in extractable ABA levels up to 32 days. Both reports suggest the increased ABA levels are the result of decreased leaf water content due to restricted water uptake by the flooded root system. However, it would appear that this needs further investigation, since the data of SIVAKUMARAN and HALL (1978a) do not show a consistent increase in water saturation deficit until 8 days after the start of flooding, while the peak in ABA levels occur after 4 days.

The source of the ABA seems to be the younger leaves of stressed plants, although the ability to produce ABA is present regardless of leaf age (SIVAKUMARAN and HALL 1978a). This does not mean that older plants are necessarily less affected by flooding, and this is illustrated by their high rates of leaf abscission under flooded conditions.

Although ABA has been found in root tissue, exudates and xylem sap (TORREY 1976), there is as yet no conclusive work that links the rapid increase of ABA in leaves to ABA synthesized in the roots. Furthermore, this seems even less likely in flooded plants which show root tip damage soon after the onset of flooding.

The role of ABA in flooded plants is not fully understood, as is the case for most hormones, but it is suggested by several authors to be responsible for increases in stomatal resistance (SIVAKUMARAN and HALL 1978, WRIGHT and HIRON 1972). Although JACKSON et al. (1978) have reported increased stomatal resistance in flooded tomato plants, these plants had high leaf water potentials at the time of stomatal closure. These authors therefore suggest that it is unlikely that closure results from increased ABA, although the levels of this hormone were not determined. JACKSON (1979) has also showed a rapid decrease in transpiration rate of flooded peas (*Pisum sativum*) which may have resulted from stomatal closure. ABA levels were not determined and the question therefore still remains open.

Preliminary work with sunflowers by one of us (RLW) has not revealed any increase in diffusive resistance of flooded plants while ABA levels appear to be increasing. Droughted sunflower plants showed a significant increase in diffusive resistance and endogenous free ABA levels. Leaves from flooded sunflower plants did respond to exogenous ABA when either applied as a spray or taken up through petioles. BRADFORD (1983b) reported the ABA treatment of nonflooded tomato plant resulted in a decrease in leaf conductance but had no effect on photosynthetic rates. BLAKE and REID (1981) did not find any evidence for a significant level of flood-induced water stress in three species of *Eucalyptus*, and although flooding did cause increases in diffusive resistance, there were no differences between flood-tolerant and flood-susceptible species. Obviously much more experimental work is needed before a convincing link could be made between flood-induced stomatal closure and flood-induced increases in ABA.

Some work on the effect of exogenous ABA treatment of flooded plants has been performed with mixed results (WAMPLE 1976). ABA stimulation of leaf epinasty in flooded sunflowers was unexpected, but may be caused by ABA stimulation of ethylene biosynthesis (GERTMAN and FUCHS 1972). ABA also reduced the level of chlorosis in flooded plants, a response which might be explained by the report that high levels of ABA inhibit pigment loss (COLQUHOUN and HILLMAN 1972). The mode of action of ABA might also involve reduced synthesis of degradative enzymes (OSBORNE 1968) or inhibition of cytokinin degradation (BLACK et al. 1972). Applied ABA has also been reported to increase endogenous GA levels to *Solanum andigena* (RAILTON and WAREING 1973; see also comments in Sect. 2.5).

It is clear that the influence of flooding on endogenous ABA levels requires further study before an accurate interpretation can be made of its contribution to the morphological responses of plants to flooding.

3.5 Gibberellins

WENT (1938a, b, 1943) postulated that shoot growth was retarded when inadequate aeration or removal of roots reduced the export of "caulocaline", a substance with gibberellin-like activity, now thought to be produced by roots (as well as other loci) and translocated in xylem to the shoots (CARR et al. 1964, PHILLIPS and JONES 1964). However, cytokinins are also translocated in the xylem (BURROWS and CARR 1969) and the possibility of ABA (HARTUNG 1977) and ethylene (JACKSON and CAMPBELL 1976) or its precursor 1-amino cyclopropane carboxylic acid (ACC) (BRADFORD and YANG 1980) also being found in xylem fluid would indicate that "caulocaline" is more complex than originally thought.

REID et al. (1969) showed that flooding caused a reduction in the GA-like activity in xylem sap of tomato plants and subsequently found that GA-like activity was reduced in roots and shoots (REID and CROZIER 1971). The latter authors reported that exogenous GA_3 could stimulate growth to a greater degree in flooded than in nonflooded plants, suggesting that flooded plants are deficient in GA. Similar results were obtained with GA_3 applications to flooded tomato (SELMAN and SANDANAM 1972) and sunflowers (WAMPLE 1976). The ability of exogenous cytokinin to overcome many of the effects of flooding (SELMAN and SANDANAM 1972, RAILTON and REID 1973), including a maintenance of endogenous GA-like activity (REID and RAILTON 1974a, b) might suggest that the level of endogenous cytokinin is the key factor, any elevation of which might alleviate most of the harmful effects of flooding. However, the addition of exogenous cytokinin could merely exert a pharmacological effect, not related to the actions of endogenous cytokinins. In any case, the limited published work shows that in flooded plants the decrease in the level of GA's precedes the decrease in cytokinins (BURROWS and CARR 1969, REID and CROZIER 1971), and therefore suggests that endogenous cytokinins do not control GA levels in flooded plants.

Adventitious roots formed by flooded plants can partially overcome some of the morphological effects of flooding (BERGMAN 1920, JACKSON 1955). Adventitious roots are also known to increase GA levels in willow (*Salix fragilis*) cuttings (MICHNIEWICZ et al. 1970). REID and CROZIER (1971) suggested a correlation between adventitious root formation and increasing gibberellin levels in flooded tomato shoots, but since there was no increase in the level of GA activity in the xylem sap, there may be yet another explanation. One possibility is the synthesis and export of GA-precursors by these new roots which would not be detected.

The disruption of GA metabolism in root apices damaged by flooding (CROZIER and REID 1971, 1972) could be a cause of reduced shoot growth. It seems clear that gibberellins play a role in plant growth and development, but whether GA produced in roots contributes significantly to this control is yet to be determined.

3.6 Auxins

Many of the symptoms exhibited by flooded plants, especially adventitious root formation and epinasty, are similar to those of plants treated with high levels of auxins (for a detailed examination of epinasty see PALMER, Chap. 5, this Vol.). KEFFORD (1962) attributed the accumulation of auxins in poorly aerated rice seedlings to high levels of phenolic inhibitors of IAA-oxidase. It has also been reported that anaerobic conditions result in reduced auxin transport in coleoptiles of other monocotyledonous plants (GOLDSMITH 1968, WILKINS and WHYTE 1968). Hence two mechanisms may be responsible for higher auxin levels in flooded plants.

KRAMER (1969) and later PHILLIPS (1964a) hypothesized that a lack of oxygen due to flooding might result in reduced auxin transport. PHILLIPS (1964b) reported that flooded sunflowers (*Helianthus annuus*) had higher endogenous auxin levels than control plants. However, the technique used did not permit him to rule out increased synthesis. Later reports (WAMPLE and REID 1978, 1979) provide evidence that the build-up of the "active" form of auxin in flooded sunflower plants is caused by both reduced transport and reduced metabolism. Endogenous auxin level of hypocotyls were examined after 24 h of flooding and were found to be significantly higher in flooded plants. Exogenous ^{14}C-IAA applied to leaves of control and flooded plants showed no difference between treatments in the level of radioactivity in the hypocotyls after 12 h. However, after 24 h, flooded plants had significantly higher levels than controls. In addition, nearly all of the radioactivity in flooded plants chromatographed with authentic IAA, while control plants had at least 50% of the radioactivity in other forms. These studies indicate that both reduced transport and reduced metabolism are contributing to higher levels of auxins in shoots of flooded plants. The mechanisms responsible are not well defined. Although high levels of phenolics may be responsible for reduced auxin catabolism, no definitive work on the effects of flooding on phenolic levels is available.

3.7 Ethylene and Its Interaction with Auxins

Ethylene production by plants is well established and for the wide range of effects of this hormone on plants, the reader is referred to ABELES (1973). The flood-induced morphological changes most commonly associated with increases in ethylene levels are: reduced shoot growth; stem, hypocotyl and/or nodal swelling (hypertrophy); leaf epinasty; lenticel formation; loss of chlorophyll; adventitious root development.

In his early papers KAWASE (1972a) reported in increased extractable ethylene levels in several different flooded species including both herbaceous and woody plants. Prior to this, he had shown that centrifugation of *Salix fragilis* cuttings in water caused the levels of endogenous ethylene to rise, presumably the result of infiltration of the tissue by water (KAWASE 1971). In subsequent papers, using a combination of ethylene extraction and treatment with ethrel, KAWASE (1972b, c, 1973, 1974) concluded that increase ethylene concentrations play a major role in the morphological changes of flooded plants. Several other groups have since reported increased ethylene content of flooded plants (JACKSON and CAMPBELL 1975a, b, 1976; CLEMENS and PEARSON 1977, CLEMENS et al. 1978, WAMPLE and REID 1979, JACKSON et al. 1978, HUNT et al. 1981, TANG and KOZLOWSKI 1982a, b, NEWSOME et al. 1982, GOMES and KOZLOWSKI 1980, BRADFORD and YANG 1980, BLAKE and REID 1981). JACKSON and CAMPBELL

(1975a, b, 1976) demonstrated that ethylene, either endogenous or supplied as a gas to the roots or directly to the shoots of tomato plants, will cause symptoms similar to those caused by flooding. Their work, using split-root systems, root removal, steam-girdling of the stem, and inhibitors of ethylene action, suggested that ethylene may play a more significant role in the morphological effects of flooding than the reductions in cytokinins and gibberellins from flooded roots discussed earlier. In these and a subsequent report (JACKSON et al. 1978) a root-produced translocatable substance is proposed which results in increased ethylene in the shoots of flooded plants. A very likely answer to this question has been provided by work performed in the laboratory of S.F. YANG. After considerable research ADAMS and YANG (1979) reported on the metabolic pathway for ethylene and identified the immediate precursor of ethylene as 1-aminocyclopropane-1-carboxylic acid (ACC). The conversion of ACC to ethylene was described as an oxygen-requiring step. In the same year LÜRSSEN et al. (1979) also reported ACC to be the immediate precursor of ethylene. These reports were followed by a study of BRADFORD and YANG (1980), which demonstrated the transport of ACC from the roots of flooded tomato plants to the shoots where it was readily converted to ethylene.

It seems clear from these studies that the roots are a source of ethylene in flooded plants, but may not be the only source. It is known that some microorganisms can produce ethylene and that the levels of ethylene production are positively correlated with the organic matter content of the soil (SMITH 1975, SMITH and RUSSELL 1969). The levels of ethylene produced by these organisms are sufficient to influence plant growth (JACKSON and CAMPBELL 1976). It was demonstrated (JACKSON and CAMPBELL 1975a) that exogenous ethylene, possibly produced by microorganisms, can move from the soil, into the roots, and then to the shoots of plants. The movement, measured by leaf epinasty and ^{14}C monitoring, apparently occurred by diffusion through the cortical tissues, since stem-girdling significantly reduced leaf epinasty and extractable ethylene levels above the girdle. ZERONI et al. (1977) questioned the levels of radioactivity in proportion to the increased levels of ethylene found in the stems of plants used by JACKSON and CAMPBELL (1975a). They suggested that some other mechanism, such as disturbance of normal diffusion patterns, was responsible for the higher ethylene levels.

Support for the idea of a translocatable substance produced in oxygen-deficient roots which moves to the shoots, is also provided by experiments in which only a small portion of the roots of sunflower plants were flooded yet caused an increase in the ethylene levels of hypocotyl-stem sections, while the roots did not show a significant rise in ethylene (WAMPLE and REID 1979). Although this does not exclude the possibility of ethylene entry from the soil, experiments using low light intensities and BA treatment show changes in ethylene of root and stem sections which are not consistent with this idea (WAMPLE and REID 1979). These results suggest a direct influence on the rate of ethylene synthesis rather than the accumulation of ethylene due to increased resistance to diffusion by water, as suggested by KAWASE (1976). Furthermore, BRADFORD and DILLEY (1978) were able to measure increased ethylene levels in both root and shoots of tomato plants grown under continuous flow of nitrogen which should facilitate more rapid diffusion. These reports point out that several sources of ethylene may contribute to the elevated levels found in flooded plants and may all contribute to flood-induced morphological changes.

It is not clear why ethylene biosynthesis increases in roots of flooded plants. Although decreased oxygen levels seem to promote prolonged synthesis of ACC (BRADFORD and YANG 1980), low oxygen does not appear to be the cause of the transient increase in ethylene in flooded but well-aerated sunflowers (WAMPLE and REID 1979). Precisely why an oxygen deficiency should cause increased synthesis in roots is not yet apparent.

Elevated auxin levels in flooded plants might contribute to increased ethylene levels (MORGAN and HALL 1962, 1964). However, based on our current information, ethylene levels increase before endogenous auxin levels increase (WAMPLE and REID 1979). However, these measurements were made in the shoots and it seems quite clear that some of the first changes must occur in the roots. What metabolic changes occur initially in the roots and would lead to increased auxin or ethylene levels are yet to be reported. In this regard, it has been shown that under waterlogged conditions the levels of malate, succinate and other organic acids rise (HOOK and CRAWFORD 1978) and the qualitative and quantitative nature of phenolics may also change (BLAKE personal communication). Both classes of compound are known to influence the activity of IAA oxidase (GOLDACRE et al. 1953, LANE and KING 1968, SCHNEIDER and WIGHTMAN 1974), thus providing a mechanism to modify endogenous auxin concentrations. Adventitious root formation in flooded plants and the established promotory effect of auxins on this process, coupled with changes in the endogenous levels of these compounds in flooded plants, seem to fit with the hypothesis that flooding causes reduced degradation of auxins. Phenolic compounds may stimulate adventitious root formation (HESS 1969, JACKSON and HARNEY 1970) and the suggestion by GASPAR et al. (1977) of a relationship between the positive effect of chlorogenic and ferulic acids on adventitious root formation in cuttings of *Pinus radiata* and on the inverse relationship of peroxidase activity and endogenous auxin level would seem to provide further support of this hypothesis.

There is a considerable amount of evidence of auxin–ethylene interaction in the control of plant form. Although it seems clear that ethylene is not directly responsible for adventitious root formation, it may indirectly contribute to its initiation (FABIJAN et al. 1981b) and subsequently speed up its development (MULLINS 1972). Following the observations that flooding stimulates ethylene production, rooting (JACKSON 1955), and auxin accumulation in hypocotyls (PHILLIPS 1964b, WAMPLE and REID 1979), and since ethylene interferes with auxin transport (BURG and BURG 1967, BEYER 1973), WAMPLE and REID (1979) suggested that the flood-induced rise in ethylene levels blocked basipetal auxins transport, causing an increase in auxin concentration in the hypocotyl, which resulted in the promotion of adventitious root formation.

WAMPLE and REID (1979) also suggested that there was an auxin–ethylene interaction with regard to flood-induced hypocotyl hypertrophy. This agrees with the hypothesis of OSBORNE (1974) and associates, which proposes that these hormones interact to loosen and cause a change in the orientation of cell-wall microfibrils in response to ethylene. This restricts longitudinal cell enlargement but in the presence of small quantities of auxin, which increases cellulase activity, the cross-linking bonds of the polysaccharide matrix in which the microfibrils are embedded are prevented from forming or are loosened. This allows for the radial expansion of these cells. RIDGE (1973) has also reported the longitudinal orientation of microfibrils in response to ethylene. Increased cellulase activity in flooded sunflower hypocotyls has been reported by KAWASE (1979), but not until 48 h after the start of flooding. This seems slow, since hypocotyl hypertrophy can be measured after 24 h of flooding (WAMPLE and REID 1978). However, the general pattern of events seems to fit the overall hypothesis for explaining the development of hypocotyl hypertrophy.

Another effect of flooding that may be regulated by an auxin–ethylene interaction is petiole epinasty. Several papers have been published which correlate epinasty with unequal distribution of auxin between the ad- and abaxial portions of the petiole (LYON 1963, 1970). It is clear that ethylene is capable of causing epinasty and in some reports there is also a strong relationship with auxin transport (for full discussion see PALMER, Chap. 5, this Vol.). A further complication is the ability of exogenous cytokinin to reduce or prevent leaf epinasty in flooded plants (RAILTON and REID 1973) but does not reduce the levels of endogenous ethylene (WAMPLE and REID 1979). This effect of cytokinin may be the result of reduced auxin transport, prevention of unequal distribution of auxin or by reducing the ability of petiole tissue to respond to auxin or ethylene. Such conflicting or confusing reports suggest only that the control of leaf epinasty is still unresolved. However, the observation that silver ions, which antagonize ethylene action (BEYER 1976), prevent flood-induced epinasty in tomato plants supports the idea that ethylene is intimately involved in the epinastic response (BRADFORD and DILLEY 1978).

Leaf epinasty has been proposed as a mechanism to reduce the heat load on a leaf by increasing the angle of the blade with respect to the incident radiation. This reaction might therefore be beneficial to flooded plants by reducing transpiration and water stress. The ability of ethylene to reduce water loss by causing stomatal closure has also been examined. Flooded tomato plants are known to have increased ethylene levels (BRADFORD and DILLEY 1978, and others) and to have reduced leaf conductances (BRADFORD and HSIAO 1982). However, on two separate occasions, treatment with ACC (BRADFORD and HSIAO 1982) or with ethylene gas (BRADFORD 1973b) had no effect on leaf conductance. Similarly JACKSON and KOWALENSKA (1983) found no effect of ACC on leaf conductance or Ψ_1 of peas. Although flooded sunflowers are known to have increased ethylene and leaf epinasty (KAWASE 1972a), and this may occur as early as 8 to 12 h after the start of flooding (WAMPLE and REID 1979), these plants do not show reduction in either leaf conductance or leaf water potential (WAMPLE, THORNTON to be published) which further supports the reports mentioned above that ethylene or ACC do not control these plant responses to flooding. Any attempt to interpret these findings is complicated by the lack of uniformity of plant water relations of flooded plants. Although the majority of the reports show an increase in leaf resistance, leaf water potential is not always reduced and in fact, several reports show increased leaf water potentials in flooded plants (JACKSON et al. 1978, SIVIKUMARIN and HALL 1978a,

BRADFORD and HSIAO 1982, TANG and KOZLOWSKI 1982a, JACKSON and KOWALEWSKA 1983).

3.8 Hormones and Photosynthate Transport and Partitioning

It is apparent from the previous sections that there are a number of changes in endogenous hormone levels in flooded plants. The possibility exists that some of these changes might influence the translocation and partitioning of photosynthates. Only a brief review will be presented here and the reader is referred to several other reviews for a more indepth look at the subject of hormone-directed transport (MARRÈ 1982, PATRICK 1976, GIAQUINTA 1983).

The number of reports which have examined the effect of flooding on transport and partitioning are few. In a general review, KOSMAKOVA and ZVEREVA (1973) stated that flooding reduced the transport of photoassimilates to roots and reproductive structures. LUXMOORE et al. (1973) showed that brief (10–15 days) periods of flooding at (15°–17° C) soil temperatures had no effect on yield but when flooding was extended to 20 to 30 days or the soil temperature was raised to 25° C that yield was reduced by approximately 20 and 70% respectively. In a field experiment these authors showed that the reduction in yield was due to a reduction in seed weight and not to fewer seed per head. A comparison of flooding and ice encasement of four cereal cultivars showed that flooding caused less ethanol accumulation and reduction in nonstructural carbohydrates than ice encasement (MCKERSIE et al. 1982). This would seem to support the findings of LUXMOORE et al. (1973). In 10-day-old corn plants, flooding caused an increase in the amount of carbohydrate (^{14}C-sucrose) found at the base of the stem and increased the percentage of carbon that was incorporated into structural components (GRINEVA and NECHIPORENKO 1977). There was also an increase in the fraction of carbon found in amino acids and this was associated with increased adventitious root formation and growth. In alfalfa (*Medicago sativa* L.) and birdsfoot trefoil (*Lotus corniculatus* L.) 7 days of flooding caused wilting, loss of chlorophyll, and shoot and root dry weight which was accompanied by reductions in nonstructural carbohydrates (BARTA 1980). These effects were greater in alfalfa than in birdsfoot trefoil, with the latter showing lower levels of ethanol (2.2 $\mu g\ g^{-1}$ fr. wt. versus 14.9 $\mu g\ g^{-1}$ fr. wt.) than alfalfa after 6 days of flooding. Recently, 4-week-old sunflowers have been shown to accumulate high levels of starch in their leaves after only 4 days of flooding (WAMPLE and DAVIS 1983). As yet unpublished work has shown that the starch begins to accumulate after only 24 h of flooding and seems to reach a maximum after 72 or 96 h (RLW recent findings). Unfortunately, none of these studies has been accompanied by an in-depth study of endogenous hormone changes, and it is therefore speculative to consider the role such changes might play in carbohydrate transport and metabolism. Besides this, our knowledge of how hormones influence such factors is meager anyway, as is pointed out by the reviewers cited above. Perhaps the best evidence to date is that provided by VREUGDENHIL (1983) and ROBINSON (1983), both of whom show reduced carbohydrate transport under exogenous or increased endogenous

ABA. VREUGDENHIL showed that pretreatment of leaf discs of four different genera (*Ricinus, Phaseolus, Petunia, Beta*) significantly reduced phloem loading of ^{14}C-labeled sucrose. ROBINSON showed that even mild water stress in flowering *Gladiolus grandiflorus*, caused by lowering the relative humidity, resulted in increased ABA levels in the inflorescence and reduced sink strength (ROBINSON et al. 1983). This was complicated, however, by the fact that application of ABA had no effect on sink strength of unstressed inflorescences. On the other hand, treatment with ethylene gave results very similar to water stress, but no change in endogenous ethylene could be found in stressed plants. As pointed out in Section 3.4, ABA levels have been shown to increase in flooded plants but, as demonstrated by the two papers just cited, the role that such changes might play in controlling carbohydrate transport in flooded plants is not established at present. Perhaps increased ABA levels in flooded plants are the cause of reduced carbon transport due to inhibition of phloem loading, but this remains to be demonstrated.

Even less can be said for the role of the other four classes of hormones with respect to their role in carbohydrate transport and partitioning. Perhaps the best approach is simply to state the need for much more research in this area of plant science for both stressed and nonstressed plants.

3.9 Summary of Section 3

This section has dealt primarily with our limited knowledge of the hormonal changes that occur in flooded plants and has made an attempt to correlate these changes with alterations in morphology associated with flooding. However, it is obvious that detailed examination of flood-induced changes in hormones in a large number of different plants has not been undertaken and that considerable research is still required.

Information on the influence of flooding on plants may be of use to agriculturists, especially as we begin to recognize that for some plants the period of flooding required to influence growth is quite brief. Flooding may also be a very useful experimental technique for examining the relationship between the roots and shoots without the necessity of dissecting the plant.

References

Abeles FB (1972) Biosynthesis and mechanism of action of ethylene. Annu Rev Plant Physiol 23:259–292

Abeles FB (1973) Ethylene in plant biology. Academic Press, New York

Abeles FB, Heggestad HE (1973) Ethylene: an urban air pollutant. J Air Pollut Control Assoc 23:517–521

Acevedo E (1975) The growth of maize (*Zea mays* L.) under field conditions as affected by its water relations. Ph.D. Thesis Univ Davis, Cal

Acevedo E, Hsiao TC, Henderson DW (1971) Immediate and subsequent growth responses of maize leaves to changes in water status. Plant Physiol 48:631–636
Ackerson RC (1980) Stomatal response of cotton to water stress and abscisic acid as affected by water stress history. Plant Physiol 65:455–459
Ackerson RC (1981) Osmoregulation in cotton in response to water stress. II. Leaf carbohydrate status in relation to osmotic adjustment. Plant Physiol 67:489–493
Ackerson RC, Hebert RR (1981) Osmoregulation in cotton in response to water stress. I. Alterations in photosynthesis, leaf conductance, translocation and ultrastructure. Plant Physiol 67:484–488
Adams DO, Yang SF (1979) Ethylene biosynthesis: Identification of 1-aminocyclopropane-1-carboxylic acid as an intermediate in the conversion of methionine to ethylene. Proc Natl Acad Sci USA 76:170–174
Adato I, Gazit S (1974) Water-deficit stress, ethylene production and ripening in avocado fruits. Plant Physiol 53:45–46
Adedipe NO, Hunt LA, Fletcher RA (1971) Effects of benzyladenine on photosynthesis, growth and senescence of the bean plant. Physiol Plant 25:151–153
Aharoni N, Blumenfeld A, Richmond AE (1977) Hormonal activity in detached lettuce leaves as affected by leaf water content. Plant Physiol 59:1169–1173
Alberta Institute of Agrologists (1971) Agriculture and the environment – a handbook. Alta Inst Agrol Edmonton, Alta pp 52–54
Alvim P (1960a) Moisture stress as a requirement for flowering of coffee. Science 132:354
Alvim P (1960b) Net assimilation rate and growth behavior of beans as affected by gibberellic acid, urea and sugar sprays. Plant Physiol 35:285–288
Anderson LWJ (1978) Abscisic acid induces formation of floating leaves in the heterophyllous aquatic angiosperm *Potamogeton nodosus*. Science 201:1135–1138
Andres J, Smith H (1976) Evidence for a rapid effect of abscisic acid on amino acid metabolism in *Lemna*. Plant Sci Lett 6:315–318
Aspinall D (1965) The effects of soil moisture stress on the growth of barley. II. Grain growth. Aust J Agric Res 16:265–275
Aspinall D (1980) Role of abscisic acid and other hormones in adaptation to water stress. In: Turner NC, Kramer PJ (eds) Adaptation of plants to water and high temperature stress. Wiley-Interscience, New York, pp 155–172
Aspinall D, Husain I (1970) The inhibition of flowering by water stress. Aust J Biol Sci 23:925–936
Aspinall D, Paleg LG (1981) Proline accumulation: physiological aspects. In Paleg LG, Aspinall D (eds) The physiology and biochemistry of drought resistance in plants. Academic Press, Sydney, pp 205–241
Aspinall D, Singh TN, Paleg LG (1973) Stress metabolism. V. Abscisic acid and nitrogen metabolism in barley and *Lolium temulentum* L. Aust J Biol Sci 26:319–327
Back A, Bittner S, Richmond AE (1972) The effect of abscisic acid on the metabolism of kinetin in detached leaves of *Rumex pulcher*. J Exp Bot 23:744–750
Barber DA, Elbert M, Evans NTS (1962) The movement of $^{15}O_2$ through barley and rice plants. J Exp Bot 13:397–403
Barley KP (1962) The effects of mechanical stress on the growth of roots. J Exp Bot 13:95–110
Barlow EWR, Munns R, Scott NS, Reisner AH (1977) Water potential, growth and polyribosome content of the stressed wheat apex. J Exp Bot 28:909–916
Barlow EWR, Munns RE, Brady CJ (1980) Drought responses of apical meristems. In: Turner NC, Kramer PJ (eds) Adaptation of plants to water and high temperature stress. Wiley-Interscience, New York, pp 191–205
Barta AL (1980) Regrowth and alcohol dehydrogenase activity in waterlogged alfalfa and birdsfoot trefoil. Agron J 72:1017–1020
Bauer R, Huber W, Sankhla N (1976) Effect of abscisic acid on photosynthesis in *Lemna minor* L. Z Pflanzenphysiol 77:237–246
Beardsell MF, Cohen D (1975) Relationships between leaf water status, abscisic-acid levels, and stomatal resistance in maize and sorghum. Plant Physiol 56:207–212
Begg JE, Turner NC (1976) Crop water deficits. Adv Agron 28:161–217

Bellandi DM, Dörffling K (1974a) Transport of abscisic acid-2-C-14 in intact pea seedlings. Physiol Plant 32:365–368

Bellandi DM, Dörffling K (1974b) Effect of abscisic acid and other plant hormones on growth of apical and lateral buds of seedlings. Physiol Plant 32:369–372

Bengtson C, Falk SO, Larsson S (1977) The after-effect of water stress on transpiration rate and changes in abscisic acid content of young wheat plants. Physiol Plant 41:149–154

Ben-Yehoshua S, Alono B (1974) Effect of water stress on ethylene production by detached leaves of Valencia orange (*Citrus sinensis* Osbeck). Plant Physiol 53:863–865

Benzioni A, Mizrahi Y, Richmond AE (1974) Effect of kinetin on plant response to salinity. New Phytol 73:315–319

Bergman HF (1920) The relation of aeration to the growth and activity of roots and its influence on the ecesis of plants in swamps. Ann Bot 34:13–33

Bergman HF (1959) Oxygen deficiency as a cause of disease in plants. Bot Rev 25:417–485

Beyer EM (1973) Abscission support for a role of ethylene modification of auxin transport. Plant Physiol 52:1–5

Beyer EM (1976) A potent inhibitor of ethylene action in plants. Plant Physiol 58:2268–2271

Beyer EM (1978) Abscission: The initial effect of ethylene is in the leaf blade. Plant Physiol 55:322–327

Beyer EM, Morgan PW (1969) Ethylene modification of an auxin pulse in cotton stem sections. Plant Physiol 44:1690–1694

Bilan MV (1960) Root development of loblolly pine seedlings in modified environments. Stephen F Austin State Coll For Bull 4

Biscoe PV, Unsworth MH, Pickney HR (1973) The effects of low concentrations of sulphur dioxide on stomatal behavior in *Vicia faba*. New Phytol 72:1299–1306

Blake J, Ferrell WK (1977) The association between soil and xylem water potential, leaf resistance, and abscisic acid content in droughted seedlings of Douglas-fir (*Pseudotsuga menziesii*). Physiol Plant 39:106–109

Blake TJ, Reid DM (1981) Ethylene, water relations and tolerance to waterlogging of three *Eucalyptus* species. Aust J Plant Physiol 8:497–505

Bolton EF, Erickson AE (1970) Ethanol concentration in tomato plants durings soil flooding. Agron J 62:220–224

Borgstrom G (1973) (1939) in Kgl Fysigr Sellsk Lund Ferh 9, 135, cited on p 148 of Abeles FB

Borohov A, Tirosh T, Halevy AH (1976) Abscisic acid content of senescing petals on cut rose flowers as affected by sucrose and water stress. Plant Physiol 58:175–178

Boussiba S, Richmond AE (1976) Abscisic acid and the after-effect of stress in tobacco plants. Planta 129:217–219

Boyer JS (1968) Relationship of water potential to growth of leaves. Plant Physiol 43:1056–1062

Boyer JS (1970) Leaf enlargement and metabolic rates in corn, soybean, and sunflower at various leaf water potentials. Plant Physiol 46:233–235

Boyer JS (1971) Nonstomatal inhibition of photosynthesis in sunflower at low leaf water potentials and high light intensities. Plant Physiol 48:532–536

Boyer JS (1976) Water deficits and photosynthesis. In: Kozlowski TT (ed) Water deficits and plant growth Vol IV. Academic Press, New York, pp 153–190

Bradbury D, Ennis WB Jr (1952) Stomatal closure in kidney bean plants treated with ammonium 2,4-dichlorophenoxyacetate. Am J Bot 39:324–328

Bradford KJ (1983a) Effects of soil flooding on leaf gas exchange of tomato plants. Plant Physiol 73:475–479

Bradford KJ (1983b) Involvement of plant growth substances in the alteration of leaf gas exchange of flooded plants. Plant Physiol 73:480–483

Bradford KJ, Dilley DR (1978) Effects of root anaerobiosis on ethylene production, epinasty and growth of tomato plants. Plant Physiol 61:506–509

Bradford KJ, Hsiao TC (1982) Stomatal behaviour and water relations of water-logged tomato plants. Plant Physiol 70:1508–1513

Bradford KJ, Yang SF (1980) Xylem transport of 1-aminocyclopropane-1-carboxylic acid, an ethylene precursor, in waterlogged tomato plants. Plant Physiol 65:322–326
Bradford KJ, Yang SF (1981) Physiological responses of plants to waterlogging. Hortic Sci 16:25–30
Brown DCW, Thorpe TA (1980) Changes in water potential and its components during shoot formation in tobacco callus. Physiol Plant 49:83–87
Brown JW (1946) Effect of 2,4-dichlorophenoxyacetic acid on the water relations, the accumulation and distribution of solid matter and the respiration of bean plants. Bot Gaz 107:332–343
Brown R (1947) The gaseous exchange between root and the shoot of the seedling of *Cucurbita pepo*. Ann Bot NS 11:417–437
Brown KW, Jordan WR, Thomas JC (1976) Water stress induced alterations of the stomatal response to decreases in leaf water potential. Physiol Plant 37:1–5
Brown WGE, Lacate DS (1961) Rooting habit of white and red pine. Can Dept For, For Res Note 108
Browning G (1973) Flower bud dormancy in *Coffea arabica* L. II. Relation of cytokinins in xylem sap and flower buds to dormancy-release. J Hortic Sci 48:297–310
Browning G (1974) 2-chloroethanephosphoric acid reduces transpiration and stomatal opening in *Coffea arabica* L. Planta 121:175–179
Bumbieris M (1976) The role of *Phytophthora cryptogea* and waterlogging in a decline of *Pinus radiata*. Aust J Bot 24:703–710
Burg SP, Burg EA (1966) The interaction between auxin and ethylene and its role in plant growth. Proc Natl Acad Sci USA 55:262–269
Burg SP, Burg EA (1967) Inhibition of polar auxin transport by ethylene. Plant Physiol 42:1224–1228
Burg SP, Burg EA (1968) Auxin-stimulated ethylene formation: its relationship to auxin-inhibited growth, root geotropism, and other plant processes. In: Wightman F, Setterfield G (eds) Biochemistry and physiology of plant growth substances. Runge, Ottawa, p 1275
Burrows WJ, Carr DJ (1969) Effects of flooding the root system of sunflower plants on the cytokinin content in the xylem sap. Physiol Plant 22:1105–1112
Butcher DN (1963) The presence of gibberellins in excised tomato roots. J Exp Biol 14:272–280
Bystrom BG, Glater RB, Scott FM, Bowler ESC (1968) Leaf surface of *Beta vulgaris* – electron microscope study. Bot Gaz 129:133–138
Caldwell MM (1976) Root extension and water absorption. In: Lange OL, Kappen L, Schulze E-D (eds) Water and plant life. Springer, Berlin Heidelberg New York, pp 59–85
Carmi A, Koller D (1978) Effects of the roots on the rate of photosynthesis in primary leaves of bean (*Phaseolus vulgaris* L.). Photosynthetica 12:178–184
Carr DJ, Reid DM (1968) The physiological significance of the synthesis of hormones in roots and of their export to the shoot system. In: Wightman F, Setterfield G (eds) Biochemistry and physiology of plant growth substances. Runge, Ottawa, pp 1169–1185
Carr DJ, Reid DM, Skene KGM (1964) The supply of gibberellins from the root to the shoot. Planta 63:382–392
Chadwick AV, Burg SP (1967) An explanation of the inhibition of root growth caused by indole-3-acetic acid. Plant Physiol 42:415–420
Chadwick AV, Burg SP (1970) Regulation of root growth by auxin-ethylene interaction. Plant Physiol 45:192–200
Childers NF, White DG (1942) Influence of submersion of the roots on transpiration, apparent photosynthesis and respiration of young apple trees. Plant Physiol 17:603–618
Chin TY, Beevers L (1970) Changes in endogenous growth regulators in *Nasturtium* leaves during senescence. Planta 92:178–188
Chin T-Y, Meyer MM, Beevers L (1969) Abscisic acid stimulated rooting of stem cuttings. Planta 88:192–196

Clarkson DT, Robards AW, Sanderson J, Peterson CA (1978) Permeability studies on epidermal-hypodermal sleeves isolated from roots of *Allium cepa* (onion). Can J Bot 56:1526–1532

Cleland R (1967) A dual role of turgor pressure in auxin-induced cell elongation in *Avena* coleoptiles. Planta 77:182–191

Clemens J, Pearson CJ (1977) The effect of waterlogging on the growth and ethylene content of *Eucalyptus robusta* SM. (Swamp Mahogany). Oecologia 29:249–255

Clemens J, Kirk A-M, Mills PD (1978) The resistance of waterlogging of three *Eucalyptus* species. Oecologia 34:125–131

Clough BF, Milthorpe FL (1975) Effects of water deficit on leaf development in tobacco. Aust J Plant Physiol 2:291–300

Collins JC, Kerrigan AP (1974) The effect of kinetin and abscisic acid on water and ion transport in isolated maize roots. New Phytol 73:309–314

Colquhoun AJ, Hillman JR (1972) The effects of abscisic acid on senescence in leaf discs of radish, *Raphanus sativus* L. Planta 105:213–224

Cooper WC, Rasmussen GK, Rogers BJ, Reece PC, Henry WH (1968) Control of abscission in agricultural crops and its physiological basis. Plant Physiol 43:1560–1576

Coulombe LJ, Paquin R (1959) Effects de l'acide gibbérellique sur le métabolisme des plantes. Can J Bot 37:897–901

Coult DA (1964) Observations on gas movement in the rhizome of *Menyanthes trifoliata* L. with comments on the role of the endodermis. J Exp Bot 15:205–218

Cracker LE, Abeles FB (1969) Abscission: Role of abscisic acid. Plant Physiol 44:1144–1149

Cram WJ, Pitman MG (1972) The actions of abscisic acid on ion uptake and water flow in plant roots. Aust J Biol Sci 25:1125–1132

Crawford RMM (1974) Tree-root survival under flooding, concrete, traffic and gas leakage. In: Wright TWJ (ed) Tree growth in the landscape. Univ London, London, pp 10–13

Crawford RMM, Tyler PD (1969) Organic acid metabolism in relation to flooding tolerance in roots. J Ecol 57:235–244

Crozier A, Reid DM (1971) Do roots synthesize gibberellins? Can J Bot 49:967–975

Crozier A, Reid DM (1972) Gibberellin metabolism in the roots of *Phaseolus coccineus* seedlings. In: Carr DJ (ed) Plant growth substances 1970. Springer, Berlin Heidelberg New York, pp 414–419

Cummins WR, Kende H, Raschke K (1971) Specificity and reversibility of the rapid stomatal response to abscisic acid. Planta 99:347–351

Currie ME, Weete JD, Peterson CM (1978) Effects of plant hormones on wax synthesis on cotton leaves. Plant Physiol (Suppl) 61:106

Cutler JM, Rains DW (1978) Effects of water stress and hardening on the internal water relations and osmotic constituents of cotton leaves. Physiol Plant 42:261–266

Cutler JM, Rains DW, Loomis RS (1977) The importance of cell size in the water relations of plants. Physiol Plant 40:255–260

Dainty J (1969) The water relations of plants. In: Wilkins MB (ed) The physiology of plant growth and development. McGraw-Hill, London, pp 421–452

Dalton FN, Raats PAC, Gardner WR (1975) Simultaneous uptake of water and solutes by plant roots. Agron J 67:334–339

Damptey HB, Aspinall D, Coombe BG (1978) Water deficit and inflorescence development in *Zea mays* L. The role of the developing tassel. Ann Bot 42:849–854

Darbyshire B (1971a) Changes in indoleacetic acid oxidase activity associated with plant water potential. Physiol Plant 25:80–84

Darbyshire B (1971b) The effect of water stress on indoleacetic acid oxidase in pea plants. Plant Physiol 47:65–67

Darbyshire B, Steer BT (1973) Dehydration of macromolecules. I. Effect of dehydration-rehydration on indoleacetic acid oxidase, ribonuclease, ribulosediphosphate carboxylase, and ketose-1-phosphate aldolase. Aust J Biol Sci 26:591–604

Das VSR, Rao IM, Raghavendra AS (1976) Reversal of abscisic acid induced stomatal closure by benzyl adenine. New Phytol 76:449–452

Davenport TL, Jordan WR, Morgan PW (1977) Movement and endogenous levels of abscisic acid during water-stress-induced abscission in cotton seedlings. Plant Physiol 59:1165–1168

Davenport TL, Morgan PW, Jordan WR (1980) Reduction of auxin transport capacity with age and internal water deficits in cotton petioles. Plant Physiol 65:1023–1025

Davies WJ (1978) Some effects of abscisic acid and water stress on stomata of *Vicia faba* L. J Exp Bot 29:175–182

Davison RM, Young H (1973) Abscisic acid content of xylem sap. Planta 109:95–98

Den Uyl D (1961) Some observation on bald cypress in Indiana. Ecology 42:841–843

Dhindsa RS, Beasley CA, Ting IP (1975) Osmoregulation in cotton fiber. Accumulation of potassium and malate during growth. Plant Physiol 56:394–398

Dittrich P, Raschke K (1977) Malate metabolism in isolated epidermis of *Commelina communis* L. in relation to stomatal functioning. Planta 134:77–81

Dörffling K, Sonka B, Tietz D (1974) Variation and metabolism of abscisic acid in pea seedlings during and after water stress. Planta 121:57–66

Dörffling K, Streich J, Kruse W, Muxfeldt B (1977) Abscisic acid and the after-effect of water stress on stomatal opening potential. Z Pflanzenphysiol 81:43–56

Dörffling K, Tietz D, Streich J, Ludewig M (1980) Studies on the role of abscisic acid in stomatal movements. In: Skoog F (ed) Plant growth substances 1979. Springer, Berlin Heidelberg New York, pp 274–285

Douglas TJ, Paleg LG (1974) Plant growth retardants as inhibitors of sterol biosynthesis in tobacco seedlings. Plant Physiol 54:238–245

Drakeford DR, Reid DM (1984) Changes in the ability of plants to alter the pH of the flooding medium as an early symptom of flooding stress in *Helianthus annuus* Can J Bot 62:2417–2422

Duniway JM, Durbin RD (1971) Detrimental effect of rust infection on the water relations of bean. Plant Physiol 48:69–72

Durley RC, Bewley JD, Railton ID, Pharis RP (1976) Effects of light, abscisic acid, and ^{6}N-benzyladenine on the metabolism of ^{3}H-gibberellin A_4 in seeds and seedlings of lettuce, cv. Grand Rapids. Plant Physiol 57:699–703

Duthion C, Mingeau M (1976) Effects of excess moisture on plant behavior and their consequences. Ann Agron 27(2):221–246

Ebell LF (1967) Cone production induced by drought in potted Douglas-fir. Bi-monthly Res. Note Can Dep For Rural Dev 23:26–27

Edelman J, Hall MA (1964) Effect of growth hormones on the development of invertase associated with cell walls. Nature 201:296–297

Eder A, Huber W (1977) About the effect of abscisic acid and kinetin on biochemical changes in *Pennisetum typhoides* during stress conditions. Z Pflanzenphysiol 84:303–311

Eder A, Huber W, Sankhla N (1977) Interaction between salinity and ethylene in nitrogen metabolism of *Pennisetum typhoides* seedlings. Biochem Physiol Pflanz 171:93–100

El-Antably HMM, Wareing PF (1966) Stimulation of flowering in certain short day plants by abscision. Nature 210:328–329

El-Beltagy AS, Hall MA (1974) Effect of water stress upon endogenous ethylene levels in *Vicia faba*. New Phytol 73:47–60

Eliasson L (1975) Effect of indoleacetic acid on the abscisic acid level in stem tissue. Physiol Plant 34:117–120

Erlandsson G, Pattersson S, Svensson S (1978) Rapid effects of abscisic acid on ion uptake in sunflower roots. Physiol Plant 43:380–384

Evans LT (1969) The induction of flowering; some case histories. MacMillan, Melbourne

Evans NTS, Elbert M (1960) Radioactive oxygen in the study of gas transport down the root of *Vicia faba* L. J Exp Bot 11:246–257

Fabijan D, Taylor JS, Reid DM (1981b) Adventitious rooting in hypocotyls of sunflower (*Helianthus annuus*) II. Action of gibberellins, cytokinins, auxins and ethylene. Physiol Plant 53:589–597

Fabijan DM, Plumb Dhindsa P, Reid DM (1981a) Effects of two growth retardants on tissue permeability in *Pisum sativum* and *Beta vulgaris*. Planta 152:481–486

Feierabend J (1969) Der Einfluß von Cytokininen auf die Bildung von Photosyntheseenzyme in Roggenkeimlingen (Influence of cytokinins on the formation of photosynthetic enzymes in rye seedlings). Planta 84:11–29
Feilding JM (1955) The seasonal and daily elongation of shoots of Monterey pine and the daily elongation of roots. Commonw Aust Timber Bur Leaflet 75
Fenton R, Mansfield TA, Wellburn AR (1976) Effects of isoprenoid alcohols or oxygen exchange of isolated chloroplasts in relation to their possible physiological effects on stomata. J Exp Bot 27:1206–1214
Fenton R, Davies WJ, Mansfield TA (1977) The role of farnesol as a regulator of stomatal opening in *Sorghum*. J Exp Bot 28:1043–1053
Ferri MG, Lex A (1948) Stomatal behavior as influenced by treatment with β-naphthoxyacetic acid. Contr Boyce Thompson Inst Plant Res 15:283–290
Fischer RA, Hsiao TL, Hagan RM (1970) After-effect of water stress on stomatal opening potential. I. Techniques and magnitudes. J Exp Bot 21:371–385
Fiscus EL (1975) The interaction between osmotic- and pressure-induced water flow in plant roots. Plant Physiol 55:917–922
Flowers TJ, Troke PF, Yeo AR (1977) The mechanism of salt tolerance in halophytes. Annu Rev Plant Physiol 28:89–121
Forrester ML, Krotkov G, Nelson CD (1966) Effect of oxygen on photosynthesis, photorespiration and respiration in detached leaves. II. Corn and other monocotyledons. Plant Physiol 41:428–531
Forsythe W, Pinghinat AM (1971) Tolerance of the bean cultivar 27-R to flooding conditions. Turrialba 21:228–231
Fry KE (1972) Inhibition of ferricyanide reduction in chloroplasts prepared from water-stressed cotton leaves. Crop Sci 12:698–701
Fujino M (1967) Role of adenosinetriphosphate and adenosinetriphosphatase in stomatal movement. Sci Bull Fac Edu Nagasaki Univ 18:1–47
Fulton JM, Erickson AE (1964) Relation between soil aeration and ethyl alcohol accumulation in xylem exudate of tomatos. Soil Sci Soc Proc 28:610–614
Funke GL, de Coeyer F, de Decker A (1938) Biol Jahrb 5:335
Gaither DH, Lutz DH, Forrence LE (1975) Abscisic acid stimulates elongation of excised pea root tips. Plant Physiol 55:948–949
Garcia-Nova F, Crawford RMM (1973) Soil aeration, nitrate reduction and flooding tolerance in higher plants. New Phytol 72:1031–1039
Gaspar T, Smith D, Thorpe TA (1977) Arguments supplémentaires en faveur d'une variation inverse du niveau auxinique au cours des deux premières phases de la rhizogenèse. CR Acad Sci Paris, 7t.285 Serie D. pp 327–330
Gates CT (1955) The response of the young tomato plant to a brief period of water shortage. I. The whole plant and its principal parts. Aust J Biol Sci 8:196–214
Gayler KR, Glasziou KT (1969) Plant enzyme synthesis: hormonal regulation of invertase and peroxidase synthesis in sugar cane. Planta 84:185–194
Gertman E, Fuchs Y (1972) Effect of abscisic acid and its interactions with other plant hormones on ethylene production in two plant systems. Plant Physiol 50:194–195
Giaquinta RT (1983) Phloem loading of sucrose. Annu Rev Plant Physiol 34:347–387
Giles KL, Beardsell MF, Cohen D (1974) Cellular and ultrastructural changes in mesophyll and bundle-sheath cells of maize in response to water stress. Plant Physiol 54:208–212
Gill CJ (1970) The flooding tolerance of woody species – a review. For Abstr 31:671–688
Gill CJ (1974) Tree pre-planting on new reservoir margins. Landscape Design 107:41–42
Glinka Z (1973) Abscisic acid effect on root exudation related to increased permeability to water. Plant Physiol 51:217–219
Glinka Z (1977) Effects of abscisic acid and of hydrostatic pressure gradient on water movement through excised sunflower roots. Plant Physiol 59:933–935
Glinka Z (1980) Abscisic acid promotes both volume flow and ion release to the xylem in sunflower roots. Plant Physiol 65:537–540
Glinka Z, Reinhold L (1971) Abscisic acid raises the permeability of plant cells to water. Plant Physiol 48:103–105

Goeschl JD, Pratt HK (1968) Regulatory roles of ethylene in the etiolated growth habit of *Pisum sativum*. In: Wightman F, Setterfield G (eds) Biochemistry and physiology of plant growth substances. Runge, Ottawa, p 1229

Goeschl JD, Rappaport L, Pratt HK (1966) Ethylene as a factor regulating the growth of pea epicotyls subjected to physical stress. Plant Physiol 41:877–884

Goldacre PL, Galston AW, Weintraub RL (1953) The effect of substituted phenols on the activity of the indoleacetic acid oxidase of peas. Arch Biochem Biophys 43:358–373

Goldsmith MHM (1968) Comparison of aerobic and anaerobic movement of 3-indoleacetic acid in coleoptiles of oats and corn. In: Wightman F, Setterfield G (eds) Biochemistry and physiology plant growth substances. Runge, Ottawa, pp 1037–1050

Gomes ARS, Kozlowski TT (1980) Effects of flooding on *Eucalyptus camaldulensis* and *Eucalyptus globulus* seedlings. Oecologia 46:139–142

Grable AR (1966) Soil aeration and plant growth. Adv Agron 18:57–106

Granatek CH, Cockerline AW (1978) Callus formation versus differentiation of cultured barley embryos: hormonal and osmotic interactions. In vitro 14:212–217

Greacen EL, Oh JS (1972) Physics of root growth. Nat New Biol 235:24–25

Greacen EL, Ponsana P, Barley KP (1976) Resistance to water flow in the roots of cereals. In: Lange OL, Kappen L, Schulze E-D (eds) Water and plant life. Springer, Berlin Heidelberg New York, pp 86–100

Greenwood DJ (1967) Studies on the transport of oxygen through the stems and roots of vegetable seedlings. New Phytol 66:337–347

Greenwood MS (1981) Reproductive development in loblolly pine. II. The effect of age, gibberellin plus water stress and out-of-phase dormancy on long shoot growth behavior. Am J Bot 9:1184–1190

Grineva GM, Nechiporenko GA (1977) Distribution and conversion of sucrose-U-C14 in corn plants under conditions of flooding. Sov Plant Physiol 24(1):32–37

Grossenbacher KA (1939) Autonomic cycle of rate of exudation of plants. Am J Bot 26:107–109

Guinn G (1976) Water deficit and ethylene evolution by young cotton bolls. Plant Physiol 57(3):403–405

Guy RD, Reid DM, Krouse HR (1980) Shifts in carbon isotope ratios of two C_3 halophytes under natural and artificial conditions. Oecologia 44:241–247

Hagan RM, Haise HR, Edminster T (1967) Irrigation of agricultural lands. Am Soc Agron, Madison, Wisc

Halevy AM, Kessler B (1963) Increased tolerance of bean plants to soil drought by means of growth-standing substances. Nature 197:310–311

Hall DM, Jones RL (1961) Physiological significance of surface wax on leaves. Nature 191:95–96

Hall MA, Kapuya JA, Sivakumaran S, John A (1977) The role of ethylene in the response of plants to stress. Pest Sci 8:217–223

Hanson AD, Nelson CE (1978) Betaine accumulation and [^{14}C] formate metabolism in water-stressed barley leaves. Plant Physiol 62:305–312

Harrison MA, Walton DC (1975) Abscisic acid metabolism in water-stressed bean leaves. Plant Physiol 56:250–254

Hartung W (1977) Der Transport von (2-^{14}C) Abscisinsäure aus dem Wurzelsystem intakter Bohnenkeimlinge in die oberirdischen Teile der Pflanze. Z Pflanzenphysiol 83:81–84

Hartung W, Fufer C (1981) Abscisic and apical dominance in *Phaseolus coccineus* L., the role of tissue age. Ann Bot 47:371–375

Hartung W, Gimmler H, Heilmann B, Kaiser G (1980) The site of abscisic acid metabolism in mesophyll cell of *Spinacia oleracea*. Plant Sci Lett 18:359–364

Hatrick AA, Bowling DJF (1973) A study of the relationship between root and shoot metabolism. J Exp Bot 24:607–613

Heide OM (1965) Interaction of temperature, auxins and kinins in the regeneration ability of *Begonia* leaf cuttings. Physiol Plant 18:891–920

Heide OM (1968) Stimulation of adventitious bud formation in *Begonia* leaves by abscisic acid. Nature 219:960–961

Heilmann B, Hartung W, Gimmler H (1980) The distribution of abscisic acid between chloroplasts and cytoplasm of leaf cells and the permeability of the chloroplast envelope for abscisic acid. Z Pflanzenphysiol 97:667–678

Heinrichs DH (1970) Flooding tolerance of legumes. Can J Plant Sci 50:435–438

Hellmers H, Horton JS, Juhren G, O'Keefe J (1955) Root systems of some chaparral plants in southern California. Ecology 36:667–678

Hemphill DD, Tukey HB (1975) Effect of intermediate mist on abscisic acid levels in *Evonymus alatus* Sieb.: Leaching vs. moisture stress. Hortscience 10:369–370

Hess CE (1969) Internal and external factors regulating root initiation. In: Whittington WJ (ed) Root growth. Whittington WJ, Buttersworth, London, pp 42–53

Hiron RWP, Wright STC (1973) The role of endogenous abscisic acid in the response of plants to stress. J Exp Bot 24:769–781

Hoad GV (1967) (+)-Abscisin II [(+)-dormin] in phloem exudate of willow. Life Sci 6:1113–1118

Hoad GV (1973) Effect of moisture stress on abscisic acid levels in *Ricinus communis* L. with particular reference to phloem exudate. Planta 113:367–372

Hoad GV (1975) Effect of osmotic stress on abscisic acid levels in xylem sap of sunflower (*Helianthus annuus* L.). Planta 124:25–29

Hong SG, Sucoff E (1976) Effects of kinetin and root tip removal on exudation and potassium (rubidium) transport in root of honey locust. Plant Physiol 57:230–236

Hook DD, Brown CL (1973) Root adaptations and relative flood tolerance of five hardwood species. For Sci 19:225–229

Hook DD, Crawford RMM (1978) Plant life in anaerobic environments. Ann Arbor Sci, Ann Arbor, Mich ISBN 0-250-40197-5

Hook DD, Brown CL, Kormanik PP (1971) Inductive flood tolerance in swamp tupelo (*Nyssa sylvatica* var biflora (Walt) Sarg). J Exp Bot 22:78–89

Horton RF (1971) Stomatal opening: the role of abscisic acid. Can J Bot 49:583–585

Horton RF (1979) Hormonal regulation of stomatal function. In: Singh DN (ed) Structure, function and ecology of stomata. (Dehra Dun, India) pp 103–210

Horton RF, Moran L (1972) Abscisic acid inhibition of potassium influx into stomatal guard cells. Z Pflanzenphysiol 66:193–196

Hsiao TC (1973) Plant responses to water stress. Annu Rev Plant Physiol 24:519–570

Hsiao TC, Acevedo E, Fereres E, Henderson DW (1976) Stress metabolism; water stress; growth and osmotic adjustment. Philos Trans R Soc Lond B. Biol Sci 273:479–500

Hsiao TC, Fereres E, Acevedo E, Henderson DW (1976) Water stress and dynamics of growth and yield of crop plants. In: Lange OL, Kappen L, Schulze E-D (eds) Water and plant life. Springer, Berlin Heidelberg New York, pp 281–306

Huber W (1974) Influence of NaCl and abscisic acid treatment on protein metabolism and some further enzymes of amino acid metabolism in seedlings of *Pennisetum typhoides*. Planta 121:225–235

Huber W, Sankhla N (1973) Eco-physiological studies on Indian arid zone plants. II. Effect of salinity and gibberellin on the activity of the enzymes of amino-acid metabolism in leaves of *Pennisetum typhoides*. Oecologia 13:271–277

Huber W, Rustagi PN, Sankhla N (1974) Eco-physiological studies on Indian arid zone plants. III. Effect of sodium chloride and gibberellin on the activity of the enzymes of carbohydrate metabolism in leaves of *Pennisetum typhoides*. Oecologia 15:77–86

Huber W, Sankhla N (1980) Effect of abscisic acid on betaine accumulation in *Pennisetum typhoides* seedlings. Z Pflanzenphysol 97:179–182

Hubick KT, Drakeford DR, Reid DM (1982) A comparison of two techniques for growing minimally water-stressed plants. Can J Bot 60:219–223

Huck MG, Klepper B, Taylor HM (1970) Diurnal variations in root diameter. Plant Physiol 45:529–530

Humphries EC, Thorne GN (1864) The effect of root formation on photosynthesis of detached leaves. Ann Bot 28:391–400

Hunt FM (1951) Effects of flooded soil on growth of pine seedlings. Plant Physiol 26:363–368
Hunt PG, Campbell RB, Sojka RE, Parsons JE (1981) Flooding-induced soil and plant ethylene accumulation and water status response of field-grown tobacco. Plant Soil 59:427–439
Husain I, Aspinall D (1970) Water stress and apical morphogenesis in barley. Ann Bot 34:393–407
Huxter TJ, Reid DM, Thorpe TA (1979) Ethylene production by tobacco (*Nicotiana tabacum*) callus. Physiol Plant 46:374–380
Ilan I (1971) Evidence for hormonal regulation of the selectivity of ion uptake by plant cells. Physiol Plant 25:230–233
Ilan I, Gilad T, Reinhold L (1971) Specific effects of kinetin on the uptake of monovalent cations by sunflower cotyledons. Physiol Plant 24:337–341
Imber D, Tal M (1970) Phenotypic reversion of Flacca, a wilty mutant of tomato, by abscisic acid. Science 169:592–593
Incol LD, Whitelam GC (1977) The effect of kinetin on stomata in the grass *Anthephora pubescens* Ness. Planta (Berl) 137:243–245
Itai C, Benzioni A (1976) Water stress and hormonal response. In: Lange OL, Kappen L, Schulze ED (eds) Water and plant life. Springer, Berlin Heidelberg New York, pp 225–242
Itai C, Vaadia Y (1965) Kinetin-like activity in root exudate of water-stressed sunflower plants. Physiol Plant 18:941–944
Itai C, Vaadia Y (1971) Cytokinin activity in water-stressed shoots. Plant Physiol 47:87–90
Itai C, Richmond AE, Vaadia Y (1968) The role of root cytokinins during water and salinity stress. Isr J Bot 17:187–195
Itai C, Weyers JDB, Hillman JR, Meidner H, Willmer C (1978) Abscisic acid and guard cells of *Commelina communis* L. Nature 271:652–653
Jackson MB (1979) Rapid injury to peas by soil waterlogging. J Sci Food Agric 30:143–152
Jackson MB, Campbell DJ (1975a) Movement of ethylene from roots to shoots, a factor in the responses of tomato plants to waterlogging soil conditions. New Phytol 74:397–406
Jackson MB, Campbell DJ (1975b) Ethylene and waterlogging effects in tomato. Ann Appl Biol 81:102–105
Jackson MB, Campbell DJ (1976) Waterlogging and petiole epinasty in tomato: the role of ethylene and low oxygen. New Phytol 76(1):21–30
Jackson MB, Harney PM (1970) Rooting cofactors, indoleacetic acid, and adventitious root initiation in mung bean cuttings (*Phaseolus aureus*). Can J Bot 48:943–946
Jackson MB, Kowalweska AKB (1983) Positive and negative messages from roots induce foliar dessication and stomatal closure in flooded pea plants. J Exp Bot 34:493–506
Jackson MB, Osborne DJ (1970) Ethylene, the natural regulator of leaf abscission. Nature 225:1019–1022
Jackson MB, Gales K, Campbell DJ (1978) Effect of waterlogged soil conditions on the production of ethylene and on water relationships in tomato plants. J Exp Bot 29(108):183–193
Jackson WT (1955) Role of adventitious roots in recovery of shoots following flooding of the original root systems. Am J Bot 42:816–819
Jackson WT (1956a) Flooding injury studied by approach-graft and split-root system techniques. Am J Bot 43:496–502
Jackson WT (1956b) The relative importance of factors causing injury to shoots of flooded tomato plants. Am J Bot 43:637–639
Jarvis PG, Jarvis MS (1963) Effects of several osmotic substances on the growth of *Lupinus albus* seedlings. Physiol Plant 16:485–500
Jewer PC, Incoll LD (1980) Promotion of stomatal opening in the grass *Anthephora pubescens* Nees by a range of natural and synthetic cytokinins. Planta 150:218–221

Jones MM, Turner NC (1978) Osmotic adjustment in leaves of sorghum in response to water deficits. Plant Physiol 61:122–126
Jones RL, Armstrong JE (1971) Evidence for osmotic regulation of hydrolytic enzyme production in germinating barley seeds. Plant Physiol 48:137–142
Jones RJ, Mansfield TA (1970) Suppression of stomatal opening in leaves treated with abscisic acid. J Exp Bot 21:714–719
Jones RL, Phillips IDJ (1966) Organs of gibberellin synthesis in light-grown sunflower plants. Plant Physiol 41:1381–1386
Jordan WR, Ritchie JT (1971) Influence of soil water stress on evaporation, root absorption and internal water status of cotton. Plant Physiol 48:783–788
Jordan WR, Morgan PW, Davenport TL (1972) Water stress enhances ethylene-mediated leaf abscission in cotton. Plant Physiol 50:756–758
Jordan WR, Brown KW, Thomas JC (1975) Leaf age as a determinant in stomatal control of water loss from cotton during water stress. Plant Physiol 56:595–599
Kamienska A, Reid DM (1978) The effect of stem girdling on levels of GA-like substances in sunflower plants. Bot Gaz 139:18–26
Kappen L, Lange OL, Schulze E-D, Evenari M, Buschbom U (1979) Ecophysiological investigations on lichens of the Negev desert. VI. Annual course of the photosynthetic production of *Ramalina maciformis* (Del) Bory. Flora 168:85–108
Kappen L, Lange OL, Schulze E-D, Buschbom U, Evenari M (1980) Ecophysiological investigations on lichens of the Negev desert. VII. The influence of the habitat exposure on dew imbibition and photosynthetic productivity. Flora 169:216–229
Karmoker JL, Van Steveninck RFM (1978) Stimulation of volume flow and ion flux by abscisic acid in excised root systems of *Phaseolus vulgaris* L. cv. Redland Pioneer. Planta 141:37–43
Karmoker JL, van Steveninck RFM (1979) The effect of abscisic acid on the uptake and distribution of ions in intact seedlings of *Phaseolus vulgaris* L. cv. Redland Pioneer. Physiol Plant 45:453–459
Kasamo K, Shimomura T (1981) Effect of cold osmotic shock on K^+ efflux from *Nicotiana tabacum* L. leaf discs induced by abscisic acid and ionophores. Plant Cell Physiol 22:939–951
Kaufmann MR (1968) Water relations of pine seedlings in relation to root and shoot growth. Plant Physiol 43:281–288
Kaufmann MR (1976) Water transport through plants: current perspectives. In: Wardlaw JF, Passioura JB (eds) Transport of transfer processes in plants. Academic, New York, pp 313–327
Kaufman PB, Ghosheh N, Ikuma H (1968) Promotion of growth and invertase activity by gibberellic acid in developing *Avena* internodes. Plant Physiol 43:29–34
Kawase M (1971) Causes of centrifugal root promotion. Physiol Plant 25:64–70
Kawase M (1972a) Effect of flooding on ethylene concentration in horticultural plants. J Am Soc Hortic Sci 97:584–588
Kawase M (1972a) Submission increases ethylene and stimulates rooting in cuttings. Proc Int Plant Propag Soc 22:360–366
Kawase M (1972b) Effect of flooding on sunflower plants. Plant Physiol (Suppl) 49:116
Kawase M (1973) Can ethylene cause flooding damage? Hortscience 8:256
Kawase M (1974) Role of ethylene in induction of flooding damage in sunflower. Physiol Plant 31:29–38
Kawase M (1976) Ethylene accumulation in flooded plants. Physiol Plant 36:236–241
Kawase M (1979) Role of cellulase in aerenchyma development in sunflower. Am J Bot 66:183–190
Kawase M, Whitmoyer RE (1980) Aerenchyma development in waterlogged plants. Am J Bot 67:18–22
Kefford NP (1962) Auxin–gibberellin interaction. Plant Physiol 37:380–386
Kemble AR, Macpherson HT (1954) Liberation of amino acids in perennial rye grass during wilting. Biochem J 58:46–49
Kende H (1965) Kinetin-like factors in the root exudate of sunflowers. Proc Natl Acad Sci USA 53:1302–1307

Kennedy HE Jr, Krinard RM (1974) Mississippi river floods impact on natural hardwood forests and plantations. US For Ser Res Note 50:177
Kinet JM, Bodson M, Jacqmard A, Bernier G (1975) The inhibitor of flowering by abscisic acid in *Sinapis alba* L. Z Pflanzenphysiol 77:70–74
King RW (1976) Implications for plant growth of the transport of regulatory compounds in phloem and xylem. In: Wardlaw JF, Passioura JB (eds) Transport and transfer processes in plants. Academic Press, New York, pp 415–431
King RW, Evans LT (1977) Inhibition of flowering in *Lolium temulentum* L. by water stress: A role for abscisic acid. Aust J Plant Physiol 4:225:233
Kirkham MB, Gardner WR, Gerloff GC (1972) Regulation of cell division and cell enlargement by turgor pressure. Plant Physiol 49:961–962
Kirkham MB, Garnder WR, Gerloff GC (1974) Internal water status of kinetin-treated, salt-stressed plants. Plant Physiol 53:241–243
Kosmakova VE, Zvereva EG (1973) In: Transport of assimilates and deposition of substances in storage in plants (in Russian). Vladivostok, p 204
Kozlowski TT (1964) Water metabolism in plants. Harper, New York
Kozlowski TT (1976) Water supply and leaf shedding. In: Kozlowski TT (ed) Water deficits and plant growth. Academic Press, New York, pp 191–231
Kozlowski TT, Pallardy SG (1979) Stomatal responses of *Fraxinus pennsylvanica* seedlings during and after flooding. Physiol Plant 46:155–158
Kramer PJ (1940) Causes of decreased absorption of water by plants in poorly aerated media. Am J Bot 27:216–220
Kramer PJ (1951) Causes of injury to plants resulting from flooding of the soil. Plant Physiol 29:722–736
Kramer PJ (1969) Plant and soil water relationships: A modern synthesis. McGraw-Hill, New York
Kramer PJ (1974) Fifty years of progress in water relations research. Plant Physiol 54:463–471
Krekule J, Horavka B (1972) The response of short-day plant *Chenopodium rubrum* L. to abscisic acid and gibberellic acid treatment applied at two levels of photoperiodic induction. Biol Plant 14:254–259
Krekule J, Ullmann J (1971) The effect of abscisic acid on flowering in *Chenopodium rubrum* L. Biol Plant 13:60–63
Kriedemann PE, Downton WJS (1981) Photosynthesis. In: Paleg LG, Aspinall D (eds) Physiology and biochemistry of drought resistance in plants. Academic Press, Sydney, pp 283–314
Kriedemann PE, Loveys BR (1974) Hormonal mediation of plant responses to environmental stress. In: Bieleski RL, Ferguson AR, Cresswell MM (eds) Mechanisms of regulation of plant growth. Bull 12, Royal Soc NZ, Wellington, pp 461–465
Kriedemann PE, Loveys BR, Fuller GL, Leopold AC (1972) Abscisic acid and stomatal regulation. Plant Physiol 49:842–847
Kulaeva ON (1962) The effect of roots on leaf metabolism in relation to the action of kinetin on leaves. Sov Plant Physiol (Engl Transl Fiziol Rast) 9:182–189
Kummerow J (1980) Adaptation of roots in water-stressed native vegetation. In: Turner NC, Kramer PJ (eds) Adaptation of plants to water and high temperature stress. Wiley-Interscience, New York, pp 57–73
Kummerow J, Krause D, Jow W (1977) Root systems of chaparral shrubs. Oecologia 29:163–177
Lamont B (1976) The effect of seasonality and waterlogging on the root systems of a number of *Hakea* species. Aust J Bot 24:691–702
Lancaster JE, Mann JD, Porter NG (1977) Ineffectiveness of abscisic acid in stomatal closure of yellow lupin, *Lupinus luteus* var. Weiko III. J Exp Bot 28:184–191
Lane HC, King EE (1968) Stimulation of indoleacetic acid oxidase and inhibition of catalase in cotton extracts by plant acids. Plant Physiol 43:1699–1702
Larque-Saavedra A, Wain RL (1974) Abscisic acid levels in relation to drought tolerance in varieties of *Zea mays* L. Nature 251:716–717
Lawlor DW (1978) Effects of water and heat stress on carbon metabolism of plants

with C_3 and C_4 photosynthesis. In: Mansele M, Staples RC (eds) Stress physiology of crop plants. Wiley-Interscience, New York

Lawlor DW, Fock H (1975) Photosynthesis and photorespiratory CO_2 evolution of water-stressed sunflower leaves. Planta 126:247–258

Lenton JR, Bowen MR, Saunders PF (1968) Detection of abscisic acid in the xylem sap of willow (*Salix viminalis* L.) by gas-liquid chromatography. Nature 220:86–87

Levitt JL (1980) Responses of plants to environmental stresses. Vol II: Water, radiation, salt and other stresses. Academic Press, New York

Leyton L, Armitage IP (1968) Cuticle structure and water relations of the needles of *Pinus radiata* (D. Don.). New Phytol 67:31–38

Lieberman M (1979) Biosynthesis and action of ethylene. Annu Rev Plant Physiol 30:533–591

Lin WT, Pool R, Wenkert W, Kriedemann PE (1978) Changes in photosynthesis, stomatal resistance and abscisic acid of *Vitis labrusca* through drought and irrigation cycles. Am J Enol Vitic 29:239–246

Lipe JA, Morgan PW (1973) Ethylene, a regulator of young fruit abscission. Plant Physiol 51:949–953

Little CHA (1975) Inhibition of cambial activity in *Abies balsamea* by internal water stress: role of abscisic acid. Can J Bot 53:3041–3050

Little CHA, Eidt DC (1968) Effect of abscisic acid on budbreak and transpiration in woody species. Nature 220:498–499

Little CHA, Heald JK, Browning G (1978) Identification and measurement of indoleacetic and abscisic acids in the cambial region of *Picea sitchensis* (Bong.) Carr by combined gas chromatography mass spectrometry. Planta 139:133–138

Livne A, Vaadia Y (1965) Stimulation of transpiration rate in barley leaves by kinetin and gibberellic acid. Physiol Plant 18:658–664

Loustalot AJ (1945) Influence of soil moisture conditions on apparent photosynthesis and transpiration of leaves. J Agr Res 71:519–532

Loveys BR (1977) The intracellular location of abscisic acid in stressed and non-stressed leaf tissue. Physiol Plant 40:6–10

Loveys BR, Kriedemann PE (1973) Rapid changes in abscisic acid-like inhibitor following alterations in vine leaf water potential. Physiol Plant 28:476–479

Luke HH, Freeman TE (1967) Rapid bioassay for phytokinins based on transpiration of excised oat leaves. Nature 215:874–875

Luke HH, Freeman TE (1968) Stimulation of transpiration by cytokinins. Nature 217:873–874

Lürssen K, Naumann K, Schröder R (1979) 1-Aminocyclopropane-1-carboxylic acid: An intermediate of the ethylene biosynthesis in higher plants. Z Pflanzenphysiol 92:285–294

Lüttge U (1969) Aktiver Transport (Kurzstreckentransport) bei Pflanzen. Protoplasmatologia VIII 76:1–146

Lüttge U, Higinbotham N (1979) Transport in plants. Springer, Berlin Heidelberg New York

Luxmoore RJ, Fischer RA, Stolzy LH (1973) Flooding and soil temperature effects on wheat during grain tilling. Agronomy J 65:361–364

Lyon CJ (1963) Auxin factor in leaf epinasty. Plant Physiol 38:567–574

Lyon CJ (1970) Ethylene inhibition of auxin transport by gravity in leaves. Plant Physiol 45:644–646

MacHardy WE, Beckman CH (1973) Water relations in American elm infected with *Ceratocystis ulmi*. Phytopathology 63:98–103

MacRobbie EAC (1981) Effects of ABA in "isolated" guard cells of *Commelina communis* L. J Exp Bot 32:563–572

Majernik O, Mansfield TA (1970) Direct effect of SO_2 pollution on the degree of opening of stomata. Nature 227:377–378

Mansfield TA (1967) Stomatal behaviour following treatment with auxin-like substances and phenylmercuric acetate. New Phytol 66:325–350

Mansfield TA (1976) Delay in the response of stomata to abscisic acid in CO_2-free air. J Exp Bot 27:559–564

Mansfield TA, Davies WJ (1981) Stomata and stomatal mechanism. In: Paleg LG, Aspinall D (eds) The Physiology and biochemistry of drought resistance in plants. Academic Press, Sydney, pp 315–346

Mansfield TA, Jones RJ (1971) Effects of abscisic acid on potassium uptake and starch content of stomatal guard cells. Planta 101:147–158

Mansfield TA, Wellburn AR, Moreira TJS (1978) The role of abscisic acid and farnesol in the alleviation of water stress. Philos Trans R Soc Lond B Biol Sci 284:471–482

Marc J, Palmer JH (1976) Relationship between water potential and leaf and inflorescence initiations in *Helianthus annuus*. Physiol Plant 36:101–104

Marrè E (1977) Physiologic implications of the hormonal control of ion transport in plants. In: Pilet PE (ed) Plant growth regulation. Springer, Berlin Heidelberg New York, pp 54–66

Marrè E (1979) Fusicoccin: A tool in plant physiology. Annu Rev Plant Physiol 30:273–288

Marrè E (1982) Hormonal regulation of transport: data and perspectives. In: Wareing PF (ed) Plant growth substances. Academic Press, New York, pp 407–417

Martin JT, Juniper BE (1970) The cuticles of plants. Edward Arnold, London

Mawson BT, Colman B, Cummins WR (1981) Abscisic acid and photosynthesis in isolated leaf mesophyll cell. Plant Physiol 67:233–236

Maximov NA (1929) The plant in relation to water. Yapp RM (ed). Allen and Unwin, London

Mayak S, Halevy AH (1972) Interrelationships of ethylene and abscisic acid in the control of rose petal senescence. Plant Physiol 50:341–346

McAlpine RG (1961) Yellow poplar seedlings intolerant to flooding. J For 59:566–568

McCree KJ (1974) Changes in the stomatal response characteristics of grain sorghum produced by water stress during growth. Crop Sci 14:273–278

McKersie BD, McDermott BM, Hunt LA, Poysa (1982) Changes in carbohydrate levels during ice encasement and floodings of winter cereals. Can J Bot 60:1822–1826

McLaren JS, Smith H (1977) Effect of abscisic acid on photosynthetic products of *Lemna minor*. Phytochemistry 16:219–221

McMichael BL, Jordan WR, Powell RD (1972) An effect of water stress on ethylene production by intact cotton petioles. Plant Physiol 49:658–660

McMinn JW, McNabb WH (1971) Early growth and development of slash pine under drought and flooding. US For Ser Res Paper SE 89, pp 1–10

McWha JA, Jackson DL (1976) Some growth promotive effects of abscisic acid. J Exp Bot 27:1004–1008

Meidner H (1954) Measurements of water intake from the atmosphere by leaves. New Phytol 53:423–426

Meidner H (1967) The effect of kinetin on stomatal opening and the rate of intake of carbon dioxide in mature primary leaves of barley. J Exp Bot 18:556–561

Meyer RF, Boyer JS (1981) Osmoregulation, solute-distribution, and growth in soybean seedlings having low water potentials. Planta 151:482–489

Michniewicz M, Kriesel K, Krassowska J (1970) Effects of adventitious roots on level of endogenous gibberellin-like substances in buds and newly formed shoots of willow cuttings (*Salix viminalis*). Bull Acad Pol Sci Ser Sci Biol 18:355–359

Milborrow BV (1974) Biosynthesis of abscisic acid by a cell-free system. Phytochemistry 13:131–136

Milborrow BV (1974) The chemistry and physiology of abscisic acid. Annu Rev Plant Physiol 25:259–307

Milborrow BV (1978) The stability of conjugated abscisic acid during wilting. J Exp Bot 29:1059–1066

Milborrow BV (1979) Antitranspirants and the regulation of abscisic acid content. Aust J Plant Physiol 6:249–254

Milborrow BV (1981) Abscisic acid and other hormones. In: Paleg LG, Aspinall D (eds) The physiology and biochemistry of drought resistance in plants. Academic Press, Sydney, pp 347–388

Milborrow BV, Noodle RC (1970) Conversion of 5-(1,2-epoxy-2,6,6-trimethylcyclohexyl)-3-methylpenta-cis-2-trans-4-dienoic acid into abscisic acid in plants. Biochem J 199:727–734

Milborrow BV, Robinson DR (1973) Factors affecting the biosynthesis of abscisic acid. J Exp Bot 24:537–548

Milburn JA (1979) Water flow in plants. Longman, London

Mills VM, Todd GW (1973) Effects of water stress on the indoleacetic acid oxidase activity in wheat leaves. Plant Physiol 51:1145–1146

Minore D (1968) Effects of artificial flooding on seedling survival and growth of six northwestern tree species. USDA For Ser Res Note PNW 92

Mittelheuser CJ, Van Stevninck RFM (1969) Stomatal closure and inhibition of transpiration induced by (RS) – abscisic acid. Nature (Lond) 221:281–282

Mizrahi Y, Blumenfeld A, Richmond AE (1970) Abscisic acid and transpiration in leaves in relation to osmotic root stress. Plant Physiol 46:169–171

Mizrahi Y, Blumenfeld A, Richmond A (1971) The role of abscisic acid and kinetin in controlling the adaptive response of plants to root stress. Isr J Bot 20:335

Morgan JM (1980) Possible role of abscisic acid in reducing seed set in water-stressed wheat plants. Nature 285:655–657

Morgan PW, Hall WC (1962) Effect of 2,4-dichlorophenoxyacetic acid on the production of ethylene by cotton and grain sorghum. Physiol Plant 15:420–427

Morgan PW, Hall WC (1964) Accelerated release of ethylene by cotton following application of indolyl-3-acetic acid. Nature 201:99

Morgan PW, Jordan WR, Davenport TL, Durham JI (1977) Abscission responses to moisture stress, auxin transport inhibitors, and ethephon. Plant Physiol 59:710–712

Most BM (1971) Bascisic acid in immature apical tissue of sugar cane and leaves of plants subjected to drought. Planta 101:61–65

Mullins MG (1970) Hormone-directed transport of assimilates in decapitated internodes of *Phaseolus vulgaris* L. Ann Bot 34:897–909

Mullins MG (1972) Auxin and ethylene in adventitious root formation in *Phaseolus aureus* (Roxb.). In: Carr DJ (ed) Plant growth substances 1970. Springer, Berlin Heidelberg New York, pp 526–533

Mullins MG, Osborne DJ (1970) Effect of abscisic acid on growth correlation in *Vitis vinifera* L. Aust J Biol Sci 23:479–493

Nadeau R, Rappaport L, Stolp CF (1972) Uptake and metabolism of ^{3}H-gibberellin A_1 by barley aleurone layers; response to abscisic acid. Planta 107:315–324

Newman EI (1970) Water movement through root systems. Philos Trans R Soc Lond B. Biol Sci 273:463–478

Newsome RE, Kozlowski TD, Tang ZC (1982) Responses of *Ulmus americana* seedlings to flooding of soil. Can J Bot 60:1688–1695

Nicholls PB, May LH (1963) Studies on the growth of the barley apex. I. Inter-relationships between primordium formation, apex, length, and spikelet formation. Aust J Biol Sci 16:561–571

Ogunkanmi AB, Tucker DJ, Mansfield TA (1973) An improved bio-assay for abscisic acid and other transpirants. New Phytol 72:277–282

Ogunkanmi AB, Wellburn AR, Mansfield TA (1976) Detection of preliminary identification and endogenous antitranspirants in water-stressed *Sorghum* plants. Planta 117:293–302

Oppenheimer HR (1960) Adaptation to drought: xerophytism. In: Plant-water relationships in semi-arid conditions: review of research. UNESCO-Paris, p 105

Osborne DJ (1968) Hormonal mechanisms regulating senescence and abscission. In: Wightman F, Setterfield G (eds) Biochemistry and physiology of plant growth substances. Runge, Ottawa, pp 815–840

Osborne DJ (1974) Auxin, ethylene and growth of cells. In: Bieleski RL, Ferguson AR, Cresswell MM (eds) Mechanisms of regulation of plant growth. Soc NZ Bull 12:645–654

Osmond CB (1978) Ion absorption and carbon metabolism in cells of higher plants. In: Lüttge N, Pitman MG (eds) Encyclopedia of plant physiol, new ser Vol IIA. Springer, Berlin Heidelberg New York, pp 347–372

Osonubi O, Davies WJ (1978) Solute accumulation in leaves and roots of woody plants subjected to water stress. Oecologia 32:323–332
O'Toole JC, Ozbun JL, Wallace DH (1977) Photosynthetic response to water stress in *Phaseolus vulgaris*. Physiol Plant 40:111–114
Overbeek J van, Loeffler JE, Mason MIR (1968) Mode of action of abscisic acid. In: Wightman F, Setterfield G (eds) Biochemistry and physiology of plant growth substances. Runge, Ottawa, pp 1593–1607
Paleg LG (1960) Physiological effects of gibberellic acid: I. On carbohydrate metabolism and amylase activity of barley endosperm. Plant Physiol 35:293–299
Pallaghy CK, Raschke K (1972) No stomatal response to ethylene. Plant Physiol 49:275–276
Pallas JE, Box JE (1970) Explanation for the stomatal response of excised leaves to kinetin. Nature 227:87–88
Pallas JE Jr, Kays SJ (1982) Inhibition of photosynthesis by ethylene – a stomatal effect. Plant Physiol 70:598–601
Parker J (1968) Drought-resistance mechanisms. In: Kozlowski TT (ed) Water deficits and plant growth Vol 1. Academic Press, New York, pp 195–234
Passioura JB (1976) The control of water movement through plants. In: Wardlaw IF, Passioura JB (eds) Transport and transfer processes in plants. Academic Press, New York, pp 373–380
Passioura JB (1981) Water collection by roots. Paleg LG, Aspinall D (eds) In: The physiology and biochemistry of drought resistance in plants. Academic Press, Sydney, pp 39–53
Patrick JW (1976) Hormone-directed transport of metabolites. In: Wardlaw IF, Passioura JB (eds) Transport and transfer process in plants. Academic Press, London New York, pp 433–446
Patrick JW, Wareing PF (1973) Auxin-promoted transport of metabolites in stems of *Phaseolus vulgaris* L.: Some characteristics of the experimental transport systems. J Exp Bot 24:1158–1171
Pemadasa MA (1979) Stomatal response to two herbicidal auxins. J Exp Bot 30:267–274
Pemadasa MA (1982) Differential abaxial and adaxial stomatal response to indole-3-acetic acid in *Commelina communis* L. New Phytol 82:69–80
Pereira JS, Kozlowski TT (1977) Variations among woody angiosperms in response to flooding. Physiol Plant 41:184–192
Pharis RP (1977) Interaction of native or exogenous plant hormones in the flowering of woody plants. In: Schütte HR, Gross D (eds) Regulation of developmental processes in plants. Proc Conf, Halle, GDR
Pharis RP (1981) Effects of temperature on strobilus production in gibberellin-treated seedlings of Western hemlock. Can For Serv Res Notes 1(3):21–22
Pharis RP, Ross SD, McMullan E (1980) Promotion of flowering in the Pinaceae by gibberellins. III. Seedlings of Douglas fir. Physiol Plant 50:119–126
Phillips IDJ (1964a) Root-shoot hormone relations. I. The importance of an aerated root system in the regulation of growth hormone levels in the shoot of *Helianthus annuus*. Ann Bot 28:17–35
Phillips IDJ (1964b) Root–shoot hormone relations. II. Changes in endogenous auxin concentration produced flooding of the root system of *Helianthus annuus*. Ann Bot 28:37–45
Phillips IDJ, Jones RL (1964) Gibberellin-like activity in bleeding sap of root systems of *Helianthus annuus* detected by a new dwarf pea epicotyl assay and other methods. Planta 63:269–278
Pickles VR, Sutcliffe JF (1955) The effects of 5-hydroxytryptamine, indole-3-acetic acid and some other substances on pigment effusion, sodium uptake and potassium efflux by slices of red beetroot, in vitro. Biochim Biophys Acta 17:244–251
Pierce M, Raschke K (1980) Correlation between loss of turgor and accumulation of abscisic acid in detached leaves. Planta 148:174–182

Pilet PE (1968) Effet de l'acide abscissique sur les racines: interactions avec l'acide-indolyacétique. CR Acad Sci Paris 267:1142–1145

Pitman MG (1972) Uptake and transport of ions in barley seedlings. III. Correlation between transport to the shoot and relative growth rate. Aust J Biol Sci 25:905–919

Pitman MG (1975) Whole plants. In: Baker DA, Hall JL (eds) Ion transport in plant cells and tissues. North Holland, Amsterdam, pp 267–308

Pitman MG (1977) Ion transport into the xylem. Annu Rev Plant Physiol 28:71–88

Pitman MG, Wellfare D (1978) Inhibition of ion transport in excised barley roots by abscisic acid; relation to water permeability by the roots. J Exp Bot 29:1125–1138

Pitman MG, Lüttge U, Läuchli A, Ball E (1974a) Action of abscisic acid on ion transport as affected by root temperature and nutrient status. J Exp Bot 25:147–155

Pitman MG, Schaefer N, Wildes RA (1974b) Effect of abscisic acid on fluxes of ions in barley roots. In: Zimmermann U, Dainty J (eds) Membrane transport in plants. Springer, Berlin Heidelberg New York, pp 391–396

Pitman MG, Lüttge U, Läuchli A, Ball E (1974c) Ion uptake to slices of barley leaves, and regulation of K content in cells of the leaves. Z Pflanzenphysiol 72:75–88

Place GA, Siddique MA, Wells BR (1971) Effects of temperature and flooding on rice growing in saline and alkaline soil. Agron J 63:62–66

Player MA (1950) Effects of some growth regulating substances on the transpiration of *Zea mays* L. and *Ricinus communis* L. Plant Physiol 25:469–477

Poovaiah BW, Wiebe HW (1973) Influence of hydrogen fluoride fumigation on the water economy of soybean plants. Plant Physiol 51:396–399

Poskuta J, Antoszewski R, Faltynowicz M (1972) Photosynthesis, photorespiration and respiration of strawberry and maize leaves as influenced by abscisic acid. Photosynthetica 6:370–374

Prisco JT, O'Leary JW (1972) Enhancement of intact bean leaf senescence by NaCl salinity. Physiol Plant 27:95–100

Prisco JT, O'Leary JW (1973) The effects of humidity and cytokinin on growth and water relations of salt-stressed bean plants. Plant Soil 39:263–276

Quarrie SA, Jones HG (1977) Effects of abscisic acid and water stress on development and morphology of wheat. J Exp Bot 28:192–203

Quarrie SA, Jones HG (1979) Genotypic variation in leaf water potential, stomatal conductance and abscisic acid concentration in spring wheat subjected to artificial drought stress. Ann Bot 44:323–332

Railton ID, Reid DM (1973) Effects of benzyladenine on the growth of waterlogged tomato plants. Planta 111:261–266

Railton ID, Wareing PG (1973) Effects of abscisic acid on the levels of endogenous gibberellin-like substances in *Solanum andigena*. Planta 112:65–69

Railton ID, Reid DM, Gasking P, MacMillan J (1974) Characterization of abscisic acid in chloroplasts of *Pisum sativum* L. cv Alaska by combined gas chromatography-mass spectrometry. Planta 117:179–182

Rajagopal V, Anderson AS (1978) Does abscisic acid influence proline accumulation in stressed leaves? Planta 143:85–88

Rajagopal V, Anderson AS (1980) Water stress and root formation in pea cuttings. I. Influence of the degree and duration of water stress on stock plants grown under two levels of irradiance. Physiol Plant 48:144–149

Raschke K (1975a) Stomatal action. Annu Rev Plant Physiol 26:309–340

Raschke K (1975b) Simultaneous requirement of carbon dioxide and abscisic acid for stomatal closing in *Xanthium strumarium* L. Planta 125:243–259

Raschke K (1979) Movements of stomata. In: Haupt W, Feinleib ME (eds) Physiology of Movements. Encyclopedia of plant physiology new ser Vol 7. Springer, Berlin Heidelberg New York, pp 383–441

Raschke K, Zeevaart JAD (1976) Abscisic acid content, transpiration and stomatal conductance as related to leaf age in plants of *Xanthium strumarium* L. Plant Physiol 58:169–174

Rasmussen OS (1976) Water stress in plants. I. Abscisic acid level in tomato leaves after a long period of wilting. Physiol Plant 36:208–212

Rawson HM (1979) Vertical wilting and photosynthesis, transpiration and water use efficiency of sunflower leaves. Aust J Plant Physiol 6:109–120

Rawson HM, Constabel GA, Howe GN (1980) Carbon production of sunflower cultivars in field and controlled environments. I. Photosynthesis and transpiration of leaves, stems and heads. Aust J Plant Physiol 7:555–573

Rayle DL (1973) Auxin-induced hydrogen-ion secretion in *Avena* coleoptiles and its implications. Planta 114:63–73

Rayle DL, Johnson KD (1973) Direct evidence that auxin-induced growth is related to hydrogen ion secretion. Plant Physiol (Suppl) 51:2

Read DJ, Bartlett EM (1972) The physiology of drought resistance in the soybean plant (*Glycine max*). I. The relationship between drought resistance and growth. J Appl Ecol 9:487–499

Regehr DL, Bazzaz FA, Boggess WR (1975) Photosynthesis, transpiration and leaf conductance of *Populus deltoides* in relation to flooding and drought. Photosynthetica 9:52–61

Rehm MM, Cline MG (1973) Rapid growth inhibition of *Avena* coleoptile segments by abscisic acid. Plant Physiol 51:93–96

Reid DM (1979) Crop responses to waterlogging. In: Gupta US (ed) Physiological aspects of crop nutrition and resistance. Atma Ram Delhi, pp 252–287

Reid DM, Carr DJ (1967) Effects of a dwarfing compound, CCC, on the production and export of gibberellin-like substances by root systems. Planta 73:1–11

Reid DM, Clements JB, Carr DJ (1968) Red light induction of gibberellin synthesis in leaves. Nature 217:580–582

Reid DM, Crozier A (1970) CCC-induced increase of gibberellin levels in pea seedlings. Planta 94:95–106

Reid DM, Crozier A (1971) Effects of waterlogging on the gibberellin content and growth of tomato plants. J Exp Bot 22:39–48

Reid DM, Railton ID (1974a) Effect of flooding on the growth of tomato plants: Involvement of cytokinins and gibberellins. In: Bieleski RL et al. (eds) Mechanisms of regulation of plant growth. R Soc NZ, Wellington, pp 789–792

Reid DM, Railton ID (1974b) The influence of benzyladenine on the growth and gibberellin content of shoots of waterlogged tomato plants. Plant Sci Lett 2:151–156

Reid DM, Watson K Ethylene: An air pollutant. In: Roberts JA, Tucker GA (eds) Ethylene and plant development. Butterworth, London (in press)

Reid DM, Crozier A, Harvey BMR (1969) The effects of flooding on the export of gibberellins from the root to the shoot. Planta 89:376–379

Reinhold L (1975) The effect of externally applied factors on the translocation of sugars in the phloem. In: Aronoff S, Dainty J, Graham PA, Srivastava LM, Swanson CA (eds) Phloem transport. Plenum, New York, pp 367–388

Richards D (1976) Root-shoot interactions: A functional equilibrium for water uptake in peach [*Prunus persica* (L). Batsch.] Ann Bot 41:279–281

Richards D, Rowe RN (1977) Effects of root restriction, root pruning and 6-benzylaminopurine on the growth of peach seedlings. Ann Bot 41:729–740

Ridge I (1973) The control of cell shape and rate of cell expansion by ethylene: effects on microfibril orientation and cell-wall extensibility in etiolated peas. Acta Bot Neerl 22:144–158

Robinson M (1983) Influence of abscisic acid and ethylene on assimilate distribution in *Gladiolus grandiflorus* Ann Bot 51:779–785

Robinson M, Halevy AH, Galili D, Plaut Z (1983) Assimilate distribution in *Gladiolus grandiflorus* as affected by water deficit. Ann Bot 51:461–468

K Ryć M, Lewak St (1980) The role of abscisic acid (ABA) in regulation of some photosynthetic enzyme activities in apple seedlings in relation to embryonal dormancy. Z Pflanzenphysiol 96:195–202

Samet JS, Sinclair TR (1980) Leaf senescence and abscisic acid in leaves of field-grown soybean. Plant Physiol 66:1164–1168

Sankhla N, Huber W (1974) Effect of abscisic acid on the activities of photosynthetic enzymes and CO_2 fixation products in leaves of *Pennisetum typhoides* seedlings. Physiol Plant 30:291–294

Sankhla N, Huber W (1979) Effects of abscisic acid on enzyme induction in *Lemna minor* L. Pflanzenphysiology 91:7–15

Sarin MN, Narayanan A (1968) Effects of soil salinity and growth regulators on germination and seedling metabolism of wheat. Physiol Plant 21:1201–1209

Sartoris GB, Belcher BA (1949) The effect of flooding on flowering and survival of sugar cane. Sugar 44:36–39

Schneider EA, Wightmann F (1974) Metabolism of auxin in higher plants. Annu Rev Plant Physiol 25:487–513

Selman IW, Sandanam S (1972) Growth responses of tomato plants in non-aerated water culture to foliar sprays of gibberellic acid and benzyladenine. Ann Bot 36:837–848

Sembdner G, Gross D, Leibisch HW, Schneider G (1980) Biosynthesis and metabolism of plant hormones. In: MacMillan J (ed) Hormonal regulation of development I. Encyclopedia of plant physiology, new series Vol 9. Springer, Berlin Heidelberg New York, pp 281–444

Setter TL, Brun WA, Brenner ML (1980) The effect of obstructed translocation of leaf abscisic acid, and associated stomatal closure and photosynthesis decline. Plant Physiol 65:1111–1115

Setter TL, Brun WA, Brenner ML (1981) Abscisic acid translocation and metabolism in soybeans following depodding and petiole girdling treatments. Plant Physiol 67:774–779

Sharpe RE, Davies WJ (1979) Solute regulation and growth by roots and shoots of water-stressed maize plants. Planta 147:43–49

Sharpe RE, Osonubi O, Wood WA, Davies WJ (1979) A simple instrument for measuring leaf extension in grasses, and its application in the study of the effects of water stress on maize and sorghum. Ann Bot 44:35–45

Sheriff DW (1972) A new apparatus for the measurement of sap flux in small shoots with the magnetohydrodynamic technique. J Exp Bot 23:1086–1095

Sheriff DW, Sinclair R (1973) Fluctuations in leaf water balance, with a period of 1 to 10 minutes. Planta 113:215–228

Shindy W, Weaver RJ (1967) Plant regulators alter translocation of photosynthetic products. Nature 214:1024–1025

Shindy WW, Kliewer WM, Weaver RJ (1973) Benzyladenine-induced movement of ^{14}C-labeled photosynthate into roots of *Vitis vinifera*. Plant Physiol 51:345–349

Simon EW (1974) Phospholipids and plant membrane permeability. New Phytol 73:377–420

Singh BN, Galson E, Dashek W, Walton DC (1979) Abscisic acid levels and metabolism in the leaf epidermal tissue of *Tulipa gesneriana* L. and *Commelina communis* L. Planta 146:135–138

Singh TN, Aspinall D, Paleg LG (1973a) Stress metabolism IV. The influence of (2-chloroethyl)trimethylammonium chloride and gibberellic acid on the growth and proline accumulation of wheat plants during water stress. Aust J Biol Sci 26:77–86

Singh TN, Aspinall D, Paleg LG, Boggess SF (1973b) Stress metabolism II. Changes in proline concentration in excised plant tissues. Aust J Biol Sci 26:57–63

Sivakumaran S, Hall MA (1978a) Effects of age and water stress on endogenous levels of plant growth regulators in *Euphorbia lathyris* L. J Exp Bot 29:195–205

Sivakumaran S, Hall MA (1978b) Effect of osmotic stress upon endogenous hormone levels in *Euphorbia lathyris* L. and *Vicia faba* L. Ann Bot 42:1403–1411

Sivakumaran S, Hall MA (1979) Hormones in relation to stress recovery in *Populus robusta* cuttings. J Exp Bot 30:53–63

Smith KA (1975) Ethylene in the soil atmosphere and its effect on root growth. Proc Assoc Applied Biol Ann Appl Biol 81:106

Smith KA, Robertson PD (1971) Effect of ethylene on root extension of cereals. Nature 234:168–169

Smith KA, Russell RS (1969) The occurrence of ethylene and its significance in anaerobic soil. Nature 222:769–771
Snaith PJ, Mansfield TA (1982) Stomatal sensitivity to abscisic acid: can it be defined? Plant Cell Environ 5:309–311
Squire GR, Mansfield TA (1972) Studies on the mechanism of action of fusicoccin, the fungal toxin that induces wilting, and its interaction with abscisic acid. Planta 105:71–78
Stålfelt MG (1955) The stomata as a hydrophotic regulator of the water deficit of the plant. Physiol Plant 8:572–593
Stansell JR, Klepper B, Browning VD, Taylor HM (1973) Plant water status in relation to clouds. Agron J 65:677–678
Steward FC (1968) Growth and organization in plants. Addison-Wesley, Reading, MA
Steward FC, Krikorian AD (1971) Plants, chemicals and growth. Academic Press, New York, pp 173–178
Stewart CR (1980) The mechanism of abscisic acid-induced proline accumulation in barley leaves. Plant Physiol 66:230–233
Stewart CR, Hanson AD (1980) Proline accumulation as a metabolic response to water stress. In: Turner NC, Kramer PJ (eds) Adaptation of plants to water and high temperature stress. Wiley, New York, pp 173–189
Stewart GR, Lee JA (1974) The role of proline accumulation in halophytes. Planta 120:279–289
Stewart CR, Morris CJ, Thompson JF (1966) Changes in amino acid content of excised leaves during incubation. II. Role of sugar in the accumulation of proline in wilted leaves. Plant Physiol 41:1585–1590
Stoker R, Weatherley PE (1971) The influence of the root system on the relationship between the rate of transpiration and depression of leaf water potential. New Phytol 70:547–554
Storey R, Jones RFW (1975) Betaine and choline levels in plants and their relationship to NaCl stress. Plant Sci Lett 4:161–168
Strack Z, Karwowska K, Kraszewska E (1975) The effect of several stress conditions and growth regulators on photosynthesis and translocation of assimilates in the bean plant. Acta Soc Bot Pol 44:567–588
Stuart DM (1973) Reduction of water permeability in potato tuber slices by cyanide, ammonia, 2,4-dinitrophenol, and oligomycin and its reversal by adenosine, 5-triphosphate and cytidine 5-triphosphate. Plant Physiol 51:485–488
Stuart KL, Coke LB (1975) The effect of vomifoliol on stomatal aperture. Planta 122:307–310
Sutcliffe JF (1962) Mineral salts: absorption in plants. Pergamon, Oxford
Taerum R (1973) Occurrence of inverted water potential gradients between soil and bean roots. Physiol Plant 28:471–475
Tal M, Imber D (1970) Abnormal stomatal behavior and hormonal imbalance in *flacca,* a wilty mutant of tomato. II. Auxin- and abscisic acid-like activity. Plant Physiol 46:373–376
Tal M, Imber D (1971) Abnormal stomatal behavior and hormonal imbalance in *flacca,* a wilty mutant of tomato. III. Hormonal effects on the water status of the plant. Plant Physiol 47:849–850
Tal M, Imber D, Erez A, Epstein E (1979) Abnormal stomatal behavior and hormonal imbalance in *flacca,* a wilty mutant of tomato. Plant Physiol 63:1044–1048
Tanada T (1968) Substances essential for a red, far-red light reversible attachment of mung bean root tips to glass. Plant Physiol 43:2070–2071
Tang ZC, Kozlowski TT (1982) Physiological, morphological, and growth responses of *Platanus occidentalis* seedlings to flooding. Plant Soil 66:243–255
Tang ZC, Kozlowski TT (1982a) Some physiological and morphological responses of *Quercus macrocarpa* seedlings to flooding. Can J For Res 12:196–202
Taylor CM, Railton ID (1977) The influence of wilting and abscisic acid application on gibberellin interconversion in etiolated seedlings of dwarf *Pisum sativum* var. Meteor. Plant Sci Lett 9:317–322

Taylor JS, Reid DM, Pharis RP (1981) Mutual antagonism of sulphur dioxide and abscisic acid in their effect on stomatal aperture in broad bean (*Vicia faba* L.) epidermal strips. Plant Physiol 68:1504–1507

Teskey RO, Hinckley TM (1981) Influence of temperature and water potential on root growth of white oak. Physiol Plant 52:363–369

Thorne GN, Evans AF (1964) Influence of tops and roots on net assimilation rate of sugar-beet and spinach beet and grafts between them. Ann Bot 28:499–508

Thorpe TA, Meier DD (1973) Effects of gibberellic acid and abscisic acid on shoot formation in tobacco callus cultures. Physiol Plant 29:121–124

Tillberg E, Eliasson L, Ericsson T (1980) Time course of uptake of ^{14}C-abscisic acid by *Lemna gibba* in relation to growth. Physiol Plant 48:584–587

Tillberg E, Dons C, Haugstad M, Nilsen S (1981) Effect of abscisic acid on CO_2 exchange in *Lemna gibba*. Physiol Plant 52:401–406

Torrey JG (1976) Root hormones and plant growth. Annu Rev Plant Physiol 27:435–459

Treharne KJ, Stoddart JL (1968) Effects of gibberellin on photosynthesis in red clover (*Trifolium pratense* L.). Nature 220:457–458

Treharne KJ, Stoddart JL, Pughe J, Paranjothy K, Wareing PF (1970) Effects of gibberellin and cytokinins on the activity of photosynthetic enzymes and plastid ribosomal RNA synthesis in *Phaseolus vulgaris* L. Nature 228:129–131

Trewavas AJ (1979) What is the molecular basis of plant hormone action? Trends Biol Sci 4:199–202

Trewavas A (1981) How do plant growth substances work? Plant Cell Environ 4:203–228

Tucker DJ (1978) Apical dominance in the tomato: The possible roles of auxin and abscisic acid. Plant Sci Lett 12:273–278

Tucker DJ, Mansfield TA (1971) A simple bioassay for detecting "antitranspirant" activity of naturally occurring compounds such as abscisic acid. Planta 98:157–163

Tucker DJ, Mansfield TA (1973) Apical dominance in *Xanthium strumarium*. J Exp Bot 24:731–740

Turner NC (1975) Concurrent comparisons of stomatal behavior, water status and evaporation of maize in soil at high or low water potential. Plant Physiol 55:932–936

Turner NC, Graniti A (1969) Fusicoccin: A fungal toxin that opens stomata. Nature 223:1070–1071

Turner NC, Jones MM (1980) Turgor maintenance by osmotic adjustments: A review and evaluation. In: Turner WC, Kramer PJ (eds) Adaptation of plants to water and high temperature stress. Wiley, New York, pp 87–103

Ungar IA (1967) Vegetation–soil relationships on saline soils in northern Kansas. Am Midl Nat 78:98–120

Unsworth MH, Biscoe PV, Pinkney HR (1972) Stomatal responses to sulphur dioxide. Nature 239:458–459

Vaadia Y (1960) Autonomic diurnal fluctuations in rate of exudation and root pressure of decapitated sunflower plants. Physiol Plant 13:701–717

Van Kirk CA, Raschke K (1978) Release of malate from epidermal strips during stomatal closure. Plant Physiol 61:474–475

Van Steveninck RFM (1972) Abscisic acid stimulation of ion transport and alteration in K^+/Na^+ selectivity. Z Pflanzenphysiol 67:282–286

Van Steveninck RFM (1976) Effect of hormones and related substances on ion transport. In: Lüttge U, Pitman MG (eds) Transport in plants Vol IIB. Tissues and organs. Encyclopedia of plant physiology, new series. Springer, Berlin Heidelberg New York, pp 307–342

Van Volkenburg E, Davies WJ (1983) Inhibition of light-stimulated leaf expansion by abscisic acid. J Exp Bot 34:835–845

Vince-Prue D (1975) Photoperiodism in plants. McGraw-Hill, London, pp 30–56

Vitagliano C (1975) Effects of ethephon on stomata, ethylene evolution and abscission in olive (*Olea europaea* L.) cv. coratina. J Am Soc Hortic Sci 100:482–485

Vreugdenhil, D (1983) Abscisic acid inhibits phloem loading of sucrose. Physiol Plant 57:463–467

Walker MA, Dumbroff EB (1981) Effects of salt stress on abscisic acid and cytokinin levels in tomato. Z Pflanzenphysiol 101:461–470
Walton DC (1980) Biochemistry and physiology of abscisic acid. Annu Rev Plant Physiol 31:453–489
Walton DC, Harrison MA, Cote P (1976) The effects of water stress on abscisic acid levels and metabolism in roots of *Phaseolus vulgaris* L. and other plants. Planta 131:141–144
Walton DC, Galson E, Harrison MA (1977) The relationship between stomatal resistance and abscisic acid levels in leaves of water-stressed bean plants. Planta 133:145–148
Wample RL (1976) Hormonal and morphological responses of *Helianthus annuus* L. to flooding. PhD Thesis, Univ Calgary, Calgary, Alberta, Canada
Wample RL, Bewley JD (1975) Proline accumulation in flooded and wilted sunflower and the effects of benzyladenine and abscisic acid. Can J Bot 53:2893–2896
Wample RL, Davies RW (1983) Effect of flooding on starch accumulation in chloroplasts of sunflower (*Helianthus annuus* L.) Plant Physiol 73:195–198
Wample RL, Reid DM (1975) Effect of aeration on the flood-induced formation of adventitious roots and other changes in sunflower (*Helianthus annuus*). Planta 127:263–270
Wample RL, Reid DM (1978) Control of adventitious root production and hypocotyl hypertrophy of sunflower (*Helianthus annuus* L.) in response to flooding. Physiol Plant 44:351–358
Wample RL, Reid DM (1979) The role of endogenous auxins and ethylene in the formation of adventitious roots and hypocotyl hypertrophy in flooded sunflower plants (*Helianthus annuus* L.). Physiol Plant 45:219–226
Wample RL, Thornton RK (1984) Differences in the response of sunflowers (*Helianthus annuus*) subjected to flooding and drought stress. Physiol Plant 61:611–616
Wangermann E (1961) The effect of water supply and humidity on growth and development. In: Ruhland W (ed) Encyclopedia of plant physiology Vol XVI. Springer, Berlin Göttingen Heidelberg, pp 618–633
Wardlaw IF (1968) The control and pattern of movement of carbohydrates in plants. Bot Rev 34:79–105
Wardlaw IF (1980) Translocation and source–sink relationships. In: Carlson PS (ed) The biology of crop productivity. Academic Press, London New York, pp 299–339
Warner HL, Leopold AC (1971) Timing of growth regulator responses in peas. Biochem Biophys Res Commun 44:989–994
Weiler EW, Schnabl H, Hornberg C (1982) Stress-related levels of abscisic acid in guard cell protoplasts of *Vicia faba* L. Planta 154:24–28
Wellburn AR, Hampp R (1976) Fluxes of gibberellic and abscisic acids, together with that of adenosine 3,5-cyclic phosphate, across plastid envelopes during development. Planta 131:95–96
Wellburn AR, Ogunkami AB, Fenton R, Mansfield TA (1974) All-trans-farnesol: A naturally occurring antitranspirant? Planta 120:255–263
Wellburn FAM, Wellburn AR, Stoddart JL, Treharne KJ (1973) Influence of gibberellic and abscisic acids and the growth retardant CCC upon plastid development. Planta 111:337–346
Went FW (1928) Wuchsstoff und Wachstum. Rec Trav Bot Neerl 25:1–116
Went FW (1938) Transplantation experiments with peas. Am J Bot 25:44–55
Went FW (1938b) Specific factors other than auxin affecting growth and root formation. Plant Physiol 13:55–80
Went FW (1943) Effect of the root system on stem growth. Plant Physiol 18:51–56
Weyers JDB, Hillman JR (1979) Uptake and distribution of abscisic acid in *Commelina* leaf epidermis. Planta 144:167–172
Weyers JDB, Hillman JR (1980) Effects of abscissic acid on $^{86}Rb^{+}$ fluxes in *Commelina communis* L. leaf epidermis. J Exp Bot 31:711–730
Wilkens MB, Whyte P (1968) Polar transport of auxin in *Zea* coleoptiles under anaerobic

conditions. In: Wightman F, Setterfied G (eds) Biochemistry and physiology of plant growth substances. Runge, Ottawa
Williamson RE (1970) Effect of soil gas composition and flooding on growth of *Nicotiana tabacum*. Agron J 62:80–83
Williamson RE, Kriz GJ (1970) Response of agricultural crops to flooding, depth of water table and soil gaseous composition. Trans ASAE 13:216–220
Winter K, Lüttge U (1976) Balance between C_3 and CAM pathway of photosynthesis. In: Lange OL, Kappen L, Schulze ED (eds) Water and plant life. Springer, Berlin Heidelberg New York, pp 323–334
Winter K, von Willert DJ (1972) NaCl-induced crassulacean acid metabolism in *Mesembryanthemum crystallinum*. Z Pflanzenphysiol 67:166–170
Wood A, Paleg LG (1972) The influence of gibberellic acid on the permeability of model membrane systems. Plant Physiol 50:103–108
Wood A, Paleg LG, Spotswood TM (1974) Hormonal–phospholipid interaction: a possible hormonal mechanism of action in the control of membrane permeability. Aust J Plant Physiol 1:167–169
Wright STC (1969) An increase in the "Inhibitor-β" content of detached wheat leaves following a period of wilting. Planta 86:10–20
Wright STC (1977) The relationship between leaf water potential (Ψ_{leaf}) and the levels of abscisic acid and ethylene in excised wheat leaves. Planta 134:183–189
Wright STC (1979) The effect of 6-benzyladenine and leaf-aging treatment on the levels of stress-induced ethylene emanating from wilted wheat leaves. Planta 144:179–188
Wright STC (1980) The effect of plant-growth regulator treatments on the levels of ethylene emanating from excised turgid and wilted wheat leaves. Planta 148:381–388
Wright STC, Hiron RWP (1969) (+)-Abscisic acid, the growth inhibitor induced in detached wheat leaves by a period of wilting. Nature 224:719–720
Wright STC, Hiron RWP (1972) The accumulation of abscisic acid in plants during wilting and under other stress conditions: In Carr DJ (ed) Plant growth substances 1970. Springer, Berlin Heidelberg New York, pp 291–298
Wyn Jones RG, Storey R (1981) Betaines. In: Paleg LG, Aspinall D (eds) The physiology and biochemistry of drought resistance in plants. Academic Press, Sydney, pp 171–204
Wyn Jones RG, Storey R, Leigh RA, Ahmad N, Pollard A (1977) A hypothesis on cytoplasmic osmoregulation. In: Marrè E, Ciferri O (eds) Regulation of cell membrane activities in plants. Elsevier/North Holland, Amsterdam, pp 121–126
Yee PT, Stolzy LM, Lettey J (1969) Survival of plants under prolonged flooded conditions. Agron J 61:844–848
Yegappan TM, Paton DM, Gates CT, Muller WJ (1980) Water stress in sunflower (*Helianthus annuus* L.): I. Effect on plant development. Ann Bot 46:61–70
Yegappan TM, Paton DM, Gates CT, Muller WJ (1982a) Water stress in sunflower (*Helianthus annuus* L.): II. Effects on leaf cells and leaf area. Ann Bot 49:63–68
Yegappan TM, Paton DM, Gates CT, Muller WJ (1982b) Water stress in sunflower (*Helianthus annuus* L.): III. Responses of cypsela size. Ann Bot 49:69–75
Yomo H (1960) Amylase-activating substance IV. Amylase-activating activity of gibberellin. Hakko Kyokaishi 18:600–602. Cited in Chem Abstr 55; 26145 (1961)
Zabadal TJ (1974) A water potential threshold for the increase of abscisic acid in leaves. Plant Physiol 53:125–127
Zahner R (1968) Water deficits and growth of trees. In: Kozlowski TT (ed) Water deficits and plant growth. Academic Press, New York, pp 191–254
Zeevaart JAD (1976) Physiology of flower formation. Ann Rev Plant Physiol 27:321–348
Zeevaart JAD (1977) Sites of abscisic acid synthesis and metabolism in *Ricinus communis* L. Plant Physiol 59:788–791
Zeevaart JAD (1980) Changes in the levels of abscisic acid and its metabolites in excised leaf blads of *Xanthium strumarium* during and after water stress. Plant Physiol 66:672–678
Zeiger E, Hepler PK (1976) Production of guard cell protoplasts from onion and tobacco. Plant Physiol 58:492–498

Zeiger E, Hepler PK (1977) Light and stomatal function: blue light stimulates swelling of guard cell protoplasts. Science 196:887–889
Zelitch I (1965) Environmental and biochemical control of stomatal movement in leaves. Biol Rev Camb Philos Soc 40:463–482
Zeroni M, Jerie PH, Hall MA (1977) Studies on the movement and distribution of ethylene in *Vicia faba* L. Planta 134:119–125
Zimmerman PW, Hitchcock AE (1933) Initiation and stimulation of adventitious roots caused by unsaturated hydrocarbon gases. Contrib Boyce Thompson Inst 5:351–369
Zimmerman RP, Kardos LT (1961) Effect of bulk density on root growth. Soil Sci 91:280–288
Zimmermann U (1978) Physics of turgor and osmoregulation. Annu Rev Plant Physiol 29:121–148

15 Pollen. Symbionts and Symbiont-Induced Structures

J.P. MASCARENHAS and D. CANARY

1 Pollen

Pollen grains, the male gametophytes of plants, are free structures for a portion of their life history. At maturity, the pollen is released from the anther in angiosperms, or the microsporangium in gymnosperms, and is carried by wind, insects, or other agents to the receptive surface of the female part of the plant, namely the stigma in flowering plants or the ovule on the scales of the female cone in gymnosperms. Pollen grains can thus be considered environmental agents which play essential roles in both fertilization and fruit development. The primary function of a pollen grain is the production of two sperm cells and their transport via a pollen tube through tissues of the stigma and style, and into the embryo sac of the ovule, where the process of double fertilization can then be completed (MASCARENHAS 1975a, b). One sperm cell fuses with the egg cell, thus forming the zygote. The other fuses with the nuclei of the central cell, forming the primary endosperm nucleus (JENSEN 1972). In a gymnosperm such as pine there is a single fusion of one male nucleus with the egg nucleus in the archegonium.

A secondary but nevertheless important function of pollen is the triggering of reactions that culminate in "fruit set", i.e., the prevention of formation of the abscission layer in the flower stalk, and the commencement of the swelling of the ovary. Plant hormones have been implicated in both these functions, viz., growth of the pollen tube and fruit set.

1.1 Effect of Hormones on Pollen Germination and Tube Growth

There have been many studies, especially in the earlier literature, on the effects of growth substances on pollen tube growth. In general, low concentrations of these substances result in varying degrees of stimulation of pollen tube germination and growth, whereas high concentrations are inhibitory. Only a few papers representative of the extensive literature in the area will be discussed here. SMITH (1942), for example, found that the inclusion of IAA, IBA, or NAA in the growth medium at concentrations of 1×10^{-4} M or lower resulted in a slight increase in both germination and pollen tube growth of *Antirrhinum majus* pollen. Concentrations of auxin greater than 10^{-4} M were inhibitory. With *Bryophyllum daigremontianum,* however, the auxin stimulation of germination and tube growth was much more dramatic, tube length after 3 h in culture with IAA being about 170% of the control (SMITH 1942). Also the auxin-treated tubes were more nearly straight, and had less tendency to coil and bend, than the control tubes. Pollen tube growth of several other species is also stimulated by auxins (BOSE 1959, RAO and CHIN 1973, SHUKLA and TEWARI 1974). Pollen of *Pinus austriaca* can be induced to germinate by IAA, no germination occurring in its absence (SMITH 1939). Concentrations of IAA, GA_3, ethylene, and ABA of 1–10 mg l^{-1} promoted pollen germination and tube growth in *Pinus roxburghii* Sarg (DHAWAN and MALIK 1981). There are also a number of other reports that auxins included in the growth medium are either without effect, or inhibitory to pollen germination and tube growth (ADDICOTT 1943, KATO 1955, LARSEN and TUNG 1950, NYGAARD 1969, PUPPO et al. 1974). It must be kept in mind, however, that all the above studies were carried out on pollen germinated in vitro in various culture media, not all of which contained the additions that most pollens seem to require, e.g., B, Ca, etc. (LINSKENS and KROH 1970) for optimum growth in vitro.

LARTER and CHAUBEY (1965) studied the growth of pollen tubes through the styles of barley plants whose stems had been treated continuously with both GA_3 and IAA, beginning 3 days before pollination. An autotetraploid barley (*Hordeum vulgare* var. Herta) was the female and the pollen came from a diploid rye (*Secale cereale* var. prolific). A high degree of incompatibility exists between these species. Significant increases in pollen tube lengths occurred with treatments of IAA ranging from 0.1 to 100 mg l^{-1}, 0.1, and 1.0 mg l^{-1} averaging 139% and 162% of controls, respectively, by 3 h after pollination. Above 1.0 mg l^{-1} the promotive effect of IAA on tube growth became less pronounced. At a concentration of 25 mg l^{-1} (optimum), GA_3 increased germination, and by h 7 after pollination caused a twofold increase in length of tubes, several having reached the base of the style. Applied together, at their individual optimum concentrations, GA_3 and IAA increased tube length by 11% over that induced by either hormone applied alone. Unfortunately, LARTER and CHAUBEY (1965) did not determine endogenous auxin or GA concentrations in the style and ovary.

In *Lilium longiflorum*, GA_3 in the growth medium at 50 mg l^{-1} stimulated pollen tube growth fivefold, relative to the control, but at 100 to 200 mg l^{-1}, elongation was inhibited (KATO 1955). A GA_3 concentration of 0.05 mg l^{-1} (optimum) in the growth medium increased length of *Pisum sativum* pollen tubes sevenfold over the control (BOSE 1959). In a more extensive study CHANDLER (1957) confirmed the growth-promoting effects of GA_3 on pollen grains of 27 species, seven of which showed an increase in germination percentage as well as in tube length.

Results with GA_3 are not always consistent or positive, pollen tube growth, but not germination of *Petunia hybrida* being greatly stimulated by GA_3 in the medium (0.9 mg l^{-1} to 9 mg l^{-1}). However, these same concentrations are inhibitory to both pollen germination and tube growth of *A. majus* (FÄHNRICH and ULLRICH 1964). For *P. sativum* pollen, where IAA and kinetin stimulate growth (although not to the extent obtained with GA_3), addition of kinetin or IAA to a medium containing GA_3 does not stimulate growth above that obtained by kinetin or IAA alone (BOSE 1959).

An optimal stimulation of pollen tube growth by ABA was observed at 3.8×10^{-8} M when other growth regulators can also influence pollen germination and tube growth. Isolated pollinia of *Calotropis procera* were incubated for 15 to 20 min in medium where ABA ranged from 1.9 to 7.6×10^{-8} M (SHUKLA and TEWARI 1973a). Growth occurred when the pollinia were subsequently transferred to ABA-free media and incubated for 24 h under constant illumination. However, although IAA, GA_3, and kinetin can all also stimulate pollen tube growth in *C. procera,* when any one is combined with ABA, tube growth falls below that of the control (SHUKLA and TEWARI 1974). Ethrel, CCC, and phosfon D retard pollen tube growth in *C. procera,* and ABA can reverse the inhibition of pollen tube elongation induced by CCC and phosfon-D, although it fails to reverse the ethrel-induced inhibition of elongation (SHUKLA and TEWARI 1973b).

Thus, there is ample evidence of interaction between growth regulators in their effects on pollen tube growth.

1.2 Hormonal Composition of Pollen

Pollen has long been known to be a rich source of growth substances. In 1909, FITTING demonstrated in Java that water extracts of pollen could stimulate the swelling of the ovary in orchids, and this stimulant was termed by him "pollen hormone". In 1932, LAIBACH, in Germany, showed that the pollen hormone of FITTING was probably identical with auxin. Pollens of several other species have also been shown to contain auxin (cf. LANG 1961).

Using the Avena curvature auxin assay MUIR (1947) found that tobacco (*Nicotiana tabacum*) pollen contained relatively small amounts of free auxin and no marked increase occurred during germination. Pollen of *Datura suaveolens* contained twice as much free auxin as pollen of tobacco (0.14 µg equivalent of IAA g^{-1} pollen), and again there was no increase after several hours of germination. There was however a three- to fourfold increase in auxin (to 0.4 µg equivalent IAA g^{-1} pollen) upon germination of *Antirrhinum majus* pollen. Dormant pollen of *Cyclamen persicum* contained the highest auxin concentration (0.24 µg IAA equivalent g^{-1} pollen), its auxin content also increasing two- to threefold after germination (MUIR 1947).

There have been some attempts to explain differences in pollen growth and germination on the basis of their growth-regulator content. Pollen from diploid varieties of apple germinates much better and has a higher pollinating value than pollen from triploid varieties (LARSEN and TUNG 1950). Neutral and acidic growth-promoting substances and a neutral growth inhibitor (in the *Avena* curvature test) were obtained from ether extracts of the apple pollen. The amounts of growth-promoting substances were less in the pollen of the diploid than in pollen of the triploid varieties. All varieties contained roughly the same amount of growth inhibitor. When IAA was added to the growth medium at 1 or 10 mg l^{-1}, it did not promote germination in either diploid or triploid varieties. Thus, differences between pollen of diploid and triploid varieties in their germinability and pollinating value could not be explained on the basis of their content of auxin-like growth promotor, or inhibitor (LARSEN and TUNG 1950). Several substances are present in the pollen of *Pinus densiflora* which inhibit growth of *Avena* coleoptiles and pollen tubes of *P. densiflora, Tradescantia,* and *Bryophyllum* (TANAKA 1960). Germinating, but not ungerminated, pollen of *Pinus radiata* contains an auxin active in the *Avena* coleoptile test (0.05 μg IAA mg^{-1} pollen), which is readily diffusible from the pollen tubes into an agar medium, and it was postulated that the auxin may enhance ovule development (SWEET and LEWIS 1969).

Purified extracts from mature pollen of *Petunia,* when assayed on *Avena* mesocotyls, contained two peaks of auxin-like activity (0.15 μg IAA equivalents g^{-1} pollen), one of which co-chromatographed on paper with IAA (BARENDSE et al. 1970). A peak of inhibitory activity was also found. Gibberellin-like substances were also extracted from immature pollen (i.e., 24 h before anthesis) and assayed after purification and chromatography, for GA's on dwarf corn. Only one peak of activity (0.006 μg GA_3 equivalent g^{-1} fresh weight) was obtained; but mature pollen of *Petunia* contained much more GA-like activity (0.057 μg) than immature pollen, possibly indicating increased synthesis or release of GA's in the pollen grain during the last 24 h, of maturation. BARENDSE et al. (1970) also extracted pollen of *Lilium henryi* 1 day prior to anthesis, and found an exceptionally high content (2.19 μg) of GA_3 equivalent g^{-1} fresh weight. Unfortunately, however, no chemical characterization was made. Mature pollen from flowers that had just opened contained 1.08 μg of GA_3 equivalents g^{-1} fresh weight. When germinated in vitro there was a large loss of GA-like activity (e.g., 0.22, 0.24, 0.17, and 0.05 μg of GA_3 equivalent g^{-1} fresh weight from pollen germinated for 1, 2.5, 3.5, and 6 h respectively). While there was some secretion of GA's into the medium (0.02, 0.03, 0.07, and 0.07 μg GA_3 equivalent g^{-1} fresh weight after 1, 2.5, 3.5, and 6 h respectively), this could not account for the 10- to 45-fold loss. This large drop in GA content during the first hour of germination suggests that GA's are important during germination, but not for further growth of the pollen tube (BARENDSE et al. 1970).

The characterization by gas chromatography-mass spectrometry of several GA's (GA_3, GA_4, GA_7 and a GA_9-like substance) from pollen of three species of *Pinus* has been accomplished (KAMIENSKA and PHARIS 1975, KAMIENSKA et al. 1976a, b). Endogenous GA's were quantified by dwarf rice cv. Tan-ginbozu (KAMIENSKA and PHARIS 1975), dormant pine pollen containing relatively high amounts of both polar and nonpolar GA's (KAMIENSKA and PHARIS 1975). During the first 15 h of germination the quantities of nonpolar GA's declined, with a parallel rise in the more polar GA's. By 48 h only slight GA activity was detected and by 72 h of germination no GA's were detectable (KAMIENSKA and PHARIS 1975). The maximum activity, mostly in more polar fractions corresponding to GA_3 (in *P. attenuata,* 250 μg of GA_3 kg^{-1} pollen), was found in all three pine species at 15 h. Since this is the period of most rapid pollen tube growth, it may possibly indicate that the increased levels of GA's present during early germination are required for pollen tube growth. It was estimated that dormant pollen of *P. attenuata* contained 5.8 μg kg^{-1} (expressed in GA_3 equivalent) pollen of nonpolar GA-like substances (including GA_9 and GA_7).

At h 15 of germination the amounts of nonpolar GA's had dropped to 3.6 μg kg^{-1}. Coupled with this decrease in nonpolar GA's was an increase in GA-like substances of a more polar nature (KAMIENSKA and PHARIS 1975), the most prominent being GA_3 (KAMIENSKA et al. 1976a). In order to test the hypothesis that there was a conversion during germination of less polar GA's (such as GA_4 and GA_7) to more polar GA's such as GA_3, pollen was germinated for 15 h in a medium containing high specific activity 1, 2- ^3H-GA_4 (KAMIENSKA et al. 1976b). Conversion of GA_4 to GA_1 and GA_{34} was noted, but not to GA_7 or GA_3. It was suggested that the predominant conversion of GA_4 to GA_{34} represents a deactivation pathway, metabolism of $GA_4 \rightarrow GA_1$ (which might be an activation pathway) being only one-tenth of the $GA_4 \rightarrow GA_{34}$ conversion. However, more complete kinetic studies with additional GA's will be necessary to establish a possible relevance of metabolic interconversion of GA's to pollen tube elongation or its cessation.

Several other as yet unidentified factors have been reported to stimulate pollen germination. Germination, but not tube growth of *Chrysanthemum morifolium* pollen has been increased three- to fivefold by diffusates, water extracts of cold ether extracts of stigma, style, ovary, and other flower parts (TSUKAMOTO and MATSUBARA 1968). Extracts of different plant parts from several other plant species were also effective in stimulating germination of chrysanthemum pollen, but this stimulation could not be duplicated by GA_3, kinetin, NAA, Ca^{2+}, boron, etc. However, until these substances are identified and studied further, their biological significance cannot be judged.

In summary, it is well documented that pollen contains auxins and GA's and several unidentified promotive and inhibitory substances. The presence of cytokinins, however, has yet to be verified. It is not clear, however, whether the primary function of these two classes of hormone is the regulation of pollen germination and tube growth, or some other purpose, such as fruit set.

1.3 Pollen, Hormones, and Fruit Set

Following pollination and growth of the pollen tube down the style, a large increase in growth of the ovary occurs. In unpollinated flowers these changes do not take place, although the physical presence of pollen and the act of double fertilization are not themselves absolute requirements for "fruit set" (AUDUS 1959). Pollen extracts or plant-growth substances can be used to stimulate the growth of the ovary in the absence of pollination, sometimes even to the point of forming mature seedless fruits (AUDUS 1959, GARCIA-MARTINEZ and GARCIA-PAPI 1979, GUSTAFSON 1961, HOLLOWAY et al. 1982, MUDGE et al. 1981, SAAVEDRA 1979). It is thus possible that the initial set of the fruit might be caused (or triggered) by growth substances liberated into the pistil tissue from the pollen. The quantities of hormones released by the pollen, however, seem inadequate to account for continued fruit growth.

In general, pollination appears to induce increased auxin levels in the pistil, and for angiosperms at least, after fertilization, the developing seeds seem to be the primary sites of auxin synthesis in the developing fruit (AUDUS 1959).

Following fertilization there is a very large increase in diffusible auxin in the ovaries of tobacco and *A. majus* (MUIR 1947). Tobacco pollen grains, tubes, and extracts of pollen grains and tubes contain an enzyme(s) that is capable of converting tryptophan into IAA. The enzyme activity increased threefold in homogenates of pollen tubes that had germinated for 10 h, and fourfold when the pollen tubes were grown with tryptophan prior to homogenization (LUND 1956). Styles and ovaries of unpollinated and pollinated flowers also contain a similar enzyme system. The enzyme activity in the ovaries remains unchanged until about 46 h after pollination, when there is an increase which continues for several days (LUND 1956). LUND (1956) has postulated that pollen tubes of tobacco contribute additional quantities of enzymes to the style and ovary which convert tryptophan to IAA, and thereby initiate fruit development. In contrast, BARENDSE et al. (1970) report no increase in concentration of either auxins or GA's in styles of *Petunia* after pollination, over the content present in the unpollinated style and the pollen.

In tomato, fruit set can be induced by pollination or treatment with several growth regulators. The changes in IAA, gibberellins, and ABA in pollinated ovaries or in tomato fruits induced parthenocarpically by several synthetic growth regulators were studied (SJUT and BANGERTH 1981). The growth regulators used were bendroquinone; methyl-2-chloro-9-hydroxyfluorene-(9)-carboxylate (CME); 4-chlorophenoxyacetic acid (4-CPA) and 5,7-dichloro-4-ethoxycarbonyl methoxy-2,1,3-benzothiadiazole. IAA was determined by the specific conversion of IAA into indole-α-pyrone and its quantitation by spectrofluorimetry. Gibberellin-like substances were purified by column and paper chromatography followed by the barley endosperm assay using GA_3 as a standard. ABA was determined by gas-liquid chromatography after methylation with diazomethane. Unpollinated ovaries were found to contain a very high concentration of ABA ($>2.5\ \mu g\ g^{-1}$ dry wt.). After pollination this level dropped to 10–30% of the initial concentration in 5-day fruits. In fruits induced parthenocarpically by the growth regulators, a similar reduction in ABA was found after 5 days. The IAA level increased with pollination and more than doubled at 5 days from ca. 80 to ca. 190 ng IAA g^{-1} dry wt. In fruits induced by some of the growth regulators, an increase in IAA was observed after 5 days, although this increase was not as great as in fruits induced by pollination (SJUT and BANGERTH 1981). With the assay procedure used, no GA-like activity was detected in extracts of unpollinated ovaries. Within 5 days after pollination there was an increase in GA-like substances to 7 ng GA_3 Eq g^{-1} dry wt. A similar increase in GA activity was seen in fruits induced parthenocarpically by the growth regulators (SJUT and BANGERTH 1981). In tomato ovaries after pollination there is a 4.7-fold increase in ornithine decarboxylase activity, and auxin, in addition to inducing fruit set, also causes a threefold increase in the activity of this enzyme (MIZAHI and HEIMER 1982). The similar changes in IAA, ABA, and gibberellins, or in ornithine decarboxylase seen in tomato fruits after pollination or after parthenocarpic induction with growth regulators, would seem to argue for a role of hormones contributed by pollen in the process of fruit set and early fruit development.

Designation of the pollen grain as an "environmental factor", may be considered by some as stretching a point. However, if germination and/or fertilization processes are indeed a trigger for subsequent fruit and seed development, and this trigger is effected by the release of hormonal substances from the pollen, then such a designation may not be so far-fetched.

Unfortunately, the possible role(s) of pollen-produced hormones in triggering fruit set, and perhaps hormone synthesis by the developing fruit, will not be resolved until accurate kinetic studies of hormone (e.g., IAA and the GA's) concentration, metabolism (including conjugation), and catabolism are accomplished for both the germinating pollen and the style and ovary. Hormone quantitation should involve accurate physical methods [e.g., selected ion monitoring by mass spectrometry (BANDURSKI 1958) and references cited therein], and turnover studies will require the use of labeled hormones and their precursors and conjugates. These methods, however, must correlate hormonal metabolism closely with morphogenetic processes. Such studies, carried out with systems known to respond to exogenous hormone application in lieu of pollination and/or fertilization, may prove most useful.

2 Symbionts and Symbiont-Induced Structures

2.1 Nitrogen-Fixing Associations[1]

Most legumes are capable of forming root nodules in response to infection by bacteria of the genus *Rhizobium*. As was done for pollen, the bacteria are considered as environmental agents interrelating with the plant via hormones. In addition to legumes, root-nodule associations that fix N_2 are known for several other orders of angiosperms, the Myricales, Fagales, Casuarinales, Rhamnales, Coriariales, and Rosales (PARKER 1977). The endophytes in the nodules of the nonlegume families are actinomycetes (BOND 1967, 1974, PARKER 1977, QUISPEL 1974, PHILLIPS 1971b, MOSER 1959). An exception are species of *Parasponia* which are nodulated by *Rhizobium* (AKKERMANS et al. 1978).

The sequence of events leading to nodule formation in legumes is summarized below. A more detailed treatment of the subject for both legume and nonlegume symbioses can be found in several recent reviews (DART 1977, GARDNER 1976, LIBBENGA and ROGERS 1974, NEWCOMB 1981, VANCE 1983). Bacteria of a *Rhizobium* strain capable of nodulating a given legume recognize and bind to the root hairs. Curling of the root hairs precedes infection. The bacteria enter the distorted root hairs at the point of attachment, and travel in a membrane-enclosed, walled, infection thread toward the interior of the root. The infection thread containing the bacteria advances from one cell through adjacent cell walls and into neighboring cells toward the vascular tissue. In successful nodule formation, meristematic activity begins for prospective nodule loci in polyploid cells of the cortex that lie ahead of the advancing infection thread. The host cells differentiate into bacteroid-containing cells. New meriste-

1 A general chapter on this topic is contained in Vol. 15A (1983), this series (Inorganic Plant Nutrition, LÄUCHLI A and BIELESKI RL eds.), Chaps. II.1–II.3 by BOTHE H et al., QUISPEL A, DÖBEREINER J, pp. 241–350

matic activity is initiated in neighboring cells and these new cells are, in turn, invaded by bacteria. The various regions of the young nodule are differentiated. There is a central bacteroid-containing tissue, surrounded by uninfected nodule cortex, a vascular connection with the root, and a meristematic zone. The exact anatomical details vary from species to species.

Hormones appear to be at least partially involved in the initial root-curling reaction, and also in the initiation and formation of the nodule (DART 1977).

2.1.1 Auxins

Various *Rhizobium* strains have the capacity to produce plant-growth substances in culture. *Rhizobia* can utilize tryptophan and convert it to IAA (ALLEN and ALLEN 1958, DULLAART 1970a, HARTMANN and GLOMBITZA 1967, RIGAUD 1970a, b, RIGAUD and TRINCHANT 1973, SABELNIKOVA et al. 1981). A large quantity of ILA is present in suspensions of washed cells of *R. meliloti*. When ILA is incubated with *R. meliloti*, IAA, indole-3-acetaldehyde, and tryptophol are produced (RIGAUD 1970a).

IAA has been conclusively identified in culture supernatants of several *Rhizobium* strains by gas chromatography-mass spectrometry. In culture supernatants of early phase cells, values of IAA from 97–711 $\mu g\ l^{-1}$ have been measured for different strains (BADENOCH-JONES et al. 1982).

Bacteroids isolated from the root nodules of *Phaseolus vulgaris* L. produce IAA and ILA from tryptophan. Low O_2 tensions strongly enhanced ILA production while lowering IAA synthesis. Production of ILA was inhibited, whereas IAA production by bacteroids containing active nitrogenase was stimulated as O_2 tension increased (TRINCHANT and RIGAUD 1977).

Nodules contain diffusible auxins in much greater concentrations than neighboring root tissues (ALLEN and ALLEN 1958, DULLAART 1970b, DULLAART and DUBA 1970, PATE 1958, THIMANN 1936). Five distinct substances active in the *Avena* coleoptile straight growth test were found in nodules of *Pisum arvense* L. and *Ulex europaeus* L. Three of the active substances were promoters and two were inhibitors of coleoptile elongation. As determined by paper chromatography and assay of fractions for activity in the *Avena* coleoptile test, the dominant auxin of *P. arvense* nodules was IAA, and it could be recovered in large amounts at all stages of nodule growth (PATE 1958). No qualitative or quantitative decrease in the growth substances of *P. arvense* nodules was observed as the nodules became senescent (PATE 1958). The nodules of the nonleguminous nitrogen-fixing *Alnus glutinosa* L. also contain high auxin levels (DULLAART 1970b). As determined by paper chromatography, two-dimensional TLC, and fluorescence spectra, IAA was the main auxin of nodule tissues, while ICA was present only in root tissues (DULLAART 1970b). No seasonal variation in auxin content either for nodules or roots was found (DULLAART 1970b).

When high specific activity ^{3}H-IAA was applied to the apical bud of pea (*Pisum sativum*) plants, radioactivity was detected in all tissues of the plant after 24 h (BADENOCH-JONES et al. 1983). More radioactivity on a fresh weight basis was accumulated in nodules than in normal root tissue. A greater proportion of the radioactivity in the nodule was as indole-3-acetylaspartic acid

(IAAsp) as compared to root tissue. Functional nodules which fix nitrogen accumulate more auxin transported from the shoot and metabolize it to IAAsp than ineffective nodules (nodules from a mutant *Rhizobium* strain that are unable to fix nitrogen). These results indicate that the nodule is markedly different in its auxin relationships from the root tissue from which it is derived, and they suggest that the metabolism of auxins received from the shoot apex may be important in the functioning of the root nodule (BADENOCH-JONES et al. 1983).

When clover (*Trifolium subterraneum* L.) was grown in distilled water for 3 weeks, tryptophan was detected in the root medium of both sterile control cultures, and in those inoculated with *R. trifolii* 3 days earlier (KEFFORD et al. 1960). Auxin with chromatographic properties of IAA, however, was detected only in the inoculated medium, suggesting that tryptophan exuded by clover roots was converted to IAA by nodule bacteria (KEFFORD et al. 1960). The addition of NAA to *Medicago sativa* L. plants grown on a mineral salts agar delayed initiation and the total number of nodules formed (KEFFORD et al. 1960). The antiauxin, p-chlorophenoxy-isobutyric acid had no effect on nodule initiation, but increased the rate of nodulation and the total number of nodules formed per plant (KEFFORD et al. 1960).

In general, added auxin usually delays the initiation of nodulation and even prevents some plants from nodulating, while antiauxins can increase nodulation (DART 1977).

Nodule roots of the nonlegumes *Myrica cerifera* and *Casuarina cunninghamiana* exhibit a negative geotropism. Non-nodulated roots show normal positive geotropism. No auxin was detectable from nodule roots when geotropism was negative, whereas in non-nodulated roots of *Myrica,* the concentration of auxin was high (SILVER et al. 1966). Root nodules of *Alnus* species, whose roots exhibit positive geotropism, contain measurable endogenous auxin (SILVER et al. 1966). It has been suggested that the negative geotropic curvature of the nodule roots of *Myrica* and *Casuarina* is due to the presence of a suboptimal auxin concentration which results from a highly active IAA oxidase system which has been shown to be present in such roots (SILVER et al. 1966), although increased formation of auxin conjugates should not be ruled out (see BANDURSKI 1958).

The infection of the root by nodule bacteria occurs through the root hair. The curling of the root hair precedes the actual entry of the bacteria. Auxins have been reported to induce curling in root hairs, this reaction being nonspecific both with respect to bacterial and host species (BECKING 1975, NUTMAN 1965).

BADENOCH-JONES et al. (1982) studied the IAA in the culture medium produced by several *Rhizobium* strains defective in nodulation and also their parent wild-type strains. Some of these defective strains were known to adhere to root hairs, but did not induce root hair curling. Similar IAA concentrations were found in the culture supernatants of both wild-type and defective strains, indicating that root hair curling is not a simple consequence of IAA production by *Rhizobium* (BADENOCH-JONES et al. 1982).

It appears, however, that it is not auxins alone which stimulate the curling reaction. Rather, macromolecules (in the presence of auxin?) from culture filtrates of *Rhizobium* induce curling (DART 1977, SOLHEIM and RAA 1973).

2.1.2 Gibberellins

The presence of GA's in nitrogen-fixing root nodules has received relatively little attention. Cultures of *R. meliloti, R. trifolii,* and *R. leguminosarum* in defined medium produce a GA-like substance which chromatographs in the GA_3 region, in amounts of 1 to 4 μg of GA_3 equivalents l^{-1} of medium after 7 days growth at 26° C (KATZNELSON and COLE 1965).

Root nodules of *Pisum sativum, Phaseolus vulgaris* and *Lupinus luteus,* and *Glycine max* contain much higher levels of GA-like substances than non-nodulated roots (DULLAART and DUBA 1970, RADLEY 1961, WILLIAMS and DEMALLORCA 1982).

Lima bean (*Phaseolus lunatus*) plants were inoculated with two different strains of *Rhizobium.* One of the strains caused plants to grow significantly taller than the other. This internode elongation in plants caused by one strain appeared to be due to the large quantities of GA-like substances in nodules which apparently contributed to the elongation of the plant (EVENSEN and BLEVINS 1981).

Levels of GA-like activity measured by the lettuce hypocotyl bioassay were higher in the root nodules of the nonlegume *Alnus glutinosa* than in normal root tissue, irrespective of whether the plants were dormant or actively growing (HENSON and WHEELER 1977a). Addition of GA_3 to agar root medium or as a leaf treatment either has no effect, or reduces nodulation of several plants (FLETCHER et al. 1959, MES 1959).

2.1.3 Cytokinins

Cytokinins are released into the medium by cultures of *R. japonicum* and *R. leguminosarum* (GIANNATTASIS and COPPOLA 1969, PHILLIPS and TORREY 1970, 1972) and the amounts increased during the logarithmic phase of growth. The cytokinin released by *R. japonicum* was identified as a zeatin-like compound on the basis of paper chromatography in four solvent systems. The predominant intracellular cytokinin found in both enzymatic hydrolysates of transfer RNA and alkaline hydrolysates of total RNA was also similar to zeatin (PHILLIPS and TORREY 1970, 1972).

Cytokinins have been detected in root nodules of both leguminous and nonleguminous plants (HENSON and WHEELER 1976, 1977b, PUPPO et al. 1974, RODRIGUEZ-BARRUECO and BERMUDEZ DE CASTRO 1973, SYONO and TORREY 1976, JAISWAL et al. 1981a, b).

At least three kinds of cytokinin were present in root nodules of *Vicia faba,* two of which had chromatographic properties similar to zeatin and its riboside. The zeatin riboside-like compound was the main cytokinin detected in the root nodules and roots, and cytokinin levels in the root nodules were 12 to 13 times those detected in the roots (HENSON and WHEELER 1976). Five cytokinin-like active zones were detected in ethanol extracts of *P. sativum* infected by *R. leguminosarum* (SYONO and TORREY 1976). The most biologically active compounds were identified as zeatin and its riboside on the basis of TLC in three solvent systems and activity in the soybean callus assay. Small amounts of zeatin ribotide, 2iP, and its riboside were also detected by TLC (SYONO and TORREY 1976). For the nonleguminous plants, *Purshia tridentata* (Pursh) D.C., *Myrica gale* L., *Alnus rubra* Bong, A. *glutinosa* (L.) Gaertn., *Hippophae rhamnoides* L. and *Colletia paradoxa* (Spreng.) escalente extracts of root nodules also contained higher levels of cytokinin activity than did extracts of roots (HENSON and WHEELER 1977b, c).

When *Phaseolus vulgaris* plants inoculated with *Rhizobium phaseoli* were grown under hydroponic conditions, substantial quantities of cytokinins were detected in the plant culture medium and their concentration remained high for the first 9 days after inoculation when nodules first appeared. Both plant and bacteria were required for this synthesis (PUPPO and RIGAUD 1978).

In an attempt to answer the question "do nodules export cytokinins to the rest of the plant?", 8-^{14}C-zeatin was applied to root nodules of *Alnus glutinosa*. Within 24 h zeatin and a number of its metabolites were detected in all parts of the plant (HENSON and WHEELER 1977c). Since transport of exogenous cytokinin applied in large doses might not reflect the endogenous movement of cytokinins, alternate methods of study were used, such as denodulation, etc., and they suggested that nodules are not likely to be a major source of shoot cytokinins (WHEELER and HENSON 1978).

2.1.4 Interaction of Growth Substances in Nodule Development

The development of root nodules reflects an intimate relationship between the bacteria and the root cells of the host plant. In *P. sativum,* the infection thread grows between and into cortical cells of the root (LIBBENGA and HARKES 1973). Nuclei of the inner cortical cells some distance ahead of the advancing infection thread undergo endoreduplication and mitosis, eventually forming a polyploid nodule meristem. The meristem itself is not infected with *Rhizobium*. Products of division of the meristem are invaded by bacteria released from the infection thread, and differentiate into nondividing, bacteroid-containing cells (LIBBENGA and HARKES 1973). As already discussed, root nodule bacteria and root nodules possess all the major classes of plant hormones. From the nature of the events occurring in the root during nodule initiation and development, one might expect that the interactions of growth substances produced by both host cell and symbiont are involved in regulating the process of nodule formation, since both auxin and cytokinin are required for cell division in plant tissue cultures (DAS et al. 1956).

The process of endoreduplication in cortical cells of the root can be duplicated by culturing pea root segments on a synthetic nutrient medium containing auxin (IAA, 2,4-D) and kinetin (LIBBENGA and TORREY 1973, TORREY 1961). In addition, pseudonodules, i.e., abnormal root outgrowths, superficially resembling true nodules, but containing no microbial endophyte, can be produced by various growth-regulator treatments (ALLEN et al. 1953, ARORA et al. 1959, BECKING 1975, RODRIGUEZ-BARRUECO and BERMUDEZ DE CASTRO 1973).

In contrast, both nodulating and non-nodulating strains of *R. leguminosarum* produce similar amounts of IAA as quantitated by gas chromatography-mass spectrometry (WANG et al. 1982). This would seem to indicate that the ability to produce IAA does not necessarily confer upon a *Rhizobium* strain the ability to nodulate pea roots.

Cytokinins are released into the medium by cultures of *R. japonicum* and *R. leguminosarum*. Calculations showed that the amount of cytokinin released could stimulate the mitotic events observed in the root cortex near the tip of the infection thread (PHILLIPS and TORREY 1970, 1972). From studies of cultured explants of root-cortex tissue from *P. sativum* plants, LIBBENGA et al. (1973) have suggested that in addition to the release of auxin and cytokinin by *Rhizobium* there is a transverse gradient of unidentified cell-division factor(s) in the root cortex which may control the induction of cell divisions in nodule initiation. Cytokinin activity was found to be highest in 2- and 3-week-old nodules of *P. sativum* roots infected with *R. leguminosarum* when the mitotic rate was high, and decreased in older nodules (SYONO et al. 1976). Moreover, cytokinin activities were positively correlated with mitotic indices, and most of the cytokinins were present in the meristematic tip and not in the infected nitrogen-fixing cells of the nodule. Based on these results, it has been proposed that cytokinins influence nodule morphogenesis by regulating the mitotic activity of the nodule meristem (SYONO et al. 1976). Uninfected pea roots contain only zeatin and its ribose derivatives (SHORT and TORREY 1972). Pea root nodules, on the other hand, contain in addition to zeatin and its riboside, small amounts of 2iP and its derivatives (SYONO and TORREY 1976). Cultures of *R. leguminosarum* produce 2ip and IPA (GIANNATTASIS and COPPOLA 1969, PHILLIPS and TORREY 1970, 1972). The initial events in nodule morphogenesis may be controlled by 2iP produced by the bacteria and later events regulated by zeatin synthesized by the root tissue (NEWCOMB et al. 1977, SYONO and TORREY 1976).

Root nodules frequently occur opposite the xylem poles in roots of pea. When one cotyledon was removed before germination, there was a promotion of nodules in the root cortex opposite the primary phloem poles (PHILLIPS 1971a). These data were interpreted in terms of an inhibitor of cortical cell division which is transported from the cotyledons via the phloem to the developing root. Since ABA inhibits nodulation at low concentrations at which neither root hair formation nor *Rhizobium* growth is inhibited, PHILLIPS (1971b) has suggested that ABA could be the host inhibitor involved.

Evidence for the presence of ABA in perennial root nodules of *Alnus glutinosa* has been obtained by chemical methods of analysis (WATTS et al. 1983). The free-ABA content of nodules fluctuates seasonally over the range 101 to 879 ng g^{-1} fresh weight. No obvious correlations were obtained between the amount of ABA detected and the rates of nodule growth. ABA pools were, however, 2.5 times higher in dormant nodules than in those from actively growing plants, indicating a possible role of ABA in dormancy of the nodules. Nodules from roots harvested during active growth showed 2.9 times more free ABA than roots (WATTS et al. 1982). Higher ABA levels were also found in nodule extracts than in root extracts of *Glycine max* (WILLIAMS and DEMALLORCA 1982).

Thus, the intriguing question of the role(s) of bacterial and host growth substances in the regulation of nodule development still remains to be clarified.

2.2 Mycorrhizae [2]

Roots, as a result of infection by symbiotic fungi, show characteristic structural modifications called mycorrhizae. Mycorrhizae are of widespred occurrence among plants (GERDEMANN 1975, MEYER 1966, MILLER 1983). Most forest trees have ectomycorrhiza (ECM), while the majority of nonwoody species, which include many agriculturally important plants, form endomycorrhizae (EDM) (SLANKIS 1974). The physiology of mycorrhizal symbioses has been recently reviewed (HAYMAN 1983).

Culture filtrates of ECM fungi induce root morphogenesis similar to that which roots undergo in the presence of the symbiont (SLANKIS 1948, 1973). Mycorrhizal fungi produce plant-growth substances in pure culture. A large number of such fungi produce IAA alone or together with IPA and/or IBA (GOGALA 1971, MOSER 1959, RUDAWSKA 1982, STRZELCZYK et al. 1977, ULLRICH 1960). A large concentration of auxins (as measured by bioassay) is present in mycorrhizal roots of pine seedlings, but only trace amounts are found in nonmycorrhizal roots (SLANKIS 1974). Mycorrhizal fungi also produce and release cytokinins (BAREA and AZCON-AGUILAR 1982, GOGALA 1971, KAMPERT and STRZELCZYK 1978, MILLER 1967, NG et al. 1982, RUDAWSKA 1982). The puffball fungus *Rhizopogon roseolus* produces and releases three cytokinins into liquid media. Two of these compounds were crystallized and found to have properties identical to those of synthetic zeatin and zeatin riboside (MILLER 1967). Fruiting bodies, mycelia, and culture medium of *Boletus edulis* are reported to contain, besides cytokinins, three GA-like compounds (GOGALA 1971).

Similarly, axenically grown *Glomus mosseae* synthesized at least two GA-like substances and four substances with properties of cytokinins (NG et al. 1982). Eight out of nine fungi isolated from mycorrhizae of *Pinus sylvestris* produced cytokinins, as assayed by the soybean callus assay (KAMPERT and STRZELCZYK 1978). Most of the fungi-synthesized compounds have Rf values identical or similar to zeatin, zeatin riboside, or 6-(γ-γ-dimethylallylamino) purine and some of the fungi produce other unidentified compounds with cytokinin activity (KAMPERT and STRZELCZYK 1978).

Several ectomycorrhizal fungi have been found to produce ethylene in media supplemented with 2.5 to 10 mM methionine (GRAHAM and LINDERMAN 1980). Ethylene was produced by aseptically grown Douglas-fir seedlings inoculated with several of the fungi and the appearance of ethylene coincided with the formation of mycorrhizae. The authors suggest that ethylene formation by mycorrhizal fungi may depend on a supply of methionine from the plant root, and that ethylene possibly has a role in symbiotic fungus–host interactions (GRAHAM and LINDERMAN 1980).

Root structures that resemble mycorrhizae have been induced with synthetic auxins both in excised and attached pine roots (SLANKIS 1949, 1950, 1958). The type of morphological modification was concentration-dependent. The my-

2 A general survey can be found in Vol. 17 (1984) of this Encyclopedia (Cellular Interactions, LINSKENS HF and HESLOP-HARRISON J eds.), Chap. 9 by HARLEY JL, pp. 148–186

corrhiza-like structures are not permanent formations, and if a periodic supply of fungus exudates or synthetic auxin is not provided, the swollen apices of the roots begin to elongate and within a few weeks the mycorrhiza-like characteristics disappear (SLANKIS 1958). SLANKIS (1958) thus suggested that the conversion of roots into mycorrhizae was induced by one or more auxin-like metabolites exuded by the symbiotic fungus.

Isolated pine roots absorb and accumulate IAA from growth media (FORTIN 1970). Since high concentrations of IAA inhibit mycelial growth, FORTIN (1970) has suggested that the absorption of IAA by roots enables the continuous production of IAA by the fungus without a build-up in concentration of IAA that might otherwise occur, thus resulting in auto-inhibition of the fungal growth.

Mycorrhiza-like roots can be induced by substances other than auxins. In *Pinus serotina,* neither IAA nor NAA induced dichotomous root branching, whereas this condition was induced by a large number of other compounds, such as amino acids, certain vitamins, and kinetin (BARNES and NAYLOR 1959a, b). Kinetin was the most effective inducer. Although present evidence suggests that growth substances produced by the symbiotic fungus influence the development of roots of the host plant, the exact extent of this influence remains to be elucidated. Hormones produced by the fungal symbiont alone, or in collaboration with other soil microorganisms, can also influence growth of the host plant. The dry weights of tomato plants inoculated with both *Glomus fasciculatus* and *Azotabacter chroococcum* and grown in phosphate-deficient soil were significantly (62%) greater than noninoculated plants (BAGYARAJ and MENGE 1978). In a study of growth and infection levels of *Lavandula spica* L., *Lycopersicum esculentum* and *Medicago sativa* plants inoculated with *G. fasciculatus,* cell-free supernatants of *R. meliloti, A. vinelandii, A. beijerinckii* and *Pseudomonas* species improved the dry weight of the plants and infection levels with mycorrhiza over those of untreated plants (AZCON et al. 1978). The bacterial cultures contain growth substances and the stimulating effects on growth and infection levels could be duplicated by a mixture of IAA, GA_3, and kinetin (AZCON et al. 1978).

Bouteloua gracilis plants were grown in defined, axenic culture with and without mycorrhizae (ALLEN et al. 1980). Total cytokinin activity was 57 and 111% greater in leaves and roots respectively in mycorrhizal over uninfected plants. The increased cytokinin levels could result from fungal production of cytokinins and their translocation into the plant, from the inhibition of cytokinin degradation by the fungus or the plant, or by the stimulation of cytokinin production by the plant as a result of fungal infection. Infection by mycorrhizal fungi resulted in significantly increased GA activity in the leaves and a decreased activity in the roots (ALLEN et al. 1982). ABA levels decreased in leaves of infected plants, but remained unchanged in roots (ALLEN et al. 1982). These altered levels of hormones in the different parts of the plants indicate that major changes in the physiology of the host plant follow mycorrhizal infection (ALLEN et al. 1980, 1982).

There are several other symbiotic systems where specialized morphological structures are formed as a result of the association of host and symbiont. Examples are the N_2-fixing association of blue-green algae (*Anabaena* species) with the aquatic fern *Azolla* species; and fungal and algal associations in lichens. One might expect from the nature of the interactions between the host plant and the symbiont that growth substances might be involved. There is, however, a paucity of information concerning the role of hormones in these and other symbiotic interrelationships.

References

Addicott FT (1943) Pollen germination and pollen tube growth as influenced by pure plant growth substances. Plant Physiol 18:270–279

Akkermans ADL, Abdulkadir S, Trininck MJ (1978) Nitrogen fixing root nodules in Ulmaceae. Nature (London) 274:190

Allen EK, Allen ON (1958) Biological aspects of symbiotic nitrogen fixation. In: Ruhland W (ed) Encyclopedia of plant physiology, vol VIII. Springer, Berlin Göttingen Heidelberg, pp 48–118

Allen EK, Allen ON, Newman AS (1953) Pseudonodulation of leguminous plants induced by 2-bromo-3,5-dichlorobenzoic acid. Am J Bot 40:429–435

Allen MF, Moore TS, Christensen M (1980) Phytohormone changes in *Bouteloua gracilis* infected by vesicular arbuscular mycorrhizae. 1. Cytokinin increases in the host plant. Can J Bot 58:371–374

Allen MF, Moore TS, Christensen M (1982) Phytohormone changes in *Bouteloua gracilis* infected by vesicular arbuscular mycorrhizae. 2. Altered levels of gibberellin-like substances and abscisic acid in the host plant. Can J Bot 60:468–471

Arora N, Skoog F, Allen ON (1959) Kinetin-induced pseudonodules on tobacco roots. Am J Bot 46:610–613

Audus LJ (1959) Plant growth substances.Hill, London, pp 164–194

Azcon R, Azcon G, de Aguilar C, Barea JM (1978) Effects of plant hormones in bacterial cultures in the formation and responses to VA endomycorrhiza. New Phytol 80:359–364

Badenoch-Jones J, Summons RE, Djordjevic MA, Shine J, Letham DS, Rolfe BG (1982) Mass spectrometric quantification of indole-3-acetic acid in *Rhizobium* culture supernatants. Relation to root hair curling and nodule initiation. Appl Environ Microbiol 44:275–280

Badenoch-Jones J, Rolfe BG, Letham DS (1983) Phytohormones, *Rhizobium* mutants, and nodulation in legumes. III. Auxin metabolism in effective and ineffective pea root nodules. Plant Physiol 73:347–352

Bagyaraj DJ, Menge JA (1978) Interaction between a VA mycorrhiza and *Azobacter* and their effects on rhizosphere microflora and plant growth. New Phytol 80:567–573

Bandurski RS (1958) Chemistry and physiology of conjugates of indole-3-acetic acid. In: Mandava NB (ed) Plant growth substances. ACS Symp Ser 111. Am Chem Soc, Washington DC, pp 1–18

Barea JM, Azcon-Aguilar C (1982) Production of plant growth regulating substances by the vesicular arbuscular mycorrhizal fungus *Glomus mosseae*. Appl. Environ Microbiol 43:810–813

Barendse GWM, Rodrigues-Pereira AS, Berkers PA, Driessen FM, Eyden-Emons van A, Linskens HF (1970) Growth hormones in pollen, styles and ovaries of *Petunia hybrida* and of *Lilium* species. Acta Bot Neerl 19:175–186

Barnes RL, Naylor AW (1959a) In vitro culture of pine roots and the use of *Pinus serotina* roots in metabolic studies. Forest Sci 5:158–168

Barnes RL, Naylor AW (1959b) Effect of various nitrogen sources on the growth of isolated roots of *Pinus serotina*. Physiol Plant 12:82–89

Becking JH (1975) Root nodules in non-legumes. In: Torrey JG, Clarkson DT (eds) The development and function of roots. Academic Press, London New York, pp 507–566

Bond G (1967) Fixation of nitrogen by higher plants other than legumes. Annu Rev Plant Physiol 18:107–126

Bond G (1974) Root-nodule symbioses with actinomycete-like organisms. In: Quispel A (ed) The biology of nitrogen fixation. North-Holland Biomedical Press, Amsterdam New York, pp 342–378

Bose N (1959) Effect of gibberellin in the growth of pollen tubes. Nature 184:1577

Chandler C (1957) The effect of gibberellic acid on germination and pollen tube growth. Contrib Boyce Thompson Inst 19:215–244

Dart P (1977) Infection and development of leguminous nodules. In: Hardy RWF, Silver WS (eds) A treatise on dinitrogen fixation, Sect III: Biology. John Wiley, New York, pp 367–472

Das NK, Pätau K, Skoog F (1956) Initiation of mitosis and cell division by kinetin and indole acetic acid in excised tobacco pith tissue. Physiol Plant 9:640–651

Dhawan AK, Malik CP (1981) Effect of growth regulators and light on pollen germination and pollen tube grown in *Pinus roxburghii* Sarg. Ann Bot 47:239–248

Dullaart J (1970a) The bioproduction of indole-3-acetic acid and related compounds in root nodules and roots of *Lupinus luteus* L. and by its rhizobial symbiont. Acta Bot Neerl 19:573–615

Dullaart J (1970b) The auxin content of root nodules and roots of *Alnus glutinosa* (L.) Vill. J Exp Bot 21:975–984

Dullaart J, Duba LI (1970) Presence of gibberellin-like substances and their possible role in auxin bioproduction in root nodules and roots of *Lupinus luteus* L. Acta Bot Neerl 19:877

Evensen KB, Blevins DG (1981) Differences in endogenous levels of gibberellin-like substances in nodules of *Phaseolus lunatus* cultivar Henderson plants inoculated with two *Rhizobium* strains. Plant Physiol 68:195–198

Fähnrich P, Ullrich H (1964) Untersuchungen zur Pollenkeimung und zum Pollenschlauchwachstum II. Die Wirkung von dreiwertigen Metallionen und von Wuchsstoffen auf Pollen von *Petunia* und *Antirrhinum*. Planta 62:39–50

Fitting H (1909) Die Beeinflussung der Orchideenblüten durch die Bestäubung und durch andere Umstände. Z Bot 1:1–86

Fletcher WW, Alcorn JWS, Raymond JC (1959) Gibberellic acid and nodulation of legumes. Nature 184:1576

Fortin JA (1970) Interaction entre basidiomycètes mycorrhizateurs et racines de pin en présence d'acide indol-3yl-acétique. Physiol Plant 23:365–371

Garcia-Martinez JL, Garcia-Papi MA (1979) The influence of gibberellic acid, 2,4-D and 6-benzylamino purine on fruit set of clementine mandarin. Sci Hortic 10:285–294

Gardner IC (1976) Ultrastructural studies of non-leguminous root nodules. In: Nutman PS (ed) Symbiotic nitrogen fixation in plants. Cambridge Univ Press, New York, pp 485–496

Gerdemann JW (1975) Vesicular-arbuscular mycorrhizae. In: Torrey JG, Clarkson DT (eds) The development and function of roots. Academic Press, London New York, pp 575–591

Giannattasis M, Coppola S (1969) Isolamento di citochinine dal *Rhizobium leguminosarum* Frank. Giorn Bot Ital 103:11–17

Gogala N (1971) Growth substances in mycorrhiza of the fungus *Boletus pinicola* Vitt. and the pine tree *Pinus sylvestris* L. Diss, Cl IV (XIV/5). Akad Sci Art Slov

Graham JH, Linderman RG (1980) Ethylene production by ectomycorrhizal fungi, *Fusarium oxysporum* f. sp. *pini* and by aseptically synthesized ectomycorrhizae and *Fusarium* infected Douglas-fir roots. Can J Microbiol 26:1340–1347

Gustafson FG, (1961) Development of fruits. In: Ruhland W (ed) Encyclopedia of plant physiology, vol XIV. Springer, Berlin Göttingen Heidelberg, pp 951–958

Hartmann T, Glombitza KW (1967) Der Tryptophanabbau bei *Rhizobium leguminosarum*. Arch Mikrobiol 56:1–8

Hayman DS (1983) The physiology of vesicular-arbuscular endomycorrhizal symbiosis. Can J Bot 61:944–963

Henson IE, Wheeler CT (1976) Hormones in plants bearing nitrogen-fixing root nodules: The distribution of cytokinins in *Vicia faba* L. New Phytol 76:433–439

Henson IE, Wheeler CT (1977a) Hormones in plants bearing nitrogen-fixing root nodules: gibberellin-like substances in *Alnus glutinosa* (L.) Gaertn. New Phytol 78:373–381

Henson IE, Wheeler CT (1977b) Hormones in plants bearing nitrogen-fixing root nodules: Cytokinin levels in roots and root nodules of some non-leguminous plants. Z Pflanzenphysiol 84:179–182

Henson IE, Wheeler CT (1977c) Hormones in plants bearing nitrogen-fixing root nodules: Cytokinin transport from the root nodules of *Alnus glutinosa* (L.) Gaertn. J Exp Bot 28:1099–1110

Holloway PS, Stushnoff C, Wildung DK (1982) Gibberellic acid induced fruiting of lingonberries. Hort Science 17:953–954

Jaiswal V, Rizvi SJH, Mukerji D, Mathur SN (1981a) Cytokinins in root nodules of *Phaseolus mungo*. Ann Bot 48:301–306

Jaiswal V, Rizvi SJH, Mukerji D, Mathur SN (1981b) Identification of cytokinins present in root nodules of *Phaseolus mungo*. Biol Plant 23:342–344

Jensen WA (1972) The embryo sac and fertilization in angiosperms. Univ Hawaii, HL. Lyon Arbor Lect 3:1–31

Kamienska A, Pharis RP (1975) Endogenous gibberellins of pine pollen II. Changes during germination of *Pinus attenuata, P. coulteri,* and *P. ponderosa*. Plant Physiol 56:655–659

Kamienska A, Durley RC, Pharis RP (1976a) Isolation of gibberellins A_3, A_4 and A_7 from *Pinus attenuata* pollen. Phytochemistry 15:421–424

Kamienska A, Durley RC, Pharis RP (1976b) Endogenous gibberellins of pine pollen. III. Conversion of 1,2-[^{3}H]GA_4 to gibberellins A_1 and A_{34} in germinating pollen of *Pinus attenuata* Lemm. Plant Physiol 58:68–70

Kampert M, Strzelczyk E (1978) Production of cytokinins by mycorrhizal fungi of pine (*Pinus silvestris* L.).Bull Acad Polon Sci Ser Sci Biol 26:499–503

Kato Y (1955) Responses of plant cells to gibberellins. Bot Gaz 117:16–24

Katznelson H, Cole SE (1965) Production of gibberellin-like substances by bacteria and actinomycetes. Can J Microbiol 11:733–741

Kefford NP, Brockwell J, Zwar JA (1960) The symbiotic synthesis of auxin by legumes and nodule bacteria and its role in nodule development. Aust J Biol Sci 13:456–467

Laibach F (1932) Pollenhormon und Wuchsstoff. Ber Dtsch Bot Ges 50:383–390

Lang A (1961) Auxins in flowering. In: Ruhland W (ed) Encyclopedia of plant physiology, vol XIV. Springer, Berlin Göttingen Heidelberg, pp 909–950

Larsen P, Tung SM (1950) Growth-promoting and growth-retarding substances in pollen from diploid and triploid apple varieties. Bot Gaz 111:436–447

Larter E, Chaubey C (1965) Use of exogenous growth substances in promoting pollen tube growth and fertilization in barley–rye crosses. Can J Genet Cytol 7:511–518

Libbenga KR, Harkes PAA (1973) Initial proliferation of cortical cells in the formation of root nodules in *Pisum sativum* L. Planta 114:17–28

Libbenga KR, Rogers RJ (1974) Root-nodule morphogenesis. In: Quispel A (ed) The biology of nitrogen fixation. North-Holland Biomedical Press, Amsterdam New York, pp 430–472

Libbenga KR, Torrey JG (1973) Hormone-induced endoreduplication prior to mitosis in cultured pea root cortex cells. Am J Bot 60:293–299

Libbenga KR, Iren van F, Bogers RJ, Schraag-Lamers MF (1973) The role of hormones and gradients in the initiation of cortex proliferation and nodule formation in *Pisum sativum* L. Planta 114:29–39

Linskens H, Kroh M (1970) Regulation of pollen tube growth. In: Moscona AA, Monroy A (eds) Current topics in developmental biology. Academic Press, London New York, pp 89–113

Lund HA (1956) The biosynthesis of indoleacetic acid in the styles and ovaries of tobacco preliminary to the setting of fruit. Plant Physiol 31:334–339

Mascarenhas JP (1975a) The biology of pollen. Module in biology No 14. Benjamin, Menlo Park, CA, pp 1–30
Mascarenhas JP (1975b) The biochemistry of angiosperm pollen development. Bot Rev 41:259–314
Mes MG (1959) Influence of gibberellic acid and photoperiod on the growth, flowering, nodulation, and nitrogen assimilation of *Vicia villosa*. Nature 184:2035–2036
Meyer FF (1966) Mycorrhiza and other plant symbioses. In: Henry SM (ed) Symbiosis, vol I. Academic Press, London New York, pp 171–255
Miller CO (1967) Zeatin and zeatin riboside from a mycorrhizal fungus. Science 157:1055–1057
Miller OK (1983) Ectomycorrhizae in the Agaricales and Gasteromycetes. Can J Bot 61:909–916
Mizahi Y, Heimer YM (1982) Increased activity of ornithine decarboxylase in tomato, *Lycopersicon esculentum* ovaries induced by auxin. Physiol Plant 54:367–368
Moser M (1959) Beiträge zur Kenntnis der Wuchsstoffbeziehungen im Bereich ectotropher Mycorrhizen. Arch Mikrobiol 34:251–269
Mudge KW, Narayanan KR, Poovaiah BW (1981) Control of strawberry *Fragaria ananassa* cultivar Ozark-Beauty fruit set and development with auxins. J Am Soc Hortic Sci 106:80–84
Muir RM (1947) The relationship of growth hormones and fruit development. Proc Natl Acad Sci USA 33:303–312
Newcomb W (1981) Nodule morphogenesis and differentiation. In: Giles KL, Atherly AG (eds) Biology of the rhizobiaceae. Academic Press, London New York, pp 247–298
Newcomb W, Syono K, Torrey JG (1977) Development of an ineffective pea root nodule: morphogenesis, fine structure, and cytokinin biosynthesis. Can J Bot 55:1891–1907
Ng PP, Cole ALJ, Jameson PE, McWah JA (1982) Cytokinin production by ectomycorrhizal fungi. New Phytol 91:57–62
Nutman PS (1965) Origin and developmental physiology of root nodules. In: Ruhland W (ed) Encyclopedia of plant physiology, vol XV, Pt 1. Springer, Berlin Göttingen Heidelberg, pp 1355–1379
Nygaard P (1969) Studies on the germination of pine pollen (*Pinus mugo*) in vitro. I. Growth conditions and effects of pH and temperature on germination, tube growth, and respiration. Physiol Planta 22:338–346
Parker CA (1977) Perspectives in biological dinitrogen fixation. In: Hardy RWF, Silver WS (eds) A treatise on dinitrogen fixation, Sect III, Biology. John Wiley, New York, pp 3–18
Pate JS (1958) Studies of the growth substances of legume nodules using paper chromatography. Aust J Biol Sci 11:516–528
Phillips DA (1971a) A cotyledonary inhibitor of root nodulation in *Pisum sativum*. Physiol Plant 25:482–487
Phillips DA (1971b) Abscisic acid inhibition of root-nodule initiation in *Pisum sativum*. Planta 100:181–190
Phillips DA, Torrey JG (1970) Cytokinin production by *Rhizobium japonicum*. Physiol Plant 23:1057–1063
Phillips DA, Torrey JG (1972) Studies on cytokinin production by *Rhizobium*. Plant Physiol 49:11–15
Pohl R (1951) Die Wirkung von Wuchsstoff und Hemmstoff auf das Wachstum der Pollenschläuche von *Petunia*. Biol Zentralbl 70:119–128
Puppo A, Rigaud J (1978) Cytokinins and morphological aspects of French bean roots in the presence of *Rhizobium phaseoli*. Physiol Plant 42:202–206
Puppo A, Rigaud J, Barthe P (1974) Sur la présence de cytokinines dans les nodules de *Phaseolus vulgaris* L. CR Acad Sci Ser D 279:2029–2032
Quispel A (1974) The endophytes of the root nodules in non-leguminous plants. In: Quispel A (ed) The biology of nitrogen fixation. North-Holland Biomedical Press, Amsterdam New York, pp 499–520
Radley M (1961) Gibberellin-like substances in plants. Nature 191:684

Rao AN, Chin LW (1973) Experimental studies in the germination of orchid pollen. Cellule 69:293–308

Rigaud J (1970a) L'acide indolyl-3-lactique et son métabolisme chez *Rhizobium*. Arch Mikrobiol 72:297–307

Rigaud J (1970b) La biosynthèse de l'acide indolyl-3-acétique en liaison avec le métabolisme du tryptophol et de l'indolyl-3-acétaldéhyde chez *Rhizobium*. Physiol Plant 23:171–178

Rigaud J, Trinchant JC (1973) Isolement d'une alcool déhydrogenase chez *Rhizobium* et rôle dans le métabolisme indolique. Physiol Plant 28:160–165

Rodriguez-Barrueco C, Bermudez de Castro F (1973) Cytokinin pseudonodules on *Alnus glutinosa*. Physiol Plant 29:277–280

Rudawska M (1982) Effect of various organic sources of nitrogen on the growth of mycelium and content of auxin and cytokinin in cultures of some mycorrhizal fungi. Acta Physiol Plant 4:11–20

Saavedra E (1979) Set and growth of *Annona cherimola* fruit obtained by hand pollination and chemical treatments. J Am Soc Hortic Sci 104:668–673

Sabelnikova VI, Voloskova MM, Brun GA (1981) *Rhizobium leguminosarum* ability to transform L-tryptophan and synthesize indolic compounds. Izv Akad Nauk SSSR Ser Biol 0(4):621–625

Short KC, Torrey JG (1972) Cytokinins in seedling roots of pea. Plant Physiol 49:155–160

Shukla SN, Tewari MN (1973a) Abscisic acid induced elongation of pollen tube in *Calotropis procera*. Ind J Exp Biol 11:258–259

Shukla SN, Tewari MN (1973b) Interaction of growth regulators in pollen tube elongation of *Calotropis procera*. Ind J Exp Biol 11:591–592

Shukla SN, Tewari MN (1974) Antagonism between plant-growth regulators in pollen tube elongation of *Calotropis procera*. Experientia 30:495

Silver WS, Bendana FE, Powell RD (1966) Root-nodule symbiosis. II. The relation of auxin to root geotropism in roots and root nodules of non-legumes. Physiol Plant 19:207–218

Sjut V, Bangerth F (1981) Effect of pollination or treatment with growth regulators on levels of extractable hormones in tomato ovaries and young fruits. Physiol Plant 53:76–78

Slankis V (1948) Einfluß von Exudaten von *Boletus variegatus* auf die dichotomische Verzweigung isolierter Kiefernwurzeln. Physiol Plant 1:390–400

Slankis V (1949) Wirkung von β-Indolylessigsäure auf die dichotomische Verzweigung isolierter Wurzeln von *Pinus sylvestris*. Sv Bot Tidskr 43:603

Slankis V (1950) Effect of α-napthelene acetic acid on dichotomous branching of isolated roots of *Pinus sylvestris*. Physiol Plant 3:40–44

Slankis V (1958) The role of auxin and other exudates in mycorrhizal symbiosis of forest trees. In: Thimann KV Physiology of forest trees. Ronald, New York, pp 427–443

Slankis V (1973) Hormonal relationships in mycorrhizal development. In: Marks GC, Kozlowski TT (eds) Ectomycorrhizae: their ecology and physiology. Academic Press, London New York, pp 231–298

Slankis V (1974) Soil factors influencing formation of mycorrhizae. Annu Rev Phytopathol 12:437–457

Smith PF (1939) The influence of 3-indole acetic acid on pollen germination. Science 90:163–164

Smith PF (1942) Studies of the growth of pollen with respect to temperature, auxins, colchicine, and vitamin B_1. Am J Bot 29:56–66

Solheim B, Raa J (1973) Characterization of the substances causing deformation of root hairs of *Trifolium repens* when inoculated with *Rhizobium trifolii*. J Gen Microbiol 77:241–247

Strzelczyk E, Sitek JM, Kowalski S (1977) Synthesis of auxins from tryptophan and tryptophan precursors by fungi isolated from mycorrhizae of pine, *Pinus sylvestris*. Acta Microbiol Pol 26:255–264

Sweet GB, Lewis PN (1969) A diffusible auxin from *Pinus radiata* pollen and its possible role in stimulating ovule development. Planta 89:380–384

Syono K, Torrey JG (1976) Identification of cytokinins of root nodules of the garden pea, *Pisum sativum* L. Plant Physiol 57:602–606

Syono K, Newcomb W, Torrey JG (1976) Cytokinin production in relation to the development of pea root nodules. Can J Bot 54:2155–2162

Tanaka K (1960) The pollen germination and pollen tube development in *Pinus densiflora* Sieb. et Zucc. V. The comparison of the growth responses of pollen tubes with those of *Avena* coleoptiles to the inhibitors from *Pinus* pollen grains. Sci Rep Tohoku Univ Ser IV Biol 26:259–268

Thimann KV (1936) On the physiology of the formation of nodules on legume roots. Proc Natl Acad Sci USA 22:511–514

Torrey JG (1961) Kinetin as trigger for mitosis in mature endomitotic plant cells. Exp Cell Res 23:281–299

Trinchant JC, Rigaud J (1977) Sur la synthèse des acides indolyl-3-lactique et indolyl-3-acétique par des bactéroides extraits de nodules de *Phaseolus vulgaris* L. et ses relations avec la fixation d'azote atmosphérique. CR Acad Sci Paris 284:301–303

Tsukamoto Y, Matsubara S (1968) Studies on germination of chrysanthemum pollen II. Occurrence of a germinating promoting substance. Plant Cell Physiol 9:237–245

Ulrich JM (1960) Auxin production by mycorrhizal fungi. Physiol Plant 13:429–443

Vance CP (1983) *Rhizobium* infection and nodulation: A beneficial plant disease. Annu Rev Microbiol 37:399–424

Wang TL, Wood EA, Brewin NJ (1982) Growth regulators, *Rhizobium* and nodulation in peas, *Pisum sativum* cultivar Wisconsin-Perfection. IAA from the culture medium of nodulating and non-nodulating strains of *Rhizobium leguminosarum*. Planta 155:345–349

Watts SH, Wheeler CT, Hillman JR, Berrie AMM, Crozier A, Math VB (1983) Abscisic acid in the nodulated root system of *Alnus glutinosa*. New Phytol 95:203–208

Wheeler CT, Henson IE (1978) Hormones in plants bearing nitrogen-fixing root nodules: The nodule as a source of cytokinins in *Alnus glutinosa* (L.) Gaertn. New Phytol 80:557–565

Williams PM, DeMallorca MS (1982) Abscisic acid and gibberellin-like substances in roots and root nodules of *Glycine max*. Plant Soil 65:19–26

16 Pathogenic and Non-pathogenic Microorganisms and Insects

G.F. PEGG

1 Hormones and Microorganisms

1.1 Ethylene

The olefinic gas ethylene (ethene) is the most unusual and perhaps powerful of the growth-regulating chemicals produced by microorganisms and by healthy and diseased plants. The earliest observations of the physiological effects of the gas by FAHNSTOCK (1858) and GIRARDIN (1864, ABELES 1973) were in relation to damage to coleus plants and lime (*Tilia vulgaris*) trees by fuel (coal) gas contaminated by ethylene. The effects of ethylene on lime trees closely simulated the disease syndrome associated with vascular wilt pathogens such as the Dutch elm disease fungus (*Ceratocystis ulmi*) on elm, but it was many years later that its biosynthesis by microorganisms (including fungal and bacterial vascular pathogens) was established (PEGG 1976b).

A principal reason why ethylene research has developed relatively recently compared with other hormones is that, prior to the middle 1960's, small quantities of the gas could only be detected by growth assays such as the tomato epinasty test or the triple response of pea (diageotropy, stunting, and thickening) in etiolated seedling epicotyls. Such tests were quantitatively insensitive and frequently difficult to apply. Recent progress on ethylene physiology is due entirely to its simple qualitative and quantitative determination by gas chromatography.

1.1.1 Fungal Production of Ethylene

The first report of suspected ethylene production by fungi was by GANE (1934), who showed that yeast produced a gas capable of causing abnormal growth in seedlings. In 1940 and 1941 it was established with certainty that *Penicillium digitatum* produced ethylene in culture (BIALE 1940, BIALE and SHEPHERD 1941, MILLER et al. 1940).

In addition to production by the pathogen, the decaying citrus fruit also produces high levels of ethylene (MILLER et al. 1940). There is, however, considerable variation in ethylene production by several single-spore isolates of *P. digitatum*, some producing none at all yet remain pathogenic (SPALDING and LIEBERMAN 1965). This variation in ethylene production appears to be true of isolates of most, if not all, microorganisms. In general, highest rates of production are associated with the log phase of growth and production falls as the mycelia mature (FERGUS 1954, PHAN 1962, SPALDING and LIEBERMAN 1965).

MEHERINK and SPENCER (1964), again working with *Penicillium,* found that maximal ethylene production occurs at sporulation.

In a survey of 228 fungal species, accounting for 81 different genera, 58 of the cultures (grown on corn-steep liquor supplemented medium) produced ethylene in the range from 0.02–500 ppm (ILAG and CURTIS 1968). It must be presumed that methionine was present in the extract from *Zea.* On this evidence it was claimed that ethylene was a general metabolic product of fungi. Abnormally high levels of ethylene production by *Fusarium oxysporum* f. sp. tulipae have been reported in vitro (DE MUNK and DE ROOY 1971), 4000 times more than in other fungi tested. The significance of this ethylene in tulip bulbs infected with the pathogen is discussed later. Saprophytes and facultative parasites all produce ethylene, although its significance, if any, in the growth and metabolism of the microorganisms is not clear. Blastomycosis pathogens have been shown to produce ethylene; *Blastomyces dermatitidis* (BERNHEIN 1942) and *B. dermatitidis, B. brasiliensis* and *Histoplasma capsulatum* (NICKERSON 1948). The microbial production of ethylene in anaerobic soils is inhibited following steam sterilization (SMITH and RESTALL 1971, SMITH and RUSSEL 1969). Although much of this soil ethylene may be of bacterial origin and produced by decomposing plant debris, following soil enrichment with glucose and methionine, *Candida vartiovaari, Trichosporon cutaneum,* and *Mucor hiemalis* all produced ethylene after isolation in pure culture (SMITH and RESTALL 1971). An ethylene-producing *Aspergillus* sp. and a *Mucor* sp. were similarly isolated (DA SILVA et al. 1974).

Reported effects of ethylene on fungi are few. Concentrations of 250 ppm have a nonspecific stimulating effect on spore germination of *Diplodia natalensis* and *Phomopsis citri* (BROOKS 1944). Ethylene appeared to substitute for a nutritional inhibition of germination in water, but ethyl acetate and methanol had the same effect. VESSELOV (1937), ABELES (1973) claimed a 10% stimulation in the respiration of *Aspergillus fumigatus, A. flavus,* and *A. niger* at 1000 ppm ethylene, a concentration which also slightly inhibited hyphal growth. In a very early paper, SHAW (1935), ABELES (1973) reported that ethylene stimulated the formation of invertase in yeast, but growth effects were variable. Ethylene, however, had no effect on hyphal growth of *Gleosporium album* except at low temperatures (3.3° C), where it was inhibitory (LOCKHART et al. 1968). Perhaps not surprisingly, high concentrations (2000 ppm) inhibited *E. oxysporum,* especially at reduced oxygen tensions. The slime mold *Physarum polycephalum* showed an increased growth rate under a 75% ethylene atmosphere, the effect appearing to be counteracted by CO_2, a well-documented phenomenon in higher plants (cf. BURG 1973). In *Agaricus bisporus* ethylene evolution was also correlated with enlargement of the fruiting body (LOCKARD and KNEEBONE 1972). Bacteria were eliminated in the casing compost or culture medium of *A. bisporus,* and the source of ethylene was conclusively shown to be of fungal origin (WOOD and HAMMOND 1977, WARD et al. 1978). Removal of the sporocarp of *A. bisporus* before maximum expansion did not change the ethylene profile (TURNER et al. 1975). This would suggest that the sporophore in some way controlled mycelial metabolism, since excised sporocarps did not release ethylene, which must therefore have been derived from the mycelium. The role of ethylene in the development of *A. bisporum* is still not clear, since WARD et al. (1978) have shown that exogeneous ethylene had no effect on sporocarp growth.

1.1.2 Bacterial Production of Ethylene

FREEBAIRN and BUDDENHAGEN (1964) were the first to demonstrate ethylene production by a bacterium (*Pseudomonas solanacearum*) and at the same time relate this phenomenon with the characteristic premature ripening of banana fruits usually associated with this wilt pathogen. Rates of ethylene as high as 0.1 μl h^{-1} were evolved from 20 ml of peptone glucose broth cultures. In a more detailed study of this organism the atmosphere over agar roll tube cultures was continuously monitored for a wild-type virulent strain and an avirulent mutant (BONN et al. 1972). The wild-type produced more than twice as much ethylene as the avirulent strain. A comparative study of two virulent strains of *P. solanacearum* and the avirulent mutant found no correlation between virulence and ethylene evolution; the two virulent strains, however, showed great and continuous variation in its evolution (PEGG and CRONSHAW 1976). Some soil bacterial cultures evolved up to 6000 μl mg^{-1} dry weight day^{-1} (PRIMROSE 1976b). A wide range of bacterial genera and species have now been shown to produce ethylene and hence most probably contribute to the soil ethylene atmosphere. These include *Acromonas hydrophila, Pseudomonas indigofera, Arthrobacter* sp., *Bacillus mycoides, Corynebacterium* sp. and *Rhizobium trifolii* (PRIMROSE 1969a, b). The production of ethylene by the latter genus is of especial interest, since OWENS et al. (1971) have described a phytotoxin – rhizobitoxin – formed by certain strains of *Rhizobium japonicum,* which inhibited the synthesis of ethylene from methionine. Rhizobitoxin, an S-containing amino compound, inhibits β-cystathionase, the enzyme which controls bacterial methionine biosynthesis.

With few exceptions (SWART and KAMERBEEK 1977, CONSIDINE 1977), all the organisms referred to have a requirement for methionine or one of its derivatives for ethylene biosynthesis. In other studies where the medium has not been defined, it is most likely that methionine was present in amino acid complexes. The availability of such precursors in the soil from decomposing organic matter, and from microorganisms, must determine the rate of ethylene production, as must the numbers of ethylene-oxidizing bacteria present (ABELES et al. 1971, CORNFORTH 1975, DE BONT 1976). In aerobic soils the levels of oxidizing bacteria will be high, and hence the level of ethylene will be lower than would be expected in waterlogged and anaerobic soils. Although high ethylene evolution by microorganisms is usually associated with high O_2 tension, anaerobes, especially in the absence of ethylene oxidizers, can produce quantities sufficient to affect plant growth. Ethylene has been claimed to be main cause of soil fungistasis (SMITH 1973, 1976, SMITH and COOK 1974). Although ARCHER (1976) could find no evidence for ethylene-induced inhibition of growth of a wide range of genera and species grown in pure cultures. It was thus argued that an ethylene-mediated fungistasis is inconsistent with the wide range of ability of microorganisms to synthesize the gas (LYNCH 1975 and ARCHER 1976). The evidence for the fungistatic effect was based on air-dried soil studies, and it was claimed (SMITH 1973) that this represented a balance between nutrient-stimulated and ethylene-inhibited germination of the spores.

1.1.3 Biosynthesis of Ethylene

It is now generally recognized that methionine is the main precursor of ethylene biosynthesis, both in higher plants and microorganisms (ABELES 1973, BURG 1973, BIALE 1940) although there are a few notable exceptions. It has, however, been proposed (YANG and BAUR 1972, YANG 1974, ADAMS and YANG 1977) that the biosynthetic pathway for bacteria and fungi differs from that for the higher plants. In general, the intermediate precursor between methionine and ethylene is thought to be 4-methyl (mercapto 2-oxobutyric acid (2-keto 4-methyl mercaptobutyric acid) (KMBA) the conversion of which is a hydrogen peroxide, peroxidase-mediated process. *Penicillium digitatum* can, however, synthesize ethylene quite independently of methionine. Glutamic acid may be the precursor in the nonmethionine pathway (CHOU and YANG 1973). This fungus, though, can also utilize methionine (CHALUTZ et al. 1977). Moreover, on *Sclerotinia fructigena*-infected apple peel, the pathogen appears to be using a different biosynthetic pathway from the host (HISLOP et al. 1973). Ethylene production was inhibited more in infected than in healthy, pre-climacteric peel, by acetate and citrate buffers. An opposite effect was obtained using a crude rhizobitoxin preparation from *Rhizobium japonicum* (HISLOP et al. 1973). Similar work on *Ceratocystis*-infected sweet potato led to the same conclusion, that two pathways were in operation (SAKAI et al. 1970).

Since 1979, it has been know that the main ethylene biosynthetic pathway in higher plants is via the adenosylation of methionine to S-adenosyl methionine (SAM) leading to the formation of the principle and rate-limiting precursor of ethylene, 1-amino cyclopropane-1-carboxylic acid (ACC) (ADAMS and YANG 1979, LÜRSSEN et al. 1979). The conversion of SAM to ACC is mediated by the enzyme ACC synthetase which is pyridoxal phosphate dependent. Rhizobitoxin a β-enol ether amino acid ($HOCH_2-CH(NH_2)-CH_2-O-CH{:}CH-CH-NH_2-COOH$) and amino ethoxyvinyl glycine (AVG) ($NH_2-CH_2-CH_2-O-CH{:}CH-CH(NH_2)-COOH$) from *Streptomyces* sp. are potent inhibitors of pyridoxal enzymes, including ACC synthetase, and treatment of plants with these compounds prevents ethylene formation through the block to precursor production. Although the pathway is now well documented for higher plants (YANG and HOFFMAN 1984) there is no such evidence for an ACC step with microorganisms. The action of rhizobitoxin and AVG in blocking ethylene production in microorganisms, cannot be taken as evidence for the existence of an ACC pathway since both block a β-cystathionase controlling the cleavage of cystathionine to homocysteine, which in turn is converted to methionine. Both enzyme stages involved in the conversion are pyridoxal phosphate dependent. This pathway of methionine synthesis has been demonstrated in *Neurospora crassa*, *E. coli* and *Aerobacter aerogenes* and is probably common to many fungi and bacteria, unlike higher plants which use homoserine as a precursor. Adenosylation of methionine involved both in anabolic and catabolic reactions of this amino acid has been demonstrated in a number of bacteria and yeasts, together with the ^{17}C-cyclopropane acid. In *Saccharamyces cerevisiae*, however, feeding experiments with SAM failed to stimulate ethylene (THOMAS and SPENCER 1977).

To date, production of ethylene by microorganisms via an ACC pathway still awaits demonstration.

1.1.4 Ethylene and Plant Pathogenesis

A major topic of interest in ethylene physiology is the production of disease-induced ethylene. This is a ubiquitous phenomenon, and since the early work of WILLIAMSON (1950) has received much attention. However, the emphasis in this chapter is on the microorganism as an "environmental factor", and not on the diseased host per se. Reviews on this latter topic have been written by HISLOP et al. (1973) and PEGG (1976a).

The areas of particular interest may be considered thus:

1. Ethylene as a phytotoxin.
2. Altered metabolic processes.
3. Ethylene-induced resistance.

It can be seen from (1) and (3) that a paradox exists in the role of this unusual growth regulator. Of primary concern here, however, is the contribution of the organism to the enhanced levels of ethylene. All known types of disease show the ethylene effect viz: those caused by fungus, bacterium, virus, and nematode. With the exception of virus and nematode diseases, it has been assumed that the invading organisms play an exclusive or substantial part in the biogenesis of the hormone. However, the evidence for this assumption is limited. In *P. solanacearum* infection of tomato, an analysis of ethylene production and bacterial cell numbers in both the infected host and in continuous culture using peptone nutrient broth was carried out (PEGG and CRONSHAW 1976). One virulent tomato strain, S-225, produced in vitro only 9.83×10^{-24} M min^{-1} $cell^{-1}$, compared with another virulent strain, K-60, which produced 1.47×10^{-22} M min^{-1} $cell^{-1}$. From a comparison of the cell numbers present in the infected plant, 1×10^6 cells in a diseased leaf contributed to 0.92 μl ethylene h_l^{-1}, but the maximum rate of the bacterium in vitro was 0.009 μl ethylene h_l^{-1} produced by 6.6×10^8 cells. Even allowing for the fact that ethylene production in vitro might be suboptimal, the shortfall in ethylene production per cell was too great to attribute to the pathogen any measurable contribution to the ethylene pool of the leaf. LUND and MAPSON (1970) similarly concluded that in *Erwinia carotovora*-infection of cauliflower florets, the pathogenic ethylene was wholly of host origin. Their conclusion was based on the failure of *E. carotovora* to synthesize ethylene when cultured on a pectate medium or cauliflower extract supplemented with methionine. They speculated that pectic enzymes such as *endo* pectate lyase and *endo* polygalacturonase liberated wall-bound oxidases which, on an available substrate, liberated H_2O_2, which in turn accelerated the synthesis of ethylene from the intermediate, KMBA. More recently ANDERSON et al. (1982) using a commercial cell wall degrading enzyme, cellulysin, presumably containing cellulose and pectolytic activity, demonstrated ethylene production in tobacco leaf discs purportedly through the activation of ACC synthetase. A more comprehensive study by VANDER MOLEN et al. (1983) showed that a crude glycosidase mixture and purified *endo* polygalacturanase from 6 day culture filtrates of *Fusarium oxysporum* f.sp. cubense each caused ethylene release when applied to leaves of a non-host *Ricinus communis*. The findings of these authors that ethylene induced vascular gel formation (interpreted as a cause of wilt symptoms) is at variance with the results of PEGG (1976) who showed that ethylene treatment of tomato plants could confer resistance to vascular wilt, colonisation and symptom development.

An interesting biological role for pathogen-produced ethylene has been demonstrated for *Fusarium oxysporum* f.sp. tulipae by BEIJERSBERGEN (1973) and BEIJERSBERGEN and LEMMERS (1978). Tulip bulbs, prior to ripening in soil are normally resistant to this organism. The chemical bases of resistance are two unsaturated lactones of the 1-acyl glucosides of α-methylene-γ-butyric acid and α-methylene-β-γ-dihydroxybutyric acid, known as tuliposides A and B. The corresponding lactones, α-methylene-γ-butyrolactone and β-hydroxy α-methylene-γ-butyrolactone are present at fungistatic levels in the outer white scale, but at maturity are metabolised to inactive substances as the white scale turns brown. During this period the bulb is highly susceptible to infection. Ethylene produced by the pathogen prolongs this period of susceptibility by inhibiting enzymic *de novo* synthesis of tuliposides A and B.

1.2 Auxins

The term auxin, as used in this section, is intended to refer only to those naturally occurring growth substances whose structure is based on the indole ring. Shortly after the first discovery in 1934 by KÖGL et al. (1934), KÖGL and KOSTERMANS (1934) of what was later to be recognized as the principal auxin – IAA, THIMANN in 1935 – isolated the same substance from culture filtrates of *Rhizopus suinus*, confirming the earlier pioneering studies of NIELSEN (1928, 1930). Many indolyl derivatives, both "free" (readily extractable in organic solvent) and "bound" (released after alkaline hydrolysis) possess apparent growth-promoting activity, including cell enlargement, root growth, lateral bud inhibition, and abscission, but it is generally believed that these compounds (such as IAN, IAAld, tryptophol, and gramine) represent precursors which are enzymatically converted to IAA, the principal substance involved in the growth response. Many bacteria and fungi, pathogens, and nonpathogens, have been shown to produce IAA readily in culture and some diseased plants infected by these organisms show typical hyperauxiny effects. It should be noted, however, that virus diseases and plants infected with root-invading nematodes similarly show abnormal growth-promoting activity, indicating a host, in addition to a possible pathogen, contribution to the enhanced auxin levels.

1.2.1 Fungal Production of Auxins

The first observation that fungi excreted substances with auxin-like activity was by NIELSEN in 1928, who showed that agar on which *Rhizopus suinus* and

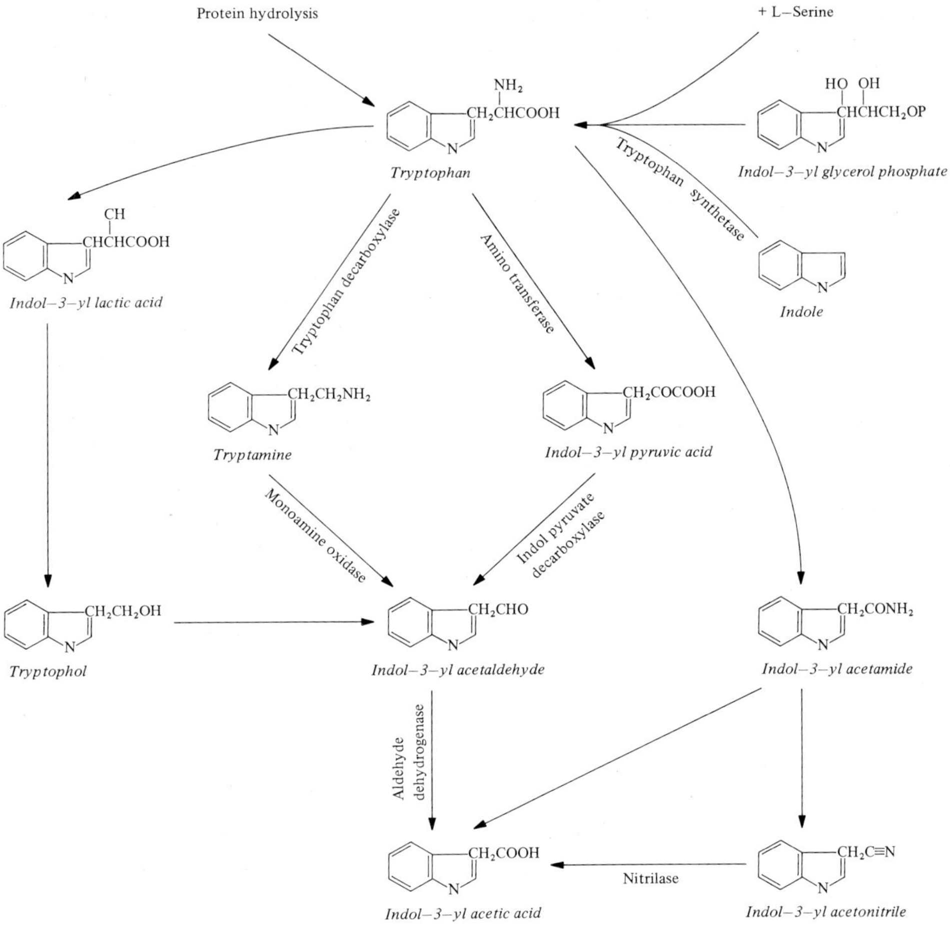

Fig. 1. The biogenesis of tryptophan and IAA showing the major pathways of microorganisms and higher plants

Absidia ramosa had grown was active in the *Avena* coleoptile test. The confirmation that NIELSEN's "rhizopin" was in fact IAA was made by THIMANN (1935). One year earlier KÖGL and KOSTERMANS (1934) had crystallized IAA from *Saccharomyces cerevisiae*. GRUEN (1959), in reviewing the production of auxins by fungi, has listed a wide range of genera and species, representing all the major fungal groups, with the ability to synthesize IAA or its derivatives. The accepted pathways of auxin biosynthesis in micro-organisms and higher plants (SOMNER 1961 b) are shown in Fig. 1. Many species of fungi have the ability to synthesize IAA in the absence of the main precursor, tryptophan, although a number showed a tryptophan dependence. A range of fungus species was tested for auxin production, and no synthesis was found when nitrate was used as a nitrogen source, whereas all isolates formed IAA when tryptophan was added to the medium (ROBERTS and ROBERTS 1939, BRIAN 1957). The possible significance of this may be seen in the interpretation of increased growth substance production in the diseased host and the availability of substrate. Tryptophan might be present in the host plant as a normal amino acid in the free pool, or an alternative source would be from proteolytic breakdown of host proteins. A further possibility, as has been demonstrated for *Neurospora crassa,* is the synthesis of tryptophan from indolyl-3-glycerol phosphate and serine, involving the enzyme tryptophan synthetase. The presumed route of IAA synthesis is via indol-3-yl pyruvic acid (IPyA) and indol-3-yl acetaldehyde (IAAld), but few studies with fungi (unlike bacteria) have confirmed this. CRADY and WOLF (1959) claimed that *Dibotryon morbosum* and *Taphrina deformans* converted tryptophan to tryptamine. PERLEY and STOWE (1966), however, could not find this precursor of IAA in tryptophan-supplemented cultures of the latter.

For obvious reasons, in vivo studies for the most part have been on saprophytes or facultative parasites. Hyperauxiny, however, is a common and well-documented feature of diseases caused by obligate parasites [cf. SEQUEIRA (1973), PEGG (1976c), DALY and KNOCHE (1976)], and it is of interest that *Avena* coleoptile growth-promoting activity on chromatograms of the acidic ether fraction of ungerminated uredospores of *Puccinia graminis tritici* was not in the IAA region (SHAW and HAWKINS 1958). WOLF (1956), working on a rare and abnormal form of *Gymnosporangium juniperi-virginianae,* found IAA in culture filtrates. These findings, however, do no more than suggest a possible contribution from biotrophic pathogens in disease-induced hyperauxiny.

1.2.2 Effects of Auxins on Fungal Growth

Extensive studies on the effect of IAA and synthetic growth regulators on the growth and metabolism of fungi (see GRUEN 1959) have yielded few conclusive results. Concentrations of IAA used were far in excess of their accepted physiological concentrations and the many reported growth inhibitory effects are perhaps not surprising. Concentrations in the range 1×10^{-3}–5×10^{-5} M IAA partly or wholly, inhibited growth of Phycomycetes (and *Phycomyces* especially), Asocomycetes, Basidiomycetes and Deuteromycetes. Growth increases, however, have been claimed for yeasts, *Saccharomyces cerevisiae* at 0.1–1.0 mg l^{-1} (TURFITT 1941), *Candida albicans* 50 mg l^{-1} (MORQUER and NYSTERAKIS 1948, see GRUEN 1959) and *Torula utilis* 1 mg l^{-1} (NAIDASCH 1955, see GRUEN 1959); *Neurospora tetrasperma* 100 mg l^{-1} (NYSTERAKIS 1954), *Agaricus campestris* f. albida 5–100 mg l^{-1} (FRASER 1953). For the last two species a significant growth promotion occurred only for the surface mycelium of aqueous cultures and not in submerged or agar cultures. Slight positive effects have been described for numerous other species, but from all the reported work it is clear that an IAA-mediated, cellular growth-promoting mechanism, similar to that in higher plants, is not present in fungi.

Increased spore germination by IAA has been claimed in a few species, *Ustilago zeae* (GUTTENBERG and STRUTZ 1952) *Gibberella zeae* (NISHIKADO and INOVE 1955, see GRUEN 1959) and *Allomyces zygotes* (TURIAN 1956, see GRUEN 1959). At 200 mg l^{-1} IAA increased germination of *Puccinia graminis* uredospores at high spore concentrations; at low spore concentrations the same

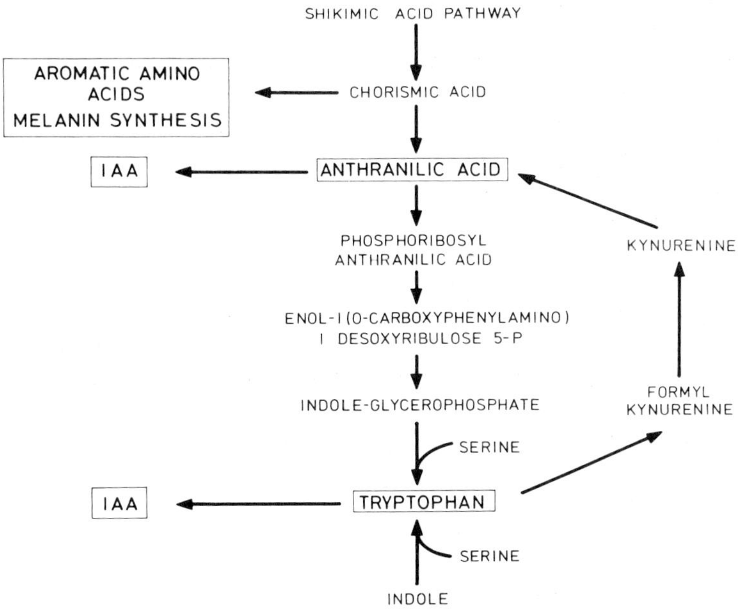

Fig. 2. Scheme to illustrate tryptophan and anthranilic acid as precursors of IAA and their relationship to phenol metabolism

amount of IAA was inhibitory (VAN SUMERE et al. 1957). It was suggested that IAA was reversing the well-known auto-inhibitory effect of rust uredospores. It is of interest that the genera *Ustilago, Gibberella,* and *Puccinia* are all associated with diseases involving hyperauxiny.

1.2.3 Bacterial Production of Auxins

The ability of bacteria, and in particular *Escherichia coli,* to metabolize tryptophan and other indole derivatives is well known (see MEISTER (1965). The biosynthesis of tryptophan in microorganisms and in selected higher plants may be intimately involved in shikimic acid metabolism, leading through the intermediate chorismic acid either to aromatic amino acid synthesis, or to anthranilic acid and the formation of tryptophan. The latter conversion would require tryptophan synthetase activity, involving terminally, the conjugation of indole, or indole glycerolphosphate with serine (Fig. 2). The chorismic synthetase catalysis of the conversion of 3-enolpyruvyl shikimic acid-5-phosphate to chorismate has been demonstrated for *E. coli* (MORELL et al. 1967), as has anthranilate synthetase (SRINIVASAN 1965). Tryptophan is a feedback inhibitor of this activity (EDWARDS et al. 1964) which would suppress IAA production from anthranilate. Protein complexes, including, among other enzymes, anthranilate synthetase and indolylglycerol phosphate synthetase, have been found in the Enterobacteriaceae and in the fungi *Saccharomyces cerevisiae* and *Neurospora crassa.* Tryptophan synthetases have also been purified from these organisms (TRUFFA-BACHI and COHEN 1968).

The relationship of tryptophan to anthranilic acid is a cyclic one, since tryptophan may also be catabolized to anthranilate via the kynurenine pathway (Fig. 2). In a virulent strain of *Pseudomonas solanacearum* (K 60) ring-labeled, but not chain-labelled, tryptophan is converted to IAA and anthranilic acid is one of the intermediates (SEQUEIRA and WILLIAMS 1964, PHELPS and SEQUEIRA 1968). The precise synthetic mechanism from anthranilic acid to IAA, however, is not known, but it is not through the conventional side chain pathway of tryptophan → indolyl pyruvic → indolyl acetaldehyde → IAA which these authors have shown for an avirulent (B1) mutant of *P. solanacearum* (Fig. 1). It has been reported in higher plants that under certain circumstances IAA may be metabolised to anthranilic acid. A reversal of this pathway in microorganisms would not therefore be unexpected. Synthesis of IAA from tryptophan in crown gall tissue by *Agrobacterium tumefaciens* may occur via indolyl pyruvate (KAPER and VELSTRA 1958). Tryptophan decarboxylase was also involved in this reaction, since ILA and tryptophol were also found.

The role of bacteria in auxin production has been considered from a different standpoint by LIBBERT and co-workers (LIBBERT et al. 1969, 1968, LIBBERT and MANTEUFFEL 1970a, LIBBERT and RISCH 1969, LIBBERT and STILHENGST 1970b). Studies on the levels of endogenous auxins in higher plants, maize, pea and cucumber, grown under aseptic conditions have shown that much of the "endogenous" auxin extracted from plants under normal nonsterile conditions can be attributed to bacterial conversion of tryptophan to IAA. LIBBERT and RISCH (1969) isolated from roots and shoots of *Pisum sativum* 58 strains of *Pseudomonas alcaligenes, Achromobacter, Bacillus* and *Flavobacterium* spp. capable of converting tryptophan to IAA and a further 15 strains with the ability to destroy IAA. Included in the latter were 13 strains with both properties. The bacterial contribution of IAA to higher plant tissue brei was further confirmed by the use of bacterial inhibitors such as chloramphenicol and streptomycin in the presence of which plant extracts were shown to have greatly reduced auxin contents compared with nonsterile controls. Similarly the level of IAA found in sterile extracts was enhanced when surface-sterilized seeds were allowed to imbibe nonsterile water or a bacterial suspension. THIMANN and GROCHOWSKA (1968) found that pea tissue grown aseptically could convert tryptamine, but not tryptophan to IAA, involving monoamine oxidase. The transformation of tryptophan by *Pisum* to IAA under nonsterile conditions was attributed solely to bacterial activity. This work implies that there is no conversion of tryptophan to tryptamine involving tryptophan decarboxylase (Fig. 1), or conversion to IPyA in sterile pea tissue. LIBBERT et al. (1968), however, claim that the tryptamine → IPyA pathway does exist in sterile pea tissue.

Bacteria may also play a role in biosynthesis of the IAA precursor, since tryptophan synthesis mediated by tryptophan synthetase from indol-3-ylglycerol-1-phosphate and L-serine has been demonstrated (SCHIEWER et al. 1970). The work of LIBBERT and co-workers, and others presents a formidable body of evidence for the involvement of bacteria in auxin production during extraction and places in question many quantitative auxin studies on higher plants and on host–parasite relations where bacterial contamination of the plant tissue was not accounted for. A re-examination of selected plants where precise IAA figures have been cited using aseptic techniques would be justified.

1.3 Cytokinins

1.3.1 Bacterial Production of Cytokinins

During the past years a number of reports of cytokinin production by microorganisms and especially bacteria have appeared. Cytokinins are N^6-substituted purine compounds with cell-division promoting properties (see SKOOG and SCHMITZ 1972 and DEKHUIJZEN 1976), and occur as natural constituents of higher plants, animals, and fungi as the free purine base, nucleoside or nucleotide and also as a nucleoside in tRNA. As with GA's, many workers have referred to "cytokinin-like" activity in cultures of microorganisms. The work cited in this chapter is restricted to chemically defined cytokinins, all N^6-substituted adenine (amino purine) with further substitution in the C4 carbon in two cases. The microbial cytokinins are: zeatin; [6-(4-hydroxy-3-methyl-trans-2-butenyl-amino)] purine; ms zeatin; 2-methyl thiopurine zeatin; 2iP, 6^2-isopentyl-amino purine; and ms-2iP. Where these compounds are found as constituents of tRNA they all occur as the nucleoside, for example, 2iPA, 2 isopentyl adenosine.

The earliest accounts of cytokinin production by bacteria was by KLAMBT et al. (1966) and HELGESON and LEONARD (1966), who isolated and identified 2iP from culture filtrates of *Corynebacterium fascians*. This bacterium is responsible for the fasciation disease of a number of dicotyledonous herbaceous plants and the hyperplastic tissue overgrowth is strongly indicative of an imbalance of cytokinins (THIMANN and SACHS 1966). 2iPA was found as a constituent of the bacterial tRNA (RATHBONE and HALL 1972) and cis-zeatin was isolated from it in culture fluids (SCARBROUGH et al. 1973). Estimates of the levels of 2iP production vary between 10 $\mu g\ l^{-1}$ and the more probable 2 $\mu g\ l^{-1}$. Under acidic conditions, hydrolysis of tRNA may be expected to release apparent "free" 2iP. Notwithstanding the ability of the bacterium to produce this powerful growth substance, the actual in vivo contribution of the pathogen may be trivial and the fasciation effect may be a stimulation of cytokinin production by the native host. More critical investigation is needed, however, to establish this point. UPPER et al. (1970) isolated 6-(3-methyl-2-butenylamino)purine from culture filtrates of *Corynebacterium fascians*.

Direct effects of cytokinins on bacteria are few. MARUZZELLA and GARNER (1963) claimed that kinetin stimulated the growth of several bacteria including the thermophil *Clostridium thermocellum* and QUINN et al. (1963) demonstrated a stimulated digestion of cellulose by this organism on a kinetin-supplemented medium. Kinetin was also claimed to induce virulent strains of *Brucella* and *Pneumococcus* from avirulent strains (BRAUN 1956). Some detailed studies have been carried out on cytokinin nucleosides in bacterial tRNA, in particular *Escherichia coli*, where 2iPA and ms 2iPA occurred in tRNA-recognizing codons commencing U (BABCOCK and MORRIS 1970, YAMADA et al. 1971). Not all tRNA's from one organism have associated cytokinins. Of four tRNA isolated from *Staphylococcus epidermis* cultures (ARMSTRONG et al. 1970), only one, $tRNA^{ser}$ UCR, contained a cytokinin. These same workers identified 2iPA and ms 2iPA from *S. epidermis*. THIMMAPPAYA and CHERAYIL (1974) identified ms zeatin riboside in tRNA from *Pseudomonas aeruginosa*. The role of tRNA cytokinins on present evidence can only be speculative. In general, it is thought that their location next to the anticodon in certain tRNA species with known base sequences may indicate a role in gene derepression at the translation stage controlling protein biosynthesis. On evidence based on *E. coli* GEFTER and RUSSELL (1969) envisage their effect to be on the binding of tRNA to the ribosomes.

CHEN and HALL (1969) have proposed that the normal catabolism of tRNA releases biologically active cytokinins and that regulation of tRNA will also regulate the levels of free cytokinins in the cell. The present evidence (see KENDE 1971) suggests that free cytokinins are key regulatory hormones in higher plants and that they are produced by a specific biosynthetic pathway rather than as a consequence of tRNA metabolism. Their metabolic role, if any, in microorganisms has yet to be demonstrated.

An unusual symbiotic relationship exists between species of Rubiaceae and Myrsinaceae and leaf nodule bacteria belonging to the genera *Chromobacterium, Bacterium, Klebsiella* and *Mycobacterium* (see LERSTEN and HORNER 1976). Of particular interest is the nitrogen-fixing capacity of the nodules; but various workers have suggested a host-dependency on bacterial hormones for normal growth. Of the major hormones examined, only cytokinins seem to have a dominant role. LAMOTTE and LERSTEN (1972) developed bacteria-free tissue cultures of *Psychotria mucronata* (Rubiaceae) which was independent for auxin but had a dependency for cytokinins at 10 $\mu g\ ml^{-1}$. RODRIGUES PEREIRA et al. (1972) showed cytokinin production by several species of symbiotic bacteria. Leaf discs from nodulated portions of young leaves of *Psychotria punctata* yielded 100 to 2000 times the level of cytokinin than non-nodulated discs, suggesting a close relationship involving cytokinins between these two organisms.

1.3.2 Fungal Production of Cytokinins

A number of fungus-induced diseases are reminiscent of uncontrolled cytokinin activity, viz.: club root, leaf galls and witches' broom. Species of *Taphrina* induce leafy galls (*T. deformans*) and excessive adventitious shoot growth (*T. cerasi*). The production of IAA in culture filtrates (HIRATA 1957, CRADY and WOLF 1959, SOMNER 1961a), though of interest, in no way accounted for the disease symptoms. JOHNSTON and TRIONE (1974) examined late log-phase culture filtrates of both pathogens and by means of soybean callus assays and paper chromatography, tentatively identified zeatin, zeatin riboside, 2iP and 2iPA from *T. cerasi* and *T. deformans*, and a third unidentified cytokinin from *T. deformans*. The total cytokinin activity expressed as kinetin equivalents was 4 $\mu g\ l^{-1}$ for *T. deformans* and 2 $\mu g\ l^{-1}$ for *T. cerasi*. BARTHE and BULARD (1974) subsequently confirmed zeatin in *T. cerasi* by GLC.

Gall formation in roots of cruciferous plants infected with *Plasmodiophora brassicae* is a result of excessive hyperplasia and hypertrophy. Uninfected cells are stimulated to divide, and nuclei and nucleoli show large increases in volume compared to healthy cells, with concomitant increases in the DNA content. Tissue cultures of infected turnip gall are largely self-sufficient in IAA and cytokinin, whereas healthy turnip tissue cultures require both for growth (WILLIAMS et al. 1969). DEKHUIJEN and OVEREEM (1971) identified zeatin in turnip club roots in levels 10 to 100 times higher than in healthy roots (0.1–1 mg 100 kg^{-1}), and at least two other cytokinins may be present (REDDY and WILLIAMS 1970).

Mycorrhizal fungi are of particular interest in relation to growth-substance production. Tree species infected by ectomycorrhizal fungi such as *Rhizopogon roseolus, Boletus elegans,* and *Boletus pinicola* frequently show a growth stimulation indicative of a growth-substance response. MILLER (1967) identified zeatin and zeatin riboside from *R. roseolus* cultures, the first record of a cytokinin from a microorganism. LALOUE and HALL (1973) found 2iP as another culture filtrate product of the same species. 2iPA has been shown to be the precursor for zeatin riboside (MIURA and HALL 1973) and 2iPA, the precursor for 2iP (MIURA and MILLER 1969). Although much has been written about growth substances and mycorrhizas, no unequivocal evidence is available to demonstrate any real growth-substance inter-relationship between host and fungus, although GOGALA (1971) stated (without unequivocal evidence) that *Pinus sylvestris* cytokinins stimulate the growth of *Boletus pinicola* mycelium. There is much need for new experimentation in this difficult field of study to reproduce with growth substances what has been elegantly demonstrated for carbon and phosphorous transport between tree and fungus.

1.4 Gibberellins

Studies on GA's in relation to microorganisms for the most part have dealt with changes in levels of these hormones in diseased plants (see PEGG 1976b). Fungi and bacteria are, however, able to synthesize GA's, but the number of genera and species with this capacity is much lower than those producing auxins and ethylene.

1.4.1 Fungal Production of GA's

CURTIS (1957) screened 1500 culture filtrates of 100 fungal and actinomycete genera for gibberellin activity, but only isolates of *Fusarium moniliforme* Sheld. the stat. conid. of *Gibberella fujikuroi* Saw. Wollen. showed activity in a bioassay. *G. fujikuroi* occupies a unique position in the fungi, since of the 52 gibberellins of known structure which have been identified, approximately 13 are produced as cultural metabolites of *F. moniliforme.* The interest in this species lies in the pathogenicity of certain strains to rice plants, one symptom of which is excessive stem elongation, giving rise to the name "bakanae" – "foolish seedling" disease. Gibberellin A_3, one of the principal culture filtrate gibberellins, can be produced on suitable media at concentrations of 1 g l^{-1}. Concentrations 1000 times less than this applied to healthy rice plants simulate part of the natural syndrome and have led to the belief that GA_3, as a fungal "toxin", is responsible for the symptoms. As with cytokinins and auxins, however, GA_3 is a native endogenous hormone of several Gramineae species (JONES et al. 1963), and there is currently no evidence to show that the pathogen in vivo makes a direct material contribution to the hormone imbalance leading to the stem overgrowth. An alternative explanation for GA production either by host or pathogen was advanced by MUROMTSEV and GLOBUS (1976). It is known that isolates of *F. moniliforme* lacking the ability to produce GA_3 in culture are still pathogenic. They found that all isolates lacking the ability to produce GA's secreted α-amylase constitutively. Similarly all isolates not secreting α-amylase synthesized GA_3. These authors speculated that the role of GA_3 or α-amylase, endogenous, or secreted into the plant, was to hydrolyse starch, indirectly or directly, thereby providing respiratory substrate for the pathogen.

AUBE and SACKSTON (1965) found GA-like activity on the dwarf maize bioassay in cultures of several pathogenic species of *Verticillium,* not generally associated in pathogenesis with stem extension effects. GOGALA (1971), in a detailed study of the growth substances of *Pinus sylvestris* and its mycorrhizal fungus *Boletus pinicola,* found three GA-like substances active on the lettuce hypocotyl bioassay in the sporophore tissue of the fungus, and in pine seedling extracts, but the fungus GA's did not appear identical to the pine GA's. PEGG (1973a) found considerable "free" and "bound" (activity released after ficin hydrolysis) GA-like activity in the dwarf maize bioassay from extracts of *Agaricus bisporus.* Highest levels of activity corresponding to 0.1 μg GA_3 equivalent per assay test seedling, were found in lamella tissue, but GA's were present also in the stipe and pileus. There was no clear correlation between sporophore age and GA content. The origin of GA's in the sporophore of a saprophyte, as with mycorrhizal species, remains a tantalizing mystery. Samples of horse straw compost extracted with ethyl acetate prior to inoculation with *A. bisporus* gave a peak of activity on the dwarf maize bioassay corresponding to the one in the fungus. The compost at this time had a large population of thermophilic microorganisms, particularly bacteria, and the possibility exists that GA's (or precursors) are translocated from compost, or compost microorganisms to the developing fungus. One argument against this is that *A. bisporus* is rich in a cytochrome involved in the pathway of mevalonate to kaurene (C. WEST personal communication), indicating that the synthetic mechanism for at least one GA precursor is present. Sporophores of basidiomycetes other than *A. bisporus* contain large quantities of GA-like substances active on the dwarf maize bioassay (PEGG 1973b): *Phallus impudicus, Boletus elegans, Grifola frondosa,* and *Phellinus pomaceous* all contained GA's and in *G. frondosa* had up to 500 μg of GA_3 equivalents kg^{-1} fresh weight. With the exception of GA_3 the only other characterised fungal gibberellin is GA_4, isolated from cultures of the pathogen *Sphaceloma manihoticola* and identified by careful mass spectrometric analysis (ZEIGLER et al. 1980).

1.4.2 Bacterial Production of GA's

There has been speculation on the possible transfer of GA's from soil-borne bacteria to the roots of higher plants. The presence of *Azotobacter chrococcum* and *Pseudomonas* spp. in the rhizospheres of tomato and cucumber was correlated with increased germination and growth of the seedling (BROWN and BURLINGHAM 1968, EKLUND 1970). Most bacterial cultures tested by KATZNELSON and COLE (1965) produced a GA-like substance on the lettuce hypocotyl bioassay in amounts varying from 1–14 μg l^{-1}. Of 29 cultures of *Pseudomonas* tested, all gave fluorescent spots on chromatograms indicative of GLS, two produced GA_3. RIVIERE (1963a, b) and SOBIESZCZANSKI (1966) also found GA-like substances active on the lettuce hypocotyl bioassay in what were predominantly *pseudomonad* cultures. Careful studies are now needed involving the isolation and unequivocal chemical characterization of GA's from pure cultures of selected rhizosphere bacteria, which together with experiments with labelled GA precursors and host plants would yield rewarding results in this interesting but difficult field.

1.5 Phytotoxins Produced by Microorganisms with Growth Hormone Properties

Most of the characterised toxins isolated from fungi and bacteria inhibit the growth of higher plants in appropriate assays, but since these act as poisons rather than growth regulators they are not considered in this review. For further detail, reference should be made to DURBIN (1981).

All classes of the recognised higher plant growth hormones are synthesised by microorganisms. Where growth-regulating substances are produced by organisms during pathogenesis and adversely affect host metabolism, they may be regarded *sensu lato* as phytotoxins. Recently ASSANTE et al. (1977) and NORMAN et al. (1983) found abscisic acid (ABA), representing the fifth class of growth-regulating substances, in cultures of *Cercospora rosicola* but to date there has been no report of pathogen-produced ABA in a diseased plant. In addition to these well-known substances, an increasing number of phytotoxic compounds produced by a diverse range of pathogenic and non-pathogenic organisms has been shown to possess hormonal properties reminiscent of the five groups.

Fusicoccin. Fusicoccin A is the principle toxin in a family of tricarbocyclic dipeptides, produced by the Mediterranean fungal, canker and vascular pathogen of peach and almond, *Fusicoccum amygdali*. Infected plants show symptoms of intense foliar necrosis following the in vivo action of Fusicoccin in causing a permanent opening of stomata, an effect opposite to ABA, leading to terminal dessication.

In bioassays, Fusicoccin mimics, IAA in inducing extension growth in etiolated pea stems (MARRÈ et al. 1974); gibberellins and cytokinins in promoting germination of light requiring seed (LADO et al. 1974) and unlike IAA, stimulates root growth at high concentration. The basic mechanism by which Fusicoccin acts as a growth hormone, thereby simulating IAA is by proton extrusion. As a result of binding to specific protein sites on the plasmalemma and tonoplast, Fusicoccin is thought to stimulate a K^+ ATPase, providing the energy for an efflux of H^+ and Na^+, with a corresponding influx of K^+, sugars, amino acids and anions in the cell. The net effect of these changes is variously to open stomata (guard cell K^+ influx) and promote cell extension via enhanced metabolism and increased osmotic potential, resulting in a cell water influx (MARRÈ 1979).

Cotylenins. These toxins produced by *Cladosporium* sp. are based on 19 deoxy-fusicoccin and exist as glucopyranose glycosides. Like fusicoccin, cotylenins are active in leaf growth assays and also induce wilting (SASSA et al. 1970; BOTALLICO et al. 1979).

Helminthosporol and Related Compounds. Helminthosporium sativum (Bipolaris sorokiniana), the conidial stage of *Cochliobolus sativus* and causal organism of seedling blight and leaf blotch of barley, produces a range of sesquiterpenoid toxins trivially known as sativanes. Two of these, Helminthosporol and Helminthosporic acid have powerful gibberellin-like properties causing: rice long sheath extension, hypocotyl extension in light grown cucumber seedlings, reversal of dwarfism in maize mutants, promotion of light sensitivity in tobacco seeds in darkness, reversal of chlorethyltrimethyl ammonium chloride (CCC) dwarfing on lettuce hypocotyls and stimulation of α-amylase in de-embryonated barley. (TAMURA and SAKURAI 1964, HASHIMOTO et al. 1967, BRIGGS 1966, OKUDA et al. 1967). The aldehyde helminthosporal is mostly active as a plant growth inhibitor but has some weak plant growth substance properties.

Victorin. This compound with an unstable and only partly resolved structure is also produced by *H. sativum*. It consists of a possible cyclic peptide moeity and a sesquiterpenoid base called Victoxinine. One of the major interests in Victorin, like the pathogen, is its strict host specificity and activity at ultra-low concentrations. In growth assays it resembles auxins and cytokinins (EVANS 1974). Experiments by SAFTNER and EVANS (1974) suggest that the action of Victorin and IAA are quite independent and act at different sites on the plasmalemma, but this awaits confirmation.

3,4-Dihydroisocoumarins. Representatives of this interesting group of fungal substances which show growth promoting activity are: Sclerotinin A, (3,4-Dihydro-3,4,5,7-tetra methyl-3,6,8-trihydroxy isocoumarin) and Sclerotinin B, (3,4-Dihydro 3,5,7-trimethyl-3,6,8-trihydroxy isocoumarin) from *Sclerotinia sclerotiorum* (SASSA et al. 1968). Mellein (ochracin) 3,4 Dihydro-3-methyl-6,8-hydroxy isocoumarin and a-3,6,8 trihydroxy form are produced variously by *Septoria nodorum* the wheat glume blotch pathogen, *Pyricularia oryzae* the cause of rice blast and *Alternaria kikuchiana* black spot disease pathogen of pear leaves and fruits.

In addition to acting as necrosis agents these substances are active growth hormones on rice leaf extension. Other similar but more complex structures, sclerin sclerolides from *Sclerotinia libertiana* may also exhibit hormone properties. Some 3,4 dihydroisocoumarins are also found as stress metabolites in higher plants.

Malformins. Some strains of *Aspergillus niger* and *A. wentii* produce cyclic pentapeptide toxins capable of inducing distorted growth of stems and petioles of *Phaseolus vulgaris* (CURTIS 1969). Malformin A – a disulphide form of cyclo-(D-cysteinyl-D-cysteinyl-n-valyl-D-leucycl-n-isoleucycl-) in addition to causing malformation, induces stem elongation (JOHN and CURTIS 1974), root curvature (CURTIS 1958) inhibits adventitious root formation (CURTIS and FELLENBURG 1972) and stimulates abscission (CURTIS 1977). Although malformin A causes increased ethylene production it is not clear whether all the effects of this toxin can be attributed solely to ethylene (CURTIS 1968).

Tyrosol. Tyrosol (p-hydroxyphenyl ethanol) has been isolated from species as diverse as *Pyricularia oryzae* (DEVYS et al. 1976), yeasts, *Gibberella fujikuroi*, *Ceratocystis fimbriata* and *Gloesporium laeticolor*. In growth assays it showed auxin-like properties. (KIMURA and TAMURA 1973).

Phenylacetic Acid. Phenylacetic acid (and a hydroxylated form) acts as a toxin in potatoes and sugar beet following infection with *Rhizoctonia solani* (KOHMOTO et al. 1973). It is also active as a growth hormone substituting for, but some seven times less active than, IAA. MOORE and TOWERS (1967) and MOORE et al. (1968) showed that ^{14}C phenylalanine could be converted to labelled phenylacetic acid in *Schizophyllum commune* and *Sporobolomyces* respectively. WIGHTMAN and LIGHTLY (1982) have subsequently shown that phenylacetic acid occurs in shoots of a number of higher plants as a naturally-occurring auxin.

Hadacidin. This substance, n-formyl hydroxyaminoacetic acid (H CON(OH) CH_2COOH) was isolated from *Penicillium purpurescens* in a screening programme for plant growth inhibitors. It is also produced by *P. frequentens* and *P. aurantio-violaceum*. Hadacidin acts as a plant growth inhibitor and antitumor agent, but unlike most other microbial toxins it does so in the absence of necrosis, thus acting like a classic growth regulator.

2 Insects and Nematodes

The relationship between *most* insects and plants is one of predator and food source respectively (except for insectivorous plants), which in the case of the plant as a food results in gross structural damage but not in a chronically disordered physiology. Thus, insects are not generally regarded as plant parasites. However, situations with insect-induced leaf galls and the histological and biochemical changes resulting from nematode injury are quite different. The plant either locally or systemically undergoes profound physiological changes and is rightly regarded as diseased.

Although much has been written about the biology of insect-induced galls, little precise information is available concerning the growth substances involved or the mechanism by which insect saliva induces the gall – an effect far greater than the zone of damage resulting from stylet or ovipositor penetration (see VIGLIERCHIO 1971). Although SCHALLER (1965) found IAA in the saliva of aphids based on a coleoptile bioassay, the aphids were unable to transform tryptophan when this was fed experimentally. The evidence to date suggests that in aphids IAA is acquired directly from plant feeding, but in the case of the plant bug *Stenodema calcaratum* and possibly other insects, by the metabolism of tryptophan (NUORTEVA 1956). Gall formation, however, cannot be induced in plants by injecting IAA.

MILES and LLOYD (1967), working with the hemipteran *Elasmolomus,* showed that its salivary apparatus could synthesize IAA. When tryptophan and phenylalanine were implanted into the body cavity, the insect induced intumescences after feeding on sunflower cotyledons. These, however, are not normally produced at the feeding sites of this insect, and the results do not support the rather tenuous evidence for a direct induction of galls by saliva. A common feature of the result of aphid leaf feeding is the retention of leaf greenness, reminiscent of the "green-island effect". Indeed, VAN STADEN (1975) has extracted cytokinin substances from wasp larvae found in *Erythrina latissima* galls. These substances co-chromatographed with zeatin and zeatin riboside on Sephadex LH-20. Larvae of this insect are essential for leaf gall formation, since if the larva is killed by a parasite the gall senesces rapidly. However, none of these studies eliminates the real possibility that the larval "cytokinins" were acquired from plant feeding. A comparison of nematode, fungal, and bacterial galls would lead to the assumption that localized increases in cytokinins are involved with the insect-initiated changes in the host genome, the latter causing increased RNA and protein synthesis. The nature (DROPKIN 1969, CHYLINSKA et al. 1972) and site of action of an inducer, however, remain unknown.

A unique parallel between insect juvenility hormones and GA's was claimed by CARLISLE et al. (1963). They noted that either the steroid ecdysone-λ obtained from 4th instar locust larvae or GA_3 shortened the length of the instar when injected into the larvae of locusts (*Locusta migratoria migratorioides*) and *Schistocera gregaria*. Ecdysone-λ from *Bombyx* and GA_3 similarly reduced the locomotory activity of locusts. They also found that ecdysone-λ significantly stimulated the growth of dwarf pea, giving a growth response 10% of the effect of an equivalent concentration of GA_3.

A number of plant parasitic nematodes in the families Heteroderidae and Tylenchidae induce symptoms strongly suggestive of growth-substance disorders; e.g., giant cell formation (*Heterodera*), hypertrophy, hyperplasia and giant cell formation (Meloidogyne) and hypertrophy and hyperplasia in grain and leaves (*Anguina* and *Ditylenchus*). Much physiological interest has centered on the formation of syncytia by *Heterodera* spp. where nuclei enlarge and cell walls partially dissolve. These are similar, but not identical, to the giant cells of *Meloidogyne* characterized by synchronous mitosis and the growth of the cell into areas where lysis of neighbouring cells has occurred. VIGLIERCHIO (1971) reviewed several accounts of auxin-containing nematodes. BIRD (1966), however, found neither GA's nor cytokinins in larvae of *M. javanica* and *M. hapla*. STANDSTEDT and SCHUSTER (1966) (see VIGLIERCHIO 1971), working on tobacco pith cultures, concluded that *M. incognita* secretes neither auxin nor cytokinins. There is little doubt, however, that galled tissue contains higher levels of hormones than healthy. Yu and Viglierchio (1964) and Viglierchio and Yu (1968) list several auxins, IAA, IAN, and IAE in galls, larvae and egg masses of a number of *Meloidogyne* spp. Serious doubt, however, should be given to their claim to have found indol-3-yl butyric acid (IBA) in eggs, larvae, and galls of *M. incognita.* This auxin (not previously reported as a naturally occurring compound) was identified only on the basis of its R_f in paper chromatography and its authenticity in the absence of other evidence, including co-chromatography with pure IBA, should be seriously questioned. Nematodes are also capable of destroying IAA under oxidative conditions in a manner analogous to IAA oxidase/peroxidase, but the kinetics of the system appears different from IAA oxidase (Viglierchio 1971). It has been shown by Giebel (1970) that solanaceous plants show a reduced IAA content mediated by phenolics as a consequence of nematode infection.

It has been suggested that increased cytokinin content of the host increases its susceptibility to nematode infection, since resistant plants increase in susceptibility after cytokinin treatment. A healthy nematode-susceptible tomato cultivar contains higher cytokinin levels than a resistant one (Van Staden 1977). Infection caused an increase in three cytokinins; zeatin, zeatin riboside, and zeatin glucoside in both S and R plants, but levels were highest in S plants. The idea that cytokinins, (induced) possibly in conjunction with auxins, are responsible for uncontrolled endomitosis in giant cells in the absence of cytokinesis is plausible (cf. the hypertrophied cells of club root) but the role of the nematode in the triggering mechanism still awaits elucidation.

References

Abeles FB (1973) Ethylene in plant biology. Academic Press, London New York

Abeles FB, Craker LE, Forrence LE, Leather GR (1971) Fate of air pollutants: removal of ethylene, sulphur dioxide and nitrogen dioxide by soil. Science 173:914–916

Adams DO, Yang SF (1977) Ethylene biogenesis: S-adenosyl methionine as an intermediate in the conversion of methionine to ethylene in apple tissue. Plant Physiol (Suppl) 59:45

Adams DO, Yang SF (1979) Ethylene biosynthesis: identification of 1-amino-cyclo-propane-1-carboxylic acid as an intermediate in the conversion of methionine to ethylene. Proc Natl Acad Sci USA 76:170–174

Anderson JD, Mattoo AK, Lieberman M (1982) Induction of ethylene biosynthesis in tobacco leaf discs by cell wall digesting enzymes. Biochem Biophys Res Comm 107:588–596

Archer SA (1976) Ethylene and fungal growth. Trans Br Mycol Soc 67(2):325–326

Armstrong DJ, Evans PK, Burrows WJ, Skoog F, Petit JF, Stewart T, Strominger J, Leonard NJ, Hecht SM, Occolowitz J (1970) Cytokinins: Activity and identification in *Staphylococcus epidermidis* transfer RNA. J Biol Chem 245:2922–2926

Arntzen CJ, Haugh MF, Bobick S (1973) Induction of stomatal closure by *Helminthosporium maydis* pathotoxin. Plant Physiol 52:569–574

Arora N, Skoog F, Allen ON (1959) Kinetin-induced pseudonodules on tobacco roots. Am J Bot 46:610–613

Assante G, Merlini L, Nasini G (1977) (+)-Abscisic acid, a metabolite of the fungus *Cercospora rosicola*. Experientia 33:1556–1557

Aubé C, Sackston WE (1965) Distribution and prevalence of *Verticillium* species producing substances with gibberellin-like biological properties. Can J Bot 43:1335–1342

Babcock DF, Morris RO (1970) V Quantitative measurement of isoprenoid nucleosides in transfer ribonucleic acid. Biochemistry 9:3701–3705

Barthe P, Bulard C (1974) Identification d'une cytokinine par chromatographie en phase gazeuse à partir de cultures pures de *Taphrina cerasi*. Can J Bot 52:1515–1518

Beijersbergen JCM (1973) Report in: Byrde RJW and Cutting CV (eds) Fungal pathogenicity and the plant's response. Academic Press, London

Beijersbergen JCM, Lemmers CBG (1978) Influence of ethylene produced by *Fusarium oxysporum* on the postulated mechanism underlying the response of *Tulipa* sp. tissues to this fungus. Proceedings of the 3rd International Congress of Plant Pathology, München 1978: p. 229

Bernheim F (1942) The effect of various substances on the oxygen uptake of *Blastomyces dermatiditis*. J. Bacteriol 44:533–539

Biale JB (1940) Effect of emanations from several species of fungi on respiration and colour development of citrus fruits. Science 91:458–459

Biale JB, Shepherd AD (1941) Respiration of citrus fruits in relation to metabolism of fungi. Effects of emanations of *Penicillium digitatum* Sacc., on lemons. Am J Bot 28:263–270

Bird AF (1966) Some observations on exudates from *Meloidogyne* larvae. Nematologica 12:471–482

Bonn WG, Sequeira L, Upper CD (1972) Determination of the growth of the rate of ethylene production by *Pseudomonas solanacearum*. Proc Can Phytopath Soc 38th Sess 26

Botallico A, Graniti A, Lerario P (1978) Further incestigations on the biological activity of some fusicoccins and cotylenins. Phytopath Mediterr 17:127–134

Braun W (1956) Cellular products affecting the establishment of bacteria of different virulence. Ann NY Acad Sci 66:348

Brian PW (1957) The effect of some microbial products on plant growth. Symp Soc Exp Biol 11:166–182

Briggs DE (1966) Gibberellin-like activity of helminthosporol and helminthosporic acid. Nature (London) 210:418–419

Brooks C (1944) Stem-end rot of oranges and factors affecting its control. J Agric Res 68:363–381

Brown ME, Burlingham SK (1968) Production of plant-growth substances by *Azotobacter chrococcum*. J Gen Microbiol 53:135–144

Burg SP (1973) Ethylene in plant growth. Proc Natl Acad Sci USA 70:591–597

Carlisle DB, Osborne DJ, Ellis PE, Moorhouse JE (1963) Reciprocal effects of insect and plant growth substances. Nature 200:1230

Chalutz E, Lieberman M, Sisler HD (1977) Methionine-induced ethylene production by *Penicillium digitatum*. Plant Physiol 60:402–406

Chen CM, Hall RH (1969) Biosynthesis of N^6-(Δ^2-isopentenyl) adenosine in the transfer ribonucleic acid of tobacco pith tissue. Phytochemistry 8:1687–1695

Chou TW, Yang SF (1973) The biogenesis of ethylene in *Penicillium digitatum*. Arch Biochem Biophys 157:73–82

Chylinska KM, Knypl JS, Brzeski MW (1972) Stimulated protein and RNA synthesis in carrot infested with the northern root-knot nematode *Meloidogyne hapla* Chitw. Bull Acad Pol Sci 20:209–212

Considine PJ, Flynn N, Patching JW (1977) Ethylene production by soil micro-organisms. Appl Environ Microbiol 33:977–979

Cornforth IS (1975) The persistence of ethylene in aerobic soils. Plant Soil 42:85–96

Crady EE, Wolf FT (1959) The production of indole acetic acid by *Taphrina deformans* and *Dibotryon morbosum*. Physiol Plant 12:526–533

Curtis RW (1957) Survey of fungi and *Actinomycetes* for compounds possessing gibberellin-like activity. Science 125:646

Curtis RW (1958) Curvatures and malformations in bean plants caused by a culture filtrate of *Aspergillus niger*. Plant Physiol 33:17–22
Curtis RW (1968) Mediation of a plant response to Malformin by ethylene. Plant Physiol 43:76–80
Curtis RW (1977) Studies on the stimulation of abscission by malformin on cuttings of *Phaseolus aureus* Roxb. Plant Cell Physiol 18:1331–1341
Curtis RW, Fellenburg G (1972) Effect of malformin on adventitious root formation and metabolism of indoleacetic acid 2-^{14}C by *Phaseolus* vulgaris. Plant Cell Physiol 13:715–726
Daly JM, Knoche HW (1976) Hormonal involvement in metabolism of host-parasite interactions. In: Friend J, Threlfall DR (eds) Biochemical aspects of plant-parasite relationships. Academic Press, London New York, pp 117–133
Da Silva EJ, Henriksson E, Henriksson LE (1974) Ethylene production by fungi. Plant Sci Lett 2:63–66
De Bont JAM (1976) Oxidation of ethylene by soil bacteria. Antonie Leeuwenhoek J Microbiol 42:59–71
Dekhuijzen HM (1976) Endogenous cytokinins in healthy and diseased plants. In: Heitefuss R, Williams PH (eds) Encyclopedia of plant physiology new series vol 4. Springer, Berlin Heidelberg New York, pp 526–559
Dekhuijzen HM, Overeem JC (1971) The role of cytokinins in clubroot formation. Physiol Plant Path 1:151–162
De Munk WJ, De Rooy M (1971) The influence of ethylene on the development of 5° C pre-cooled "Apeldoorn" tulips during forcing. Hortic Sci 6:40–41
Devys M, Bousquet JF, Barbier M (1976) Le tyrosol (p-hydroxyphényl éthanol) inhibiteur de la germination isolé du milieu du culture de *Pyricularia oryzae*. Phytopath Z 85:176–178
Dropkin VH (1969) Cellular responses of plants to nematode infections. Ann Rev Phytopath 7:101–122
Durbin RD (1981) Toxins in plant disease. Academic Press New York, London, Toronto, Sydney, San Francisco pp. 515
Edwards JM, Gibson F, Jackman LM, Shannon JS (1964) The source of the nitrogen atom for the biosynthesis of anthranilic acid. Biochim Biophys Acta 93:78–84
Eklund E (1970) Secondary effects of some pseudomonads in the rhizoplane of peat-grown cucumber plants. Acta Agric Scand Suppl 17:57
Evans ML (1974) Rapid responses to plant hormones. Ann Rev Plant Physiol 25:195–223
Fahnestock GW (1858) Memoranda of the effects of carburetted hydrogen gas upon a collection of exotic plants. Proc Acad Nat Sci Phila 1:118–134
Fergus CL (1954) The production of ethylene by *Penicillium digitatum*. Mycology 46:543–555
Fraser LM (1953) Effects of indole acetic acid on growth of *Agaricus*. Aust J Biol Sci 6:379–395
Freebairn HT, Buddenhagen IW (1964) Ethylene production by *Pseudomonas solanacearum*. Nature 202:313–314
Gane R (1934) Production of ethylene by some ripening fruits. Nature 134:1008
Gefter ML, Russel RL (1969) Role of modifications in tyrosine transfer RNA: A modified base affecting ribosome binding. J Mol Biol 39:145–157
Giebel J (1970) Phenolic content in roots of some Solanaceae and its influence on IAA-oxidase activity as an indicator of resistance to *Heterodera rostochiensis*. Nematologica 16:22–32
Gogala N (1971) Growth substances in mycorrhiza of the fungus *Boletus pinicola* Vitt. and the pine tree, *Pinus sylvestris* L. Razprave XIV/5:123–202
Gruen H (1959) Auxins and fungi. Annu Rev Plant Physiol 10:405–440
Guttenberg H von, Strutz I (1952) Zur Keimungsphysiologie von *Ustilago zeae*. Arch Mikrobiol 17:189–198
Hashimoto T, Sakurai A, Tamura S (1967) Physiological activities of helminthosporol and helminthosporic acid. 1. Effects on growth of intact plants. Plant Cell Physiol 8:23–34

Helgeson JP, Leonard NJ (1966) Cytokinins: Identification of compounds isolated from *Corynebacterium fasciens*. Proc Natl Acad Sci USA 56:60–63
Hirata S (1957) Studies on the phytohormone in the malformed portion of the diseased plants. III. Auxin formation on culture-grown *Exobasidium*, *Taphrina*, and *Ustilago* spp. Ann Phytopath Soc J 22:153–158
Hislop EC, Archer SA, Hoad GV (1973) Ethylene production by healthy and *Sclerotinia fructigena*-infected apple peel. Phytochemistry 12:1281–1286
Ilag L, Curtis RW (1968) Production of ethylene by fungi. Science 159:1357–1358
John WW, Curtis RW (1974) Stimulation of plant growth by malformin A. Experientia 30:1392–1393
Johnston JC, Trione EJ (1974) Cytokinin production by the fungi *Taphrina cerani* and *Taphrina deformans*. Can J Bot 52:1583–1589
Jones DF, MacMillan J, Radley M (1963) Plant hormones III. Identification of gibberellic acid in immature barley and immature grass. Phytochemistry 2:307–314
Kaper JM, Veldstra H (1958) On the metabolism of tryptophan by *Agrobacterium tumefaciens*. Biochem Biophys Acta 30:401–420
Katznelson H, Cole SE (1965) Production of gibberellin-like substances by bacteria and actinomycetes. Can J Microbiol 11:733–741
Kende H (1971) The cytokinins. Int Rev Cytol 31:301–338
Khomoto K, Fukui R, Mizuno M, Nishimura S (1973) Pathochemical studies on *Rhizoctonia* disease III. Characteristic production of *Rhizoctonia* toxins and their *vivo* toxicity. J Fac Agric Tottori Univ 8:21–31
Kimura Y, Tamura S (1973) Isolation of L-β-phenyllactic acid and tyrosol as plant growth regulators from *Gloeosporium laeticolor*. Agric Biol Chem 37:2925
Klambt D, Thies G, Skoog F (1966) Isolation of cytokinins from *Corynebacterium fasciens*. Proc Natl Acad Sci USA 56:52–59
Kögl F, Kostermans DRFG (1934) Heteroauxin als Stoffwechselprodukt niederer pflanzlicher Organismen. Isolierung aus Hefe. XIII Mitt Z Physiol Chem 228:113–121
Kögl F, Haagensmit AJ, Erxleben H (1934) Über ein neues Auxin („Herteroauxin") aus Harn. XI. Mitt Z Physiol Chem 228:104–112
Kono Y, Knoche HW, Daly JM (1982) Structure: Fungal host-specific. In: Durbin RD (Ed.) Toxins in plant disease, pp. 221–257. Academic Press
Lado P, Rasi-Caldogono F, Colombo R (1974) Promoting effects of fusicoccin on seed germination. Physiologia Plantarum 31:149–152
Laloue M, Hall RH (1973) Cytokinins in *Rhizopogon roseolus*-secretion of N-[9-(β-D-Ribofuranosyl-9H) purin-6-yl-carbomoyl] threonine. Plant Physiol 51:559–562
Lamotte CE, Lersten NR (1972) Attempts to obtain bacteria-free plants of *Psychotria punctata* (Rubiaceae): Growth and root formation in callus cultures. Am J Bot 59:89–96
Lersten NR, Horner HT (1976) Bacterial leaf nodule symbiosis in angiosperms with emphasis on Rubiaceae and Myrsinaceae. Bot Rev 42(2):145–214
Libbert E, Manteuffel R (1970a) Interactions between plants and epiphytic bacteria regarding their auxin metabolism. VII Physiol Plant 23:93–98
Libbert E, Risch H (1969) Interactions between plants and epiphytic bacteria regarding their auxin metabolism. V Physiol Plant 22:51–58
Libbert E, Silhengst P (1970b) Interactions between plants and epiphytic bacteria regarding their auxin metabolism. VIII Physiol Plant 23:480–487
Libbert E, Wichner S, Duerst E, Kaiser W, Kunert R, Manicki A, Manteuffel R, Riecke E, Schroder R (1968) Auxin content and auxin synthesis in sterile and non-sterile plants, with special regard to the influence of epiphytic bacteria. In: Wightman F, Setterfield G (eds) Biochemistry and physiology of plant growth substances. Runge, Ottawa, pp 213–230
Libbert E, Kaiser W, Kunert R (1969) Interactions between plants and epiphytic bacteria regarding their auxin metabolism. IV Physiol Plant 22:432–439
Lockard JE, Kneebone OR (1972) Investigation of the metabolic gases produced by *Agaricus bisporus* (Lange) Sing. Mushroom Sci 5:281–299

Lockhart CL, Forsyth FR, Eaves CA (1968) Effect of ethylene on development of *Gloeosporium album* in apple and on growth of the fungus in culture. Can J Plant Sci 48:557–559

Lund BM, Mapson LW (1970) Stimulation by *Erwinia carotovora* of the synthesis of ethylene in cauliflower tissue. Biochem J 119:251–263

Lürssen K, Naumann K, Schröder R (1979) 1-Amino cyclopropane-1-carboxylic acid – an intermediate of the ethylene biosynthesis in higher plants. Z Pflanzenphysiol 92:285–294

Lynch JM (1975) Ethylene in soil. Nature 256:576–577

Marrè E (1979) Fusicoccin: a tool in plant physiology, Annual Review of Plant Physiology 30:273–288

Marrè E, Colombo R, Lado P, Rasi-Caldogno F (1974) Correlation between proton extrusion and simulation of cell enlargement. Effects of fusicoccin and of cytokinins on leaf fragments and isolated cotyledons. Plant Science Letters 2:139–150

Maruzzella JC, Garner JG (1963) Effect of kinetin on bacteria. Nature 200:385

Meherink M, Spencer M (1964) Ethylene production during germination of oat seeds and *Penicillium digitatum* spores. An J Bot 42:337–340

Meister A (1965) Biochemistry of the amino acids vol 2. Academic Press, London New York

Miles PW, Lloyd J (1967) Synthesis of a plant hormone by the salivary apparatus of plant-sucking Hemiptera. Nature 213:801–802

Miller CO (1967) Zeatin and zeatin riboside from a mycorrhizal fungus. Science 157:1055–1056

Miller EV, Winston JR, Fisher DF (1940) Production of epinasty by emanations from normal and decaying citrus fruits from *Penicillium digitatum*. J Agric Res 60:269–277

Miura GA, Hall RH (1973) Trans-ribosyl zeatin. Its biosynthesis in *Zea mays* endosperm and the mycorrhizal fungus *Rhizopogon roseolus*. Plant Physiol 51:563–569

Miura GA, Miller CO (1969) 6-(γ,γ-dimethyl-allylamino) purine as a precursor of zeatin. Plant Physiol 44:372–376

Moore K, Towers GHN (1967) Degradation of aromatic amino acids by fungi. I Fate of L. phenylalanine in *Schizophyllum commune*. Can J Biochem 45:1959–1965

Moore K, Subba Rao PV, Towers GHN (1968) Degradation of phenylalanine and tyrosine in *Sporobolomyces roseus*, Biochem J 106:507–514

Morell H, Clark MJ, Knowles PF, Sprinson DB (1967) The enzymic synthesis of chorismic and prephenic acids from 3-enolpyruvyl shikimic acid 5-phosphate. J Biol Chem 242:82–90

Muromtsev GS, Globus GA (1976) Adaptive significance of the ability to synthesize gibberellins for the phytopathogenic fungus. Dokl Akad Nauk SSSR vol 226 no 1:204–206

Nickerson WJ (1948) Ethylene as a metabolic product of the pathogenic fungus, *Blastomyces dermatiditis*. Arch Biochem 17:225–233

Nielsen N (1928) Materials which hasten the growth of *Avena* coleoptiles. Planta 6(2):376–378

Nielsen N (1930) Untersuchungen über einen neuen wachstumregulierenden Stoff: Rhizopin. Jahrb Wiss Bot 73:125

Norman SM, Bennett RD, Maier VP, Poling SM (1983) Cytokinins inhibit abscisic acid biosynthesis in *Cercospora rosicola*. Plant Science Letters 28:255–263

Nuorteva P (1956) Studies on the effect of the salivary secretions of some *Heteroptera* and *Homoptera* on plant growth. Ann Ent Fenn 22:108–117

Nysterakis R (1954) Résistance de *Neurospora tetrasperma* aux fortes doses d'auxine et relation avec l'hypothèse de l'existence des métabolites antiauxines. C R Acad Sci 238:143–145

Okuda M, Kato J, Tamura S (1967) Effects of helminthosporol and helminthosporic acid on activation of α-amylase production in barley endosperm. Planta 72:289–291

Owens LD, Lieberman M, Kunishi A (1971) Inhibition of ethylene production by rhizobitoxine. Plant Physiol 48:1–4

Pegg GF (1973a) Gibberellin-like substances in the sporophores of *Agaricus bisporus* (Lange) Imbach. J Exp Bot 24:675–688
Pegg GF (1973b) Occurrence of gibberellin-like growth substances in basidiomycete sporophores. Trans Brit Mycol Soc 61:277–286
Pegg GF (1976) The response of ethylene-treated tomato plants to infection by *Verticillium albo-atrum*. Physiological Plant Pathology 9:215–216
Pegg GF (1976a) Endogenous gibberellins in healthy and diseased plants. In: Heitefuss R, Williams PH (eds) Encyclopedia of plant physiology new ser vol 4. Springer, Berlin Heidelberg New York, pp 592–606
Pegg GF (1976b) The involvement of ethylene in plant pathogenesis. In: Heitefuss R, Williams PH (eds) Encyclopedia of plant physiology new ser vol 4 Springer, Berlin Heidelberg New York, pp 582–591
Pegg GF (1976c) Endogenous auxins in healthy and diseased plants. In: Heitefuss R, Williams PH (eds) Encyclopedia of plant physiology new ser vol 4. Springer, Berlin Heidelberg New York, pp 560–581
Pegg GF, Cronshaw DK (1976) Ethylene production in tomato plants infected with *Verticillium albo-atrum*. Physiol Plant Path 8:279–295
Perley JE, Stowe BB (1966) On the ability of *Taphrina deformans* to produce indoleacetic acid from tryptophan by way of tryptamine. Plant Physiol 41:234–237
Phan CT (1962) Contribution a l'étude de la production de l'éthylène par le *Penicillium digitatum* Sacc. Rev Gen Bot 69:505–593
Phelps RH, Sequeira L (1968) Auxin biosynthesis in a host-parasite complex. In: Wightman F, Setterfield G (eds) Biochemistry and physiology of plant growth substances. Runge, Ottawa, pp 197–212
Phillips DA, Torrey JG (1972) Studies on cytokinin production by *Rhizobium*. Plant Physiol 49:11–15
Primrose SB (1976a) Formation of ethylene by *Escherichia coli*. J Gen Microbiol 95:159–165
Primrose SB (1976b) Ethylene-forming bacteria from soil and water. J Gen Microbiol 97:343–346
Quinn LY, Oates RP, Beers TS (1963) Support of cellulose digestion by *Clostridium thermocellum* in a kinetin-supplemented basal medium. J Bacteriol 86:1359
Rathbone MP, Hall RH (1972) Concerning the presence of the cytokinin $N^6(\Delta^2$-isopentenyl) adenine in cultures of *Corynebacterium fasciens*. Planta 108:93–102
Reddy MN, Williams PH (1970) Cytokinin activity in *Plasmodiophora brassicae*-infected cabbage tissue cultures. Phytopathology 60:1463–1465
Rivière J (1963a) Action des micro-organismes de la rhizosphère sur la croissance du blé. II. Isolement et charactérisation des bactéries produisant des phytohormones. Ann Inst Pasteur 105:303–314
Rivière J (1963b) Rhizosphère et croissance du blé. Ann Agron 14:619–653
Roberts JL, Roberts E (1939) Auxin production by soil micro-organisms. Soil Sci 48:135–139
Rodrigues Pereira AS, Houwen PJW, Deurenberg-Vos H-WJ, Dey EBF (1972) Cytokinins and the bacterial symbiosis of *Ardisia* species. Z Pflanzenphysiol 68:170–177
Saftner RA, Evans ML (1974) Selective effects of Victorin on growth and the auxin response in *Avena*. Plant Physiol 53:382–383
Sakai S, Imaseki H, Uritani I (1970) Biosynthesis of ethylene in sweet potato root tissue. Plant Cell Physiol 11:737–745
Sandstedt R, Schuster ML (1966) The role of auxins in root-knot nematode-induced growth on excised tobacco stem sections. Physiol Plant 19:960–967
Sassa T (1970) Promotion and enlargement of cotyledons by cotylenins A and B. Agric Biol Chem 34:1588–1589
Sassa T, Tojyo T, Munakata K (1970) Isolation of a new plant growth substance with cytokinin activity. Nature (London) 227:379
Scarbrough E, Armstrong DJ, Skoog F, Frihart CR, Leonard NJ (1973) Isolation of cis-zeatin from *Corynebacterium fascians* cultures. Proc Natl Acad Sci USA 70:3825–3829

Schaller von G (1965) Untersuchungen über den β-Indolylessigsäure-Gehalt des Speichels von Aphidenarten mit unterschiedlicher Phytopathogenität. Zool Jb Physiol 71:385–392

Schiewer V, Erdmann N, Libbert E (1970) Conversion of indolylglycerol phosphate into tryptophan by extracts of *Pisum*. Physiol Plant 23:473–479

Sequeira L (1973) Hormone metabolism in diseased plants. Annu Rev Plant Physiol 24:353–380

Sequeira L, Williams PH (1964) Synthesis of indoleacetic acid by *Pseudomonas solanacearum*. Phytopathology 54:1240–1246

Shaw M, Hawkins AR (1958) The physiology of host – parasite relations. I. A preliminary examination of the level of free endogenous indoleacetic acid in rusted and mildewed cereal leaves and their ability to decarboxylate exogenously supplied radioactive indoleacetic acid. Can J Bot 36:1–16

Skoog F, Schmitz RY (1972) Cytokinins. In: Steward FC (ed) Plant physiology. Physiology of Development: The Hormones. Vol VIB. Academic Press, London New York, pp 181–213

Smith AM (1973) Ethylene as a cause of soil fungistasis. Nature 246:311–313

Smith AM (1976) Ethylene in soil biology. Ann Rev Phytopathol 14:53–73

Smith AM, Cook RJ (1974) Implications of ethylene production by bacteria for biological balance of soil. Nature 252:703–705

Smith KA, Restall SWF (1971) The occurrence of ethylene in anaerobic soil. J Soil Sci 22:430–443

Smith KA, Russel RS (1969) Occurrence of ethylene, and its significance in anaerobic soil. Nature 222:769–771

Sobieszczanski (1966) Studies on the role of micro-organisms in the life of cultivated plants. III. Origin of the bacterial substances stimulating the growth of plants. Acta Microbiol Poloni 15:67–84

Somner NF (1961a) Production by *Taphrina deformans* of substances stimulating cell elongation and division. Physiol Plant 14:460–469

Somner NF (1961b) Longitudinal and lateral response of etiolated pea sections to indoleacetic acid, gibberellin, kinetin, sucrose and cobaltous chloride. Physiol Plant 14:741–749

Spalding DH, Lieberman M (1965) Factors affecting the production of ethylene by *Penicillium digitatum*. Plant Physiol 40:645–648

Srinivasan PR (1965) The biosynthesis of anthranilate from (3,4-C^{14}) glucose in *Escherichia coli*. Biochemistry 4:2860–2865

Swart A, Kamerbeek GA (1977) Ethylene production and mycelium growth of the tulip strain of *Fusarium oxysporum* as influenced by shaking of and oxygen supply to the culture medium. Physiol Plant 39:38–44

Tamura S, Sakura A (1964) Syntheses of several compounds related to helminthosporol and their plant growth promoting activities. Agric Biol Chem 29:595–596

Thimann KV (1935) On the plant-growth hormone produced by *Rhizopus suinus*. J Biol Chem 109:179–291

Thimann KV, Grochowska MV (1968) The role of tryptophan and tryptamine as IAA precursors. In: Wightman F, Setterfield G (eds) Biochem physiol plant growth substances. Runge, Ottawa, pp 231–242

Thimann KV, Sachs T (1966) The role of cytokinins in the "fasciation" disease caused by *Corynebacterium fasciens*. Am J Bot 53:731–739

Thimmappaya B, Cherayil JD (1974) Unique presence of 2-methylthio-ribosylzeatin in the transfer ribonucleic acid of the bacterium *Pseudomonas aeruginosa*. Biochem Biophys Res Commun 60:665–672

Thomas KC, Spencer M (1977) L-Methionine as an ethylene precursor in *Saccharomyces cerevisiae*. Can J Microbiol 23:1669–1674

Truffa-Bachi P, Cohen GN (1968) Some aspects of amino acid biosynthesis in microorganisms. Ann Rev Biochem 37:79–108

Turfitt GF (1941) A new method for the determination of phytohormone activity. Biochem J 35:237–244

Turner EM, Wright M, Ward T, Osborne DJ, Self R (1975) Production of ethylene and other volatiles and changes in cellulose and laccase activities during the life cycle of the cultivated mushroom, *Agaricus bisporus*. J Gen Microbiol 91:167–176

Upper CD, Helgeson JP, Kemp JD, Schmidt CJ (1970) Gas liquid chromatographic isolation of cytokinins from natural sources. Plant Physiol 45:543–547

Vander Molen GE, Labavitch JM, Strand LL, Devay JE (1983) Pathogen-induced vascular gels: ethylene as a host intermediate. Physiol Plantarum 59:573–580

Van Staden J (1975) Cytokinins from larvae in *Erythrina latissima* galls. Plant Sci Lett 5:227–230

Van Staden J, Dimalia GG (1977) A comparison of the endogenous cytokinins in the roots and xylem exudate of nematode-resistant and susceptible tomato cultivars. J Exp Bot 28:1351–1356

Van Sumere CF, Van Sumere-De Preter C, Vining LL, Ledingham GA (1957) Coumarins and phenolic acids in the uredospores of wheat stem rust. Can J Microbiol 3:847–862

Viglierchio DR (1971) Nematodes and other pathogens in auxin-related plant growth disorders. Bot Rev 37:1–21

Viglierchio DR, Yu PK (1968) Plant growth substances and parasitic nematodes. II. Host influence on auxin content. Exp Parasitol 23:88–95

Ward T, Turner EM, Osborne DJ (1978) Evidence for the production of ethylene by the mycelium of *Agaricus bisporus* and its relationship to sporocarp development. J Gen Microbiol 104:23–30

Wightman F, Lighty DL (1982) Identification of phenylacetic acid as a natural auxin in the shoots of higher plants. Physiol Plant 55:17–24

Williams PH, Reddy MN, Strandberg JO (1969) Growth of non-infected and *Plasmodiophora brassicae*-infected cabbage callus in culture. Can J Bot 47:1217–1221

Williamson CE (1950) Ethylene, a metabolic product of diseased or injured plants. Phytopathology 40:205–208

Wolf FT (1956) The production of indolacetic acid by cedar apple rust fungus and its identification by paper chromatography. Phytopathol Z 26:219–233

Wood DA, Hammond JBW (1977) Ethylene production by axenic fruiting cultures of *Agaricus bisporus*. Appl Environ Microbiol 34:228–229

Yamada Y, Nishimura S, Ishikura H (1971) The presence of 2 methylthio-N^6 (Δ^2-isopentenyl) adenosine in leucine, tryptophan and cysteine tRNAs from *Escherichia coli*. Biochim Biophys Acta 247:170–174

Yang SF (1974) Biochemistry of ethylene: biosynthesis and metabolism. In: Runeckles VC, Sondheimer E, Walton DC (eds) Recent advances in phytochemistry Vol 7. Academic Press, London New York, pp 131–164

Yang SF, Baur AH (1972) Biosynthesis of ethylene in fruit tissues. In: Carr DC (ed) Plant growth substances 1970. Springer, Berlin Heidelberg New York, pp 510–517

Yang SF, Hoffman NE (1984) Ethylene biosynthesis and its regulation in higher plants. Ann Rev Plant Physiol 35:155–189

Yu PK, Viglierchio DR (1964) Plant growth substances and parasitic nematodes. I. Root knot nematodes and tomato. Exp Parasitol 15:242–248

Zeigler RS, Powell LE, Thurston HD (1980) Gibberellin A_4 production by *Sphaceloma manihoticola*, causal agent of Cassava superelongation disease. Phytopathology 70:589–593

17 Electrical and Magnetic Stimuli

D.S. FENSOM

1 Introduction

A current part of our thinking is that matter is comprised of particles or waves which have electromagnetic properties. Therefore it is to be expected that externally applied electrical and magnetic fields will cause perturbations in atoms or subatomic particles. Bonding, enzyme action to associate or dissociate molecules, or to alter their configuration, must all be influenced, however slightly, by electromagnetic stimuli. This is taken for granted, but the major question which concerns us in this chapter is whether electromagnetic stimuli applied externally (usually artificially) can cause changes in plant hormone location, concentration, or action, which will in turn affect growth. Pure magnetic effects may also occur, but are more difficult to demonstrate experimentally, and discussion of them will be reserved until later, but it is quite easy to show that electrical effects cause reactions in plants and conversely, that hormonal effects can cause detectable changes in bioelectricity. Therefore it seems desirable to discuss the interaction of electrical fields and hormone activity, in as far as we can presently detect them with readily available apparatus. As noted above, there will be some response at deeper levels to every electrical change in the environment; but this response may be subatomic and will not be considered if we cannot detect it. Rather, we will first review the present state of knowledge; then, we will consider when electrical stimuli may lead to detectable changes in hormones or responses which may be suspected of having a hormonal basis. In the last part we will briefly discuss what little is known about magnetic effects.

2 Electrical Stimuli

2.1 Historical Background

Bioelectric measurements have now been made on plants for at least a century. Indeed, if KUNKEL's publication in 1878 can be considered the first, we have celebrated our centennial of the detection of bioelectric phenomena in the Plant Kingdom. Historical reviews of the subject have been published by ROSENE and LUND (1953), UMRATH (1956, 1959), by SCOTT (1967, by MEYLAN (1971), and by PICKARD (1973); but it is not the purpose of this chapter to deal with bioelectric phenomena as a whole. Rather, we will review work where a *close association* seems probable between electric fields and plant hormones, whether the electric field occurs naturally (as it does everywhere in living tissue), or is applied artificially.

If an electrical potential difference is applied at high voltage across plant tissue, several things may happen. If the voltage is too high, the tissue will produce heat and may actually burn, as in the case with lightning. Somewhat lower voltages (<50 V cm^{-1}) result in less serious damage, such as membrane rupture (BEILBY and COSTER 1980). At even less intense voltage (<1.5 V cm^{-1}), disturbances to ions can occur along apoplast, symplast, and membrane. In tissue it is the voltage gradient which is important, and since both apoplast and symplast may be relatively good conductors, the main resistance is apt to occur at membranes, particularly the plasmalemmae. Across these nm-thin barriers with their high resistance and capacitance properties (of the order of 1 μ farad cm^{-2} and 10^4 Ω cm^2 WILLIAMS et al. 1964), the voltage gradients will be greatest and may then build up to many kV cm^{-1}. All of the above will affect polar compounds, their properties as dielectrics, and their alignment or movement in small electric fields which are established as the current follows its tortuous path between or through the living cells. Every living cell will, of course, possess its own natural bioelectric potential between symplast and apoplast across the plasma membrane, or between vacuolar fluid and cytoplasm across the tonoplast, or indeed across any membrane or membrane-like inclusion found within the cell. These natural biopotentials can be changed by processes of oxidation reduction, or photochemistry, or electrogenic ion pumping of various types. Bioelectric potentials also occur due to natural charge accumulation on the surface of living matter, are highly complex, and are capable of continued flux or alteration as the life of the cell proceeds. It follows, then, that alterations in such natural charge effects must lead to concomitant alterations in any charged molecules (e.g., proteins, other macromolecules, polar hormones or ions) which are in the vicinity. Therefore, the effects of even small amounts of applied electrical voltage may produce measurable effects inside the living cell, depending upon the amount of current (amperage), or the potential gradients which are induced.

Natural gradients of some 130 V m^{-1} normally exist in the atmosphere above the earth's surface. These voltages, probably electrostatic in origin, fluctuate somewhat in magnitude, but are normally electropositive when compared with the earth. Thus, plants exist continually in this electric field, and indeed a tree usually has its top electropositive to its base (except for the growing tips which are electronegative to the rest), so that positive ions tend to move basipetally in it (FENSOM 1963) unless carried by the upward transpiration stream.

In plant tissues DC passes mainly through the apoplast if there are many cell membranes in the path (cf. FENSOM 1966, HAYDEN et al. 1969). Since in multicellular tissue, or in xylem tissue, the current path will largely be apoplastic, only small subsidiary effects will occur across membranes, or inside cells, unless the cells are near the electrodes supplying the current. On the other hand, for cells which have a large amount of cytoplasm, or are continuous with high ion content and few membrane barriers (e.g., members of the Characeae, phloem sieve tubes or cambium tissue in many trees), the electric current will mainly travel in symplastic regions. The amount going via the symplast (or the apoplast) can often be calculated (FENSOM and DAINTY 1963, SPANSWICK 1972, 1976). When DC is used to pass electricity through living plant tissue, the anode may become highly oxidizing and the tissue around it may be damaged as a result of the release of elemental oxygen or chlorine at this terminal. The cathode, on the other hand, may become highly reducing, and although the damage is seldom as obvious, it can nonetheless produce nearby effects. Therefore these major electrical disturbances can induce damages or distortions to tissue without involving the intermediate action of plant hormones at all. When the current is AC, however, sufficiently small (e.g., at physiological levels of 1 μamp cm^{-2} or less), or pulsed DC of short duration (as in the "Shigometer" used by BLANCHARD et al. 1983) and nonpolarizing electrodes are used, damage to living cells near electrodes or in the tissue as a whole may be trivial.

Plant membranes can act as small electrical capacitors, therefore under AC they charge and discharge in such a manner that with high frequency AC, the current *appears* to pass completely through the cell. In fact, however, it is charging and discharging the membranes so quickly that leakage through the condensers (e.g., plasmalemma and tonoplast) is not terribly important. At low frequencies AC acts in a manner somewhat similar to DC, and nonappreciable amounts will pass through the membrane as built-up charges on either side attempt to leak past. Thus, the low-frequency AC will follow apoplastic pathways along ion channels in the cell wall, or pass through extracellular free space unless cell membranes between electrodes are few in number, as is likely in cambial tissue (ZIMMERMANN and BROWN 1971). If a plant hormone is polar or its production is activated by an applied electric field, this activity could alter normal plant hormone action, but very little work of a quantitative nature has been pusblished. Therefore, we must restrict our present consideration to the work which has been done, using – as we have said before – measurable electrical stimuli and measurable hormonal responses.

Electric potential differences are, then, a "fact of life" in plants as well as in other systems. They occur on all surfaces and within solutions wherever there are differences in ion adsorption or mobility. In particular they occur across membranes, for the living membrane is not only a site of charge accumulation and a barrier to ion mobility, it is often the site of active charge production, or as we say, of electrogenic pumps, batteries, as it were, where positive and negative charges are separated and released with the insulating membrane barrier in between (cf. P. MITCHELL's concept, reviewed by RAVEN and SMITH 1980). The role of the bioelectric potential in most processes, and especially in hormone-mediated processes, is still largely unknown and only inferred. Indeed, there are few processes where it can be proven unequivocally that differences in bioelectric phenomena are causal, and not just effects of other activity.

SCHWAN (1977), summarizing his studies of electric field interaction with biological matter, notes that "rather strong field values of the order of kV cm^{-1} are probably needed to orient polar macromolecules"; "classical membrane excitation phenomena occur at field levels of the order of 1 V cm^{-1} ... but weak field effects have been reported corresponding to tissue field gradients of the order of 10^{-7} V cm^{-1}." In this context it should be noted that membrane potentials in plants are frequently of the order of 10–200 mV. This would correspond to field strengths across a 10-nm membrane of 10^3–10^4 V cm^{-1}, which is strong enough to alter the orientation of polar macromolecules, yet is below the level required to induce membrane breakdown.

2.2 Measurement Techniques

It is convenient to divide electrical measurements into those concerned with biopotentials, resistances, currents, capacitances, and wave function. However, since the effects of high frequency AC are in some ways contradictory, it will be better to discuss here only the simplest electrical properties of plant tissue and the measurements of biopotential, resistances, currents, and surface charges; that is, properties altered by application of DC or direct field rather than alternating ones.

2.2.1 Biopotential Differences

Biopotentials must generally be measured by devices which do not draw appreciable current, using either potentiometers (where the measuring current is ultimately nil), or high impedance amplifiers (where very small amounts of charge are magnified by appropriate electronics) to read the existing biopotential without appreciably altering it. For the latter method we use electrometers which have more than 10^8 ohm input impedance. However, there are so many sources of biopotential, including surface charge, concentration effects, thermocouple effects etc., that special precautions need to be taken before or during measurements. The use, for instance of salt bridges and nonpolarizing electrodes is particularly important (cf. SUCKLING 1961, SPANNER 1964, NOBEL 1974, HIGINBOTHAM 1973, MEYLAN 1971, FINDLAY and HOPE 1976).

2.2.2 Resistance

Electrical resistances are in many ways simpler to measure, but, if done by a continuously applied DC may cause distortion or damage to the tissue unless the currents are very close to physiological level. Therefore, the applied current should either be not much higher than 1 μamp cm^{-2}, or should be in tiny pulses with pauses between, so that polarizing effects at the electrode are reduced (cf. HAYDEN et al. 1969, DIXON et al. 1978, SKUTT et al. 1972, FENSOM 1966, GREENHAM et al. 1972, TATTER and BLANCHARD 1976, WILNER and BRACH 1979).

When electrical resistances are measured by AC, they are generally reported as impedances. Here the measuring device is an AC bridge, but convenient devices which can lead to continuous recording have been worked out for the monitoring of water potential in stress, and could be useful in other situations (DIXON et al. 1978). Some work with electrical impedance has also been done using an oscilloscope, the square-wave input pulses being monitored further down the tissue (cf. WILLIAMS et al. 1964, FINDLAY and HOPE 1976). Of course, it should be realized that conductance measurements can be made by the same pieces of apparatus and essentially are the reciprocal of the resistance or impedance, as the case may be.

2.2.3 Current

Some electrical current measurements have been made directly on trees (FENSOM 1963, LEVENGOOD 1970, 1973) and in Characeae (PICKARD 1972, FENSOM et al. 1973), but these imply the draining of the natural charge from the plant. This may be disturbing, since an ammeter can hardly measure the actual current which is inside the tissue, but rather measures a short circuit part of it by draining off some accumulated charge. Nonetheless, streaming currents can be useful in some translocation situations (cf. TYREE and FENSOM 1968, FARQUAHAR and FIELD 1971), and it is quite possible that close feedback relations could exist with processes under hormone control or with hormonal movement (JAFFE 1980), see Section 2.7 below.

2.2.4 Capacitance

Until recently, capacitance measurements were difficult to make without complex high-frequency or time-scanning apparatus. This approach required a measure of decay of charge or phase change in the circuit (WILLIAMS et al. 1964, FINDLAY and HOPE 1976). Detectable capacitance changes have been associated with membrane properties, tissue structure, temperature, and other parameters in plant tissue, therefore this technique offers considerable promise (GREENHAM 1966a, b, GREENHAM et al. 1972, TATTAR et al. 1974, GREENHAM et al. 1978, GREENHAM et al. 1980, SMITH and COSTER 1980). Now that cheap, portable capacitance meters operating at frequencies of 1 to 10 MHz are available, changes in capacitance in leaves can be related to water of hydration or even water potential (BIDGOOD 1982), both of which are likely to be sensitive to hormonal regulation. A portable oscilloscope has also been developed to attempt to detect the breaking of dormancy in trees (RIETVELD and WILLIAMS, personal communication), and a portable AC capacitance meter has recently been used on balsam fir by FENSOM et al. (1983).

2.2.5 Dielectric Constant

The dielectric constant is normally measured using fixed plates with high-frequency AC. This is essentially a capacitance technique, and since the dielectric constant of water is so high compared with lipids or other plant material (about 80 compared to 5–10), dielectric measurements primarily pick up the amount of water between the measurement plates. Membranes, ions and polar compounds will also contribute to the total dielectric, therefore changes in this parameter are of physiological interest and may now be used to study molecular and hormonal effects (TATTAR and SYLVIA 1977, STARTSEVA et al. 1975, C.G. GREENHAM, personal communication, SCHWAN 1977, SCHWARZ 1977, MARCHEL 1977, BIDGOOD 1982, CARERI, GIANSANTI, FENSOM and CÔTÉ, personal communication).

2.3 Seed or Tissue Viability

Seed viability, as expressed by germination, is controlled by a number of processes, including some of a hormonal nature. The relationship (if any) between seed germination and bioelectric potential is not well substantiated, and between these phenomena in seeds and hormone-mediated or -induced processes such a relationship is purely speculative. However, seeds of maize (*Zea mays* L.), when touched by a nonpolarizing electrode system, are reported to give two potentials, that which can be read a few seconds after the moment of touch, and another after several minutes (BURR 1947). The first of these is claimed by BURR (1947) to be associated with the viability of seeds. While this work has not been independently confirmed, the technique has been attempted on seeds of *Pinus resinosa* Ait. The readings are, however erratic, and with conifer seeds small in magnitude (FENSOM, personal communication). Other methods (e.g., tetrazolium chloride) look more promising, but as it is possible that there may be some hormone activity correlative with electrical activity, it seems worth noting that an electrical approach may some day become useful (such as RIETVELD and WILLIAM'S, personal communication, use of a portable oscilloscope to detect viability in trees). In seaweeds, Na^+, K^+ or Cl^- ion uptake is much greater in living cells than in damaged or dead tissue and can be followed by tracer studies (MACDONALD et al. 1974).

2.4 Photoelectric Responses

It is well known that light induces changes in biopotentials across chloroplast membranes of different kinds (cf. MEYLAN 1971, VREDENBERG 1976). In situations involving light-induced tropisms the question is whether auxin is synthesized (or released) as a response to electrical stimulus, or is directly induced by photochemical reactions. At the moment it is impossible to answer this question. Similar bioelectric responses to light and to auxin have been reported (Figs. 1 and 2, see also FELLE and BENTRUP 1974, JAFFE 1970) but these bioelectric responses are complex, highly variable between tissues, and often hard to interpret. Even so, it is well accepted that when light is shone unilaterally onto a stem or hypocotyl, a biopotential gradient may be detected across the axis of the stem (ROSENE and LUND 1953, NOVAK et al. 1975, PILET 1961, JOHNSSON 1965a, 1967). Biopotential responses to light and dark in stomatal guard cells are reviewed by SIMONS (1981) and WILLMER (1983). The biopotential gradient (Fig. 1) is usually set up quite rapidly and is similar in time-course to that produced by auxin additions (BENTRUP et al. 1977). It seems probable that light stimulates auxin production/movement/release, which then is accompanied by a change in biopotential, but this is far from certain; the biopotential change could occur first. Where chloroplasts are present, light will alter the charge on the surface of membranes (VREDENBERG 1974, 1976), which will in itself change the biopotential of the plant tissue as a whole (e.g., ways in which this might occur are suggested by SPANSWICK 1974, CLELAND and LOMAX 1977, RAVEN and SMITH 1977). Examples of other light-stimulated biopotential

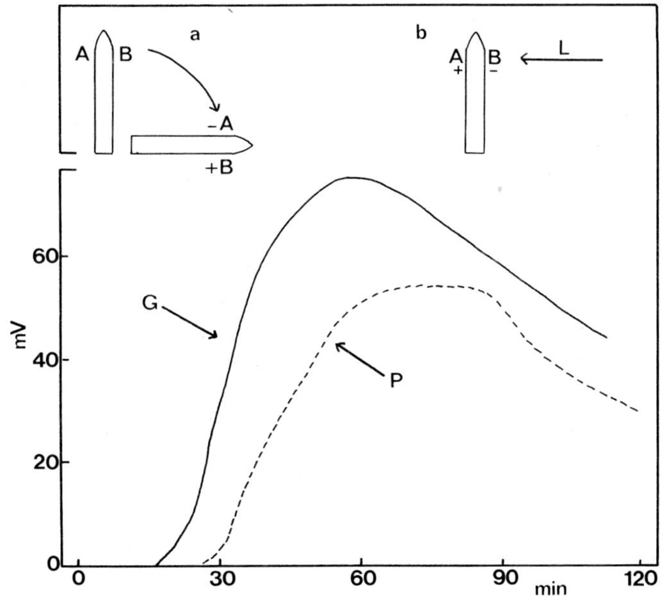

Fig. 1. Definition of geoelectric effect *a*, and photoelectric effect *b*. Development in the coleoptile of *Zea mays*. Geoelectrical effect: *G* at time 0 set at the horizontal position for 1 h. Photoelectric effect: *P* at time 0 exposed to continuous light from one side of 1.5×10^{-2} mJ cm^{-2} s^{-1} (28 lx). *Abscissa* is time; *ordinate* is biopotential. (MEYLAN 1971, after JOHNSSON 1965a)

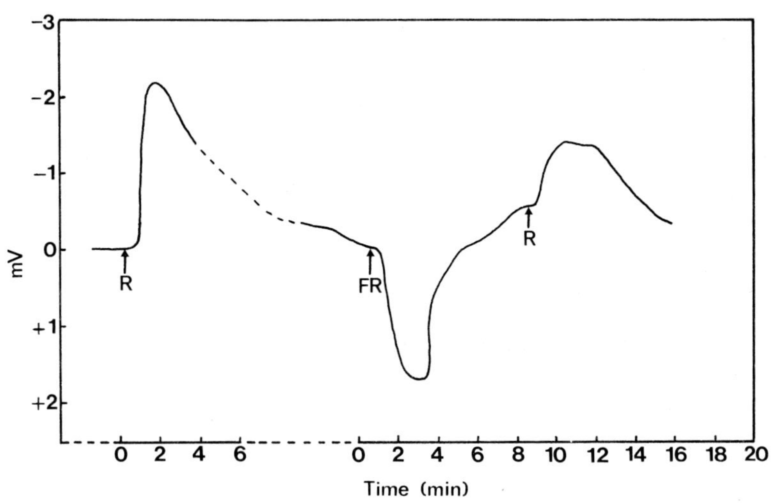

Fig. 2. Electric response of an oat cell to red and far-red light. These representative vacuolar potential changes of one parenchyma cell of an intact etiolated coleoptile were measured with an inserted microelectrode. The peak in the response occurs at about 80 s after the start of the initial red light, which lasted 2 s. *R* treatment by 50 mJ cm^{-2} red light; *FR* treatment by 50 mJ cm^{-2} far-red light. Note the reversal of effects. (After NEWMAN and SULLIVAN 1976)

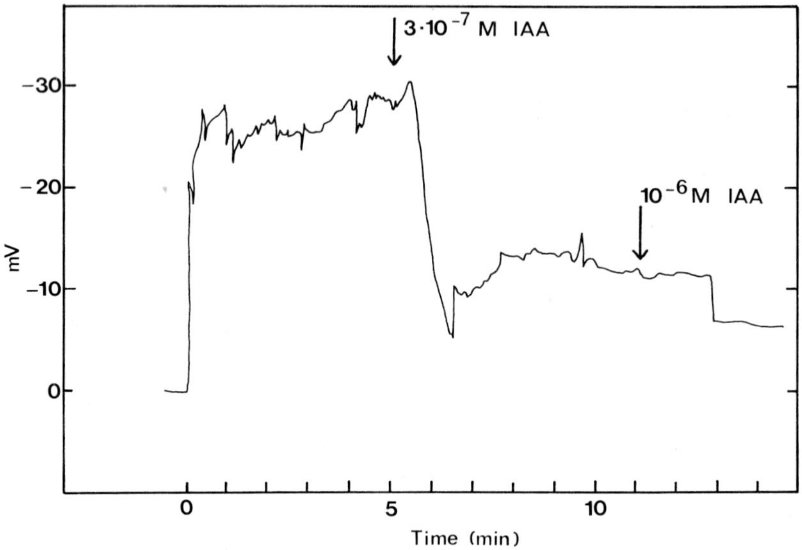

Fig. 3. Time course of the membrane potential, in mV, of a *Petroselinum* cell upon addition of IAA as indicated. Impalement was commenced at t=0 and finished at t=13 min, indicating that the time of stimulus response between application of auxin and change in membrane potential is short. (After BENTRUP et al. 1977)

changes where hormones could be involved are shown in Figs. 2 and 3. Some of the responses to light are slow, e.g., 5 min (Fig. 1), but other responses are much faster, ≈ 1 min (Figs. 2, 3), the latter being as fast as auxin acts in effecting biopotential change (Fig. 3). JOHNSSON (1965a, 1967) suggests that light causes charge separation or electrogenic pump activity to occur across membranes, probably through the chlorophyll mechanism as MITCHELL proposed (1961). Changes in auxin level can be expected to follow, indicating auxin biosynthesis, release, and/or movement, but not necessarily implying that biopotential change produces auxin per se.

RACUSEN and SATTER (1975) have produced evidence of a biological clock and two different phytochrome systems which interact to regulate the membrane biopotential on the pulvini of the nyctinastic plant *Samanea*. Both white light and the membrane localized chromoprotein phytochrome regulated the membrane properties (Fig. 2a, b). Light may also, upon occasion, induce action potentials as in *Phaseolus* (STOLAREK et al. 1980).

2.5 Geoelectric Responses

Biopotential changes resulting from a change in gravitational direction have been reported since 1907 (BOSE), both auxin movement and bioelectric effects occurring very close together in time (JOHNSSON 1965b, 1967, MEYLAN 1971, p. 65). Both may be the result of pressure, stress, or other gravitational effects

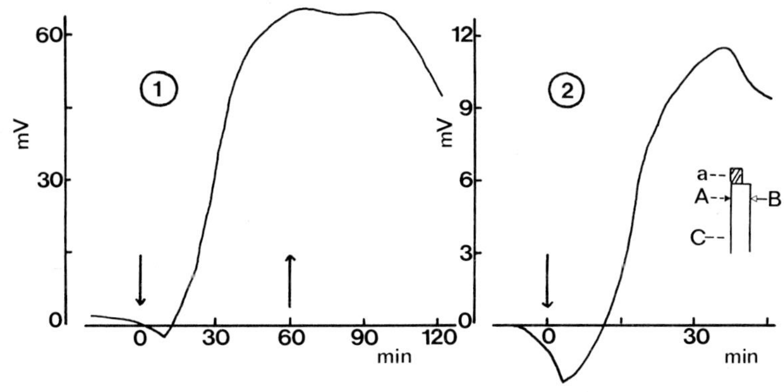

Fig. 4. Development of a geoelectric effect (*1*) and the effect of auxin dissymmetry (*2*) in the coleoptile of *Zea mays*. *Abscissa* is time; *ordinate* is bioelectric potential. *1* Taken in horizontal position at time 0 for 60 min; the potential measured by the method of HERTZ (1960) in his Fig. 1. (Adapted from GRAHM and HERTZ 1962 by MEYLAN 1971). *2* application of IAA at 0.02% in lanolin at time 0. *a* on the decapitated coleoptile *C*; *A* and *B* contacts with continually replenished solution at 4 mm below the section to measure the transverse biopotential according to the classical method. (After WILKINS and WOODCOCK 1965 by MEYLAN 1971)

acting on the cell wall or plasma membrane (AUDUS 1972, CLIFFORD et al. 1982), the correlation being only coincidental (cf. Figs. 1 and 4).

A considerable amount of work has been done on geoelectric responses in plants (see Sect. 2.2 above), and it is now certain that an electrical change does occur as a result of the gravity change in the tissue. Again, however, there are conflicting views as to whether the electrical change precedes the hormonal change or the reverse, since movement of auxin has been detected at about the same time as other gravitational-induced movements (e.g., starch grains, AUDUS 1972). It now seems likely that starch grains and protoplasm inside a cell slightly distort the membrane on the bottom side of any tissue, and by this distortion (or ion accumulation) affect the biopotential (ZIMMERMANN and BENZ 1980, ZIMMERMANN and STEUDLE 1980), as well as the hormone distribution (AUDUS 1972, PICKARD 1973, HERTZ 1960, GRAHM 1964, GRAHM and HERTZ 1964). Alternatively it is conceivable that tension in cell walls will alter the biopotential in a different manner than compression, thus acting as piezoelectric devices (CLIFFORD et al. 1982).

There is, however, little evidence to prove that auxin increases are a result of a primary bioelectric stimulus; indeed the reverse seems much more probable at the moment. Nonetheless, feedback mechanisms (e.g., Sect. 2.8) might be involved.

2.6 Hormone Transport

Direct measurement of hormone transport has been accomplished (NEWMAN 1959, 1963, PILET 1964, PILET and MEYLAN 1953, SCOTT 1967). In the case

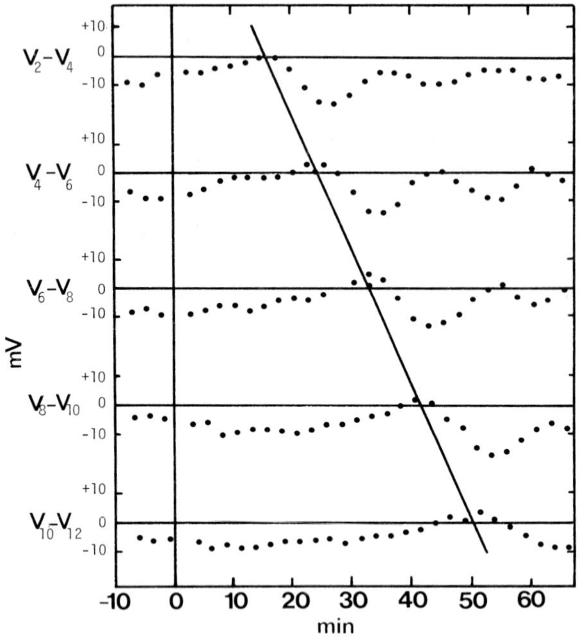

Fig. 5. Time-course of biopotential change in intact dark-grown coleoptiles upon illumination with white light. The illumination was limited to the apical 2 mm with an intensity of 2×10^{-1} J cm^{-2} s^{-1}. The subscript on the biopotentials V_2–V_{12} indicates mm from the tip. Light has apparently induced movement of IAA from the coleoptile apex, as observed by the auxin's electrical effect on the tissue. Previous experiments showed that IAA from an agar block was moving at the rate of 12.4 mm h^{-1} and the present observations indicate that the electrical front was moving at the rate of 14 mm h^{-1}. It is suggested that the potential electrical gradients down a coleoptile are an effect of the distinct advancing from of the stream of IAA from the agar block. (After NEWMAN 1959)

of NEWMAN's work (1959, 1963), the movement of IAA down a root was shown to be accompanied by bioelectric response at the same speed, the hormone apparently inducing the bioelectric change as it passed down the tissue at about 14 mm h^{-1} (Fig. 5). A much faster movement of from 6–21 cm min^{-1} has been reported (PICKARD 1973, VAN SAMBEEK et al. 1976, VAN SAMBEEK and PICKARD 1976, see Fig. 6a, b). CHEESEMAN and PICKARD (1977) think that injury effects release a separate hormone, which they call Ricca's factor, and that this substance is capable of moving rapidly in the tissue and inducing an action or variation potential as it moves. On the other hand, LEBEDEV et al. (1975) interpret the moving of the injury effect at 6 cm min^{-1} as indicating that free radicals are released with the passing of an action potential. This view, the opposite of that held by CHEESEMAN and PICKARD, postulates that the electrical effect is the stimulus, the hormone (or free radical release) the response, and is held also by others (PASZEWSKI and ZAWADZKI 1973, MASLOBROD et al. 1975, OPRITOV et al. 1974, KALININ et al. 1970).

Further evidence in favor both of a high-speed electrical signal and of some sort of fast chemical transmission is given by UMRATH and KASTBERGER (1983). These signals are clearly influenced by hormones (WATANABE and UMRATH 1983). Recent experiments indicate the fast transport of ions in seaweeds or *Tradescantia* (PENOT and PENOT 1979, PENOT 1976).

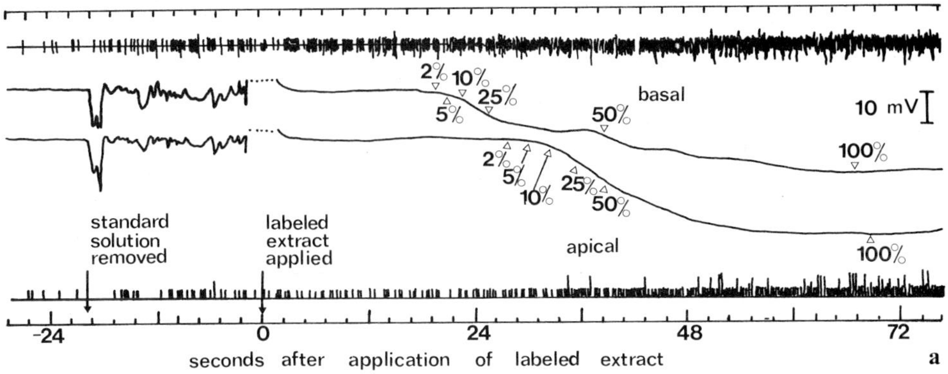

Fig. 6a. Recordings of the progress of radioactively labeled extract ($^{32}PO_4$) and of the variation potential (10 mV scale shown on *right*) up the petiole-rachis of an excised leaf of tomato. *Outer trace* Geiger-Müller counter output; *middle traces* voltage record. The count rate and the variation potential can be seen to increase together both in the paired basal recordings and in the paired apical recordings. The detectors were centered on the rachis 105 and 140 mm above the cut base of the petiole. The interwindow velocity of the variation potential is 4.0 mm s^{-1}. **b** *Upper plots* are the counts s^{-1} measured on the accelerated recordings corresponding to the outer recordings of **a**. *Lower plots* illustrate the fitting of regression lines to ascertain their intercepts as estimates of the times of arrival of the labeled extract at the two counter window slits. The interwindow velocity of the radioactivity is 4.4 mm s^{-1}, corresponding closely with the velocity of the variation potential. (After VAN SAMBEEK et al. 1976)

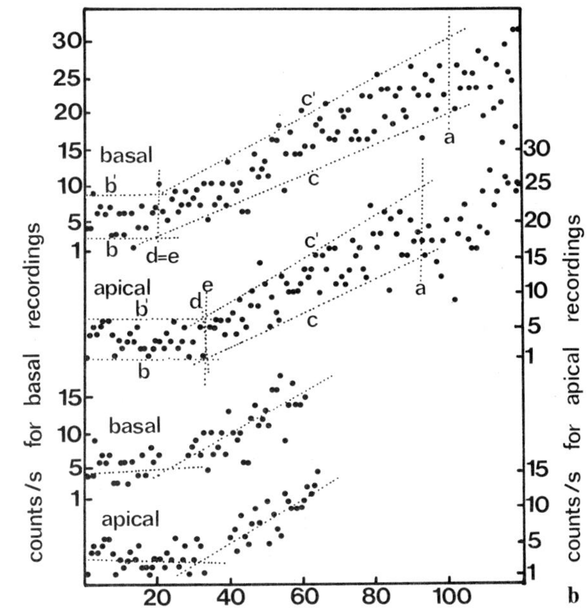

2.7 Growth-Regulating Compounds in Relation to the Membrane or Cell Walls

The close association in time and space between bioelectric activity and hormone action has been touched upon in the previous sections with regard to plant tissue. It can also be shown that the association persists at the level of individual cell membranes and cell walls.

Considerable work has been done upon large algal cells, particularly members of the Characeae. Additions of IAA and other auxins have been shown to alter the transmembrane potential (plasmalemma and tonoplast; BENTRUP et al. 1973, 1977, LELKES et al. 1980), but no reports exist which prove that an applied biopotential stimulus alters auxin concentration or activity. Nonetheless, it is thought that a possible feedback relationship may exist (see Sect. 2.9). Other than by using electron microscopy, electrical measurements are recognized as providing important evidence about the structure of membranes in living cells. For instance, measurements of capacitance and conductance of synthetic and animal membranes (EISENBERG et al. 1977) have shown that many membranes appear labile in nature with pores forming and disappearing in small fluctuating bursts of ms duration. Antibiotics such as valinomycin or nonactin facilitate the passive diffusion of ions like K^+ across membranes via a carrier mechanism. The ion complexes with the carrier molecule and shuttles across the membrane down an electrochemical gradient. This, in effect, alters the permeability of the membrane to univalent ions and might be one way in which some growth-regulating compounds could act in plants. Another possibility arises in connection with similar experiments. HAYDON and HLADKY (1972), EISENBERG et al. (1977) also point out that other membrane-modifying molecules exist (e.g., gramicidin A, alamethicin), which facilitate the passive diffusion of ions across membranes via the formation of structures which can bridge the membranes and create a pore or pathway through which the ions can move down the electrochemical gradient. Plant hormones might also act on membranes by altering the number or size of these pathways and electro-osmotic experiments supports this view. In electro-osmotic studies, the simultaneous measurement of water flow and ion flow have shown that the membrane–wall complex in *Nitella* is normally electronegative in surface charge (zeta potential). The flow of water and univalent cations appear to be coupled when very small electric currents (<3 µamp cm^{-2}) are caused to pass through *Nitella* cells (FENSOM and DAINTY 1963, MACROBBIE and FENSOM 1969, FENSOM and WANLESS 1967, WANLESS et al. 1973, FENSOM 1977). While all the coupling of 80 to 500 H_2O (univalent+ion)$^{-1}$, called E-O efficiency, is probably not entirely through pores (BARRY and HOPE 1969), much of it must be so, and this itself is strong evidence that small pores under 1 nm exist in these plant membranes, passages which on the average exclude anions (WANLESS et al. 1973, WILLIAMS et al. 1972). Electro-osmosis has been used not only to probe for the presence of water-filled pores capable of conducting ions down an electrochemical gradient, but also to explore the effect of growth-regulating compounds including hormones on water-ion coupling. Indeed the same experiments simultaneously measure a transcellular water flow (TWF) at the membrane which seems to be related to probable electrogenic pump activity, particularly to active fluxes of H^+, HCO_3^- or OH^- and to water movements which are independent of true electro-osmotic coupling (i.e., passive movement through pores) and of the hydraulic permeability of the membrane to water alone (FENSOM 1977). Plant hormones or membrane conditioners can alter these three parameters,

Table 1. Relative effect of some growth regulatory compounds on electroosmotic efficiency (EO)[a], transcellular water flow (TWF)[b] in *Nitella* after 1 h in APW[c]

Treatment	Concentration mM	EO %	TWF %	Reference
Control (APW[c] alone)	—	100	100	1
IAA	10^{-2}	230[d]	60[e], 120[d]	1
Light-degraded IAA	10^{-2}	100	190[d]	1
Skatol	10^{-2}	100	200[d]	1
IAA + GA_3	$10^{-2}+10^{-2}$	360[d]	42[e], 130	1
IAA + ABA	10^{-2} mM 4×10^{-5}	200[d]	65	1
ABA	3.8×10^{-2}	88	140[d]	1
	0.4×10^{-2}	120	130	1
NAA	10^{-2}	100	123[d]	1
FC	10^{-2}	194[d]	120[d]	2
Valinomycin	10^{-4}	100	not known	2
Decylsuccinamide	10^{-1} to 10^{-3}	100	100	1, 3
Ethylene	4 to 10^{-3}	100	100	3
Diquatdibromide	10^{-4}	77[d]	9[d]	1
2,4-D	10^{-3}	135	119[d]	4
TIBA	10^{-3}	83	121[d]	4
Kinetin	10^{-3}	135	106[d]	4
CHA (cyclohexane carboxylic acid)	10^{-3}	216[d]	96	4
Polymixine sulphate	10^{-2}	122[d]	145[d]	5
Chloramphenicol	10^{-2}	104	127[d]	5

[a] Electroosmotic efficiency measured in moles H_2O F^{-1} or cm^3 $Coulomb^{-1}$.
[b] Transcellular water flow, measured in pico-litre s^{-1}.
[c] APW or artificial pond water is 1 mM NaCl + 0.1 mM KCl + 0.1 mM $CaCl_2$.
[d] Significantly different than control at 5% level or below.
[e] Initial measurement in 10 min or less followed by a second effect after at least 30 min.
References: 1 WANLESS et al. (1973), FENSOM et al. (1973). 2 TAZAWA and FENSOM unpublished. 3 FENSOM and ROSS (1977). 4 REY-BELLET and FENSOM unpublished. 5 GILLET et al. (1971)

passive pore water-coupling, uncoupled water (hydraulic) permeability, and water movements associated with active ion pumps, severally or together, as Table 1 shows. Thus ABA alters hydraulic permeability but not, unless with IAA, the other two; IAA and FC, and particularly IAA + GA increase water-ion coupling in pores. Valinomycin, on the other hand, had no effect on water-ion coupling (E-O efficiency) nor on TWF, while NAA stimulated TWF but not the E-O efficiency. When hormones were added, a measurable change occurred within 5 min after the addition. Therefore it seems likely that IAA, IAA + GA and NAA all act upon the membrane itself, the first two (and FC) by altering pore number or size, the last (and skatol) by exciting electrogenic pump activity. Diquat, on the other hand, must eventually destroy the negative charge on the membrane pores as well as the electrogenic pump activity (FENSOM et al. 1973), while valinomycin, which increases K^+ permeability, must act upon binding molecules in the membrane rather than on pores where water would be coupled (cf. LELKES et al. 1980, SPANSWICK 1980, GREENLEAF and FERRIER 1980, PALTA and STADELMANN 1980).

Table 2. Similarities between the effects of IAA and FC in pea internode segments. (After MARRÈ 1977, VREDENBERG 1976)

	Controls	IAA	FC
Cell enlargement (% Increase in fresh weight in 2 h)	5	15	28
K^+ uptake[a] (1 mM) $\mu Eq\ g^{-1}\ h^{-1}$	0.48	0.96	2.20
Na^+ uptake[a] (1 mM) $\mu Eq\ g^{-1}\ h^{-1}$	0.13	0.16	0.26
Cl^- uptake[a] (1 mM) $\mu Eq\ g^{-1}\ h^{-1}$	1.7	2.6	3.9
SO_4^{-2} uptake[a] (1 mM) $\mu Eq\ g^{-1}\ h^{-1}$	0.59	0.67	1.31
Titratable H^+ extrusion (1 mM KCl) $\mu Eq\ g^{-1}\ h^{-1}$	0.16	0.40	0.89
Transmembrane potential, mV	−60	−77	−83

[a] H^+ extrusion and uptake of ions in $\mu Eq\ g^{-1}\ h^{-1}$

In cell walls there is some evidence that auxins alter the H^+ concentration, either by direct dissociation or by acting to stimulate an H^+ extrusion pump on the plasmalemma. This, then, is thought to lead to a relaxation of cell-wall elasticity, thus permitting cell enlargement to occur (CLELAND and LOMAX 1977, SMITH and RAVEN 1976).

For multicellular tissues there is also evidence of biopotential changes closely associated with or caused by auxin additions, a number of electrical effects being altered by addition of IAA and fusicoccin FC (MARRÈ 1977, 1980, COCUCCI et al. 1980).

Much work on higher plant cells has been done on the guard cells of stomata. As these cells change size, there is a movement of ions (H^+ efflux, K^+ influx), followed by a water influx (RASCHKE 1977, JESCHKE 1976, WILLMER 1983). Plant hormones can alter these movements, ABA preventing stomatal opening, FC preventing stomatal closure. In these situations it is clear that addition of hormone provides the stimulus which leads to the electrical response, and later to the change in size (MACROBBIE 1980). It is also possible that electrical gradients may alter hormone production or migration in roots (KRASAVINA and KTITOROVA 1975, ANDERSON 1976), or in formation of secondary meristems etc. (JEFFS and NORTHCOTE 1966, FENSOM 1960), but these possibilities have never been established with certainty.

Before leaving this subject, it should also be mentioned that electrical resistance of parenchyma tissue is altered by hormone action (MEYLAN 1976). This change in resistance has been interpreted as being due both to change in cell wall *and* membrane properties in these tissues (cf. CLELAND and LOMAX 1977, LÜTTGE 1980).

2.8 Alterations of Growth by Applied Electric Fields

From time to time there have been reports of alterations in growth or development by applied electric fields (cf review by BLACK et al. 1971), although the evidence for hormone interaction is mainly inferential. Certainly applied electric currents can alter the growth rates of tomatoes (and probably other plants), usually in a detrimental fashion, but occasionally in a stimulating way (BLACK et al. 1971). There have been reports of increased abcission by applied bioelectric effects in citrus crops, and also that fruit can be ripened more rapidly (MOLITERITZ 1965, GUNN et al. 1972). The difficulty about these reports is that they have not been substantiated by *consistent* evidence published in professionally recognized scientific journals. Therefore one gathers that while they seem to have occurred, their reproducibility has been *inconsistent*, or at least not thoroughly substantiated. In view of this, it is currently easier to interpret these results as indicating that applied electrical impulses probably distort membrane fields and endogenous electrical activity, upsetting hormone action where polar compounds are involved.

2.9 Electrogenesis, Membranes, ATPases and Contractile Proteins

It is generally agreed that electrogenesis in plants is primarily localized across membranes. Of these, the plasmalemma and tonoplast have been extensively studied, followed by membranes of organelles. There is now considerable evidence that in the Characeae an electrogenic pump extruding H^+ exists at the plasmalemma. This pump is vanadate-inhibited (and by oligomycin) but not by NO_3^- (SHIMMEN and TAZAWA 1982, BENNETT and SPANSWICK 1984, COCUCCI et al. 1980). The tonoplast, on the other hand, is vanadate-insensitive, but its ATPase is inhibited by NO_3^- and stimulated (in red beet vacuoles) by Cl^- (BENNETT and SPANSWICK 1984, TAZAWA and SHIMMEN 1980). Thus it seems likely that plant hormones or growth-regulating compounds could disturb the functioning of the ATPases specific to plasmalemma in some cases or to tonoplasts or plastids in others. For instance, all herbicides seem to cause depolarization of membranes (cf. SZYNKIER 1984), but since membrane excitability and cytoplasmic streaming are insensitive to vanadate in Characeae (SHIMMEN and TAZAWA 1982), it seems that the ATPases of different cell processes are distinct from each other, at least to the extent that they respond differently – or not at all – to different hormones, ions, and metabolic stimulants or inhibitors. Proton pump activity is certainly influenced by plant hormones, e.g., fusicoccin and IAA both stimulate it as described above, therefore a hormone–electric interaction seems to be mediated at membranes by specific ATPase responses.

In addition to membrane electrogenesis via ion pumps, electrical changes are known to accompany action potentials (GRADMANN and MUMMERT 1980, TAZAWA and SHIMMEN 1980), variation potentials and changes in cytoplasmic streaming. For instance, in the Characeae cytoplasmic streaming ceases during an action potential. This is thought to be caused by the release of Ca^{2+} into the cytoplasm near the actin-myosin. Actin-myosin normally produces movement through the mediation of Mg^{2+}, ATP, and ATPase. An electrical stimulus, and probably local change of ion or charge accompanies contractile protein activity, and it is this which can be upset by small concentrations of Ca^{2+}. Cytochalasin B or phalloidin stop streaming by interfering with the actin, normal-ethylmaleimide (NEM) by interfering with the myosin functioning. Therefore these contractile protein systems, which now seem to be widely occurring in plants (KAMIYA 1981, ALLEN 1980) will also be subject to hormone–electric interactions (WATANABE and UMRATH 1983).

For a review of the role of electrical activity in relation to movements in plants, the reader is referred to SIMONS (1981). One topic omitted from this review is the possible mode of operation of plasmadesmata. If these intercellular connections are merely inert pipes which permit transport through diffusion or pressure forces, hormone–electric interactions can be expected to have minimal effects on them. On the other hand, if contractile proteins or even some cytoskeletal or reticulum strands are needed through them to maintain active transport, electrical and hormonal controls can be expected (BARCLAY and FENSOM, unpublished).

Brief mention must be made of electrical effects which have been reported to occur in the guard cells of stomata (reviewed by SIMONS 1981, WILLMER 1983). It now seems likely that electrical changes in the membranes of guard cells which accompany switches from light to darkness or which accompany opening and closing of each stoma are produced by H^+, K^+ exchanges, OH^-, HCO_3^-, or malate$^-$ movements (MACROBBIE 1980, LÜTTGE 1980), through the plasma membranes. Therefore the movements in these cells too are tied in with ion pumps, ATPase activity and turgor regulation, all likely to be influenced by hormonal changes (cf. RASCHKE 1977, RAVEN and SMITH 1980).

In similar theme is the current view that loading of ions into roots (BOWLING 1980), or sugars into transporting cells like sieve tubes (GIAQUINTA 1980, KOMOR and TANNER 1980, FELLE and BENTRUP 1980, BAKER 1980), are all active processes mediated through electrogenic H^+ pumps and thus will be sensitive to hormone–bioelectric interactions.

It is possible, however, that small currents applied close to the natural or physiological flux may be found to *assist* in normal hormone activity by increasing growth, cell elongation, or fruiting (JAFFE 1980). For example, there have been reports from the USSR (TARCHEVSKY, personal communication from the Kazan Group) that electrical resistance measurements on wheat can enhance viability for early planting. If this is correct, experiments should be done elsewhere for independent confirmation. In single cells of the Characeae there is no evidence at all that such applied currents produce hormone changes or alter growth. Indeed at present, currents under 1 μamp cm^{-2} seem to produce only passive flux changes (WILLIAMS et al. 1972).

2.10 Hypothesis of Hormone–Electric Interaction

A number of hormone–electric interaction hypotheses have been put forward; one is that plant hormones activate nucleic acid which in turn changes metabolism, alters turgor and then as a secondary or tertiary effect, causes a change in the bioelectric potential of the membrane. Changes in turgor per se may alter the pressure exerted on the membrane, thereby slightly changing membrane capacitance and resistance (cf. ZIMMERMANN et al. 1976, ZIMMERMANN and STEUDLE 1980). One argument against this rather tortuous explanation is that

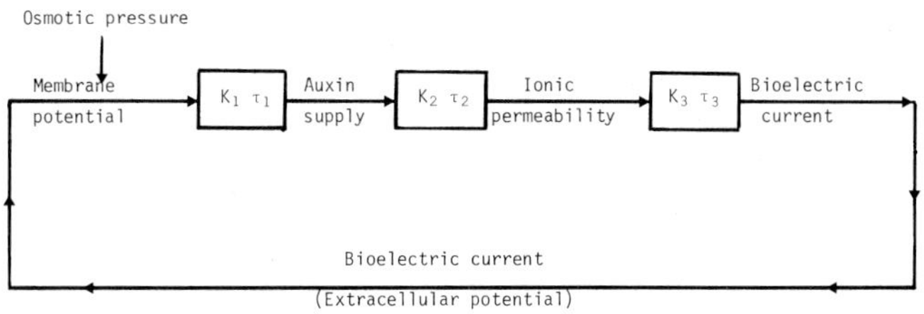

Fig. 7. Proposed feedback loop in which auxin, Ca ions, and other effects can alter membrane potential in the presence of three exponential delays. (After JENKINSON and SCOTT 1961, republished by VAN STEVENINCK 1976)

some hormone effects occur within a very short time, within 5 min of application (e.g. see CLELAND and LOMAX 1977, FENSOM et al. 1973, BENTRUP et al. 1977, JAFFE, Chap. 12, this Vol.).

One could, however, explain a fast action by postulating that auxin acts on the membrane itself or on protoplasm very close to it. We now know that auxin can rapidly alter the passive pore size or frequency, and thereby the ion-water coupling entering the cell (see Sect. 2.7 above). We also know that plant hormones or FC may rapidly alter the biopotential across the membrane and TWF, probably by acting directly on electrogenic pump sites. It has been postulated that increased electrogenic pump activity increases turgor both by electro-osmotic and osmotic mechanisms (FENSOM 1977). On the other hand, it has been found that increased turgor will increase the biopotential gradient at the membrane (COSTER et al. 1976, ZIMMERMANN and BENZ 1980), and since electric field gradients across membranes are of the order of kV cm^{-1} (a change of 10 mV across a plasmalemma becomes 10 kV cm^{-1}), the strength of this electric field is large enough to cause configurational changes in polar macromolecules such as bound proteins or enzymes. Thus it seems very likely that auxin or growth-regulating substances do act first directly on the membrane to alter electrogenic ion pumps, membrane potential *and* then passive ion fluxes as well as turgor and water flows mediated both by electrical and metabolic forces. Such changes would be mediated by a change in active ion pumping (probably bicarbonate and other ion influx), H^+ efflux, and a change in the dissociation constant of the auxin itself (SMITH and RAVEN 1976, FENSOM 1977, ZIMMERMANN and STEUDLE 1977). All or any of the above could lead then to secondary changes in turgor pressure as well as to action at a later time on nucleic acids. A more complete discussion of these theories is presented in MARRÈ and CIFERRI (1977), or by LUCAS (1984), OKAMOTO (1984), VAN STEVENINCK (1984), MARRÈ (1980).

However, in spite of the large amount of work accomplished with auxins in the last two decades, the precise ways in which these plant hormones act remain matters for speculation. Therefore, the relationship between auxin and electrical responses (or the reverse) is also mostly hypothetical. Nonetheless, since feedback systems (metabolic, ionic, pressure?) exist which regulate cell metabolism and membrane activity, it is entirely likely that similar systems operate so that *either* electrical stimuli *or* auxin stimuli could cause alterations

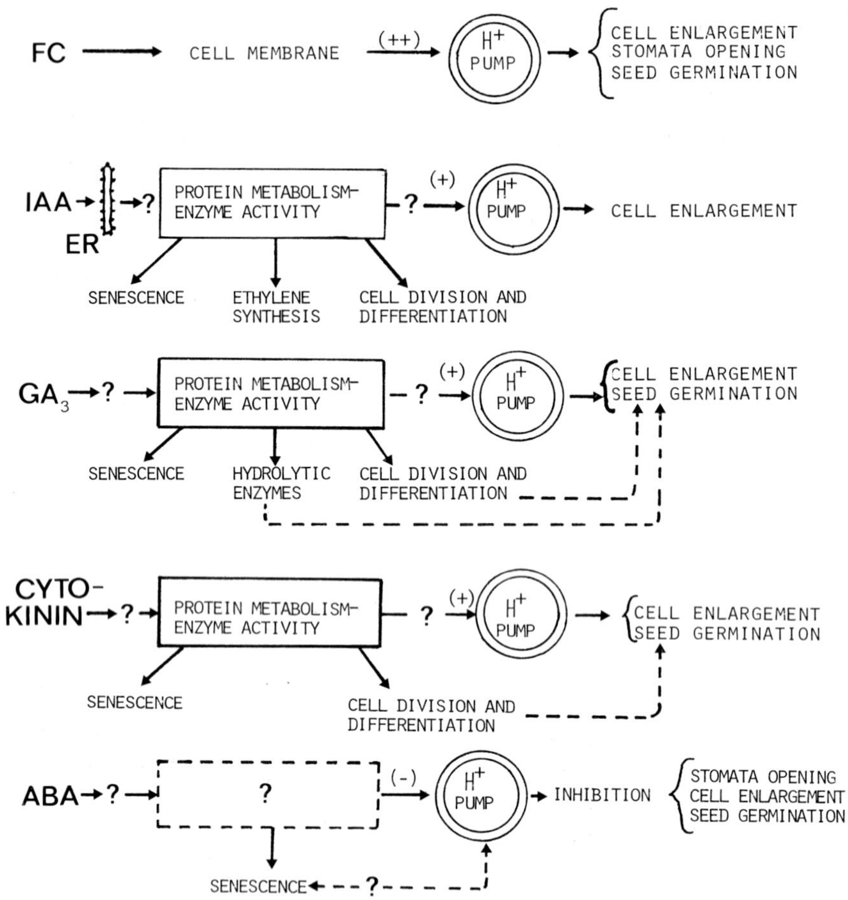

Fig. 8. Model for the different modes of action of FC, IAA, and other growth substance on the H^+/K^+ exchange system and associated responses. Among the hormonal effects only those are considered in the scheme which: (a) may be interpreted as direct consequences of H^+/K^+ exchange activation or inhibition; or (b) may be induced in at least some materials without a simultaneous detectable effect on H^+/K^+ exchange. (After MARRÈ 1977)

of enzyme or protein configuration, charge on membrane, ion migration rate, or change in pH. Such changes by auxin or electrical stimuli would be reflected in altered cell/organ metabolism. Several models have been proposed to explain this interaction, for instance those by JENKINSON (1962). Figures 7, 8 and 9 are included here as possible systems which should be considered. Other models (not shown) have been put forward by POLEVOY and SALAMATOVA (1975, 1977), MASLOBROD et al. (1975), PILET (1964), PITMAN et al. (1977), KRASAVINA and KTITOROVA (1975), VAN STEVENINCK (1976) and CRAM (1975, 1976), SPANSWICK and MILLER (1978), and are discussed in the proceedings of the 1983 *Symposium on Plant Membrane Transport* held in Prague (to be published 1984).

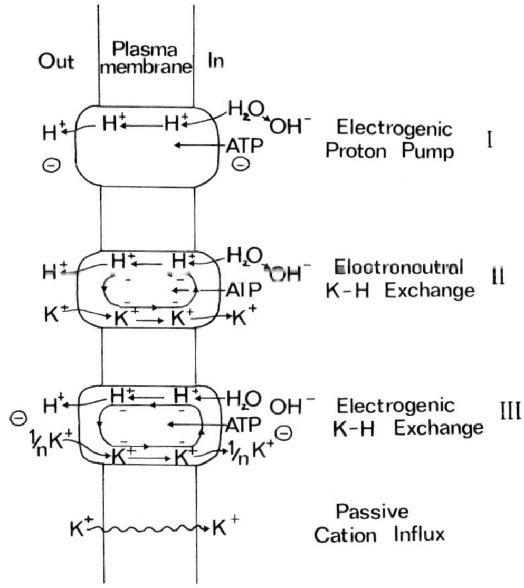

Fig. 9. Three models as possible models for H^+ ion excretion. In each case the effects of auxin or related substances on the biopotential needs to be known, as does the uptake of potassium ion and the potassium/hydrogen ratio. It is assumed that auxin or FC acts on one of these three parameters and that one or more of these mechanisms for ion excretion is then set in operation. (After CLELAND et al. 1977)

3 Magnetic Stimuli and Hormones

There have been reports from time to time that magnetic treatment of plants and seeds can cause specific growth responses, or accelerate germination. For a review of the recent literature on this subject the reader is referred to PITTMAN (1972, 1977), PITTMAN and ORMROD (1970, 1971), LEBEDEV et al. (1975). It is known that work has been done in the U.S., in the USSR (KLEIMENOV 1975), the UK., Europe, and Canada, attempting to carry out experiments with magnetic fields. However, these experiments are invariably empirical in nature, and duplication of results has not been consistent or even feasible under different conditions or in different locations. In experiments with cytoplasmic streaming the results have consistently been negative (PICKARD 1972). In this respect the problems are similar to those of electrical stimuli on growth, but unlike work with electrical stimuli, hormone changes accompanying magnetic effects have not been reported to date, even though many of the systems under observation are readily controlled or affected by hormones. L.G. AUDUS (formerly at Bedford College, University of London, personal communication), after extensive review in this area, indicated that he knew "of no evidence in the current literature which indicates that hormones can be involved in any response to magnetic fields". This does not mean that some responses may not occur, for the interrelation between electrical effects and magnetic effects, particularly for changing magnetic fields, is well established. Thus, in spite of some commercial claims to the contrary, there is no firm evidence that hormone responses to magnetic stimuli occur; one must merely remain open-minded about the possibility. Another comment is pertinent here. As KLEIMENOV (1975) points out in a recent review, one should use "not the value of induction, but the density of magnetic field energy and time" – i.e., the density rate-of-change of fields in dealing with plant responses. Even then the changing magnetic field should be strong enough when acting on moving ions to induce 10 kV cm^{-1} across membranes, which is probably why the magnetic effects observed to date have been undetectable or at best marginal.

4 Conclusion

Evidence for the interrelationships of bioelectric phenomena and the known plant hormones in plant growth and differentiation processes is thus far largely inferential in nature. There is little doubt that movement of certain plant hormones is closely *associated* with changes in electrical fields, as are a number of growth processes which are known to be induced or affected by plant hormones. The association, at this time at least, is correlative, increases in hormone levels (auxin) probably causing bioelectric changes, rather than the reverse! The very movement of plant hormones (at least auxins) is accompanied by a change in bioelectric potential, although opinion is split as to which is cause and which effect.

The close association in time and space between bioelectric activity and hormone action persists to the level of individual cell membranes/walls, growth-regulating compounds being able rapidly to affect electro-osmosis and electrogenic pump activity at the membrane. Here the change in bioelectric potential appears to the caused by the presence of the hormone (auxin), but any change in cell metabolism or turgor is potentially able to produce electric field alterations of magnitude great enough to cause changes in configurations of enzymes or polar molecules (or auxin) at a membrane. These changes in trans-membrane potential, electro-osmosis and electrogenic pump activity may lead to changes in H^+ concentration, simultaneous with an increase in turgor, hence in some cases to a relaxation of cell-wall elasticity, and subsequently an increase in cell enlargement. While this is now quite likely, there is, alas, no general agreement that this is indeed the mechanism by which auxin acts.

With all of these close associations between growth processes/hormone levels/hormone movement and bioelectric potential changes, one would suppose that applied electric fields would induce at least some significant growth alteration (ideally of a positive nature). Unfortunately, the summation of such reports is inconsistent. They do little to further substantiate the theories that bioelectric potential differences could cause hormone movement or increased hormone production/release, or even that growth which is induced by hormones acts via a mechanism(s) involving bioelectric potential as an inducer! In the Characeae small electric currents below the action potential level have not produced detectable hormone responses.

The close interaction between hormones and changes in bioelectric potential is thought to be a result of hormone-induced changes to the membrane, or the nearby protoplasm, thereby engendering changes in activity of passive ion fluxes, of electrogenic ion pumps, enzyme action, and charges on fixed structures. Any, or all, could lead to secondary changes on cell turgor, or at the nucleic acid level, and a number of models have been put forward to explain just how either auxin, or electrical stimuli, could be working at the subcellular and membrane levels.

The relationship between magnetic stimuli and plant hormones has not been proven, and indeed there is much conflict and a dearth of readily repeatable results as to whether treatment with a magnetic field actually invokes specific growth responses, including increases in germination. Nonetheless, if electrical fields affect ion movements, as they must, magnetic stimuli, particularly changing magnetic fields, must also do so, and in so doing may cause interaction with hormones.

Note Added in Proof. Since writing this chapter it has come to my attention that changes in electrical protonic properties of root tips have been reported in relation to plant hormone distribution and axial growth (P.E. Pilet, J.M. Versel and G. Mayor, 1983, Planta 398–402, also P.E. Pilet, Abstracts of Fed. of European Soc. of Plant Physiol., Strasbourg 1984, p. 153). Also recent experiments indicate that differential changes in root gravi-reactions can be induced by applied magnetic fields to capped and decapped roots (Pilet, unpublished).

References

Allen NS (1980) Cytoplasmic streaming and transport in the characean algae *Nitella*. Can J Bot 58:786–796

Anderson WP (1976) Transport through roots. In: Lüttge U, Pittman MG (eds) Transport in plants. Encyclopedia of plant physiology., new ser Vol 2/A. Springer, Berlin Heidelberg New York, p 146

Audus LJ (1972) Plant growth substances, vol I. Chemistry and physiology. Harper & Row, New York

Baker D (1980) Solute loading and the control of phloem turgor in *Ricinus communis* L. In: Spanswick RM, Lucas WJ, Dainty J (eds) Plant membrane transport: current conceptual issues. Elsevier/North-Holland Biomedical Press, Amsterdam New York, pp 543–544

Barry PH, Hope AB (1969) Electroosmosis in *Chara* and *Nitella* cells. Biochim Biophys Acta 193:124

Beilby MJ, Coster HGL (1980) Electromechanical effects in membranes: their role in the dependence of punch-through on temperature. In: Spanswick RM, Lucas WJ, Dainty J (eds) Plant membrane transport: current conceptual issues. Elsevier/North-Holland Biomedical Press, Amsterdam New York, pp 611–612

Bennett AB, Spanswick RM (1984) H^+/ATP stoichiometry of an ion sensitive H^+-ATPase from red beet. Plant Membr Transp Symp, Prague

Bentrup FW, Gratz HJ, Unbehauen H (1973) Membrane potentials in leaf cells. In: Anderson W (ed) Ion transport in plants. Academic Press, London New York

Bentrup FW, Gutknecht W, Pfrüner H (1977) Auxin effects on the ionic relations of *Petroselinum* cell structure. In: Marrè E, Ciferri O (eds) Regulation of cell membrane activities in plants. Elsevier/North Holland Biomedical Press, Amsterdam New York, p 203

Bidgood PT (1982) A determination of water potential using dielectric/capacitance measurements of plants in situ. Honours Thes, Mount Allison Univ

Black JD, Forsyth FR, Fensom DS, Ross RB (1971) Electrical stimulation and its effects on growth and ion accumulation in tomato plants. Can J Bot 49:1809–1815

Blanchard RO, Shortle WC, Davis W (1983) Mechanism relating cambial electrical resistance to periodic growth rate of balsam fir. Can J For Res 13:472–480
Bose JC (1907) Comparative electro-physiology. Longmans, Green & Co, London New York
Bowling DJF (1980) An electrogenic phosphate pump in sunflower roots. In: Spanswick RM, Lucas WJ, Dainty J (eds) Plant membrane transport: current conceptual issues. Elsevier/North-Holland Biomedical Press, Amsterdam New York, pp 405–406
Burr HS (1947) Field theory in biology. Sci Mon 64(3):217–225
Cheeseman JM, Pickard BG (1977) Depolarization of cell membranes in leaves of *Lycopersicon* by extract containing Ricca's factor. Can J Bot 55:511
Cleland RE, Lomax T (1977) Hormonal control of H^+-excretion from oat cells. In: Marrè E, Ciferri O (eds) Regulation of cell membrane activities in plants. Elsevier/North-Holland Biomedical Press, Amsterdam New York, pp 161–172
Clifford PE, Fensom DS, Munt BI, McDowell WD (1982) Lateral stress initiates bending responses in dandelion peduncles: a clue to geotropism? Can J Bot 60:2671–2673
Cocucci M, Ballarin-Denti A, Marrè MT (1980) Effects of orthovanadate on H^+ and K^+ transport, transmembrane potential and plasmalemma ATPase activity in plant tissues. In: Spanswick RM, Lucas WJ, Dainty J (eds) Plant membrane transport: current conceptual issues. Elsevier/North-Holland Biomedical Press, Amsterdam New York, pp 505–506
Coster HGL, Steudle E, Zimmermann U (1976) Turgor pressure sensing in plant cell membranes. Plant Physiol 58:636–643
Cram WJ (1975) Systems analysis of ion transport in plant cells. Proc XIIth Int Bot Congr. Nauka, Leningrad (1979)
Cram WJ (1976) Negative feedback regulation of transport in cells. The maintenance of turgor, volume and nutrient supply. Lüttge U, Pitman MG (eds) Transport in plants II. Encyclopedia of plant physiology, new ser Vol 2A. Springer, Berlin Heidelberg New York, p 284–316
Dixon MA, Thompson RG, Fensom DS (1978) Electrical resistance measurements of water potential in avocado and *Picea*. Can J For Res 8:73–80
Eisenberg M, Kleinberg ME, Shaper JH (1977) Channels across black lipid membranes. In: Takashima S, Fishman HM (eds) Ann NY Acad Sci 303:201–291
Farquahar GD, Field CD (1971) Transpiration-linked short-circuit currents in the xylem of a liana. J Exp Bot 22:818–829
Felle H, Bentrup FW (1974) Light-dependent changes of membrane potential and conductance in *Riccia fluitans*. In: Zimmermann MH, Dainty J (eds) Membrane transport in plants. Springer, Berlin Heidelberg New York, pp 120–125
Felle H, Bentrup FW (1980) Electrogenic hexose transport in *Riccia fluitans*. In: Spanswick RM, Lucas WJ, Dainty J (eds) Plant membrane transport: current conceptual issues. Elsevier/North-Holland Biomedical Press, Amsterdam New York, pp 549–550
Fensom DS (1960) The bio-electric potentials of plants and their functional significance. III. The production of continuous potentials across membranes in plant tissue by the circulation of the hydrogen ion. Can J Bot 37:1003–1026
Fensom DS (1963) The bio-electric potentials of plants and their functional significance. V. Some daily and seasonal changes in the electric potential and resistance of living trees. Can J Bot 41:831
Fensom DS (1966) On measuring electrical resistance in situ in higher plants. Can J Plant Sci 46:169–175
Fensom DS (1977) A role of electrogenic pumps in producing turgor in *Nitella*. In: Marrè E, Ciferri (eds) Regulation of cell membrane activities in plants. Elsevier/North-Holland Biomedical Press, Amsterdam New York, p 91
Fensom DS, Dainty J (1963) Electro-osmosis in *Nitella*. Can J Bot 41:685–691
Fensom DS, Ross SM (1977) Note on lack of effect of ethylene on water permeability, electro-osmotic efficiency or transcellular water flow of the plasma membranes in *Nitella*. Can J Bot 55:615–616
Fensom DS, Wanless IR (1967) Further studies of electro-osmosis in *Nitella* in relation to pores in membranes. J Exp Bot 18:563–577

Fensom DS, Barclay SL, Law S, Thompson RG (1973) A transcellular water flux induced by light in *Nitella*. Can J Bot 51(5):1045–1053

Fensom DS, Thompson RG, Alexander K (1983) Proceedings of CANUSA workshop on bud-worm damage and control. Serv Environ Can Bangor (Maine)

Findlay GP, Hope AB (1976) Electrical properties of plant cells: methods and findings. In: Lüttge U, Pitman MG (eds) Transport in plants II. Encyclopedia of plant physiology, new ser Vol 2A, Springer, Berlin Heidelberg New York, pp 53–92

Giaquinta R (1980) Sucrose/proton cotransport during phloem loading and its possible control by internal sucrose concentration. In: Spanswick RM, Lucas WJ, Dainty J (eds) Plant membrane transport: current conceptual issues. Elsevier/North-Holland Biomedical Press, Amsterdam New York, pp 273–282

Gillet C, Fensom DS, Lefèbre J (1971) Effects immédiats de quelques antibiotiques sur l'éléctro-osmose chez *Nitella flexilis*. Experientia 27:853–854

Gradmann D, Mummert H (1980) Plant action potentials. In: Spanswick RM, Lucas WJ, Dainty J (eds) Plant membrane transport: current conceptual issues. Elsevier/North-Holland Biomedical Press, Amsterdam New York, pp 333–344

Graham L (1964) Measurements of geoelectric and auxin-induced potentials in coleoptiles with the refined vibrating electrode technique. Physiol Plant 17:231

Graham L, Hertz CH (1964) Measurement of the geoelectric effect in coleoptiles. Physiol Plant 17:186

Greenham CG (1966a) The stages at which frost injury occurs in alfalfa. Can J Bot 44:1471–1483

Greenham CG (1966b) Bruise and pressure injury in apple fruits. J Exp Bot 17:404–409

Greenham CG, Randall PJ, Ward MM (1972) Impedance parameters in relation to phosphorus and calcium deficiencies in subterranean clover (*Trifolium subterraneum* L.). J Exp Bot 23:197–209

Greenham CG, Helms K, Muller WJ (1978) Influence of virus infections on impedance parameters. J Exp Bot 29:867–877

Greenham CG, Groves RH, Muller WJ (1980) Variation between populations of one form of Skeleton Weed (*Chondrilla juncea* L.) shown by electrical parameters. J Exp Bot 31:967–974

Greenleaf CRJ, Ferrier JM (1980) Electrical consequences of acid/alkaline banding in *Chara corallina*. In: Spanswick RM, Lucas WJ, Dainty J (eds) Plant membrane transport: current conceptual issues. Elsevier/North-Holland Biomedical Press, Amsterdam New York, pp 593–594

Gunn RF, Craig DL, Fensom DS (1972) Induction of early ripening by electricity on *Vitis vinifera*: Rep Can Soc Hortic, Charlottetown, PEI

Hayden RI, Moyse CA, Calder FW, Crawford DP, Fensom DS (1969) Electrical impedance studies on potato and alfalfa tissue. J Exp Bot 20:177–200

Haydon DA, Hladky SB (1972) Ion transport across thin lipid membranes: a critical discussion of mechanisms in selected systems. Q Rev Biophys 5:187–282

Hertz CH (1960) Electrostatic measurement of the geoelectric effect. Nature 187:320

Higinbotham N (1973) Electropotentials of plant cells. Annu Rev Plant Physiol 24:25–46

Jaffe L (1980) Control of plant development by steady ionic currents. In: Spanswick RM, Lucas WJ, Dainty J (eds) Plant membrane transport: current conceptual issues. Elsevier/North-Holland Biomedical Press, Amsterdam, New York, pp 381–388

Jaffe MJ (1970) Evidence for the regulation of phytochrome-mediated processes in bean roots by the neurohormone, acetylcholine. Plant Physiol 46:768–777

Jeffs RA, Northcote DH (1966) Experimental induction of vascular tissue in an undifferentiated plant callus. Biochem J 101:146–152

Jenkinson IS (1962) Bioelectric oscillations of bean roots: further evidence for a feedback oscillator. III. Excitation and inhibition of oscillation by osmotic pressure, auxins, and antiauxins. Aust J Biol Sci 15:115–125

Jenkinson IS, Scott BH (1961) Bioelectric oscillations of bean roots: further evidence for a feedback oscillator. Aust J Biol Sci 14:231–243

Jeschke WD (1976) Ionic relations of leaf cell. In: Lüttge U, Pitman MG (eds) Transport in plants II. Encyclopedia of plant physiology, new ser Vol 2B. Springer, Berlin Heidelberg New York, pp 160–194

Johnsson A (1965a) Photoinduced lateral potentials in *Zea mays*. Physiol Plant 18: 574

Johnsson A (1965b) Investigations of the reciprocity rule by means of geotropic and geoelectric measurements. Physiol Plant 18:945

Johnsson A (1967) Relationships between photoinduced and gravity-induced electrical potentials in *Zea mays*. Physiol Plant 20:562

Kalinin VA, Opritov VA, Khudyakov VA (1970) Formation of free radicals in the spread of an excitation wave in the vascular system of higher plants. Fiziol Rast 17(2):309–313

Kamiya N (1981) Physical and chemical basis of cytoplasmic streaming. Annu Rev Plant Physiol 32:205–236

Kleimenov EV (1975) Effect of a magnetic field on the correlation of some physiological properties of plants. Fiziol Rast 22(4):852–853

Komor E, Tanner W (1980) Proton-cotransport of sugars in plants. In: Spanswick RM, Lucas WJ, Dainty J (eds) Plant membrane transport: current conceptual issues. Elsevier/North-Holland Biomedical Press, Amsterdam New York, pp 247–257

Krasavina MS, Ktitorova IN (1975) Effects of adenine and imidazole derivatives and IAA on membrane potential of root hair cells in *Trianea bogotensis*. Fiziol Rast 22(6):1156–1161

Kunkel AJ (1878) Über elektromotorische Wirkungen an unverletzten lebenden Pflanzentheilen. Arb Bot Inst Würzburg 2:1

Lebedev SI, Baranskii PI, Litvinenko LG, Shiyan LT (1975) Physiological and biochemical characteristics of plants after pre-sowing treatment with a constant magnetic field. Fiziol Rast 22(1):103–109

Lelkes PI, Bach D, Miller IR (1980) Interactions of auxin (IAA) with lipid membranes. In: Spanswick RM, Lucas WJ, Dainty J (eds) Plant membrane transport: current conceptual issues. Elsevier/North-Holland Biomedical Press, Amsterdam New York, pp 539–540

Levengood WC (1970) Redox currents associated with ion mobility in stems of plants. Can J Bot 48:1099–1108

Levengood WC (1973) Bioelectric currents and oxidant levels in plant systems. J Exp Bot 24:626–640

Lucas WJ (1984) How are the transport processes of the *Chara* plasmalemma regulated? Plant Membr Transp Symp, Prague

Lüttge U (1980) Malic acid transport across the tonoplast of *Kalanchoe* leaf cells: tonoplast biophysics and biochemistry in relation to crassulacean acid metabolism (CAM). In: Spanswick RM, Lucas WJ, Dainty J (eds) Plant membrane transport: current conceptual issues. Elsevier/North-Holland Biomedical Press, Amsterdam New York, pp 49–60

MacDonald MA, Fensom DS, Taylor ARA (1974) Electrical impedance in *Ascophyllum nodosum* and *Fucus vesiculosus* in relation to cooling, freezing and desiccation. J Phycol 10:462–469

MacRobbie EAC (1980) Stomatal ionic relations. In: Spanswick RM, Lucas WJ, Dainty J (eds) Plant membrane transport: current conceptual issues. Elsevier/North-Holland Biomedical Press, Amsterdam New York, pp 97–107

MacRobbie EAC, Fensom DS (1969) Measurements of electro-osmosis in *Nitella translucens*. J Exp Bot 20:466–484

Marchel E (1977) Aggregation and conformational changes in solution studied by dielectric absorption of polypeptides and substituted polysaccharides. In: Takashima S, Fishman HM (eds). Ann NY Acad Sci 303:190–197

Marrè E (1977) Effects of fusicoccin and hormones on plant cell membrane activities: Observations and hypotheses. In: Marrè E, Ciferri O (eds) Regulation of cell membrane activities in plants. Elsevier/North-Holland Biomedical Press, Amsterdam New York, pp 175–202

Marrè E (1980) Fusicoccin: Mechanism of action on electrogenic H^+ extrusion. In: Spanswick RM, Lucas WJ and Dainty J (eds) Plant membrane transport: current conceptual issues. Elsevier/North-Holland Biomedical Press, Amsterdam New York, pp 227–240

Marrè E, Ciferri O (eds) (1977) Regulation of cell membrane activities in plants. Elsevier/North-Holland Biomedical Press, Amsterdam New York

Maslobrod SA, Zemshman AYa, Stepanov KI, Lysikov VN, Semin VS (1975) Functional role of photoinduced electric response of the plant during uptake of phosphorus from soil in light. Fiziol Rast 22(6):1162–1167

Meylan S (1971) Bio-éléctricité: Quelque problèmes. Masson & Cie, Paris

Meylan S (1976) Étude conductimétrique des effets de trois auxines sur un parenchyme de réserve. Physiol Plant 38P:267–272

Mitchell P (1961) Coupling of phosphorylation to electron and hydrogen transfer by a chemi-osmotic type of mechanism. Nature 191:144–148

Moliteritz J (1965) Effect of electric current on citrus trees. Rep 37th Annu Rural Electr Conf, sponsored by the Dep Agric Eng, Univ Cal, Davis

Newman IA (1959) Electrical determination of transport of 3-indole-acetic acid in *Avena*. Nature 184:1728–1729

Newman IA (1963) Electric potentials and auxin translocation in *Avena*. Aust J Biol Sci 16:629

Newman IA, Sullivan JK (1976) Auxin transport in oats: A model for the electric changes. In: Wardlaw IF, Passioura JB (eds) Transport and transfer processes in plants. Academic Press, London, New York, pp 153–159

Nobel PS (1974) Biophysical plant physiology. Freeman, San Francisco

Novak VA, Ivankina NG (1975) Comparative study of light-induced changes in electric potentials in plants. Fiziol Rast 22(1):49–54

Okamoto AH (1984) Hormonal and osmotic control of electrogenic ion pumps and growth by xylem perfusion of bean hypocotyl segments. Plant Membr Transp Symp, Prague

Opritov VA, Kalinin VA, Yarchenkova IM (1974) Participation of free radicals in the mechanism of action potential propagation in vascular system cells of higher plants. Fiziol Rast 21(3):545–553

Palta JP, Stadelmann EJ (1980) Simultaneous transport of water and solutes through plant cell membranes. In: Spanswick RM, Lucas WJ, Dainty J (eds) Plant membrane transport: current conceptual issues. Ed. Elsevier/North-Holland Biomedical Press, Amsterdam New York, pp 457–458

Paszewski A, Zawadzki T (1973) Action potentials in *Lupinus angustifolius* L. shoots. J Exp Bot 24:804–809

Penot M (1976) Mise en évidence de déplacements de ^{86}Rb à grande vitesse, dans le phloème de *Tradescantia viridis*. Planta 132:183–187

Penot Ma, Penot Mi (1979) High speed translocation of ions in seaweeds. Z Pflanzenphysiologie 95:265–273

Pickard BG (1973) Action potentials in higher plants. Bot Rev 39:172–201

Pickard WF (1972) Further observations on cytoplasmic streaming in *Chara braunii*. Can J Bot 50:703–711

Pilet PE (1961) Les phytohormones de croissance. Masson & Cie, Paris

Pilet PE (1964) Auxines et polarité morphologique des tissus de carotte cultivés in vitro. Rev Cytol Biol Vég 27:269

Pilet PE, Meylan S (1953) Polarité électrique, auxines et physiologie des racines du *Lens culinaris* Medikus. Bull Soc Bot Sui 63:430

Pitman MG, Anderson WP, Schaefer N (1977) H^+ ion transport in plant roots. In: Marrè E, Ciferri O (eds) Regulation of cell membrane activities in plants. Elsevier/North-Holland Biomedical Press, Amsterdam New York, pp 147–160

Pittman UJ (1972) Biomagnetic responses in potato. Can J Plant Sci 52:727–733

Pittman UJ (1977) Effects of magnetic seed treatment of yields of barley, wheat and oats in southern Alberta. Can J Plant Sci 57:37–46

Pittman UJ, Ormrod DP (1970) Physiological and chemical features of magnetically treated winter seeds and resultant seedlings. Can J Plant Sci 50:211–217

Pittman UJ, Ormrod DP (1971) Biomagnetic responses in germinating malting barley. Can J Plant Sci 51:64–65

Polevoi VV, Salamatova TS (1975) Mechanism of the auxin effect on membrane transport of hydrogen ions. Fiziol Rast 22(3):519–526

Polevoi VV, Salamatova TS (1977) Auxin, proton pump and cell trophics. In: Marrè E, Ciferri O (eds) Regulation of cell membrane activities in plants. Elsevier/North-Holland Biomedical Press, Amsterdam New York, pp 209–216

Racusen R, Satter RL (1975) Rhythmic and phytochrome-regulated changes in transmembrane potential in *Samanea* pulvini. Nature (London) 255:408–410

Raschke K (1977) The stomatal turgor mechanism and its responses to CO_2 and abscisic acid: Observations and a hypothesis. In: Marrè E, Ciferri O (eds) Regulation of cell membrane activities in plants. Elsevier North-Holland Biomedical Press, Amsterdam New York, pp 173–184

Raven JA, Smith FA (1977) Characteristics, function and regulation of active proton extrusion. In: Marrè E, Ciferri O (eds) From regulation of cell membrane activities in plants. Elsevier/North-Holland Biomedical Press, Amsterdam New York, p 25

Raven JA, Smith FA (1980) The chemiosmotic viewpoint. In: Spanswick RM, Lucas WJ, Dainty J (eds) Plant membrane transport: current conceptual issues. Elsevier/North-Holland Biomedical Press. Amsterdam New York, pp 161–174

Rosene HF, Lund EJ (1953) Bioelectric fields and correlation in plants. In: Loomis WE (ed) Growth and differentiation in plants. Iowa State College Press, Des Moines, pp 219–252

Sambeek van JW, Pickard BG (1976) Mediation of rapid electrical, metabolic, transpirational and photosynthetic changes by factors released from wounds. I. Variation potentials and putative action potentials in intact plants. Can J Bot 54:2642–2650

Sambeek van JW, Pickard BG, Ulbright CE (1976) Mediation of rapid electrical, metabolic, transpirational and photosynthetic changes by factors released from wounds. II. Mediation of the variation potential by Ricca's factor. Can J Bot 54:2651–2661

Schwan HP (1977) Field interactions with biological matter. In: Takashima S, Fishman HM (eds) Electrical properties of biological polymers, water and membranes. Ann NY Acad Sci 303:198–213

Schwarz G (1977) Chemical transitions of biopolymers induced by an electric field and their effects on dielectrics and birefringence. In: Takashima S, Fishman HM (eds) Ann NY Acad Sci 303:190–197

Scott BIH (1967) Electric fields in plants. Annu Rev Plant Physiol 18:409–418

Shimmen T, Tazawa M (1982) Effects of intracellular vanadate on electrogenesis, excitability and cytoplasmic streaming in *Nitellopsis obtusa*. Plant Cell Physiol 23:669–677

Simons PJ (1981) The role of electricity in plant movements. New Phytol 87:11–37

Skutt HR, Shigo AL, Lessard RA (1972) Detection of discoloured and decayed wood in living trees using a pulsed electric current. Can J For Res 2:54–56

Smith FA, Raven JA (1976) H^+ transport and regulation of cell pH. In: Lüttge U, Pittman MG (eds) Transport in plants II. Encyclopedia of plant physiology, new ser Vol 2A. Springer, Berlin Heidelberg New York, p 338

Smith JR, Coster HGL (1980) Frequency dependence of the AC membrane impedance of *Chara:* the effect of temperature. In: Spanswick RM, Lucas WJ, Dainty J (eds) Plant membrane transport: current conceptual issues. Elsevier/North-Holland Biomedical Press, Amsterdam New York, pp 609–610

Spanner DC (1964) Introduction to thermodynamics. Academic Press, London New York

Spanswick RM (1972) Electrical coupling between cells of higher plants: A direct demonstration of intercellular communication. Planta 102:215–227

Spanswick RM (1974) Symplastic transport in plants. Symp Soc Exp Biol 28:125–135

Spanswick RM (1976) Symplastic transport in tissues. In: Lüttge U, Pittman UJ (eds) Transport in plants II. Encyclopedia of plant physiology. new ser Vol 2A. Springer, Berlin Heidelberg New York, pp 34–53

Spanswick RM (1980) Biophysical control of electrogenicity in the Characeae. In: Spanswick RM, Lucas WJ, Dainty J (eds) Plant membrane transport: current conceptual issues. Elsevier/North-Holland Biomedical Press, Amsterdam New York, pp 305–313

Spanswick RM, Miller AG (1977) The effect of CO_2 and Cl^- influx and electrogenic pump in *Nitella translucens*. Échanges ioniques transmembranes chez les végétaux. Colloque du CNRS 258:239–245

Startseva AV, Burtaeva TE, Ruban NF (1975) Effect of a weak electric current on the state of water in wheat leaves. Fiziol Rast 22(4):853–856

Steveninck van RFM (1976) Effect of hormones and related substances on ion transport. In: Lüttge U, Pittman UJ (eds) Transport in plants II. Encyclopedia of plant physiology new ser Vol 2B. Springer, Berlin Heidelberg New York, pp 307–342

Steveninck van RFM (1984) Role of abscisic acid in nutrient transport and osmoregulation. Plant Membr Transp Symp, Prague

Stolarek J, Jablonska B, Pazurkiewicz-Kocot K (1980) Light-induced action potentials in *Phaseolus vulgaris* L. In: Spanswick RM, Lucas WJ, Dainty J (eds) Plant membrane transport: current conceptual issues. Elsevier/North-Holland Biomedical Press, Amsterdam New York, pp 619–620

Suckling EE (1961) Bioelectricity. McGraw-Hill, New York

Szynkier K (1984) Plant membrane transport symposium, Prague (in press)

Tattar TA, Blanchard RO (1976) Electrophysiological research in plant pathology. Annu Rev Phytopathol 14:309–325

Tattar RA, Sylvia DM (1977) Frequency spectrum analysis of plant storage tissue during deterioration. Can J Bot 55:2437–2438

Tattar TA, Blanchard RO, Sanfley GC (1974) Relationship between electrical resistance and capacitance of wood in progressive states of discoloration and decay. J Exp Bot 25:658–662

Tazawa M, Shimmen T (1980) Action potentials in Characeae: some characteristics revealed by internal perfusion studies. In: Spanswick RM, Lucas WJ, Dainty J (eds) Plant membrane transport: current conceptual issues. Elsevier/North-Holland Biomedical Press, Amsterdam New York, pp 349–362

Tyree MT, Fensom DS (1968) Methods of measuring hydrokinetic pressure gradients in the xylem of plants in situ. Can J Bot 46:310–314

Umrath K (1956) Electrophysiologische Phänomene. In: Ruhland W (ed) Encyclopedia of plant physiology vol II. Springer, Berlin Göttingen Heidelberg, pp 747–778

Umrath K (1959) Der Erregungsvorgang. In: Ruhland W (ed) Encyclopedia of plant physiology Vol XVII. Springer, Berlin Göttingen Heidelberg, pp 44–110

Umrath K, Kastberger G (1983) Action potentials of the high-speed conduction in *Mimosa pudica* and *Neptunia plena*. Phyton (Horn, Austria) 23:65–78

Vahey M, Scordilis SP (1980) Contractile proteins from the tomato. Can J Bot 58:779–801

Vredenberg WJ (1974) Changes in transport determining electrical parameters of cell and chloroplast membranes associated with primary and associated photosynthetic reactions. In: Zimmermann MH, Dainty J (eds) Membrane transport in plants. Springer, Berlin Heidelberg New York, pp 126–130

Vredenberg WJ (1976) Electrical interactions and gradients between chloroplast compartments and cytoplasm In: Barber J (ed) The intact chloroplast. Elsevier/North-Holland Biomedical Press, Amsterdam New York

Wanless IR, Bryniak N, Fensom DS (1973) The effect of some growth-regulating compounds upon electro-osmotic movements, transcellular water flows and Na, K, and Cl influxes in *Nitella flexilis*. Can J Bot 51:1055–1070

Watanabe S, Umrath K (1983) The influence of plant hormones on leaf movements of *Mimosa pudica*. Phyton (Horn, Austria) 23:49–54

Wilkins MB, Woodcock AER (1965) Origin of the geoelectric effect. Nature 208:990

Williams EJ, Johnston RJ, Dainty J (1964) The electrical resistance and capacitance of the membrane of *Nitella translucens*. J Exp Bot 15:1–14

Williams EJ, Munro C, Fensom DS (1972) The influence of small applied electrical currents on Na, K and Cl fluxes in *Nitella translucens*. Can J Bot 50:2255–2263

Willmer CM (1983) Stomata. Longman, London New York, pp 126–130

Wilner J, Brach EJ (1979) Utilization of bioelectric tests in biological research. Rep 1–139, Eng Statist Res Inst, Agric, Ottawa Canada

Zimmermann MH, Brown CL (1971) Trees, structure and function. Springer, Berlin Heidelberg New York, p 78
Zimmermann U, Benz R (1980) Electromechanical properties of cell membranes and lipid bilayers. In: Spanswick RM, Lucas WJ, Dainty J (eds) Plant membrane transport: current conceptual issues. Elsevier/North-Holland Biomedical Press, Amsterdam New York, pp 467–468
Zimmermann U, Steudle E (1977) Action of indoleacetic acid on membrane structure and transport. In: Marrè E, Ciferri O (eds) Regulation of cell membrane activities in plants. Elsevier/North-Holland Biomedical Press, Amsterdam New York, pp 231–242
Zimmermann U, Steudle E (1980) Fundamental water relations parameters. In: Spanswick RM, Lucas WJ, Dainty J (eds) Plant membrane transport: current conceptual issues. Elsevier/North-Holland Biomedical Press, Amsterdam New York, pp 113–127
Zimmermann U, Steudle E, Lelkes PI (1976) Turgor pressure regulation in *Valonia utricularis*: effect of cell-wall elasticity and auxin. Plant Physiol 58:608–613

18 Regulators of Plant Reproduction, Growth and Differentiation in the Environment

D.E. BILDERBACK

1 Introduction

Biological substances in the environment have profound effects upon the growth and reproduction of many plants. A variety of lower plants release substances or pheromones into the environment which alter the development or behavior of other closely related plants. With many algae, fungi, and ferns, these pheromones may play an essential role in regulating sexual differentiation and maturation, act as sex attractants, or modify asexual or vegetative development. Since these regulatory substances are small molecules and are active at low concentrations, they are hormones in the broadest sense and constitute a significant part of the environment influencing plant growth and development.

The behavior and physiology of many plants may also be influenced by regulatory substances in the environment that are not pheromones. Motile asexual propagules of many fungi are attracted chemotactically to new areas of colonization by substances produced by higher plants. Substances from higher plants and even nematode worms can control the differentiation of several fungi. On the other hand, soil microflora and epiphytic bacteria produce growth regulators which may alter the hormonal metabolism and growth of higher plants. Finally, many vascular plants release growth regulators or toxic allelopathic compounds which profoundly affect the development of other associated plants.

2 Factors Regulating Sexual Differentiation

Sexual reproduction provides a considerable advantage for the evolutionary survival of organisms. With the recombination of genetic material, there is an increased probability that organisms will adapt to ever-changing environments. To enhance this advantage, many organisms have evolved mating strains which are sexually self-incompatible. However, self-incompatibility itself can reduce sexual reproduction, since sexual reactions only occur when different mating strains are in close proximity. To overcome this difficulty, lower plants have evolved mechanisms which enhance sexual differentiation and coordinate the sexual reaction between mating strains. The release of specific diffusible substances into the environment have become an integral component of the sexual process. With lower plants the coordinating systems can be simple or highly complex and the chemical nature of the signals quite diverse. The sexual reactions of algae, fungi, and ferns, and the factors regulating sexual differentiation in those groups of plants are described below.

2.1 Algae

Of all the algae, only the sexual reactions of *Volvox* and *Oedogonium* are known in some detail. Since the sexual reaction of *Oedogonium* contains several chemotropic components, this alga will be described in Section 3.1.1.

2.1.1 Sexual Induction

The green alga, *Volvox,* is a spherical colonial organism of biflagellate vegetative cells embedded in a gelatinous matrix. The colony reproduces asexually by producing specialized cells called gonidia which undergo complex developmental changes involving enlargement, division, and inversion of the young multicellular spheroid to form a new colony within the parental colony. Under certain conditions, male initials differentiate and divide to form packets of sperm, and female initials differentiate as eggs within a colony. In some species of *Volvox,* undivided gonidia may function as eggs in female strains or may divide to form sperm in male strains (DARDEN 1966, STARR 1968, VAN DE BERG and STARR 1971).

With many strains of *Volvox,* a few colonies within a population differentiate spontaneously as males, and only frequent transfer to fresh medium prevents sexual development (DARDEN 1966, 1970, KOCHERT 1968, MCCRACKEN and STARR 1970, VAN DE BERG and STARR 1971). These male colonies produce a substance which is species- and strain-specific and induces all male colonies of the subsequent generation to differentiate sexually (DARDEN 1966, MCCRACKEN and STARR 1970, STARR 1969, VAN DE BERG and STARR 1971). Levels of the substance increase sharply in the aqueous environment as sperm packets are released from disintegrating colonies (MCCRACKEN and STARR 1970, STARR 1969, VAN DE BERG and STARR 1971), and conditioned medium containing the inducing substance can be diluted from 10^3- to 10^6-fold and still cause sexual differentiation of presumptive male colonies (KARN et al. 1974, KOCHERT 1968, MCCRACKEN and STARR 1970, STARR 1968). When purified, the sexual inducer from *V. carteri f. nagariensis* IYENGAR is active at concentrations as low as 10^{-10} gm 1^{-1} or about 3×10^{-16} M (STARR and JAENICKE 1974). Apparently, two molecules of the inducer substance are sufficient to cause a young colony to develop sexually (PALL 1973).

Female colonies do not spontaneously form eggs. However, if male and female strains are placed together, or if female colonies are placed in conditioned medium from sexually differentiated males, the females will produce eggs (KARN et al. 1974, KOCHERT 1968, MCCRACKEN and STARR 1970, STARR 1968, 1969, VAN DE BERG and STARR 1971). In most species of *Volvox,* there is no evidence that female colonies produce any inducing substances (MCCRACKEN and STARR 1970, VAN DE BERG and STARR 1971; however, STARR (1972) noted that female colonies of *V. dissipatrix* SHAW produce both self- and male-inductive substances.

The colony is sensitive to inducer only during a particular phase in development. In some species of *Volvox,* inducer must be present before gonidia within the parental colony begin to divide (KARN et al. 1974, KOCHERT 1968, STARR 1968). For other species, sexual induction occurs when gonidia are dividing and organizing a new colony (DARDEN 1966, 1970, KOCHERT 1968, STARR 1969). Finally, colonies of a few species can be induced even after they are released from the parental colony. In the latter case, the inducer either influences the development of reproductive initials already present in the colony or causes vegetative cells to differentiate sexually (MCCRACKEN and STARR 1970, STARR 1968, VAN DE BERG and STARR 1971).

For most species of *Volvox,* inducer must be present for a relatively long time (15 h) during the most sensitive phase of colony development (MCCRACKEN and STARR 1970). On the other hand, DARDEN (1966) reported that only a 5-min exposure to inducer during the sensitive phase was sufficient to induce a significant number of male colonies of *V. aureus* EHRENBERG.

Inducer is relatively stable after heating and destroyed by protease (DARDEN 1966, 1970, KOCHERT 1968, MCCRACKEN and STARR 1970, STARR 1969, STARR and JAENICKE 1974, VAN DE BERG and STARR 1971). STARR and JAENICKE (1974) have purified and characterized the inducer from *V. carteri* as an elongated asymmetrical glycoprotein with molecular weight of 27,500 to 30,000 containing a single polypeptide. The carbohydrate moiety, which accounts for about 45% of the weight, consists of pentoses, hexoses, and aminohexoses. The inducer can be distinguished from the colonial matrix glycoproteins which are released upon colony disintegration and sperm release (KOCHERT and YATES 1974).

The glycoprotein is produced and released from the sperm packets and not from vegetative cells or gonidia (GILLES et al. 1981). It first appears 8 to 11 h after the formation of the packets, and the concentration of the inducer increases exponentially for 10 h thereafter until the packets begin to separate into individual sperm. One male colony releases 6×10^{-11} g of inducer which corresponds to 1.6×10^{-13} g per sperm packet or 5.5×10^4 molecules per sperm cell. The protein moiety is synthesized for 35 h prior to secretion, stored within the sperm cells and glycosylated immediately prior to secretion.

STARR et al. (1980) have found that low concentrations of the amino acid, L-glutamic acid, induce sexual development of *Volvox capensis* RICH et POCOCK grown in the light. Other amino acids, various analogs of glutamic acid, compounds with similar structure to glutamic acid, and intermediates in biochemical pathways involving glutamic acid have no inductive effect. The glutamic-acid inducer is released from the colonial matrix by digestion of the glycoprotein.

2.2 Fungi

Sexual reactions of at least one genus from each of the four major groups of fungi have been investigated to a varying degree. With the myxomycetes, an understanding of the sexual reaction of cellular slime molds has occurred only recently, whereas the sexual reactions of phycomycete genera, *Achlya* and *Mucor,* are well known, having been studied since 1939 and 1960, respectively.

For the ascomycetes only the sexual reactions of *Ascobolus* and the economic yeasts have been studied in any detail. Very little is known about the sexual reactions of the basidiomycetes.

2.2.1 Myxomycetes

2.2.1.1 Macrocyst Formation

When opposite mating types of the cellular slime molds *Dictyostelium* or *Polysphondylium* are combined, the free-living amoeboid cells aggregate to form a macrocyst, enclosed in a fibrillar sheath (BLASKOVICS and RAPER 1957, CLARK et al. 1973, ERDOS et al. 1972, 1973, 1975). Two cells fuse to form a zygote and increase in size by phagocytoses of the other cells of the aggregation (ERDOS et al. 1972, FILOSA and DENGLER 1972). After ingestion is complete, the giant cell forms a rigid cell wall and undergoes meiosis (ERDOS et al. 1972). Presumably, only one of the meiotic products survives and divides to form a new population of haploid amoebae (ERDOS et al. 1972, 1975, MACINNES and FRANCIS 1974).

Macrocyst formation is controlled by a substance of low molecular weight which is released into the environment (MACHAC and BONNER 1975, O'DAY and LEWIS 1975). Only one mating type produces the substance, and only its opposite mating type can respond by migrating and fusing to form a giant zygote cell. Contact between mating types is not essential, since inducer diffusing through a dialysis membrane will cause migration (MACHAC and BONNER 1975, O'DAY and LEWIS 1975). The inducer is a volatile, heat-labile, species-specific compound with a molecular weight less than 12,000, and although its chemical nature is unknown, it definitely is not cyclic AMP (LEWIS and O'DAY 1976, 1977, O'DAY and LEWIS 1975). Macrocyst formation may actually be a two-step phenomenon. A low level of substance may induce migration and fusion, while a higher level of substance or perhaps an unidentified second substance may be required for macrocyst formation (LEWIS and O'DAY 1977). The giant zygote cell acts as an aggregation center and attracts adjacent amoebae by secreting 3′, 5′-cyclic AMP (O'DAY 1979).

2.2.2 Phycomycetes

2.2.2.1 Antheridial and Oogonial Formation

The sexuality of the saprophytic water mold, *Achlya,* is coordinated by diffusible chemicals (BARKSDALE 1963b, MCMORRIS 1978, RAPER 1939a, b, 1940a, b, 1942a, b, 1950, 1951, RAPER and HAAGEN-SMIT 1942). During vegetative growth, female strains continuously produce two substances, A and A^2, which are released into the aqueous environment. These substances intitiate the sexual reaction by inducing nearby hyphae of male strains to form many twisted and highly ramified lateral branches which are immature male sex organs or antheridial hyphae. Substance A is a steroid called antheridiol (Fig. 1) (ARSENAULT et al. 1968, MCMORRIS and BARKSDALE 1967), which induces antheridial hyphae at a concentration as low as 0.01 ng ml^{-1}. The steroid is removed from the

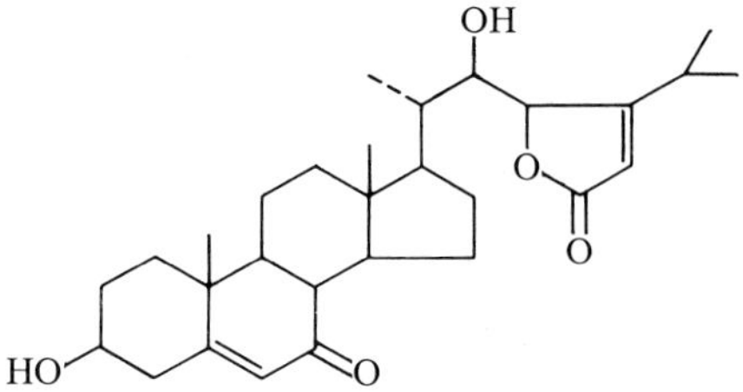

Fig. 1. Antheridiol

environment by the male strain but cannot be extracted from male hyphae (BARKSDALE 1963b, 1970). In fact, many male strains rapidly convert antheridiol into an inactive metabolite (MUSGRAVE and NIEUWENHUIS 1975). Since this inactivation requires both RNA and protein synthesis, antheridiol apparently induces the synthesis of its own inactivating enzyme (MUSGRAVE and NIEUWENHUIS 1975, MUSGRAVE et al. 1978).

The induction and growth of antheridial hyphae also requires RNA and protein synthesis (HOROWITZ and RUSSELL 1974, KANE et al. 1973, TIMBERLAKE 1976). More specifically, a messenger-like fraction of RNA accumulates (SILVER and HORGEN 1974), and one new cytoplasmic protein, as well as a number of new extra-cellular proteins, appears after treatment with antheridiol (GRONER et al. 1976). Since cellulase activity increases dramatically prior to and during the initiation of antheridial hyphae, the extra-cellular proteins are probably cellulases (KANE et al. 1973, MULLINS 1973, THOMAS 1970, THOMAS and MULLINS 1967, 1969). THOMAS (1970) have proposed that the induced cellulase causes a localized softening of the hyphal cell wall which develops as antheridial hyphae when the cytoplasm is forced by turgor pressure through the weakened wall. Localized groups of vesicles occur in the hyphal cytoplasm adjacent to thin portions of the wall which fuse with the plasma membrane, presumably releasing cellulase into the wall (MULLINS and ELLIS 1974). Vesicle formation could provide a method of concentrating cellulase at the presumptive initiation site of antheridial hyphae (MULLINS 1979). However, little is known about the factors regulating the localization of the vesicles at specific loci.

The other substance (A^2) produced by females can be separated from A and can induce antheridial hyphae alone (RAPER 1950). However, substances A and A^2 have a synergistic effect upon the induction of antheridial hyphae (RAPER 1950). The chemical nature of substance A^2 is unknown.

RAPER (1942b, 1950) found that vegetative hyphae of the male strain release two factors, A^1 and A^3, which alter the response of males to the substances A and A^2 from females. Substance A^1 augments the inductive effect of the female substances but cannot alone induce antheridial hyphae. The chemical nature of substance A^1 has not been determined; however, it may be a nutritional factor (BARKSDALE 1970). Substance A^3 limits the number of antheridial hyphae produced in response to the female substances, but little is known about its chemical nature. However, there is an isomer of antheridiol which can inhibit completely the induction of antheridial hyphae by antheridiol (BARKSDALE et al. 1974).

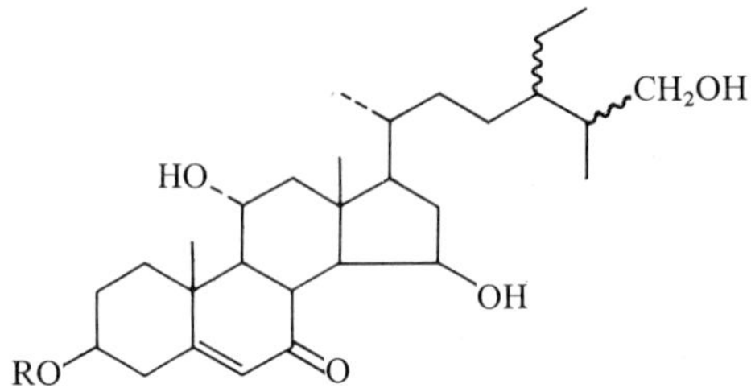

1. R = $(CH_3)_2CHC{=}O$
2. R = $CH_3CH_2C{=}O$
3. R = $CH_3C{=}O$

Fig. 2. Oogoniols

Antheridial hyphae produce substance B, which induces lateral branches on the female hyphae, the tips of which enlarge to form spherical immature female sex organs, oogonial initials (BARKSDALE and LASURE 1973, 1974, RAPER 1939a, b, 1940a, b, 1951). Substance B is actually three steroids which have been named oogoniols (Fig. 2) (MCMORRIS et al. 1975).

Antheridial hyphae grow toward and touch oogonial initials, become flattened, branch, and form multinucleate antheridia. RAPER (1939a, 1940a) thought that antheridial hyphae had to make contact with an oogonial initial or an abiotic surface before delimitation of antheridia would occur and believed that a new substance (C) regulated their chemotropic response. However, BARKSDALE (1963b) conclusively showed that antheridial hyphae are attracted to substance A and that contact is not required for antheridial differentiation.

Antheridia also produce a substance D of unknown chemical nature which controls the maturation of the oogonium initials into separate oogonia with eggs (RAPER 1939a, b, 1951). Conjugation tubes grow from antheridia into the oogonium and make contact with the eggs. Finally, a single male nucleus is transferred to the egg, and the zygote secretes a thick wall and enters dormancy. This directed growth of the conjugation tube may be regulated by a chemotropic substance produced by the egg (RAPER 1951).

Sexual reproduction of heterothallic species of the fungus *Phytophthora* is also controlled by diffusible substances called α-hormones (KO 1978). Oospore formation by A^1 isolates depends on α^2 sex hormone secreted by the complementary A^2 isolates, and vice versa. The sexuality of homothallic species is also under hormonal regulation (KO 1980), and oospores of *Phytophthora* can be induced by exogenous sterols (ELLIOTT 1979).

2.2.2.2 Induction of Progametangia

The mating strains of the genera *Mucor* or *Blakeslea* are morphologically similar and are designated as plus or minus mating types. Each mating type releases into the environment small amounts of specific substances called progamones which are derived from β-carotene (AUSTIN et al. 1969, 1970, BU'LOCK et al. 1972, 1974, BU'LOCK and HARDY 1979, PLEMPEL 1963, SUTTER et al. 1973, VAN DEN ENDE 1968, VAN DEN ENDE et al. 1972, WERKMAN and VAN DEN ENDE

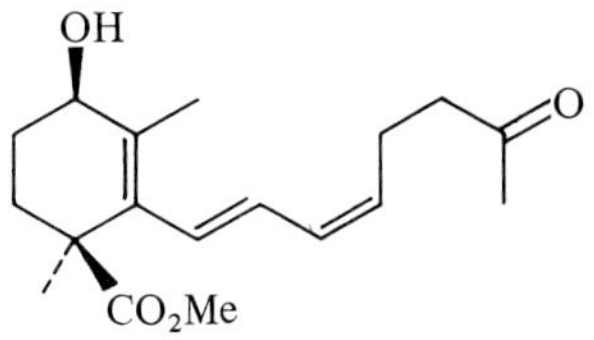

Fig. 3. Hydroxymethyltrisporate B

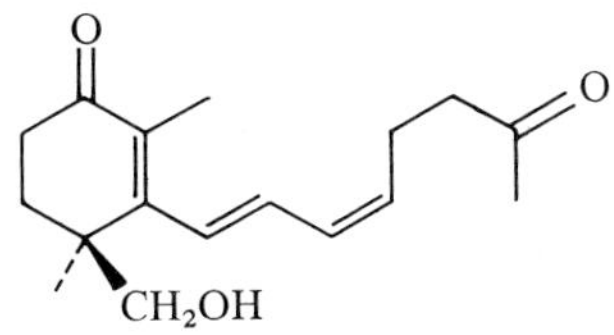

Fig. 4. Trisporal B

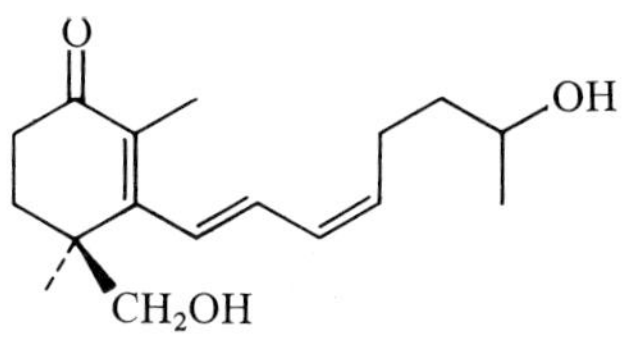

Fig. 5. Trisporal C

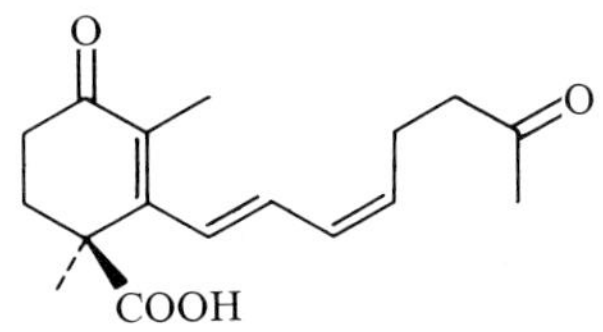

Fig. 6. Trisporic acid B

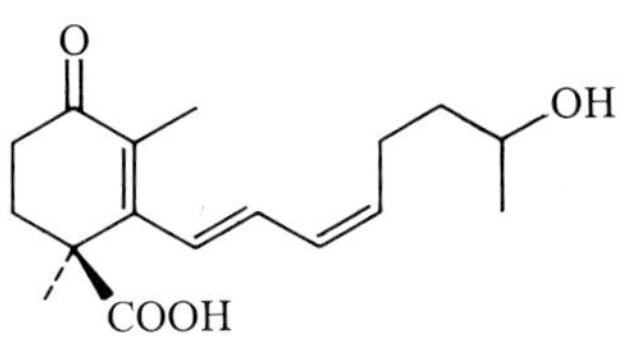

Fig. 7. Trisporic acid C

1973). Progamone synthesis by opposite mating types may be regulated in turn by unknown metabolites from each mating type (SUTTER et al. 1973, VAN DEN ENDE et al. 1972). The progamone from the plus strain is hydroxymethyltrisphosphate B (Fig. 3) (BU'LOCK et al. 1973, 1974a, NIEUWENHUIS and VAN DEN ENDE 1975, WERKMAN and VAN DEN ENDE 1973). The minus strain produces the progamones, trisporals B and C, or trisporins B and C (Figs. 4, 5) (BU'LOCK et al. 1973, 1974a, NIEUWENHUIS and VAN DEN ENDE 1975). The progamones produced by each strain are then transformed by the opposite mating type into trisporic acids B and C (Figs. 6, 7) (BU'LOCK et al. 1972, SUTTER 1970, SUTTER et al. 1973, 1974, VAN DEN ENDE et al. 1972, WERKMAN and VAN DEN ENDE 1973), with few or no trisporic acids being synthesized by isolated mating strains (VAN DEN ENDE et al. 1970). Only the minus strain has specific enzymes which transform the progamone produced by the plus strain into trisporic acids B and C (WERKMAN 1976). Furthermore, the end products, trisporic acids, will stimulate the further synthesis of these enzymes (WERKMAN 1976). Presumably, the plus strain also has enzymes which convert the progamones produced by the minus strain into trisporic acids, since the plus strain of *Blakeslea* has a greater capacity to oxidize trisporal B and C to trisporic acids B and C than the minus strain (BU'LOCK et al. 1972). The trisporic acids induce the sexual differentiation of negatively geotropic progametangia in both strains (VAN DEN ENDE 1968).

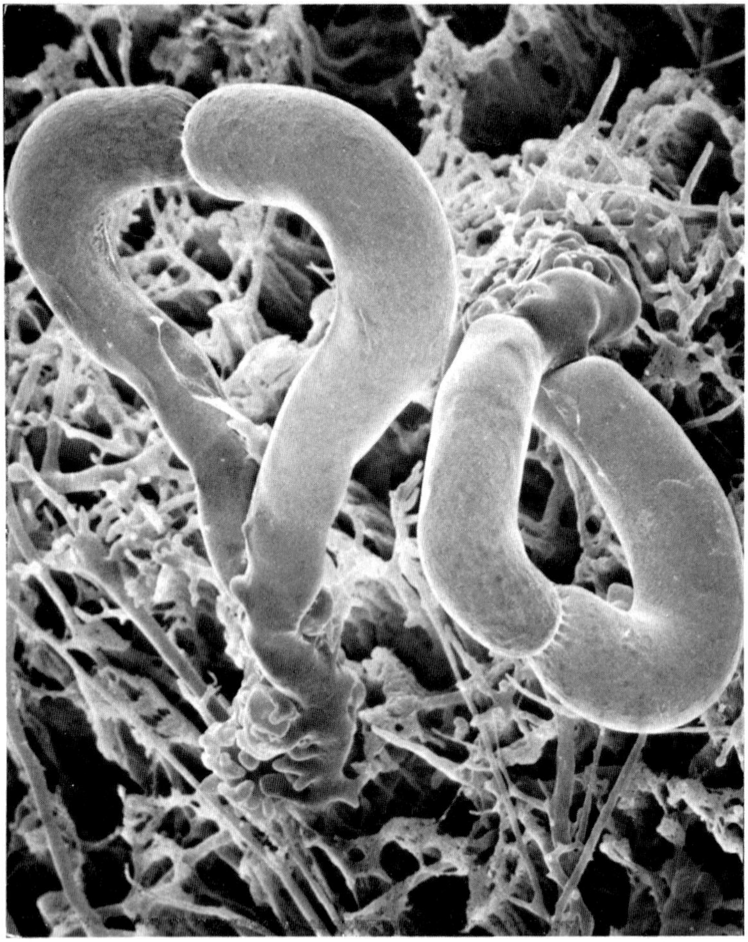

Fig. 8. Fusion of progametangia of *Phycomyces nitens*. (From O'DONNELL et al. 1978)

The erect progametangia of each strain bend, grow toward the progametangia of the opposite strain, and make physical contact (Fig. 8). A gradient of volatile compounds produced by the progametangia of each strain is responsible for this chemotropic reaction (BANBURY 1954, PLEMPEL 1960, 1962, 1963, PLEMPEL and DAWID 1961). The chemical nature of these substances is not known; however, the progamones are volatile (BU'LOCK et al. 1974a, NIEUWENHUIS and VAN DEN ENDE 1975), and fractions containing progamones (but not trisporic acids) cause the progametangial reaction (MESLAND et al. 1974). Genetic experimentation also indicates that the progamones control the chemotropic reaction. WURTZ and JOCKUSCH (1975) were able to isolate several mutations in *Mucor* which affected progametangial development but not the chemotropic reaction. Mutants with just the chemotropic reaction blocked would not be expected if progametangial induction and chemotropism are controlled by the same compounds.

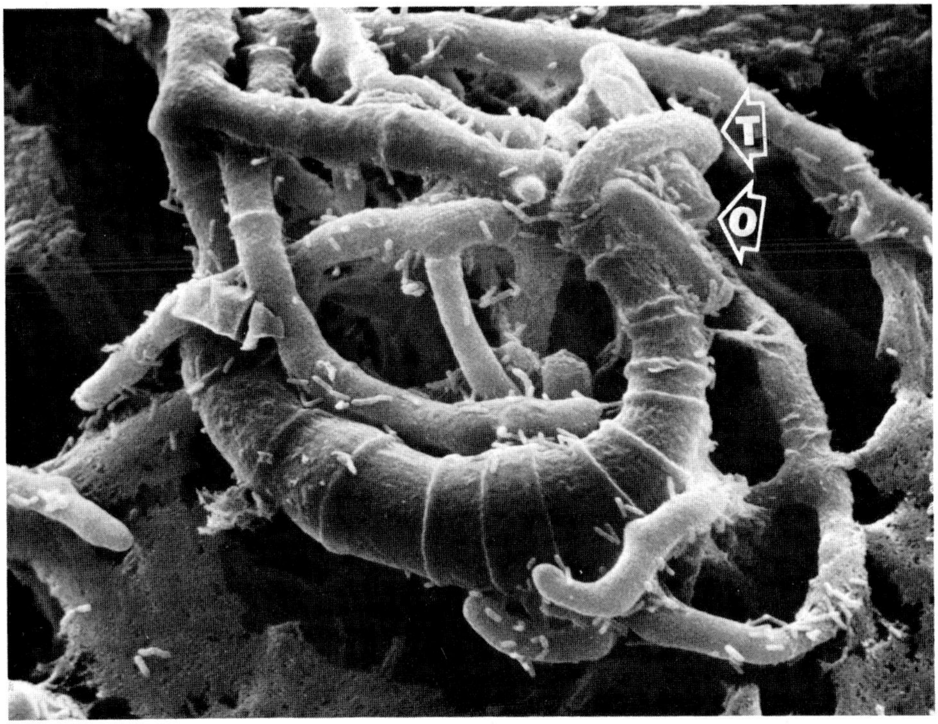

Fig. 9. Fusion of the tip (*T*) of the ascogonium with an oidial spore (*O*) of *Ascobolus furfuraceus*. (From O'DONNELL et al. 1974)

2.2.3 Ascomycetes

2.2.3.1 Ascogonial Induction

When vegetative hyphae of opposite mating types of the genus, *Ascobolus*, grow in close proximity, hyphae of the donor or fertilizing strain form highly branched filaments which then differentiate antheridia at their tips (BISTIS 1956, BISTIS and RAPER 1963). Other portions of the fungus, such as ungerminated asexual oidial spores, oidial germ tubes, and ascospore germ tubes may also form antheridia (BISTIS and RAPER 1963). Antheridia then produce a substance which induces the formation of ascogonial hyphae by the opposite strain (BISTIS 1956). If this interaction is interrupted, the young ascogonium ceases growth, reverts to vegetative growth or branches dichotomously (BISTIS 1957). However, if the sexual interaction is uninterrupted, the tip of the ascogonium is attracted chemotropically toward the antheridium and fuses with it (Fig. 9; BISTIS 1956, 1957). If plasmogamy is continuously postponed, the ascogonium will lose its ability to respond chemotropically (BISTIS 1957). Finally, an investing sheath of vegetative hyphae are attracted chemotropically to the ascogonium and form a fruiting body. The chemical nature of the antheridial- and ascogonial-inducing substances or the chemotropic agents is not known.

2.2.3.2 Factors Affecting Sexual Fusion

Saccharomyces are unicellular, heterothallic yeasts with morphologically similar a- and α-mating strains. When haploid cells of opposite mating types are mixed together, they agglutinate and form conjugation tubes. The walls separating the two mating pairs dissolve, plasma membranes fuse, nuclear fusion occurs and diploid buds emerge at the original point of contact (OSUMI et al. 1974).

Isolated mating strains reproduce asexually by producing buds. Within a population of cells there is no synchrony of nuclear division and bud formation. Budding cells are incapable of fusing sexually (HARTWELL 1973), but each mating type produces continuously diffusible factors which inhibit DNA synthesis and budding of the other mating type (BILIŃSKI et al. 1974, DUNTZE et al. 1970, MC CULLOUGH and HERSKOWITZ 1979, SHIMODA and YANAGISHIMA 1972, 1973, WILKINSON and PRINGLE 1974, YANAGISHIMA 1978). Much more is known about the proteinaceous α-factor produced by α-cells than the a-factor released from a-cells (DUNTZE et al. 1970, 1973). Actually, STÖTZLER and DUNTZE (1973, 1976) have been able to isolate four related peptides with α-factor activity. The primary peptide has an amino acid sequence, (H_2N) Trp-His-Trp-Leu-Gln-Leu-Lys-Pro-Gly-Gln-Pro-Met-Tyr (COOH) (STÖTZLER et al. 1976). A second peptide is similar to the primary one but lacks the N-terminal tryptophan which may have been removed by an aminopeptidase (STÖTZLER and DUNTZE 1976). These two oligopeptides have been synthesized and shown to inhibit DNA synthesis and promote sexual agglutination (CIEJEK et al. 1977). The remaining two peptides are methionine sulfoxide oxidation products the former α-peptides. However, TANAKA et al. (1977) have been able to isolate only the primary peptide from their α-cell cultures. The primary peptide appears to be derived from a larger precursor peptide with 23 amino acids by the removal of 11 amino acids from the acid terminal of the molecule (TANAKA and KITA 1977). The primary α-factor is initially degraded by a surface-bound endopeptidase into two inactive peptides by cleavage of the peptide bond between leucine and lysine (CIEJEK and THORNER 1979, TANAKA and KITA 1977). Hydrolysis of the α-factor releases the a-cells from cell cycle arrest (CHAN 1977, CIEJEK and THORNER 1979, FINKELSTEIN and STRAUSBERG 1979) and is not necessary for α-factor action (CIEJEK and THORNER 1979, FINKELSTEIN and STRAUSBERG 1979) as suggested by MANESS and EDELMAN (1978). The minimum amino acid sequence required for activity is (H_2N) His-Trp-Leu-Gln-Leu-Lys-Pro-Gly-Gln (COOH) (MASUI et al. 1977). The N-terminal tryptophan is not required for biological activity but substitution of this amino acid by D-tryptophan, tyrosine or leucine results in a significant reduction or loss of activity (MASUI et al. 1979). Histidine and the second tryptophan is also an essential portion of the peptide. However, the second tryptophan can be replaced by other amino acids with hydrophobic side chains with only a slight reduction in activity. Apparently, even the slightest change in conformation resulting from the substitution of other amino acids of the peptide results in a reduction in biological activity. Mating type a-cells will bind 10^4 α-factor molecules per cell but naked a-spheroplasts do not bind the factor (MANESS and EDELMANN 1978). When α-factor is added to a-cells, asexual budding and DNA synthesis are arrested, but RNA and protein synthesis remain unaffected (BÜCKING-THROM et al. 1973, THROM and DUNTZE 1970). Treatment of a-cells with α-factor results in a localized alteration of the cell walls (TKACZ and MACKAY 1979). There is also an increased rate of synthesis and activation of chitin synthase in the plasma membrane resulting in the localized deposition of chitin on the cell surface (SCHEKMAN and BRAWLEY 1979). A portion of the wall of a-cells then loses its characteristic trilaminar morphological structure, becomes thinner and develops a diffuse surface layer (LIPKE et al. 1976). The a-factor presumably affects α-cells in a similar manner (SHIMODA and YANAGISHIMA 1973). When the wall of α-cells is removed, the α-spheroplasts can inactivate a-factor and respond morphologically to α-factor resembling the response of a-cells (MANESS and EDELMAN 1978).

Another diffusible component of the sexual reaction is produced by the a-cells and reduces the effectiveness of the α-factor (HICKS and HERSKOWITZ 1976). Once all of the α-factor is inhibited, removed or inactivated after contact with the binding sites on a-cells, the cells begin to synthesize DNA and bud again (CHAN 1977). Apparently, the sexual reaction is delicately balanced between inducers and inhibitors.

When cells of opposite mating types of *Saccharomyces* are mixed, not only is DNA synthesis inhibited but the cells agglutinate after a lag period (SAKAI and YANAGISHIMA 1971). The α-cells continuously release some substance which induces an alteration of the wall of a-cells so that they will agglutinate with α-cells. Induction of agglutination requires protein synthesis (SAKAI and YANAGISHIMA 1972), and the substance (α-substance-I_A) has been characterized by SAKURAI et al. (1976, 1977a, b) as being a peptide with the same amino acid sequence as the second α-factor peptide of STÖTZLER and DUNTZE (1976). However, only minute amounts (0.001–0.01 ng ml^{-1}) of the substance are required to induce agglutination whereas higher concentrations (0.1–1 ng ml^{-1}) are required to inhibit DNA synthesis (TAKAO et al. 1970). Even when a-cells were treated with 1 ng ml^{-1}, agglutination is detected after 20 min, but DNA synthesis is inhibited only by 8.7% after 1 h (SHIMODA et al. 1978). The primary α-factor peptide apparently induces agglutination in some strains (BETZ et al. 1978) but not in others (TANAKA et al. 1977). Using fluorescein isothiocyanate-conjugated α-mating factor, TANAKA and KITA (1978) have shown that the factor binds to the cell surface and is translocated to the nucleus of a-cells. The peptide enhances sexual agglutination by inducing the synthesis of wall glycoprotein by the a-cells (BURKE et al. 1980, FEHRENBACHER et al. 1978, SHIMODA et al. 1976, TOHOYAMA et al. 1979, YOSHIDA et al. 1976). On the other hand, type a-cells release a specific substance into the environment which binds to the agglutination inducing substance and inactivates it (YANAGISHIMA et al. 1977). This inactivation is nonenzymatic, and the ability to induce agglutination can be regained by heating the inactivated complex (YANAGISHIMA et al. 1977). A second substance (α-substance -I_B) with the amino acid sequence, (H_2N) Arg-Gly-Pro-Phe-Pro-Ile (COOH), has been isolated which also induces sexual agglutination of a-cells (AOYAGI et al. 1977).

Type a-cells also produce a peptide which enhances the ability of α-cells to agglutinate (YANAGISHIMA et al. 1976). BETZ and DUNTZE (1979) have isolated three compounds exhibiting the a-factor activity of arresting DNA synthesis in α-cells. The most active compound is a hydrophobic undecapeptide with the following amino acid composition:

$$(H_2N)\ \text{Try}\ (\text{Asx}_1, \text{Gly}_1, \text{Ala}_1, \text{Val}_1, \text{Ile}_1, \text{Phe}_1, \text{Lys}_1, \text{Trp}_1, \text{Pro}_1)$$

It is now apparent that both mating strains produce peptides which specifically either induce or enhance the sexual agglutination of the opposite mating type.

Yeast cells elongate significantly during conjugation, and LEVI (1956) first suggested that diffusible factors from α-cells caused the elongation of a-cells. In fact, a-cells do elongate extensively in response to the α-factor peptides (DUNTZE et al. 1970, THROM and DUNTZE 1970). However, other substances produced continuously by the mating strains cause the opposite strain to elongate or expand (YANAGISHIMA 1969). These substances have been tentatively identified as steroids (TAKAO et al. 1970). Although the female animal steroid, estradiol, causes a-cells to elongate and the animal male steroid, testosterone, will cause α-cells to elongate, the yeast steroids are chemically different from these steroids (YANAGISHIMA et al. 1970). The reproductive system may be even more complex with the a-steroid inducing the production of a non-steroidal substance by the α-cell which with the α-steroid, can cause the a-cells to elongate (YANAGISHIMA 1971).

2.2.4 Basidiomycetes

2.2.4.1 Induction of Conjugation Tubes

Tremella mesenterica FR. is a single-celled organism which reproduces asexually by budding, but when cells of opposite mating type are mixed together, they form filamentous conjugation tubes. Each mating type produces diffusible substances which cause the opposite type to produce the tubes (BANDONI 1965). Three conjugation substances from one mating type have been isolated which appear to be nonpolar amino acids or small peptides (REID 1974). A peptidyl hormone, tremerogen A-10, has been isolated from a culture filtrate of the A-mating type which can induce conjugation tubes by a-cells at the concentration of 1–2 ng ml^{-1} (SAKAGAMI et al. 1978a, b). Tremerogen A-10 has an amino acid sequence, H_2N-Glu-His-Asp-Pro-Ser-Ala-Pro-Gly-Asn-Gly-Tyr-Cys (S-isoprenoid)-OCH_3 (SAKAGAMI et al. 1979). For biological activity, the peptide must possess the lipophilic isoprenoid moiety, a blocked C-terminus and the total peptide sequence (FUJINO et al. 1980).

Rhodosporidium also has morphologically similar A- and a-mating strains (ABE et al. 1975). The A-cells constitutively produce A-factor which induces the a-cells to form conjugation tubes. The a-cells then secrete a-factor which causes A-cells to form mating tubes. The tips of the tubes recognize the opposite mating-type cells or their tubes and fuse with them to form zygotes. The A-factor, rhodotorucine A, has been determined to have the amino acid sequence (H_2N) Tyr-Pro-Glu-Ile-Ser-Trp-Thr-Arg-Asn-Gly-Cys (S-farnesyl)-(COOH) (KAMIYA et al. 1978). The lipophilic amino acid, S-farnesyl cysteine, as well as the entire amino acid sequence, is required for biological activity (TSUCHIYA et al. 1978).

Thus, for many algae and fungi a number of diverse substances are released into the external environment, where they act as inductive stimuli for sexual differentiation or as attractants. Without these major environmental hormone-like factors, the sexual interactions of many lower plants would be severely restricted.

2.3 Ferns

2.3.1 Antheridia Formation

Many fern gametophytes produce a hormone-like pheromone which induces the precocious formation of sperm-producing structures, antheridia. The pheromone is called an antheridiogen. Some ferns possess their own native antheridiogen systems (DÖPP 1950, HAUFLER and GASTONY 1978, NÄF 1956, 1959, NÄF et al. 1969, SCHEDLBAUER and KLEKOWSKI 1972, TRYON and VITALE 1977, VOELLER 1971), while others respond to gibberellins or native antheridiogens (SCHRAUDOLF 1962, VOELLER 1964a, b, 1966).

When spores of the fern genera *Pteridium* and *Ceratopteris* are sown on a nutrient medium, some of the gametophytes or prothalli differentiate notched meristems more rapidly than others (NÄF 1958, SCHEDLBAUER 1976a). These rapidly growing prothalli synthesize and release antheridiogen into the environment, which induces the more slowly growing prothalli to differentiate precociously antheridia (DÖPP 1950, 1959, NÄF 1956, 1958, NÄF et al. 1975, SCHEDLBAUER 1976a, VOELLER 1966). For *Pteridium,* the antheridiogen appears in the external environment after the organization of a notch meristem and formation of a heart-shaped prothallus (NÄF 1958). The notch meristem is the site of antheridiogen synthesis (DÖPP 1959, 1962, VOELLER and WEINBERG 1969). However, before synthesis begins the prothallus becomes insensitive to the antheridiogen and never differentiates antheridia but eventually forms female sex organs, archegonia (DÖPP 1959, 1962, NÄF 1958, 1962b). If the notch meristem is surgically removed from an antheridiogen-insensitive prothallus, the remaining portion will differentiate antheridia when exposed to antheridiogen (DÖPP 1959). Apparently, the notch meristem produces a substance which renders the prothallus insensitive to antheridiogen (DÖPP 1959) or to spontaneous antheridium formation (NÄF 1961). Slower growing prothalli lack a notch meristem, do not synthesize antheridiogen, but do respond to it by initiating antheridia (NÄF 1958). Eventually, some of these slower-growing prothalli form notch meristems; however, they stop forming antheridia and initiate archegonia. Only a small number of the slower-growing prothalli never form notch meristems and continue to produce antheridia.

When a *Pteridium* prothallus is grown in isolation, it becomes insensitive to antheridiogen prior to synthesizing it, and antheridia do not differentiate. However, if a young isolated prothallus is treated with antheridiogen before notch meristem formation, many antheridia differentiate (NÄF 1958).

For *Ceratopteris* the more rapidly growing prothalli also become heart-shaped and only these synthesize antheridiogen (SCHEDLBAUER 1976a). However, unlike *Pteridium,* these prothalli do not become insensitive to antheridiogen.

They form antheridia before and after the notch meristem has been organized and eventually differentiating archegonia. Antheridiogen produced by these more rapidly growing prothalli induce more slowly growing prothalli to form only antheridia (SCHEDLBAUER 1976a). The production of antheridia is dependent on the continuous presence of the antheridiogen; if antheridiogen is absent or removed, prothalli become hermaphroditic (SCHEDLBAUER and KLEKOWSKI 1972).

Although prothalli of *Onoclea* can be grown so that they will not form antheridia spontaneously, they can be induced to do so by *Pteridium* antheridiogen (NÄF 1958). However, this must be done before antheridiogen sensitivity ends which is shortly after prothalli become heart-shaped. If prothalli are exposed to the antheridiogen at an early age, the period of sensitivity can be prolonged considerably (NÄF 1962a). During the sensitive phase, there is a light-dependent block to antheridium formation which can be overcome by *Pteridium* antheridiogen, although the block becomes irreversible once the meristem has formed (NÄF 1962a, NÄF et al. 1974).

Species of the schizaeaceous ferns in the genera *Lygodium* and *Anemia* also produce antheridiogens. With *Lygodium,* the most rapidly growing individuals form archegonia, sometimes followed by antheridia (NÄF 1960, TYRON and VITALE 1977). However, with slower-growing prothalli, the sequence of organ formation becomes reversed in the presence of antheridiogen from rapidly growing prothalli. With *Anemia* more rapidly growing prothalli also become heart-shaped and initially lack antheridia (VOELLER 1966). For both ferns, the more slowly growing prothalli forming antheridia become heart-shaped, and the transition from antheridiogen-sensitive phase to an insensitive phase is not abrupt. Apparently, the number of prothallial cells sensitive to antheridiogen is reduced gradually (NÄF 1959, 1960, SCHRAUDOLF 1962), and in *Lygodium* a restricted area of the prothallus remains sensitive indefinitely (NÄF 1960). Since both gametophyte morphologies possess notch meristems, neither sensitivity to, nor production of, antheridogen can be absolutely correlated to the presence of this characteristic.

Antheridiogens are distinct chemical entities and can be separated chromatographically (BURCKY et al. 1978, NÄF 1959, 1960, 1968, NÄF et al. 1969, SCHEDLBAUER 1974, 1976b, VOELLER 1964a). The *Pteridium* antheridiogen is nonspecific and will induce antheridia on prothalli of ferns of various families (NÄF et al. 1975). It has a low molecular weight, a dissociation constant near 5, an unsaturated carbon-to-carbon bond and a carboxyl group (PRINGLE 1961, PRINGLE et al. 1960) linking it with the growth substances, gibberellins (VOELLER 1964a). The *Ceratopteris* antheridiogen is genus-specific and will not induce antheridia on *Pteridium* or *Anemia* prothalli (SCHEDLBAUER 1974). It also has a low molecular weight and a dissociation constant near 5 (SCHEDLBAUER 1976b). The antheridiogen of *Lygodium* also appears to be genus-specific and will not induce antheridia on prothalli of *Pteridium, Onoclea,* or *Anemia* (NÄF 1960); it has been chemically characterized as the methyl ester of gibberellin A_9 (YAMANE et al. 1979).

Finally, the *Anemia* antheridiogen is family-specific, being active only in the Schizaeaceae (NÄF 1959), and has been isolated and chemically characterized

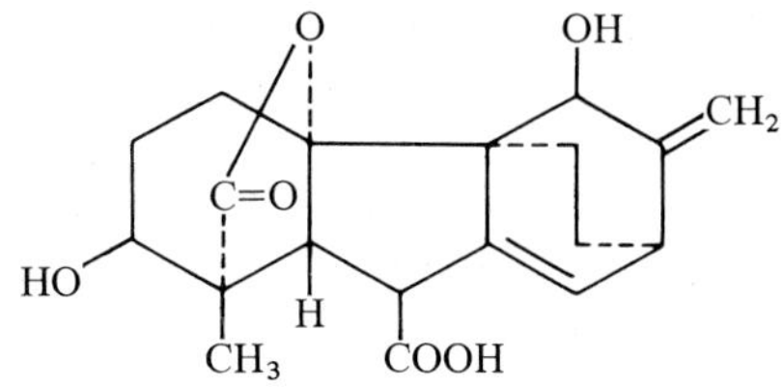

Fig. 10. *Anemia* antheridiogen

(Fig. 10) (ENDO et al. 1972, NAKANISHI et al. 1971). *Anemia* antheridiogen is structurally related to gibberellins, and gibberellins can replace it in the induction of antheridia on prothalli of various members of Schizaeaceae (SCHRAUDOLF 1962, 1964, VOELLER 1964a, b). Thus, fern antheridiogens constitute hormone-like factors produced by certain laboratory-grown prothalli which influence the sexual differentiation of neighboring prothalli. Even though the evidence is not as conclusive as with the algae and fungi, a substance enhancing the sexual exchange of genetic material in ferns could be a major regulatory factor in the natural environment of these plants (TRYON and VITALE 1977).

3 Sex Attractants

For the most part, the union of animal gametes and fertilization are purely random events with the egg having no effect on the locomotor activity of swimming sperm. With the lower plants, female gametes release sexual attractants into the environment which elicit a chemotactic response by motile male gametes. When male gametes enter the diffusion field of the chemotactic substance, they swim in a directed manner toward the female gamete. As a result of this chemotactic reaction, gamete fusion becomes a non-random event, and fertilization is enhanced. Sex attractants are significant environmental factors affecting the directional movement of sperm of algae, fungi, and ferns. Since they operate at some distance from the site of synthesis and are active at low concentrations, these substances are hormone-like pheromones.

3.1 Algae

3.1.1 Chlorophyta

Oedogonium is a filamentous freshwater alga. In all species some cells of the filament form oogonia, each containing an egg. Antheridia are formed either on the same filament with oogonia or on separate male plants. After release from an antheridium, multiflagellate sperm are chemotropically attracted to oogonia (HOFFMAN 1960) by a substance with an estimated molecular weight between 500 and 1500, which is highly soluble in water and destroyed by dilute acids or bases and heat (MACHLIS et al. 1974).

With other species of *Oedogonium,* male plants form androsporangia which release motile androspores. These androspores become attached to the oogonial mother cell or oogonium and then develop into dwarf male plants. Dwarf male plants differentiate antheridia which ultimately release motile sperm.

Sexual reproduction of these species of *Oedogonium* is regulated by three or four unidentified substances released in the aqueous environment (RAWITCHER-KUNKEL and MACHLIS 1962). The oogonial mother cells release a compound which attracts androspores. If a cotton thread is soaked first in a dish with fertile female filaments and then introduced into a suspension of androspores, the androspores immediately swarm to the thread. The attraction to the oogonial mother cell continues for as long as 24 d or until the mother cell divides to form an oogonium. After attachment to the oogonial mother cell, the androspore grows asymmetrically with those androspores attached to the upper portion of the oogonial mother cell bending down toward the future oogonium, and those attached to the lower portion bending upward. When androspores attach to oogonial mother cells which are enclosed within a thin sheath of agar, no asymmetrical growth occurs. This asymmetrical growth may be controlled by the androspore attractant or by another substance produced by the oogonial mother cell. Attachment of the androspores for at least 8 h is required to induce the oogonial mother cell to differentiate an oogonium.

Sperm released from dwarf male plants move randomly for a time within a massive gelatinous oogonial sheath, eventually becoming attracted to a particular area of the oogonium. An oogonial pore develops in that area, and a cytoplasmic papilla appears to which all sperm in the area attach. Fusion with one sperm occurs and the papilla is withdrawn, and the remaining sperm disperse.

3.1.2 Phaeophyta

Chemotaxis of male gametes plays a central role in the reproduction of the marine brown algae, *Ectocarpus, Cutleria* and *Fucus* (JAENICKE 1977a, b). *Ectocarpus siliculosus* (DILLW.) LYNGB. produces biflagellate, isomorphic gametes which can be differentiated physiologically as female and male gametes. Female gametes produce a chemotactic substance called ectocarpen which attracts male gametes. The attracted male gametes become attached to the female gamete by the tips of their anterior flagella. When male gametes are exposed to tritiated ectocarpen, radioactivity appears first on the longer anterior flagella and then later on the sperm body but never on the shorter, posterior flagellum (JAENICKE 1977b). In the presence of ectocarpen, the male gametes no longer swim in straight paths but begin to swim in circular paths toward the egg (MÜLLER 1978). Once fertilization has been achieved by one male gamete, the other attached males disperse, and the zygote loses its capacity to produce ectocarpen (MÜLLER 1968, 1972, MÜLLER et al. 1971). Ectocarpen is allo-cis-1-(cycloheptadien-2′, 5′ yl) but-1-ene (Fig. 11) (MÜLLER et al. 1971). It is derived from highly unsaturated fatty acids and possesses a high specificity, since no chemical homologs or sterically modified compounds will attract male gametes (JAENICKE 1972). Ectocarpen is not species-specific (MÜLLER and GASSMANN 1980).

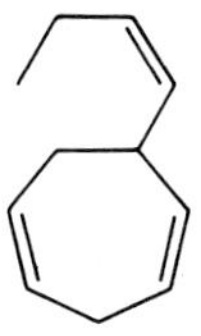

Fig. 11. Ectocarpen

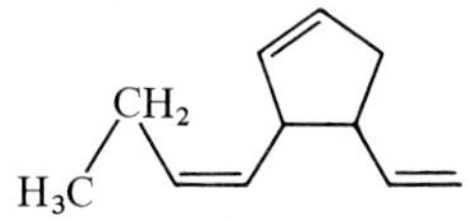

Fig. 12. Multifiden

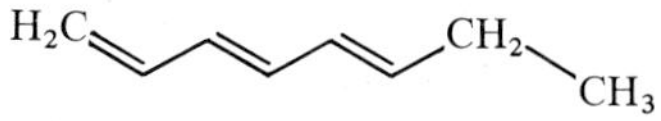

Fig. 13. Fucoserraten

Cutleria multifida (SMITH) GREY, releases into the sea biflagellate, anisogamous gametes with the female being much larger than the male. Male gametes are attracted by compounds produced by the more slowly swimming females (MÜLLER 1974). The major attractant is multifiden or cis-4-vinyl-5-(cis-1′ butenyl) cyclopentene (Fig. 12); however, its by-product, aucanten or trans-4-vinyl-5-(trans-1′-propenyl) cyclohexene and ectocarpen are also released in small amounts (JAENICKE 1977b, JAENICKE et al. 1974). Multifiden and its isomers have been synthesized (BOLAND and JAENICKE 1978). The threshold concentration of multifiden is 60 ng l^{-1} (DERENBACH et al. 1980). In the absence of multifiden, the male gametes swim in a straight line, but in the presence of the substance they are stimulated to swim in loops and circles (DERENBACH et al. 1980). Male gametes of *Cutleria* are ten times less sensitive to ectocarpen than *Ectocarpus* gametes (MÜLLER 1976). Nevertheless, female gametes of both genera attract male gametes of the other genus, but fusions never occur (MÜLLER 1976).

Eggs of *Dictyota dichotoma* (HUDSON) LAMOUR secrete a substance which attracts sperm (MÜLLER et al. 1981). The substance is chemically similar to ectocarpen and has the absolute configuration of (−)-(R)-6-butylcyclohepta-1,4-diene (MÜLLER et al. 1981).

Mature female gametophytes of the kelp, *Laminaria digitata* (HUDS) LAMOUR, release a substance which not only attracts sperm but will also cause the explosive discharge of sperm from antheridia of the male gametophytes (LÜNING and MÜLLER 1978). The compound has been isolated and partially identified as a $C_{11}H_{14}O$ compound of high polarity (MÜLLER and GASSMANN 1980, MÜLLER et al. 1979). Seawater in which the female gametophytes of three species of *Laminaria* or one species of *Alaria, Pteryophora,* or *Macrocystis* have separately matured cause mass release and attraction of sperm in all 36 combinations with male gametophytes (LÜNING and MÜLLER 1978). Apparently, the substance is not genus-or species-specific among these selected laminarian algae (LÜNING and MÜLLER 1978, MÜLLER and GASSMANN 1980).

Fucus serratus L. produces large immotile eggs and small motile biflagellate sperm. Eggs release a sperm attractant called fucoserraten which has been identified as a 1, 3-trans, 5-cis-octatriene (Fig. 13) (MÜLLER 1972, MÜLLER and GASSMANN 1978, MÜLLER and JAENICKE 1973). The attractant is not species-specific, but most other conjugated octatrienes are less effective than fucoserraten (MÜLLER and SEFERIADIS 1977).

The sperm attractants from *Sargassum* have been determined to be fucoserraten and its isomers, 1, trans-3, trans-5-octatriene, 1, cis-3, trans-5-octatriene and cis-2, cis-4, trans-6-octatriene (KAJIWARA et al. 1980). All four compounds strongly attract sperm. To a lesser degree, other conjugated triene isomers, as well as ectocarpen, also will attract *Sargassum* sperm.

3.2 Fungi

The filamentous water mold, *Allomyces,* has isomorphic haploid and diploid generations. The haploid gametophytic plant produces female and male gametangia. Released male gametes not only gather around the female gametangia containing differentiating gametes but also around large, slowly swimming uniflagellate female gametes (MACHLIS 1958a). Female gametangia and individual female gametes release into the surrounding aqueous medium a chemotropic agent called sirenin, which is a heat-stable sesquiterpene (Fig. 14) (MACHLIS 1958a, b, c, MACHLIS et al. 1968, NUTTING et al. 1968). Its specificity is high; no analogs are active, and the side chain allylic alcohol, as well as the specific steriochemistry of the side chain, is required for activity (MACHLIS 1973, PLATTNER and RAPOPORT 1971). When large numbers of male gametes are present, chemotaxis can be detected at concentrations as low as 10^{-10} M, but is more apparent at higher concentrations with an optimal response at 10^{-6} M. Chemotaxis is diminished at 10^{-5} M and is very weak at 10^{-4} M sirenin (CARLILE and MACHLIS 1965a).

There must be a concentration gradient of sirenin for male gametes to respond chemotactically (CARLILE and MACHLIS 1965a). If sirenin is added to a suspension of male gametes, the chemotactic activity rapidly disappears from the medium (CARLILE and MACHLIS 1965a), although no enzymes are released into the medium (MACHLIS 1973). Apparently, the male gametes take up sirenin, inactivate it (CARLILE and MACHLIS 1965a, MACHLIS 1973) and become fully responsive to a second addition of sirenin only after some time has elapsed (MACHLIS 1973). In the absence of sirenin, sperm exhibit a smooth nonhelical swimming behavior with numerous rapid turns in direction (POMMERVILLE 1978). When exposed to sirenin, sperm swimming behavior changes, and they swim helically to the source of the attractant without undertaking very many rapid directional changes (POMMERVILLE 1978). Male gametes produce a chemical factor which inhibits the chemotactic response to sirenin (KLAPPER and KLAPPER 1977). This factor is a polar heat-stable molecule with an estimated molecular weight between 300 and 400 and has been named keerosin (KLAPPER and KLAPPER 1977).

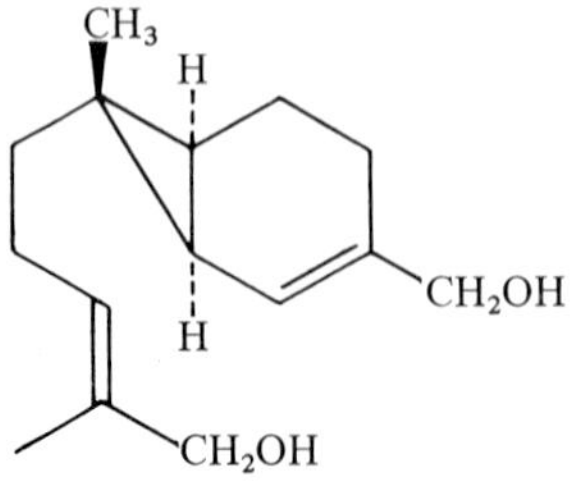

Fig. 14. Sirenin

3.3 Ferns

Sperm of ferns will respond chemotactically to gradients created by the diffusion of sodium malate from the open tip of a small glass capillary tube (PFEFFER 1884, ROTHSCHILD 1952). In a more detailed study, BROKAW (1958) noted that the bimalate form of malate reversibly binds with surface sites at the anterior end of the sperm, apparently is not metabolized, and alters the locomotor activity of the sperm's flagella. At least 10^7 to 10^8 bimalate ions are required to turn the sperm toward the chemotactic source (BROKAW 1958).

4 Factors Regulating Vegetative Development

Algae and fungi also release pheromones or hormone-like substances into the environment, which regulate various aspects of their own vegetative development. Among various algae, these factors control the formation of resistant spores, rhizoid orientation, and filament repair. With the alga *Ulva* inhibitory substances prevent the multicellular vegetative thallus from forming gametes or zoospores. Among the fungi, the aggregation of cellular slime molds and the dimorphism of *Mucor* are regulated by substances produced by the respective organisms and released into the environment.

4.1 Algae

4.1.1 Spore Differentiation

The filamentous blue-green alga, *Cylindrospermum licheniforme* KÜTZ is a simple organism consisting of only three cell types: photosynthetic vegetative cells, nitrogen-fixing heterocysts, and spores resistant to various stressful conditions. The filaments release a substance into their environment which stimulates the differentiation of vegetative cells into spores (FISHER and WOLK 1976). Conditioned medium in which filaments have been grown induce the immediate differentiation of spores. Induction does not result from nutrient depletion, nor does it alter the growth of filaments. Spores can be induced by only 0.3 μM of the substance (HIROSAWA and WOLK 1979). The compound probably consists of two fused five-membered rings, one of which is a lactam, while the other has a thioketone group (Fig. 15) (HIROSAWA and WOLK 1979). In nature, this

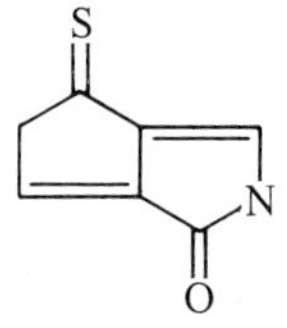

Fig. 15. Spore-inducing substance from a blue-green alga

substance may regulate sporulation during times of stress. During desiccation or stagnation of a pond containing blue-green algae, the substance would increase in concentration and induce the differentiation of resting spores (FISHER and WOLK 1976).

4.1.2 Thallus Stability

Ulva mutabilis FØYN, a marine green alga, exists as isomorphic haploid and diploid plants. When laboratory-grown thalli are transferred to fresh medium, almost all of the cells of the gametophyte plant divide mitotically to form gametes, while the cells of the sporophyte divide meiotically to form zoospores (FÖYN 1959, NILSON and NORDBY 1975, NORDBY 1974, NORDBY and HOXMARK 1972, THIADENS and ZEUTHERN 1967). Both haploid and diploid plants release at least two substances which inhibit these cell divisions (NILSEN and NORDBY 1975). However, meiosis and zoospore formation appear to be more easily inhibited than mitosis and gamete formation. During zoospore production, about 40 h are required for the cells to prepare for meiosis. Inhibitors must be absent for the first 35 h of this time or meiosis and subsequent zoospore formation will not occur, but after 39 h the inhibitors have no effect. The chemical identities of the inhibitors are unknown, but it is apparent that the vegetative multicellular state of *Ulva* can be sustained only if inhibitors of reproductive processes are present in the environment.

4.1.3 Rhizoid Initiation

The zygote of the brown alga *Fucus* is initially radially symmetrical, but after a short period of time, it becomes polarized and forms an anchoring rhizoid. When two or more zygotes are in close proximity, there is a mutual orientation of rhizoids or a group effect (HURD 1920). Slight acidity of the seawater and large numbers of zygotes favor the growth of rhizoids toward each other, while slight alkalinity and low numbers of zygotes favor the growth of rhizoids away from each other (JAFFE and NEUSCHELER 1969, WHITTAKER 1940). These complex group interactions are regulated by two diffusible substances (JAFFE and NEUSCHELER 1969). One substance, rhizin, stimulates rhizoid initiation, while the other, antirhizin, inhibits rhizoid formation. Rhizin appears to be a relatively small molecule which diffuses over greater distances than the larger, more stable antirhizin.

Slightly acid conditions favor a group interaction regulated by rhizin. Rhizin is released from a localized region of the zygote that will become the rhizoid, and is not produced synchronously by all zygotes at the same time. A rapidly differentiating zygote will release rhizin and polarize a more slowly growing neighbor toward it before the neighbor produces its own rhizin. Then the more slowly developing zygote releases rhizin, which causes the elongating rhizoid of the more rapidly developing zygote to grow toward it.

Alkaline conditions enhance an interaction regulated primarily by antirhizin, which is produced by all zygotes at the same time and diffuses uniformly from them. Rhizoidal growth is directed away from areas containing high concentrations of antirhizin where diffusion gradients overlap. Even in slightly acidic conditions, some inhibitory influence of antirhizin is apparent when zygotes are in very close proximity. This inhibition decreases rapidly when the distance between the zygotes is increased.

The chemical nature of rhizin and antirhizin is unknown. However, it has been suggested that rhizin may be an auxin (JAFFE 1968, OLSON and DU BUY 1937), but after careful study, KLEMKE and BENTRUP (1973) have concluded that rhizin is not IAA.

4.1.4 Filament Repair

For a filamentous organism, damage to individual cells can result in its fragmentation. *Griffithsia pacifica* KYLIN, a red alga, is capable of replacing dead cells and restoring the filament integrity (WAALAND and CLELAND 1974). The cell above a dead cell divides to form a repair rhizoid, which elongates through the lumen of the dead cell, and the cell below divides to form a dome-shaped repair-shoot cell. As this cell elongates, it develops the morphology of a rhizoid. The repair rhizoid and repair-shoot cell grow toward each other and fuse to form a new cell which expands to fill the cavity of the dead cell.

Filament repair is regulated by a species-specific substance called rhodomorphin (WAALAND 1975, WAALAND and CLELAND 1974, WAALAND and WATSON 1980). Rhodomorphin is released from the repair rhizoid and regulates not only the induction but the change in morphology of the repair-shoot cell. In the presence of the repair rhizoid, the cell division forming the repair-shoot cell occurs more rapidly than in its absence. If the repair rhizoid is removed after the repair-shoot cell has assumed its rhizoidal morphology, the repair-shoot cell stops growing, expands laterally and organizes a new apical cell. Furthermore, rhodomorphin elicits a chemotropic response from the repair-shoot cell since it bends and grows toward the repair rhizoid prior to the fusion of the two cells.

4.2 Fungi

4.2.1 Aggregation of Amoebae

The cellular slime molds *Dictyostelium* and *Polysphondylium* exist as free-living amoebae in the soil by feeding on bacteria and increase in number by binary fission. When the food becomes exhausted, the amoebae differentiate and chemotactically aggregate (Fig. 16). Multicellular masses called pseudoplasmodia form which subsequently differentiate into stalked fruiting bodies (BONNER 1967).

During preaggregation differentiation, amoebae of *Dictyostelium discoideum* RAPER develop a series of competencies essential for aggregation. Initially, they become competent to respond to a chemotactic gradient after about 4 h (BONNER et al. 1969, GERISCH 1968, ROBERTSON et al. 1972). By 6 h, they are competent to synthesize and relay their own chemotactic signal after first being stimulated

Fig. 16. Aggregation of *Dictyostelium discoideum*. (From TREFFRY and WATTS 1976)

by the chemotactic agent (GINGLE and ROBERTSON 1976, ROBERTSON and DRAGE 1975, SHAFFER 1957, 1962). After 6 to 8 h, about 1% of the amoebae begin to produce autonomously a pulse of the chemotactic substance about every 300 s (KONIJN 1972, KONIJN and RAPER 1961, RAMAN 1976, RAMAN et al. 1976, SHAFFER 1957). Finally, amoebae become competent to make contact and adhere to each other (COHEN and ROBERTSON 1971a, GERISCH 1968, GREGG and NESOM 1973, ROBERTSON et al. 1972, SHAFFER 1957, 1962).

The chemotactic attractant produced by many species of *Dictyostelium* with large fruiting bodies is 3′, 5′-cyclic AMP (Fig. 17) (BARKLEY 1969, KONIJN et al. 1967, 1968, 1969). Since the role of cyclic AMP as a chemotactic attractant in *Dictyostelium* has been studied extensively, more so than any other attractant, a discussion of its synthesis, release, attachment, and subsequent metabolism may have relevance for the future studies of other attractants.

Cyclic AMP is synthesized from ATP by the enzyme, adenyl cyclase, the

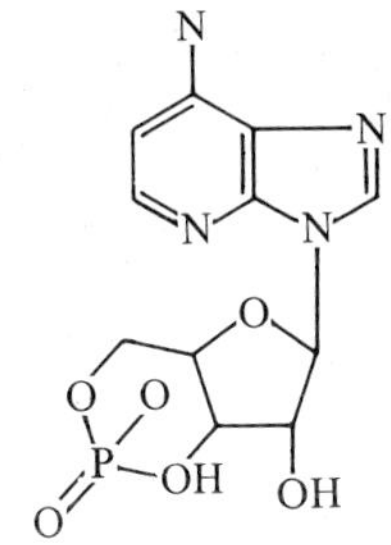

Fig. 17. Adenosine 3′, 5′-monophosphate

specific activity of which increases as amoebae begin to develop aggregative competency (KLEIN 1976, KLEIN and DARMON 1977, ROSSOMANDO and SUSSMAN 1972). However, the enzyme does appear to be stable and associated with the inner portion of the plasma membrane (CUTLER and ROSSOMANDO 1975, FARNHAM 1975, KLEIN 1976, ROSSOMANDO and CUTLER 1975, ROSSOMANDO and SUSSMAN 1972) or with small intercellular vesicles (HINTERMANN and PARISH 1979, MAEDA and GERISCH 1977).

Cyclic AMP is degraded to 5′ AMP by the extracellular enzyme, phosphodiesterase, which is secreted by feeding and dividing amoebae (CHANG 1968, CHASSY 1972). At the onset of differentiation, amoebae secrete a proteinaceous inhibitor which binds tightly to the extracellular phosphodiesterase (GERISCH et al. 1972, RIEDEL and GERISCH 1971, RIEDEL et al. 1972, 1973), and this inhibition of phosphodiesterase can be correlated with the attainment of aggregative competence (GERISCH et al. 1972). The inhibitor may trigger differentiation by contributing to an increase in the external concentration of cyclic AMP (BONNER et al. 1969, MALKINSON and ASHWORTH 1972, 1973, RIEDEL and GERISCH 1971). Cyclic AMP in turn induces aggregation competence of the amoebae (DARMON et al. 1975, KLEIN and DARMON 1975, 1977).

When the concentration of extracellular cyclic AMP increases above a threshold level of 10^{-9} to 10^{-8} M, competent amoebae move toward a natural or artificial source of the chemotactic agent (KONIJN et al. 1968, MATO et al. 1975, ROBERTSON et al. 1972). If a pulsating source of cyclic AMP is present, many amoebae respond to the first pulse and most have responded by the time of the third pulse (ROBERTSON and DRAGE 1975). Apparently, there is no refractory period during which amoebae are chemotactically insensitive to cyclic AMP (ROBERTSON and DRAGE 1975), the reaction being a simple, all or nothing, response and not dependent upon a continuously maintained concentration gradient of cyclic AMP (COHEN and ROBERTSON 1971 b).

Cyclic AMP binds in an oscillatory manner with specific receptor sites on the surface of the amoebae (KING and FRAZIER 1978, MALCHOW and GERISCH 1973, 1974, MATO and KONIJN 1975, MULLENS and NEWELL 1978) and does not have to penetrate through the plasma membrane to exert its chemotactic effect (MOENS and KONIJN 1974). Binding of cyclic AMP is a calcium-dependent, noncovalent interaction which increases during the attainment of aggregation competence so that about 5×10^5 binding sites ultimately form on the surface of each amoeba (MALCHOW and GERISCH 1974). Cyclic AMP is bound to the receptor by three hydrogen bonds at the 3′-oxygen of the ribose and the 6-amino and the 7-nitrogen of the base and possibly by one ionic interaction with the negatively charged phosphate group (MATO et al. 1978). There may also be hydrophobic interactions between the adenine moiety and the binding site. An amoeba is sensitive to a concentration difference of cyclic AMP as low as 3.6×10^{-11} M from one end to the other of the cell, which represents a difference of about 12 occupied receptor sites (MATO et al. 1975). When exposed to cyclic AMP, the amoebae move a distance of about 20 µm toward the source of the chemoattractant at a constant rate for about 100 s (COHEN and ROBERTSON 1971 b).

Cyclic AMP is released from the receptor sites and hydrolyzed to nonbinding 5′ AMP by phosphodiesterase and then to adenosine by a 5′ nucleotidase (DARMON et al. 1978, MALCHOW and GERISCH 1974, MALCHOW et al. 1972). Both enzymes are found externally to the plasma membrane of amoebae (CUTLER and ROSSOMANDO 1975, FARNHAM 1975, MALCHOW et al. 1972, ROSSOMANDO and CUTLER 1975). The activity of membrane-bound phosphodiesterase increases significantly during differentiation, appearing about 1 to 2 h after an increase in cyclic AMP binding capacity (MALCHOW and GERISCH 1974, MALCHOW et al. 1972, MATO and KONIJN 1975, PANNBACKER and BRAVARD 1972) and is induced by cyclic AMP (KLEIN 1975, KLEIN and DARMON 1975, 1977). An aggregating amoeba has enough enzyme to hydrolyze 10^8 molecules of cyclic AMP per min (MALCHOW et al. 1972), and membrane-bound phosphodiesterase is not sensitive to the inhibitor secreted by amoebae at the beginning of differentiation (MALCHOW et al. 1972).

Receptor sites for cyclic AMP are different from the phosphodiesterase catalytic sites, and the cyclic AMP-receptor complex is not a better substrate for phosphodiesterase than is free cyclic AMP (MALCHOW and GERISCH 1973, 1974, MALCHOW et al. 1975). In all likelihood, the receptor site and the membrane-bound phosphodiesterase react independently with free cyclic AMP (MALCHOW and GERISCH 1974). The end result of the binding and hydrolysis of cyclic AMP is a chemotactic response by amoebae, followed by a rapid drop in the concentration of the extracellular cyclic AMP below threshold concentrations until a new pulse of cyclic AMP appears in the environment of the amoebae.

When the extracellular concentration of cyclic AMP increases to 10^{-5} M, amoebae not only respond chemotactically but are stimulated to produce a pulsatile relay signal of cyclic AMP after a delay of about 15 s (COHEN and ROBERTSON 1971 a, DURSTON 1974, GERISCH and WICK 1975, ROBERTSON and DRAGE 1975, SHAFFER 1956, 1957). After release of the signal the amoebae enter a refractory period of about 7 to 10 min, during which the amoebae can no longer be stimulated to secrete a chemotactic signal (COHEN and ROBERTSON 1971 a, DURSTON 1974, GERISCH 1965, 1968, ROBERTSON and DRAGE 1975, SHAFFER 1957, 1962). This refractory period becomes progressively shorter as aggregation proceeds (DURSTON 1974, ROBERTSON and DRAGE 1975). Since amoebae relaying a signal are unable to induce a retrograde signaling among the cells that signaled them, a unidirectional wave propagation is ensured (ROBERTSON and DRAGE 1975).

Chemotaxis, phosphodiesterase activity, and signal induction results in the outward propagation of chemotactic signals and inward waves of cell movement toward the autonomous center of aggregation.

Cyclic AMP is not the chemoattractant for *Dictyostelium lacteum* VAN TIEGHEM and *minutum* RAPER (KAKEBEEKE et al. 1978, MATO et al. 1977). The amoebae of these slime molds are not capable of relaying signals or pulsating movement and form small fruiting bodies (KAKEBEEKE et al. 1978). An attractant from yeast extract is a small, heat-stable species-specific molecule containing glycine which is inactivated by a non-species-specific enzyme (KAKEBEEKE et al. 1978, MATO et al. 1977). It may be a small peptide with an aromatic moiety bound to the amino terminal end of the molecule (MATO et al. 1977). This substance may not be the natural attractant of these species of *Dictyostelium*.

Cross-attraction experiments between amoebae of the slime mold genera *Dictyostelium* and *Polysphondylium* indicate that amoebae of *Polysphondylium* are not attracted to cyclic AMP and do not bind it (KONIJN et al. 1969, MATO and KONIJN 1975, RAPER and THOM 1941, SHAFFER 1958). During differentiation, amoeba of this slime mold become increasingly sensitive to the attractant (FRANCIS 1965), and aggregation centers are initiated by autonomous founder cells which release a chemotactic agent attracting amoebae in the immediate vicinity (FRANCIS 1965, SHAFFER 1961). It is not clear whether amoebae other than founder cells release the chemotactic agent, but no pulses of cellular movement have been observed (FRANCIS 1965). The chemotactic substance of *Polysphondylium violaceum* BREFELD has a molecular weight of less than 1500 and is heat-stable (WURSTER et al. 1976). It might be a peptide with a blocked terminal amino group and is inactivated by a specific enzyme (WURSTER et al. 1976).

4.2.2 Morphological Transformation

The fungi *Mucor racemosus* FRESENIUS and *Mucor rouxii* (CALMETTE) WEHMER can exist either as single cells which multiply by budding or as coenocytic hyphae with extensive branching. When *Mucor* is grown anaerobically under nitrogen or argon in a nutrient medium, both growth forms of the fungus occur (MOONEY and SYPHERD 1976). If gas flow rates over or through the medium are low, mycelia develop. When gas flow rates are more rapid, the fungi grow as single cells but are rapidly transformed into hyphae when gas flow rates are decreased. However, the transition from mycelia back to single cells is more complex and accomplished after a greater length of time. The mycelial growth form is regulated by one or more volatile compounds which are produced by the fungus in aerobic growth conditions. The volatile regulators can be collected from one culture and can affect the morphology of fungi growing in other cultures.

5 Chemotactic Agents

Substances released from the roots of legumes chemotactically attract nitrogen-fixing bacteria to the site of nodule formation. Small molecules released into the environment also may play an important role in directing fungal plasmodia, hyphae, or zoospores toward new areas where food supplies are adequate or the probability of pathogenic infection is enhanced. For this reason, these factors strongly influence not only the distribution but also the growth processes of many fungi; without these chemotactic substances, the fungi would either become quiescent and die, or encyst and enter dormancy.

5.1 Attraction of Nitrogen-Fixing Bacteria

Nitrogen-fixing bacteria of the genus *Rhizobium* are attracted to the root exudate of various legumes (CURRIER and STROBEL 1976). For *Lotus corniculatus* L. the attractant is a glycoprotein with an approximate molecular weight of 65,000 and has been given the name of chemotactin (CURRIER and STROBEL 1977, 1981). The purified substance will attract rhizobia at concentrations of 10^{-16} to 10^{-6} M with maximum attraction occurring at 10^{-14} M (CURRIER and STROBEL 1981). The carbohydrate portion of the molecule is 38% of the molecule, contains glucose, galactose, and mannose and is linked to serine and threonine of the peptide (CURRIER and STROBEL 1977, 1981). Forty-three of 109 residues of the peptide are hydrophobic amino acids (CURRIER and STROBEL 1981).

Rhizobium is also attracted to simple sugars (CURRIER 1980). Mannose, ribose, and maltose are strong attractants; fructose, galactose, and sucrose are weak ones; and glucose, rhamnose, arabinose, xylose, and deoxyribose do not attract the bacterium.

5.2 Attraction of Plasmodia

The syncytial plasmodium of *Physarum polycephalum* SCHWEINITZ is attracted chemotactically toward a source of the sugars, fucose, glucose, lactose, maltose mannitol, or galactose; the amino acids, leucine, serine, or phenylalanine, and the nucleotide uracil (CARLILE 1976, CHET et al. 1977, KNOWLES and CARLILE 1978). Most of the carbohydrates which attract the plasmodium also support its growth (KINCAID and MANSOUR 1978, KNOWLES and CARLILE 1978). However, fructose which can be metabolized will not attract or will only weakly attract *Physarum*, and 2-deoxyglucose which cannot be metabolized will attract the plasmodium (KINCAID and MANSOUR 1978, KNOWLES and CARLILE 1978). Attraction to a compound depends on a dynamic conformational change of the proteins of the plasma membrane (ISHIDA et al. 1978) and an increase in the level of calcium in the cytosol of the plasmodium (ISHIDA et al. 1979).

5.3 Attraction of Zygotes, Mitospores, and Meiospores

After fusion of the male and female gametes of the filamentous water mold *Allomyces*, the biflagellate zygote swims around, finally coming to rest, where it then develops into a sporophytic plant with two kinds of sporangia. Colorless sporangia release motile mitospores that develop into new sporophyte plants, whereas brown thick-walled sporangia release meiospores which develop into new gametophyte plants. Zygotes, mitospores, and meiospores are all attracted to a mixture of amino acids found in casein hydrolysate but are not attracted to sirenin. Conversely, the gametes do not respond to casein hydrolysate (CARLILE and MACHLIS 1965b). When individual amino acids are tested, zygotes exhibit a strong chemotactic reaction toward the amino acids, proline and cysteine (CARLILE and MACHLIS 1965b). Mitospores or meiospores do not respond to a diffusion gradient of individual amino acids, but rather to a gradient of the amino acid pair leucine and lysine (MACHLIS 1969). This chemotropic response to the two amino acids can be enhanced further by the presence of proline, which by itself has no effect (MACHLIS 1969).

Hyphae of another water mold, *Saprolegnia,* will also grow toward a high concentration of a mixture of amino acids (FISHER and WERNER 1955). The mixture must contain at least five to seven amino acids for the chemotactic reaction to occur; when tested individually, the amino acids are inactive.

Asexual zoospores of the pathogenic fungi *Pythium* and *Phytophthora* are attracted to exudate from intact or wounded plant roots, become encysted on the root near the source of the exudate, and subsequently germinate and infect the host (CUNNINGHAM and HAGEDORN 1972, DUKES and APPLE 1961, HICKMAN 1970, ROYLE and HICKMAN 1964a, ZENTMYER 1961, 1970). The exudates and root extracts have been fractionated into various components and tested for chemotactic activity. Zoospores are attracted to sugars, organic acids, and amino acids especially glutamic and aspartic acids (BIMPONG and CLERK 1970, CARLILE 1966, CHANG-HO and HICKMAN 1970, HICKMAN and HO 1966, KHEW and ZENTMYER 1973, RAI and STROBEL 1966, ROYLE and HICKMAN 1964b, ZENTMYER 1966). However, care must be taken to differentiate between a true chemotactic reaction and simple entrapment of zoospores, since various vitamins, nucleotides, organic acids, and growth regulators will cause the accumulation of zoospores by entrapment without evoking directional movement of the zoospores (HALSALL 1976, KHEW and ZENTMYER 1973).

KHEW and ZENTMYER (1973) found that attraction of zoospores to glutamic and aspartic acids increases with increasing concentration of these amino acids, but varies with different species of *Phytophthora*. Characteristically, the zoospores possess a negative charge and amino acids must have a positive charge for maximal chemotactic activity (KHEW and ZENTMYER 1974). The chemoreceptors of the zoospores are nonspecific for the two amino acids, and metabolic inhibitors as well as surface-active agents do not prevent chemotaxis (KHEW and ZENTMYER 1973).

With Australian isolates of *Phytophthora,* the amino acids with neutral charges, glutamine and asparagine are active chemotactic agents (HALSALL 1976). Compounds structurally related to these amino acids do not attract zoospores, and both amino and amide groups are necessary for activity. Other compounds can interact in the absence of amino acids to increase the complexity of the chemotactic reaction (HALSALL 1976). A gradient of sugars or ethanol will attract zoospores (ALLEN and NEWHOOK 1973, HALSALL 1976), and ethanol or sucrose will act synergistically with amino acids to increase the chemotactic attraction of zoospores (HALSALL 1976).

6 Factors Regulating Fungal Morphogenesis

Factors released into the environment by higher plants as well as nematode worms can affect profoundly the morphogenesis of certain fungi. Rust fungi are induced to form infection structures by compounds released from the leaves of host plants. Nematode-trapping fungi are induced to form traps by substances released by the nematode prey. Since these regulating substances are produced by organisms unrelated to the fungi, they are not pheromones.

6.1 Induction of Infection Structures

Rust fungi, *Puccinia graminis* PERS., form the infection structures, appressoria, and substomatal vesicles prior to the formation of intercellular hyphae and haustoria in the leaf of the host plant. Differentiation of these structures is regulated by volatile and nonvolatile factors in the environment at the entrance of stomata (GRAMBOW and REISENER 1976, GRAMBOW and RIEDEL 1977). In the presence of an uncharacterized fraction of leaf extract, volatile hexanols as well as 6-methyl-5-hepten-2-1,2-methyl-2-hexen, tert-butanol and chloroform will induce rust fungi to form appressoria (GRAMBOW 1977).

6.2 Nematode Trap Formation

The fungus *Arthrobotrys oligospora* FRES. is normally saprophytic but can become predacious on nematodes by differentiating special trapping hyphae (PRAMER 1964). Trap formation is induced by a soluble substance from nematodes called nemin which has the characteristics of a peptide (PRAMER and STOLL 1959), but appears to be more than one specific substance (PRAMER and KUYAMA 1963).

When individual amino acids are added to the fungal culture, only valine, isoleucine, and leucine will induce trap formation (NORDBRING-HERTZ 1973, WOOTTON and PRAMER 1966), although a large number of synthetic dipeptides containing valine are even more active (NORDBRING-HERTZ 1973). Since both valine-phenylalanine and phenylalanine-valine dipeptides have an equal effect on trap formation, the terminal amino position of valine is not required for activity. When the valyl peptides are increased in length, the inductive capacity is enhanced further.

Peptides isolated from enzyme-hydrolyzed casein can also induce trap formation (NORDBRING-HERTZ 1973, NORDBRING-HERTZ and BRINCK 1974). The active peptides are small with only two to seven amino acids, with nonpolar and aromatic amino acids being essential for trap induction. Single active peptides were difficult to isolate, but all active peptide preparations contained leucine, isoleucine, valine, proline, and tyrosine, and some contained phenylalanine, glycine, and alanine. The amino terminal of the active peptides can be occupied by tryosine, valine, or phenylalanine, and the carboxyl terminal by leucine or isoleucine. Thus, many different peptides can induce fungal trap formation.

7 Factors Regulating the Growth of Higher Plants

Many factors found within the environment are well-known regulators of the growth and development of higher plants. These hormonal substances originate from diverse sources. Bacteria and fungi from the soil rhizosphere or root surfaces, as well as bacteria on the surface of leaves, are capable of synthesizing and releasing hormones which affect the growth of the closely associated higher plants. Hormones can even be leached from leaves or released from roots of higher plants, and thereby alter the growth of other individuals in the plant community.

7.1 Bacteria

7.1.1 Inhabitants of Soil

Microorganisms from the soil, rhizosphere, and root surfaces, when isolated and grown in axenic culture, produce regulators of growth and differentiation in higher plants. These bacteria can synthesize IAA and other auxin-like compounds (KATZNELSON and SIROIS 1961), GA-like substances (GAIROLA et al. 1972, KATZNELSON and COLE 1965, KATZNELSON et al. 1962, VANČURA 1961), and cytokinins (BLONDEAU 1970, KAMPERT and STRZELCZYK 1978, VAN STADEN and DIMALLA 1976).

Bacteria from rhizosphere soils and the surface of roots of maize (*Zea mays* L.) and wheat (*Triticum aestivum* L.) significantly enhance the growth of those plants in sterile soils after bacterial inoculation (BROWN 1972, HUSSAIN and VANCŮRA 1970). For maize, supernatants from various bacterial cultures enhance seed germination, primary root growth and lateral root production by plants grown in sterile conditions (HUSSAIN and VANCŮRA 1970). Bacterial strains produce both IAA- and GA-like substances, one of these having a chromatographic mobility similar to GA_3 (HUSSAIN and VANČURA 1970). For wheat, seedlings grown aseptically in soil treated with GA_3 and IAA or bacteria develop in a similar manner but differ morphologically from those plants grown only under sterile conditions (BROWN 1972). These bacteria also produce IAA- and GA-like substances in culture (BROWN 1972).

Rhizosphere bacteria which solubilize phosphate also are capable of synthesizing IAA-, GA-like substances, and cytokinins (BAREA et al. 1976). The bacteria belong to several genera, and growth regulator production is not specific to any particular genus. BAREA et al. (1976) have suggested that increased plant growth caused by the bacteria may be due to the growth regulators synthesized by these bacteria as well as the solubilized phosphate.

Bacteria of the genus *Azotobacter* promote the growth of tomatoes (AZCÓN and BAREA 1975, BAREA and BROWN 1974, BROWN and BURLINGHAM 1968, JACKSON et al. 1964). Inoculation of seed or roots of young tomato seedlings with *Azotobacter* accelerates early stem and leaf growth, shortens the time between the appearance of flower buds and petal fall and causes an increase in fruit yield. Similar responses are obtained with supernatants from cultures of *Azotobacter* which synthesize and release auxins, cytokinins, and GA-like substances, the latter appearing to be the primary substance controlling the enhanced growth of tomato.

Cultured strains of bacteria isolated from the seedling root zone of Scots pine (*Pinus sylvestris* L.) produce IAA and other auxins when tryptophan is added to the medium (KAMPERT et al. 1975). When this amino acid is excluded from the medium, only trace amounts of auxins are synthesized by the bacteria. Generally, bacteria isolated from the root surfaces produce more auxin than those isolated at random from the soil. However, soil did contain more strains of bacteria capable of independently synthesizing auxin in the absence of tryptophan.

The interaction of specific soil bacteria and legumes results in a morphological alteration of the root and the formation of a nodule. Young plants of clover, *Trifolium subterraneum* L., release tryptophan, which is converted to IAA by the nodulating bacterium, *Rhizobium* (KEFFORD et al. 1960). Auxin may play a role during the initiation of nodulation by causing the curling of root hairs around the infecting bacteria (BRIAN 1957, NUTMAN 1956). However, strains of *Rhizobium* which infect soybean do not convert tryptophan to IAA, and it has not yet been determined whether auxin plays a critical role during the nodulation of soybean roots (KEFFORD et al. 1960).

In addition to producing substances which promote the growth of higher plants, some bacterial strains also produce plant growth inhibitors (BAREA et al. 1976, BROWN 1972, HUSSAIN and VANČURA 1970).

Even though bacterial strains isolated from roots and the rhizosphere produce plant growth regulators under ideal culture conditions, little is known about the production of these regulators in more natural field settings. Since roots are covered with microorganisms, it is assumed that the growth regulators produced by them are absorbed by the roots and subsequently alter the growth of the plant. However, this assumption may be an oversimplification. BROWN (1972) has shown that the population structure of bacteria in the root region of a wheat plant changes as the plant matures, whereas the population structure of bacteria from the soil remains unchanged. As plants age and root microflora change, the number of bacterial isolates utilizing exogenous tryptophan for IAA synthesis decreases. Also, bacteria isolated from the surface of young wheat roots produce more inhibitory substances to the growth of wheat plants than those from older plant roots. It is thus apparent that different populations of bacteria could be expected to have quite different effects on the growth of higher plants.

Soil bacteria not in the rhizosphere may play only a minor role in regulating plant growth through the production of plant growth substances. Gibberellin is rapidly inactivated when added to soil (KATZNELSON and COLE 1965), and a number of strains of bacteria can decompose auxin (BAREA et al. 1976, STRZELCZYK et al. 1973). Thus, to avoid rapid degradation the growth regulator would have to be synthesized by bacteria near the root surface and be taken up quickly by the plant.

7.1.2 Epiphytes on Leaves

Epiphytic bacteria also can significantly influence the growth of higher plants by producing auxin (BANERJEE and CHANDRA 1978, MANTEUFFEL et al. 1973). Forty-two strains of bacteria have been isolated from the aerial portions of the pea (*Pisum sativum*), 26 of which produce IAA (LIBBERT and RISCH 1969). With maize (*Zea mays*) 29 epiphytic bacterial strains have been isolated, of which 18 could synthesize IAA (MANTEUFFEL et al. 1972). These epiphytic bacteria probably have a profound influence on the auxin metabolism of plants. In fact, those plants with epiphytic bacteria have much more extractable and actively transported auxin than those without epiphytes (LIBBERT et al. 1969, LIBBERT and MANTEUFFEL 1970, MANTEUFFEL et al. 1972). When sterile plants are inoculated with epiphytic bacteria, the amount of auxin increases to levels found in the nonsterile plants. When bacteria fed on radioactive tryptophan are applied to the tips of sterile maize coleoptiles, much more radioactive auxin is transported to the base than by sterile coleoptiles just treated with radioactive tryptophan (LIBBERT and SILHENGST 1970). It has been shown that corn plants exude amino acids including tryptophan and sugars (KUNERT and LIBBERT 1972), which are utilized by the bacteria for growth (MANTEUFFEL et al. 1973b). The bacteria not only utilize tryptophan to synthesize IAA, but can even synthesize auxin when given an artificial exudate lacking tryptophan (MANTEUFFEL et al. 1973b).

7.2 Fungi

Soil fungi also produce auxin and GA-like substances (AUBE and SAKSTON 1965, CHAILAKHYAN et al. 1958, CURTIS 1957, KATZNELSON and COLE 1965, PURUSHOTHAMAN et al. 1974). YOUSSEF (1974) has isolated and grown in liquid culture rhizosphere fungi from the broad bean (*Vicia faba*) and cotton (*Gossypium barbadense*). When irrigated with fungal filtrates, the growth of plants in sterile sand is appreciably stimulated relative to plants irrigated with just pure medium (YOUSSEF 1974), presumably due to the growth regulators in the fungal filtrate (YOUSSEF and MANKARIOS 1975).

Many soil fungi enter into a symbiotic relationship with roots of higher plants to form mycorrhizae. The mycelia of ectomycorrhizae which do not penetrate the roots of the higher plant have been shown to produce auxins (HORAK 1964, MOSER 1959, STRZELCZYK et al. 1977, ULRICH 1960), GA-like substances (STRZELCZYK et al. 1975) and cytokinins (CRAFTS and MILLER 1967, KAMPERT and STRZELCZYK 1980, MILLER 1967). Ectomycorrhizal roots of conifers lack root hairs and are usually short, radially expanded and dichotomously branched. When the lateral roots of several species of *Pinus* are treated with IAA, they are transformed morphologically into structures which clearly resemble mycorrhizae (SLANKIS 1973), but this auxin effect may not be universal for all *Pinus* species, since a cytokinin induces dichotomous root formation in *Pinus serotina* MICHX. and not IAA (BARNES and NAYLOR 1959a, b).

Ethylene gas is also found in soil and litter (DASILVA et al. 1974, LILL 1979, LILL and McWHA 1976, LILL et al. 1979, LYNCH 1975, SMITH and RUSSELL 1969). Since no ethylene evolves from sterilized soil, the production of the gas is dependent on biological processes (SMITH and RESTALL 1971). Substantially more ethylene evolves from anaerobic soils than from aerobic soils, the concentration often reaching 10 to 20 ppm or more in waterlogged or compacted soils (SMITH and COOK 1974, SMITH and RESTALL 1971, SMITH and RUSSELL 1969), more than enough to inhibit the elongation of intact roots (CROSSETT and CAMBELL 1975, SMITH and ROBERTSON 1971, SMITH and RUSSELL 1969).

Ethylene is produced primarily by spore-forming bacteria in anaerobic microsites (SMITH and COOK 1974), although a common soil fungus, *Mucor hiemalis* WEHMER and two yeasts have been isolated which produce ethylene in pure culture from methionine or ethionine (LYNCH 1972, LYNCH and HARPER 1980). In culture, ethylene production by *Mucor* is enhanced by oxygen (LYNCH and HARPER 1974a), although it has been suggested that aerobic soils may have inadequate substrates for ethylene production (LYNCH and HARPER 1974a). Ethylene production by fungi also is favored by low growth rates, a situation analogous to that found in more natural conditions (LYNCH and HARPER 1974b).

The concentration of ethylene in the soil can be altered by other factors. Ethylene is metabolized by a number of organisms in aerobic soils (ABELES et al. 1971, CORNFORTH 1975), and there is a volatile substance within soils which inhibits ethylene evolution (SMITH and RESTALL 1971).

7.3 Higher Plants

Roots of higher plants may release growth regulators into the soil. The cytokinins zeatin and zeatin riboside are released from the roots of maize (*Zea mays*) (VAN STADEN 1976) whereas the roots of tomatoes (*Lycopersicon esculentum*) in waterlogged soils produce ethylene gas, some of which escapes into the rhizosphere (JACKSON and CAMPBELL 1975, JACKSON et al. 1978, also see REID and WAMPLE, Chap. 14, this Vol.). Hormones may also be leached from the aerial portions of higher plants into the soil. Gibberellin-like substances are leached from leaves and stems of slow-growing *Chrysanthemum* plants but not from rapidly growing ones (KOZEL and TUKEY 1968). The growth inhibitor abscisic acid can also be leached from the leaves of *Euonymus* (HEMPHILL and TUKEY 1973, 1975, IL and TUKEY 1972).

8 Allelopathic Substances

Vascular plants release diverse substances into the environment (TUKEY 1970). Some of these compounds have inhibitory or allelopathic effects on the germination and growth of other plants (HORSLEY 1977, RICE 1974, 1977).

Not all of the inhibitory effects on plant growth result from interactions with hormones or hormone-mediated processes. However, certain allelopathic compounds (phenolics) may alter the IAA concentrations of a plant by stimulating or inhibiting the activity of IAA oxidase (ANDREAE 1952, TOMASZEWSKI and THIMANN 1966, ZENK and MÜLLER 1963). Other allelopathic compounds inhibit GA-induced growth of plants, and this inhibition can be overcome by exogenously supplied GA_3 (CORCORAN et al. 1972).

8.1 Ferns

8.1.1 Inhibitors from Leaves

The leaves of the fern, *Pteridium aquilinum* (L.) KUHN, contain phytotoxic compounds (GLASS and BOHM 1969) which enter the rhizosphere by leaching or decomposition (GLIESSMAN and MULLER 1972, 1978). Soils around *Pteridium* contain phenolic compounds at concentrations ranging from 0.4 to 4.9×10^{-5} M, (WHITEHEAD 1964) and the growth of barley (*Hordeum vulgare*) is inhibited when they are grown in a solution containing phenolics at concentrations approximating those found in the soil (GLASS 1976).

8.1.2 Inhibitors of Gametophyte Cell Division from Roots

With the fern *Thelypteris normalis* (C. CHR.) MOXLEY gametophytes are not found around established sporophyte plants even though spore production is prolific. This conspicuous absence of gametophytes is caused by inhibitors released from the roots (DAVIDONIS and RUDDAT 1973). Two of these, thelypterin A and B, have been isolated, the former being tentatively identified as an indole derivative (DAVIDONIS 1972). Thelypterin A and B are not species-specific and will inhibit gametophytes of other ferns (DAVIDONIS and RUDDAT 1973). They do not affect spore germination, but subsequent cell divisions within the gametophyte are inhibited to such a degree that the gametophyte remains as a single cell (DAVIDSON and RUDDAT 1974).

8.2 Tree Leaves and Litter

8.2.1 Conifers

Unequivocal studies of allelopathy in conifers are few. LEE and MONSI (1963) observed that herbaceous vegetation beneath the red pine, *Pinus densiflora* SIEB. & ZUCC, of Korea and Japan is very sparse and reproductively depauperate. The forest soil contains p-coumaric acid and a tannin which inhibit the growth of those plants not normally found in the forest but not of those characteristic of red pine forests. These inhibitors presumably leach from the needles and litter of the trees into the soil. JAMESON (1966, 1968) found that the litter from *Pinus edulis* ENGELM. and *Juniperus osteosperma* (TORR.) LITTLE growing in the arid southwestern United States inhibits the growth of the blue gramma grass, *Bouteloua gracilis* (H.B.K.) LAG. The soil beneath the litter contains inhibitory phenolic substances as well (JAMESON 1970). Other species of *Pinus* as well as *Abies, Picea, Thuja,* and *Taxus* are suspected of having allelopathic effects on herbaceous undergrowth (DEL MORAL and CATES 1971).

8.2.2 Angiosperms

Eucalyptus camaldulensis DEHNHARDT leaves produce large quantities of volatile terpenes which are absorbed by dry soil during the summer and inhibit the germination and growth of herbaceous plants during the subsequent rainy season (DEL MORAL and MULLER 1970). Leaf litter from this tree may also have allelopathic effects on herbaceous vegetation (DEL MORAL and MULLER 1970). An aqueous extract of the litter contains ten phytotoxins, five of which have been identified as phenolic compounds (DEL MORAL and MULLER 1970).

Several trees also produce nonvolatile allelopathic compounds which leach from leaves or litter into the soil, thereby preventing the growth of the understory vegetation. The inhibitory substances produced by leaves of *Eucalyptus globulus* LABILL., *Platanus occidentalis* L., *Celtis laevigata* WILLD., *Celtis occidentalis* L., *Quercus borealis* MICHX. and *Quercus alba* L. have been characterized as tannins, coumarins, phenolic, benzoic acid and cinnamic acid derivatives (AL-NAIB and RICE 1971, DEL MORAL and MULLER 1969, LODHI 1975, 1976, 1978, LODHI and RICE 1971).

8.3 Woody Shrubs

Leaves of several species of *Salvia,* as well as many other shrubs of the Californian chaparral community, produce volatile compounds which inhibit the germination and growth of native annual grasses as well as the laboratory test plant, the garden cucumber, *Cucumis sativus* (MULLER 1966, MULLER et al. 1964). Other portions of the *Salvia* plant have little effect on the growth of cucumber (MULLER et al. 1964). Air around *Salvia* leaves contains two terpenes, cineole and camphor (MULLER 1965), both of which severely inhibit the root growth of cucumber and onion (LORBER and MULLER 1980a, MULLER and MULLER 1964). The volatile terpenes apparently either interact directly with the aboveground portions of sensitive plants or inhibit germination and root growth after being absorbed by the soil (MULLER 1965, MULLER and DEL MORAL 1966). The terpenes suppress cell division and elongation, cause excessive cutin deposition by epidermal cells, and inhibit the differentiation of vascular tissue of cucumber (MULLER and HAUGE 1967). Cells exposed to the terpenes exhibit severe ultrastructural damage, a reduction in the number of organelles, and an increase in the number of lipid bodies, presumably as a result of organelle degradation (LORBER and MULLER 1976). Physiologically, the terpenes inhibit O_2 uptake, interfere with the citric acid cycle and decrease the permeability of cell membranes (LORBER and MULLER 1980b, MULLER et al. 1968, 1969).

Various species of the shrub, *Artemisia,* also produce volatile terpenes which inhibit growth of herbaceous plants (MCCOHEN et al. 1973, MULLER et al. 1964).

Furthermore, *Artemisia* produces nonvolatile sesquiterpene lactones and a coumarin, which inhibit growth and increase the respiration of cucumber (McCOHEN et al. 1973).

Leaves of the Californian chaparral plants, *Adenostoma fasciculatum* H. & A. and *Arctostaphylos glandulosa* EASTW. produce nonvolatile allelopathic compounds which leach from the surface of leaves by rain or fog, dripping into the soil where they inhibit germination and early growth of many herbaceous plants (CHOU and MULLER 1972, MCPHERSON and MULLER 1969). Nine inhibitors from the leachate of *Adenostoma* leaves have been identified as phenolic compounds (MCPHERSON et al. 1971), and *Arctostaphylos* produces benzoid acid derivatives and tannins (CHOU and MULLER 1972).

8.4 Herbaceous Angiosperm Leaves

Herbaceous plant leaves are also known to produce allelopathic substances. The leachate from the leaves of the sunflower, *Helianthus annuus,* contain substances including a suspected α-naphthol derivative and scopolin, which not only selectively inhibit the germination of a number of associated plants but also the germination and growth of sunflower seedlings (WILSON and RICE 1968).

8.5 Woody and Herbaceous Angiosperm Roots

Roots of many plants release allelopathic compounds into the soil. Walnut (*Juglans nigra* and *J. cinerea*) roots release juglone (5-hydroxy-α-naphthaquinone) (DAVIS 1928) which causes wilting and death of nearby alfalfa and tomato plants (MASSEY 1925). Roots of guayule, *Parthenium argentatum* GRAY release cinnamic acid which severely inhibits the growth of guayule seedlings (BONNER and GALSTON 1944) at concentrations as low as 10 mg l^{-1} in the soil (BONNER 1946). The exudate from roots of sunflower, *Helianthus annuus* contains chlorogenic and isochlorogenic acids which inhibit the growth of some associated herbaceous plants as well as sunflower seedlings (WILSON and RICE 1968). Ragweed, *Ambrosia artemisiifolia,* roots also produce phenolics which inhibit the germination and growth of roots and shoots of their own seedlings (JACKSON and WILLEMSEN 1976).

8.6 Monocot Roots

Rhizomes and/or roots of several grasses release allelopathic substances into the soil which inhibit seed germination. Johnson-grass, *Sorghum halepense,* produces chlorogenic acid, p-coumaric acid, HCN and p-hydroxybenzaldehyde (ABDUL-WAHAB and RICE 1967). Chlorogenic, isochlorogenic and sulfosalicylic acids are produced by crabgrass, *Digitaria sanguinalis* (L.) SCOP. (PARENTI and RICE 1969), and dropseed or rush grass, *Sporobolus pyramidatus* (LAM.) HITCHC., probably exudes p-coumaric and ferulic acids (RASMUSSEN and RICE 1971).

8.7 Seeds of Angiosperms

Seeds of the cow-parsnip, *Heracleum laciniatum* HORN, release allelopathic furanocoumarins which inhibit the germination and root growth of lettuce (*Lactuca sativa* L.) and the willow (*Salix pentandra*) but have no effect on the germination of spruce seeds (*Picea abies*) (JUNTTILA 1975, 1976).

9 Conclusions

9.1 Sexual Reactions

Factors released into the environment play an essential regulatory role during the sexual reactions of many algae, fungi and ferns. With the alga, *Volvox* male colonies produce a glycoprotein which not only induces other male colonies to produce sperm but causes female colonies to form eggs. With the cellular slime mold *Dictyostelium* one mating strain produces a volatile substance (as yet unidentified) which induces the formation of special reproductive structures called macrocysts by the opposite mating strain. The sexual reaction of *Achlya* is coordinated by several steroids. First, female hyphae release antheridiol, which induces and attracts male antheridial filaments. These antheridial filaments produce in turn a group of steroids, which induce the formation of the female sex organs, oogonia. Finally, the antheridia release a substance which controls the final maturation of the oogonium. For *Mucor* and *Blakeslea,* each mating strain releases specific progamones into the environment, which are converted to trisporic acids by the opposite strain. The trisporic acid induces the formation of progametangia, which are chemotactically attracted to each other and fuse to form zygospores. The ascomycete, *Ascobolus,* has a sexual reaction which is similar to that of *Achlya,* the female strain releasing a substance which induces the formation of antheridia by the male strain. The antheridia in turn induce the formation of ascogonial hyphae which are attracted chemotactically to the antheridia. With the yeast *Saccharomyces* mating types release peptide factors which inhibit DNA synthesis and asexual budding in the opposite strain and regulate cellular agglutination. Upon contact, the cells elongate to form conjugation tubes. This elongation is controlled by peptide as well as by steroidal factors. Conjugation tube formation by the basidiomycetes *Tremella* and *Rhodosporidium* are also controlled by a peptide factor released into the environment. With ferns, more rapidly growing prothalli release a gibberellin-like compound, antheridiogen, which induces the formation of antheridia by more slowly growing prothalli.

Factors found in the environment also serve as chemotactic sex attractants. Various substances produced by female sex organs or gametes of several algae, fungi, and ferns are capable of altering the behavior of androspores or male gametes in such a way that they swim toward the female sex organs or gametes.

9.2 Vegetative Growth and Development

Various aspects of vegetative development of many plants are controlled by growth regulators in the environment. Among the algae, such factors control the formation of resistant spores, rhizoid orientation, and filament repair. With one alga, *Ulva,* certain factors maintain the multicellular thallus by inhibiting gamete and zoospore formation. With the cellular slime molds, factors such as cyclic AMP regulate the aggregation of individual amoeboid cells into multicellular structures prior to sporulation. Cellular dimorphism of the fungus *Mucor* is regulated by volatile substances in the environment. Environmental substances such as glycoproteins and amino acids attract nitrogen-fixing bacteria and asexual propagules of fungi toward new areas of nodulation, colonization, or pathogenic infection. Other substances released from the leaves of higher plants control the morphogenesis of pathogenic rust fungus, *Puccinia.* A few fungi are predacious on nematodes and special hyphal traps are induced by peptides released by the nemotode prey.

Microflora found in the rhizosphere and on roots produce auxin, gibberellin-like compounds, cytokinins, and ethylene which influence the growth and development of higher plants. Epiphytic bacteria on the surface of leaves can modify the hormonal metabolism of higher plants by synthesizing and releasing auxin.

Finally a large number of allelopathic substances exuded or leached from leaves, litter, roots, and seeds of higher plants are environmental factors which affect other plants in the immediate vicinity by inhibiting germination and growth. In many cases, these allelopathic substances act by inhibiting hormonally controlled growth processes, although it is not clear whether these substances actually inhibit plant-growth hormones.

References

Abdul-Wahab AS, Rice EL (1967) Plant inhibition by Johnson grass and its possible significance in old-field succession. Bull Torrey Bot Club 94:486–497

Abe K, Kusaka I, Fukui S (1975) Morphological change in the early stages of the mating process of *Rhodosporidium toruloides*. J Bacteriol 122:710–718

Abeles FB, Craker LE, Forrence LE, Leather GR (1971) Fate of air pollutants: removal of ethylene, sulfur dioxide and nitrogen dioxide by soil. Science 173:914–916

Allen RN, Newhook FJ (1973) Chemotaxis of zoospores of *Phytophthora cinnamomi* to ethanol in capillaries of soil pore dimensions. Trans Brit Mycol Soc 61:287–302

Al-Naib FA, Rice EL (1971) Allelopathic effect of *Platanus occidentalis*. Bull Torrey Bot Club 98:75–82

Andreae WA (1952) Effect of scopoletin on indoleacetic acid metabolism. Nature 170:83–84

Aoyagi H, Izumiya N, Sakurai A, Sakata K, Tamura S (1977) Synthetic identification as a hexapeptide of α substance -I_B inducing sexual agglutination in *Saccharomyces cerevisiae*. Experientia 339:870

Arsenault GP, Biemann K, Barksdale AW, McMorris TC (1968) The structure of antheridiol, a sex hormone in *Achyla bisexualis*. J Am Chem Soc 90:5635–5636

Aube C, Sakston WE (1965) Distribution and prevalence of *Verticillum* species producing substances with gibberellin-like properties. Can J Botany 43:1335–1342

Austin DJ, Bu'Lock JD, Winstanley DJ (1969) Trisporic acid biosynthesis and carotenogenesis in *Blakeslea trispora*. Biochem J 113:34

Austin DJ, Bu'Lock JD, Drake D (1970) The biosynthesis of trisporic acids from β-carotene via retinol and trisporal. Experientia 36:347–349

Azcón R, Barea JM (1975) Synthesis of auxins, gibberellins and cytokinins by *Azotobacter vinelandii* and *Azotobacter beijerinckii* related to effects produced on tomato plants. Plant Soil 43:609–619

Banbury GH (1954) Processes controlling zygospore formation and zygotropism in *Mucor mucedo* Brefeld. Nature 173:499–500

Bandoni RJ (1965) Secondary control of conjugation in *Tremella mesenterica*. Can J Botany 43:627–630

Banerjee M, Chandra AK (1978) Auxin production potentiality of nitrogen fixers isolated from the phyllosphere of crop plants. Curr Sci 47:962–963

Barea JM, Brown ME (1974) Effects on plant growth produced by *Azotobacter paspali* related to synthesis of plant-growth-regulating substances. J Appl Bacteriol 37:583–593

Barea JM, Navarro E, Montoya E (1976) Production on plant growth regulators by rhizosphere phosphate-solubilizing bacteria. J Appl Bacteriol 40:129–134

Barkley DS (1969) Adenosine-3′,5′ phosphate: identification as acrasin in a species of cellular slime mold. Science 165:1133–1134

Barksdale AW (1963a) The uptake of exogenous hormone A by certain strains of *Achlya*. Mycologia 55:164–171

Barksdale AW (1963b) The role of hormone A during sexual conjugation in *Achlya ambisexualis*. Mycologia 55:627–632

Barksdale AW (1970) Nutrition and antheridiol-induced branching in *Achlya ambisexualis*. Mycologia 62:411–420

Barksdale AW, Lasure LL (1973) I. Induction of gametangial phenotypes in *Achlya*. Bull Torrey Bot Club 100:199–202

Barksdale AW, Lasure LL (1974) Production of hormone B by *Achlya heterosexualis*. Appl Microbiol 28:544–546

Barksdale AW, McMorris TC, Seshadri R, Arunachalam T, Edwards JA, Sundeen J, Green DM (1974) Response of *Achyla* (sic) *ambisexualis* E87 to the hormone antheridiol and certain other steroids. J Gen Microbiol 82:295–299

Barnes RL, Naylor AW (1959a) In vitro culture of pine roots and the use of *Pinus serotina* roots in metabolic studies. For Sci 5:158–168

Barnes RL, Naylor AW (1959b) Effect of various nitrogen sources on the growth of isolated roots of *Pinus serotina*. Physiol Plant 12:82–89

Betz R, Duntze W (1979) Purification and partial characterization of a factor, a mating hormone produced by mating-type-a cells from *Saccharomyces cerevisiae*. Eur J Biochem 95:469–475

Betz R, Duntze W, Manney TR (1978) Mating-factor-mediated sexual agglutination in *Saccharomyces cerevisiae*. FEMS Microbiol Lett 4:107–110

Biliński T, Jachymczyk W, Litwińska J, Żuk J, Gajewski W (1974) Mutual inhibition of DNA synthesis in a and α-cells of *Saccharomyces cerevisiae* during conjugation. J Gen Microbiol 82:97–101

Bimpong CE, Clerk GC (1970) Motility and chemotaxis in zoospores in *Phytophthora palmivora* (Butl) Butl. Ann Bot 34:617–624

Bistis G (1956) Sexuality in *Ascobolus stercorarius*. I. Morphology of the ascogonium; plasmogamy; evidence for a sexual hormonal mechanism. Am J Bot 43:389–394

Bistis G (1957) Sexuality in *Ascobolus stercorarius*. II. Preliminary experiments on various aspects of the sexual process. Am J Bot 44:436–443

Bistis GN, Raper JR (1963) Heterothallism and sexuality in *Ascobolus stercorarius*. Am J Bot 50:880–891

Blaskovics JC, Raper KB (1957) Encystment stages of *Dictyostelium*. Biol Bull 113:58–88

Blondeau R (1970) Production d'une substance de type cytokinine par des *Arthrobacter* d'origine rhizosphérique. C R Ser D 270:3158–3161

Boland DW, Jaenicke L (1978) Synthesis of multifiden, the gamete-attractant of the phaeophyte *Cutleria multifida*, and some of its isomers. Chem Ber 111:3262–3275
Bonner J (1946) Further investigation of toxic substances which arise from guayule plants: Relation of toxic substances to the growth of guayule in soil. Bot Gaz 107:343–351
Bonner J, Galston AW (1944) Toxic substances from the culture media of guayule which may inhibit growth. Bot Gaz 106:185–198
Bonner JT (1967) The cellular slime molds. Princeton Univ Press, Princeton NJ
Bonner JT, Barkley DS, Hall EM, Konijn TM, Mason JW, O'Keefe III G, Wolfe PB (1969) Acrasin, acrasinase and the sensitivity to acrasin in *Dictyostelium discoideum*. Dev Biol 20:72–87
Brian PW (1957) The effects of some microbial metabolic products on plant growth. Symp Soc Exp Biol 11:166–182
Brokaw CJ (1958) Chemotaxis of bracken spermatozoids. Implications of electrochemical orientation. Exp Biol 35:197–212
Brown ME (1972) Plant growth substances produced by micro-organisms of soil and rhizosphere. J Appl Bacteriol 35:443–451
Brown ME, Burlingham SK (1968) Production of plant-growth substances by *Azotobacter chroococcum*. J Gen Microbiol 53:135–144
Bücking-Throm E, Duntze W, Hartwell LH, Manney TR (1973) Reversible arrest of haploid yeast cells at the initiation of DNA synthesis by a diffusible sex factor. Exp Cell Res 76:99–110
Bu'Lock JD, Hardy TM (1979) Mating reaction between sexually altered strains of *Mucor pusillus*. Exp Mycol 3:194–196
Bu'Lock JD, Drake D, Winstanley DJ (1972) Specificity and transformations of the trisporic acid series of fungal sex hormones. Phytochemistry 11:2011–2018
Bu'Lock JD, Jones BE, Quarrie SA, Winskill N (1973) The biochemical basis of sexuality in Mucorales. Naturwissenschaften 60:550–551
Bu'Lock JD, Jones BE, Winskill N (1974a) Structures of the mating-type-specific prohormones of Mucorales. J Chem Soc Chem Commun 17:708–709
Bu'Lock JD, Jones BE, Taylor D, Winskill N, Quarrie SA (1974b) Sex hormones in Mucorales. The incorporation of C_{20} and C_8 precursors into trisporic acids. J Gen Microbiol 80:301–306
Burcky K, Gemmrich A, Schraudolf H (1978) Gas chromotographic evidence for antheridiogen of *Lygodium japonicum* (A_{LY}) (Schizaeaceae) Experientia 34:718
Burke D, Mendonça-Previato L, Ballou CE (1980) Cell-cell recognition in yeast: purification of *Hansenula wingii* 21-cell sexual agglutination factor and comparison of the factors from three genera. Proc Natl Acad Sci USA 77:318–322
Carlile MJ (1966) The orientation of zoospores and germ tubes. In: Madelin MF (ed) The fungus spore. Butterworth London, pp 175–187
Carlile MJ (1976) Nutrition and chemotaxis in the myxomycete *Physarum polycephalum*: the effect of carbohydrate on the plasmodium. J Gen Microbiol 63:221–226
Carlile MJ, Machlis (1965a) The response of male gametes of *Allomyces* to the sexual hormone sirenin. Am J Bot 52:478–483
Carlile MJ, Machlis L (1965b) A comparative study of the chemotaxis of the motile phases of *Allomyces*. Am J Bot 52:484–486
Chailakhyan MKH, Aseeva IJ, Khlopenkova LP (1958) Gibberellin-like substances formed by soil yeasts. Dokl Akad Nauk USSR 123:274–276
Chan RK (1977) Recovery of *Saccharomyces cerevisiae* mating type a cells from Gl arrest by α factor. J Bacteriol 130:766–774
Chang YY (1968) Cyclic 3′, 5′-adenosine monophosphate phosphodiesterase produced by the slime mold *Dictyostelium discoideum*. Science 161:57–59
Chang-Ho Y, Hickman CJ (1970) Some factors involved in the accumulation of phycomycete zoospores on plant roots In: Toussoun TA, Bega RV, Nelson PE (eds) Root diseases and soil-borne pathogens. Univ Calif Press, Berkeley, pp 103–108
Chassy BM (1972) Cyclic nucleotide phosphodiesterase in *Dictyostelium discoideum*: interconversion of two enzyme forms. Science 175:1016–1018
Chet I, Naveh A, Henis Y (1977) Chemotaxis of *Physarum polycephalum* towards carbohydrates, amino acids and nucleotides. J Gen Microbiol 102:145–148

Chou CH, Muller CH (1972) Allelopathic mechanisms of *Arctostaphylos glandulosa* var. *zacaensis*. Am Midl Nat 88:324–347
Ciejek E, Thorner J (1979) Recovery of *S. cerevisiae* a cells from Gl arrest by α factor pheromone requires endopeptidase action. Cell 18:623–635
Ciejek E, Thorner J, Geier M (1977) Solid phase peptide synthesis of α-factor, a yeast mating hormone. Biochem Biophys Res Commun 78:952–961
Clark MA, Francis D, Eisenberg R (1973) Mating types in cellular slime molds. Biochem Biophys Res Commun 52:672–678
Cohen MH, Robertson A (1971a) Wave propagation in the early stages of aggregation of cellular slime molds. J Theor Biol 31:101–118
Cohen MH, Robertson A (1971b) Chemotaxis and the early stages of aggregation of cellular slime molds. J Theor Biol 31:119–130
Corcoran MR, Geissman TA, Phinney BO (1972) Tannins as gibberellin antagonists. Plant Physiol 49:323–330
Cornforth IS (1975) The persistance of ethylene in aerobic soils. Plant Soil 42:85–96
Crafts CB, Miller CO (1974) Detection and identification of cytokinins produced by mycorrhizal fungi. Plant Physiol 54:586–588
Crossett RN, Campbell DJ (1975) The effects of ethylene in the root environment upon the development of barley. Plant Soil 42:453–464
Cunningham JL, Hagedorn DT (1962) Attraction of *Aplanomyces euteiches* zoospores to pea and other plant roots. Phytopathology 52:616–618
Currier WW (1980) Chemotaxis of a birdsfoot trefoil strain of *Rhizobium* to simple sugars. FEBS Microbiol Lett 8:43–46
Currier WW, Strobel GA (1976) Chemotaxis of *Rhizobium* spp. to plant root exudates. Plant Physiol 57:820–823
Currier WW, Strobel GA (1977) Chemotaxis of *Rhizobium* spp. to a glycoprotein produced by birdsfoot trefoil roots. Science 197:434–436
Currier WW, Strobel GA (1981) Characterization and biological activity of trefoil chemotactin. Plant Sci Lett 21:159–165
Curtis RW (1957) Survey of fungi and actinomycetes for compounds possessing gibberellin-like activity. Science 125:646
Cutler LS, Rossomando EF (1975) Localization of adenylate cyclase in *Dictyostelium discoideum*. II. Cytochemical studies on whole cells and isolated plasma membrane vesicles. Exp Cell Res 95:79–87
Darden WH Jr (1966) Sexual differentiation in *Volvox aureus*. J Protozool 13:239–255
Darden WH Jr (1970) Hormonal control of sexuality in the genus *Volvox*. Ann NY Acad Sci 175:757–763
Darmon M, Brachet P, Pereira Da Silva LH (1975) Chemotactic signals induce cell differentiation in *Dictyostelium discoideum*. Proc Natl Acad Sci USA 73:3163–3166
Darmon M, Barra J, Brachet P (1978) The role of phosphodiesterase in aggregation of *Dictyostelium discoideum*. J Cell Sci 34:233–243
Dasilva EJ, Henriksson E, Henriksson LE (1974) Ethylene production by fungi. Plant Sci Lett 2:63–66
Davidonis G (1972) The regulation of fern gametophyte growth by substances released from the sporophyte. Ph D Thesis, Univ Chicago
Davidonis G, Ruddat M (1973) Allelopathic compounds, thelypterin A and B in the fern *Thelypteris normalis*. Planta 111:23–32
Davidonis G, Ruddat M (1974) Growth inhibition in gametophytes and oat coleoptiles by thelypterin A and B released from roots of the fern *Thelypteris normalis*. Am J Bot 61:925–930
Davis EF (1928) The toxic principle of *Juglans nigra* as identified with synthetic juglone and its toxic effects on tomato and alfalfa plants. Am J Bot 15:620
Del Moral R, Cates RG (1971) Allelopathic potential of the dominant vegetation of Western Washington. Ecology 52:1030–1037
Del Moral R, Muller CH (1969) Fog drip: A mechanism of toxin transport from *Eucalyptus globulus*. Bull Torrey Bot Club 96:467–475
Del Moral R, Muller CH (1970) The allelopathic effects of *Eucalyptus camaldulensis*. Am Midl Nat 83:254–282

Derenbach JB, Boland W, Fölster E, Müller DG (1980) Interference tests with the pheromone system of the brown alga *Cutleria multifida*. Mar Ecol Prog Ser 3:357–361
Döpp W (1950) Eine die Antheridienbildung bei Farnen fördernde Substanz in den Prothallien von *Pteridium aquilinum*. L. Kuhn. Ber Dtsch Bot Ges 63:139–147
Döpp W (1959) Über eine hemmende und eine fördernde Substanz bei der Antheridienbildung in den Prothallien von *Pteridium aquilinum*. Ber Dtsch Bot Ges 72:11–24
Döpp W (1962) Weitere Untersuchungen über die Physiologie der Antheridienbildung bei *Pteridium aquilinum*. Planta 58:483–508
Dukes PD, Apple JL (1961) Chemotaxis of zoospores of *Phytophthora parasitica* var. nicotianae by plant roots and certain chemical solutions. Phytopathology 51:195–197
Duntze W, MacKay V, Manney TR (1970) *Saccharomyces cerevisiae:* a diffusible sex factor. Science 169:1472–1473
Duntze W, Stötzler D, Bücking-Throm E, Kalbitzer S (1973) Purification and partial characterization of α-factor, a mating-type-specific inhibitor of cell reproduction from *Saccharomyces cerevisiae*. Eur J Biochem 35:357–365
Durston AJ (1974) Pacemaker activity during aggregation in *Dictyostelium discoideum*. Dev Biol 37:225–235
Elliott CG (1979) Influence of the structure of the sterol molecule on sterol-induced reproduction in *Phytophthora cactorum*. J Gen Microbiol 115:117–126
Endo M, Nakanishi K, Näf U, McKeon W, Walker R (1972) Isolation of the antheridiogen of *Anemia phyllitidis*. Physiol Plant 26:183–185
Erdos GW, Nickerson AW, Raper KB (1972) The fine structure of macrocysts in *Polysphondylium violaceum*. Cytobiol 6:351–366
Erdos GW, Raper KB, Vogen LK (1973) Mating types and macrocyst formation in *Dictyostelium discoideum*. Proc Natl Acad Sci USA 70:1828–1830
Erdos GW, Raper K, Vogen LK (1975) Sexuality in the cellular slime mold *Dictyostelium giganteum*. Proc Natl Acad Sci USA 72:970–973
Farnham CJM (1975) Cytochemical localization of adenylate cyclase and 3′,5′-nucleotide phosphodiesterase in *Dictyostelium*. Exp Cell Res 91:36–46
Fehrenbacher G, Perry K, Thorner J (1978) Cell-cell recognition in *Saccharomyces cerevisiae:* regulation of mating-specific adhesion. J Bacteriol 134:893–901
Filosa MF, Dengler RE (1972) Ultrastructure of macrocyst formation in the cellular slime mold *Dictyostelium mucoroides:* extensive phagocytosis of amoebae by a specialized cell. Dev Biol 29:1–16
Finkelstein DB, Strausberg S (1979) Metabolism of α-factor by a mating type cells of *Saccharomyces cerevisiae*. J Biol Chem 254:796–803
Fischer FG, Werner G (1955) Eine Analyse des Chemotropismus einiger Pilze, insbesondere der Saprolegniaceen. Hoppe-Seylers Z Physiol Chem 300:211–236
Fisher RW, Wolk CP (1976) Substances stimulating the differentiation of spores of the blue-green alga *Cylindrospermum licheniforme*. Nature 259:394–395
Föyn B (1959) Geschlechtskontrollierte Vererbung bei der marinen Grünalge *Ulva mutabilis*. Arch Protistenkd 104:236–253
Francis D (1965) Acrasin and the development of *Polysphondylium pallidum*. Dev Biol 12:329–346
Fujino M, Kitada C, Sakagami Y, Isogai A, Tamura S, Suzuki A (1980) Biological activity of synthetic analogs of tremerogen A-10, a mating hormone of heterobasidiomycetous yeast, *Tremella mesenterica*. Naturwissenschaften 67:406–407
Gairola C, Bhalla PR, Sabharwal PS, Alleem MIH (1972) Production of gibberellin-like substances by an autotrophically grown *Thiobacillus*. Planta 106:177–180
Gerisch G (1965) Stadienspezifische Aggregationsmuster von *Dictyostelium discoideum* Roux. W Roux Arch Entwicklungsmech Org 116:127–144
Gerisch G (1968) Cell aggregation and differentiation in *Dictyostelium*. Curr Top Dev Biol 3:157–197
Gerisch G, Wick U (1975) Intracellular oscillations and release of cyclic AMP from *Dictyostelium* cells. Biochem Biophys Res Commun 65:364–370
Gerisch G, Malchow D, Riedel V, Müller E, Every E (1972) Cyclic AMP phosphodiesterase and its inhibitors in slime mold development. Nature New Biol 235:90–92

Gilles R, Bittner C, Jaenicke L (1981) Site and time of formation of the sex-inducing glycoprotein in *Volvox carteri*. FEBS Lett 124:57–61
Gingle AR, Robertson A (1976) The development of the relaying competence in *Dictyostelium discoideum*. J Cell Sci 20:21–27
Glass ADM (1976) The allelopathic potential of phenolic acids associated with the rhizosphere of *Pteridium aquilinum* (L.) Kuhn: Can J Bot 54:2440–2444
Glass ADM, Bohm BA (1969) The accumulation of cinnamic and benzoic acid derivatives in *Pteridium aquilinum* and *Athyrium felix-femina*. Phytochemistry 8:371–377
Gliessman SR, Muller CH (1972) The phytotoxic potential of bracken, *Pteridium aquilinum* (L.) Kuhn. Madrõno 21:299–304
Gliessman SR, Muller CH (1978) The allelopathic mechanisms of dominance in bracken (*Pteridium aquilinum*) in southern California. J Chem Ecol 4:337–362
Grambow HJ (1977) The influence of volatile leaf constituents on the in vitro differentiation and growth of *Puccinia graminis* f. sp. tritici. Z Pflanzenphysiol 85:361–372
Grambow HJ, Reisener HJ (1976) Differenzierung und Wachstum von *Puccinia graminis* f. sp. tritici in vitro. Ber Dtsch Bot Ges 89:555–561
Grambow HJ, Riedel S (1977) The effect of morphogenetically active factors from host and nonhost plants on the in vitro differentiation of infection structures of *Puccinia graminis* f. sp. tritici. Physiol Plant Pathol 11:213–234
Gregg JH, Nesom MG (1973) Response of *Dictyostelium* plasma membranes to adenosine 3′,5′ cyclic monophosphate. Proc Natl Acad Sci USA 70:1630–1633
Groner B, Hynes N, Sippel AE, Schutz G (1976) Induction of specific proteins in hyphae of *Achlya ambisexualis* by the steroid hormone antheridiol. Nature 261:599–601
Halsall DM (1976) Zoospore chemotaxis in Australian isolates of *Phytophthora* species. Can J Microbiol 22:409–422
Hartwell LH (1973) Synchronization of haploid yeast cell cycles, a prelude to conjugation. Exp Cell Res 76:111–117
Haufler CH, Gastony GJ (1978) Antheridiogen and breeding system in the fern genus *Bommeria*. Can J Bot 56:1594–1601
Hemphill DD Jr, Tukey HB Jr (1973) The effect of intermittent mist on abscission and content of *Euonymus alatus* Sieb 'compactus'. J Am Soc Hort Sci 98:416–442
Hemphill DD Jr, Tukey JB Jr (1975) Effect of intermittent mist on abscisic acid levels in *Euonymus alatus* Sieb: Leaching vs. moisture stress. Hor Science 10:369–370
Hickman CJ (1970) Biology of *Phytophthora* zoospores. Phytopathology 60:1128–1135
Hickman CJ, Ho HH (1966) Behavior of zoospores in plant-pathogenic Phycomycetes. Annu Rev Phytopathol 4:195–220
Hicks JB, Herskowitz I (1976) Evidence for a new diffusible element of mating pheromones in yeast. Nature 260:246–248
Hintermann R, Parish RW (1979) The intracellular location of adenyl cyclase in the cellular slime molds *Dictyostelium discoideum* and *Polysphondylium pallidum*. Exp Cell Res 123:429–434
Hirosawa T, Wolk CP (1979) Isolation and characterization of a substance which stimulates the formation of akinetes in the cyanobacterium *Cylindrospermum licheniforme* Kutz. J Gen Microbiol 114:433–434
Hoffman LR (1960) Chemotaxis of *Oedogonium* sperms. Southwest Nat 5:111–116
Horak E (1964) Die Bildung von IES-Derivaten durch ectotrophe Mykorrhizapilze (*Phlegmacium* spp.) von *Picea abies* Karsten. Phytopathol Z 51:491–515
Horowitz DK, Russell PJ (1974) Hormone-induced differentiation of antheridial branches in *Achyla ambisexualis*: dependence on ribonucleic acid synthesis. Can J Microbiol 20:977–980
Horsley SB (1977) Allelopathic interference among plants II. Physiological modes of action. In: Wilcox HE, Hamer AF (eds) Proceedings of the 4th North American For Biol Workshop. School Cont Ed Coll Environ Sci For, Syracuse New York, pp 93–136
Hurd AM (1920) Effect of unilateral monochromatic light and group orientation on the polarity of germinating *Fucus* spores. Bot Gaz 70:25–50

Hussain A, Vančura V (1970) Formation of biologically active substances by rhizosphere bacteria and their effect on plant growth. Folia Microbiol 15:468–478

Il CL, Tukey HB Jr (1972) Effect of intermittent mist on development of fall color in foliage of *Euonymus alatus* Sieb 'compactus'. J Am Soc Hortic Sci 97:97–101

Ishida N, Kurihara K, Kobatake Y (1978) Dynamic conformational changes of surface membrane of true slime mold in response to chemoreception revealed by reactivity with chemical agents. Comp Biochem Physiol 60A:241–246

Ishida N, Kurihara K, Kobatake Y (1979) Selective modification of positive chemotaxis in the true slime mold *Physarum polycephalum* by ethylenediaminetetraacetic acid treatment. Biochem Biophys Acta 587:89–98

Jackson JR, Willemsen RW (1978) Allelopathy in the first stages of secondary succession on the piedmont of New Jersey. Am J Bot 63:1015–1023

Jackson MB, Campbell DJ (1975) Movement of ethylene from roots to shoots, a factor in the responses of tomato plants to waterlogged soil conditions. New Phytol 74:397–406

Jackson MB, Gales K, Campbell DJ (1978) Effect on waterlogged soil conditions on the production of ethylene and on water relationships in tomato plants. J Exp Bot 29:183–193

Jackson RM, Brown ME, Burlingham SK (1964) Similar effects on tomato plants of *Azotobacter* inoculation and application of gibberellins. Nature 203:851–852

Jaenicke L (1972) Biochemistry of the sex attractants of marine plants. Biochem J 128:38

Jaenicke L (1977a) Sex hormones of known algae. Naturwissenschaften 64:69–75

Jaenicke L (1977b) Sex and sex-attraction in seaweed. Trends Biochem Sci 2:152–154

Jaenicke L, Müller DG, Moore RE (1974) Multifidene and Aucantene, C_{11} hydrocarbons in the male-attracting essential oil from the gynogametes of *Cutleria multifida* (Smith) Grev. (Phaeophyta). J Am Chem Soc 86:3324–3325

Jaffe LF (1968) Localization in the developing *Fucus* egg and the general role of localizing currents. Adv Morphog 7:295

Jaffe LF, Neuscheler W (1969) On the mutual polarization of nearby pairs of fucaceous eggs. Dev Biol 19:549–565

Jameson DA (1966) Pinyon-juniper litter reduces growth of blue gramma. J Range Manage 19:214–217

Jameson DA (1968) Species interaction of growth inhibitors in the native plants of northern Arizona. US For Ser Rocky Mt For Range Exp Sta Res Note RM-113, p 2

Jameson DA (1970) Degradation and accumulation of inhibitory substances from *Juniperus osteosperma* (Torr) Little. Plant Soil 33:213–224

Junttila O (1975) Allelopathy in *Heracleum laciniatum*. Inhibition of lettuce seed germination and root growth. Physiol Plant 33:22–27

Junttila O (1976) Allelopathic inhibitors in seeds of *Heracleum laciniatum*. Physiol Plant 36:374–378

Kajiwara T, Kodama K, Hatanaka A (1980) Male-attracting substance in a marine brown alga *Sargassum horneri*. Naturwissenschaften 67:612–613

Kakebeeke PIJ, Mato JM, Konijn TM (1978) Purification and preliminary characterization of an aggregation-sensitive chemoattractant of *Dictyostelium minutum*. J Bacteriol 133:403–405

Kamiya Y, Sakurai A, Tamura S, Takahashi N, Abe K, Tsuchiya E, Fukui S, Kitada C, Fujino M (1978) Structure of rhodotorucine *A*, a novel lipopeptide, inducing mating tube in *Rhodosporidium toruloides*. Biochem Biophys Res Commun 83:1077–1083

Kampert M, Strzelczyk E (1978) Production of cytokinins by mycorrhizal fungi of pine (*Pinus silvestris* L.). Bull Acad Polon Sci 26:499–503

Kampert M, Strzelczyk E (1980) The synthesis of cytokinin-like substances by coryneform-bacteria isolated from the roots of pine seedlings (*Pinus silvestris* L.). Acta Microbiol Polon 29:117–124

Kampert M, Strzelczyk E, Pokojska A (1975) Production of auxins by bacteria isolated from the roots of pine seedlings (*Pinus silvestris* L.). Acta Microbiol Polon Ser B 7:135–143

Kane BE, Reiskind JB, Mullins JT (1973) Hormone control of sexual morphogenesis in *Achlya:* dependence on protein and ribonucleic acid synthesis. Science 180:1192–1193

Karn RC, Starr RC, Hudock GA (1974) Sexual and asexual differentiation in *Volvox obversus* (Shaw) Printz, strains WD3 and WD7. Arch Protistenkd 116:142–148

Katznelson H, Cole SE (1965) Production of gibberellin-like substances by bacteria and actinomycetes. Can J Microbiol 11:733–741

Katznelson H, Sirois JC (1961) Auxin production by species of *Arthrobacter*. Nature 191:1323–1324

Katznelson H, Sirois JC, Cole SE (1962) Production of a gibberellin-like substance by *Arthrobacter globiformis*. Nature 196:1012–1013

Kefford NP, Brockwell J, Zwar JA (1960) The symbiotic synthesis of auxin by legumes and nodule bacteria and its role in nodule development. Aust J Biol Sci 13:456–467

Khew KL, Zentmyer GA (1973) Chemotactic response of zoospores of five species of *Phytophthora*. Phytopathology 63:1511–1517

Khew KL, Zentmyer GA (1974) Electrotactic response of zoospores of seven species of *Phytophthora*. Phytopathology 64:500–507

Kincaid RL, Mansour TE (1978) Measurement of chemotaxis in the slime mold *Physarum polycephalum*. Exp Cell Res 116:365–375

King AC, Frazier WA (1978) Reciprocal periodicity in cyclic AMP binding and phosphorylation of differentiating *Dictyostelium discoideum* cells. Biochem Biophys Res Commun 78:1093–1099

Klapper BF, Klapper MH (1977) A natural inhibitor of sexual attraction in the water mold *Allomyces*. Exp Mycol 1:352–355

Klein C (1975) Induction of phosphodiesterase by cyclic adenosine 3′, 5′ monophosphate in differentiating *Dictyostelium discoideum* amoebae. J Biol Chem 250:6134–6138

Klein C (1976) Adenylate cyclase activity in *Dictyostelium discoideum* and its changes during differentiation. FEBS Lett 68:125–128

Klein C, Darmon M (1975) The relationship of phosphodiesterase to the developmental cycle of *Dictyostelium discoideum*. Biochem Biophys Res Commun 67:440–447

Klein C, Darmon M (1977) Effects of cyclic AMP pulses on adenylate cyclase and the phosphodiesterase inhibitor of *D. discoideum*. Nature 268:76–78

Klemke I, Bentrup FW (1973) Zur Frage nach den Signalstoffen beim Gruppeneffekt keimender Fucuszygoten. Plant Sci Lett 1:315–320

Knowles DJC, Carlile MJ (1978) The chemotactic response of plasmodia of the myxomycete *Physarum polycephalum* to sugars and related compounds. J Gen Microbiol 108:17–25

Ko WH (1978) Heterothallic *Phytophthora:* evidence for hormonal regulation of sexual reproduction. J Gen Microbiol 107:15–18

Ko WH (1980) Hormonal regulation of sexual reproduction in *Phytophthora*. J Gen Microbiol 116:459–463

Kochert G (1968) Differentiation of reproductive cells in *Volvox carteri*. J Protozool 15:438–452

Kochert G, Yates I (1974) Purification and partial characterization of a glycoprotein sexual inducer from *Volvox carteri*. Proc Natl Acad Sci USA 71:1211–1214

Konijn TM (1972) Cyclic AMP as a first messenger. In: Greengard P, Robinson GA, Paoletti R (eds) Advances in cyclic nucleotide research Vol 1 Raven, New York, pp 17–31

Konijn TM, Raper K (1961) Cell aggregation in *Dictyostelium discoideum*. Dev Biol 3:725–756

Konijn TM, Van De Meene JGC, Bonner JT, Barkley DS (1967) The acrasin activity of adenosine 3′,5′-phosphate. Proc Natl Acad Sci USA 58:1152–1154

Konijn TM, Barkley DS, Chang YY, Bonner JT (1968) Cyclic AMP: a naturally occurring acrasin in cellular slime molds. Am Nat 102:225–233

Konijn TM, Chang YY, Bonner JT (1969) Cyclic AMP synthesis by *Dictyostelium discoideum* and *Polysphondylium pallidum*. Nature 224:1211–1212

Kozel PC, Tukey HB Jr (1968) Loss of gibberellins by leaching from stems and foliage of *Chrysanthemum morifolium* 'Princess Anne'. Am J Bot 55:1184–1189

Kunert R, Libbert E (1972) Beziehungen zwischen Pflanzen und epiphytischen Bakterien hinsichtlich ihres Auxinstoffwechsels. X. Die Exsudation von Aminosäuren und Kohlenhydraten durch Maissprosse als Ernährungsgrundlage für epiphytische Bakterien. Biochem Physiol Pflanzen 163:524–535

Lee IK, Monsi M (1963) Ecological studies on *Pinus densiflora* forest. I. Effects of plant substances on the floristic comparison of the undergrowth. Bot Mag 76:400–413

Levi JD (1956) Mating reaction in yeast. Nature 177:753–754

Lewis KE, O'Day DH (1976) Sexual hormone in the cellular slime mold *Dictyostelium purpureum*. Can J Microbiol 22:1269–1273

Lewis KE, O'Day DH (1977) Sex hormone of *Dictyostelium discoideum* is volatile. Nature 268:730–731

Libbert E, Manteuffel R (1970) Interactions between plants and epiphytic bacteria regarding their auxin metabolism. VII. The influence of the epiphytic bacteria on the amount of diffusible auxin from corn coleoptiles. Physiol Plant 23:93–98

Libbert E, Risch H (1969) Interactions between plants and epiphytic bacteria regarding their auxin metabolism. V. Isolation of IAA-producing and -destroying bacteria from pea plants. Physiol Plant 22:51–58

Libbert E, Silhengst P (1970) Interactions between plants and epiphytic bacteria regarding their auxin metabolism. VIII. Transfer of ^{14}C indole acetic acid from epiphytic bacteria to corn coleoptiles. Physiol Plant 23:480–487

Libbert E, Kaiser W, Kunert R (1969) Interactions between plants and epiphytic bacteria regarding their auxin metabolism. VI. The influence of epiphytic bacteria on the content of extractable auxin in plants. Physiol Plant 22:432–439

Lill RE (1979) Conditions influencing the production of ethylene by incubated litter of *Pinus radiata*. N Z J Sci 22:21–24

Lill RE, McWha JA (1976) Production of ethylene by incubated litter of *Pinus radiata*. Soil Biol Biochem 8:61–63

Lill RE, McWha JA, Cole ALJ (1979) The influence of volatile substances from incubated litter of *Pinus radiata* on seed germination. Ann Bot 43:81–85

Lipke PN, Taylor A, Ballou CE (1976) Morphogenic effects of α-factor on *Saccharomyces cerevisiae* a cells. J Bacteriol 127:610–618

Lodhi MAK (1975) Allelopathic effects of hackberry in a bottomland forest community. J Chem Ecol 1:171–182

Lodhi MAK (1976) Role of allelopathy as expressed by dominating trees in a lowland forest in controlling the productivity and pattern of herbaceous growth. Am J Bot 63:1–8

Lodhi MAK (1978) Allelopathic effects of decaying litter of dominant trees and their associated soil in a lowland forest community. Am J Bot 65:340–344

Lodhi MAK, Rice EL (1971) Allelopathic effects of *Celtis laevigata*. Bull Torrey Bot Club 98:83–89

Lorber P, Muller WH (1976) Volatile growth inhibitors produced by *Salvia leucophylla*: Effects on seedling root tip ultrastructure. Am J Bot 63:196–200

Lorber P, Muller WH (1980a) Volatile growth inhibitors produced by *Salvia leucophylla*: effects on cytological activity in *Allium cepa*. Comp Physiol Ecol 5:60–67

Lorber P, Muller WH (1980b) Volatile growth inhibitors produced by *Salvia leucophylla*: effects on metabolic activity in mitochondrial suspensions. Comp Physiol Ecol 5:68–75

Lüning K, Müller DG (1978) Chemical interaction in sexual reproduction of several Laminariales (Phaeophyceae): release and attraction of spermatozoids. Z Pflanzenphysiol 89:333–341

Lynch JM (1972) Identification of substrates and isolation of microorganisms responsible for ethylene production in the soil. Nature 240:45–46

Lynch JM (1975) Ethylene formation by soil microorganisms. Ann Appl Biol 81:114–115

Lynch JM, Harper SHT (1974a) Formation of ethylene by a soil fungus. J Gen Microbiol 80:187–195

Lynch JM, Harper SHT (1974b) Fungal growth rate and the formation ethylene in soil. J Gen Microbiol 85:91–96

Lynch JM, Harper SHT (1980) Role of substrates and anoxia in the accumulation of soil ethylene. Soil Biol Biochem 12:363–367

Machac MA, Bonner JT (1975) Evidence of a sex hormone in *Dictyostelium discoideum*. J Bacteriol 124:1624–1625

Machlis L (1958a) Evidence for a sexual hormone in *Allomyces*. Physiol Plant 11:185–192

Machlis L (1958b) A study of sirenin, the chemotactic sexual hormone from the water mold *Allomyces*. Physiol Plant 11:845–854

Machlis L (1958c) A procedure for the purification of sirenin. Nature 181:1790–1791

Machlis L (1969) Zoospore chemotaxis in the water mold *Allomyces*. Physiol Plant 22:126–139

Machlis L (1973) The chemotactic activity of various sirenins and analogues and the uptake of sirenin by the sperm of *Allomyces*. Plant Physiol 52:527–530

Machlis L, Nutting WH, Rapoport H (1968) The structure of sirenin. J Am Chem Soc 90:1674–1676

Machlis L, Hill GJC, Steinback KE, Reed W (1974) Some characteristics of the sperm attractant from *Oedogonium cardiacum*. J Phycol 10:199–203

MacInnes M, Francis D (1974) Meiosis in *Dictyostelium mucoroides*. Nature 251:321–324

Maeda Y, Gerisch G (1977) Vesicle formation in *Dictyostelium discoideum* cells during oscillations of cAMP synthesis and release. Exp Cell Res 110:119–126

Malchow D, Gerisch G (1973) Cyclic AMP binding to living cells of *Dictyostelium discoideum* in presence of excess cyclic GMP. Biochem Biophys Res Commun 55:200–204

Malchow D, Gerisch G (1974) Short-term binding and hydrolysis of cyclic 3′, 5′ adenosine monophosphate by aggregating *Dictyostelium* cells. Proc Natl Acad Sci USA 71:2423–2427

Malchow D, Nägele B, Schwarz H, Gerisch G (1972) Membrane-bound cyclic AMP-phosphodiesterase in chemotactically responding cells of *Dictyostelium discoideum*. Eur J Biochem 28:136–141

Malchow D, Fuchila J, Nanjundiah V (1975) A plausible role for a membrane-bound cyclic AMP phosphodiesterase in cellular slime mold chemotaxis. Biochim Biophys Acta 385:421–428

Malkinson AM, Ashworth JM (1972) Extracellular concentrations of adenosine 3′,5′ cyclic monophosphate during axenic growth of myxamoebae of the cellular slime mold *Dictyostelium discoideum*. Biochem J 127:611–612

Malkinson AM, Ashworth JM (1973) Adenosine 3′,5′-cyclic monophosphate concentrations and phosphodiesterase activities during axenic growth and differentiation of cells of the cellular slime mold *Dictyostelium*. Biochem J 134:311–319

Maness PF, Edelman GM (1978) Inactivation and chemical alteration of mating factor α by cells and spheroplasts of yeast. Proc Natl Acad Sci USA 75:1304–1308

Manteuffel R, Siegl E, Libbert E (1972) Beziehungen zwischen Pflanzen und epiphytischen Bakterien hinsichtlich ihres Auxinstoffwechsels. XI. Der Einfluß von IES-bildenden, IES-abbauenden und IES-neutralen Bakterien auf Auxingehalt der Maispflanze im Vergleich zur Wirkung von Zucker, Aminosäure oder Vitaminapplikationen. Biochem Physiol Pflanz 163:586–595

Manteuffel R, Siegl E, Kunert R, Libbert E (1973a) Beziehungen zwischen Pflanzen und epiphytischen Bakterien hinsichtlich ihres Auxinstoffwechsels. XIII. Bakterielle Auxinsynthese in Pflanzenexsudaten. Biochem Physiol Pflanz 164:303–314

Manteuffel R, Siegl E, Kunert R, Libbert E (1973b) Beziehungen zwischen Pflanzen und epiphytischen Bakterien hinsichtlich ihres Auxinstoffwechsels. XIV. Der Einfluß der epiphytischen Bakterien auf das Wachstum, den Geotropismus und die Adventivwurzelbildung der Pflanzen. Biochem Physiol Pflanz 164:338–348

Massey AB (1925) Antagonism of the walnuts (*Juglans nigra* L. and *J. cinerea* L.) in certain plant associations. Phytopathology 15:773–784

Masui Y, Chino N, Sakakibara S, Tanaka T, Murakami T, Kita H (1977) Synthesis of the mating factor of *Saccharomyces cerevisiae* and its truncated peptides: the structure-activity relationship. Biochem Biophys Res Commun 78:534–546

Masui Y, Tanaka T, Chino N, Kita H, Sakakibara S (1979) Amino acid substitution of mating factor of *Saccharomyces cerevisiae*: structure-activity relationship. Biochem Biophys Res Commun 86:982–987
Mato JM, Konijn TM (1975) Chemotaxis and binding of cyclic AMP in cellular slime molds. Biochem Biophys Acta 385:173–179
Mato JM, Losada A, Nanjundiah V, Konijn TM (1975) Signal input for a chemotactic response in the cellular slime mold *Dictyostelium discoideum* Proc Natl Acad Sci USA 72:4991–4993
Mato JM, Van Haastert PJM, Krens FA, Konijn TM (1977) An acrasin-like attractant from yeast extract specific for *Dictyostelium lacteum*. Dev Biol 57:450–453
Mato JM, Jastorff B, Morr M, Konijn TM (1978) A model for cyclic AMP-chemoreceptor interaction in *Dictyostelium discoideum*. Biochem Biophys Acta 544:309–314
McCohen CB, Kelsey RG, Sheridan RP, Shafizadeh F (1973) Physiological effects of compounds extracted from sagebrush. Bull Torrey Bot Club 100:23–28
McCracken MD, Starr RC (1970) Induction and development of reproductive cells in the K32 strains of *Volvox rosseletii*. Arch Protistenkd 112:262–282
McCullough J, Herskowitz I (1979) Mating pheromones of *Saccharomyces kluyveri*: pheromone interactions between *Saccharomyces kluyveri* and *Saccharomyces cerevisiae*. J Bacteriol 138:146–154
McMorris TC (1978) Sex hormones of the aquatic fungus *Achlya*. Lipids 13:716–722
McMorris TC, Barksdale AW (1967) Isolation of a sex hormone from the water mold *Achlya bisexualis*. Nature 215:320–321
McMorris TC, Seshadri R, Weihe GR, Arsenault GP, Barksdale AW (1975) Structures of oogonial −1, −2, and −3, steroidal sex hormones of the water mold *Achlya*. J Am Chem Soc 97:2544–2545
McPherson JK, Muller CH (1969) Allelopathic effects of *Adenostoma fasciculatum*. "chamise," in the California chaparral. Ecol Mon 39:177–197
McPherson JK, chou CH, Muller CH (1971) Allelopathic constituents of the chaparral shrub *Adenostoma fasciculatum*. Phytochemistry 10:2925–2933
Mesland DAM, Huisman JG, Van den Ende H (1974) Volatile sex hormones in *Mucor mucedo*. J Gen Microbiol 80:111–117
Miller CO (1967) Zeatin and zeatin riboside from a mycorrhizal fungus. Science 157:1055–1057
Moens PB, Konijn TM (1974) Cyclic AMP as a cell surface-activating agent in *Dictyostelium discoideum*. FEBS Lett 45:44–46
Mooney DT, Sypherd PS (1976) Volatile factor involved in the dimorphism of *Mucor racemosus*. J Bacteriol 126:1266–1270
Moser M (1959) Beiträge zur Kenntnis der Wuchstoffbeziehungen im Bereich ektotropher Mykorrhizen I. Arch Mikrobiol 34:251–269
Mullens IA, Newell PC (1978) cAMP binding to cell surface receptors of *Dictyostelium*. Differentiation 10:171–176
Muller CH (1965) Inhibitory terpenes volatilized from *Salvia* shrubs. Bull Torrey Bot Club 92:38–45
Muller CH (1966) The role of chemical inhibition (allelopathy) in vegetational composition. Bull Torrey Bot Club 93:332–351
Muller CH, Del Moral R (1966) Soil toxicity induced by terpenes from *Salvia leucophylla*. Bull Torrey Bot Club 93:130–137
Muller CH, Muller WH, Haines BL (1964) Volatile growth inhibitors produced by shrubs. Science 143:471–473
Müller DG (1968) Versuche zur Charakterisierung eines Sexual-Lockstoffes bei der Braunalge *Ectocarpus siliculosus*. Planta 81:160–168
Müller DG (1972) Befruchtungsstoffe bei Braunalgen. Ber Dtsch Bot Ges 85:363–369
Müller DG (1974) Sexual reproduction and isolation of a sex attractant in *Cutleria multifida* (Smith) Grev. (Phaeophyta). Biochem Physiol Pflanzen 165:212–215
Müller DG (1976) Quantitative evaluation of sexual chemotaxis in two marine brown algae. Z Pflanzenphysiol 80:120–130
Müller DG (1978) Locomotive responses of male gametes to species-specific sex attractant in *Ectocarpus siliculosus* (Phaeophyta). Arch Protistenkd 120:371–377

Müller DG, Gassmann G (1978) Identification of the sexual attractant in the marine brown alga *Fucus vesiculosus*. Naturwissenschaften 65:389–390
Müller DG, Gassmann G (1980) Sexual hormone specificity in *Ectocarpus* and *Laminaria*. Naturwissenschaften 67:462–463
Müller DG, Jaenicke L (1973) Fucoserratin, the female sex attractant of *Fucus serratus*. FEBS Lett 30:137–139
Müller DG, Seferiadis K (1977) Specificity of sexual chemotaxis in *Fucus serratus* and *Fucus vesiculosus* (Phaeophyceae). Z Pflanzenphysiol 84:85–94
Müller DG, Jaenicke L, Donike M, Akintobi T (1971) Sex attractants in a brown alga: chemical structure. Science 171:815–817
Müller DG, Gassmann G, Lüning K (1979) Isolation of a spermatozoid-releasing and attracting substance from female gametophytes of *Laminaria digitata*. Nature 279:430–431
Müller DG, Gassmann G, Boland W, Marner F, Jaenicke L (1981) *Dictyota dichotoma* (Phaeophyceae): identification of the sperm attractant. Science 212:1040–1041
Muller WH, Hauge R (1967) Volatile growth inhibitors produced by *Salvia leucophylla:* Effect on seedling anatomy. Bull Torrey Bot Club 94:182–191
Muller WH, Muller CH (1964) Volatile growth inhibitors produced by *Salvia* species. Bull Torrey Bot Club 91:327–330
Muller WH, Lorber P, Haley B (1968) Volatile growth inhibitors produced by *Salvia leucophylla:* Effect on seedling growth and respiration. Bull Torrey Bot Club 95:415–422
Muller WH, Lorber P, Haley B, Johnson K (1969) Volatile growth inhibitors produced by *Salvia leucophylla:* Effect on oxygen uptake by mitochondrial suspensions. Bull Torrey Bot Club 96:89–96
Mullins JT (1973) Lateral branch formation and cellulose production in the water molds. Mycologia 65:1007–1014
Mullins JT (1979) A freeze-fracture study of hormone-induced branching in the fungus *Achlya*. Tissue Cell 11:585–595
Mullins JT, Ellis EA (1974) Sexual morphogenesis in *Achlya:* ultrastructural basis for the hormonal induction of antheridial hyphae. Proc Natl Acad Sci USA 71:1347–1350
Musgrave A, Nieuwenhuis D (1975) Metabolism of radioactive antheridiol by *Achlya* species. Arch Microbiol 105:313–318
Musgrave A, Ero L, Schaffer R (1978) The selfinduced metabolism of antheridiol in water moulds. Acta Bot Neerl 27:397–404
Näf U (1956) The demonstration of a factor concerned with the initiation of antheridia in Polypodiaceous ferns. Growth 20:91–105
Näf U (1958) On the physiology of antheridium formation in the bracken fern. [*Pteridium aquilinum* (L.) Kuhn]. Physiol Plant 11:728–746
Näf U (1959) Control of antheridium formation in the fern species *Anemia phyllitidis*. Nature 184:798–800
Näf U (1960) On the control of antheridium formation in the fern species *Lygodium japonicum*. Soc Exp Biol Med 105:82–86
Näf U (1961) Mode of action of an antheridium-inducing substance in ferns. Nature 189:900–903
Näf U (1962a) Loss of sensitivity to the antheridial factor in maturing gametophytes of the fern *Onoclea sensibilis*. Phyton 18:173–182
Näf U (1962b) Developmental physiology of lower archegoniates. Annu Rev Plant Physiol 13:507–532
Näf U (1968) On separation and identity of fern antheridiogens. Plant Cell Physiol 9:27–33
Näf U, Sullivan J, Cummins M (1969) New antheridiogen from the fern *Onoclea sensibilis*. Science 163:1357–1358
Näf U, Sullivan J, Cummins M (1974) Fern antheridiogen: cancellation of a light-dependent, block to antheridium formation. Dev Biol 40:355–365

Näf U, Nakanishi K, Endo M (1975) On the physiology and chemistry of fern antheridiogen. Bot Rev 41:315–359

Nakanishi K, Endo M, Näf U (1971) Structure of the antheridium-inducing substance of the fern *Anemia phyllitidis*. J Am Chem Soc 93:5579–5581

Nieuwenhuis M, Van Den Ende H (1975) Sex specificity of hormone synthesis in *Mucor mucedo*. Arch Microbiol 102:167–170

Nilsen G, Nordby Ø (1975) A sporulation-inhibiting substance from vegetative thalli of the green alga *Ulva mutabilis*, Føyn. Planta 125:127–139

Nordbring-Hertz B (1973) Peptide-induced morphogenesis in the nematode-trapping fungus *Arthrobotrys oligospora*. Physiol Plant 29:223–233

Nordbring-Hertz B, Brinck C (1974) Qualitative characterization of some peptides inducing morphogenesis in the nematode-trapping fungus *Arthrobotrys oligospora*. Physiol Plant 31:59–63

Nordby Ø (1974) Light microscopy of meiotic zoosporogenesis and mitotic gametogenesis in *Ulva mutabilis*. J Cell Sci 15:443–455

Nordby Ø, Hoxmark R (1972) Changes in cellular parameters during synchronous meiosis in *Ulva mutabilis* Føyn. Exp Cell Res 75:321–328

Nutman PS (1956) The influence of the legume in root-nodule symbiosis. Biol Rev 31:109–151

Nutting WH, Rapoport H, Machlis L (1968) The structure of sirenin. J Am Chem Soc 90:6434–6438

O'Day DH (1979) Aggregation during sexual development in *Dictyostelium discoideum*. Can J Microbiol 25:1416–1426

O'Day DH, Lewis KE (1975) Diffusible mating-type factors induce macrocyst development in *Dictyostelium discoideum*. Nature 254:431–432

O'Donnell KL, Fields WG, Hooper GR (1974) Scanning ultrastructural ontogeny of cleistohymenial apothecia in the operculate discomycete *Ascobolus furfuraceus*. Can J Bot 52:1653–1656

O'Donnell KL, Flegler SL, Hooper GR (1978) Zygosporangium and zygospore formation in *Phycomyces nitens*. Can J Bot 56:91–100

Olson RA, Du Buy HG (1937) The role of growth substance in the polarity and morphogenesis of *Fucus*. Am J Bot 24:611–615

Osumi M, Shimoda C, Yanagishima N (1974) Mating reaction in *Saccharomyces cerevisiae*. V. Changes in the fine structure during the mating reaction. Arch Microbiol 97:27–38

Pall ML (1973) Sexual induction in *Volvox carteri:* a quantitative study. J Cell Biol 59:238–241

Pannbacker RG, Bravard LJ (1972) Phosphodiesterase in *Dictyostelium discoideum* and the chemotactic response to cyclic AMP. Science 175:1014–1015

Parenti RL, Rice EL (1969) Inhibitional effects of *Digitaria sanguinalis* and possible role in old-field succession. Bull Torrey Bot Club 96:70–78

Pfeffer W (1884) Locomotorische Richtungsbewegungen durch chemische Reize. Cent Bot Inst Tübingen 1:304–382

Plattner JJ, Rapoport H (1971) The synthesis of *d* and *l* sirenin and their absolute configurations. J Am Chem Soc 93:1758–1761

Plempel M (1960) Die zygotropische Reaktion bei *Mucorineen* I. Planta 55:258

Plempel M (1962) Die zygotropische Reaktion bei *Mucorineen* III. Planta 58:509–520

Plempel M (1963) Die chemischen Grundlagen der Sexualreaktion bei Zygomyceten. Planta 59:492–508

Plempel M, Dawid W (1961) Die zygotropische Reaktion bei *Mucorineen* II. Planta 56:438–446

Pommerville J (1978) Analysis of gamete and zygote motility in *Allomyces*. Exp Cell Res 113:161–172

Pramer D (1964) Nematode-trapping fungi. Science 144:382–388

Pramer D, Kuyama S (1963) Symposium on biochemical base of morphogenesis in fungi. II. Nemin and the nematode-trapping fungi. Bacteriol Rev 27:282–292

Pramer D, Stoll NR (1959) Nemin, a morphogenetic substance causing trap formation by predacious fungi. Science 129:966–967
Pringle RB (1961) Chemical nature of antheridogen-A, a specific inducer of the male sex organ in certain fern species. Science 133:284
Pringle RB, Näf U, Braun AG (1960) Purification of a specific inducer of the male sex organ in certain fern species. Nature 186:1066–1067
Purushothaman D, Marimuthu T, Venkataraman CV, Kesavan R (1974) Role of actinomycetes in the biosynthesis of indole acetic acid in soil. Curr Sci 43:413–414
Rai PV, Strobel GA (1966) Chemotaxis of zoospores of *Aphanomyces cochlioides* to sugar beet seedling. Phytopathology 56:1365–1369
Raman RK (1976) The autonomous cell in *Dictyostelium discoideum*. J Theor Biol 59:491–495
Raman RK, Hashimoto Y, Cohen MH, Robertson A (1976) Differentiation for aggregation in the cellular slime molds: the emergence of autonomously signalling cells in *Dictyostelium discoideum*. J Cell Sci 21:243–259
Raper JR (1939a) Sexual hormones in *Achlya*. I. Indicative evidence for a hormonal coordinating mechanism. Am J Bot 26:639–650
Raper JR (1939b) Role of hormones in the sexual reaction of heterothallic *Achlyas*. Science 89:321–322
Raper JR (1940a) Sexual hormones in *Achlya*. II. Distance reactions, conclusive evidence for a hormonal coordinating mechanism. Am J Bot 27:162–173
Raper JR (1940b) Sexuality in *Achlya ambisexualis*. Mycologia 32:710–727
Raper JR (1942a) Sexual hormones in *Achlya*. III. Hormone A and the initial male reaction. Am J Bot 29:159–166
Raper JR (1942b) Sexual hormones in *Achlya*. V. Hormone A, a male secreted augmenter or activiator of hormone A. Proc Natl Acad Sci USA 28:509–515
Raper JR (1950) Sexual hormones of *Achlya*. VI. The hormones of the A complex. Proc Natl Acad Sci USA 36:524–533
Raper JR (1951) Sexual hormones in *Achlya*. Am Sci 39:110–120, 130
Raper JR, Haagen-Smit AJ (1942) Sexual hormones in *Achlya*. IV. Properties of hormone A of *A. bisexualis*. J Biol Chem 143:311–320
Raper KB, Thom C (1941) Interspecific mixtures in the Dictyosteliaceae. Am J Bot 28:69–78
Rasmussen JA, Rice EL (1971) Allelopathic effects of *Sporobolus pyramidatus* on vegetational pattern. Am Midl Nat 86:309–326
Rawitcher-Kunkel E, Machlis L (1962) The hormonal integration of sexual reproduction in *Oedogonium*. Am J Bot 49:177–183
Reid ID (1974) Properties of conjugation hormones (erogens) from the basidiomycete *Tremella mesenterica*. Can J Bot 52:521–524
Rice EL (1974) Allelopathy. Academic Press, London New York
Rice EL (1977) Some roles of allelopathic compounds in plant communities. Biochem Syst Ecol 5:201–206
Riedel V, Gerisch G (1971) Regulation of extracellular cyclic-AMP-phosphodiesterase activity during development of *Dictyostelium discoideum*. Biochem Biophys Res Commun 42:119–123
Riedel V, Malchow D, Gerisch G, Nägele B (1972) Cyclic AMP phosphodiesterase interaction with its inhibition of the slime mold *Dictyostelium discoideum*. Biochem Biophys Res Commun 46:279–287
Riedel V, Gerisch G, Müller E, Beug H (1973) Defective cyclic adenosine-3′, 5′-phosphate-phosphodiesterase regulation in morphogenetic mutants of *Dictyostelium discoideum*. J Mol Biol 74:573–585
Robertson A, Drage DJ (1975) Stimulation of late interphase *Dictyostelium discoideum* amoebae with an external cyclic AMP signal. Biophys J 15:765–776
Robertson A, Drage DJ, Cohen MH (1972) Control of aggregation in *Dictyostelium discoidium* by an externally applied periodic pulse of cyclic AMP. Science 175:333–335
Rossomando EF, Cutler LS (1975) Localization of adenylate cyclase in *Dictyostelium discoideum*. I. Preparation and biochemical characterization of cell fractions and isolated plasma membrane vesicles. Exp Cell Res 95:67–78

Rossomando EF, Sussman M (1972) Adenyl cyclase in *Dictyostelium discoideum*, a possible control element of the chemotactic system. Biochem Biophys Res Commun 47:604–610

Rothschild L (1952) The behavior of spermatozoa in the neighborhood of eggs. Int Rev Cytol 1:257–263

Royle DS, Hickman CJ (1964a) Analyses of factors governing in vitro accumulation of zoospores of *Pythium aphonidermatum* on roots. I. Behavior of zoospores. Can J Microbiol 10:151–162

Royle D, Hickman CJ (1964b) Analyses of factors governing in vitro accumulation of zoospores of *Pythium aphonidermatum* on roots. II. Substances causing response. Can J Microbiol 10:201–219

Sakagami Y, Isogai A, Suzuki A, Tamura S, Tsuchiya E, Fukui S (1978a) Isolation of a novel sex hormone, tremerogen A-10, controlling conjugation tube formation in *Tremella mesenterica* Fr. Agric Biol Chem 42:1093–1094

Sakagami Y, Isogai A, Suzuki A, Tamura S, Tsuchiya E, Fukui S (1978b) Amino acid sequence of tremerogen A-10, a peptidal hormone, inducing conjugation tube formation in *Tremella mesenterica* Fr. Agric Biol Chem 42:1301–1302

Sakagami Y, Isogai A, Suzuki A, Tamura S, Kitada C, Fujino M (1979) Structure of tremerogen *A*-10, a peptidal hormone-inducing conjugation tube formation in *Tremella mesenterica*. Agric Biol Chem 43:2643–2645

Sakai K, Yanagishima N (1971) Mating reaction in *Saccharomyces cerevisiae*. I. Cell agglutination related to mating. Arch Mikrobiol 75:260–265

Sakai K, Yanagishima N (1972) Mating reaction in *Saccharomyces cerevisiae*. II. Hormonal regulation of agglutinability of a type cells. Arch Mikrobiol 84:191–198

Sakurai A, Tamura S, Yanagishima N, Shimoda C (1976) Structure of the peptidyl factor inducing sexual agglutination in *Saccharomyces cerevisiae*. Agric Biol Chem 40:1057–1058

Sakurai A, Tamura S, Yanagishima N, Shimoda C (1977a) Isolation of a peptidyl factor, α substance-I_A, inducing sexual agglutinability in *Saccharomyces cerevisiae*. Agric Biol Chem 41:389–393

Sakurai A, Tamura S, Yanagishima N, Shimoda C (1977b) Structure of a peptidyl factor, α-substance-I_A, inducing sexual agglutinability in *Saccharomyces cerevisiae*. Agric Biol Chem 41:395–398

Schedlbauer MD (1974) Biological specificity of the antheridogen from *Ceratopteris thalictroides* (L.) Brongn. Planta 116:39–43

Schedlbauer MD (1976a) Fern gametophyte development: controls of dimorphism in *Ceratopteris thalictroides*. Am J Bot 63:1080–1087

Schedlbauer MD (1976b) Specificity of the antheridogen from *Ceratopteris thalictroides* (L.) Brongn. Plant Physiol 57:666–669

Schedlbauer MD, Klekowski EJ Jr (1972) Antheridogen activity in the fern *Ceratopteris thalictroides* (L) Brongn. Bot J Linn Soc 65:399–413

Schekman R, Brawley V (1979) Localized deposition of chitin on the yeast cell surface in response to mating pheromone. Proc Natl Acad Sci USA 76:645–649

Schraudolf H (1962) Die Wirkung von Phytohormonen auf Keimung und Entwicklung von Farnprothalien. I. Auslösung der Antheridienbildung und Dunkelkeimung bei Schizaeaceen durch Gibberellinsäure. Biol Zentralbl 81:731–740

Schraudolf H (1964) Relative activity of the gibberellins in the antheridium induction in *Anemia phyllitidis*. Nature 201:98–99

Shaffer BM (1956) Acrasin, the chemotactic agent in cellular slime moulds. J Exp Biol 33:645–657

Shaffer BM (1957) Aspects of aggregation in cellular slime moulds. Am Nat 91:19–35

Shaffer BM (1958) Integration in aggregating cellular slime moulds. Quart J Microsc Sci 99:103–121

Shaffer BM (1961) The cells founding aggregation centers in the slime mould *Polysphondylium violaceum*. J Exp Biol 38:833–849

Shaffer BM (1962) The acrasina Part I. Adv Morphol 2:109–182

Shimoda C, Yanagishima N (1972) Mating reaction in *Saccharomyces cerevisiae*. III. Changes in autolytic activity. Arch Mikrobiol 85:310–318

Shimoda C, Yanagishima N (1973) Mating reaction in *Saccharomyces cerevisiae.* IV. Retardation of deoxyribonucleic acid synthesis. Physiol Plant 29:54–59

Shimoda C, Yanagishima N, Sakurai A, Tamura S (1976) Mating reaction in *Saccharomyces cerevisiae.* IX. Regulation of sexual cell agglutinability of *a* type cells by a sex factor produced by α-type cells. Arch Microbiol 180:27–33

Shimoda C, Yanagishima N, Sakurai A, Tamura S (1978) Induction of sexual agglutinability of α mating-type cells as the primary action of the peptidyl sex factor from α mating-type cells in *Saccharomyces cerevisiae.* Plant Cell Physiol 19:513–517

Silver JC, Horgen PA (1974) Hormonal regulation of presumptive mRNA in the fungus *Achlya ambisexualis.* Nature 249:252–254

Slankis V (1973) Hormonal relationships in mycorrhiza. In: Marks GC, Kozlowski TT (eds) Ectomycorrhizae. Academic Press, London New York, pp 231–298

Smith AM, Cook RJ (1974) Implications of ethylene production by bacteria for biological balance of soil. Nature 252:703–705

Smith KA, Restall SWF (1971) The occurrence of ethylene in anaerobic soil. J Soil Sci 22:430–443

Smith KA, Robertson PD (1971) Effect of ethylene on root extension of cereals. Nature 234:148–149

Smith KA, Russell RS (1969) Occurrence of ethylene and its significance, in anaerobic soil. Nature 222:769–771

Starr RC (1968) Cellular differentiation in *Volvox.* Proc Natl Acad Sci USA 59:1082–1088

Starr RC (1969) Structure, reproduction and differentiation in *Volvox carteri* f. nagariensis Iyengar, strains HK 9 and 10. Arch Protistenkd 111:204–222

Starr RC (1972) Sexual reproduction in *Volvox dissipatrix.* Abst Br Phycol J 7:284

Starr RC, Jaenicke L (1974) Purification and characterization of the hormone initiating sexual morphogenesis in *Volvox carteri* f. nagariensis Iyengar, Proc Natl Acad Sci USA 71:1050–1054

Starr RC, O'Neil RM, Miller III CE (1980) L-glutamic acid as a mediator of sexual morphogenesis in *Volvox capensis.* Proc Natl Acad Sci USA 77:1025–1028

Stötzler D, Duntze W (1973) Charakterisierung eines regulatorischen Peptids aus *Saccharomyces cerevisiae.* Hoppe Seyler's Z Physiol Chem 354:1247

Stötzler D, Duntze W (1976) Isolation and characterization of four related peptides exhibiting α-factor activity from *Saccharomyces cerevisiae.* Eur J Biochem 65:257–262

Stötzler D, Kiltz HH, Duntze W (1976) Primary structure of α-factor peptides from *Saccharomyces cerevisiae.* Eur J Biochem 69:297–400

Strzelczyk E, Kampert M, Dahm H (1973) Production and decomposition of indole acetic acid (IAA) by microorganisms isolated from the root zone of two crop plants. Acta Microbiol Polon Ser B 5:71–79

Strzelczyk E, Sitek J, Kowalski S (1975) Production of gibberellin-like substances by fungi isolated from mycorrhizae of pine (*Pinus silvestris* L.). Acta Microbiol Polon Ser B 7:145–153

Strzelczyk E, Sitek J, Kowalski S (1977) Synthesis of auxins from tryptophan and tryptophan precursors by fungi isolated from mycorrhizae of pine (*Pinus silvestris* L.) Acta Microbiol Polon Ser B 25:255–264

Sutter RP (1970) Trisporic acid synthesis in *Blakeslea trispora.* Science 168:1590–1592

Sutter RP, Capage DA, Harrison TL, Keen WA (1973) Trisporic acid biosynthesis in separate plus and minus cultures of *Blakeslea trispora:* Identification by *Mucor* assay of two mating type-specific components. J Bacteriol 114:1074–1082

Sutter RP, Harrison TL, Galasko G (1974) Trisporic acid biosynthesis in *Blakeslea trispora* via mating type-specific precursors. J Biol Chem 249:2282–2284

Takao N, Shimoda C, Yanagishima N (1970) Chemical nature of yeast sexual hormones. Dev Growth Differ 12:199–205

Tanaka T, Kita H (1977) Degradation of mating factor by α-mating type cells of *Saccharomyces cerevisiae.* J Biochem 82:1689–1693

Tanaka T, Kita H (1978) Site of action of mating factor in *a*-mating cell of *Saccharomyces cerevisiae.* Biochem Biophys Res Commun 83:1319–1324

Tanaka T, Kita H, Murakami T, Narita K (1977) Purification and amino acid sequence of mating factor from *Saccharomyces cerevisiae.* J Biochem 82:1681–1687

Thiadens AJH, Zeuthern E (1967) Meiosis and sporulation induced in sporophytes of *Ulva mutabilis* ("Slender") with synchronous mitosis. Planta 72:60–65
Thomas D Des S (1970) Effects of water stress on induction and secretion of cellulose and on branching in the water mold *Achlya ambisexualis*. Can J Bot 48:977–979
Thomas D Des S, Mullins JT (1967) Role of enzymatic wall softening in plant morphogenesis: hormonal induction in *Achlya*. Science 156:84–85
Thomas D Des S, Mullins JT (1969) Cellulose induction and wall extension in the water mold *Achlya ambisexualis*. Physiol Plant 22:347–353
Throm E, Duntze W (1970) Mating-type-dependent inhibition of deoxyribonucleic acid synthesis in *Saccharomyces cerevisiae*. J Bacteriol 104:1388–1390
Timberlake WE (1976) Alterations in RNA and protein synthesis associated with steriod hormone-induced sexual morphogenesis in the water mold, *Achlya*. Dev Biol 51:202–214
Tkacz JS, Mackay VL (1979) Sexual conjugation in yeast. Cell surface changes in response to the action of mating hormones. J Cell Biol 80:326–333
Tohoyama H, Hagiya M, Yoshida K, Yanagishima N (1979) Regulation of the production of the agglutination substances responsible for sexual agglutination in *Saccharomyces cerevisiae:* changes associated with conjugation and temperature shift. Molec Gen Genet 174:269–280
Tomaszewski M, Thimann KV (1966) Interactions of phenolic acids, metallic ions and chelating agents on auxin-induced growth. Plant Physiol 41:1443–1454
Treffry TE, Watts DJ (1976) Development of *Dictyostelium discoideum:* A study by scanning electron microscopy. Microsc 7:11–20
Tryon RM, Vitale G (1977) Evidence for antheridogen production and its mediation of a mating system in natural populations of fern gametophytes. Bot J Linn Soc 74:243–249
Tsuchiya E, Fukui S, Kamiya Y, Sakagami Y, Fujino M (1978) Requirements of chemical structure for hormonal activity of lipopeptidyl factors inducing sexual differentiation in vegetative cells of heterobasidiomycetous yeasts. Biochem Biophys Res Commun 85:459–463
Tukey HB Jr (1970) The leaching of substances from plants. Annu Rev Plant Physiol 21:305–324
Ulrich JM (1960) Auxin production by mycorrhizal fungi. Physiol Plant 13:429–443
Vančura V (1961) Detection of gibberellic acid in *Azotobacter* cultures. Nature 192: 88–89
Van De Berg WJ, Starr RC (1971) Structure, reproduction and differentiation in *Volvox gigas* and *Volvox powersii*. Arch Protistenkd 113:195–219
Van den Ende H (1968) Relationship between sexuality and carotene synthesis in *Blakeslea trispora*. J Bacteriol 96:1298–1303
Van Den Ende H, Wiechmann AHCA, Reyngoud RJ, Henriks T (1970) Hormonal interactions in *Mucor* and *Blakeslea trispora*. J Bacteriol 101:423–428
Van Den Ende H, Werkman BA, Van Den Briel ML (1972) Trisporic acid synthesis in mated cultures of the fungus *Blakeslea trispora*. Arch Mikrobiol 86:174–184
Van Staden J (1976) The release of cytokinins by maize roots. Plant Sci Lett 7:279–283
Van Staden J, Dimalla GG (1976) Cytokinins from soils. Planta 130:85–87
Voeller BR (1964a) Antheridogens in ferns. In: Régulateurs naturels de la croissance végétale. Coll Int Cent Natl Rech Sci Paris 123:665–684
Voeller BR (1964b) Gibberellins: their effect on antheridium formation in fern gametophytes. Science 143:373–375
Voeller BR (1966) Gibberellins and growth in ferns. In: Proc Int Symp Plant Stimulation. Bulgarian Acad Sci, Sophia, pp 247–258
Voeller B (1971) Developmental physiology of fern gametophytes: relevance for biology. Bioscience 21:266–270
Voeller BR, Weinberg ES (1969) Evolutionary and physiological aspects of antheridium induction in ferns. In: Gunckel JE (ed) Current topics in plant science. Academic Press, London New York, pp 77–93
Waaland SD (1975) Evidence for a species-specific cell fusion hormone in red algae. Protoplasma 86:253–261

Waaland SD, Cleland RE (1974) Cell repair through cell fusion in the red alga *Griffithsia pacifica*. Protoplasma 79:185–196
Waaland SD, Watson BA (1980) Isolation of a cell-fusion hormone from *Griffithsia pacifica* Kylin, a red alga. Planta 149:493–497
Werkman BA, Van Den Ende H (1973) Trisporic acid synthesis in *Blakeslea trispora*: interactions between plus and minus types. Arch Mikrobiol 90:365–374
Werkman GA (1976) Localization and partial characterization of a sex-specific enzyme in homothallic and heterothallic mucorales. Arch Microbiol 109:209–214
Whitehead DC (1964) Identification of p-hydroxybenzoic, vanillic, p-coumaric and ferulic acids in soil. Nature 202:417–418
Whittaker DM (1940) The effect of alkalinity upon mutual influences determining the developmental axis in *Fucus* eggs. Biol Bull 78:407–411
Wilkinson LE, Pringle JR (1974) Transient G_1 arrest of *S. cerevisiae* cells of mating type α by a fraction produced by cells of mating type *a*. Exp Cell Res 89:175–187
Wilson RE, Rice EL (1968) Allelopathy as expressed by *Helianthus annuus* and its role in old-field succession. Bull Torrey Bot Club 95:432–448
Wootton LMO, Pramer D (1966) Valine-induced morphogenesis in *Arthrobotrys conoides*. Bacteriol Proc p 75
Wurster B, Pan P, Tyan GG, Bonner JT (1976) Preliminary characterization of the acrasin of the cellular slime mold *Polyspondylium violaceum*. Proc Natl Acad Sci USA 73:795–799
Wurtz T, Jockusch H (1975) Sexual differentiation in *Mucor*: Trisporic and response mutants and mutants blocked in zygospore development. Dev Biol 43:213–220
Yamane H, Takahashi N, Takeno K, Furuya M (1979) Identification of gibberellin A_9 methyl ester as a natural substance regulating formation of reproductive organs in *Lygodium japonicum*. Planta 147:351–356
Yanagishima N (1969) Sexual hormones in yeast. Planta 87:110–118
Yanagishima N (1971) Induced production of a sexual hormone in yeast. Physiol Plant 24:260–263
Yanagishima N (1978) Sexual cell agglutination in *Saccharomyces cerevisiae*: Sexual cell recognition and its regulation. Bot Mag (Tokyo) Spec Issue 1:61–81
Yanagishima N, Shimoda C, Takahashi T, Takao N (1970) Responsiveness of yeast cells to auxin, animal sex hormones and yeast sexual hormones. Dev Growth Differ 11:277–286
Yanagishima N, Yoshida K, Hamada K, Hagiya M, Kawanabe Y, Sakurai A, Tamura S (1976) Regulation of sexual agglutinability in *Saccharomyces cerevisiae* of *a* and α types of sex-specific factors producted by their respective opposite mating types. Plant Cell Physiol 17:439–450
Yanagishima N, Shimizu T, Hagiya M, Yoshida K (1977) Physiological detection of a binding substance for the agglutinability inducing pheromone, α substance I, in *Saccharomyces cerevisiae*. Plant Cell Physiol 18:1181–1192
Yoshida K, Hagiya M, Yanagishima N (1976) Isolation and purification of the sexual agglutination substance of mating type *a* cells in *Saccharomyces cerevisiae*. Biochem Biophys Res Commun 71:1085–1094
Youssef YA (1974) Studies on the rhizosphere mycoflora on broad bean and cotton. Effect of fungal filtrate on plant growth. Folia Microbiol 19:381–385
Youssef YA, Mankarios AT (1975) Production of plant growth substances by rhizosphere mycoflora of broad bean and cotton. Biol Plant 17:175–181
Zenk MH, Müller G (1963) In vivo destruction of exogenously applied indolyl-3-acid as influenced by naturally occurring phenolic acids. Nature 200:761–776
Zentmyer GA (1961) Chemotaxis zoospores for root exudates. Science 133:1595–1596
Zentmyer GA (1966) Role of amino acids in chemotaxis of zoospores of three species of *Phytophthora*. Phytopathology 56:907
Zentmyer GA (1970) Tactic responses of zoospores of *Phytophthora*. In: Toussoun TA, Bega RV, Nelson PE (eds) Root Diseases and Soil-borne Pathogens. Univ California Press, Berkeley, pp 109–111

19 The Ecological Role of Plant Growth Substances

F.B. SALISBURY and N.G. MARINOS

1 Introduction

An ecophysiologist might rephrase our title as a question: "In what ways do events in the life cycle of a plant that are influenced by growth substances contribute to the plant's adaptation in a particular environment?" Adaptation means survival, and survival often means maintaining a favorable energy balance in the plant. So, what roles do growth substances play in survival (often, maintenance of a favorable energy balance) of plants in natural ecosystems?

We have found few papers specifically concerned with the ecological implications of phenomena influenced by growth substances. Apparently, relatively few investigators have studied the comparative growth-substance physiology of a given species in distinct environments, or of two closely related plants adapted to different environments. Yet there is a vast literature on how various environmental factors influence growth substances – and on how applied growth substances influence plant responses to environment. Thus our assignment was to interpret existing information in ecological terms.

Early in our discussions, we found it essential to define certain terms and to delineate some concepts. It is fundamental, for example, to understand the nature of *environment* and to define the terms *growth substance* and *hormone*. A discussion of *stress* was particularly valuable. We could relate stress to virtually all the interactions between a plant and its environment, including those that involve growth substances. Thus, nearly the first fourth of this review is an introduction to these crucial broad ideas.

2 Growth Substances and Ecology

We recognize that the traditional concept of plant hormones is presently being questioned (HANSON and TREWAVAS 1982, TREWAVAS 1981), and we are sympathetic to the need for some re-evaluation of older definitions. The classical concept of a *plant hormone* is that of a natural plant product that controls or influences plant functions (especially growth and development), being synthesized in one location and translocated to another, where its control or influence is exercised via its changing concentration. Hormones are active in extremely small amounts and are thus not acting as energy sources; they are *messengers*.

There seem to be at least two problems with this venerable concept (TREWAVAS 1981): First, it has become increasingly evident that many plant growth substances are synthesized in the same cells where they are active in controlling or influencing plant response. The messenger concept was strongly suggested by work in animal physiology, since animal hormones are often synthesized in a particular gland and then transported to a target tissue where control is exercised. This is probably much less common in plants, although there is still excellent evidence for transport of a few compounds such as the flowering substance (florigen) – although in this case, at least, the substance remains to be isolated and identified. Because the term *hormone* has strong connotations of being a transported messenger, we will avoid its use here unless there really is evidence for such translocation. Hopefully, these matters will be clarified in the not-too-distant future.

Second, it is possible that growth substances do not always control or influence plant responses via changes in their concentration. Control might sometimes be exercised via changes in sensitivity of the target cells, while growth-substance concentrations remain relatively constant. Growth rates of coleoptile segments, for example, are closely correlated with sensitivity of the segments to auxin at a constant concentration; rates are not correlated with tissue concentrations of auxin, which are highest near the tip where there is virtually no growth, and sensitivity changes with time after excision (VESPER and EVANS 1978). Many of the reports that we cite have the implicit assumption that growth-substance concentrations are controlling the process being studied. Only future research can settle the question in most cases, but the traditional approach works fairly well in much of our discussion. We will avoid use of the term *growth regulator* most of the time, because it seems to imply that growth is being regulated by concentration of the substance in question. We do use the term with reference to synthetic substances, a usage that is well established.

Growth substance is perhaps the term least loaded with connotations that need to be tested by future research, although it is more cumbersome than hormone. We define *growth substances* as compounds that are natural plant products active in extremely small amounts, controlling (presumably via concentration) or influencing (with control perhaps being via changes in cell sensitivity to the growth substance) various plant processes, most of which are examples of growth and/or development. We are concerned mostly with the auxins, GA's, cytokinins, ethylene, and inhibitors including ABA, although a few other compounds are also mentioned.

Ecology is the study of the interactions between organisms and their total environment. We shall be concerned with the way processes that are influenced by growth substances contribute to the successful establishment and survival of a plant in particular environmental situations.

3 The Concept of the Environment

In the broadest sense, the *environment* of an organism includes all the surrounding energy and matter in the universe. But certainly that is an unwieldy concept. An alternative is to limit one's attention to those environmental parameters that presently have demonstrable effects on plants, realizing that other environmental factors (e.g., cosmic radiation) may sometime be proven to have important effects. This is the view advanced by MASON and LANGENHEIM (1957), who defined *operational environmental factors* as those that actually influence a particular organism. A corollary is that operational factors must be defined in terms of a particular organism. A cricket's chirp, for example, may be an important aspect of another cricket's environment, but is probably of no consequence to a nearby plant. As far as we know, neither the cricket nor the plant respond to long radio waves, but both certainly respond to heat.

In a thermodynamic sense, a plant is a system, the environment provides the surroundings, and the plant's surfaces constitute the boundary between system and surroundings. SPOMER (1973) has pointed out that operational factors transfer heat or energy or do work across the boundary; such transfer is an *interaction*. Many traditional factors fit this definition nicely: water, light, heat (thermal energy), various gases, mineral elements, organic materials, etc. Many other common terms do not describe operational factors, but instead describe flux measurements or potentials of the operational factors. They include water potential, temperature, pH, partial pressures of gases, concentrations, chemical potential, etc. These measurements indicate the potential for exchange of such operational factors as water, thermal energy, hydrogen ions, gases, solutes, chemicals, etc., across the boundaries of intracellular compartments, cell to cell, one part of the plant to another, and finally between the plant and its surroundings.

Complications arise since plants in a natural environment are influenced by complexes of environmental factors that often vary independently in terms of their flux, direction, and gradients. Gravity is a constant operational factor, but most factors change with time. Three changes are regular and predictable: diurnal, lunar, and annual. Other changes experienced by a plant are irregular, relatively unpredictable, and sometimes totally random. Clearly, a plant must perceive changes in its environment and respond in ways that assure its survival. These responses could well be mediated by growth substances.

4 The Concepts of Stress and Strain

The term *stress* is widely used in biology, but often it is not carefully defined; it may well have different meanings for different biologists.

LEVITT (1972) has defined stress and the resulting strain in a biological system by analogy to their definitions in Newtonian physics. *Stress* is a force applied to an object. If the force exceeds the mechanical resistance in the object, the object is changed in size or shape, and this deformation is called the *strain*. If, when the stress is removed, the original shape is restored (i.e., the strain is completely reversible), the strain is said to be *elastic*. If the original shape is not restored (i.e., the strain is not reversible), the strain is said to be *plastic*.

Biological stress is any environmental factor that limits plant growth or yield below its maximum genetic potential. Environmental stress factors can be expressed in energy terms (as LEVITT suggests), but it is more convenient to express them in the potential units of temperature, pH, water potential, etc. *Biological strain* is the reduced growth or other manifestation induced by the stress. The strain may be elastic as, for example, a plant that stops growing when placed at a low temperature but, after restoration to the initial temperature, resumes its previous growth rate. Such elastic strains in plants have been largely ignored, although most study in animal stress physiology has emphasized this aspect. *Plastic strains* have been much more extensively studied in plants. A plastic strain begins at the *yield point* where the strain becomes permanent: reduced growth rate or death. Growth substances are often involved in plant responses to stress; that is, in influencing the extent of the resulting strains.

For some authors, the term stress has carried a connotation of permanent injury or death (plastic strain), while other authors speak of stress in the sense of elastic strain. LEVITT's careful definition avoids this problem.

As LEVITT (1972) points out, biological systems are more complex than physical ones in at least three ways: First, plants may acclimate to stresses. Second, the time factor is almost always important in biological systems; with prolonged stress, an elastic strain may eventually become plastic (a phenomenon that sometimes also appears in physical systems). We would add that circadian and other internal timing mechanisms may influence the extent to which an organism experiences an applied stress as a strain. For example, a plant's metabolism may be less influenced by low temperature at midnight than at noon. Third, plastic strains in a biological system may be repaired.

4.1 The Concept of a Stressful Environment

What constitutes a stressful environment? Let us approach the question from the other end and first clarify our ideas about a *nonstressful environment*. We suggest that in such an environment plants should grow and reproduce at their maximum genetic potential. The concept of a nonstressful environment has three corollaries:

1. Organisms experience stress and consequently are strained when any environmental parameter limits their growth, reproduction, or some other function to a level below their maximum genetic potential. A temperature response curve provides a good example. Growth (or some other measured response) is usually maximal over a relatively narrow range of temperatures (i.e., optimal temperature, nonstressful condition) and progressively drops off on either side (increasing stress and strain). As the temperature progressively deviates from the optimal range, elastic strains increase until a yield point is reached beyond which growth does not resume its maximum rate even if an optimal temperature is restored.

An example based on a single environmental parameter is unrealistic, however, because several conditions may limit the expression of maximum genetic potential. Furthermore, the apparent optimal (nonstressful) temperature may shift as other conditions change.

A key point is that the exact stress conditions are determined by the individual organism. In some cases, conditions stressful enough to cause death of one organism are essential for the growth of another. (Think of certain salt- and freshwater organisms, the Arctic tundra and the tropical rainforest, etc.) Stress conditions are also determined by the response being studied. Cell enlargement, for example, is much more sensitive to water stress than is photosynthesis (BOYER 1970, BRADFORD and HSIAO 1982).

2. There may be no nonstressful environments. It may be virtually impossible to be certain that a nonstressful environment has been achieved or exists in nature. This corollary focuses on an important practical limitation to the application of LEVITT's concept. Since there are no truly objective standards of comparison, how could one be certain that a plant is growing at its maximum genetic potential – that all limiting elements of the environment have been removed?

One approach to achieving a nonstressful environment is to try all combinations and permutations of environmental factors in an artificial environment. Clearly, the number of potential combinations becomes astronomical, so such an approach remains technologically impractical. This is particularly so because many plants are adjusted to fluctuating environments (diurnal and annual). To try all possible environmental combinations, they must be tried in all possible cycling sequences!

We like LEVITT's definitions in spite of difficulties of applying them in a manner that is completely satisfying. As physicists were long able to consider GALILEO's principles of falling bodies, although these principles could logically only be tested in an unachievable perfect vacuum, so we can perhaps make some reasonable assumptions that allow us to side-step the question of completely nonstressful environments.

We can begin by assuming that a given species must be well adapted to the environmental conditions where it occurs naturally and appears to flourish. Then we can consider to what extent these conditions can be adjusted to produce even better performance. If this always proves to be possible, then according to LEVITT's definitions, plants are always experiencing some level of stress.

3. An organism is most likely to be found in locations where the environment is least stressful for that organism. This must be true if the organism has had the opportunity to get there in the first place. Thus, alpine and desert plants are growing where the environment is least stressful for them, which is not to say that they might not grow better in other environments (e.g., as established by an experimenter); only that such other environments are normally not available to most of them. Competition is an ever-present stress that must eliminate organisms from environments that would otherwise be acceptable or even less stressful. The stresses, and thus the limitations in unsuitable environments, may well be the competition imposed by other plants that can grow larger and/or faster (thereby changing the operational environment), the presence of predators, and so on.

It may be reasonable to assume that the strains experienced by plants growing in their natural environments are mostly of the elastic type; if a completely nonstressful environment could be provided, the plant would immediately begin to grow at its maximum potential. This potential may be approached during the "best years", which must be defined for each plant. Perhaps a criterion for "best years" is the maximum biomass production per unit land area for a population: the highest photosynthetic conversion of light energy to chemical bond energy. From the standpoint of population survival, we might characterize "best years" as those that bring forth the maximum reproductive performance of the population, but in horticulture it is well known that various stresses (e.g., drought, mineral deficiencies, mutilations) lead to increased flowering and seed production in many plants. In such situations, reproductive effort might be thought of as a strain. Nevertheless, abundant water (which is thought to increase cytokinin production by roots) promotes femaleness and high seed production in various monoecious species, and water stress (which is thought to increase gibberellin production) promotes maleness and thus low seed production (FREEMAN et al. 1981).

We do not mean to imply that organisms that usually thrive in a given environment never experience plastic strains. Extreme fluctuations that sometimes occur ("worst years" caused by drought, extremes of temperatures, etc.) limit plant growth to levels well below the optimum, and sometimes individual plants may die, or even whole species may become extinct in a given location. As a rule, however, at least some members of a population normally survive stressful conditions.

4.2 A Model of Plant Responses to Stress

Consider the physiological responses of plants under two kinds of environmental conditions: short-term shifts and the normal seasonal pattern of environmental changes. In the latter case the plant becomes seasonally acclimated to the changing conditions, while in the former case there is no time for acclimation to occur. The total physiology of a plant may be visualized as a complex of distinct and yet interacting processes with rates that oscillate in the course of time in response to both internal controls and changing operational environmental factors. Under relatively nonstressful conditions, when the environmental parameters fluctuate within relatively narrow limits, the oscillations of the rates of the various processes will remain well within the elastic strain limits of the plant (Fig. 1A). In a well-regulated system, although the amplitude and oscillation period will be different for different processes, there will be no violent fluctuations. Feedback will remain unimpaired.

If environmental conditions become temporarily more severe (Fig. 1B), then the rate of one or more processes may oscillate close to the elastic strain limits (the yield points), and certain functions of the plant will operate less efficiently (e.g., gas exchange if the stomates remain closed). Once the environmental conditions return to their original level, then the "deformed" processes will recover their normal oscillation.

Finally, under extreme conditions (Fig. 1C), the rates of most or all of the essential processes will escape the control of feedback loops and will pass the yield points into the state of plastic strain from which there can be no complete recovery.

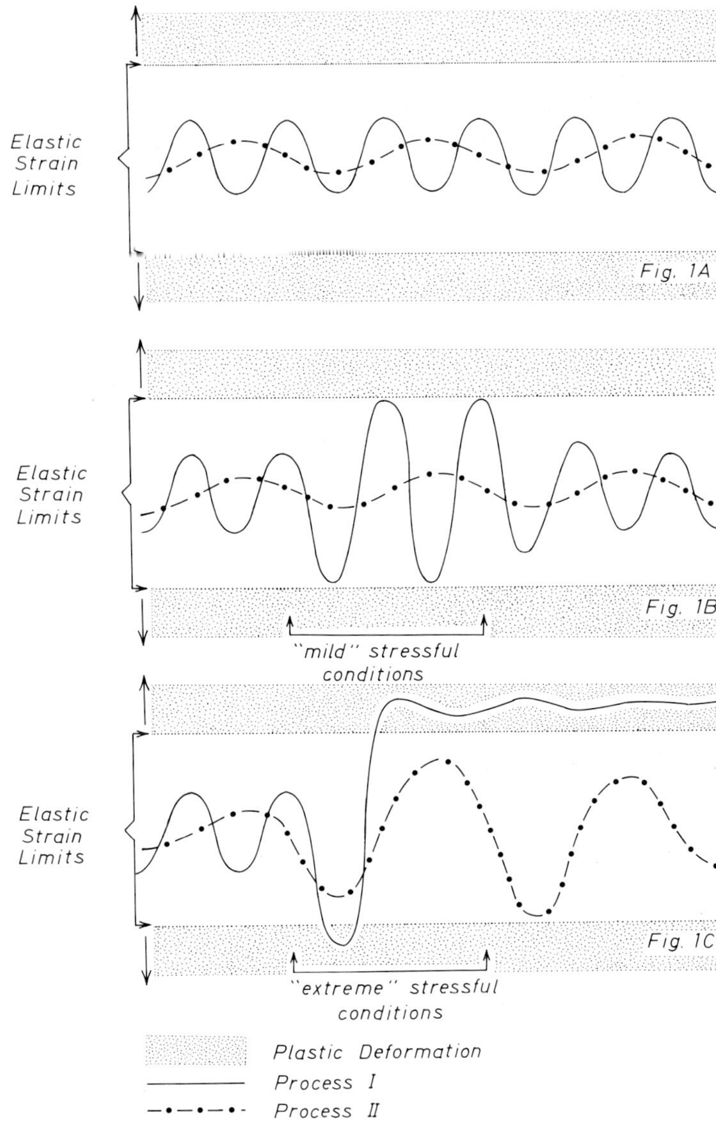

Fig. 1 A–C. Some possible modes of fluctuation of plant response (many plant functions such as photosynthesis, cell elongation, etc.) within or beyond the elastic strain limits of the plant as determined by its genetics and previous experience. When the plant activity fluctuates beyond the strain limits (**C**), the plant may perish

Plants are exposed to *predictable* drifts in seasonal conditions and become acclimated to them. As a consequence, the low temperatures (for example) to which plants are exposed with the approach of winter induce frost hardiness and thus need not be considered stressful, although these same temperatures,

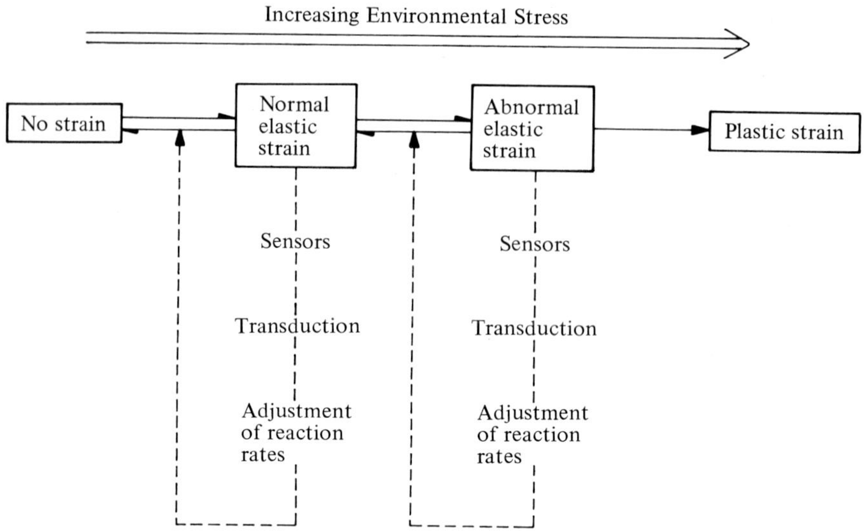

Fig. 2. Illustrating feedback processes that might homeostatically maintain plant functions within tolerable limits during increasing stress

if experienced in the middle of summer, could cause strain that might even exceed the yield points and result in permanent damage (LEVITT 1972).

Although plants acclimate to the seasonal drifts of environment, they are still subject to short-term environmental extremes. If the intensity of any environmental factor does not cause strains that exceed the elastic limits set by the acclimated state of the plant at the time when an environmental extreme occurs, the plant will be able to adjust and survive. If, however, the transient environmental stress pushes the plant past the limits of elastic strain, then the plant will perish (Fig. 1C).

The range within which the limits of elastic strain may shift in response to environmental drifts is determined by the genetic make-up of the plant. For example, a species from the northern temperate region may be able to adjust its temperature elastic strain limits from, say, the range of 20° to 30° C in the summer to −30° C in the winter, whereas a subtropical species may be able to shift only to the range of 5° to 40° C, or not shift at all, in which case if exposed to temperatures below 5° C it would be injured.

These observations suggest that the survival capacity of a plant is a function of two things: First, the range of its elastic-strain yield-limits at a given time (i.e., the wider the limits, the greater the range of environmental conditions it will be able to cope with), and second, the regulated adjustment of its functions so they operate well within the elastic yield limits set by its genetic constitution and state of acclimation.

Very few organisms are adapted to relatively constant environments (e.g., organisms at abysmal depths in the oceans). Most environments are in a state of flux, so plants have been selected for their ability to cope with the kinds and magnitudes of such fluctuations in a given environment. Since no natural environment is likely to allow full expression of the genetic potential of a plant, some level of strain must be normal. One can visualize, then, that the normal physiological status of a plant will oscillate around a theoretical nonstrained condition and within the elastic strain limits. Figure 2 is an attempt to visualize the concept of feedback systems that maintain a plant within its elastic yield limits. Response mechanisms, many of them involving growth substances, maintain the homeostatic status of a plant and thus promote survival.

5 An Approach to the Ecophysiological Role of Plant Growth Substances

We recognize that plants respond to their environment by mechanisms that may not involve growth substances. For example, light is absorbed by photosynthetic pigments, and photosynthesis indirectly affects various aspects of growth and development; enzymes are directly influenced by temperature; and mineral deficiencies affect the physiology of a plant in various ways (sometimes via growth substances!). It is, however, also clear that growth substances could be involved in responses to a wide range of environmental conditions. This is true for the immediate and direct responses to stress that may result in elastic or plastic strains and for the kinds of stresses that are "anticipated" as plants gradually acclimate by responding to changing seasonal environments (e.g., cold hardening during the autumn). A change in some environmental factor is perceived, which might result in a change in the internal concentration of one or more growth substances and/or their translocation. An environmental factor is thus *transduced* to a *messenger* or stimulus, which could be a chemical or chemicals. Since our central theme depends on the demonstration that environmental conditions are transduced in the plant via plant growth substances, it is appropriate to establish that substance levels are, in fact, influenced by a wide range of environmental conditions. Many papers referred to later in this discussion collectively establish this point.

An important and little-appreciated aspect of the acclimation responses of plants is the role of biological rhythms. Plants, like other organisms, have internal timing mechanisms that apparently adjust their metabolic systems so that they function optimally under changing environmental conditions (BÜNNING 1973, HILLMAN 1976, SALISBURY and ROSS 1985, SWEENEY 1969). For example, leaves of many flowering plants are horizontal during the day and vertical at night. That is, they are in an ideal position to photosynthesize during the day and in a position less likely to absorb moonlight at night, light that could affect the flowering of photoperiodically sensitive species (BÜNNING 1969). These manifestations are examples of circadian (approximately 24-h) rhythms. Probably growth substances are involved, perhaps as messengers between the biological clock and the observed response. Indeed, WATANABE and SIBAOKA (1983) found that light of suitable wavelengths (726 and 403 nm but not 585 or 656 nm) would cause detached pinnae of *Mimosa pudica* to open, but only from 06:00 to 11:00 h (demonstrating participation of the biological clock). Applied IAA, NAA, or 2,4-D would cause opening any time.

We considered a number of possible ways to organize our sample of papers from the over 1000 available to us: according to environmental factors, specific growth substances, or the phenological events that occur during the annual cycle of a "model" flowering plant. Although each of these approaches has its merits, the annual-cycle scheme is perhaps the most ecological. Yet in practice we were not able to apply it rigorously. We have modified it into the following four topics, considering in each case how growth-substance-mediated adaptive/acclimation responses might have survival value:

1. They regulate the *timing* of certain phenological events so that the events occur at times when environmental conditions are likely to be most favorable (e.g., the timing of germination and flowering).
2. They modify the *morphology* of a plant and thus reduce the impact of stressful environmental conditions (e.g., leaf abscission in autumn or during drought).
3. They conserve some particular *resource* of the plant during periods of stress (e.g., water and energy conservation).
4. They *reduce metabolic and growth activities* and thus render the plant less vulnerable to environmental stresses (e.g., winter and summer dormancy, frost hardiness).

In addition to responses that can be accommodated into at least one of the above groups, there are some ecophysiologically important responses that do not easily fit this scheme. They will be considered separately.

Our discussion is rather general, emphasizing the broad role that growth substances might play in an ecological sense. Our intention is to concentrate on cases where growth-substance involvement is firmly established or strongly indicated, but it would not take much imagination to link every environmental response to some chemical mechanism! We will think mostly in terms of natural ecosystems, but much work has been done with agricultural plants, and we will not hesitate to refer to it. Although synthetic growth regulators have many long-term implications for ecosystems, they fall outside the scope of this review.

6 The Adaptive Importance of Timing

It is axiomatic that no adaptive response is of any value unless it occurs at the right time in relation to prevailing or anticipated environmental conditions. Precise timing is important in the control of short-term events, for example in the regulation of water balance of a plant, where a sensitive and rapidly responding perception and transduction system must operate that is essentailly devoid of anticipatory properties and functions. It is a "here-and-now" type of response. The importance of timing becomes even more striking when we consider seasonal events where gradual adjustment of metabolic, physiological, and structural features of a plant are so timed that when environmental conditions are likely to be either for the process (germination, early seedling growth, flowering, and fruit development) or highly stressful (winter cold or summer drought), they have been anticipated.

Although the notion of timing pervades all the adaptive events that we shall consider in this paper, reproductive development and germination are particularly well suited for a more general discussion and illustration of timing mechanisms that involve growth-substance controls.

6.1 Reproductive Development

The timing of sexual reproduction is crucial for the survival of a species. In most plants it is triggered or at least influenced by environment, especially temperature and photoperiod (BERNIER et al. 1981, CHAILAKHYAN 1968, EVANS 1969, 1971, LANG 1965, SALISBURY 1981, 1982, SCHWABE 1971, VINCE-PRUE 1975, ZEEVAART 1976). That flowering-promoting hormones must be involved has seemed fairly clear since the late 1930's, although it has often been suggested that combinations of known growth substances might be involved rather than some special florigen (CHAILAKHYAN's term 1968), and recently the importance of assimilates (nutrients) has again been emphasized (see discussion below). The role of inhibiting substances has been much more controversial. Alternative explanations, such as environmental effects on translocation, could almost always be considered (ZEEVAART 1976). Finally, in a series of grafting experiments with SD, LD, and day-neutral tobacco plants, LANG et al. (1977) clearly showed that inhibiting as well as promoting substances were a normal part of the flowering process. Yet the chemical natures of both florigens and flower inhibitors remain as elusive as ever. It has not been possible to reliably extract florigen(s) and flower inhibitor(s). A good bioassay is needed. Numerous solvents and application techniques have been tried, and sporadic successes have been reported. The best-known method (LINCOLN et al. 1966) extracts quick-frozen induced *Xanthium* leaves with organic solvents, followed by application to noninduced plants. Extracts made from vegetative plants have no effect, but extracts from a day-neutral sunflower and even from a fungus (*Calonectria*) have sometimes been effective. Only a low level of flowering is induced in test plants, and GA applied with the extracts sometimes promotes flower development. Activity in the extracts follows the acid fraction and has been called florigenic acid. The experiment has sometimes succeeded (CARR 1967, HODSON and HAMNER 1970); it has often failed, even when carried out exactly according to instructions (CLELAND 1978, SALISBURY 1982).

CLELAND and AJAMI (1974) collected honeydew from aphids feeding on induced *Xanthium strumarium* (cocklebur) and other species. Using a LD duckweed as a test plant, strong florigenic activity was found. Two active substances proved to be salicylic acid and acetylsalicylic acid (common aspirin!). Unfortunately, salicylic acid and aspirin do not induce flowering in vegetative cocklebur or other tested plants. Furthermore, salicylic acid concentrations are equal in the honeydew of both vegetative and induced cocklebur plants.

Few attempts have been made to extract floral inhibitors, although gallic acid appears to be a specific flower-inhibiting substance in vegetative plants of the SD *Kalanchoe blossfeldiana* growing in LD (PRYCE 1972).

In the early part of this century, there was much discussion about the balance in carbohydrate and nitrogen in plants, suggesting that a high carbohydrate/nitrogen ratio promoted reproductive development, while a low ratio promoted vegetative growth. During recent years, SACHS (1977, BERNIER et al. 1981) has proposed that reproductive development does indeed involve a diversion of sugars and perhaps other nutrients to the meristems that will form flowers. He emphasizes that there must be much validity to the old idea that treatments promoting flowering (e.g., pruning, ringing of fruit trees) do so by inhibiting vegetative growth, and that this is because nutrients are diverted to the reproductive meristems. SACHS and colleagues have studied *Bougainvillea,* a SDP, showing that SD increases photosynthesis and that this is reversed by GA treatment, which also inhibits flowering (EVEN-CHEZ and SACHS 1980). A cytokinin (benzyladenine) promotes flowering, perhaps by diverting nutrients to the meristems, but GA overcomes the cytokinin effects, apparently by promoting conversion of the benzyladenine to the riboside (EVEN-CHEZ et al. 1979). VANSTADEN (1981) found increasing cytokinins in *Bougainvillea* under SD.

Numerous experiments have been performed in which known plant growth substances and growth regulators have been applied to plants to test their effects upon flower induction or development. We will summarize information presented in much more detail by VINCE-PRUE (1975) and add some recent papers.

6.1.1 Auxins and Ethylene

More often than not, auxins inhibit flowering (VINCE-PRUE 1975, LOZHNIKOVA 1977). In SD plants, this occurs before translocation of florigen from the leaf is complete, after which there may be marginal promotive effects. Promotions have also been observed in LD plants held under days just too short for induction, and auxin clearly promotes flowering in some bromeliads, including the pineapple. Applied auxins cause a production of ethylene, which itself influences flowering the same as auxins (inhibition in SD plants, promotion in bromeliads). In the bromeliads, IAA is relatively ineffective, since it is apparently broken down by the plant's enzymes, but synthetic auxins such as NAA or 2,4-D are more effective. Auxin concentrations required to inhibit flowering usually produce severe epinasties and other responses, and measured plant auxin levels seldom correlate with flowering in any meaningful way. Hence, although endogenous auxins may influence flowering to some extent, they probably do not control it.

6.1.2 Gibberellins

GA's will substitute for the LD requirement of many LDP's and for the cold requirement in many plants that require vernalization (COTTRELL et al. 1982, PARUPS 1980, VINCE-PRUE 1975). GA's seldom substitute for both a LD and a vernalization requirement, however. A few LDP's fail to respond, and several SDP's do respond to GA's (e.g., cosmos and rice, VINCE-PRUE 1975; *Impatiens*, SOOD and NANDA 1979, SAWHNEY 1979; *Sorghum*, WILLIAMS and MORGAN 1979). Sometimes, GA inhibits flowering of SDP's (e.g., strawberry, TAFAZOLI and VINCE-PRUE 1978), and sometimes GA inhibitors promote flowering of SDP's under LD (CCC promotes flowering of *Salix*, JUNTTILA 1980). In many species, GA's increase under LD, so it seemed reasonable to assume that a build-up in GA's might be part of floral initiation in LDP's and cold-requiring plants; but the situation is not so simple. In LDP's, for example, LD's increase both the rate of synthesis and the rate of loss of GA's, and many studies have shown a lack of correlation between extractable GA's and flowering (HILLER et al. 1974, ZEEVAART 1976). ZEEVAART stated in a 1976 review that there was "conclusive evidence" that flower formation and stem elongation were separate processes, with GA promoting stem elongation only (see also discussion below on GA's in vernalization). Nevertheless, ZEEVAART and co-workers (BENGSTON et al. 1979, JONES and ZEEVAART 1982, METZGER and ZEEVAART 1982, PROEBSTING and HEFTMANN 1980) have continued to study the metabolism of the various GA's in LDP's. They have evidence that certain GA's (e.g., GA_{20}, GA_{29}, polar GA's) play a much more important role than others.

KAREGE et al. (1982) have studied spinach (*Spinacia*), in which flowering can be induced by LD (red light promotion), low temperatures, and GA_3. They find a specific change in isoperoxidase activity in the leaves in response to all three treatments. Effects of red light were reversed by far-red.

GA's are often effective in shortening the juvenile period and inducing reproduction in conifers. KATO et al. (1958), PHARIS and KUO (1977), and others have pioneered in this work. PHARIS and KUO induced formation of staminate strobili on Arizona cypress (*Cupressus arizonica*) when plants were only 55 days old by spraying with GA_3, but the less polar GA's (GA_4, GA_5, GA_7, GA_9) are most effective in achieving cone formation in members of the Pinaceae. Several researchers have used $GA_{4/7}$ with various results: increased male-cones on *Picea sitchensis* when combined with GA_3 (TOMPSETT and FLECHTER 1979); increased female cones on *Pinus taeda* when combined with water stress (GREENWOOD 1981); increased strobili on *Tsuga heterophylla* when combined with drying cycles (BRIX and PORTLOCK 1982); no promotion of cone initiation on *Pinus contorta* although there were interesting clonal differences and promotions by SD and cool temperatures (LONGMAN 1982) and increased cones of both sexes on *Picea sitchensis* when combined with heat or water stress (PHILLIPSON 1983).

6.1.3 Cytokinins

Cytokinins promote flower formation in a few SD and LD plants (VINCE-PRUE 1975), although effects are usually marginal and often complex (e.g., a promotion when plants receive near-threshold dark periods (OGAWA and ZEEVAART 1967); a promotion when combined with red-light interruption in *Pharbitis nil* (OGAWA and KING 1979). A combination of cytokinin (benzyladenine) and GA's clearly induced flowering of one SD variety of chrysanthemum (PHARIS

1972), but most work (unpublished) suggests that cytokinins do not affect flowering in most LDP's and SDP's.

6.1.4 Abscisic Acid

When certain SDP's (e.g., *Chenopodium rubrum*, KREKULE and KOHLI 1981) are already slightly induced and then treated with ABA, flowering is promoted. But no conclusions can be drawn, since ABA inhibits flowering in other SDP's and LDP's (e.g., *Spinacia*) and is completely innocuous in still others (VINCE-PRUE 1975). ABA is generally but not always higher in shoots under LD's.

6.1.5 Sterols

A substance called TDEAP [tris-(2-diethylaminoethyl)-phosphate trihydrochloride], which inhibits synthesis of cholesterol in animal systems, inhibits flowering of *Xanthium* and *Pharbitis*. Does this imply that florigen is a sterol related to cholesterol? Probably not (ZEEVAART 1976), since application of various sterols does not overcome the effect of TDEAP, nor can any significant changes in sterol fractions be detected during induction. Sterol biosynthesis can be inhibited by compounds other than TDEAP without influencing flowering. Flowering can be induced by application of sitosterol and lanosterol to *Chrysanthemum morifolium* cv. Princess Anne and estradiol to *Callistephus chinensis*, both SD plants (PARUPS 1980).

6.1.6 γ-Tocopherol

BATTLE et al. (1976, 1977) measured γ-tocopherol (a form of vitamin E) in leaves of cocklebur plants at various times during an inductive dark period and following 20-min light interruptions given at various times during such a dark period. By 10 to 12 h of continuous darkness, the γ-tocopherol had increased in the leaves to levels five times those at the beginning of the dark period. Red light interruptions at various times up to and including 8 h after the beginning of darkness maintained the level of γ-tocopherol at about that measured at the beginning, and the effect was completely reversed by FR. Furthermore, when the light interruption was given 9 h after the beginning of darkness, the substance was almost as high as in dark controls. Just as floral induction reaches a point where it can no longer be reversed by light, so did the level of γ-tocopherol, and at the same time. This excellent correlative study suggests that γ-tocopherol might play some critical role in the induction process or even act as a hormone. Further studies are needed.

6.1.7 Some Complications

Such environmental parameters as temperature, photoperiod, and nutrient supply could influence the establishment and survival of a species in a given environment by affecting not only the timing of reproduction but also sex expression and fertility. And these parameters could be acting via effects on growth substances. All the common growth substances have influenced sex expression in certain species (VINCE-PRUE 1975). The most consistently effective substances appear to be auxins and ethylene, both of which favor femaleness in several different plants (e.g., *Cannabis sativa, Cucumis sativus, Cucurbita pepo)*. Often this is correlated with SD treatments, which also increase natural auxin and ethylene levels. GA's typically act the same as LD, promoting either maleness or femaleness. As is so typical in this field, however, there are many exceptions and complications. For example, TAKAHASHI et al. (1983) found that femaleness was promoted by SD in a SD-sensitive cultivar and by LD in a LD-sensitive cultivar of cucumber. But both GA and ethylene were higher in both cultivars under SD than under LD. Thus, effects of photoperiod could be separated from effects of GA or ethylene on sex expression. GREGG (1982) found more ethylene and more female flowers when *Catasetum* and *Cycnoches* plants were grown in the shade than when they were grown in the sun. Here, then, is an effect of light level rather than daylength on growth substances and on sex expression (although a cause/effect relationship was not demonstrated).

The responsiveness of different species to environmental cues like photoperiod varies widely. For example, the morphological transformation of the apex from vegetative to reproductive may take place within 2 or 3 days from the time the photoperiodic cue has been perceived, e.g., in the case of plants that respond to a single inductive cycle (*Xanthium strumarium, Lolium temulentum, Pharbitis nil*). Most photoperiodically sensitive plants, however, respond only after repeated exposure to inductive cycles, and the morphological transformation of the apex is not expressed until a considerable time has elapsed. Such a protracted period of induction is also common to such plants as winter cereals and biennials that require exposure to cold early in their life cycle before they eventually flower.

Some plants have complex induction requirements (VINCE-PRUE 1975). For example, *Trifolium repens* will flower under LD provided it has previously experienced SD, while the reverse sequence of photoperiodic regimes is required by *Kalanchoe laxiflora*. The situation becomes even more complex when both photoperiodic and temperature requirements must be satisfied before floral induction can occur.

6.1.8 Vernalization

Low temperatures have a rather special effect on the flowering of some species (*vernalization*), although the subject has not been researched much in recent years. For example, sugar beet, celery, and biennial henbane have an absolute low temperature requirement, while other plants such as winter cereals show a quantitative response in that their flowering is accelerated by low temperatures (EVANS 1969). Do growth substances play a role in this fundamentally ecological response? If a "hormone" is involved, it is not transportable within the plant. With a few exceptions (LANG 1965, PURVIS 1961), it has been found that the meristematic regions that eventually become flowers must be exposed directly to low temperature. However, the work by WELLENSIEK (summarized in WELLENSIEK 1962, ZEEVAART 1976) suggests that young expanding leaves may act as perceptive sites. It seems that the fundamental requirement for the perception of the low temperature stimulus by an organ is that it be in a phase of active cell division, or at least DNA synthesis.

The nature of the low temperature stimulus is as elusive as that of the agent responsible for photoperiodic induction. The "vernalin" postulated by MELCHERS has about the same status as "florigen". Of the known growth substances, only GA has been shown to replace the cold requirement of certain rosette plants (LANG 1965, PURVIS 1961, WELLENSIEK 1962). Furthermore, in some cases, small increases in GA level occur during cold treatment (EVANS 1971, LANG 1965), but there is great reluctance to equate GA with "vernalin". As with LD plants, there is an important difference in the developmental behavior of plants that have been induced by low temperature or by GA: When plants are induced to flower by exposure to cold, floral differentiation is apparent before the shoot begins to elongate, but when GA's are applied to a cold-requiring rosette plant, the first observable response is the elongation of a vegetative shoot followed by the transformation of the apex to the floral state (NITSCH 1966, ZEEVAART 1976).

Could GA's induce changes that in turn lead to flowering or to the production of "vernalin"? Could several substances influence the morphogenetic program in different but coordinated ways so that the end result is the same? Various hypotheses have been proposed to explain a great diversity of observations, for example CHAILAKHYAN's (1968) view that two hormones are involved (GA and "anthesin"), but such ideas remain speculative.

There are a number of ecophysiological implications of vernalization that have not been explored to any extent. For example, since the vernalization effect is apparently localized in the region of the plant that has actually been exposed to low temperature, and since there is an optimal vernalization temperature (around 5° C for most species, PURVIS 1961), then the overwintering habit and microclimate in the immediate vicinity of the meristematic regions is a crucial determinant of flowering behavior.

The vast majority of plants in the alpine tundra, for example, are perennial, overwintering mostly underground and with perennating buds, and completing their reproductive phase during a short period of time when environmental conditions are least severe (BILLINGS and MOONEY 1968). The timing of floral inception and subsequent development are crucial to survival and could be closely linked to the low temperatures that the meristematic regions experience, although there is little evidence that such plants require vernalization. In any case, plants strictly adapted to the alpine tundra (e.g., *Geum turbinatum*) have failed to initiate flowers away from their natural environment (SPOMER and SALISBURY 1968). In *Geum,* flowers first appear within 1 to 3 weeks after the onset of spring growth. These flowers were formed as primordia during the previous July. We have unpublished data suggesting that soil temperatures could be important at that time.

Clearly, plants perceive complex environmental cues that set in motion morphogenetic mechanisms that are timed so that flowering occurs at an optimal time of the year when physiological strains imposed by environmental stresses are likely to remain within elastic limits. Any process that depends on environmental cues requires some insurance against responding to aberrant cues that may occur "out of season". Although changes in photoperiod are reliable as a timing cue, temperature, particularly in its short-term fluctuations, is quite unpredictable. Some temperature-dependent mechanisms such as flowering in fruit trees (GUSTA et al. 1982) operate on a cumulative basis, in which an event occurs only after the plant has experienced some definite number of hours within a specified range of temperatures (LOMBARD and RICHARDSON 1979).

Another way the effect of unseasonal temperature cues can be minimized involves a "two-step" timing mechanism. Many species are known (e.g., henbane, the winter cereals) that have a vernalization requirement followed by a LD requirement (PURVIS 1961, VINCE-PRUE 1975), for example, and a chrysanthemum variety and several grasses have a cold requirement followed by a SD requirement.

Our understanding of the ecophysiological role of plant growth substances in the timing of reproductive events might be greatly advanced by studies of plants adapted to naturally harsh environments. The chances of correlating ecophysiologically significant cues and growth-substance physiology could be good in such a situation. There are limitations to how far laboratory studies of a few favored plants can be extrapolated to the field.

6.2 Germination in General

Germination must occur at a time when environmental conditions are conducive to the establishment of the seedlings and development and completion (annuals) of the life cycle during the less stressful times of the year. Such timing requirements are particularly stringent in such marginal environments as deserts and tundras.

Chemical mechanisms are known to prevent germination even when the embryo is morphologically and physiologically mature. Such mechanisms have been extensively discussed in various reviews and books (e.g., BEWLEY and BLACK 1978, 1982, KOLLER 1969, MAYER and SHAIN 1974, ROBERTS 1972).

The initiation of a physiological event like germination could come about either through the promotive activity of some substance or the disappearance of some inhibiting substance that maintained the system in a suppressed state. The two mechanisms are not mutually exclusive, and the event could depend on the balance of promoting and inhibiting substances, a suggestion that is encountered frequently in the literature. It seems, however, that the release of a system from an inhibited condition by the leaching out of some inhibitor(s) is more likely than the synthesis of some promoter(s) when the seed is dry and the metabolic system of the seed is clearly at a low ebb. Indeed, most changes in growth substances that have been observed occurred in seeds that were at least slightly hydrated or in seeds that were involved in the germination process. Removal of an inhibitor in a single rainstorm heavy enough to wet the soil to a considerable depth would ensure water for survival of the growing seedling (KOLLER and HADAS 1982, WENT 1957).

One important point that emerges from the wide range of germination studies is that seed dormancy is often under control of a series of mechanisms. Some seeds may contain leachable inhibitors, for example, have a cold requirement, and be light-sensitive. From the ecophysiological point of view, such a multiple back-up system might be ideal for survival, particularly in unpredictable environments.

There have been many studies of growth substances in seeds and of effects of applied growth substances on germination of seeds, especially dormant ones. Changes in substances during early germination could be secondary events instead of causitive, and thus not critical from the ecophysiological point of view, but such changes could also be responses to the environmental conditions that lead to germination. Consider a few recent examples from the literature:

Ethylene is known to promote germination of many seeds. For example, germination of cocklebur (*Xanthium*) and redroot pigweed (*Amaranthus*) seeds was strongly promoted by injecting ethylene into the soil (Egley 1980). Schonbeck and Egley (1981) found greatly enhanced sensitivity of pigweed to ethylene after seeds had been buried in the soil, and suggested that this could be important in nature, but Adkins and Ross (1981) concluded that endogenous ethylene production was not likely to be a main cause of dormancy breakage in *Avena fatua*.

Many authors have reported a promotion of germination by GA application and/or an increase in endogenous GA during the early stages of germination. For example, Taylor and Wareing (1979) found a promotion of sitka spruce (*Picea sitchensis*) germination in response to a 10-min exposure to red light or to chilling. Both treatments led to rapid increases in cytokinins and GA, but GA declined during the subsequent dark period. It is not unusual for GA_3 to be less effective than $GA_{4/7}$ (e.g. Nelson and Sharples 1980). Carpita and Nabors (1981) found that both red light and GA induced a water potential decrease in the axes of lettuce (*Lactuca*) embryos resulting in germination of intact seeds or growth of the axes of isolated embryos. The decreased water potential would lead to increased water uptake and hence higher pressures in cells of the swelling axes. Germination of *Phacelia tanacetifolia* seeds is inhibited by light and promoted by darkness, opposite to the situation with most light-sensitive seeds (see next section), yet germination is accompanied by an increase of GA-like substances (Cocucci et al. 1981).

Several workers have observed a clearly defined but transient peak of endogenous cytokininlike substances during germination of both dormant and nondormant species (Julin-Tegelman and Pinfield 1983 and references cited therein). The seeds typically required a cold period to break dormancy, raising the possibility that cytokinins are intermediates in the transduction of the cold stimulus to the breaking of dormancy. There are other subtle effects. For example, treatment of lettuce seeds with kinetin raised their upper temperature limit for germination in the light from 22.5 to 30.5° C (Gray and Steckel 1977).

In contrast to the growth substances mentioned so far, ABA typically inhibits germination of seeds (Daines and Minocha 1983). Among other things, this is because of reduced growth potential of the embryo (Takeba and Matsubara

1979). (Increased growth potential was indicated by water absorption into embryos from more concentrated mannitol solutions; i.e., from solutions with increasingly negative osmotic potentials.) Again, there are subtle effects of possible ecophysiological importance such as an increased drought resistance induced by seed pretreatment with ABA (GOLLI and LEVI 1982).

6.3 Germination and Phytochrome

Germination of light-sensitive seeds, along with many other phytochrome-dependent responses, presents an ecophysiologically interesting situation, because the phenological timing of one species may affect the life cycle of another species that is a member of the same community. The control of germination by phytochrome (ROLLIN 1972, SATTER and GALSTON 1976) has special ecological significance in situations where there is a seasonal change in the spectral composition of the light at ground level (MORGAN and SMITH 1981). The canopy of a dense stand of herbaceous plants, or of a deciduous forest, is relatively transparent to FR but not to R or blue light. Under these conditions, the phytochrome system of any plant or seed under the canopy is shifted toward the inactive P_r form. Early in the spring, before the leaf canopy forms, and with the coming of autumn as chlorophyll disappears and leaves abscise, the spectral composition at ground level, even under dead leaves, will be richer in R, so the phytochrome system will shift toward the physiologically active P_{fr} form (GORSKI 1975, HOLMES 1976, KASPERBAUER 1971), and germination may occur (see also Sect. 7.5).

GORSKI et al. (1978) placed seeds of 110 herbaceous species in dishes under a dense canopy of rhubarb leaves that permitted no sunflecks. Light under the canopy was greatly enriched in FR compared to normal sunlight. Seeds of 88 of the species were more or less inhibited in their germination, exhibiting what the authors called FR dormancy. Altogether, these authors have studied 240 species, finding that FR dormancy is much more common among wild plants than among cultivated ones.

How deep can seeds be buried and still respond to the light? WOOLEY and STOLLER (1978) observed some germination of light-sensitive lettuce seeds 2 mm below the soil surface but not 6 mm.

There is evidence that phytochrome-dependent germination is mediated by plant-growth substances. We noted above that phytochrome has been linked to the production of various substances in seeds and seedlings (see also KANG and BURG 1973, REID et al. 1968, VAN STADEN 1972), and application of GA and cytokinins may replace the light requirement in several species. In spite of these leads, there is a large gap in our understanding of any mechanism whereby phytochrome can control growth-substance levels. SCHOPFER (1977) expresses the view that on current evidence growth substances do not appear to serve "as causal links in P_{fr}-mediated enzyme control" and presumably metabolic and developmental events.

7 Morphological Adaptations and Responses

We suspect that the overall morphology and anatomy of a plant reflect an optimal adaptive configuration for a particular environment, but this is certainly not simple or obvious. Consider plants of greatly differing morphology flourishing side by side, such as succulents sharing a desert environment with annual mesophytes and nonsucculent perennials.

7.1 Vegetative Morphology

Morphology is often a dynamic feature that is subject to growth-substance influence, responds to a range of environment stresses, and contributes to the maintenance of a favorable energy balance. Examples of environmentally dependent changes in the morphology of a plant would be the relative amounts of leaf area carried by deciduous plants in the summer and winter, or plants that lose their foliage during drought. One of the most ubiquitous responses of plants to photoperiod is stem elongation in response to LD (VINCE-PRUE 1975); such a response is virtually independent of flowering response-type (i.e., whether the plant has a LD, SD, or DN flowering response). Of course, stem elongation is considered to be controlled by growth substances. Other situations are more subtle. For example, hybrid vigor (heterosis) in maize has been correlated with increased GA-like substances (ROOD et al. 1983), and GA_3 apparently influences leaf and plant form of the heterophyllous *Proserpinaca intermedia;* photoperiod (phytochrome) and temperature are also critical (KANE and ALBERT 1982). We should not lose sight, however, of the possibility that some features might simply indicate evolutionary trends that have not been eliminated by selection but confer no particular advantage to a plant in a given environment.

7.2 Senescence and Abscission

Development of the *autumn syndrome* (formation of dormant buds, development of frost hardiness, leaf abscission, etc.) is one of the most universal responses of temperate-zone plants to SD. Virtually always, plants also develop the autumn syndrome in response to low temperature (VINCE-PRUE 1975); in nature, the response is usually to both conditions.

Senescence and abscission of leaves and other organs is not simply a degradative process, since during its occurrence and particularly in the early stages there is active synthesis of enzymes that mobilize reserve materials and, in the case of abscission, separate cells in the abscission zone (ABELES and LEATHER 1971).

When leaves senesce and abscise in response to extended drought or short photoperiods and temperatures that signal the coming of winter, the morphology of a plant is changed to an energetic configuration that matches the circumstances. Water resources are conserved, since transpirational losses are nearly abolished, and the loss of leaves before the time of killing frost is life-preserving.

In addition, some energy, as well as minerals and carbon chains, are conserved by a redistribution from the senescing leaves to storage sites in the stem, roots, or specialized organs.

The role of growth substances and particularly of IAA, ABA, and ethylene in the control of senescense and abscission have been repeatedly reviewed (e.g., ADDICOTT 1970, BEEVERS 1976). The discovery in recent years that ABA and ethylene increase under the stress conditions that induce senescence and abscission has infused new interest in this field (e.g., DARENPAT et al. 1977, JORDON et al. 1972, MIZRAHI et al. 1971, MORGAN et al. 1977, PORPIGLIA and BARDEN 1980). Interpretations based on a single-substance (IAA) control system, which dominated thinking until the 1960's, became inadequate.

The current picture may be summarized as follows: A progressive lowering of water potential (or SD's or low temperatures) leads to physiological strains that initiate the formation of ABA and ethylene, which in turn accelerate the processes of senescence and abscission. These are physiological facts, but our understanding of how environmental signals direct the production or destruction of specific substances remains elusive. Again, phytochrome must play a pivotal role in the perception and transduction of photoperiodic stimuli (see Sect. 6.3).

The conclusions based on correlations between endogenous growth-substance levels and the occurrence of senescence and abscission are reinforced by observations on the effects of applying various substances singly or in combinations. Active substances fall into two groups: those that accelerate senescence and abscission (i.e., ABA and ethylene) and those that under certain conditions retard them (i.e., cytokinins, GA's, and IAA). Various interactions complicate interpretations, however. For example, effects of ABA may be caused by an induced production of ethylene (CRACKER and ABELES 1969), and the stimulation of abscission by the delayed application of IAA could also be the result of ethylene production (LIPE and MORGAN 1972). Furthermore, ADDICOTT (1970) discusses cases where abscission was stimulated by exogenous GA, and RAILTON and WAREING (1973c) found that ABA increases the endogenous level of GA. PETERSON et al. (1980) found no relationship between ABA content of *Ficus benjamina* leaves and abscission.

Apparently, the substances involved in the control of these phenomena influence various physiological processes that accelerate or retard senescence and abscission. For example, ethylene stimulates the activity of various enzymes, possibly by releasing the enzyme from a bound form (ABELES and LEATHER 1971, VARNER and HO 1976). Cytokinins direct metabolic transport to sites of high cytokinin content, and their absence allows degradation processes to dominate the overall metabolic status of the plant. In their presence, senescence is retarded (KAO 1980). Membranes are probably involved (ELLIOTT 1982). However, goes of these processes discussion beyond the scope of this review.

7.3 Apical Dominance

Apical dominance contributes to the overall shape of the shoot and root systems of a plant. For example, a plant in which apical dominance is weak is more bushy and sprawling, so the scaffolding of its branches can support a relatively large leaf area – perhaps an advantage in the interception of light. On the other side of the coin, strong apical dominance at the expense of a profusion of laterals may be a distinct advantage for plants growing in crowded conditions that restrict the light available to each plant.

Apical dominance is one of the earliest phenomena in which the correlative action of auxins was implicated. In view of subsequent information on the possible involvement of other growth substances (e.g., BURG and BURG 1968, DÖRFFLING 1966), the original, perhaps simplistic interpretation based on the growth-inhibiting effect of apically derived auxin upon lateral buds has proven inadequate. Older ideas involving the flow of nutrients (GREGORY and VEALE 1957) and the importance of the vascular relationships of the lateral buds to the main conducting system (SOROKIN and THIMANN 1964) have re-entered the arena. For this review the mechanism of apical dominance is of secondary importance to the manner in which environmental factors influence it via the plant's growth-substance physiology. It is, after all, the modifying influence of factors like light and temperature on apical dominance that contributes to the development of a particular growth habit adapted to a given environment.

Examples of work on environmental influences on apical dominance are rather scarce, but the following illustrate at least the beginning of an ecophysiological approach to the phenomenon.

Light has been explored by JACKSON and FIELD (1972) and SHEIN and JACKSON (1972), who studied many complex interactions between light, other environmental factors, known plant-growth substances, and apical dominance. No meaningful relationships were found, and they concluded that growth of laterals and main stems was dependent on a hormone balance that could be critically modified by a wide range of internal and external factors, the nature of which remains to be determined. WOOLLEY and WAREING (1972) also concluded that "lateral shoot development is controlled by a delicate balance between auxins, GA's, and root-produced cytokinins." Two other papers are illuminating: PILLAY and RAILTON (1983) were able to completely release *Pisum sativum* buds from apical dominance, producing rapidly elongating, leafy shoots, by application of two cytokinins: zeatin or 6-benzyladenine. Kinetin or isopentenyladenine caused only limited outgrowth that stopped 6 days after application. DAMPTEY et al. (1978) note that in *Zea* the lower axillary inflorescences are promoted to grow by water stress during early tassel development. They observed increased ABA in response to the water stress and were able to promote the axillary influorescence development with applied ABA.

TUCKER and MANSFIELD (1972) showed that lateral buds of cocklebur were completely inhibited when 30 min of illumination from tungsten lamps were given after a 16-h photoperiod from fluorescent tubes, whereas plants given an identical photoperiod (16.5 h), but only from fluorescent tubes (no FR from tungsten lamps), branched profusely. The cytokinin content was much higher in inhibited than released buds, but so were inhibitors resembling ABA. Inhibitor concentrations in inhibited buds were 50 to 250 times those occurring in all other plant parts examined. It appears that end-of-day FR, by removing the FR-absorbing form of phytochrome (P_{fr}), in some way mediates a sharp increase in inhibitor concentration in this species. Strong apical dominance is also a feature of plants grown in heavy leaf shade, as discussed in Section 7.5.

7.4 Etiolation and Plumular Hook Opening

The capacity of plants to sustain rapid growth in the dark can be considered an adaptive response that brings the leaves to the light as soon as possible. This is accomplished by rapid growth of epicotyls and hypocotyls and inhibition of unfolding of the plumular hook and embryonic leaves until the dicot seedling emerges. The compact and mechanically strong configuration of the arched plumule bears the brunt of the physical forces of pressure and friction in the soil. The coleoptile serves this function in grasses. These features of seedlings grown in the dark constitute the *etiolation syndrome*.

It has long been known that the symptoms of the etiolation syndrome are reversed by R, the effects of which can be overcome by FR, implicating the phytochrome system. Are these light effects mediated by changes in levels of growth substances, or does light influence the sensitivity of the tissue to the growth substances already present? There is evidence supporting both mechanisms. For example, inhibition of plumular hook opening is linked to ethylene (KANG et al. 1973), which ties the response to thigmomorphogenesis (GOESCHL et al. 1966; see Section 7.8). Furthermore, treatment with R caused a 45% decrease in ethylene production of the hypocotyls of dark-grown soybeans (SAMIMY 1978). The treatment promoted hypocotyl elongation and prevented radial swelling, suggesting that ethylene acts as a growth regulator in this case. But ethylene is apparently not involved in red-light-induced stimulation of chlorophyll formation in etiolated cucumber cotyledons (DEI 1981), and ethylene increased in etiolated pea seedlings following irradiation with a pulse of R (ROHWER and SCHIERLE 1982). Blue and red light increased cytokinins in *Amaranthus* seedlings, apparently accounting for increases in the pigment, amaranthin (KOHLER et al. 1980), and applied cytokinins suppressed hypocotyl elongation and promoted anthocyanin and protochlorophyll synthesis in *Raphanus* seedlings (BUSCHMANN and LICHTENTHALER 1982). IINO (1982) reports that R inhibits mesocotyl growth in maize partially by reducing the auxin supply from the coleoptile.

De-etiolation could also involve production of inhibitors in response to light. Ethylene could be the inhibitor, as noted above, but HASEGAWA et al. (1982) have chemically identified a neutral growth inhibitor from light-exposed *Raphanus* seedlings. The inhibitor, which they called raphanasanin, increased under R and produced the symptoms of de-etiolation when applied at concentrations above 1.5 μM.

Sensitivity may also play a role. IINO (1982) found that only part of the growth inhibition of maize mesocotyls caused by light (see above) could be accounted for by reduced IAA. Another part of the inhibition was not related to IAA concentration and could not be relieved by applied IAA. WALTON and RAY (1981) found that auxin-binding sites decreased in dark-grown maize seedlings after 4 h of irradiation with R. GA overcomes the red inhibition of lettuce-hypocotyl elongation, and THOMAS et al. (1980) observed that the optimum concentration for applied GA_3 decreased during de-etiolation from 100 to 10 micromolar. SCHNEIDER and LEGOCKA (1981) also found a rapid increase in transcriptional activity of the chromatin of etiolated barley leaves in response to irradiation with R. The response (also to applied cytokinins) depended strongly upon the age of the leaves. KELLER and COULTER (1982) found the usual inhibition of stem elongation with R in etiolated pea seedlings, but they found much higher levels of GA's (mostly GA_1 and GA_5) in the light-grown plants. Clearly, GA's were not controlling (promoting) the rate of stem elongation; sensitivity changes could have been responsible.

7.5 Plants Grown in Leaf Shade

Sometimes plants grown in the shade are said to be etiolated, but they differ greatly from etiolated seedlings, and this use of the term should probably be avoided. Etiolated seedlings and shade-grown plants both have rapid stem elongation, but shade leaves, rather than being reduced, are large, dark green, and thin. Stems grown in leaf shade also typically exhibit strong apical dominance. The features of shade-grown plants are apparently mediated by phytochrome responding to the enriched FR in light that has passed through a leaf canopy (rather than inhibition of elongation by R). Growth substances could be involved. VEZINA and BOULTER (1966) demonstrated that light passing through a maple (*Acer saccharum*) canopy is indeed much richer in FR. Trees without leaves in early spring and late fall have no significant effect on the light composition, nor do pine (*Pinus resinosa*) needles. (See also the discussion in Sect. 6.3.)

By far the most work in this field has been done by SMITH, HOLMES, and coworkers (MORGAN and SMITH 1981), who in 1977 published a series of four papers on the function of phytochrome in natural environments. In the first two (HOLMES and SMITH 1977a, b), they again document the enriched FR under leaf canopies (see also HOLMES 1976, HOLMES and SMITH 1975). In the third paper (HOLMES and SMITH 1977c), they use hypocotyl hook sections of *Cucurbita pepo* and *Phaseolus vulgaris* to calculate values of P_{fr}/P_{total}. Values under clear or cloudy skies were as high as 0.54, but under a wheat canopy values varied from 0.20 to 0.54; they were as low as 0.04 under sugar beet leaves. In the fourth paper (SMITH and HOLMES 1977), they grew *Cucurbita pepo* and *Chenopodium album* under simulated wheat-canopy shade at equivalent photosynthetically active radiation levels, showing that the greatest stem growth occurred in light most enriched in FR (deepest shade). Thus, phytochrome is clearly implicated in plant detection of shade. In a fifth paper, TASKER and SMITH (1977) note seasonal changes in the radiant energy quality in birch, beech, and oak woodlands.

MORGAN and SMITH (1978) further showed that the logarithmic rate constant of stem elongation (and also leaf/stem dry weights) in *Chenopodium album* were linearly related to the estimated P_{fr}/P_{total}, with high P_{fr} producing the least growth. This is further impressive evidence that phytochrome detects shade. MORGAN and SMITH (1979) extended the work on phytochrome-controlled development to species habitats for several species. They found that true shade plants were the least sensitive to P_{fr}/P_{total}. They concluded that "stem extension rate increases as the depth of shade (in light quality terms) increases, and that the magnitude of this increase depends upon the natural habitat of the species concerned." In short, shade plants are adapted to tolerate the shade, while sun plants are adapted to escape the shade.

Are growth substances involved in the response to leaf shade? MUIR and ZHU (1983) observed an increased synthesis of auxin in leaves of *Helianthus* irradiated with R and increased auxin inhibitor under FR contrary to expectations based on the above discussion. GA's could play a role (especially GA_3 and $GA_{4/7}$, JUNTTILLA 1980), as could brassinosteroid (KRIZEK and MANDARA 1983).

7.6 Phototropism Including Sun-Tracking

The ability of a plant to intercept an optimal (not necessarily maximal) amount of light is an important aspect of its survival in a given environment. The light intercepted will depend both on fixed morphological features such as leaf area and absorbance and on such variable characteristics as the angle of leaves in relation to the direction of incoming radiation. Adjustments of the orientation of the entire plant or its leaves in relation to light direction are examples of phototropism (DENNISON 1979).

Charles DARWIN assisted by his son Francis suggested in 1881 that the phototropic response of grass coleoptiles was mediated by some transported messenger, since the coleoptile tips were much more sensitive in detecting the light than was the tissue below (DARWIN and DARWIN 1881). The so-called Cholodny-Went theory, formulated independently by the two authors in 1926, stated that phototropic bending occurs in coleoptiles (and presumably in hypocotyls and stems) as auxin is transported from the light to the dark side of a unilaterally irradiated organ (BRIGGS 1963). The idea was accepted as standard doctrine for half a century, but it has been questioned in recent years (FIRN and DIGBY 1980; cf. PICKARD, Chap. 10, this vol.).

For one thing, growth stops almost immediately on the irradiated side of a coleoptile or hypocotyl (FRANSSEN et al. 1982), and it has been argued that such a rapid response could not occur if it depended on auxin transport out of the irradiated side. Indeed, FRANSSEN and BRUINSMA (1981) could measure no gradient in auxin across a unilaterally irradiated sunflower hypocotyl, although the inhibitor, xanthoxin, accumulated on the lighted side. The hypocotyl itself was sensitive to the light, while the tip and cotyledons were insensitive (true also for coleoptiles; FRANSSEN et al. 1982). Actually, the DARWIN's (1881) and many workers since (BRENNAN et al. 1976, DENNISON 1979) observed that tissue below the tip would respond directly to light; in many coleoptiles, however, the tip is about 1000 times as sensitive as the more basal tissue. It will require future research to determine the role of growth substances in the phototropic response.

Action spectra for phototropism have long implicated flavoproteins and/or β-carotene as the photoreceptor pigment(s). Most recent work favors the flavoproteins (DENNISON 1979, MANOZ and BUTLER 1975, VIERSTRA and POFF 1981).

If one half of a *Xanthium* leaf is covered with aluminum foil, two movements are observed: the petiole side corresponding to the shaded half of the leaf elongates, displacing the blade away from the covered side, and the shaded portion of the blade bends upward (SALISBURY and ROSS 1985). These two responses, perhaps mediated by hormones, would tend to move a leaf in a natural environment toward less shaded regions and could account for leaf mosaics.

Solar tracking by leaves is one of the more intricate phototropic responses, because it involves complex movements and constant readjustment of position in relation to a moving light source. The phenomenon was studied by the DARWINS (1881) and then virtually neglected until some recent interest by physiological ecologists (EHLERINGER and FORSETH 1980, SHACKEL and HALL 1979, WAINWRIGHT 1977). Most of our physiological understanding is based upon work by SCHWARTZ and KOLLER (1978, 1980), who have studied *Lavatera cretica,* a small herbaceous mallow that grows along roadsides in Israel. Its slightly cup-shaped leaves track the sun across the sky much as a moving satellite is tracked by a radar antenna. At sunset, the leaves are almost vertical, facing the point on the western horizon where the sun set. An hour or two later, they have assumed a "resting" horizontal position. An hour or two before sunrise, the leaves change their orientation so that they are facing the point on the eastern horizon where the sun came up the day before (SALISBURY and ROSS 1985). Clearly, the biological clock is implicated. Leaf orientation is controlled by motor cells in a pulvinus where the blade joints the petiole. Movement of water in and out of these cells is almost certainly controlled by osmotic solutes, especially K^+. Hence solar tracking, although a true tropism, is not a growth phenomenon as is phototropism of stems and coleoptiles.

SCHWARTZ and KOLLER (1978, 1980) found that this response was most sensitive to blue light (as is stem phototropism). The major veins in the leaves detect the direction of the sun's rays. There is evidence that auxin plays a role in the message sent from the major veins to the pulvinus, but the situation is unexpectedly complex, because the message can be either positive or negative. If the rays come from toward the tip, the exported message is negative, so cells in the pulvinus lose water, causing the leaf to turn toward the sun. Rays coming from toward the leaf base produce a positive signal that causes the cells in the pulvinus below the responding vein to gain water, again reorienting the blade toward the sun.

7.7 Gravitropism[1] (Geotropism)

The ability of plants to adjust their morphology in response to the all-pervasive gravitational force is of obvious ecophysiological significance. Gravi-responses

1 Traditionally, the response has been called *geotropism,* which means response to the *earth,* but the response is to *gravity* (would work well on the moon or Mars if other conditions for growth were suitable; see VOLKMANN and SIEVERS 1979). The term *heliotropism* (response to the sun) was changed decades ago to *phototropism* (response to light).

of both the aerial and underground parts of a plant determine to a large extent its general habit and therefore some of its adaptive characteristics.

For example, the horizontal (diagravitropic or plagiotropic[2]) growth of stolons, rhizomes, and lateral roots, when compared with the main tap root, are adaptive expressions in the sense that they enable an individual plant to propagate vegetatively and explore the soil in its immediate vicinity, while the positively gravitropic tap root grows toward the water table. Furthermore, plagiogropic underground organs effectively anchor the plant, particularly in unstable substrates such as sand dunes and environments subjected to severe wind and water erosion.

The literature on gravitropism and the role of growth substances is voluminous and goes back to the early days of plant physiology. The experimental evidence and development of ideas is traced in a number of reviews (e.g., HOSHIZAKI 1973, JUNIPER 1976, SHEN-MILLER and HINCHMANN 1974, VOLKMANN and SIEVERS 1979, WILKINS 1979). Current ideas on the mechanisms of perception and transduction of gravity seem to be better crystallized for positively orthotropic (vertically growing) roots than for negatively orthotropic stems or especially for plagiotropic organs.

The picture that emerges regarding roots (WILKINS 1979) can be summarized along these lines. A group of cells (*statocytes*) in the core of the root cap and adjacent to the calyptrogen layer contains *statoliths* that sediment at the lower end of the cell. In roots, there is little doubt that the statoliths are amyloplasts (HILLMAN and WILKINS 1982, SHEN-MILLER and HINCHMAN 1974). When the orientation of the root is changed, the statoliths are shifted against the "new" lower part of the cell and initiate a sequence of events that result in the asymmetric distribution within the root cap of growth inhibitors. These substances are then translocated to the lower side of the growing region of the root, where they inhibit growth and produce the positively gravitropic (downward) bending of the root. In stems, there is evidence that movement of the amyloplasts acts as a kind of mechanical stress within the cells; mechanical stress applied to one side of dandelion peduncles causes bending (CLIFFORD et al. 1982). In general, the action of the statoliths remains a mystery, since we have only theories as to how their movement might result in movement of hormones or other actions that lead to bending (AUDUS 1979). There are some interesting ideas that consider the greater number of plasmodesmata at the statocyte end walls compared with lateral walls and possible effects of the endoplasmic reticulum, blocking the transport of materials through plasmodesmata (JUNIPER 1976, VOLKMANN and SIEVERS 1979).

As in phototropism, the Cholodny-Went hypothesis has been accepted for half a century as the explanation of gravitropism. The hypothesis suggests that auxin moves from the top toward the bottom of horizontal coleoptiles, hypocotyls, stems, or roots. In shoots, the effect is a promotion of growth on the bottom and an upward bending; in roots, there is an inhibition and downward bending. The hypothesis could be modified to suggest other promoters or inhibitors besides auxin; in its broadest sense, it could be retained if there is any

2 Plagiotropic = growth at some determined angle other than vertical; diagravitropic = horizontal growth (transverse to the gravitational force).

movement of promoters or inhibitors (hormones in such a case) in response to gravity. In roots, perception of gravity is clearly localized in the root cap, but in stems gravity is perceived in the same tissues that respond. This was noted by the DARWINS (1881) and many workers since then (VOLKMANN and SIEVERS 1979, SLIWINSKI and SALISBURY 1984). Peripheral cell layers may be especially important in stems (FIRN and DIGBY 1980). In recent years the Cholodny-Went hypothesis has come under fire, particularly from DIGBY and FIRN and their coworkers (DIGBY and FIRN 1976, FIRN and DIGBY 1980, HALL et al. 1980), who call some important problems to our attention. Consider three points:

1. Growth rates do not correlate with measured auxin (or other growth substance) concentration gradients; differences in growth rates are extremely high while growth-substance concentration gradients are low or nonexistent.

2. Measurements are usually made after the latent period; observed growth-substance gradients could be the result of bending rather than its cause.

3. Exogenously applied auxin does not always produce effects predicted by the Cholodny-Went theory, although few data relate to this point. In one study (PHILLIPS and HARTUNG 1976), auxin was applied to sunflower hypocotyls such that there was four times more in the upper than the lower half, yet the shoot curved upward.

One thing is clear: When a stem or coleoptile is laid on its side, the cells on top almost immediately stop or nearly stop growing (CARRINGTON et al. 1983a, b, SALISBURY et al. 1982). In some cases cells on top shrink slightly. Growth on the bottom may continue at the same rate as growth in comparable parts of vertical control plants, or it may be accelerated (the more typical situation). When growth stops on top and continues at the control rate on the bottom, if the Cholodny-Went mechanism is functioning, there could be no build-up of auxin on the bottom; only an almost instantaneous depletion of auxin on top.

In a recent study of IAA, ABA, and GA_3 and GA_7 by extremely sensitive immunoassay techniques, MERTENS and WEILER (1983) could not detect any significant lateral asymmetry of any of the growth regulators in horizontal root tips of *Zea mays,* nor in hypocotyls of *Helianthus annuus,* although 65% of the IAA in maize coleoptiles was found in the lower side. Measurements were made as a function of time after gravistimulation.

On the other hand, several recent studies support at least a modified Cholodny-Went hypothesis. SUZUKI et al. (1979) found increased IAA, ABA, and an unidentified inhibitor in the bottom halves of maize roots; they suggest that the unidentified substance might be more important in root gravitropism than IAA or ABA. They worked with a variety of maize in which light irradiation of the roots is necessary for gravitropic sensitivity, and they found increased ABA and inhibitor in irradiated roots. MULKEY and EVANS (1982) eliminated gravitropic bending of maize roots by treatment with several inhibitors of auxin transport and with orthovanadate, which inhibits efflux of H^+ in response to auxins. The inhibitors also stopped the previously observed efflux of H^+ on the top of horizontal roots. Other workers have also implicated ABA, IAA, and other inhibitors in gravitropism of roots (e.g., CHANSON and PILET 1981, FELDMAN 1981, PILET and ELLIOTT 1981). WRIGHT and RAYLE (1983) also used auxin transport inhibitors and other approaches to study the role of auxin in gravitropism of sunflower hypocotyls. They found no asymmetry in acid-growth capacity (sensitivity) but did find evidence in support of the Cholodny-Went hypothesis. BANDURSKI et al. (1984) measured an IAA asymmetry with 56–57% of the IAA on the bottoms of horizontal *Zea mays* mesocotyls within 15 min after seedlings were placed on their sides.

There are numerous experiments indicating that auxin plays some role in shoot gravitropism, even if the role is not a hormonal one. For example, BRAUNER and HAGER (1958) were able to separate perception of gravity from actual manifestation by depleting sunflower seedlings of auxin (by removing cotyledons) and laying plants horizontally. Hypocotyls bent after plants were placed upright, only providing they were supplied with auxin. Furthermore, when stems of cocklebur or castor bean plants are removed above the portion that ultimately will bend the most sharply, bending is reduced compared to control plants, but IAA (1% in lanolin) applied to the cut stump restores most of the bending (SLIWINSKI and SALISBURY 1984). Auxin gradients may not account for gravitropic bending, but auxin seems to be essential if such bending is to occur; and sensitivity to auxin might change (VESPER and EVANS 1978).

WHEELER and SALISBURY (1980, 1981) present evidence, based upon four different inhibitors of ethylene synthesis and action, that ethylene is positively involved in the gravitropic response of some species. CLIFFORD et al. (1983) present evidence that ethylene is not directly involved in response of dandelion peduncles to gravity, although ethylene might modify the response. It would seem logical that ethylene might inhibit growth of top cells of a gravistimulated stem, but increasing ethylene is observed in the *bottoms* of such stems during bending rather than the top (CLIFFORD et al. 1983, SALISBURY et al. 1982).

There is currently considerable interest in a possible role for calcium ions in the gravitropic response. Application of calcium-chelating agents (EDTA or EGTA) to the tips of maize roots caused a loss of gravitropic sensitivity, and sensitivity could be restored by applying calcium (LEE et al. 1983). Application of calcium to the upper surface of a horizontal root caused upward bending of the root, although control roots bent strongly downward.

There has been a recent revival of interest in the possibility that the gravitropic signal is somehow transduced electrically in plants. Vibrating electrodes have shown that current flow around test roots changes during gravitropic stimulation (BEHRENS et al. 1982).

It is especially difficult to understand the mechanisms of plagiotropic growth in roots, stolons, rhizomes, and runners. There are many complications. For example, BENNET-CLARK and BALL (1951) found that rhizomes of *Aegopodium podagraria* are sensitive to red light and CO_2 concentrations. Short exposures to R result in a change of orientation from the normal diagravitropic in darkness to positively plagiotropic (downward), while high CO_2 concentrations result in negative plagiotropic (upward) orientation. A similar response has been reported for the stolons of some grasses, which change their orientation from diagravitropic to negatively gravitropic if the parent plant is darkened (PALMER 1956). Other examples are found in the work of ZIV et al. (1973), who studied growth substances in erect and runner branches of peanut plants (*Arachis hypogaea*). Side branches are erect when plants are held in the dark, but become plagiotropic when plants are returned to the light. Auxins and GA's were present under both conditions, but runner branches had exceptionally high levels of an inhibitor. The effect of light is particularly interesting and suggests that the phytochrome system and its effect on growth-substance balance may be involved, as in so many other responses to light.

MOORE and PASIENIUK (1984) propose that the plagiotropic response of lateral roots of *Ricinus communis* is a result of their small size. If half of a root cap is removed, the roots bend toward the side with the intact half, suggesting that the root cap is capable of producing inhibitors. It appears that the small size prevents establishment of a concentration gradient of inhibitor (and also somewhat less inhibitor). Although much remains to be learned about plant gravitropism, growth substances – and perhaps hormones in the transport sense of the word – are clearly involved.

7.8 Responses to Mechanical Stresses [3]

In recent years, it has been realized that plants exhibit marked morphological responses to such mechanical stimuli as shaking, bending, rubbing, blowing, and the impact of raindrops. Mechanical stresses must be universally present in any natural situation and undoubtedly influence the habit of plants adapted to a given environment (see also Sect. 7.9). For example, the stunted, dense, and spreading growth habit of perennial cushion plants in places like the alpine fellfields (faeldmark) that are almost constantly buffeted by strong winds and are bombarded by wind-driven gravel particles may well be, at least partly, morphogenetic responses to mechanical stresses. It may be appreciated that such responses can be noted only by comparison with plants that are protected in a greenhouse or in a growth chamber from the inducing stresses. Such responses to touch have been called *thigmomorphogenesis* (JAFFE 1976a); responses to shaking, *seismomorphogenesis* (MITCHELL 1977).

By far the most common observation under laboratory conditions when plants are mechanically stressed is reduced stem elongation with concurrent thickening and often reduction in leaf growth, fruit yield, fresh and dry weights, etc. (JAFFE 1976a, b, MITCHELL 1977, NEEL and HARRIS 1970, SUGE 1978, TURGEON and WEBB 1977, WHEELER and SALISBURY 1979). Typically, stem elongation is retarded 40% or more when plants are mechanically stressed (shaken, sprayed with water, rubbed, etc.) for 10 s once or twice a day. A striking feature of these response is the rapidity with which they occur. For example, growth inhibitions have been demonstrated within 6 to 30 min from the time of stimulation with full recovery of the normal rate of growth after 2 to 3 d.

Are the responses to mechanical stresses mediated through growth substances? GOESCHL et al. (1966) studied pea seedlings growing through glass beads and observed an evolution of ethylene. Several workers have since observed ethylene production, usually about 20 to 30 min after the beginning of mechani-

3 See also JAFFE, Chap. 12, this vol.

cal stress (HIRAKI and OTA 1975, IRVINE and OSBORNE 1973, KAYS et al. 1974, LEOPOLD et al. 1972, MITCHELL 1977, ROBITAILLE and LEOPOLD 1974, VICTOR and VANDERHOEF 1975). JAFFE and BIRO (1977) reported that they could mimic responses to mechanical stresses with applied ethylene. On the other hand, JAFFE (1976b) emphasizes that the immediate responses to mechanical stresses (a decrease in electrical resistance of tissue, a sudden burst of growth lasting about 3 min, and then cessation of stem elongation) occur in the first few minutes after mechanical stress and before ethylene can be detected. Thus the primary responses to mechanical stresses might not involve ethylene.

On the basis of available information, it certainly appears that ethylene evolution is involved in the control of thigmomorphogenetic responses, if only indirectly, but it would be surprising if ethylene alone amongst all the known growth substances holds the complete answer. Several workers have looked for auxin and other growth substance interactions in mechanical stress responses, but these have not yet been clearly demonstrated (BOYER 1967, MITCHELL 1977, VICTOR and VANDERHOEF 1975, SUGE 1978). SUGE (1978) observed retarded growth and gibberellin production in response to stroking the upper part of common bean plants.

The preceding comments indicate the need for comparative studies on plants that are maintained in their natural environment but that are or are not protected by some means from the impact of such environmental stresses as wind and rain, while keeping all other conditions identical. We hazard a guess that such studies will be particularly productive if they deal with plants that are normally adapted to environments where mechanical stresses are exceptionally severe such as the alpine tundra and coastlines.

7.9 Anatomical Responses

It is not surprising that the metabolic, physiological, and gross morphological effects resulting from exposure to various environmental stresses are also reflected at the anatomical level. For example, QUARRIE and JONES (1977) studied several responses to drought and found that both ABA and water stress decreased mean cell size, reduced the number of stomata per leaf, and increased the production of trichomes in wheat leaves. The development of such xerophytic characteristics under conditions of water stress have long been known, but the involvement of ABA (known to increase under water stress) is especially interesting (see Sect. 8.1).

Another excellent example of anatomical modifications induced by environment is the formation of reaction wood, characterized by conspicuous development of supportive tissues (fibers, tracheids) when woody stems are forcefully maintained in unnatural orientations (SCURFIELD 1973, WILSON and ARCHER 1977). This is a manifestation of gravitropism (Sect. 7.7). A similar response is seen when plants are exposed to some unilateral mechanical stress as when strong winds come from a prevailing direction. Trees growing along windy coastlines or mountain ridges are not only bent, but the sides of the stem in line with the direction of the wind develop mechanical elements in the wood that impart greater directional strength. Clearly, such responses are related to thigmomorphogenetic phenomena (Sect. 7.8).

Probably growth substances are involved in the formation of reaction wood, but unequivocal experimental evidence is lacking. Auxin is implicated by the importance of the stem tip in reaction wood formation (WARDROP 1956) and by the successful induction of the response by the unilateral application of the growth substance. However, in the latter case, reaction wood does not necessarily form on the side of higher concentration. In conifers (WERSHING and BAILEY 1942), reaction wood forms on the side of the stem with high auxin, while in some angiosperms (CASPERSEN 1965) reaction wood forms on the low-auxin side. The possible role of auxin is clouded even more by the failure of some workers (ZIMMERMANN et al. 1968) to demonstrate any effect by exogenous applications or asymmetric distribution of endogenous auxin in horizontally maintained stems (SMITH and WAREING 1964). Other work suggests that growth substances other than auxin may be involved (LEACH and WAREING 1967). In normal tree growth, formation of thick-walled tracheids was associated with a build-up of an ABA-like inhibitor in the cambium and nearby phloem (WODZICKI and WODZICKI 1980).

8 Conservation of Resources

Some resources must be conserved during short-term fluctuations in the environment (e.g., water during the diurnal cycle), while other resources must be conserved over relatively long periods of time (e.g., storage energy and molecules). There is a plethora of publications on the subject.

8.1 Too Little or Too Much: Water Stress and Flooding

Perhaps the most striking observation in this field, first made by WRIGHT in 1969, is that ABA accumulates to high levels in leaves subjected to water stress (BRADFORD 1982, WALTON 1980). Furthermore, it was soon discovered that applied ABA caused stomatal closure in many plants (MITTELHEUSER and VAN STEVENINCK 1969, RASCHKE 1975, 1976). In addition, high ABA inhibits growth, further conserving water. By now, many complications have been encountered; only a few examples can be given.

Gymnosperms respond as do flowering plants (BACO and BACHELARD 1979, FEDOROVA 1977). Related compounds such as trans-farnesol may play the same role as ABA (OGUNKANMI et al. 1974). Changes in other growth substances may accompany ABA build-up: proline may accumulate (BENGSTON et al. 1978, RAJAGOPAL and ANDERSON 1978), cytokinins may decrease and applied cytokinins may antagonize ABA effects (BENGSTON et al. 1979, HALEVY and MAYAK 1975), GA's may antagonize ABA effects (KOCH and BERGMANN 1979), and ethylene often increases along with ABA (APELBAUM and YANG 1981, PALLAS and KAYS 1982, WRIGHT 1981) but not always (SIVAKUMARAN and HALL 1978). Applied ABA acts as an antitranspirant (because stomates close on many species; e.g., DAVIES and KOZLOWSKI 1975), but yield of apples was not increased by spraying with ABA during drought (GOODE et al. 1978). In cotton (*Gossypium hirsutum*), ABA increase appeared sooner under long day than under short day (HUBAC and LE PAGE-DEGIVRY 1981). ABA built up in young cotton balls in response to water stress, probably contributing to abscission of many of the bolls (GUINN 1982). Root conductance is decreased as well as stomatal conductance, providing another adaptation to water stress (MARKHART et al. 1979). The flacca mutant of tomato is normally wilted because stomates remain open; it can be phenotypically reverted to the wild type by application of ABA, which causes stomates to close (BRADFORD 1983). Finally, ABA often increases in plants in response to stresses other than lowered water potential (e.g., cold: RIKIND and RICHMOND 1975, see review by WALTON 1980).

Build-up of ABA in response to water stress is often influenced by various factors besides lowered water potential. In one study, ABA levels were lower in response to water stress when the plants had previously experienced a period of water stress (HENSON 1983); but not in another study (ACKERSON 1980). That is, there is sometimes a conditioning effect. RASMUSSEN (1976) found a rapid rise in ABA in water-stressed tomato plants when detached leaves wilted, but plants conditioned by 18 days of water stress did not have elevated ABA levels. WALKER and DUMBROFF (1981) also observed a drop in ABA in tomatoes back to normal levels after 2 days of osmotic stress. When the stress was removed, growth returned to normal: an elastic strain response. Young leaves were much more responsive to water stress by synthesizing ABA than older, especially senescing leaves (EZE et al. 1981, SIVAKUMARAN and HALL 1978), and adequate nitrogen nutrition made plants more resistant to water stress [i.e., ABA content increased at higher water potentials in N-deficient plants than in normal plants (RADIN and ACKERSON 1981)].

An important question from the ecological standpoint concerns the role of ABA in plants that are normally resistant to water stress. Are ABA levels especially elevated in such plants as stress develops? The answer for many mesophytes is apparently yes. Most studies have been done with drought sensitive and drought-resistant cultivars of crop plants (e.g., LARQUE-SAAVEDRA and WAIN 1974, 1976, QUARRIE 1980, QUARRIE and HENSON 1982, QUARRIE and JONES 1979). In one study, applied ABA converted a drought-sensitive cultivar of wheat to a drought-resistant form (KIRKHAM 1983). On the other hand, some plants are remarkably insensitive to applied ABA (e.g., a yellow lupin, *Lupinus luteus*, LANCASTER and MANN 1977) or do not show changes in ABA correlated with water stress (e.g., *Arbutus unedo*, BURSCHKA et al. 1983). It would not be surprising if many or most true xerophytes have features that help resist drought besides or instead of stomatal control via ABA.

In a few cases, rapid water stress caused stomatal closure *before* significant accumulation of ABA (e.g., HENSON 1981). This observation along with other more subtle features such as lack of correlation of ABA levels across physiologically diverse tissues such as leaves of different ages (RASCHKE and ZEEVAART 1976) have caused some to question the growth-regulator role of ABA in stomatal closure. TREWAVAS (1981) suggests that sensitivity of guard cells to endogenous ABA might be more important than changes in ABA levels, and HENSON et al. (1982) present evidence that sensitivity of stomata to ABA does indeed change during the day. The considerable body of correlative evidence that ABA levels control stomatal apertures is quite compelling, however, and RADIN and ACKERSON (1982) point out that there are still ways to reconcile the apparently conflicting observations. For example, stomatal closure before an increase in ABA levels (usually observed with detached and rapidly drying leaves) could be the result of evaporative water loss from the guard cells themselves. It is also necessary to consider such complications as the amount of ABA in the apoplast of leaf tissue and if it might be released to move to stomates only in response to water stress (ACKERSON 1982). All in all, the case for participation of growth substances (hormones, in this case) in plant response to environment is probably nowhere stronger than it is with ABA and water stress.

Another example is that of ethylene induced by flooding – a stress situation opposite to that of water stress (CRAWFORD 1982, JORDON et al. 1972). Again, many papers report that ethylene production increases in many species in response to flooding. A signal is sent from flooded roots to shoots, and this signal is apparently 1-aminocyclopropane-1-carboxylic acid (ACC), the immediate precursor of ethylene (BRADFORD and YANG 1980). Leaves on the shoots frequently become epinastic. In *Euphorbia lathyris,* both ABA and ethylene increased in response to waterlogging (SIVAKUMARAN and HALL 1978). Cytokinin (benzyladenine) and GA countered the effects of ethylene on waterlogged tomato plants, although they did not decrease the amount of ethylene produced (JACKSON and CAMPBELL 1979).

Does ethylene play a role in plant resistance to flooding? BLAKE and REID (1981) compared three species of *Eucalyptus* and found that ethylene production was high in the flood-tolerant species in response to flooding, and that the ethylene caused tissue hypertrophy (opening channels for air movement to the roots) and basal stem thickening. Applied ethylene caused the same responses. The flood-susceptible species did not show these responses. Thus ethylene may well be important in a positive way in the responses of tolerant species to flooding.

8.2 Long-Term Energy Storage

The survival of many plants adapted to desert, arctic, or alpine tundra environments, where net energy input can range from high to negative for long periods of time and seed production can also be highly variable, depends largely on their ability to store energy in such specialized organs as rhizomes, stolons, tubers, bulbs, and roots. This energy can then be used to maintain the living state during periods of stress or dormancy and to provide adequate energy for rapid growth early in the season, in many cases even before the snow and ice have receded (SALISBURY 1976). Such an early start may be an essential timing adaptation, because it enables plants to complete their life cycle during the short growing seasons normally associated with such environments.

The formation of storage organs is a major morphogenetic event that involves more than the diversion of metabolites into specific, pre-existing sites. The enlargement of the bases of bud scales during the development of bulbs and the transformation of a stolon tip into a tuber are just two examples of how development is affected. The initiation of such organs is in many cases controlled by temperature and/or photoperiod (VINCE-PRUE 1975). Furthermore, such inductive responses can be studied on explants (EWING and WAREING 1978).

Storage organs are induced by a variety of conditions. Tuberization in potato, begonia, some dahlias, *Helianthus tuberosus,* and *Phaseolus coccineus,* for example, is induced by SD (ALLARD and GARNER 1940, ESASHI 1960, ESASHI and LEOPOLD 1968, MOSER and HESS 1968), while the induction of bulbs in onions, garlic and certain rhizomes as in Chinese radish and iris, is a response to LD (ALLARD and GARNER 1940, VINCE-PRUE 1975). Photoperiodic effects on storage organ formation often show the familiar response to night interruption. For example, potato tuber induction is inhibited if inductive long nights are interrupted by a brief exposure to light (VINCE-PRUE 1975). In common with the photoperiodic effects on flowering, the site of perception is the leaf (ESASHI 1960, HEALTH and HOLDWORTH 1943), and the resultant messenger is transported to the target organ (GREGORY 1965), which then embarks on a defined developmental path. Thus hormonal action is implied.

The involvement of all known growth substances, acting in a promoting or inhibiting capacity in storage organ production, is indicated by many investigations (e.g., BIRAN et al. 1972, CATCHPOLE and HILLMAN 1969, CHAPMAN 1958, EL-ANTABLY et al. 1967, ESASHI et al. 1964, NAGAO and OKAGAMI 1966, OKAZAWA et al. 1967, PALMER and SMITH 1969, SANO and NAGAO 1970, SELMAN and KALASEGARAM 1967, TIZIA 1966, VAN SCHREVEN 1956, VINCE-PRUE 1975).

Changes in growth-substance balance may precede the visible morphological initiation of storage organs. For example, in potatoes cytokinin level increases to a peak 8 to 10 d before tuber initiation (FORSLINE and LANGILLE 1974, LANGILLE and FORSLINE 1974, MAUK and LANGILLE 1978). A similar pattern, but involving IAA, has been reported for onions (CLARK and HEATH 1959, 1962), where bulbing is induced by LD. They found a large increase in IAA content during the first 7 d following exposure to inductive conditions before any visible swelling of the leaf bases. It is even more interesting that subsequently the level of auxin drops. Hence, high auxin levels are apparently not essential for the continuation of the growth processes that lead to bulb formation; it is as if IAA acts as a trigger. Of course, subsequent events may be controlled by other growth substances, but we have no information on this point.

GA_3 might promote leaf and stem growth at the expense of storage organ formation. Induction is sometimes correlated with a reduction of GA, and exogenous GA application inhibits induction (CHO 1970, OKAZAWA 1960, RAILTON and WAREING 1973c). It is likely that high temperatures (MENZEL 1980) and long days (GOLENIOWSKI et al. 1980) inhibit tuberization by promoting GA synthesis.

Temperature typically interacts with daylength. For example, corm formation in *Freesia* is accelerated by low temperatures, and in some plants, e.g.,

Begonia evansiana and potato, tuber formation may occur even in LD provided night temperatures are low enough (ESASHI 1964, GREGORY 1965). These temperature effects are correlated with a reduction of stem growth, suggesting (as in the case of GA) that a competitive situation may exist between stem elongation and development of storage organs.

Understandably, all the work on storage organ initiation to date is concerned with plants of agricultural importance. It is important that such studies be extended to plants adapted to extreme environments where storage organ formation is crucial to survival.

9 Metabolic Acclimation and Associated Changes

Many plants respond to and survive transient or long-term, seasonal, stress conditions by going into dormancy or otherwise suspending growth and by developing hardiness. Both changes might well be accompanied by changes at the ultrastructural level of organization and, in the context of this review, linked to changes in growth substances.

9.1 Ultrastructural Changes

Ultrastructural changes in response to various stressful conditions have been reported for various plants, tissues, and organelles (e.g., FELLOWS and BAYER 1978, KIMBALL and SALISBURY 1973, 1974, MARINAS and FIFE 1972, NIR et al. 1970b), but the role of growth substances in the evocation of such changes has received less attention. GILES et al. (1976) reported that the increased ABA levels that occur during water stress are correlated with the disappearance of starch, swelling of the outer chloroplast membrane, and the breakdown of the large central vacuole into small vesicles. However, one cannot be sure that ultrastructural changes are caused by increased ABA levels and not by the reduced water status of the plant. Chilling in cotton apparently damaged the microtubular network (RIKIN et al. 1983; see Section 9.3 below).

Studies on ultrastructural changes influenced by growth substances will certainly yield important information on the nature of dormancy and hardiness, but the separation of growth-substance effects from other metabolic and physical changes in plant cells (WEISER 1970) during entry into dormancy and increased hardiness will be experimentally difficult.

9.2 Metabolic Acclimation Responses

A wide spectrum of metabolic changes brought about by various environmental conditions may not be mediated by growth substances and thus do not fall within the ambit of this review. The direct effect of temperature on the rate of some enzymatic reaction is a simple illustration. However, some environmentally dependent metabolic changes can be linked to the transducing function of one or more growth substances.

The question of enzyme activity regulated by growth substances has attracted the attention of plant biochemists for a long time. The classical case of GA-mediated amylase synthesis (JONES 1973) is supported by the demonstration of other growth-substance/enzyme interactions, particularly in the area of nucleic acid metabolism (SMITH 1978). Studies that correlate environmental stress, endogenous growth-substance levels, and some metabolic event have concentrated on the large increases of proline in various plants under conditions of water stress, plus associated increases in ABA levels (e.g., ASPINALL et al. 1973, MCMICHAEL and ELMORE 1977, WAMPLE and BEWLEY 1975). Reduced cytokinin supplies to shoots from roots exposed to saline conditions are held responsible for the decline in protein synthesis in leaves and the acceleration of senescence (BENZIONI et al. 1974, ITAI et al. 1968), but salinity symptoms are not always overcome by exogenous application of cytokinins (PRISCO and O'LEARY 1973). Such metabolic changes related to growth substances could have survival value under natural conditions, but this aspect has not been experimentally established.

9.3 Dormancy, Chilling, and Cold Hardiness

Metabolic acclimation assumes particular significance in view of changes that must take place during development of plant hardiness toward heat, cold, or drought (or when plants enter dormancy). The interesting point is that all these hardiness conditions involve similar growth substance changes.

Although dormancy and hardiness are related conceptually and physiologically, a useful distinction can be made between them. Hardening involves a spectrum of changes in the physiology and structure of plant cells that enable them to withstand stressful environmental conditions while still maintaining a reasonable level of metabolic and growth activity. Dormancy is, in a sense, an extension of the hardened state and implies a virtually complete "shut-down" of all energy-requiring processes except those that are essential for the maintainance of the living state. In spite of this distinction between the two states, the associated physiological changes have common features including the role of growth substance controls.

The voluminous information on dormancy and hardiness and the effects of growth substances (ALDEN and HERMANN 1971, TAYLORSON and HENDRICKS 1977, WAREING and SAUNDERS 1971) leads us to a few general statements:

1. Dormancy and hardiness are induced, depending on species, by high or low temperatures, water stress, and/or short photoperiods.

2. There is much support for the belief, originally suggested by HEMBERG (1949) and subsequently pursued by many workers, that the onset of dormancy is marked by increased ABA and other inhibitor concentrations and a concurrent decrease in levels of such promoters as IAA and GA; these trends are reversed during emergence from dormancy (MOUSDALE 1983, PHILLIPS and WAR-

EING 1954, RAPPAPORT 1969, RIKIN et al. 1978, SEELEY and POWELL 1981, WALSER et al. 1981, WEBBER et al. 1979). Similar changes have been observed in seed dormancy. There is also evidence that ethylene and cytokinins are produced when dormancy is broken (DOMANSKI and KOZLOWSKI 1968, HEWETT and WAREING 1973, POAPST et al. 1968, VAN STADEN and DIMALLA 1981). On the other hand, such a correlation between increasing ABA and developing dormancy could be coincidental rather than a cause-effect relationship. For example, LD treatment of willow (*Salix viminalis*) prevented the development of dormancy, but ABA levels were higher than in plants kept under natural daylength (ALVIM et al. 1978). MIELKE and DENNIS (1978) prevented the normal increase in ABA that occurs in the fall in sour cherry by mechanical defoliation of trees, but the intensity of dormancy was not affected. FREEMAN and MARTIN (1981) influenced the termination of dormancy in peach floral buds with various environmental treatments, again without changing the ABA levels in the buds.

WRIGHT (1975) found that dormancy in blackcurrant and beech was correlated with a high free ABA/bound ABA ratio, while the reverse relation obtained during bud burst. This relationship was less clear in *Freesia* corms (GILBERTSON-FERRISS et al. 1981).

The normal requirement for the elapse of a fairly well-defined period of time before emergence from dormancy implies that, although the plant operates at a very low metabolic level, changes must occur in the metabolism of growth substances or other compounds. Furthermore, the fact that the duration of dormancy can be modified to a certain extent by such environmental conditions as photoperiod and temperature implies that receptive sites for the detection of such environmental cues remain active throughout dormancy.

3. Dormancy in bulbs, tubers, and other storage organs also seems to involve growth substances, although the situation is again complex. Typically, ABA is high in dormant organs and drops to much lower levels as dormancy breaks; for example: in potato tubers (KORABLEEVA et al. 1980), tubers of yellow nutsedge, *Cyperus esculentus* (FERNANDEZ and SAUCHEZ 1982), and corms of *Freesia hybrida* (GILBERTSON-FERRISS et al. 1981). Cytokinins and GA's increase as storage organs become active: in bulbs of daffodil, *Narcissus pseudonarcissus* (VAN STADEN 1978) or *Tulipa alberti* (RAKHIMBAER 1978). Applied $GA_{4/7}$ appeared to substitute for an otherwise necessary cold period during storage of tulip bulbs (HANKS 1982).

4. Some plants are damaged by chilling (temperatures a few degrees above freezing), and these responses often involve growth substances. Chilling of common bean (*Phaseolus vulgaris*) at 5° C for 18 h led to a threefold increase in ABA – but only when humidity was low (EAMUS and WILSON 1983). DAIE and coworkers (1981) also measured increased ABA in several warm-season crops exposed to 10° C. Ethylene release from cucumbers (*Cucumis sativus*) occurred only upon warming following chilling; when chilling was prolonged (4 days or longer), there was no ethylene release, suggesting that the ethylene-synthesizing mechanism had been damaged (WANG and ADAMS 1982). Chilling in cotton apparently involved damage to the microtubular network, and this damage was reduced by applied ABA, which also decreased chilling injury

(RIKIN et al. 1983). Cotton plants were more sensitive to chilling when they were deficient in photosynthate (RIKIN et al. 1981).

5. Many plants are capable of being hardened against freezing (often to several tens of degrees below the freezing point), and again growth substances appear to play a role. Hardening is typically associated with increases in such inhibitors as ABA (BRYANTSEVA 1983, EAMUS and WILSON 1983, VOLKOVA et al. 1981). Auxins and other promoters sometimes decrease (VOLKOVA et al. 1981).

6. High temperatures (e.g., above 40° C) have also led to increased ABA levels (DAIE et al. 1981, MAKUS and SHANNON 1979).

Such generalizations may cause us to overlook important species, variety, and cultivar differences. For example, a comparative study of northern and southern races of red maple (*Acer rubrum*) of their responses to short and long photoperiods, exogenous applications of ABA, and changes in endogenous levels of ABA led PERRY and HELLMERS (1973) to conclude that ABA levels were not causally related to cold hardiness. Similar negative results were obtained for bud dormancy of birch, maple and sycamore (LENTON et al. 1972).

Information on the comparative hormonal physiology of related plants with different cold-hardiness capacity is rather scarce. Examples of this approach are shown by the work of WALDMAN et al. (1975) and RIKIN et al. (1975), RIKIN and RICHMOND (1975), who conducted a comparative study of two cultivars of alfalfa (*Medicago sativa*) that differ in their cold-hardiness and found, among other things, that under SD and low temperatures that induce hardiness, GA_3-like activity was practically absent in the hardy variety, although it increased in the nonhardy. ABA levels were not significantly affected, and the authors conclude that ABA/GA balance determines the capacity of alfalfa to become cold-hardy. Such comparative studies will provide a better perspective of the ecophysiological role of growth substances in these responses.

10 Miscellaneous Environmental Stresses and Plant Responses

This section deals with responses to environmental stresses, some adaptive, that are not easily classified into one of the main groups of survival mechanisms that have provided the framework of this review.

10.1 Atmospheric Pollutants

Industrial atmospheric pollutants such as fluoride provide a new environmental stress. One suspects that responses are pathological rather than adaptive, since there has not been time enough for such conditions to make an impact on the genetic composition of plant populations. More ancient pollutants, as from fire and volcanoes, were sporadic, again giving little time for genetic adjustment.

Although early work emphasized morphological and metabolic responses (DUGGER and TING 1970), changes in growth substances in response to atmospheric pollutants have been reported. For example, ethylene evolution increases in response to ozone and SO_2 (BUCHER 1981), in the latter case, before symptoms of damage appear (on forest trees).

As shown by ROSEN et al. (1978), damage to grapevines depended upon entry of SO_2 or O_3 into the leaves via stomates. FLETCHER et al. (1972) were able to reduce O_3 toxicity symptoms by treating bean plants with ABA. The protective effect was no doubt caused by stomatal closure (Sect. 8.1) and the consequent restrictions on O_3 penetration into the leaf. But do plants respond to atmospheric pollution by producing more ABA and thus protecting themselves against the pollution? KONDO and SUGAHARA (1978) found such a response in peanut and tomato plants, which were resistant to SO_2 (again no doubt because of stomatal closure), but ABA did not increase in radish, perilla, or spinach, which were sensitive to SO_2. Incidentally, the fact that peanut and tomato respond to SO_2 by increasing ABA levels does not mean that we can conclude that their response is a genetic adaptation achieved in the distant past. It might well be coincidental.

TAYLOR et al. (1981) found that SO_2 (as sulfurous acid in solution) in low concentrations (10^{-10} to 10^{-7} molar) led to opening of stomates in *Vicia faba* epidermal strips floated on the solutions. A higher concentration (10^{-5} M) of SO_2, or ABA between 10^{-10} and 10^{-5} M, caused stomatal closure. There was a direct antagonism between low SO_2 and ABA.

10.2 Mineral Nutrients and Salinity

Plants may experience transient or chronic mineral deprivation or high salt concentrations in the vicinity of their roots. Although some plants are obviously well adapted to salinity conditions, it is not known to what extent, if any, growth substances contribute to their salt tolerance. There is evidence, however, that mineral deprivation changes endogenous growth-substance levels. For example, cytokinin activity in *Coleus* exudates was reduced by a number of environmental stresses including low nutrition (BANKO and BOE 1975), and an increase in the content of leaf ABA in tobacco plants occurred after transferring the plants from complete nutrient solution to distilled water (MIZRAHI and RICHMOND 1972b). In an expansion of these studies, RIKIN et al. (1973) subjected tobacco plants to water stress, 5° C air temperatures, and mineral deprivation. Plants from all treatments exhibited improved resistance to subzero temperatures as well as markedly increased levels of ABA in the leaves. Their experimental design makes it possible to attribute the observed increases in ABA to any one of the experimental conditions, but the important observation is that improved resistance to environmental stresses was correlated with increases in ABA.

ABA features prominently in studies on nutritional or salinity stress, as it does in so many other stress situations. It is difficult to separate water and salt stress, since both involve lowered water potential; certainly salinity (specific ion effects?) causes effects similar to those observed with water stress: Tobacco plants grown in salinated solutions had increased ABA levels and were at the same time considerably more resistant to anaerobic conditions around the root system (MIZRAHI et al. 1972). Grapevine leaves also increased in ABA (and, later, proline) when subjected to NaCl solutions (DOWNTON and LOREYS 1981). This was also true for cold and drought resistance (RIKIN and RICHMOND 1975).

A study of the recovery of plants subjected to mineral deprivation or salination showed that ABA and the extent of stomatal opening were inversely related during recovery (BOUSSIBA and RICHMOND 1976); it was suggested that "the after-effect results from the delay in resumption of the pre-stress hormonal balance in the plant, particularly with regard to ABA, after termination of the stress".

Other investigations have linked cytokinins to the response of plants to salinity. For example, BENZIONI et al. (1974) found "that exogenous application of kinetin to tobacco plants prevented salinity damage by retarding cellular senescence." The growth inhibition that is normally observed under salinity conditions has been attributed to impaired cytokinin metabolism and/or transport, but it is more likely that any growth effects involve various growth substances acting on several steps of the growth process (LOMBARD and RICHARDSON 1979). Similar conclusions were reached with respect to kinetin and GA in lettuce seeds (BRAUN and KHAN 1976), and IAA levels dropped in sunflower, bean, and maize seedlings exposed to boron deficiency – and sensitive to such deficiency; wheat, a tolerant species, showed no change in IAA (SMIRNOV et al. 1977). Seeds of the halophytic genus *Suaeda* exposed to saline conditions showed that the levels of cytokinins and GA's were reduced, but only the exogenous application of GA reversed the dormancy that was induced by saline solutions, whereas kinetin had no effect (BOUCAUD and UNGAR 1976). Unfortunately, in this report the distinction between salinity and osmotic effects is not made clear. Applied GA_3, kinetin, and ascorbic acid all ameliorated effects of salinity on chickpea (*Cicer arietinum;* SINGH and SINGH 1980).

An interesting approach to the ecophysiological aspects of salinity stress is illustrated by the work of KHAN et al. (1976), who studied the growth-regulator composition of *Zea mays* and *Suaeda fructicosa,* which differ in salt tolerance. Although their finding that there was no significanct difference in inhibitor content between the two when exposed to saline conditions did not account for differences in behavior, their work points to the direction that needs to be explored.

10.3 Plant Diseases

Plant diseases are stressful factors that can place the survival of a plant at risk. Even closely related plants can differ widely in their ability to resist various diseases, but the possible role that growth substances may play as defensive agents has not been explored to any great extent. Most of the work to date has concerned pathogen-induced changes in growth-substance metabolism and growth (SEQUEIRA 1973, STROBEL 1974). For example, PUPPO and RIGAUD (1978) found that the morphological response that follows infection of bean roots with *Rhizobium* is accompanied by increases in cytokinin level. Also, a significant fall in cytokinin levels has been reported (MISAGHI et al. 1972) in cotton plants in response to infection with *Verticillium* wilt disease. Other work indicates that disturbances of the normal balance of GA and ABA are responsible for the suppression of growth that follows viral infections (AHARONI et al. 1977a, b, FERNANDEZ and GABORJAYI 1976). Formation of N_2-fixing nodules on a plant like *Alnus glutinosa* also has more than nutritional implications, because the distribution and seasonal levels of cytokinins are also affected (HENSON and WHEELER 1977).

Hormonal balances may be disturbed by various pathogenic agents, but as BRIAN (1978) has pointed out, the mechanisms and reasons for such changes are not clear. Furthermore, pathogens may produce toxins that cause effects similar to those caused by growth substances, although they bear no structural relationship to known growth substances (MOSER 1968).

Some attempts have been made to study the possible link between hormonal status and tolerance or susceptibility toward pathogenic agents. For example, VAN STADEN and DIMALLA (1977) compared the cytokinin activity in roots and xylem exudates of nematode-resistant and susceptible tomato cultivars and found that cytokinin levels increased in both, but the increase was greater in the susceptible variety and seemed to be related to the formation of giant cells. Other workers have attempted to modify susceptibility to pathogens by exogenous treatment with hormones and related substances. For example, SAWHNEY and WEBSTER (1975) obtained some resistance to a root-knot nematode by treating tomato plants with a combination of an auxin (NAA) and kinetin. It is not surprising, in view of such growth-substance disturbances, that infections may lead to changes in winter hardiness and drought tolerance (FRIMMEL 1977).

11 Growth Substances as Allelopathic Agents

Intra- and interspecific competition and other population interactions are at the heart of community ecology. Considerable information is available on the role of a variety of substances produced and secreted by plants (TUKEY 1970) and soil micro-organisms on the establishment of seedlings of the same or different species, thereby controlling plant spacing (ANDERSON et al. 1978, KRUL 1978, PUTNAM and DUKE 1978, WHITAKER and FEENY 1971). However, information on the possible role of growth substances in influencing the interactions of plants under natural conditions is extremely meager. The substances that are leached or secreted by plants are predominantly inhibitors of some kind with little chemical relationship to growth substances, but there is enough information to indicate that known growth substances may be important allelopathic agents in certain situations. Most of the work published today involves the mediation of soil micro-organisms in such interactions rather than the direct release of some growth substance by the plant. For example, ethylene has been shown to be produced by various soil micro-organisms (LYNCH 1975b, 1977, SMITH and COOK 1974, SMITH 1976). BROWN (1972) described micro-organisms isolated from rhizospheres of wheat plants and from root-free soil that produced growth-regulating substances with the properties of GA's and IAA. Inhibitors were also produced, especially by bacteria from the root region of 6-day-old seedlings. Seedlings grown aseptically differed from those grown with a soil inoculum or aseptically with added GA_3 and IAA.

Clearly, studies on possible roles of growth substances acting as allelopathic agents must involve rigorous studies of soil microflora, because soil organisms not only produce such substances, but undoubtedly influence the destruction of any growth substances released into the soil by a plant. Soil composition could also be important.

Leachates from coniferous leaf litter have been shown to contain water-soluble growth regulating factors (BLASCHKE 1977) that could play an important role as allelochemic agents in forests.

12 Epilog

In this review, we attempted to interpret various aspects of the growth-substance and hormonal physiology of plants from a particular point of view: the way growth substances and hormones act as transducers of perceived environmental cues and control a wide range of structural and functional adaptations that define the limits of elastic strain and enable a plant to survive environmental challenges.

There seem to be two main approaches that call for a concerted effort before we can reach some understanding of the complete sequence of events responsible for the survival of plants in various environments, and particularly the part that growth substances and hormones play:

1. Precise definition of the complex edaphic and atmospheric operational factors that are in a constant state of flux and impinge on a plant under natural conditions. Many studies have measured environmental parameters, but the value of such studies can be increased by wise selection of the micro- and macro-environments and ecosystems that are likely to yield the most information, and by the thoroughness of the environmental profiles that emerge.

2. Co-ordinated with the definition of environmental parameters, there should be comparative studies on the growth-substance and hormonal history of related plants adapted to definably different environments. This objective will encounter problems, both in the collection of informations and its interpretation. Perhaps the most serious problem is the multiplicity of responses to growth substances and hormones and the complex ways in which growth substance and hormone levels are affected by environmental conditions. In the present stage of our knowledge, it is impossible to bring together all the threads of evidence into a unifying hypothesis. How can we possibly interpret on a common basis the effects of a given substance, say GA, on such diverse phenom-

ena as cell elongation, flowering, dormancy, cold-hardiness, germination, expression of sexuality, and parthenocarpy, to mention only a few, and relate these to the superimposed influence of various environmental conditions on growth substance and hormone levels?

The difficulties we have just outlined are confounded even further by the fact that just about every known phenomenon is affected by more than one growth substance or hormone, which can act independently, synergistically, or antagonistically. In the face of this complexity, we wonder if growth-substance and hormone systems and responses might be redundant. It might be that plants evolved more or less parallel systems of chemical controls, each of which proved successful in the transduction of a wide range of environmental cues and the regulation of a wide spectrum of processes essential to survival. In fact, they could have been so successful that they were never really subjected to any strong selective pressures, so they remained functional to the present day, keeping us largely mystified!

Acknowledgements. The participation of F.B. Salisbury in preparation of this review was supported by the Utah Agricultural Experiment Station, which also supported Deleyne Wentz in a search of the literature and Mary Jo Hansen, Elaine Ewing, and Linda Gillespie during work on the manuscript. N.G. Marinos, while on sabbatical leave, received support for work on this review from the following organizations: The Flinders University of South Australia, the Utah Agricultural Experiment Station, the National Parks Foundation, and the Rocky Mountain Nature Association. This is Utah Agricultural Experiment Station Journal Article number 2667.

References

Abeles FB, Leather GR (1971) Abscission: Control of cellulase secretion by ethylene. Planta 97:87–91

Ackerson RC (1980) Stomatal response of cotton to water stress and abscisic acid as affected by water-stress history. Plant Physiol 65:455–459

Ackerson RC (1982) Synthesis and movement of abscisic acid in water-stressed cotton leaves. Plant Physiol 69:609–613

Addicott FT (1970) Plant hormones in the control of abscission. Biol Rev 45:485–524

Adkins SW, Ross JD (1981) Studies in wild oat seed dormancy. I. The role of ethylene in dormancy breakage and germination of wild oat seeds (*Avena fatua* L.) Plant Physiol 67:358–362

Aharoni N, Blumenfeld A, Richmond AE (1976) Hormonal activity in detached lettuce leaves as affected by leaf water content. Plant Physiol 59:1169–1173

Aharoni N, Marco S, Levi D (1977b) Involvement of gibberellins and abscisic acid in the supression of hypocotyl elongation in CMV-infected cucumbers. Physiol Plant Pathol 11:189–194

Alden J, Hermann RK (1971) Aspects of the cold-hardiness mechanism in plants. Bot Rev 37:37–142

Allard HA, Garner WW (1940) Observations on responses to length of day. USDA Tech Bull 727

Alvim R, Thomas S, Saunders PF (1978) Seasonal variation in the hormone content of willow II. Effect of photoperiod on growth and abscisic acid content of trees under field conditions. Plant Physiol 62:779–780

Anderson RC, Katz AJ, Anderson MR (1978) Allelopathy as a factor in the success of *Helianthus*. J Chem Ecol 4:9–16
Apelbaum A, Yang SF (1981) Biosynthesis of stress ethylene induced by water deficit. Plant Physiol 68:594–596
Aspinall D, Singh TN, Paleg LG (1973) Stress metabolism. 5. Abscisic acid and nitrogen metabolism in barley and *Lolium temulum* L. Aust J Biol Sci 26:319–327
Audus LJ (1979) Plant geosensors. J Exp Bot 30:1051–1073
Bacon GJ, Bachelard EP (1974) Changes in growth substance levels associated with the conditioning of *Pinus caribaea* Mor. var. *hondurensis* B and G seedlings to water stress. Aust For Res 9:241–254
Bandurski RS, Schulze A, Dayanandan P, Kaufman PB (1984) Response to gravity by *Zea mays* seedlings. I. Time course of the response. Plant Physiol 74:284–288
Banko RJ, Boe AA (1975) Effects of pH, temperature, nutrition, ethephon, and chlormequat on endogenous cytokinin levels of *Coleus blumei* Benth J Am Soc Hortic Sci 100:168–172
Battle RW, Gaunt JK, Laidman DL (1976) The effect of photoperiod on endogenous γ-tocopherol and plastochromanol in leaves of *Xanthium strumarium* L (cocklebur). Biochem Soc (Lond) Trans 4:484–486
Battle RW, Laidman DL, Gaunt JK (1977) The relationship between floral induction and γ-tocopherol concentrations in leaves of *Xanthium strumarium* L. Biochem Soc (Lond) Trans 5:322–324
Beevers L (1976) Senescence. In: Bonner J, Varner JE (eds) Plant Biochemistry. Academic Press, New York, pp 771–794
Behrens HM, Weisenseel MH, Sievers A (1982) Rapid changes in the pattern of electric current around the root tip of *Lepidium sativum* L. following gravistimulation. Plant Physiol 70:1079–1083
Bengston C, Klockare B, Klockare R, Larsson S, Sundqvist C (1978) The after-effect of water stress on chlorophyll formation during greening and the levels of abscisic acid and proline in dark grown wheat seedlings. Physiol Plant 43:205–212
Bengtson C, Falk SO, Larsson S (1979) Effects of kinetin on transpiration rate and abscisic acid content of water stressed young wheat plants. Physiol Plant 45:183–188
Bennet-Clark TA, Ball NG (1951) The diageotropic behavior of rhizomes. J Exp Bot 2:169–203
Benzioni A, Mizrahi Y, Richmond AE (1974) Effect of kinetin of plant-response to salinity. New Phytol 73:315–319
Bernier G, Kinet JM, Sachs RM (1981) The physiology of flowering Vol I: The initiation of flowers. Vol II. Transition of reproductive growth. CRC, Boca Raton
Bewley JD, Black M (1978, 1982) Physiology and biochemistry of seeds in relation to germination. In 2 Volumes. Springer, Berlin Heidelberg New York
Billings WD, Mooney HA (1968) The ecology of arctic and alpine plants. Biol Rev 43:481–529
Biran I, Gur I, Halevy AH (1972) The relationship between exogenous growth inhibitors and endogenous levels of ethylene and tuberization of dahlias. Physiol Plant 27:226–230
Blake TJ, Reid DM (1981) Ethylene, water relations and tolerance to waterlogging of three *Eucalyptus* species. Aust J Plant Physiol 8:497–505
Blaschke H (1977) Water soluble growth regulating factors occurring in coniferous leaf litter leachates. Flora 166:537–545
Boucaud J, Ungar IA (1976) Hormonal control of germination under saline conditions of three halophytic taxa in the genus *Suaeda*. Physiol Plant 37:143–148
Boussiba S, Richmond AE (1976) Abscisic acid and the after-effect of stress in tobacco plants. Planta 129:217–219
Boyer JS (1970) Leaf enlargement and metabolic rates in corn, soybean, and sunflower at various leaf water potentials. Plant Physiol 46:233–235
Boyer N (1967) Modifications de la croissance de la tige de bryone (*Bryonia dioica*) à la suite d'irritations tactiles. CR Acad Sci Paris 264:2114–2117

Bradford KJ (1983) Water relations and growth of the *flacca* tomato mutant in relation to abscisic acid. Plant Physiol 72:251–255

Bradford KJ, Hsiao TC (1982) Physiological responses to moderate water stress. In: Lange OL, Nobel PS, Osmond CB, Ziegler H (eds) Physiological plant ecology II. Water relations and carbon assimilation. Encyclopedia of plant physiology, new ser. Vol 12B. Springer, Berlin Heidelberg New York, pp 263–324

Bradford KJ, Yang SF (1980) Xylem transport of 1-aminocyclopropane-1-carboxylic acid, an ethylene precursor, in waterlogged tomato plants. Plant Physiol 65:322–326

Braun JW, Khan AA (1976) Alleviation of salinity and high temperature stress by plant growth regulators permeated into lettuce seeds via acetone. J Am Soc Hortic Sci 101:716–721

Brauner L, Hager A (1958) Versuche zur Analyse der geotropischen Perzeption I. Planta 51:115–1147

Brennan T, Gunckel JE, Frenkel C (1976) Stem sensitivity and ethylene involvement in phototropism of mung bean. Plant Physiol 57:286–289

Brian RW (1978) Hormones in healthy and diseased plants. Proc R Soc Lond Biol Sci 200:231–243

Briggs WR (1963) Mediation of phototropic responses of corn coleoptiles by lateral transport of auxin. Plant Physiol 38:237–247

Brix H, Portlock FT (1982) Flowering response of western hemlock seedlings to gibberellin and water-stress treatments. Can J For Res 12:76–82

Brown ME (1972) Plant growth substances produced by microorganisms of soil and rhizosphere. J Appl Bacteriol 35:443–451

Bryantseva ZN (1983) Effect of chlorocholine chloride on functional activity of wheat roots in connection with cold resistance. Sov Plant Physiol 29:579–584

Bucher JB (1981) SO_2-induced ethylene evolution of forest tree foliage, and its potential use as stress-indicator. (Short commun.) Eur J For Pathol 11:369–373

Bünning E (1969) Die Bedeutung tagesperiodischer Blattbewegungen für die Präzision der Tageslängenmessung. Planta 86:209–217

Bünning E (1973) The physiological clock (3rd edn) Academic Press, New York

Burg SP, Burg EA (1968) Ethylene formation in pea seedings: its relation to the inhibition of bud growth caused by indoleacetic acid. Plant Physiol 43:1069–1074

Burschka C, Tenhunen JD, Hartung W (1983) Diurnal variations in abscisic acid content and stomatal response to applied abscisic acid in leaves of irradiated and non-irradiated *Arbutus unedo* plants under naturally fluctuating environmental conditions. Oecologia 58:128–131

Buschmann C, Lichtenthaler HK (1982) The effect of cytokinins on growth and pigment accumulation of radish seedlings (*Raphanus sativus* L.) grown in the dark and at different light quanta fluence rates. Photochem Photobiol 35:217–221

Carpita NC, Nabors MW (1981) Growth physics and water relations of red-light-induced germination in lettuce seeds. V. Promotion of elongation in the embryonic axes by gibberellins and phytochrome. Planta 152:131–136

Carr DJ (1967) The relationship between florigen and the flower hormones. In: Fredrick JF, Weyer EM (eds) Plant growth regulators. Ann N Y Acad Sci Pub No 144:305–312

Carrington CM Sean, Firn RD (1983a) Cell enlargement during gravicurvature of sunflower hypocotyls. J Exp Bot 34:283–290

Carrington CM Sean, Firn RD (1983b) Solute production and net wall synthesis in the growing and non-growing cells of gravistimulated sunflower hypocotyls. Planta 157:524–530

Casperson G (1965) Über endogene Faktoren der Reaktionsholzbildung. Planta 64: 225–240

Catchpole AH, Hillman J (1969) Effect of ethylene on tuber formation in *Solanum tuberosum*. Nature 223:1387

Chailakhyan MKH (1968) Internal factors of plant flowering. Ann Rev Plant Physiol 19:1–36

Chanson A, Pilet PE (1981) Effect of abscisic acid on maize root gravireaction. Plant Sci Lett 22:1–5

Chapman HW (1958) Tuberization in the potato plant. Physiol Plant 11:215–224

Cho SC (1970) Response to gibberellic acid of the sterile cultured buds of *Begonia*. Sci Rep Tohoku Uni 25:139–148

Clark JE, Heath OVS (1959) Auxin and the bulbing of onions. Nature 184:345–347

Clark JE, Heath OVS (1962) Studies on the physiology of the onion plant V. Growth substance content. J Exp Bot 12:227–249

Cleland CF (1978) The flowering enigma. Bio Science 28:265–269

Cleland CF, Ajami A (1974) Identification of flower-inducting factor isolated from aphid honeydew as being salicylic acid. Plant Physiol 54:904–906 (See also 54:899–903)

Clifford PE, Fensom DS, Munt BI, McDowell WD (1982) Lateral stress initiates bending responses in dandelion peduncles: a clue to geotropism? Can J Bot 60:2671–2673

Clifford PE, Reid DM, Pharis RP (1983) Endogenous ethylene does not initiate but may modify geobending – a role for ethylene in autotropism. Plant Cell Envir 6:433–436

Cocucci S, Ranieri AM, Morgutti S, Ciroli F (1981) The role of darkness, GA and fusicoccin (FC) in breaking photodormancy in *Phacelia tanacetifolia* seeds. Physiol Plant 52:177–180

Cottrell JE, Dale JE, Jeffcoat B (1982) The effects of daylength and treatment with gibberellic acid on spikelet initiation and development in Clipper barley. Ann Bot 50:57–68

Cracker LE, Abeles FB (1969) Abscission: role of abscisic acid. Plant Physiol 44:1144–1149

Crawford RMM (1982) Physiological responses to flooding. In: Lange OL, Nobel PL, Osmond CB, Ziegler H (eds) Physiological plant ecology II. Water relations and carbon assimilation. Encyclopedia of plant physiology, new ser Vol 12B. Springer, Berlin Heidelberg New York, pp 453–477

Daie J, Campbell WF, Seeley SD (1981) Temperature-stress-induced production of abscisic acid and dihydrophaseic acid in warm- and cool-season crops. J Am Soc Hortic Sci 106:11–13

Daines RJ, Minocha SC (1983) Regulation of phenylalanine ammonia-lyase in germinating lettuce seeds (*Lactuca sativa* L. cv. Grand Rapids): Effects of abscisic acid and water stress. Z Pflanzenphysiol 110:69–76

Damptey HB, Coombe BG, Aspinall D (1978) Apical dominance, water deficit and axillary inflorescence growth in *Zea mays:* the role of abscisic acid. Ann Bot 42:1447–1458

Darwin C, Darwin F (1966) The power of movement in plants. Appleton New York "Authorized edition," 1896, Appleton D, New York, reprinted by De Capo, New York

Davenport TL, Jordan WR, Morgan PW (1977) Movement and endogenous levels of abscisic acid during water-stress-induced abscission in cotton seedlings. Plant Physiol 59:1165–1168

Davies WJ, Kozlowski TT (1975) Effects of applied abscisic acid and plant water stress on transpiration of woody angiosperms. For Sci 21:191–195

Dei M (1981) Evidence that ethylene is not involved in red-light-induced stimulation of chlorophyll formation in etiolated cucumber cotyledons. Plant Cell Physiol 22:699–707

Dennison DS (1974) Phototropism. In: Haupt W, Feinleib ME (eds) Physiology of movements. Encyclopedia of plant physiology, new ser Vol 7. Springer, Berlin Heidelberg New York, pp 506–566

Digby J, Firn RD (1976) A critical assessment of the Cholodny-Went theory of shoot geotropism. Curr Adv Plant Sci 8:953–960

Domanski R, Kozlowski TT (1968) Variations in kinetin-like activity in *Betula* and *Populus* during release from dormancy. Can J Bot 46:397–403

Dörffling K (1966) Weitere Untersuchungen über korrelative Knospenhemmung. Planta 70:257–274

Downton WJS, Loveys BR (1981) Abscisic acid content and osmotic relations of salt-stressed grapevine leaves. Aust J Plant Physiol 8:443–452

Dugger WM, Ting IP (1970) Air pollution oxidants – their effects on metabolic processes in plants. Annu Rev Plant Physiol 21:215–234
Eamus D, Wilson JM (1983) ABA levels and effects in chilled and hardened *Phaseolus vulgaris*. J Exp Bot 34:1000–1006
Egley GH (1980) Stimulation of common cocklebur (*Xanthium pensylvanicum*) and redroot pigweed (*Amaranthus retroflexus*) seed germination by injections of ethylene into soil. Weed Sci 28:510–514
Ehleringer J, Forseth I (1980) Solar tracking by plants. Science 210:1094–1098
El-Antably HMM, Wareing PF, Hillman J (1967) Some physiological responses to DL abscisin (dormin). Planta 73:74–90
Elliott DC (1982) Inhibition of cytokinin action and of heat aging induced potential for cytokinin action by inhibitors of membrane synthesis and function. Plant Physiol 69:1169–1172
Esashi Y (1960) Studies on the formation and sprouting of aerial tubers in *Begonia evansiana* IV. Photoperiodic conditions for tuberization and sprouting in the cutting plants. Sci Rep Tohoku Univ 27:101–112
Esashi Y (1964) Studies on the formation and sprouting of aerial tubers in *Begonia*. X. Tuberization under long days and in darkness. Plant Cell Physiol 5:101–107
Esashi Y, Leopold AC (1968) Regulation of tuber development in *Begonia* by cytokinin. In: Wightman P, Setterfield G (eds) Biochemistry and physiology of plant growth substances. Runge, Ottawa, pp 923–941
Esashi Y, Eguchi T, Nagao M (1964) The role of auxin in the photoperiodic tuberization in *Begonia evansiana*. Plant Cell Physiol 5:413–427
Evans LT (1969) The induction of flowering, some case histories. Macmillan, South Melbourne, Victoria
Evans LT (1971) Flower induction and the florigen concept. Annu Rev Plant Physiol 22:365–394
Even-Chez Z, Sachs RM (1980) Photosynthesis as a function of short day induction and gibberellic acid treatment in *Bougainvillea* San Diego Red. Plant Physiol 65:65–68
Even-Chez Z, Sachs RM, Hackett WP (1979) Control of flowering in *Bougainvillea* San Diego Red. Metabolism of benzyladenine and the action of gibberellic acid in relation to short day induction. Plant Physiol 64:646–651
Ewing EE, Wareing PF (1978) Shoots, stolon and tuber formation on potato (*Solanum tuberosum* L.) cuttings in response to photoperiod. Plant Physiol 61:348–353
Eze JMO, Dumbroff EB, Thompson JE (1981) Effects of moisture stress and senescence on the synthesis of absicisic acid in the primary leaves of bean. Physiol Plant 51:418–422
Fedorova AI (1977) Effects of unfavorable environmental conditions on the level of abscisic acid in Siberina larch. Sov Plant Physiol 24:985–988
Feldman LJ (1981) Light-induced inhibitors from intact and cultured caps of *Zea* roots. Planta 153:471–475
Fellows RJ, Boyer JS (1978) Altered ultrastructure of cells of sunflower leaves having low water potentials. Protoplasma 93:381–395
Fernandez MB, Sanchez TR (1982) Seasonal pattern of growth regulators in tubers of yellow nutsedge (*Cyperus esculentus*). Weed Sci 30:83–86
Fernandez TF, Gaborjanyi R (1976) Reversion of dwarfing induced by virus infection: effect of polyacrylic and gibberellic acids. Acta Phytopathol Acad Sci Hung 11:271–276
Firn RD, Digby J (1980) The establishment of tropic curvatures in plants. Ann Rev of Plant Physiol 31:131–148
Fischer H (1905) Über die Blütenbildung in ihrer Abhängigkeit vom Licht und über die blütenbildenden Substanzen. Flora. 94:478–490
Fletcher RA, Adedipe NO, Ormrod DP (1972) Abscisic acid protects bean leaves from ozone-induced phytotoxicity. Can J Bot 50:2389
Forsline PL, Langille AR (1974) The influence of temperature and photoperiod on cytokinin activity in the katahdin potato. Am Potato J 51:309
Franssen JM, Bruinsma J (1981) Relationships between xanthoxin, phototropism, and elongation growth in the sunflower seedling *Helianthus annuus* L. Planta 151:365–370

Franssen JM, Firn RD, Digby J (1982) The role of the apex in the phototropic curvature of *Avena* coleoptiles: positive curvature under conditions of continuous illumination. Planta 155:281–286

Freeman BC, McArthur ED, Harper K, Blauer AC (1981) Influence of environment on the floral sex ratios of monoecious plants. Evolution 35:194–197

Freeman MW, Martin GC (1981) Peach floral bud break and abscisic acid content as affected by mist, light, and temperature treatments during rest. J Am Soc Hortic Sci 106:333–336

Frimmel G (1977) The effect of mildew infection on winter hardiness and drought resistance. Z Pflanzenzücht 79:256–260

Galli MG, Levi M (1982) Increased drought resistance induced by pretreatment with abscisic acid in germinating embryos of *Haplopappus gracilis*. Physiol Plant 54:425–430

Gianfagna T, Zeevaart JAD, Lusk WJ (1983) Effect of photoperiod on the metabolism of deuterium-labeled gibberellin A_{53} in spinach. Plant Physiol 72:86–89

Gilbertson-Ferriss TL, Brenner ML, Wilkins HF (1981) Localization of endogenous growth substances in *Freesia hybrida* Moya corms before and after a heat treatment. J Am Soc Hortic Sci 106:460–463

Giles KL, Cohen D, Beardsell MF (1976) Effects of water stress on the ultrastructure of leaf cells of *Sorghum bicolor*. Plant Physiol 57:11–14

Goeschl JD, Pratt HK, Bonner BA (1967) An effect of light on the production of ethylene and the growth of the plumular portion of etiolated pea seedlings. Plant Physiol 42:1077–1080

Goeschl JD, Rappaport L, Pratt HK (1966) Ethylene as a factor regulating the growth of pea epicotyls subjected to physical stress. Plant Physiol 41:877–884

Goleniowski M, DeBottini GA, Tizio R (1980) The effect of CCC (2-chloroethyl) trimethylammonium chloride and its interaction with gibberellic acid (GA_3) on quali- and quantitative evolution of gibberellin-like substances in potato plants (*Solanum tuberosum* L.) grown under unfavorable photoperiodic conditions for tuberization. Phyton 38:13–22

Goode JE, Higgs KH, Hyrycz KJ (1978) Abscisic acid applied to orchard trees of Golden Delicious apple to control water stress. J Hortic Sci 53:99–103

Gorski T (1975) Germination of seeds in the shadow of plants. Physiol Plant 34:342–346

Gorski T, Gorska K, Rybicki J (1978) Studies on germination of seeds under leaf canopy, Flora 167, 289–299

Gray D, Steckel JRA (1977) Pre-sowing seed treatment with cytokinin to prevent high temperature dormancy in lettuce (*Lactuca sativa*) seeds. Seed Sci Technol 5:473–477

Greenwood MS (1981) Reproductive development in loblolly pine. II. The effect of age, gibberellin plus water stress and out-of-phase dormancy on long shoot growth behavior. Am J Bot 68:1184–1190

Gregg KB (1982) Sunlight-enhanced ethylene evolution by developing inflorescence of *Catasetum* and *Cycnoches* and its relation to female flower production. Bot Gaz 143:466–475

Gregory FG, Veale JA (1957) A reassessment of the problem of apical dominance. Soc Exp Biol Symp 11:1–20

Gregory LE (1956) Some factors for tuberization in the potato plant. Am J Bot 43:281–288

Gregory LE (1965) Physiology of tuberization in plants. In: Ruhland W (ed) Encycl of Plant Physiol. Vol 15. Springer, Berlin Göttingen Heidelberg, pp 1328–1356

Guinn G (1982) Abscisic acid and abscission of young cotton bolls in relation to water availability and boll load. Crop Sci 22:580–583

Gusta LV, Fowler DB, Tyler NJ (1982) The effect of abscisic acid and cytokinins on the cold hardiness of winter wheat. Can J Bot 60:301–305

Halevy AH, Mayak S (1975) Interrelationship of several phytohormones in the regulation of rose petal senescence. Acta Hortic 41:103–116

Hall AB, Firn RD, Digby J (1980) Auxins and shoot-tropisms – a tenuous connection? J Biol Educ 14:195–199

Hanks GR (1982) The response of tulips to gibberellins following different durations of cold storage. J Hortic Sci 57:109–119
Hanson JB, Trewavas AJ (1982) Regulation of plant cell growth: the changing perspective. New Phytol 90:1–18
Hasegawa K, Shiihara S, Iwagawa T, Hase T (1982) Isolation and identification of a new growth inhibitor, raphanusanin, from radish seedlings and its role in light inhibition of hypocotyl growth. Plant Physiol 70:626–628
Health OVS, Holdsworth M (1943) Bulb formation and flower production in onion. Nature 152:334–335
Hemberg T (1949) Significance of growth-inhibiting substances and auxins for the rest period of potato. Physiol Plant 2:24–36
Henson IE (1981) Changes in abscisic acid content during stomatal closure in pearl millet (*Pennisetum americanum* (L.) Leeke) Plant Sci Lett 21:121–127
Henson IE (1982) Abscisic acid and water relations of rice (*Oryza sativa* L.): sequential responses to water stress in the leaf. Ann Bot 50:9–24
Henson IE (1983) Abscisic acid and water relations of rice (*Oryza sativa* L.): Effects of drought conditioning on abscisic acid accumulation in the leaf and stomatal response. Ann Bot 52:247–255
Henson IE, Wheeler CT (1977) Hormones in plants bearing nitrogen fixing root nodules: distribution and seasonal changes in levels of cytokinins in *Alnus glutinosa*. J Exp Bot 28:205–214
Henson IE, Alagarswamy G, Mahalakshmi V, Bidinger FR (1982) Diurnal changes in endogenous abscisic acid in leaves of pearl millet (*Pennisetum americanum* (L.) Leeke) under field conditions. J Exp Bot 33:416–425
Hewett EW, Wareing PF (1973) Cytokinins in *Populus* × *robusta:* Changes during chilling and bud burst. Physiol Plant 28:393–399
Hiller LK, Kelly WC, Powell LE (1979) Temperature interactions with growth regulators and endogenous gibberellin-like activity during seedstalk elongation in carrots. Plant Physiol 63:1055–1061
Hillman SK, Wilkins MB (1982) Gravity perception decapped roots of *Zea mays*. Planta 155:267–271
Hillman W (1976) Biological rhythms and physiological timing. Ann Rev Plant Physiol 27:159–179
Hiraki Y, Ota Y (1975) The relationship between growth inhibition and ethylene production by mechanical stimulation in *Lilium longiflorum*. Plant Cell Physiol 16:185–189
Hodson HK, Hamner KC (1970) Floral inducing extract from *Xanthium*. Science 167:384–385
Holmes MG (1976) Spectral energy distribution in the natural environment and its implications for phytochrome function. In: Smith H (ed) Light and plant development. Butterworth, London, pp 407–476
Holmes MG, Smith H (1975) The function of phytochrome in plants growing in the natural environment. Nature 254:512–514
Holmes MG, Smith H (1977a) The function of phytochrome in the natural environment. I. Characterization of daylight for studies in photomorphogenesis and photoperiodism. Photochem Photobiol 25:533–538
Holmes MG, Smith H (1977b) The function of phytochrome in the natural environment. II. The influence of vegetation canopies on the spectral energy distribution of natural daylight. Photochem Photobiol 25:539–545
Holmes MG, Smith H (1977c) The function of phytochrome in the natural environment. IV. Light quality and plant development. Photochem Photobiol 25:551–557
Hoshizaki T (1973) Influence of gravitational forces on plants. Environ Biol Med 2:47–79
Hubac C, LePage-Degivry MT (1981) Abscisic acid content evolution during drought in cotton plants under different photoperiodic conditions. Physiol Vég 19:87–97
Iino M (1982) Inhibitory action of red light on the growth of the maize mesocotyl: Evaluation of the auxin hypothesis. Planta 156:388–395
Irvine RF, Osborne DJ (1973) The effect of ethylene on (1-^{14}C) glycerol incorporation into phospholipids of etiolated pea stems. Biochem J 136:1133–1135

Itai C, Richmond C, Vaadia Y (1968) The role of root cytokinins during water and salinity stress. Isr J Bot 17:187–195

Jackson DI, Field RJ (1972) Light and hormone interaction in apical dominance in *Phaseolus vulgaris* L. Ann Bot 36:525–532

Jackson MB, Campbell DJ (1979) Effects of benzyladenine and gibberellic acid on the responses of tomato plants to anaerobic root environments and to ethylene. New Phytol 82:331–340

Jaffe MJ (1976a) Thigmomorphogenesis: a detailed characterization of the response of beans (*Phaseolus vulgaris* L) to mechanical stimulation. Z Pflanzenphysiol 77:437–453

Jaffe MJ (1976b) Thigmomorphogenesis: electrical resistance and mechanical correlates of the early events of growth retardation due to mechanical stimulation in beans. Z Pflanzenphysiol 78:24–32

Jaffe MJ, Biro R (1979) Thigmomorphogenesis: the effect of mechanical perturbation on the growth of plants, with special reference to anatomical changes, the role of ethylene, and interaction with other environmental stresses. In: Mussell H, Staples RC (eds) Stress physiology in crop plants. John Wiley & Sons, New York, p 25–59

Jones MG, Zeevaart JAD (1982) Effect of photoperiod on metabolism of ^{3}H gibberellins A_1, 3-epi-A_1, and A_{20} in *Agrostemma githago* L. Plant Physiol 69:660–662

Jones RL (1973) Gibberellins: their physiological role. Annu Rev Plant Physiol 24:571–598

Jordon WR, Morgan PW, Davenport TL (1972) Water stress enhances ethylene-mediated leaf abscission in cotton. Plant Physiol 50:756–758

Julin-Tegelman A, Pinfield N (1983) Changes in the levels of endogenous cytokinin-like substances during cold-induced germination of *Acer platanoides* L. seeds. Z Pflanzenphysiol 110:89–95

Juniper BE (1976) Geotropism. Ann Rev Plant Physiol 27:385–406

Junttila O (1980) Flower bud differentiation in *Salix pentandra* as affected by photoperiod, temperature and growth regulators. Physiol Plant 49:127–134

Junttila O (1981) Effects of different gibberellins on elongation growth under short day conditions in seedlings of *Salix pentandra*. Physiol Plant 53:315–318

Kane ME, Albert LS (1982) Environmental and growth regulator effects on heterophylly and growth of *Proserpinaca intermedia* (*Haloragaceae*). Aquat Bot 13:1 (Special issue) 73–86

Kang BG, Burg SP (1973) Role of ethylene in phytochrome-induced anthocyanin synthesis. Planta 110:227–235

Kang BG, Yokum CS, Burg SP, Ray PM (1967) Ethylene and carbon dioxide: mediation of hypocotyl hook opening response. Science 156:958–959

Kannangara T, Durley RC, Simpson GM, Stout DG (1982) Drought resistance of *Sorghum bicolor* 4. Hormonal changes in relation to drought stress in field-grown plants. Can J Plant Sci 62:317–330

Kao CH (1980) Retardation of senescence by low temperature and benzyladenine in intact primary leaves of soybean. Plant Cell Physiol 21:339–344

Karege F, Penel E, Greppin H (1982) Floral induction in spinach leaves by light, temperature and gibberellic acid: use of the photocontrol of basic peroxidase activity as biochemical marker. Z Pflanzenphysiol 107:357–365

Kasperbauer MJ (1971) Spectral distribution of light in a tobacco canopy and effect of end-of-day light quality on growth and development. Plant Physiol 47:775–778

Kato Y, Fukunharu N, Kobayashi R (1958) Stimulation of flower bud differentiation of conifers by gibberellin. In: Abstracts of the 2nd Meeting of the Japan Gibberellin Research Association. pp 67–68

Kawase M (1974) Role of ethylene in induction of flooding damage in sunflower. Physiol Plant 31:29–38

Kays SJ, Nicklow CW, Simons DH (1974) Ethylene in relation to the response of roots to physical impedance. Plant Soil 40:565–571

Keller PL, Coulter MW (1982) The relationship of endogenous gibberellins to light-regulated stem elongation rates in dwarf and normal cultivars of *Pisum sativum* L. Plant Cell Physiol 23:409–416

Khan MI, Khan MA, Khizar T (1976) Plant-growth regulators from species differing in salt tolerance as affected by soil salinity. Plant Soil 45:267–271

Kimball SL, Salisbury FB (1973) Ultrastructural changes of plants exposed to low temperatures. Am J Bot 60:1028–1033

Kimball SL, Salisbury FB (1974) Plant development under snow. Bot Gaz 135:147–149

Kirkham MB (1983) Effect of ABA on the water relations of winter-wheat cultivars varying in drought resistance. Physiol Plant 59:153–157

Koch J, Bergmann H (1979) Effect of phytohormones on water use efficiency (WUE) in *Hordeum vulgare* L. and *Triticum aestivum* L. Biochem Physiol Pflanz 174:486–490

Kohler KH, Dorfler M, Goring H (1980) The influence of light on the cytokinin content of *Amaranthus* seedlings. Biol Plant 22:128–134

Koller D (1969) The physiology of dormancy and survival of plants in desert environments. In: Woolhouse HW (ed) Dormancy and Survival. Symp Soc Exp Biol No 23, pp 449–469

Koller D, Hadas A (1982) Water relations in the germination of seeds. In: Lange OL, Osmond CB, Ziegler H (eds) Physiological plant ecology II. Water relations and carbon assimilation. Encyclopedia of plant physiology, new ser Vol 12B, Springer, Berlin Heidelberg New York, pp 401–432

Kondo N, Sugahara K (1978) Changes in transpiration rate of SO_2-resistant and -sensitive plants with SO_2 fumigation and the participation of abscisic acid. Plant Cell Physiol 19:365–373

Korableva NP, Karavaeva KA, Metlitskii LV (1980) Changes of abscisic acid content in potato tuber tissues in the period of deep dormancy and during germination. Sov Plant Physiol 27:441–446

Krekule J, Kohli RK (1981) The condition of the apical meristem of seedlings response to a promotive effect of abscisic acid on flowering in the short-day plant, *Chenopodium rubrum*. Z Pflanzenphysiol 103:45–51

Krizek DT, Mandava NB (1983) Influence of spectral quality on the growth response of intact bean plants to brassinosteroid, a growth-promoting steroidal lactone. I. Stem elongation and morphogenesis. Physiol Plant 57:317–323

Krul WR (1978) Diffusible inhibitors of imbibition from senescent soybean pods. Hortic Sci 13:41–42

Lancaster JE, Mann JD (1977) Extrastomatal control of transpiration in leaves of yellow lupin. J Exp Bot 28:1373–1379

Lang A (1965) Physiology of flower initiation. In: Ruhland W (ed) Encyclopedia of plant physiology Vol 15. Springer, Berlin Göttingen Heidelberg, pp 1380–1535

Lang A, Chailakhyan MKH, Frolova IA (1977) Promotion and inhibition of flower formation in a day-neutral plant in grafts with a short-day plant and a long-day plant. Proc Natl Acad Sci 74:2412–2416

Langille AR, Forsline PL (1974) Influence of temperature and photoperiod on cytokinin pools in the potato *Solanum tuberosum* L. Plant Sci Lett 2:189–191

Larque-Saavedra A, Wain RL (1974) Abscisic acid levels in relation to drought tolerance in varieties of *Zea mays* L. Nature 251:716–717

Larque-Saavedra A, Wain RL (1976) Studies on plant growth-regulating substances. XLII. Abscisic acid as a genetic character related to drought tolerance. Ann Appl Biol 83:291–297

Leach RWA, Wareing PF (1967) Distribution of auxin in horizontal woody stems in relation to gravimorphism. Nature 214:1025–1027

Lee JS, Mulkey TJ, Evans ML (1983) Reversible loss of gravitropic sensitivity in maize roots after tip application of calcium chelators. Science 220:1375–1376

Lenton JR, Perry VM, Saunders PF (1972) Endogenous abscisic acid in relation to photoperiodically induced bud dormancy. Planta 106:13–22

Leopold AC, Brown KM, Emerson FH (1972) Ethylene in the wood of stressed trees. Hortic Sci 7:175

Levitt J (1972) Responses of plants to environmental stresses. Academic Press, London New York

Lincoln RG, Cunningham A, Carpenter BH, Alexander J, Mayfield DL (1966) Florigenic acid from fungal cultures. Plant Physiol 41:1079–1080
Lipe JA, Morgan PW (1972) Ethylene role in fruit abscission and dehiscence processes. Plant Physiol 50:759–764
Lombard P, Richardson EA (1979) Physical principles involved in controlling phenological development. In: Barfield BJ, Gerber JF (eds) Modification of the aerial environment of plants. St Joseph, Michigan, Am Soc Agric Eng, pp 429–440
Longman KA (1982) Effects of gibberellin, clone and environment on cone initiation, shoot growth and branching in *Pinus contorta*. Ann Bot 50:247–257
Lozhnikova VN, Krekule J, Saidlova F, Bavrina TV, Chailakhyan MK (1977) Effects of growth substances on flowering of *Chenopodium rubrum* under different conditions of photoperiodic induction. Sov Plant Physiol 24:810–813
Lynch JM (1975b) Ethylene in soil. Nature 256:576–577
Lynch JM, Harper SHT (1977) Effects of microorganisms and their products on seed germination. Letcombe Lab Ann Rep 50–52
Makus DJ, Shannon C (1979) Temperature and photoperiod effects on abscisic acid content in Earliwax snapbean seed. Hort Science 14:732–733
Marinos NG, Fife DN (1972) Ultrastructural changes in wheat embryos during a "presowing drought hardening" treatment. Protoplasma 74:381–396
Markhart III AH, Fiscus EL, Naylor AW, Kramer PJ (1979) Effect of abscisic acid on root hydraulic conductivity. Plant Physiol 64:611–614
Mason HL, Langenheim JH (1957) Language analysis and the concept environment. Ecology 38:325–340
Mauk CS, Langille AR (1978) Physiology of tuberization in *Solanum tuberosum* L. Cis-zeatin riboside in the potato plant: its identification and changes in endogenous levels as influenced by temperature and photoperiod. Plant Physiol 62:438–442
Mayer AM, Poljakoff-Mayber A (1975) The germination of seeds. 2nd edn. Pergamon, New York
Mayer AM, Shain Y (1974) Control of seed germination. Annu Rev Plant Physiol 25:167–193
McMichael BL, Elmore CD (1977) Proline accumulation in water-stressed cotton leaves. Crop Sci 17:905–908
Menzel CM (1980) Tuberization in potato at high temperatures: responses to gibberellin and growth inhibitors. Ann Bot 46:259–265
Mertens R, Weiler EW (1983) Kinetic studies on the redistribution of endogenous growth regulators in gravireacting plant organs. Planta 158:339–348
Metzger JD, Zeevaart JAD (1982) Photoperiodic control of gibberellin metabolism in spinach. Plant Physiol 69:287–291
Mielke EA, Dennis FG (1978) Hormonal control of flower bud dormancy in sour cherry (*Prunus cerasus* L.). III. Effects of leaves, defoliation and temperature on levels of abscisic acid in flower primordia. J Am Soc Hortic Sci 103:446–449
Misaghi I, Devay JE, Kosuge T (1972) Changes in cytokinin activity associated with the development of *Verticillium* wilt and water stress in cotton plants. Physiol Plant Pathol 2:187–196
Mitchell CA (1977) Influence of mechanical stress on auxin-stimulated growth of excised pea stem sections. Physiol Plant 41:129–134
Mittelheuser CJ, Van Steveninck RFM (1969) Stomatal closure and inhibition of transpiration induced by (RS)-abscisic acid. Nature 221:281–282
Mizrahi Y, Blumenfeld A, Bittner S, Richmond AE (1971) Abscisic acid and cytokinin contents of leaves in relation to salinity and relative humidity. Plant Physiol 48:752–755
Mirzahi Y, Blumenfeld R, Richmond A (1972) The role of abscisic acid and salination in the adaptive response of plants to reduced root aeration. Plant Cell Physiol 13:15–21
Mizrahi Y, Richmond AE (1972b) Mineral deficiency and abscisic acid. Isr J Bot 21:123
Moore PD (1978) Adaptation to waterlogged environments. Nature 271:209

Moore R, Pasieniuk J (1984) Graviresponsiveness and the development of columella tissue in primary and lateral roots of *Ricinus communis*. Plant Physiol 74:529–533

Morgan DC, Smith H (1978) The relationship between phytochrome photoequilibrium and development in light-grown *Chenopodium album*. Planta 142:187–194

Morgan DC, Smith H (1979) A systematic relationship between phytochrome-controlled development and species habitat, for plants grown in simulated natural radiation. Planta 145:253–258

Morgan DC, Smith H (1981) Non-photosynthetic responses to light quality. In: Lange OL, Nobel PS, Osmond CB, Ziegler H (eds) Physiological plant ecology I. Responses to the physical environment. Encyclopedia of plant physiology, new ser Vol 12A. Springer, Berlin Heidelberg New York, pp 109–134

Morgan PW, Jordan WR, Davenport TL, Durham JI (1977) Abscission responses to moisture stress, auxin transport inhibitors, and ethephon. Plant Physiol 59:710–712

Moser BC, Hess CE (1968) The physiology of tuberous root development in *Dahlia*. Proc Am Soc Hortic Sci 93:595

Mousdale DM (1983) Seasonal variation and metabolism of abscisic acid in shoot bark and lateral buds of apple (*Malus domestica* Borkh.) Biochem Physiol Pflanz 178:503–510

Muir RM, Zhu L (1983) Effect of light in the control of growth by auxin and its inhibitor(s) in the sunflower. Physiol Plant 57:407–410

Mulkey TJ, Evans ML (1982) Suppression of asymmetric acid efflux and gravitropism in maize roots treated with auxin transport inhibitors of sodium orthovanadate. J Plant Growth Regul 1:259–265

Munoz V, Butler WL (1975) Photoreceptor pigments for blue light in *Neurospora crassa*. Plant Physiol 55:421–426

Nagao M, Okagami N (1966) Effect of 2-chlorethyl-trimethyl-ammonium-chloride on the formation and dormancy of aerial tubers of *Begonia*. Bot Mag 79:687–692

Neel PL, Harris RW (1971) Motion-induced inhibition of elongation and induction of dormancy in *Liquidambar*. Science 173:58–59

Nelson JM, Sharples GC (1980) Effect of growth regulators on germination of cucumber and other cucurbit seeds at suboptimal temperatures. Hort Science 15:253–254

Nir I, Klein S, Poljakoff-Mayber A (1970b) The effect of water stress on mitochondria of root cells. Plant Physiol 45:173–177

Nitsch JP (1966) Photoperiodism et tubérisation. Bull Soc Fr Physiol Vég 12:233–246

Ogawa Y, King RW (1979) Establishment of photoperiodic sensitivity by benzyladenine and a brief red irradiation in dark grown seedlings of *Pharbitis nil* Chois. Plant Cell Physiol 20:115–122

Ogawa Y, Zeevaart JAD: The relation of growth regulators to flowering. In: Imamura S (ed) Physiology of flowering in pharbitis nil. Jpn Soc Plant Physiol, pp 107–119

Ogunkanmi AB, Wellburn AR, Mansfield TA (1974) Detection and preliminary identification of endogenous antitranspirants in water-stressed sorghum plants. Planta 117:293–302

Okazawa Y (1960) Studies on the relation between tuber formation of potato and its natural gibberellin content. Proc Crop Sci Soc Jpn 29:121–124

Okazawa Y, Katsura N, Tagawa T (1967) Effects of auxin and kinetin on the development and differentiation of potato tissue cultured in vitro. Physiol Plant 20:862–869

Pallas Jr JE, Kays SJ (1982) Inhibition of photosynthesis by ethylene – a stomatal effect. Plant Physiol 70:598–601

Palmer CE, Smith OE (1964) Effect of abscisic acid on elongation and kinetin-induced tuberization of isolated stolons of *Solanum tuberosum*. Plant Cell Physiol 10:657–664

Palmer JH (1956) The nature of the growth response to sunlight shown by certain stoloniferous and prostrate tropical plants. New Phytol 55:346–355

Parups EV (1980) Gibberellins in photoperiodically treated *Chrysanthemum* cv. Improved Indianapolis White. Phyton 39:121–126

Perry TO, Hellmers H (1973) Effects of abscisic acid on growth and growth and dormancy of two races of red maple. Bot Gaz 134:783–789

Peterson JC, Sacalis jN, Durkin DJ (1980) Alterations in abscisic acid content of *Ficus benjamina* leaves resulting from exposure to water stress and its relationship to leaf abscission. J Am Soc Hortic Sci 105:793–798

Pharis RP (1972) Flowering of *Chrysanthemum* under non-inductive long days by gibberellins and N^6-benzyladenine. Planta 105:205–212

Pharis RP, Kuo CG (1977) Physiology of gibberellins in conifers. Can J For Res 7:299–325

Phillips IDJ (1975) Apical dominance. Annu Rev Plant Physiol 26:341–367

Phillips IDJ, Hartung W (1976) Longitudinal and lateral transport of [3-4-^{3}H] gibberellin A 1 and 3-indolyl (acetic acid-2-^{14}C) in upright and geotropically responding green internode stems of *Helianthus annuus*. New Phytol 76:1–9

Phillips IDJ, Wareing PF (1959) Studies in dormancy of sycamore. II. The effect of daylength on the natural growth inhibitor content of the shoot. J Exp Bot 10:504–514

Phillipson JJ (1983) The role of gibberellin A4/7, heat, and drought in the induction of flowering in Sitka spruce. J Exp Bot 34:291–302

Pilet PE, Elliott MC (1981) Some aspects of the control of root growth and georeaction: the involvement of indoleacetic acid and abscisic acid. Plant Physiol 67:1047–1050

Pillay I, Railton ID (1983) Complete release of axillary buds from apical dominance in intact, light-grown seedlings of *Pisum sativum* L. following a single application of cytokinin. Plant Physiol 71:972–974

Poapst PA, Durkee AB, McGugan WA, Johnston FB (1968) Identification of ethylene in gibberellic acid-treated potatoes. Sci Food Agric 19:325

Porpiglia PJ, Barden JA (1980) Peach leaf abscission following CGA-15281 and CGA-17856 applications as affected by temperature. J Am Soc Hortic Sci 105:227–229

Prisco JT, O'Leary JW (1973) The effects of humidity and cytokinin on growth and water relations of salt-stressed bean plants. Plant Soil 39:263–276

Proebsting WM, Heftmann E (1980) The relationship of ^{3}H-GA$_9$ metabolism to photoperiod-induced flowering in *Pisum sativum* L. Z Pflanzenphysiol 98:305–309

Pryce RJ (1972) Gallic acid as a natural inhibitor of flowering in *Kalanchoe blossfeldiana*. Phytochemistry 11:1911–1918

Puppo A, Rigaud J (1978) Cytokinins and morphological aspects of French bean roots in the presence of *Rhizobium*. Physiol Plant 42:202–206

Purvis ON (1961) The physiological analysis of vernalization. In: Ruhland W (ed) Encyclopedia of plant physiology. Springer, Berlin Göttingen Heidelberg, Vol 16, pp 76–122

Putnam AR, Duke WB (1978) Allelopathy in agroecosystems. Annu Rev Phytopathol 16:431–451

Quarrie SA (1980) Genotypic differences in leaf water potential, abscisic acid and proline concentrations in spring wheat during drought stress. Ann Bot 46:383–394

Quarrie SA, Henson IE (1982) Biparental inheritance of drought-induced accumulation of abscisic acid in wheat and pearl millet (Short commun.) Ann Bot 49:265–268

Quarrie SA, Jones HG (1977) Effects of abscisic acid and water stress on development and morphology of wheat. J Exp Bot 28:192–203

Quarrie SA, Jones HG (1979) Genotypic variation in leaf water potential, stomatal conductance and abscisic acid concentration in spring wheat subjected to artificial drought stress. Ann Bot 44:323–332

Radin JW, Ackerson RC (1981) Water relations of cotton plants under nitrogen deficiency. III. Stomatal conductance, photosynthesis, and abscisic acid accumulation during drought. Plant Physiol 67:115–119

Radin JW, Ackerson RC (1982) Does abscisic acid control stomatal closure during water stress? What's New In Plant Physiol 13(3):9–12

Railton ID, Wareing PF (1973c) Effects of daylength on endogenous gibberellins in *Solanum andigena*. III. Gibberellin production by the leaves. Physiol Plant 29:430–433

Rajagopal V, Anderson AS (1978) Does abscisic acid influence proline accumulation in stressed leaves? Planta 143:85–88

Rakhimbaev IR, Syrtanova GA, Solomina VF (1978) Effect of cold treatment on the level of biological activity of endogenous growth regulators in tulip bulbs. Sov Plant Physiol 25:197–200

Rappaport L, Wolf N (1969) The problem of dormancy in potato tubers and related structures. Soc Exp Biol Symp 23:219–240

Raschke K (1975) Stomatal action. Annu Rev Plant Physiol 26:309–340

Raschke K (1976) How stomata resolve the dilemma of opposing priorities. Philos Trans R Acad Lon Biol Sci 273:551–560

Raschke K, Zeevaart JAD (1976) Abscisic acid content, transpiration, and stomatal conductance as related to leaf age in plants of *Xanthium strumarium* L. Plant Physiol 58:169–174

Rasmussen OS (1976) Water stress in plants. I. Abscisic acid level in tomato leaves after a long period of wilting. Physiol Plant 36:208–212

Reid DM, Clements JB, Carr D (1968) Red light induction of gibberellin synthesis in leaves. Nature 217:580–582

Rikin A, Atsmon D, Gitler C (1983) Quantitation of chill-induced release of a tubulin-like factor and its prevention by abscisic acid in *Gossypium hirsutum* L. Plant Physiol 71:747–748

Rikin A, Boussiba S, Mizrahi Y, Richmond AE (1973) The role of abscisic acid in the interrelationship between plant response and environmental stresses. Isr J Bot 22:265–266

Rikin A, Gitler C, Atsmon D (1981) Chilling injury in cotton (*Gossypium hirsutum* L.): light requirement for the reduction of injury and for the protective effect of abscisic acid. Plant Cell Physiol 22:453–460

Rikin A, Richmond AE (1975) Abscisic acid as a common factor in hormonal regulation of resistance to cold and drought in plants. Isr J Bot 24:54

Rikin A, Richmond AE, Waldman M, Dovrat A (1978) The role of abscisic acid and gibberellic acid in regulation of morphogenesis and cold resistance of alfalfa seedlings (*Medicago sativa* L). Isr J Bot 24:49

Rikin A, Waldman M, Richmond AE, Dovrat A (1975) Hormonal regulation of morphogenesis and cold resistance. J Exp Bot 26:175–183

Roberts EH (1972) Dormancy: a factor affecting seed survival in the soil. In: Roberts EH (ed) Viability of Seeds. Syracuse Univ Press, Syracuse, pp 321–359

Robitaille HA, Leopold AC (1974) Ethylene and the regulation of apple stem growth under stress. Physiol Plant 32:301–304

Rohwer F, Schierle J (1982) Effect of light on ethylene production: red light enhancement of 1-aminocyclopropane-1-carboxylic acid concentration in etiolated pea shoots. Z Pflanzenphysiol 107:295–300

Rollin P (1972) Phytochrome control of seed germination. In: Mitrakos K, Shropshire Jr W (eds) Phytochrome. Academic Press, London New York, pp 229–254

Rood SB, Pharis RP, Koshioka M, Major DJ (1983) Gibberellins and heterosis in maize. I. Endogenous gibberellin-like substances. Plant Physiol 71:639–644

Rosen PM, Musselman RC, Kender WJ (1978) Relationship of stomatal resistance to SO_2 and O_3 injury in grapevines. Sci Hortic 8:137–142

Sachs RM (1977) Nutrient diversion: an hypothesis to explain the chemical control of flowering. Hortic Sci 12:220–222

Salisbury FB (1976) Snow flowers. Utah Sci 37:35–41

Salisbury FB (1981) Response to photoperiod. In: Lange OL, Nobel PS, Osmond CB, Ziegler H (eds) Physiological plant ecology I. Vol 12A. Encyclopedia of plant physiology, new ser. Springer, Berlin Heidelberg New York, pp 135–167

Salisbury FB (1982) Photoperiodism. Hortic Rev 4:66–105

Salisbury FB, Ross CW (1985) Plant physiology, 3rd edn. Wadsworth, Belmont, Cal

Salisbury FB, Sliwinski JE, Mueller WJ, Harris CS (1982) How stems bend up. Utah Sci 43:42–49

Samimy C (1978) Effect on light on ethylene production and hypocotyl growth of soybean seedlings. Plant Physiol 61:772–774

Sano H, Nagao M (1970) Change in the indoleacetic acid oxidase levels in leaves of *Begonia evansiana* under short day conditions. Plant Cell Physiol 11:849–856

Satter RL, Galston AW (1976) The physiological functions of phytochrome. In: Goodwin TW (ed) Chemistry and Biochemistry. Academic Press, London New York, pp 680–735

Sawhney R, Webster JM (1975) The role of plant growth hormones in determining the resistance of tomato plants to the root-knot nematode, *Meloidogyne incognita*. Nematologica 21:95–103

Sawhney SS, Sawhney N, Kumar S, Nanda KK (1979) Enzyme activity and electrophoretic pattern of isoenzymes of amylase, catalase, and peroxidase in photo- and gibberellin-induced plants of *Impatiens balsamina* L. var. rose. New Phytol 82:41–47

Schneider J, Legocka J (1981) The role of cytokinins in the development and metabolism of barley leaves. 5. Effect of light and kinetin on the transcriptional activity of chromatin in etiolated leaves. Biochem Physiol Pflanz 176:614–624

Schonbeck MW, Egley GH (1981) Changes in sensitivity of *Amaranthus retroflexus* L. seeds to ethylene during preincubation. Plant Cell Envir 4:237–242

Schopfer P (1977) Phytochrome control of enzymes. Annu Rev Plant Physiol 28:223–252

Schwabe WW (1971) Physiology of vegetative reproduction and flowering. In: Steward FC (ed) Plant physiology – a treatise. Academic Press, London New York, pp 233–411

Schwartz A, Koller D (1978) Phototropic response to vectorial light in leaves of *Lavatera cretica* L. Plant Physiol 61:924–928

Schwartz A, Koller D (1980) Role of the cotyledons in the phototropic response of *Lavatera cretica* seedlings. Plant Physiol 66:82–87

Scurfield G (1973) Reaction wood: its structure and function. Science 179:647–655

Seeley SD, Powell Jr LE (1981) Seasonal changes of free and hydrolyzable abscisic acid in vegetative apple buds. J Am Soc Hortic Sci 106:405–409

Selman IW, Kulasegaram S (1967) Development of the stem tuber in kohlrabi. J Exp Bot 18:471–490

Sequeira L (1973) Hormone metabolism in diseased plants. Annu Rev Plant Physiol 24:353–380

Shackel KA, Hall AE (1979) Reversible leaflet movements in relation to drought adaptation of cowpeas, *Vigna unguiculata* L. Walp Aust J Plant Physiol 6:265–276

Shein T, Jackson DI (1972) Interaction between hormones, light, and nutrition on extension of lateral buds in *Phaseolus vulgaris* L. Ann Bot 36:791–800

Shen-Miller J, Hinchman RR (1974) Gravity sensing in plants: a critique of the statolith theory. Bio Science 24:643–651

Singh G, Singh H (1980) Effect of growth regulators on the growth parameters of chickpea grown under different salinity levels. Indian J Agric Sci 50:23–30

Sivakumaran S, Hall MA (1978) Effects of age and water stress on endogenous levels of plant-growth regulators in *Euphorbia lathyris* L. J Exp Bot 29:195–205

Sliwinski JE, Salisbury FB (1984) Gravitropism in higher plant shoots III. Cell dimensions during gravitropic bending; perception of gravity. Plant Physiol 76:1000–1008

Smirnov YS, Krupnikova TA, Shkolnik MY (1977) Contents of IAA in plants with different sensitivity to boron deficits. Sov Plant Physiol 24:270–276

Smith AM (1976) Ethylene in soil biology. Annu Rev Phytopathol 14:53–74

Smith AM, Cook RJ (1974) Implications of ethylene production by bacteria for biological balance of soil. Nature 252:703–705

Smith H (1978) Molecular biology of plant cells. Blackwell, Oxford

Smith H, Holmes MG (1977) The function of phytochrome in the natural environment. III. Measurement and calculation of phytochrome photoequilibrium. Photochem Photobiol 25:547–550

Smith H, Wareing PF (1964) Gravimorphism in trees. III. The possible implication of a root factor in the growth and dominance relationships of the shoots. Ann Bot 28:297–309

Sood V, Nanda KK (1979) Effect of gibberellic acid and monophenols on flowering of *Impatiens balsamina* in relation to the number of inductive and non-inductive photoperiodic cycles. Physiol Plant 45:250–254

Sorokin HP, Thimann KV (1964) The histological basis for inhibition of axillary buds in *Pisum*. Protoplasma 59:326–350

Spomer GG (1973) The concepts of "interaction" and "operational environment" in environmental analyses. Ecology 54:200–204

Spomer GG, Salisbury FB (1968) Eco-physiology of *Geum turbinatum* and implications concerning alpine environments. Bot Gaz 129:33–49

Stan HJ, Schicker S, Kassner H (1981) Stress ethylene evolution of bean plants – a parameter indicating ozone pollution. Atmos Environ 15:391–395

Strobel GA (1974) Phytotoxins produced by plant parasites. Annu Rev Plant Physiol 25:541–566

Suge H (1978) Growth and gibberellin production in *Phaseolus vulgaris* as affected by mechanical stress (Short commun.) Plant Cell Physiol 19:1557–1560

Suzuki T, Kondo N, Fujii T (1974) Distribution of growth regulators in relation to the light-induced geotropic responsiveness in *Zea* roots. Planta 145:323–329

Sweeney BM (1979) Rhythmic phenomena in plants. Academic Press, London New York

Tafazoli E, Vince-Prue D (1978) A comparison of the effects of long days and exogenous growth regulators on growth and flowering in strawberry, *Fragaria* × *ananassa* Duch. J Hortic Sci 53:255–259

Takahashi H, Saito T, Suge H (1983) Separation of the effects of photoperiod and hormones on sex expression in cucumber. Plant Cell Physiol 24:147–154

Takeba G, Matsubara S (1979) Measurement of growth potential of the embryo in New York lettuce seed under various combinations of temperature, red light and hormones. Plant Cell Physiol 20:51–61

Tasker R, Smith H (1977) The function of phytochrome in the natural environment. V. Seasonal changes in the radiant energy quality in woodlands. Photochem Photobiol 26:487–491

Taylor JS, Reid DM, Pharis RP (1981) Mutual antagonism of sulfur dioxide and abscisic acid in their effect on stomatal aperature in broad bean (*Vicia faba* L.) epidermal strips. Plant Physiol 68:1504–1507

Taylor JS, Wareing PF (1979) The effect of light on the endogenous levels of cytokinins and gibberellins in seeds of sitka spruce (*Picea sitchensis* Carriere). Plant Cell Envir 2:173–179

Taylorson RR, Hendricks SB (1977) Dormancy in seeds. Annu Rev Plant Physiol 28:331–354

Thomas B, Tull SE, Warner TJ (1980) Light-dependent gibberellin responses in hypocotyls of *Lactuca sativa* L. Plant Sci Lett 19:355–362

Tizio R (1966) Présence de kinines dans le périderme de tubercules de pomme de terre. C R Acad Sci 262:868–869

Tompsett PB, Fletcher AM (1979) Promotion of flowering on mature *Picea sitchensis* by gibberellin and environmental treatments. The influence of timing and hormonal concentration. Physiol Plant 45:112–116

Trewavas AJ (1981) How do plant growth substances work? Plant Cell Envir 4:203–228

Tucker DJ, Mansfield TA (1972) Effects of light quality on apical dominance in *Xanthium strumarium* and the associated changes in endogenous levels of abscisic acid and cytokinins. Planta 102:140–151

Tukey HB Jr (1970) The leaching of substances from plants. Annu Rev Plant Physiol 21:305–324

Turgeon R, Webb JA (1971) Growth inhibition by mechanical stress. Science 174:961–962

van Schreven DA (1956) On the physiology of tuber formation in potatoes. I and II. Plant Soil 8:49–86

Van Staden J (1978) The levels of endogenous cytokinins in daffodil bulbs stored under different environmental conditions. Z Pflanzenphysiol 86:323–330

Van Staden J (1981) Effect of photoperiod and gibberellic acid on flowering and cytokinin levels in bougainvillea. S Afr J Sci 77:327

Van Staden J, Dimalla GG (1977) A comparison of the endogenous cytokinins in roots and xylem exudate of nematode resistant and susceptible tomato cultivars. J Exp Bot 28:1351–1356

Van Staden J, Dimalla GG (1981) The production and utilization of cytokinins in rootless, dormant almond shoots maintained at low temperature. Z Pflanzenphysiol 103:121–129

Van Staden J, Wareing PF (1972) The effect of light on endogenous cytokinin levels in seeds of *Rumex obtusifolius*. Planta 104:126–133

Varner JE, Ho DT (1976) Hormones. In: Bonner J, Varner JE (eds) Plant biochemistry. Academic Press, London New York, pp 713–770
Vegis A (1964) Dormancy in higher plants. Annu Rev Plant Physiol 15:185–224
Vesper MJ, Evans ML (1978) Time-dependent changes in the auxin sensitivity of coleoptile segments. Plant Physiol 61:204–208
Vezina PE, Boulter DWK (1966) The spectral composition of near UV and visible radiation beneath forest canopies. Can J Bot 44:1267–1284
Victor TS, Vanderhoef LN (1975) Mechanical inhibition of hypocotyl elongation induces radial enlargement. Plant Physiol 56:845–846
Vierstra RD, Poff KL (1981) Role of carotenoids in the phototropic response of corn seedlings. Plant Physiol 68:798–801
Vince-Prue D (1975) Photoperiodism in plants. McGraw-Hill, New York
Volkmann D, Sievers A (1979) Graviperception in multicellular organs. In: Haupt W, Feinleib ME (eds) Encyclopedia of Plant Physiology, new ser Vol 7. Springer, Berlin Heidelberg New York, pp 573–600
Volkova RI, Drozdov SN, Sycheya ZF, Balagurova NI (1981) On the regulatory function of auxins in actively vegetating plants subjected to temperature influence. Sov Plant Physiol 28:446–451
Wainwright CM (1977) Sun-tracking and related leaf movements in a desert lupin (*Lupinus arizonicus*). Am J Bot 64:1032–1041
Waldman M, Rikin A, Dourat A, Richmond AE (1975) Hormonal regulation of morphogenesis and cold resistance. II. Effect of cold-acclimation and of exogenous abscisic acid on gibberellic acid and abscisic acid activities in alfalfa (*Medicago sativa* L.) seedlings. J Exp Bot 26:853–859
Walker MA, Dumbroff EB (1981) Effects of salt stress on abscisic acid and cytokinin levels in tomato. Z Pflanzenphysiol 101:461–470
Walser RH, Walker DR, Seeley SD (1981) Effect of temperature, fall defoliation, and gibberellic acid on the rest period of peach leaf buds. J Am Soc Hortic Sci 106:91–94
Walton DC (1980) Biochemistry and physiology of abscisic acid. Annu Rev Plant Physiol 31:453–489
Walton JD, Ray PM (1981) Evidence for receptor function of auxin-binding sites in maize. Red light inhibition of mesocotyl elongation and auxin binding. Plant Physiol 68:1334–1338
Wample RL, Bewley JD (1975) Proline accumulation in flooded and wilted sunflower and the effects on benzyladenine and abscisic acid. Can J Bot 53:2893–2896
Wang CY, Adams DO (1982) Chilling-induced ethylene production in cucumbers (*Cucumis sativus* L.) Plant Physiol 69:424–427
Wardrop AB (1956) The nature of reaction wood. V. The distribution and formation of tension wood in some species of *Eucalyptus*. Aust J Bot 4:152–166
Wareing PF, Saunders PF (1971) Hormones and dormancy. Annu Rev Plant Physiol 22:261–288
Watanabe S, Sibaoka T (1983) Light- and auxin-induced leaflet opening in detached pinnae of *Mimosa pudica*. Plant Cell Physiol 24:641–647
Webber JE, Laver ML, Zaerr JB, Lavender DP (1979) Seasonal variation of abscisic acid in the dormant shoots of Douglas-fir. Can J Bot 57:534–538
Weiser CJ (1970) Cold resistance and injury in woody plants. Science 169:1269–1278
Wellensiek SJ (1962) Dividing cells as the locus for vernalization. Nature 195:307–308
Wellensiek SJ (1968) Identity of floral hormones in *Xanthium strumarium* L., and *Silene armeria* L. Acta Bot Neerl 17:197–198
Went F (1957) The experimental control of plant growth. Chron Bot Waltham, Mass
Wershing HF, Bailey IW (1942) Seedlings as experimental material in the study of redwood in conifers. J For 40:411–414
Wheeler RM, Salisbury FB (1978) Water spray as a convenient means of imparting mechanical stimulation to plants. Hort Science 14:270–271
Wheeler RM, Salisbury FB (1980) Gravitropism in plant stems may require ethylene. Science 299:1126–1127

Wheeler RM, Salisbury FB (1981) Gravitropism in higher plant shoots. I. A role for ethylene. Plant Physiol 67:686–690
Whitaker RH, Feeny PP (1971) Allelochemics: chemical interaction between species. Science 171:757–770
Wilkins MB (1979) Growth-control mechanisms in gravitropism. In: Haupt W, Feinleib ME (eds) Encycl of Plant Physiol. Springer, Berlin Heidelberg New York, Vol 7, pp 601–626
Williams EA, Morgan PW (1979) Floral initiation in sorghum hastened by gibberellic acid and far-red light. Planta 145:269–272
Wilson BF, Archer RR (1977) Reaction wood: induction and mechanical action. Annu Rev Plant Physiol 28:23–43
Wodzicki TJ, Wodzicki AB (1980) Seasonal abscisic acid accumulation in stem cambial region of *Pinus silvestris*, and its contribution to the hypothesis of a late-wood control system in conifers. Physiol Plant 48:443–447
Woolley DJ, Wareing PF (1972) The interaction between growth promoters in apical dominance. II. Environmental effects on endogenous cytokinin and gibberellin levels in *Solanum andigena*. New Phytol 71:1015–1025
Woolley JT, Stoller EW (1978) Light penetration and light-induced seed germination in soil. Plant Physiol 61:597–600
Wright LZ, Rayle DL (1983) Evidence for a relationship between H^+ excretion and auxin in shoot gravitropism. Plant Physiol 72:99–104
Wright STC (1969) An increase in the "inhibitor-beta" content of detached leaves following a period of wilting. Planta 86:10–20
Wright STC (1975) Seasonal changes in the levels of free and bound abscisic acid in blackcurrant (*Ribes nigrum*) buds and beech (*Fagus sylvatica*) buds. J Exp Bot 26:161–174
Wright STC (1981) The effect of light and dark periods on the production of ethylene from water-stressed wheat leaves. Planta 153:172–180
Zeevaart JAD (1976) Physiology of flower formation. Annu Rev Plant Physiol 27:321–348
Zimmermann MH, Wandrop AB, Tomlinson PB (1968) Tension wood in aerial roots of *Ficus benjamina*. Wood Sci Tech 2:95–104
Ziv M, Halevy AH, Ashri A (1973) Phytohormones and light regulation of the growth habit in peanuts (*Arachis hypogaea* L.). Plant Cell Physiol 14:727–735

Author Index

Page numbers in *italics* refer to the references

Subject Index

ABA see Abscisic acid
Abbreviations of chemicals
  ABA = Abscisic acid
  ACC = 1-aminocyclopropane-1-carboxylic acid
  Alar (see Daminozide)
  AMO-1618 = Amonium (5-hydroxycarvacryl) trimethyl choride piperidine carboxlate
  ancymidol = -cyclopropyl- -(4-methoxyphenyl)
  ATP = Adenosine triphosphate
  AVG = aminoethoxyvinylglycine
  BA = Benzyladenine = BAP
  BAP = 6-Benzylaminopurine = BA
  BOA = benzothiazole-2-oxyacetic acid
  cAMP = Cyclic adenosine monophosphate
  CBBP = tributyl (2,4 dichlorobenzyl) phosphonium chloride
  CCC = 2-Chlorethyltrimethylammonium chloride
    = Chlorocholine chloride
    = chlormequat
  CEPA (ethrel; ethephon) = 2-(chloroethyl) phosphonic acid
  CK = cytokinins
  CPA = 4-chlorophenoxyacetic acid
  2,4-D = 2,4-dichlorophenoxyacetic acid
  2,6-D = 2,6-dichlorophenoxyacetic acid
  Daminozide = N-dimethylaminosuccinic acid = Butanedioic acid mono (2,2-dimethylhydrazide)
  DIHB = 3,5-diiode-5-hydrobenzoic acid
  DMSO = dimethylsulfoxide
  DNA = Deoxyribonucleic acid
  DNP = 2,4-dinitrophenol
  EDTA = ethylene diaminetraacetic acid
  EGTA = ethyleneglycol-bis-(-aminoethyl ether) N,N[1]-tetraacetic acid
  ET = ethylene
  ethephon (see CEPA)
  FC = fusicoccin
  GA = Gibberellin
  GMP = guanadine monophosphate
  9-HFCA = 9-hydroxyfluorenecarboxylic acid
  IAA = Indoleacetic acid
  IAAsp = IAA-aspartate
  IBA = Indolebutyric acid
Abbreviations of chemicals
  ILA = indole-3-lactic acid
  IPA = isopentenyl adenosine
  2ip = isopentenyladenine
  kinetin = 6-furfurylamino purine
  MCPA = [(4-chloro-o-tolyl)] acetic acid
  mefluidide = [(N-(2,4-dimethyl-5-(trifluoremethyl) amino) phenyl] acetamide
  NAA = Naphthaleneacetic acid
  NADH = nicotinamide adenine dinucleotide reduced
  NOA = -naphthoxyacetic acid
  NPA = N-1-Naphthylphthalamic acid
  PAL = phenylalanine ammonia-lyase
  PBA = 6-(benzylamino)-9-(2-tetrahydropyranyl)-9H-purine
  PCIB = 4-chlorophenoxy-isobutyric acid
  picloram = 4-amino-3,5,6-trichloropicolinic acid
  RNA = ribonucleic acid
  RUDPase = ribulosediphosphate carboxylase
  SADH (see Daminozide)
  SAM = S-adenosylmethionine
  2,3,5-T = 2,3,5-trichlorophenoxyacetic acid
  TIBA = Triiodobenzoic acid
Abelmoschus 263
Abies 176, 225, 523
Abrosia 687
Abscisic acid (ABA) 7, 17, 20, 26, 27, 29, 30, 56, 57, 92, 104, 130, 423–425, 435, 437, 440, 515–544, 581, 637–638, 642, 684
  and
    abscission 538
    age of leaf 515
    apical dominance 535
    assimilate movement 17, 26, 27, 530
    auxin metabolism 535
    cambial growth 174–176, 533
    carbohydrate transport 555
    cell expansion 533, 642
    chilling (0°–12° C) resistance 435
    ^{36}Cl accumulation in roots 530
    CO_2 and stomatal opening 526
    cold acclimation or dormancy 423–427, 440
      deacclimation 437
    cytokinins (interaction) 532, 538